Sequential Counting Principle (SCP)

If one event can occur in m ways, a second event can occur in n ways, a third event can occur in p ways, and so on, then the sequence of events can occur in $m \times n \times p \ldots$ ways.

$$P(n, r) = \frac{n!}{(n-r)!}$$

The number of permutations of n objects taken r at a time.

$$C(n, r) = \frac{n!}{r!(n-r)!}$$

The number of combinations of n objects taken r at a time.

$$P(T') = 1 - P(T)$$

The probability of the complement of an event.

$$P(A \cup B) = P(A) + P(B) - P(A \cap B)$$

The probability of event A **or** event B.

$$P(A \mid B) = \frac{P(A \cap B)}{P(B)}$$

The probability of event A **given** event B.

$$P(A \cap B) = P(A) \cdot P(B)$$

The probability of event A **and** event B where A and B are independent.

Odds $f : u$ in favor of an event

Where f and u, respectively, represent the number of favorable and unfavorable ways that an event can occur.

$$E = a_1 p_1 + a_2 p_2 + \cdots + a_n p_n$$

The expected value E for an event where $a_1, a_2, \ldots, a_n$ are the values that occur with probability $p_1, p_2, \ldots, p_n$.

Mean $\bar{x}$ — The sum of a set of data values divided by the number of values.

Median — The middle value when data values are ranked in order of magnitude. If there is an even number of values, it is the mean of the two middle values.

Mode — The value(s) that occur most often in a set of data.

Range — The difference between the greatest and least values in a set of data.

$$s = \sqrt{\frac{(x_1 - \bar{x})^2 + (x_2 - \bar{x})^2 + \cdots + (x_n - \bar{x})^2}{n - 1}}$$

The standard deviation of a set of data values $x_1, x_2, \ldots, x_n$.

$$z = \frac{x - \mu}{\sigma}$$

The z-score associated with the raw data value x where μ is the mean and σ is the standard deviation.

NINTH EDITION

Topics in
Contemporary Mathematics

Ignacio Bello
University of South Florida

Jack R. Britton
Late of University of South Florida

Anton Kaul
California Polytechnic State University

HOUGHTON MIFFLIN COMPANY
Boston   New York

Publisher: Richard Stratton
Senior Sponsoring Editor: Lynn Cox
Senior Marketing Manager: Katherine Greig
Marketing Associate: Naveen Hariprasad
Development Editor: Lisa Collette
Editorial Assistant: Laura Ricci
Project Editor: Kathleen Deselle
Art and Design Manager: Gary Crespo
Cover Design Manager: Anne S. Katzeff
Senior Photo Editor: Jennifer Meyer Dare
Composition Buyer: Chuck Dutton

Cover image: José Mijares, *Lo Concreto en Rojo,* 1954, oil on wood. Copyright © José Mijares/Cernuda Arte, Coral Gables, Florida. "José María Mijares left Cuba but Cuba never left Mijares."—http://www.mokagallery.com/artists/mijares/mijares.html

Acknowledgments appear on page A48.
Photo Credits appear on page A49.

URLs cited herein may have changed since publication of this book.

Printed in the U.S.A.

Library of Congress Control Number: 2006935527

Instructor's examination copy
   ISBN-10: 0-618-83303-X
   ISBN-13: 978-0-618-83303-0
For orders, use student text ISBNs
   ISBN-10: 0-618-77524-2
   ISBN-13: 978-0-618-77524-8

123456789-VH-10 09 08 07 06

Contents

Chapter 4   Numeration Systems   177

Chapter 5   Number Theory and the Real Numbers   223

Chapter 9   Mathematical Systems and Matrices   583

Chapter 10   Counting Techniques   656

Web: The following chapters and appendix are available on this textbook's Online Study Center. Visit **college.hmco.com/PIC/bello9e** and click on the Online Study Center icon.

Chapter 14   Voting and Apportionment

Chapter 15   Graph Theory

Appendix   The Metric System

Preface

In this ninth edition we continue with our goal of introducing students to the many interesting mathematical concepts used in our contemporary world. We bring out the basic ideas and techniques as simply and clearly as possible and relate these ideas to other areas—such as sociology, psychology, business, and technology—that will be attractive to the reader. Elementary applications are given whenever feasible. These applications can be found in the *Getting Started* feature, throughout the text discussions and examples, and in the lesson problem sets. As suggested by the *Crossroads in Mathematics: Standards for Introductory College Mathematics of AMATYC,* we deemphasized the more abstract and theoretical aspects of the subject matter and, instead, placed emphasis on promoting the understanding and use of the various concepts introduced. Important aids to reaching this goal are the exercise sets, which include problems ranging from routine drills to more challenging exercises for advanced students. Readers will find considerable support and explanation in the worked-out examples.

A Word About Problem Solving

Problem solving has become a fixture in mathematics textbooks. Inspired by the teachings of George Polya, and following the recommendations of the NCTM and the MAA, most mathematics books at this level cover the topic. Many texts front-load much of the presentation in the first chapter. We have chosen to integrate problem solving where it is needed and, consequently, where it can be taught and learned most effectively.

For example, a few of the strategies suggested by Polya himself call for making a table, writing an equation, making a diagram, and accounting for all possibilities. We introduce these ideas in Chapter 1 and present these techniques in the chapters dealing with truth tables, algebra, geometry, and counting, respectively, where the pertinent methods can be effectively displayed.

As an ongoing theme of this text, problem solving is presented purposefully not only in Chapter 1 but in meaningful and appropriate contexts where students can best understand and appreciate its methods. Above all, we hope that this integrated approach will help students learn how to apply problem-solving techniques in the real world once they have completed the course.

New to the Ninth Edition

At the request of the reviewers, we have updated the data in most of the application problems throughout the textbook. In addition, we have made the following changes:

- **Chapter 1:** Many new applications have been added to Section 1.3 including those pertaining to sports, taxes, and poverty rates.

- **Chapter 2:** We updated many of the applications throughout the chapter. New *Collaborative Learning* exercises, in Section 2.2, and *Using Your Knowledge* exercises, in Section 2.3, were also added.

- **Chapter 3:** Examples were added and updated, with corresponding exercises in the exercise sets, throughout the chapter. Conditional and equivalent statements are discussed in the Section 3.3 exercise set. Section 3.7 (Switching Networks: A Problem Solving Tool) now appears on this textbook's Online Study Center.

- **Chapter 4:** New application problems have been added to Section 4.2 and new *Collaborative Learning* exercises were added to Section 4.5.

- **Chapter 5:** Additional coverage, including new examples and new exercises on the Order of Operations, was added to this chapter.

- **Chapter 6:** Examples were added and updated throughout the chapter, with corresponding exercises in the exercise sets, including those pertaining to beverages and insurance. Some of the *In Other Words* exercises have been expanded to cover additional concepts.

- **Chapter 7:** Sections 7.2, 7.3, and 7.5 contain many new exercises and applications. To make the procedures for Solving Systems of Equations easier, we included a summary of the techniques involved for quick reference. Section 7.8 (Linear Programming) now appears on this textbook's Online Study Center.

- **Chapter 8:** Application examples were added and updated, with corresponding exercises in the exercise sets, throughout the chapter including those pertaining to the distance traveled by Venus, crop circles, and hurricanes. A new summary of formulas for the area of different polygons is provided in Section 8.4. Section 8.7 is a new section on Right Triangle Trigonometry. Section 8.7 (Right Triangle Trigonometry) and Section 8.8 (Chaos and Fractals) are available on this textbook's Online Study Center.

- **Chapter 9:** This chapter now includes a discussion of the Luhn algorithm with corresponding exercises in the exercise set. Research questions about Hans Peter Luhn and his algorithm are presented as well.

- **Chapter 10:** An expanded interpretation of the formula for the number of combinations of n objects taken r at a time, $C(n, r)$, is given in Section 10.3.

- **Chapter 11:** This chapter has been expanded to cover empirical probability and expands upon the concept of independent events. Also, examples were added and updated, with corresponding exercises in the exercise sets, throughout this chapter.

- **Chapter 12:** We have extensively revised this chapter to include a discussion of surveys (sample and target populations) and how to summarize the data obtained from these surveys. We expanded the problems dealing with frequency distributions and problems dealing with misuse of statistics. We provide additional examples dealing with mean, median, and mode. Also, examples were added and updated, with corresponding exercises in the exercise sets, throughout this chapter.

- **Chapter 13:** We have extensively revised this chapter to include new formulas to find the APY and discuss how to use them to compare investments, FHA maximum loan values by states for both conventional and FHA loans, and a new formula to calculate monthly payments. Section 13.5 (Investing in Stocks, Bonds, and Mutual Funds) is a new section that contains five new examples and sixty-two new exercises.

- **Chapters 14, 15, and Appendix:** Chapter 14 (Voting and Apportionment), Chapter 15 (Graph Theory), and the Appendix (The Metric System), now appear on this textbook's Online Study Center.

Key Continuing Features

The text has been carefully written in order to promote student success. Pedagogical themes such as *problem solving*, *motivation*, *connections*, and *assessment and review* are incorporated throughout the text. Important pedagogical features have been retained in this new edition in order to continue to promote student success.

The ninth edition continues to incorporate the "Standards of Introductory College Mathematics" set forth by AMATYC. Accordingly, deductive proof (Chapters 1 and 2), numeracy (Chapters 4 and 5), symbolism and algebra (Chapter 6), functions (Chapter 7), geometry and measurement (Chapter 8), and probability (Chapter 11) and statistics (Chapter 12) are among the topics covered in the book.

We also made many significant efforts to address the NCTM curriculum recommendations regarding communication (*In Other Words*), problem solving (Chapter 1 and as a pedagogical theme throughout the book), reasoning (Chapter 3), connections (*Discovery* feature and mathematical systems), algebra (Chapters 6 and 7), geometry (Chapter 8), mathematical structures (Chapter 9), probability (Chapters 10 and 11), and statistics (Chapter 12).

Problem Solving

Problem solving is a pedagogical theme that is incorporated throughout the text to promote student success. Specific *Problem Solving* examples are formatted using the RSTUV method (Read, Select, Think, Use, and Verify) to guide the reader through the problem. The solution is carefully developed to the right of the problem-solving steps. Students are encouraged to cover the solution, write their own solutions, and check their work in order to build problem-solving skills. This feature also includes references to similar problems in the exercise sets.

Motivation

Each chapter begins with a preview that details the material to be covered in the chapter and the ways in which the topics are related to each other throughout the chapter. Each chapter opening photo is connected to a *Getting Started* feature from the chapter.

Each section begins with a *Getting Started* vignette. The applications in this feature are drawn from a vast array of fields and offer a motivating introduction for the techniques and ideas covered in the chapter.

Each chapter presents the *Human Side of Math,* a margin feature, which is a brief biography of a person who devised or contributed to the development of the mathematical topics covered in the chapter. *Looking Ahead* links the biography to the upcoming material.

Connections

Real-life applications, from many disciplines, are included throughout the book to show students how they can apply the material covered in the text. These applications usually conclude the examples in a section and appear in the exercise sets.

In Other Words exercises give students the opportunity to use writing to clarify and express ideas, concepts, and procedures.

Using Your Knowledge exercises help students to generalize concepts and apply them immediately to similar real-life situations.

Discovery exercises are more challenging problems that help students to make connections between concepts and to develop better critical thinking and problem-solving skills.

Calculator Corner exercises provide essential background on how to solve problems using a calculator.

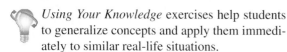 *Collaborative Learning* exercises give students the opportunity to express their thoughts verbally and to become accustomed to working in groups. The type of work involved in these exercises includes *problem solving, modeling, reasoning,* and *connecting* to other disciplines.

Skill Checker exercises, included in relevant sections, help students check their mastery of the skills that they will need in order to proceed to the next section.

Graph It examples appear in the margin next to corresponding examples. The *Graph Its* illustrate an alternative method of solving a problem or checking a solution with the use of a graphing utility.

Online Study Center The *Web It* feature directs students to annotated links on the text-specific Web site. The links correlate to material covered in the text and provide students with access to additional information on a topic, practice problems, tutorials, or downloadable software. In later chapters, selected exercise sets conclude with optional *Web It* exercises that require the use of the Internet.

Research Questions are included at the end of each chapter and at the end of selected exercise sets. Many of these are correlated to specific Web sites to help students master research techniques as well as to explore how the topics under discussion were developed. An expanded research bibliography, detailing sources for researching these questions, is provided at the end of the book.

Assessment and Review

The *Chapter Summary* provides brief definitions and examples for key topics within a given chapter. It also contains section references to encourage students to reread sections.

The *Practice Test* at the end of the chapter is followed by *Answers to Practice Test.* The answers to each question are keyed to the specific section, example, and page that the students should reference if they miss a problem. This provides students with a means to diagnose skills and concepts that they have mastered and identifies those that require further work. This tool also assists students in taking responsibility for their own learning.

Courses for Which This Book Is Intended

This textbook contains a large selection of topics and is suitable for use in various courses. The entire book can be covered easily in a full year's course, while many alternative choices can be made for a two-quarter or a one-semester course. Here are some of the courses for which the book is suggested:

- General education or Liberal arts mathematics [the text follows most of the CUPM (Committee on the Undergraduate Program in Mathematics) and AMATYC recommendations for liberal arts mathematics]

- Topics in contemporary mathematics

- College mathematics or Survey of mathematics

- Introduction to mathematics or Applications of mathematics

A few more advanced topics may be included or omitted at the instructor's discretion. These choices will not affect the continuity of any chapter presentation or syllabus as a whole. The topics include the following sections: 2.5 Infinite Sets; 3.7 Switching Networks; 7.8 Linear Programming; 8.6 Networks, Non-Euclidean Geometry, and Topology; 8.8 Chaos and Fractals; and 9.5 Game Theory. Some of these sections (3.7, 7.8, and 8.8) are available on this textbook's Online Study Center.

Supplements for the Instructor

Topics in Contemporary Mathematics, Ninth Edition, has an extensive support package for the instructor that includes:

- *Online Teaching Center* This free companion Web site contains an abundance of instructor resources such as solutions to all exercises in the text, digital art and tables, suggested course syllabi, and Chapter Tests. Visit **college.hmco.com/ PIC/bello9e** and click on the Online Teaching Center icon.

- **Online Instructor's Answer Key:** The *Online Instructor's Answer Key* offers answers to *all* of the exercises in each exercise set.

- **HM Testing™ CD-ROM (powered by Diploma™):** *HM Testing™* (powered by *Diploma™*) offers instructors a flexible and powerful tool for test generation and test management. Now supported by the Brownstone Research Group's market-leading *Diploma* software, this new version of *HM Testing™* significantly improves the functionality and ease of use by offering all the tools needed to create, author, deliver, and customize multiple types of tests—including authoring and editing algorithmic questions.

- **Eduspace®:** Eduspace®, powered by Blackboard®, is Houghton Mifflin's customizable and interactive online learning tool. Eduspace® provides instructors with online courses and content. By pairing the widely recognized tools of Blackboard® with quality, text-specific content from Houghton Mifflin Company, Eduspace® makes it easy for instructors to create all or part of a course online. This online learning tool also contains ready-to-use homework exercises, quizzes, tests, tutorials, and supplemental study materials. Visit **eduspace.com** for more information.

Supplements for the Student

Topics in Contemporary Mathematics, Ninth Edition, has an extensive support package for students that includes:

- **Student Solutions Manual:** The *Student Solutions Manual* contains study tips, worked-out solutions to *all* odd-numbered exercises, and *all* the solutions to the Chapter Practice Tests in the text.

- *Online Study Center* This free companion Web site contains an abundance of student resources. A comprehensive set of annotative *Web It* links, related to concepts covered in each chapter, provides students with access to additional information, practice problems, tutorials, or downloadable software. Material supported on the Online Study Center is denoted by the Online Study Center icon. Visit **college.hmco.com/PIC/ bello9e** and click on the Online Study Center icon.

- **Online Preparing for the CLAST Guide:** The *Preparing for the CLAST Mathematics Guide* is a competency-based study guide that reviews and offers preparatory material for the CLAST (College Level Academic Skills Test) objectives required by the State of Florida for mathematics.

- **CLAST Videos:** A set of five videotapes (prepared by Fran Hopf) covers key topics on the CLAST test and correlates to the *Preparing for the CLAST Mathematics Guide* with definitions, explanations, and examples.

- **Instructional DVDs:** These text-specific DVDs, professionally produced by Dana Mosely, provide explanations of key concepts, examples, and exercises in a lecture-based format. They offer students a valuable resource for further instruction and review. They also provide support for students in online courses.

- **Eduspace®:** Eduspace®, powered by Blackboard®, is Houghton Mifflin's customizable and interactive online learning tool for instructors and students. Eduspace® is a text-specific Web-based learning environment that your instructor can use to offer students a combination of practice exercises, multimedia tutorials, video explanations, online algorithmic homework, and more. Specific content is available 24 hours a day to help you succeed in your course.

- **SMARTHINKING® Live, Online Tutoring:** Houghton Mifflin has partnered with SMARTHINKING® to provide an easy-to-use, effective, online tutorial service. Through state-of-the-art tools and a two-way whiteboard, students communicate in real-time with qualified e-structors, who can help the students understand difficult concepts and guide them through the problem-solving process while studying or completing homework.

  Three levels of service are offered to the students.

  Live Tutorial Help provides real-time, one-on-one instruction.

  Question submission allows students to submit questions to the tutor outside the scheduled hours and receive a response within 24 hours.

  Independent Study Resources connects students around-the-clock to additional educational resources, ranging from interactive Web sites to Frequently Asked Questions.

  Visit **smarthinking.com** for more information.

  Limits apply; terms and hours of SMARTHINKING® service are subject to change.

Acknowledgments

We wish to thank the reviewers of this ninth edition for their many valuable suggestions and constructive criticism:

Robin L. Ayers, *Western Kentucky University;* Yungchen Cheng, *Missouri State University;* John C. D. Diamantopoulos, Ph. D, *Northeastern State University;* John D. Gieringer, *Alvernia College;* Raymond E. Griffith, *South Piedmont Community College;* Donald R. Goral, Ph. D, *Northern Virginia Community College;* Harriet Higgins Kiser, *Floyd College;* Bernard Omolo, *University of South Carolina Upstate;* Kathy Pinchback, *University of Memphis;* Nancy Ressler, *Oakton Community College;* Daniel M. Seaton, *University of Maryland Eastern Shore;* Sankara N. Sethuraman, *Augusta State University;* William H. Shaw, Jr., *Coppin State University;* Nader Vakil, *Western Illinois University;* and Dr. Judith Wood, *Central Florida Community College.*

We also wish to thank Professor Stephen Suen, of the University of South Florida, who shared the information on his Web site and offered valuable comments regarding Chapters 1, 2, 10, 11, and 12. Professor Manoug Manougian, of the University of South Florida, pointed out ways to improve some of the material regarding sets, lines, and probability. Also, Professor John Davis provided valuable suggestions for the probability chapter.

For their help on earlier editions, we wish to express our appreciation to Dr. Heriberto Hernandez, Bill Albrecht, Barbara Burrows, Joe Clemente, Gary Etgen, Josephine Rinaldo, and Prakash Sach. We thank the following colleagues for all their helpful criticisms and suggestions: Diana Fernandez, James Gard, George Kosan, Chester Miles, Donald Clayton Rose, Donald Clayton Rose II, and, especially, Charles Osborne. Thanks also to the people involved in the preparation of the ancillary materials for previous editions: Alex Ambrioso, Marcus McWaters, Mark Oglesby, Rose Reyes, Robert Schatzow, T. Tran, and William Wilder.

Among the other people involved with the previous editions, we would like to thank Philip Lanza and Debbie Gruetzmacher.

For the ninth edition, we would like to thank Kathleen Deselle, Project Editor; Lisa Collette, Developmental Editor; and Lauri Semarne, our final accuracy checker. In addition, Debbie Garrison checked answers in the *Instructor's Answer Key* and helped in the statistics chapter. Fran Hopf checked some of the *Student's Solutions Manual,* Louis Camara and Karol McIntosh checked the statistics chapter, and Jolene Rhodes reviewed the graph theory chapter. We also thank Mile Krajcesvki and Scott Rimbey of the University of South Florida.

We wish to express our sincere thanks to the many users—students and instructors alike—of the previous editions. We hope that this ninth edition will please them even more. We always welcome comments and suggestions from students, professors, and readers. You may send them to us at the following address: Ignacio Bello, Mathematics Department, University of South Florida (Phy 342), 4202 Fowler Avenue, Tampa, FL 33620, or e-mail Ignacio Bello at cubanmath@aol.com.

Finally, we would like to acknowledge the passing of our colleague Jack R. Britton. His dedication to the study of mathematics provided a foundation for the current edition of this textbook.

I. B.
A. K.

In leaner economic times, families are forced to adhere to more strict household budgets. Families must carefully allot the amount of money that they spend on necessity and luxury items in order to live within their means. Graphs are a useful tool for creating and analyzing a budget. In Section 1.3 you will learn how to interpret different types of graphs.

Problem Solving

It's not where you're from; it's where you're going. It's not what you drive; it's what drives you. It's not what's on you; it's what's in you. It's not what you think; it's what you know.

GATORADE COMMERCIAL

The question really is, Do you know how to solve problems? If you are not sure, we will help you right here, right now. As René Descartes said: "It is not enough to have a good mind. The main thing is to use it well." How? We start this chapter by giving you a procedure that you can use to solve any type of problem. Why?

Quite simply, students cannot solve word problems reliably because they are presented with inconsistent models of problem solving that contradict the logical processes they have learned in other courses and in everyday life. *

Together, we can fix that!

Our RSTUV procedure is based on a concept developed by George Polya, a Hungarian mathematician and professor featured in this chapter's Human Side of Math. Along with our procedure, we discuss an important concept in problem solving: inductive reasoning.

The rest of the chapter is devoted to problem-solving techniques that will help you solve any problem: patterns (Section 1.1), estimations (Section 1.2), or a picture in the form of a graph (Section 1.3). Along the way, we will try to have some fun and remember two important ideas:

The value of a problem is not so much in coming up with the answer as in the ideas and attempted ideas it forces on the would-be solver.

ISRAEL NATHAN HERSTEIN

Math is fun, and you can do it!

IGNACIO BELLO

*Source: www.hawaii.edu/suremath/why1.html.

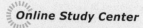

Online Study Center

For links to Internet sites related to Chapter 1, please access **college.hmco.com/PIC/bello9e** and click on the Online Study Center icon.

1

1.1   Inductive Reasoning

HUMAN SIDE OF MATH

George Polya was awarded a Ph.D. in mathematics with a minor in physics in 1912. His mathematical (1887–1985) output was broad and included papers in number theory, combinatorics, and voting systems.

   After a brief stay at Brown University in 1942. Polya moved to Standford University in 1942. He published many articles and books; however, his book *How to Solve It* became a resounding success, selling more than 1 million copies. The book outlined his famous problem-solving strategy.

   Polya also contributed greatly in the classroom. He regularly visited schools and inspired many students to pursue a career in mathematics.

Looking Ahead
In this chapter we will look at the concepts developed by George Polya and use them to learn how to solve problems.

Knowledge of mathematics is power. As a student, mathematics gives you the power to pursue many careers. As a citizen, it gives you the power to reach informed decisions. As a nation, it gives us the power to compete in a technological world. Many students today are not prepared for the jobs of tomorrow or even for the jobs of today! In fact, three out of four Americans stop studying mathematics before completing career or job prerequisites. The mathematics you learn today may have an impact on the job you get tomorrow. Over 75% of all jobs require proficiency in simple algebra and geometry, either as a prerequisite to enter a training program or as part of an examination required to be licensed in a specific field. In addition, earning a college degree requires taking at least a minimal amount of mathematics courses. Unfortunately, many students are burdened by unpleasant experiences in mathematics. They convince themselves that they can't do mathematics, so they won't. They believe that they never liked mathematics, so they don't. Now it is time to change all that. This time try mathematics with an attitude—a positive one. The way to start is by learning how to solve problems. ▶

A. Problem-Solving Procedure

One of the first problem-solving models was developed by George Polya, a Hungarian-born mathematician and researcher. A problem-solving strategy is *your plan for action.* First, you must choose your problem-solving strategy. What strategies did Polya recommend? Here are a few.

Look for a pattern.	Make a table.
Draw a picture or diagram.	Use logical reasoning.
Make a model.	Work a simpler problem first.
Use a formula.	Make a list.

Polya's original procedure as it appeared in *How to Solve It* (Princeton, NJ: Princeton University Press, 1973) consisted of four parts.

1. Understand the problem.

2. Devise a plan.

3. Carry out the plan.

4. Look back.

We expand this procedure so that you can use it as a model to solve *any* problem. Here are the five steps we shall use.

RSTUV Problem-Solving Procedure

1. **R**ead the problem, not once or twice but until you understand it.

2. **S**elect the unknown; that is, find out what the problem asks for.

3. **T**hink of a plan to solve the problem.

4. **U**se the techniques you are studying to carry out the plan.

5. **V**erify your answer.

PROBLEM SOLVING

Hints and Tips

Our problem-solving procedure (RSTUV) contains five steps. The steps are given in the left column, and hints and tips in the right.

❶ Read the problem.

Mathematics is a language, so it is important to learn how to read it. If you do not understand or even get through reading the problem the first time, read it again and pay attention to key words or instructions such as *compute, draw, write, construct, make, show, identify, state, simplify, solve,* and *graph.*

❷ Select the unknown.

How can you answer a question if you do not understand the question and cannot determine the unknown? One good way to look for the unknown is to look for the question mark "?" and carefully read the preceding material. Try to determine what information is given and what is missing from the problem.

❸ Think of a plan.

Problem solving requires many skills and strategies. Some strategies are *look for a pattern; examine a related problem; make tables, pictures, and diagrams; write an equation; work backward;* and *make a guess.*

❹ Use the techniques you are studying to carry out the plan.

If you are studying a mathematical technique, use your knowledge about the technique to solve the given problem. Look for procedures that can be used to solve each problem. Then carry out the plan and check each step.

❺ Verify the answer.

Look back and check the result of the original problem. Is the answer reasonable? Can you find the answer some other way?

Online Study Center

To further explore problem solving, go to the Web site **college.hmco.com/PIC/bello9e**, click on the Online Study Center icon, select Chapter 1 from the drop-down menu, and then select "Web It" Links. Finally, select Chapter 1, and click on link 1.1.1. Bookmark the Online Study Center site for your convenience; you will need it to access all the "Web It" Links.

Look at the first letter in each sentence. To help you remember the steps, we call them the **RSTUV** procedure.

Problem solving will be presented in a two-column format. Cover the answers in the right column (a 3-by-5 index card will do), and write *your own answers* as you practice the problems. After you complete the problems, uncover the answers and check if your answers are correct. You will then be given a similar example and its solution.

B. Inductive Reasoning

As we mentioned, one of the strategies used in problem solving is to find a pattern. Reasoning based on examining a variety of cases, discovering patterns, and forming conclusions is called **inductive reasoning.** For example, you want to make sure that you have a good instructor in your next mathematics course. You ask several of your friends about different instructors. They all say that Professor X is a good instructor. You conclude that Professor X is indeed a good instructor. This is an example of inductive reasoning, but it may be a hasty generalization. Now consider this example: Suppose you become a doctor and one of your patients comes to you and says,

> It hurts when I press here (pressing his side).
> And when I press here (pressing his other side).

And here (pressing his leg).

And here, here, and here (pressing his other leg and both arms).

You examine the patient all over, trying to invoke your inductive powers. What can you conclude? (The answer is on page 9.)

Inductive Reasoning

Inductive reasoning is the process of arriving at a general conclusion on the basis of repeated observations of specific examples.

Look at panels A, B, and C of Figure 1.1. Panel A consists of 1 square. Panel B consists of 1 large square and 4 smaller squares. Panel C consists of 1 large square, 4 medium squares, and 9 smaller squares. How many squares are there in panels D and E? Draw the next figure in the pattern and write an expression for the total number of squares.

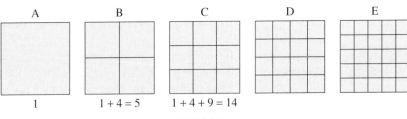

FIGURE 1.1

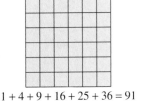

$1 + 4 + 9 + 16 + 25 + 36 = 91$

FIGURE 1.2

Panel D consists of $1 + 4 + 9 + 16 = 30$ squares, and panel E consists of $1 + 4 + 9 + 16 + 25 = 55$ squares. The next square in the pattern (see Figure 1.2) consists of $1 + 4 + 9 + 16 + 25 + 36 = 91$ squares. An arrangement of numbers according to a pattern is called a **sequence,** and each number in the sequence is called a **term.** Thus, the sequence of the number of squares is 1, 5, 14, 30, 55, and 91. Here, the first term is 1, the second term is 5, and so on.

EXAMPLE 1 ▶ Identifying Patterns and Finding Terms

Identify the pattern in each sequence and find the next three terms.

(a) 1, 4, 7, 10, _____, _____, _____   (b) 1, 2, 4, 8, _____, _____, _____

Solution

The strategy is to examine the difference between successive terms to discover the pattern.

(a) The difference between adjacent numbers is 3. Thus, the pattern is *add 3.* The next three terms are $10 + 3 = 13$, $13 + 3 = 16$, and $16 + 3 = 19$.

(b) Each number after the first is *2 times* the preceding number. The pattern is *multiply by 2,* and the next three terms are $2 \times 8 = 16$, $2 \times 16 = 32$, and $2 \times 32 = 64$.   ∎

EXAMPLE 2 ▶ Identifying Patterns and Finding Terms

Identify the pattern in each sequence and find the next term.

(a) 4, 9, 3, 8, 2, _____   (b) 1, 2, 8, 22, 47, _____

Solution

(a) Examine the differences between successive terms.

The pattern is *add 5, subtract 6.*

Add 5: $2 + 5 = 7$
Subtract 6: $7 - 6 = 1$

Thus, the term after 2 is ⑦ and the term after 7 is ①.

Another way of looking at this pattern is to concentrate on the alternate terms, 4, 3, 2, __?__ and 9, 8, __?__, as shown. The answer is the same!

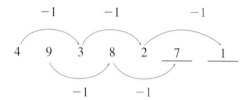

Moral: There may be more than one way to identify patterns!

(b) Examine the differences between successive terms.

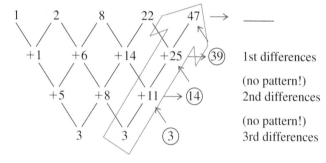

Now, follow the red arrows to get *3 + 11 = 14, 14 + 25 = 39,* and $39 + 47 = $ __86__. Do you see that there is a quicker way of doing this? Simply add 3 + 11 + 25 + 47 and you get the __86__. The technique here is to get the first differences, second differences, third differences, and so on, until you get a constant, and then add up the diagonal as shown. ∎

A Word of Warning: Some sequences follow more than one pattern. Thus, the next three terms in the sequence

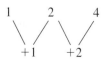

are 7, 11, and 16 if the pattern is *add 1, add 2, add 3,* and so on. On the other hand, if you view the sequence 1, 2, 4 as *doubling* the preceding number, the next three terms in the sequence are 8, 16, and 32! Sometimes it is necessary to examine a large number of cases before realizing that the conjectured pattern does not continue, as the next example shows. If there is just *one* case in which the conjecture does not work, you have a *counterexample* and the conjecture is false.

EXAMPLE 3 ► Predicting the Number of Regions

Choose points on a circle and connect them to form distinct, nonoverlapping regions as shown in Figure 1.3. Two points determine $2 = 2^1$ regions, three points determine $4 = 2 \times 2 = 2^2$ regions, and four points determine a total of $8 = 2 \times 2 \times 2 = 2^3$ regions. These results are entered in the table. How many regions would you predict for

(a) five points?   (b) six points?

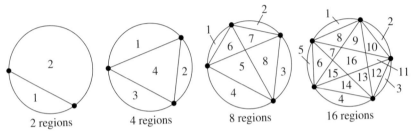

FIGURE 1.3

Solution

(a) $16 = 2 \times 2 \times 2 \times 2 = 2^4$ regions, as shown in Figure 1.3.

(b) The formula predicts $2^5 = 32$. However, if you choose six unequally spaced points on the circle and count the regions correctly, you get only 31 regions. How many regions do you get if the points are equally spaced? Try it!

Number of Points	2	3	4	5	6
Number of Regions	$2 = 2^1$	$4 = 2^2$	$8 = 2^3$	$16 = 2^4$	?

What would happen for the seventh point? As it turns out, adding a seventh point would yield 57 regions. The pattern is

$$1, 2, 4, 8, 16, 31, 57$$

and the formula to find the number of regions is

$$\frac{n^4 - 6n^3 + 23n^2 - 18n + 24}{24}$$

How do we know that this is right? We can test it by substituting $n = 2$, $n = 3$, and so on. We can also use a grapher, as shown in the margin. However, we cannot be sure about a conjecture until a general formula or pattern is proved because just *one* counterexample can make the conjecture false. Unfortunately, we are not in a position to prove or disprove this formula because the result uses combinatorics, which will be discussed in Chapter 10. (See the Research Questions at the end of this section for further information.) The next example illustrates the use of inductive as well as deductive reasoning, the process of proving a conclusion from one or more general statements.

GRAPH IT

In order to find the number of regions using a grapher, let y_1 be the number of regions. Then go to $\boxed{Y=}$ and enter the formula

$$y_1 = \frac{(x^4 - 6x^3 + 23x^2 - 18x + 24)}{24}$$

using x's instead of n's, pressing the $\boxed{\land}$ key to enter exponents, and inserting parentheses as shown. Now press $\boxed{\text{2nd}}$ TABLE, and the values for the number of points and the corresponding number of regions appear as shown.

EXAMPLE 4 ▶ Making and Proving Conjectures

Consider this procedure. Select a number. Multiply the number by 9. Add 6 to the product and divide the sum by 3. Subtract 2 from the quotient.

(a) Follow the procedure for four different numbers and make a conjecture about the relationship between the original number and the final result.

(b) Represent the original number by n and prove the conjecture in part (a).

Solution

Let us select four different numbers—say, 5, 10, 21, and 100—and apply the procedure given in the example.

(a)

Select a number.	5	10	21	100
Multiply the number by 9.	$9 \cdot 5 = 45$	$9 \cdot 10 = 90$	$9 \cdot 21 = 189$	$9 \cdot 100 = 900$
Add 6 to the product.	$45 + 6 = 51$	$90 + 6 = 96$	$189 + 6 = 195$	$900 + 6 = 906$
Divide the sum by 3.	$\dfrac{51}{3} = 17$	$\dfrac{96}{3} = 32$	$\dfrac{195}{3} = 65$	$\dfrac{906}{3} = 302$
Subtract 2 from the quotient.	$17 - 2 = 15$	$32 - 2 = 30$	$65 - 2 = 63$	$302 - 2 = 300$

Since we have to make a conjecture relating the original number and the final result, let us look at the original numbers and the final results.

5	10	21	100
15	30	63	300

Do you see the pattern? Using inductive reasoning, our conjecture is that the final result is three times the original number. But can we prove it? Let us go to part (b) and repeat the process using $\boldsymbol{n}$ as the original number.

(b) Select a number. $\qquad$ n

Multiply the number by 9. $\qquad$ $9 \cdot n$

Add 6 to the product. $\qquad$ $9n + 6$

Divide the sum by 3. $\qquad$ $\dfrac{9n + 6}{3} = 3n + 2$

> It gets a little tricky here. You have to divide $9n$ by 3 and get $3n$ and 6 by 3 and get 2.

Subtract 2 from the quotient. $\quad 3n + 2 - 2 = 3n$

The final result $3n$ is indeed three times the original number n; this proves our original conjecture. ■

Here is one that will amaze your friends. Try it with your own age and the amount of change in your pocket and make a conjecture about what the final result means. For example, if your final result is 2015, what does that mean?

1. Take your age. _____
2. Multiply it by 2. _____
3. Add 5. _____
4. Multiply this sum by 50. _____
5. Subtract 365. _____
6. Add the amount of loose change in your pocket. _____ (Must be less than $1!)
7. Add 115. _____

C. Applications

Did you know that sequences, patterns, and induction were used to find some of the planets of our solar system (see Figure 1.4)? In 1772, the German astronomer Johann Bode discovered a pattern in the distances of the planets from the Sun. His sequence is shown in the next example.

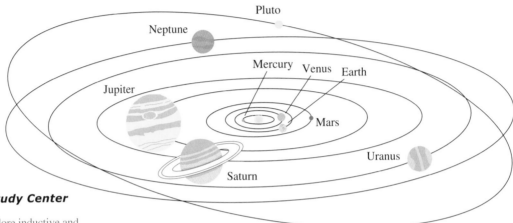

FIGURE 1.4

Online Study Center

To further explore inductive and deductive reasoning, use one of the search engines on the Internet and look under "Inductive and Deductive Reasoning," or access links 1.1.2 and 1.1.3 on this textbook's Online Study Center. Look for "Introduction to Inductive and Deductive Reasoning," and try some practice problems (with answers)! To learn more about Polya and problem solving, access link 1.1.4.

EXAMPLE 5 ▶ The Number of a Planet

P1	P2	P3	P4	P5	P6	P7	P8
Mercury	Venus	Earth	Mars	___	Jupiter	Saturn	___
↓	↓	↓	↓	↓	↓	↓	↓
0 + 4	3 + 4	6 + 4	12 + 4	___	48 + 4	96 + 4	___

Use the RSTUV procedure to find which number corresponds to

(a) the fifth planet.   (b) the eighth planet.

PROBLEM SOLVING

Discovering Planets Using Induction

❶ Read the problem.

You are asked to find the numbers corresponding to the *fifth* and *eighth* missing planets.

❷ Select the unknown.

The unknowns are the missing numbers in the pattern.

❸ Think of a plan. Is there a pattern? Can you find it?

If you can find a pattern, you can find the missing numbers. All the numbers are of the form _____ + 4. The numbers in front of the 4 for the first four planets are 0, 3, 6, and 12. The numbers in this sequence after the 3 are obtained by *doubling*, so the numbers will be

0 + 4   3 + 4   6 + 4   12 + 4   <u>24 + 4</u>   48 + 4   96 + 4   <u>192 + 4</u>

$$\overset{\times 2}{\frown}\ \overset{\times 2}{\frown}\ \overset{\times 2}{\frown}\ \overset{\times 2}{\frown}\ \overset{\times 2}{\frown}\ \overset{\times 2}{\frown}$$

↓	↓	↓	↓	↓	↓	↓	↓
4	7	10	16	28	52	100	196

continued

④ Use inductive reasoning to find the pattern.

(a) The number for the fifth planet should be $2 \times 12 + 4 = 28$. (This "planet" is really Ceres, a planetoid or asteroid.)

(b) The number for the eighth planet is $2 \times 96 + 4 = 196$. (This corresponds to Uranus, discovered by William Herschel in 1781.)

⑤ Verify the solution.

The differences between successive terms in the sequence follow the doubling pattern shown below.

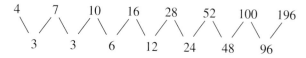

Cover the solution, write your own solution, and then check your work.

The numbers in this sequence—4, 7, 10, 16, and so on—are very important because they correspond to 10 times the distance of the planets from the Sun measured in astronomical units, where 1 astronomical unit is the average distance of the Earth from the Sun. ∎

What about the answer to the doctor question on page 4? There are many, but here is one: The patient had a broken finger!

EXERCISES 1.1

Ⓐ Problem-Solving Procedure

1. What are the four steps in Polya's problem-solving procedure?

2. What do the letters in the RSTUV procedure stand for?

3. When solving a problem, what is the first thing you should try to determine?

The first step in the RSTUV procedure is to **R**ead the problem: Read the following information and do problems 4–10.

Many search engines or "Web crawlers" on the Internet or the World Wide Web (WWW) can help you find information. One such service is illustrated here.

WebQuery

WebQuery is a research service that makes finding information easy and fast. You can search WWW pages, online newspapers, magazines, press releases, medical and health information, several databases, and much more. To enroll, select one of our membership plans:

Plan	Monthly Fee	Free Inquiries/Month	Transaction Fee
Occasional	Free	0	20 cents
Light Use	$1.95	10	15 cents
Standard	$9.95	100	10 cents

4. What is WebQuery? How many membership plans are available for a customer?

5. If you are in the Occasional Plan and make 15 transactions in a given month, what is your total cost for the month?

6. If you are in the Light Use Plan and make 15 transactions in a given month, what is your total cost for the month?

7. On the basis of your answers to problems 5 and 6, if you are planning to make about 15 transactions per month, which plan should you select?

8. What is the maximum number of calls you can make in the Occasional Plan and come out spending less than in the Light Use Plan?

9. After how many calls would it save you money to change from a Light Use Plan to a Standard Plan?

10. If you expect to have 50 transactions in a month, which plan should you select?

B **Inductive Reasoning**

In problems 11–20, identify the pattern and find the next three terms.

11. 1, 2, 4, 7, _____, _____, _____

12. 2, 5, 10, 17, _____, _____, _____

13. 1, 5, 1, 10, 1, 15, _____, _____, _____

14. 3, 3^2, 3^4, 3^8, _____, _____, _____

15.

16.

17. 1, $\frac{1}{2}$, $\frac{1}{4}$, $\frac{1}{8}$, _____, _____, _____

18. $\frac{1}{14}$, $\frac{1}{9}$, $\frac{1}{12}$, $\frac{1}{7}$, $\frac{1}{10}$, _____, _____, _____

19. 1, 5, 2, 6, 3, _____, _____, _____

20. 6, 1, 9, 5, 12, 9, 15, _____, _____, _____

21. The figures represent the *triangular numbers*.

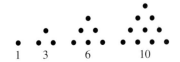

a. Draw the next triangular number.
b. Describe the pattern and list the next three triangular numbers.
c. What is the tenth triangular number?

22. The first 10 triangular numbers are

1   3   6   10   15   21   28   36   45   55

a. Find the difference between adjacent triangular numbers. For example, $3 - 1 = 2$ and $6 - 3 = 3$. What numbers do you get?
b. Find the sum between adjacent triangle numbers. For example, $1 + 3 = 4$. What numbers do you get?
c. Follow the pattern and find the sum of the ninth and tenth triangular numbers.
d. Find the sum of the fourteenth and fifteenth triangular numbers.

23. The fourth triangular number is 10 and the sum $1 + 2 + 3 + 4 = 10$.
a. What is the eighth triangular number?
b. What is $1 + 2 + 3 + 4 + 5 + 6 + 7 + 8$?
c. What is the twelfth triangular number?
d. What is $1 + 2 + 3 + \cdots + 12$?
e. Make a conjecture regarding the nth triangular number and the sum of the first n counting numbers.
f. Carl Friedrich Gauss, a German mathematician born in 1777, was confronted with a similar problem when he was seven years old. His teacher wanted Gauss to find the sum

$$1 + 2 + 3 + \cdots + 99 + 100$$

Gauss noticed that $1 + 100 = 101$, $2 + 99 = 101$, $3 + 98 = 101, \ldots$; thus, you have 50 pairs of numbers, each pair summing to 101. What is the result? Use the idea in parts **(a)–(e)**.

24. The figures represent the *square numbers*.

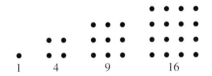

a. Draw the next square number.
b. Describe the pattern and list the next three square numbers.
c. What is the twelfth square number?

25. The numbers in problems 21 and 24 are examples of *figurate numbers*. Another type of figurate number is the *pentagonal number,* shown below.

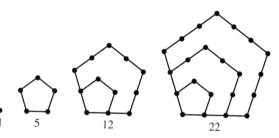

1   5   12   22

    a. Draw the next pentagonal number.
    b. Describe the pattern.
    c. What is the sixth pentagonal number?

26. The figures below show the number of line segments that can be drawn between two points and between three, four, and five noncollinear points. How many line segments can be drawn between
    a. six noncollinear points?
    b. nine noncollinear points?

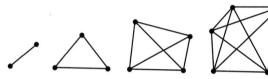

27. The figures below show all the diagonals that can be drawn from one vertex of a quadrilateral (four sides), pentagon (five sides), hexagon (six sides), and heptagon (seven sides). How many diagonals can be drawn from one vertex of a decagon (ten sides)?

28. Follow this procedure.

    Select a number.
    Add 7 to it.
    Multiply the sum by 3.
    Subtract 6 from the product.
    Divide by 3.
    Subtract 5 from the quotient.

    a. Follow the procedure above for four different numbers and make a conjecture about the relationship between the original number and the final result.
    b. Represent the original number by n and prove the conjecture in part (**a**).

29. Follow this procedure.

    Select a number.
    Add 7 to it.
    Multiply the sum by 3.
    Subtract 6 from the product.
    Divide by 3.
    Subtract the original number from the quotient.

    a. Follow the procedure above for four different numbers and make a conjecture about the relationship between the original number and the final result.
    b. Represent the original number by n and prove the conjecture in part (**a**).

30. Follow this procedure.

    Select a number.
    Add 5 to it.
    Multiply the sum by 4.
    Divide the product by 2.
    Subtract 10 from the quotient.

    a. Follow the procedure above for four different numbers and make a conjecture about the relationship between the original number and the final result.
    b. Represent the original number by n and prove the conjecture in part (**a**).

31. Follow this procedure.

    Select a number.
    Add 5 to it.
    Multiply the sum by 4.
    Divide the product by 2
    Subtract twice the original number from the quotient.

    a. Follow the procedure above for four different numbers and make a conjecture about the relationship between the original number and the final result.
    b. Represent the original number by n and prove the conjecture in part (**a**).

32. Pick a number and follow this pattern.
    1. If the number is even, divide by 2.
    2. If the number is odd, multiply by 3, then add 1.
Whatever answer you get, follow rules 1 and 2 again and proceed until you get a 1. For example, start with 13 for good luck. Here is the pattern.

$$13 \rightarrow 40 \rightarrow 20 \rightarrow 10 \rightarrow 5 \rightarrow 16 \rightarrow 8 \rightarrow 4 \rightarrow 2 \rightarrow 1$$

a. What is the pattern if you start with 22?
b. What is the pattern if you start with 15?
c. What do you notice about the last three numbers in each pattern?
d. Can you find any number so that the last three numbers in the pattern are different from the last three numbers you obtained in parts (a) and (b)?

33. Pick a number and write it in words; then do the following:

1. Write the number of letters in the words.
2. Write the number obtained in step 1 in words.
3. Repeat steps 1 and 2.

For example, if you pick the number 24, the pattern is

twenty-four, 10, ten, 3, three, 5, five, 4, four, 4

a. Pick a different number and follow the pattern. What is the last number in the pattern?
b. Can you explain why this works?

34. Consider the pattern

$$1= 1^2$$
$$1 + 3= 2^2$$
$$1 + 3 + 5= 3^2$$
$$1 + 3 + 5 + 7= 4^2$$

a. What are the next three lines in this pattern?
b. The pattern suggests that if you add the first two odd numbers, you get 2^2; if you add the first three odd numbers, you get 3^2; and so on. What would the answer be if you added the first 10 odd numbers?

35. Consider the pattern

$$1^2= 1^3$$
$$(1 + 2)^2= 1^3 + 2^3$$
$$(1 + 2 + 3)^2= 1^3 + 2^3 + 3^3$$

a. What are the next three lines in this pattern?
b. What does this pattern suggest?

36. Consider the pattern

$$3^2 + 4^2 = \underline{\qquad}^2$$
$$3^3 + 4^3 + 5^3 = \underline{\qquad}^3$$

a. What do you think is the next line in this pattern?
b. Is the result you get in part (a) a true statement?

C Applications

37. At the age of 19, Galileo Galilei, an Italian astronomer, mathematician, and physicist, made discoveries that led to the invention of the pendulum clock. The following table lists the lengths of a series of pendulums having different swing times. Find the pattern that relates the length of the pendulum to the time of the swing.

Time of Swing	Length of Pendulum
1 sec	1 unit
2 sec	4 units
3 sec	9 units
4 sec	16 units

38. There is a pattern relating length of foot and shoe size. For men, the pattern is

Foot Length	9 in.	10 in.	11 in.	12 in.	13 in.	14 in.
Shoe Size	5	8	11			

a. Fill in the table with the appropriate numbers.
b. Matthew McGrory has an 18-in.-long foot. (He is in the *Guinness Book of Records*.) Follow the pattern and find his shoe size.

39. The pattern relating foot length and shoe size for women is

Foot Length	8 in.	9 in.	10 in.	11 in.	12 in.	13 in.
Shoe Size	3	6	9			

a. Fill in the table with the appropriate numbers.
b. Suppose a woman wears size 8 shoes. What is the length of her foot?

40. According to the Health Insurance Association of America, the average daily room charge by U.S. hospitals is as shown in the following table:

Year	1980	1985	1990	1995	2000	2005	2010
Daily Cost	$127	$212	$297				

a. Fill in the table with the appropriate numbers.
b. What is the pattern?
c. Do you think the pattern will continue after 1995?

In Other Words

41. Use a dictionary to find the definition of *deduction* and then explain in your own words the difference between inductive and deductive reasoning.

42. Explain in your own words the definition of the word *problem.*

43. Briefly describe an instance in which you used induction as a problem-solving strategy to solve a problem.

Using Your Knowledge

Leonardo Fibonacci, an Italian mathematician, wrote a book dealing with arithmetic and algebra in which he proposed the following problem: A pair of rabbits 1 month old is too young to produce more rabbits, but suppose that each month from its second month on, it produces a new pair of rabbits. If each pair produces a pair of rabbits and none of the rabbits die, how many pairs of rabbits will there be at the beginning of each month?

Beginning month 1 (1 pair)	1
End month 1 (1 pair)	1
End month 2 (2 pairs)	$1 + 1 = 2$
End month 3 (3 pairs)	$1 + 2 = 3$
End month 4 (5 pairs)	$2 + 3 = 5$

Online Study Center

To further explore drawing, access link 1.1.5 on this textbook's Online Study Center.

The resulting sequence 1, 1, 2, 3, 5, and so on, in which the first two terms are 1s and each succeeding term is the sum of the previous two terms, is called the *Fibonacci sequence.* Use your knowledge of patterns to do the following problems.

44. Write the first 12 terms of the Fibonacci sequence.

45. Find the sum of the first five terms of the sequence. How does this sum compare with the *seventh* term of the sequence?

46. Find the sum of the first eight terms of the sequence. How does this sum compare with the *tenth* term of the sequence?

47. Use induction to predict the sum of the first 12 terms.

Collaborative Learning

The Fibonacci numbers are related to the number of petals in certain flowers, the leaf arrangements in certain plants, the number of spirals in pine cones, and the arrangement of seeds on flower heads.

1. Have each student select one of the examples listed, examine it, and report to the rest of the class on his or her findings.

A bee colony consists of the *queen* ♀, *worker bees* (females who produce no eggs), and *drone bees* ♂ (males who do no work). Male bees are produced by the queen's unfertilized eggs, so male bees have a mother but no father! All the females are produced when the queen mates with a male, so females have two parents. The family tree of a female bee and that of a male drone bee are shown in the figures.

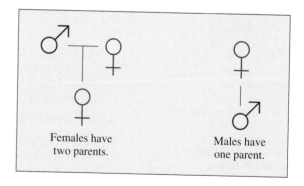

Females have two parents.   Males have one parent.

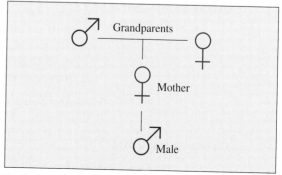

Family tree of a male (drone) bee.

2. Have one of the students in the class draw the next generation for a male bee, starting with the family tree of a male bee shown in the diagram. Then have another student draw the next generation, and so on.

3. Count the number of bees in the first generation (1), the second generation (1), the third generation (2), and so on. Compare the numbers you get with the Fibonacci numbers. What is your conclusion?

4. Here are the first 15 Fibonacci numbers.

$$1 \quad 1 \quad 2 \quad 3 \quad 5 \quad 8 \quad 13 \quad 21 \quad 34 \quad 55$$
$$89 \quad 144 \quad 233 \quad 377 \quad 610$$

Note: Every *third* Fibonacci number

$$2, 8, 34, 144, 610$$

is a multiple of 2.

a. Have a member of the class examine all the multiples of *3* in the list. Is there a similar conclusion regarding multiples of *3*?

b. Have another member of the class examine all the multiples of *5* in the list. What conclusion can be reached regarding multiples of *5*?

c. Have another member of the class examine all the multiples of *8* in the list. What conclusion can be reached regarding multiples of *8*?

On the basis of the observations made by the members of the class, what could be a general rule regarding the numbers in the Fibonacci sequence and their multiplicity?

A model of Galileo's escapement and pendulum made in 1883 following a design given by Galileo. The invention of the pendulum is generally credited to Galileo.

Online Study Center

 Research Questions

1. The formula

$$\frac{n^4 - 6n^3 + 23n^2 - 18n + 24}{24}$$

used in Example 3 to obtain the number of regions into which the interior of a circle is divided by a family of lines is discussed in the January 1988 issue of the *Mathematics Teacher* in an article entitled "Counting Pizza Pieces and Other Combinatorial Problems." Describe the techniques the authors used and how they arrived at the formula.

2. Galileo Galilei (mentioned in problem 37) is so well known that Rice University devotes a complete Web page to his life. Access link 1.1.6 on this textbook's Online Study Center and answer the following questions:

a. When and where was Galileo born?

b. According to legend, the inspiration for the discovery of the pendulum came to Galileo while in the Cathedral of Pisa. Describe the events and circumstances that led to the discovery of the pendulum, and state the year in which he made "his most notable discovery about the pendulum."

c. Name one more invention made by Galileo.

3. Another famous Italian mathematician was Leonardo Fibonacci (see Collaborative Learning).

a. What was Fibonacci's real name, where was he born, and where did he die?

b. Access link 1.1.7 on this textbook's Online Study Center, and name several instances in which the Fibonacci numbers appear in nature.

Online Study Center

1.2 Estimation: A Problem-Solving Tool

Have you attended a football game or other sporting event lately? How does the event staff estimate attendance? Does it matter? Read on and see.

The Million Man March took place on Monday, October 16, 1995, along the Mall in Washington, D.C. The National Park Service announced that 400,000 people took part in the event. The Nation of Islam, organizers of the march, vehemently objected, called it a gross underestimate of what was counted on-site to be a crowd 1.5 to 2 million strong, and threatened to sue the National Park Service. (Source: www.eomonline.com/ Common/Archives/February96/baz.htm.)

The crowd at the Million Man March and estimates of the number of persons in several different areas.

How was the controversy settled? Here are the problem-solving steps they used.

1. Ten color photographs of the crowd were collected.

2. The area of the mall in the photos was divided into square grids.

3. The crowd density was estimated on the basis of different degrees of packing in each square meter, ranging from six people per square meter to one person per five square meters.

4. The number of people per square meter was multiplied by the number of square meters (grids), and an estimate of *878,587* was reached, with an estimated error margin of 25% (about 219,647).

5. Therefore, the Million Man March had between 658,940 and 1,098,234 participants. Why?

Here are some more facts that would result in a better estimate. At what time were the photos taken (peak attendance was between 12 and 2 P.M.)? Some areas were more densely populated (six people per square meter) than others (one person per 10 square meters). What was the exact area of the mall? To see the answers to some of these questions and a revised crowd estimate, read the article cited in the photo credits at the end of this book. ▶

A. Estimation

As you can see from Getting Started, an important step in solving problems is to be able to estimate to make sure the answer you arrive at makes sense. One way to estimate, or approximate, an answer is to use round numbers. For example, suppose you want to buy 2 soft drinks (at $.99 each) and 2 bags of popcorn at $1.75 each. To estimate the cost of the items, you can round like this.

$.99 → $1   So the drinks are about 2 × $1 = $2.
$1.75 → $2   So the popcorn is about 2 × $2 = $4.
Your total purchase is about $6.

Some people prefer to write the procedure using the symbol ≈ to indicate an approximation. Thus, we can also write

$.99 cents ≈ $1   So the drinks are about $2.
$1.75 ≈ $2   So the popcorn is about $4.
The total purchase is still about $6.

Below is the procedure we use to round off numbers (left) with a worked out example on the right.

Rule for Rounding Numbers

Rule	*Example:*
	Round 258.34 to the nearest hundred.
1. **Underline** the place to which you are rounding.	1. **Underline** the 2: 2̲58.34.
2. If the first number to the *right* of the underlined place is 5 *or more*, **add** one to the underlined number. Otherwise, *do not change* the underlined number.	2. The first number to the *right* of 2 is 5, so we **add** 1 to the underlined digit 2 to get 3.
3. *Change* all the numbers to the *right* of the underlined number to zeros.	3. *Change* all the digits to the right of 3 to zeros, obtaining 300.00 or 300. Note that if you count by hundreds (100, 200, and so on), 258.34 is closer to 300 than to 200. The procedure is written as
	2̲58.34 → 300

GRAPH IT

Your grapher can round numbers for you; however, you have to know how to tell it what to do. Suppose you want to round the answer $31.84 of Example 1 to the nearest dollar. Tell the grapher you want to do math with a number by pressing MATH ► 2. Enter the 31.84 ' and tell the grapher you want no decimals by entering a 0. Press ENTER . You get 32! (See below.)

```
7.99+2.29+3.79+1
.89+8.69+7.19
              31.84
round(31.84,0)
                 32
```

EXAMPLE 1 ► Estimating the Amount of a Purchase

A student bought perfume for $7.99, nail enamel for $2.29, candy for $3.79, adhesive paper for $1.89, a curling iron for $8.69, and sunglasses for $7.19. Find a reasonable estimate of the total amount spent.

Solution

If we round each of the amounts to the nearest dollar, we have

| 7.99 → 8 | 3.79 → 4 | $8.69 → 9 |
| 2.29 → 2 | $1.89 → 2 | $7.19 → 7 |

Since $8 + 2 + 4 + 2 + 9 + 7 = 32$, a reasonable estimate for the total amount spent is $32. ∎

Note that in Example 1 we could have decided to round to the nearest dime, obtaining $8 + $2.30 + $3.80 + $1.90 + $8.70 + $7.20, or $31.90. This is a better estimate because the true cost is $31.84. However, *estimates are not supposed to give exact answers but rather tell us if the answers we are getting are reasonable.* Next, we look at a specialized type of rounding: electric meters.

How to Read Your Electric Meter

• Stand directly in front of your meter. Looking at dials from an angle can distort the reading.

• Read your meter dials from *right* to *left.*

• If the dial hand is between numbers, use the smaller of the two numbers.

• If the dial hand is positioned *exactly* on a number, look at the dial to the right to determine the correct reading. *Has the dial to the right passed zero?*

• If *no,* use the smaller number on the dial you're reading.

• If *yes,* use the number the hand is pointing to on the dial you're reading.

FIGURE 1.5

Do you see why the reading on the first dial of **B** in Figure 1.5 is 9 and not 0? It is because the pointer is on the number 0, which means that you must look at the dial to the right of it. That dial is before 9 and has not passed 0, so we use the smaller number on the dial we are reading—in this case 9 (not 10 or 0).

EXAMPLE 2 ▶ Reading Your Light Meter

(a) Read the meter dials for today as shown in Figure 1.6.

FIGURE 1.6

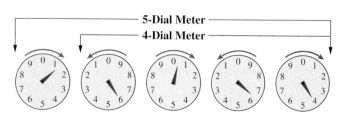

(b) If the reading yesterday was 16,003 kWh, how many kilowatt-hours have been consumed since yesterday?

(c) If a kilowatt-hour costs $.08, how much is your 1-day bill?

(d) If you estimate the same consumption each day for a 30-day period, what is your estimated monthly electricity bill?

Solution

(a) The reading is 1 6 0 6 4.

(b) The reading today is 1 6 0 6 4, and yesterday it was 1 6 0 0 3, so we have consumed 16,064 − 16,003 = 61 kWh.

(c) Your 1-day bill is 0.08 × 61 = $4.88.

(d) For 30 days it would be 30 × 4.88 = $146.40 because your 1-day bill is $4.88. ■

One of the most important estimations you can make is the estimation of your taxes. This estimation depends on your earnings and marital status. Let us use our problem-solving procedure to see how it works.

EXAMPLE 3 ▶ Estimating Taxable Income

Suppose you are single and estimate that the taxable income on line 5 of your tax return is $30,000. How much would you estimate your taxes to be?

Solution
We are going to use our problem-solving procedure to find the answer.

PROBLEM SOLVING

Calculating Estimated Taxes

❶ **Read** the problem.

We are looking for your estimated taxes.

❷ **Select** the unknown.

The unknown is the estimated amount of taxes you have to pay.

❸ **Think** of a plan.

We can use an estimated tax table from the Internal Revenue Service (IRS). The table is available on the Internet at www.irs.gov/pub/irs-pdf/f1040e01.pdf. (This table changes every year. Enter e04 for 2004, e05 for 2005, and so on.) It is called Form 1040 ES, Estimated Tax for Individuals. Select the table for singles.*

Single –Schedule X

Schedule X–Use if your **2005** filing status is **Single**			
If line 5 is:		**The tax is:**	*of the amount over–*
Over–	*But not over–*		
$0	$7,300	---------- 10%	$0
7,300	29,700	$730.00 + 15%	7,300
29,700	71,950	4,090.00 + 25%	29,700
71,950	150,150	14,652.50 + 28%	71,950
150,150	326,450	36,548.50 + 33%	150,150
326,450	---------	94,727.50 + 35%	326,450

continued

*For the latest table, go to www.irs.gov/pub/irs-pdf/f1040es.pdf.

❹ **Use** the techniques you are studying to carry out the plan.

The problem says that line 5 is $30,000, which is over $29,700 but not over $71,950 (second row). This means that the tax is $4090 + 25% of the amount over $29,700.

❺ **Verify** the answer.

Since your taxable income is $30,000, the amount over $29,700 is $30,000 − $29,700 = $300.

Thus, your tax is $4090 + 0.25($300)

$$= \$4090 + \$75$$
$$= \$4165$$

You can use an online calculator to verify this. (Caution: The $30,000 figure already includes a deduction for singles.)

Cover the solution, write your own solution, and then check your work.

Can you estimate the height of a person with only a bone as a clue? Anthropological detectives do! Suppose a detective finds a 17.9-in. femur bone (that is the one connecting the hip bone to the knee bone) from a male. To find the height H of its owner, use the formula $H = 1.88f + 32$, where f is the length of the femur bone. If the bone is 17.9 in. long, the owner's height H (to the nearest inch) must have been $H = 1.88(17.9) + 32 \approx 66$ in. ($\approx$ means "approximately equal to").

EXAMPLE 4 ▶ Estimating Height from Femur Length

A detective found a 17.9-in. femur bone. He is looking for a missing female 66 in. tall. The formula indicating the relationship between the femur of a female and her height is $H = 1.94f + 28$.

(a) Determine if the femur could belong to the missing female.

(b) Estimate a person's height (to the nearest inch) with a 17.9-in. femur bone.

(c) How much difference would there be in the height if you round off the measurements in the original formulas to the nearest whole number?

Solution

(a) Substituting 17.9 in $H = 1.94f + 28$, we obtain
$$H = 1.94(17.9) + 28$$
$$\approx 63 \text{ in.}$$

(b) The height of a person depends on whether the person is a male or a female.

If the person is a male, use $H = 1.88(17.9) + 32 \approx 66$ in.

If the person is a female, use $H = 1.94(17.9) + 28 \approx 63$ in.

Online Study Center

To further explore inferring height from bone lengths for various races and in the metric system, access link 1.2.1 on this textbook's Online Study Center.

(c) When rounded to the nearest whole numbers, the formulas would be:

Male: $H = 2f + 32 = 2(17.9) + 32 \approx 68$ in.
Female: $H = 2f + 28 = 2(17.9) + 28 \approx 64$ in.

Thus, the difference in the male's height with the new formula is 2 in. (68 instead of 66) and for the female is 1 in. (64 instead of 63). ∎

Now that we know that there is a relationship between the length of your bones and your height, is there a relationship between the length of your bones and your weight? Of course there is! However, a better measurement for your healthy weight is your body mass index (BMI). According to the National Heart, Lung and Blood Institute, you can figure out your BMI using this formula.

$$\text{BMI} = \frac{705W}{H^2}$$

where W is your weight (in pounds) and H is your height (in inches). In order to interpret your BMI, you use the following table.

If Your *BMI* Is	You Are
18 or below	Underweight
19–24	Normal
25–29	Overweight
30 or more	Obese

EXAMPLE 5 ▶ Estimating Your Body Mass Index

Suppose you weigh 162 lb and are 69 in. tall.

(a) Find your BMI to two decimal places.

(b) Refer to the table and interpret the results.

(c) Round the height, weight, and answer to the nearest 10, and use the formula

$$\text{BMI} = \frac{700W}{H^2}$$

to estimate your BMI.

Solution

(a) $\text{BMI} = \dfrac{705W}{H^2}$, where $W = 162$ and $H = 69$.

$$= \frac{705 \cdot 162}{69^2} \approx 23.99$$

(b) According to the table, if your BMI is between 19 and 24, your weight is "normal" for your height.

(c) When the height and weight are rounded to the nearest 10, $162 \rightarrow 160$ and $69 \rightarrow 70$ and

$$\text{BMI} = \frac{700 \cdot 160}{70 \cdot 70} = \frac{10 \cdot 16}{7} \approx 22.9, \text{ or } 20 \text{ (to the nearest 10)}$$

which is still "normal" (see the table).   ■

We have examined the weight of humans based on their BMI, but what is the "normal" weight for a horse? First, we have to figure out how to weigh a horse! According to Lon Lewis, author of *Feeding and Care of the Horse,*

> The importance of accurate equine weight estimates is unquestionable. Veterinarians, equine management facilities, stables, and individual horse owners rely on accurate weight information to determine proper

medication dosage, feed and nutrition considerations, racing performance, and transportation requirements.

Here is the formula you use to estimate the weight of a horse.

$$W \text{ (in pounds)} = \frac{G^2 \cdot L}{330}$$

where G (inches) is the horse's girth, the circumference of the horse's body about 4 in. behind its front legs, and L (inches) is its length (see Figure 1.7).

EXAMPLE 6 ▶ Estimating the Weight of a Horse

(a) Estimate the weight of a 65-in.-long horse with a 70-in. girth.

(b) A horse requires about 0.6 gal of water, 1 lb of hay, and $\frac{1}{2}$ lb of grain for each 100 lb of body weight daily. Estimate how much water, hay, and grain this horse needs.

Solution

(a) Substitute 70 for G and 65 for L in

$$W = \frac{G^2 \cdot L}{330}$$

obtaining

$$\frac{70 \cdot 70 \cdot 65}{330} \approx 965 \text{ lb}$$

Check this with a horse-weight calculator at link 1.2.2 on this textbook's Online Study Center.

(b) We have to estimate how many 100 lb of body weight the horse has. It is about $\frac{965}{100} = 9.65$. We can approximate the 9.65 to 10 to be safe. Thus, our horse needs 0.6 gal of water for every 100 lb of body weight; that is, $10 \times 0.6 = 6$ gal per day. The horse also needs 1 lb of hay for each 100 lb of body weight—that is, $10 \times 1 = 10$ lb of hay per day—and, finally, it needs $\frac{1}{2}$ lb of grain for each 100 lb of body weight—that is, $10 \times \frac{1}{2} = 5$ lb of grain per day.

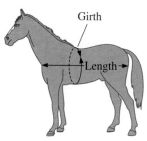

Girth

Length

Source: http://horses.about.com/cs/horsecare/a/eqweight2075.htm.

FIGURE 1.7

Online Study Center

To further explore horse feeding, access links 1.2.3 and 1.2.4 on this textbook's Online Study Center.

EXERCISES 1.2

Ⓐ Estimation

1. An investor owns 416.38 shares of a mutual fund valued at $30.28 per share. Find a reasonable estimate of the value of the investor's stock (to the nearest hundred dollars) by rounding 416.38 to the nearest 100 and $30.28 to the nearest dollar.

2. Water is sold in thousand-gallon units. If a unit of water costs $1.88 and 50.439 units were used, find a reasonable estimate of the bill to the nearest hundred dollars.

3. A student bought artichokes for $7.80, cucumbers for $2.29, lettuce for $3.75, tomatoes for $1.85, and broccoli for $2.90. Find a reasonable estimate of the total amount the student spent on vegetables by rounding each price to the nearest dollar.

4. A student bought a towel for $8.99, soap for $2.39, toothpaste for $3.79, shampoo for $1.79, a pair of shorts for $8.79, and a hat for $9.99. Find a reasonable estimate of the total purchases by rounding each quantity to the nearest dollar.

5. A herbicide is to be applied at the rate of 5.75 gal per acre. Find, to the nearest hundred gallons, a reasonable estimate for the amount of herbicide needed for $154\frac{1}{2}$ acres.

6. A bag of bahia grass covers 1.75 acres. What is a reasonable estimate of the number of acres that could be covered with $158\frac{1}{2}$ bags of seed? Answer to the nearest hundred acres.

7. To find the batting average A of a player (to three decimal places), divide the number of hits H by the number of times the player has been at bat (AB). Pete Rose holds the record for most career hits with 4256 in 14,053 at bats.
   a. What is Rose's career batting average?
   b. Estimate his average by rounding 4256 and 14,053 to the nearest hundred.

8. The highest batting average for a season belongs to Hugh Duffy, who played for Boston. He had 236 hits in 539 at bats.
   a. What was Duffy's batting average?
   b. Estimate his batting average by rounding the hits and the at bats to the nearest ten.

9. The *Guinness Book of Records* reports that the lowest earned run average (ERA) for a season belongs to Ferdinand M. Schupp, who in 1916 pitched 140 innings (IP) for New York and only allowed 14 earned runs (ER). If

$$ERA = \frac{9 \times ER}{IP}$$

what was Schupp's ERA rounded to three decimal places?

10. In 1914, Hubert "Dutch" Leonard of the Boston Red Sox gave up 25 earned runs in $222\frac{2}{3}$ innings. Find Leonard's ERA rounded to three decimal places. Is it lower than Schupp's?

The following information, from cooking.com, will be used in problems 11–16. As a rough average, one medium-sized, medium-hungry person could consume what is listed below.

4 to 8 hors d'oeuvres	$\frac{1}{4}$ lb vegetables
1 cup soup	$\frac{1}{3}$ lb rice, beans, etc.
$\frac{1}{4}$ head of lettuce	$\frac{1}{4}$ lb raw pasta
$\frac{1}{3}$ lb boneless meat or fish	$\frac{1}{4}$ cup gravy

Suppose you are planning a banquet for 100 people.

11. Estimate how many hors d'oeuvres you will need for the banquet.

12. Estimate how many cups of soup you will need for the banquet. If 1 gal is 16 cups, how many gallons of soup do you need?

13. Estimate how many pounds (to the nearest pound) of boneless meat or fish you will need for the banquet.

14. Estimate how many pounds (to the nearest pound) of rice, beans, and so on you will need for the banquet.

15. Estimate how many pounds of raw pasta you will need for the banquet.

16. Estimate how many cups of gravy you will need for the banquet. How many gallons is that?

17. a. Read the meter for today.
    b. If the reading yesterday was 5102, how many kilowatt-hours have been used?
    c. If electricity costs 8 cents per kilowatt-hour, how much is your 1-day bill?
    d. If you estimate the same consumption each day for a 30-day period, what is your estimated monthly electricity bill?

18. a. Read the meter for today.
    b. If the reading yesterday was 5501, how many kilowatt-hours have been used?
    c. If electricity costs 8 cents per kilowatt-hour, how much is your 1-day bill?
    d. If you estimate the same consumption each day for a 30-day period, what is your estimated monthly electricity bill?

19. a. Read the meter for today.
    b. If the reading yesterday was 6951, how many kilowatt-hours have been used?

c. If electricity costs 8 cents per kilowatt-hour, how much is your 1-day bill?

d. If you estimate the same consumption each day for a 30-day period, what is your estimated monthly electricity bill?

20. a. Read the meter for today.

b. If the reading yesterday was 6100, how many kilowatt-hours have been used?

c. If electricity costs 8 cents per kilowatt hour, how much is your 1-day bill?

d. If you estimate the same consumption each day for a 30-day period, what is your estimated monthly electricity bill?

21. On the basis of the table in Example 3, what would your estimated tax be if your taxable income were $40,000?

22. On the basis of the table in Example 3, what would your estimated tax be if your taxable income were $50,000?

Use the following information in problems 23–24.

The relationship between the length h of a humerus bone (the one connecting your shoulder to your elbow) and the height H of a person is given by

Male: $H = 2.89h + 27.81$
Female: $H = 2.75h + 28.14$

23. A detective found a 15-in. humerus bone belonging to a male.

a. How tall was the male?

b. What would the difference in height be if you rounded the measurements in the original formula to the nearest whole number?

24. Suppose the 15-in. humerus bone belonged to a female.

a. How tall was she?

b. What would the difference in height be if you rounded the measurements in the original formula to the nearest whole number?

25. Use the information in Example 5 to find the BMI for a person 68 in. tall and weighing 150 lb. What can you conclude from your answer?

26. Repeat problem 25 if the person is 70 in. tall and weighs 170 lb.

27. Use the information in Example 6 to estimate the weight of a 66-in.-long horse with a 70-in. girth. How much hay, grain, and water should this horse consume daily?

28. Repeat problem 27 for a 70-in.-long horse with a 70-in. girth. How much hay, grain, and water should this horse consume daily? Answers to the nearest whole number.

29. The Ohio Turnpike estimates that the annual cost C of routine maintenance per lane-mile is given by $C = 596 + 0.0019V + 21.7A$, where C is the annual cost of routine maintenance per lane-mile (in 1967 dollars), V is the volume of traffic on the roadway (measured in equivalent standard axle loads, ESAL, so that a heavy truck is represented as equivalent to many automobiles), and A is the age of the pavement in years since the last resurfacing. Estimate C (to the nearest dollar) when $V = 500,300$ ESAL and $A = 5$ years.

30. Repeat problem 29 for $V = 500,000$ and $A = 10$ years.

31. Can you estimate the age of your dog? One way to do it is to assume that if your dog is 1 year old, it would be the equivalent of 15 years old in human years. If your dog is 2 years old, it would be the equivalent of 24 years old in human years. After the second year, you add 4 dog-years for every actual year.

a. Estimate the equivalent human age of a 5-year-old dog.

b. Estimate the equivalent human age of a 10-year-old dog.

32. What about cats? According to the *Daily Cat*, here is the conversion:

Cat Years	Human Years
1	16–18
2	21–25
3	29

After the third year, add 4 cat years for every actual year.
   a. Estimate the equivalent human age of a 5-year-old cat.
   b. Estimate the equivalent human age of a 10-year-old cat.

You can estimate the distance between two points on a map by using a scale. In the accompanying map, each inch represents approximately 15 mi. Thus, the distance from the beginning of Interstate 90 to its intersection with Route 128 (about 1 in. on the map) represents an actual distance of 15 mi.

33. Estimate the distance between the intersection of 90 and 128 and the intersection of 90 and 495.

34. Estimate the distance between the intersection of 90 and 495 and the intersection of 90 and 290.

35. Estimate the distance between the intersection of 90 and 290 and the intersection of 90 and 86.

36. Estimate the distance between the intersection of 90 and 86 and the intersection of 90 and 32.

37. Estimate the distance between the intersection of 90 and 32 and the intersection of 90 and 91.

Use the following assumptions in problems 38–42. Your car makes about 20 mi/gal, and gasoline costs about $2.40/gal.

38. How much does it cost to travel the distance in problem 33?

39. How much does it cost to travel the distance in problem 34?

40. How much does it cost to travel the distance in problem 35?

41. How much does it cost to travel the distance in problem 36?

42. How much does it cost to travel the distance in problem 37?

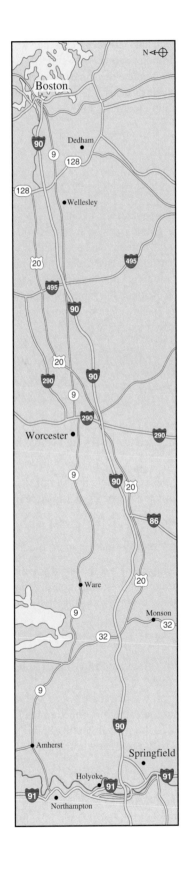

43. A gray whale eats about 268,000 lbs of amphipods (a marine crustacean) per month. In a month with 30 days, estimate how many pounds of amphipods a gray whale consumes in a day.

44. If the average weight of an amphipod is 0.004 oz, estimate how many amphipods a gray whale might eat in a day.

45. A 12,000-lb killer whale might eat as much as 14,000 lb of herring per month. In a month with 30 days, estimate to the nearest hundredth how many pounds of herring a killer whale would eat in a day.

46. If the average weight of a herring is 3.2 oz, estimate how many herring a killer whale would eat in a day. (Source: Oregon Coast Aquarium: www.aquarium.org/.)

Use the following information for problems 47–48. Your basal metabolic rate (BMR) is the amount of energy your body needs to maintain body temperature, breathe, and make your heart beat. To estimate the BMR, use the following formulas:

Male	Multiply the body weight by 10; add twice the body weight to this value.
Female	Multiply body weight by 10; add the body weight to this value.

47. What is the BMR for a 150-lb male?

48. What is the BMR for a 120-lb female?

 In Other Words

49. Write the procedure that you use to round the number 157 to the nearest hundred.

50. Explain, in your own words why the reading on the first dial is 9 and not 0.

 Using Your Knowledge

In Getting Started, we discussed how the number of participants in the Million Man March was estimated. How does the event staff estimate the number of people in the Tournament of Roses Parade? Below is the method Robert Gillette, a reporter for the *Los Angeles Times,* uses to calculate the number of people in a specific area.

He measures the depth of the standing-room area, which is 23 ft. Then he multiplies by the 5.5-mi parade route and doubles the amount because there are spectators on both sides of the street, obtaining 23 ft · 5.5 mi · 2. Unfortunately, the answer will be in feet · miles. 1 mi = 5280 ft.

51. Find 23 ft · 5.5 · 5280 ft · 2.

52. Assume that each spectator occupies 2 ft^2 of space (2 ft thick and 1 ft wide). How many spectators fit in the space you obtained in problem 51? That's the estimate Mr. Gillette provides!

 Research Questions

1. We have mentioned the estimation of crowds in relation to the Million Man March and the Tournament of Roses Parade. But there is another way of estimating crowds. Looking at the garbage they leave behind! To further explore how to estimate a crowd, access link 1.2.5 on this textbook's Online Study Center. Then list the top five "garbage" parades of all time in New York, according to Casey Kasem.

2. How many people are there in the United States for each square kilometer? Outline how to answer this question. To learn how to make the outline, access link 1.2.6 on this textbook's Online Study Center.

Collaborative Learning

Don't Eat the Beans!

How many fish in the lake? How many deer in the forest? We cannot let you go fishing, capture every fish in the lake, and count them, let alone catch deer in the forest. In this Collaborative Learning, we will discuss an **estimation** method called **capture-recapture** and simulate the fishing and hunting. You need several bags of lima beans, one for each group (1-lb bags are ideal), and a nontoxic washable marker. In this investigation, the bag of beans represents the lake or the forest. The objective is to find out the total number of beans (fish or deer) in the bag. Here are the steps for each group:

Step 1 Reach into the lake, remove T fish, and tag them by marking them with the marker. Return the fish to the lake (meaning: put beans back in the bag!).

Step 2 Allow the fish to mingle (shake the bag!), remove a handful of fish H and count them. Count the number of tagged (marked) fish M in the handful.

Step 3 Make a table like the one below.

Group	Number (T) of Tagged	Number (H) in Handful
1		
2		
.		
.		
.		

The method assumes that the ratio of the actual population P to the sample size T is the same as the ratio of the number of marked fish H to the number marked in the recapture sample M; hence, $PM = TH$. Thus, the formula for the population is

$$\text{Population} = \frac{TH}{M}$$

This method of estimation is called the *Lincoln Index*. If you marked the beans with a nontoxic marker, you may reuse them; otherwise, *don't eat the beans*.

You can calculate your percent of error by counting the beans in the bag (B), recording by how many beans your count is off (O), and calculating the percent O/B.

Online Study Center

To further explore sampling, access link 1.2.7 on this textbook's Online Study Center.

1.3 Graph Interpretation: A Problem-Solving Tool

Do you have a budget? What are your main expenses? The **circle graph (pie chart)** on the next page (Figure 1.8) gives general guidelines to establish your budget. In a circle graph, a circle is divided into sectors (wedges) that are *proportional* to the size of the category. The information can also be presented using a **bar graph** (Figure 1.9), in which the size of the categories is proportional to the *length* of the bars.

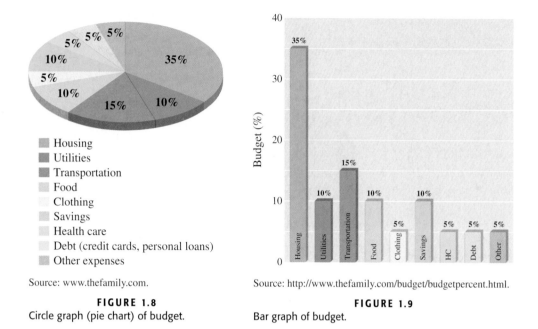

Source: www.thefamily.com.

FIGURE 1.8
Circle graph (pie chart) of budget.

Source: http://www.thefamily.com/budget/budgetpercent.html.

FIGURE 1.9
Bar graph of budget.

Now, which category in the budget above is the main expense? From either the circle or the bar graph, you can see that the greatest percentage of the money will go toward housing. What about the category that receives the least percentage of the money? There are four of them. Can you name them?

In this section we will concentrate on solving problems that involve the interpretation of different graphs: circle (pie), bar, and line. ▶

A. Interpreting Circle Graphs (Pie Charts)

Can you predict the weather using a circle graph? The Bureau of Meteorology does this in Australia! The following charts (Figure 1.10) tell us how often Australia will have low (dry), normal, or high (wet) weather, depending on the type of year (El Niño, Normal, or La Niña). Now, if you know that you are having an El Niño year, what can you say about rainfall? The probability that it will be dry is about 50% (half the circle), wet 17%, and normal 33%. Can you predict what will happen if you know that you are having a La Niña year?

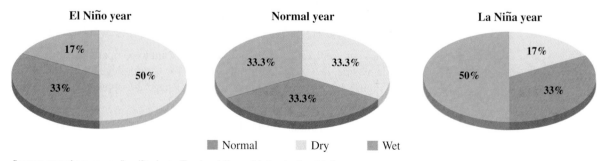

Source: www.bom.gov.au/lam/Students_Teachers/climprob/rainprbprim.shtml.

FIGURE 1.10

EXAMPLE 1 ▶ Interpreting Circle Graphs

The two circle graphs (Figure 1.11) show the ethnic makeup of schools in California for fiscal years 1981–1982 and 2003–2004.

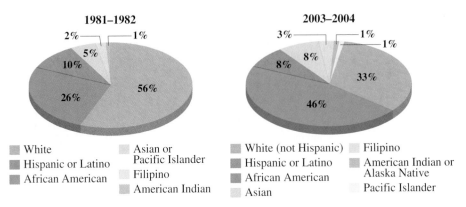

1981–1982

2% — 1%
5%
10%
26%
56%

■ White
■ Hispanic or Latino
■ African American
 Asian or Pacific Islander
 Filipino
 American Indian

2003–2004

3% — 1%
1%
8% 8%
33%
46%

■ White (not Hispanic)
■ Hispanic or Latino
■ African American
 Asian
 Filipino
 American Indian or Alaska Native
 Pacific Islander

FIGURE 1.11

(a) Which category has grown the most?

(b) Which category has grown the least?

(c) Which category has decreased the most?

(d) If you were in charge of the English as a Second Language (ESL) or English for Academic Purposes (EAP) program, which type of language proficiency would you look for in your teachers?
(Source: www.cde.ca.gov/re/pn/fb/documents/factbook2005.pdf.)

Solution

Remember, you can follow the RSTUV procedure, so Read carefully, and Select the unknowns, which are certain types of categories. The plan is to inspect the chart.

(a) Hispanic or Latino has grown the most (from 26% to 46%).

(b) American Indian or Alaska Native (American Indians were 1% in 1981 and American Indians *plus* Alaska Natives are still 1%, so it can be inferred that American Indians grew less than 1%. See how you can use deduction!).

(c) White (56% to 33%).

(d) You should look for teachers proficient in Spanish. ■

B. Interpreting Bar Graphs

Sometimes bar graphs show categories that you want to compare by drawing the bars side by side. Did your family save money for your college education? How much? Figure 1.12, the bar graph at the top of page 29, shows three different household categories:

    Saved less than $5000   Saved $5000 to $19,999   Saved $20,000 or more

Which is your household category? What is the percent difference between the households that saved less than $5000 and those that saved $20,000 or more?

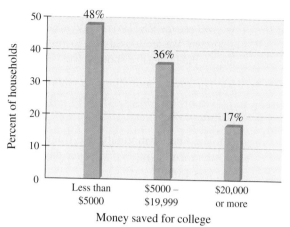

By Darryl Haralson and Jeff Dionise, *USA TODAY*. Source: State Farm.

FIGURE 1.12

What is the percent difference between the households that saved the most and those that saved the least?

Let us interpret more bar graphs. Does drinking shorten your life span? A 12-year-long study of over 200,000 men found that subjects who had consumed alcohol in moderation were less likely to die than those who abstained from alcohol. Let us see how this can be deduced.

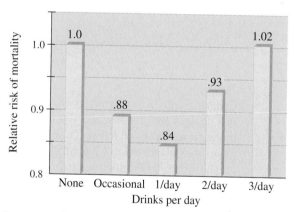

Source: www2.potsdam.edu/alcohol-info/AlcoholAnd.html.

FIGURE 1.13

EXAMPLE 2 ▶ Interpreting the Risk of Mortality

The bar graph (Figure 1.13) shows the relative risk of mortality for 200,000 men.

(a) Which group has the lowest mortality rate, and what is this mortality rate?

(b) Which group has the highest mortality rate, and what is this mortality rate?

(c) What is the numerical difference between the lowest and highest relative risk of mortality?

Solution

We will use the RSTUV method. Read the problem carefully, and then Select the unknown. We are looking for lowest or highest mortality rates, so we are looking for the longest or shortest bars. The plan: Look at the graph.

(a) The lowest mortality rate (shortest bar) corresponds to the third category, the men having 1 drink per day (1/day). The mortality rate for this group was 0.84.

(b) The highest mortality rate (longest bar) corresponds to the fifth category, the men having 3 drinks per day (3/day). The mortality rate for this group was 1.02.

(c) The numerical difference between the lowest (0.84) and highest (1.02) relative risk of mortality is $1.02 - 0.84 = 0.18$.   ■

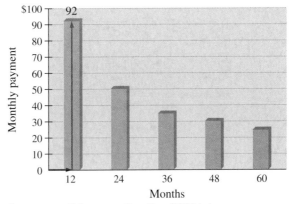

Source: www.dinkytown.net/java/PayoffCC.html.

FIGURE 1.14

Can bar graphs save you money? Suppose you have a $1000 credit card balance that charges 18% interest and you only make the minimum $25 payment each month. It will take you forever (well, actually 5 years) to pay it off. If you decide to pay the balance off in 12 months, how much will your payments be? The bar graph (Figure 1.14) tells you, provided you know how to read it!

First, start at the 0 point and move right horizontally until you get to the category labeled 12 months (blue arrow), and then go up vertically to the end of the bar (red arrow). According to the vertical scale labeled Monthly payment, the arrow is 92 units long, which means that the monthly payment will be $92 per month. How much money will you save? In 60 months, making the minimum $25 payment, your total payments would be $60 \times \$25 = \$1500$. In 12 months, making $92 payments, you will pay $12 \times 92 = \$1104$. Thus, the savings are $\$1500 - \$1104$, or $396.

EXAMPLE 3 ▶ Estimating Savings on Your Credit Card

Find the savings if you decide to pay the balance in 24 months.

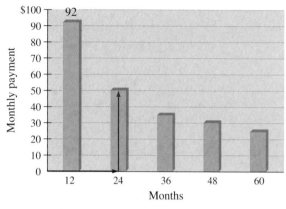

FIGURE 1.15

Solution

Review Figure 1.15 to find the savings! To find the payment corresponding to 24 months on the horizontal axis, move *right* to the category *labeled* 24 months and then *vertically* to the end of the bar. According to the vertical scale, the monthly payment will be $50. If you pay $50 for 24 months, you would pay 24 × $50 = $1200 and have savings of $300 ($1500 − $1200). Remember, when paying off any debts, the faster you pay, the more interest you save!   ▪

C. Interpreting Line Graphs

We have already discussed how to interpret circle and bar graphs. Now we will learn how to interpret line graphs. Look at the graph in Figure 1.16.

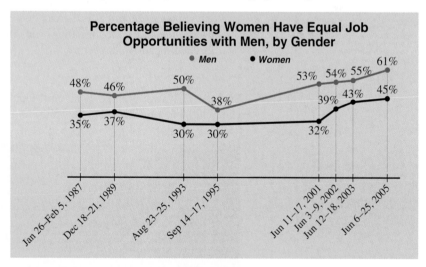

FIGURE 1.16

Do women have equal job opportunities? It depends on who you ask and when! For example, in 1987, not even 50% of men or women thought so, but in 2005, 61% of the men and 45% of the women did.

EXAMPLE 4 ► Interpreting Equal Opportunity

(a) In what year did the largest percent of women believe that they had achieved equality of opportunity? What about men?

(b) In what year did the smallest percent of women believe that they had achieved equality of opportunity? What about men?

(c) In what year was the difference of opinion greatest? smallest?

(d) Which gender (men or women) is more likely to believe that women have achieved equality of opportunity in the work force?

Source: The Gallup Organization, "Gender Difference in Views of Job Opportunities," by Jeffrey Jones.

Solution

(a) First, note that the graph for women is at the bottom in black. Follow the black line until you encounter the *largest* percent, which is 45%. The 45% is *over* the date June 6–25, **2005.** Thus, the year in which the *largest* percent of women believed that they had achieved equality of opportunity is **2005.** For men, you have to follow the green line. The *largest* percent on the green line is 61%, also occurring in **2005.**

(b) Follow the same procedure as in part (a), except we are looking for the *smallest* percent on the black line, which is 30% (occurring in **1993 and 1995**). For men (green line), the smallest percent is 38% in **1995.**

(c) Now we are looking for the *greatest* gap (distance) between the black and green lines. This occurs in **2001.** You can make this precise by looking at the percent difference at each date. For **2001,** the difference is 53% − 32%, or 21 percentage points. The *smallest* difference occurs in **1995** (38% − 30%, or 8 percentage points).

(d) Men are more likely to believe that women have achieved equality of opportunity in the work force. The green line (representing men) is always higher than the black line (representing women), so the percentage of males with this belief is always higher.   ∎

EXAMPLE 5 ▶ Chill Out!

Figure 1.17 shows the new wind-chill temperatures (top, yellow) and the old wind-chill temperatures (bottom, white) for different wind speeds.

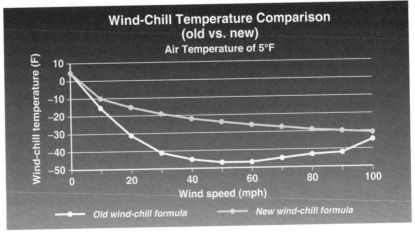

Source: www.erh.noaa.gov/er/iln/tables.htm.

FIGURE 1.17

(a) If the wind speed is 90 mph, what is the approximate new wind-chill factor?

(b) If the wind speed is 90 mph, what is the approximate old wind-chill factor?

(c) When the wind is 90 mph, what is the wind-chill temperature difference between the new and the old?

Solution

(a) Go to 90 on the horizontal axis and up vertically until you meet the yellow graph. This occurs at about $-30°F$.

(b) The old wind chill is slightly under $-40°F$ (be careful here, the numbers go from $-40°F$ to $-50°F$), so we estimate the answer as $-42°F$.

(c) The difference between the two is about $12°F$ (from $-30°F$ to $-42°F$). ∎

EXERCISES 1.3

A Interpreting Circle Graphs (Pie Charts)

In problems 1–10, answer the questions about the circle graph.

1. Do you have a job? Which method of transportation do you use to get to work? The chart shows the different modes of transportation used by people going to work in England.

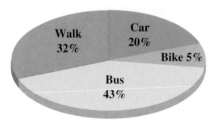

Source: www.learn.co.uk/default.asp? WCI=Unit&WCU=307.

a. What is the most preferred mode of transportation?

b. What is the least preferred mode of transportation?

c. In Dallas–Fort Worth, about 91% of the people drive to work. What is the percent difference between people driving to work in Dallas–Fort Worth and in England?

2. The following circle graph is divided into 12 equal parts (slices) that illustrate daily activities.

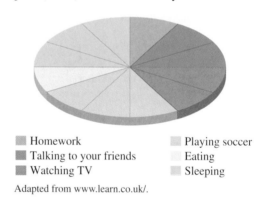

Homework   Playing soccer
Talking to your friends   Eating
Watching TV   Sleeping

Adapted from www.learn.co.uk/.

a. How many parts (slices) were spent sleeping?

b. How many parts (slices) were spent watching TV?

c. Which activities took the most time?

d. Which activity took the least time?

e. What fraction of the time was spent eating? Remember that the pie has 12 equal parts (slices).

f. What fraction of the time was spent doing homework?

3. The circle graph that follows shows the percent of different types of cheese produced.

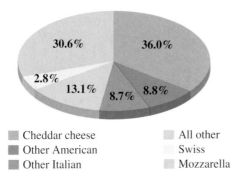

- ■ Cheddar cheese
- ■ Other American
- ■ Other Italian
- ▨ All other
- ▨ Swiss
- ▨ Mozzarella

Source: U.S. Department of Agriculture.

a. Which type of cheese was produced the most?
b. Which type of cheese was produced the least?
c. If you assume that the cheese that is produced the most is also the most popular, which is the second most popular cheese?

4. What does a stellar sea lion eat? The circle graph illustrates the answer. Suppose you are in charge of feeding the stellar sea lions in the zoo.

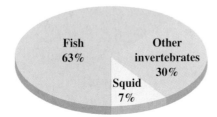

Source: www.saburchill.com/facts/facts0014.html.

a. Which food would you stock the most?
b. If you buy 100 lb of sea lion feed, how many pounds should be squid?
c. If you buy 200 lb of stellar sea lion feed, how many pounds should be squid?
d. If octopuses, shrimp, and crabs, preferably in the same amounts, fall in the category of "other invertebrates" and you buy 300 lb of stellar sea lion feed, how many pounds of crab should it contain?
e. A male stellar sea lion weighs about 2200 lb and eats about 200 lb of food each day. How many pounds of fish does he eat?
f. A female stellar seal, on the other hand, weighs about 600 lb and eats 50 lb of food each day. How many pounds of squid does she eat each day?
g. How many pounds of shrimp does a female stellar seal eat every day?

Online Study Center

To further explore stellar sea lions, access link 1.3.1 on this textbook's Online Study Center.

5. The graph shows the average indoor water use in the United States.

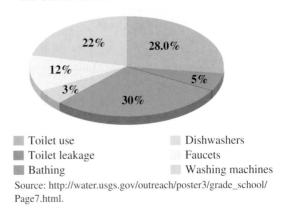

- ■ Toilet use
- ■ Toilet leakage
- ■ Bathing
- ▨ Dishwashers
- ▨ Faucets
- ▨ Washing machines

Source: http://water.usgs.gov/outreach/poster3/grade_school/Page7.html.

a. Where is water used the most?
b. If you use 500 gal of water, how much would you use for bathing?
c. Which uses more water, the dishwasher or a toilet leak?
d. If your dishwasher used 5 gal of water, how much water would be used by the faucets?

6. The pie chart shows the fraction of each ingredient (by weight) used in making a sausage pizza.

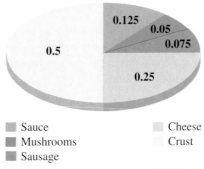

- ■ Sauce
- ■ Mushrooms
- ■ Sausage
- ▨ Cheese
- ▨ Crust

Source: www.mathleague.com/help/data/data.htm.

a. What fraction of the pizza is crust?

b. What fraction of the pizza is cheese?

c. Which ingredient makes the smallest part of the pizza by weight?

d. If you estimated that a pizza weighs 4 lb, how many pounds would be crust and how many pounds cheese?

e. If you were to make 100 of these 4-lb pizzas, how many pounds of cheese would you need?

7. Have you looked in your trash lately? You have an average trash can if your trash divides into the same percents as those shown.

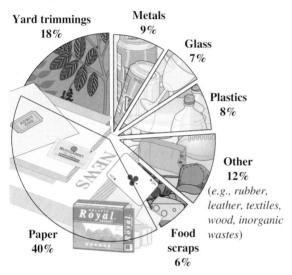

Source: www.epa.gov/epaoswer/non-hw/reduce/catbook/what.htm.

a. What is the most prevalent item in average trash?

b. Which is the second most prevalent item in average trash?

c. If you have an average trash load weighing 50 lb, how many pounds of paper would it contain? How many pounds of yard trimmings? Actually, you probably recycle and do not have as much paper!

8. What is your favorite sport to watch? The charts show the results of a Gallup Poll comparing viewership in December 2004 and November 1998.

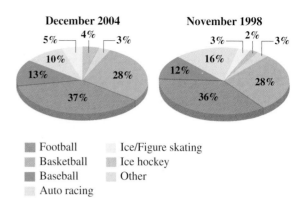

Football	Ice/Figure skating
Basketball	Ice hockey
Baseball	Other
Auto racing	

In which sports did the percent of viewers

a. stay the same?

b. increase? By what percent?

c. decrease? By what percent?

d. If you were the programming director of a sports network and each percent of increase or decrease in viewership was worth $1 million when renewing contracts, how much would the contract of each of the sports change?

9. The circle graph shows the breakdown of how the world produces its energy.

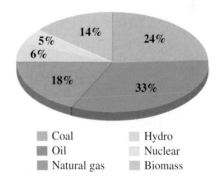

Coal	Hydro
Oil	Nuclear
Natural gas	Biomass

Source: www. envirolink.org/orgs/edf/sitemap.html.

a. Which energy source produces the most energy?

b. Which energy source produces the least energy?

c. Fossil fuels (coal, oil, and natural gas) emit greenhouse gases when burned. Which of the three fossil fuels produces the least energy?

10. Suppose you paid **$10,000** in federal income taxes. The chart shows where the money went!

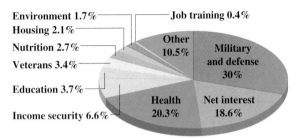

Source: www.nationalpriorities.org/auxiliary/ interactive taxchart/taxchart.html.

a. Where did most of the money go?
b. What percent of the money went to health?
c. How much money went to health?
d. Which category received the least money?
e. How much money went to education?
f. What is the difference between the amount of money spent for military and defense and the amount of money spent for education?

B Interpreting Bar Graphs

In problems 11–22, answer the questions about the bar graphs.

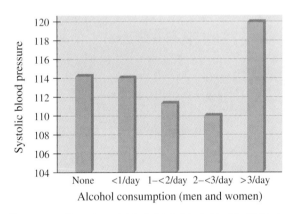

Source: www2.potsdam.edu/alcohol-info/AlcoholAndHealth.html.

11. Can moderate alcohol consumption (yes, we said *moderate*) reduce blood pressure? Judge for yourself by examining the bar graph that illustrates the average systolic blood pressure among young adults. Find the average systolic blood pressure for young adults consuming

a. no drinks per day.
b. less than 1 drink per day ($<$1/day).
c. at least 1 but less than 2 drinks per day (1–$<$2/day).
d. Which category has the lowest blood pressure? What is the measure of the blood pressure?
e. Which category has the highest blood pressure? What is the measure of the blood pressure?

12. People who abstain from drinking have double the risk of a stroke as drinkers. How can we deduce this from the bar graph? Look at the vertical scale (0 to 3.5).

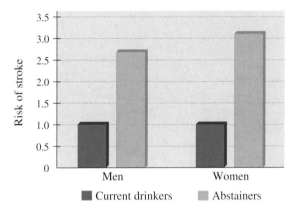

Source: www2.potsdam.edu/alcohol-info/AlcoholAndHealth.html.

a. What is the risk of stroke for current drinking men?
b. What is the risk of stroke for men who abstain?
c. Estimate how many more times the risk of stroke is for abstaining men than for current drinking men?
d. What is the risk of stroke for current drinking women?
e. What is the risk of stroke for abstaining women?
f. Estimate how many more times is the risk of stroke for abstaining women than for current drinking women.
g. Which category has the highest risk of stroke?

13. The bar graph shows the number of traffic accident victims who died at the scene of the accident and the blood alcohol level (BAL) of the driver.

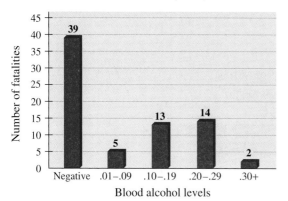

Source: www.metrokc.gov/health/examiner/2003report/accident.pdf.

a. What was the number of fatalities with a negative (no) blood alcohol level?

b. In many states, a person is legally drunk if his or her BAL is 0.10 or more. How many people were legally drunk? (In some states, 0.08 or more is legally drunk.)

c. What was the most prevalent BAL range for the people who were legally drunk? How many persons had that BAL?

14. At what time do fatal accidents occur? Refer to the graph below.

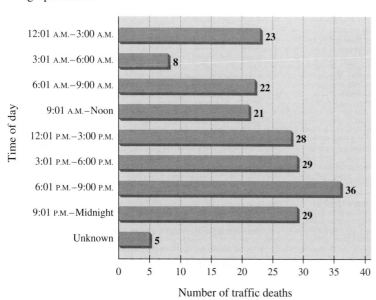

Number of traffic deaths

Source: www.metrokc.gov/health/examiner/2003report/accident.pdf.

a. Find the number of fatalities between 12:01 and 3:00 A.M.

b. Find the number of fatalities between 3:01 and 6:00 A.M.

c. What is the most likely time period for a fatal traffic accident to occur?

d. Aside from "unknown," what is the least likely time period for a fatal traffic accident to occur?

15. The graph at the top of page 38 indicates the number of traffic fatalities and age of the decedent.

a. In which age group are most of the fatalities?

b. Which age group has the least (nonzero) fatalities? Why do you think that is?

c. Are there more fatalities involving people who are less than 50 years old or more than 50 years old?

d. Which age group had only two fatalities? Why?

Graph for problem 15 on page 37.

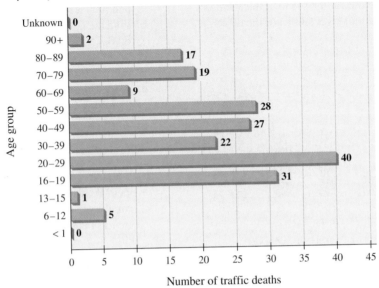

Number of traffic deaths

Source: www.metrokc.gov/health/examiner/2003report/accidents.pdf.

16. Do you have a computer at home? The bar graph below shows the percent of U.S. households with a computer in 1998 and in 2000.

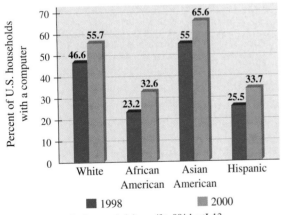

■ 1998   ■ 2000

Source: www.ntia.doc.gov/ntiahome/fttn00/chartI-13. htm#f13. NTIA and ESA, U.S. Department of Commerce, using U.S. Bureau of the Census Current Population Survey supplement.

a. Which ethnic group had the most households with a computer? In what year?

b. Which ethnic group had the fewest households with a computer? In what year?

c. If you were the marketing director for a computer manufacturer, to which ethnic group would you direct your advertisements?

17. The graph below shows the percent of U.S. households with Internet access by income in thousands.

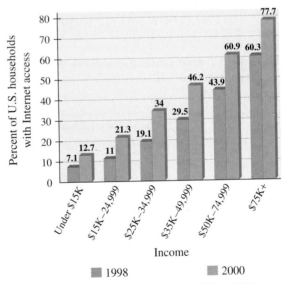

Income

■ 1998   ■ 2000

Source: http://www.ntia.doc.gov/ntiahome/fttn00/chartI-6.htm.

a. Which income group had the highest percent of households with Internet access? What percent of that group had Internet access in 2000? In 1998?

b. Which income group had the lowest percent of households with Internet access? What percent of that group had Internet access in 2000? In 1998?

c. What was the difference in income (in thousands) between the category with the highest percent of U.S. households with Internet access and the category with the lowest percent of U.S. households with Internet access?

18. Which has more calories, ice cream or yogurt? The graph shows the number of calories for $\frac{1}{2}$ cup of ice cream or yogurt.

Source: http://journalism.smcvt.edu/echo/03-07-02/yogurticecream.htm.

a. How many calories are in the Cherry Garcia ice cream?

b. How many calories are in the Cherry Garcia yogurt?

c. How many calories are in the Cookie Dough ice cream?

d. How many calories are in the Cookie Dough yogurt?

e. Which is the product with the least calories?

f. Which is the product with the most calories?

19. Refer to the graph in problem 18.

a. How many calories would you save if you ate a whole cup of the product with the least calories instead of a cup of the product with the most calories?

b. How many extra calories would you consume when you have a pint of the product with the most calories in the graph instead of the product with the least calories? *Hint:* 2 cups = 1 pint.

20. What is the most popular spectator sport in Japan? A survey of 3000 Japanese aged 20 or older says that it is baseball! The bar graph for the survey is shown below.

a. What is the second most popular sport? What percent of the respondents said they preferred high school baseball?

b. What was the least popular spectator sport in the survey? What percent of the people preferred this sport?

c. Name the three sports that enjoyed about the same popularity in the survey.

d. How many more people preferred Japanese professional baseball than major league baseball?

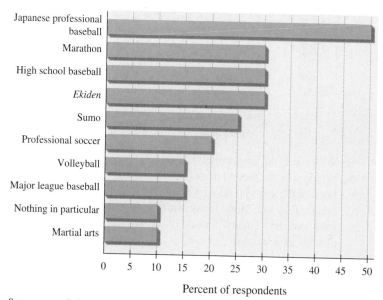

Source: www.jinjapan.org.

21. La Cubanita Cafe has the following breakfast menu, and the items sold are displayed in the graph.

Breakfast 7 am to 11 am

Cuban Toast
Cheese Toast
Plain Omelette
Western Omelette
Ham Omelette
Bacon Omelette
Sausage Omelette
Cheese Omelette
Turkey Omelette
Cafe Con Leche   12 oz sm
                        16 oz Lg
Hot Chocolate   12 oz sm
                        16 oz Lg
Espresso
All Omelettes served on Cuban Toast
Side items
Bacon
Add cheese
Seasoned Home Fries

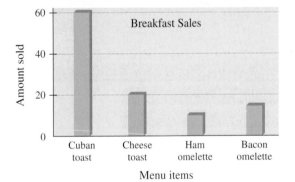

a. Which was the most popular item sold?
b. Which was the second most popular item?
c. Each of the items is served with $\frac{1}{4}$ loaf of Cuban bread. Estimate how many breakfasts were sold and how many loaves of bread were needed for breakfast.

22. La Cubanita Cafe has the following sandwich lunch menu, and the items sold are displayed in the graph.

Sandwiches	Lunch
Cuban	3.59
Cuban special (includes croquettes)	4.19
Media noche	3.49
Steak sandwich	4.59
Breaded steak sandwich	4.59
Turkey	4.29
Turkey club	4.59
BLT	3.99
Chicken	4.59
Vegetarian	3.50
Ham & cheese	3.59
"ALEX" Cuban special	
(regular Cuban w/fresh roast pork)	4.59
Pork sandwich	4.25
add lettuce & tomato	.45
add cheese	.55

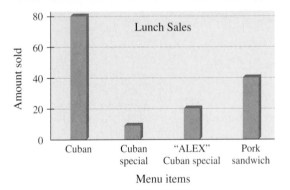

a. Which was the most popular sandwich sold?
b. Which was the least popular sandwich sold?
c. Each sandwich requires $\frac{1}{4}$ loaf of Cuban bread. Estimate how many sandwiches were sold and how many loaves of bread were needed to make those sandwiches.

C Interpreting Line Graphs

In problems 23–36, answer the questions about the line graph.

23. Are you poor? If you are a family of one, the U.S. Department of Health and Human Services says that you are poor (below poverty level) if your annual income is less than $8350. The graph shows the percent of persons below the poverty level.

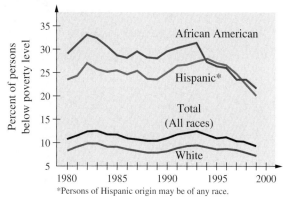

*Persons of Hispanic origin may be of any race.

Source: U.S. Census Bureau.

a. Which group has the highest percent of persons below the poverty level?
b. Which group has the lowest percent of persons below the poverty level?
c. Estimate the percent difference between the highest and lowest percents for persons below the poverty level in 1980. In 1999.
d. By examining the graph, are the poverty levels getting better or worse? Why?

24. The graph shows the annual percent change in consumer price indexes.

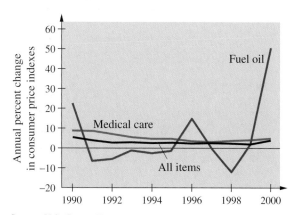

Source: U.S. Census Bureau.

a. Which item has the highest percent change in the year 2000? What is the percent change?
b. Estimate the percent decline in medical care from 1990 to 2000.

25. The graph shows the years of school completed by persons 25 years old and over.

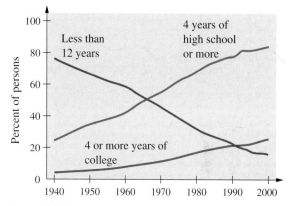

Source: U.S. Department of Commerce.

a. What percent of the people had 4 years of high school or more in 1940? In 2000? What is the percent difference?
b. What percent of the people had 4 or more years of college in 1940? In 2000? What is the percent difference?
c. What are the trends for each of the three categories shown in the graph?

26. Do you want to be a teacher? The graph shows the average annual salary for public elementary and secondary school teachers.

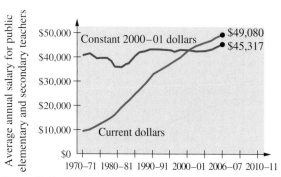

Source: National Education Association, NCES.

a. What was the average salary (current dollars) in 1970–1971? In 2006–2007? What is the dollar difference?
b. What was the average salary (constant dollars) in 1970–1971? In 2006–2007? What is the dollar difference?
c. Are salaries going up? Explain.

27. According to the Census Bureau, a family of one is poor if it makes less than $10,000 a year. The chart shows the poverty rate by age.

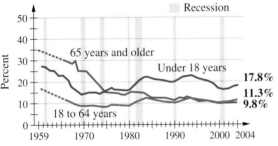

Poverty Rates by Age: 1959 to 2004

Note: The data points are placed at the midpoints of the respective years. Data for people 18 to 64 and 65 and older are not available from 1960 to 1965.
Source: U.S. Census Bureau, Current Population Survey: 1960–2005 Annual Social and Economic Supplements.

a. Which age group had the highest percent of poor in 2004?

b. Which age group had the lowest percent of poor in 2004?

c. What percent of people 65 years of age and older were poor in 1966?

d. From 1966 to 2004, which age group showed a decrease in their poverty rate?

e. From 1966 to 2004, which age group showed an increase in their poverty rate?

f. From 1966 to 2004, which age group stayed about the same in their poverty rate?

28. The graph shows the evolution of VCRs, DVD players, and DVD recorders for 2000–2006.

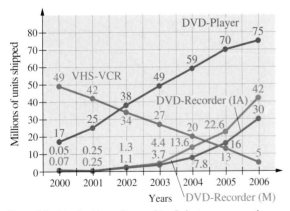

Notes: M = Matsushita estimates; IA = Industry average estimates. MMIS data posted January 2004; graph 2004-DVDV-100.
Source: http://mmislueck.com/Archives/011204.htm.

a. What was the peak and lowest years for VCR player sales?

b. What about for DVD player sales?

c. What was the difference in units shipped between DVD and VCR players in 2006?

d. What was the difference in units shipped between M and IA estimates in 2006?

29. How old is your dog? The graph gives the relationship between a dog's age (in human years) and a dog's age in dog years. Thus, if a dog is 1 year in human years, it is about 12 years in dog years.

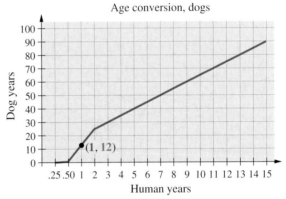

Age conversion, dogs

Source: www.k9clips.com/age_chart.html.

a. If a dog is 9 years old in human years, how old is it in dog years?

b. If retirement age is 65 for humans, what is the equivalent retirement age for dogs in human years? (Answer to the nearest whole number.)

c. If the drinking age for humans is 21, what is the equivalent drinking age for dogs in human years? (Answer to the nearest whole number.)

30. We need equal time for cats! The graph gives the relationship between a cat's age (in human years) and a cat's age in cat years.

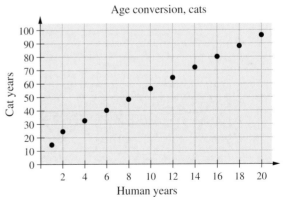

Age conversion, cats

Source:www.cnn.com/US/9802/03/fringe/old.cat. From www.cats.alpha.pl/catsage.htm.

a. If a cat is 6 years old in human years, how old is it in cat years?

b. If retirement age is 65 for humans, what is the equivalent retirement age for cats in human years? (Answer to the nearest whole number.)

c. If the drinking age for humans is 21, what is the equivalent drinking age for cats in human years? (Answer to the nearest whole number.)

d. Comparing your answers to problem 29, who would retire first (in human years), a cat or a dog?

31. Do you know what a badger is? It is a small burrowing animal that looks like a weasel.

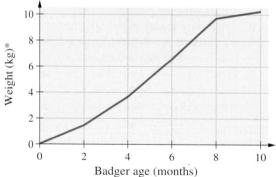

*1 kilogram ≈ 2.2 lb

Source: www.knockonthedoor.com.

a. To the nearest kilogram, how much does a badger weigh when it is 4 months old? (1 kg ≈ 2.2 lb)

b. To the nearest kilogram, how much does a badger weigh when it is 8 months old?

c. At approximately what age do badgers stop growing?

Use the following graph for problems 32–36. The graph shows the amount owed on a $1000 debt at an 18% interest rate when the minimum $25 payment is

made (blue) or when a new monthly payment of $92 is made (red). Thus, at the current monthly payment of $25, it will take 60 months to pay off the $1000 balance. On the other hand, with a new $92 monthly payment, you pay off the $1000 balance in 12 months.

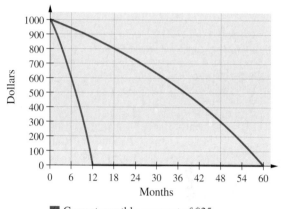

 Current monthly payment of $25
New monthly payment of $92

Source: www.dinkytown.net/java/PayoffCC.html.

32. What is your balance after 6 months if you are paying $25 a month?

33. What is your balance after 6 months if you are paying $92 per month?

34. What is your balance after 18 months if you are paying $25 a month?

35. What is your balance after 48 months if you are paying $25 a month?

36. What is your balance after 60 months if you are paying $25 a month?

In Other Words

37. Describe in your own words a circle, a bar, and a line graph.

38. Write in your own words situations in which it is more advantageous to use one type of graph (circle, bar, or line) than another.

39. What type of graph would you use if you had to show

a. a relationship between groups that do not affect each other?

b. continuing data?

c. how parts of a whole relate to each other?

d. In each case, discuss why you would use the type of graph you indicated.

Using Your Knowledge

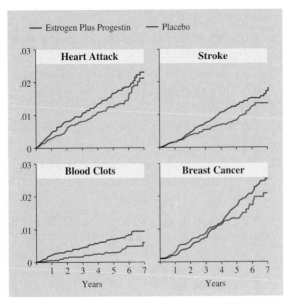

HORMONE REPLACEMENT THERAPY

The most popular prescription for relieving the effects of menopause has more risks than benefits, and the 6 million women in the United States who take the estrogen-plus-progestin preparation should consult their doctors right away, a national study determined. The research found these numbers of illnesses per 10,000 women annually:

■ Estrogen Plus Progestin   ■ Placebo

The *Tampa Tribune* graphic. Sources: Knight Ridder/Tribune, *Journal of the American Medical Association*.

40. A recent milestone study regarding hormone replacement therapy for relieving the effects of menopause showed that there were more risks than benefits in the treatment. How did they persuade doctors to stop the experiments? By studying the statistics using bar and line graphs. Look at the bar graphs on the left. The red bars show the number of women (per 10,000) suffering heart attacks, strokes, breast cancer, and blood clots when taking the estrogen plus progestin (medicine), while the blue bars show the comparable numbers for women taking a placebo (fake medicine).

  a. Estimate the total number of women (per 10,000) who had heart attacks, strokes, breast cancer, and blood clots while taking the estrogen-plus-progestin medicine

  b. Estimate the total number of women (per 10,000) who had heart attacks, strokes, breast cancer, and blood clots while taking the placebo.

  c. Estimate the numerical difference in the total number (per 10,000) of women taking the estrogen-plus-progestin medicine, having heart attacks, strokes, breast cancer, and blood clots as compared with the number (per thousand) taking the placebo.

  d. There were some benefits associated with taking the medicine. What were those benefits and how could you measure them?

  e. There were some areas that were neutral (not much difference between taking the medicine and taking the placebo). Which were those areas?

41. The four line graphs relate the *risk* of some illnesses (*vertical* scale, from 0 to 0.03) and the *length* of time the women have taken the medicine (*horizontal* scale, from 0 to 7 years). Study the graphs and determine in which years (to the nearest year) the placebo group (blue) had

  a. fewer heart attacks than the medicine group (red).

  b. fewer strokes than the medicine group.

  c. fewer blood clots than the medicine group.

  d. less breast cancer than the medicine group.

  e. In what year and from what condition was the difference between taking medicine and taking a placebo greatest? What was the numerical difference in the risk?

f. Which was the only condition in which the patients taking the medicine fared better than the ones taking the placebo? In which years were the patients with this condition and taking the medicine better off than the ones taking the placebo?

 Research Questions

 *Online Study Center*

1. List three different verbs you can use to indicate changes on a graph and explain what meaning they would have if you were referring to a line graph. For help, access link 1.3.2 on this textbook's Online Study Center.
  a. Explain the meaning of the verbs if you are referring to a circle graph.
  b. Explain the meaning of the verbs if you are referring to a bar graph.

 *Online Study Center*

2. List at least three ways in which you can portray misleading information using a graph. Try looking up "How to Lie with Statistics" on the Internet. For help, access link 1.3.2 on this textbook's Online Study Center.

Chapter **1 Summary**

Section	Item	Meaning	Example
1.1	RSTUV	A five-step problem-solving procedure	Read the problem. Select the unknown. Think of a plan. Use a strategy to carry out the plan. Verify your answer.
1.1	Inductive reasoning	The process of arriving at a general conclusion on the basis of repeated observations of specific examples	The pattern 1, 3, 6, 10, ... is obtained by adding 2, adding 3, adding 4, and so on.
1.1	Sequence	An arrangement of numbers according to a pattern	1, 3, 6, 10, ... is a sequence.
1.1	Term	Each number or item in a sequence	1, 3, 6, 10, and so on, are terms in the sequence 1, 3, 6, 10, ...
1.2	Estimation	The process of arriving at an approximate answer to a question	
1.2	Rounding	The process by which numbers are approximated to a certain level of precision	To round the number 46.27 to one decimal or the nearest unit, write $46.\underline{2}7 \rightarrow 46.3$ $4\underline{6}.27 \rightarrow 46$

Section	Item	Meaning	Example
1.3A	Circle (pie) chart	A circle (pie) chart is a circle graph divided into sectors, each displaying the size of some related piece of information.	

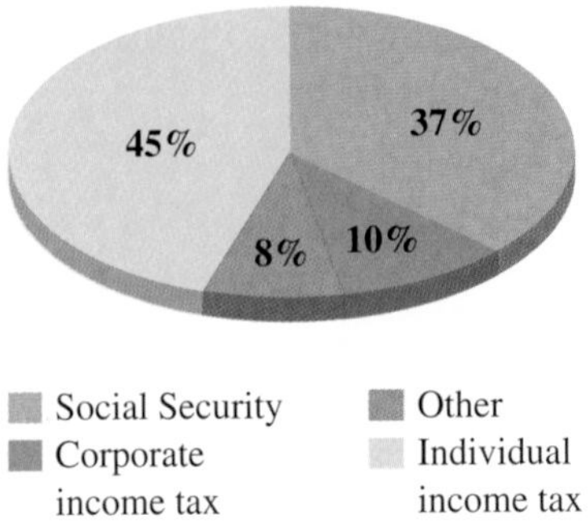

37% • 45% • 8% • 10%

■ Social Security   ■ Other
■ Corporate   ■ Individual
  income tax   income tax

The circle graph shows the revenue sources of the federal government.

Source: http://cstl.syr.edu/fipse/TabBar/ReadCirc/REVCIRCL.HTM.

Section	Item	Meaning	Example
1.3B	Bar graph	Bar graphs consist of an axis with horizontal or vertical bars that represent different values for each bar. The set of numbers along a side of the bar graph is called the scale.	

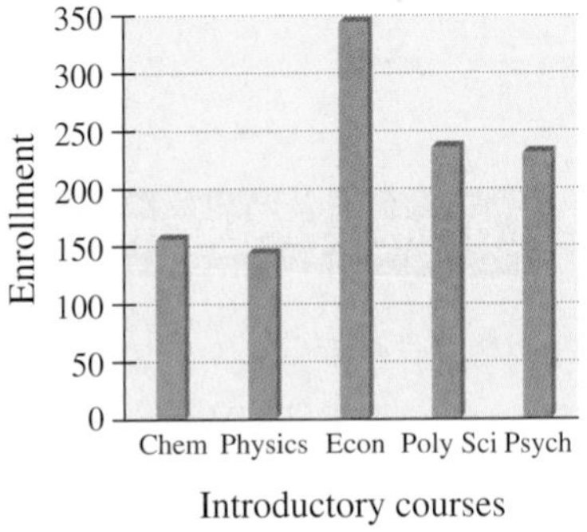

The bar graph represents the enrollment in different courses at a university.

Source: http://cstl.syr.edu/fipse/TabBar/RevBar/REVBAR.HTM.

Section	Item	Meaning	Example
1.3C	Line graph	A line graph is a tool used to represent information and summarize how the information is related and varies, depending on one another.	

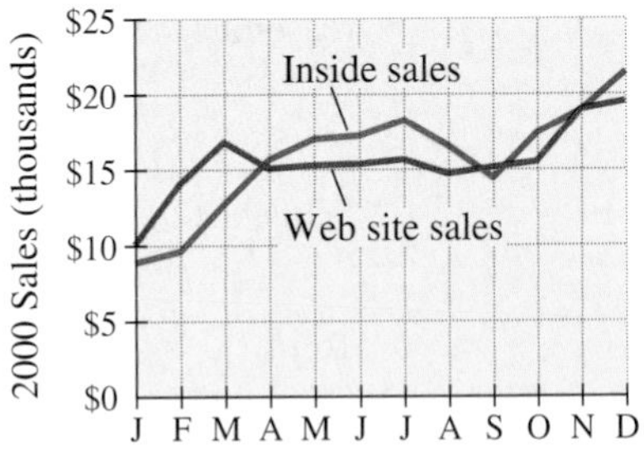

The graph shows how Web site sales and inside sales compare to each other.

Source: www.visualmining.com/examples/styles/line-graphs.html.

Research Questions

1. Access link 1.1.6 on this textbook's Online Study Center and contrast induction and deduction. Include a diagram of how each method works.

2. In a short paragraph, give three reasons why estimation is important in the study of mathematics.

Chapter 1 Practice Test

1. What does RSTUV mean?

2. What is inductive reasoning?

3. Identify the pattern and find the next three terms.

$$1, 2, 7, 19, 41, 76, \underline{\hspace{1cm}}, \underline{\hspace{1cm}}, \underline{\hspace{1cm}}$$

4. **a.** Follow this procedure: Select a number, multiply it by 4, add 6 to the product, divide the sum by 2, and subtract 3 from the quotient.
   b. Follow the same steps as in part (**a**) using the numbers 1, 10, and 100. What results do you get?
   c. Make a conjecture about the relationship between the original number and the final results.

5. Round 319.26 to
   a. the nearest tenth.   **b.** the nearest hundred.

6. **a.** Read the meter for today.

   b. If the reading for yesterday was 6002 kilowatt-hours, how many kilowatt-hours have you consumed since yesterday?
   c. If a kilowatt-hour costs $.10, how much was your 1-day bill?
   d. If you estimate the same consumption each day for a 30-day period (a month), what is your estimated monthly bill?

7. The relationship between the height H of a person and the length t of their tibia (the bone connecting the knee to the ankle) is
   Male: $H = 32.2 + 2.4t$ Female: $H = 28.6 + 2.5t$
   a. Estimate the height of a person (to the nearest inch) with a 15-in. tibia if the person is a female.
   b. Do the same as in part (**a**) if the person is a male.

8. The circle graphs at the top of page 48 show the budget requirements and resources for the city of St. Petersburg.

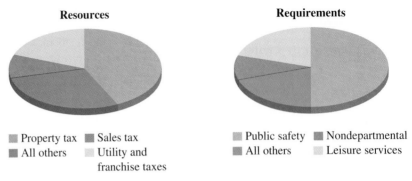

Resources

■ Property tax ■ Sales tax
■ All others   Utility and
                      franchise taxes

Requirements

■ Public safety ■ Nondepartmental
■ All others   Leisure services

Source: www.stpete.org/2002charts.htm.

  a. What is the biggest category in the requirements chart?
  b. About what fraction of the requirements is for public safety?
  c. What is the smallest category in the resources chart?

9. The bar graph shows the property tax (orange) versus police and fire expenses (green) for the city of St. Petersburg.

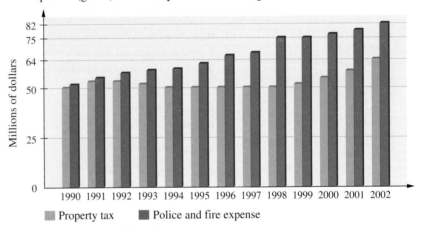

■ Property tax   ■ Police and fire expense

Source: www.stpete.org/2002charts.htm.

  a. How much were the property taxes in 1990?
  b. What were the police and fire expenses in 1990?
  c. Estimate the police and fire expenses for 2002.
  d. Estimate the property taxes for 2002.
  e. What is the difference in property taxes in 1998 and in 2002?

10. The line graph shows the 30-year fixed mortgage rates on different dates.
  a. What was the rate on 5/15?   **b.** What was the rate on 6/5?
  c. What was the difference in rates between 5/15 and 6/5?
  d. What seems to be the trend for 30-year fixed-rate mortgages?

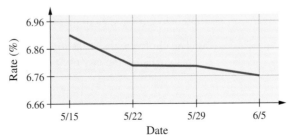

Source: www.bankrate.com/kip/subhome/mtg_m1.asp.

Answers to Practice Test

ANSWER	IF YOU MISSED	REVIEW		
	Question	Section	Example(s)	Page(s)

1. Read the problem.
   Select the unknown.
   Think of a plan.
   Use the techniques you are studying to carry out the plan.
   Verify the answer.

2. The process of arriving at a general conclusion on the basis of repeated observations of specific examples

3. Look at the following differences:

$$1\ 2\ 7\ \ 19\ \ 41\ \ 76 \rightarrow \boxed{127}\ \boxed{197}\ \boxed{289}\ \cdots$$
$$1\ 5\ 12\ \ 22\ \ 35 \rightarrow 51\ \ \ \ 70\ \ \ \ 92\ \cdots$$
$$4\ 7\ 10\ \ 13 \rightarrow 16\ \ \ \ 19\ \ \ \ 22\ \cdots$$
$$3\ 3\ \ 3\ \ \ 3\ \ \ 3\ \ \ \ \ \ 3\ \cdots$$

The third differences are constant, so the next numbers in each line can be constructed by addition. For example,
   $3 + 13 = 16$
   $16 + 35 = 51$
   $51 + 76 = 127$, or simply $3 + 13 + 35 + 76$
The next three terms are 127, 197, and 289.

4. a. $n, 4n, 4n + 6, 2n + 3, 2n$
   b. Using 1, the final result is 2.
      Using 10, the final result is 20.
      Using 100, the final result is 200.
   c. The conjecture is that the final result is twice the original number.

5. a. $319.\underline{2}6 \rightarrow 319.3$
   b. $\underline{3}19.26 \rightarrow 300$

6. a. 6064 kWh
   b. 62 kWh
   c. $6.20
   d. $186.00

7. a. Female: $H = 28.6 + 2.5(15) \approx 66$ in.
   b. Male: $H = 32.2 + 2.4(15) \approx 68$ in.

8. a. Public safety
   b. $\frac{1}{2}$
   c. Sales tax

9. a. About $50 million
   b. About $52 million
   c. About $82 million
   d. About $64 million
   e. About $14 million

10. a. 6.91
   b. 6.76
   c. 0.15
   d. They seem to be decreasing.

	Question	Section	Example(s)	Page(s)
	1	1.1A	Problem-Solving Box	2–3
	2	1.1B	Definition of Inductive Reasoning	4
	3	1.1B	1–3	4–6
	4	1.1B	4	7
	5	1.2	1	16–17
	6	1.2	2	17–18
	7	1.2	4–6	19–21
	8	1.3	1	28
	9	1.3	2–3	29–31
	10	1.3	4–5	31–33

Before diagnosing a disease, doctors must recognize the symptoms that differentiate one disease from another. When presented with symptoms, doctors can use concepts of set operations in order to offer a diagnosis to a patient. In Section 2.2, you will use Venn diagrams to solve other application problems.

Sets

The modern mathematical theory of sets is one of the most remarkable creations of the human mind. Because of the unusual boldness of some of the ideas found in its study, and because of some of the singular methods of proof to which it has given rise, the theory of sets is indescribably fascinating. But above this, the theory has assumed tremendous importance for almost the whole of mathematics.

HOWARD EVES

According to David M. Burton's *History of Mathematics,* "the birth of set theory can be marked by Cantor's paper 'On a Property of the System of all the Real Algebraic Numbers.'" Many mathematicians studied finite sets, but Georg Ferdinand Ludwig Philip Cantor was one of the first to solve some of the problems in the theory of infinite sets. You can read about Cantor in the Human Side of Math as you study sets. First, we consider **finite sets** and their characteristics. Second, we discuss **subsets** and the operations that can be performed with them and visualize these operations by using **Venn diagrams,** as shown in Section 2.3. Third, we will examine how to count the number of elements in a set and use this idea to analyze statistical surveys. Finally, we end the chapter with an optional section dealing with **cardinal numbers** and the "indescribably fascinating" **infinite sets.**

Online Study Center

For links to Internet sites related to Chapter 2, please access
college.hmco.com/PIC/bello9e
and click on the Online Study Center icon.

2.1 Sets: A Problem-Solving Tool

Packaging, Recycling, and Sets

When the clerk at the supermarket asks, "Paper or plastic?" most people assume that the correct answer is *paper*. That is not necessarily correct. (If you want to see why, read *The Green Pages,* published by Random House.) Table 2.1 lists some kitchen product packaging.

TABLE 2.1 Kitchen Product Packaging

Packaging	Recycled	Recyclable
Cellophane	No	No
Glass bottles	Yes	Yes
Plastic	No	Yes

HUMAN SIDE OF MATH

Georg Cantor made the first successful attempts to answer questions concerning infinite sets. His most important contributions appeared in papers published between 1874 and 1884. These papers attacked the basic questions of infinite sets.

(1845–1919)

Unfortunately, Cantor did not receive much recognition during this period and was rewarded by ridicule from many of his most famous contemporaries.

It was only later in his life that Cantor's ideas gained a measure of recognition from his colleagues. Today, we know that much of the foundation of set theory rests directly on Cantor's work.

Looking Ahead
In this chapter we look at the ideas of sets and infinite sets, ideas that were the main focus of Cantor's research and writings.

Which set of products is recycled? Which set of products is recyclable? Which set of products is neither recycled nor recyclable? Can you think of a product that is recycled but not recyclable? Is the set of recycled products the same as the set of recyclable products? You will answer more questions like these later in this section in Exercises 2.1, problems 25 and 26. ▶

The idea of a **set** is familiar in everyday life. Do you have a set of dishes, a set of tools, or a set of books? Each of these sets is regarded as a unit.

Sets, however, need not consist of physical objects; they can consist of abstract ideas. For instance, the Ten Commandments is a set of moral laws. The Constitution is the basic set of laws of the United States.

A. Well-Defined Sets and Notation

We study sets in this book not only because much of elementary mathematics can be stated and developed by using this concept but also because many mathematical ideas can be stated most simply in the language of sets.

> **Sets**
>
> A **set** is a well-defined collection of objects, called **elements** or **members** of the set.

The main property of a set in mathematics is that it is **well defined.** This means that given any object, it must be clear whether that object is a member (element) of the set. Thus, if we consider the set of even whole numbers, we know that every even whole number, such as 0, 2, 4, 6, and so on, is an element of this set. Thus, the set of even whole numbers is well defined. On the other hand, the set of funny comic strips in the daily newspaper is *not* well defined, because what one person thinks is funny may not be the same as what another person thinks is funny.

EXAMPLE 1 ▶ Well-Defined Sets

Which of the following descriptions define sets?

(a) Interesting numbers   (b) Multiples of 2

(c) Good writers   (d) Current directors of General Motors

(e) Numbers that can be substituted for x so that $x + 4 = 5$

Solution

Descriptions (b), (d), and (e) are well defined and therefore define sets. Descriptions (a) and (c) are *not* well defined because people do not agree on what is "interesting" or what is "good." Descriptions (a) and (c) therefore do not define sets.

We use capital letters, such as A, B, C, X, Y, and Z, to denote sets and use lowercase letters, such as a, b, c, x, y, and z, to denote **elements** (members) of sets. It is customary, when practical, to list the elements of a set in braces and to separate these elements with commas. Thus, $A = \{1, 2, 3, 4\}$ means that "A is the set consisting of the elements 1, 2, 3, and 4." To indicate the fact that "4 is an element of the set A," or "4 is in A," we write $4 \in A$. To indicate that "6 is not an element of A," we write $6 \notin A$. ∎

EXAMPLE 2 ▶ Set Inclusion

Let $X = \{$Eva, Mida, Jack, Janice$\}$. Which of the following are correct statements?

(a) Mida $\in X$ (b) Jack $\notin X$ (c) Janice $\in \{$Eva, Mida, Jack, Janice$\}$

(d) $E \in X$ (e) $X \in X$

Solution

Statements (a) and (c) are the only correct statements. ∎

B. Describing Sets

Sets can be described in three ways.

Online Study Center

To learn more about Cantor, access link 2.1.1 on this textbook's Online Study Center.

Three Methods to Describe Sets
1. By giving a **verbal or written description** of the set
2. By **listing** the elements of the set within braces (**roster** method)
3. By using **set-builder notation** (shown on p. 53)

Below are some examples of sets described by words and by lists.

Description	*List*
The set of counting numbers less than 5	$\{1, 2, 3, 4\}$
The set of natural Earth satellites	$\{$Moon$\}$
The set of counting numbers	$\{1, 2, 3, \ldots \}$
	The three dots, called an **ellipsis,** mean that the list goes on in the same pattern without end.

The set of odd counting numbers less than 15	$\{1, 3, 5, \ldots, 13\}$
	The three dots mean the odd numbers after 5 and before 13 are in the set but are not listed.
The set of whole numbers less than or equal to 3	$\{0, 1, 2, 3\}$

In *set-builder notation*, we use a defining property to describe the set. A vertical bar (|) is used to mean "such that," and the words to the right of the bar describe the rule. Thus, the preceding sets can be written as follows:

Set-Builder Notation	*Read*
$\{x \mid x$ is a counting number less than 5$\}$	The set of all elements x such that x is a counting number less than 5
$\{x \mid x$ is a natural Earth satellite$\}$	The set of all elements x such that x is a natural Earth satellite
$\{x \mid x$ is a counting number$\}$	The set of all elements x such that x is a counting number
$\{x \mid x$ is an odd counting number less than 15$\}$	The set of all elements x such that x is an odd counting number less than 15
$\{x \mid x$ is a whole number less than or equal to 3$\}$	The set of all elements x such that x is a whole number less than or equal to 3

EXAMPLE 3 ▶ **Writing Descriptions of Sets**

Write descriptions for the following sets:

(a) $\{a, b, c, \ldots, z\}$ (b) $\{1, 3, 5, \ldots\}$ (c) $\{3, 6, 9, \ldots, 27\}$

Solution

(a) The set of letters in the English alphabet

(b) The set of odd counting numbers

(c) The set of counting numbers that are multiples of 3 and less than or equal to 27   ■

EXAMPLE 4 ▶ **Using Roster and Set-Builder Notation**

Describe the following sets using the listing (roster) method and using set-builder notation:

(a) The set of digits in the number 1896

(b) The set of odd counting numbers greater than 6

(c) The set of counting numbers greater than 0 and less than 1

(d) The set of counting numbers that are multiples of 4

Solution

List (Roster)	*Set-Builder Notation*
(a) $\{1, 8, 9, 6\}$	$\{x \mid x \text{ is a digit in the number } 1896\}$
(b) $\{7, 9, 11, \ldots\}$	$\{x \mid x \text{ is an odd counting number and } x > 6\}$ The symbol $>$ means "greater than."
(c) $\{\}$	$\{x \mid x \text{ is a counting number and } 0 < x < 1\}$ The symbol $<$ means "less than," and $0 < x < 1$ can be read as "x is between 0 and 1" or "x is greater than 0 and is less than 1."
(d) $\{4, 8, 12, \ldots\}$	$\{x \mid x \text{ is a counting number that is a multiple of } 4\}$ ■

A set with *no* elements, as in part (c) of Example 4, can be denoted by the symbol $\{\}$ or $\varnothing$. *Note:* The notation $\varnothing$ is preferred.

Notation for the Empty Set

The symbol $\{\}$ or $\varnothing$ represents the **empty,** or **null,** set.

EXAMPLE 5 ▶ From Set-Builder to Roster Notation

Write the following sets using the listing (roster) method:

(a) $\{x \mid x \text{ is a counting number less than } 10 \text{ and } x \text{ is divisible by } 4\}$

(b) $\{x \mid x \text{ is a counting number between } 8 \text{ and } 13 \text{ and } x \text{ is divisible by } 7\}$

Solution

(a) The counting numbers less than 10 that are divisible by 4 are 4 and 8. Hence, the required set is $\{4, 8\}$.

(b) None of the numbers between 8 and 13 is divisible by 7. Thus, the required set is empty, and the answer can be written as $\{\}$ or $\varnothing$. No roster!

Note: It would *not* be correct to write $\{\varnothing\}$ for the answer to part (b) because the set $\{\varnothing\}$ is not empty, it contains the element $\varnothing$. ■

We now have used three different types of notation to write sets: description, roster, and set builder. Why do we have three types of notation? Because some sets can only be written using one of the notations.

In order to illustrate this fact, we will write some of the sets of numbers that will be used later in the book. Do not worry if you do not know about these sets of numbers; we are only illustrating the fact that sets can be written different ways!

Written Description	*Roster Notation*	*Set-Builder Notation*
The set of natural numbers	$\{1, 2, 3, \ldots\}$	$\{x \mid x \text{ is a natural number}\}$
The set of whole numbers	$\{0, 1, 2, 3, \ldots\}$	$\{x \mid x \text{ is a whole number}\}$

Written Description	Roster Notation	Set-Builder Notation
The set of integers	$\{\ldots, -2, -1, 0, 1, 2, \ldots\}$	$\{x \mid x \text{ is an integer}\}$
The set of rational numbers	*Impossible* to give roster. $\frac{5}{3}, -\frac{3}{7}, 8,$ and 0 are rational. Any rational number may be written as a terminating decimal $(0.8, -3.4)$ or a repeating decimal $(0.333\ldots, -2.666\ldots)$.	$\{x \mid x = a/b, a \text{ and } b$ integers and $b \neq 0.$
The set of irrational numbers	*Impossible* to give roster. Nonterminating, nonrepeating decimals $(0.101001000\ldots,$ and numbers like $\sqrt{2},$ $-\sqrt{7})$ are irrationals.	$\{x \mid x \text{ is not rational}\}$ or $\{x \mid x \text{ cannot be written as a quotient of integers}\}$
The set of real numbers	*Impossible* to give roster. Any rational or irrational number is a real number.	$\{x \mid x \text{ is a rational or an irrational number}\}$ or $\{x \mid x \text{ is a number that can be written as a decimal}\}$

C. Equality of Sets

Note that the order in which the elements of a set are listed does not affect membership in the set. Thus, if we are asked to write the set of digits in the year in which Columbus discovered America, we may write the set as $\{1, 4, 9, 2\}$. Someone else may write the set as $\{1, 2, 4, 9\}$. Both are correct! Thus, we see that $\{1, 4, 9, 2\} = \{1, 2, 4, 9\}$. Similarly, $\{a, b, c, d, e\} = \{e, d, c, b, a\}$.

Definition of Equal Sets

In general, two sets A and B are **equal,** denoted by $A = B$, if they have the same elements (members) not necessarily listed in the same order.

For example, $\{1, 3, 2\} = \{1, 2, 3\}$ and $\{20, \frac{1}{2}\} = \{\frac{1}{2}, 20\}$.

Notice also that repeated listings do not affect membership. For example, the set of digits in the year in which the Declaration of Independence was signed is $\{1, 7, 7, 6\}$. This set can also be written as $\{1, 6, 7\}$. Therefore, the set $\{1, 7, 7, 6\} = \{1, 6, 7\}$. In the same way, $\{a, a, b, b, c, c\} = \{a, b, c\}$ because the two sets have the same elements. By convention, we do not list an element of a set more than once.

D. Subsets and Problem Solving

Sometimes all the elements of a set A are also elements of another set B. For example, if A is the set of all students in your class, and B is the set of all students in your school, every element of A is also in B (because every student in your class is a student in your school). In such cases, we say that the set A is a **subset** of the set B. We denote this by writing $A \subseteq B$.

Definition of Subset

The set A is a **subset** of B (denoted by $A \subseteq B$) if every element of A is also an element of B.

Thus, if $A = \{a, b\}$, $B = \{a, b, c\}$, and $C = \{b\}$, then $A \subseteq B$, $C \subseteq A$, and $C \subseteq B$.

It is a consequence of the definition of a subset that $A = B$ when both $A \subseteq B$ and $B \subseteq A$. Furthermore, for any set A, since every element of A is an element of A, according to the definition of a subset, $A \subseteq A$.

The definition of a subset may be restated in the following form:

Alternative Definition of a Subset

The set A is a **subset** of B if there is no element of A that is not an element of B.

From this definition it follows that $\varnothing \subseteq A$ because there is no element of $\varnothing$ that is not in A. This reasoning holds for every set, so *the empty set is a subset of every set.*

Definition of Proper Subset

A set A is said to be a **proper subset** of B, denoted by $A \subset B$, if A is a subset of B but $A \neq B$ (A is not equal to B).

In other words, $A \subset B$ means that all elements of A are also in B, but B contains at least one element that is not in A. For example, if $B = \{1, 2\}$, the proper subsets of B are $\varnothing$, $\{1\}$, and $\{2\}$, but the set $\{1, 2\}$ itself is not a *proper* subset of B.

In everyday discussion, we are usually aware of the "universe of discourse," that is, the set of all things we are talking about. In dealing with sets, the universe of discourse is called the **universal set.**

Definition of a Universal Set

The **universal set** $\mathcal{U}$ is the set of all elements under discussion.

Thus, if we agree to discuss all the letters in the English alphabet, then the set $\mathcal{U} = \{a, b, c, \ldots, z\}$ is our universal set. On the other hand, if we are to discuss the counting numbers, our universal set is $\mathcal{U} = \{1, 2, 3, \ldots\}$.

EXAMPLE 6 ▶ Finding Subsets of a Set

Find all the subsets of the set $\mathcal{U} = \{a, b, c\}$.

Solution

Form subsets of the set $\mathcal{U}$ by assigning some, none, or all of the elements of $\mathcal{U}$ to these subsets. Organize the work as follows:

Form all the subsets with no elements.	$\varnothing$
Form all the subsets with 1 element.	$\{a\}, \{b\}, \{c\}$
Form all the subsets with 2 elements.	$\{a, b\}, \{a, c\}, \{b, c\}$
Form all the subsets with 3 elements.	$\{a, b, c\}$ ∎

The set in this example has 3 elements and $2^3 = 2 \times 2 \times 2 = 8$ subsets. Similarly, a set such as $\{a, b\}$, containing 2 elements, has $2^2 = 2 \times 2 = 4$ subsets, namely, $\varnothing, \{a\}, \{b\}$, and $\{a, b\}$. The Discovery section of Exercises 2.4 shows another way of constructing all the subsets of a given set. It also illustrates the following:

The Number of Subsets of a Set

A set of n elements has 2^n subsets.

EXAMPLE 7 ▶ Counting the Subsets of a Set

If $A = \{a_1, a_2, a_3, \ldots, a_8\}$, how many subsets does A have?

Solution

Because the set A has 8 elements, A has $2^8 = 256$ subsets. ∎

EXAMPLE 8 ▶ Listing Proper Subsets

List all the proper subsets of the set $\mathcal{U} = \{1, 2, 3\}$.

Solution

The proper subsets are $\varnothing, \{1\}, \{2\}, \{3\}, \{1, 2\}, \{1, 3\}$, and $\{2, 3\}$. By the definition of a proper subset, the set $\{1, 2, 3\}$ is *not* a proper subset of $\mathcal{U}$. Thus, a set with n elements has $2^n - 1$ proper subsets. ∎

The ideas in the preceding examples can be applied to many practical problems. You have heard the slogans "Have it your way" and "I'm loving it." Now, suppose that you have a small hamburger place and you want to advertise that you have a great variety of burgers. If you have three condiments (catsup, mustard, and onions), how many different types of hamburgers can you prepare? (By the way, Wendy's did something similar some time ago. See problem 71 of Exercises 2.1.) To solve this problem, use the RSTUV procedure. The strategy here is to *make a list* of the subsets, as shown next.

PROBLEM SOLVING

Finding Subsets

❶ **Read** the problem.

List the different types of hamburgers that can be prepared if catsup (c), mustard (m), and onions (o) are available as condiments.

Some problems must be read two or three times. Make sure you understand the problem before you attempt a solution.

❷ **Select** the unknown.

Look for the *different* types of hamburgers that can be prepared if catsup, mustard, and onions are available.

❸ **Think** of a plan.
What is given?
What do you do?

Given the universal set

$$\mathcal{U} = \{c, m, o\}$$

find all the subsets of $\mathcal{U}$. The strategy here is to make a list of the elements in each subset. Organize the work! Start with subsets of 0 elements, then 1 element, and so on.

❹ **Use** your knowledge to carry out the plan. Find the
subset of 0 elements
subsets of 1 element
subsets of 2 elements
subsets of 3 elements

$\varnothing$. There is 1 type of hamburger with no condiments.
$\{c\}, \{m\}, \{o\}$. There are 3 types of hamburgers with only 1 condiment.
$\{c, m\}, \{c, o\}, \{m, o\}$. There are 3 types of hamburgers with 2 condiments.
$\{c, m, o\}$. There is 1 type of hamburger with everything.

❺ **Verify** the answer.

The set $\mathcal{U} = \{c, m, o\}$ has 3 elements and $2^3 = 8$ different subsets, and you have found them all!

TRY EXAMPLE 9 NOW.

Cover the solution, write your own solution, and then check your work.

EXAMPLE 9 ▶ Police Options

A police dispatcher has four patrol cars a, b, c, and d available. List all the options for handling a particular call.

Solution

Send no cars.	$\varnothing$
Send 1 car.	$\{a\}, \{b\}, \{c\}, \{d\}$
Send 2 cars.	$\{a, b\}, \{a, c\}, \{a, d\}, \{b, c\}, \{b, d\}, \{c, d\}$
Send 3 cars.	$\{a, b, c\}, \{a, c, d\}, \{a, b, d\}, \{b, c, d\}$
Send 4 cars.	$\{a, b, c, d\}$

The total number of choices is $2^4 = 16$, and you included all of them. ■

EXAMPLE 10 ▶ **Picking Coins and Sums**

If you have to select *at least one* coin from the set of coins shown, how many different sums of money can you make? What is the largest sum you can make if you decide to use three coins?

Solution

Since a set of n elements has $2^n - 1$ nonempty subsets and we have $n = 5$ coins, you can have $2^5 - 1 = 32 - 1 = 31$ different sums of money. The largest sum you can make with three coins would include the half-dollar, the quarter, and the dime, a total of 85 cents. ■

EXERCISES 2.1

Ⓐ Well-Defined Sets and Notation

In problems 1–8, which of the descriptions describe a set?

1. Grouchy people

2. Good tennis players in the United States

3. Retired baseball players with lifetime batting averages of .400 or better

4. Students taking mathematics courses at Yale University at the present moment

5. $\{x \mid x \text{ is an odd counting number}\}$

6. $\{x \mid x \text{ is an even counting number}\}$

7. $\{x \mid x \text{ is a good college course}\}$

8. $\{x \mid x \text{ is a bad instructor}\}$

9. Let $A = \{$Desi, Gidget, Jane, Dora$\}$. Which of the following are correct statements?
   a. $D \in A$ **b.** Desi $\in A$
   c. $A \in$ Jane   **d.** $D \notin A$
   e. Jane $\notin A$

In problems 10–14, let $X = \{a, b, x, y\}$. Fill in the blank with $\in$ or $\notin$ to make each statement correct.

10. a _____ X 11. x _____ X

12. X _____ X 13. A _____ X

14. $\{bay\}$ _____ X

Ⓑ Describing Sets (with Verbal Descriptions)

In problems 15–24, write a verbal description.

15. $\{a, z\}$

16. $\{m, a, n\}$

17. $\{$Adam, Eve$\}$

18. $\{$Christopher Columbus$\}$

19. $\{7, 2, 6, 3, 5, 4, 1\}$

20. $\{2, 6, 12, 20, 30\}$

21. $\{1, 3, 5, \ldots, 51\}$

22. $\{3, 6, 9, 12, \ldots, 36\}$

23. $\{1, 4, 7, 10, \ldots, 25\}$

24. $\{1, 6, 11, \ldots, 31\}$

The following table, which lists toxic substances found in the fat tissue of humans, was compiled by the National Adipose Tissue Survey of the Public Health Service and will be used in problems 25 and 26.

Compound	Possible Sources of Exposure	Frequency in Test Subjects
Chloroform	Drinking water	76%
Dioxin	Wood treatment, herbicides, auto exhaust	100%
Heptachlor	Termite control	67%
Toluene	Gasoline	91%
Xylene	Gasoline, paints	100%

25. Which set of compounds was found in everybody's tissue?

26. Which set of compounds was found in less than 90% of the people?

In problems 27–36, write the sets using the listing (roster) method.

27. $\{x \mid x$ is a counting number less than 8$\}$

28. $\{x \mid x$ is a counting number less than 2$\}$

29. $\{n \mid n$ is a whole number less than $7\frac{1}{2}\}$

30. $\{n \mid n$ is a whole number less than $8\frac{1}{4}\}$

31. $\{x \mid x$ is a counting number between 3 and 8$\}$

32. $\{x \mid x$ is a counting number between 2 and 7$\}$

33. $\{n \mid n$ is a counting number between 6 and 7$\}$

34. $\{n \mid n$ is a counting number between 8 and 10$\}$

35. $\{x \mid x$ is a counting number greater than 3$\}$

36. $\{x \mid x$ is a counting number greater than 0$\}$

Listed below are the most fuel-efficient cars for this model year. Use the numbers from 1 to 10 at the beginning of the column to identify the car. For example, the set of cars made by Honda is $\{2, 3, 4, 5, 6\}$.

Gas-Efficient Cars

Make/Model	City mpg	Highway mpg	Trans-mission	Type	Annual Fuel Costs Estimate
1. Toyota Prius	60	51	M	Hybrid	$491.00
2. Honda Insight	61	66	M	Hybrid	$483.00
3. Honda Civic Hybrid	48	47	M	Hybrid	$562.00
4. Honda Civic GX NGV	30	34	A	CNG	$491.00
5. Honda Civic HX	36	44	M	Gasoline	$729.00
6. Honda Accord Hybrid	29	37	A	Hybrid	NA
7. Ford Escape HEV 2WD	36	31	A	Hybrid	$818.00
8. Volkswagen Jetta Wagon	36	47	M	Diesel	$646.00
9. Volkswagen Passat TDI	27	38	A	Diesel	$751.00
10. Volkswagen Golf TDI	38	46	M	Diesel	$646.00

Source: www.gasefficientcars.com/?source=google.

Use this idea to find the set of cars

37. made by Volkswagen.

38. with automatic transmissions.

39. with manual transmissions.

40. that use diesel fuel.

41. with the lowest fuel costs.

42. with the highest fuel costs.

43. making more than 50 highway mpg.

C Equality of Sets

In problems 44–47, state whether the sets A and B are equal.

44. $A = \{2n + 1 \mid n$ is a counting number$\}$
$B = \{2n - 1 \mid n$ is a counting number$\}$

45. $A = \{4n \mid n$ is a counting number$\}$
$B = \{2n \mid n$ is a counting number$\}$

46. $A = \{1, 1, 2, 2, 3\}$, $B = \{1, 2, 3\}$

47. $A = \{x \mid x$ is a cow that has jumped over the moon$\}$
$B = \{x \mid x$ is an astronaut who has landed on Pluto$\}$

48. Let $A = \{5\}$, $B = \{f, i, v, e\}$, $C = \{e, f, v, i\}$, and D be the set of letters in the word *repeat*. Find the following:
  a. The set containing five elements
  b. The set equal to B
  c. The set of letters in the word *five*

49. Let $A = \{1, 2, 3, 4\}$, $B = \{4, 3, 2, 1\}$, and $C = \{4, 3, 2, 1, 0\}$. Fill in the blanks with $=$ or $\neq$ to make true statements.
  a. A _____ B **b.** A _____ C
  c. B _____ C

50. Let $A = \{x \mid x$ is a counting number between 4 and 5$\}$, $B = \varnothing$, and $C = \{\varnothing\}$. Fill in the blanks with $=$ or $\neq$ to make true statements.
  a. A _____ B **b.** A _____ C
  c. B _____ C

D Subsets and Problem Solving

In problems 51–56, list all the subsets and indicate which are proper subsets of the given set.

51. $\mathcal{U} = \{a, b\}$

52. $\mathcal{U} = \{1, 2, 3\}$

53. $\mathcal{U} = \{1, 2, 3, 4\}$

54. $\mathcal{U} = \{\varnothing\}$

55. $\{1, 2\}$

56. $\{x, y, z\}$

57. How many subsets does the set $A = \{a, b, c, d\}$ have?

58. How many proper subsets does the set $\{1, 2, 3, 4\}$ have?

59. If $A = \{\frac{1}{1}, \frac{1}{2}, \frac{1}{3}, \ldots, \frac{1}{10}\}$, how many subsets does A have?

60. How many proper subsets does set A of problem 57 have?

61. A set has 32 subsets. How many elements are there in the set?

62. A set has 31 proper subsets. How many elements are there in the set?

63. A set has 64 subsets. How many elements are there in the set?

64. A set has 63 proper subsets. How many elements are there in the set?

65. Is $\varnothing$ a subset of $\varnothing$? Explain.

66. Is $\varnothing$ a proper subset of $\varnothing$? Explain.

67. If A is the set of numbers that are divisible by 2 and B is the set of numbers that are divisible by 4, is $A \subseteq B$? Is $B \subseteq A$?

68. Give an example of a set P and a set Q such that $P \in Q$ and $P \subseteq Q$.

69. Gino's Pizza offers the following set of toppings: $\{C, M, O, P, S\}$, where $C, M, O, P,$ and S mean cheese, mushrooms, onions, pepperoni, and sausage. How many types of pizza can you order with
 a. one topping?   **b.** two toppings?
 c. three toppings?

70. Referring to problem 69, how many different kinds of pizza with at least one topping can you order?

71. Some time ago, Wendy's Hamburger claimed that it could prepare your hamburger 256 ways. How many condiments do you need in order to be able to prepare 256 different hamburgers?

72. If Gino's Pizza decides to top Wendy's claim and advertises that it has 500 different types of pizza, what is the minimum number of toppings it must carry?

In Other Words

73. Find the definition of the word *set* in a dictionary. Does the definition contain the word *collection* or the word *thing?* Now, find the definitions of the words *collection* and *thing.* Why do you think it is almost impossible to give a formal definition of the word *set?*

74. Is the set of all good students in your class well defined? Why or why not? If it is not well defined, can you make it well defined? How?

75. Explain why
 a. $\varnothing \notin \varnothing$.   **b.** $\varnothing \in \{\varnothing\}$.
 c. $\varnothing \neq \{0\}$.   **d.** $\varnothing = \{\}$.

76. Explain why for any nonempty set A and universal set $\mathcal{U}$,
 a. $A \subseteq A$.   **b.** $A \not\subseteq A$.
 c. $\varnothing \subset A$.   **d.** $\varnothing \subseteq A$.
 e. $A \subseteq \mathcal{U}$.

Using Your Knowledge

Gepetto Scissore, a barber in the small town of Sevilla, who was naturally called the Barber of Sevilla, decided that as a public service he would shave all those men and only those men of the village who did not shave themselves. Let $S = \{x \mid x$ is a man of the village who shaves himself$\}$ and $D = \{x \mid x$ is a man of the village who does not shave himself$\}$.

77. If g represents Gepetto,
 a. is $g \in S$?
 b. is $g \in D$?

The preceding problem is a popularization of the *Russell paradox,* named after its discoverer, Bertrand Russell. In studying sets, it seems that one can classify sets as those that are members of themselves and those that are not members of themselves. Suppose that we consider the two sets of sets

$$M = \{X \mid X \in X, X \text{ is a set}\}$$

and

$$N = \{X \mid X \notin X, X \text{ is a set}\}$$

78. a. Is $N \in M$?
 b. Is $N \in N$?
 Think about the consequences of your answers!

 **Discovery**

You should find the following paradox amusing, puzzling, and perhaps even thought provoking. Define a *self-descriptive word* to be a word that makes good sense when put into both blanks of the sentence "_____ is a(n) _____ word." Two simple examples of self-descriptive words are "English" and "short." Just try them out!

Now define a *non-self-descriptive word* to be a word that is not self-descriptive. Most words will fit into this category. Try it out again. Now consider the following question.

79. Let *S* be the set of self-descriptive words, and let *S'* be the set of non-self-descriptive words. How would you classify the word *non-self-descriptive?* Is it an element of *S*? Or is it an element of *S'*? You should get into difficulty no matter how you answer these questions. Think about it!

80. *Russell's Paradox on the Web.* Suppose you construct a Web page (Paradox.html) that has a link to every Web page that does not link to itself. Does Paradox.html have a link to itself?

 Collaborative Learning

This group activity is designed to determine the number of subsets of a set of *n* elements that have to be selected before we find a pair with the property that *one is a subset of the other.* Form three groups. One group will work with the set {1, 2}, the second group with the set {+, −}, and the third group with the set {*a*, *b*}. Each of the groups will do the following.

1. Find all subsets of the set you are working with.

The object is to find a pair of subsets so that *one is a subset of the other.*

2. How many subsets does your group have to select before you find two of the selected subsets are such that *one is a subset of the other?*

3. Answer the same question when the given sets for each of the groups are {1, 2, 3}, {+, −, ×}, and {*a*, *b*, *c*}.

4. Answer the same question when the given sets for each of the groups are {1, 2, 3, 4}, {+, −, ×, /}, and {*a*, *b*, *c*, *d*}.

5. Each of the groups fill in a table like the one below.

Try to answer this question: How many subsets of a set of *n* elements have to be selected so that two of the selected subsets have the property that *one is a subset of the other.* Compare answers and report to the rest of the class.

Number of Elements in the Set	Number of Subsets	Number of Subsets That Have to Be Selected Before Finding Two of the Selected Subsets So That One is a Subset of the Other
1		
2		
3		
4		

To see a proof of the result obtained, go to www.cut-the-knot.org/pigeonhole/subsets.shtml.

Online Study Center

To further explore Cantor's history, access links 2.1.2 and 2.1.3 on this textbook's Online Study Center.

Online Study Center

To further explore Russell's paradox, access links 2.1.4 and 2.1.5 on this textbook's Online Study Center.

 Research Questions

1. Cantor's professor and one of his most acerbic critics was described as "a tiny man, who was increasingly self-conscious of his size with age." Who was this professor, and what were his objections to Cantor's work? (See link 2.1.2.)

2. In 1872, Cantor traveled to Switzerland, where he met Richard Dedekind. What were Dedekind's influences on Cantor's work? (See link 2.1.3.)

3. Write a short paragraph describing Russell's paradox, when it was discovered, its significance, and Russell's response to the paradox.

2.2 Set Operations

Diagnosis and Set Operations

In diagnosing diseases, it is extremely important to recognize the symptoms that distinguish one disease from another (usually called the *differential diagnosis*). For example, some symptoms for hypoglycemia (too little sugar in the blood) and hyperglycemia (too much sugar in the blood) are identical. Short of a blood test, how does a doctor determine whether a person is hypoglycemic or hyperglycemic? First, doctors are aware that a certain universal set of symptoms may indicate the presence of either hypoglycemia or hyperglycemia. Next, they discard symptoms that are common to both conditions. Table 2.2 shows that nausea and headaches are associated with both diseases. Thus, those two symptoms will not help to make a diagnosis and should be disregarded. The emphasis should be on visual disturbances, trembling, stomach cramps, and rapid breathing. Do you know which diagnosis to make when presented with the symptoms in Table 2.2? You will encounter similar questions in Exercises 2.2, problems 40–54, and in Using Your Knowledge, problems 81–87. ▶

It is often important to ascertain which elements two given sets have in common. For example, the sets of symptoms exhibited by patients with too little sugar in the blood (hypoglycemia) or too much sugar in the blood (hyperglycemia) are as given in Table 2.2.

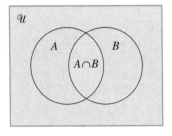

Dr. Margaret Colleran looks for some of these symptoms to diagnose diabetes.

TABLE 2.2

Too Little Sugar in the Blood	Too Much Sugar in the Blood
Nausea (n)	Headache (h)
Visual disturbances (v)	Stomach cramps (s)
Trembling (t)	Nausea (n)
Headache (h)	Rapid breathing (r)

A. Intersections and Unions

We can clearly see that the set of symptoms *common* to both sets listed in Table 2.2 is $\{n, h\}$. The set $\{n, h\}$ is called the **intersection** of the two given sets.

FIGURE 2.1
If A and B are represented by circles and the universal set $\mathcal{U}$ by a rectangle, $A \cap B$ is the shaded area common to both sets.

Definition of the Intersection of Sets

If A and B are sets, the **intersection** of A and B, denoted by $A \cap B$ (read "A intersection B"), is the set of all elements that are common to both A and B. That is,

$$A \cap B = \{x \mid x \in A \text{ and } x \in B\}$$

Thus, if $A = \{a, b, c, d\}$ and $B = \{b, d, e\}$, then $A \cap B = \{b, d\}$. See Figure 2.1.

If we list all the symptoms mentioned in Table 2.2, we obtain the set $\{n, v, t, h, s, r\}$. This set is called the **union** of the two given sets.

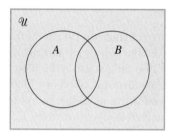

FIGURE 2.2
If A and B are represented by circles and the universal set $\mathcal{U}$ by a rectangle, $A \cup B$ is the shaded area.

> ### Definition of the Union of Sets
>
> If A and B are sets, the **union** of A and B, denoted by $A \cup B$ (read "A union B"), is the set of all elements that are either in A or in B or in both A and B. That is,
>
> $$A \cup B = \{x \mid x \in A \textbf{ or } x \in B\}$$

Note that we use the *inclusive or;* that is, $x \in A$, $x \in B$, or x is in both A and B.

Hence, if $A = \{1, 3, 4, 6\}$ and $B = \{3, 6, 7\}$, then $A \cup B = \{1, 3, 4, 6, 7\}$. See Figure 2.2.

EXAMPLE 1 ▶ Finding Unions and Intersections

Let $A = \{a, b, c, d, e\}$ and $B = \{a, c, e, f\}$. Find the following:

(a) $A \cap B$ (b) $A \cup B$

Solution

(a) $A \cap B$ is the set of all elements common to both A and B. That is, $A \cap B = \{a, c, e\}$.

(b) $A \cup B$ is the set of all elements in A or B or both. That is, $A \cup B = \{a, b, c, d, e, f\}$. Note that a, c, and e occur in both sets, yet we list each of these elements only once. ▪

Two sets with no elements in common are said to be **disjoint.** If sets A and B are disjoint, then $A \cap B = \varnothing$. For example, the set $A = \{1, 2, 3\}$ and the set $B = \{4, 5\}$ are disjoint; they have no elements in common, or $A \cap B = \varnothing$.

B. Complement of a Set

We are often interested in the set of elements in the universal set $\mathcal{U}$ under discussion that are *not* in some specified subset A of $\mathcal{U}$.

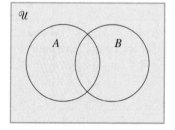

FIGURE 2.3
The set $A' = \mathcal{U} - A$ consists of all the elements in $\mathcal{U}$ but not in A.

> ### Definition of the Complement of a Set
>
> Let $\mathcal{U}$ be the universal set, and let A be a subset of $\mathcal{U}$. The **complement** of A, denoted by A' (read "A prime" or "A complement"), is the set of elements in $\mathcal{U}$ that are not in A. That is, $A' = \{x \mid x \in \mathcal{U} \text{ and } x \notin A\}$. This set is also symbolized by $\mathcal{U} - A$.

For example, if $\mathcal{U}$ is the set of students in your school, and A is the set of those students who have taken algebra, then A' is the set of those students in your school who have not taken algebra. Similarly, if $\mathcal{U}$ is the set of all letters in the English alphabet, and $A = \{s, k, y\}$, then A' is the set of all letters in the alphabet except s, k, and y. Refer to Figure 2.3.

EXAMPLE 2 ▶ **Finding Complements**

Let $\mathscr{U} = \{1, 2, 3, 4, 5, 6\}$, $A = \{1, 3, 5\}$, and $B = \{2, 4\}$. Find the following:

(a) A'

(b) B'

(c) $A' \cap B'$

(d) $A \cap B$

Solution

(a) $A' = \{2, 4, 6\}$

(b) $B' = \{1, 3, 5, 6\}$

(c) $A' \cap B' = \{6\}$

(d) $A \cap B = \emptyset$, because there are no elements common to these two sets. ■

Note that, because $\emptyset \subseteq \mathscr{U}$, and no element of $\mathscr{U}$ is an element of $\emptyset$, it follows that $\emptyset' = \mathscr{U}$ and $\mathscr{U}' = \emptyset$.

EXAMPLE 3 ▶ **Finding Complements of Unions and Intersections**

If A, B, and $\mathscr{U}$ are the same as in Example 2, find the following:

(a) $(A \cup B)'$

(b) $(A \cap B)'$

(c) $A' \cup B'$

(d) $A \cup (A \cup B)'$

Solution

(a) $A \cup B = \{1, 3, 5\} \cup \{2, 4\} = \{1, 2, 3, 4, 5\}$
Hence, $(A \cup B)' = \{6\}$.
[Note that in order to find $(A \cup B)'$, you must first find $A \cup B$ and then take its complement. Always do the operations inside parentheses *first*.]

(b) $A \cap B = \{1, 3, 5\} \cap \{2, 4\} = \emptyset$
Hence, $(A \cap B)' = \emptyset' = \{1, 2, 3, 4, 5, 6\} = \mathscr{U}$.

(c) $A' \cup B' = \{1, 3, 5\}' \cup \{2, 4\}'$
$= \{2, 4, 6\} \cup \{1, 3, 5, 6\}$
$= \{1, 2, 3, 4, 5, 6\} = \mathscr{U}$

(d) Since $(A \cup B)' = \{6\}$ [see part (a)],

$A \cup (A \cup B)' = \{1, 3, 5\} \cup \{6\} = \{1, 3, 5, 6\}$ ■

Notice that the answers to parts (b) and (c) are identical; that is, $(A \cap B)' = A' \cup B'$. Also notice that the answers to Example 2(c) and Example 3(a) are identical; that is, $(A \cup B)' = A' \cap B'$. In general, we have

De Morgan's Law
For any sets A and B, $(A \cap B)' = A' \cup B'$ and $(A \cup B)' = A' \cap B'$.

[See Exercises 2.3, problem 34(a) and (b).]
It is also possible to form intersections, unions, and complements using more than two sets, as in the next example.

EXAMPLE 4 ▶ Unions, Intersections, and Complements

Let $\mathcal{U} = \{a, b, c, d, e, f\}$, $A = \{a, c, e\}$, $B = \{b, e\}$, and $C = \{a, b, d\}$. Find $(A \cup B) \cap C'$.

Solution

Since $A \cup B$ is in parentheses, find $A \cup B$ first.

$$A \cup B = \{a, c, e\} \cup \{b, e\} = \{a, b, c, e\}$$

Then

$$C' = \{c, e, f\}$$

Hence

$$\begin{aligned}(A \cup B) \cap C' &= \{a, b, c, e\} \cap \{c, e, f\} \\ &= \{c, e\}\end{aligned}$$ ■

C. Difference of Two Sets

In some cases, we may be interested in only part of a given set. For example, we may want to consider the set of all nonpoisonous snakes. If we let S be the set of all snakes and P be the set of all poisonous snakes, we are interested in the set of all snakes except (excluding) the poisonous ones. This set will be denoted by $S - P$.

Definition of the Difference of Sets

If A and B are two sets, the **difference** of A and B, denoted by $A - B$ (read "A minus B"), is the set of all elements that are in A and not in B. That is, $A - B = \{x \mid x \in A \text{ and } x \notin B\}$.

Notice that

1. the definition of A' is a special case of the preceding definition because $\mathcal{U} - A = A'$ (see the definition of the complement of set A).

2. $A \cap B' = A - B$ because $A \cap B'$ is the set of all elements in A and not in B, and this is precisely the definition of $A - B$.

EXAMPLE 5 ▶ Finding Differences of Sets

Let $\mathcal{U} = \{1, 2, 3, 4, 5, 6\}$, $A = \{1, 2, 3, 4\}$, and $B = \{1, 2, 5\}$. Find

(a) $\mathcal{U} - A$.   (b) A'.   (c) $A - B$.   (d) $B - A$.

Solution

(a) $\mathcal{U} - A$ is the set of all elements in $\mathcal{U}$ and not in A, that is, $\{5, 6\}$.

(b) A' is the set of all elements in $\mathcal{U}$ and not in A, that is, $\{5, 6\}$.

(c) $A - B$ is the set of all elements in A and not in B, that is, $\{3, 4\}$.

(d) $B - A$ is the set of all elements in B and not in A, that is, $\{5\}$. ■

D. Applications

EXAMPLE 6 ▶ Classifying Employees

A small company has 10 employees, listed by number as 01, 02, 03, . . . , 10. The company classifies these employees according to the work they do.

P = the set of part-time employees
F = the set of full-time employees
S = the set of employees who do shop work
O = the set of employees who do outdoor field work
I = the set of employees who do indoor office work

The payroll department lists these employees as follows:

$\mathcal{U}$ = {01, 02, 03, 04, 05, 06, 07, 08, 09, 10}
P = {01, 02, 05, 07}
F = {03, 04, 06, 08, 09, 10}
S = {01, 04, 05, 08}
O = {03, 04, 06, 09}
I = {02, 05, 07, 10}

Find and describe each of the following sets:

(a) $P \cap S$ (b) $O \cup S$ (c) $F \cap I$ (d) P'

(e) $F \cap (S \cup O)$ (f) $S' \cup (O' \cap I')$

Solution

(a) $P \cap S$ = {01, 05}, the set of part-time employees who do shop work.

(b) $O \cup S$ = {01, 03, 04, 05, 06, 08, 09}, the set of employees who do outside field work or shop work.

(c) $F \cap I$ = {10}, the set of full-time employees who do indoor office work.

(d) P' = {03, 04, 06, 08, 09, 10}, the set of employees who are not part time, that is, the set of full-time employees.

(e) $F \cap (S \cup O)$ = {03, 04, 06, 08, 09}, the set of full-time employees who do shop work or outside field work.

(f) $S' \cup (O' \cap I')$ = {01, 02, 03, 06, 07, 08, 09, 10}, the set of employees who do not do shop work combined with the set of employees who do neither outdoor field work nor indoor office work. ■

EXAMPLE 7 ▶ Parents' and Children's Blood Types

In 1900, Karl Landsteiner identified four blood groups as A, B, AB, and O (which is neither A nor B). Figure 2.4 shows how blood types are passed from parents to a child.

(a) If one parent has A blood and the other O blood, which blood type could their child have?

(b) Blood types can be used to settle paternity cases. Suppose a child has B blood. What are the possible blood types for the parents?

(c) What blood type in a child would be the most difficult to use to identify paternity?

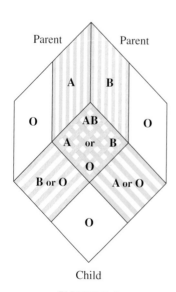

FIGURE 2.4

Solution

(a) Select blood type A at the top left (blue lines) and O at the top right (white). Follow the columns to their intersection. The child could have A or O blood.

(b) Look at the bottom part of the cube. The B appears twice: where O and B intersect and where A and B intersect. Thus, one parent can be O and the other B or one parent can be A and the other B.

(c) Again, look at the bottom of the cube and note that type O appears four times; this means that there are four different sets of parental combinations. Type O blood in a child is the most difficult to use to identify paternity. ■

EXAMPLE 8 ► Movies, Directors, and Intersections

What is your favorite movie? The lists give the 10 best movies of all time according to the critics (set *C*) and according to readers of a movie magazine (set *R*). (Source: www.filmsite.org/mrshowbz.html.)

Critics' Picks (set C)

1. *Casablanca* (1942), director Michael Curtiz
2. *The Godfather Part II* (1974), director Francis Ford Coppola
3. *North by Northwest* (1959), director Alfred Hitchcock
4. *Citizen Kane* (1941), director Orson Welles
5. *Lawrence of Arabia* (1962), director David Lean
6. *Manhattan* (1979), director Woody Allen
7. *Gone With the Wind* (1939), director Victor Fleming
8. *Chinatown* (1974), director Roman Polanski
9. *The Man Who Shot Liberty Valance* (1962), director John Ford
10. *City Lights* (1931), director Charles Chaplin

Readers' Picks (set R)

1. *Star Wars* (1977), director George Lucas
2. *The Godfather* (1972), director Francis Ford Coppola
3. *Pulp Fiction* (1994), director Quentin Tarantino
4. *Casablanca* (1942), director Michael Curtiz
5. *Gone With the Wind* (1939), director Victor Fleming
6. *Raiders of the Lost Ark* (1981), director Steven Spielberg
7. *Schindler's List* (1993), director Steven Spielberg
8. *Citizen Kane* (1941), director Orson Welles
9. *The Empire Strikes Back* (1980), director Irvin Kershner
10. *It's a Wonderful Life* (1946), director Frank Capra

Online Study Center

To further explore questions from students, access links 2.2.1 and 2.2.2 on this textbook's Online Study Center. To learn more about blood and parentage, access link 2.2.3.

(a) Find $C \cap R$. (b) How many movies are in both lists?

(c) Find the set of directors in the intersection of both lists.

Solution

(a) $C \cap R$ is the set of all elements common to *C* and *R*, that is,

   {*Casablanca, Citizen Kane, Gone With the Wind*}

(b) There are three movies that are in both lists.

(c) The set of directors in the intersection of both lists is

   {Michael Curtiz, Francis Ford Coppola, Orson Welles, Victor Fleming} ■

EXERCISES 2.2

Ⓐ Intersections and Unions

In problems 1–8, let set $A = \{1, 2, 3, 4, 5\}$, $B = \{1, 3, 4, 6\}$, and $C = \{1, 6, 7\}$. Find the following:

1. a. $A \cap B$ **b.** $A \cap C$ **c.** $B \cap C$

2. a. $A \cup B$ **b.** $A \cup C$ **c.** $B \cup C$

3. a. $A \cap (B \cup C)$ **b.** $A \cup (B \cap C)$

4. a. $(A \cap B) \cup C$ **b.** $(A \cap B) \cup (A \cap C)$

5. $A \cup (B \cup C)$ **6.** $(A \cup B) \cap (A \cup C)$

7. $A \cap (B \cap C)$ **8.** $(A \cup B) \cap C$

In problems 9 and 10, let $A = \{\{a, b\}, c\}$, $B = \{a, b, c\}$, and $C = \{a, b\}$. Find the following:

9. a. $A \cap B$ **b.** $A \cap C$

10. a. $A \cup B$ **b.** $A \cup C$

In problems 11–14, let $A = \{\{a, b\}, \{a, b, c\}, a, b\}$ and $B = \{\{a, b\}, a, b, c, \{b, c\}\}$. Which of the given statements are correct?

11. a. $\{b\} \subseteq (A \cap B)$ **b.** $\{b\} \in (A \cap B)$

12. a. $\{a, b\} \subseteq (A \cap B)$ **b.** $\{a, b\} \in (A \cap B)$

13. a. $\{a, b, c\} \subseteq (A \cup B)$ **b.** $\{a, b, c\} \in (A \cup B)$

14. a. $3 \subseteq (A \cap B)$ **b.** $3 \in (A \cap B)$

Ⓑ Complement of a Set

The sets $\mathcal{U} = \{a, b, c, d, e, f\}$, $A = \{a, c, e\}$, $B = \{b, d, e, f\}$, and $C = \{a, b, d, f\}$ will be used in problems 15–26. Find each specified set.

15. a. A' **b.** B'

16. a. $A' \cap B'$ **b.** $(A \cap B)'$

17. a. $(A \cup B)'$ **b.** $A' \cup B'$

18. a. $(A \cup B) \cap C'$ **b.** $(A \cup B)' \cap C$

19. a. $(A \cap B) \cup C'$ **b.** $C \cup (A \cap B)'$

20. a. $A' \cup B$ **b.** $A \cup B'$

21. a. $A' \cap B$ **b.** $A \cap B'$

22. a. $A' \cap (A \cup B')$ **b.** $A \cup (A \cap B')$

23. a. $C' \cup (A \cap B)'$ **b.** $C' \cup (A \cup B)'$

24. a. $(C \cup B)' \cap A$ **b.** $(C \cup B) \cap A'$

25. a. $\mathcal{U} - A$ **b.** $\mathcal{U} - B$

26. a. $A - B$ **b.** $B - A$

Ⓒ Difference of Two Sets

In problems 27 and 28, let $\mathcal{U} = \{1, 2, 3, 4, 5\}$, $A = \{2, 3, 4\}$, and $B = \{1, 4, 5\}$. Find each specified set.

27. a. B' **b.** $\mathcal{U} - B$

28. a. $A - B$ **b.** $B - A$

In problems 29–38, $\mathcal{U}$ is some universal set of which A is a subset. In each case find the indicated set in terms of A, $\mathcal{U}$, or $\varnothing$ alone.

29. $\varnothing'$ **30.** $\mathcal{U}'$ **31.** $A \cap \varnothing$

32. $A \cap A$ **33.** $A \cap \mathcal{U}$ **34.** $A \cup \varnothing$

35. $A \cap A'$ **36.** $A \cup A'$ **37.** $(A')'$

38. $A \cup A$

39. If $A = \{1, 2, 3\}$, $B = \{2, 3, 4\}$, and $C = \{1, 3, 5\}$, find the smallest set that will serve as a universal set for A, B, and C.

Ⓓ Applications

Problems 40–45 refer to the following data, the results of an attempt to analyze factors in popularity between members of the same sex and between members of opposite sexes. A psychologist asked 676 college men and women to consider a few persons whom they liked and to tell why they liked those persons.

Traits Men Liked in Women (M_w)	Traits Women Liked in Men (W_m)
Beauty Intelligence Cheerfulness Congeniality	Intelligence Consideration Kindliness Cheerfulness

Traits Men Liked in Men (M_m)	Traits Women Liked in Women (W_w)
Intelligence Cheerfulness Friendliness Congeniality	Intelligence Cheerfulness Helpfulness Loyalty

40. Find the smallest set that will serve as a universal set for M_w, W_m, M_m, and W_w.

41. Find the set of traits that are mentioned only once.

42. Find $M_w \cap W_w$.   **43.** Find $M_w \cap M_m$.

44. What set of traits is common to M_m and M_w?

45. Name the traits that are common to all four of the sets; that is, find $M_w \cap M_m \cap W_w \cap W_m$.

In an article in the *Harvard Business Review,* 606 participants reported on 17 specific changes in their bosses' behaviors from one year to the next. Here are some of the traits that were most frequently mentioned in each of these years:

First Year	Second Year
Encourages suggestions.	Is self-aware.
Sets goals with me.	Listens carefully.
Gets me to have high goals.	Follows up on action.
Listens carefully.	Gets me to have high goals.
Is aware of others.	Encourages suggestions.
Is self-aware.	Sets goals with me.

Let S_1 be the set of traits mentioned in the first year; let S_2 be the set of traits mentioned in the second year.

46. Find $S_1 \cap S_2$, the set of traits mentioned in both years.

47. What traits were mentioned only once?

48. Find the smallest set that will serve as a universal set for S_1 and S_2.

49. Find S_1' relative to the universal set found in problem 48.

50. Find S_2' relative to the universal set found in problem 48.

In problems 51–57, let

$$\mathcal{U} = \text{set of employees of a company}$$
$$M = \text{set of males who are employees}$$
$$F = \text{set of females who are employees}$$
$$D = \text{set of employees who work in the data-processing department}$$
$$T = \text{set of employees who are under 21}$$
$$S = \text{set of employees who are over 65}$$

In problems 51 and 52, find a single letter to represent each specified set.

51. a. M' **b.** F'

52. a. $M \cup F$ **b.** $M \cap F$ **c.** $T \cap S$

In problems 53 and 54, verbally describe each specified set.

53. a. $M \cap D$ **b.** $F \cap T$

54. a. $M \cap T'$ **b.** $(T \cup S)'$ **c.** $(D \cap T)'$

In problems 55–57, find a set representation for each set.

55. Employees in data processing who are over 65

56. Female employees who are under 21

57. Male employees who work in data processing

In problems 58–60, write out in words the complement of the set in each specified problem.

58. Problem 55   **59.** Problem 56   **60.** Problem 57

For problems 61 and 62, refer to the data in Example 6. Find and describe the sets in each problem.

61. a. $F \cap S$ **b.** $P \cap (O \cup I)$

62. a. $P \cap I$ **b.** $P \cap O' \cap S'$

Problems 63–66 refer to the following data. (Sources: www.vh1.com/shows/dyn/the_greatest/series.html www.tcotrel.tripod.com/100rockartists.html http://tcotrel.tripod.com/100hardrocks.html; and http://www.answers.com/topic/100-greatest-songs-of-rock-n-roll.) The set on the left shows the ten greatest artists of rock and roll (RR); the set on the right lists the ten greatest artists of hard rock (HR).

Ten Greatest Artists of Rock and Roll	Ten Greatest Artists of Hard Rock
1. The Beatles	1. AC/DC
2. David Bowie	2. Black Sabbath
3. James Brown	3. Guns N' Roses
4. Bob Dylan	4. Jimi Hendrix
5. Jimi Hendrix	5. KISS
6. Led Zeppelin	6. Led Zeppelin
7. The Police	7. Metallica
8. Elvis Presley	8. Nirvana
9. The Rolling Stones	9. Van Halen
10. The Who	10. The Who

63. Find the intersection of the two sets.

64. Find the set of all names with exactly four letters.

65. How many artists appear on both lists?

66. Find the set of artists that appear on only one list.

Problems 67–76 refer to the following data. Use the numbers from 1 to 10 at the beginning of the column to identify the car. For example, the set of cars made by Ford is {5, 9}. Use this idea to find the indicated sets.

Sales Rank	Make/ Model/ Trim	Fixed Costs	Operating Costs	Total Annual Costs
1.	Toyota Camry LE V6	$4,741	$1,238	$5,979
2.	Honda Accord EX	$4,673	$1,142	$5,815
3.	Honda Civic EX	$4,233	$1,017	$5,250
4.	Chevrolet Impala LS	$5,496	$1,389	$6,885
5.	Ford Taurus SES	$5,358	$1,547	$6,905
6.	Nissan Altima 2.5S	$4,485	$1,301	$5,786
7.	Chevrolet Cavalier LS	$4,771	$1,244	$6,015
8.	Toyota Corolla S	$3,629	$960	$4,589
9.	Ford Focus ZTS	$4,535	$1,317	$5,852
10.	Pontiac Grand Prix GT2	$5,242	$1,373	$6,615

Source: http://www.cars.com/carsapp/boston/?srv=parser&act=display&tf=/advice/bestworst/opcosts/04opcosts.tmpl.

67. The set of cars made by Chevrolet

68. The set of cars made by Honda

69. The set of cars made by Honda with the lowest total annual cost

70. The set of cars made by Toyota

71. The set of cars made by Toyota with the lowest operating costs

72. A customer wants to purchase a car based on total annual costs. Find the set of cars with annual costs under $8000.

73. A customer wants to purchase a car based on fixed costs. Find the set of cars with fixed costs under $5000.

74. Find the set of cars with total annual costs under $7000.

75. Find the set of cars with total annual costs under $8000 and operating costs under $2500.

76. Find the set of cars with total annual costs under $8000, operating costs under $2500, and fixed costs under $5000.

 In Other Words

In problems 77–80, diagrams are given showing certain relationships between sets A and B represented by the circular areas. State these relationships in your own words, and find sets A and B that satisfy the conditions shown in the diagrams. *Hint:* In problem 77, A may be a set of cats and B a set of dogs.

77.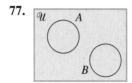

78.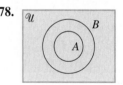

79.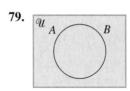

80.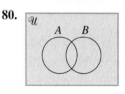

 Using Your Knowledge

The ideas of sets, subsets, unions, intersections, and complements are used in zoology and in other branches of science. Here are some typical applications.

81. A zoology book lists the following characteristics of giraffes and okapis:

Giraffes	Okapis
Tall	Short
Long neck	Short neck
Long tongue	Long tongue
Skin-covered horns	Skin-covered horns
Native to Africa	Native to Africa

Let G be the set of characteristics of giraffes, and let O be the set of characteristics of okapis.
a. Find $G \cap O$.
b. What set of characteristics is common to okapis and giraffes?
c. Find the smallest set $\mathcal{U}$ that will serve as a universal set for G and O.
d. Find G'.
e. Find O'.

Giraffes and okapis have many common characteristics.

Problems 82–87 refer to the table below. The table shows the number of heavy alcohol users (five or more drinks per occasion on five or more days in the past month) by age group.

Age	Male	Female
12–17	224,000	51,000
18–25	2,192,000	921,000
26–34	2,174,000	603,000
35 and older	3,293,000	685,000

Source: U.S. Department of Health and Human Services.

Let M be the set of males, F be the set of females, A be the set of persons 12–17 years old, B be the set of persons 18–25 years old, C be the set of persons 26–34 years old, and D be the set of persons 35 and older.

82. How many elements are in the set $M \cap D$?

83. How many elements are in the set $F \cap D$?

84. How many elements are in the set $(F \cup M) \cap D$?

85. Which category has the fewest members? Which set corresponds to that category?

86. Describe the set $M \cup F$.

87. Describe the set $M \cap F$.

Is it worth it to go to college? The following table shows the average annual earnings by degree (high school, associate, bachelor's).

Education	Male (M)	Female (F)
High school	$32,521	$21,893
Associate degree	$39,873	$28,403
Bachelor's degree	$63,127	$41,339

Let M be the set of males, F be the set of females, H be the set of average earnings of high school graduates, A be the set of average earnings of persons with an associate degree, and B be the set of average earnings of persons with a bachelor's degree. In problems 88–94, find the average earnings for the persons in the following sets:

88. $M \cap B$ **89.** $F \cap B$

90. $A \cap M$ **91.** $A \cap F$

92. Which set has the lowest earnings?

93. Describe the set $H \cap M$.

94. Describe the set $M \cup F$.

Online Study Center

To further explore unions and intersections, access link 2.2.4 on this textbook's Online Study Center.

Collaborative Learning

Form three groups of students: the Asteroids, the Comets, and the Intergalactics. Now examine the table below very carefully. Cut out each of the properties (labels) from the table. The Asteroids are to select the properties that pertain to asteroids only, the Comets select the properties that pertain to comets only, and the Intergalactics select the properties that pertain to both comets and asteroids.

To get some information about the topics in the table, consult an online encyclopedia (MSN Encarta) or try http://tinyurl.com/dt9jg.

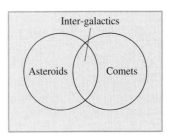

Take a large piece of paper and draw a diagram like the one above. The Asteroids should paste (glue) the properties pertaining to asteroids only in the proper region, the Comets should paste the properties pertaining to comets only in the proper region, and the Intergalactics should paste the properties common to both in the proper region.

Want to see the answer? Go to www.enchantedlearning.com/subjects/astronomy/activities/venn/cometasteroid/answers.shtml.

Made of frozen ice, gas and dust	Have a long gas tail	Also known as planetoids
Made of rock and/or metal	Have a long dust tail	Surrounded by hydrogen cloud
Orbit the Sun	Have a long ion tail	Have no atmosphere
Ceres is the biggest	Have no tail	Part of our Solar System
Halley is one	Highly elliptical orbit	Sungrazers are ones that crash into the Sun
Some come close to the Earth	Most orbit between Mars and Jupiter	Some have hit the Earth

2.3 Venn Diagrams

Blood Types and Venn Diagrams

Did you know that there are many different types and groups of human blood? Blood is classified by indicating which of three particular antigens A, B, and Rh are present. If we represent the set of persons carrying antigens A, B, and Rh by circles enclosed in a rectangle representing the set $\mathcal{U}$ of all persons, the result shown is called a **Venn diagram, Euler circle,** or **Euler diagram.** A person can be A or B or AB depending on which antigens the person has; a type O person has neither A nor B antigen. A person is Rh positive if the person has the Rh antigen and Rh negative otherwise. Plus and minus signs are used to indicate

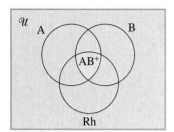

FIGURE 2.5

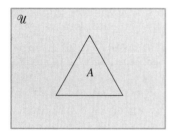

FIGURE 2.6

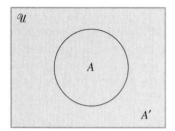

FIGURE 2.7

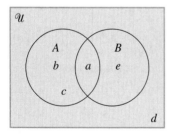

FIGURE 2.8

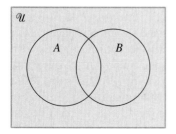

FIGURE 2.9

positive and negative blood types. Thus, AB^+ means that a person has all three antigens, as shown in Figure 2.5. You can learn more about the distribution of blood types in the Using Your Knowledge section.

What else can we do with Venn diagrams? We can show that the *commutative, associative,* and *distributive laws,* familiar from arithmetic, apply to sets by verifying that

$$\left.\begin{array}{l} A \cup B = B \cup A \\ A \cap B = B \cap A \end{array}\right\} \quad \text{Commutative laws}$$

$$\left.\begin{array}{l} A \cup (B \cup C) = (A \cup B) \cup C \\ A \cap (B \cap C) = (A \cap B) \cap C \end{array}\right\} \quad \text{Associative laws}$$

$$\left.\begin{array}{l} A \cup (B \cap C) = (A \cup B) \cap (A \cup C) \\ A \cap (B \cup C) = (A \cap B) \cup (A \cap C) \end{array}\right\} \quad \text{Distributive laws}$$

We verify these laws in problems 30–32 of Exercises 2.3. ▶

The ideas of sets, subsets, and the operations used to combine sets can be illustrated graphically by the use of diagrams called *Venn diagrams,* after John Venn (1834–1923), the English mathematician and logician who invented them. In these diagrams, we represent the universal set $\mathcal{U}$ by a rectangle, and we use regions enclosed by simple curves (usually circles) drawn inside the rectangle to represent the sets being considered. For example, if A is a subset of a universal set $\mathcal{U}$, we can represent this universal set by the set of points in the interior of the rectangle shown in Figure 2.6. The interior of the circle represents the set of points in A, whereas the set of points inside the rectangle and outside the circle represents the set A'. Obviously, closed figures other than circles can be used to represent the points of the set A. Figure 2.7 shows a Venn diagram in which A is represented by the points inside a triangle.

A. Drawing Venn Diagrams

Venn diagrams are illustrated in the following examples:

EXAMPLE 1 ▶ Drawing Venn Diagrams

Let $\mathcal{U} = \{a, b, c, d, e\}$, $A = \{a, b, c\}$, and $B = \{a, e\}$. Draw a Venn diagram to illustrate this situation.

Solution

Draw a rectangle whose interior points represent the set $\mathcal{U}$ and two circles whose interior points represent the points in A and B. The completed diagram appears in Figure 2.8 at the left. Note that a is in both A and B because $A \cap B = \{a\}$. Also, d is the only element that is in $\mathcal{U}$, but not in A or in B, so $(A \cup B)' = \{d\}$. ■

Intersections and unions of sets can be represented by Venn diagrams. For example, given two sets A and B, you can draw a Venn diagram to represent the region corresponding to $A \cap B$. Proceed as follows:

1. As usual, the points inside the rectangle represent $\mathcal{U}$, and the points inside the two circles represent A and B (Figure 2.9). Note that A and B overlap to allow for the possibility that A and B have points in common.

2. Shade the set A using vertical lines (Figure 2.10).

3. Shade the set B using horizontal lines (Figure 2.11). The region in which the lines *intersect* (cross-hatched in diagram) corresponds to $A \cap B$.

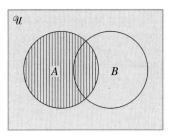

FIGURE 2.10

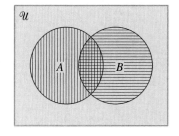

FIGURE 2.11

EXAMPLE 2 ▶ **Venn Diagrams for Complements and Intersections**

Draw a Venn diagram to represent the set $A' \cap B$.

Solution

Proceed as in Example 1.

1. Draw a rectangle and two circles as in Figure 2.12.

2. Shade the points of A' (the points in $\mathcal{U}$ and outside A) with vertical lines.

3. Shade the points of B with horizontal lines. Then $A' \cap B$ is represented by the cross-hatched region in Figure 2.12. ■

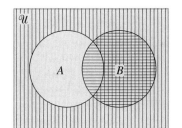

FIGURE 2.12 $A' \cap B$

We may associate with the formation of $A \cap B$ a command to shade the region representing the set to the left of the symbol $\cap$ one way and the region representing the set to the right of the symbol another way. For example, in finding $A \cap B$, because A is to the left of $\cap$, we shade region A vertically, and because B is to the right of the symbol $\cap$, we shade region B horizontally. As before, $A \cap B$ is represented by the region in which the lines intersect. (Recall that if $A \cap B = \varnothing$, A and B are said to be *disjoint*.) On the other hand, the operation $\cup$ may be thought of as a command to shade the regions representing the sets to the left and right of the symbol $\cup$ with the same type lines (horizontal or vertical). Thus, in finding $A \cup B$, we shade A with, say, horizontal lines and shade B in the same way. The union of A and B will be represented by the entire shaded region (see Figure 2.13).

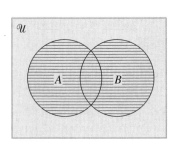

FIGURE 2.13 $A \cup B$

A somewhat simpler procedure can be adopted in the construction of Venn diagrams. As before, we start with a rectangle and two circles representing the sets $\mathcal{U}$, A, and B, respectively. We then number the four regions into which the universal set is divided (see Figure 2.14). *Note:* The numbering of the regions is completely arbitrary. By referring to the figure, we can identify the various sets as follows:

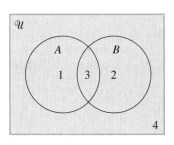

FIGURE 2.14

$A \cap B$ is the set of elements common to A and B—represented by region 3.
$A' \cap B$ is the set of elements not in A and in B—represented by region 2.
$A \cap B'$ is the set of elements in A and not in B—represented by region 1.
$A \cup B$ is the set of elements in A or in B or in both A and B—represented by regions 1, 2, and 3.
A' is the set of elements that are not in A—represented by regions 2 and 4.
B' is the set of elements that are not in B—represented by regions 1 and 4.

B. Verifying Equality

Venn diagrams are convenient for analyzing problems involving sets as long as there are not too many subsets of $\mathcal{U}$ to be considered. For example, referring to Figure 2.14, we note that $A \cap B$ is the set of points in A and in B (region 3), but $B \cap A$ is the set of points in B and in A (region 3). Hence, these two sets refer to the same region, and we can verify that $A \cap B = B \cap A$.

We now use this technique to verify one of De Morgan's Laws:

$$(A \cap B)' = A' \cup B'$$

PROBLEM SOLVING

Verifying One of De Morgan's Laws

❶ **Read** the problem.

Verify that $(A \cap B)' = A' \cup B'$.

❷ **Select** the unknown.

We actually have two unknowns: $(A \cap B)'$ and $A' \cup B'$.

❸ **Think** of a plan.

The technique used to verify equality consists of showing that the regions corresponding to the sets $(A \cap B)'$ and $A' \cup B'$, respectively, are the same. To do this, we start with a diagram similar to the one in Figure 2.14.

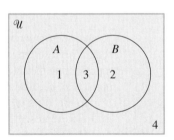

❹ **Use** the techniques you are studying to carry out the plan.

The region corresponding to $A \cap B$ is region 3. Thus, the regions corresponding to $(A \cap B)'$, the complement of $(A \cap B)$, are all the regions that *are not* region 3, that is, regions 1, 2, and 4. Thus, $(A \cap B)'$ is represented by regions 1, 2, and 4.

Now, let us find the regions for $A' \cup B'$.

A' consists of the regions outside of A, that is, 2 and 4.
B' consists of the regions outside of B, that is, 1 and 4.
The union of regions 2, 4 and 1, 4, is $A' \cup B'$ and consists of regions 1, 2, and 4. Since $(A \cap B)'$ and $A' \cup B'$ both consist of regions 1, 2, and 4, we have $(A \cap B)' = A' \cup B'$.

❺ **Verify** the answer.

You can try some examples to verify that the result is true.

TRY EXAMPLE 3 NOW.

Cover the solution, write your own solution, and then check your work.

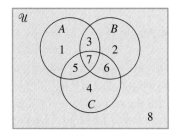

FIGURE 2.15

Venn diagrams can be used to illustrate three or more sets where the diagrams are drawn so the three sets overlap, creating eight different regions, as shown in Figure 2.15. Here are some sets and their corresponding regions in Figure 2.15.

$A \cap B \cap C$	Region 7		$A \cap B$	Regions 3 and 7
$A \cap C$	Regions 5 and 7		$B \cap C$	Regions 6 and 7
$A \cup B$	Regions 1, 2, 3, 5, 6, and 7		$A \cup C$	Regions 1, 3, 4, 5, 6, and 7
$B \cup C$	Regions 2, 3, 4, 5, 6, and 7			

EXAMPLE 3 ▶ Using Venn Diagrams to Verify Equality

If A, B, and C are subsets of $\mathcal{U}$, use the preceding method to verify the **distributive law,** $A \cap (B \cup C) = (A \cap B) \cup (A \cap C)$.

Solution

We have drawn the rectangle and the circles representing the sets $\mathcal{U}$, A, B, and C and numbered regions 1, 2, 3, 4, 5, 6, 7, and 8 in Figure 2.15. Note that when you had 2 sets, you used $2^2 = 4$ regions. In this example, you have 3 sets; hence, you need $2^3 = 8$ regions.

First consider $A \cap (B \cup C)$.

(a) A is represented by regions 1, 3, 5, and 7.

(b) $B \cup C$ is represented by regions 2, 3, 4, 5, 6, and 7.

(c) $A \cap (B \cup C)$ is therefore represented by the regions common to the two sets in parts (a) and (b), that is, by regions 3, 5, and 7.

Next consider $(A \cap B) \cup (A \cap C)$.

(d) $A \cap B$ is represented by the regions common to the circles representing A and B, that is, by regions 3 and 7.

(e) $A \cap C$ is represented by the regions common to the circles representing A and C, that is, by regions 5 and 7.

(f) $(A \cap B) \cup (A \cap C)$ is therefore represented by all the regions found in parts (d) and (e), that is, by regions 3, 5, and 7.

Because $A \cap (B \cup C)$ and $(A \cap B) \cup (A \cap C)$ are both represented by regions 3, 5, and 7, you see that

$$A \cap (B \cup C) = (A \cap B) \cup (A \cap C) \qquad ■$$

In problems 30, 31, and 34, you will verify the *commutative,* the *associative,* and *De Morgan's laws.*

C. Applications

As we saw in Getting Started, human blood is grouped according to the presence of three antigens, A, B, and Rh. Suppose that we want to use a Venn diagram to visualize all the different blood groups. We begin with three circles representing the sets of persons having antigens A, B, and Rh, respectively (Figure 2.16). As we can see, there are eight different regions, so there are eight different blood groups. Blood groups inside the set Rh will carry a plus sign (+), and those outside Rh will carry a minus sign (−). Blood with neither the A nor the B antigen will be labeled type O.

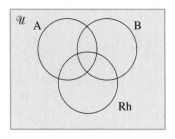

FIGURE 2.16

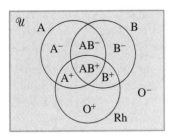

FIGURE 2.17
Human blood types.

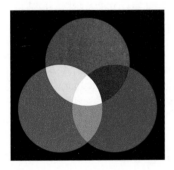

FIGURE 2.18

⁝⁝⁝*Online Study Center*

By the way, do you know why the diagram uses a black background? To produce the white color in the intersection, you have to start with no color at all (black) and add the red, green, and blue. To further explore mixing colors, access link 2.3.1 on this textbook's Online Study Center.

EXAMPLE 4 ▶ **Identifying Blood Groups**

Draw a Venn diagram to identify all possible blood groups.

Solution

The eight possible groupings of blood are A^-, A^+, B^-, B^+, AB^-, AB^+, O^-, and O^+, where A^+ means that the person has both A and Rh antigens, A^- means that the person has antigen A but not Rh, and similarly for the remaining groupings. Note in Figure 2.17 that the circle labeled A represents the set of persons having the A antigen and A^+, A^-, AB^+, or AB^- blood, and likewise for the other two circles. Read more about blood types in the Using Your Knowledge section. ∎

EXAMPLE 5 ▶ **Colors and Venn Diagrams**

Do you have an RGB monitor or television set? What does RGB mean, anyway? It means "Red, Green, and Blue," the additive primary colors. Suppose that you make a Venn diagram consisting of three circles, one red, one green, and one blue, as shown in Figure 2.18.

(a) Which color is the intersection of the three circles?

(b) The secondary colors are found by blending two of the primary colors. Name the secondary colors of light and their primary components.

Solution

(a) The intersection of the three circles is white.

(b) The secondary colors are yellow (red and green), magenta (red and blue), and cyan (green and blue). ∎

EXAMPLE 6 ▶ **Weather Diagrams**

The ideas presented in this section have been used in recent years by forecasters in the National Weather Service. For example, on the map in Figure 2.19, rain, showers, snow, and flurries are indicated with different types of shadings.

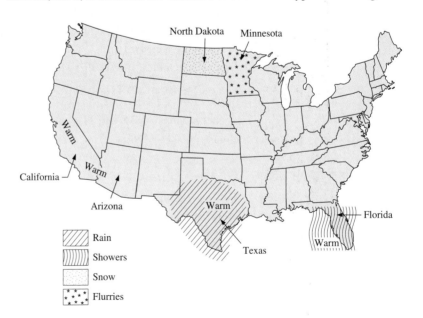

FIGURE 2.19

Find the states in which

(a) it is warm and there are showers. (b) it is warm and raining.

(c) it is warm only. (d) it is snowing. (e) there are snow flurries.

Solution

(a) Florida (b) Texas (c) California, Arizona

(d) North Dakota, Minnesota (e) Minnesota ■

Have you done an Internet search lately? The computer uses concepts of set operations in order to complete the search. For example, suppose you want to buy a car. First, you must go to the Web and select a *search engine* (a *program* that searches documents for specified *keywords* and returns a list of the documents that contain the keywords). In this illustration, we use Google. We write the word *car* as shown.

Press Google Search.

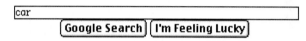

The bar shows how many results (hits) you have: 448,000,000. Too many! In order to decrease the number of results, we can state the size of the car that you want to purchase. We indicate *midsize* by typing the word *midsize* next to the word car and then pressing the Return key. The results have decreased to 1,570,000.

There are still too many results, so we add the category *American*. We indicate *American* by writing *American* next to *car* and *midsize* and pressing the Return key. Now we are down to a mere 404,000 (not cars, but Web pages or documents that mention *car midsize American*).

What does this have to do with Venn diagrams? This is what is happening.

EXAMPLE 7 ▶ Venn Diagrams and the Web

Make a Venn diagram of the sets we obtained at each of the steps in the process we just described. The first set contains the Web pages that mention *cars,* as shown in yellow. The second set contains the Web pages that mention *midsize* cars (blue). Finally, the third category is *American* (purple) and the *intersection* of the three regions has 404,000 Web pages that mention *car midsize American.* ■

Online Study Center

To further explore Venn diagrams, access link 2.3.2 on this textbook's Online Study Center. To further explore sets and Venn diagrams, access link 2.3.3.

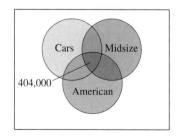

EXERCISES 2.3

Ⓐ Drawing Venn Diagrams

1. Draw a Venn diagram to illustrate the relationships among the sets $\mathcal{U} = \{1, 2, 3, 4, 5\}$, $A = \{1, 2\}$, and $B = \{1, 3, 5\}$.

2. Do the same as in problem 1 for the sets $\mathcal{U} = \{a, b, c, d, e, f\}$, $A = \{a, b, c\}$, and $B = \{d, e, f\}$.

In problems 3–8, draw a Venn diagram (see Figure 2.14) and shade the region representing each specified set.

3. $A \cap B'$ **4.** $A' \cup B'$ **5.** $(A \cup B) - (A \cap B)$

6. $A \cup B'$ **7.** $A' \cap B'$ **8.** $(A \cup B) - A$

In problems 9–17, use the numbered regions of the diagram below to identify each specified set.

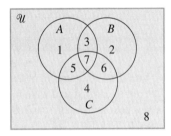

9. $A - (B \cup C)$

10. $C \cap (A \cup B)$

11. $(A \cap B \cap C) - (A \cap B)$

12. $(A \cap B') \cup (A \cap C')$

13. $(A \cup B') \cap C$

14. $(A \cup B) - C$

15. $(A \cap B') \cup C$

16. $B \cap C' \cap A$

17. $(A \cup B \cup C)'$

In problems 18–25, draw a Venn diagram to illustrate each specified set.

18. $\{x \mid x \in A \text{ or } x \in B\}$

19. $\{x \mid x \notin A \text{ or } x \notin B\}$

20. $\{x \mid x \in A \text{ and } x \notin B\}$

21. $\{x \mid x \in A \text{ and } (x \in B \text{ and } x \notin C)\}$

22. $\{x \mid x \in A \text{ and } x \in B\}$

23. $\{x \mid x \notin A \text{ and } x \in B\}$

24. $\{x \mid x \in A \text{ or } (x \in B \text{ or } x \in C)\}$

25. $\{x \mid x \notin A \text{ and } (x \in B \text{ and } x \in C)\}$

Ⓑ Verifying Equality

In problems 26–29, draw a Venn diagram that satisfies each equation.

26. $A \cap B = \varnothing$ **27.** $A \cap B = B$

28. $(A \cup B) \cap C = \varnothing$ **29.** $A \cap (A \cap B) = A$

In problems 30–34, use the numbered regions in the diagram for problems 9–17 to verify each equality.

30. a. $A \cup B = B \cup A$
    b. $A \cap B = B \cap A$
    (These two equations are called the **commutative laws** for set operations.)

31. a. $A \cup (B \cup C) = (A \cup B) \cup C$
    b. $A \cap (B \cap C) = (A \cap B) \cap C$
    (These two equations are called the **associative laws** for set operations.)

32. $A \cup (B \cap C) = (A \cup B) \cap (A \cup C)$
    [This equation and the equation $A \cap (B \cup C) = (A \cap B) \cup (A \cap C)$, which was verified in Example 3, are known as the **distributive laws** for set operations.]

33. a. $A \cup A' = \mathcal{U}$
    b. $A \cap A' = \varnothing$
    c. $A - B = A \cap B'$

34. a. $(A \cup B)' = A' \cap B'$
    b. $(A \cap B)' = A' \cup B'$
    (These two equations are known as **De Morgan's laws**.)

35. Referring to the diagram for problems 9–17, the set of regions {3, 7} represents which of the following?
    a. $A \cap B$ **b.** $A \cap B \cap C$
    c. $(A \cup B) \cap C$ **d.** $(A \cap B) \cup C$
    e. None of these

36. Referring to the diagram for problems 9–17, the set of regions {1, 2, 3} represents which of the following?
    a. $(A \cup B) \cap C$ **b.** $(A \cup B) \cap C'$
    c. $(A \cap B) \cup C$ **d.** $(A \cap B) \cup C'$
    e. None of these

37. Given $A \cap B = \{a, b\}$, $A \cap B' = \{c, e\}$, $A' \cap B = \{g, h\}$, and $(A \cup B)' = \{d, f\}$, use a Venn diagram to find the following:
 a. A, B, and $\mathcal{U}$ **b.** $A \cup B$
 c. $(A \cap B)'$

38. Given $A \cap B = \{b, d\}$, $A \cup B = \{b, c, d, e\}$, $A \cap C = \{b, c\}$, and $A \cup C = \{a, b, c, d\}$, use a Venn diagram to find the following:
 a. A, B, and C **b.** $A \cap B \cap C$
 c. $A \cup B \cup C$

39. Draw a Venn diagram representing the most general situation for four sets A, B, C, and D. (*Hint:* There should be $2^4 = 16$ regions, but *do not* use circles!)

40. Referring to Figure 2.19, find the states in which it is warm.

41. Referring to Figure 2.19, find the states in which it is warm and/or it is raining.

42. Referring to Figure 2.19, find the states in which it is warm and it is raining or there are showers.

C Applications

Do you own any stocks or bonds? To analyze your holdings (portfolio), Morning Star employs a nine-square grid that uses the idea of a Venn diagram "that provides a graphical representation of the 'investment style' of stocks and mutual funds. It classifies securities according to market capitalization, the total dollar value of all outstanding shares (shown on the vertical axis) and growth, and value factors (shown on the horizontal axis)." For example, the Vanguard Long Term Treasury Fund is classified as a high-capitalization fund.

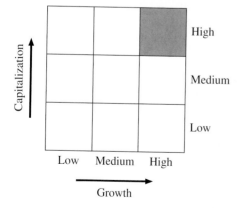

This means that the growth and value factors (horizontal axis) of the fund are high, and the capitalization (vertical axis) is also high. The intersection of the two (black area) is the analysis for the fund.

Draw a diagram for a fund with

43. low capitalization and high growth.

44. high capitalization and low growth.

45. medium capitalization and high growth.

46. low capitalization and medium growth.

47. medium capitalization and low growth.

48. high capitalization and medium growth.

Are you ready for a Web game? You need a computer for this one! Following the procedure of Example 7, go to Google.com. When we entered the word *car* (a query), we got 448,000,000 responses. There is a game called "Googlewhack" whose rules are at (where else?) www.googlewhack.com. The goal: Enter two words; get one response. For example, as of this printing, if you enter the words

Traumatizing monkiness

you should get one response. (To make the game more interesting, you can now make a question whose answer is the two words you entered!) Here is our mathematical variation of the game: Mathegoogle.

49. Can you enter two words and get two results?

50. Can you enter three words and get three results?

51. Can you enter four words and get four results?

52. The generalization: Enter n words and get n results. (Do as many as you want.)

In Other Words

In problems 53–56, write in words the set represented by the shaded region of each Venn diagram.

53.

54.
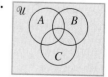

55.
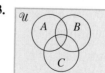

56.

In problems 57–60, refer to the following diagram and assume that none of the eight regions is empty. Answer true or false.

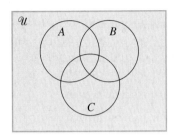

57. Any element in *A* is also in *B*.

58. No element is a member of *A*, *B*, and *C*.

59. Any element in 𝒰 is also in *C*.

60. None of the statements in 57–59 are true.

 Using Your Knowledge

In Example 4 of this section, blood was classified into eight different types. In blood transfusions, the recipient (the person receiving the blood) must have all or more of the antigens present in the donor's blood. For instance, an A^+ person cannot donate blood to an A^- person, because the recipient does not have the Rh antigen, but an A^- person can donate to an A^+ person. Refer to Figure 2.17.

61. Identify the blood type of universal recipients.

62. Identify the blood type of universal donors.

63. Can an AB^- person give blood to a B^- person?

64. Can a B^- person give blood to an AB^- person?

65. Can an O^+ person give blood to an O^- person?

66. Can an O^- person give blood to an O^+ person?

The approximate distribution of blood types in the U.S. population is as follows. (Distribution may be different for specific racial and ethnic groups.)

O Rh-positive	38%	B Rh-positive	9%
O Rh-negative	7%	B Rh-negative	2%
A Rh-positive	34%	AB Rh-positive	3%
A Rh-negative	6%	AB Rh-negative	1%

Source: www.aabb.org/Content/Donate_Blood/
Blood_Donation_FAQs/donatefaqs.htm. You can read all about blood on this site!

Can you use your knowledge to find what percent of the people have antigen

67. A?   **68.** B?

69. O? *Hint:* See Figure 2.17 to see which people have O blood.

70. Based on your answers to 67–69, which blood types (A, B, or O) are the most and least common?

 Discovery

John Venn invented the diagrams introduced in this section to illustrate his work, *Symbolic Logic.* The Swiss mathematician Leonhard Euler (1707–1783) used similar diagrams to illustrate his work. For this reason, Venn diagrams are sometimes called *Euler circles* or *Euler diagrams.*

We have seen in the preceding examples that if we have one set, the corresponding Venn diagram divides the universe into two regions. Two sets divide the universe into four regions, and three sets divide it into eight regions.

71. What is the maximum number of regions into which four sets will divide the universe?

72. The diagram for a division of the universal set into the 16 regions corresponding to four given sets may look like the figure that follows. Can you guess the maximum number of regions into which *n* sets will divide the universe?

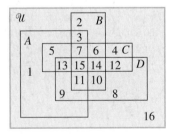

73. Referring to the diagram for problem 72, find the regions corresponding to the following:
a. $A \cap B \cap C' \cap D$ **b.** $(A \cup B \cup C)'$

 Collaborative Learning

This group activity is designed to determine how to mix the *additive* primary colors, *red, green,* and *blue,* to obtain *white* and how to mix the *subtractive* primary

colors, *yellow, cyan* (light blue), and *magenta* (purple), to obtain *black*.

Form two groups of students. One group should have a piece of black paper and three different-colored markers or cans of paint (red, green, and blue) or three plastic disks similar to those in Example 5. (You can use plastic report cover protectors to make the disks.) The other group should have a piece of white paper and three different-colored markers or cans of paint—cyan (light blue), magenta (purple), and yellow—or three plastic disks.

Each of the groups should make a Venn diagram similar to the one in Example 5 consisting of three circles of different colors.

1. Which color is the intersection of the three additive primary colors?

2. Which color is the intersection of the three subtractive secondary colors?

3. Which other intersection of three colors (if any) would produce white?

4. Which other intersection of three colors (if any) would produce black?

Additive secondary colors are composed of two of the primary colors.

5. Which color do you get if you mix red and green?

6. Which color do you get if you mix red and blue?

7. Which color do you get if you mix green and blue?

 Online Study Center

To further explore mixing colors, access links 2.3.4 and 2.3.5 on this textbook's Online Study Center.

 *Research Questions*

As we have mentioned before, John Venn invented the diagrams bearing his name, but Leonhard Euler used similar diagrams to illustrate his own work. Euler also contributed to modern notation in mathematics.

1. Find at least three different types of notation that Euler developed.

2. Catherine the Great once asked for Euler's help to quiet the famous French philosopher Diderot, who was attempting to convert her subjects to atheism. One day in the court, Diderot was informed that someone had a mathematical proof of the existence of God. Write a short paragraph indicating how Euler proceeded to quiet Diderot.

2.4 The Number of Elements in a Set: A Problem-Solving Tool

Test Registration and Counting

When registering for the Test of English as a Foreign Language (TOEFL) or the Test of Spoken English (TSE), students must indicate which test they plan to take by checking one of the boxes on the envelope (see Figure 2.20 on the following page). Suppose that a total of 800 students are registered for the TOEFL exam and 500 students are registered for the TSE; 200 of these students indicate that they will take both tests. How many students are participating; that is, how many students are registered? To answer this question, we have to develop a notation indicating how many elements we have in each set. If F is the set of students registered for TOEFL and S is the set of students registered for TSE, we are asked to find the total number of registered students. (They can be taking TOEFL only,

TSE only, or both.) We answer this question and more in Example 1. Similar questions are presented in Exercises 2.4, problems 7–9.

FIGURE 2.20   ▶

One of the simplest counting techniques involves the counting of elements in a given set. If A is any set, the number of elements in A is denoted by $n(A)$. The number $n(A)$ is frequently called the **cardinal number** of A. If $n(A)$ is a whole number, the set A is a **finite** set. For example, if $A = \{g, i, r, l\}$, then $n(A) = 4$. Likewise, if $B = \{@, \#, \$\}$, then $n(B) = 3$, and of course, A and B are finite sets. We shall be interested here in counting the number of elements in sets formed by the operations of union, intersection, and taking complements.

A. Counting the Elements of a Set

EXAMPLE 1 ▶ **TOEFL and TSE Survey**

In a particular school, 800 students are registered for TOEFL, 500 are registered for TSE, and 200 are registered for both tests.

(a) What is the total number of registered students?

(b) How many students are taking TOEFL only?

(c) How many students are taking TSE only?

Solution

Let F be the set of students registered for TOEFL and S be the set of students registered for TSE. First draw a Venn diagram with overlapping regions to show the information (Figure 2.21).

1. $S \cap F$ has 200 students. Write 200 in the region corresponding to $S \cap F$.

2. Since $n(F) = 800$ and 200 students are in the intersection of F and S, the number of students taking TOEFL only is $800 - 200 = 600$, as shown in Figure 2.21.

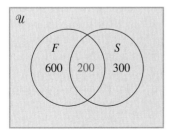

FIGURE 2.21

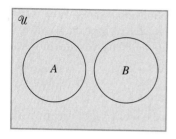

FIGURE 2.22
$A \cap B = \varnothing$

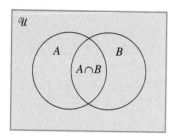

FIGURE 2.23
$A \cap B \neq \varnothing$

3. The number of students taking TSE only is $500 - 200 = 300$.

(a) The number of registered students is

$$n(F \cup S) = 600 + 200 + 300 = 1100$$

(b) The number of students taking TOEFL only is 600.

(c) The number of students taking TSE only is 300.

Note that $n(F) = 600 + 200 = 800$ and $n(S) = 200 + 300 = 500$. Thus,

$$n(F \cup S) = n(F) + n(S) - n(S \cap F)$$

that is,

$$1100 \quad = 800 + 500 - 200$$

We now examine the problem of finding the number of elements in the union of two sets in a more general way. Let us assume that A and B are any two given sets. We must consider two possibilities.

1. $A \cap B = \varnothing$ (see Figure 2.22): In this case, $n(A \cup B) = n(A) + n(B)$.

2. $A \cap B \neq \varnothing$ (see Figure 2.23): In this case, $A \cup B$ includes all the elements in A and all the elements in B, but each element is counted only once. It is thus clear that $n(A) + n(B)$ counts the elements in $A \cap B$ twice and so exceeds $n(A \cup B)$ by $n(A \cap B)$. Therefore,

$$n(A \cup B) = [n(A) + n(B)] - n(A \cap B)$$

or

Number of Elements in $A \cup B$	
$n(A \cup B) = n(A) + n(B) - n(A \cap B)$	(1)

Online Study Center

To further explore Venn diagrams, access links 2.4.1 and 2.4.2 on this textbook's Online Study Center.

Notice that equation (1) is correct even if $A \cap B = \varnothing$ because in that case $n(A \cap B) = 0$.

EXAMPLE 2 ▶ Finding $n(A \cup B)$

If $n(A) = 20$, $n(B) = 30$, and $n(A \cap B) = 10$, find $n(A \cup B)$.

Solution
Using equation (1), we have

$$n(A \cup B) = n(A) + n(B) - n(A \cap B) = 20 + 30 - 10 = 40$$

It is possible to develop a formula similar to equation (1) for the case in which three or more sets are considered. However, we will rely on the use of Venn diagrams to solve such problems. For example, suppose we want to buy a pair of running shoes on the Internet using a search engine like Google. How many Web pages include information about running shoes? Go to Google.com, type the word *shoes* and press Google Search (or Enter). The blue bar tells us that

shoes
Google Search
Results 1 - 10 of about 9,510,000.

shoes running
Google Search
Results 1 - 10 of about 1,090,000.

there are about 9,510,000 Web pages with information about shoes. How many pages do we have to choose from if we want *running shoes?* Type the word *running* after shoes and press Enter. Now, you have 1,090,000 pages with information about running shoes.*

EXAMPLE 3 ▶ Buying Shoes and Venn Diagrams

If S is the set of pages dealing with shoes and R is the set of pages dealing with running,

(a) what is $n(S)$?   (b) what is $n(S \cap R)$?

(c) Draw a Venn diagram and find the number of pages that deal with nonrunning shoes.

Solution

(a) Since there are about 9,510,000 pages dealing with shoes, $n(S) = 9,510,000$.

(b) Since there are 1,090,000 pages dealing with shoes that are running shoes, $n(S \cap R) = 1,090,000$.

(c) Draw two overlapping circles labeled S and R. We know that there are 1,090,000 elements in the intersection of S and R, so we enter 1,090,000 as shown in Figure 2.24. The elements that represent nonrunning shoes are in S and not in R, that is, in $S \cap R'$. Since we have 9,510,000 in S and 1,090,000 in $S \cap R$, the number of elements in $S \cap R'$ is $9,510,000 - 1,090,000 = 8,420,000$. ■

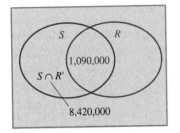

FIGURE 2.24

B. Applications

Venn diagrams can also be used to study surveys, as shown next. The strategy used to solve these problems is to make a diagram and assign to each region the correct number of elements.

PROBLEM SOLVING

❶ **Read** the problem.

Surveys

To estimate the number of persons interested in recycling aluminum cans, glass, and newspapers, a company conducts a survey of 1000 people and finds that

200 recycle glass (G)	300 recycle paper (P)
450 recycle cans (C)	50 recycle cans and glass
15 recycle paper and glass	60 recycle cans and paper
10 recycle all three	

(a) How many people do not recycle at all?

(b) How many people recycle cans only?

❷ **Select** the unknown.

We want to find the number of people who do not recycle at all and the number of people who recycle cans only.

continued

*These numbers change daily, so your numbers may be different.

❸ Think of a plan. The strategy is to draw a Venn diagram.

❹ Use the information to carry out the plan.

We draw a Venn diagram with three overlapping circles labeled C, G, and P (Figure 2.25).

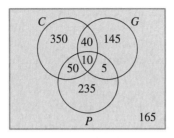

FIGURE 2.25

We have to start with the 10 persons who recycle all three and place them in $C \cap G \cap P$. Since 15 recycle paper and glass, the intersection of P and G must have 15 people. We already have 10 in $C \cap G \cap P$, so we add 5 in the remainder of $P \cap G$.

Do we know how to distribute the 200 people who recycle glass inside the circle G?

Similarly, we place 50 more in the remainder of $C \cap P$ and 40 in the remainder of $C \cap G$. We now have $10 + 50 + 40 = 100$ persons in C, so we place $450 - 100 = 350$ in the region corresponding to C only. Similarly, we place $200 - 55 = 145$ in the region corresponding to G only and $300 - 65 = 235$ in the region corresponding to P only.

We have $350 + 40 + 145 + 50 + 10 + 5 + 235 = 835$ people inside the circles.

(a) Since 1000 persons were surveyed, $1000 - 835 = 165$ persons are outside the three circles and do not recycle at all.

(b) 350 recycle cans only.

❺ Verify the answer.

The sum of all the numbers in the diagram is 1000.

TRY EXAMPLE 4 NOW.

Cover the solution, write your own solution, and then check your work.

EXAMPLE 4 ▶ A Music Survey

A survey of students at Prince Tom University shows that

> 29 like jazz.
> 23 like rock.
> 40 like classical music.
> 10 like classical music and jazz.
> 13 like classical music and rock.
> 5 like rock and jazz.
> 3 like rock, jazz, and classical music.

If there are a total of 70 students in the survey, find

(a) the number of students who like classical music only.

(b) the number of students who like jazz and rock but not classical music.

(c) the number of students who do not like jazz, rock, or classical music.

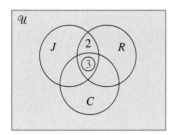

Step 1

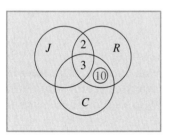

Step 2

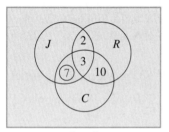

Step 3

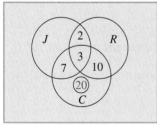

Step 4

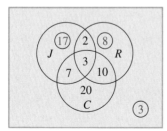

Step 5

FIGURE 2.26

Solution

Read the problem carefully; there are three questions that are to be answered.

Select the letters J, R, and C to represent the sets of students who like jazz, rock, and classical music, respectively.

Think of a plan to solve the problem. Since the data indicate some overlap, draw a Venn diagram (Figure 2.26, step 1).

Use the Venn diagram to fill in the numbers as follows:

Since 3 students like all three types of music, the region common to J, R, and C ($J \cap R \cap C$) must contain the number 3. Next, note that 5 students like jazz and rock, so the region common to J and R ($J \cap R$) must contain a total of 5. But there are already 3 in $J \cap R \cap C$, a portion of $J \cap R$, so put 2 in the remainder of $J \cap R$ (Figure 2.26, step 1).

Similarly, since 13 students like rock and classical music, the region common to R and C ($R \cap C$) must contain a total of 13. There are already 3 in $J \cap R \cap C$, a portion of $R \cap C$, so put 10 in the remainder of $R \cap C$ (Figure 2.26, step 2).

Since 10 students like classical music and jazz, the region common to C and J ($J \cap C$) must contain a total of 10. There are already 3 in a portion of this region ($J \cap R \cap C$), so the remainder of the region must contain $10 - 3 = 7$ (Figure 2.26, step 3).

Next, note that 40 students like classical music, and $7 + 3 + 10 = 20$ of these are already accounted for. Thus, the remainder of region C, outside of regions J and R, must contain $40 - 20 = 20$ (Figure 2.26, step 4).

Proceeding in the same way, fill in the numbers for the remaining regions in the figure. From the completed diagram (Figure 2.26, step 5), note that

(a) 20 students like classical music only.

(b) 2 students like jazz and rock but not classical music.

(c) since the numbers inside the circles add up to 67, 3 students do not like jazz, rock, or classical music.   ∎

Verification is done by counting all the elements in Figure 2.26, step 5, and making sure they add up to 70, the total number of students in the survey. Note that the problem-solving strategy is to draw a Venn diagram and fill it in, starting with the innermost region and then working step by step toward the outside, as done in Figure 2.26, steps 1–5.

EXAMPLE 5 ▶ Surveying Courses

In a survey of 100 students, the numbers taking algebra (A), English (E), and philosophy (P) are shown in Figure 2.27.

(a) How many students are taking algebra or English but not both?

(b) How many students are taking algebra or English but not philosophy?

(c) How many students are taking one or two of these courses but not all three?

(d) How many students are taking at least two of these courses?

(e) How many students are taking at least one of these courses?

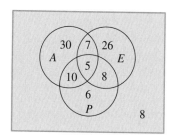

FIGURE 2.27

Solution

(a) Here, we want all the elements in A or E but not in both. We read the numbers from the Venn diagram to get

$$30 + 10 + 26 + 8 = 74$$

Note that we have taken all the numbers in $A \cup E$ except those in $A \cap E$.

(b) Here, we need all the numbers in A or E that are not in P. From the diagram, we find

$$30 + 7 + 26 = 63$$

(c) The required number here is the number in the entire universal set, 100, minus the number taking all three courses, 5, or none of these courses, 8. Thus, the result is

$$100 - 5 - 8 = 87$$

(d) The required number here is the number in

$$(A \cap E) \cup (A \cap P) \cup (E \cap P)$$

which is

$$7 + 5 + 10 + 8 = 30$$

(e) We can get the required number by taking the number in the universal set minus the number taking none of these courses. We find the result to be

$$100 - 8 = 92$$

■

EXAMPLE 6 ▶ **Reading Tables**

An insurance company has classified a group of drivers as indicated in Table 2.3 below. Find the number of persons who

(a) are low risks under age 21.

(b) are not high risks and are over age 35.

(c) are under age 21.

(d) are low risks.

TABLE 2.3

Risk	Under Age 21	Age 21–35	Over Age 35
Low risk	15	20	35
Average risk	25	15	10
High risk	50	10	30

Solution

(a) We see that there are 15 persons in the column labeled "Under Age 21" *and* in the row labeled "Low risk."

(b) There are 30 persons who are over age 35 and high risks, so the rest of the persons in the column labeled "Over Age 35" are not high risks. They number $10 + 35 = 45$.

(c) We count all the persons in the column labeled "Under Age 21." The sum is
$15 + 25 + 50 = 90$.

(d) We count the persons in the row labeled "Low risk." The sum is
$15 + 20 + 35 = 70$.   ■

Results of polls are often shown using a table. Here are the basic relationships
between each of the three habits being considered (smoking, drinking, overweight) in this analysis and self-reported "excellent" health.

	Excellent Physical Health	*Excellent Mental Health*
0 Habits	42%	47%
1 Habit	31%	45%
2 Habits	27%	40%
3 Habits	17%	42%

Source: The Gallup Organization: Three Deadly Habits.

EXAMPLE 7 ▶ Interpreting a Gallup Poll

Referring to the table,

(a) what percent of the people have 0 habits and claim to be in excellent physical
health?

(b) what percent of the people have 3 habits and claim to be in excellent physical
health?

(c) is there a relationship between the number of habits and excellent physical
health?

(d) what percent of the people who are not overweight and do not drink or
smoke report excellent physical health?

(e) what percent of the people who are not overweight and do not drink or
smoke report excellent mental health?

Solution

Note that the number of habits refers to the **row**, and the health (physical or mental) refers to the **column**. Column **2** for physical and column **3** for mental.

(a) 0 habits (row 1) and excellent physical health (column **2**) lists 42%. Thus,
42% of the people have 0 habits and claim to be in excellent physical health.

(b) The percent of people with 3 habits (row **4**) in excellent physical health (column **2**) is **17%**.

(c) The table suggests that as the number of habits increases, physical health
decreases.

(d) If you are not overweight and do not drink or smoke, then you have 0 habits
(row **1**). If you are in excellent physical health, you are in column **2**. The
number in row **1**, column **2** is 42%. Thus, the percent of people who are not
overweight, do not drink or smoke, and claim to be in excellent physical
health is **42%.**

(e) The percent of people who are not overweight, do not drink or smoke, and
claim to be in excellent mental health is **47%** (row **1**, column **3**).

EXERCISES 2.4

Ⓐ Counting the Elements of a Set

1. Suppose that $n(A) = 15$, $n(B) = 20$, and $n(A \cap B) = 5$. Find $n(A \cup B)$.

2. Suppose that $n(A) = 12$, $n(B) = 6$, and $n(A \cup B) = 14$. Find $n(A \cap B)$.

3. Suppose that $n(A) = 15$, $n(A \cap B) = 5$, and $n(A \cup B) = 30$. Find $n(B)$.

4. There are 50 students in an algebra (A) class and 30 students in a chemistry (C) class. Find
  a. $n(A)$.   **b.** $n(C)$.
  c. the total number of students taking either algebra or chemistry if it is known that none of the students are taking both courses.
  d. the number of students taking algebra and/or chemistry if it is known that 10 students are taking both courses.

5. On checking with 100 families, it was found that 75 families subscribe to *Time*, 55 to *Newsweek*, and 10 to neither magazine. How many subscribe to both?

6. If, on checking with 100 families, it was found that 83 subscribe to *Time*, 40 to *Newsweek*, and 30 to both magazines, how many subscribe to neither?

Ⓑ Applications

7. In a survey of 100 students, the numbers taking various courses were found to be English, 60; mathematics, 40; chemistry, 50; English and mathematics, 30; English and chemistry, 35; mathematics and chemistry, 35; and courses in all three areas, 25.
  a. How many students were taking mathematics, but neither English nor chemistry?
  b. How many were taking mathematics and chemistry, but not English?
  c. How many were taking English and chemistry, but not mathematics?

8. Mr. N. Roll, the registrar at a university, has observed that, of the students,

    45% have a 9 A.M. class.
    45% have a 10 A.M. class.
    40% have an 11 A.M. class.
    20% have a 9 and a 10 A.M. class.
    10% have a 9 and an 11 A.M. class.
    15% have a 10 and an 11 A.M. class.
    5% have a 9, a 10, and an 11 A.M. class.

  a. What percent of the students have only a 9 A.M. class?
  b. What percent of the students have no classes at these times?

9. The following table shows the distribution of employees at the Taste-T Noodle Company.

Personnel Distribution

Department*	(A)	(C)	(O)	(SK)	(SS)	(U)
Purchasing (P)	1	14	7	0	0	0
Quality control (Q)	11	7	6	21	53	11
Sales (S)	8	8	40	0	0	0
Manufacturing (M)	5	7	0	9	23	37
Janitorial (J)	3	0	0	6	8	11

*(A) administrative, (C) clerical, (O) other, (SK) skilled, (SS) semiskilled, (U) unskilled.

  a. How many employees are there in the Purchasing Department?
  b. How many skilled employees are there in the factory?
  c. How many of the skilled employees are in the Janitorial Department?

10. Use the table in problem 9 to find the number of people in the following sets:
  a. $A \cap S$ **b.** $S \cup P$
  c. $M \cap A' \cap SK'$ **d.** $S \cap A' \cap C'$

11. The following table gives the estimated costs for a proposed computer system in 2-year intervals, projected over 10 years (figures in thousands of dollars).

Cost Projections

Item	Year				
	1–2	*3–4*	*5–6*	*7–8*	*9–10*
Data-processing equipment	215	240	260	295	295
Personnel	85	85	95	95	105
Materials	120	65	35	35	40
All others	90	85	90	95	120

a. How much money would be spent on materials in the first 2 years?

b. What would be the total cost at the end of the second year?

c. What would be the cost of the data-processing equipment over the 10-year period?

12. In a survey of 100 investors, it was found that

> 5 owned utilities stock only.
> 15 owned transportation stock only.
> 70 owned bonds.
> 13 owned utilities and transportation stock.
> 23 owned transportation stock and bonds.
> 10 owned utilities stock and bonds.
> 3 owned all three kinds.

a. How many investors owned bonds only?

b. How many investors owned utilities and/or transportation stock?

c. How many investors owned neither bonds nor utilities stock?

13. In a recent survey of readers of the *Times* and/or the *Tribune,* it was found that 50 persons read both the *Times* and the *Tribune.* If it is known that 130 persons read the *Times* and 120 read the *Tribune,* how many people were surveyed?

14. In a survey conducted in a certain U.S. city, the data in the following table were collected:

Income	White (W)	Black (B)	Other (O)
Over $10,000 (H)	50	15	10
$7000–$10,000 (M)	40	25	15
Under $7000 (L)	30	35	20

Find the number of people in the following sets:
a. M **b.** M'
c. $(O \cup B) \cap W'$ **d.** $L \cup O'$
e. $H \cap B'$

15. In a survey of 100 customers at the Royal Hassle Restaurant, it was found that

> 40 had onions on their hamburgers.
> 35 had mustard on their hamburgers.
> 50 had catsup on their hamburgers.
> 15 had onions and mustard on their hamburgers.
> 20 had mustard and catsup on their hamburgers.
> 25 had onions and catsup on their hamburgers.
> 5 had onions, mustard, and catsup on their hamburgers.

a. How many customers had hamburgers with onions only?

b. How many customers had plain hamburgers (no condiments)?

c. How many customers had only one condiment on their hamburgers?

16. A survey of 900 workers in a plant indicated that 500 owned houses, 600 owned cars, 345 owned boats, 300 owned cars and houses, 250 owned houses and boats, 270 owned cars and boats, and 200 owned all three.

a. How many of the workers did not own any of the three items?

b. How many of the workers owned only two of the items?

17. A coffee company was willing to pay $1 to each person interviewed about his or her likes and dislikes on types of coffee. Of the persons interviewed, 200 liked ground coffee, 270 liked instant coffee, 70 liked both, and 50 did not like coffee at all. What was the total amount of money the company had to pay?

18. In a recent survey, a statistician reported the following data:

> 15 persons liked brand A.
> 18 persons liked brand B.
> 12 persons liked brand C.
> 8 persons liked brands A and B.
> 6 persons liked brands A and C.
> 7 persons liked brands B and C.
> 2 persons liked all three brands.
> 2 persons liked none of the three brands.

When the statistician claimed to have interviewed 30 persons, he was fired. Can you explain why?

19. In problem 18, a truthful statistician was asked to find out how many people were interviewed. What was this statistician's answer?

20. In an experiment, it was found that a certain substance could be of type x or type y (not both). In addition, it could have one, both, or neither of the characteristics m and n. The following table gives the results of testing several samples of the substance. Let M and N be the sets with characteristics m and n, respectively, and let X and Y be the sets of type x and y, respectively. How many samples are in each of the following sets?

a. $M \cap X$

b. $(X \cup Y) \cap (M \cup N)$

c. $(Y \cap M) - (Y \cap N')$

d. $(X \cup Y) \cap (M \cup N')$

	m Only	*n* Only	*m and n*	*Neither m nor n*
x	6	9	10	20
y	7	11	15	9

21. The number of students taking algebra (A) or chemistry (C) is shown in the diagram below. Find the following:

a. $n(A)$

b. $n(C)$

c. $n(A \cap C)$

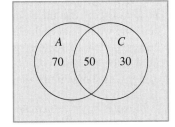

22. Referring to the diagram for problem 21, find $n(A \cup C)$.

23. If the total number of students surveyed to obtain the data for problem 21 is 200, find

a. $n(A')$.   **b.** $n(C')$.   **c.** $n(A' \cap C')$.

24. With the total number of students as in problem 23, find

a. $n(A' \cup C)$.   **b.** $n(A \cup C')$.

25. On checking with 100 investors to see who owned electric company stock (E), transportation stock (T), or municipal bonds (M), the numbers shown in the diagram below were found.

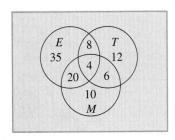

a. How many investors owned electric company or transportation stock but not both?

b. How many owned electric company or transportation stock but not municipal bonds?

c. How many had one or two of these types of investments but not all three?

d. How many had at least two of these types of investments?

e. How many had none of these types of investments?

26. A number of people were interviewed to find out who buys products A, B, and C regularly. The results are shown in the diagram below.

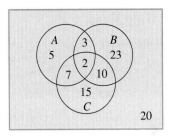

a. How many buy product A?

b. How many buy product A but not B?

c. How many buy product B or C, but not A?

d. How many do not buy product C?

e. How many people were interviewed?

A book titled *Are you Normal?* by Bernice Kanner offers many amusing but important pieces of information gleaned from actual surveys.

Solve each of problems 27–29 using a Venn diagram.

27. *What do you eat first, the frosting or the cake?* Astonishingly, only about 3% of us eat them together. Most people (69%) take Marie Antoinette's infamous words literally: They eat cake first (C). Just shy of 30%—more with younger people—pluck off the frosting first (F). On the basis of this information, what percent of the people do not eat cake?

28. An overwhelming majority of us (91%) write our return address on the front left corner of an envelope (L), avoiding the back flap (B) as if it were a uranium mine. Just 7% of letter writers put the return address on the back flap, and 2% do it either way. What percent of the people do not write a return address on their envelopes?

29. *Is the glass half-full or half-empty?* Some 46% of men and 52% of women describe themselves as optimists (O), whereas 8% of men and 11% of women see themselves as pessimists (P). The rest say they are neither. What percent of the men are neither? *Hint:* Draw a diagram for men *only*.

30. On the basis of the information in exercise 29, what percent of the women are neither?

 In Other Words

Classify the following statements as true or false. If true, explain why. If false, give a counterexample.

31. If $n(A) = n(B)$, then $A = B$.

32. If $A = B$, then $n(A) = n(B)$.

33. If $A - B = \varnothing$, then $n(A) = n(B)$.

34. If $n(A) = n(B)$, then $A - B = \varnothing$.

 Using Your Knowledge

The cartoon below seems to indicate that it is impossible to have the following morale statistics:

> 58% want out (*WO*).
> 14% hate his guts (*HG*).
> 56% plan to desert (*PD*).
> 8% are undecided (*UD*) (do not plan to do any of the above).

However, a new statistician is hired and finds that in addition to the original information, the following statements are also true:

> 12% want to do only one thing—hate his guts.
> 36% want to do exactly two things. Of these,
>   34% want out and plan to desert, and
>   2% hate his guts and want out.

Of course, nobody in his right mind would do all three things.

35. On the basis of all the information, both old and new, draw a Venn diagram and show that it is possible to have the statistics quoted in the cartoon.

 **Discovery**

In Section 2.1 we discussed the subsets of a given set. It is interesting to diagram the formation of such subsets. We imagine that the elements of the given set are listed, and we look at each element in turn and decide whether to include it in the subset. For example, suppose that the given set is $\{a, b\}$. Then our diagram has two steps, as shown in the diagram below.

Include a?	Include b?	Resulting subset
Yes	Yes	$\{a, b\}$
	No	$\{a\}$
No	Yes	$\{b\}$
	No	$\varnothing$

This diagram makes it clear that there are 2×2, or 4, subsets in all.

Reprinted with special permission of North America Syndicate, Inc.

The following diagram is for the three-element set $\{a, b, c\}$.

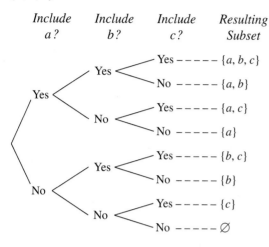

| Include
a? | Include
b? | Include
c? | Resulting
Subset |

Diagrams like these are called **tree diagrams,** and we shall discuss them in more detail in Chapter 10. The second tree diagram shows that a three-element set has $2 \times 2 \times 2$, or 8, subsets. Can you discover an easy way to explain this?

36. Can you discover the tree diagram for a four-element set and its subsets?

37. Can you discover how to count the number of subsets of a four-element set?

38. Can you now discover how to explain why an n-element set has

$$\underbrace{2 \times 2 \times 2 \times \cdots \times 2}_{n \text{ twos}}$$

or 2^n, subsets?

Online Study Center

Look under the topic "Controversial Venn Diagrams" on the Web. To further explore controversial Venn diagrams, access link 2.4.4 on this textbook's Online Study Center.

 Research Question

1. We have already mentioned that Venn diagrams are sometimes called Euler circles. Find out why this is so and in what work Euler first used "Venn" diagrams.

2.5 Infinite Sets

Infinite Sets and One-to-One Correspondences

Is the number of stars infinite? Mathematically speaking, if S is the set of all stars, what is $n(S)$? We do know that if $A = \{1, 2, 3\}$, then $n(A) = 3$. Now, suppose that $N = \{1, 2, 3, \ldots\}$; what is $n(N)$? To find $n(N)$ is to determine how many natural (counting) numbers there are! Georg Cantor studied this problem and assigned the **transfinite cardinal** $\aleph_0$ (read "aleph null"; $\aleph$ is the first letter of the Hebrew alphabet) to $n(N)$. Thus, $n(N) = \aleph_0$. Now, which set has more elements, $N = \{1, 2, 3, \ldots\}$ or $E = \{2, 4, 6, \ldots\}$?

At first, it seems that there should be twice as many natural numbers as there are even numbers. But consider the **one-to-one correspondence**

$$N = \{1, 2, 3, \ldots, n, \ldots\}$$
$$\updownarrow \updownarrow \updownarrow \quad \updownarrow$$
$$E = \{2, 4, 6, \ldots, 2n, \ldots\}$$

where every natural number n in N is paired with the even number $2n$ in E. Can you see that there are as many even numbers as there are natural numbers? The two sets are said to be **equivalent.** Moreover, there are as many natural numbers as there are fractions, as you will discover later in the book. For now, you will learn how to use one-to-one correspondences to determine whether sets are

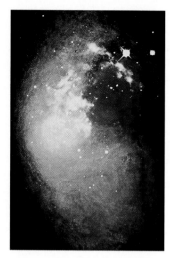

Is the number of stars infinite? The infrared map of the Orion nebula shown above unveils new star formation.

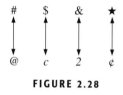

FIGURE 2.28

equivalent (see problems 1–5 in Exercises 2.5). In problems 16–20, you will even determine whether a set is infinite! ▶

What is an infinite set? In the years 1871–1884, Georg Cantor created a completely new mathematical discipline, **set theory.** Cantor asked himself: "What do we mean when we say of two finite sets that they consist of equally many things, that they have the same number, that they are equivalent?" The tools he used to answer this question were *cardinal numbers* and *one-to-one correspondences.*

A. Equivalent Sets and Cardinal Numbers

In Section 2.4 we noted that the cardinal number of a set A is the number of elements in A, denoted by $n(A)$ as before. Thus, if $A = \{a, b, c\}$, the cardinal number of A is $n(A) = 3$. To ascertain whether two sets have the same cardinal number, we determine whether each element of A can be matched with a unique element of B and vice versa. That is, we check whether there is a one-to-one correspondence between A and B.

EXAMPLE 1 ▶ Showing that A and B Have the Same Cardinality

Show that the sets $A = \{\#, \$, \&, \bigstar\}$ and $B = \{@, c, 2, ¢\}$ have the same cardinal number.

Solution
You have to show that there is a one-to-one correspondence between the elements of A and those of B. One such correspondence is given in Figure 2.28. If, as before, $n(A)$ represents the number of elements in A, $n(A) = n(B) = 4$. Can you find a different correspondence?   ■

B. Equivalent Sets

Definition of Equivalent Sets
If two sets A and B can be placed into a one-to-one correspondence with each other, the two sets are said to be **equivalent,** denoted by $A \sim B$.

Thus, the set of vowels V in the English alphabet is equivalent to the set of the first five counting numbers F, as can be seen by the correspondence

a	e	i	o	u
$\updownarrow$	$\updownarrow$	$\updownarrow$	$\updownarrow$	$\updownarrow$
1	2	3	4	5

We then say that $V \sim F$.

EXAMPLE 2 ▶ Showing that N and E Are Equivalent

Show that the set N of counting numbers and the set E of even counting numbers are equivalent.

Solution
Set up the correspondence

1	2	3	4	$\cdots$	n	$n+1$	$\cdots$
$\updownarrow$	$\updownarrow$	$\updownarrow$	$\updownarrow$		$\updownarrow$	$\updownarrow$	
2	4	6	8	$\cdots$	$2n$	$2(n+1)$	$\cdots$

Since the two sets can be placed into a one-to-one correspondence, $N \sim E$. In mathematics, when a set S is equivalent to the set N of counting numbers, S is said to be **denumerable.**

C. Infinite Sets

At first sight, it may seem strange that the set of counting numbers and a proper subset of itself (the set of even counting numbers) can be put into a one-to-one correspondence. This apparent paradox puzzled mathematicians for years until it was resolved by Georg Cantor. Cantor defined an infinite set as follows:

> **Definition of an Infinite Set**
>
> A set is **infinite** if it is equivalent to one of its proper subsets.

In his theory, he assigned the cardinal number $\aleph_0$ to the set of counting numbers; that is, if $N = \{1, 2, 3, \ldots \}$, then $n(N) = \aleph_0$.

With this convention it is possible to find the cardinal number of certain infinite sets. For example, because the set E of even counting numbers is equivalent to N, we must have $n(N) = n(E) = \aleph_0$.

To show that N is infinite, we note that N has the proper subset E that can be put into a one-to-one correspondence with N. Thus, we conclude that N is an infinite set. In contrast to the cardinal number of a finite set, that of an infinite set is usually called a **transfinite cardinal number.**

EXAMPLE 3 ▶ **Showing Equivalency and Cardinality**
Consider the set $S = \{1^2 = 1, 2^2 = 4, 3^2 = 9, \ldots, n^2, \ldots \}$.

(a) Show that S is equivalent to N.

(b) Find the cardinality of S.

(c) Show that S is infinite.

Solution

(a) To show that S and N are equivalent, set up a one-to-one correspondence between S and N as follows:

1	4	9	$\cdots$	n^2	$\cdots$
$\updownarrow$	$\updownarrow$	$\updownarrow$		$\updownarrow$	
1	2	3	$\cdots$	n	$\cdots$

Thus, S and N are equivalent; that is, $S \sim N$.

(b) Since S and N are equivalent and $n(N) = \aleph_0$, $n(N) = n(S) = \aleph_0$.

(c) To show that S is infinite, you must place S into a one-to-one correspondence with one of its proper subsets. Here is such a correspondence.

$$
\begin{array}{cccccc}
1 & 4 & 9 & 16 & \cdots & n^2 & \cdots \\
\updownarrow & \updownarrow & \updownarrow & \updownarrow & & \updownarrow & \\
4 & 9 & 16 & 25 & \cdots & (n+1)^2 & \cdots
\end{array}
$$

Note that

$$\{4, 9, 16, 25, \ldots, (n+1)^2, \ldots\} \subset \{1, 4, 9, 16, \ldots, n^2, \ldots\} \quad \blacksquare$$

EXERCISES 2.5

Ⓐ Equivalent Sets and Cardinal Numbers

In problems 1–5 show that the sets are equivalent by setting up a one-to-one correspondence between the two given sets.

1. The set N of counting numbers and the set O of odd counting numbers

2. The set N of counting numbers and the set F of positive multiples of 5

3. The set E of even counting numbers and the set G of even counting numbers greater than 100

4. The set O of odd counting numbers and the set E of even counting numbers

5. The set G of even counting numbers greater than 200 and the set T of even counting numbers greater than 300

Ⓑ Equivalent Sets

In problems 6–10, show that the two sets are equivalent.

6. $A = \{1, 2, 3, 4, 5\}$ and $B = \{a, b, c, d, e\}$

7. $P = \{2, 4, 8, 12\}$ and $Q = \{6, 12, 24, 36\}$

8. $W = \{0, 1, 2, 3, \ldots\}$ and $N = \{1, 2, 3, \ldots\}$

9. $I^- = \{-1, -2, -3, \ldots\}$ and $N = \{1, 2, 3, \ldots\}$

10. $N = \{1, 2, 3, \ldots\}$ and $F = \{\frac{1}{1}, \frac{1}{2}, \frac{1}{3}, \ldots\}$

In problems 11–15, find the cardinality of each set.

11. $A = \{a, b, c, \ldots, z\}$

12. $B = \{x \mid x \text{ is one of the Ten Commandments}\}$

13. $C = \{x \mid x \text{ is a star on the American flag}\}$

14. $D = \{\frac{1}{2}, \frac{1}{4}, \frac{1}{6}, \ldots\}$

15. $E = \{\frac{1}{1}, \frac{1}{4}, \frac{1}{9}, \ldots\}$

Ⓒ Infinite Sets

In problems 16–20, determine whether each set is finite or infinite.

16. $\{1, 2, 3, \ldots, 999{,}999\}$

17. $\{100, 200, 300, \ldots\}$

18. $\{5, 10, 15, \ldots\}$

19. $\{\frac{1}{3}, \frac{2}{3}, \frac{3}{3}, \ldots\}$

20. $\{2^{64}, 2^{32}, 2^{16}, \ldots, 2\}$

Use the sets $A = \{1, 2, 3, 4, 5, 6\}$, $B = \{a, b, c, d\}$, $C = \{w, x, y, z\}$, and $D = \{d, c, b, a\}$ to answer problems 21–23.

21. Which set(s) are equal and equivalent?

22. Which set(s) are equivalent but not equal?

23. Which set(s) are not equivalent and not equal?

In Other Words

As you recall, $n(N) = \aleph_0$ and $\aleph_0$ is called a *transfinite cardinal*. Can you perform arithmetic operations with these cardinals? Fill in the blanks and justify your answers.

24. $\aleph_0 + 1 = $ ____

25. $\aleph_0 + \aleph_0 = $ ____

26. $2 \times \aleph_0 = $ ____

27. $\aleph_0 \times \aleph_0 = $ ____

28. Let $A = \{1, 3, 5, \ldots\}$ and $B = \{2, 4, 6, \ldots\}$. As you recall, $n(A \cup B) = n(A) + n(B)$ if A and B are disjoint. Substitute A and B in the equation and state your result.

29. Consider the line segment shown. It is 1 unit long.

Draw an identical segment on a sheet of paper. Cut off the middle piece, the piece between $\frac{1}{3}$ and $\frac{2}{3}$, and paste it on a second sheet of paper. Then divide the piece between 0 and $\frac{1}{3}$ into three equal parts, each of length $\frac{1}{9}$. Cut off the middle piece, the piece between $\frac{1}{9}$ and $\frac{2}{9}$, and paste it next to the first piece you pasted on the paper. Repeat the process with the piece between $\frac{2}{3}$ and 1. The middle piece you will cut off is the piece between $\frac{7}{9}$ and $\frac{8}{9}$. Paste this next to the second piece on the paper. Imagine that this process is continued.

a. The points $\frac{1}{3}, \frac{2}{3}, \frac{1}{9},$ and $\frac{2}{9}$ are the first four points of the Cantor set. What are the next two points?

b. If you continue the pasting process, what do you think will be the total length of the pieces you pasted?

30. Do you think the Cantor set and the set of all points on the line segment in problem 29 are equivalent? Explain.

Online Study Center

To further explore Cantor sets, access link 2.5.1 on this textbook's Online Study Center. You may find other sites on your own.

Using Your Knowledge

Have you ever heard of the Infinity Hotel? It is a peculiar establishment indeed. The only prerequisite for employment is a thorough knowledge of infinite sets. In fact, the *Employment Handbook* consists entirely of the section you have just read. Georg was hired as manager, and his first day on the job was a cinch.

The hotel soon filled all of its rooms, 1, 2, 3, ..., $n, n + 1$, and so on. Trouble started on the second day with the arrival of a new guest. Where would Georg put this new guest? He thought about it for a split second, and then up went a neatly handwritten sign.

> If you are presently in room n, please move next door to room $n + 1$.

Where would the nice family in room 222 go? And what about the newcomer?

The third day things got more involved, for a group of eager customers arrived, in a brand-new Infinity no less. Tensions were high at the hotel. Could Georg accommodate them all? No time was wasted. Without hesitation, the next sign went up.

> If you are presently in room n, please move to room $2n$.

What an odd arrangement of rooms that would leave!

Use your knowledge to answer the following questions:

31. On the second day, where would the family in room 222 go?

32. On the second day, in which room would the newcomer go?

33. Which rooms were vacated for the people in the Infinity?

34. Where would the guest of problem 32 go on the third day?

35. On the third day, where would the family in room 333 go?

Online Study Center

To read more about Hotel Infinity go to 2.5.2 on this textbook's Online Study Center.

Discovery

Here is one of the most striking results obtained by Cantor while studying the theory of *nondenumerable* sets (sets that cannot be put into a one-to-one correspondence with the set of counting numbers). This result may appear incredible to you, but if you have mastered the idea of a one-to-one correspondence, you should be able to prove it!

Two unequal line segments contain the same number of points!

Cantor reasoned in the following manner: Two sets contain the same number of elements if and only if the elements can be paired off one to one. Then he diagrammed the two unequal line segments. (Call the line segments $\overline{AB}$ and $\overline{CD}$ as in the diagram below.) Notice that line segment $\overline{OQ}$ cuts $\overline{CD}$ at P, and $\overline{AB}$ at Q. You may regard P and Q as corresponding points.

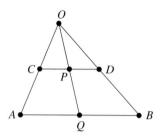

36. Can you discover the proof that there are as many points in $\overline{AB}$ as in $\overline{CD}$?

 Research Questions

Is infinity a number? How big is it? What is infinity plus 1? What about infinity plus infinity?

1. Which set has more elements, the set of counting numbers or the set of counting numbers with a shoe thrown in? Prove it!

2. While on this journey, you may consider staying at the Hotel Infinity (Web branch). Get some background information, familiarize yourself with the big ideas and concepts like infinity and transfinite arithmetic, and write a diary containing a summary of the topics involved.

3. Give two sources where you can get more information about the topics you listed.

Online Study Center

To further explore the concept of infinity, access link 2.5.3 on this textbook's Online Study Center.

When you are finished with the research questions, you can revisit our very own Infinity Hotel in the Using Your Knowledge section.

Chapter 2 Summary

Section	Item	Meaning	Example
2.1A	{}	Set braces	$\{1, 2, 3\}$ is a set.
2.1A	$\in$	Is an element of	$2 \in \{1, 2, 3\}$
2.1A	$\notin$	Is not an element of	$4 \notin \{1, 2, 3\}$
2.1B	$\{1, 2\}$	List (roster) notation	
2.1B	$\{x \mid x \text{ has property } P\}$	Set-builder notation	$A = \{x \mid x \text{ is a counting number}\}$
2.1B	{} or $\varnothing$	The empty, or null, set	The set of words that rhyme with *orange* is an empty set.

Section	Item	Meaning	Example
2.1C	$=$	Equals	$\{1, 2\} = \{2, 1\}$
2.1D	$\subseteq$	Is a subset of	$\{1, 2\} \subseteq \{1, 2, 3\}$
2.1D	$\subset$	Is a proper subset of	$\{a\} \subset \{a, b\}$
2.1D	$\mathcal{U}$	The universal set	$\mathcal{U}$ is the set of all elements under discussion.
2.2A	$\cap$	Intersection	$\{1, 2, 3\} \cap \{2, 3, 4\} = \{2, 3\}$
2.2A	$\cup$	Union	$\{1, 2, 3, 4\} \cup \{2, 3, 4\}$ $= \{1, 2, 3, 4\}$
2.2B	A'	Complement of set A	If $\mathcal{U} = \{1, 2, 3, 4, 5\}$ and $A = \{1, 2\}$, then $A' = \{3, 4, 5\}$.
2.2B	De Morgan's Laws	$(A \cap B)' = A' \cup B'$ $(A \cup B)' = A' \cap B'$	If $\mathcal{U} = \{1, 2, 3, 4, 5\}$, $A = \{1, 2, 3, 4, 5\}$, and $B = \{1, 2\}$, then $A \cap B = \{1, 2\}$, and $(A \cap B)' = \{3, 4, 5\}$, $A' = \varnothing, B' = \{3, 4, 5\}$, and $A' \cup B' = \{3, 4, 5\}$.
2.2C	$A - B$	Set difference	$A - B = \{3, 4, 5\}$
2.4	$n(A)$	Cardinal number of set A	If $A = \{a, b, c\}$ and $B = \{d, e\}$, then $n(A) = 3$ and $n(B) = 2$.
2.4	Finite set A	A set A for which $n(A)$ is a whole number	If $A = \{1, 5, 17\}$, $n(A) = 3$ and A is finite.
2.5B	$A \sim B$	Set equivalence	If $A = \{a, b\}$ and $B = \{1, 2\}$, A and B are equivalent.
2.5C	Infinite set	A set that can be placed into a one-to-one correspondence with one of its proper subsets	N is infinite since it can be placed into a one-to-one correspondence with its proper subset E, the even numbers.
2.5C	$\aleph_0$	The cardinal number for the set N of counting numbers	$n(N) = \aleph_0$

Research Questions

Sources of information for these questions can be found in the Bibliography at the end of the book or on the Web.

1. We have used the symbols $\{\ \}$, $\in$, $\cap$, $\cup$, $\varnothing$, and $\aleph_0$ in this chapter. Find out who invented these symbols and where (book, journal, article, etc.) they were first used.

2. Venn diagrams are attributed to John Venn (see Discovery 2.3). Write a paragraph about Venn's life, and find out the name of the work in which these diagrams were first used. What is another name for Venn diagrams?

Chapter 2 Practice Test

1. Which of the following descriptions define a set?
   a. Brilliant students   b. Students with math SAT scores over 400
   c. Natural numbers greater than 5   d. Natural numbers less than 1

2. List the elements of the set

   $\{x \mid x$ is a counting number between 2 and 10$\}$

3. Describe the following sets verbally and in set-builder notation:
   a. $\{a, e, i, o, u\}$ b. $\{2, 4, 6, 8\}$

4. List all the proper subsets of the set $\{\$, \cent, \% \}$.

5. Complete the following definitions by filling in the blanks with the symbol $\in$ or $\notin$.
   a. $A \cup B = \{x \mid x \underline{\hspace{1cm}} A$ or $x \underline{\hspace{1cm}} B\}$
   b. $A \cap B' = \{x \mid x \underline{\hspace{1cm}} A$ and $x \underline{\hspace{1cm}} B\}$
   c. $A' = \{x \mid x \underline{\hspace{1cm}} \mathcal{U}$ and $x \underline{\hspace{1cm}} A\}$
   d. $A - B = \{x \mid x \underline{\hspace{1cm}} A$ and $x \underline{\hspace{1cm}} B'\}$

6. Let $\mathcal{U} = \{$Ace, King, Queen, Jack$\}$, $A = \{$Ace, Queen, Jack$\}$, and $B = \{$King, Queen$\}$. Find the following:
   a. A' b. $(A \cup B)'$ c. $A \cap B$ d. $\mathcal{U} - (A \cap B)'$

7. If, in addition to the sets in problem 6, $C = \{$Ace, Jack$\}$, find the following:
   a. $(A \cap B) \cup C$ b. $(A' \cup C) \cap B$

8. Draw a pair of Venn diagrams to show that $A - B = A \cap B'$.

9. Draw a Venn diagram to illustrate the set $A \cap B \cap C'$.

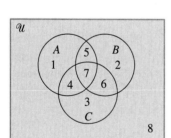

10. Find the sets of numbered regions in the diagram at the left that represent the following sets:
    a. $(A \cup B) \cap C'$ b. $A' \cup (B' \cap C)$

11. Use the numbered regions in the diagram of problem 10 to verify that $(A \cap B) \cup C = (A \cup C) \cap (B \cup C)$.

12. Use the numbered regions in the diagram of problem 10 to verify that $(A \cap B)' = A' \cup B'$.

13. Refer to the diagram of problem 10 and determine which of the following sets (if any) is represented by regions 5, 6, and 7.
    a. $B - (A \cup C)$ b. $(A \cup C) \cap B$
    c. $(A \cap B \cap C) - (A \cap C)$ d. $(A \cup C') \cap B$

14. Refer to the diagram of problem 10 and determine which of the following sets (if any) is represented by regions 6 and 7.
    a. $A \cap B \cap C$ b. $(A \cup C) \cap B$
    c. $A \cap C$ d. $(A \cap C) \cup B$

15. Let $\mathcal{U} = \{1, 2, 3, 4, 5, 6, 7, 8\}$, $A = \{1, 3, 5, 7\}$, $B = \{2, 4, 6, 8\}$, and $C = \{1, 4, 5, 8\}$. Fill in the blanks with = or ≠ to make correct statements.

    a. $n(A \cup C)$ _____ 6   **b.** $n(B \cap C)$ _____ 3

16. Let $n(A) = 25$ and $n(B) = 35$. Find $n(A \cup B)$ if

    a. $A \cap B = \varnothing$.   **b.** $n(A \cap B) = 5$.

17. Let $n(A) = 15$, $n(B) = 25$, and $n(A \cup B) = 35$. Find the following:

    a. $n(A \cap B)$ **b.** $n(\mathcal{U})$ if $n(A' \cap B') = 8$

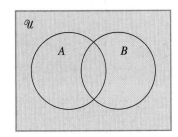

18. In the diagram at the left the rectangular region represents the universal set $\mathcal{U}$, and the circular regions represent the subsets A and B of $\mathcal{U}$. Find an expression for the shaded region in the diagram.

19. On checking 200 students, it is found that 70 are taking French, 40 are taking German, 75 are taking Spanish, 10 are taking French and German, 30 are taking French and Spanish, 15 are taking German and Spanish, and 70 are taking no language. If it is known that no students are taking all three languages, draw a Venn diagram to determine the answers to the following questions.

    a. How many are taking two languages?

    b. How many are taking Spanish and no other language?

    c. How many are taking Spanish and not French?

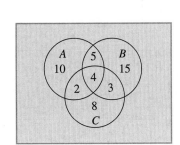

20. A survey of people to determine who regularly buys products A, B, and C gave the numbers shown in the diagram at the left.

    a. How many people were surveyed?

    b. How many buy product A but not B?

    c. How many buy product B or C but not A?

    d. How many buy both products B and C but not A?

    e. How many do not buy either product B or product C?

21. Show that the sets $\{1, 3, \ldots, 2n - 1, \ldots\}$ and $\{2, 4, \ldots, 2n, \ldots\}$ have the same cardinal number.

22. Show that the sets $\{1, 2, \ldots, n, \ldots\}$ and $\{4, 16, \ldots, (2n)^2, \ldots\}$ are equivalent.

23. Find the cardinality of the set $\{1, 4, \ldots, n^2, \ldots, 144\}$.

24. Find the cardinality of the set $\{1, 4, \ldots, n^2, \ldots\}$.

25. Show that the set of all proper fractions with numerator 1 is infinite.

Answers to Practice Test

ANSWER	IF YOU MISSED	REVIEW		
	Question	Section	Example(s)	Page(s)
1. b, c, and **d**	1	2.1	1	52
2. $\{3, 4, 5, 6, 7, 8, 9\}$	2	2.1	2–3	52, 53
3. a. The set of vowels in the English alphabet: $\{x \mid x \text{ is a vowel in the English alphabet}\}$	3	2.1	4, 5	53, 54
b. The set of even counting numbers less than 10: $\{x \mid x \text{ is an even counting number less than 10}\}$				
4. The proper subsets are $\varnothing$, $\{\$\}$, $\{\cent\}$, $\{\%\}$, $\{\$, \cent\}$, $\{\$, \%\}$, and $\{\cent, \%\}$.	4	2.1	6, 8, 9	57, 58
5. a. Both blanks take the symbol $\in$.	5	2.2	2	65
b. First blank takes $\in$; second blank takes $\notin$.				
c. First blank takes $\in$; second blank takes $\notin$.				
d. Both blanks take the symbol $\in$.				
6. a. $\{$King$\}$ **b.** $\varnothing$ **c.** $\{$Queen$\}$ **d.** $\{$Queen$\}$	6	2.2	1–3	64, 65
7. a. $\{$Ace, Queen, Jack$\}$ **b.** $\{$King$\}$	7	2.2	4, 5	66
8. The shaded region in the diagram on the left corresponds to $A - B$. The darkest region in the diagram on the right corresponds to $A \cap B'$. This shows that $A - B = A \cap B'$. 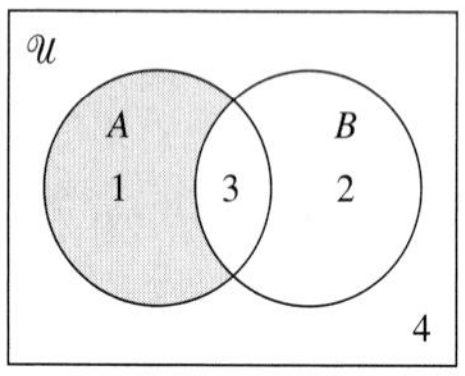 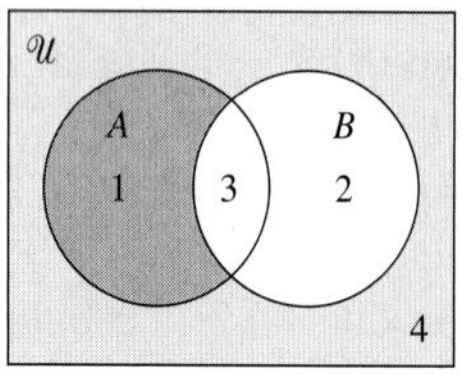	8	2.3	1, 2	74, 75

	IF YOU MISSED	REVIEW		
9. The darkest region in the diagram corresponds to the set $A \cap B \cap C'$. 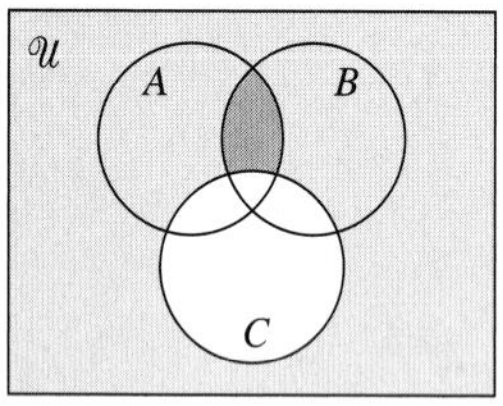	9	2.3	2	75

10. a. Regions 1, 2, and 5	10	2.3	2	75
b. Regions 2, 3, 4, 6, and 8				
11. $(A \cap B) \cup C$ corresponds to regions 3, 4, 5, 6, and 7, and $(A \cup C) \cap (B \cup C)$ corresponds to the same regions. This verifies the equation $(A \cap B) \cup C = (A \cup C) \cap (B \cup C)$.	11	2.3	3	77
12. $(A \cap B)'$ and $A' \cup B'$ both correspond to regions 1, 2, 3, 4, 6, and 8. This verifies the equation $(A \cap B)' = A' \cup B'$.	12	2.3	3	77
13. Part **(b)** is correct.	13	2.3	5	78
14. None of these	14	2.3	5	78
15. a. $=$ **b.** $\neq$	15	2.4	1, 2	84, 85

ANSWER	IF YOU MISSED	REVIEW		
	Question	Section	Example(s)	Page(s)
16. a. 60 **b.** 55	16	2.4	2	85
17. a. 5 **b.** 43	17	2.4	2, 3	85, 86
18. $B - A$ or $B \cap A'$ or $A' \cap B$	18	2.4	4	87–88
19. a. 55 **b.** 30 **c.** 45	19	2.4	4, 5	87–89
20. a. 47 **b.** 12 **c.** 26 **d.** 3 **e.** 10	20	2.4	4, 5	87–89
21. The following correspondence shows that the two sets have the same cardinal number: $$\begin{array}{ccccc} 1 & 3 & \cdots & 2n-1 & \cdots \\ \updownarrow & \updownarrow & & \updownarrow & \\ 2 & 4 & \cdots & 2n & \cdots \end{array}$$	21	2.5	1	96
22. The following correspondence shows that the two sets are equivalent: $$\begin{array}{ccccc} 1 & 2 & \cdots & n & \cdots \\ \updownarrow & \updownarrow & & \updownarrow & \\ 4 & 16 & \cdots & (2n)^2 & \cdots \end{array}$$	22	2.5	2	96–97
23. 12	23	2.5	3	97–98
24. $\aleph_0$	24	2.5	3	97–98
25. A one-to-one correspondence can be set up between the given set $\{\frac{1}{2}, \frac{1}{3}, \frac{1}{4}, \ldots, \frac{1}{n}, \ldots\}$ and a subset of itself $\{\frac{1}{3}, \frac{1}{4}, \frac{1}{5}, \ldots, \frac{1}{n+1}, \ldots\}$, so the given set is infinite.	25	2.5	3	97–98

The Massachusetts Institute of Technology is an important center for computer research and innovation. Building on applications pioneered at MIT by Claude Shannon, today's computer scientists still use logic gates to build circuits and networks of circuits.

Logic

One of our most precious possessions is our ability to think and reason. Logic, the methods and principles used in distinguishing correct from incorrect thinking, logical thinking, and correct reasoning are used in many fields—law, insurance, science, and mathematics, to name a few. The study of logic dates back to the Greek philosopher Aristotle (ca. 384–322 B.C.), who systematized the principles of reasoning and laws of logic in his *Organon*.

We start this chapter with a basic concept in logic, *statements*. We learn how to write statements and how to determine the conditions under which they are true or false. This technique dates back to the German mathematician and logician Gottlob Frege (1848–1925), who first tried to rewrite the established body of mathematics in logical symbolism.

In order to determine the truth or falsity of more complex statements, we develop a problem-solving tool called a **truth table** (Section 3.2) and then turn our attention to conditional and biconditional statements and their variations, as well as the related idea of implication. How do we determine whether an argument is valid (Section 3.6)? We do this by using Euler diagrams in Section 3.5 and truth tables Section 3.6.

Online Study Center

For links to Internet sites related to Chapter 3, please access **college.hmco.com/PIC/bello9e** and click on the Online Study Center icon.

3.1 Statements

George Boole was born in Lincoln, England, on November 2, 1815.

(1815–1864)

Throughout the years, Boole had some of his writings publlished. At the age of 34, Boole was appointed professor of mathematics at Queen's College in Cork, Ireland.

At the age of 39, he published his masterpiece, *An Investigation of the Laws of Thought, on Which Are Founded the Mathematical Theories of Logic and Probabilities.* Boole was primarily responsible for bringing the study of formal logic from the field of philosophy into that of mathematics.

Looking Ahead
Boole's work on logic and thought was instrumental in establishing modern symbolic logic, which is the focus of this chapter.

Online Study Center

To further explore the meaning, forms, and faulty reasoning associated with syllogisms, access link 3.1.1 on this textbook's Online Study Center.

Making a Statement

Consider the following directions:

Lessee shall not paint, paper, or otherwise redecorate or make alterations to the premises without the prior written consent of Lessor.
Checking "Yes" will not change your tax or reduce your refund.
Make sure your application is signed on page 3 and in the appropriate place on page 4.

Can you guess where these directions come from? Do you understand what they mean? In this section you study **statements,** that is, sentences that can be classified as true or false. The preceding statements were constructed using **conjunctions, disjunctions,** and **negations.** You will study these types of statements and practice how to write them in symbols in Exercises 3.1, problems 9–16.

Now look at an excerpt from a menu, given below.

---

6. Country Fried Steak Breakfast

Includes 2 Eggs (any style), Country Fried Steak and Gravy, Grits or Breakfast Potatoes, Toast or Biscuits, and Fruit Jelly $6.99

Can you get Eggs, Country Fried Steak and Gravy, Grits and Breakfast Potatoes, and Toast and Biscuit and Jelly? Not unless you pay extra! When translating statements containing commas into symbolic form, the commas indicate which simple statements are grouped together. Thus, using the symbols $\wedge$ for *and,* and $\vee$ for *or,* your breakfast menu can be translated as

$$e \wedge (s \wedge a) \wedge (g \vee p) \wedge (t \vee b) \wedge j$$

You will learn how to translate sentences into symbols and vice versa in Exercises 3.1, problems 9–16. ▶

The word *logic* is derived from the Greek word *logos,* which can be interpreted as "reason" or "discourse." The principles discovered by the Greeks were first systematized by Aristotle, and Aristotle's type of reasoning constitutes the traditional logic that has been studied and taught from his era to the present day. A simple illustration of Aristotelian logic goes as follows:

All men are mortal.
Socrates was a man.
Therefore, Socrates was mortal.

This is a typical argument that is known as a **syllogism.**

Whether we are trying to solve a problem, taking part in a debate, or working a crossword puzzle, we are engaging in a mental activity called *logical reasoning.* This reasoning is usually expressed in declarative sentences. We now turn our attention to the study of these sentences.

Is this a statement? Is it true or false?

Are these statements or commands?

Online Study Center

To further explore paradoxes involving "self-references," access link 3.1.2 on this textbook's Online Study Center.

A. Recognizing Statements

In this and the following sections we will examine a certain type of declarative sentence called a *statement* and the manner in which we can combine such sentences and arrive at valid conclusions.

Definition of Statement

A **statement** is a declarative sentence that can be classified as true or false but not both simultaneously.

This capability of being classified as true or false makes statements different from questions, commands, or exclamations. Questions can be asked, commands given, and exclamations shouted, but only statements can be classified as true or false. The sentences in Example 1 are illustrations of statements.

EXAMPLE 1 ▶ Sentences That Are Statements

(a) Boston is the capital of Massachusetts.

(b) The number 2 is even and less than 20.

(c) There are 5 trillion grains of sand in Florida.

(d) Either you study daily or you get an F in this course.

(e) If 2 is even, then 2 + 2 is even. ■

Note that the truth or falsity of the first statement in Example 1 can be determined by a direct check. The third statement is also true or false, even though there are no immediate or practical methods to determine its truth or falsity.

In contrast to the statements in Example 1, the following are illustrations of **nonstatements.**

EXAMPLE 2 ▶ Sentences That Are Not Statements

(a) What time is it?

(b) Elvis for president!

(c) Good grief, Charlie Brown!

(d) Close the door.

(e) This statement is false. ■

The sentences in Example 2 are not statements. Notice that if we assume that sentence (e) is true, then it is false, and if we assume that it is false, then it is true. Hence, the sentence cannot be classified as either true or false, so it is not a statement. A self-contradictory sentence of this type is called a **paradox.**

B. Conjunction, Disjunction, and Negation

Having explained what is meant by a statement, we now turn our attention to various combinations of statements. In Example 1, for instance, statements (a) and

(c) have only one component each (that is, each says only one thing), whereas statement (b) is a combination of two components, namely, "2 is even" and "2 is less than 20." Statements (a) and (c) are **simple;** statements (b), (d), and (e) are **compound.**

As a further example, "John is 6 ft tall" is a simple statement. On the other hand, "John is 6 ft tall *and* he plays basketball" is a compound statement because it is a combination of the two simple statements "John is 6 ft tall" and "he plays basketball."

There are many ways in which simple statements can be combined to form compound statements. Different combinations are formed by using words called **connectives** to join the statements. Two of the most common connectives are the words *and* and *or.* Suppose we use the letters p and q to represent the statements.

> p: It is 100° today.
> q: The air conditioner in this room is broken.

Then we can form the following compound sentences:

> p **and** q: It is 100° today *and* the air conditioner in this room is broken.
> p **or** q: It is 100° today *or* the air conditioner in this room is broken.

In the study of logic, the word *and* is symbolized by $\wedge$ and the word *or* by $\vee$ as mentioned earlier. Thus,

> p **and** q is written $p \wedge q$
> p **or** q is written $p \vee q$

Using these ideas, we make the following definitions:

Definition of Conjunction

If two statements are combined by the word **and** (or an equivalent word such as **but**), the resulting statement is called a **conjunction.** If the two statements are symbolized by p and q, respectively, then the conjunction is symbolized by $p \wedge q$.

Definition of Disjunction

If two statements are combined by the word **or** (or an equivalent word such as **otherwise**), the result is called a **disjunction.** If the two statements are symbolized by p and q, respectively, then the disjunction is symbolized by $p \vee q$.

EXAMPLE 3 ▶ Writing Conjunctions in Symbols

Symbolize the following conjunctions:

(a) Tom is taking a math course and Mary is taking a physics course.

(b) Ann is passing math but she is failing English.

Solution

(a) Let m stand for "Tom is taking a math course" and p stand for "Mary is taking a physics course." Then the given conjunction can be symbolized by $m \wedge p$.

(b) Let p stand for "Ann is passing math" and f stand for "she is failing English." The given conjunction can then be symbolized by $p \wedge f$. Here, the word *but* is used in place of *and*. ∎

EXAMPLE 4 ▶ Writing Disjunctions in Symbols

Symbolize the disjunction "We stop inflation or we increase wages."

Solution

Letting p stand for "We stop inflation" and q stand for "we increase wages," we can symbolize the disjunction by $p \vee q$. ∎

Another important construction in logic is negating a given statement.

Definition of Negation

The **negation** of a given statement is a statement that is false whenever the given statement is true and true whenever the given statement is false. If the given statement is denoted by p, its negation is denoted by $\sim p$. (The symbol $\sim$ is called a *tilde*.)

The negation of a statement can always be written by prefixing it with a phrase such as "It is not the case that." Sometimes the negation can be obtained simply by inserting the word *not* in the given statement. For example, the negation of the statement "Today is Friday" can be written as

"*It is not the case that* today is Friday"

or as

"Today is *not* Friday."

Similarly, if p stands for "It is hot today," then $\sim p$ (read "not p") can be written either as "It is not the case that it is hot today" or as "It is not hot today."

In the preceding illustrations, we have negated simple statements. We often have to consider the negation of compound statements, as in the next example.

EXAMPLE 5 ▶ Translating Symbolic Statements into English

Let p be "the sky is blue," and let q be "it is raining." Translate the following statements into English:

(a) $\sim(p \wedge q)$ (b) $\sim p \wedge \sim q$ (c) $\sim q \wedge p$

Solution

(a) It is not the case that the sky is blue and it is raining. Another form of the negation is "The sky is not blue or it is not raining."

(b) The sky is not blue and it is not raining.

(c) It is not raining and the sky is blue. ∎

The two forms of the solution to part (a) of Example 5 illustrate the fact that the negation of $p \wedge q$ can be written either as $\sim(p \wedge q)$ or as $\sim p \vee \sim q$.

Negation of the Conjunction p and q

$\sim(p \wedge q)$ means $\sim p \vee \sim q$

Similarly,

Negation of the Disjunction p or q

$\sim(p \vee q)$ means $\sim p \wedge \sim q$

Thus, the statement "p or q" is false when and only when p and q are *both* false. These two facts are known as **De Morgan's laws.**

De Morgan's Laws

For any statement p and q,

$\sim(p \wedge q)$ means $\sim p \vee \sim q$

and

$\sim(p \vee q)$ means $\sim p \wedge \sim q$

You will be asked to verify De Morgan's laws in Exercises 3.2, problems 45 and 46. Compare the preceding two laws with De Morgan's laws for sets, $(A \cap B)' = A' \cup B'$ and $(A \cup B)' = A' \cap B'$.

EXAMPLE 6 ▶ Translating from English into Symbols

Consider the two statements

    p: Sherlock Holmes is alive.
    q: Sherlock Holmes lives in London.

Write the following statements in symbolic form:

(a) Sherlock Holmes is alive and he lives in London.

(b) Either Sherlock Holmes is alive or he lives in London.

(c) Sherlock Holmes is neither alive nor does he live in London.

(d) It is not the case that Sherlock Holmes is alive and he lives in London.

Solution

(a) $p \wedge q$ (b) $p \vee q$ (c) $\sim p \wedge \sim q$ (d) $\sim(p \wedge q)$ ■

Note the use of parentheses in symbolic statements to indicate which items are to be taken as a unit. Thus, in part (a) of Example 5 and in part (d) of Example 6, $\sim(p \wedge q)$ means the negation of the entire statement $p \wedge q$. It is important to distinguish $\sim(p \wedge q)$ from $\sim p \wedge q$. The latter means that only statement p is negated. For example, if p is "John likes Mary" and q is "Mary likes John," then $\sim(p \wedge q)$ is "It is not true that John and Mary like each other." But $\sim p \wedge q$ is "John does not like Mary, but Mary likes John."

Note in statements written in words the use of commas to indicate which simple statements are grouped together, as shown in the next example. This must be watched carefully when translating verbal statements containing commas into symbolic form.

EXAMPLE 7 ▶ Writing Statements in Symbols

Write the following in symbolic form:

(a) "You are a full-time student (f) or over 21 (o), and a resident of the state (r)."

(b) "You are a full-time student, or over 21 and a resident of the state."

Solution

(a) $(f \vee o) \wedge r$ (b) $f \vee (o \wedge r)$ ■

Statements (a) and (b) do not have the same meaning!

EXAMPLE 8 ▶ Writing Negations, Disjunctions, and Conjunctions

Let p be the statement "Tarzan likes Jane," and let q be the statement "Jane likes Tarzan." Symbolize and write the following in words:

(a) The negation of the conjunction of p and q

(b) The disjunction of the negations of p and q

(c) The conjunction of the negations of p and q

(d) What can you conclude about the statements in (a), (b), and (c) according to De Morgan's laws?

Solution

(a) The conjunction of p and q is $p \wedge q$. Thus, the negation of the conjunction of p and q is $\sim(p \wedge q)$. In words, it is not the case that Tarzan likes Jane and Jane likes Tarzan. That is, it is not the case that Tarzan and Jane like each other.

(b) The negations of p and q are $\sim p$ and $\sim q$, respectively. Thus, the disjunction of the negations of p and q is $\sim p \vee \sim q$. In words, either Tarzan does not like Jane or Jane does not like Tarzan.

(c) $\sim p \wedge \sim q$. In words, Tarzan does not like Jane and Jane does not like Tarzan. That is, Tarzan and Jane do not like each other.

(d) By De Morgan's laws, the statements $\sim(p \wedge q)$ and $\sim p \vee \sim q$ in parts (a) and (b), respectively, have the same meaning.   ■

What is the negation of this statement?

Statements involving the **universal quantifiers**—*all, no,* and *every*—or the **existential quantifiers**—*some* and *there exists at least one*—are more complicated to negate. Table 3.1 may help you.

TABLE 3.1

Statement	Negation
All a's are b's. **No** a's are b's.	**Some** a's **are not** b's. **Some** a's **are** b's.

All ——— No (none)

Negations Negations

Some are. Some are not.
(There is one.)

To help you remember how to negate statements, use this diagram.

Online Study Center

To further explore conjunctions, disjunctions, negations, and quantifiers, access link 3.1.3 on this textbook's Online Study Center.

Thus, the following statements p and $\sim p$ are negations of each other, as are q and $\sim q$:

- p: **All** homeowners participate in recycling.
- $\sim p$: **Some** homeowners **do not** participate in recycling, or "**Not** all homeowners participate in recycling."
- q: **Some** of us will graduate.
- $\sim q$: **None** of us will graduate.

Keep in mind that the definition of a negation requires that the negation of a statement must be false whenever the statement is true and true whenever the statement is false. You can check this by looking at statements q and $\sim q$ above.

EXAMPLE 9 ▶ Writing Negations Involving All, None, and Some

Write the negation of each of the following:

(a) All of us like pistachio nuts.

(b) Nobody likes freezing weather.

(c) Some students work part-time.

Solution

(a) Some of us do not like pistachio nuts. An alternative form is "Not all of us like pistachio nuts."

(b) Somebody likes freezing weather.

(c) No student works part-time. ∎

Note: Remember that the negation of *all* is *some do not* and the negation of *some* is *none*.

EXAMPLE 10 ▶ Translating English into Symbols

Write the indicated statements in symbols.

(a) Do you know what happens when you sign an application for a credit card? Here is the fine print:

> I request that a Visa account be opened (o) and cards be issued as indicated (i), and I authorize the bank to receive (r) and exchange (e) information and investigate (n) the references and data (d) collected pertinent to my creditworthiness.

(b) Here is a tip from a software manual.

> Some of the things you can use the Control Panel for are changing your screen colors, installing or changing settings for hardware and software, and setting up or changing settings for a network.

Write a statement that will indicate how you can use the Control Panel.

Solution

(a) $(o \wedge i) \wedge (r \wedge e) \wedge (n \wedge d)$

(b) First, we have to determine how many statements are present. This can be done by assigning a letter to each of the statements.

> Changing your screen colors is p.
> Installing settings for hardware is q.
> Installing settings for software is r.
> Changing settings for hardware is s.
> Changing settings for software is t.
> Setting up settings for a network is u.
> Changing settings for a network is v.

The translation is $p \wedge (q \vee s) \wedge (r \vee t) \wedge (u \vee v)$.   ■

In Section 1.3 we learned how to interpret bar graphs. Now we are ready to use that knowledge to determine if statements about those graphs are true or false.

EXAMPLE 11 ▶ Credit Card Surveys and Statements

The bar graph shows the results of a survey regarding the number of credit cards you own (None, 1, 2, 3, 4, or 5 or more). Determine if the given statements are true or false.

(a) The most common number of credit cards owned was **none (0).**

(b) The least number of credit cards owned was **1.**

(c) There were fewer people owning **4** cards than people owning **5 or more.**

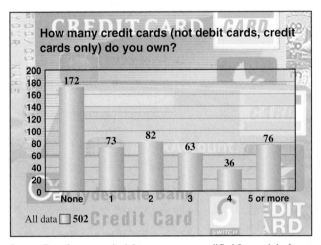

Source: Data from www.insightexpress.com, modified from original content.

Solution

To determine the truth or falsity of the statements, look at the length of the bars (bars are labeled at the top with the number of people answering).

(a) **True** (the longest bar is the one labeled None: 172 people)

(b) **False** (the shortest bar is the one labeled 4: 36 people)

(c) **True** (the number of people owning 4 cards was 36, which is less than the number of people owning 5 or more cards: 76)   ■

EXERCISES 3.1

A Recognizing Statements

In problems 1–8, determine whether each sentence is a statement. Classify each sentence that is a statement as simple or compound. If it is compound, give its components.

1. Circles are dreamy.

2. Lemons and oranges are citrus fruits.

3. Jane is taking an English course, and she has four themes to write.

4. Apples are citrus fruits.

5. Do you like mathematics?

6. Walk a mile.

7. Students at Ohio State University are required to take either a course in history or a course in economics.

8. Today is Sunday, and tomorrow is Monday.

B Conjunction, Disjunction, and Negation

In problems 9–16, write each statement in symbolic form using the indicated letters to represent the corresponding components.

9. This is April (a), and income tax returns must be filed (f).

10. Logic is a required subject for lawyers (r) but not for most engineers ($\sim e$).

11. Dick Tracy is a detective (d) or a fictitious character in the newspaper (f).

12. Snoopy is not an aviator ($\sim a$), or the Sopwith Camel is an airplane (p).

13. Violets are blue (b), but roses are pink (p).

14. The stock market goes up (u); nevertheless, my stocks stay down (d).

15. I will take art (a) or music (m) next term.

16. I will not drive to New York ($\sim d$); however, I shall go by train (t) or by plane (p).

In problems 17–20, let p be "Robin can type" and let q be "Robin takes shorthand." Write each statement in symbolic form.

17. Robin can type and take shorthand.

18. Robin can type but does not take shorthand.

19. Robin can neither type nor take shorthand.

20. It is not the case that Robin can type and take shorthand.

In problems 21–25, let p be "Ricky loves Lucy" and let q be "Lucy loves Ricky." Give a verbal translation of each statement.

21. $p \vee \sim q$ 22. $\sim(p \vee q)$ 23. $p \wedge \sim q$

24. $\sim p \wedge \sim q$ 25. $\sim(p \wedge q)$

In problems 26–31, write the negation of each sentence.

26. It is a long time before the end of the term.

27. Bill's store is making a good profit.

28. The number 10 is a round number.

29. My dog is a spaniel.

30. Your cat is not a Siamese.

31. I do not like to work overtime.

In problems 32–34, determine whether the statements p and q are negations of each other.

32. p: Sally is a very tall girl.
    q: Sally is a very short girl.

33. p: All squares are rectangles.
    q: Some squares are not rectangles.

34. p: All whole numbers are even.
    q: At least one whole number is not even.

In problems 35–49, give the negation of each statement.

35. All men are mortal.

36. Some women are teachers.

37. Some basketball players are not 6 ft tall.

38. Some things are not what they appear to be.

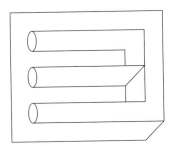

Some things are not what they appear to be.

39. Either he is bald or he has a 10-in. forehead.

40. Nobody does not like Sara Lee.

41. Some circles are round.

42. Some men earn less than $5 per hour, and some men earn more than $50 per hour.

43. Somebody up there loves me.

44. Nothing is certain but death and taxes.

45. Everybody likes to go on a trip.

46. No one can sue us under this coverage.

47. All persons occupying your covered auto are insured.

48. None of your contributions are deductible.

49. Some expenses are not subject to the 2% limit.

50. The statement "Not all people are awkward" is directly transformed into the statement "Some people are not awkward" by which one of the following logical equivalences?
   a. "If p, then q" means the same as "If not q, then not p."
   b. "All are not p" means the same as "None are p."
   c. "Not all are p" means the same as "Some are not p."
   d. "Not (not p)" means the same as "p."

51. The chairperson of the city council told the members that if they declared a holiday, at least one of the 10 banks in the city would remain open. The chairperson was mistaken. Which of the following statements (if any) is consistent with this situation?
   a. The council did not declare a holiday and all the banks remained closed.
   b. The council did not declare a holiday and all the banks remained open.
   c. The council declared a holiday and none of the banks remained closed.

   d. The council declared a holiday and none of the banks remained open.

Write the statements in problems 52–55 in symbolic form.

52. You are legally married (m) or divorced (d), and currently enrolled (e).

53. I am a dependent person (d) and my parents are residents (p), or I have maintained legal residence in the state (r).

54. "You must be a U.S. citizen (c) or an eligible noncitizen (n), and be enrolled as a degree-seeking student (e)."—*Excerpt from a Pell Grant brochure*

55. "You must be a resident (r) and be ranked in the top 10% of your graduating class (t) or GED scores (g)."—*Excerpt from the Paul Douglas Teacher Scholarship*

 In Other Words

56. Let p be "Today is Friday." Let q be "Tomorrow is Saturday." Write the negation of each of the following in words:
   a. $p \wedge q$ **b.** $p \wedge \sim q$ **c.** $p \vee q$

57. Let p be "The diagram is a square." Let q be "The diagram is a rectangle." Write the negation of each of the following in words:
   a. $p \vee q$ **b.** $\sim p \wedge q$ **c.** $\sim p \vee \sim q$

58. In Example 2 we discussed the paradoxical statement "This sentence is false." Now, take a 3-by-5 card. On one side of the card write

   The sentence on the other side of this card is true.

On the other side of the card write

   The sentence on the other side of this card is false.

Is the first sentence true or false? Explain.

59. On another 3-by-5 card write these three sentences:
   (1) This sentence contains five words.
   (2) This sentence contains eight words.
   (3) Exactly one sentence on this card is true.

Which statement(s) are true, and which false?

60. Consider the sentence "This sentence is true." Is this sentence a statement? Explain.

Online Study Center

To further explore interactive paradoxes, access link 3.1.4 on this textbook's Online Study Center.

Using Your Knowledge

In Section 1.3 we learned how to interpret circle graphs. Use your knowledge to determine some facts regarding average costs at public and private universities. In problems 61–65, refer to the diagram below to determine if the statements are true or false.

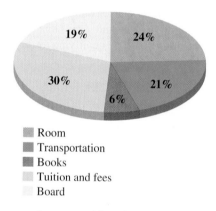

■ Room
■ Transportation
■ Books
■ Tuition and fees
■ Board

Average annual costs at public universities.
Source: www.moneyopolis.com/town_answers_school.asp.

61. The biggest expense is transportation.

62. The smallest expense is board.

63. The expenses for transportation and room are greater than the expenses for tuition and fees and board.

64. The biggest expense is not tuition and fees.

65. The cost of room and books is the same as that of tuition and fees.

In problems 66–70, refer to the diagram below to determine if the statements are true or false.

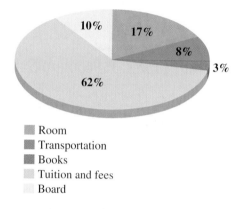

■ Room
■ Transportation
■ Books
■ Tuition and fees
■ Board

Average annual costs at private universities.

66. The biggest expense is tuition and fees.

67. The smallest expense is books.

68. The expenses for room and books are greater than the expenses for transportation and board.

69. The percent of total expenses allocated for tuition and fees at private universities is more than double the percent allocated at public universities.

70. The percent of total expenses allocated for transportation at a private university are less than a third of the percent allocated at a public university.

Research Questions

1. In this section we have indicated that a statement can be true or false, but not both. In 1910, Charles S. Peirce developed a three-valued logic. Write a short paragraph describing how this three-valued logic worked and the truth tables that were associated with it.

2. Who invented logic? What is the definition of logic? How many kinds of logic are there? What is critical thinking?

Online Study Center

To further explore logic and critical thinking, access link 3.1.5 on this textbook's Online Study Center.

3. In the Human Side of Math, we discussed George Boole, who died in England in December 1864. Recount the tragic circumstances of Boole's death as described by A. Macfarlane in *Lectures on Ten British*

Mathematicians of the Nineteenth Century (New York, 1916), or you can go to one of the many sites on the Web dealing with the history of mathematics for the information.

4. We have mentioned the *universal quantifiers* and the *existential quantifiers.* What symbols are used in logic for these quantifiers, who invented these symbols, and where were they used first?

3.2 Truth Tables: A Problem-Solving Tool

Tax Forms and Truth Values

Have you ever *really* read the instructions for your 1040 tax form? For example, suppose you are interested in knowing whether a person qualifies as your dependent. You turn to page 10 of the instruction booklet, and it says that for someone to qualify as your dependent, he or she must pass five qualifying tests, including the following:

Test 3—Citizen or Resident

The person must have been a U.S. citizen [u] or resident alien [r], a resident of Canada [c] or Mexico [m], or your adopted child [a] who is not a U.S. citizen [$\sim u$], but who lived with you all year in a foreign country [l].

Here is some bad news. This is only *one* of the *five* tests required to qualify a person as a dependent. For present purposes, just consider when a person qualifies under this test. First, translate the statement to symbolic form as follows:

$$(u \vee r) \vee (c \vee m) \vee (a \wedge \sim u \wedge l)$$

As you recall, a disjunction is true when any of its components is true; thus, if any of the statements in parentheses is true, the test is satisfied. Can you think of three different conditions under which the test is satisfied? How can you make the third statement in parentheses true? In this section you will learn to identify circumstances under which compound statements are true or false. Then you will practice doing this in problems 17–25 of Exercises 3.2. ▶

We regard T (for true) and F (for false) as the possible **truth values** of a statement. Thus, the statement "George Washington was the first president of the United States" has the truth value T, and the statement "The moon is made of green cheese" has the truth value F.

One of the principal problems in logic is that of determining the truth value of a compound statement when the truth values of its components are known. In order to attack this problem, we must first assign appropriate truth values to such statements as $p \wedge q$, $p \vee q$, and $\sim p$. Although the symbols $\wedge$, $\vee$, and $\sim$ were introduced in Section 3.1, they were not completely defined. We shall complete their definitions by assigning appropriate truth values to statements involving these symbols.

TABLE 3.4
Negation (~)

1	2
p	$\sim p$
T	F
F	T

Table 3.4 defines the negation symbol $\sim$. This table expresses the definition:

> **Truth Value of the Negation, $\sim p$**
>
> $\sim p$ is false whenever p is true, and $\sim p$ is true whenever p is false.

EXAMPLE 4 ▶ Driver's License Renewals

Your driver's license will be renewed under the following conditions: You are a safe driver, have no physical disability, and are not addicted to drugs or intoxicants. Let s be the statement "You are a safe driver," p be the statement "You have a physical disability," and q be the statement "You are addicted to drugs or intoxicants." Write a statement in symbolic form whose truth will guarantee that your driver's license will be renewed.

Solution

$$s \wedge \sim p \wedge \sim q \qquad\blacksquare$$

In many cases it is convenient to construct truth tables to determine the truth values of certain compound statements involving the symbols $\wedge$ (and), $\vee$ (or), and $\sim$ (not). It is important to keep the following in mind:

1. A conjunction $p \wedge q$ is true when p and q are both true and is false otherwise.

2. A disjunction $p \vee q$ is false when p and q are both false and is true otherwise.

3. If p and $\sim p$ are negations of each other, then $\sim p$ is false whenever p is true, and $\sim p$ is true whenever p is false.

EXAMPLE 5 ▶ Singles and Taxes

When filing your Form 1040A, you will be classified as *single* if any *one* of the following is true:

> You were not married ($\sim m$).
> You were legally separated under a decree of divorce (d) or of separate maintenance (s).
> You were widowed before January 1 (w) and did not remarry ($\sim r$).

(a) Write in symbols the conditions under which you will be classified as *single*.

(b) Write the conditions under which you will be classified as *not single*.

Solution

(a) We take the disjunction of the three given statements.

$$\sim m \vee (d \vee s) \vee (w \wedge \sim r)$$

(b) We have to make the disjunction $\sim m \vee (d \vee s) \vee (w \wedge \sim r)$ false; this means that each of the components must be false.

> For $\sim m$ to be false, m must be true.
> For $d \vee s$ to be false, both d must be false and s must be false.
> For $(w \wedge \sim r)$ to be false, either w has to be false or $\sim r$ must be false; that is, r must be true.

Thus, for you to be classified as *not single,* you must be *all* of the following:

> Married (*m*)
> Not legally separated under a decree of divorce ($\sim d$) *and* not legally separated under a decree of separate maintenance ($\sim s$)
> Either a widower (*w*) *or* remarried (*r*)

Note that the statement $m \wedge \sim d \wedge \sim s \wedge (\sim w \vee r)$ is the negation of $\sim m \vee (d \vee s) \vee (w \wedge \sim r)$; that is, *not single* is the negation of *single.* ∎

D. Making Truth Tables

We have already constructed truth tables for the basic statements, conjunctions, disjunctions, and negations. We now give you a procedure that can be used to help you construct any truth table containing compound statements.

Procedure for Constructing Truth Tables

1. Examine the statement and determine if the final result is a negation, conjunction, or disjunction. For example, $\sim p \vee q$ is a disjunction (you are connecting the negation $\sim p$ and the statement q with the disjunction $\vee$); the statement $\sim(p \wedge \sim q)$ is a negation (you are negating $p \wedge \sim q$); the statement $p \wedge (q \vee r) \Leftrightarrow (p \wedge q) \vee (p \wedge r)$ is called an *equivalency* (you have to show that the statements $p \wedge (q \vee r)$ and $(p \wedge q) \vee (p \wedge r)$ have identical truth tables).

2. Complete the columns under the simple statements *p, q,* and *r.*

3. Complete the columns under the connectives with parentheses.

4. Complete the column under any remaining statements and their negations.

5. Complete the column under any remaining connectives keeping in mind that the answer will appear under the final column determined in step 1. It is a good idea to highlight or circle the column containing the final answer and number the columns in the order they were completed.

TABLE 3.5

1	*2*	*3*	*4*	
p	*q*	$\sim p$	$\vee$	*q*
T	T	F	T	
T	F	F	F	
F	T	T	T	
F	F	T	T	

Note: You do *not* have to write the truth values for *q* in the far-right column; simply refer to column 2.

Some students prefer to write the columns like this.

1	*2*	*3*	*4*
p	*q*	$\sim p$	$\sim p \vee q$

Note: The procedure and the end result under column 4 will be the same! (And you save some work!)

EXAMPLE 6 ▶ Truth Tables with Negations and Disjunctions

Construct the truth table for the statement $\sim p \vee q$.

Solution

1. We write the four possible pairs of truth values for *p* and *q* in columns 1 and 2 of Table 3.5.

2. Using column 1 as a reference, we negate statement *p* to get the entries in column 3. (We simply write *F* in the rows where we wrote *T* for *p* and *T* in the rows where we wrote *F* for *p*.)

3. To finish Table 3.5, we look at column 3 ($\sim p$) and column 2 (*q*). We combine columns 3 ($\sim p$) and 2 (*q*) using the disjunction *or,* denoted by $\vee$, to get the truth value for $\sim p \vee q$ and write the result in column 4. We recall (see Table 3.3) that a disjunction is false only when both its components are false; it is true otherwise. Thus, we write *F* in the second row of column 4 where both components, $\sim p$ and *q*, are false, and we write *T* in the other rows. This completes Table 3.5. ∎

EXAMPLE 7 ▶ Finding When a Disjunction Is False

Let p be "I lie," and let q be "I would tell you." When will the statement "I do not lie or I would tell you" be false?

Solution

The statement under consideration can be symbolized as $\sim p \vee q$. From Table 3.5, we see that $\sim p \vee q$ is false when p is true and q is false. Hence, the given statement is false when "I lie" and "I would not tell you." ■

EXAMPLE 8 ▶ Truth Tables Involving Negating Conjunctions

Construct the truth table for the statement $\sim(p \wedge \sim q)$.

Solution

As in Example 6, the plan is first to break down the given statement into its primitive components. The statement $\sim(p \wedge \sim q)$ is the negation of $p \wedge \sim q$, which is the conjunction of the components p and $\sim q$. We can write the truth values of $\sim q$ from those of q, and p is a primitive component. Thus, we use p, q, and $\sim(p \wedge \sim q)$ as headings for the truth table.

To find the truth values of $\sim(p \wedge \sim q)$, we must have the truth values of p, q, $\sim q$, and $p \wedge \sim q$, so we number the columns in that order. Table 3.6 is filled out in the following steps:

TABLE 3.6

Remember, you can also write the columns like this.

1	2	5	4	3
p	q	$\sim(p$	$\wedge$	$\sim q)$
T	T	T	F	F
T	F	F	T	T
F	T	T	F	F
F	F	T	F	T

1	2	3	4	5
p	q	$\sim q$	$p \wedge \sim q$	$\sim(p \wedge \sim q)$

1. We write the four possible pairs of truth values for p and q in columns 1 and 2.

2. Using column 2 (q) as our reference, we negate q to get $\sim q$ (column 3).

3. To get column 4 ($p \wedge \sim q$), we combine columns 1 (p) and 3 ($\sim q$), using the conjunction *and,* denoted by $\wedge$. We recall (see Table 3.2) that a conjunction is true only when both its components are true and is false otherwise. Thus, we write T in the second row of column 4, where p and $\sim q$ are both true, and we write F in the other rows.

4. We negate the truth values in column 4 to get those of $\sim(p \wedge \sim q)$ in column 5. The statement $\sim(p \wedge \sim q)$ has the truth values shown in column 5 of the table. ■

Online Study Center

Want to learn more about constructing a truth table? Access link 3.2.1 on this textbook's Online Study Center.

E. Equivalent Statements

Notice that the statements given in Examples 6 and 8 have exactly the same truth values, *TFTT*; hence, the two statements must have the same meaning.

Definition of Equivalent

Two statements p and q that have identical truth values are said to be **equivalent** (denoted by $p \Leftrightarrow q$).

Accordingly, from Examples 6 and 8, we can write an equivalent statement in symbolic form.

An Example of Equivalence

$$\sim p \lor q \Leftrightarrow \sim(p \land \sim q)$$

PROBLEM SOLVING

Equivalent Statements

Show that $p \land (q \lor r) \Leftrightarrow (p \land q) \lor (p \land r)$.

❶ Read the problem and **select** the unknown.
Show that two statements are equivalent. What does this mean?

To show that two statements are equivalent, you have to show that the statements have identical truth tables.

❷ Think of a plan.
Make a truth table for each of the statements. How many lines will each truth table have?

Because there are three statements, p, q, and r, each of which has two possible truth values, there are $2 \times 2 \times 2 = 8$ possible cases. Thus, the truth table must have 8 lines.

❸ Use truth tables to carry out the plan.
Break the statement $p \land (q \lor r)$ into components and label the columns accordingly. Then proceed as in the previous examples to make the truth table.

The components of $p \land (q \lor r)$ are p and $(q \lor r)$. Thus, the truth table (Table 3.7) must have columns for p, q, r, $q \lor r$, and $p \land (q \lor r)$. Table 3.7 is filled out in the same manner as the tables for the previous examples—by steps.

1. In the first three columns, write the possible truth values for p, q, and r. In column 1, enter four T's and four F's; in column 2, enter two T's, two F's, two T's, and two F's; in column 3, enter alternately one T and one F. This gives all the possible combinations of T's and F's for the three statements.

2. To obtain the truth values of $q \lor r$, combine columns 2 and 3 with a disjunction $\lor$, as shown in column 4. Because $q \lor r$ is false only when both q and r (rows 4 and 8) are false, enter F's in rows 4 and 8 and T's in the remaining rows.

3. To obtain the truth values of $p \land (q \lor r)$, combine columns 1 (p) and 4 ($q \lor r$) with a conjunction $\land$. Since a conjunction is true only when both components are true, complete column 5 by writing T's in the first three rows and F's in the other rows. The given statement has the truth values shown in column 5 of Table 3.7.

continued

TABLE 3.7

1	2	3	5	4
p	q	r	$p \wedge$	$(q \vee r)$
T	T	T	T	T
T	T	F	T	T
T	F	T	T	T
T	F	F	F	F
F	T	T	F	T
F	T	F	F	T
F	F	T	F	T
F	F	F	F	F

Break the statement $(p \wedge q) \vee (p \wedge r)$ into components.

The components of $(p \wedge q) \vee (p \wedge r)$ are $(p \wedge q)$ and $(p \wedge r)$, and these two components have components p, q, and r. The truth table (Table 3.8) is filled out in the same manner as Table 3.7; that is, do columns 4, 5, and then 6.

TABLE 3.8

1	2	3	4	6	5
p	q	r	$(p \wedge q)$	$\vee$	$(p \wedge r)$
T	T	T	T	T	T
T	T	F	T	T	F
T	F	T	F	T	T
T	F	F	F	F	F
F	T	T	F	F	F
F	T	F	F	F	F
F	F	T	F	F	F
F	F	F	F	F	F

④ **Verify** the answer. Are the results of the two tables identical?

TRY EXAMPLE 9 NOW.

Since the truth values in the final columns of the two tables (5 and 6, respectively) are identical, the two statements are equivalent.

Cover the solution, write your own solution, and then check your work.

EXAMPLE 9 ▶ Showing Equivalency

(a) Show that $(\sim p \vee q) \wedge r \Leftrightarrow (\sim p \wedge r) \vee (q \wedge r)$.

(b) Let p be the statement "You recycle paper," let q be the statement "You recycle glass," and let r be the statement "You are an environmentalist." If we know that you recycle paper, under what condition(s) will the statement $(\sim p \vee q) \wedge r$ be true?

Solution

(a) To show that $(\sim p \vee q) \wedge r \Leftrightarrow (\sim p \wedge r) \vee (q \wedge r)$, we have to show that the truth tables for both statements are identical. We first construct the table for $(\sim p \vee q) \wedge r$ (Table 3.9). The components for this statement are p, q, r, $\sim p$, and $(\sim p \vee q)$, as shown in Table 3.9. The statement $\sim p \vee q$ is false only when $\sim p$ and q are both false. Thus, in column 5 we enter F's in rows 3 and 4 and T's in the rest of the rows. Finally, $(\sim p \vee q) \wedge r$ is true only when both $(\sim p \vee q)$ in column 5 and r in column 6 are true. Thus, in column 7 we enter T's in rows 1, 5, and 7 and F's in the rest of the rows. The final truth values appear in column 7.

TABLE 3.9

1	2	3	4	5	7	6
p	q	r	$\sim p$	$(\sim p \vee q)$	$\wedge$	r
T	T	T	F	T	T	T
T	T	F	F	T	F	F
T	F	T	F	F	F	T
T	F	F	F	F	F	F
F	T	T	T	T	T	T
F	T	F	T	T	F	F
F	F	T	T	T	T	T
F	F	F	T	T	F	F

TABLE 3.10

1	2	3	4	5	7	6
p	q	r	$\sim p$	$(\sim p \wedge r)$	$\vee$	$(q \wedge r)$
T	T	T	F	F	T	T
T	T	F	F	F	F	F
T	F	T	F	F	F	F
T	F	F	F	F	F	F
F	T	T	T	T	T	T
F	T	F	T	F	F	F
F	F	T	T	T	T	F
F	F	F	T	F	F	F

Online Study Center

To further explore truth tables and the truth table constructor, access link 3.2.2 on this textbook's Online Study Center.

The components for $(\sim p \wedge r) \vee (q \wedge r)$ are p, q, r, $\sim p$, $(\sim p \wedge r)$, and $(q \wedge r)$, as shown in Table 3.10. Thus, $(\sim p \wedge r)$ (column 5) is true only when $\sim p$ and r are both true (rows 5 and 7 in column 5), and $(q \wedge r)$ (column 6) is true only when both q and r are true (rows 1 and 5 in column 6). Finally, $(\sim p \wedge r) \vee (q \wedge r)$ in column 7 is false only when both $(\sim p \wedge r)$ and $(q \wedge r)$ are false (rows 2, 3, 4, 6, and 8 in column 7). The final truth values are in column 7. Since columns 7 in Tables 3.9 and 3.10 are identical, the statements are equivalent.

(b) Since p has truth value T (first four rows), $(\sim p \vee q) \wedge r$ is true only when q and r are both true (row 1), that is, when the components q "You recycle glass" and r "You are an environmentalist" are both true. ∎

EXERCISES 3.2

A The Conjunction
B The Disjunction
C Negations

In problems 1–4, let p be "Today is Friday" and let q be "Today is Monday."

1. Write in words the disjunction of the two statements.

2. Write in words the conjunction of the two statements.

3. Write in words the negation of statement p.

4. Which of the statements in problems 1–3 always has the truth value F?

In problems 5–7, let g be "He is a gentleman" and let s be "He is a scholar." Write the following in words:

5. The disjunction of the two statements

6. The negation of the statement g

7. The conjunction of the two statements

In problems 8–10, use statements *g* and *s* of problems 5–7 and write each statement in symbolic form.

8. He is not either a gentleman or a scholar.

9. He is a gentleman and a scholar.

10. He is neither a gentleman nor a scholar.

11. Consider the statements *p* and *q*.

> *p*: It is raining.
> *q*: I will go to the beach.

Write the statements in parts (**a**) and (**b**) in symbolic form.
a. It is raining but I will go to the beach.
b. It is raining or I will go to the beach.
c. Assume that *p* is true and *q* is false. Find the truth values of the statements given in parts (**a**) and (**b**).

In problems 12–16, let *p* be "Mida is cooperative" and let *q* be "Desi is uncooperative." Write each statement in symbolic form.

12. Mida and Desi are both cooperative.

13. Neither Desi nor Mida is uncooperative.

14. It is not the case that Mida and Desi are both uncooperative.

15. Either Mida is cooperative or Desi is uncooperative.

16. Assume that Mida is cooperative and Desi is uncooperative. Which of the statements in problems 12–15 are true?

In problems 17–21, suppose that *p* is true and *q* is false. Write each statement in symbolic form, and find its truth value.

17. Either *p* or *q*

18. Either *p* or not *q*

19. Neither *p* nor *q*

20. *p* or *q*, but not both

21. Not *q* and not *p*

In problems 22–26, consider the following statements:

> *g*: I go to college.
> *j*: I join the army.

Suppose that *g* is false and *j* is true. Write each statement in symbolic form, and find its truth value.

22. Either I go to college or I join the army.

23. I go to college or I do not join the army.

24. I neither go to college nor join the army.

25. I go to college or I join the army, but not both.

26. I do not go to college and I do not join the army.

D **Making Truth Tables**

In problems 27–40, construct a truth table for each statement.

27. $p \vee \sim q$

28. $\sim(p \vee q)$

29. $\sim p \wedge q$

30. $\sim p \vee \sim q$

31. $\sim(p \vee \sim q)$

32. $\sim(\sim p \vee \sim q)$

33. $\sim(\sim p \wedge \sim q)$

34. $(p \vee q) \wedge \sim(p \wedge q)$

35. $(p \wedge q) \vee (\sim p \wedge q)$

36. $(p \wedge \sim q) \wedge (\sim p \wedge q)$

37. $p \wedge (q \vee r)$

38. $p \vee (q \wedge r)$

39. $(p \vee q) \vee (r \wedge \sim q)$

40. $[(p \wedge q) \vee (q \wedge \sim r)] \vee (r \wedge \sim s)$

41. Let *p* be "Mida is blonde," and let *q* be "Mida is 6 ft tall."
a. Under which conditions is the statement "Mida is blonde and 6 ft tall" true?
b. Under which conditions is the statement "Mida is blonde and 6 ft tall" false?
c. Under which conditions is the statement "Mida is blonde or 6 ft tall" true?
d. Under which conditions is the statement "Mida is blonde or 6 ft tall" false?

42. Let *p* be "Eva is a high school graduate," and let *q* be "Eva is over 16 years old."
a. Under which conditions is the statement "Eva is neither a high school graduate nor over 16 years old" true?
b. Under which conditions is the statement in part (**a**) false?
c. Under which conditions is the statement "Either Eva is a high school graduate or she is over 16 years old" true?
d. Under which conditions is the statement in part (**c**) false?

E **Equivalent Statements**

In problems 43–46, use truth tables to show that the two given statements are equivalent.

43. $p \vee (q \wedge r)$ and $(p \vee q) \wedge (p \vee r)$

44. $p \wedge (q \vee r)$ and $(p \wedge q) \vee (p \wedge r)$

45. $\sim(p \vee q)$ and $\sim p \wedge \sim q$

46. $\sim(p \wedge q)$ and $\sim p \vee \sim q$

In problems 47 and 48, use truth tables to show the equivalence.

47. $(p \wedge q) \vee \sim p \Leftrightarrow q \vee \sim p$

48. $(p \vee q) \wedge (\sim p \vee \sim q) \Leftrightarrow (p \wedge \sim q) \vee (\sim p \wedge q)$

49. a. Verify the entries in the following table:

p	q	$p \wedge q$	$p \wedge \sim q$	$\sim p \wedge q$	$\sim p \wedge \sim q$
T	T	T	F	F	F
T	F	F	T	F	F
F	T	F	F	T	F
F	F	F	F	F	T

b. Look at the last four columns of this table. Each of these columns has one T and three F's, and there is exactly one T on each line. This T occurs on the line where both components of the corresponding column heading are true. The headings of these last four columns are called **basic conjunctions.** By using these conjunctions, you can write statements having any given four-entry truth table. For instance, a statement with the truth table $TFFT$ is $(p \wedge q) \vee (\sim p \wedge \sim q)$. Explain this.

c. By forming disjunctions of the basic conjunctions, you can write statements with given truth tables as noted in part (**b**). Write a statement having truth table $FTTF$. Do the same for $FTTT$. Can you write a simpler statement for the truth table $FTTT$?

50. The ideas in problem 49 can be generalized to include statements with any given number of components. Thus, the statement $p \wedge \sim q \wedge \sim r$ would have a truth table with a T on the line corresponding to p true, q false, and r false and F's on the other seven lines. What would be the truth table for $(p \wedge \sim q \wedge r) \vee (\sim p \wedge \sim q \wedge \sim r)$?

51. To be eligible for a position in a banking firm, an applicant must be at least 25 years old (t), have a college degree in business administration (d), and be married (m). Assume that t, d, and m are true statements. Which of the following

three applicants (if any) is eligible for the above position?

> Joe is married, has a college degree in fine arts, and is 26 years old.
> Mary is married, has a degree in business administration, and is 22 years old.
> Ellen has a degree in business administration, is 25 years old, and is single.

52. To qualify for a $40,000 loan, an applicant must have a gross income of $30,000 if single ($50,000 combined income if married) and assets of at least $10,000. Which of the following three applicants (if any) would qualify for the loan?

> Mr. Perez is married, has two children, and makes $35,000 at his job. His wife does not work.
> Ms. Jefferson and her husband have assets of $50,000. One makes $22,000; the other makes $19,000.
> Tran Quang is a bachelor and works at two jobs. He makes $28,000 at one job and $5000 at the other; his only asset is a $7000 Toyota.

 In Other Words

Mathematics has a specialized vocabulary. For example, the negation of "9 is less than 3" is "9 is not less than 3," but it can also be "9 is greater than or equal to 3."

In problems 53–57, negate each statement without using the word *not*.

53. 7 is less than 5.

54. 8 is more than 9.

55. 0 is greater than 3.

56. $\frac{1}{3}$ is greater than or equal to 1.

57. $\frac{1}{2}$ is less than or equal to $\frac{1}{8}$.

58. Let f be "I will go fishing." Let g be "The sun is shining." Write the following in words:
a. The conjunction of f and g
b. The negation of g
c. The disjunction of f and g

59. For the statements given in problem 58, write in words the negation of the conjunction of f and g. State under what circumstances this negation would be true.

Using Your Knowledge

The Higher Education Act states that any student is eligible to apply for a loan, provided the student fulfills the following requirements:

1. Enrolled (*e*) and in good standing (*g*), or accepted for enrollment (*a*) at an eligible school
2. Registered for at least one-half the normal full-time work load as determined by the school (*h*)
3. A citizen (*c*) or national (*n*) of the United States, or in the United States for other than a temporary purpose (~*t*)

60. Translate requirements 1, 2, and 3 into symbolic form.

61. What is the general symbolic compound statement whose truth implies that the student can apply for a loan?

62. A three-component statement whose truth implies that the student can apply for a loan is $a \land h \land c$. What are two others?

 **Discovery**

Four men, Mr. Baker, Mr. Carpenter, Mr. Draper, and Mr. Smith, live in Logictown. One is a baker, one a carpenter, one a draper, and one a smith, but none follows the vocation corresponding to his name. A logician tries to find out who is who and obtains the following partially correct information:

Mr. Baker is the smith.
Mr. Carpenter is the baker.
Mr. Draper is not the smith.
Mr. Smith is not the draper.

63. If it is known that three of the four statements are false, who is the carpenter? (*Hint:* Consider the four possible sets of truth values given for the statements in the table below.)

Statement	*I*	*II*	*III*	*IV*
1	*T*	*F*	*F*	*F*
2	*F*	*T*	*F*	*F*
3	*F*	*F*	*T*	*F*
4	*F*	*F*	*F*	*T*

Online Study Center

For a more contemporary puzzle, access link 3.2.3 on this textbook's Online Study Center.

3.3 The Conditional and the Biconditional

On What Condition?

Have you read your insurance policy or your income tax instructions lately? They contain many of the connectives we have studied and more. Here is an excerpt from an automobile insurance policy.

> If the final recomputed premium exceeds the premium stated on the declarations page, you must pay the excess to Allstate.

(If you do not understand this, it may cost you money.) Here is another statement.

> If an injured person unreasonably refuses to take the examination, we are not required to pay any subsequent personal injury protection benefits.

(Failure to understand this one may cost you your benefits!)

What about income taxes? Here is an excerpt from an income tax form.

> If you checked NO to any of the above questions, you may not take the earned income credit.

(Here you may make a mistake on your return, resulting in an audit.)

The field of law requires the ability to apply logic to reasoning and argument.

TABLE 3.11

p	q	$p \rightarrow q$
T	T	T
T	F	F
F	T	T
F	F	T

```
IF YOU STOP
HERE — YOUR
PAIN WILL TOO
```

A chiropractor's sign makes use of a conditional statement.

Or what about this one?

> If your return is more than 60 days late, the minimum penalty will be $100 or the amount of any tax you owe, whichever is smaller.

Do you know what the penalty will be?

Finally, look at a hypothetical application of conditional statements in the field of law. A man was being tried for participation in a robbery. Here is a partial transcript of the proceedings.

Prosecutor:	"If the defendant is guilty, then he had an accomplice."
Defense attorney:	"That is not true!"
Judge:	"In that case, I declare a mistrial. The defendant needs a new attorney."

Why? The attorney said that the statement "If the defendant is guilty (g), then he had an accomplice (a)," or symbolically $g \rightarrow a$, was not true; that is, $g \rightarrow a$ was false. Now, look at the second row of Table 3.11. When is a conditional false? When the antecedent (g in this case) is true and the consequent (a in this case) is false. What does that mean in terms of the defendant? ▶

It is sometimes necessary to specify the conditions under which a given event will be true. For example, one might say, "If the weather is nice, then I will go to the beach." If we let p stand for "the weather is nice" and q stand for "I will go to the beach," then the preceding compound statement is of the form "If p, then q." Statements of this kind are called **conditional statements** and are symbolized by $p \rightarrow q$. (Read "if p, then q" or "p arrow q" or "p conditional q.") In $p \rightarrow q$, statement p is called the **antecedent** and q the **consequent.**

A. The Conditional

To understand the truth table for the conditional, consider the sign in the margin. It promises "If you stop here, your pain will too." Under which circumstances is this promise broken? Only if you *do* stop here and your pain *does not* stop. Thus, we should write F for $p \rightarrow q$ if p is true and q is false; otherwise, we should write T.

Table 3.11 expresses these facts. It shows that if p and q are both true, then $p \rightarrow q$ is true, and if p is true and q is false, then $p \rightarrow q$ is false. In the last two lines of Table 3.11, p is false, so it would be incorrect to say that $p \rightarrow q$ is false. Since we want a complete truth table, we have assigned the value T to $p \rightarrow q$ in these two lines. See problem 52, Exercises 3.3.

Note that "All p's are q's" is translated as "If it is a p, then it is a q." This idea will be used in Exercises 3.3, problems 35–39.

Truth Value of the Conditional Statement $p \rightarrow q$

The **conditional statement** $p \rightarrow q$ ("if p, then q") is false only when p is true and q is false; otherwise, it is true.

A gas pump sign also uses a conditional statement.

EXAMPLE 1 ▶ Diesel Logic

(a) Write the statement in the photograph in symbolic form using

v: the pump runs very slowly r: release nozzle trigger
c: count to 20 t: try again

(b) When is the statement false?

Solution

(a) The statement is written as the conditional $v \rightarrow (r \wedge c \wedge t)$.

(b) A conditional is false when the antecedent v is true and the consequent $(r \wedge c \wedge t)$ is *false*, that is, when the pump is running very slowly and you did not release the trigger or count to 20 or try again. ∎

B. The Biconditional

In certain statements the conditional is used twice; in the second conditional, the antecedent and the consequent of the first conditional are reversed. For example, the statement "If money is plentiful, then interest rates are low, and if interest rates are low, then money is plentiful" uses the conditional twice in this manner. It is for this reason that such statements, which can be written in the form $(p \rightarrow q) \wedge (q \rightarrow p)$, are called **biconditionals.** The biconditional is usually symbolized by the shorter form $p \leftrightarrow q$ (read as *p if and only if q* or *p biconditional q*).

Definition of the Biconditional Statement $p \leftrightarrow q$

$$p \leftrightarrow q \Leftrightarrow (p \rightarrow q) \wedge (q \rightarrow p)$$

Table 3.12 is the truth table for the statement $(p \rightarrow q) \wedge (q \rightarrow p)$. It is filled out in the usual way.

1. In columns 1 and 2, we write the four possible pairs of truth values for p and q.

TABLE 3.12

1	*2*	*3*	*5*	*4*
p	q	$(p \rightarrow q)$	$\wedge$	$(q \rightarrow p)$
T	T	T	T	T
T	F	F	F	T
F	T	T	F	F
F	F	T	T	T

2. We combine columns 1 (p) and 2 (q) with the conditional ($\rightarrow$) to form column 3 ($p \rightarrow q$). Since $p \rightarrow q$ is false only when p is true and q is false, we write F in the second row and T in the other rows.

3. We combine columns 2 (q) and 1 (p)—*in that order*—with the conditional ($\rightarrow$) to form column 4 ($q \rightarrow p$). Because $q \rightarrow p$ is false only when q is true and p is false, we write F in the third row and T in the other rows.

4. We combine columns 3 ($p \rightarrow q$) and 4 ($q \rightarrow p$) with the conjunction $\wedge$ to form column 5 [$(p \rightarrow q) \wedge (q \rightarrow p)$]. Since the conjunction is true only when both components are true, we write T in the first and fourth rows and F in the other two rows. This completes Table 3.12.

Table 3.12 shows the following:

Truth Value of the Biconditional Statement $p \leftrightarrow q$

The biconditional $p \leftrightarrow q$ is true when and only when p and q have the same truth values; it is false otherwise.

EXAMPLE 2 ▶ Finding the Truth Value of Conditionals

Give the truth value of each of the following, assuming that the week starts on Sunday.

(a) If Tuesday is the last day of the week, then the next day is Sunday.

(b) If Tuesday is the third day of the week, then the next day is Sunday.

(c) If Tuesday is the third day of the week, then Wednesday is the fourth day of the week.

(d) If Tuesday is the last day of the week, then the next day is Wednesday.

Solution

All these statements are of the form $p \rightarrow q$, where p is the antecedent and q is the consequent. (Recall that the antecedent is the "if" part, and the consequent is the "then" part.)

(a) Because p is false, the statement $p \rightarrow q$ is true.

(b) Because p is true and q is false, the statement $p \rightarrow q$ is false.

(c) Because p and q are both true, the statement $p \rightarrow q$ is true.

(d) Because p is false, the statement $p \rightarrow q$ is true.

Note the results in (a) and (d). The moral is that if you start off with a false assumption, then you can prove anything!   ■

EXAMPLE 3 ▶ Finding the Truth Value of Biconditionals

Is the statement $(3 + 5 = 35) \leftrightarrow (2 + 7 = 10)$ true or false?

Solution

This statement is of the form $p \leftrightarrow q$, where p is "$3 + 5 = 35$" and q is "$2 + 7 = 10$." Since the biconditional $p \leftrightarrow q$ is true when p and q have the same truth value, and p and q in this example are both false, the given biconditional statement is true.   ■

EXAMPLE 4 ▶ Finding When Conditionals Are False

Let p be "x is a fruit," and let q be "x is ripe." Under what conditions is the statement $p \rightarrow q$ false?

Solution

The statement $p \rightarrow q$ is a conditional statement and thus is false only when p is true and q is false; hence, the given statement is false if x is a fruit that is not ripe. ◼

EXAMPLE 5 ▶ Showing the Equivalency of $p \rightarrow q$ and $\sim p \lor q$

Show that the statements $p \rightarrow q$ and $\sim p \lor q$ are equivalent; that is, show that $(p \rightarrow q) \Leftrightarrow (\sim p \lor q)$. Use the equivalence to write the statement "If you are under 18, then you must register" as a disjunction.

Solution

To show that two statements are equivalent, we must show that they have identical truth tables. The statement $p \rightarrow q$ is false only when p is true and q is false (see Table 3.13).

To make a truth table for $\sim p \lor q$ (see Table 3.14), we write the components p, q, $\sim p$, and $\sim p \lor q$ as shown. We negate statement p by writing truth values in column 3 opposite to those shown in column 1.

The statement $\sim p \lor q$, shown in column 4, is false only when both components, $\sim p$ and q, are false (row 2); otherwise it is true. Since the final results $TFTT$ are identical, the statements are equivalent; that is, $(p \rightarrow q) \Leftrightarrow (\sim p \lor q)$.

Now, let p be the statement "You are under 18" and q be the statement "You must register." Since $(p \rightarrow q) \Leftrightarrow (\sim p \lor q)$, the statement "If you are under 18, then you must register" can be written as the disjunction "You are not under 18 or you must register." ◼

The equivalence in Example 5 $(p \rightarrow q) \Leftrightarrow (\sim p \lor q)$, is of great importance because it allows us to handle a conditional statement in terms of the logical symbols for negation and disjunction. This equivalence will be used in several of the problems in Exercises 3.3 and later in this chapter.

TABLE 3.13

1	2	3
p	q	$p \rightarrow q$
T	T	T
T	F	F
F	T	T
F	F	T

TABLE 3.14

1	2	3	4
p	q	$\sim p$	$\sim p \lor q$
T	T	F	T
T	F	F	F
F	T	T	T
F	F	T	T

Conditional Equivalence

$$(p \rightarrow q) \Leftrightarrow (\sim p \lor q)$$

Note that since $(p \rightarrow q) \Leftrightarrow (\sim p \lor q)$, the negation of $p \rightarrow q$ is equivalent to the negation of $\sim p \lor q$; that is, the negation of a conditional statement can be written as follows:

Negation of a Conditional Statement

$$\sim(p \rightarrow q) \Leftrightarrow \sim(\sim p \lor q)$$

See problem 40 in Exercises 3.3 for the simplification of the right-hand side of this equivalence.

Sometimes, conditional statements are implicit, as shown in the excerpt of a brochure dealing with stress given in Example 6.

EXAMPLE 6 ▶ Writing Statements in Symbolic Form

Frustration mounts (f) when equipment (e) or materials (m) necessary to do an assigned task are insufficient, work schedules are overly demanding (o), the workload too great (w), or work demands regularly spill over into time reserved for personal life (s).

Use the indicated letters to represent the corresponding components and write the statement in symbolic form.

Solution

Note that in this case, the conclusion reached is (f). We translate the statement as $[(e \vee m) \vee o \vee w \vee s] \to f$. (We hope that frustration did not mount when doing this problem!) ▪

Online Study Center

To further explore conditionals, access link 3.3.1 on this textbook's Online Study Center.

EXERCISES 3.3

A The Conditional

1. Show that the statement $\sim q \to \sim p$ is equivalent to $p \to q$.

In problems 2 and 3, use truth tables to show the equivalences.

2. $p \to \sim q \Leftrightarrow \sim(p \wedge q)$

3. $\sim p \to q \Leftrightarrow p \vee q$

In problems 4–7, find the truth value of each statement.

4. If $2 + 2 = 22$, then $22 = 4$.

5. If $2 + 2 = 4$, then $8 = 5$.

6. If $2 + 2 = 22$, then $8 = 4 + 4$.

7. If $2 + 2 = 22$, then $4 = 26$.

In problems 8–11, find all the number replacements for x that make each sentence true.

8. If $2 + 2 = 4$, then $x - 2 = 5$.

9. If $2 + 2 = 22$, then $x - 2 = 5$.

10. If $x + 2 = 6$, then $3 + 2 = 5$.

11. If $x + 2 = 6$, then $2 + 2 = 32$.

12. Let p be "I kiss you once," and let q be "I kiss you again." Under which condition is the statement $p \to q$ false?

13. Under which condition is the statement "If you've got the time, we've got the beer" false?

In problems 14–16, construct a truth table for each statement. Note the importance of the parentheses and the brackets to indicate the order in which items are grouped.

14. $[(p \to q) \to p] \to q$ **15.** $(p \to q) \leftrightarrow (p \vee r)$

16. $(p \to q) \leftrightarrow (p \to \sim q)$

B The Biconditional

In problems 17 and 18, construct a truth table for each statement.

17. $p \to (q \wedge r)$ **18.** $(p \to q) \wedge (p \to r)$

19. Are the statements in problems 17 and 18 equivalent?

In problems 20–25, let p be "I will buy it" and let q be "It is a poodle." Translate each statement into symbolic form.

20. If it is a poodle, then I will buy it.

21. If I will buy it, then it is a poodle.

22. It is a poodle if and only if I will buy it.

23. If it is not a poodle, then I will not buy it.

24. If I will not buy it, then it is not a poodle.

25. If it is a poodle, then I will not buy it.

In problems 26–28, let $\sim s$ be "You are out of Schlitz" and let $\sim b$ be "You are out of beer." Translate each statement into symbolic form.

26. If you are out of Schlitz, you are out of beer.

27. If you are out of beer, you are out of Schlitz.

28. Having beer is equivalent to having Schlitz.

In problems 29–31, write each statement in symbolic form using $\sim$ and $\vee$. Also write the corresponding statements in words. (*Hint:* $p \rightarrow q$ is equivalent to $\sim p \vee q$.)

29. If the temperature is above 80° (a), then I will go to the beach (b).

30. If Mida is home by 5 (h), then dinner will be ready by 6 (r).

31. If Eva has a day off (a), then she will go to the beach (g).

In Example 5 it was shown that $p \rightarrow q$ is equivalent to $\sim p \vee q$. In problems 32–34, use this equivalence to write each statement as a disjunction.

32. If you work, you have to pay taxes.

33. If you've got the time, we've got the beer.

34. If you find a better one, then you buy it.

The statement "All even numbers are divisible by 2" can be translated as "If it is an even number, then it is divisible by 2." In general, the statement "All ____ are ____" can be translated as "If it is a ____, then it is a ____." In problems 35–39, use this idea to write each statement in the if-then form.

35. All dogs are mammals.

36. All cats are felines.

37. All men are created equal.

38. All prime numbers greater than 2 are odd numbers.

39. All rectangles whose diagonals are perpendicular to each other are squares.

40. Because $p \rightarrow q$ is equivalent to $\sim p \vee q$ (see Example 5), the negation of $p \rightarrow q$ should be equivalent to the negation of $\sim p \vee q$. Show that the negation of $\sim p \vee q$ is $p \wedge \sim q$; that is, show that $\sim(\sim p \vee q)$ is equivalent to $p \wedge \sim q$.

41. From problem 40 it is clear that the negation of $p \rightarrow q$ is equivalent to $p \wedge \sim q$. Verify this by means of a truth table.

Problem 41 verified that the negation of $p \rightarrow q$ is $p \wedge \sim q$. This means that to negate an "if ____, then ____" statement, we simply assert the *if clause* and deny the *then clause*. For instance, the negation of the statement "If you are out of Schlitz, you are out of beer" is the statement "You are out of Schlitz, but you are not out of beer." In problems 42–47, write in words the negation of each statement.

42. If you earn a lot of money, then you pay heavy taxes.

43. If Johnny does not play quarterback, then his team loses.

44. If Alice passes the test, then she gets the job.

45. If I kiss you once, I kiss you again.

46. If Saturday is a hot day, I will go to the beach.

47. Evel Knievel will lose his life if he is careless.

48. From problem 41 you can see that $p \wedge \sim q$ has truth values *FTFF*. If you know that $p \wedge \sim q$ is the negation of $p \rightarrow q$, how can you define the truth table for $p \rightarrow q$?

In problems 49–51, write each statement in the if-then form.

49. Johnny does not play quarterback or his team wins.

50. Alice fails the test or she gets the job.

51. Joe had an accident or he would be able to get car insurance.

52. In defining $p \rightarrow q$, it is easy to agree that if p is true and q is true, then $p \rightarrow q$ is true; also, if p is true and q is false, then $p \rightarrow q$ is false. Assuming that the entries in the first two rows in the following table are *TF*, respectively, we have four possible definitions for $p \rightarrow q$, as listed in the table.

		Definition of $p \rightarrow q$			
p	q	1	2	3	4
T	T	T	T	T	T
T	F	F	F	F	F
F	T	F	F	T	T
F	F	F	T	F	T

a. Show that if we use definition 1, then $p \rightarrow q$ and $p \wedge q$ have the same truth table.

b. Show that if we use definition 2, then $p \rightarrow q$ and $p \leftrightarrow q$ have the same truth table.

c. Show that if we use definition 3, then $p \rightarrow q$ and q have the same truth table.

Thus, the table shows that if we wish $p \rightarrow q$ to be different from $p \wedge q$, $p \leftrightarrow q$, and q, then we must use definition 4.

53. A mother promises her child "If you eat the spinach and the liver, then you may go out to play." The child eats only the spinach, but the mother lets him go out to play. Has she broken her original promise?

54. Now suppose the mother in problem 53 says "If you do not eat the spinach and liver, then you may not go out to play." If the child then eats only the spinach and the mother lets him go out to play, has she broken her promise?

55. Which of the following statements is logically equivalent to "If Mary is in Tampa, then she is in Florida"?
a. Mary is in Florida, or she is in Tampa.
b. If Mary is not in Tampa, then she is not in Florida.
c. If Mary is in Florida, then she is in Tampa.
d. If Mary is not in Florida, then she is not in Tampa.

56. Which of the following statements is logically equivalent to "If you want to buy organic food, you have to let your grocer know"?
a. You want to buy organic food, or you let your grocer know.
b. If you do not let your grocer know, then you do not want to buy organic food.
c. If you let your grocer know, then you want to buy organic food.
d. If you do not want to buy organic food, then you do not let your grocer know.

57. Which statement is the logical negation of "If you studied hard, you passed the course"?
a. If you didn't study hard, you didn't pass the course.
b. If you didn't pass the course, you didn't study hard.
c. You didn't study hard, but you passed the course.
d. You studied hard, and you did not pass the course.

58. Which statement is the logical negation of "If it rains, we will not go to the beach"?
a. If it doesn't rain, then we will go to the beach.
b. It is raining, and we do not go to the beach.
c. If we go to the beach, then it will not rain.
d. It is raining, and we will go to the beach.

 In Other Words

59. Assume that the chiropractor's advertisement "If you stop here, your pain will too," found on page 130, is a true statement. Does it follow logically that if you do not stop, your pain will not stop either? Explain.

60. Here is an excerpt from an automobile insurance policy. "If the loss is $50 or less, we will not make any payment." A policyholder suffers a $75 loss. What are the insurance company's options? Explain.

61. A sophomore college student reads the following statement: "If you are entering college for the first time, you are required to take the placement examination." Does the student have to take the placement examination? Explain.

62. Write in your own words why the conditional statement "If p then q" is **false** only when p is true is q is false, and it is true otherwise.

63. Write in your own words the conditions under which the statement $p \leftrightarrow q$ must be true and the conditions under which it must be false.

64. What does it mean to you when we say that two statements are **equivalent?**

65. What is the difference between $\leftrightarrow$ and $\Leftrightarrow$?

 Using Your Knowledge

A certain credit union issues the following memorandum with its monthly statement of account: "Please carefully examine the enclosed memorandum. Report all differences to the Auditing Division. If no differences are reported in 10 days, we shall understand that the balance is correct as shown."

Let d be "A difference is found."
Let r be "A report is made in 10 days."
Let a be "The credit union makes the adjustment."

66. Write in symbols "If a difference is found, then a report is made in 10 days."

67. Write in symbols "If a report is made in 10 days, the credit union makes the adjustment."

68. Write in symbols "If a difference is found, then a report is made in 10 days and the credit union makes the adjustment."

69. Does the statement in problem 68 indicate that the credit union will make no adjustment if a late report of differences is made?

 **Discovery**

A logician is captured by a tribe of savages, whose chief makes the following offer: "One of these two roads leads to certain death and the other to freedom. You may select either road after asking any one question of one of these two warriors. I must warn you, however, that one of them is always truthful and the other always lies."

Let p be "The first road leads to freedom."
Let q be "You are telling the truth."

70. What should the question be? (*Hint:* Construct a question so that if p is true, the answer is "Yes" and, if p is false, the answer is "No." Complete the table below, and then refer to problem 49, Exercises 3.2, to find the desired question.)

p	q	Answer	Truth Table of Question to Be Asked
T	T	Yes	
T	F	Yes	
F	T	No	
F	F	No	

Online Study Center

Problems 71–73 are taken from the Web. Some are attributed to Raymond Smullyan, a well-known mathematician and logician, and Oscar Ewing, professor emeritus of philosophy at Indiana University and professor emeritus of the City University of New York–Lehman College and Graduate Center. To read them and explore their solutions, access link 3.3.2 on this textbook's Online Study Center. For more puzzles (some dealing with logic, some not), access link 3.3.3.

71. Three logicians, A, B, and C, are wearing hats, which they know are either black or white, but not all three are white. A can see the hats of B and C, B can see the hats of A and C, and C is blind. They are asked in turn if they know the color of their own hat. The answers are A, no; B, no; and C, yes. What color is C's hat and how does she know?

72. Two women stand at a fork in the road. One fork leads to Someplaceorother; the other fork leads to Nowheresville. One of these people always answers the truth to any yes/no question that is asked of her. The other always lies when asked any yes/no question. By asking one yes/no question, can you determine the road to Someplaceorother?

73. While three logicians were sleeping under a tree, a malicious child painted their heads red. On waking, each logician spies the child's handiwork as it applied to the heads of the other two. Naturally they start laughing. Suddenly one falls silent. Why?

 Collaborative Learning

Go to your library, Post Office, IRS office, or link 3.3.4 on this textbook's Online Study Center and get Form 1040EZ. Here is a statement appearing at the top of page 2.

Use this form (u) if you had *only* wages (w), salaries (s), tips (t), taxable scholarships (ts) or fellowship grants (g), unemployment compensations (c), or Alaska Permanent Fund dividends (d), and your taxable interest income was not over \$400 ($\sim o$).

1. Write this statement in symbols.

2. Ask the members of the group if they qualify to use Form 1040EZ, and discuss why or why not.

3. If your taxable interest income was \$400, can you use Form 1040EZ? Discuss why or why not.

4. Suppose that a person in the group received a Form 1099 indicating that the person earned some miscellaneous income. Can that person use Form 1040EZ? Discuss why or why not.

5. The form has the following warning:

Remember, you must report all wages, salaries, and tips even if you do not get a W-2 form from your employer.

Let r be the statement "You must report all wages, salaries, and tips," and let w be "You get a W-2 form from your employer." If a **tautology** is defined as *a statement that is always true* and a **contradiction** as *a statement that is always false,* according to the warning, which statement does the IRS want you to believe is a tautology and which statement a contradiction? Give reasons for your decision.

 Skill Checker

The Skill Checker exercises sometimes appear at the end of exercise sets and help you to maintain previously studied skills that you may need in the next section.

The Using Your Knowledge feature in Section 3.4 shows a relationship between logic and sets. To understand this relationship better, complete the truth table below.

p	q	$\sim p$	$p \vee q$	$p \wedge q$
T	T			
T	F			
F	T			
F	F			

3.4   Variations of the Conditional and Implications

This lawyer's sign wisely emphasizes winning, not paying.

Conditional Statements in Advertising

Can lawyers advertise in your state? Suppose that you are a lawyer and you want to subtly convey the message that you are a contingency lawyer; that is, you get paid a certain percentage of the money awarded in the cases you win. How would you word your message? You could say either of the following:

1. I win if you pay.

2. Pay only if I win.

Do the messages give you the impression that if you have enough money, you can buy the wins? Here are two other wordings.

3. Paying is sufficient for winning.

4. Winning is a necessary condition for paying.

These two statements still place a lot of importance on paying. Probably the best message is the one in the photo.

5. If I do not win, you do not pay.

Here the emphasis is on winning rather than paying. Do you realize that the five statements are equivalent? You will see why in this section when the different forms in which a conditional statement can be expressed are discussed. ▶

In the preceding section we observed that equivalent statements have identical truth tables and can be considered different forms of the same statement. In this section we shall be concerned with some of the different forms in which the conditional statement $p \rightarrow q$ can be expressed.

A. Converse, Inverse, and Contrapositive

The conditional differs from conjunctions, disjunctions, and biconditionals in that the two components may *not* be interchanged to give an equivalent statement.

Write the converse, the inverse, and the contrapositive of the statement. Which one is equivalent to "If you think you had reservations you're in the wrong place"?

Thus, $p \vee q \Leftrightarrow q \vee p$ and $p \wedge q \Leftrightarrow q \wedge p$, but $p \to q$ is *not* equivalent to $q \to p$.* If we attempt to discover a statement that is equivalent to $p \to q$ (that is, that has an identical truth table) and involves p and q and the conditional or $\sim p$ and $\sim q$ and the conditional, we find the following possibilities:

Statements Related to the Conditional Statement

$q \to p$	**Converse** of $p \to q$
$\sim p \to \sim q$	**Inverse** of $p \to q$
$\sim q \to \sim p$	**Contrapositive** of $p \to q$

Table 3.15 shows the truth tables for these statements. Notice that $p \to q$ is equivalent to its contrapositive, $\sim q \to \sim p$ (because they have identical truth tables).

TABLE 3.15

Equivalent

Equivalent

		Conditional	Converse	Inverse	Contrapositive
p	q	$p \to q$	$q \to p$	$\sim p \to \sim q$	$\sim q \to \sim p$
T	T	T	T	T	T
T	F	F	T	T	F
F	T	T	F	F	T
F	F	T	T	T	T

The contrapositive of a statement is used in proving theorems for which a direct proof is difficult but the proof of the contrapositive is easy. Thus, to prove that "If n^2 is odd, then n is odd," we can prove the contrapositive "If n is not odd, then n^2 is not odd." That is, "If n is even, then n^2 is even."

EXAMPLE 1 ▶ Writing the Contrapositive of a Statement

Write in words the contrapositive of the statement "If n is an even integer, then n^2 is an even integer."

Solution

The given statement is of the form $p \to q$, where p is "n is an even integer" and q is "n^2 is an even integer." The contrapositive is $\sim q \to \sim p$, which translates into "If n^2 is not an even integer, then n is not an even integer." ■

The contrapositive is also helpful in giving a different perspective to statements containing negatives. For example, the contrapositive of the statement "If you were not at the party, then you were not invited" is "If you were invited, then you were at the party." The contrapositive says that everyone that was invited was at the party. This may not be clear from the original statement.

Have you seen this sign in the back of trucks on the highway? Is there another way of telling you that for the driver to see you, you have to see his mirrors? There is: the contrapositive!

*Recall that p and q are equivalent (denoted by $p \Leftrightarrow q$) if p and q have identical truth tables.

EXAMPLE 2 ▶ Writing Contrapositives

Write the contrapositive of "If *you* can't see my mirrors (~*m*), *I* can't see *you* (~*y*)."

Solution

The statement is of the form ~*m* → ~*y*. The contrapositive of *p* → *q* is ~*q* → ~*p*. Thus, the contrapositive of ~*m* → ~*y* is ~(~*y*) → ~(~*m*) or, in simple terms, *y* → *m*. This means that "If I can see you, then you can see my mirrors." ∎

EXAMPLE 3 ▶ Writing Conditionals, Converses, Inverses, and Contrapositives

Let *t* be "I will tip" and let *s* be "Service is good." Write in symbols and in words:

(a) The conditional "If *s*, then *t*"   (b) The converse of *s* → *t*

(c) The inverse of *s* → *t*   (d) The contrapositive of *s* → *t*

Solution

(a) The conditional "If *s*, then *t*" is symbolized by *s* → *t*. In words, "If service is good, then I will tip."

(b) The converse of *s* → *t* is *t* → *s*. In words, "If I will tip, then service is good."

(c) The inverse of *s* → *t* is ~*s* → ~*t*. In words, "If service is not good, then I will not tip."

(d) The contrapositive of *s* → *t* is ~*t* → ~*s*. In words, "If I do not tip, service is not good." ∎

B. Conditional Equivalents

Frequently in mathematics the words **necessary** and **sufficient** are used in conditional statements. To say that *p* is sufficient for *q* means that when *p* happens (is true), *q* will also happen (will also be true). Hence, "*p* is sufficient for *q*" is equivalent to "If *p*, then *q*."

Similarly, the sentence "*q* is necessary for *p*" means that if *q* does not happen, neither will *p*. That is, ~*q* → ~*p*. The statement ~*q* → ~*p* is equivalent to *p* → *q*, so the sentence "*q* is necessary for *p*" is equivalent to "If *p*, then *q*."

Finally, "*p* only if *q*" also means that if *q* does not happen, neither will *p*; that is, ~*q* → ~*p*. The statement ~*q* → ~*p* is equivalent to *p* → *q*, so the sentence "*p* only if *q*" is equivalent to *p* → *q*. The equivalences discussed, together with the variation "*q* if *p*," are summarized in Table 3.16. To aid you in understanding this table, notice that in the statement *p* → *q*, *p* is the sufficient condition (the antecedent) and *q* is the necessary condition (the consequent).

From Table 3.16, you can see that the statements

p is necessary and sufficient for *q*
q is necessary and sufficient for *p*
q if and only if *p*

are all equivalent to the statement "*p* if and only if *q*" and can be symbolized by *p* ↔ *q*.

TABLE 3.16

Statement	Equivalent Forms
If *p*, then *q*	*p* is sufficient for *q*; *q* is necessary for *p*; *p* only if *q*; *q* if *p*

EXAMPLE 4 ▶ Writing Variations of the Conditional

Let *s* be "You study regularly" and let *p* be "You pass this course." Translate the following statements into symbolic form.

(a) You pass this course only if you study regularly.

(b) Studying regularly is a sufficient condition for passing this course.

(c) To pass this course, it is necessary that you study regularly.

(d) Studying regularly is a necessary and sufficient condition for passing this course.

(e) You do not pass this course unless you study regularly. (*Hint: a* unless *b* means $\sim b \rightarrow a$.)

Solution

(a) $p \rightarrow s$

(b) Because *s*, studying regularly, is the sufficient condition, write $s \rightarrow p$.

(c) Since *s* is the necessary condition, write $p \rightarrow s$.

(d) $p \leftrightarrow s$ or $s \leftrightarrow p$.

(e) "You do not pass this course unless you study regularly" can be written as $\sim p$ unless *s*, which means $\sim s \rightarrow \sim p$. ■

Two statements *p* and *q* were defined to be equivalent (symbolized by $p \Leftrightarrow q$) if they had identical truth tables. An alternative definition states that *p* is equivalent to *q* ($p \Leftrightarrow q$) if the biconditional $p \leftrightarrow q$ is always true. (Can you see why these definitions are the same?)

C. Tautologies and Contradictions

> **Tautology and Contradiction**
>
> A statement that is always true is called a **tautology.** A statement that is always false is called a **contradiction.**

Is the statement "Objects in mirror are closer than they appear" a tautology?

TABLE 3.17

p	$\sim p$	$p \vee \sim p$
T	*F*	*T*
F	*T*	*T*

TABLE 3.18

p	$\sim p$	$p \wedge \sim p$
T	*F*	*F*
F	*T*	*F*

EXAMPLE 5 ▶ Using Truth Tables to Show Tautologies

Show the following by means of a truth table:

(a) The statement $p \vee \sim p$ is a tautology.

(b) The statement $p \wedge \sim p$ is a contradiction.

Solution

(a) Table 3.17 gives the truth table for $p \vee \sim p$. Note that in every possible case, $p \vee \sim p$ is true; therefore, the statement $p \vee \sim p$ is a tautology.

(b) Table 3.18 gives the truth table for $p \wedge \sim p$. Note that in every possible case, $p \wedge \sim p$ is false; therefore, $p \wedge \sim p$ is a contradiction (always false). ■

It is easy to restate the definition of equivalence in terms of a tautology.

Definition of Equivalent

The statement p is **equivalent** to the statement q ($p \Leftrightarrow q$) if and only if the **biconditional** $p \leftrightarrow q$ is a **tautology.**

EXAMPLE 6 ▶ Using Truth Tables to Show Tautologies

Show that the biconditional $\sim(p \wedge q) \leftrightarrow (\sim p \vee \sim q)$ is a tautology.

Solution

Do this by showing that $\sim(p \wedge q)$ and $\sim p \vee \sim q$ have the same truth table. First, do the truth table for $\sim(p \wedge q)$. The statement $(p \wedge q)$ is true when both p and q are true (row 1, Table 3.19) and false otherwise. Thus, in column 1 the first row is true; all other rows are false. To negate $(p \wedge q)$, write the opposite values of those shown in column 1 in column 2, the truth table of $\sim(p \wedge q)$. To get the truth table for $\sim p \vee \sim q$, start by negating p and then negate q, as shown in columns 3 and 4, respectively. Then use the results of columns 3 and 4 as guides to complete column 5. Remember that $\sim p \vee \sim q$ is false only when both components $\sim p$ and $\sim q$ are false. This occurs in row 1 only, so in column 5 enter F for the first row and T's for the rest of the rows. Since columns 2 and 5 are identical and the biconditional is true when both statements have the same truth values, $\sim(p \wedge q) \leftrightarrow (\sim p \vee \sim q)$ is a tautology.

TABLE 3.19

		2	1	3	5	4
p	q	$\sim$	$(p \wedge q)$	$\sim p$	$\vee$	$\sim q$
T	T	F	T	F	F	F
T	F	T	F	F	T	T
F	T	T	F	T	T	F
F	F	T	F	T	T	T

Same truth values: Equivalent

D. Implications

Another relationship between statements that is used a great deal by logicians and mathematicians is that of **implication**. Note that an **implication** is a *relationship* between the statements p and q, while a **conditional** is simply a *connective.*

Definition of Implication

The statement p is said to **imply** the statement q (symbolized by $p \Rightarrow q$) if and only if the conditional $p \rightarrow q$ is a tautology.

EXAMPLE 7 ▶ **Showing Implications**

Show that $[(p \rightarrow q) \wedge p] \Rightarrow q$.

Solution

First method By the definition of *implication* we must show that $[(p \rightarrow q) \wedge p] \rightarrow q$ is a tautology. A conditional is true whenever the antecedent is false, so we need to check only the cases in which the antecedent is true. Thus, if $(p \rightarrow q) \wedge p$ is true, then $p \rightarrow q$ is true and p is true. But if p is true, then q is also true (why?), so both sides of the conditional are true. This shows that the conditional is a tautology, and thus, $(p \rightarrow q) \wedge p$ implies q.

Second method A different procedure, which some people prefer, uses truth tables to show an implication. In order to show that $a \Rightarrow b$, we need to show that $a \rightarrow b$ is a tautology. But this means only that the truth tables for a and b (in this order) must not have a row with the values *TF* (in the same order), because this is the only case in which $a \rightarrow b$ is false. Thus, we may simply examine Table 3.20 for $(p \rightarrow q) \wedge p$ and q to show the implication. Notice that in columns 2 and 3 of Table 3.20 there is no row with *TF* (in this order). Thus, $[(p \rightarrow q) \wedge p] \Rightarrow q$.

TABLE 3.20

p	q	1 $p \rightarrow q$	2 $(p \rightarrow q) \wedge p$	3 q
T	T	T	T	T
T	F	F	F	F
F	T	T	F	T
F	F	T	F	F

We can convince ourselves of this fact by making a truth table for the statement $[(p \rightarrow q) \wedge p] \rightarrow q$. The result is a tautology (all true). ■

EXERCISES 3.4

Ⓐ Converse, Inverse, and Contrapositive

1. Write in words the contrapositive of the statement "If n is not an even number, then n is not divisible by 2."

2. Let p be "You brush your teeth with Clean," and let q be "You have no cavities." Write the converse, contrapositive, and inverse of the statement $p \rightarrow q$, "If you brush your teeth with Clean, then you have no cavities."

Ⓑ Conditional Equivalents

In problems 3–5, let p and q be defined as in problem 2. Translate each statement into symbolic form.

3. You have no cavities only if you brush your teeth with Clean.

4. Having no cavities is a sufficient condition for brushing your teeth with Clean.

5. To have no cavities, it is necessary that you brush your teeth with Clean.

In problems 6–14, write each statement in if-then form.

6. If I kissed you once, I will kiss you again.

7. To be a mathematics major, it is necessary to take calculus.

8. A good argument is necessary to convince Eva.

9. A two-thirds vote is sufficient for a measure to carry.

10. To have rain, it is necessary to have clouds.

11. A necessary condition for a stable economy is that we have low unemployment.

12. A sufficient condition for joining a women's club is being a woman.

13. Birds of a feather flock together.

14. All dogs are canines.

15. Use a truth table to show that, in general, the converse and the inverse of the statement $p \rightarrow q$ are equivalent (have identical truth tables).

In problems 16–23, let h be "honk" and u be "you love Ultimate." Write the following in symbolic form:

HONK IF YOU ♥
Ultimate

16. Honk if you love Ultimate.

17. If you love Ultimate, honk.

18. Honk only if you love Ultimate.

19. A necessary condition for loving Ultimate is to honk.

20. A sufficient condition for loving Ultimate is to honk.

21. To honk is a necessary condition for loving Ultimate.

22. To love Ultimate is a sufficient condition for honking.

23. To love Ultimate, it is sufficient and necessary that you honk.

24. Write in symbols the converse, inverse, and contrapositive of "Honk if you love Ultimate."

25. Write in symbols the converse, inverse, and contrapositive of "Honk only if you love Ultimate."

In problems 26–28, let p be "I will pass this course" and let s be "I will study daily." Write each statement in symbolic form.

26. Studying daily is necessary for my passing this course.

27. A necessary and sufficient condition for my passing this course is studying daily.

28. I will pass this course if and only if I study daily.

29. Write the converse, inverse, and contrapositive of each of the following statements:
 a. If you do not eat your spinach, you will not be strong.
 b. If you eat your spinach, you will be strong.
 c. You will be strong only if you eat your spinach.

30. Which statements in problem 29 are equivalent?

In problems 31–35, write the converse of each statement. State whether each converse is always true.

31. If an integer is even, then its square is divisible by 4.

32. If it is raining, then there are clouds in the sky.

33. In order to get a date, I must be neat and well dressed.

34. If M is elected to office, then all our problems are over.

35. In order to pass this course, it is sufficient to get passing grades on all the tests.

In problems 36–40, write the contrapositive of each statement.

36. In an equilateral triangle, the three angles are equal.

37. If the research is adequately funded, we can find a cure for cancer.

38. Black is beautiful.

39. All radicals want to improve the world.

40. Everyone wants to be rich.

41. The statement "If n is even, then $3n$ is even" is directly transformed into the statement "If $3n$ is not even, then n is not even" by which one of the following logical equivalences?
 a. "Not (p and q)" is equivalent to "not p or not q."
 b. "If p, then q" is equivalent to "(not p) or q."
 c. "If p, then q" is equivalent to "if not q, then not p."

42. The statement "If it can be recycled, place it in this container" is directly transformed into the statement "If it is not placed in this container, then it cannot be recycled" by which one of the following logical equivalences?
  a. "If p, then q" is equivalent to "(not p) or q."
  b. "If p, then q" is equivalent to "if not q, then not p."
  c. "Not (p and q)" is equivalent to "not p or not q."

43. Which of the following statements is *not* logically equivalent to "If the day is cool, I will go fishing"?
  a. I will go fishing or the day is not cool.
  b. If I go fishing, the day is cool.
  c. If I do not go fishing, then the day is not cool.
  d. The day is not cool or I go fishing.

44. Select the statement that is *not* logically equivalent to "If the class is canceled, Mary will go to the library."
  a. Mary will go to the library or the class is not canceled.
  b. It is not true that Mary will not go to the library and the class is canceled.
  c. If Mary does not go to the library, the class is canceled.
  d. If Mary does not go to the library, the class will not be canceled.

C **Tautologies and Contradictions**

45. Show by means of a truth table that the statement $(p \wedge q) \to p$ is a tautology. This demonstrates that $(p \wedge q) \Rightarrow p$.

46. Show by means of a truth table that the statement $[(p \to q) \wedge (q \to r)] \to (p \to r)$ is a tautology. This demonstrates that $[(p \to q) \wedge (q \to r)] \Rightarrow (p \to r)$.

47. Show by means of a truth table that the statement $p \leftrightarrow \sim p$ is a contradiction.

D **Implications**

In problems 48–54, two statements are given. Determine whether they are equivalent, one implies the other, or neither is the case.

48. $\sim(p \vee q); \sim p \wedge \sim q$ **49.** $\sim p \wedge q; p \to q$

50. $\sim p \to \sim q; \sim p \to q$ **51.** $p \vee (p \wedge q); p$

52. $p \wedge (p \vee q); p$ **53.** $\sim p \vee \sim q; p \wedge \sim q$

54. $(p \wedge q) \to r; \sim p \vee \sim q \vee r$

 In Other Words

55. Explain the difference between the words *necessary* and *sufficient*.

56. Write in words two statements of the form "If p, then q," using the word *sufficient*.

57. Write in words two statements of the form "If p, then q," using the word *necessary*.

58. Write in words two statements of the form "If p, then q," using the words *only if*.

59. Explain the difference between $p \to q$ and $p \Rightarrow q$.

60. Explain why a true statement is implied by any statement.

61. Explain why a false statement implies any statement.

 Using Your Knowledge

In describing sets using set-builder notation, the close connection between a set and the statement used to define that set is apparent. If we are given a universal set $\mathcal{U}$, there is often a simple way in which to select a subset of $\mathcal{U}$ corresponding to a statement about the elements of $\mathcal{U}$. For example, if $\mathcal{U} = \{1, 2, 3, 4, 5, 6\}$ and p is the statement "The number is even," the set corresponding to this statement is $P = \{2, 4, 6\}$; that is, P is the subset of $\mathcal{U}$ for which the statement p is true. The set P is called the **truth set** of p. Similarly, P' is the truth set of $\sim p$.

Let $\mathcal{U} = \{a, b, c, d, e\}$. Then, let p be the statement "The letter is a vowel," let q be the statement "The letter is a consonant," and let r be the statement "The letter is the first letter in the English alphabet."

  P, the truth set of p, is $\{a, e\}$.
  Q, the truth set of q, is $\{b, c, d\}$.

Find the following:

  R, the truth set of r
  R', the truth set of $\sim r$

Because p and q are statements, $p \vee q$ and $p \wedge q$ are also statements; hence, they must have truth sets. To find the truth set of $p \vee q$, select all the elements of $\mathcal{U}$ for which $p \vee q$ is true (that is, the elements that are vowels or consonants). Thus, the truth set of $p \vee q$

is $P \cup Q = \{a, b, c, d, e\} = \mathcal{U}$. Similarly, the truth set of $p \wedge q$ is the set of all elements of $\mathcal{U}$ that are vowels and consonants; that is, the truth set of $p \wedge q$ is $P \cap Q = \varnothing$.

62. With this information, complete the following table:

Statement Language	Set Language
p	P
q	Q
$\sim p$	P'
$\sim q$	
$p \vee q$	
$p \wedge q$	
$p \Rightarrow q$	
$p \Leftrightarrow q$	
t, a tautology	
c, a contradiction	

63. If P, Q, and R are the truth sets of p, q, and r, respectively, find the truth sets of the following statements:
  a. $q \wedge \sim r$ **b.** $(p \wedge q) \wedge \sim r$

64. As in problem 63, find the truth sets of the following statements:
  a. $p \wedge \sim (q \vee r)$ **b.** $(p \vee q) \wedge \sim (q \vee r)$

 **Discovery**

The following properties are used by mathematicians and logicians. In problems 65–69, express each statement in symbols and explain why it is true.

65. The contrapositive of the statement $\sim q \rightarrow \sim p$ is equivalent to $p \rightarrow q$.

66. The inverse of the inverse of $p \rightarrow q$ is equivalent to $p \rightarrow q$.

67. The contrapositive of the inverse of $p \rightarrow q$ is equivalent to $q \rightarrow p$.

68. The statement $r \vee s \vee \sim p \vee \sim q$ is equivalent to the contrapositive of $(p \wedge q) \rightarrow (r \vee s)$.

69. The statement $(\sim r \wedge \sim s) \vee (p \vee q)$ is equivalent to the converse of $(p \vee q) \rightarrow (r \vee s)$.

Some of the properties in problems 65–69 are summarized in the following table, where d stands for direct statement, c for converse, p for contrapositive, and i for inverse.

	d	c	p	i
d	d	c	p	i
c	c	d	i	p
p	p	i	d	c
i	i	p	c	d

Using the table, find the following:

70. The contrapositive of the contrapositive

71. The inverse of the inverse

72. The converse of the contrapositive

73. The inverse of the converse

74. The inverse of the contrapositive

 Collaborative Learning

1. Assign several members in your group to find a logical implication in a newspaper or magazine or on the Internet. Copy the article and reference its location by giving the name and date of the publication. Provide each member of the group with a copy and let each member identify the antecedent and the consequent in the implication. State the inverse, converse, and contrapositive of the conditional associated with the implication. Does the contrapositive sound more convincing than the original implication? Why or why not?

2. As you recall, the statement $p \rightarrow q$ is *true* if p is *false*. Can your group find an example in a newspaper or magazine of such an occurrence?

3. The idea mentioned in item 2 has been used to create many logic puzzles. Here is one.

  If this sentence is true, then Santa Claus exists.

Can you explain why this proves (implies) that Santa Claus does exist? (Puzzle 240 taken from *What Is the Name of This Book* by Raymond Smullyan.)

4. As we have mentioned, many statements can be restated in the if-then form. Have the members of your group rewrite the following sentences symbolically in the if-then form.

  a. Time is the best medicine.

  b. The ideal doctor is patient.

  c. The mark of a true doctor is usually illegible.

  d. Wealth is the product of man's ability to think.

  e. Rich people are just poor people with money.

3.5 Euler Diagrams: A Problem-Solving Tool

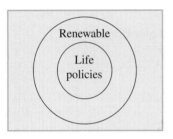

FIGURE 3.1

Insurance Policies and Euler Diagrams

Sometimes statements in logic involve relationships between sets. Thus, a renewal provision of a life insurance policy states "All life policies are renewable for additional term periods." This statement is equivalent to the following two statements:

1. If it is a life policy, then it is renewable for additional term periods.

2. The set of life policies is a subset of the set of all policies renewable for additional term periods.

Statements 1 and 2 can be visually represented by an Euler diagram (another name for a type of Venn diagram), as shown in Figure 3.1.

   In this section you will learn the techniques used to diagram statements and then use these diagrams to determine whether given arguments are valid. ▶

A. Drawing Euler Diagrams

We shall now study the analysis of arguments by using Euler diagrams, a method that is most useful for arguments containing the words *all, some,* or *none.* In order to proceed, we must define what is meant by an **argument.**

Definition of an Argument

An **argument** is a set of statements, the **premises,** and a claim that another statement, the **conclusion,** follows from the premises.

   We can represent four basic types of statements in Euler diagrams; these are illustrated in Figures 3.2–3.5.

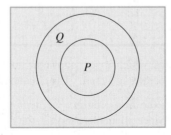

FIGURE 3.2
All *P*'s are *Q*'s. The set *P* is inside the set *Q*.

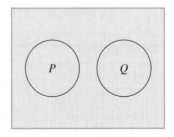

FIGURE 3.3
No *P*'s are *Q*'s. The set *P* and the set *Q* have no common elements.

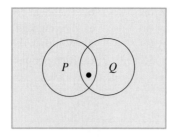

FIGURE 3.4
Some *P*'s are *Q*'s. There is an element (represented by the dot) in *P* and also in *Q*.

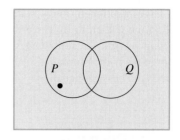

FIGURE 3.5
Some *P*'s are not *Q*'s. There is an element (represented by the dot) in *P* but not in *Q*.

The following examples discuss some simple arguments. Note that the premises are written on individual lines above a horizontal line; the conclusion is written below this line. This conclusion is preceded by the symbol ∴, which is read "therefore."

EXAMPLE 1 ▶ Premises, Conclusions, and Euler Diagrams

Consider the following argument:

> All men are mortal.
> Socrates is a man.
> ∴ Socrates is mortal.

(a) Identify the premises and the conclusion.

(b) Make an Euler diagram for the premises.

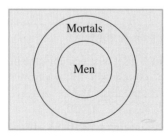

FIGURE 3.6

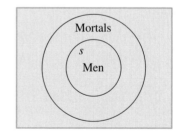

FIGURE 3.7

Online Study Center

A well-illustrated exposition dealing with Euler diagrams can be found by accessing link 3.5.1 on this textbook's Online Study Center. To further explore syllogisms, access link 3.5.2.

Solution

(a) The premises, which appear above the horizontal line, are "All men are mortal" and "Socrates is a man." The conclusion is "Socrates is mortal."

(b) To diagram the first premise, we begin by drawing a region to represent "mortals." Since all men are mortal, the region for "men" appears inside the region for mortals, as shown in Figure 3.6. The second premise, "Socrates is a man," indicates that "Socrates" goes inside the region representing "men." If *s* represents "Socrates," the diagram showing both premises appears in Figure 3.7. ∎

B. Valid Arguments

If we use the premises in an argument to reach a conclusion, we would like the resulting argument to be **valid.**

> **Validity of an Argument**
>
> An argument is **valid** if the conclusion is true whenever the premises are assumed to be true. If an argument is not valid, it is said to be **invalid.**

Thus, the argument shown in Figure 3.7 is valid because if Socrates is in the set of all men and the set of all men is inside the set of mortals, it must follow that Socrates is mortal.

EXAMPLE 2 ▶ Using Euler Diagrams to Test Validity

Use an Euler diagram to test the validity of the following argument:

All foreign cars are expensive.

My car is expensive.

∴ My car is a foreign car.

Solution

The first premise means that the set of all foreign cars is a subset of the set of expensive cars; the second premise places "my car" (represented by m) within the set of expensive cars without specifying exactly where, as shown in Figure 3.8.

Since the information given is not enough to determine which alternative is correct—that is, m can be foreign or not—we say the argument is invalid. ■

EXAMPLE 3 ▶ Using Euler Diagrams to Test Validity

Use an Euler diagram to test the validity of the following argument:

Some students are dangerous.

All dangerous people are crazy.

∴ Some students are not crazy.

Solution

The diagram in Figure 3.9 shows the premise "Some students are dangerous" by two intersecting circles and includes at least one student (represented by s) in both circles. The second premise, "All dangerous people are crazy," can be shown by enclosing the set of dangerous people inside the set of crazy people, as in Figure 3.10 or as in Figure 3.11. Since we do not know which of these two drawings is correct, we cannot conclude that "some students are not crazy." Thus, the argument is invalid. ■

EXAMPLE 4 ▶ Using Euler Diagrams to Test Validity

Use an Euler diagram to test the validity of the following argument:

Some students are intelligent.

All intelligent people are snobs.

∴ Some students are snobs.

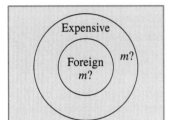

FIGURE 3.8

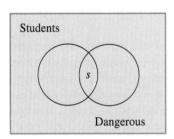

FIGURE 3.9

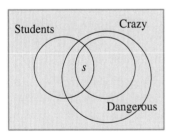

FIGURE 3.10

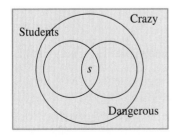

FIGURE 3.11

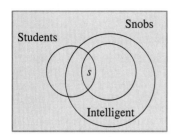

FIGURE 3.12

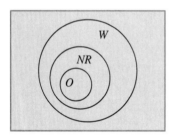

FIGURE 3.13

Online Study Center

To further explore testing validity with Euler diagrams, access link 3.5.3 on this textbook's Online Study Center.

Solution

The first premise indicates that the set of students and the set of intelligent people have at least one element in common, represented by s (see Figure 3.12). Since all intelligent people are snobs, the set of intelligent people appears inside the circle of snobs. In this case, we can conclude that some students are snobs (that is, there is at least one student s who is also a snob). The argument is valid.

EXAMPLE 5 ▶ Using Euler Diagrams to Test Validity

Use an Euler diagram to determine the validity of the following argument:

> All persons taking the test outside the United States should not register.
> All persons not registered should write to the company.
> ∴ Persons taking the test outside the United States should write to the company.

Solution

Let O be the set of all persons taking the test outside the United States, let NR be the set of all persons not registered for the test, and let W be the set of all persons who should write to the company. The first premise is diagrammed by placing O inside NR; the second premise indicates that NR is inside the set W (see Figure 3.13). It follows from the diagram that the set O is inside the set W. Thus, the argument is valid.

Do you know who Lewis Carroll (Charles Lutwidge Dodgson) is? He is the author of *Alice In Wonderland, Through the Looking Glass,* and many other mathematical works, stories, and poems. Sometimes, Carroll presents several premises and asks about valid conclusions that could be derived from these premises. But he did not use Euler diagrams! He used what he called "Triliteral diagrams." Can we reach valid conclusions using our familiar Euler diagrams? Let us try one of Carroll's problems where two premises are given, and we need to find a valid conclusion. Here are the premises.

1. No son of mine is dishonest.

2. People always treat an honest man with respect.

To find a valid conclusion, Carroll translated the first statement as

1'. All sons of mine are honest.

and the second statement as

2'. All honest men are treated with respect.

Now, can we do an Euler diagram and find a valid conclusion using premises (1') and (2')? We will proceed as before.

1'. All sons of mine are honest.

This means that the set S (all sons of mine) is inside the set H (honest men) as shown in Figure 3.14.

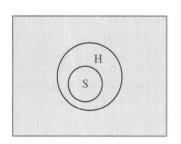

FIGURE 3.14

2'. All honest men are treated with respect.

This means that the set H must be inside the set R (respect) (Figure 3.15). Now what can we conclude? Since the set S is completely inside the set R, one possible valid conclusion is

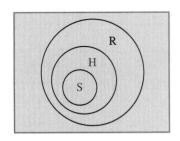

FIGURE 3.15

Online Study Center

To read more about Lewis Carroll, access link 3.5.4 on this textbook's Online Study Center.

All sons of mine are treated with respect (see Figure 3.15).

Note: Carroll translated this conclusion as the equivalent

No son of mine ever fails to be treated with respect.

Let us do a more contemporary example based on Carroll's work.

EXAMPLE 6 ▶ Supplying Valid Conclusions

Use Euler diagrams to supply a valid conclusion based on the following premises:

1. No professor sings rap songs.

2. Some math teachers sing rap songs.

Solution
We translate each premise using an Euler diagram (see Figure 3.16).

1. No professor sings rap songs.

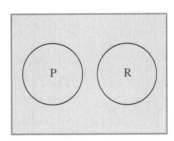

FIGURE 3.16

This means that the set P (professors) and the set R (people who sing rap songs) do not have any elements in common (they are disjoint), as shown in Figure 3.16.

2. Some math teachers (M) sing rap songs (R).

This means that the set M of math teachers and the set R have at least one element in common, represented by the little *m* (a math teacher who sings rap songs!), as shown in Figure 3.17.

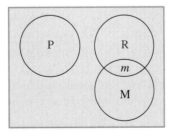

FIGURE 3.17

What can we conclude? Since *m* is outside the set P, we can conclude that *Some math teachers are not professors.* ◼

EXERCISES 3.5

A Drawing Euler Diagrams: Premises and Conclusions

In problems 1–6, state the premises and the conclusion for each argument.

1. No misers are generous.
   Some old persons are not generous.
   ∴ Some old persons are misers.

2. No thieves are honest.
   Some dishonest people are convicted.
   ∴ Some thieves are convicted.

3. All diligent students get A's.
   All lazy students are not successful.
   ∴ All diligent students are lazy.

4. All students like logic.
   Robin likes logic.
   ∴ Robin is a student.

5. No kitten that loves fish is unteachable.
   No kitten without a tail will play with a gorilla.
   ∴ No unteachable kitten will play with a gorilla.

6. No birds are proud of their tails.
   Some birds cannot sing.
   ∴ Peacocks cannot sing.

B Valid Arguments

In problems 7–20, use Euler diagrams to determine the validity of each argument.

7. All professors are wise.
   Ms. Brown is a professor.
   ∴ Ms. Brown is wise.

8. All students are studious.
   Mr. Smith is studious.
   ∴ Mr. Smith is a student.

9. No drinkers are healthy.
   No joggers drink.
   ∴ No joggers are healthy.

10. All students are dedicated.
    All wealthy people are students.
    ∴ All wealthy people are dedicated.

11. All men are funny.
    Joey is a man.
    ∴ Joey is funny.

12. All football players are muscular.
    Jack is muscular.
    ∴ Jack is a football player.

13. All felines are mammals.
    No dog is a feline.
    ∴ No dogs are mammals.

14. Some students drink beer.
    All beer drinkers are dangerous.
    ∴ All students are dangerous.

15. No mathematics teacher is wealthy.
    No panthers teach mathematics.
    ∴ No panthers are wealthy.

16. All hippies have long hair.
    Some athletes are hippies.
    ∴ Some athletes have long hair.

17. All mathematics teachers have publications.
    Some Ph.D.'s have publications.
    ∴ Some Ph.D.'s are mathematics teachers.

18. All beer lovers like Schlitz.
    All people who like Schlitz get drunk.
    ∴ All beer lovers get drunk.

19. All heavy cars are comfortable to ride in.
    No car that is comfortable to ride in is poorly built.
    ∴ No heavy car is poorly built.

20. Some Nissan owners save money.
    Some fast drivers save money.
    ∴ Some fast drivers are Nissan owners.

In problems 21–26, use Euler diagrams to determine which (if any) of the given arguments are valid.

21. All bulldogs are ugly. This dog is ugly. So it must be a bulldog.

22. All peacocks are proud birds. This bird is not proud. Therefore, it is not a peacock.

23. All students who get A's in mathematics are smart. This student got a B in mathematics. Hence, this student is not smart.

24. No southerners like freezing weather. Joe likes freezing weather. Therefore, Joe is not a southerner.

25. Some people who fish are lucky. Fred is unlucky. Therefore, Fred is not a person who fishes.

26. Some students do well in history. Bobby failed history. Therefore, Bobby is not a student.

In problems 27–36, use an Euler diagram to determine which conclusion (a, b, or neither) can be logically deduced.

27. (1) All math teachers are salseros.
(2) Some math teachers do not sing hip-hop.

    a. Some hip-hop singers are not salseros.
    b. Some salseros do not sing hip-hop.

28. (1) All math teachers can teach calculus.
(2) Some students cannot teach calculus.

    a. Some students are not math teachers.
    b. Some math teachers are not students.

29. (1) No doctors are enthusiastic.
(2) You are enthusiastic.

    a. You are not a doctor.
    b. You are a doctor.

30. (1) No Frenchmen like plum pudding.
(2) All Englishmen like plum pudding.

    a. Some Frenchmen do not like plum pudding.
    b. Englishmen are not Frenchmen.

31. (1) No old misers are cheerful.
(2) Some old misers are thin.

    a. Some thin people are cheerful.
    b. Some thin people are not cheerful.

32. (1) No professor can be crossed in love.
(2) A student may be crossed in love.

    a. Some students are not professors.
    b. Some professors are not students.

33. (1) No wasps are unfriendly.
(2) No puppies are unfriendly.

    a. No puppies are wasps.
    b. All wasps are not puppies.

34. (1) All pigs are fat.
(2) Nothing that is vegetarian is fat.

    a. Pigs are not vegetarian.
    b. Some pigs are vegetarian.

35. (1) Some math problems are difficult.
(2) All difficult problems need attention.

    a. Some math problems need attention.
    b. Some math problems are not difficult.

36. (1) No professors are salseros.
(2) Some professors are rappers.

    a. Some rappers are not salseros.
    b. Some professors are not rappers.

37. Given the following:

All highway patrol officers direct traffic.
Persons who direct traffic must be obeyed.

Use an Euler diagram to determine which conclusion(s) can be logically deduced.
    a. Persons who direct traffic are highway patrol officers.
    b. All highway patrol officers must be obeyed.
    c. Some persons who direct traffic are not highway patrol officers.
    d. None of the above.

38. All four of the following arguments have true conclusions, but one of the arguments is *not* valid. Use Euler diagrams to determine which argument is *not* valid.
    a. All fish have gills and all trout are fish. Therefore, all trout have gills.
    b. All trout have gills and all fish have gills. Therefore, all trout are fish.
    c. All fish have tails and all trout are fish. Therefore, all trout have tails.
    d. Every bird has a beak and the robin is a bird. Therefore, the robin has a beak.

In Other Words

39. Can a valid argument reach
    a. a true conclusion from true premises? Explain.
    b. a false conclusion from true premises? Explain.

40. Can an invalid argument reach
    a. a true conclusion from true premises? Explain.
    b. a false conclusion from true premises? Explain.

41. If an argument is invalid, does it have to have false premises? Explain and give examples.

42. Suppose an argument has a false conclusion. Which of the following statements should be true? Give reasons.
   a. All premises are false.
   b. Some premises are false.
   c. The argument must be invalid.

43. Suppose all premises in an argument are true. If the argument is valid, what can you say about the conclusion?

44. Suppose all premises in an argument are false. If the argument is valid, what can you say about the conclusion?

 Using Your Knowledge

At the beginning of this chapter we mentioned a type of argument called a *syllogism*. The validity of this type of argument can be tested by using the information about Venn diagrams presented in Chapter 2. We shall first diagram the four types of statements involved in these syllogisms.

Recall that when we make a Venn diagram for two sets that are subsets of some universal set, we divide the region representing the universal set into four different regions, as shown in diagrams A–D.

Diagram A represents the statement "All *P*'s are *Q*'s." Note that region 1 must be empty, because all *P*'s are *Q*'s.

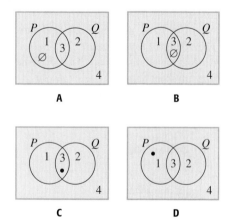

A. All *P*'s are *Q*'s. Since all the *P*'s are in *Q*, there are no elements in region 1.
B. No *P*'s are *Q*'s. Since no *P*'s are *Q*'s, there are no elements in region 3.

C. Some *P*'s are *Q*'s. Since some *P*'s are *Q*'s, there must be at least one element in *P* and *Q* (region 3).
D. Some *P*'s are not *Q*'s. Since some *P*'s are not *Q*'s, there must be at least one element in *P* and not in *Q* (region 1).

Diagram B represents the statement "No *P*'s are *Q*'s." Note that region 3 must be empty because no *P*'s are *Q*'s.

Diagram C represents the statement "Some *P*'s are *Q*'s." The dot in region 3 indicates that there is at least one *P* that is a *Q*.

Diagram D represents the statement "Some *P*'s are not *Q*'s." The dot in region 1 indicates that there is at least one *P* that is not a *Q*.

Similar considerations govern diagrams involving three sets, as shown when diagramming the following syllogism:

> All kangaroos are marsupials.
> All marsupials are mammals.
> Therefore, all kangaroos are mammals.

In order to diagram this argument, we draw a rectangle and three circles and label the circles *K* (kangaroos), *M* (marsupials), and *Ma* (mammals). As before, these circles divide the rectangle (the universal set) into eight regions. The statement "All kangaroos are marsupials" makes regions 1 and 5 empty (see the following diagram); the statement "All marsupials are mammals" makes regions 2 and 3 empty (again, see the diagram). In order for the argument to be valid, regions 1 and 3 must be empty because we wish to conclude that all kangaroos are mammals. Since the diagram shows that this is the case, the conclusion follows and the argument is valid.

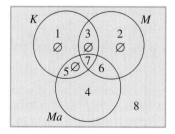

Using a similar technique and the diagram on page 155, we can show that the following argument is invalid:

> Some intelligent people are attractive.
> All models are attractive.
> Therefore, some intelligent people are models.

In order to diagram the statement "Some intelligent people are attractive," we put dots in both regions 3 and 7 and join them with a curved arrow symbol. This indicates that a dot may be in region 3 or in region 7 or dots may be in both regions, but we do not know which is the case. The second premise makes regions 4 and 5 empty. There is no statement in the argument that prevents region 7 from being empty. Thus, the conclusion could be false, so the argument is invalid.

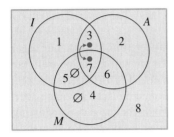

45–54. Use the ideas just discussed to examine the validity of the arguments in problems 11–20 of Exercises 3.5. Use the same ideas to solve problems 55 and 56.

55. Given the following premises:

No student who does not study will get an A in this course.
Some students in this course do not study.

Which of the following conclusions (if any) are valid?
a. Some students in this course will not get A's.
b. No students in this course will get A's.
c. All students in this course who do study will get A's.

56. Given the following premises:

Unless it rains, the grass will not grow.
If the grass grows, I will cut it.
It rains.

Which of the following conclusions (if any) are valid?
a. I will cut the grass.
b. The grass grows.
c. If the grass does not grow, I will not cut it.

57. Draw an Euler diagram to represent the two premises

All x's are y's.
All x's are z's.

Can you conclude that "Some y's are z's"? Explain why or why not.

58. Draw an Euler diagram to represent the two premises

All x's are y's.
No z is a y.

What can you conclude from the diagram? Explain.

59. Draw an Euler diagram to represent the two premises

All x's are y's.
Some z's are x's.

What can you conclude from the diagram? Explain.

60. On the basis of the knowledge you gained in solving problems 57–59, can you derive a valid conclusion involving an *existential* quantifier (some) from two premises involving *universal* quantifiers (all, no)? Explain.

✓ Skill Checker

In the next section we are going to use truth tables to determine if the conclusion of an argument is true whenever the premises are true. To help you succeed, construct and complete the following truth table:

1.

p	q	$\sim p$	$\sim q$	$p \to \sim q$	$p \vee \sim q$	$\sim p \wedge q$
T	T					
T	F					
F	T					
F	F					

2. You also have to remember how to form the contrapositive of a statement. As practice now, find the contrapositive of
a. $p \to q$.   **b.** $\sim q \to p$.
c. $\sim q \to \sim p$.   **d.** $p \to \sim q$.

3.6 Truth Tables and Validity of Arguments

From Premises to Valid Conclusions

Here is an excerpt from the IRS instructions for Form 1040.

> If you do not file your return by the due date, the penalty is usually 5% of the amount due for each month or part of a month your return is late, unless you have a reasonable explanation.

Now suppose you filed your return 1 month late. Will there always be a penalty involved? Will it be 5% of the amount due? The answer is no in both cases. Do you see why?

In this section we shall study how to determine whether an argument is valid by writing the argument in symbolic form and then using truth tables to determine its validity (Exercises 3.6, problems 1–16). Moreover, we shall discuss how to reach valid conclusions from given premises. Of course, we must be careful about the conclusions we reach. A sixth-century B.C. paradox involving arguments claims that Epimenides, the poet and prophet of Crete, made the statement

All Cretans are liars.

What is paradoxical about this statement? First, rewrite the situation as

1. If it is a statement made by a Cretan, then it is not true.

2. The statement was made by a Cretan (Epimenides).

3. Thus statement 1 is not true.

But statement 3 makes statement 1 true! Do you see the problem? We shall stay away from such paradoxical arguments! ▶

In the preceding section an argument was defined as **valid** if the conclusion is true whenever all the premises are assumed to be true. If an argument is not valid, it is said to be **invalid.** This definition suggests that a truth table can be used to check the validity of an argument. In order to construct such a truth table efficiently, the argument must be in symbolic form. The following examples illustrate converting arguments into symbolic form.

EXAMPLE 1 ▶ Writing Arguments in Symbolic Form

Write the following argument in symbolic form:

> If today is Sunday, then I will go to church.
> Today is Sunday.
> ∴ I will go to church.

Solution

Let s be "Today is Sunday" and let c be "I will go to church." Then the argument is symbolized as

$$s \rightarrow c$$
$$\underline{s}$$
$$\therefore c$$

■

EXAMPLE 2 ▶ Symbolizing an Argument

Symbolize the following argument:

> A whole number is even or odd.
> This whole number is not even.
> ∴ This whole number is odd.

Solution

Let e be "A whole number is even" and let o be "A whole number is odd." Then the argument is symbolized as

$$e \vee o$$
$$\underline{\sim e}$$
$$\therefore o$$

■

A. Problem Solving and Arguments

To determine whether an argument is valid, we use the following procedure:

Determining Validity

1. Write each premise on a separate line.

2. Write the conclusion after the premises and separate it by a horizontal line.

3. Make a truth table using a column for each premise and a column for the conclusion.

4. Check *only* the rows in which *all* the premises are *true*. For the argument to be *valid*, the conclusion must also be *true*.

PROBLEM SOLVING

❶ Read the problem and **select** the unknown.
Determine whether the given argument is valid.

❷ Think of a plan.

❸ Use the four-step procedure.
  1. Write each premise on a separate line.
  2. Draw a horizontal line after the last premise. Write the conclusion after the premises.

Determining Validity

Determine the validity of the following argument: If today is Sunday (s), then I will not go to school ($\sim g$). Today is not Sunday, so I will go to school.

Follow the four-step procedure given.

$$s \rightarrow \sim g \qquad \text{First premise}$$
$$\underline{\sim s} \qquad\qquad \text{Second premise}$$
$$\therefore g \qquad\qquad \text{Conclusion}$$

3. Make a truth table
    (Table 3.21).

TABLE 3.21

			Premise	*Premise*	*Conclusion*
s	*g*	$\sim g$	$s \rightarrow \sim g$	$\sim s$	*g*
T	T	F	F	F	T
T	F	T	T	F	F
F	T	F	T	T	T
F	F	T	T	T	F	← True premises / False conclusion

❹ Check the rows in which the premises $s \rightarrow \sim g$ and $\sim s$ are true.

Check rows 3 and 4. In the fourth row, both premises are true, but the conclusion *g* is false, so the argument is *invalid*.

TRY EXAMPLE 3 NOW.

Cover the solution, write your own solution, and then check your work.

EXAMPLE 3 ▶ Using Truth Tables to Determine Validity

Use a truth table to determine the validity of the argument in Example 2.

Solution

The argument is symbolized as

$$e \vee o$$
$$\underline{\sim e}$$
$$\therefore o$$

Make a truth table (Table 3.22) for this argument with a column for each premise and a column for the conclusion.

According to the definition of a valid argument, you need to examine only those rows of Table 3.22 where all the premises are true. Consequently, you need to check only the third row. (In rows 1, 2, and 4, at least one of the premises is false, and so these rows are crossed out.) For the argument to be valid, the remaining items in the conclusion column must all be *T*'s. There is only one item, and it is *T*. Thus, Table 3.22 shows that the argument is valid.

TABLE 3.22

			Premise	*Premise*	*Conclusion*
e	*o*	$e \vee o$	$\sim e$	*o*
~~T~~	~~T~~	~~T~~	Ⓕ	~~T~~
~~T~~	~~F~~	~~T~~	Ⓕ	~~F~~
F	T	T	T	T	← True premises / True conclusion
~~F~~	~~F~~	Ⓕ	~~T~~	~~F~~

■

EXAMPLE 4 ▶ **Using Truth Tables to Determine Validity**

Use a truth table to determine the validity of the following argument:

> Either the puppy is cute or I will not buy it.
> The puppy is cute.
> ∴ I will buy it.

Solution

By writing p for "The puppy is cute" and b for "I will buy it," we can symbolize the argument in the form

$$p \vee \sim b$$
$$p$$
$$\therefore b$$

Now, we construct Table 3.23 for this argument.

TABLE 3.23

			Premise	Premise	Conclusion	
p	b	$\sim b$	$p \vee \sim b$	p	b	
T	T	F	T	T	T	
T	F	T	T	T	F	← True premises / False conclusion
F	T	F	F	F	T	
F	F	T	T	F	F	

Next, we cross out all the rows where an F occurs in a premise column (rows 3 and 4 in Table 3.23). In the remaining rows, the premises are all true, so the remaining items in the conclusion column must all be T's if the argument is valid. Since there is an F in the second row of this column, the argument is invalid. ∎

EXAMPLE 5 ▶ **Arguments and Overtime Pay**

Use a truth table to determine the validity of the following argument: You will not get overtime pay unless you work more than 40 hr per week. You work more than 40 hr per week. So you will get overtime pay.

Solution

To symbolize the argument, let p be "You get overtime pay" and let m be "You work more than 40 hr per week." Then, the argument is as follows:

$$p \rightarrow m \qquad \textit{Note: } \sim a \textit{ unless } b \textit{ means } a \rightarrow b. \textit{ Why?}$$
$$m$$
$$\therefore p$$

TABLE 3.24

		Premise	*Premise*	*Conclusion*
p	*m*	$p \to m$	*m*	*p*
T	T	T	T	T
~~T~~	~~F~~	(F)	(F)	~~T~~
F	T	T	T	F	←
~~F~~	~~F~~	~~T~~	(F)	~~F~~

True premises
False conclusion

Notice that "You will not get overtime pay unless you work more than 40 hours per week" means that "If you get overtime pay, then you work more than 40 hours per week." Table 3.24 is the truth table for this argument.

In the first and third rows of Table 3.24, both premises are true, but in the third row, the conclusion is false. Hence, the argument is invalid. ■

EXAMPLE 6 ► Using Truth Tables to Determine Validity

Use a truth table to determine the validity of the following argument:

> All dictionaries are useful books.
> All useful books are valuable.
> ∴ All dictionaries are valuable.

Solution

We first write the argument in symbolic form. The statement "All dictionaries are useful books" is translated as "If the book is a dictionary (*d*), then it is a useful book (*u*)." "All useful books are valuable" means "If the book is a useful book (*u*), then it is valuable (*v*)." "All dictionaries are valuable" is translated as "If the book is a dictionary (*d*), then it is valuable (*v*)." Thus, the argument is symbolized as

$$d \to u$$
$$\underline{u \to v}$$
$$\therefore d \to v$$

From Table 3.25 we see that whenever all the premises are true (rows 1, 5, 7, and 8), the conclusion is also true. Hence, the argument is valid.

TABLE 3.25

			Premise	*Premise*	*Conclusion*
d	*u*	*v*	$d \to u$	$u \to v$	$d \to v$
T	T	T	T	T	T	←
~~T~~	~~T~~	~~F~~	~~T~~	(F)	~~F~~
~~T~~	~~F~~	~~T~~	(F)	~~T~~	~~T~~
~~T~~	~~F~~	~~F~~	(F)	~~T~~	~~F~~
F	T	T	T	T	T	←
~~F~~	~~T~~	~~F~~	~~T~~	(F)	~~T~~
F	F	T	T	T	T	←
F	F	F	T	T	T	←

True premises
True conclusion

The results of this problem can also be determined by using an Euler diagram. Try it and compare the results. ∎

   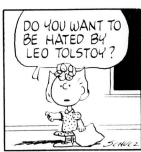

Peanuts. Reprinted by permission of United Feature Syndicate, Inc.

EXAMPLE 7 ▶ **Translating Cartoons and Checking Validity**

Translate the argument in the cartoon into symbolic form and check its validity.

Solution

Let r be "You do not know how to read."
Let w be "You cannot read *War and Peace*."
Let h be "Leo Tolstoy will hate you."

Then the argument can be translated as follows:

$$r \to w$$
$$\underline{w \to h}$$
$$\therefore r \to h$$

Since the form of this argument is identical to that in Example 6, this argument is also valid. ∎

Examples 6 and 7 are illustrations of the fact that an argument of the form

$$p \to q$$
$$\underline{q \to r}$$
$$\therefore p \to r$$

is always valid.

In using a truth table to check the validity of an argument, we need to examine only those rows where the premises are all true. This points out a basic logical principle: **If there is no case in which the premises are all true, then the argument is valid regardless of the conclusion.**

This principle is used in the next example.

EXAMPLE 8 ▶ **Determining the Validity of an Argument**

Determine the validity of the following argument:

It is raining now.

It is not raining now.

∴ It is raining now.

TABLE 3.26

p	$\sim p$	$p \wedge \sim p$
T	F	F
F	T	F

Solution

We let p be "It is raining now" and symbolize the argument, getting

$$p$$
$$\underline{\sim p}$$
$$\therefore p$$

From Table 3.26 we see that there is no case in which the conjunction of the premises, $p \wedge \sim p$, is true. Thus, the argument is valid. ∎

   In many cases we are confronted with arguments of the following form:

1. $p \rightarrow q$

2. $q \rightarrow r$

3. $\underline{r \rightarrow s}$

4. $\therefore p \rightarrow s$

Four statements are involved in this argument, so the truth table to determine its validity will have $2^4 = 16$ rows. To construct such a table would be a tedious task, indeed. For this reason, we develop an alternative method for establishing the validity of such arguments. We first consider the premises $p \rightarrow q$ and $q \rightarrow r$. From these we can conclude (see Example 6) that $p \rightarrow r$. Hence, substituting our conclusion in the given argument, we have the following:

2′. $p \rightarrow r$

3. $\underline{r \rightarrow s}$

4. $\therefore p \rightarrow s$

But this new argument again has the form of the one in Example 6, so we know it is valid.

   In many cases, instead of determining the validity of an argument, we have to supply a valid conclusion for a given set of premises. Using the idea in Example 6, we can see that if the premises are $(p \rightarrow q)$, $(q \rightarrow r)$, . . . , $(x \rightarrow y)$, $(y \rightarrow z)$, a valid conclusion is $p \rightarrow z$. We illustrate this idea in Examples 9–11.

EXAMPLE 9 ▶ Supplying Conclusions Using All Premises

Supply a valid conclusion using all the following premises:

1. $p \rightarrow q$

2. $q \rightarrow r$

3. $\sim s \rightarrow \sim r$

Solution

The third premise can be rewritten as $r \rightarrow s$ because a statement and its contrapositive are equivalent. Hence, the entire argument can be written as follows:

1. $p \rightarrow q$

2. $q \rightarrow r$

3′. $r \rightarrow s$

Thus, a valid conclusion using all the premises is $p \rightarrow s$. ∎

EXAMPLE 10 ▶ Showing Conclusions Involving TV Watching

Suppose you know the following to be true:

1. If Alice watches TV, then Ben watches TV.

2. Carol watches TV if and only if Ben watches.

3. Don never watches TV if Carol is watching.

4. Don always watches TV if Ed is watching.

Show that Alice never watches TV if Ed is watching.

Solution

> Let *a* be "Alice watches TV."
> Let *b* be "Ben watches TV."
> Let *c* be "Carol watches TV."
> Let *d* be "Don watches TV."
> Let *e* be "Ed watches TV."

The preceding argument can be symbolized as follows:

1. $a \to b$

2. $c \leftrightarrow b$, or equivalently, 2′. $b \leftrightarrow c$

3. $c \to \sim d$

4. $e \to d$, or equivalently, 4′. $\sim d \to \sim e$

If these premises are arranged in the order 1, 2′, 3, and 4′, we obtain the following:

1. $a \to b$

2′. $b \leftrightarrow c$

3. $c \to \sim d$

4′. $\sim d \to \sim e$

Consequently, we can conclude that $a \to \sim e$; that is, "If Alice watches TV, then Ed does not watch TV," or equivalently, "Alice never watches TV if Ed is watching." ■

In the preceding examples we have supplied a valid conclusion using all the given premises. Sometimes it may be necessary to supply an additional (hidden) premise to reach a valid conclusion. Here is one such instance. Look at the statement on the shampoo bottle. It says

> If seal is broken (b), do not use ($\sim$u).

In symbols,

> $b \to \sim u$

Can you ever use the shampoo? How would you use it without breaking the seal? Here is the hidden premise. If you use the shampoo, you have to break the seal. In symbols,

> $u \to b$

You do *not* have to break the seal to use the shampoo.

![Online Study Center]
Online Study Center

Today, these types of puzzles are usually described as paradoxes. To further explore them, access link 3.6.1 on this textbook's Online Study Center.

EXAMPLE 11 ▶ Finding a Valid Conclusion

Supply a valid conclusion using the premises

1. $b \rightarrow \sim u$

2. $u \rightarrow b$ (hidden premise)

Solution

We arrange the premises so that the consequent of one statement corresponds to the antecedent of the next; that is, we write

   $u \rightarrow b$ and then $b \rightarrow \sim u$

or simply

   $u \rightarrow b \rightarrow \sim u$

Thus, the conclusion is

   If you use the shampoo, then do not use it.

This means that we will not use the shampoo! For a dissenting opinion, see the photo on the left!   ■

In our day-to-day reasoning, we do not make truth tables or check our arguments in any formal fashion. Instead, we (perhaps unconsciously) learn a few argument forms, which we use as we need them. The most commonly used of these argument forms are as follows:

Modus Ponens (*Law of Detachment*)	*Modus Tollens* (*Law of Contraposition*)	*Hypothetical Syllogism*	*Disjunctive Syllogism*
$p \rightarrow q$	$p \rightarrow q$	$p \rightarrow q$	$p \vee q$
$\underline{p}$	$\underline{\sim q}$	$\underline{q \rightarrow r}$	$\underline{\sim p}$
$\therefore q$	$\therefore \sim p$	$\therefore p \rightarrow r$	$\therefore q$

The names of the first two of these are derived from the Latin and mean, respectively, "a manner of affirming" and "a manner of denying" (the parts of a conditional). The modus ponens and the two types of syllogisms have already been discussed in this section; see Examples 1 and 2, the problem-solving example, and Example 6. It is left for you to show the validity of the modus tollens in problem 20 of Exercises 3.6.

Let us discuss a more immediate problem. Have you been attending all your math class sessions, or do you have excessive absences? If you have been absent a number of times, here is an argument you and/or your instructor may consider.

   1 absence does not constitute excessive absences.

   If 1 absence does not constitute excessive absences, then 2 absences do not.

   If 2 absences do not constitute excessive absences, then 3 absences do not.
      .
      .
      .

If 99 absences do not constitute excessive absences, then 100 absences do not.
_____

∴ 100 absences do not constitute excessive absences.

This type of argument form is called a *sorite* from the word *soros,* meaning "heap." (Can you find out why?) Does the argument seem valid to you? See the Collaborative Learning for more information.

EXERCISES 3.6

A Problem Solving and Arguments

In problems 1–16, symbolize each argument using the suggested abbreviations. In each case, determine the validity of the given argument.

1. If you eat your spinach (e), you can go out and play (p).
You did not eat your spinach.
Therefore, you cannot go out and play.

2. If you eat your spinach (e), you can go out and play (p).
You cannot go out and play.
Therefore, you did not eat your spinach.

3. If you study logic (s), mathematics is easy (e).
Mathematics is not easy.
Therefore, you did not study logic.

4. I will learn this mathematics (m), or I will eat my hat (e).
I will not eat my hat.
Therefore, I will learn this mathematics.

5. The Good Taste Restaurant has good food (g).
Hence, the Good Taste Restaurant has good food, and I will recommend it to everyone (r).

6. If prices go up (u), management will scream (s).
If management screams, then supervisors will get tough (t).
Hence, if prices go up, supervisors will get tough.

7. If I work (w), then I have money (m).
If I don't work, I have a good time (g).
Therefore, I have money or a good time.

8. Babies (b) are illogical (i).
Nobody is despised (d) who can manage a crocodile (m).
Illogical persons are despised.
Hence, babies cannot manage crocodiles.

9. If you have the time (t), we've got the beer (b).
You have the time.
So we've got the beer.

10. Bill did not go to class this morning ($\sim g$), because he wore a red shirt (r), and he never wears a red shirt to class.

11. Where there is smoke (s) there is fire (f).
There is smoke.
Hence, there is fire.

12. If you are enrolled (e) or have been accepted half-time at a college (h), you can apply for a loan (a).
You have not been accepted half-time at a college.
Hence, you cannot apply for a loan.

13. You will be eligible for a grant (e) if you meet all the criteria (m).
You do not meet all the criteria.
So you are not eligible for a grant.

14. We will pay for collision loss (p) only if collision coverage is afforded (a).
Collision coverage is not afforded.
Hence, we will not pay for collision loss.

15. If spouse is also filing (f), give spouse's Social Security number (s).
Spouse is not filing.
Hence, do not give spouse's Social Security number.

16. Additional sheets of paper are not attached ($\sim a$) unless more space is needed (m).
More space is needed.
Hence, additional sheets of paper are attached.

Determine whether each argument in problems 17–21 is valid.

17. $p \lor q$ **18.** $p \to q$ **19.** $p \to q$

$\underline{\sim p}$ $\underline{p}$ $\underline{\sim p}$

$\therefore q$ $\therefore q$ $\therefore q$

20. $p \to q$ **21.** $p \to q$

$\underline{\sim q}$ $\underline{q \to r}$

$\therefore \sim p$ $\therefore \sim r \to \sim p$

In problems 22–27, find valid conclusions using all the premises.

22. $p \to q$ **23.** $p \to q$

$q \to r$ $\underline{\sim q \lor r}$

$\underline{r \to \sim s}$ [*Hint:* $(\sim q \lor r) \Leftrightarrow (q \to r)$]

24. $p \to q$ **25.** $p \to q$

$s \to \sim r$ $q \to r$

$t \to r$ $\sim s \to \sim r$

$q \to u$ $\underline{p}$

$\underline{\sim u \lor t}$

26. $p \to \sim q$ **27.** $\sim p \to q$

$r \to q$ $\sim p \lor r$

$\underline{r}$ $\underline{\sim r}$

In problems 28–35, find valid conclusions using all the premises.

28. No people who assign work are lovable.
All supervisors assign work.

29. All politicians run for office.
No people who run for office are reliable.

30. No employee who doesn't arrive for work on time will be promoted.
Some employees don't arrive for work on time.

31. Some students are intelligent.
All intelligent people are snobs.

32. If it rains, then the grass will grow.
A sufficient condition for cutting the grass is that it will grow.
The grass is cut only if it is higher than 8 in.

33. The only books in this library that I do not recommend are unhealthy.
All bound books are well written.
All romances are healthy in tone.
I do not recommend any unbound books.

34. All ducks can fly.
No land bird eats shrimp.
Only flightless birds do not eat shrimp.

35. If you are not patriotic, then you do not vote.
Aardvarks have no emotions.
You cannot be patriotic if you have no emotions.

36. Given that

No music student who doesn't practice will learn to play well.
Some music students don't practice.

Which conclusion can be logically deduced?
a. Some music students won't learn to play well.
b. No music student will learn to play well.
c. All music students who practice will learn to play well.
d. None of these answers.

37. Given that

No college student who doesn't go to class will pass the course.
Some college students don't go to class.

Which conclusion can be logically deduced?
a. No college student will pass the course.
b. All college students who go to class will pass the course.
c. Some college students will not pass the course.
d. None of these answers.

38. Given that

No student who works late can be a party animal.
All supervisors work late.

Which conclusion can be logically deduced?
a. All supervisors are party animals.
b. Some supervisors are party animals.
c. No supervisors are party animals.
d. None of these answers.

39. Given that

No people who teach classes are dumb.
All teachers teach classes.

Which conclusion can be logically deduced?
a. All teachers are dumb.
b. No teacher is dumb.
c. Some teachers are dumb.
d. None of these answers.

40. Classify the arguments in problems 2, 4, 6, and 11 as modus ponens, modus tollens, hypothetical syllogism, or disjunctive syllogism.

In problems 41–44, two premises are given in each problem. Select the conclusion that will make each entire argument valid.

41. If I drive to work, then I will not be late.
If I am not late, then I do not lose any pay.

  a. If I am not late, then I drive to work.
  b. If I do not lose any pay, then I drive to work.
  c. If I drive to work, then I do not lose any pay.
  d. If I do not drive to work, then I lose some pay.

42. If the Bears win the final game, then they will play in the NFL playoffs.
If they play in the NFL playoffs, their owners will make a good profit.

  a. If their owners made a good profit, then the Bears played in the NFL playoffs.
  b. If their owners made a good profit, then the Bears won the final game.
  c. If the Bears do not win the final game, then their owners will not make a good profit.
  d. If the Bears win the final game, then their owners will make a good profit.

43. If all persons pay their bills on time, then no collection agencies are needed.
Unfortunately, some collection agencies are needed.

  a. Some people pay their bills on time.
  b. Some people do not pay their bills on time.
  c. If there are no collection agencies, then all persons pay their bills on time.
  d. All people pay their bills on time.

44. If all students learn from their books alone, then no teachers are needed.
However, some teachers are needed.

  a. No students learn from their books alone.
  b. Some students learn from their books alone.
  c. If no teachers are needed, then all students learn from their books alone.
  d. Some students do not learn from their books alone.

45. If Bill studies economics, he will make good money. If he studies business procedures, he will make good money. Bill studies economics but not business procedures. Which of the following (if any) is a logical conclusion?
  a. Bill does not make good money.
  b. Bill makes good money.
  c. Bill does not get a college degree.

46. All college graduates are educated. All educated people dress neatly. Jackie dresses neatly. Which of the following (if any) is a logical conclusion?
  a. Jackie is a college graduate.
  b. Jackie is not a college graduate.
  c. Jackie is educated.

 Online Study Center

If you want some more practice, access link 3.6.2 on this textbook's Online Study Center.

 **In Other Words**

The job of advertisers is to convince people to buy their products. In problems 47 and 48, write the conclusions you think the advertisers want you to reach.

47. If you read *X* magazine, then you will not make bad financial decisions. You certainly do not want to make bad financial decisions, so _____.

48. If you do not join our club, you will not be popular. You did join, so _____.

49. The name of the fallacy in problem 47 is "affirming the consequent." Explain what this means.

50. The name of the fallacy in problem 48 is "denying the antecedent." Explain what this means.

Discovery

Consider the following premises taken from *Symbolic Logic,* a book written by Lewis Carroll (logician, mathematician, and author of *Alice's Adventures in Wonderland*):

No kitten that loves fish is unteachable.
No kitten without a tail will play with a gorilla.

Kittens with whiskers always love fish.
No teachable kitten has green eyes.
Kittens that have no whiskers have no tails.

51. Find a valid conclusion using all the premises.
(*Hint:* "No _____ is _____" is translated as "If it is a _____, then it is not a _____.")

Collaborative Learning

Form three groups called *Library, Internet,* and *Other.*
Let each of the groups find the answers to the following questions, compare the groups' answers, and write a report using information from all of the three groups:

1. Who has been attributed with the discovery of the sorites?

2. How many sorites were there originally?

3. There is another variation of this argument form that uses subtraction (taking just 1 away is not going to make a difference!) rather than addition to reach its conclusion. It goes like this:

A person with **10,000** hairs on his or her head is not bald,
If a person with **10,000** hairs on his or her head is not bald,
then a person with **9999** hairs on his or her head is not bald.
And so on.

Can you prove from these premises that a person with 1 hair on his or her head is not bald?

4. If you assume that your test can be postponed just 1 day, can you construct a sorite arguing that your test can be postponed forever?

5. Sometimes students argue that their lowest test score should be dropped. Can you construct a sorite showing that all your test grades (or at least the undesirable ones) should be dropped?

Chapter 3 Summary

Section	Item	Meaning	Example
3.1A	p, q, r, etc.	Statements	p: Today is Monday. q: The sky is blue.
3.1B	$p \wedge q$	Conjunction (p **and** q)	Today is Monday **and** the sky is blue.
3.1B	$p \vee q$	Disjunction (p **or** q)	Today is Monday **or** the sky is blue.
3.1B	$\sim p$	Negation (**not** p)	Today is **not** Monday.
3.2A, B, C	(truth table below)	Truth tables for the statements $p \wedge q$, $p \vee q$, and $\sim p$	
3.2E	$p \Leftrightarrow q$	The statements p and q are equivalent; that is, they have identical truth tables.	$\sim(p \vee q) \Leftrightarrow \sim p \wedge \sim q$

p	q	$p \wedge q$	$p \vee q$	$\sim p$
T	T	T	T	F
T	F	F	T	F
F	T	F	T	T
F	F	F	F	T

Section	*Item*	*Meaning*	*Example*
3.3A	$p \rightarrow q$	p conditional q	**If** today is Monday, **then** I will go to school.
3.3B	$p \leftrightarrow q$	p biconditional q	Today is Monday **if and only if** I go to school.
3.3A, B	(truth table below)	Truth tables for the conditional and the biconditional	
3.4A	$q \rightarrow p$ $\sim p \rightarrow \sim q$ $\sim q \rightarrow \sim p$	Converse of $p \rightarrow q$ Inverse of $p \rightarrow q$ Contrapositive of $p \rightarrow q$	
3.4B	p is **sufficient** for q. q is **necessary** for p. p **only if** q q **if** p	Statements equivalent to "If p, then q"	
3.4C	A tautology	A statement that is always true	2 is even. $p \vee \sim p$
3.4C	A contradiction	A statement that is always false	2 is odd. $p \wedge \sim p$
3.4D	$p \Rightarrow q$	p implies q; used when the conditional $p \rightarrow q$ is a tautology.	The animal is a dog (d) implies that the animal is a mammal (m), since $d \rightarrow m$ is a tautology.
3.5A	$p \rightarrow q$ $\underline{\quad p \quad}$ $\therefore q$	An argument (a set of statements, the premises, and a claim that another statement, the conclusion, follows from the premises)	If it rains, then I get wet. $\underline{\text{It rains.}}$ $\therefore$ I get wet.
3.5B	Valid argument	An argument is valid if, whenever all the premises are true, the conclusion is also true.	All men are mortal. $\underline{\text{Socrates is a man.}}$ $\therefore$ Socrates is mortal.
3.6	Modus ponens	$p \rightarrow q$ $\underline{\quad p \quad}$ $\therefore q$	
3.6	Modus tollens	$p \rightarrow q$ $\underline{\quad \sim q \quad}$ $\therefore \sim p$	

Truth table for 3.3A, B:

p	q	$p \rightarrow q$	$p \leftrightarrow q$
T	T	T	T
T	F	F	F
F	T	T	F
F	F	T	T

Section	*Item*	*Meaning*	*Example*
3.6	Hypothetical syllogism	$p \rightarrow q$ $q \rightarrow r$ $\therefore p \rightarrow r$	
3.6	Disjunctive syllogism	$p \vee q$ $\sim p$ $\therefore q$	

Research Questions

Sources of information for these questions can be found in the Bibliography at the end of the book.

1. Go to a logic book and find and discuss at least five different fallacies in logic. Give examples.

2. What is a paradox? Find at least two famous paradoxes in logic.

3. Find out what Boolean algebra is and how it relates to logic.

4. Find some newspaper or magazine articles that use logic to persuade the reader.

5. Find some newspaper editorials that use logic to persuade readers about important issues.

6. Write a paragraph about the author of the *Organon* and the material discussed in this book.

7. Write a report about the life and work of Gottlob Frege.

8. Are you doing well in this course? We hope so, but if not, read on. Here is a comment made about Augustus De Morgan, the discoverer of De Morgan's laws, studied in the text.

   > At school De Morgan did not excel and, because of his physical disability, he did not join in the sports of other boys, and he was even made the victim of cruel practical jokes by some schoolfellows.

   Write a report indicating where De Morgan was born, where he went to school, who his teachers were, and some of the publications he authored during his career.

Online Study Center

To further explore fuzzy logic, access links 3.7.4 and 3.7.5 on this textbook's Online Study Center.

9. At the beginning of this chapter a statement was defined as a declarative sentence that can be classified as true or false, but not both simultaneously. There is a type of logic, called *fuzzy logic,* in which this is not the case.
   a. Find the definition of *fuzzy logic.*
   b. Find at least three applications of fuzzy logic.
   c. We have shown that logic and sets are related. Find out what a fuzzy set is.

Chapter 3 Practice Test

1. Which of the following are statements?
   a. Green apples taste good.
   b. 1991 was a leap year.
   c. No fish can live without water.
   d. Some birds cannot fly.
   e. If it rains today, my lawn will get wet.
   f. Can anyone answer this question?

2. Identify the components and the logical connective or modifier in each of the following statements. Write each statement in symbolic form using the suggested abbreviations.
   a. If the number of a year is divisible by 4 (d), then the year is a presidential election year (p).
   b. I love Bill (b), but Bill does not love me ($\sim m$).
   c. A candidate is elected president of the United States (e) if and only if he or she receives a majority of the electoral college votes (m).
   d. Janet can make sense out of symbolic logic (s) or fail this course (f).
   e. Janet cannot make sense out of symbolic logic ($\sim s$).

3. Let g be "He is a gentleman" and let s be "He is a scholar." Write the following in words:
   a. $\sim(g \wedge s)$ b. $\sim g \wedge s$

4. Write the negation of each of the following statements:
   a. I will go to the beach or to the movies.
   b. I will stay in my room and do my homework.
   c. Pluto is not a planet.

5. Write the negation of each of the following statements:
   a. All cats are felines.
   b. Some dogs are well trained.
   c. No dog is afraid of a mouse.

6. Write the negation of each of the following statements:
   a. If Joey does not study, he will fail this course.
   b. If Sally studies hard, she will make an A in this course.

7. In the table below, which of the truth tables under (a), (b), (c), (d), and (e) matches each of the following statements: $\sim p$, $p \rightarrow q$, $p \wedge q$, $p \vee q$, and $p \leftrightarrow q$?

p	q	(a)	(b)	(c)	(d)	(e)
T	T	T	T	F	T	T
T	F	F	F	F	F	T
F	T	F	F	T	T	T
F	F	T	F	T	T	F

8. Construct a truth table for the statement $(p \vee q) \wedge (\sim p \vee \sim q)$.

9. Construct a truth table for the statement $(p \vee q) \rightarrow \sim p$.

10. Which of the following statements is equivalent to $\sim p \vee q$?
    a. $\sim p \wedge \sim q$ b. $\sim(p \wedge \sim q)$

11. Under what condition is the following statement true? "Sally is naturally beautiful, or she knows how to use makeup."

12. Is the following statement true or false? If $2 + 2 = 5$, then $2 \times 3 = 6$.

13. Construct the truth table for the statement $(p \rightarrow q) \leftrightarrow (q \vee \sim p)$.

14. Write the following for the statement "If you make a golf score of 62 once, you will make it again:"
   a. The converse   **b.** The inverse   **c.** The contrapositive

15. Let p be "You get overtime pay" and let m be "You work more than 40 hours per week." Symbolize the following statements.
   a. If you work more than 40 hr per week, then you get overtime pay.
   b. You get overtime pay only if you work more than 40 hr per week.
   c. You get overtime pay if and only if you work more than 40 hr per week.

16. Let b be "You get a bank loan" and let c be "You have a good credit record." Symbolize the following statements:
   a. For you to get a bank loan, it is necessary that you have a good credit record.
   b. Your having a good credit record is sufficient for you to get a bank loan.
   c. A necessary and sufficient condition for you to get a bank loan is that you have a good credit record.

a	b	c
T	T	T
T	F	T
F	F	F
T	F	F

17. The table at the left gives the truth values for three statements a, b, and c. Find all the implications among these statements.

18. Which of the following statements are tautologies?
   a. $p \wedge \sim p$ **b.** $p \vee \sim p$ **c.** $(p \rightarrow q) \leftrightarrow (\sim q \vee p)$

19. Use an Euler diagram to check the validity of the following argument:

   All students study hard.
   <u>John is not a student.</u>
   ∴ John does not study hard.

20. Use an Euler diagram to check the validity of the following argument:

   No loafers work hard.
   <u>Sally does not work hard.</u>
   ∴ Sally is a loafer.

21. Use a truth table to check the validity of the argument in problem 19.

22. Use a truth table to check the validity of the argument in problem 20.

23. Use a truth table to check the validity of the following argument:

   If you win the race, you are a good runner.
   <u>You win the race.</u>
   ∴ You are a good runner.

24. Supply a valid conclusion using all three premises: (1) $p \rightarrow q$, (2) $\sim p \rightarrow \sim r$, and (3) $s \rightarrow \sim q$.

Answers to Practice Test

ANSWER	IF YOU MISSED	REVIEW		
	Question	Section	Example(s)	Page(s)
1. Parts (**b**), (**c**), (**d**), and (**e**) are statements. Each is either true or false. Part (**a**) is not a statement, because people do not agree on what is good. Part (**f**) is a question, so is neither true nor false.	1	3.1	1, 2	108
2. a. d: The number of years is divisible by 4. p: The year is a presidential election year. The logical connective is *if-then*. $d \rightarrow p$ **b.** b: I love Bill. $\sim m$: Bill does not love me. The logical connective is *and*. $b \wedge \sim m$ **c.** e: A candidate is elected president of the United States. m: He or she receives a majority of the electoral college votes. The logical connective is *if and only if*. $e \leftrightarrow m$ **d.** s: Janet can make sense out of symbolic logic. f: She fails this course. The logical connective is *or*. $s \vee f$ **e.** s: Janet can make sense out of symbolic logic. The logical modifier is *not*. $\sim s$	2	3.1	3, 4	109–110
3. a. It is not the case that he is a gentleman and a scholar. **b.** He is not a gentleman, but he is a scholar.	3	3.1	5	110
4. a. I will go neither to the beach nor to the movies. **b.** I will either not stay in my room or not do my homework. **c.** Pluto is a planet.	4	3.1	6d, 8a	111, 112
5. a. Some cats are not felines. **b.** No dog is well trained. **c.** Some dogs are afraid of a mouse.	5	3.1	9	113
6. a. Joey does not study, but he will not fail this course. **b.** Sally studies hard, but she does not get an A in this course.	6	3.2	3	120
7. (a) $p \leftrightarrow q$ (b) $p \wedge q$ (c) $\sim p$ (d) $p \rightarrow q$ (e) $p \vee q$	7	3.1–3.3	Tables 3.2 3.3, 3.4, 3.11, and 3.12	119, 120, 121, 130 and 131
8.	8	3.2	6–9	122–123 and 125–126

8.

1	2	3	7	4	6	5
p	q	$(p \vee q)$	$\wedge$	$(\sim p$	$\vee$	$\sim q)$
T	T	T	F	F	F	F
T	F	T	T	F	T	T
F	T	T	T	T	T	F
F	F	F	F	T	T	T

ANSWER	IF YOU MISSED	REVIEW		
	Question	Section	Example(s)	Page(s)

9.

1	2	3	5	4
p	q	$(p \vee q)$	$\rightarrow$	$\sim p$
T	T	T	F	F
T	F	T	F	F
F	T	T	T	T
F	F	F	T	T

9	3.2	6, 8	122, 123

10. Statement (**b**)

| 10 | 3.2 | 9 | 125–126 |

11. When at least one of the statements "Sally is naturally beautiful" and "Sally knows how to use makeup" is true.

| 11 | 3.3 | 2 | 132 |

12. The premise "$2 + 2 = 5$" is false, so the statement (a conditional) is true.

| 12 | 3.3 | 3 | 132 |

13.

1	2	3	6	5	4
p	q	$(p \rightarrow q)$	$\leftrightarrow$	$(q \vee$	$\sim p)$
T	T	T	T	T	F
T	F	F	T	F	F
F	T	T	T	T	T
F	F	T	T	T	T

| 13 | 3.3 | 5 | 133 |

14. a. If you make a golf score of 62 again, then you made it once.
   b. If you do not make a golf score of 62 once, then you will not make it again.
   c. If you do not make a golf score of 62 again, then you did not make it once.

| 14 | 3.4 | 1–3 | 139–140 |

15. a. $m \rightarrow p$ **b.** $p \rightarrow m$ **c.** $p \leftrightarrow m$

| 15 | 3.4 | 4 | 141 |

16. a. $b \rightarrow c$ **b.** $c \rightarrow b$ **c.** $b \leftrightarrow c$

| 16 | 3.4 | 4 | 141 |

17. Statement (b) implies (a); (b) implies (c); (c) implies (a).

| 17 | 3.4 | 7 | 143 |

18. Only statement (**b**).

| 18 | 3.4 | 5, 6 | 141, 142 |

19. Nothing in the premises tells whether the J (for John) is inside or outside of the circle H. Thus, the argument is invalid.

| 19 | 3.5 | 1–4 | 148–150 |

ANSWER	IF YOU MISSED	REVIEW		
	Question	Section	Example(s)	Page(s)
20	20	3.5	1–5	148–150
21	21	3.6	3–5	158–160
22	22	3.6	3–5	158–160

20. Nothing in the premises tells whether S (for Sally) goes inside the circle L or not. Thus, the argument is invalid.

21. With s for "He is a student" and h for "He studies hard," the argument can be symbolized and a truth table constructed as follows:

$$\begin{array}{l} s \to h \\ \underline{\sim s} \\ \therefore \sim h \end{array}$$

		Prem.	*Prem.*	*Concl.*
s	h	$s \to h$	$\sim s$	$\sim h$
~~*T*~~	~~*T*~~	~~*T*~~	~~*F*~~	~~*F*~~
~~*T*~~	~~*F*~~	~~*F*~~	~~*F*~~	~~*T*~~
F	*T*	*T*	*T*	*F*
F	*F*	*T*	*T*	*T*

In the third row of the table, both premises are true and the conclusion is false, so the argument is invalid.

22. With f for "She is a loafer" and h for "She works hard," the argument can be symbolized and a truth table constructed as follows:

$$\begin{array}{l} f \to \sim h \\ \underline{\sim h} \\ \therefore f \end{array}$$

		Prem.	*Prem.*	*Concl.*
f	h	$f \to \sim h$	$\sim h$	f
~~*T*~~	~~*T*~~	~~*F*~~	~~*F*~~	~~*T*~~
T	*F*	*T*	*T*	*T*
~~*F*~~	~~*T*~~	~~*T*~~	~~*F*~~	~~*F*~~
F	*F*	*T*	*T*	*F*

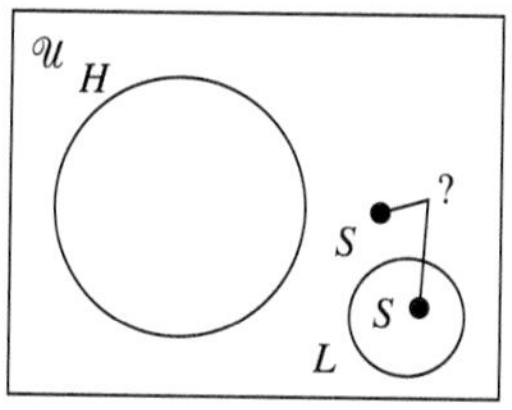

ANSWER	IF YOU MISSED	REVIEW			
	Question	Section	Example(s)	Page(s)	

In the fourth row of the truth table, the premises are both true and the conclusion is false, so the argument is invalid.

23. With w for "You win the race" and r for "You are a good runner," the argument can be symbolized and a truth table constructed as follows:

| 23 | 3.6 | 6–8 | 160–162 |

$$w \to r$$
$$\underline{w }$$
$$\therefore r$$

		Prem.	Prem.	Concl.
r	w	$w \to r$	w	r
T	T	T	T	T
T	F	T	F	T
F	T	F	T	F
F	F	T	F	F

The first row of the table is the only row where the premises are both true. In this row the conclusion is also true, so the argument is valid.

| 24 | 3.6 | 9, 11 | 162, 164 |

24. $\sim s$.

$$p \to q$$
$$\sim p \to \sim r$$
$$s \to \sim q$$

is equivalent to

$$p \to q$$
$$r \to p$$
$$q \to \sim s$$

which is equivalent to

$$r \to \sim s$$

Numeration Systems

Imagine yourself in Egypt 5500 years ago, at a time when Egyptians used their fingers as standard counting units. To keep a record of counted objects, the Egyptians used short, straight strokes called **tally marks.** Representing the numbers from 1 to 9 by the proper number of tally marks resulted in an additive system. In this system, the position of a numeral has no significance; ∩ | and | ∩ both represent 11. Number values are simply added.

In Babylon, you would find the oldest system known in which the value of a numeral depends on its placement. Using 60 as the base, ❚❚ means $1 + 1 = 2$, but ❚ ❚ means $60 + 1 = 61$.

The Roman numeral system uses base 10, as the Egyptians did, and constructs numerals by means of simple addition and subtraction. XI means $10 + 1 = 11$, but IX is $10 - 1 = 9$.

Then you study our present decimal system, also called the *Hindu-Arabic system.* Exponents play a major role in this system when we write numbers in expanded form. You will notice that the base commonly used is 10, but you will look at number systems using bases other than 10 as well.

You will be surprised to know that **binary** systems, using 2 as a base, were among the earliest systems invented. Even today, some South Sea Islanders have a binary system in which counting by 2 means repeating the words for 1 and 2. For these islanders, *urapun* is 1, *okosa* is 2, *okosa urapun* 3, *okosa okosa* 4, and so on. How does the computer count? Well, for a computer, 1_2 (one base two) is 1, 10_2 (one zero base two) is 2, 11_2 (one one base two) is 3, and so on. Do you see the similarity?

From antiquity to the present, the ideas are the same!

Online Study Center

For links to Internet sites related to Chapter 4, please access **college.hmco.com/PIC/bello9e** and click on the Online Study Center icon.

4.1 Egyptian, Babylonian, and Roman Numeration Systems

One of the most influential mathematicians of the early Arab empire was Mohammed (ca. 780–850) ibn Musa al-Khowarizmi. He wrote two books, one on algebra and one on the Hindu numeral system.

Our present-day symbols for the digits 1–9 originated with the Hindus. These numerals were designed for a decimal (base 10) system of counting.

It was not until about the year 500 that the Hindus devised a positional notation for the decimal system. They discarded the separate symbols for numbers greater than 9 and standardized the symbols for the digits from 1 through 9. The symbol for zero came into use much later.

Looking Ahead
In this chapter, we will look at ancient numeration systems and then at our familiar Hindu-Arabic system that al-Khowarizmi championed over 1000 years ago, centuries before it became widely used.

From Egyptians to Tolstoy

The Egyptian system of numeration is additive; that is, the numeral ‖ represents $1 + 1 = 2$ and ‖| ∩ represents $1 + 1 + 1 + 10 = 13$. An additive system found in Leo Tolstoy's novel *War and Peace* gives each letter of the French alphabet the number shown in Table 4.1. (Note there is no j.) Thus, the number associated with the name Sue is $90 + 110 + 5 = 205$. Similarly, the number associated with the name Pierre, a character in the novel, is $60 + 9 + 5 + 80 + 80 + 5 = 239$. This character, Pierre Bezukhof, wondered what the number corresponding to the Emperor would be. So he wrote

Le empereur Napoleon

TABLE 4.1

a	b	c	d	e	f	g	h	i	k	l	m	n
1	2	3	4	5	6	7	8	9	10	20	30	40

o	p	q	r	s	t	u	v	w	x	y	z	
50	60	70	80	90	100	110	120	130	140	150	160	

and added the numbers for these letters, 20 for l, 5 for e, 30 for m, and so on. Can you find his answer? It was 666, the number of the beast prophesied in the Bible. Thus, Pierre reasoned that Napoleon must die. But who would vanquish the feared beast? He had some ideas.

l'empereur Alexandre? La nation russe?

One was too low, the other too high. Can you find which is which? In desperation, Pierre wrote

Comte Pierre Bezukhov

Now the number was far away from what he wanted. He changed the spelling and added his nationality to get

Le russe Besuhof

See whether you can find his number. He knew that he was now near the coveted answer, and he finally wrote

l'russe Besuhof

This gave him the desired answer: He himself would take care of Napoleon! Can you find the final number that Pierre obtained and determine its relationship to Napoleon's number?

In this section we will examine some relationships among the Egyptian, Babylonian, and Roman numeral systems and our own decimal system. ▶

During the period of recorded history (beginning about 4000 B.C.), people began to think about numbers as abstract concepts. That is, they recognized that two

Napoleon Bonaparte, French military leader and emperor of France from 1804–1815.

TABLE 4.2 Ancient Numerals

	Egyptian	Sumerian	Mayan		Egyptian	Sumerian	Mayan
1	I	I	•	7	IIII III	III III I	•• ▬
2	II	II	••				
3	III	III	•••	8	IIII IIII II	III III II	••• ▬
4	IIII	III I	••••				
5	III II	III II	▬	9	III III III	III III III	•••• ▬
6	III III	III III	•̣				

fruits and two arrows have something in common—a quantity called *two*—which is independent of the objects. The perception of this quantity was probably aided by the process of tallying. Different tallying methods have been found in different civilizations. For example, the Incas of Peru used knots in a string or rope to take the census, the Chinese used pebbles or sticks for computations, and the English used tally sticks as tax receipts. As a result of human efforts to keep records of numbers, the first numerals, reflecting the process of tallying, were developed. (A **numeral** is a symbol that represents a number. For example, the numeral 2 represents the number two.) Table 4.2 shows three ancient sets of numerals. The property shared by these three numeration systems is that they are **additive;** that is, the values of the written symbols are added to obtain the number represented.

A. The Egyptian System

Let us look at the Egyptian system in greater detail. The Egyptians used hieroglyphics (sacred picture writing) for their numerals. The first line shown in Table 4.3 shows these symbols and their probable numerical values. Convenient names for the hieroglyphics are given in the margin.

As we can see from Table 4.3, the Egyptians used a **base of 10.** That is, when the tallies were added and they reached 10, the 10 tallies were replaced by the symbol ∩. The Egyptians generally wrote their numbers from right to left, although they sometimes wrote from left to right or even from top to bottom! Thus, the number 12 could have been represented by ‖ ∩ or by ∩ ‖. For this reason, we say that the Egyptian system is *not* a positional system. In contrast, our own **decimal system** (called *decimal* because we use a base of 10) is a **positional system.** In our system, the numerals 12 and 21 represent different numbers.

Computation in the Egyptian system was based on the **additive principle.** For example, to add 24 to 48, the Egyptians proceeded as follows:

```
Exchange                          ────────────────→  ∩
   24          ││││ ∩ ∩                              ∩ ∩
 + 48        ││││││││ ∩ ∩ ∩ ∩                   ‖ ∩ ∩ ∩ ∩
 ────                                           ─────────────
   72                                           ‖ ∩ ∩ ∩ ∩ ∩ ∩ ∩
```

	Stroke
∩	Heel bone
ꝰ	Scroll
⚶	Lotus flower
⟋	Pointing finger
⌔	Fish
ⵅ	Astonished man

TABLE 4.3 Numerals

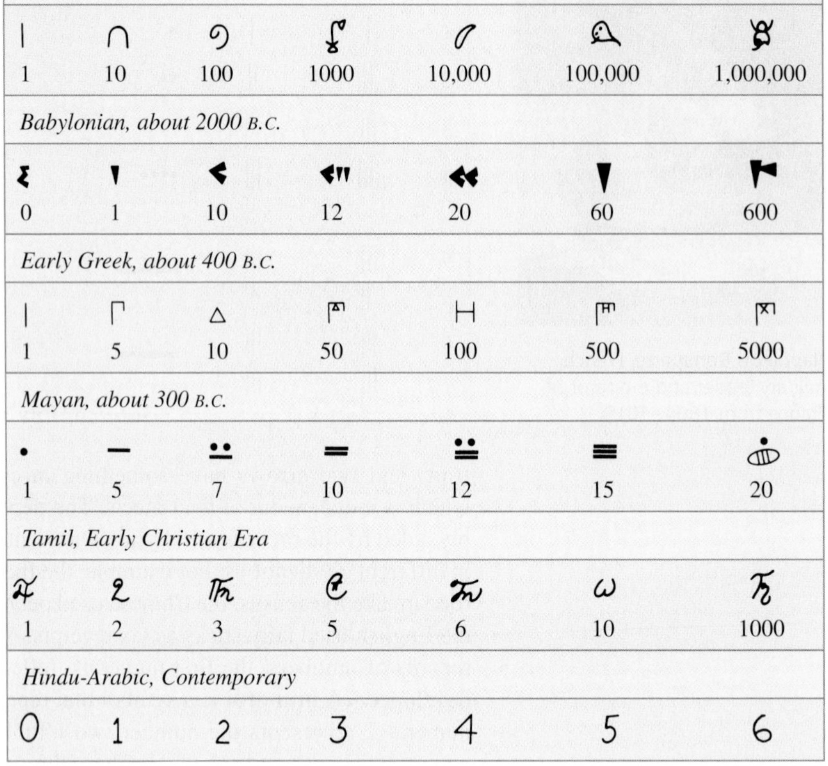

Egyptian, about 3000 B.C.						
|	∩	ꝰ	⚶	⟋	⌔	ⵅ
1	10	100	1000	10,000	100,000	1,000,000

Babylonian, about 2000 B.C.						
⟨	▼	◀	◀▼▼	◀◀	▼	⊢
0	1	10	12	20	60	600

Early Greek, about 400 B.C.						
|	Γ	△	Γᴬ	H	Γᴴ	Γˣ
1	5	10	50	100	500	5000

Mayan, about 300 B.C.						
•	—	••	=	••	≡	⬭
1	5	7	10	12	15	20

Tamil, Early Christian Era						
ℱ	ℓ	ℼ	ℓ	𝒯𝓃	ω	𝒦
1	2	3	5	6	10	1000

Hindu-Arabic, Contemporary						
0	1	2	3	4	5	6

Online Study Center

To further explore Egyptian calculators or converters, access link 4.1.1 on this textbook's, Online Study Center.

As you can see, before the computation was done, 10 of the strokes were replaced by a heel bone (see the names given in the margin next to Table 4.3).

The Egyptians performed subtraction in a similar manner. Thus, to subtract 13 from 22, they proceeded as follows:

Exchange

$$
\begin{array}{r}
22 \\
-\,13 \\
\hline
9
\end{array}
$$

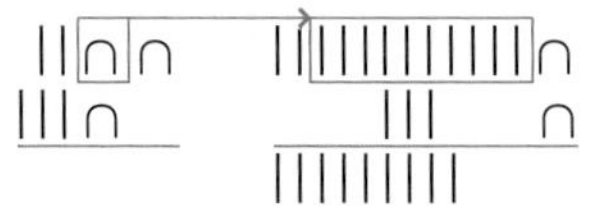

The procedure for Egyptian multiplication is explained in the Rhind papyrus (an ancient document found at Thebes and bought in Egypt by A. Henry Rhind and named in his honor). The operation was performed by **successive duplications.** Multiplying 19 by 7, for example, the Egyptians would take 19, double it, and then double the result. Then they would add the three numbers; thus

	\1	19
	\2	38
	\4	76
Total	7	133

The symbol \ is used to designate the submultipliers that add up to the total multiplier, in this case 7. In the Rhind papyrus, the problem 22 times 27 looks like this:

$$
\begin{array}{rr}
1 & 27 \\
\backslash\,2 & 54 \\
\backslash\,4 & 108 \\
8 & 216 \\
\backslash16 & \underline{432} \\
\text{Total}\quad 22 & 594 \quad \leftarrow \text{Answer}
\end{array}
$$

Again, the numbers to be added are only those in the lines starting with the symbol \ (54 + 108 + 432 = 594).

EXAMPLE 1 ▶ Successive Duplications and the Rhind Papyrus

Problem 79 of the Rhind papyrus states, "Sum the geometrical progression of five terms, of which the first term is 7 and the multiplier 7." It can be shown that the solution of the problem is obtained by multiplying 2801 by 7. Use the method of successive duplications to find the answer.

Solution

$$
\begin{array}{rr}
\backslash\,1 & 2\,,801 \\
\backslash\,2 & 5,602 \\
\backslash\,4 & \underline{11,204} \\
\text{Total}\quad 7 & 19,607
\end{array}
$$

■

It is said that in later years the Egyptians adopted another multiplication technique generally known as **mediation and duplation.** This system consists of halving the first factor and doubling the second. For example, to find the product 19×7, we successively halve 19, discarding remainders at each step, and successively double 7. The process is completed when a 1 appears in the left-hand column.

	Half	*Double*
19 is odd. →	19	⑦
9 is odd. →	9	⑭
	4	28
	2	56
1 is odd. →	1	⑪⑫

Notice that half of 19 is regarded as 9 because all the remainders are discarded. Opposite each number in the left-hand column there is a corresponding number in the column of numbers being doubled. The product 19×7 is found by adding the circled numbers—those opposite the odd numbers in the column of halves. (Can you see why this works?) Thus, $19 \times 7 = 7 + 14 + 112 = 133$.

EXAMPLE 2 ▶ Multiplication by Mediation and Duplation

Use the method of mediation and duplation to find the product 18×43.

Solution

	Half	*Double*	
	18	43	
9 is odd. →	9	86	
	4	172	$18 \times 43 = 86 + 688 = 774$
	2	344	
1 is odd. →	1	688	

Additive numeral systems were devised to keep records of large numbers. As these numbers became larger and larger, it became evident that tallying them was difficult and awkward. Thus, numbers began to be arranged in groups and exchanged for larger units, as in the Egyptian system in which 10 tallies were exchanged for a heel bone (∩). The scale used to determine the size of the group to be exchanged (10 in the case of the Egyptians) was the **base** for the system. These additive systems were advantageous for record-keeping operations, but computation in these systems was extremely complicated. As problems became even more complex, a new concept evolved to help with computations, that of a **positional numeral system.** In such a system, a numeral is selected as the base; then symbols ranging from 1 to the numeral that is 1 less than the base are also selected. Numbers are represented by placing the symbols in a specified order. For example, the Babylonians used a sexagesimal (base 60) system. In their system, a vertical wedge **▼** was used to represent 1, and the symbol **◀** represented 10. [These symbols first appeared on the clay tablets of the Sumerians (see the photograph) and Chaldeans but were later adopted by the Babylonians.]

Sumerian clay tablets and inscriptions contain numbers dating back as far as 4000 B.C.

B. The Babylonian System

The Babylonian numerals, which may look odd to you, were simply wedge marks in clay. Figure 4.1 shows a few numerals. Notice that the same symbols

1	2	3	4	5	10	20	60
▼	▼▼	▼▼▼	▼▼▼▼	▼▼▼▼▼	◀	◀◀	▼

FIGURE 4.1
Babylonian numerals.

are used for the numerals 1 and 60. To distinguish between them, a wider space was left between the characters. Thus, **▼▼▼** represents the numeral 3, whereas **▼ ▼▼** is 62.

What does 231 mean in base 10 notation? It means 2 hundreds, 3 tens, and 1 unit; that is,

$10^2 = 100$s	10s	Units
2	3	1

Since the Babylonians used base 60, the number 231 would be written as

▼▼▼   ◄ ◄ ◄ ◄ ◄▼
60s   Units

which means three sixties (▼▼▼ or 180) and fifty-one (◄ ◄ ◄ ◄ ◄▼) units. Note that

$$180 + 51 = 231$$

EXAMPLE 3 ▶ Writing Decimal Numerals in Babylonian Notation
Write the numbers 82, 733, and 4443 in Babylonian notation.

Solution

$82 =$	$1 \times 60 + 22 =$	▼	◄ ◄▼▼
$733 =$	$12 \times 60 + 13 =$	◄▼▼	◄▼▼▼
$4443 = 1 \times 3600 + 14 \times 60 +$	$3 =$	▼ ◄▼▼▼▼	▼▼▼

EXAMPLE 4 ▶ Writing Babylonian Numerals in Decimal Notation
Write the given Babylonian numerals in decimal notation.

(a) ▼▼   ◄▼▼▼   (b) ◄   ▼▼▼▼▼ (c) ◄▼▼▼▼▼ (d) ▼▼   ◄▼   ◄▼▼

Solution

(a) ▼▼   ◄▼▼▼ $= 2 \times 60 + 13 = 133$

(b) ◄   ▼▼▼▼▼ $= 10 \times 60 + 5 = 605$

(c) ◄▼▼▼▼▼ $= 10 + 5 = 15$

(d) ▼▼   ◄▼   ◄▼▼ $= 2 \times 3600 + 11 \times 60 + 12 = 7872$

Note that the symbols in (b) and (c) are the same, but the spacing is different. The lack of a symbol for zero in the Babylonian system was a significant shortcoming that made it difficult to distinguish between numbers such as 605 and 15.

Our degree-minute-second system of measuring angles undoubtedly stems from the Babylonian division of a circle into 360 equal parts. Addition of angles in this system is essentially Babylonian-style addition.

The next example illustrates the method used by the Babylonians to add numbers.

EXAMPLE 5 ▶ Adding Numbers in Babylonian Notation
Write in Babylonian notation and add $64 + 127$.

Solution

64	▼	▼▼▼▼		◄
127	▼▼	▼▼▼▼▼▼ ▼	→	
191	▼▼▼	◄▼		$191 = 3(60) + 11$

TABLE 4.4

Number	Roman Numeral
1	I
5	V
10	X
50	L
100	C
500	D
1000	M

C. The Roman System

The Roman numeral system is still used today—for example, on the faces of clocks, for chapter numbers in books, on cornerstones of buildings, and for copyright dates on films and television shows. How does the Roman system work? The Roman symbol for the number 1 is I, which is repeated for 2 and 3. Thus, II is 2, and III is 3. Though similar to the Egyptian system, the Romans introduced a special symbol for the number 5. They then used another special symbol for 10 and repeated this symbol for 20 and again for 30. Other special symbols are used for 50, 100, 500, and 1000, as shown in Table 4.4.

Although both the Roman and Egyptian systems used the addition principle, the Romans went one step further and used the **subtraction principle** as well. For instance, instead of writing IIII for the number 4, the Romans wrote IV with the understanding that the I (1) was to be *subtracted* from the V (5). In the Roman system, the value of a numeral is found by starting at the left and adding the values of the succeeding symbols to the right, unless the value of a symbol is less than that of the symbol to its right. In that case, the smaller value is *subtracted* from the larger one. Thus, XI = 10 + 1 = 11, but IX = 10 − 1 = 9. Only the numbers 1, 10, and 100 are allowed to be subtracted, and these only from numbers not more than two steps larger. For example, I can be subtracted from V to give IV = 4 or from X to give IX = 9 but cannot be subtracted from C or L. Thus, 99 is written as LXLIX *not* IC. Here are some other examples.

Addition Principle	Subtraction Principle
LX = 50 + 10 = 60	XL = 50 − 10 = 40
CX = 100 + 10 = 110	XC = 100 − 10 = 90
MC = 1000 + 100 = 1100	CM = 1000 − 100 = 900

EXAMPLE 6 ▶ Writing Roman Numerals in Decimal Notation

Write the following Roman numerals in decimal notation:

(a) DCXII   (b) MCMXLIX

Solution

(a) Since the value of each symbol is larger than that of the one to its right, we simply add these values.

$$DCXII = 500 + 100 + 10 + 1 + 1 = 612$$

(b) In this case the values of some symbols are less than the values of the symbols to their right, so we use the subtraction principle and write

$$M(CM)(XL)(IX) = 1000 + (1000 − 100) + (50 − 10) + (10 − 1)$$
$$= 1000 + 900 + 40 + 9$$
$$= 1949 \quad \blacksquare$$

Another way in which the Roman system goes further than the Egyptian system is the use of a **multiplication principle** for writing larger numbers. A multiplication by 1000 is indicated by placing a bar over the entire numeral. Thus,

$$\overline{X} = 10 × 1000 = 10{,}000$$
$$\overline{LI} = 51 × 1000 = 51{,}000$$
$$\overline{DC} = 600 × 1000 = 600{,}000$$
$$\overline{M} = 1000 × 1000 = 1{,}000{,}000$$

The largest number that can be written using Roman numerals without using either the bar or the subtraction principle is

MMMDCCCLXXXVIII

What is this number in decimal notation?

EXAMPLE 7 ▶ Writing Decimal Numerals as Roman Numerals

Write the following in Roman numerals:

(a) 33,008

(b) 42,120

Solution

(a) 33 is written as XXXIII in Roman numerals, so $33,000 = \overline{\text{XXXIII}}$. To write 33,008, 8 must be added. Thus

$$33,008 = 33,000 + 8 = \overline{\text{XXXIII}}\text{VIII}$$

(b) $42 = \text{XLII}$, so $42,000 = \overline{\text{XLII}}$. Since $120 = \text{CXX}$,

$$42,120 = \overline{\text{XLII}}\text{CXX}$$ ■

EXAMPLE 8 ▶ From False to True in One Move

Now that you know about Roman numerals, we can have some fun with them. Get some toothpicks and construct the four false equations shown in Figure 4.2. Can you make them true by moving only one stick? *Note:* You can use the symbol for any mathematical operation or notation to make the equation true. The square represents 0.

Solution

We do not want to spoil all your fun, so we will only give you hints. On the first one, you have to move the equal sign; on the second one, you need a multiplication sign; the third one takes the $\pm$ sign; and the last one involves the $\sqrt{}$ sign. Now can you do it? ■

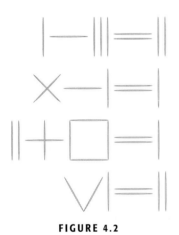

FIGURE 4.2

Online Study Center

To further explore ancient number systems, access link 4.1.2 on this textbook's Online Study Center.

EXERCISES 4.1

A The Egyptian System

In problems 1–6, use the symbols given in Tables 4.2 and 4.3 to write each number in Egyptian notation.

1. 24   **2.** 54   **3.** 142

4. 1247   **5.** 835   **6.** 11,209

In problems 7–12, translate each Egyptian numeral into decimal notation.

7. ꟼ∩|||

8. ẞꟼꟼꟼꟼꟼꟼꟼꟼꟼꟼ||||

9. ꟼꟼꟼ∩∩||

10. ∩∩∩∩|||||

11. ℒ ẞ ꟼꟼꟼ∩∩∩||

12. ꟼꟼꟼ∩∩∩|||
    ꟼꟼꟼ∩∩ ||

In problems 13–16, write each number in Egyptian notation and perform the indicated operation.

13.   34
       $+ 23$

14.   148
       $+ 45$

15.   432
       $- 143$

16.   1203
       $- 502$

In problems 17–20, use the Egyptian method of successive duplications to find each product.

17. 15×40 **18.** 25×15

19. 22×51 **20.** 21×63

In problems 21–24, use the Egyptian method of mediation and duplation to find each product.

21. 18×32 **22.** 15×32

23. 12×51 **24.** 40×61

B The Babylonian System

In problems 25–34, write each number in Babylonian notation.

25. 6   **26.** 24

27. 32   **28.** 64

29. 123   **30.** 144

31. 258   **32.** 192

33. 3733   **34.** 3883

In problems 35–40, write each Babylonian number in decimal notation.

35. ▼ ◀ ◀ ◀ ▼▼   **36.** ◀ ▼▼▼ ◀ ▼▼

37. ▼▼▼ ◀ ▼▼   **38.** ◀ ▼▼▼▼ ▼▼▼

39. ▼ ◀ ▼▼ ▼▼   **40.** ▼▼ ◀ ▼ ▼▼▼▼

In problems 41–44, write each number in Babylonian notation and perform the addition using the Babylonian system.

41. $\begin{array}{r} 32 \\ +43 \\ \hline \end{array}$ **42.** $\begin{array}{r} 63 \\ +81 \\ \hline \end{array}$

43. $\begin{array}{r} 133 \\ +68 \\ \hline \end{array}$ **44.** $\begin{array}{r} 242 \\ +181 \\ \hline \end{array}$

C The Roman System

In problems 45–50, write each number in decimal notation.

45. CXXVI   **46.** DCXVII

47. $\overline{\text{XLII}}$ **48.** $\overline{\text{XXX}}$DCI

49. $\overline{\text{XCC}}$DV   **50.** $\overline{\text{LDDC}}$

In problems 51–56, write each number in Roman numerals.

51. 72   **52.** 631

53. 145   **54.** 1709

55. 32,503   **56.** 49,231

Numerology, like astrology, is a pseudoscience concerning itself with birth dates, names, and other personal characteristics. A popular scheme used by numerologist Juno Jordan gives each letter the value of the number above it as shown. Thus, A is 1, M is 4, and Z is 8.

1	2	3	4	5	6	7	8	9
A	B	C	D	E	F	G	H	I
J	K	L	M	N	O	P	Q	R
S	T	U	V	W	X	Y	Z	

The number associated with James is

$$1 + 1 + 4 + 5 + 1 = 12 = 1 + 2 = \mathbf{3}$$

For the surname Brown, we have

$$2 + 9 + 6 + 5 + 5 = 27 = 2 + 7 = \mathbf{9}$$

Thus, the final number associated with James Brown is **3** + **9** $= 12 = 1 + 2 = $ **3**. What does this mean? It means that he has the characteristics corresponding to the number 3 in the following list:

1. Creative, inventive, positive
2. Gentle, imaginative, romantic
3. Ambitious, proud, independent
4. Rebels, unconventional
5. Mercurial, high-strung, risk-taker
6. Magnetic, romantic, artistic
7. Independent, individualistic
8. Lonely, misunderstood
9. Fighter, determined, leader

Thus, James Brown should be ambitious, proud, and independent! Do problems 57–60 just for fun. Do not take the results seriously!

57. Find the number for John Fitzgerald Kennedy.

58. Find the number for Sonya Kovalevski.

59. Find the number for Ringo Starr.

60. Find your own number.

In Other Words

61. Can you write 99 as IC in Roman numerals? Explain.

62. What are the differences between the Egyptian and our decimal systems of numeration?

63. What are the differences between the Babylonian numeration system and our decimal system of numeration?

64. What are the differences between the Roman numeration system and our decimal system of numeration?

65. What are the differences between the Egyptian and the Babylonian systems of numeration?

66. Explain why it is difficult to distinguish between 605 and 15 in the Babylonian system.

 Using Your Knowledge

The Rhind papyrus is a document that was found in the ruins of a small ancient building in Thebes. The papyrus was bought in 1858 by a Scottish antiquarian, A. Henry Rhind, and most of it is preserved in the British Museum, where it was named in Rhind's honor.

The scroll was a handbook of Egyptian mathematics containing mathematical exercises and practical examples. Many of the problems were solved by the **method of false position.**

For example, one of the simple problems states "A number and its one-fourth added together become 15. Find the number." The solution by false position goes like this: Assume that the number is 4. A number (4) and its one-fourth ($\frac{1}{4}$ of 4) added become 15; that is,

$$4 + \tfrac{1}{4}(4) \text{ must equal } 15$$

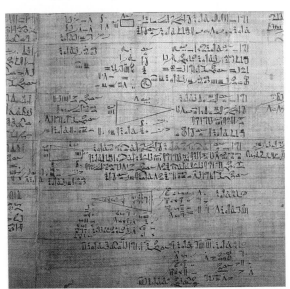

A portion of the Rhind papyrus dating back to ca. 1650 B.C. It is the most extensive mathematical document from ancient Egypt.

But

$$4 + \tfrac{1}{4}(4) = 5$$

and we need 15, which is 3 times the 5 we obtained. Therefore, the correct answer must be 3 times the assumed answer, that is, 3×4, or 12.

See if you can use the method of false position to solve the following problems:

67. A number and its one-sixth added together become 21. What is the number?

68. A number, its one-half, and its one-quarter add up to 28. Find the number.

69. If a number and its two-thirds are added and from the sum one-third of the sum is subtracted, then 10 remains. What is the number?

 Discovery

A Babylonian tablet giving the values of $n^3 + n^2$ for $n = 1$ to 30 was discovered a few years ago. The decimal equivalents of the first few entries in the table can be found as follows:

For $n = 1$, we have $1^3 + 1^2 = 2$.
For $n = 2$, we have $2^3 + 2^2 = 12$.
(Recall that $2^3 = 2 \times 2 \times 2 = 8$ and
$2^2 = 2 \times 2 = 4$.)
For $n = 3$, we have $3^3 + 3^2 = 36$.

Complete the following:

$4^3 + 4^2 = $ _____ $5^3 + 5^2 = $ _____
$6^3 + 6^2 = $ _____ $7^3 + 7^2 = $ _____
$8^3 + 8^2 = $ _____ $9^3 + 9^2 = $ _____
$10^3 + 10^2 = $ _____

Using the preceding information, find the solution of the following.

70. $n^3 + n^2 - 810 = 0$

71. $n^3 + n^2 - 576 = 0$

72. There are many equations for which this method does not seem to work. For example, $n^3 + 2n^2 - 3136 = 0$. However, a simple transformation will reduce the sum of the first two terms to the familiar form $(\ )^3 + (\ )^2$. For example, let $n = 2x$. Now try to solve the following equation:

$$n^3 + 2n^2 - 3136 = 0$$

Research Questions

1. When were Roman numerals developed, and how were the symbols for 1, 5, 10, 50, 100, and 1000 derived?

2. Find out the historians' opinion of why Roman numerals have been replaced by the Arabic system. What do you think?

3. Have you noticed that the copyright year of movies is written using Roman numerals? How would you write 1999 (there's lots of controversy here; see link 4.1.3 on this textbook's Online Study Center.)

4. We have mentioned the numeration systems of the Egyptians, the Greeks, and the Arabs. Discuss the Mayan numeration system, making sure that you answer these questions: How did they write the number 0? What symbols did they use to express their numbers? Make a table showing the numbers from 1 to 19. Then translate the numbers in the table below.

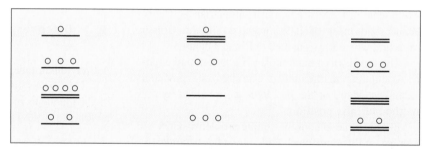

Mayan Numbers for You to Translate.
Source: *Mayan Numbers for You to Translate,* by Rhonda Robinson.

Online Study Center

To further explore replacing Roman numerals with Arabic, as well as adding Mayan numerals, access links 4.1.3, 4.1.4, and 4.1.5 on this textbook's, Online Study Center.

4.2 The Hindu-Arabic (Decimal) System

GETTING STARTED

Packaging, Garbage, and Exponents

Do you know that in just one day Americans throw out 1.5×10^5 tons of packaging material? How many pounds is that? Since 1 ton $= 2 \times 10^3$ lb, the amount of packaging material thrown out is

$$\underbrace{(1.5 \times 10^5)}_{\text{tons}} \times \underbrace{(2 \times 10^3)}_{\text{lb/ton}} \text{ lb}$$

How do you find this number? First, $1.5 \times 2 = 3$, so you need to find

$$3 \times 10^5 \times 10^3 \text{ lb}$$

The environmental stress created by landfills, such as this one, necessitates a greater awareness of waste management and recycling.

In this section you will learn the laws of exponents for multiplication. These laws state that $a^m \times a^n = a^{m+n}$ and $a^m \div a^n = a^{m-n}$. Using the law for multiplication, you get

$$3 \times 10^5 \times 10^3 = 3 \times 10^{5+3} = 3 \times 10^8 = 300,000,000 \text{ lb}$$

Thus, 300 million lb of packaging material is thrown out daily.

Now suppose a tractor trailer can carry a 15-ton load. How many tractor trailers do you need to carry out the packaging material? The answer is

$$\frac{\text{Total weight of material}}{\text{Weight per load}} = \frac{3 \times 10^8 \text{ lb}}{15 \text{ tons/load}}$$

Since 1 ton = 2000 lb, 15 tons = 30,000 lb = 3×10^4 lb. The answer is

$$\frac{3 \times 10^8 \text{ lb}}{3 \times 10^4 \text{ lb/load}} = \frac{10^8}{10^4} \text{ loads}$$

To find this answer, you need to divide 10^8 by 10^4. Using the law of exponents for division,

$$\frac{10^8}{10^4} = 10^{8-4} = 10^4$$

Thus, the number of tractor trailers needed is $1 \times 10^4 = 10,000$!

In this section you will work with exponents and write standard numbers in expanded form containing exponents and vice versa.

A few more consumption problems will be addressed in Exercises 4.2, problems 50–55. ▶

In this section, we will study our familiar **decimal system,** which is also called the **Hindu-Arabic system.** This numeration system is a positional system with 10 as its base, and it uses the symbols (called **digits**) 0, 1, 2, 3, 4, 5, 6, 7, 8, and 9. Furthermore, each symbol in this system has a **place value;** that is, the value represented by a digit depends on the position of that digit in the numeral. For instance, the digit 2 in the numeral 312 represents 2 ones, but in the numeral 321 the digit 2 represents 2 tens.

A. Expanded and Decimal Forms

To illustrate the idea of place value further, we can write both numbers in **expanded form.**

$$312 = 3 \text{ hundreds} + 1 \text{ ten} \quad + 2 \text{ ones}$$
$$= (3 \times 100) + (1 \times 10) + (2 \times 1)$$
$$321 = 3 \text{ hundreds} + 2 \text{ tens} \quad + 1 \text{ one}$$
$$= (3 \times 100) + (2 \times 10) + (1 \times 1)$$

These numbers can also be written using exponential form, a notation introduced by the French mathematician René Descartes. As the name indicates, **exponential form** uses the idea of exponents. An **exponent** is a number that indicates how many times another number, called the **base,** is a factor in a product. Thus, in 5^3 (read "5 cubed" or "5 to the third power"), the exponent is 3, the base is 5, and $5^3 = 5 \times 5 \times 5 = 125$. Similarly, in 2^4, 4 is the exponent, 2 is the base, and

$2^4 = 2 \times 2 \times 2 \times 2 = 16$. On the basis of this discussion, we state the following definition:

Definition of a^n

If a is any number and n is any counting number, then a^n (read "a to the nth power") is the product obtained by using a as a factor n times; that is,

$$a^n = \underbrace{a \times a \times \cdots \times a}_{n \ a's}$$

For any nonzero number a, we define $a^0 = 1$.

Using this definition, we can write $1 = 10^0$, $10 = 10^1$, $100 = 10^2$, $1000 = 10^3$, $10{,}000 = 10^4$, and so on. Thus, in expanded form,

$$312 = (3 \times 100) + (1 \times 10) + (2 \times 1)$$
$$= (3 \times 10^2) + (1 \times 10^1) + (2 \times 10^0)$$

The exponent 1 usually is not written; we understand that $10^1 = 10$ and, in general, $a^1 = a$. Also, we used $10^0 = 1$ in the last term of the equation above. With these conventions, we can write any number in expanded form.

EXAMPLE 1 ▶ Writing Decimal Numerals in Expanded Form

Write 3406 in expanded form.

Solution

$$3406 = (3 \times 10^3) + (4 \times 10^2) + (0 \times 10^1) + (6 \times 10^0)$$

Notice that we could have omitted the term (0×10^1), because $0 \times 10 = 0$. Using the same ideas, we can convert any number from expanded form into our familiar decimal form. ■

EXAMPLE 2 ▶ Converting from Expanded to Decimal Form

Write $(5 \times 10^3) + (2 \times 10^1) + (3 \times 10^0)$ in ordinary decimal form.

Solution

$$(5 \times 10^3) + (2 \times 10^1) + (3 \times 10^0) = (5 \times 1000) + (2 \times 10) + 3$$
$$= 5023$$ ■

B. Operations in Expanded Form

The ideas of expanded form and place value can greatly simplify computations involving addition, subtraction, multiplication, and division. In the following examples, on the left we present the usual way in which these operations are performed, and on the right we present the expanded form depending on place value.

EXAMPLE 3 ▶ **Adding Using Decimal Numerals and Using Expanded Form**

Add 38 and 61 in the usual way and using expanded form.

Solution

$$
\begin{array}{r}
38 \\
+\ 61 \\
\hline
99
\end{array}
\qquad
\begin{array}{l}
(3 \times 10^1) + (8 \times 10^0) \\
(6 \times 10^1) + (1 \times 10^0) \\
\hline
(9 \times 10^1) + (9 \times 10^0)
\end{array}
$$

◼

EXAMPLE 4 ▶ **Subtracting Using Decimal Numerals and Using Expanded Form**

Subtract 32 from 48 in the usual way and using expanded form.

Solution

$$
\begin{array}{r}
48 \\
-\ 32 \\
\hline
16
\end{array}
\qquad
\begin{array}{l}
(4 \times 10^1) + (8 \times 10^0) \\
-\ (3 \times 10^1) + (2 \times 10^0) \\
\hline
(1 \times 10^1) + (6 \times 10^0)
\end{array}
$$

◼

Before illustrating multiplication and division, we need to determine how to multiply and divide numbers involving exponents. For example, $2^2 \times 2^3 = (2 \times 2) \times (2 \times 2 \times 2) = 2^5$, and in general,

$$a^m \times a^n = \underbrace{(a \times a \times a \times \cdots \times a)}_{m\ a\text{'s}} \times \underbrace{(a \times a \times a \times \cdots \times a)}_{n\ a\text{'s}}$$

$$= a^{m+n}$$

In order to divide 2^5 by 2^2, we proceed as follows:

$$\frac{2^5}{2^2} = \frac{\cancel{2} \times \cancel{2} \times 2 \times 2 \times 2}{\cancel{2} \times \cancel{2}} = 2 \times 2 \times 2 = 2^3$$

In general, if $m > n$,

$$a^m \div a^n = \frac{a^m}{a^n} = a^{m-n}$$

Product and Quotient Laws for Exponents

$$a^m \times a^n = a^{m+n} \qquad \text{and} \qquad a^m \div a^n = \frac{a^m}{a^n} = a^{m-n}, \qquad m > n$$

EXAMPLE 5 ▶ **Using the Law of Exponents**

Perform the following indicated operations, and leave the answers in exponential form.

(a) $4^5 \times 4^7$ (b) $3^{10} \div 3^4$

Solution

(a) $4^5 \times 4^7 = 4^{5+7} = 4^{12}$ (b) $3^{10} \div 3^4 = 3^{10-4} = 3^6$

◼

Note that by the definition of a^n, we have, for example,

$$(2^3)^2 = 2^3 \times 2^3 = 2^{3+3} = 2^{3 \times 2} = 2^6$$

and

$$(5^2)^4 = 5^2 \times 5^2 \times 5^2 \times 5^2 = 5^{2+2+2+2} = 5^{2 \times 4} = 5^8$$

In general,

Power Law for Exponents

$$(a^m)^n = a^{m \times n}$$

EXAMPLE 6 ▶ Show Me the Result

Show that $32^2 = 2^{10}$.

Solution

Direct computation shows that $32 = 2^5$. Thus

$$32^2 = (2^5)^2 = 2^{5 \times 2} = 2^{10}$$

∎

EXAMPLE 7 ▶ Multiplying Using Decimal Numbers and Using Expanded Form

Multiply 32 and 21 in the usual way and using expanded form.

Solution

$$
\begin{array}{r}
32 \\
\times\,21 \\
\hline
32 \\
64 \\
\hline
672
\end{array}
\qquad
\begin{array}{r}
(3 \times 10) + (2 \times 10^0) \\
\times\,(2 \times 10) + (1 \times 10^0) \\
\hline
(3 \times 10) + (2 \times 10^0) \\
(6 \times 10^2) + (4 \times 10) \\
\hline
(6 \times 10^2) + (7 \times 10) + (2 \times 10^0)
\end{array}
$$

∎

EXAMPLE 8 ▶ Dividing Using Decimal Numbers and Using Expanded Form

Divide 63 by 3 in the usual way and using expanded form.

Solution

$$
3\overline{)63} = 21
\qquad
3\overline{)(6 \times 10) + (3 \times 10^0)} = (2 \times 10) + (1 \times 10^0)
$$

∎

C. Applications

A computer uses a collection of ones (1s) and zeros (0s) to store information. The smallest piece of information is defined as a **bit.** As information gets more complex, larger units are needed, as shown in the accompanying table. As you can see,

$$1 \text{ byte} = 8 = 2^3 \text{ bits}$$

and

$$1 \text{ kilobyte (kB)} = 1024 = 2^{10} \text{ bytes}$$

UNIT	DEFINITION
1 bit	The smallest piece of information two possible values: "0" or "1"
1 byte	8 bits
1 kB (Kilobyte)	1,024 bytes
1 MB (Megabyte)	1,048,576 bytes
1 GB (Gigabyte)	1,073,741,824 bytes
1 TB (Terabyte)	1,099,511,627,776 bytes
1 PB (Petabyte)	1,125,899,906,842,624 bytes

EXAMPLE 9 ▶ Using Your Powers

Write 1 MB and 1 GB as powers of 2.

Solution

Since 1 kB $= 2^{10}$ bytes, we may suspect that 1 MB is 1000 times as big. Since 1000 is about $2^{10} = 1024$,

$$1 \text{ MB} = 2^{10} \cdot 2^{10} = 2^{20} \text{ bytes}$$

and

$$1 \text{ GB} = 2^{10} \cdot 2^{10} \cdot 2^{10} = 2^{30} \text{ bytes}$$

Check this out with a calculator! ■

EXERCISES 4.2

Ⓐ Expanded and Decimal Forms

In problems 1–6, write each number in expanded form.

1. 432 **2.** 549 **3.** 2307

4. 3047 **5.** 12,349 **6.** 10,950

In problems 7–15, write each number in decimal form.

7. 5^0

8. $(3 \times 10) + (4 \times 10^0)$

9. $(4 \times 10) + (5 \times 10^0)$

10. $(4 \times 10^2) + (3 \times 10) + (2 \times 10^0)$

11. $(9 \times 10^3) + (7 \times 10) + (1 \times 10^0)$

12. $(7 \times 10^4) + (2 \times 10^0)$

13. $(7 \times 10^5) + (4 \times 10^4) + (8 \times 10^3) + (3 \times 10^2) + (8 \times 10^0)$

14. $(8 \times 10^9) + (3 \times 10^5) + (2 \times 10^2) + (4 \times 10^0)$

15. $(4 \times 10^6) + (3 \times 10) + (1 \times 10^0)$

Ⓑ Operations in Expanded Form

In problems 16–19, add in the usual way and using expanded form.

16. $32 + 15$ **17.** $23 + 13$

18. $21 + 34$ **19.** $71 + 23$

In problems 20–23, subtract using decimal numerals and using expanded form.

20. $34 - 21$ **21.** $76 - 54$

22. $45 - 22$ **23.** $84 - 31$

In problems 24–35, perform the indicated operation, and leave each answer in exponential form.

24. $3^5 \times 3^9$ **25.** $7^8 \times 7^3$ **26.** $4^5 \times 4^2$

27. $6^{19} \times 6^{21}$ **28.** $5^8 \div 5^3$ **29.** $6^{10} \div 6^3$

30. $7^{15} \div 7^3$ **31.** $6^{12} \div 6^0$ **32.** $(3^2)^4$

33. $(5^4)^3$ **34.** $(7^3)^5$ **35.** $(10^3)^{10}$

In problems 36–39, multiply in the usual way and using expanded form.

36. 41×23 **37.** 25×51

38. 91×24 **39.** 62×25

In problems 40–43, divide in the usual way and using expanded form.

40. $48 \div 4$ **41.** $64 \div 8$

42. $93 \div 3$ **43.** $72 \div 6$

In problems 44 and 45, write each answer in exponential form.

44. Each American produces about 5 lb of garbage every day. If the U.S. population is assumed to be 3×10^8, how many pounds of garbage per day are produced nationwide?

45. If the *New York Times* printed one Sunday edition on recycled paper instead of new paper, 75,000 trees would be saved. How many trees would be saved if the *New York Times* used recycled paper for one month (4 Sundays)?

C Applications

In problems 46–56, write the answer in scientific notation.

46. How many pet reptiles are there in the United States? According to the American Pet Products Manufacturers Association, there are 4.4×10^6 pet households, each owning 2.5 reptiles on average. How many pet reptiles is that? Source: www.infoplease.com/ipa/A0931712.html.

47. What about pet dogs? There are 45.5×10^6 pet households, each owning 1.7 dogs on average. How many dogs is that?

48. According to Nielsen Net, about 7.2×10^7 unique searches were made using Google in a given month (30 days). How many unique searches per day is that? (Source: www.infoplease.com/ipa/A-0902390.html.)

49. About 4.8×10^7 unique searches were made using Yahoo in a given month (30 days). How many unique searches per day is that?

50. A typical American consumes almost 300 lb of vegetables per year. If there are 300 million Americans, how many pounds of vegetables a year is that?

51. Of the 300 lb of vegetables consumed per person per year, 170 lb are fresh. If there are 300 million Americans, what is the total fresh vegetable consumption in the United States? (Source: www.usda.gov/nass/pubs/agr05/05_ch4.PDF.)

52. Americans consume 16.7 lb of ice cream per person per year. If there are 300 million Americans, how many pounds of ice cream are consumed each year? (Source: www.usda.gov/nass/pubs/agr05/05_ch8.PDF.)

53. Americans eat 102 lb of chicken per person per year. If there are 300 million Americans, how many pounds of chicken are consumed each year?

54. Turkey consumption is down to 16.6 lb per person per year. If there are 300 million Americans, how many pounds of turkey are consumed each year? (Source: www.usda.gov/nass/pubs/agr05/05_ch8.PDF.)

55. The average American drinks 24 gal of coffee each year. If there are 300 million Americans, how many gallons of coffee is that? (Source: U.S. Dept. of Agriculture, Economic Research Service.)

56. Americans drink 50 gal of carbonated soft drinks each year. How many gallons of soft drinks are consumed each year? (Source: *Statistical Abstract of the United States.*)

57. The United States produces about 270 million tons of garbage each year. Since a ton is 2000 lb and there are about 360 days in a year, the number of pounds of garbage produced each day of the year for each man, woman, and child in the United States is

$$\frac{(2.7 \times 10^8) \times (2 \times 10^3)}{(3 \times 10^8) \times (3.6 \times 10^2)}$$

How many pounds is that? Answer to the nearest pound.

58. Oil reserves in the United States are estimated to be 3.5×10^{10} barrels (bbl). Production amounts to 3.2×10^9 bbl per year. At this rate, how long will U.S. oil reserves last? Answer to the nearest year.

59. The world's oil reserves are estimated to be 6.28×10^{11} bbl. Production is 2.0×10^{10} bbl per year. At this rate, how long will the world's oil reserves last? Answer to the nearest year.

60. Scientists have estimated that the total energy received from the Sun each minute is 1.02×10^{19} cal. Since the area of the Earth is 5.1×10^8 km^2, and 1 km^2 is 10^{10} cm^2, the amount of energy received per square centimeter of earth surface per minute (the **solar constant**) is

$$\frac{1.02 \times 10^{19}}{(5.1 \times 10^8) \times 10^{10}} \text{ cal}$$

Write the answer as a whole number.

 In Other Words

61. Explain what you must do to multiply a^m by a^n.

62. Explain what you must do to divide a^m by a^n.

63. Explain what you must do to raise a^m to the nth power.

64. By the laws of exponents,

$$\frac{a^m}{a^n} = a^{m-n}, \quad m > n$$

What does

$$\frac{a^m}{a^n} = \frac{a^m}{a^m}, \quad m = n$$

equal in exponential form and as a number? Explain. On the basis of your answer, how would you define a^0?

 Discovery

The Rhind papyrus contains a problem that deals with exponents. Problem 79 is very difficult to translate, but historian Moritz Cantor formulates it as follows:

An estate consisted of seven houses; each house had seven cats; each cat ate seven mice; each mouse ate seven heads of wheat; and each head of wheat was capable of yielding seven hekat measures of grain: Houses, cats, mice, heads of wheat, and hekat measures of grain, how many of these in all were in the estate?

Here is the solution.

Houses	$7 = 7^1$
Cats	$49 = 7^2$
Mice	$343 = 7^3$
Heads of wheat	$2{,}401 = 7^4$
Hekat measures	$\underline{16{,}807 = 7^5}$
Total	$19{,}607$

Because the items in the problem correspond to the first five powers of 7, it was at first thought that the writer was introducing the terminology houses, cats, mice, and so on, for first power, second power, third power, and so on!

65. A similar problem can be found in *Liber Abaci,* by Leonardo Fibonacci (A.D. 1170–1250). Can you find the answer?

There are seven old women on the road to Rome. Each woman has seven mules; each mule carries seven sacks; each sack contains seven loaves; with each loaf are seven knives; and each knife is in seven sheaths. Women, mules, sacks, loaves, knives, and sheaths, how many are there in all on the road to Rome?

66. A later version of the same problem reads as follows:

As I was going to St. Ives
I met a man with seven wives;
Every wife had seven sacks;
Every sack had seven cats;
Every cat had seven kits.
Kits, cats, sacks, and wives,
How many were going to St. Ives?

Hint: The answer is not 2801. If you think it is, then you did not read the first line carefully.

 Online Study Center

To further explore Arabic numerals and the contributions of Indian mathematicians, access link 4.2.1 on this textbook's Online Study Center..

Research Questions

1. Trace the evolution of the Arabic numerals starting with the earliest preserved examples, found on ancient columns in India believed to have been erected by King Ashoka.

2. Explore the contributions made by Indian mathematicians such as Brahmagupta (seventh century), Mahavira (ninth century), and Bhaskara (twelfth century).

4.3   Number Systems with Bases Other Than 10

Binary Card Magic

Have you heard of binary cards? The five-card set shown below has a series of numbers on each card, the largest of which is 31. Say you have a smaller set of three cards A, B, and C that have numbers on them as follows:

A	5	3	1	7
B	6	3	2	7
C	6	7	4	5

Now, pick a number, any number, from 1 to 7—say, 3.

Is it on card A? Yes. Note that the lowest number on A is 1.
Is it on card B? Yes. Note that the lowest number on B is 2.
Is it on card C? No.

The answer (that is, the number you picked) is $1 + 2$, the sum of the lowest numbers on the cards containing the number you picked. How does this work and why? It will be much easier to understand the trick if you know a little bit more about **binary** numbers, which you examine in this section. As a matter of fact, the Using Your Knowledge section in Exercises 4.3 explains how you can make your own set of cards and make the trick work.

A set of five binary cards.   ▶

As you learned in Section 4.1, in a positional number system a number is selected as the base, and objects are grouped and counted using that base. In the decimal system, the base chosen was 10, probably because the fingers are a convenient aid in counting.

A. Other Number Bases

As we saw earlier, it is possible to use other numbers as bases for numeration systems. For example, if we decide to use 5 as our base (that is, count in groups of five), then we can count the 17 asterisks that follow in this way:

We write 32_{five}, where the subscript "five" indicates that we are grouping by fives.

If we select 8 as our base, the asterisks are grouped this way:

(********) (********)   *

2 eights   + 1 one

We write 21_{eight}. Thus, if we use subscripts to indicate the manner in which we are grouping the objects, we can write the number 17 as

$$17_{\text{ten}} = 32_{\text{five}} = 21_{\text{eight}}$$

Using groups of seven, we can indicate the same number of asterisks by

$$23_{\text{seven}} = 2 \text{ sevens} + 3 \text{ ones}$$

Note: In the following material, when no subscript is used, it will be understood that the number is expressed in base 10.

EXAMPLE 1 ▶ Writing Decimals in Bases 8, 5, and 7

Arrange 13 asterisks in groups of eight, five, and seven, and write the number 13 in the following:

(a) Base 8   (b) Base 5   (c) Base 7

Solution

(a) (********)   ***** $= 15_{\text{eight}}$

1 eight   + 5 ones

(b) (*****) (*****)   *** $= 23_{\text{five}}$

2 fives   + 3 ones

(c) (******)   ****** $= 16_{\text{seven}}$

1 seven   + 6 ones

B. Changing to Base 10

Now consider the problem of "translating" numbers written in bases other than 10 into our decimal system. For example, what number in our decimal system corresponds to 43_{five}? First, recall that in the decimal system, numbers can be written in expanded form. For example, $342 = (3 \times 10^2) + (4 \times 10) + (2 \times 10^0)$. As you can see, when written in expanded form, each digit in 342 is multiplied by the proper power of 10 (the base being used). Similarly, when written in expanded form, each digit in the numeral 43_{five} must be multiplied by the proper power of 5 (the base being used). Thus

> Recall that $a^0 = 1$ for $a \neq 0$. So, $10^0 = 1$ and $2 \times 10^0 = 2$.

$$43_{\text{five}} = (4 \times 5^1) + (3 \times 5^0) = 23$$

EXAMPLE 2 ▶ Converting from Bases 5 and 8 to Decimal

Write the following numbers in decimal notation:

(a) 432_{five} (b) 312_{eight}

Solution

(a) $432_{\text{five}} = (4 \times 5^2) + (3 \times 5^1) + (2 \times 5^0) = 117$

(b) $312_{\text{eight}} = (3 \times 8^2) + (1 \times 8^1) + (2 \times 8^0) = 202$

In the seventeenth century, the German mathematician Gottfried Wilhelm Leibniz advocated use of the **binary system** (base 2). This system uses only the digits 0 and 1, and the grouping is by twos. The advantage of the binary system is that each position in a numeral contains one of just two values (0 or 1). Thus, electric switches, which have only two possible states, *off* or *on,* can be used to designate the value of each position. Computers using the binary system have revolutionized technology and the sciences by speedily performing calculations that would take humans years to complete. Handheld calculators operate internally on the binary system.

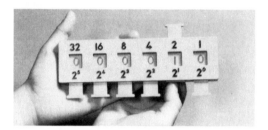

A binary counter.

Now consider the problem of converting a number from base 2 to decimal notation. It will help you to keep in mind that in the binary system, numbers are built up by using blocks that are powers of the base 2.

$$2^0 = 1, \qquad 2^1 = 2, \qquad 2^2 = 4, \qquad 2^3 = 8, \qquad 2^4 = 16, \qquad \ldots$$

When you write 1101 in the binary system, you are saying in the yes/no language of the computer, "a block of 8, yes; a block of 4, yes; a block of 2, no; a block of 1, yes." Thus,

$$\begin{aligned}
1101_{\text{two}} &= (1 \times 2^3) + (1 \times 2^2) + (0 \times 2^1) + (1 \times 2^0) \\
&= 8 + 4 + 0 + 1 \\
&= 13
\end{aligned}$$

EXAMPLE 3 ▶ Converting from Binary to Decimal

Write the number 10101_{two} in decimal notation.

Solution We write

$$\begin{aligned}
10101_{\text{two}} &= (1 \times 2^4) + (0 \times 2^3) + (1 \times 2^2) + (0 \times 2^1) + (1 \times 2^0) \\
&= 16 + 4 + 1 \\
&= 21
\end{aligned}$$

So far we have used only bases less than 10, of which the most important ones are base 2 (**binary**) and base 8 (**octal**), because they are used by computers. Bases greater than 10 are also possible, but then new symbols are needed for the digits greater than 9. For example, base 16 (**hexadecimal**) is also used by computers, with the "digits" A, B, C, D, E, and F used to correspond to the decimal numbers 10, 11, 12, 13, 14, and 15, respectively. We can change numbers from hexadecimal to decimal notation in the same way that we did for bases less than 10.

Some scientific calculators convert between decimal, binary, and octal. For example, enter 625.

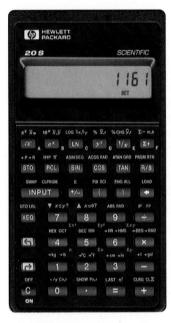

Press ▶ HEX OCT. You get the answer 1161_{eight}.

EXAMPLE 4 ▶ **Converting from Hexadecimal to Decimal**

Write the number $5AC_{\text{sixteen}}$ in decimal notation.

Solution

$$
\begin{aligned}
5AC_{\text{sixteen}} &= (5 \times 16^2) + (10 \times 16^1) + (12 \times 16^0) \\
&= (5 \times 256) + (160) \qquad + 12 \\
&= 1280 \qquad + 160 \qquad + 12 \\
&= 1452
\end{aligned}
$$

C. Changing from Base 10

Up to this point we have changed numbers from bases other than 10 to base 10. Now we shall change numbers from base 10 to other bases. A good method for doing this depends on successive divisions. For example, to change 625 to base 8, we start by dividing 625 by 8 and obtain 78 with a remainder of 1. We then divide 78 by 8 and get 9 with 6 remaining. Next, we divide 9 by 8 and obtain a quotient of 1 and a remainder of 1. We diagram these divisions as follows:

$$
\begin{array}{r|l}
8 & 625 \\
8 & 78 \quad\; 1 \qquad \text{Remainder} \\
8 & 9 \quad\;\; 6 \qquad \text{Remainder} \\
& 1 \quad\;\; 1 \qquad \text{Remainder}
\end{array}
$$

The answer is read upward as 1161_{eight} (see the arrow).

Why does this method work? Suppose we wish to find how many eights there are in 625. To find out, we divide 625 by 8. The quotient 78 tells us that there are 78 eights in 625, and the remainder tells us that there is 1 left over. Dividing the 78 by 8 (which is the same as dividing the 625 by $8 \times 8 = 64$) tells us that there are 9 sixty-fours in 625, and the remainder tells us that there are 6 eights left over. Finally, dividing the quotient 9 by 8 gives a new quotient of 1 and a remainder of 1. This tells us that there is 1 five hundred twelve ($8 \times 8 \times 8$) contained in 625 with 1 sixty-four left over. Thus, we see that

$$
625 = (1 \times 8^3) + (1 \times 8^2) + (6 \times 8) + (1 \times 8^0) = 1161_{\text{eight}}
$$

EXAMPLE 5 ▶ **Changing Decimals to Bases 2 and 5**

Change the number 33 to the following:

(a) Base 2   (b) Base 5

Solution

(a)
$$
\begin{array}{r|l}
2 & 33 \\
2 & 16 \quad 1 \\
2 & 8 \quad\; 0 \\
2 & 4 \quad\; 0 \\
2 & 2 \quad\; 0 \\
& 1 \quad\; 0
\end{array}
$$

Thus, $33 = 100001_{\text{two}}$.

(b)
$$
\begin{array}{r|l}
5 & 33 \\
5 & 6 \quad 3 \\
& 1 \quad 1
\end{array}
$$

Thus, $33 = 113_{\text{five}}$.

EXAMPLE 6 ▶ Changing Decimal Numerals to Bases 8 and 6

Change the number 4923 to the following:

(a) Octal notation   (b) Hexadecimal notation

Solution

(a) 8 | 4923
        8 | 615   3
          8 | 76   7
            8 | 9   4
              1   1

Thus, $4923 = 11473_{eight}$.

(b) 16 | 4923
         16 | 307   11
            16 | 19   3
                1   3

Thus, $4923 = 133B_{sixteen}$.

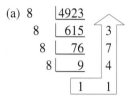

Online Study Center

To verify the conversions in Examples 1–6 [except 1(c)], access link 4.3.2 on this textbook's Online Study Center.

D. Applications

How does your computer work? A computer uses a collection of ones (1s) and zeros (0s), known as binary digits or bits, to store or read information. In particular, all color information (monitor and color printer) is stored numerically. One bit uses a 0 or a 1 to represent each pixel on a screen. A color depth of 1 bit can only be a black-and-white picture.

EXAMPLE 7 ▶ Following Patterns

Complete the following table:

> 1 bit allows $2^1 = 2$ colors, black and white.
> 4 bits allow _____ different color combinations.
> 8 bits allow _____ different color combinations.
> 16 bits allow _____ color combinations.
> 24 bits allow _____ color combinations.

Solution

The pattern here is $2^1 = 2$, $2^4 = 16$, $2^8 = 256$, $2^{16} = 65,536$, and $2^{24} = 16,777,216$. Thus, a monitor that supports 24 bits can show more than 16 million color combinations!

24 bits   8 bits   6 bits   4 bits   1 bit

Have you seen an envelope with symbols like the ones at the bottom of the address in Figure 4.3? What are these symbols and what do they mean? Because the symbols are composed of bars, they are called **bar codes.** One of the simplest bar codes is the **Postnet code** used by the U.S. Postal Service to represent the ZIP code + 4 for addresses. In this case, the bars represent the numbers **33675-5096.** The tenth digit is a *checking* digit (see Exercises 4.3, problems 41 and 42).

I. Bello
Ibello Hcc Usf
Ybor Campus POB 5096
Tampa FL 33675-5096

|ııllıııllııllıılııı:lıllıılıılıllııılııılıııllıııllııl|

FIGURE 4.3

The code is made up of 52 bars starting with a long bar, followed by ten groups of five bars each containing two long and three short bars and ending with a long bar. The long bars represent 1s, and the short bars 0s. The relationships between decimal digits, bar codes, and Postnet binary codes are shown in Table 4.5.

TABLE 4.5

Decimal	Bar Code	Binary Code
1	ıııll	00011
2	ıılıl	00101
3	ııllı	00110
4	ılııl	01001
5	ılılı	01010
6	ıllıı	01100
7	lıııl	10001
8	lıılı	10010
9	lılıı	10100
0	llııı	11000

Do you see the relationship between the binary codes for the numbers 1–6 in Postnet code and in regular binary? Try omitting the last digit! Now, look at the binary codes for 7–10. All the numbers start with 1. Do you see how to form the codes for these numbers?

EXAMPLE 8 ▶ Zipping to Postnet

Write the ZIP code 33657 using Postnet binary code.

Solution

Start and end with a long bar, and copy the binary code for the digits 33657 like this:

Start with a long bar.	3	3	6	5	7	End with a long bar.	■

| | ııllı | ııllı | ıllıı | ılılı | lıııl | |

Online Study Center

For a tutorial on decoding, access link 4.3.3 on this textbook's Online Study Center.

EXERCISES 4.3

Ⓐ Other Number Bases

In problems 1–4, write numerals in the bases indicated by the manners of grouping.

1. (✱✱✱) (✱✱✱) ✱✱

2. (✱✱✱✱✱) (✱✱✱✱✱) (✱✱✱✱✱) ✱✱✱

3. (✱✱✱✱) (✱✱✱✱) (✱✱✱✱) ✱

4. (✱✱✱✱✱✱✱✱✱) (✱✱✱✱✱✱✱✱✱) (✱✱✱✱✱✱✱✱✱) ✱✱✱✱✱✱

In problems 5–8, draw diagrams as shown in problems 1–4, and then write each decimal number in the given base.

5. 15 in base 8
6. 15 in base 5
7. 15 in base 7
8. 15 in base 12

Ⓑ Changing to Base 10

In problems 9–16, write each number in decimal notation.

9. 42_{five}
10. 31_{five}
11. 213_{eight}
12. 563_{eight}
13. 11011_{two}
14. 101001_{two}
15. $123_{sixteen}$
16. $ACE_{sixteen}$

Ⓒ Changing from Base 10

In problems 17–32, use the method of successive divisions.

17. Write the number 15 in base 5 notation.
18. Write the number 27 in base 5 notation.
19. Write the number 28 in binary notation.
20. Write the number 43 in binary notation.
21. Write the number 25 in hexadecimal notation.
22. Write the number 121 in hexadecimal notation.
23. Write the number 25 in base 6 notation.
24. Write the number 38 in base 6 notation.

25. Write the number 64 in base 7 notation.
26. Write the number 123 in base 7 notation.
27. Write the number 38 in octal notation.
28. Write the number 135 in octal notation.
29. Write the number 1467 in hexadecimal notation.
30. Write the number 145,263 in hexadecimal notation.
31. Write the number 73 in binary and in octal notation.
32. Write the number 87 in octal and in hexadecimal notation.

Ⓓ Applications

In problems 33–36, write each ZIP code in binary code and as a decimal.

33. | ₁₁ll₁ ₁l₁l₁ ₁l₁l ₁₁ll ₁ll₁₁ |
34. | ₁l₁l₁ ₁₁ll₁ ₁l₁l ₁₁ll ₁l₁l₁ |
35. | ₁l₁l ₁l₁l ₁₁ll ₁ll₁ ₁l₁l₁ |
36. | l₁l₁ ll₁₁ l₁₁l ₁l₁l ₁ll₁ |

In problems 37–40, write each ZIP code in bar code.

37. 02218
38. 70605
39. 95472
40. 15744

 In Other Words

41. The ZIP code + 4 shown at the beginning of this section contains 52 bars and starts and ends with a long bar. If each digit in a ZIP code + 4 number consists of 5 bars, is this consistent with the fact that the complete ZIP code + 4 number has 52 bars? Explain.

42. As it turns out, there is an extra digit in the ZIP code + 4 shown at the beginning of this section. The digit represented by the five bars *before* the last long bar is called the *checking digit*.
    a. What is this checking digit?
    b. The first part of the procedure used to find the checking digit for 33675-5096 consists of adding the digits in 33675. Write the rest of the procedure.

You may need a dictionary to answer questions 43–46.

43. What does *binary* mean? What does the prefix *bi-* indicate?

44. What does *octal* mean? What does the prefix *oct-* indicate?

45. What does *hexadecimal* mean? What does the prefix *hexa-* indicate?

46. The binary system uses the numbers 0 and 1; the octal uses 0, 1, 2, 3, 4, 5, 6, and 7; and the hexadecimal uses 0, 1, 2, 3, 4, 5, 6, 7, 8, 9, A, B, C, D, E, and F. Why are the A, B, C, D, E, and F needed?

 Using Your Knowledge

Here is a trick that you can use to amaze your friends. Write the numbers from 1 to 7 in binary notation. They look like this:

Decimal	Binary
1	1
2	10
3	11
4	100
5	101
6	110
7	111

Now label three columns *A*, *B*, and *C*. In column *A*, write the numbers that have a 1 in the units place when written in binary notation; in column *B*, write the numbers that have a 1 in the second position from the right when written in binary notation; and then in column *C*, write the numbers with a 1 in the third position from the right when written in binary notation.

A	B	C
1	2	4
3	3	5
5	6	6
7	7	7

Ask someone to think of a number between 1 and 7 and tell you in which columns the number appears. Say that the number is 6 (which appears in columns *B* and *C*). You find the sum of the numbers at the top of columns *B* and *C*. This sum is $2 + 4 = 6$, and you have the desired number!

47. Can you explain why this works?

48. If you extend this trick to cover the first 15 numbers, how many columns do you need?

49. Can you discover how to do the trick with 31 numbers?

 Calculator Corner

Converting from Base b to Base 10 For convenience and to save space, we write the base *b* as an ordinary decimal rather than spelling it out. Thus, for example, 47_8 means exactly the same thing as 47_{eight}. Suppose that 4735_8 is to be converted to base 10. We know that

$$4735_8 = (4 \times 8^3) + (7 \times 8^2) + (3 \times 8) + 5$$

Since the three quantities in parentheses are all divisible by 8, we can rewrite the expression to get

$$4735_8 = 8 \times [(4 \times 8^2) + (7 \times 8) + 3] + 5$$

Next, we see that the two quantities in parentheses inside the square brackets are divisible by 8, so we can rewrite again to get

$$4735_8 = 8 \times \{8 \times [(4 \times 8) + 7] + 3\} + 5$$

Now we can evaluate the last expression by a simple step-by-step procedure. Start with the innermost parentheses, multiply 4 by the base 8, and add 7 to the result. Then multiply the last result by 8 and add 3 to the product. Finally, multiply the preceding result by 8 and add 5 to the product. The final sum is the required answer.

The arithmetic can be done on a calculator by keying in the following:

4	×	8	+	7	=	×	8	+	3	=

×	8	+	5	=

The calculator will show the result 2525.

Notice that we multiplied the first octal digit, 4, by the base, 8, and added the next octal digit, 7. Then we multiplied the result by the base and added the next octal digit, 3. Finally, we multiplied by the base again and added the last octal digit, 5.

This procedure holds for any base. Thus, to convert from any base *b* to base 10, do the following:

a. Multiply the first digit of the number by *b*, and add the second digit to the result.

b. Multiply the preceding result by *b*, and add the third digit of the given numeral.

c. Continue the same procedure until you have added the last digit of the given numeral.

The calculator will show the final answer. Be sure that you work from left to right when using the digits of the given numeral.

As another example, let us convert the hexadecimal numeral $BF3_{16}$ to base 10. Recall that the hexadecimal digits B and F are the numbers 11 and 15, respectively, in decimal notation, so we key in the following:

Our calculator will show the result 3059.

Convert each of the following numerals to decimal notation:

1. 1101_2 **2.** 231_4 **3.** 423_5

4. 752_8 **5.** 3572_8 **6.** 873_9

Online Study Center

To verify these answers and further explore base conversion, access link 4.3.4 on this textbook's Online Study Center.

Collaborative Learning

We already know that we can change a number from a given base to decimal notation by writing the number in expanded notation as we did in Example 4. Now, we are going a step further; we are going to translate words from a given base (16 in this case) to numbers. Take the word $FACE_{16}$. As you recall (see Example 4),

F = 15, A = 10, C = 12, and E = 14. Thus, $FACE_{16}$ = $15 \cdot 16^3 + 10 \cdot 16^2 + 12 \cdot 16^1 + 14$. Now your base converter or a simple calculator will tell you that $FACE_{16}$ = 64,206.

1. When the base of a number system is greater than 10, the missing digits come from the alphabet. Thus, when using base 12, 11 = A and 12 = B. Using this idea, discuss letters we use as numbers in different bases (such as 13, 16, and so on). If we follow this scheme, what is the highest base we can use?

2. $ADD_{36} = (10 \cdot 36^2 + 13 \cdot 36^1 + 13) = 13,441$. The number 13,441 is prime (it is only divisible by itself and 1).

Online Study Center

To further explore base conversion and primes, access link 4.3.5 on this textbook's Online Study Center.

We will call the word ADD_{36} a prime word. Can you find other prime words in base 36?

3. Assign group members to find prime words in base 36. Can you find a prime three-letter word starting with B? What about with Z?

4. Collect all prime words from the members in your group. Write prime sentences using the words. An example: TWIN TOBOGGANIST TOURED TANZANIAN TOWN (discovered by David Gogomolov). What is the longest sentence your group can create? *Hint:* Go to link 4.3.5 on this textbook's Online Study Center.

4.4 Binary Arithmetic

Modems and ASCII Code

Do you know what a modem is? It is a device that transmits data to or from a computer via telephone. How does one work? Suppose you are writing a message starting with the letter A on your computer. A stream of digital bits with 0 volts for binary 0 and a constant voltage for binary 1 flows from the computer into the modem carrying the American Standard Code for Information Interchange (ASCII) code for the letter A: 01000001. At the receiving end, the signal is demodulated—that is, changed back—and resumes its original form, a series of pulses representing 0s and 1s resulting in 01000001, the letter A. You can learn how these letters are encoded in different bases by working problems 27–32 in Exercises 4.4. ▶

TABLE 4.6 Binary Addition

+	0	1
0	0	1
1	1	10

TABLE 4.7 Binary Multiplication

×	0	1
0	0	0
1	0	1

Now that we know how to represent numbers in the binary system, we look at how computations are done in that system. First, we can construct addition and multiplication tables like the ones for base 10 arithmetic. See Tables 4.6 and 4.7.

The only entry that looks peculiar is the 10 in the addition table, but recall that 10_2 means 2_{10}, which is exactly the result of adding $1 + 1$. (Be sure to read 10 as "one zero" not as "ten.")

A. Addition

Binary addition is done in the same manner as addition in base 10. We line up the corresponding digits and add column by column.

EXAMPLE 1 ► Adding Two Terms in Binary

Add 1010_2 and 1111_2.

Solution

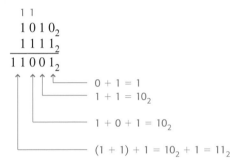

$$
\begin{array}{c}
1\ 1 \\
1\ 0\ 1\ 0_2 \\
1\ 1\ 1\ 1_2 \\
\hline
1\ 1\ 0\ 0\ 1_2
\end{array}
$$

$0 + 1 = 1$	
$1 + 1 = 10_2$	Write 0 and carry 1 to the next column.
$1 + 0 + 1 = 10_2$	Write 0 and carry 1 to the next column.
$(1 + 1) + 1 = 10_2 + 1 = 11_2$	

Thus, $1010_2 + 1111_2 = 11001_2$. ■

EXAMPLE 2 ► Adding Three Terms in Binary

Perform the addition $1101_2 + 110_2 + 11_2$.

Solution

We shall omit the subscript 2 in the computation, but keep in mind that all numerals are binary.

$$
\begin{array}{c}
1\ 1\ 1 \\
1\ 1\ 0\ 1 \\
1\ 1\ 0 \\
1\ 1 \\
\hline
1\ 0\ 1\ 1\ 0
\end{array}
$$

$1 + 0 + 1 = 10$	Write 0 and carry 1 to the next column.
$(1 + 1) + 0 + 1 = 10 + 1 = 11$	Write 1 and carry 1 to the next column.
$(1 + 1) + 1 = 10 + 1 = 11$	Write 1 and carry 1 to the next column.
$1 + 1 = 10$	

Thus, $1101_2 + 110_2 + 11_2 = 10110_2$. ■

B. Subtraction

To subtract in the binary system, we line up corresponding digits and subtract column by column, "borrowing" as necessary.

EXAMPLE 3 ▶ Subtracting in Binary

Perform the subtraction $1111_2 - 110_2$.

Solution

Again, we omit the subscript 2 in the computation.

```
    1 1 1 1
  −   1 1 0
    1 0 0 1
```

```
          1 − 0 = 1
          1 − 1 = 0
          1 − 1 = 0
          1 is left in this column.
```

This shows that $1111_2 - 110_2 = 1001_2$. Check this by adding $1001_2 + 110_2$ to get 1111_2.   ■

Example 3 did not require "borrowing," but the next example does.

EXAMPLE 4 ▶ More Subtraction in Binary

Subtract 101_2 from 1010_2.

Solution

```
  0 10 0 10
    1 0 1 0
  −   1 0 1
      1 0 1
```

```
          0 − 1 requires borrowing; then, 10 − 1 = 1.
          0 − 0 = 0
          0 − 1 requires borrowing; then, 10 − 1 = 1.
```

The result is $1010_2 - 101_2 = 101_2$. You can check this answer by adding $101_2 + 101_2$ to get 1010_2.   ■

C. Multiplication

Multiplication in the binary system is done in a manner similar to that for base 10, as the following examples illustrate:

EXAMPLE 5 ▶ Multiplication in Binary

Multiply $101_2 \times 110_2$.

Solution

```
      1 1 0
  ×   1 0 1
      1 1 0      ← Start from the right and multiply by 1.
    0 0 0        ← Indent and multiply by 0.
  1 1 0          ← Indent and multiply by the left-hand 1.
  1 1 1 1 0      ← Add the above products.
```

The result shows that $101_2 \times 110_2 = 11110_2$. Note that we could abbreviate a little by omitting the two leftmost 0s in the second partial product and writing the third product on the same line.   ■

EXAMPLE 6 ▶ More Multiplication in Binary

Multiply 1110_2 by 110_2.

Solution

```
        1 1 1 0
      × 1 1 0
      1 1 1 0 0       ← Multiply by 0, and write one 0. Then multiply by 1, using the
                        same line.
    1 1 1 0           ← Indent two places and multiply by 1.
  1 0 1 0 1 0 0       ← Add the partial products.
```

The computation shows that $110_2 \times 1110_2 = 1010100_2$. ▪

D. Division

The procedure used for division in the base 10 system can also be used in the binary system. This is illustrated next.

EXAMPLE 7 ▶ Division in Binary

Divide 1011_2 by 10_2.

Solution

```
                  1 0 1
         1 0 ) 1 0 1 1
Step 1 ──────→ − 1 0
Step 2 ──────→     1 1
Step 3 ──────→   − 1 0
                     1    ←─────── Remainder
```

Step 1 $1 \times 10 = 10$. Write 1 in the quotient above the 10. Then subtract.

Step 2 Bring down the next digit. 10 does not go into 1, so write 0 in the quotient and bring down the final digit.

Step 3 $1 \times 10 = 10$. Write 1 in the quotient and then subtract to get the remainder.

The computation shows that $1011_2 \div 10_2 = 101_2$ with remainder 1. You can check this answer by multiplying 101_2 by 10_2 and adding 1 to the product. Thus

$$(10_2 \times 101_2) + 1 = 1010_2 + 1 = 1011_2$$

This shows that the answer is correct. ▪

EXAMPLE 8 ▶ More Division in Binary

Divide 110111_2 by 101_2.

Solution

```
                      1 0 1 1
         1 0 1 ) 1 1 0 1 1 1
Step 1 ──────→ − 1 0 1
Step 2 ──────→     1 1 1
Step 3 ──────→   − 1 0 1
Step 4 ──────→       1 0 1
Step 5 ──────→     − 1 0 1
                         0    ←─────── Remainder
```

Step 1 $1 \times 101 = 101$. Write 1 in the quotient above the 110 and then subtract.

Step 2 Bring down the next digit. 101 does not go into 11, so write 0 in the quotient and bring down the next digit.

Step 3 $1 \times 101 = 101$. Write 1 in the quotient and then subtract.

Step 4 Bring down the final digit.

Step 5 $1 \times 101 = 101$. Write 1 in the quotient and then subtract to get the remainder.

The division here is exact; the quotient is 1011_2, and the remainder is 0.

You can check the answer by multiplying the quotient 1011_2 by the divisor 101_2. Thus

$$
\begin{array}{r}
1\ 0\ 1\ 1 \\
\times\quad 1\ 0\ 1 \\
\hline
1\ 0\ 1\ 1 \\
1\ 0\ 1\ 1\ 0 \\
\hline
1\ 1\ 0\ 1\ 1\ 1 \\
\end{array}
$$

Therefore, the answer is correct.   ■

Online Study Center

To further explore addition and multiplication tables for bases 2 to 36, access links 4.4.1 and 4.4.2 on this textbook's Online Study Center. To explore the conversion of binary numbers and decimals, access link 4.4.3.

E. Applications

Binary numbers can be used to transmit, print, and encode pictures or images. (More about this later in the Discovery section.) How would you "translate," or code, a picture to binary numbers? Look at the next example (Figure 4.4).

EXAMPLE 9 ▶ Say Hi in Binary

Translate, or code, a white square with a 0 and a black square with a 1.
On the first line, all squares are white, so the first line corresponds to

0000000000000000

The second line is

0000000001100110

What are the third, fourth, and last lines?

Solution
The third line is

0000000001100110

The fourth line is

0000000000000110

The last line is

0000000000000000

You have encoded the information into a 16-bit binary code!
Can you answer these questions? Which line(s) have the most 1s? Which line(s) have the same number of 1s? What about the least number of 1s?   ■

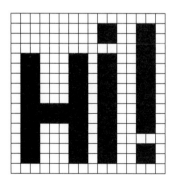

FIGURE 4.4

EXERCISES 4.4

A Addition

In problems 1–6, perform the indicated additions (base 2).

1. 111_2
$+ \ 10_2$

2. 111_2
$+ 101_2$

3. 1101_2
$+ \ 110_2$

4. 1111_2
$+ 1101_2$

5. 110_2
101_2
$+ 111_2$

6. 1101_2
1110_2
$+ \ 101_2$

B Subtraction

In problems 7–12, perform the indicated subtractions (base 2).

7. 111_2
$- \ 10_2$

8. 110_2
$- \ 11_2$

9. 1000_2
$- 111_2$

10. 1101_2
$- 111_2$

11. 1111_2
$- 101_2$

12. 1010_2
$- 101_2$

C Multiplication

In problems 13–18, multiply as indicated (base 2).

13. 110_2
$\times \ 11_2$

14. 101_2
$\times \ 10_2$

15. 1111_2
$\times \ 11_2$

16. 1110_2
$\times 111_2$

17. 1011_2
$\times 101_2$

18. 1011_2
$\times 111_2$

D Division

In problems 19–24, divide as indicated (base 2).

19. $10_2 \overline{)1101_2}$

20. $11_2 \overline{)1101_2}$

21. $11_2 \overline{)1110_2}$

22. $11_2 \overline{)11011_2}$

23. $101_2 \overline{)111011_2}$

24. $111_2 \overline{)1110111_2}$

 In Other Words

In ASCII (American Standard Code for Information Interchange), characters are numbered in **binary notation.** The characters A–O are prefixed by 0100 and are numbered in order, starting with A = 0001, B = 0010, C = 0011, D = 0100, and so on. The characters P–Z are numbered in order starting with P = 0000 and pre-fixed by 0101. In problems 25 and 26, write the messages in words.

25. 01001000 01000101 01001100
01001100 01001111
What decimal number corresponds to each of the letters?

26. 01000011 01000001 01001100
01001100 01001101 01000101
What decimal number corresponds to each of the letters?

 Using Your Knowledge

In problems 27–32, use ASCII to write the binary, decimal, and hexadecimal codes for the letters. Remember that in the hexadecimal system the "digits" A, B, C, D, E, and F correspond to the numbers 10, 11, 12, 13, 14, and 15, respectively.

		Binary	Decimal	Hexadecimal
27.	A			
28.	I			
29.	Q			
30.	V			
31.	X			
32.	Z			

In problems 33 and 34, change the hexadecimal numerals representing letters to binary and then use ASCII to write the messages.

33. 48 45 4C 50

34. 49
44 4F
4E 4F 54
47 45 54
49 54

 **Online Study Center**

To further explore binary conversions, access link 4.4.4 on this textbook's Online Study Center.

Discovery

A camera installed on the spacecraft *Mariner IV* took the first pictures of the planet Mars and sent them back to Earth by radio signals on July 14, 1965. On Earth, a computer received the pictures in the form of binary numerals consisting of 6 bits. (A **bit** is a binary digit.) The shade of each dot in the final picture was determined by 6 bits.

The numeral 000000_2 (0 in base 10) indicated a white dot, and the numeral 111111_2 (63 in base 10) indicated a black dot. The 62 numerals between represented various shades of gray between white and black. To make a complete picture, 40,000 dots, each described by 6 bits, were needed!

35. If one of the numerals received was 110111_2, what was the corresponding decimal numeral?

36. Does the dot corresponding to the numeral received in problem 35 represent a shade of gray closer to white or to black?

37. What binary numeral represents the lightest shade of gray that is not white?

38. What binary numeral represents the darkest shade of gray that is not black?

39. What binary numeral represents the shade numbered 31?

Calculator Corner

Some scientific calculators will perform operations in different bases. Thus, to add $1101_2 + 110_2 + 11_2$ (Example 2), we place the calculator in the binary mode by pressing [2nd] [mode] [bin] and proceed as in regular addition by entering $1101 + 110 + 11 =$. The answer will appear on the screen as 10110. We can do addition, subtraction, and multiplication in the same manner, but we must be careful with division. In Example 7 we divided 1011_2 by 10_2. The answer was 101_2 with a remainder of 1. To do this on our calculator, we enter [2nd] [mode] [bin] (now we are in binary) and $1011 \div 10 =$. The calculator will give us the quotient 101 but not the remainder. To discover that there is a remainder, we must check our division by multi-

plying the quotient (101_2) by the divisor (10_2). The result is 1010_2, not 1011_2, so the remainder must be 1. The moral of the story is that when doing division problems in different bases, we must check the problem by multiplication and, during this process, find the remainder.

1. Use your calculator to find the remainder (if any) in problems 19, 21, and 23 of Exercises 4.4.

Collaborative Learning

Form two (or more) groups of students and provide each group with a 16-bit binary code chart like the ones shown below. Each group should construct a diagram similar to the one shown in Example 9 but using the word *LO*! Write the 16 lines of code on a 3-by-5 card.

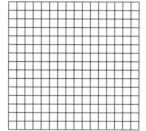 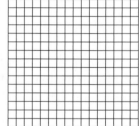

1. Are the numbers on the first and last lines of code identical? If so, does that mean that the rest of the numbers will be identical? Discuss this.

2. The number of bits (the simplest pieces of information, a 0 or a 1) to write each line can be shortened. For example, the fourth line in Example 9 can be written as

    13 0   2 1   1 0

Is there a problem with this notation? What is the problem? Can you fix this problem?

3. To encode a character, say, the letter A, takes 8 bits = 1 byte. How many bits would it take to write *supercalifragilisticexpialidocious?* Each of the groups can find the number of bits it would take to write what they consider the longest word in the English language. Is the answer the same for both groups?

4. If you assume that an average-sized novel has 500,000 characters, then, according to the table below, it would take about 500 kB to write all the characters (not in the novel but the letters and symbols in the book itself!). An encyclopedia is equivalent to 2500 average novels. How many bytes is that? How many bits?

UNIT	Value (byte)	Value (bit)
1 character	1 byte	8 bits
1 novel	500 kB	4 Mbits
1 encyclopedia	1.25 GB	10 Gbits
1 library	6.25 TB	50 Tbits

Online Study Center

To further explore bytes and bits, access link 4.4.5 on this textbook's Online Study Center.

4.5 Octal and Hexadecimal Arithmetic

Native American Arithmetic

When counting in the decimal system, you can use 10 fingers. In the binary system you can use 2 fingers. What could you use in an octal system? The answer is not at your fingertips but *between* your fingers! The Yuki Indians of California use the spaces between their fingers for counting. Thus, some of their counting is in base 4 and some is in base 8 (octal or octonary). In this section we shall study octal and hexadecimal arithmetic. ▶

Arithmetic in the base 8 and base 16 systems can be done with addition and multiplication tables in much the same way as in the base 10 system. We consider the octal system first and construct the required tables (Tables 4.8 and 4.9 on p. 212).

TABLE 4.8 Octal Addition Table

+	0	1	2	3	4	5	6	7
0	0	1	2	3	4	5	6	7
1	1	2	3	4	5	6	7	10
2	2	3	4	5	6	7	10	11
3	3	4	5	6	7	10	11	12
4	4	5	6	7	10	11	12	13
5	5	6	7	10	11	12	13	14
6	6	7	10	11	12	13	14	15
7	7	10	11	12	13	14	15	16

Yuki woman in mourning.

TABLE 4.9 Octal Multiplication Table

×	0	1	2	3	4	5	6	7
0	0	0	0	0	0	0	0	0
1	0	1	2	3	4	5	6	7
2	0	2	4	6	10	12	14	16
3	0	3	6	11	14	17	22	25
4	0	4	10	14	20	24	30	34
5	0	5	12	17	24	31	36	43
6	0	6	14	22	30	36	44	52
7	0	7	16	25	34	43	52	61

A. Octal Addition

To add octal numbers, we align corresponding digits and add column by column, carrying over from column to column as necessary.

As in the preceding section, we shall omit the subscripts in the computations, but keep in mind the system with which we are working.

EXAMPLE 1 ▶ Octal Addition

Add $673_8 + 52_8$.

Solution

$$
\begin{array}{r}
\overset{1}{} \\
6\ 7\ 3 \\
+\ \ 5\ 2 \\
\hline
7\ 4\ 5
\end{array}
$$

$\longleftarrow$ 3 + 2 = 5

7 + 5 = 14 (Table 4.8)   Write 4 and carry 1. (Read 14 as "one four" not as "fourteen.")

1 + 6 = 7

Thus, $673_8 + 52_8 = 745_8$.   ∎

EXAMPLE 2 ▶ More Octal Addition

Add $705_8 + 374_8$.

Solution

$$
\begin{array}{r}
1\ \ 1 \\
7\ 0\ 5 \\
+\ 3\ 7\ 4 \\
\hline
1\ 3\ 0\ 1
\end{array}
$$

$\longleftarrow$ 5 + 4 = 11 (Table 4.8)   Write 1 and carry 1.

1 + 0 + 7 = 10 (Table 4.8)   Write 0 and carry 1.

1 + (7 + 3) = 1 + 12 = 13 (Table 4.8)

The required sum is 1301_8.   ∎

B. Octal Multiplication

EXAMPLE 3 ▶ Octal Multiplication

Multiply 46_8 by 5_8.

Solution

$$
\begin{array}{r}
4\ 6 \\
\times\ \ \ 5 \\
\hline
2\ 7\ 6
\end{array}
$$

← 5 × 6 = 36 (Table 4.9)   Write 6 and carry 3.

5 × 4 = 24 (Table 4.9)   Add the 3 to get 27.

The answer is 276_8. (*Note:* You can check the answer by converting all the numbers to base 10. Thus, since $46_8 = 38$ and $5_8 = 5$, in base 10 we have $5 \times 38 = 190$. We also find that $276_8 = 190$, so the answer checks.)   ∎

EXAMPLE 4 ▶ More Octal Multiplication

Multiply 237_8 by 14_8.

Solution

$$
\begin{array}{r}
2\ 3\ 7 \\
\times\ \ \ 1\ 4 \\
\hline
1\ 1\ 7\ 4 \\
2\ 3\ 7\ \ \ \\
\hline
3\ 5\ 6\ 4
\end{array}
$$

4 × 7 = 34 (Table 4.9)   Write 4 and carry 3.

4 × 3 = 14 (Table 4.9); 14 + 3 = 17   Write 7 and carry 1.

4 × 2 = 10 (Table 4.9); 10 + 1 = 11

1 × 237 = 237

Addition of the partial products gives the final answer, 3564_8. The check is left to you.   ∎

C. Octal Subtraction and Division

Subtraction and division also can be done by using tables as in the examples that follow.

EXAMPLE 5 ▶ Octal Subtraction

Subtract 56_8 from 747_8.

Solution

$$
\begin{array}{r}
^{6}\ ^{14} \\
7\ 4\ 7 \\
-\ \ \ 5\ 6 \\
\hline
6\ 7\ 1
\end{array}
$$

← 7 − 6 = 1 (Table 4.8 shows 6 + 1 = 7.)

4 − 5 requires borrowing. Change the leftmost 7 to 6 and add 10 to the 4. Then subtract. To get this result, refer to Table 4.8, go down the left-hand column under + to the 5, and read across to 14. The 7 in the top line above the 14 is the required number.

Bring down the 6.

The final answer is 671_8. You can check this answer by adding $56_8 + 671_8$ to get 747_8.   ∎

EXAMPLE 6 ▶ More Octal Subtraction

Subtract $643_8 - 45_8$.

Solution

```
      5 13
        3 13
    6 4 3
  −   4 5
    5 7 6   ←── 3 − 5 requires borrowing. Change the 4 to 3 and add 10 to the first
                 3 to give 13. Then subtract 13 − 5 = 6 (Table 4.8).
                 3 − 4 requires borrowing. Change the 6 to 5 and add 10 to the 3 to
                 give 13. Then subtract 13 − 4 = 7 (Table 4.8).
                 Bring down the 5.
```

The result shows that $643_8 - 45_8 = 576_8$. This can be checked by addition as in Example 5. ■

EXAMPLE 7 ▶ Octal Division

Divide 765_8 by 24_8.

Solution

```
        3 1
   2 4)7 6 5
       7 4   ←── 3 × 24 = 74 (Table 4.9). Write 3 in the quotient above the 76.
         2 5 ←── Subtract and bring down the 5.
         2 4 ←── 1 × 24 = 24
           1 ←── Subtract to get the remainder.
```

This computation shows that $765_8 \div 24_8 = 31_8$ with a remainder of 1_8. This can be checked by multiplying 31_8 by 24_8 and adding 1 to the product. ■

EXAMPLE 8 ▶ More Octal Division

Divide 4357_8 by 21_8.

Solution

```
          2 0 6
   2 1)4 3 5 7
       4 2       ←── 2 × 21 = 42. Write 2 in the quotient above the 43.
         1 5 7   ←── Subtract and bring down the 5. 21 does not go into 15, so
         1 4 6   ←  write 0 in the quotient and bring down the 7.
             1 1 ←  6 × 21 = 146 (Table 4.9). Write 6 in the quotient.
                    Subtract to get the remainder.
```

The computation shows that the quotient is 206_8 and the remainder is 11_8.

As before, the answer can be checked by multiplying the quotient by the divisor and adding the remainder. Thus,

```
        206
   ×     21
        206
       414
      4346
   +     11
      4357
```

which shows the answer is correct. ■

D. Hexadecimal Addition and Multiplication

In order to perform arithmetic in the hexadecimal system (base 16), we first construct the addition and multiplication tables (Tables 4.10 and 4.11). These tables furnish the "number facts" we need to perform the computations.

TABLE 4.10 Hexadecimal Addition Table (base 16)

+	0	1	2	3	4	5	6	7	8	9	A	B	C	D	E	F
0	0	1	2	3	4	5	6	7	8	9	A	B	C	D	E	F
1	1	2	3	4	5	6	7	8	9	A	B	C	D	E	F	10
2	2	3	4	5	6	7	8	9	A	B	C	D	E	F	10	11
3	3	4	5	6	7	8	9	A	B	C	D	E	F	10	11	12
4	4	5	6	7	8	9	A	B	C	D	E	F	10	11	12	13
5	5	6	7	8	9	A	B	C	D	E	F	10	11	12	13	14
6	6	7	8	9	A	B	C	D	E	F	10	11	12	13	14	15
7	7	8	9	A	B	C	D	E	F	10	11	12	13	14	15	16
8	8	9	A	B	C	D	E	F	10	11	12	13	14	15	16	17
9	9	A	B	C	D	E	F	10	11	12	13	14	15	16	17	18
A	A	B	C	D	E	F	10	11	12	13	14	15	16	17	18	19
B	B	C	D	E	F	10	11	12	13	14	15	16	17	18	19	1A
C	C	D	E	F	10	11	12	13	14	15	16	17	18	19	1A	1B
D	D	E	F	10	11	12	13	14	15	16	17	18	19	1A	1B	1C
E	E	F	10	11	12	13	14	15	16	17	18	19	1A	1B	1C	1D
F	F	10	11	12	13	14	15	16	17	18	19	1A	1B	1C	1D	1E

TABLE 4.11 Hexadecimal Multiplication Table (base 16)

×	0	1	2	3	4	5	6	7	8	9	A	B	C	D	E	F
0	0	0	0	0	0	0	0	0	0	0	0	0	0	0	0	0
1	0	1	2	3	4	5	6	7	8	9	A	B	C	D	E	F
2	0	2	4	6	8	A	C	E	10	12	14	16	18	1A	1C	1E
3	0	3	6	9	C	F	12	15	18	1B	1E	21	24	27	2A	2D
4	0	4	8	C	10	14	18	1C	20	24	28	2C	30	34	38	3C
5	0	5	A	F	14	19	1E	23	28	2D	32	37	3C	41	46	4B
6	0	6	C	12	18	1E	24	2A	30	36	3C	42	48	4E	54	5A
7	0	7	E	15	1C	23	2A	31	38	3F	46	4D	54	5B	62	69
8	0	8	10	18	20	28	30	38	40	48	50	58	60	68	70	78
9	0	9	12	1B	24	2D	36	3F	48	51	5A	63	6C	75	7E	87
A	0	A	14	1E	28	32	3C	46	50	5A	64	6E	78	82	8C	96
B	0	B	16	21	2C	37	42	4D	58	63	6E	79	84	8F	9A	A5
C	0	C	18	24	30	3C	48	54	60	6C	78	84	90	9C	A8	B4
D	0	D	1A	27	34	41	4E	5B	68	75	82	8F	9C	A9	B6	C3
E	0	E	1C	2A	38	46	54	62	70	7E	8C	9A	A8	B6	C4	D2
F	0	F	1E	2D	3C	4B	5A	69	78	87	96	A5	B4	C3	D2	E1

EXAMPLE 9 ▶ **Hexadecimal Addition**

Add $2B4_{16} + A1_{16}$.

Solution

```
    1
   2 B 4
 +   A 1
   3 5 5  ←— 4 + 1 = 5
   ↑ ↑——————— B + A = 15 (Table 4.10)     Write 5 and carry 1.
   └——————————— 1 + 2 = 3
```

The answer is 355_{16}. (You can check this by converting to base 10.) ∎

EXAMPLE 10 ▶ **More Hexadecimal Addition**

Add $1AB2_{16} + 2CD3_{16}$.

Solution

```
    1 1
   1 A B 2
 + 2 C D 3
   4 7 8 5  ←— 2 + 3 = 5
   ↑ ↑ ↑——————— B + D = 18 (Table 4.10)          Write 8 and
   │ │ └——————— (A + C) + 1 = 16 + 1 = 17 (Table 4.10)   carry 1.
   │ │                                           Write 7 and
   │ └——————————— 1 + 1 + 2 = 4                  carry 1.
```

The answer is 4785_{16}. (You can check this by converting to base 10.) ∎

EXAMPLE 11 ▶ **Hexadecimal Multiplication**

Multiply $1A2_{16}$ by B_{16}.

Solution

```
   1 A 2
 ×     B
 1 1 F 6
     ↑ ↑——————— B × 2 = 16 (Table 4.11)     Write 6 and carry the 1.
     │ └——————— B × A = 6E (Table 4.11)
     │————————— 6E + 1 = 6F (Table 4.10)     Write F and carry the 6.
     └————————— B × 1 = B; B + 6 = 11 (Table 4.10)
```

The answer is $11F6_{16}$. ∎

EXAMPLE 12 ▶ **More Hexadecimal Multiplication**

Multiply $2B4_{16}$ by $B1_{16}$.

Solution

```
     2 B 4
   ×   B 1
     2 B 4  ←— 1 × 2B4 = 2B4
   1 D B C  ←— B × 4 = 2C (Table 4.11); write C and carry the 2. Then,
   1 D E 7 4      B × B = 79 (Table 4.11); 79 + 2 = 7B (Table 4.10); write B
       ↑ ↑        and carry the 7. Then B × 2 = 16 (Table 4.11);
       │ │        16 + 7 = 1D (Table 4.10).
       │ └——————— B + C = 17 (Table 4.10); write 7 and carry the 1.
       └————————— (2 + B) + 1 = D + 1 = E (Table 4.10)
```

The answer is $1DE74_{16}$. ∎

Online Study Center

To further explore the decimal, binary, octal, and hexadecimal systems, access links 4.5.1, 4.5.2, and 4.5.3 on this textbook's Online Study Center.

EXERCISES 4.5

A **Octal Addition**

In problems 1–4, perform the indicated additions (base 8).

1. $531_8 + 47_8$

2. $425_8 + 364_8$

3. $7256_8 + 634_8$

4. $5732_8 + 747_8$

B **Octal Multiplication**

In problems 5–8, perform the indicated multiplications (base 8).

5. $57_8 \times 6_8$

6. $45_8 \times 7_8$

7. $216_8 \times 32_8$

8. $312_8 \times 65_8$

C **Octal Subtraction and Division**

In problems 9–12, perform the indicated subtractions (base 8).

9. $534_8 - 25_8$

10. $617_8 - 47_8$

11. $3264_8 - 756_8$

12. $4763_8 - 654_8$

In problems 13–16, perform the indicated divisions (base 8).

13. $317_8 \div 7_8$

14. $4355_8 \div 5_8$

15. $4215_8 \div 15_8$

16. $7342_8 \div 31_8$

D **Hexadecimal Addition and Multiplication**

In problems 17–20, perform the indicated additions (base 16).

17. $3CB_{16} + 4C_{16}$

18. $4FE_{16} + 35_{16}$

19. $98D_{16} + 2B_{16}$

20. $CBD_{16} + AF_{16}$

In problems 21–24, perform the indicated multiplications (base 16).

21. $2C5_{16} \times 3B_{16}$

22. $4DE_{16} \times 12_{16}$

23. $6F3_{16} \times AB_{16}$

24. $29A_{16} \times E0F_{16}$

 In Other Words

25. Explain why it is easier to add and multiply in the binary system (base 2) than in the hexadecimal system (base 16).

26. Why is it easier to write a number in expanded form in the decimal system than in the binary (base 2) or hexadecimal (base 16) system?

 Using Your Knowledge

You know that we used decimals in the base 10 system with the understanding that the place values to the right of the decimal point are

$$\frac{1}{10^1}, \quad \frac{1}{10^2}, \quad \frac{1}{10^3}, \quad \cdots$$

For example, the decimal numeral 23.759 stands for

$$(2 \times 10) + 3 + \frac{7}{10^1}, + \frac{5}{10^2}, + \frac{9}{10^3}$$

The same idea is used in other number systems. Thus, in the binary system (base 2), the numeral

$$11.101_2 = (1 \times 2) + 1 + \frac{1}{2^1} + \frac{0}{2^2} + \frac{1}{2^3}$$

In the octal system (base 8), the numeral

$$73.524_8 = (7 \times 8) + 3 + \frac{5}{8^1} + \frac{2}{8^2} + \frac{4}{8^3}$$

To convert numbers from binary (base 2) or octal (base 8) form to decimal form, it is convenient to have the values

$$\frac{1}{2} = 0.5, \quad \frac{1}{4} = 0.25, \quad \frac{1}{8} = 0.125, \quad \cdots$$

For instance,

$$1.111_2 = 1 + 0.5 + 0.25 + 0.125 = 1.875$$
$$3.5_8 = 3 + (5 \times 0.125) = 3.625$$

Change the following to decimal form:

27. 10.101_2

28. 11.011_2

29. 10.001_2

30. 21.4_8

31. 72.6_8

32. 31.7_8

Calculator Corner

Some scientific calculators are able to perform arithmetic in different bases. (You need a key displaying the desired base.) Thus, to add $673_8 + 52_8$ (Example 1), we place the calculator in the octal mode by pressing $\boxed{\text{2nd}}$ $\boxed{\text{mode}}$ $\boxed{\text{oct}}$ and entering the indicated operation 673 + 52 =, as in regular addition. The answer 745 will appear on the calculator screen. Keep in mind that doing division on the calculator will not provide the remainder. Thus, if you divide 765_8 by 24_8 (Example 7), the calculator gives the quotient as 31. Is there a remainder? If you multiply the quotient 31_8 by the divisor 24_8, the result is 764_8 and not 765_8. Thus, the remainder must be 1.

What about the hexadecimal system? First, you need a $\boxed{\text{hex}}$ key and A, B, C, D, E, and F buttons to represent the numbers 10, 11, 12, 13, 14, and 15, respectively. You can then multiply $1A2_{16}$ by B_{16} (Example 11) by entering $\boxed{\text{2nd}}$ $\boxed{\text{mode}}$ $\boxed{\text{hex}}$ $\boxed{\text{1A2}}$ $\boxed{\times}$ $\boxed{\text{B}}$ $\boxed{=}$. The calculator gives the answer 11F6.

1. Use your calculator to find the remainder (if any) in problems 13 and 15, Exercises 4.5.

Online Study Center

For a calculatory that will do octal, hexal, and binary arithmetic, try link 4.54 on this textbook's Online Study Center.

Collaborative Learning

This chapter is about numeration systems, and we are about to expand on the topic. Form three groups of students. Each of the groups will be assigned one of the numbers **7, 11,** or **18.**

1. Write your number in binary, octal, and hexadecimal base.

2. Your number is the _____ Lucas number. (What is a Lucas number, anyway?)

3. Your number is the sum of the first _____ Fibonacci numbers. (What is a Fibonacci number?)

4. Your number is the sum of the _____ and _____ prime numbers.

5. Your number is the sum of the _____ and _____ triangular number. (What is a triangular number?)

6. Numerology converts letters to numbers with $A = 1, B = 2, C = 3, \ldots$, as shown below.

1	2	3	4	5	6	7	8	9
A	B	C	D	E	F	G	H	I
J	K	L	M	N	O	P	Q	R
S	T	U	V	W	X	Y	Z	

   a. In numerology, the number **4** corresponds to the word **HE** (H = 8, E = 5 because 8 + 5 = 13 and 1 + 3 = **4**). It also corresponds to **AL** (1 + 3 = **4**) Can you find some two-letter words corresponding to 7? to 11? to 18?

   b. What about some three-letter words corresponding to 7, 11, and 18?

7. Here are some questions that pertain to one and only one of the numbers (7, 11, or 18). Which one?
   a. Number of orifices (holes) in the human head
   b. Atomic number of sodium
   c. One, Allah, Brahma (numerology)
   d. Number of dwarfs in Snow White tale
   e. It, my, no, jar (numerology)
   f. Number of stars in the Big Dipper
   g. Atomic number of argon
   h. Fox, hen, red (numerology)
   i. Atomic number of nitrogen
   j. James Polk, Andrew Jackson, Ulysses S. Grant

Chapter 4 Summary

Section	Item			Meaning	Example
	Egyptian	*Babylonian*	*Roman*		
4.1	|	▼	I	1	
4.1	|||||	▼▼▼▼▼	V	5	
4.1	∩	◄	X	10	
4.1	∩∩∩∩∩	◄◄◄◄◄	L	50	
4.1	☌		C	100	
4.1	☌☌☌☌☌		D	500	
4.1	(glyph)		M	1000	
4.1	(glyph)		$\overline{\text{X}}$	10,000	
4.1	(glyph)		$\overline{\text{C}}$	100,000	
4.1	(glyph)		$\overline{\text{M}}$	1 million	
4.2A	a^n			$\underbrace{a \times a \times \cdots \times a}_{n\ a\text{'s}}$	$10^3 = 10 \times 10 \times 10$
4.2A	$(2 \times 10^2) + (4 \times 10^1) + (5 \times 10^0)$			Expanded form of 245	
4.2B	$a^m \times a^n = a^{m+n}$			Law of exponents	$5^3 \times 5^6 = 5^9$
4.2B	$a^m \div a^n = a^{m-n}$			Law of exponents	$4^7 \div 4^2 = 4^5$
4.3A	43_{five}			$(4 \times 5^1) + (3 \times 5^0)$	
4.3B	10001_{two}			$(1 \times 2^4) + (1 \times 2^0)$	
4.3B	$A_{16}, B_{16}, C_{16}, D_{16}, E_{16}, F_{16}$			10, 11, 12, 13, 14, 15	
4.3B	17_{eight}			$(1 \times 8^1) + (7 \times 8^0)$	

Research Questions

Sources of information for these questions can be found in the Research Bibliography at the end of the book.

1. Write a report on the Egyptian numeration system.

2. Trace the development of the Babylonian numeration system, with special emphasis on the base used.

3. Write a report on the Sumerian numeration system.

4. Write a report on the Mayan calendar. Include the *glyph* (graphic symbol or character) used for each day and the lengths of its months and year. Find information on how to correlate the Gregorian and Mayan calendars.

5. Write a report on the life and works of Mohammed al-Khowarizmi, with special emphasis on the books he wrote.

6. Write a report on A. Henry Rhind and the Rhind papyrus.

7. Write a report on the uses of binary arithmetic in computers.

8. Write a report on Leonardo Fibonacci and his book the *Liber Abaci*.

9. Write a report about the development and use of ASCII.

Chapter 4 Practice Test

1. Write the following in Egyptian numerals:
   a. 63   **b.** 735

2. Write the following in decimal notation:
   a. ∩∩||| **b.** ɔ∩∩|

3. Write the following in Babylonian numerals:
   a. 63   **b.** 735

4. Write the following in decimal notation:
   a. ▼ ◄◄▼▼   **b.** ▼▼ ◄▼

5. Do the multiplication 23×21 using the following:
   a. The Egyptian method of successive duplication
   b. The Egyptian method of mediation and duplation

6. Write the following in decimal notation:
   a. LXVII
   b. $\overline{\text{XLVIII}}$

7. Write the following in Roman numerals:
   a. 53   **b.** 42   **c.** 22,000

8. Write the following in expanded form:
   a. 2507   **b.** 189

9. Write the following in decimal notation:
   a. $(3 \times 10^3) + (7 \times 10^2) + (2 \times 10^0)$ **b.** $(5 \times 10^4) + (9 \times 10^3) + (4 \times 10)$

10. Do the following computations in the usual way and in expanded form:
    a. $75 + 32$ **b.** $56 - 24$

11. Perform the indicated operations, leaving the answers in exponential form:
    a. $3^4 \times 3^8$ **b.** $2^9 \div 2^3$

12. Do the following computations in the usual way and in expanded form:
    a. 83×21 **b.** $54 \div 7$

13. Change the following to decimal notation:
    a. 203_4 **b.** 143_5 **c.** 1101_2

14. Change the following to decimal notation:
    a. 152_8 **b.** $A2C_{16}$

15. Convert the number 33 to the following:
    a. Base 5   **b.** Base 6

16. Convert the following to binary notation:
    a. 39   **b.** 527

17. Convert the number 47 to the following:
    a. Base 8   **b.** Base 16

18. Perform the indicated computations in the binary system.
    a. $\begin{array}{r} 1101_2 \\ + \ 101_2 \end{array}$ **b.** $\begin{array}{r} 1101_2 \\ - \ 111_2 \end{array}$

19. Perform the indicated computations in the binary system.
    a. $1101_2 \times 11_2$ **b.** $10110_2 \times 101_2$

20. Do the indicated computations in the binary system.
    a. $10110_2 \div 11_2$ **b.** $110111_2 \div 110_2$

21. Add $632_8 + 46_8$ in the octal system.

22. Multiply 37_8 by 5_8 in the octal system.

23. Subtract 46_8 from 632_8 in the octal system.

24. Divide 572_8 by 6_8 in the octal system.

25. Do the indicated computations in the hexadecimal system:

a. $2BC_{16} + 5D_{16}$ b. $3C4_{16} \times 2B_{16}$

Answers to Practice Test

ANSWER	IF YOU MISSED	REVIEW		
	Question	Section	Example(s)	Page(s)
1. a. ∩∩∩‖‖ ∩∩∩ **b.** ꝯꝯꝯꝯ∩∩‖‖‖ ꝯꝯꝯ∩‖‖	1	4.1	Table 4.3	180
2. a. 23 **b.** 121	2	4.1	Table 4.3	180
3. a. ▼ ▼▼▼ **b.** ◀▼▼ ◀▼▼▼▼▼	3	4.1	3	183
4. a. 82 **b.** 131	4	4.1	4	183
5. a. \1 21 **b.** 23 ㉑	5	4.1	1, 2	181, 182
\2 42 11 ㊷				
\4 84 5 ㊷				
8 168 2 168				
\16 336 1 ⟨336⟩				
483 483				
6. a. 67 **b.** 48,000	6	4.1	6	184
7. a. LIII **b.** XLII **c.** $\overline{XXII}$	7	4.1	7	185
8. a. $(2 \times 10^3) + (5 \times 10^2) + (0 \times 10) + (7 \times 10^0)$	8	4.2	1	190
b. $(1 \times 10^2) + (8 \times 10) + (9 \times 10^0)$				
9. a. 3702 **b.** 59,040	9	4.2	2	190
10. a. 75 $(7 \times 10) + 5$	10	4.2	3, 4	191
$+ 32$ $+ (3 \times 10) + 2$				
107 $(10 \times 10) + 7$				
$= (1 \times 10^2) + 7$				
$= 107$				
b. 56 $(5 \times 10) + 6$				
$- 24$ $(-)(2 \times 10) + 4$				
32 $(3 \times 10) + 2$				
$= 32$				
11. a. $3^4 \times 3^8 = 3^{4+8} = 3^{12}$	11	4.2	5	191
b. $2^9 \div 2^3 = 2^{9-3} = 2^6$				
12. a. 83 $(8 \times 10) + 3$	12	4.2	7, 8	192
$\times 21$ $\times (2 \times 10) + 1$				
83 $(8 \times 10) + 3$				
166 $(16 \times 10^2) + (6 \times 10)$				
1743 $(16 \times 10^2) + (14 \times 10) + 3$				
$= (1 \times 10^3) + (6 \times 10^2)$				
$+ (1 \times 10^2) + (4 \times 10) + 3$				
$= (1 \times 10^3) + (7 \times 10^2)$				
$+ (4 \times 10) + 3 = 1743$				

ANSWER	IF YOU MISSED	REVIEW		
	Question	Section	Example(s)	Page(s)
b. $\dfrac{\quad 7R5}{7\overline{)54}}$ $\dfrac{49}{5}$	12	4.2	7, 8	192
$\dfrac{(7 \times 10^0)\,R\,(5 \times 10^0)}{7 \times 10^0\overline{)(5 \times 10) + (4 \times 10^0)}}$ $\dfrac{(4 \times 10) + (9 \times 10^0)}{(5 \times 10^0)}$				
13. a. 35 **b.** 48 **c.** 13	13	4.3	2, 3	197, 198
14. a. 106 **b.** 2604	14	4.3	4	199
15. a. 113_5 **b.** 53_6	15	4.3	5	199
16. a. 100111_2 **b.** 1000001111_2	16	4.3	5	199
17. a. 57_8 **b.** $2F_{16}$	17	4.3	6	200
18. a. 10010_2 **b.** 110_2	18	4.4	1–4	205–206
19. a. 100111_2 **b.** 1101110_2	19	4.4	5, 6	206, 207
20. a. $111_2\,R\,1_2$ **b.** $1001_2\,R\,1_2$	20	4.4	7, 8	207–208
21. 700_8 **22.** 233_8	21, 22	4.5	1–4	212–213
23. 564_8 **24.** 77_8	23, 24	4.5	5–8	213–214
25. a. 319_{16} **b.** $A1EC_{16}$	25	4.5	9–12	216

Throughout its history, there have been several changes made to the game of basketball to increase its pace. After an extensive study and the use of ratios, the 24-second clock was added to the game. In Section 5.3, you will study ratios and their many uses as a problem-solving tool.

Number Theory and the Real Numbers

In Chapter 4 we took a journey through time and the numbers of antiquity. Now we shall explore more recent developments in the history of mathematics by studying the **natural** or **counting** numbers, their properties, and their uses. We will also look closely at some special natural numbers: the **primes.** Since ancient times, mathematicians have been fascinated by the divisors of numbers and on that basis classified the numbers as abundant, perfect, or deficient. Prime numbers have only two divisors, themselves and 1. The search for prime numbers continues today using supercomputers, and the results are used in such sophisticated areas as cryptography, the art or science dealing with making information unintelligible and for restoring the encrypted information to intelligible form.

As useful as they are, the natural numbers are not enough for the needs of everyday life. We need numbers to measure such things as subzero temperatures, yardage losses in a football game, and financial losses in the stock market. In other words, we need **negative** numbers, which we shall consider when we study the **integers.**

Division of one integer by another yields a new number, a **rational** number. Rational numbers can also be written as decimals. We might imagine that the rationals would be the end of this numerical journey. However, the Pythagoreans, a secret society of scholars in ancient Greece (ca. 540–500 B.C.), made a stunning discovery. The **irrational** numbers defied their knowledge of number properties, because they could not write them as the ratio of two whole numbers.

The union of the rationals and the irrationals yields the **real** numbers. One of the most interesting real numbers is π, and we shall trace different attempts to find its value. We shall then learn how to operate with irrational numbers written using radical notation. Finally, we shall see how to find the sums of certain sequences of numbers. In Chapter 6 we shall find out how all these numbers are used in a new world: algebra.

Online Study Center

For links to Internet sites related to Chapter 5, please access **college.hmco.com/PIC/bello9e** and click on the Online Study Center icon.

223

5.1 Number Theory: Primes and Composites

HUMAN SIDE OF MATH

Leopold Kronecker

(1823–1891)

was born on December 7, 1823, in Liegnitz, Prussia. Throughout his schooling, Kronecker's performance was diverse and brilliant; however, his greatest talent was in mathematics.

He finished an outstanding memoir on the theory of equations, which was published in 1853. Throughout his mathematical work, much of it on number theory and the theory of equations, he tried to make concise and expressive formulas tell the whole story. Most of Kronecker's papers have a strong arithmetic flavor; he wanted to explain everything in terms of the whole numbers and in a finite number of steps, a style that has had a great influence on modern mathematics.

Looking Ahead
In this chapter we will look at number systems from the natural numbers to the real numbers.

Emirps and Primes

Do you know what an *emirp* is? It is a prime number that turns into another prime number when its digits are reversed; 13 and 31, 37 and 73 are emirps. A **prime number** is a number that has exactly two distinct factors, itself and 1. Prime numbers are used in cryptography, the encoding of secret messages. The lowest prime is 2 and the largest known prime at this time is $2^{32,582,657} - 1$, discovered by Dr. Curtis Cooper, Dr. Steven Boone, and the Great Internet Research Prime Search (GIMPS) on September 11, 2006. Here you will learn about prime and composite (not prime) numbers, factors, and divisors and their uses.

On September 11, 2006, Dr. Curtis Cooper and Dr. Steven Boone, professors at Central Missouri State University, discovered the largest prime known at this time, $2^{32,582,657} - 1$.

The proper divisors of a number have fascinated mathematicians as far back as the Pythagoreans, who classified numbers according to the sum of their proper divisors (all the divisors, except the number itself) as follows:

Sum of Number's Proper Divisors	*Classification of Number*
More than the number	*Abundant* (12 and 20)
Equal to the number	*Perfect* (6 and 28)
Less than the number	*Deficient* (14 and 15)

You will learn more about prime, perfect, abundant, and deficient numbers in the Discovery section of Exercises 5.1. Can all numbers be written as a sum of primes: $10 = 5 + 5$, $18 = 11 + 7$, and $21 = 19 + 2$? (See problem 71, Exercises 5.1.) Here is another problem. 17 is prime and 19 is prime; their difference is 2. Also, 43 is prime and 41 is prime; their difference is 2. These pairs of numbers are called **twin primes.** Are there infinitely many twin primes? Nobody knows!

How can you find whether the number of primes is infinite? (See problem 15, Exercises 5.1.) One way is to find a formula that generates primes. Here is one such formula: $n^2 - n + 17$. When $n = 1$, $1 - 1 + 17 = 17$, a prime number. When $n = 2$, $2^2 - 2 + 17 = 4 - 2 + 17 = 19$, a prime number. Will this formula generate primes all the time? Can you find another formula? ▶

Look at the numbers on the keys of your calculator. They can be used to form a special set of numbers, called the set N of **natural** or **counting numbers** $\{1, 2, 3, \ldots\}$. In this section you will see some uses and properties of the set of natural numbers and one of its important subsets: the set of prime numbers.

A. Using the Natural Numbers

The natural numbers can be used in three different ways: as cardinals (for counting or showing how many), as ordinals (for indicating order in a series), or for identification. Thus, the number of elements in $A = \{a, b, c, d\}$ is 4. We then say that the **cardinal number of** A is 4 and write $n(A) = 4$. Read as "n of A is 4." Here, the number 4 is used as a **cardinal number.**

Numbers can also be used to assign an *order,* or position, to the elements of a set, that is, to indicate which element is *first, second, third,* and so on. We then refer to these numbers as **ordinal numbers.** Finally, numbers can be used for **identification.** Our Social Security numbers, passport numbers, and savings account numbers are used for identification purposes.

EXAMPLE 1 ▶ Classifying Numbers

Determine whether the underlined word is used as a cardinal or ordinal number or for identification.

(a) If she kissed you once, would she kiss you two times?

(b) My account number is 123456.

(c) This is my first and last warning.

Solution

(a) Cardinal   (b) Identification   (c) Ordinal   ■

Now consider some relationships among natural numbers and the terminology used to express these relationships. Suppose there are 40 students in your class and you wish to divide them into 5 equal groups. Each group will have 8 students, and no student will be left out. Thus, 40 divided by 5 is exactly 8 with no remainder. In other words, 5 is an **exact divisor** of 40, or more briefly, 5 is a **divisor** or **factor** of 40. Similarly, 63 divided by 7 is exactly 9, so 7 is a divisor, or factor, of 63. However, 7 is not a divisor, or factor, of 60 because 60 is not exactly divisible by 7.

B. Prime and Composite Numbers

If 40 is written as a product of factors, the product—for example, 5×8—is called a **factorization** of 40. The 5 and 8 are factors (divisors) of 40, and the number 40 is a **multiple** of each of these factors. Note that 8 itself has divisors other than 8 and 1 (2 and 4), but the only divisors of 5 are itself and 1, so 5 is a prime number.

Definition of Prime and Composite Numbers

A natural number with exactly two distinct divisors (1 and itself) is called a **prime number** or is said to be **prime.**

Any number with more than two distinct divisors is called a **composite number** or is said to be **composite.**

The number 1 is by definition neither prime nor composite, because it has only one divisor (itself).

According to the definition, 5 is a prime number, and 40 and 8 are composite.

In the third century B.C., the Greek mathematician Eratosthenes devised the following procedure to find prime numbers: He listed all the numbers from 2

	2	3	~~4~~	5	~~6~~	7	~~8~~	~~9~~	~~10~~
11	~~12~~	13	~~14~~	~~15~~	~~16~~	17	~~18~~	19	~~20~~
~~21~~	~~22~~	23	~~24~~	~~25~~	~~26~~	~~27~~	~~28~~	29	~~30~~
31	~~32~~	~~33~~	~~34~~	~~35~~	~~36~~	37	~~38~~	~~39~~	~~40~~
41	~~42~~	43	~~44~~	~~45~~	~~46~~	47	~~48~~	~~49~~	~~50~~

FIGURE 5.1

up to a given number (say, from 2 through 50, as in Figure 5.1). He concluded that 2 is prime, but every multiple of 2 (4, 6, 8, and so on) is composite because it is divisible by 2. He then circled 2 as a prime and crossed out all multiples of 2. Next, because 3 is prime, he circled it and crossed out all following multiples of 3. Similarly, he circled 5 (the next prime) and crossed out its multiples. He continued this process until he reached 11, the first prime whose square exceeds 50. We note that all the multiples of 11 ($11 \cdot 2$, $11 \cdot 3$, and $11 \cdot 4$) were eliminated earlier when the multiples of 2 and 3 were crossed out. In the same way, the multiples of any prime larger than 11 were also eliminated. Hence, all the numbers left in the table were primes. Eratosthenes then circled those numbers. If the table had gone as far as 200, he would have had to continue his stepwise exclusion of composite numbers until he came to 17, the first prime whose square exceeds 200. (Can you explain why?) Because it sifts out prime numbers, the preceding method is called the **Sieve of Eratosthenes.**

The set of primes is not finite; that is, there is no largest prime. This fact is proven in problem 15 of Exercises 5.1.

The applications of prime numbers that we shall consider depend on a theorem which we now state without proof.

> **THEOREM 5.1 Fundamental Theorem of Arithmetic**
>
> Every composite number can be expressed as a unique product of primes (disregarding the order of the factors).

Thus,

$$180 = 2 \cdot 2 \cdot 3 \cdot 3 \cdot 5 = 2^2 \cdot 3^2 \cdot 5$$

and

$$92 = 2 \cdot 2 \cdot 23 = 2^2 \cdot 23$$

C. Factorization

The prime factorization of a composite number can be found by "dividing out" the prime factors of the number, starting with the smallest factor. Thus,

$$40 = 2 \cdot 20 = 2 \cdot 2 \cdot 10 = 2 \cdot 2 \cdot 2 \cdot 5 = 2^3 \cdot 5$$

and

$$63 = 3 \cdot 21 = 3 \cdot 3 \cdot 7 = 3^2 \cdot 7$$

Some students prefer to keep track of the factors by using a factor tree. Thus, to find the prime factorization of 40 and 63, we make two trees as shown.

Divide 40 by the smallest prime, 2.	$40 = 2 \times 20$
Divide 20 by 2 to find the factors of 20.	$20 = 2 \times 10$
Divide 10 by 2 to find the prime factors of 10.	$10 = 2 \times 5$

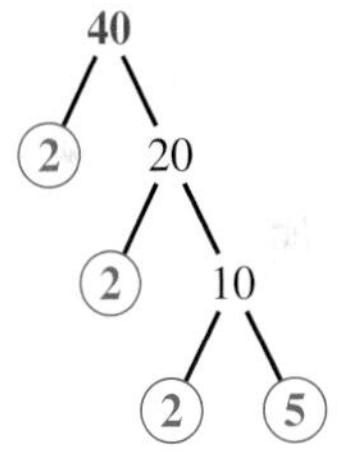

Thus,

$$40 = ②\times②\times②\times⑤ = 2^3 \times 5$$

Similarly,

Divide 63 by the smallest prime that divides 63, that is, by 3.	$63 = 3 \times 21$
Now, divide 21 by 3 to find the prime factors of 21.	$21 = 3 \times 7$

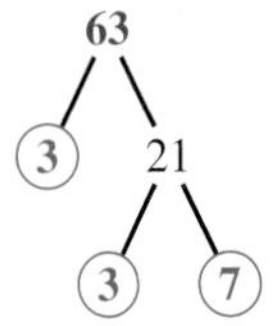

Online Study Center

To further explore techniques for factoring primes, access link 5.1.2 on this textbook's Online Study Center.

Thus,

$$63 = ③ \times ③ \times ⑦ = 3^2 \times 7$$

To save time and space, we can write these computations using repeated divisions as follows:

Divide by 2 →	2	40	Divide by 3 →	3	63
Divide by 2 →	2	20	Divide by 3 →	3	21
Divide by 2 →	2	10	Divide by 7 →	7	7
Divide by 5 →	5	5			1
		1			

Reading downward, as indicated by the arrows, we get

$$40 = 2 \cdot 2 \cdot 2 \cdot 5 = 2^3 \cdot 5 \qquad \text{and} \qquad 63 = 3 \cdot 3 \cdot 7 = 3^2 \cdot 7$$

as before.

EXAMPLE 2 ▶ Writing a Number as a Product of Primes

Write 1440 as a product of primes.

Solution

```
2 | 1440
2 |  720
2 |  360
2 |  180
2 |   90
3 |   45
3 |   15
5 |    5
          1
```

Hence, $1440 = 2 \cdot 2 \cdot 2 \cdot 2 \cdot 2 \cdot 3 \cdot 3 \cdot 5 = 2^5 \cdot 3^2 \cdot 5$. ▪

EXAMPLE 3 ▶ Prime or Composite?

Is the number 197 prime or composite?

Solution

To answer this question, we must determine whether 197 has factors other than 1 and itself. Thus, we try dividing by the consecutive prime numbers 2, 3, 5, 7, 11, 13, ... Since 197 is not even, it is not divisible by 2. It is also not divisible by 3 or 5 or 7 or 11 or 13, as we can find by trial. Where do we stop?

In trying to divide 197 by 3, 5, 7, 11, and 13, we notice that the quotient is greater than the trial divisor each time (for example, dividing 197 by 3 gives a quotient of 65). The Sieve of Eratosthenes (Figure 5.1) gives the next prime as 17. If we try to divide 197 by 17, we get a quotient (11) that is *less* than 17 and a remainder showing that 17 is not a factor of 197. Consequently, we need go no further (can you see why?), and we have shown that 197 is prime. ▪

Online Study Center

To determine if a number is prime, access link 5.1.3 on this textbook's Online Study Center.

D. Divisibility Rules

Finding the prime factorization of a number or determining whether the number is prime or composite can require a number of trial divisions. Some of these divisions can be avoided if we know the **divisibility rules** demonstrated in Table 5.1.

TABLE 5.1

Divisible by	Test	Example
2	The number ends in 0, 2, 4, 6, or 8.	345,678 ends in 8, so it is divisible by 2.
3	The sum of the number's digits is divisible by 3.	258 is divisible by 3, since $2 + 5 + 8 = 15$, which is divisible by 3.
5	The number ends in 0 or 5.	365 ends in 5, so it is divisible by 5.

There are other rules for divisibility by larger primes such as 7 and 11, but most people feel that these rules are too complicated to bother with; it is easier to check by direct division. To see other divisibility rules (for 4, 6, 8, 9, 10, and 12), see the table in the Using Your Knowledge section of Exercises 5.1.

EXAMPLE 4 ▶ **Determining Divisibility by 2, 3, or 5**

Which of the following numbers is divisible by 2, 3, or 5?

(a) 8925
(b) 39,120
(c) 2553

Solution

(a) 8925 is not divisible by 2. The sum of the digits $8 + 9 + 2 + 5 = 24$ is divisible by 3; hence, 8925 is divisible by 3. Since 8925 ends in 5, it is divisible by 5.

(b) 39,120 is divisible by 2, 3, and 5. Note that $3 + 9 + 1 + 2 = 15$ is divisible by 3, so 39,120 is divisible by 3. Since 39,120 ends in a 0, the number 39,120 is divisible by 5.

(c) 2553 is not divisible by 2 or 5, but is divisible by 3 because the sum of its digits, $2 + 5 + 5 + 3 = 15$, is divisible by 3. ■

PROBLEM SOLVING

Divisibility

① **Read** the problem. What do *remainder* and *divided into* mean?

② **Select** the unknown.

③ **Think** of a plan. What can you say when a number leaves a remainder of 2 when divided into 50? What about when the remainder is 3 when divided into 63?

④ **Use** the information to carry out the plan. You need the divisors of 48 and 60 larger than 3.

⑤ **Verify** the answer.

TRY EXAMPLE 5 NOW.

How many numbers leave a remainder of 2 when divided into 50 and a remainder of 3 when divided into 63?

3 divided into 50 means $\frac{50}{3}$. The answer is 16 with a remainder of 2.

You need to find all numbers with the property stated in the problem.

If the remainder is 2 when the number is divided into 50, then $50 - 2 = 48$ is divisible by the number.

Similarly, $63 - 3 = 60$ is divisible by the number. Moreover, the number must be larger than 3, the larger of the two remainders.

The divisors of 48 larger than 3 are

48,   24,   16,   12,   8,   6,   and   4

Of these, the divisors of 48 *and* 60 larger than 3 are 12, 6, and 4. Thus, there are exactly three numbers that meet the given conditions: 12, 6, and 4.

If 12, 6, and 4, respectively, are divided into 50, the remainders are 2. If 12, 6, and 4 are divided into 63, the remainders are 3. Try it!

Cover the solution, write your own solution, and then check your work.

EXAMPLE 5 ▶ Finding Multiples

Find the smallest multiple of 5 that leaves a remainder of 2 when divided by 3 and a remainder of 6 when divided by 7.

Solution

Write all the multiples of 5 larger than 6, the larger of the two remainders, and select the multiples that satisfy the two given conditions.

Multiples of 5	10	15	20	25	30	35	· · ·
Remainder when divided by 3	1	0	2	1	0	2	· · ·
Remainder when divided by 7	3	1	6	4	2	0	· · ·

The desired number is 20.   ■

E. Using Prime Factorization: GCF and LCM

The number 8 is the **greatest common factor** of 16 and 24. We abbreviate greatest common factor by **GCF** and write GCF(16, 24) = 8. A fraction a/b is said to be in **simplest form** if GCF(a, b) $= 1$. In this case we say that a and b are **relatively prime.** If GCF(a, b) $\neq 1$, the fraction is reduced to lowest terms by dividing a and b by their GCF. The most commonly used way of finding the GCF of a set of numbers is as follows:

How to Find the GCF

1. Write the prime factorization of each number.

2. Select *all* the primes that are common to *all* the factorizations and apply to each such prime the smallest exponent to which it occurs.

3. The product of the factors selected in step 2 is the GCF of all the numbers.

The GCF is easier to find if you write the same primes in a column, as in the next example.

EXAMPLE 6 ▶ **Finding GCFs to Reduce Fractions**

Find the GCF(216, 234), and then reduce $\frac{216}{234}$ to lowest terms.

Solution

Step 1 We write the prime factorization of each number.

Pick the one with the smallest exponent in each column.

$$216 = 2^3 \cdot 3^3$$
$$234 = 2 \cdot 3^2 \cdot 13$$

Step 2 We select the *common* prime factors with their *smallest* exponents; these factors are 2 and 3^2.

Step 3 GCF(216, 234) $= 2 \cdot 3^2 = 18$.

To reduce the fraction $\frac{216}{234}$ to lowest terms, we divide out the GCF(216, 234), which we found to be 18. Thus, we write

$$\frac{216}{234} = \frac{18 \cdot 12}{18 \cdot 13} = \frac{12}{13}$$

Notice that we can obtain the 12 and the 13 from the respective prime factorizations. We just use all the factors except those in the GCF. ∎

A second use of prime factorization occurs in the addition of fractions, where we need to find a common denominator. For example, to add $\frac{3}{8}$ and $\frac{5}{12}$, we must first select a common denominator for the two fractions. Such a denominator is any natural number that is exactly divisible by both 8 and 12. Thus, we could use $8 \cdot 12 = 96$. However, it is usually most efficient to use the **least** (smallest) **common denominator (LCD).** In our case, we note that

$$8 = 2^3 \qquad \text{and} \qquad 12 = 2^2 \cdot 3$$

GRAPH IT

The numerical calculations in this section can be done with a grapher (graphing calculator), but be aware that calculator procedures vary. (When in doubt, read the manual!) Start at the home screen by pressing [ON] [2ND] [MODE]. To find the LCM of 8 and 12, tell the grapher you want a calculation by pressing [MATH] [▶]. Specify which type by pressing [8], which calculates LCMs; press 8 [.] 12 [)] [ENTER] to get the answer 24 as shown in the window. To get the GCD (or gcd) follow the same steps but press [9] instead of [8]. Can you find the LCM of three numbers with your grapher? Moral: You have to know your math, even with a great grapher!

```
lcm(8,12)
                    24
gcd(8,12)
                     4
```

To have a common multiple of 8 and 12, we must include 2^3 and 3 (remember we are looking for a multiple of *both* numbers), at the very least. Thus, we see that the **least common multiple (LCM)** of 8 and 12 is $2^3 \cdot 3 = 24$ (see margin), so the LCD of $\frac{3}{8}$ and $\frac{5}{12}$ is 24.

The procedure that we used to find the LCM(8, 12) in the Graph It can be generalized to find the LCM of any two or more natural numbers.

How to Find the LCM

1. Write the prime factorization of each number.

2. Select every prime that occurs, raised to the *highest* power to which it occurs, in these factorizations.

3. The product of the factors selected in step 2 is the LCM.

EXAMPLE 7 ▶ **Finding LCMs to Add Fractions**

Find the LCM(18, 21, 28), and use it to add $\frac{1}{18} + \frac{1}{21} + \frac{1}{28}$.

Solution

Step 1

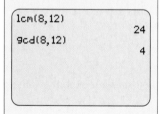

Pick the one with the greatest exponent in each column.

$$18 = 2 \cdot 3^2$$
$$21 = \quad 3 \cdot 7$$
$$28 = 2^2 \quad \cdot 7$$

Step 2 We select every prime factor (not just the common factors) with the greatest exponent to which it occurs and obtain 2^2, 3^2, and 7.

Step 3 LCM(18, 21, 28) $= 2^2 \cdot 3^2 \cdot 7 = 252$

We now use the LCM 252 to replace the given fractions by equivalent fractions with 252 as their common denominator. To obtain these equivalents, we refer to the factored forms of the denominators and the LCM. Since the LCM is

$$252 = 2^2 \cdot 3^2 \cdot 7 \qquad \text{and} \qquad 18 = 2 \cdot 3^2$$

we have to multiply 18 by $2 \cdot 7 = 14$ to get 252. Hence, we multiply the fraction $\frac{1}{18}$ by 1 in the form $\frac{14}{14}$ to get

$$\frac{1 \times 14}{18 \times 14} = \frac{14}{252}$$

Note: $\frac{1}{18}$ is *not* changed; it is written as the equivalent fraction $\frac{14}{252}$.

The same procedure can be used for the other two fractions.

$$\frac{1}{21} = \frac{1 \times 12}{21 \times 12} = \frac{12}{252} \qquad \frac{1}{28} = \frac{1 \times 9}{28 \times 9} = \frac{9}{252}$$

Then we add.

$$\frac{1}{18} + \frac{1}{21} + \frac{1}{28} = \frac{14}{252} + \frac{12}{252} + \frac{9}{252} = \frac{14 + 12 + 9}{252} = \frac{35}{252}$$

Same denominators

Since $35 = 5 \times 7$ and $252 = 36 \times 7$, we can reduce the last fraction by dividing out the 7. This gives a final answer of $\frac{5}{36}$. ∎

EXAMPLE 8 ▶ Subtraction of Fractions

Do the subtraction $\frac{7}{24} - \frac{5}{84}$.

Solution

To do this subtraction, we have to change the fractions to equivalent fractions with a common denominator. We factor the denominators to get

$$24 = 2^3 \cdot 3 \qquad \text{and} \qquad 84 = 2^2 \cdot 3 \cdot 7$$

Thus, the LCM of 24 and 84 is $2^3 \cdot 3 \cdot 7 = 168$. Consequently, we multiply the first fraction by $\frac{7}{7}$ and the second by $\frac{2}{2}$ to write each fraction with 168 as a denominator. Thus,

$$\frac{7}{24} - \frac{5}{84} = \frac{7 \cdot 7}{24 \cdot 7} - \frac{5 \cdot 2}{84 \cdot 2}$$

$$= \frac{49 - 10}{168} = \frac{39}{168} = \frac{3 \cdot 13}{3 \cdot 56} = \frac{13}{56}$$ ∎

PROBLEM SOLVING

Hot Dogs and Buns

❶ Read the problem.

Can you ever apply these ideas to anything besides adding and subtracting fractions? You bet your buns! Have you noticed that hot dogs come in packages of 10, but buns come in packages of 8 (or 12)? What is the smallest number of packages of hot dogs (10 to a package) and buns (8 to a package) you must buy so that you have as many hot dogs as you have buns?

❷ Select the unknown.

We want to find the smallest number of packages of hot dogs and the smallest number of packages of buns you have to buy so that you have the same number of hot dogs and buns.

❸ Think of a plan.

What we need is the same number of hot dogs and buns; that is, we need the smallest multiple (LCM) of 10 and 8.

❹ Use the LCM procedure.

$$10 = 2 \times 5 \qquad \text{and} \qquad 8 = 2^3$$

Thus, the LCM is $2^3 \times 5 = 40$. But 40 *is not* the answer. We were asked how many packages of hot dogs (10 to a package) and buns (8 to a package) we needed. Since we need 40 hot dogs and 40 buns, we need *4* packages of hot dogs and *5* packages of buns.

❺ **Verify** the solution.

4 packages of hot dogs contain 40 hot dogs.
5 packages of buns contain 40 buns.

TRY EXAMPLE 9 NOW.

Cover the solution, write your own solution, and then check your work.

George wonders why there are buns left!

EXAMPLE 9 ▶ Hot Dogs and Buns

What is the smallest number of packages of hot dogs (10 to a package) and buns (12 to a package) you must buy so that you have as many hot dogs as you have buns?

Solution
As before, we write 10 and 12 in factored form.

$$10 = 2 \times 5 \quad \text{and} \quad 12 = 2^2 \times 3$$

The LCM is $2^2 \times 3 \times 5 = 60$. Thus, we need $\frac{60}{10} = 6$ packages of hot dogs and $\frac{60}{12} = 5$ packages of buns. ∎

EXAMPLE 10 ▶ Reduced Distances

A motorist on a 2500-mi trip drove 500 mi the first day, 600 mi the second day, and 750 mi the third day. What (reduced) fraction of the total distance did he drive each of these days?

Solution

First day $\quad \dfrac{500}{2500} = \dfrac{1 \times 500}{5 \times 500} = \dfrac{1}{5}$

Second day $\quad \dfrac{600}{2500} = \dfrac{6 \times 100}{25 \times 100} = \dfrac{6}{25}$

Third day $\quad \dfrac{750}{2500} = \dfrac{3 \times 250}{10 \times 250} = \dfrac{3}{10}$ ∎

The procedures used to find the GCF and LCM of two or more numbers are very similar. If you are asked to find the GCF *and* LCM of two numbers, you can follow the procedure used below to find the GCF and LCM of 42, 28, and 210.

		42	28	210
2		21	14	105
7		3	②	15
3				

No number divides → 21, 14, and 105 (except 1).

1　　2　　5

The LCM is $2 \times 7 \times 3 \times 1 \times 2 \times 5 = 420$.

> **Finding the GCF and LCM**
>
> 1. Write the numbers over a horizontal line and divide them by a prime divisor that is common to them all.
>
> 2. Repeat the procedure with the quotients until there is no longer any common divisor.
>
> 3. The product of all the divisors in steps 1 and 2 is the GCF (in the example, $2 \times 7 = 14$).
>
> 4. Divide by a prime factor common to two or more numbers. If any of the numbers is not divisible by this prime, circle the number and carry it down to the next line.
>
> 5. The LCM is the product of all the divisors in steps 1, 2, and 4 and the numbers in the final row.

EXERCISES 5.1

A Using the Natural Numbers

In problems 1–5, identify the underlined items as cardinal numbers, as ordinal numbers, or for identification only.

1. My telephone number is <u>123-7643</u>.

2. This is the <u>second</u> problem in this exercise.

3. It takes <u>two</u> to tango.

4. <u>One</u>, <u>two</u>, <u>three</u>, go!

5. <u>First</u> National Bank is number <u>one</u>.

6. Three numbers (14, 5696, and 0166) are circled on the check. Identify each of these numbers as a cardinal number, an ordinal number, or for identification only.

B Prime and Composite Numbers

7. Continue the Sieve of Eratosthenes (Figure 5.1 on page 226), and find all the primes between 50 and 100.

In problems 8–11, use Figure 5.1 and the results of problem 7 to find how many primes there are between the given numbers.

8. 1 and 25

9. 25 and 50

10. 50 and 75

11. 75 and 100

12. Refer to Figure 5.1, and find
   a. the smallest prime.   **b.** an even prime.

13. Refer to Figure 5.1.
   a. Find a pair of primes that are consecutive counting numbers.
   b. Can there be a second pair of primes that are consecutive counting numbers? Why?

14. Primes that differ by 2 are called **twin primes.** The smallest twin primes are 3 and 5. Refer to Figure 5.1 and find two other pairs of twin primes.

15. To show that there is no largest prime, Euclid (in about 300 B.C.) gave the following *proof by contradiction:* Assume that there is a largest prime; call it P. Now form a number—say, m—by taking the product of all the primes from 2 through P and adding 1 to the result. This gives $m = (2 \cdot 3 \cdot 5 \cdot 7 \cdot \cdots \cdot P) + 1$.
   a. m is not divisible by 2. Why?
   b. m is not divisible by 3. Why?
   c. m is not divisible by 5. Why?
   d. m is not divisible by any of the primes from 2 through P. Why?

e. m is greater than P, so it cannot be a prime. Why?

f. m cannot be a composite number. Why?

Now we have a contradiction! Since m is a natural number greater than 1, it must be either prime or composite. But if our assumption that P is the largest prime is correct, then m cannot be either prime or composite. Therefore, the assumption is invalid, and there is no largest prime.

Ⓒ Factorization

In problems 16–21, find all the factors (divisors) of each number.

16. 28   **17.** 50   **18.** 119

19. 128   **20.** 1365   **21.** 1001

In problems 22–27, find the prime factorization of each number or state if the number is prime.

22. 24   **23.** 41   **24.** 82

25. 91   **26.** 191   **27.** 148

In problems 28–31, find the natural number whose prime factorization is given.

28. $2 \cdot 3^2 \cdot 5^2$ **29.** $2 \cdot 5 \cdot 7^2$

30. $2 \cdot 3 \cdot 5 \cdot 11$ **31.** $2^4 \cdot 3 \cdot 5^2$

Ⓓ Divisibility Rules

In problems 32 and 33, determine whether each number is divisible by 2, 3, or 5.

32. a. 468   **b.** 580   **c.** 795   **d.** 3942

33. a. 6345   **b.** 8280   **c.** 11,469,390

34. Find the smallest whole number multiple of 4 that leaves a remainder of 1 when divided by either 5 or 7.

35. How many whole numbers leave a remainder of 1 when divided into either 23 or 45?

Ⓔ Using Prime Factorization: GCF and LCM

In problems 36–45, find the GCF of the given numbers or state that the numbers are relatively prime.

36. 14 and 210   **37.** 135 and 351

38. 315 and 350   **39.** 147 and 260

40. 368 and 80   **41.** 282 and 329

42. 12, 18, and 30   **43.** 12, 15, and 20

44. 285, 315, and 588   **45.** 100, 200, and 320

In problems 46–52, reduce each fraction to lowest terms.

46. $\frac{80}{92}$ **47.** $\frac{62}{88}$ **48.** $\frac{140}{280}$ **49.** $\frac{156}{728}$

50. $\frac{315}{420}$ **51.** $\frac{96}{384}$ **52.** $\frac{716}{4235}$

In problems 53–63, find the LCM of the given numbers and use it to add the fractions.

53. 15 and 55; $\frac{1}{15} + \frac{1}{55}$

54. 17 and 136; $\frac{1}{17} + \frac{1}{136}$

55. 32 and 124; $\frac{3}{32} + \frac{1}{124}$

56. 124 and 155; $\frac{3}{124} + \frac{1}{155}$

57. 180 and 240; $\frac{1}{180} + \frac{1}{240}$

58. 284 and 568; $\frac{3}{284} + \frac{1}{568}$

59. 12, 18, and 30; $\frac{1}{12} + \frac{1}{18} + \frac{1}{30}$

60. 12, 15, and 20; $\frac{1}{12} + \frac{1}{15} + \frac{1}{20}$

61. 285, 315, and 588; $\frac{1}{285} + \frac{1}{315} + \frac{1}{588}$

62. 100, 200, and 320; $\frac{1}{100} + \frac{1}{200} + \frac{1}{320}$

63. 200, 300, and 420; $\frac{1}{200} + \frac{1}{300} + \frac{1}{420}$

Applications

64. In a recent year, \$1340 billion was spent on health care in the United States. Of this amount, $\frac{1}{10}$ was used for drugs, eyeglasses, and necessities; $\frac{1}{4}$ for professional services; $\frac{2}{25}$ for nursing homes; $\frac{3}{20}$ for miscellaneous; and the rest for hospital costs. What fraction of the money was used for hospital costs? (Source: U.S. Centers for Medicare and Medicaid Services.)

65. A car depreciates an average of $\frac{1}{4}$ of its value the first year, $\frac{3}{20}$ the second, and $\frac{1}{10}$ the third. What fraction of its original value is the car worth after the third year?

66. A bookstore owner spent $\frac{3}{5}$ of her initial investment on bookcases, $\frac{1}{10}$ on office machines, and the rest on books. What proportion of her initial investment was spent on books?

67. The operating ratios (budgets) for hardware stores in the United States suggest that $\frac{3}{5}$ of their sales go to pay for merchandise, $\frac{1}{4}$ for expenses, and the rest is profit. What proportion of sales is profit?

68. Do you know what is in your trash? $\frac{1}{5}$ is yard waste; $\frac{1}{10}$ is food waste; $\frac{1}{4}$ is metal, glass, and plastics; $\frac{1}{8}$ is wood, rubber, and miscellaneous. The rest is paper and cardboard, which can be recycled. What fraction of the garbage is paper and cardboard? (Source: www.epa.gov/epaoswer/non-hw/reduce/catbook/what.htm.)

69. How do you use your water indoors? On the average, $\frac{3}{10}$ is used for showering and bathing, $\frac{1}{5}$ is used for laundry, $\frac{1}{20}$ is lost to leaky toilets, $\frac{3}{20}$ is used for dish washing and faucets. Where does the rest go? Toilet flushing! What fraction of the water is used for toilet flushing? (Source: *Healthy Environment–Healthy Me* [New Brunswick, N.J.: Rutgers University, 1991].)

70. Mathematicians dream of finding a formula that will yield an infinite number of primes only when natural numbers are substituted into the formula. No such formula has ever been found, but there are formulas that give a large number of primes. One such formula is $n^2 - n + 41$. This formula gives primes for all natural numbers n less than 41. The following are examples:

For $n = 1$, $1^2 - 1 + 41 = 41$, a prime.
For $n = 2$, $2^2 - 2 + 41 = 4 - 2 + 41 = 43$, a prime.
For $n = 3$, $3^2 - 3 + 41 = 9 - 3 + 41 = 47$, a prime.

  a. What prime is obtained if $n = 4$?
  b. What prime is obtained if $n = 5$?
  c. What number is obtained if $n = 41$?
  d. Is the number obtained in part (**c**) prime?

71. There are many **conjectures** (unproved theories) regarding primes. One of these was transmitted to Leonhard Euler in 1742 by C. Goldbach. Goldbach conjectured that every even natural number except 2 could be written as the sum of two primes. For example, $4 = 2 + 2$, $6 = 3 + 3$, $8 = 3 + 5$, $10 = 5 + 5$, and so on.
  a. Write 100 as the sum of two primes.
  b. Write 200 as the sum of two primes. (*Hint:* Look at the given examples.)

72. In 1931, the Russian mathematician Lev Schnirelmann proved that every natural number can be written as the sum of not more than 300,000 primes. Another mathematician, I. M. Vinogradoff, has proved that every sufficiently large natural number can be expressed as the sum of at most four primes!
  a. Try to write 20 as the sum of three primes.
  b. Try to write 43 as the sum of four primes.

 In Other Words

73. Look at the definition of prime and composite and explain why the number 1 cannot be classified either as a prime number or as a composite number.

74. In view of your explanation in problem 73, how would you classify the number 1?

75. To determine that 211 is prime, start by trying to divide it by 2, 3, 5, and 7. What is the highest prime you will have to try in order to determine that 211 is prime? Explain.

76. Explain why Euclid's proof that there is no largest prime (problem 15) also shows that the number of primes is infinite.

 Using Your Knowledge

The validity of the divisibility by 3 rule can be shown by using some of the properties of the natural numbers. For example, consider the number 2853. Since $10 = 9 + 1$, $100 = 99 + 1$, and $1000 = 999 + 1$, you can write

$$2853 = 2(999 + 1) + 8(99 + 1) + 5(9 + 1) + 3$$
$$= 2 \cdot 999 + 2 + 8 \cdot 99 + 8 + 5 \cdot 9 + 5 + 3$$
$$= 2 \cdot 999 + 8 \cdot 99 + 5 \cdot 9 + (2 + 8 + 5 + 3)$$

Now, as you can see, the first three terms of the last expression are all divisible by 3 because 999, 99, and 9 are all divisible by 3. Thus, the number is divisible by 3 if and only if the sum of the numbers in parentheses is divisible by 3. But the numbers in parentheses are exactly the digits 2, 8, 5, and 3 of the number 2853. Since the sum of these digits is 18, which is exactly divisible by 3, the number 2853 is also divisible by 3.

The reasoning used for 2853 applies to any natural number. Hence, the divisibility by 3 rule is valid.

77. If some of the digits of a number are 3s, 6s, or 9s, you can omit them in figuring the sum of the digits to check for divisibility by 3. For example, 2,963,396,607 is divisible by 3, because $2 + 7 = 9$ is divisible by 3. Explain this.

78. Is 5,376,906,391 divisible by 3?

79. The way in which we wrote

$$2853 = 2(999 + 1) + 8(99 + 1) + 5(9 + 1) + 3$$

shows that 2853 is divisible by 9 if and only if the sum of its digits is divisible by 9. Explain this.

The following table shows some other simple divisibility rules for the natural numbers:

Divisible by	If and Only If
4	The last two digits of the number form a number divisible by 4.
6	The number is divisible by both 2 and 3.
8	The last three digits of the number form a number divisible by 8.
9	The sum of the digits of the number is divisible by 9.
10	The number ends in 0.
12	The number is divisible by both 3 and 4.

The rule for divisibility by 8 can be obtained as follows: Consider the number 2,573,649,336. This can be written in the form 2,573,649 × 1000 + 336. Since 1000 is divisible by 8, the entire number is divisible by 8 if and only if 336 is. Since 336 is divisible by 8, the given number is also divisible by 8. This reasoning applies to the natural numbers in general. Therefore, the rule is valid.

80. Which of the following numbers are divisible by 9? by 6?
a. 405 b. 676
c. 7488 d. 309,907,452

81. Which of the following numbers are divisible by 4? by 8?
a. 1436 b. 21,408
c. 347,712 d. 40,924

82. Which of the following numbers are divisible by 10? by 12?
a. 4920 b. 943
c. 52,341,120 d. 60,210

 **Discovery**

A natural number is said to be **perfect** if the number is the sum of its proper divisors. (The **proper divisors** of a number include all its divisors except the number itself.) For example, 6 is a perfect number because the proper divisors of the number 6 are 1, 2, and 3 and $1 + 2 + 3 = 6$.

Some historians believe that the Pythagoreans were the first to define perfect numbers. In any event, it is certain that the Pythagoreans knew about these numbers and endowed them with mystical properties. In ancient Greek numerology, 6 was regarded as the most beautiful of all numbers; it is not only a perfect number (equal to the sum of its proper divisors), but it is also the product of all its proper divisors: $6 = 1 \times 2 \times 3$.

83. Can you discover why 6 is the smallest perfect number? This fact may be what led St. Augustine in about the year A.D. 400 to assert, "God created all things in 6 days because 6 is a perfect number."

84. The next perfect number after 6 is a number between 25 and 30. Can you discover what number this is?

85. The third perfect number is 496. Find its proper divisors and so prove that 496 is a perfect number.

86. In about the year A.D. 800, Alcuin remarked that the whole human race descended from the 8 souls of Noah's Ark, and he regarded this as imperfect. Can you discover why?

When the sum of the proper divisors of a number is less than the number, the number is called **deficient.** For example, 4 is a deficient number because its proper divisors are 1 and 2, and $1 + 2 = 3$, which is less than 4. Similarly, 7 is a deficient number, because it has only 1 as a proper divisor.

87. If n is any prime number, can you discover whether n is a deficient number?

88. It is known that there are infinitely many prime numbers. Can you use this fact to prove that there are infinitely many deficient numbers?

89. In problems 83–85 we found the perfect numbers 6, 28, and 496. The next perfect number is 8128. Aside from the fact that they are perfect numbers, what do you notice about these four numbers?

90. Look at the first four perfect numbers in binary that follow:

110

11100

111110000

1111111000000

If you continue this pattern, what do you think the next perfect number is? Can it be odd?

 Collaborative Learning

How do you discover prime numbers? For small primes, the Sieve of Eratosthenes is acceptable, or you can go to a table of primes.

 *Online Study Center*

For a table of primes, access link 5.1.5 on this textbook's Online Study Center.

To discover larger primes you need a new method. For many years mathematicians felt that numbers of the form $2^n - 1$ were always prime. Form two or more study groups and determine whether the following numbers are prime:

1. a. $2^{11} - 1$ **b.** $2^{17} - 1$ **c.** $2^{19} - 1$

2. In 1603; Pietro Cataldi conjectured that $2^n - 1$ was prime when $n = 23, 29, 31,$ and 37. Let some members of the group check the result for 23 and 29 and other members check it for 31 and 37. Discuss your finding with the rest of the group.

3. In 1644, Marin Mersenne finally showed that $2^n - 1$ was prime for certain values of n. Find what values do make $2^n - 1$ prime.

4. Divide into several groups and prepare reports about the following topics:
   a. Abundant numbers   **b.** Frugal numbers
   c. Equidigital numbers   **d.** Extravagant numbers
   e. Economical numbers

 Online Study Center

To further explore kinds of numbers, access link 5.1.6 on this textbook's Online Study Center. Try to find other sites on your own.

Research Questions

At this writing, the largest known prime is the Mersenne prime $2^{32,582,657} - 1$, discovered by Dr. Curtis Cooper and Dr. Steven Boone. The second largest prime number is $2^{30,402,457} - 1$, and was discovered by Cooper and Boone.

1. Write a short paragraph about Mersenne primes. What are they? Who discovered them? When? Which is the largest known one at this time?

2. Write a short biography about each of the persons involved in the discovery of the largest known prime, as of this writing.

3. What is GIMPS?

4. Write a short paragraph describing how the largest known prime was discovered.

5. Finally, a trivia question. If you were to print all the 4,053,946 decimal digits of $2^{13,466,917} - 1$ in a single line of type, using a particular font size, how long would the line be? A mile? Two miles? Find your answers for the following font sizes:
   a. 10   **b.** 12   **c.** 14

5.2   Whole Numbers, Integers, and Order of Operations

Those Absurd Numbers!

Look at the thermometer in the photo. It uses *zero, positive* numbers, and *negative* numbers to measure temperature. The set $\{\ldots, -2, -1, 0, 1, 2, \ldots\}$ is called the set of *integers* and will be studied in this section. In the twelfth century, the Chinese invented negative numbers for counting purposes and used red rods for positive quantities and black rods for negatives. However, most mathematicians of the sixteenth and seventeenth centuries refused to accept negative numbers. In fact, Nicolas Chuquet and Michael Stifel both referred to negative numbers as *absurd* numbers.

Here are some other examples that use integers.

The product of the first seven primes: 510,510
The lowest temperature ever recorded on Earth: $-129°$F (Vostock, Antarctica)
The highest temperature ever recorded on Earth: 136°F (Libyan desert)
The lowest temperature ever recorded in the atmosphere: $-225°$F
The income from the movie *Heaven's Gate:* $-\$57$ million
Twenty seconds before liftoff: -20

What about operations with integers? We can add, subtract, and multiply them and get more integers. Note in particular that the set of integers is closed under subtraction; that is, the difference of two integers is always an integer. This is not true of the natural numbers that we studied earlier; for instance, $2 - 5 = -3$ is not a natural number.

The solutions of the next three problems involve operations with integers. The greatest temperature ranges are in Vekrhoyanks, near Siberia, where temperatures vary from 98 to $-94°$F. How many degrees difference is that? The greatest temperature variation in one day occurred in Browning, Montana, on January 23–24, 1916, when the temperature went from 44 to $-56°$F. But this is not as spectacular as the most freakish rise that occurred in Spearfish, South Dakota, on January 22, 1943. On this day, the temperature rose from -40 to 45°F in 2 min. In each of these cases, to find the range in temperature requires the subtraction of a negative number. This section will discuss such operations with integers. ▶

The thermometer shows a very familiar application of negative numbers.

The set of natural numbers N and the operations of addition, subtraction, multiplication, and division form a **mathematical system.** We have already discussed many properties of this system, but there is one more property that leads to some interesting and important ideas.

A.   Properties of the Whole Numbers

The number 1 has the unique property that multiplication of any natural number *a* by 1 gives the number *a* again. Because the *identity* of the number *a* is preserved under multiplication by 1, the number 1 is called the **multiplicative identity,** and this property is called the **identity property for multiplication.**

Identity Property for Multiplication

If a is any natural number, then

$$a \cdot 1 = 1 \cdot a = a$$

It can be shown (problem 63, Exercises 5.2) that 1 is the only element with this property.

Is there an **additive identity?** That is, is there an element z such that $a + z = a$ for every a in N? There is no such element in N. This lack of an additive identity spoils the usefulness of N in many everyday applications of arithmetic. The Babylonians realized this difficulty, as we noted in Chapter 4, and simply used a space between digits as a zero placeholder, but it was not until about A.D. 1400 that the Hindu-Arabic system popularized the idea of zero. The Hindus used the word *sunya* (meaning "void"), which was later adopted by the Arabs as *sifr,* or "vacant." This word passed into Latin as *zephirum* and became, over the years, *zero.*

The number zero (0) provides us with an **identity for addition.** Therefore, we enlarge the set N by adjoining this new element to it. The set consisting of all the natural numbers and the number 0—that is, the set $\{0, 1, 2, 3, \ldots\}$—is called the set of **whole numbers** and is denoted by the letter W. It is an important basic assumption that the set W obeys the same fundamental laws of arithmetic as the natural numbers do. Moreover, adding 0 to a whole number does not change the identity of the whole number. Thus, we have the following:

Identity Property for Addition

If a is any whole number, then

$$0 + a = a + 0 = a$$

As with the number 1, we can prove (problem 64, Exercises 5.2) that the number 0 is unique. It also can be shown that if a is any natural number, then

---

Properties of 0

$$a - 0 = a, \qquad 0 \cdot a = 0, \qquad \text{and} \qquad 0 \div a = \frac{0}{a} = 0$$

The proofs for these facts are in problems 65–67, Exercises 5.2. Note that $a \div 0$ is *not* defined. The importance of 0 is also evident in algebra. When solving an equation such as

$$(x - 1)(x - 2) = 0$$

we argue that either $x - 1 = 0$ or $x - 2 = 0$ and conclude that $x = 1$ or $x = 2$. This argument is based on the theorem that follows.

> **THEOREM 5.2 Zero Product Property**
>
> If $a \cdot b = 0$,   then $a = 0$ or $b = 0$

Proof Assume that $a \neq 0$. Divide both sides by a.

$$\frac{a \cdot b}{a} = \frac{0}{a} = 0$$

Since

$$\frac{a \cdot b}{a} = b$$

then $b = 0$.

A similar argument shows that if $b \neq 0$, then $a = 0$. Of course, if a and b are both 0, the theorem is obviously true. ∎

B. The Set of Integers

The set W is extremely useful when the idea of "How many?" is involved. However, the whole numbers are inadequate even for some simple everyday problems. For instance, the below-zero temperature on a winter's day cannot be described by a whole number, and the simple equation $x + 3 = 0$ has no solution in W. Because similar problems occur repeatedly in the applications of mathematics, the set of whole numbers is extended to include the **negative numbers,** $-1, -2, -3, \ldots$. This new set, called the set of **integers,** is denoted by the letter I, so $I = \{\ldots, -3, -2, -1, 0, 1, 2, 3, \ldots\}$. The set I consists of the following three subsets:

1. The **positive integers:** $1, 2, 3, \ldots$

2. The number **0**

3. The **negative integers:** $-1, -2, -3, \ldots$

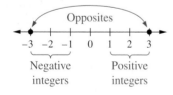

Negative   Positive
integers   integers

FIGURE 5.2

For every positive integer n, there is a negative integer $-n$. To see this, we represent the integers using a **number line.** Draw a horizontal straight line, choose any point on the line, and label it 0. Then measure successive equal intervals to the *right* of 0, and mark the endpoints with the *positive* integers 1, 2, 3, and so on. Mark the endpoints to the *left* of 0 with the *negative* integers. The result is a number line, as shown in Figure 5.2. We graph the integers 3 and -3 by adding dots to the number line. Note that 3 and -3 are the *same distance* from 0 on the number line but in *opposite directions*. This is why 3 and -3 are **opposites.** The following are some other numbers and their opposites:

Number	*Opposite (Additive Inverse)*
5	-5
-8	$-(-8) = 8$
2	-2

Moreover, since 5 and -5 are the same distance from 0, we say that their *absolute values* are the same. The **absolute value** of a number a is defined as the distance between the number a and 0 on the number line and is denoted by $|a|$

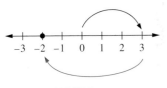

FIGURE 5.3

(read "the absolute value of a"). Thus, $|5| = 5$ and $|-5| = 5$, and $|-8| = 8$ and $|8| = 8$. Since $|a|$ represents a distance, $|a|$ is *never* negative. Also, since 0 is 0 units from 0, $|0| = 0$.

We can give a physical interpretation of addition using the number line. For example, $3 + 2$ can be regarded as a move of 3 units from 0 in the positive direction (right) followed by a move of 2 units more in the positive direction (right). The terminal point is 5. Similarly, $3 + (-5)$ can be interpreted as a move of 3 units in the positive direction followed by a move of 5 units in the negative direction (left). The terminal point is -2 (see Figure 5.3). Thus

$$3 + (-5) = -2$$

Using the same idea,

$$3 + (-3) = 0$$

We can always verify that two numbers are opposites by addition. Thus, 5 and -5 and -12 and 12 are opposites because $5 + (-5) = 0$ and $(-12) + 12 = 0$. Because of this, 5 and -5, and -12 and 12 are also known as **additive inverses.** This discussion can be summarized as follows:

Additive Inverse Property

If n is any integer, then there exists a unique integer $-n$ such that

$$n + (-n) = (-n) + n = 0$$

The numbers n and $-n$ are said to be **additive inverses (opposites)** *of each other.*

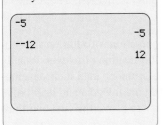
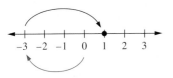
It can be proven (see problem 68, Exercises 5.2) that each integer has exactly one additive inverse.

The operation of subtraction can also be represented on the number line. For instance, $3 - 5$ can be thought of as a move of 3 units from 0 in the positive (right) direction, followed by a move of 5 units in the negative (left) direction. The terminal point is -2. Thus, we have $3 - 5 = -2$. Notice that this is the same result we found for $3 + (-5)$ (see Figure 5.3); that is,

$$3 - 5 = 3 + (-5)$$

The minus sign is used in two different ways in this last equation: On the left side of the equal sign, it means to subtract 5 from 3; on the right, it means that -5 is the integer 5 units in the negative direction from 0.

EXAMPLE 1 ▶ Subtraction on the Number Line

Illustrate the subtraction $-3 - (-4)$ on the number line.

Solution

This subtraction can be considered as a move of 3 units to the left of 0, followed by a move of 4 units to the right (in the opposite direction from that indicated by the -4). The terminal point is 1 unit to the right of 0, as indicated in Figure 5.4, so we have $-3 - (-4) = 1$. Notice that this is the same result as $-3 + 4 = 1$. ■

FIGURE 5.4

This discussion should convince you that subtraction can be viewed as the opposite of addition, and any subtraction problem can be transformed into an addition problem as follows:

Subtraction Problem	Equivalent Addition Problem
$3 - 1$	$3 + (-1)$
$10 - 7$	$10 + (-7)$
$-3 - 5$	$-3 + (-5)$
$2 - (-3)$	$2 + 3$

This motivates the following definition:

Definition of Subtraction

If a and b are any integers, the **subtraction** of b from a is defined as

$$a - b = a + (-b)$$

This means that subtracting b from a is the same as adding the inverse (opposite) of b to a.

EXAMPLE 2 ▶ Subtraction and Additive Inverses

Change the subtraction problem $4 - 3$ to an equivalent addition problem and illustrate on the number line.

Solution

By definition, $4 - 3 = 4 + (-3)$. On the number line, we think of this as a move of 4 units to the right, followed by a move of 3 units to the left. The result, as shown in Figure 5.5, is 1. ∎

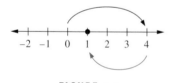

FIGURE 5.5

We assume that the reader is familiar with the operations of addition and subtraction of integers. However, the rules for the multiplication of integers usually have been memorized without any attempt to see or understand their development. To remedy this, we state the rules for multiplying integers and later prove the second rule using Theorem 5.3. You should try to prove the first and third rules on your own!

1. The product of two positive integers is positive.

2. The product of a positive integer and a negative integer is negative.

3. The product of two negative integers is positive.

Table 5.2 summarizes these results. Note that the product of integers with the same (*like*) signs is *positive* and the product of integers with different (*unlike*) signs is *negative*.

TABLE 5.2

Rules of Signs in Multiplication	Examples
1. Positive $\times$ positive = positive	$4 \times 3 = 12$
2. Positive $\times$ negative = negative	$4 \times (-3) = -12$
3. Negative $\times$ negative = positive	$(-4) \times (-3) = 12$

Signed Number Properties

If n is a negative odd integer, what kind of number is $(n - 2)^3$?

① **Read** the problem.
What are the negative odd integers?

The negative odd integers are $-1, -3, -5, \ldots$.

② **Select** the unknown.

We want to know what type number $(n - 2)^3$ is.

③ **Think** of a plan.
What kind of number is $n - 2$?
Try n equal to $-1, -3, -5$, and so on.
What does $(n - 2)^3$ mean?

If n is a negative odd integer, n can be $-1, -3, -5, \ldots$. Thus, $n - 2$ can be $-3, -5, -7, \ldots$; that is, $n - 2$ is also a negative odd integer.

$(n - 2)^3$ means $(n - 2)(n - 2)(n - 2)$. Is $(n - 2)^3$ odd or even? Try $n = -1$, $n = 1$, and so on.

④ **Use** the laws of signs to carry out the plan.

If $(n - 2)$ is negative, $(n - 2)(n - 2)$ is positive $[(-) \times (-) = (+)]$ and $(n - 2)(n - 2)(n - 2)$ is negative $[(-) \times (-) \times (-) = (-)]$. Thus, $(n - 2)^3$ is a negative odd integer.

⑤ **Verify** the answer.
Take some examples and see if it works.

If $n = -1$, then $n - 2 = -1 - 2 = -3$ and $(-3)^3 = (-3)(-3)(-3) = -27$, which is a negative odd integer. Of course, this is *not* a proof, only a verification!

TRY EXAMPLE 3 NOW.

Cover the solution, write your own solution, and then check your work.

EXAMPLE 3 ▶ **Cubing Integers**

If n is a negative even integer, what kind of number is $(n + 1)^3$?

Solution
If n is a negative even integer, n can be $-2, -4, -6, \ldots$. Thus, $n + 1$ can be $-1, -3, -5, \ldots$; that is, $n + 1$ is a negative odd integer. Now,

$$(n + 1)^3 = (n + 1)(n + 1)(n + 1)$$
$$\qquad\qquad\;\; (-) \qquad (-) \qquad (-)$$

which is a negative odd integer. So $(n + 1)^3$ is a negative odd integer. ▪

Theorem 5.3 illustrates the way in which we can prove the second rule of signs.

THEOREM 5.3   A Positive Number Multiplied by a Negative Number

If a and b are positive, then

$$\underset{\text{Positive}}{(a)} \;\times\; \underset{\text{Negative}}{(-b)} \;=\; \underset{\text{Negative}}{-(ab)}$$

Before proceeding with the proof, we note that in Theorem 5.3, a is a positive number, $-b$ is negative (because b is positive), $a \cdot b$ is positive (by rule of signs 1), and hence, $-(ab)$ is negative. Thus, Theorem 5.3 states that a positive number (a) times a negative number ($-b$) yields a negative number $-(ab)$.

Proof

1. $b + (-b) = 0$ Definition of inverse

2. $a[b + (-b)] = a \cdot 0$ Multiplying both sides of the equation by a

3. $a \cdot b + a(-b) = 0$ Multiplying each term of the equation by a and recalling that $a \cdot 0 = 0$

4. By step 3, the additive inverse of $a \cdot b$ is $a(-b)$, but because $a \cdot b$ is positive, its additive inverse is also $-(ab)$.

5. However, the additive inverse of a number is unique, so we conclude that $a(-b) = -(ab)$.   ■

We now summarize the properties of the integers under the operations of addition and multiplication.

Properties of Addition

1. The set of integers is closed with respect to addition.
   The sum of any two integers a and b is always an integer; that is, $a + b$ is an integer.

2. The set of integers has the commutative property of addition.
   For any two integers a and b, $a + b = b + a$.

3. The set of integers has the associative property of addition.
   For any three integers a, b, and c, $a + (b + c) = (a + b) + c$.

4. The set of integers has a unique identity (the number 0) for addition.
   $a + 0 = a$ and $0 + a = a$.

5. Each integer a has a unique additive inverse $(-a)$ in the set of integers.
   $a + (-a) = 0$.

Properties of Multiplication

1. The set of integers is closed with respect to multiplication.
   The product of two integers a and b is always an integer.

2. The set of integers has the commutative property of multiplication.
   For any two integers a and b, $a \cdot b = b \cdot a$.

3. The set of integers has the associative property of multiplication.
   For any three integers a, b, and c, $a \cdot (b \cdot c) = (a \cdot b) \cdot c$.

4. The set of integers has a unique identity (number 1) for multiplication.
   $a \cdot 1 = a$ and $1 \cdot a = a$.

5. Not all the integers have multiplicative inverses in the set of integers.
   For example, the multiplicative inverse for 3 would be a number b such that $3 \cdot b = 1$. But b is not an integer.

Besides these properties, the set of integers has the distributive property of multiplication over addition.

Distributive Property

$$a(b + c) = ab + ac \qquad \text{and} \qquad (a + b)c = ac + bc$$

Now that we have discussed the properties of the set of integers, how do we use them? We give an example next.

FIGURE 5.6

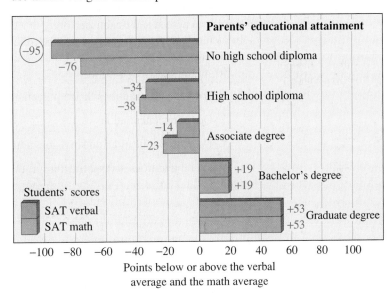

Source: The College Board.

EXAMPLE 4 ▶ Interpreting Graphs Using Integers

Is there a relationship between how much education your parents received (educational attainment) and your scores on the SAT? Figure 5.6 shows that the verbal scores for test takers with parents that had no high school diploma is −95, that is, 95 points *below* average. (By the way, the average score in math was 514 and in the verbal 506.)

(a) What integer corresponds to the verbal scores of test takers whose parents had a high school diploma? What does the integer mean?

(b) What integer corresponds to the math scores of test takers whose parents had a graduate degree? What does the integer mean?

Solution

(a) The verbal scores for test takers whose parents had a high school diploma appear in red to the left of the row labeled "High school diploma." The score is −34. It means that the verbal score of the test taker was 34 points *below* the verbal average (506). Thus, the verbal score for the test taker was 506 − 34 = 472.

(b) The math scores for test takers whose parents had a graduate degree appear in blue to the right of the row labeled "Graduate degree." The score is +53.

It means that the math score of the test taker was 53 points *above* the math average (514). Thus, the math score for the test taker was $514 + 53 = 567$.

◼

C. Order of Operations

Suppose we wish to evaluate the expression $8 - 3 \times 5$. Which evaluation is correct?

$$8 - 3 \times 5 = 5 \times 5 = 25 \qquad \text{or} \qquad 8 - 3 \times 5 = 8 - 15 = -7$$

In this example, -7 is the correct evaluation. To avoid this sort of ambiguity, there is an agreed and established order in which operations are to be performed. In such calculations, we always use the following order of operations:

You can remember the order of operations if you remember
Please
   Excuse
      My **D**ear
         Aunt **S**ally

Order of Operations (PEMDAS)

1. Evaluate all quantities inside **p**arentheses. (P)

2. Do all **e**xponentiations. (E)

3. Do all **m**ultiplications and **d**ivisions as they occur from left to right. (MD)

4. Do all **a**dditions and **s**ubtractions as they occur from left to right. (AS)

Note: $8 \div 4 \times 2 = 2 \times 2 = 4$ Division done first (It *occurs* first.)

$8 \times 4 \div 2 = 32 \div 2 = 16$ Multiplication done first (It *occurs* first.)

EXAMPLE 5 ▶ Evaluating Expressions

Evaluate $7 \times 8 \div 4 \times 10^2 - 3(-5 + 2) \times 10^3$.

Solution

$$7 \times 8 \div 4 \times 10^2 - 3(-5 + 2) \times 10^3$$

Step 1 $= 7 \times 8 \div 4 \times 10^2 - 3(-3) \times 10^3$ $(-5 + 2) = -3$

Step 2 $= 7 \times 8 \div 4 \times 100 - 3(-3) \times 1000$ $10^2 = 100, 10^3 = 1000$

Step 3 $= 56 \div 4 \times 100 - (-9) \times 1000$ $7 \times 8 = 56, 3(-3) = -9$

$= 14 \times 100 - (-9000)$ $56 \div 4 = 14,$
$(-9) \times 1000 = -9000$
$14 \times 100 = 1400$

$= 1400 - (-9000)$

Step 4 $= 1400 + 9000 = 10,400$ $1400 - (-9000) =$
$1400 + 9000 = 10,400$

◢◣ **GRAPH IT**

If you enter the operations from left to right as given, keeping in mind that exponents are obtained by using the $\boxed{\wedge}$ key, subtraction uses the blue key $\boxed{-}$, and additive inverses need the gray key $\boxed{(-)}$, the grapher takes care of the order of operations.

```
7*8/4*10^2-3(-5+2)
*10^3
                10400
```

Note: Sometimes we use the centered dot symbol "·" to indicate multiplication.

EXAMPLE 6 ▶ Simplifying Expressions

Simplify $8 \div 2^2 \cdot 2 + 3(5 - 2) - 3 \cdot 2$.

Given

$$8 \div 2^2 \cdot 2 + 3(5 - 2) - 3 \cdot 2$$

$$= 8 \div 2^2 \cdot 2 + 3(3) - 3 \cdot 2$$

$$= 8 \div 4 \cdot 2 + 3(3) - 3 \cdot 2$$

$$= 2 \cdot 2 \quad\quad + 3(3) - 3 \cdot 2$$

$$= 4 \quad\quad + 3(3) - 3 \cdot 2$$

$$= 4 \quad\quad + 9 \quad - 3 \cdot 2$$

$$= 4 \quad\quad + 9 \quad - 6$$

$$= \quad\quad\quad 13 \quad - 6$$

$$= 7$$

Solution

1. Evaluate quantities inside parentheses. $(5 - 2) = 3$

2. Do exponentiations. $2^2 = 4$

3. Do multiplications and divisions as they occur.
   Divide: $8 \div 4 = 2$
   Multiply: $2 \times 2 = 4$
   Multiply: $3 \times 3 = 9$
   Multiply: $3 \times 2 = 6$

4. Do additions and subtractions as they occur.
   Add: $4 + 9 = 13$
   Subtract: $13 - 6 = 7$ ∎

Sometimes we may use more than one type of grouping symbols, such as brackets [ ] and braces { }. When grouping symbols occur **within** other grouping symbols, computations in the innermost symbols are done *first*. Thus, to simplify

$$[2 \cdot (14 + 88)] + 12$$

we *first* add 14 and 88 (the computation inside the innermost grouping symbol), then multiply the result 102 by 2, and finally add 12. The procedure looks like this:

$$[2 \cdot (14 + 88)] + 12 = [2 \cdot (102)] + 12$$
Add inside parenthesis: $14 + 88 = \mathbf{102.}$

$$= 204 + 12$$
Multiply inside brackets: $2 \cdot 102 = \mathbf{204.}$

$$= 216$$
Add $204 + 12 = \mathbf{216.}$

EXAMPLE 7 ▶ Simplifying Expressions

Simplify $20 \div 4 + \{2 \cdot 3^2 - [3 + (6 - 2)]\}$.

$$20 \div 4 + \{2 \cdot 3^2 - [3 + (6 - 2)]\}$$

$$= 20 \div 4 + \{2 \cdot 3^2 - [3 + 4]\}$$
Subtract inside parentheses $(6 - 2 = \mathbf{4})$.

$$= 20 \div 4 + \{2 \cdot 3^2 - 7\}$$
Add inside brackets $[3 + 4 = \mathbf{7}]$.

$$= 20 \div 4 + \{2 \cdot 9 - 7\}$$
Do exponents $(3^2 = \mathbf{9})$.

$$= 20 \div 4 + \{18 - 7\}$$
Multiply *first* inside braces $(2 \cdot 9 = \mathbf{18})$.

$$= 20 \div 4 + 11$$
Then subtract inside braces $(18 - 7 = \mathbf{11})$.

$$= 5 + 11$$
Next, divide 20 by 4 $(20 \div 4 = \mathbf{5})$.

$$= 16$$
Finally, add $5 + 11 = \mathbf{16.}$

EXERCISES 5.2

Ⓐ Properties of the Whole Numbers

In problems 1–4, graph each indicated set of numbers on the number line.

1. The integers between 4 and 9, inclusive

2. The integers between -5 and 6, inclusive

3. The whole numbers between -2 and 5, inclusive

4. The natural (counting) numbers between -2 and 4, inclusive

In problems 5–8, find the additive inverse of each number.

5. 3   6. 47   7. -8 8. -1492

Ⓑ The Set of Integers

In problems 9–16, illustrate each indicated operation on the number line.

9. $4 - 5$ 10. $-3 - 2$

11. $4 - 2$ 12. $3 + 4$

13. $-5 + 4$ 14. $4 - 4$

15. $3 + (-5)$ 16. $5 + (-2)$

In problems 17–28, write each problem as an equivalent addition problem and find the answer.

17. $3 - 8$ 18. $8 - 3$

19. $3 - 4$ 20. $-3 - 4$

21. $-5 - 2$ 22. $-3 - 5$

23. $5 - (-6)$ 24. $6 - (-3)$

25. $-3 - (-4)$ 26. $-5 - (-6)$

27. $-5 - (-3)$ 28. $-10 - (-5)$

In problems 29–34, find each product.

29. **a.** $(-5) \times 3$ **b.** $(-8) \times 9$

30. **a.** $(-3) \times (-4)$ **b.** $(-9) \times (-3)$

31. **a.** $4 \times (-5)$ **b.** $3 \times (-13)$

32. **a.** $0 \times (-4)$ **b.** $(-0) \times 0$

33. **a.** $3 \times (4) \times (-5)$ **b.** $5 \times (-4) \times (3)$

34. **a.** $-2 \times (-3) \times (-1)$ **b.** $-4 \times (-5) \times (-7)$

35. If n is a negative odd integer, what kind of a number is $(n - 1)^3$?

36. If n is a positive odd integer, what kind of a number is $(n + 1)^5$?

In problems 37–46, refer to the chart in Example 4.

37. What integer corresponds to the verbal score of test takers whose parents earned an associate degree? What was their score?

38. What integer corresponds to the math score of test takers whose parents earned an associate degree? What was their score?

39. What integer corresponds to the verbal score of test takers whose parents earned a bachelor's degree? What was their score?

40. What integer corresponds to the math score of test takers whose parents earned a bachelor's degree? What was their score?

41. What integer corresponds to the verbal score of test takers whose parents earned a graduate degree? What was their score?

42. Find the difference between the integers corresponding to the verbal (-95) and math (-76) scores for a test taker whose parents did not earn a high school diploma.

43. Find the difference between the integers corresponding to the verbal and math scores for a test taker whose parents earned a high school diploma.

44. Find the difference between the integers corresponding to the verbal and math scores for a test taker whose parents earned an associate degree.

45. Find the difference between the integers corresponding to the verbal and math scores for a test taker whose parents earned a bachelor's degree.

46. Find the difference between the integers corresponding to the verbal and math scores for a test taker whose parents earned a graduate degree.

Ⓒ Order of Operations

In each of problems 47–56, use the order of operations to simplify the expression.

47. a. $-3(4 + 5)$ **b.** $-4(4 - 5)$

48. a. $-2(-3 + 1)$ **b.** $-5(-4 + 2)$

49. a. $-5 + (-5 + 1)$ **b.** $-8 + (-2 + 5)$

50. a. $-2(4 - 8) - 9$ **b.** $-3(5 - 7) - 11$

51. $(-2 - 4)(-3) - 8(5 - 4)$

52. $(-3 - 5)(-2) + 8(3 + 4 - 5)$

53. $6 \times 2 \div 3 + 6 \div 2 \times (-3)$

54. $8 \div 2 \times 4 - 8 \times 2 \div 4$

55. $4 \times 9 \div 3 \times 10^3 - 2 \times 10^2$

56. $5 \times (-2) \times 3^2 + 6 \div 3 \times 5 \times 3^2$

57. $20 \div 5 + \{3 \cdot 4 - [4 + (5 - 3)]\}$

58. $30 \div 6 + \{4 \div 2 \cdot 3 - [3 + (5 - 4)]\}$

59. $(20 - 15) \cdot [20 \div 2 - (2 \cdot 2 + 2)]$

60. $(30 - 10) \cdot [52 \div 4 - (3 \cdot 3 + 3)]$

61. The highest point on Earth is Mount Everest, 9 km above sea level. The lowest point is the Marianas Trench, 11 km below sea level. Use signed numbers to find the difference in altitude between these two points.

Mount Everest, the highest mountain in the world, rises 9 km above sea level.

62. Julius Caesar died in 44 B.C. Nero is said to have set Rome afire in A.D. 64. Use signed numbers to find how many years elapsed between Caesar's death and the burning of Rome.

63. Supply reasons for steps 1, 3, and 4 in the following proof by contradiction that 1 is the unique multiplicative identity. Suppose there is another multiplicative identity—say, m—different from 1.

Then, for every element a of N, you have the following:

Step 1 $m \cdot a = a$

Step 2 $m \cdot 1 = 1$ Substituting 1 for a

Step 3 $m \cdot 1 = m$

Step 4 Therefore, $m = 1$

64. Fill in the blanks in the following proof that 0 is the unique additive identity. Assume that there are two additive identity elements—say, 0 and z.

a. $0 + z = z$ Because 0 is an additive _____

b. $0 + z = 0$ Because z is an additive _____

c. Therefore, $z = 0$ Because they are both equal to _____

This shows that the assumed two identity elements are the same; that is, 0 is the unique additive identity.

65. Supply reasons for steps 1, 3, and 5 in the following proof that $a - 0 = a$.

Step 1 $a - 0 = a + (-0)$

Step 2 $\quad = [a + (-0)] + 0$ Because 0 is the additive identity

Step 3 $\quad = a + [(-0) + 0]$

Step 4 $\quad = a + 0$ Because 0 and -0 are additive inverses of each other

Step 5 $\quad = a$

66. Fill in the blanks in the following proof that $0 \cdot a = 0$.

a. $0 \cdot a = [1 + (-1)] \cdot a$ Because 1 and -1 are _____ inverses

b. $\quad = 1 \cdot a + (-1) \cdot a$ By the _____ property

$\quad = a + (-a)$ Multiplying and using the rule of signs

c. $\quad = 0$ Because a and $-a$ are _____

67. Fill in the blanks in the following proof that for $a \neq 0$, $0/a = 0$.

a. Let the quotient $0/a = q$. Then

$q \cdot a = 0$	By the definition of division

b. $q \cdot a + a = 0 + a$ By _____ to both sides
c. $q \cdot a + a = a$ Because 0 is the additive _____
d. $q \cdot 1 + 1 = 1$ By dividing both sides by a
e. Therefore, $q + 1 = 1$ Because $q \cdot 1 = $ _____
f. Consequently, $q = 0$ Because the additive _____ is _____

68. Fill in the blanks in the following proof that each integer has only one additive inverse, $-n$.

Suppose that the integer n has another such inverse—say, x. Then because $n + (-n) = 0$,

a. $x = x + [n + (-n)]$ Because 0 is the _____ identity
b. $= (x + n) + (-n)$ By the _____ property of addition
c. $= (n + x) + (-n)$ By the _____ property of addition
d. $= 0 + (-n)$ Because x was assumed to be an additive inverse of n
e. $= -n$ Because 0 is the _____

69. Supply reasons for steps 1, 3, and 5 in the following alternative proof that $a \cdot 0 = 0$.

Step 1 $0 + 0 = 0$

Step 2 $a \cdot (0 + 0) = a \cdot 0$ Multiplying both sides by a

Step 3 $a \cdot (0 + 0) = a \cdot 0 + a \cdot 0$

Step 4 So, $a \cdot 0 + a \cdot 0 = a \cdot 0$ Substituting from step 3 into step 2

Step 5   Hence, $a \cdot 0 = 0$

70. Supply reasons for steps 1, 3, 4, and 6 in the following proof that $(-a)(-b) = ab$.

Step 1 $a + (-a) = 0$

Step 2 $[a + (-a)](-b) = 0$ Multiplying both sides by $(-b)$ and using $0 \cdot (-b) = 0$

Step 3 $(a)(-b) + (-a)(-b) = 0$

Step 4 $(a)(-b) = -ab$

Step 5 $-ab + (-a)(-b) = 0$ Substituting from step 4 into step 3

Step 6 Hence, $(-a)(-b)$ is the additive inverse of $-ab$.

Step 7 Therefore, $(-a)(-b) = ab$ Because the additive inverse of a number is unique

 In Other Words

71. Write in your own words the following rule of signs:

Positive $\times$ positive = positive

72. Write in your own words the following rule of signs:

Positive $\times$ negative = negative

73. Write in your own words the following rule of signs:

Negative $\times$ negative = positive

74. Two numbers having the *same* sign are said to have *like* signs. If two numbers have *different* signs, they are said to have *unlike* signs. Use the words *like* and *unlike* to summarize the rules of signs given in Table 5.2.

Using Your Knowledge

The oxidation number (or valence) of a molecule is found by using the oxidation numbers of the atoms present in the molecule. For example, the oxidation number of hydrogen (H) is $+1$, that of sulfur (S) is $+6$, and that of oxygen (O) is -2. Thus, we can get the oxidation number of sulfuric acid (H_2SO_4) as follows:

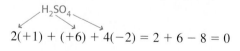

$$2(+1) + (+6) + 4(-2) = 2 + 6 - 8 = 0$$

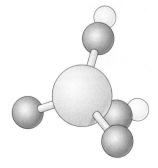

A sulfuric acid molecule.

Use this idea to find the oxidation number of the following:

75. Phosphate, PO_4, if the oxidation number of phosphorus (P) is $+5$ and that of oxygen (O) is -2

76. Sodium dichromate, $Na_2Cr_2O_7$, if the oxidation number of sodium (Na) is $+1$, that of chromium (Cr) is $+6$, and that of oxygen (O) is -2

77. Baking soda, $NaHCO_3$, if the oxidation number of sodium (Na) is $+1$, that of hydrogen (H) is $+1$, that of carbon (C) is $+4$, and that of oxygen (O) is -2

 **Discovery**

Nomographs are graphs that can be drawn to perform various numerical operations. For example, the nomograph in the diagram below can be used to do addition and subtraction. To add any two numbers, we locate one of the numbers on the lower scale and the other number on the upper scale. We then connect these by a straight line. The point where the line crosses the middle scale gives the result of the addition. The diagram shows the sum $-2 + 4 = 2$.

78. Can you discover how to subtract on a nomograph?

 Calculator Corner

Most scientific calculators perform operations in **algebraic order;** that is, they perform multiplications and divisions first, followed by additions and subtractions. Thus, if you enter ③ ⊞ ⑤ ⊠ ② ⊟ in an algebraic calculator, the answer is 13. (If you do not get this answer, you have to multiply 5×2 first and

then add 3 to the result.) Of course, if you have an expression such as $-3(-5 + 2)$, remember that the parentheses represent an indicated multiplication. Thus, you must enter this expression as

Practice this concept by evaluating the expression from Examples 4 and 5.

$$7 \times 8 \div 4 \times 10^2 - 3(-5 + 2) \times 10^3$$

on your calculator. (Recall that $10^2 = 10 \times 10$ and $10^3 = 10 \times 10 \times 10$.)

1. Use your calculator to check the answers to problems 53, 55, and 56 in Exercises 5.2.

Collaborative Learning

Form two or more groups.

1. Let each group select three integers from 0 to 9, and subtract its *reversal* (the *reversal* of 856 is 658). Then, if the difference is *positive,* add its reversal. If the difference is *negative,* subtract its reversal. Here are three possibilities for three numbers.

$$
\begin{array}{r} 856 \\ -658 \\ \hline 198 \\ +891 \end{array}
\qquad
\begin{array}{r} 159 \\ -951 \\ \hline -792 \\ -297 \end{array}
\qquad
\begin{array}{r} 872 \\ -278 \\ \hline 594 \\ +495 \end{array}
$$

What answer did your group get? Did all groups get the same answer? Why do you think this works?

Online Study Center

To further explore reversals, access link 5.2.3 on this textbook's Online Study Center.

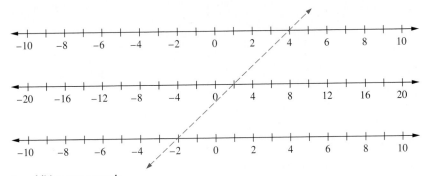

An addition nomograph.

2. Look at these patterns:

$$99 + 100 + 101 + 102 + 103 = 505$$
$$99^2 + 100^2 + 101^2 + 102^2 + 103^2 = 51{,}015$$

The answers 505 and 51,015 are *palindromic* numbers. Do not get alarmed; it only means that 505 and 51,015 read the same from left to right as from right to left. Will the pattern continue if six numbers are used instead of five? Let the members of your group decide.

3. Look at the next patterns.

$$10{,}099 + 10{,}100 + 10{,}101$$
$$+ 10{,}102 + 10{,}103 = 50{,}505$$
$$10{,}099^2 + 10{,}100^2 + 10{,}101^2$$
$$+ 10{,}102^2 + 10{,}103^2 = 510{,}151{,}015$$

5.3 The Rational Numbers

On Pyramids and Basketball

The photo below shows the great pyramid at Giza, erected about 2600 B.C. by Khufu, whom the Greeks called Cheops. The Greek historian Herodotus claimed that the ratio b/a, where b is one-half the base and a is the altitude of one of the face triangles, is the Golden Ratio

$$\frac{\sqrt{5} - 1}{2}$$

an irrational number that can be approximated by $0.6180339 \cdots$. How close were the Egyptians to this ratio? Actual measurements show that $b = 377.89$ ft and $a = 611.85$ ft. Substituting in b/a will tell you how close! In addition, the Golden Ratio is the only positive number that is 1 less than its reciprocal!

Here is an example of a more modern application of ratios. During the 1953–1954 basketball season, the NBA had a problem with the game: It was boring! Fans wanted plenty of action, shooting, and scoring, but if a team had an early lead, their best ball handler would dribble in the backcourt until he was

The Great Pyramid at Giza near Cairo, Egypt.

fouled. How could the game be speeded up? Danny Biasone, the owner of the Syracuse Nationals, an obscure team from upstate New York, thought that more shots could be encouraged by limiting the time a team could have the ball. But how many seconds should be allowed between shots? He figured out that in a fast-paced game, a team would take about 60 shots during the 48 minutes the game lasted (4 quarters of 12 minutes each). He then looked at the ratio

$$\frac{\text{Seconds}}{\text{Shots}}$$

Now, the number of shots for both teams combined would be 2×60, and the number of seconds in each game is 48×60. Thus, the ratio becomes

$$\frac{\text{Seconds}}{\text{Shots}} = \frac{48 \times 60}{2 \times 60} = \frac{24}{1}$$

and now you know how the 24-second shot clock was born!

In this section you will study the rational numbers, which are numbers that can be written in the form *a/b*, where *a* and *b* are integers, and the operations that can be performed using these numbers. ▶

In Section 5.2 we extended the set of natural numbers and obtained the set of integers. As we saw in studying the multiplication of integers, not all the integers have multiplicative inverses; for instance, there is no integer *b* such that $3 \cdot b = 1$. Just as the lack of additive inverses impairs the usefulness of the set of natural numbers, so does the lack of multiplicative inverses impair the usefulness of the set of integers.

A. Rational Numbers

The difficulty lies in the fact that the set of integers is not closed under division; division of two integers can produce fractions such as $\frac{3}{4}, \frac{5}{7}, \frac{8}{3}$, and so on. Thus, we extend the system of integers to include such fractions and call the resulting system the set of **rational numbers.** The set of rational numbers is symbolized by the letter *Q*. (Note that a common fraction is often called a *ratio;* hence the name rational numbers.)

Definition of a Rational Number

A **rational number** is a number that can be put in the form

$$\frac{a}{b}, \qquad \text{where } a \text{ and } b \text{ are integers and } b \neq 0$$

Thou shalt not divide by zero

Cartoon by Paul Kicklighter.

The following are some important facts about rational numbers:

1. Every integer is a rational number because the integers can be written in the form $\ldots, -\frac{2}{1}, -\frac{1}{1}, \frac{0}{1}, \frac{1}{1}, \frac{2}{1}, \ldots$.

2. The integer *a* in the definition of a rational number is called the **numerator,** and the integer *b* is called the **denominator.**

3. The restriction $b \neq 0$ is necessary. By the definition of division, if $a/b = c$, then $a = bc$. So, if $b = 0$, this means that $a = 0$. Thus, if $a \neq 0$, then the

attempted division by 0 leads to a contradiction. If $a = 0$, then the equation $0 = 0 \cdot c$ is true for every number c; that is, the quotient c is not uniquely defined. The only way to avoid this dilemma is to **forbid division by 0.**

4. The definition of a rational number includes the words "a number that *can* [emphasis here] be put in the form a/b, where a and b are integers and $b \neq 0$." Why could it not say "a number of the form a/b, where a and b are integers and $b \neq 0$"? To answer this question, consider the number $0.333 \cdots$. This number is *not* in the form a/b, but because $0.333 \cdots = \frac{1}{3}$, it *can* be put in the form a/b, where $a = 1$ and $b = 3$. Similarly, $1\frac{3}{4}$ is not in the form a/b. But because $1\frac{3}{4} = \frac{7}{4}$, the number *can* be put in the form a/b, where $a = 7$ and $b = 4$. From this discussion we conclude that $0.333 \cdots$ and $1\frac{3}{4}$ are rational numbers.

How do we recognize the fact that $\frac{1}{3}$ and $\frac{3}{9}$ represent the same rational number? To do this, we need to define the equality of rational numbers.

Definition of the Equality of Rational Numbers

$$\frac{a}{b} = \frac{c}{d} \qquad \text{if and only if} \qquad ad = bc$$

Thus, $\frac{1}{3} = \frac{3}{9}$ because $1 \cdot 9 = 3 \cdot 3$, and $\frac{1}{8} = \frac{4}{32}$ because $1 \cdot 32 = 8 \cdot 4$. Using the definition of the equality of rational numbers, we can prove the following useful result:

THEOREM 5.4 Fundamental Property of Rational Numbers

If a, b, and k are integers, where $b \neq 0$ and $k \neq 0$, then

$$\frac{a}{b} = \frac{ak}{bk}$$

Proof

$$\frac{a}{b} = \frac{ak}{bk} \qquad \text{if and only if} \qquad a(bk) = b(ak) \qquad \text{By the equality of rational numbers}$$

But this equality is true because we can use the associative and commutative properties of multiplication for the integers to show that $a(bk) = b(ak)$. ∎

Theorem 5.4 enables us to reduce $\frac{4}{6}$ by writing

$$\frac{4}{6} = \frac{2 \cdot 2}{3 \cdot 2}$$

and "canceling" the 2s to obtain $\frac{4}{6} = \frac{2}{3}$. The theorem also assures us that if we multiply the numerator and denominator of a fraction by the same nonzero number k, the fraction is unchanged in value. Thus,

$$\frac{1}{2} = \frac{1 \cdot 2}{2 \cdot 2} = \frac{2}{4} \qquad \text{and} \qquad \frac{1}{8} = \frac{1 \cdot 3}{8 \cdot 3} = \frac{3}{24}$$

EXAMPLE 1 ▶ **Reducing Fractions**

Use Theorem 5.4 to reduce $\frac{10}{30}$.

Solution

$$\frac{10}{30} = \frac{1 \cdot 10}{3 \cdot 10} = \frac{1}{3}$$

∎

EXAMPLE 2 ▶ **Fractions with Specified Denominators**

Using Theorem 5.4, find a rational number with a denominator of 12 and equal to $\frac{5}{6}$.

Solution

$$\frac{5}{6} = \frac{5 \cdot 2}{6 \cdot 2} = \frac{10}{12}$$

∎

B. Operations with Rational Numbers

We now define the operations of addition, subtraction, multiplication, and division of rational numbers.

Definition of the Product of Two Rational Numbers

The **product** of two rational numbers a/b and c/d is defined by

$$\frac{a}{b} \cdot \frac{c}{d} = \frac{ac}{bd}$$

Thus,

$$\frac{2}{7} \cdot \frac{3}{5} = \frac{2 \cdot 3}{7 \cdot 5} = \frac{6}{35}$$

Definition of the Sum of Two Rational Numbers

The **sum** of two rational numbers a/b and c/d is defined as

$$\frac{a}{b} + \frac{c}{d} = \frac{ad + bc}{bd}$$

For example,

$$\frac{2}{7} + \frac{3}{5} = \frac{(2 \cdot 5) + (7 \cdot 3)}{7 \cdot 5} = \frac{10 + 21}{35} = \frac{31}{35}$$

We can arrive at this definition by noting that

$$\frac{a}{b} = \frac{ad}{bd} \qquad \text{and} \qquad \frac{c}{d} = \frac{bc}{bd}$$

We then write

$$\frac{a}{b} + \frac{c}{d} = \frac{ad}{bd} + \frac{bc}{bd}$$

$$= \frac{ad + bc}{bd} \qquad \text{The fractions have a common denominator.}$$

Note: If $d = b$, it is easy to show that the definition of the sum of two rational numbers gives

$$\frac{a}{b} + \frac{c}{b} = \frac{a + c}{b}$$

as we should expect.

Definition of the Difference of Two Rational Numbers

The **difference** of two rational numbers a/b and c/d is defined as

$$\frac{a}{b} - \frac{c}{d} = \frac{a}{b} + \frac{-c}{d} = \frac{ad - bc}{bd}$$

Thus,

$$\frac{1}{3} - \frac{1}{4} = \frac{1}{3} + \frac{-1}{4} = \frac{(1 \cdot 4) - (3 \cdot 1)}{3 \cdot 4} = \frac{4 - 3}{12} = \frac{1}{12}$$

Now suppose that we want to do the addition $\frac{3}{4} + \frac{1}{16}$. Using the definition of the sum of two rational numbers, we proceed as follows:

$$\frac{3}{4} + \frac{1}{16} = \frac{(3 \cdot 16) + (4 \cdot 1)}{4 \cdot 16} = \frac{48 + 4}{64} = \frac{52}{64} = \frac{13}{16}$$

However, it is easier to use the LCD of 4 and 16 (instead of 64).

$$\frac{3}{4} = \frac{3 \cdot 4}{4 \cdot 4} = \frac{12}{16}$$

and thus,

$$\frac{3}{4} + \frac{1}{16} = \frac{12}{16} + \frac{1}{16} = \frac{13}{16}$$

The number 16 is the *least common denominator* (LCD) of $\frac{3}{4}$ and $\frac{1}{16}$. Recall from Section 5.1 that this number is the LCM (least common multiple) of the given denominators, so it can be obtained by following the procedure discussed in Section 5.1. The next example illustrates the use of this idea.

EXAMPLE 3 ▶ Adding Fractions

Perform the addition $\frac{3}{8} + \frac{7}{36}$.

Solution

First find the LCM of 8 and 36 by using the procedure given in Section 5.1.

$$8 = 2^3$$
$$\underline{36 = 2^2 \cdot 3^2}$$
$$\text{LCM} = 2^3 \cdot 3^2 = 72$$

As you can see, LCM(8, 36) = 72. Thus,

$$\frac{3}{8} = \frac{3 \cdot 9}{8 \cdot 9} = \frac{27}{72} \quad \text{and} \quad \frac{7}{36} = \frac{7 \cdot 2}{36 \cdot 2} = \frac{14}{72}$$

so that

$$\frac{3}{8} + \frac{7}{36} = \frac{27}{72} + \frac{14}{72} = \frac{41}{72}$$

using 72 as the LCD.   ■

To define division, we introduce the idea of a *reciprocal.*

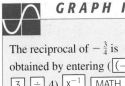

GRAPH IT

The reciprocal of $-\frac{3}{4}$ is obtained by entering ($(-)$ 3 $\div$ 4) x^{-1} MATH 1 ENTER .

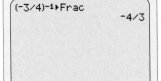

```
(-3/4)⁻¹►Frac
                -4/3
```

How would you find the reciprocal of $1\frac{3}{8}$?

Definition of the Reciprocal of a Rational Number

The **reciprocal** of a rational number

$$\frac{a}{b} \quad \text{is} \quad \frac{b}{a}, \qquad a \neq 0 \text{ and } b \neq 0$$

Note that

$$\frac{a}{b} \cdot \frac{b}{a} = 1$$

Thus, the product of any nonzero rational number and its reciprocal is 1. For this reason, the reciprocal of a rational number is its **multiplicative inverse.**

EXAMPLE 4 ▶ Finding Reciprocals

Find the reciprocals of the following: (a) $\frac{1}{3}$ (b) $-\frac{3}{4}$ (c) $1\frac{3}{8}$

Solution

(a) The reciprocal of $\frac{1}{3}$ is $\frac{3}{1} = 3$.

(b) The reciprocal of $-\frac{3}{4}$ is $-\frac{4}{3}$. Note that $\left(-\frac{3}{4}\right)\left(-\frac{4}{3}\right) = (-1)(-1)\left(\frac{3}{4}\right)\left(\frac{4}{3}\right) = 1$.

(c) We first write $1\frac{3}{8}$ in the form a/b. Because $1\frac{3}{8} = \frac{11}{8}$, the reciprocal of $1\frac{3}{8}$ is $\frac{8}{11}$.   ■

Note: The reciprocal of any negative rational number is always negative, as shown in part (b) of Example 4.

Definition of the Quotient of Two Rational Numbers

The **quotient** of two rational numbers a/b and c/d is defined as

$$\frac{a}{b} \div \frac{c}{d} = \frac{a}{b} \cdot \frac{d}{c}, \qquad c \neq 0$$

Briefly, we say "to *divide* by a number, *multiply* by its reciprocal."

$$\frac{1}{3} \div \frac{2}{7} = \frac{1}{3} \cdot \frac{7}{2} = \frac{7}{6} \quad \text{and} \quad \frac{4}{5} \div \frac{3}{5} = \frac{4}{5} \cdot \frac{5}{3} = \frac{4 \cdot 5}{3 \cdot 5} = \frac{4}{3}$$

$\uparrow$ Reciprocal $\uparrow$ Reciprocal

We can check the definition of the quotient of two rational numbers by multiplying the quotient $(a/b) \cdot (d/c)$ by the divisor c/d to obtain the dividend a/b. Thus,

$$\left(\frac{a}{b} \cdot \frac{d}{c}\right) \cdot \frac{c}{d} = \frac{a}{b} \cdot \left(\frac{d}{c} \cdot \frac{c}{d}\right) = \frac{a}{b} \cdot 1 = \frac{a}{b}$$

as required.

As a consequence of these definitions of the basic operations, the set of rational numbers under addition and multiplication is closed and has the associative, commutative, and distributive properties. It also has an additive identity (0) and additive inverses. With respect to multiplication, the set of rational numbers has a multiplicative identity (1), and every rational number except 0 has a multiplicative inverse (its reciprocal).

It can be shown that the definitions we have given are the only possible ones if we require that the rational numbers obey the same basic laws of arithmetic as the integers do. $\left(\text{Just imagine how unpleasant it would be if } \frac{1}{2} + \frac{1}{4} \neq \frac{1}{4} + \frac{1}{2}.\right)$

From the preceding definitions and Theorem 5.4, we can obtain some additional important results. By the definition of the rational number -1, we know that $-1 = \frac{-1}{1}$, and by Theorem 5.4,

$$\frac{-1}{1} = \frac{(-1)(-1)}{(1)(-1)} = \frac{1}{-1}$$

Therefore,

$$-\frac{a}{b} = -1 \cdot \frac{a}{b} = \frac{-1}{1} \cdot \frac{a}{b} = \frac{-1 \cdot a}{1 \cdot b} = \frac{-a}{b}$$

Similarly,

$$-\frac{a}{b} = -1 \cdot \frac{a}{b} = \frac{1}{-1} \cdot \frac{a}{b} = \frac{a}{-b}$$

Thus, we have the important conclusion that

$$-\frac{a}{b} = \frac{-a}{b} = \frac{a}{-b}$$

Furthermore, because $1 = \frac{1}{1} \cdot \frac{-1}{-1} = \frac{-1}{-1}$, we see that

$$\frac{a}{b} = \frac{-1}{-1} \cdot \frac{a}{b} = \frac{-a}{-b}$$

If you think of a and b as positive numbers, you can see that the rules of signs in division are exactly the same as those in multiplication: The quotient of two numbers with like signs is positive and of two numbers with unlike signs is negative. For example,

$$\frac{9}{3} = \frac{-9}{-3} = 3 \quad \text{and} \quad \frac{-9}{3} = \frac{9}{-3} = -3$$

If the operations under discussion involve **mixed numbers,** first write the mixed numbers as fractions. Thus, to add $2\frac{3}{4} + \frac{5}{6}$, note that $2\frac{3}{4} = 2 + \frac{3}{4} = \frac{8}{4} + \frac{3}{4} = \frac{11}{4}$. A simpler way to do this (in just one step) is

$$2\frac{3}{4} = \frac{2 \cdot 4 + 3}{4} = \frac{11}{4}$$

Same denominator

Thus,

$$2\frac{3}{4} + \frac{5}{6} = \frac{11}{4} + \frac{5}{6} = \frac{33}{12} + \frac{10}{12} = \frac{43}{12} = 3\frac{7}{12}$$

or, separating the whole and fractional parts,

$$2\frac{3}{4} + \frac{5}{6} = 2 + \frac{3}{4} + \frac{5}{6} = 2 + \frac{9}{12} + \frac{10}{12} = 2 + \frac{19}{12} = \frac{24}{12} + \frac{19}{12} = \frac{43}{12} = 3\frac{7}{12}$$

Note that to convert the fraction $\frac{43}{12}$ to a mixed number, we divide 43 by 12 (the answer is 3) and write the remainder 7 as the numerator of the remaining fraction, with the denominator unchanged.

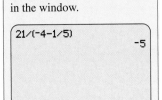

GRAPH IT

To verify part (d), enter
21 ÷ ((−) 4 − 1 ÷ 5)
ENTER to get −5 as shown
in the window.

21/(-4-1/5)
                    -5

Did you notice that $-4\frac{1}{5}$ was entered as $(-4 - \frac{1}{5})$?

EXAMPLE 5 ▶ Operations with Fractions

Perform the following indicated operations:

(a) $3\frac{1}{4} + 4\frac{1}{6}$ (b) $5 - 2\frac{1}{7}$

(c) $3\frac{1}{4} \times (-8)$ (d) $21 \div \left(-4\frac{1}{5}\right)$

Solution

(a) We first change the mixed numbers to fractions.

$$3\frac{1}{4} = \frac{3 \times 4 + 1}{4} = \frac{13}{4} \quad \text{and} \quad 4\frac{1}{6} = \frac{4 \times 6 + 1}{6} = \frac{25}{6}$$

We then obtain the LCD, which is 12, and change the fractions to equivalent ones with 12 as a denominator.

$$3\frac{1}{4} + 4\frac{1}{6} = \frac{13}{4} + \frac{25}{6} = \frac{39}{12} + \frac{50}{12} = \frac{89}{12} = 7\frac{5}{12}$$

or, separating the whole and fractional parts,

$$3\frac{1}{4} + 4\frac{1}{6} = (3 + 4) + \frac{1}{4} + \frac{1}{6}$$

$$= 7 + \frac{3}{12} + \frac{2}{12} = 7 + \frac{5}{12} = 7\frac{5}{12}$$

(b) Changing the $2\frac{1}{7}$ to $\frac{15}{7}$, and 5 to $\frac{35}{7}$, we write

$$5 - 2\frac{1}{7} = \frac{35}{7} - \frac{15}{7} = \frac{20}{7} = 2\frac{6}{7}$$

or

$$5 - 2 - \frac{1}{7} = 3 - \frac{1}{7} = \frac{21}{7} - \frac{1}{7} = \frac{20}{7} = 2\frac{6}{7}$$

(c) Write $3\frac{1}{4}$ as $\frac{13}{4}$, and recall that the product of two numbers with unlike signs is negative.

$$3\frac{1}{4} \times (-8) = \frac{13}{\overset{\cancel{4}}{\underset{1}{}}} \times (-\overset{-2}{\cancel{8}}) = -26$$

(d) Change the $-4\frac{1}{5}$ to $-\frac{21}{5}$, invert, and then multiply to get

$$21 \times \left(-\frac{5}{21}\right) = -5$$

Note that the answer is negative.   ■

If there are several operations involved, we follow the order of operations given on page 248.

EXAMPLE 6 ▶ Simplifying Several Operations

Simplify $(\frac{1}{2})^3 \div \frac{1}{4} \cdot \frac{1}{2} + \frac{1}{3}(\frac{5}{2} - \frac{1}{2}) - \frac{1}{3} \cdot \frac{1}{2}$.

$$\left(\frac{1}{2}\right)^3 \div \frac{1}{4} \cdot \frac{1}{2} + \frac{1}{3} \left(\frac{5}{2} - \frac{1}{2}\right) - \frac{1}{3} \cdot \frac{1}{2}$$

$$= \left(\frac{1}{2}\right)^3 \div \frac{1}{4} \cdot \frac{1}{2} + \frac{1}{3}(2) - \frac{1}{3} \cdot \frac{1}{2}$$

Do operations inside parentheses: $\left(\dfrac{5}{2} - \dfrac{1}{2}\right) = \left(\dfrac{4}{2}\right) = (2)$

$$= \frac{1}{8} \div \frac{1}{4} \cdot \frac{1}{2} + \frac{1}{3}(2) - \frac{1}{3} \cdot \frac{1}{2}$$

Do exponents: $\left(\dfrac{1}{2}\right)^3 = \dfrac{1}{8}$

$$= \frac{1}{2} \cdot \frac{1}{2} + \frac{1}{3}(2) - \frac{1}{3} \cdot \frac{1}{2}$$

Do division: $\dfrac{1}{8} \div \dfrac{1}{4} = \dfrac{1}{8} \cdot \dfrac{4}{1} = \dfrac{1}{2}$

$$= \frac{1}{4} + \frac{2}{3} - \frac{1}{6}$$

Do multiplication:

$\dfrac{1}{2} \cdot \dfrac{1}{2} = \dfrac{1}{4}; \quad \dfrac{1}{3}(2) = \dfrac{2}{3}; \quad \dfrac{1}{3} \cdot \dfrac{1}{2} = \dfrac{1}{6}$

$$= \frac{11}{12} - \frac{1}{6}$$

Do addition: $\dfrac{1}{4} + \dfrac{2}{3} = \dfrac{3}{12} + \dfrac{8}{12} = \dfrac{11}{12}$

$$= \frac{9}{12}$$

Do subtraction: $\dfrac{11}{12} - \dfrac{1}{6} = \dfrac{11}{12} - \dfrac{2}{12} = \dfrac{9}{12}$

$$= \frac{3}{4}$$

Reduce: $\dfrac{9}{12}$ to $\dfrac{3}{4}$

C. Applications

Have you seen the nutrition labeling attached to most food products? It gives the information (per serving) about the product you are about to consume. Let us look at an example.

EXAMPLE 7 ▶ Nutrition and Fractions

The nutrition panel shown in Figure 5.7 came from a bag of potato chips. For one serving (about 20 chips),

(a) how many calories come from fat?

(b) what fraction of the calories are from fat?

Nutrition Facts

Serving size 1 oz. (28g/About 20 chips)
Servings Per Container About 8

Amount Per Serving

Calories 150   Calories from Fat 90

	% Daily Value*
Total Fat 10g	**15%**
Saturated Fat 3g	**15%**
Cholesterol 0mg	**0%**
Sodium 180mg	**8%**
Total Carbohydrate 15g	**5%**
Dietary Fiber 1g	**2%**
Sugars 0g	
Protein 2g	

Vitamin A 0%	•	Vitamin C 10%
Calcium 0%	•	Iron 0%

* Percent Daily Values are based on a 2,000 calorie diet. Your daily values may be higher or lower depending on your calorie needs:

	Calories:	2,000	2,500
Total Fat	Less than	65g	80g
Sat Fat	Less than	20g	25g
Cholesterol	Less than	300mg	300mg
Sodium	Less than	2,400mg	2,400mg
Total Carbohydrate		300g	375g
Dietary Fiber		25g	30g

Calories per gram:

Fat 9 • Carbohydrates 4 • Protein 4

FIGURE 5.7

(c) what fraction of the daily value of total fat would you consume in one serving?

(d) how many calories per chip?

Solution

(a) 90 calories come from fat.

(b) 90 out of the 150, or

$$\frac{90}{150} = \frac{30 \cdot 3}{30 \cdot 5} = \frac{3}{5}$$

of the calories, are from fat.

(c) The daily value of total fat is 65 g and the total fat in one serving is 10 g; thus the fraction of the daily value of total fat is $\frac{10}{65} = \frac{2}{13}$.

(d) One serving has 20 chips and there are 150 calories in one serving; thus there are $\frac{150}{20} = 7\frac{1}{2}$ calories per chip. ■

EXERCISES 5.3

Ⓐ Rational Numbers

In problems 1–4, identify the numerator and denominator of each rational number.

1. $\frac{3}{4}$ **2.** $\frac{4}{5}$ **3.** $\frac{3}{-5}$ **4.** $\frac{-4}{5}$

In problems 5–7, identify the rational numbers that are equal by using the definition of equality.

5. $\frac{17}{41}, \frac{289}{697}, \frac{714}{1682}$ **6.** $\frac{438}{529}, \frac{19}{23}, \frac{323}{391}$

7. $\frac{11}{91}, \frac{111}{911}, \frac{253}{2093}$

In problems 8–16, reduce each rational number.

8. $\frac{14}{21}$ **9.** $\frac{95}{38}$ **10.** $\frac{42}{86}$

11. $\frac{21}{48}$ **12.** $\frac{15}{12}$ **13.** $\frac{30}{28}$

14. $\frac{22}{33}$ **15.** $\frac{52}{78}$ **16.** $\frac{224}{84}$

Ⓑ Operations with Rational Numbers

It is possible to add and subtract rational numbers by converting them to equivalent rational numbers with the same denominator. In problems 17–19, express each sum as a sum of rational numbers with a denominator of 18.

17. $\frac{2}{9} + \frac{1}{6} + \frac{7}{18}$ **18.** $\frac{7}{3} + \frac{7}{9} + \frac{5}{6}$

19. $\frac{1}{3} + \frac{1}{6} + \frac{1}{9}$

In problems 20–90, perform the indicated operations and reduce each answer (if possible).

20. $\frac{1}{7} + \frac{1}{3}$ **21.** $\frac{1}{7} + \frac{1}{9}$

22. $\frac{2}{7} + \frac{3}{11}$ **23.** $\frac{3}{4} + \frac{5}{6}$

24. $\frac{1}{12} + \frac{7}{18}$ **25.** $\frac{3}{17} + \frac{7}{19}$

26. $\frac{1}{3} - \frac{1}{7}$ **27.** $\frac{1}{7} - \frac{1}{9}$

28. $\frac{2}{7} - \frac{3}{11}$ **29.** $\frac{3}{4} - \frac{5}{6}$

30. $\frac{7}{18} - \frac{1}{12}$ **31.** $\frac{7}{19} - \frac{3}{17}$

32. $\frac{3}{4} \times \frac{2}{7}$ **33.** $\frac{2}{5} \times \frac{5}{3}$

34. $\frac{7}{9} \times \frac{3}{8}$ **35.** $\frac{3}{4} \div \frac{2}{7}$

36. $\frac{2}{5} \div \frac{5}{3}$ **37.** $\frac{7}{9} \div \frac{3}{8}$

38. $\left(\frac{-2}{5}\right) \times \frac{4}{9}$ **39.** $\left(-\frac{6}{7}\right) \times \left(-\frac{3}{11}\right)$

40. $\frac{4}{5} \div \left(\frac{-7}{9}\right)$ **41.** $\left(-\frac{3}{4}\right) \div \left(-\frac{7}{6}\right)$

42. $\frac{3}{4} \div \left(-\frac{1}{5}\right)$ **43.** $\frac{1}{8} \div \left(-\frac{3}{4}\right)$

44. $\left(-\frac{1}{4}\right) + \left(-\frac{1}{7}\right)$ **45.** $\left(-\frac{1}{8}\right) + \left(\frac{1}{4}\right)$

46. $\left(\frac{1}{3} + \frac{1}{4}\right) + \frac{7}{8}$ **47.** $\frac{3}{8} - \left(\frac{1}{4} - \frac{1}{8}\right)$

48. $\left(\frac{1}{5} \times \frac{1}{4}\right) \times \frac{3}{7}$ **49.** $\frac{1}{2} \times \left(\frac{7}{8} \times \frac{7}{5}\right)$

50. $\frac{1}{2} \div \left(\frac{1}{8} \div \frac{1}{4}\right)$ **51.** $\left(\frac{1}{2} \div \frac{1}{8}\right) \div \frac{1}{4}$

52. $\frac{3}{4} + \frac{1}{2}\left(\frac{3}{2} + \frac{1}{4}\right)$ **53.** $\frac{2}{3}\left(\frac{1}{2} + \frac{3}{4}\right) + \frac{2}{3}$

54. $\frac{1}{2}\left(\frac{3}{4} - \frac{1}{2}\right) - \frac{1}{12}$ **55.** $\frac{1}{3}\left(\frac{3}{5} - \frac{1}{5}\right) - \frac{1}{30}$

56. $\frac{1}{2}\left(\frac{5}{2} - \frac{1}{3}\right) - \frac{5}{12}$

57. $1\frac{1}{2} + \frac{1}{7}$

58. $5 - 1\frac{1}{3}$

59. $\frac{1}{4} \times 1\frac{1}{7}$

60. $5 \div \left(-2\frac{1}{2}\right)$

61. $3\frac{1}{4} + \frac{1}{6}$

62. $4 - 2\frac{1}{4}$

63. $\frac{1}{5} \times 2\frac{1}{7}$

64. $6 \div \left(-1\frac{1}{5}\right)$

65. $-3 + 2\frac{1}{4}$

66. $-\frac{2}{3} - (-2)$

67. $(-8) \times 2\frac{1}{4}$

68. $7 \div \left(-2\frac{1}{3}\right)$

69. $-2 + 1\frac{1}{5}$

70. $-\frac{3}{4} - (-3)$

71. $(-9) \times 3\frac{1}{3}$

72. $\left(-\frac{1}{6}\right) \div \left(-\frac{5}{7}\right)$

73. $7\frac{1}{4} + \left(-\frac{1}{8}\right)$

74. $-3\frac{1}{8} - (-2)$

75. $\left(-1\frac{1}{4}\right) \times \left(-2\frac{1}{10}\right)$

76. $\left(-1\frac{1}{8}\right) \div \left(-2\frac{1}{4}\right)$

77. $\frac{1}{2} \times \frac{1}{6} - \frac{1}{3} + \frac{1}{4}$

78. $\frac{3}{8} - 6\left(\frac{1}{4} - \frac{1}{8}\right)$

79. $\frac{1}{3} - \frac{1}{3} \times \frac{2}{3} \div \frac{2}{5}$

80. $\frac{1}{2} \div \frac{1}{4} - \frac{3}{4}$

81. $\left(2\frac{1}{2}\right) \times \left(-3\frac{1}{4}\right) - \left(-7\frac{1}{8}\right) \div 3$

82. $\left(-6\frac{2}{5}\right) \div (-4) + \left(2\frac{1}{10}\right) \times (-2)$

83. $12 \div 6 - \left(\frac{1}{3} + \frac{1}{2}\right)$

84. $18 \div 9 - \left(\frac{1}{4} + \frac{1}{6}\right)$

85. $\frac{1}{3} \cdot \frac{1}{4} \div \frac{1}{2} + \left(\frac{5}{6} - \frac{1}{2}\right)$

86. $\frac{1}{3} \cdot \frac{1}{6} \div \frac{1}{2} + \left(\frac{4}{5} - \frac{1}{2}\right)$

87. $\frac{1}{6} \div \frac{1}{3} \cdot \frac{1}{3} \cdot \frac{1}{3} + \left(\frac{1}{4} - \frac{1}{9}\right)$

88. $\frac{1}{10} \div \frac{1}{2} \cdot \frac{1}{2} \cdot \frac{1}{2} + \left(\frac{2}{3} - \frac{1}{2}\right)$

89. $8 \div \frac{1}{2} \cdot \frac{1}{2} \cdot \frac{1}{2} - \left(\frac{1}{3} + \frac{1}{5}\right)$

90. $6 \div \frac{1}{3} \cdot \frac{1}{3} \cdot \frac{1}{3} - \left(\frac{1}{3} + \frac{1}{5}\right)$

C Applications

91. The normal body temperature is $98\frac{6}{10}°$F. Carlos had the flu and his temperature was $101\frac{6}{10}°$F. How many degrees above normal is that?

92. An average human brain weighs approximately $3\frac{1}{8}$ lb. The brain of the writer Anatole France weighed only $2\frac{1}{4}$ lb. How much weight under the average is that?

93. As stated in problem 92, the average human brain weighs approximately $3\frac{1}{8}$ lb. The heaviest brain ever recorded was that of Ivan Sergeyevich Turgenev, a Russian author. His brain weighed approximately $4\frac{7}{16}$ lb. How much above the average is this weight?

94. A board $\frac{3}{4}$ in. thick is glued to another board $\frac{5}{8}$ in. thick. If the glue is $\frac{1}{32}$ in. thick, how thick is the result?

95. A recent survey found that $\frac{3}{10}$ of the U.S. population work long hours and smoke. The survey also found that $\frac{1}{5}$ of the people are overweight non-smokers who do not work long hours. What fraction of the people are not overweight, do not smoke, and do not work long hours?

96. Human bones are $\frac{1}{4}$ water, $\frac{9}{20}$ minerals, and the rest living tissue. What fraction of human bone is living tissue?

97. In a recent year, Americans spent $\$3\frac{1}{2}$ billion on daily newspapers and $\$1\frac{2}{5}$ billion on Sunday newspapers. How many billions of their dollars were spent on newspapers?

98. Americans work an average of $46\frac{3}{5}$ hours per week, whereas Canadians work $38\frac{9}{10}$ hours. How many more hours per week do Americans work on average?

99. Do husbands help with the household chores? A recent survey estimated that husbands spend about $7\frac{1}{2}$ hours during weekdays helping around the house, $2\frac{3}{5}$ hours on Saturday, and 2 hours on Sunday. How many hours do husbands work around the house during the entire week?

Problems 100–109 refer to the diagram below.

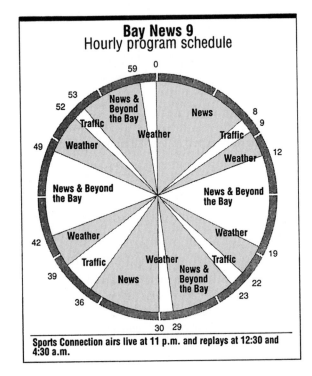

What fraction of the hour is devoted to

100. weather?   **101.** traffic?

102. news and Beyond the Bay?

103. only news?

104. Which of the features uses the most time? What fraction of an hour does it use?

105. Which of the features uses the least time? What fraction of an hour does it use?

106. What is the time difference (as a fraction) between the features that use the least and the most time each hour?

107. Is the amount of time devoted to a particular feature the same at the top and bottom of the hour? If not, in which part of the hour (top or bottom) does the feature use the most time?

108. What is the time difference (as a fraction) between the news at the top of the hour and news at the bottom of the hour?

109. Which features use the same amount of time?

 In Other Words

110. Why do you think the set of rational numbers is symbolized by the letter Q?

111. You know that $\frac{3}{3} = 1$, $\frac{2}{2} = 1$, and $\frac{1}{1} = 1$. Does $\frac{0}{0} = 1$? Why or why not?

112. The Golden Ratio mentioned in the Getting Started section is

$$\frac{b}{a} = \frac{\sqrt{5} - 1}{2}$$

but it does not satisfy the definition of a rational number. Why?

 Using Your Knowledge

Road maps and other maps are drawn to **scale**. The scale on a map shows what distance is represented by a certain measurement on the map. For example, the scale on a certain map is

1 in. = 36 mi

Thus, if the distance on the map is $2\frac{1}{4}$ in. from Indianapolis to Dayton, the actual distance is

$$36 \times 2\frac{1}{4} = 36 \times \frac{9}{4} = (9 \times \cancel{4}) \times \frac{9}{\cancel{4}} = 81 \text{ mi}$$

113. Find the actual distance from Indianapolis to Cincinnati, a distance of $3\frac{1}{2}$ in. on the map.

114. Find the actual distance from Indianapolis to Terre Haute, a distance of $3\frac{1}{4}$ in. on the map.

115. If the actual distance between two cities is 108 mi, what is the distance on the map?

116. If the actual distance between two cities is 162 mi, what is the distance on the map?

 Discovery

Let us try to discover how many positive rational numbers there are. We arrange the first few fractions as shown below.

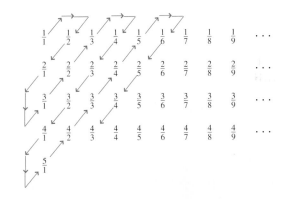

117. Can you discover the next two rows in this array?

118. Notice that the number $\frac{1}{5}$ appears in the first row, fifth column. Can you discover where the number $\frac{37}{43}$ will be?

119. By following the path indicated in the array, we can set up a correspondence between the set of natural numbers N and the set of positive rational numbers. The first few terms look like this.

1 2 3 4 5 6 7 8 9 10 11 $\cdots$

$\updownarrow$ $\updownarrow$ $\updownarrow$ $\updownarrow$ $\updownarrow$ $\updownarrow$ $\updownarrow$ $\updownarrow$ $\updownarrow$ $\updownarrow$ $\updownarrow$

1 $\frac{1}{2}$ $\frac{2}{1}$ $\frac{3}{1}$ $\frac{1}{3}$ $\frac{1}{4}$ $\frac{2}{3}$ $\frac{3}{2}$ $\frac{4}{1}$ $\frac{5}{1}$ $\frac{1}{5}$ $\cdots$

Notice that in this pairing we skipped $\frac{2}{2}$, $\frac{4}{2}$, $\frac{3}{3}$, and $\frac{2}{4}$. Can you discover why?

120. The correspondence indicated in problem 119 can be continued indefinitely. If we assign a cardinal number to the set N of natural numbers, say, $n(N) = \aleph_0$ (read *aleph null*), so that we can speak of the "number of natural numbers" as being $\aleph_0$, what would $n(Q^+)$ be? That is, how many positive rational numbers are there?

121. On the basis of your answer to problem 120, would you say that there are more or fewer rational numbers than natural numbers?

Collaborative Learning

Do you remember the *Fibonacci sequence* in Section 2.1? The sequence is

$$1, 1, 2, 3, 5, 8, 13, 21, 34, 55, 89, 144, 233, \ldots$$

Form two groups.

1. One group calculates the ratios of numbers that precede each other in the sequence, that is, 1/1, 2/1, 3/2, and so on.

2. The other group calculates the ratios of the numbers following each other in the sequence 1/1, 1/2, 2/3, and so on. (These are the multiplicative inverses of the numbers obtained in 1.)

3. Compare results. Are the ratios multiplicative inverses?

4. What conjecture would you make for the ratios of the numbers obtained in part (1)? What about for the ratio of the numbers obtained in part (2)? In the next section, we convert rational numbers of the form *a/b* to decimals by dividing *a* by *b*. Do you notice anything about the decimal part in each of the ratios?

5. How close is the answer you get for the ratios 1/1, 1/2, 2/3, and so on, to the Golden Ratio $\dfrac{(\sqrt{5} - 1)}{2}$ we mentioned at the beginning of the section?

5.4 Decimals, Scientific Notation, and Significant Digits

Adverbs, Adjectives, and Decimals

How fast does a snail move? You can express the answer as a decimal. Many measurements involving decimals are so small or so large that they must be written in a special way called **scientific notation,** which you will study in part B of this section. The following are some speeds in decimal form and in scientific notation:

Speed at which a snail moves	$0.0310 = 3.10 \times 10^{-2}$ mph
Brisk walking speed of a person	$5.72 = 5.72 \times 10^{0}$ mph
Fast running speed of a man	$16.70 = 1.67 \times 10^{1}$ mph
Speed of a peregrine falcon	$168 = 1.68 \times 10^{2}$ mph
Speed of an SR-71A plane	$2193 = 2.193 \times 10^{3}$ mph

Decimals are used in many ways. For example, social scientists often try to quantify different characteristics of people or things. Thus, psychologists and linguists have a numerical way to indicate the difference between *nice* and *very nice* or between *unpleasant* and *very unpleasant.* Suppose you assign a positive number (+2, for example) to the adjective *nice* and a negative number (say −2) to *unpleasant* and a positive number greater than 1 (say +1.25) to *very.* Then, *very nice* is quantified as

Very nice
↓   ↓
$(1.25) \cdot (2) = 2.50$

That is, 1.25 and 2 are multiplied. Similarly, *very unpleasant* is quantified as

Very unpleasant
↓   ↓
$(1.25) \cdot (-2) = -2.50$

Table 5.3 gives some adverbs and adjectives and their average numerical values, as rated by several panels of college students. (Values differ from one panel to another.)

TABLE 5.3

Adverbs		Adjectives	
Slightly	0.54	Wicked	−2.5
Rather	0.84	Disgusting	−2.1
Decidedly	1.16	Average	−0.8
Very	1.25	Good	3.1
Extremely	1.45	Lovable	2.4

Now you are able to quantify phrases such as *extremely disgusting* and *very wicked*. Which of the two phrases is worse? If you do the multiplication, you will know, and if you get the correct answer, you are 4.495! ▶

As you probably know, the number $\frac{1}{2}$ can be written in decimal form, that is, as 0.5. Because all rational numbers can be written in the form a/b, it is always possible to change a rational number a/b to its decimal form simply by dividing a by b. For example, $\frac{1}{2} = 0.5$, $\frac{1}{8} = 0.125$, and $\frac{1}{4} = 0.25$. Of course, if the denominator of a fraction is already a power of 10 (10, 100, 1000, and so on), then the fraction can easily be written as a decimal. Thus,

$$\frac{7}{10} = 0.7, \qquad \frac{19}{100} = 0.19, \qquad \text{and} \qquad \frac{17}{1000} = 0.017$$

You should keep in mind that the successive places to the right of the decimal point have the place values

$$\frac{1}{10^1}, \qquad \frac{1}{10^2}, \qquad \frac{1}{10^3}, \qquad \frac{1}{10^4}, \qquad \dots$$

The exponent on the 10 is the number of the place. For example, the underlined digit in 2.56794 has the place value $1/10^4$, because the 9 is in the fourth decimal place.

A. Decimals in Expanded Form

You learned in Section 4.2 that a positive integer can always be written in expanded form as, for instance,

$$361 = (3 \times 10^2) + (6 \times 10^1) + (1 \times 10^0)$$

Can you do something like this for a decimal, say, 3.52? You know that

$$0.5 = \frac{5}{10} \qquad \text{and} \qquad 0.02 = \frac{2}{100}$$

so

$$3.52 = (3 \times 10^0) + \left(5 \times \frac{1}{10^1}\right) + \left(2 \times \frac{1}{10^2}\right)$$

This is somewhat awkward, so to make it more convenient, define **negative exponents** as follows:

$$10^{-1} = \frac{1}{10^1} = \frac{1}{10}, \quad 10^{-2} = \frac{1}{10^2} = \frac{1}{100}, \quad 10^{-3} = \frac{1}{10^3} = \frac{1}{1000}, \quad \cdots$$

Definition of Negative Exponents

If n is a positive integer and $a \neq 0$,

$$a^{-n} = \frac{1}{a^n}$$

The definition of negative exponents yields a good pattern.

$10^3 = 1000$
$10^2 = 100$
$10^1 = 10$
$10^0 = 1$
$10^{-1} = \frac{1}{10}$
$10^{-2} = \frac{1}{100}$
$10^{-3} = \frac{1}{1000}$

And so on

It can be shown that **negative exponents obey exactly the same laws as positive exponents.**

Using these negative exponents, you can write decimals in expanded form. For example,

$$3.52 = (3 \times 10^0) + (5 \times 10^{-1}) + (2 \times 10^{-2})$$

and

$$362.754 = (3 \times 10^2) + (6 \times 10^1) + (2 \times 10^0)$$
$$+ (7 \times 10^{-1}) + (5 \times 10^{-2}) + (4 \times 10^{-3})$$

These look like very natural generalizations of the expanded form that you used for positive exponents.

EXAMPLE 1 ▶ Writing Decimals in Expanded Form

(a) Write 25.603 in expanded form.

(b) Write $(7 \times 10^2) + (2 \times 10^0) + (6 \times 10^{-1}) + (9 \times 10^{-4})$ in standard decimal form.

Solution

(a) $25.603 = (2 \times 10^1) + (5 \times 10^0) + (6 \times 10^{-1}) + (0 \times 10^{-2}) + (3 \times 10^{-3})$
$= (2 \times 10^1) + (5 \times 10^0) + (6 \times 10^{-1}) + (3 \times 10^{-3})$

(b) Notice that the exponent in 10^{-n} tells you that the first nonzero digit comes in the nth decimal place. With this in mind,

$$(7 \times 10^2) + (2 \times 10^0) + (6 \times 10^{-1}) + (9 \times 10^{-4}) = 702.6009 \qquad ■$$

B. Scientific Notation

In science, very large or very small numbers frequently occur. For example, a red cell of human blood contains 270,000,000 hemoglobin molecules, and the mass of a single carbon atom is 0.000 000 000 000 000 000 000 019 9 g. Numbers in this form are difficult to write and to work with, so they are written in *scientific notation*.

GRAPH IT

Your grapher has to be in the mode to do scientific notation! Press $\boxed{\text{MODE}}$ $\boxed{\blacktriangleright}$ $\boxed{\text{ENTER}}$ and then $\boxed{\text{2ND}}$ $\boxed{\text{MODE}}$ to go to the home screen. For Example 2(a), enter 270,000,000 $\boxed{\text{ENTER}}$ to obtain 2.7E8, which means 2.7×10^8 as shown in the window.

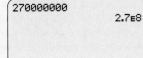

Go ahead and try Example 2(b)!

Definition of Scientific Notation

A number is said to be in **scientific notation** if it is written in the form

$$m \times 10^n$$

where m is a number greater than or equal to 1 and less than 10, and n is an integer.

For any given number, the m is obtained by placing the decimal point so that there is exactly one nonzero digit to its left. The n is then the number of places that the decimal point must be moved from its position in m to its original position; it is positive if the point must be moved to the right and negative if the point must be moved to the left. The following are examples:

$5.3 = 5.3 \times 10^0$	Decimal point in 5.3 must be moved 0 places.
$87 = 8.7 \times 10^1 = 8.7 \times 10$	Decimal point in 8.7 must be moved 1 place to the right to get 87.
$68,000 = 6.8 \times 10^4$	Decimal point in 6.8 must be moved 4 places to the right to get 68,000.
$0.49 = 4.9 \times 10^{-1}$	Decimal point in 4.9 must be moved 1 place to the left to get 0.49.
$0.072 = 7.2 \times 10^{-2}$	Decimal point in 7.2 must be moved 2 places to the left to get 0.072.
$0.0003875 = 3.875 \times 10^{-4}$	Decimal point in 3.875 must be moved 4 places to the left to get 0.0003875.

EXAMPLE 2 ▶ Writing Numbers in Scientific Notation

Write the following in scientific notation:

(a) 270,000,000

(b) 0.000 000 000 000 000 000 000 019 9

Solution

(a) $270,000,000 = 2.7 \times 10^8$

(b) $0.000\ 000\ 000\ 000\ 000\ 000\ 000\ 019\ 9 = 1.99 \times 10^{-23}$ ■

EXAMPLE 3 ▶ From Scientific Notation to Decimals

Write the following in standard decimal notation:

(a) 2.5×10^{10} (b) 7.4×10^{-6}

Solution

(a) $2.5 \times 10^{10} = 25,000,000,000$

(b) $7.4 \times 10^{-6} = 0.0000074$ ■

We noted earlier in this section that negative exponents obey the same laws as positive exponents. We can verify this quite easily in simple cases. For example,

$$10^4 \times 10^{-2} = 10,000 \times \frac{1}{100} = 100 = 10^2 = 10^{4+(-2)}$$

$$10^{-2} \times 10^{-3} = 0.01 \times 0.001 = 0.00001 = 10^{-5} = 10^{-2+(-3)}$$

The laws of exponents can be used to do calculations with numbers in scientific notation.

EXAMPLE 4 ▶ Calculations in Scientific Notation

Do the following calculation, and write the answer in scientific notation:

$$(5 \times 10^4) \times (9 \times 10^{-7})$$

Solution

$$
\begin{aligned}
(5 \times 10^4) \times (9 \times 10^{-7}) &= (5 \times 9) \times (10^4 \times 10^{-7}) \\
&= 45 \times 10^{4-7} \\
&= 45 \times 10^{-3} \\
&= (4.5 \times 10^1) \times 10^{-3} \\
&= 4.5 \times 10^{1-3} \\
&= 4.5 \times 10^{-2}
\end{aligned}
$$

■

C. Operations with Decimals

Can you buy 100 King Supremes for under $1?

We have already studied the operations of addition, subtraction, multiplication, and division with whole numbers, integers, and fractions. We now consider the same operations with decimals. To add or subtract decimals, we proceed as with whole numbers, making sure that the numbers to be added have the same number of decimal places. Thus, to add 5.1 and 2.81, we attach a 0 to 5.1 so that it has two decimal places and then add. To see why this works, we write the numbers in expanded form and add as follows:

$$
\begin{aligned}
5.10 &= 5 + \tfrac{1}{10} + \tfrac{0}{100} \\
\underline{2.81} &= \underline{2 + \tfrac{8}{10} + \tfrac{1}{100}} \\
7.91 &= 7 + \tfrac{9}{10} + \tfrac{1}{100}
\end{aligned}
$$

In practice, we place the numbers to be added in a vertical column *with the decimal points aligned,* and then add or subtract as required. The rules for "carrying" in addition and "borrowing" in subtraction are justified here also, as shown by the expanded forms in the next examples.

EXAMPLE 5 ▶ Adding and Subtracting Decimals

(a) Add 4.81 and 3.7.   (b) Subtract 6.53 from 8.71.

Solution

(a) We attach a 0 to the 3.7, align the decimal points, and add.

Short Form

$$
\begin{array}{r}
4.81 \\
+\ 3.70 \quad \leftarrow \text{Attach a 0.} \\
\hline
8.51
\end{array}
$$

Expanded Form

$$
\begin{aligned}
& 4 + \tfrac{8}{10} + \tfrac{1}{100} \\
+\ & 3 + \tfrac{7}{10} + \tfrac{0}{100} \\
\hline
& 7 + \tfrac{15}{10} + \tfrac{1}{100} \\
=\ & 7 + \tfrac{10}{10} + \tfrac{5}{10} + \tfrac{1}{100} \\
=\ & 8 + \tfrac{5}{10} + \tfrac{1}{100} \\
=\ & 8.51
\end{aligned}
$$

(b) *Short Form*   *Expanded Form*

$$\begin{array}{r} 8.71 \\ -6.53 \\ \hline 2.18 \end{array}$$

$$\begin{array}{r} 8 + \frac{7}{10} + \frac{1}{100} = 8 + \frac{6}{10} + \frac{11}{100} \\ -6 + \frac{5}{10} + \frac{3}{100} = 6 + \frac{5}{10} + \frac{3}{100} \\ \hline 2 + \frac{1}{10} + \frac{8}{100} = 2.18 \end{array}$$

Note: In the first line of the expanded form, we wrote

$$\frac{7}{10} = \frac{6}{10} + \frac{1}{10} = \frac{6}{10} + \frac{10}{100}$$

and then combined the hundredths.   ◾

To understand the rule for multiplying decimals, look at the following example, in which the decimals are first replaced by equivalent fractions:

EXAMPLE 6 ▶ Multiplying Decimals

Multiply 0.37×7.2.

Solution

$$0.37 \times 7.2 = \frac{37}{100} \times \frac{72}{10} = \frac{37 \times 72}{1000}$$

The numerator 37×72 tells us that the 37 and the 72 are to be multiplied as usual for whole numbers. The 1000 in the denominator tells us that the final answer will have three decimal places, the sum of the number of decimal places in the two factors. Thus, we can multiply the 0.37 and the 7.2 just as if they were whole numbers and then place the decimal point so that there are three decimal places in the product. The short form of the multiplication follows:

$$\begin{array}{r} 0.37 \\ \times \quad 7.2 \\ \hline 74 \\ 259 \quad\;\; \\ \hline 2.664 \end{array}$$

← 2 decimal places
← 1 decimal place

← 2 + 1 =
3 decimal places   ◾

In general, we multiply two decimals as if they were whole numbers and then place the decimal point so that the number of decimal places is equal to the sum of the number of decimal places in the two factors. If signed numbers are involved, the rules of signs apply. Thus,

$$(-0.37) \times 7.2 = -2.664$$
$$0.37 \times (-7.2) = -2.664$$
$$(-0.37) \times (-7.2) = 2.664$$

A division involving decimals is easier to understand if the divisor is a whole number. If the divisor is not a whole number, we can make it so by moving the decimal point to the right the same number of places in both the divisor and the dividend. This procedure can be justified as in the next example.

EXAMPLE 7 ▶ **Applications to Unit Cost**

A $6\frac{1}{2}$-oz can of tuna fish is on sale for 55¢. What is the cost per ounce to the nearest tenth of a cent?

Solution

The cost per ounce can be obtained by dividing the total cost by the number of ounces. Thus, we have to find $55 \div 6.5$, and we can make the divisor a whole number by writing the division in fractional form.

$$\frac{55}{6.5} = \frac{550}{65} \qquad \text{Multiply numerator and denominator by 10.}$$

This is equivalent to moving the decimal point one place to the right in both the divisor and the dividend. Then we divide in the usual way.

$$
\begin{array}{r}
8.46 \\
65\overline{)550.00} \\
\underline{520} \\
30\,0 \\
\underline{26\,0} \\
4\,00 \\
\underline{3\,90} \\
10
\end{array}
$$

To two decimal places, the cost per ounce is 8.46¢. Therefore, to the nearest tenth of a cent, the answer is 8.5¢. ∎

Again, if signed numbers are involved, the rules of signs apply. For example,

$$\frac{-14}{2.5} = -5.6, \qquad \frac{14}{-2.5} = -5.6, \qquad \text{and} \qquad \frac{-14}{-2.5} = 5.6$$

PROBLEM SOLVING

Applications to Decimals

❶ Read the problem.

Bank A offers a checking account costing $3 per month plus $.20 per check after the first 10 checks. Bank B offers unlimited checking for a $6 monthly charge.

(a) Which bank is cheaper if you write 20 checks per month?

(b) How many checks do you have to write so that the cost is the same in both banks?

❷ Select the unknown.

Find out (a) which bank is cheaper if you write 20 checks per month and (b) when the cost is the same for both banks.

❸ Think of a plan.
Find the cost of writing 20 checks in bank A and in bank B.

Bank A costs $3 plus $.20 for each check over 10. For 20 checks, 10 are free and 10 cost $.20 each.

continued

④ **Use** the information to find the answers.

Which bank is cheaper if you write 20 checks?

The cost of writing 20 checks in bank A is $3 + 10 · $.20 = $3 + $2 = $5. The cost in bank B is $6. Bank A is cheaper.

When is the cost the same?

The cost is $6 in bank A when you pay the $3 monthly charge and you spend an additional $3 on checks. Since checks are $.20 each, you can write $3/$.20 = 15 checks plus the 10 free ones for a total of 25 checks. Thus the cost is the same ($6) when you write 25 checks.

⑤ **Verify** the answer.

The cost of writing 25 checks in bank A is $3 + 15 · $.20 = $3 + $3 = $6, the same as in bank B.

TRY EXAMPLE 8 NOW.

Cover the solution, write your own solution, and then check your work.

EXAMPLE 8 ▶ Car Rental and Decimals

Suppose you need to rent a car for a 3-day, 400-mi trip. Your choice of rental companies is as follows:

Company A $50 per day, with unlimited free mileage
Company B $25 per day, plus $.20 per mile

(a) Which company offers the best deal for your trip?

(b) How many miles do you have to travel for the 3-day trip for the cost to be the same with both companies?

Solution
The cost for each company is as follows:

Company A costs $50 each day for 3 days, or $3 \times \$50 = \$150$.
Company B costs $3 \times \$25 + \$.20 \times 400 = \$75 + \$80 = \$155$.

(a) It is less expensive to rent from company A.

(b) The 3-day charge is $150 for company A and $75 for company B. The $75 difference ($150 − $75) can be used to pay the mileage with company B. Since the cost is $.20 per mile, the number of miles you can travel is

$$\frac{75}{0.20} = 375 \text{ mi}$$

Verifying that the cost is the same can be done by looking at the cost for company B, $75 + $.20 × 375 = $75 + $75 = $150; it is the same as for company A.   ∎

D. Rounding

In Example 7, we **rounded off** the quotient 8.46 to one decimal place to get 8.5. This type of procedure is necessary in many practical problems and is done in the following way:

Rules for Rounding Off Numbers

1. Underline the digit in the place to which you are rounding.

2. If the first digit to the right of the underlined digit is 5 or more, add 1 to the underlined digit. Otherwise, do not change the underlined digit.

3. Drop all digits to the right of the underlined digit and attach 0s to fill in the place values if necessary.

For instance, to round off 8.46 to one decimal place, that is, to the tenths place, do the following:

1. Underline the 4: 8.4̲6.

2. The digit to the right of the underlined digit is 6 (greater than 5). So add 1 to the underlined digit.

3. Drop all digits to the right of the underlined digit. The result is 8.5.

If it is necessary to round off 2632 to the hundreds place, you should underline the 6 to show 2632. Since the next digit is 3 (less than 5), the underlined 6 is unchanged and the 32 is replaced by 00 to give 2600.

In many applications, the numbers are obtained by measurement and are usually only approximate numbers correct to the stated number of decimal places. When calculations are performed with approximate numbers, two rules are customarily used to avoid presenting results with a false appearance of accuracy. The rule for addition and subtraction is as follows:

Addition and Subtraction Rule for Approximate Numbers

The result of an addition or subtraction of approximate numbers should be written with the same number of decimal places possessed by the approximate number with the *least* number of decimal places.

EXAMPLE 9 ▶ Rounding Numbers

The three sides of a triangle are measured to be approximately 26.3, 7.41, and 20.64 cm long, respectively. What is the approximate perimeter of the triangle?

Solution
The perimeter is the total distance around the triangle, so we must add the lengths of the three sides. We get

$$
\begin{array}{r}
26.3 \\
7.41 \\
+\ 20.64 \\
\hline
54.35
\end{array}
$$

However, the least precise of the measurements is 26.3, with just one decimal place. Therefore, we round off the answer to 54.4 cm.　■

Before giving the rule that is used when multiplying or dividing approximate numbers, we must explain what is meant by the **significant digits** of a number.

Significant Digits

1. The digits 1, 2, 3, 4, 5, 6, 7, 8, and 9 are always significant.

2. The digit 0 is significant if it is preceded and followed by other significant digits.

3. The digit 0 is not significant if its only purpose is to place the decimal point.

Thus, all the digits in the numbers 2.54 and 1.73205 are significant. However, the number 0.00721 has only three significant digits: 7, 2, and 1; the 0s in this number do nothing except place the decimal point. There is one ambiguous case left. For example, how many significant digits does the number 73,200 have? We cannot answer this question unless we know more about the final 0s. To avoid this type of difficulty, we write 73,200 in the following forms:

7.32×10^4 If three digits are significant
7.320×10^4 If four digits are significant
7.3200×10^4 If all five digits are significant

EXAMPLE 10 ▶ Number of Significant Digits

State the number of significant digits in each of the following:

(a) 0.50 in.   (b) 0.05 lb   (c) 8.2×10^3 L   (d) 8.200×10^3 km

Solution

(a) 0.50 has two significant digits.

(b) 0.05 has one significant digit. The 0 before the 5 only places the decimal point.

(c) 8.2×10^3 has two significant digits.

(d) 8.200×10^3 has four significant digits.   ■

We can now state the multiplication and division rule for approximate numbers as follows:

Multiplication and Division Rule for Approximate Numbers

The result of a multiplication or division of approximate numbers should be given with the same number of significant digits as are possessed by the approximate number with the *fewest* significant digits.

For instance, if we wish to convert 6 cm to inches, we note that 1 cm is approximately 0.394 in. Therefore, we multiply

$6 \times 0.394 = 2.364$

However, 0.394 has only three significant digits, so we must round our answer to three significant digits. Our final result is

6 cm ≈ 2.36 in.   ≈ means "approximately equal to."

Notice carefully that the 6 in 6 cm is an *exact,* not an approximate, number.

EXAMPLE 11 ▶ Area of Rectangles

Two adjacent sides of a rectangular lot are measured to be approximately 91.5 and 226.6 ft. Find the area of this lot.

Solution

The area of a rectangle of sides a and b is $A = ab$. Thus, we have

$$A = 91.5 \times 226.6$$
$$= 20{,}733.90$$

However, the 91.5 has only three significant digits, so our answer must be rounded to three significant digits. This gives $20{,}700$ ft^2 for the area. A better form for the answer is 2.07×10^4 ft^2. Use of scientific notation frequently helps to display the significant digits of the answer. ■

EXAMPLE 12 ▶ Scientific Notation and Rounding

A computer uses a collection of ones (1s) and zeros (0s) to store information. The units used are shown in the table below.

(a) Write the number of bytes in a kilobyte (kB) and a megabyte (MB) in scientific notation.

(b) Round the decimal part to two decimal places and write the answers in (a) in scientific notation.

Solution

(a) $1024 = 1.024 \times 10^3$
   $1{,}048{,}576 = 1.048576 \times 10^6$

(b) To two decimal places, $1.024 = 1.02$, so

   $1024 \approx 1.02 \times 10^3$

   To two decimal places, $1.048576 = 1.05$, so

   $1{,}048{,}576 \approx 1.05 \times 10^6$ ■

By the way, do you recognize the symbol GB? It is usually called a "gig," and it measures the capacity of information that can be stored on the hard drive of your computer.

UNIT	DEFINITION
1 bit	The smallest piece of information two possible values: "0" or "1"
1 byte	8 bits
1 kB (Kilobyte)	1,024 bytes
1 MB (Megabyte)	1,048,576 bytes
1 GB (Gigabyte)	1,073,741,824 bytes
1 TB (Terabyte)	1,099,511,627,776 bytes
1 PB (Petabyte)	1,125,899,906,842,624 bytes

EXERCISES 5.4

A Decimals in Expanded Form

In problems 1–4, write each number in expanded notation.

1. 692.087

2. 30.2959

3. 0.00107

4. 4.30008

In problems 5–8, write each number in standard decimal notation.

5. $(5 \times 10^3) + (2 \times 10^1) + (3 \times 10^{-1}) + (9 \times 10^{-2})$

6. $(4 \times 10^2) + (5 \times 10^0) + (6 \times 10^{-2}) + (9 \times 10^{-4})$

7. $(4 \times 10^{-3}) + (7 \times 10^{-4}) + (2 \times 10^{-6})$

8. $(2 \times 10^{-1}) + (5 \times 10^{-2}) + (4 \times 10^{-4})$

B Scientific Notation

In problems 9–12, write each number in scientific notation.

9. 935

10. 0.372

11. 0.0012

12. 3,453,000

In problems 13–16, write each number in standard notation.

13. 8.64×10^4

14. 9.01×10^7

15. 6.71×10^{-3}

16. 4.02×10^{-7}

In problems 17–22, simplify and write each answer in scientific notation.

17. $0.0346 \div 1,730,000$

18. $0.00741 \times 225,000$

19. $(3.1 \times 10^5) \times (2.2 \times 10^{-6})$

20. $(4.9 \times 10^{-2}) \times (3.5 \times 10^{-1})$

21. $\dfrac{(2 \times 10^6)(6 \times 10^{-5})}{4 \times 10^3}$

22. $\dfrac{(8 \times 10^2)(3 \times 10^{-2})}{24 \times 10^{-3}}$

23. The most plentiful form of sea life known is the nematode sea worm. It is estimated that there are

40,000,000,000,000,000,000,000,000

of these worms in the world's oceans. Write this number in scientific notation.

24. Sir Arthur Eddington claimed that the total number of electrons in the universe is 136×2^{256}. It can be shown that 2^{256} is approximately 1.16×10^{77}. Use this result to write Eddington's number in scientific notation.

In problems 25–30, write each answer in scientific notation.

25. The width of the asteroid belt is 2.8×10^8 km. The speed of *Pioneer 10* in passing through this belt was 1.4×10^5 km/hr. Thus, *Pioneer 10* took

$$\frac{2.8 \times 10^8}{1.4 \times 10^5} \text{ hr}$$

to go through the belt. How many hours was that?

26. The mass of Earth is 6×10^{21} tons. The Sun is about 300,000 times as massive. Thus, the mass of the Sun is $(6 \times 10^{21}) \times 300,000$ tons. How many tons is that?

27. The velocity of light can be measured using the distance from the Sun to Earth (1.47×10^{11} m) and the time it takes for sunlight to reach Earth (490 sec). Thus, the velocity of light is

$$\frac{1.47 \times 10^{11}}{490} \text{ m/sec}$$

How many meters per second is that?

28. U.S. oil reserves are estimated to be 2.2×10^{10} barrels (42 gal). We use about 7.2×10^9 barrels per year. At this rate, how long will U.S. oil reserves last? (Give the answer to the nearest year.) (Source: eia.doe.gov.)

29. The world's oil reserves are estimated to be 1.03×10^{12} barrels. We use 2.67×10^{10} per year. At this rate, how long will the world's oil reserves last? (Give the answer to the nearest year.) (Source: eia.doe.gov.)

30. Scientists have estimated that the total energy received from the Sun each minute is 1.02×10^{19} cal. Since the area of Earth is 5.1×10^8 km^2, the amount of energy received per square centimeter of Earth's surface per minute (the solar constant) is

$$\frac{1.02 \times 10^{19}}{(5.1 \times 10^8) \times 10^{10}}$$

Note: 1 km^2 = 10^{10} cm^2

How many calories per square centimeter is that?

C Operations with Decimals

In problems 31–38, perform the indicated operations.

31. a. $3.81 + 0.93$ **b.** $-3.81 + (-0.93)$

32. a. $18.64 - 0.983$ **b.** $-18.64 - 0.983$

33. a. $2.08 - 6.238$ **b.** $3.07 - 8.934$

34. a. 2.48×2.7 **b.** $(-2.48) \times (-2.7)$

35. a. $(-0.03) \times (-1.5)$ **b.** $(-3.2) \times (-0.04)$

36. a. $10.25 \div 0.05$ **b.** $2.16 \div 0.06$

37. a. $(-0.07) \div 1.4$ **b.** $(-0.09) \div (-4.5)$

38. a. $(-1.8) \div (0.09)$ **b.** $3.6 \div (-0.012)$

39. The revenues (in millions) of City Investing were derived from the following sources: manufacturing, $926.30; printing, $721.21; international, $488.34; housing, $674; food services, $792.45; and insurance, $229.58. Find the total revenue.

40. Ann rented a car at a cost of $25 per day plus $.15 per mile traveled. What was her cost for a 2-day trip of 700 mi?

41. A 6-oz can of orange juice concentrate costs $.48 and a 12-oz can costs $.89. How much is saved by buying 36 oz of the more economical size?

42. A worker was paid $4.50 per hour for a 40-hour week. The worker was paid an additional $1.50 per hour for each hour over 40 hours per week. How much did the worker earn in 3 weeks if he worked 43 hours the first week, 42 hours the second week, and 45 hours the third week?

43. George, Harry, and Joe played poker one evening. Joe won $25.75, which was the total amount lost by George and Harry. If George lost $7.25 more than Harry, how much did each lose?

44. A high school library charged for an overdue book at the rate of 10¢ for the first day and 5¢ for each additional day. If Mabel paid a fine of 75¢ for an overdue book, how many days overdue was the book?

45. A bookstore ordered 30 books for a certain course. Each book, which cost the bookstore $25, was to be sold for $31.50. A service charge of $3 was to be paid for each unsold book returned. If the bookstore sold 28 of the books and returned the other 2, what was its gross profit?

D Rounding

46. Round the measure of 45.49 m to the nearest meter.

47. Round 3265 oz to the nearest hundred ounces.

48. The sides of a triangular lot are measured to be 392.1, 307.25, and 507 ft. Find the perimeter of the lot to the correct number of decimal places.

49. The mileage indicator on a car read 18,327.2 mi at the beginning of a trip. It read 18,719.7 mi at the end. How far did the car go?

50. A business reported annual sales of $8.5 million. If the expenses were reported as $6.52 million, find the profit to the correct number of decimal places.

51. The average weekly circulation of a magazine is 19,230,000 copies. If each copy is sold for 75¢, how much money (on the average) do sales amount to each week?

52. On October 17, 1965, the *New York Times* contained a total of 946 pages. If the thickness of one sheet of newspaper is 0.0040 in., find (to the correct number of decimal places) the thickness of this edition of the paper.

53. A light-year is about 5.878×10^{12} mi. The *Great Galaxy* in *Andromeda* is about 2.15 million light-years away. How many miles is the *Great Galaxy* from us? (Write your answer in scientific notation.) By the way, this galaxy is the most remote heavenly body visible to the naked eye.

54. Lionel Harrison drove the 1900 mi from Oxford, England, to Moscow in a Morris Minor fitted with a 62-gal tank. If he used all the 62 gal of gas, how many miles per gallon did he get? (Round the answer to the nearest tenth.)

55. Stuart Bladon holds the record for the greatest distance driven without refueling, 1150.3 mi on 19.41 gal of fuel. To the correct number of decimal places, how many miles per gallon is that?

56. The highest average yardage gain for a football season belongs to Beattie Feathers. He gained 1004 yd in 101 carries. What was his average number of yards per carry? (Give the answer to the nearest tenth of a yard.)

57. In 1925, a 3600-rpm (revolution-per-minute) motor was attached to a gramophone with a 46:1 gear ratio. The resulting speed was $\frac{3600}{46}$ rpm. What is this ratio to the nearest whole number? This ratio gave birth to a new kind of record, and if you find the answer, you will find what kind of record.

58. The longest and heaviest freight train on record was about 21,120 ft in length and consisted of 500 coal cars. To the nearest foot, how long was each coal car?

 In Other Words

59. Explain in your own words some of the advantages of decimals over ordinary fractions.

60. Explain why, when multiplying two decimals, the number of decimal places in the answer is equal to the sum of the number of decimal places in the two factors.

 Using Your Knowledge

In doing numerical computations, it is often worth making an estimate of the answer to avoid bad errors. (Even if you use a calculator, you might hit an incorrect key!) In many instances, a good estimate may be all that is needed. For example, suppose you want to find the area of a rectangular lot that is measured to be 98.2 ft wide by 347 ft deep. The area of a rectangle of sides a and b is $A = ab$. If you round 98.2 to 100, and 347 to 340, then you have a quick estimate of 34,000 ft^2 for the area of the lot.

Using a calculator to multiply the given dimensions yields an answer of 34,075.4. Rounding to three significant digits (to agree with the accuracy of the dimensions) gives 34,100 ft^2. Thus, the estimate is off by only 100 ft^2.

61. Edie goes to the grocery store and picks up four items that cost $1.98, $2.35, $2.85, and $4.20. Before going to the checkout counter, Edie wants an estimate of her total purchases. Which of the following is the best estimate?
a. $9   b. $11   c. $13   d. $8

62. Jasper has a room that is 15 ft wide and 24.5 ft long. He wants to put in flooring that costs $10.25 per square yard. Which of the following is the best estimate of the total cost?
a. $300   b. $400   c. $350   d. $3500

 Calculator Corner

If your calculator has a ⌈sci⌉ key or its equivalent, then you can convert numbers to scientific notation automatically. Thus, to write 270,000 in scientific notation, enter ⌈2⌉⌈7⌉⌈0⌉⌈0⌉⌈0⌉⌈0⌉⌈=⌉ and ⌈2nd⌉⌈sci⌉ (or ⌈MODE⌉⌈sci⌉). If you do not have a ⌈sci⌉ key, you must know how to enter numbers in scientific notation by using the ⌈exp⌉ or the ⌈EE⌉ key. Thus, to enter 270,000, you have to know that

$$270,000 = 2.7 \times 10^5$$

and enter ⌈2⌉⌈.⌉⌈7⌉⌈exp⌉⌈5⌉. The display will show 2.7 05.

In order to perform operations in scientific notation (without a ⌈sci⌉ key), enter the numbers as discussed. Thus, to find $(5 \times 10^4) \times (9 \times 10^{-7})$ (as in Example 4), enter ⌈(⌉⌈5⌉⌈exp⌉⌈4⌉⌈)⌉⌈×⌉⌈(⌉⌈9⌉⌈exp⌉ ⌈7⌉⌈±⌉⌈)⌉⌈=⌉. The display will give the answer as 4.5 −02, that is, 4.5×10^{-2}. If you have a ⌈sci⌉ key or if you can place the calculator in scientific mode, then enter ⌈5⌉⌈exp⌉⌈4⌉⌈×⌉⌈9⌉⌈exp⌉⌈7⌉⌈±⌉⌈=⌉ (no need to use parentheses), and the same result as before appears in the display.

1. Use your calculator to check problems 13, 15, 17, and 19 in Exercises 5.4.

The operations of addition, subtraction, multiplication, and division involving decimals are easy to perform with a calculator. You just press the appropriate numbers and indicated operations, and the calculator does the rest! However, you must know how to round off answers using the appropriate rules. Thus, in Example 9, you enter 26.3 + 7.41 + 20.64, but you have to know that the answer must be rounded off to 54.4.

 Collaborative Learning

Do you know what unit pricing is? You should if you want to save money. We are ready for the supermarket scavenger hunt. Form two or more groups. The purpose is to determine whether when buying groceries, bigger is better and/or generic is cheaper. Many supermarket items carry a label stating the *unit price* of the item The *unit price* is the price of the item divided by the number of units (ounces, grams, etc.). Your first task is to determine whether the unit price given by the supermarket is correct. To find the *best buy,* find the unit price for each item and select the best unit price.

1. Make a list of ten different items you usually buy at the supermarket. Have one group select brand name items, and another group generic items, each containing the same amount of units. Record the results. Are brand name items always more expensive than generic?

2. For the same ten items, have a group select one size, and another group a different size. Compare unit prices. Is the bigger size always cheaper? *Hint:* We found two items for which this was not always the case: Sun Maid Raisins and Goya Sardines.

5.5 Rationals as Decimals: Percents

The New York Stock Exchange (NYSE), the world's largest, converted to decimals. Price increases are now reported in dollars and cents instead of fractions. Announcements will now say that a particular stock, for example, is "up $1.75" above its previous value instead of "up 1 and $\frac{3}{4}$."

Online Study Center

To get today's data, access link 5.5.1 on this textbook's Online Study Center.

Stock Prices and Fractions

Before April 9, 2001, stock prices were quoted using fractions. Thus, a stock costing $16\frac{31}{32}$ would now cost $16.97. Is there a difference in the price? Absolutely not, but the fractional part $\frac{31}{32}$ will now be written as .97. The pricing of stocks using fractions goes back hundreds of years.

The tradition of pricing stocks in fractions with 16 as the denominator takes its roots from the fact that Spanish traders some 400 years ago quoted prices in fractions of Spanish gold doubloons. A doubloon could be cut into 2, 4, or even 16 pieces. Presumably, it was too difficult to split those $\frac{1}{16}$ wedges any further. (Source: http://invest-faq.com/; Trivia.) Quotes were even priced in 32nds! Can you imagine the problems that would arise if prices were in $\frac{3}{7}$s or $\frac{7}{11}$s? (Try dividing 3 by 7, or 7 by 11.)

What do you think the denominators of all the fractions that are easy to convert to decimals (10, 2, 4, and 8) have in common? The answer is in Theorem 5.5 of this section. As a matter of fact, after you read and understand Theorem 5.5, you will be able to tell whether a fraction has a terminating or a nonterminating decimal representation (problems 25–30, Exercises 5.5). You will also learn to write this representation and even change a terminating or nonterminating repeating decimal to a fraction (problems 43–58, Exercises 5.5). You finish this section by studying a special type of decimal: *percents.* ▶

As we mentioned in Section 5.4, a rational number can always be written in decimal form. If the number is a fraction, we divide the numerator by the denominator and obtain either a terminating or a nonterminating decimal.

A. Terminating and Nonterminating Decimals

Numbers such as $\frac{1}{2}$, $\frac{1}{8}$, and $\frac{1}{5}$ are said to have **terminating decimal representations** because division of the numerator by the denominator terminates (ends). However, some rational numbers—for example, $\frac{1}{3}$—have **infinite repeating decimal representations.** Such a representation is obtained by dividing the numerator of the fraction by its denominator. In the case of $\frac{1}{3}$, we obtain

GRAPH IT

Can you verify that $\frac{1}{3} = 0.\overline{3}$ and $\frac{1}{7} = 0.\overline{142857}$? Enter 1 ÷ 3; you get a terminating (not an infinite) decimal approximation as shown. For $\frac{1}{7}$, the approximation is 0.1428571429 (not infinite either). Again, the moral is: You have to know your math!

```
1/3
            .3333333333
1/7
            .1428571429
```

$$
\begin{array}{r}
0.333\cdots \\
3\overline{)1.0} \\
\underline{9} \\
10 \\
\underline{9} \\
10 \\
\underline{9} \\
1\cdots
\end{array}
$$

For convenience, we shall write $0.333\cdots$ as $0.\overline{3}$. The bar over the 3 indicates that the 3 repeats indefinitely. Similarly,

$$\tfrac{1}{7} = 0.142857142857\cdots = 0.\overline{142857}$$

EXAMPLE 1 ▶ Fractions to Decimals

Write the following as decimals, and state the value in the second decimal place.

(a) $\frac{3}{4}$ (b) $\frac{2}{3}$

Solution

(a) Dividing 3 by 4, we obtain

$$
\begin{array}{r}
0.75 \\
4\overline{)3.0} \\
\underline{28} \\
20 \\
\underline{20} \\
0
\end{array}
$$

Thus, $\frac{3}{4} = 0.75$, and the value in the second decimal place is $\frac{5}{100}$.

(b) Dividing 2 by 3, we obtain

$$
\begin{array}{r}
0.666\cdots \\
3\overline{)2.0} \\
\underline{1\ 8} \\
20 \\
\underline{18} \\
20 \\
\underline{18} \\
2\cdots
\end{array}
$$

Thus, $\frac{2}{3} = 0.666\cdots = 0.\overline{6}$, and the value in the second decimal place is $\frac{6}{100}$. ∎

You should be able to convince yourself of the truth of Theorem 5.5, which indicates which rational numbers have terminating decimal representations.

THEOREM 5.5 Criterion for Terminating Decimals

A rational number a/b (in lowest terms) has a terminating decimal expansion if and only if b has no prime factors other than 2 and 5.

Notice that b does not have to have 2 *and* 5 as factors; it can have only one of them as a factor, or perhaps neither, as in the following illustrations:

$$\frac{1}{25} = 0.04$$

Since $25 = 5 \times 5$, $\frac{1}{25}$ has a terminating decimal expansion.

$$\frac{1}{4} = 0.25$$

Since $4 = 2 \times 2$, $\frac{1}{4}$ has a terminating decimal expansion.

$$\frac{8}{1} = 8.0$$

The denominator is 1, which has no prime factors, so $\frac{8}{1}$ has a terminating decimal expansion. Of course, if the rational number is given as a mixed number—for example, $5\frac{1}{2}$—you can first convert it to $\frac{11}{2}$ and notice that the denominator has only 2 as a factor. Thus $5\frac{1}{2} = \frac{11}{2}$ has a terminating decimal expansion.

The stock markets used the fractions with denominators 2, 4, 8, 16, 32, and 64 because these fractions have simple terminating decimal forms. The denominators are powers of 2 and

$$\frac{1}{2^n} = (0.5)^n$$

so there will be exactly n decimal places.

It is easy to see that every rational number has an infinite repeating decimal representation. In the case of a terminating decimal, you can simply adjoin an infinite string of 0s. For example, $\frac{3}{4} = 0.75\overline{0}$, $\frac{1}{20} = 0.05\overline{0}$, and so on. If the rational number a/b has no terminating decimal representation, then it must have a repeating decimal representation, as you can see by carrying out the division of a by b. The only possible remainders are $1, 2, 3, \ldots$, and $b - 1$. Therefore, after at most $(b - 1)$ steps of the division, a remainder must occur for the second time. Thereafter, the digits of the quotient must repeat. The following division illustrates the idea:

$$
\begin{array}{r}
1.692307 \\
13\overline{)22} \\
\underline{13} \\
90 \\
\underline{78} \\
120 \\
\underline{117} \\
30 \\
\underline{26} \\
40 \\
\underline{39} \\
100 \\
\underline{91} \\
9
\end{array}
$$

Notice that the remainder 9 occurs just before the digit 6 appears in the quotient and again just after the digit 7 appears in the quotient. Thus, the repeating part of the decimal must be 692307, and $\frac{22}{13} = 1.\overline{692307}$.

Of the first 20 counting numbers, only 7, 17, and 19 have reciprocals with the maximum possible number of digits in the repeating part of their decimal representations:

$\frac{1}{7} = 0.\overline{142857}$
$\frac{1}{17} = 0.\overline{0588235294117647}$
$\frac{1}{19} = 0.\overline{052631578947368421}$

B. Changing Infinite Repeating Decimals to Fractions

The preceding discussion shows that **every rational number can be written as an infinite repeating decimal.** Is the converse of this statement true? That is, does every infinite repeating decimal represent a rational number? If the repeating part is simply a string of 0s so that the decimal is actually terminating, it can be written as a rational number with a power of 10 as the denominator. For example, $0.73 = \frac{73}{100}$, $0.7 = \frac{7}{10}$, and $0.013 = \frac{13}{1000}$. If the decimal is repeating but not terminating, then we can proceed as in the next example.

EXAMPLE 2 ▶ Repeating Decimals to Fractions

Write $0.\overline{23}$ as a quotient of integers.

Solution

If $x = 0.232323\cdots$, then $100x = 23.232323\cdots$. Now we can remove the repeating part by subtraction as follows:

$$\begin{array}{r} 100x = 23.232323\cdots \\ (-)\quad x = 0.232323\cdots \\ \hline 99x = 23 \end{array}$$

Then, dividing by 99, we get

$$x = \frac{23}{99}$$ ∎

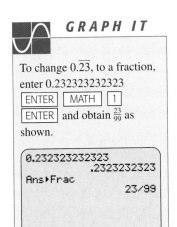

As you can see, every infinite repeating decimal represents a rational number. Here is the procedure for obtaining this representation.

Changing an Infinite Repeating Decimal to a Fraction

1. Let $x =$ the given decimal.

2. Multiply by a power of 10 to move the decimal point to the right of the first sequence of digits that repeats.

3. If the decimal point is not at the left of the first repeating sequence of digits, multiply by a power of 10 to place it there.

4. Subtract the result of step 3 from that of step 2.

5. Divide by the multiplier of x in the result of step 4 to get the desired fraction.

The idea in steps 2 and 3 is to line up the repeating parts so that they drop out in the subtraction in step 4.

EXAMPLE 3 ▶ More Repeating Decimals to Fractions

Write $3.5212121\cdots$ as a quotient of two integers.

Solution

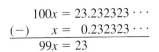

1. Let $x = 3.5212121\cdots$.   We want the decimal point here so that the "21" repeats.

2. Since we want the decimal point to the right of the first 21, we multiply by 1000.

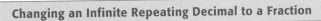

$$1000x = 3521.212121\cdots$$

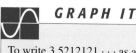

3. In this step, we want the decimal point to the left of the first 21 in

$$3.5212121 \cdots \qquad \text{Here.}$$

so we multiply x by 10.

$$10x = 35.212121 \cdots$$

4.
$$
\begin{array}{r}
1000x = 3521.212121 \cdots \\
(-)\quad 10x = 35.212121 \cdots \\
\hline
990x = 3486
\end{array}
$$
     The decimal parts drop out.

5. We divide by 990 to get

$$x = \frac{3486}{990} = \frac{581 \cdot 6}{165 \cdot 6} = \frac{581}{165}$$ ∎

This discussion can be summarized as follows:

> Every rational number has a repeating decimal representation, and every repeating decimal represents a rational number.

If you use the procedure of Examples 2 and 3 for the repeating decimal $0.999 \cdots$, you come out with the result $0.999 \cdots = 1$.

If you are bothered by this result, the following examples may convince you of its truth:

1. $\frac{1}{3} = 0.333 \cdots$

   $\frac{2}{3} = 0.666 \cdots$

   What is the result when you add these two equations?

2. $\frac{4}{9} = 0.444 \cdots$

   $\frac{5}{9} = 0.555 \cdots$

   What is the result when you add these two equations?

3. $\frac{1}{9} = 0.111 \cdots$

   What do you get by multiplying both sides of this equation by 9? (You can try it on your calculator! Begin by entering 1; then press $\boxed{÷}$ $\boxed{9}$ $\boxed{×}$ $\boxed{9}$ $\boxed{=}$ and watch the display.)

C. Percent

In many of the daily applications of decimals, information is given in terms of **percents**. The interest rate on a mortgage may be 12% (read "12 percent"), the Dow-Jones stock average may increase by 2%, your savings account may earn interest at 6%, and so on. The word *percent* comes from the Latin words *per* and *centum* and means "by the hundred." Thus, 12% is the same as $\frac{12}{100}$ or the decimal 0.12.

> ### Changing a Percent to a Decimal
>
> Move the decimal point in the number two places to the left and omit the % symbol.

EXAMPLE 4 ▶ Percents to Decimals

Write the following as decimals:

(a) 18%   (b) 11.5%   (c) 0.5%

Solution

(a) 18% = 0.18   (b) 11.5% = 0.115   (c) 0.5% = 0.005   ∎

To change a decimal to a percent, just reverse the procedure.

Changing a Decimal to a Percent

Move the decimal point two places to the right and affix the percent sign.

EXAMPLE 5 ▶ Decimals to Percents

Change the following to percents:

(a) 0.25   (b) 1.989

Solution

(a) 0.25 = 25%   (b) 1.989 = 198.9%   ∎

Changing a Fraction to a Percent

Divide the numerator by the denominator, and then convert the resulting decimal to a percent.

EXAMPLE 6 ▶ Fractions to Percents

Write the following as percents. (Give the answers to one decimal place.)

(a) $\frac{2}{5}$ (b) $\frac{3}{7}$

Solution

(a) $\frac{2}{5}$ = 0.40 = 40%

(b) $\frac{3}{7}$ ≈ 0.42857   By ordinary division

    ≈ 0.429   Rounded to three decimal places

    = 42.9%   ∎

EXAMPLE 7 ▶ Popcorn and Percent

When you buy popcorn at the theater, you get popcorn, butter substitute, and a bucket. Which of these is the most expensive? The bucket! Here are the theater's approximate costs: popcorn, 5¢; butter substitute, 2¢; and bucket, 25¢.

(a) Find, to two decimal places, the percent of the total cost for each of these components.

(b) If the average profit on this popcorn is 86% of the cost, how much is the profit?

Solution

(a) The cost of the popcorn is $\frac{5}{32} = 0.15625$, or 15.63%, of the total cost.

The cost of the butter substitute is $\frac{2}{32} = 0.0625$, or 6.25%, of the total cost.

The cost of the bucket is $\frac{25}{32} = 0.78125$, or 78.13%, of the total cost.

(b) 86% of 32 = 0.86 × 32 = 27.52, or nearly 28¢. ■

EXAMPLE 8 ▶ Concessions and Percent

Theater concession stands make 25% of all the popcorn sales in the United States. If these stands make about 250 million sales annually, about how many popcorn sales are made per year in the United States?

Solution

We know that 25% of all sales is about 250 million. If we let s be the total number of sales, then

$$0.25s = 250 \text{ million}$$

Thus, dividing both sides by 0.25, we get

$$s = \frac{250 \text{ million}}{0.25} = 1000 \text{ million}$$

or about 1 billion sales per year. ■

⟨⟨⟨Online Study Center

To further explore the current debt and the interest paid on it, access links 5.5.2 and 5.5.3 on this textbook's Online Study Center.

Do you owe some money? Most of us do, but not as much as the federal government. As of this writing, the national debt amounts to $8,414,376,610,438.13, about $8.4 × 10^{12}$ or about $8.4 trillion. How much interest do we have to pay on that debt? Go to the next example and see.

EXAMPLE 9 ▶ Finding the Interest

The simple interest I for 1 year on a principal P at a rate r is given by $I = Pr$. If the rate r on the national debt is 6%, what is the annual interest I paid on the debt?

Solution

For the national debt, $P = 8.4 \times 10^{12}$ and $r = .06 = 6 \times 10^{-2}$. Thus, $I = 8.4 \times 10^{12} \times 6 \times 10^{-2} = 50.4 \times 10^{12-2} = 50.4 \times 10^{10} = 5.04 \times 10^{11}$, or $504,000,000,000! (504 billion!) ■

D. Percent Increase and Decrease

EXAMPLE 10 ▶ Engineering Salaries

(a) In a certain company, male engineers' salaries began at $46,000, whereas female engineers' salaries began at $31,000. To the nearest percent, how much higher were male engineers' salaries?

(b) According to the *Almanac of Jobs and Salaries,* the average starting salary offered to bachelor's degree candidates in engineering was $30,300, whereas to those in computer sciences it was $27,600. To the nearest percent, how much more was offered to engineers?

Solution

(a) The difference in salaries is $46,000 − $31,000 = $15,000. Thus, the male engineers earned

Difference → $\frac{15,000}{31,000} \approx 0.484$ $\boxed{1}\boxed{5}\boxed{÷}\boxed{3}\boxed{1}\boxed{0}\boxed{=}$
Original →

or about a whopping 48% more. (Note that the female engineers' salary is in the denominator of the fraction because that salary is the basis for comparison.)

(b) Here, the salary difference is $30,300 − $27,600 = $2700. Thus, the engineers were offered

$\frac{2700}{27,600} \approx 0.0978$ $\boxed{2}\boxed{7}\boxed{÷}\boxed{2}\boxed{7}\boxed{6}$

or approximately 9.78%, more than the computer scientists.   ■

EXERCISES 5.5

A Terminating and Nonterminating Decimals

In problems 1–24, write each number in decimal form.

1. $\frac{9}{10}$ 2. $\frac{3}{10}$ 3. $\frac{11}{10}$ 4. $\frac{27}{10}$

5. $\frac{17}{100}$ 6. $\frac{38}{100}$ 7. $\frac{121}{100}$ 8. $\frac{3520}{100}$

9. $\frac{3}{1000}$ 10. $\frac{143}{1000}$ 11. $\frac{1243}{1000}$ 12. $\frac{25,360}{1000}$

13. $\frac{3}{5}$ 14. $\frac{7}{8}$ 15. $\frac{9}{16}$ 16. $\frac{15}{32}$

17. $\frac{5}{8}$ 18. $\frac{5}{4}$ 19. $\frac{5}{7}$ 20. $\frac{7}{6}$

21. $\frac{4}{15}$ 22. $6\frac{1}{4}$ 23. $7\frac{1}{7}$ 24. $3\frac{2}{3}$

In problems 25–30, determine whether each number has a terminating decimal expansion. If it does, give the expansion.

25. $\frac{3}{16}$ 26. $\frac{3}{14}$ 27. $\frac{1}{64}$

28. $\frac{4}{28}$ 29. $\frac{31}{3125}$ 30. $\frac{9}{250}$

In problems 31–42, rewrite each repeating decimal, using a bar and as few digits as possible.

31. $0.555555\cdots$ 32. $0.777777\cdots$

33. $0.646464\cdots$ 34. $0.737373\cdots$

35. $0.235235\cdots$ 36. $0.930930\cdots$

37. $0.215555\cdots$ 38. $0.7132222\cdots$

39. $0.079353535\cdots$ 40. $0.23515151\cdots$

41. $5.070707\cdots$ 42. $9.23373737\cdots$

B Changing Infinite Repeating Decimals to Fractions

In problems 43–58, write each number as a fraction (a quotient of two integers). Reduce if possible.

43. $0.\overline{8}$ 44. $0.\overline{6}$ 45. $0.\overline{31}$

46. $0.\overline{21}$ 47. $0.\overline{114}$ 48. $0.\overline{102}$

49. $2.\overline{31}$ 50. $5.\overline{672}$ 51. $1.\overline{234}$

52. $0.0\overline{17}$ 53. $1.\overline{27}$ 54. $2.4\overline{8}$

55. $0.45\overline{75}$ 56. $0.23\overline{15}$ 57. $0.2\overline{016}$

58. $0.201\overline{6}$

C Percent

In problems 59–67, write each percent as a decimal.

59. 29%   60. 23.4%   61. 0.9%

62. 56.9%   63. 45.69%   64. 0.008%

65. 34.15%   66. 93.56%   67. 0.0234%

In problems 68–76, write each decimal as a percent.

68. 0.38   69. 3.45   70. 9.998

71. 0.567   72. 0.00452   73. 9.003

74. 0.0004   75. 0.0045   76. 0.0008

In problems 77–80, write each fraction as a percent.

77. $\frac{3}{5}$

78. $\frac{4}{7}$ (to one decimal place)

79. $\frac{5}{6}$ (to one decimal place)

80. $\frac{7}{8}$

81. Find 13% of 70.

82. What is 110% of 90?

83. 24 is what percent of 72?

84. 15 is 30% of what number?

85. 100 is what percent of 80?

86. Find $12\frac{1}{4}$% of 320.

87. In a recent year, about 16,300,000 cars were sold by the U.S. auto industry. Of these, 8,214,671 were made in the United States. What percent is that? (Answer to one decimal place.)

88. One of the most expensive British standard cars is the Rolls-Royce Phantom VI, quoted at $312,954. An armor-plated version was quoted at $560,000. What percent more does the armor-plated version cost?

89. By weight, the average adult is composed of 43% muscle, 26% skin, 17.5% bone, 7% blood, and 6.5% organs. Assume a person weighs 150 lb.
 a. How many pounds of muscle does the person have?
 b. How many pounds of skin does the person have?

90. Refer to problem 89.
 a. How many pounds of bone does the person have?
 b. How many pounds of organs does the person have?

91. A portable paint compressor is priced at $196.50. If the sales tax rate is 5.5%, what is the tax?

92. The highest recorded shorthand speed was 300 words per minute for 5 min with 99.64% accuracy. How many errors were made?

93. In a recent year, the United States had 236,760,000 cars. This represented about 40% of the total world stock of cars. To the nearest million, how many cars were in the world stock? (Source: www.bts.gov/publications/national_transportation_statistics/2005/html/table_01_11.html.)

94. In a recent year, 41.2 million households with televisions watched the Superbowl. This represented 47.1% of homes with televisions in major cities. How many homes with televisions are there in major cities? (Source: www.nielsnmedia. com/newsreleases/2003/presuperbowl_2003_html.)

The data for problems 95–98 come from the College Board.

95. The average annual cost of tuition, room, and board at 2-year public colleges is about $4800. This represents 30% of comparable costs at 2-year private colleges. How much is the cost of tuition, room, and board at 2-year private colleges?

96. The average annual cost of tuition, room, and board at 4-year public colleges is about $8800. This represents 40% of the comparable cost at 2-year public colleges. How much is the cost of tuition, room, and board at 2-year public colleges?

97. Dorm rooms at 4-year public colleges cost about $2500. At 2-year public colleges, they cost only 60% of that amount. What is the cost of a dorm room at 2-year public colleges?

98. Dorm rooms at 4-year private colleges cost about $3200. At 2-year public colleges, they cost 90% as much as at the 4-year private colleges. What is the cost of a dorm room at 2-year public colleges?

Problems 99–104 refer to the graph below and assume a total annual cost of $12,000 for in-state fees at a 4-year public university.

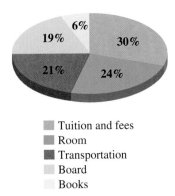

- Tuition and fees
- Room
- Transportation
- Board
- Books

Source: www.moneyopolis.com/teachers/answers/teachers_lped.asp.

99. What is the cost of tuition and fees?

100. What is the cost of transportation?

101. Tracy paid $3000 for a room. Is that more than, less than, or the same as the estimate on the graph?

102. Farah paid $720 for books. Is that more than, less than, or the same as the percent estimated on the graph?

103. Kalin spent $2400 for board. Is that a higher, a lower, or the same percent as is in the graph?

104. Youmbi spent $2520 for transportation. Is that a higher, a lower, or the same percent as is in the graph?

D Percent Increase and Decrease

105. If 25 is increased to 30, what is the percent increase?

106. If 32 is decreased to 28, what is the percent decrease?

107. If you increase 28 by 25% of itself, what is the result?

108. If you decrease 24 by $16\frac{1}{2}$% of itself, what is the result?

109. Joe Long is a computer programmer. His salary was increased by $8000. If his previous salary was $28,000, what was his percent increase? (Answer to the nearest percent.)

110. Felicia Perez received a $1750 annual raise from the state of Florida. If her salary was $25,000 before the raise, what percent raise did she receive?

111. Joseph Clemons had a salary of $20,000 last year. This year, his salary was increased to $22,000. What percent increase is this?

112. Here are some professions and the average salaries earned by males and females. To the nearest percent, how much more did males make in each profession? (Source: http://www.consumersunion.org/products/kitchen_802.htm.)

	Male	Female
a. Management supervisor	$71,000	$68,000
b. Chief technology officer	$73,000	$50,000
c. CEO	$127,000	$95,000

113. Andy was told that if he switched to the next higher grade of gasoline, his mileage would increase to 115% of his present mileage of 22 mi/gal. What would be his increased mileage?

114. Gail learned that a new hairdresser's fees were only 80% of her present hairdresser's fees. If her present fees are $25 per month, what would be her monthly savings if she changed to the new hairdresser?

115. A furniture dealer sold two small sets of furniture for $391 each. On one set he made 15% and on the other he lost 15%. What was his net profit or loss on the two sets?

 In Other Words

116. Answer true (T) or false (F) for each of the following statements and give reasons to support each answer:
  a. Some repeating decimals are not rational numbers.
  b. All counting numbers are rational numbers.
  c. Some integers are not rational numbers.
  d. 0.20200200020000 · · · is a repeating decimal.
  e. All terminating decimals are rational numbers.

117. When stocks are valued in $\frac{1}{16}$ increments, price increases are measured in 6.25¢ increments. Why?

118. Discuss the type of units you will use when measuring increases in the decimal stock market.

119. Discuss the benefits of decimal pricing.

120. How do you think decimalization of the stock market changes the way people invest?

 Using Your Knowledge

121. The **spread** is the smallest value that a stock can change using fractional increments. What is the smallest spread you can have when using $\frac{1}{8}$ increments?

122. What is the smallest spread you can have if you use $\frac{1}{16}$ increments?

123. What is the smallest spread you can have if you use $\frac{1}{64}$ increments?

124. How many shares of stock at $4\frac{3}{8}$ can you purchase with \$4375?

125. How many shares of stock at $5\frac{1}{16}$ can you purchase with \$506.25?

126. How many shares of stock at $7\frac{5}{8}$ can you purchase with \$7625?

 **Discovery**

In this section we developed a procedure for expressing an infinite repeating decimal as a quotient of two integers. For example,

$$0.\overline{3} = 0.333\cdots = \tfrac{3}{9}$$
$$0.\overline{6} = 0.666\cdots = \tfrac{6}{9}$$
$$0.\overline{21} = 0.212121\cdots = \tfrac{21}{99}$$
$$0.\overline{314} = 0.314314314\cdots = \tfrac{314}{999}$$

127. From these examples, can you discover how to express $0.\overline{4} = 0.444\cdots$ as a quotient of integers?

128. Can you express $0.\overline{4321}$ as a quotient of integers?

If we have a repeating decimal with one digit as a **repetend** (the part that repeats), the number can be written as a fraction by dividing the repetend by 9. For example, $0.\overline{7} = \tfrac{7}{9}$. If the number has two digits in the repetend, then the number can be written as a fraction by dividing the repetend by 99. For example, $0.\overline{21} = \tfrac{21}{99}$. Use this idea to check your results in problems 43–48, Exercises 5.5.

 Calculator Corner

A calculator can perform most operations in this section. For example, to change a fraction to a decimal, divide the numerator by the denominator. But can you find the decimal expansion for $\tfrac{3}{14}$? If the calculator has ten decimal places, then when you divide 3 by 14 the answer is given as 0.214285714, and you must know that the repeating part is 142857.

Note that you can check the decimal expansion of a fraction by simply leaving your answer in the calculator and multiplying by the denominator.

Some calculators have a ⬚% key that changes percents to decimals. You can determine whether your calculator does this by entering 11.5 ⬚2nd ⬚%. The answer should appear as 0.115 [Example 4(b)].

1. Use your calculator to check problems 59–67 in Exercises 5.5.

2. Use your calculator to show that a 10% increase followed by a 10% decrease does not result in the original amount.

 Research Questions

In the last sections we have used the horizontal bar to write common fractions. Sometimes, instead of a horizontal bar, we use the diagonal fraction bar like this: a/b. Do you know how they wrote the decimal point many years ago or who invented the percent symbol? Here are some questions for you.

1. Who invented the horizontal bar to write common fractions, and who was the first European mathematician to use the horizontal fraction bar?

2. Who were the first users of the diagonal fraction bar? From what did the diagonal bar evolve?

3. What symbols were used to write the decimal point and by whom?

4. Who invented the percent symbol and when?

Online Study Center

To further explore the number π, access link 5.6.1 on this textbook's Online Study Center.

GRAPH IT

When you enter the number π on your grapher by pressing [2nd] [π] [ENTER], you get the number shown. Is that number equal to π? Is it rational or irrational? Is π rational or irrational?

π	
	3.141592654

Remember, a rational number has a repeating or terminating decimal representation. Irrationals are nonrepeating and nonterminating. See the definition of an irrational number on page 293.

Pi (π) Through the Ages

One of the most unusual and interesting real numbers is π (the Greek letter *pi*, pronounced "pie"), which is the ratio of the circumference to the diameter of a circle; that is,

$$\pi = \frac{C}{d}$$

What is the numerical value of π? There have been many versions. In 1800–1600 B.C., the Babylonians gave π a value of 3. The Hebrews used the same approximation in the Old Testament (1 Kings 7:23, written in 650 B.C. but dating back to 900 B.C.). It reads, "And he made a molten sea, 10 cubits from one brim to the other: it was round all about . . . : and a line of 30 cubits did compass it round about." More recent findings (a tablet in Susa) indicate that the Babylonians might have used $3\frac{1}{8} = 3.125$ as the value for π. The Egyptians used the approximation $4 \times \left(\frac{8}{9}\right)^2 = 3.1605$ in the Rhind papyrus, written in 1650 B.C.

In the first century A.D., Liu Hsin used 3.1547, and Chang Hen (A.D. 78–139) used the value $\sqrt{10}$, which is approximately 3.1623, or the fraction $\frac{92}{29}$, which is about 3.1724. In the third century B.C., Archimedes approximated π by drawing a circle between two polygons of 6, 12, 24, 48, and 96 sides. As the number of sides increased, the perimeter of the polygons approximated the circumference of the circle and showed that π was between $3\frac{10}{71}$ and $3\frac{1}{7}$. This same method was employed by Liu Hui to find his best value for π, namely, 3.14159. Tsu Chung-Chi (A.D. 430–501) refined the method and obtained boundaries of 3.1415926 and 3.1415727, giving the fraction $\frac{22}{7}$ as the "inaccurate" value for π and $\frac{355}{113}$, a value correct to six decimals, as the "accurate" value. (No fraction with a denominator less than 113 gives a closer approximation for π.)

How close can you get to π? Get a soda can, measure around it, and divide by the distance across the top. What do you get? What about using computers to find a more accurate approximation? Using a Hitachi supercomputer, the latest approximation for 3.141592654 · · · has 1.24 trillion decimal places! ▶

In Section 5.2 we presented the usefulness and beauty of the set of integers. The Pythagoreans were so certain that the entire universe was made up of the whole numbers that they classified them into categories such as "perfect" and "amicable." They labeled the even numbers "feminine" and the odd numbers "masculine," except 1, which was the generator of all other numbers. (At that time, the symbol for marriage was the number 5, the sum of the first feminine number, 2, and the first masculine number, 3.) In the midst of these charming fantasies, the discovery of a new type of number was made—a type of number so unexpected that the brotherhood tried to suppress its discovery. It had found the numbers that we call *irrational numbers* today.

Here is a general idea of what happened. Suppose we draw a number line 2 units long (see Figure 5.8). We divide the unit interval into two equal parts and graph $\frac{1}{2}, \frac{2}{2} = 1, \frac{3}{2},$ and $\frac{4}{2} = 2$. We then proceed in the same way and divide the unit interval into three parts, marking $\frac{1}{3}, \frac{2}{3}, \frac{3}{3} = 1, \frac{4}{3}, \frac{5}{3},$ and $\frac{6}{3} = 2$ as shown in Figure 5.8. For any whole number q, we can divide the unit interval into q equal parts and then graph $1/q, 2/q,$ and so on. It seems reasonable to assume that this

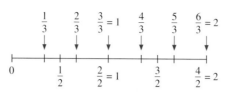

FIGURE 5.8

process continued indefinitely would assign a rational number to every point on the line.

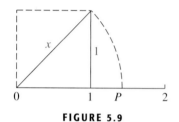

FIGURE 5.9

Now suppose we construct a unit square and draw its diagonal as shown in Figure 5.9. Turn this diagonal (called the **hypotenuse** of the resulting triangle) clockwise to coincide with the number line extending from 0 to a point marked *P*. Which rational number corresponds to *P*? None! Why? Because the required number is not obtainable by dividing any whole number by another whole number; it is not a rational number.

Ironically enough, Pythagoras himself had proved the famous theorem bearing his name: *The square of the hypotenuse of a right triangle is equal to the sum of the squares of the other two sides.* If we let *x* be the length of this hypotenuse (the diagonal of the square in Figure 5.9), the theorem says that

$$x^2 = 1^2 + 1^2 = 2$$

so that

$$x = \sqrt{2}$$ Square root of 2, a positive number that when squared yields 2

The Pythagoreans were able to prove that $\sqrt{2}$ is not obtainable by dividing any whole number by another, that is, that $\sqrt{2}$ is not a rational number. To do so, they used a method of proof called *reductio ad absurdum,* meaning "reduction to the absurd" (we now call this **proof by contradiction**). Their proof may have gone as follows:

Proof
Assume that $\sqrt{2}$ is a rational number, say, *a/b*, in lowest terms; that is, *a* and *b* have no common factor other than 1. Then we have the following:

1. $$\sqrt{2} = \frac{a}{b}$$

2. Multiply by *b*. $\sqrt{2}b = a$

3. Square both sides. $(\sqrt{2}b)^2 = a^2$

4. Simplify. $2b^2 = a^2$

5. Thus, a^2 is an even number, and since only even numbers have squares that are even, *a* is an even number. Suppose $a = 2c$.

6. Substitute $a = 2c$ in step 4. $2b^2 = (2c)^2 = 4c^2$

7. Divide by 2. $b^2 = 2c^2$

8. This last equation says that b^2 is an even number, so that *b* is an even number. (If *b* were odd, b^2 would also be odd.)

9. Now, we have a contradiction. Step 5 says that *a* is even, and step 8 says that *b* is even; this means that *a* and *b* would have the common factor 2, contrary

to our assumption. Hence, our assumption is invalid, and $\sqrt{2}$ is not a rational number. ∎

Definition of an Irrational Number

Numbers that are not rational are called **irrational.**

A. Square Roots

The preceding proof can be generalized to show that the only numbers with rational square roots are the **perfect squares,** as follows:

$$\sqrt{1} = 1 \quad \text{because} \quad 1^2 = 1$$
$$\sqrt{4} = 2 \quad \text{because} \quad 2^2 = 4$$
$$\sqrt{9} = 3 \quad \text{because} \quad 3^2 = 9$$
$$\vdots$$

The square roots of all natural numbers that are not perfect squares are irrational.

EXAMPLE 1 ▶ **Classifying Square Roots**

Classify the following as rational or irrational:

(a) $\sqrt{36}$ (b) $\sqrt{44}$ (c) $\sqrt{81}$

Solution

(a) $\sqrt{36} = 6$ is rational.   (b) $\sqrt{44}$ is irrational.   (c) $\sqrt{81} = 9$ is rational. ∎

B. Irrational Numbers and Decimal Numbers

Irrational numbers such as $\sqrt{2}$ can be approximated to any finite number of decimal places. But these decimal numbers can never repeat (as in $\frac{1}{7}$) or terminate (as in $\frac{1}{2}$) because if they did, they would be rational numbers. For example,

$$\sqrt{2} = 1.4142 \cdots \quad \text{or} \quad \sqrt{2} = 1.414213 \cdots$$

We use this idea to define irrational numbers.

Representation of an Irrational Number

An **irrational number** is a number that has a **nonterminating, nonrepeating** decimal representation.

For example, $0.909009000 \cdots$ (the successive sets of digits are 90, 900, 9000, and so on) is nonterminating and nonrepeating and thus is irrational. Another irrational number is the decimal $1.23456789101112 \cdots$, where we continue writing the digits of the successive counting numbers. Here again, although there is a definite pattern, the decimal is nonrepeating and nonterminating. The set consisting of all decimals is called the set R of **real numbers.** This set includes all the numbers we have studied: the natural numbers, the whole numbers, the integers, the rational numbers, and the irrational numbers. (Keep in mind that rational numbers can be written as quotients of two integers and irrationals cannot.) The rationals and the irrationals completely cover the number line: To each point on the line, there corresponds a unique real number, and to each real number, there corresponds a unique point on the line.

EXAMPLE 2 ▶ Classifying Decimals

Classify the following numbers as rational or irrational:

(a) $0.35626262\cdots$ (b) $0.305300530005\cdots$ (c) 0.12345678

(d) $-\frac{1}{3}$ (e) $\sqrt{65}$ (f) $\sqrt{144}$

Solution

(a) A repeating decimal, therefore rational

(b) A nonrepeating, nonterminating decimal, therefore irrational

(c) A terminating decimal, therefore rational

(d) A fraction, therefore rational

(e) Irrational

(f) $\sqrt{144} = 12$, therefore rational   ∎

Looking at the decimal approximations of $\sqrt{2}$ given earlier, we can see that 1.4 is less than $\sqrt{2}$ but 1.5 is greater; that is, $\sqrt{2}$ is between 1.4 and 1.5. Can we always find an irrational number between any two rational numbers? In order to answer this question, we must first make the ideas of **less than** ($<$) and **greater than** ($>$) more precise.

Definition of "Less Than" and "Greater Than"

If a and b are real numbers, then

$\quad a < b$ if and only if there is a positive number c such that
$\qquad\qquad a + c = b$

$\quad b > a$ if and only if $a < b$

Thus, $3 < 5$ because $3 + 2 = 5$, $-4 < -1$ because $-4 + 3 = -1$, $5 > 3$ because $3 < 5$, and $\frac{1}{3} = 0.333\cdots < \frac{1}{2} = 0.5$ because $\frac{1}{3} + \frac{1}{6} = \frac{1}{2}$.

A basic property of the real numbers is given by the following statement:

The Trichotomy Law

If a and b are any real numbers, then exactly one of the following relations must occur:

$\quad a = b$ (1)

$\quad a < b$ (2)

$\quad a > b$ (3)

Thus, if $a \not> b$ (a is not greater than b), then $a \leq b$ (a is less than or equal to b). If $a \not< b$ (a is not less than b), then $a \geq b$ (a is greater than or equal to b).

It is not difficult to compare two rational numbers, but what about comparing an irrational number such as $\sqrt{48}$ and a rational number such as 6.9? We know that the numbers are close because $\sqrt{49} = 7$, which is close to 6.9, but can we fill in the blank with $<$ or $>$ to obtain a correct statement in the following expression?

$$\sqrt{48} \underline{\quad\quad} 6.9$$

By squaring both sides, we get the equivalent comparison

$$48 \underline{\quad\quad} (6.9)^2 = 47.61$$

which shows that the $>$ symbol is the correct choice. Thus,

$$\sqrt{48} > 6.9$$

EXAMPLE 3 ▶ Using the Trichotomy Law

Insert $<$, $>$, or $=$ in the following to make correct statements:

(a) $\sqrt{60} \underline{\quad\quad} 7.7$ (b) $\sqrt{30} \underline{\quad\quad} 5.5$

Solution

(a) As before, we write $\sqrt{60} \underline{\quad\quad} 7.7$

    Squaring both sides, $60 \underline{\quad\quad} (7.7)^2 = 59.29$

    Since $60 > 59.29$, we have $\sqrt{60} > 7.7$

(b) Squaring both sides, we get $30 \underline{\quad\quad} (5.5)^2 = 30.25$

    Since $30 < (5.5)^2 = 30.25$, $\sqrt{30} < 5.5$ ∎

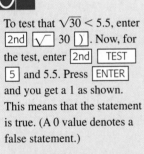

GRAPH IT

To test that $\sqrt{30} < 5.5$, enter [2nd] [√] 30 [ )]. Now, for the test, enter [2nd] [TEST] [5] and 5.5. Press [ENTER] and you get a 1 as shown. This means that the statement is true. (A 0 value denotes a false statement.)

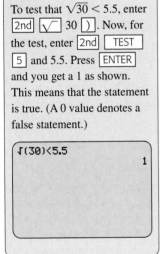

$\sqrt{(30)}{<}5.5$

                     1

We now return to the question "Can we always find an irrational number between any two rational numbers?" The answer is affirmative. For example, to find an irrational number between $\frac{1}{3}$ and $\frac{1}{2}$, we first write $\frac{1}{3}$ and $\frac{1}{2}$ as decimals.

$$\frac{1}{2} = 0.5$$
$$\frac{1}{3} = 0.333 \cdots$$

Obviously, the number 0.4 is between 0.5 and $0.333 \cdots$; however, this number is not irrational. We now add a nonrepeating, nonterminating part to this number as shown.

$$\frac{1}{2} = 0.5$$
$$\quad\quad 0.4101001000 \cdots \quad\quad \text{Nonterminating, nonrepeating}$$
$$\frac{1}{3} = 0.333 \cdots$$

The number $0.4101001000 \cdots$ is bigger than $0.333 \cdots$, smaller than 0.5, and irrational. We could have found infinitely many other numbers using a similar technique. Can you find two more?

EXAMPLE 4 ▶ Finding Numbers Between Decimals

Find the following:

(a) A rational number between 0.121 and 0.122

(b) An irrational number between 0.121 and 0.122

Solution

(a) The rational number 0.1215 is between 0.121 and 0.122 as shown.

0.121

0.1215

0.122

(b) The irrational number 0.121567891011 · · · is between 0.121 and 0.122 as shown.

0.121

0.121567891011 · · ·   Nonterminating, nonrepeating

0.122   ■

In the preceding discussion we saw that we are able to find an irrational number between any two given rational numbers. We can also show that it is possible to find a rational number between any two given rational numbers. For example, given the rational numbers $\frac{4}{7}$ and $\frac{5}{7}$, we write

$$\frac{4}{7} = \frac{4 \cdot 2}{7 \cdot 2} = \frac{8}{14} \quad \text{and} \quad \frac{5}{7} = \frac{5 \cdot 2}{7 \cdot 2} = \frac{10}{14}$$

We can now see by inspection that one rational number between $\frac{4}{7}$ and $\frac{5}{7}$ is $\frac{9}{14}$. If the two given rationals do not have the same denominator, then we can proceed as in the next example.

EXAMPLE 5 ▶ **Finding Numbers Between Decimals and Fractions**

Find a rational number between $0.\overline{4}$, and $\frac{6}{13}$.

Solution

Since $0.\overline{4} = \frac{4}{9}$, the problem is equivalent to that of finding a rational number between $\frac{4}{9}$ and $\frac{6}{13}$. We can do this easily by changing to fractions with common denominators. Thus,

$$\frac{4}{9} = \frac{4 \cdot 13}{9 \cdot 13} = \frac{52}{117} \quad \text{and} \quad \frac{6}{13} = \frac{6 \cdot 9}{13 \cdot 9} = \frac{54}{117}$$

Therefore, an obvious choice for the number we seek is $\frac{53}{117}$.   ■

C. The Number π

There is one more important irrational number that caused the Pythagoreans further difficulties. The number π introduced in Getting Started as the ratio of the circumference of a circle C to its diameter d is itself an irrational number, 3.14159 · · · . Use of the symbol π was probably inspired by the first letter in the Greek word *periphereia,* meaning "periphery." Thus,

$$\frac{C}{d} = \pi$$

or, solving for the circumference,

$$C = \pi d$$

Proving that π is irrational is a very difficult mathematical problem, and we shall simply accept the fact. As we have already noted, the value of π has been approximated by various people throughout the ages, and different methods have been used to calculate the value of π to a large number of decimal places. Today, this value is known to millions of decimal places. Here is the value to a mere 32 places.

$$\pi = 3.141\ 592\ 653\ 589\ 793\ 238\ 462\ 643\ 383\ 279\ 50$$

In most applications, however, we use only two decimal places, 3.14; if greater accuracy is desired, we customarily use the value 3.1416.

EXAMPLE 6 ▶ Applications of Circumference

A manufacturer of cylindrical tanks wants to put a reinforcing steel strap around a 2-ft-diameter tank. The strap is to be 2 in. longer than the circumference to allow for riveting the overlap. What is the ideal length of the strap? (Answer to the nearest 0.1 in.)

Solution
We use the formula $C = \pi d$, taking π as 3.14. Thus,

$$
\begin{aligned}
C &= 3.14 \times 2 \text{ ft} \\
&= 3.14 \times 2 \times 12 \text{ in.} \\
&\approx 75.4 \text{ in.}
\end{aligned}
$$

Adding the 2-in. overlap, we get the required length to be 77.4 in. ∎

PROBLEM SOLVING

Circumferences and Diameters

❶ Read the problem.

A bullring in Spain has a fence that is 450 ft long and uses a 20-ft circular gap as its gate. What is the diameter of this bullring?

❷ Select the unknown.

We want to find the diameter of the bullring.

❸ Think of a plan.
We need to find the diameter of a circle (the bullring) when we know its circumference.

Since the fence is 450 ft long and there is a 20-ft gate, the circumference C must be $450 + 20$, or 470, ft. We also know that $C = \pi d$.

❹ Use $C = \pi d$ and solve for $d = C/\pi$ to find the diameter.

Solve for $d = \dfrac{C}{\pi}$, where $C = 470$. Thus

$$d = \frac{C}{\pi} = \frac{470}{3.14} \approx 149.68$$

Since the value for π has three significant digits, we round the answer 149.68 to three significant digits and obtain $d = 150$ ft.

continued

⑤ Verify the answer.

TRY EXAMPLE 7 NOW.

The circumference of a circle with a 150-ft diameter is $C = 3.14 \times 150 = 471$, or about 470 ft, as required.

Cover the solution, write your own solution, and then check your work.

EXAMPLE 7 ▶ More Circumference Applications

A rectangular piece of silver approximately 0.5 by 6.75 in. is to be formed into a circular bracelet. If the band is bent so that a half-inch circular gap is left, what will be the diameter of the bracelet? (Use the approximate value 3.14 for π.)

Solution

The circumference C of a circle is given by $C = \pi d$, where d is the diameter. The total circumference of the bracelet will be $6.75 + 0.5$ in. (the length of the piece of silver plus the 0.5-in. gap). Since $C = \pi d$, solving for d gives

$$d = \frac{C}{\pi} = \frac{6.75 + 0.5}{3.14} = \frac{7.25}{3.14} \approx 2.3089$$

Since the 7.25 and 3.14 both have only three significant digits, we round our answer to three significant digits and get $d = 2.31$ in.   ■

EXERCISES 5.6

Ⓐ Square Roots

Ⓑ Irrational Numbers and Decimal Numbers

In problems 1–20, classify each number as rational or irrational.

1. $\sqrt{120}$ **2.** $\sqrt{121}$

3. $\sqrt{125}$ **4.** $\sqrt{169}$

5. $\sqrt{\frac{9}{16}}$ **6.** $\sqrt{\frac{9}{15}}$

7. $\frac{3}{5}$ **8.** $-\frac{22}{7}$

9. $-\frac{5}{3}$ **10.** -0

11. $0.232323\cdots$ **12.** $0.023002300023\cdots$

13. $0.121231234\cdots$ **14.** 0.121231234

15. $6\frac{1}{4}$ **16.** $\sqrt{6\frac{1}{4}}$

17. $0.24681012\cdots$ **18.** 0.1122334455

19. 3.1415 **20.** π

In problems 21–32, evaluate each expression.

21. $\sqrt{16}$ **22.** $\sqrt{49}$ **23.** $\sqrt{64}$

24. $-\sqrt{144}$ **25.** $\sqrt{81}$ **26.** $\sqrt{256}$

27. $-\sqrt{169}$ **28.** $-\sqrt{25}$ **29.** $\sqrt{196}$

30. $\sqrt{225}$ **31.** $-\sqrt{81}$ **32.** $-\sqrt{121}$

In problems 33–46 insert $<$, $>$, or $=$, as appropriate.

33. 3 _____ 4 **34.** 17 _____ 11

35. $\frac{1}{5}$ _____ $\frac{1}{4}$ **36.** $\frac{12}{19}$ _____ $\frac{11}{17}$

37. $\frac{5}{7}$ _____ $\frac{10}{14}$ **38.** $1\frac{2}{3}$ _____ $\frac{8}{6}$

39. $\sqrt{20}$ _____ 4.5

40. $3.777\cdots$ _____ $\sqrt{15}$

41. $0.333\cdots$ _____ $0.333444\cdots$

42. 0.101001000 _____ $0.1101001000\cdots$

43. $0.999\cdots$ _____ 1

44. $0.333\cdots + 0.666\cdots$ _____ 1

45. $3(0.333\cdots)$ _____ 1

46. 0.112233 _____ $0.111222333\cdots$

47. Find a rational number between 0.31 and 0.32.

48. Find a rational number between 0.28 and 0.285.

49. Find an irrational number between 0.31 and 0.32.

50. Find an irrational number between 0.28 and 0.285.

51. Find a rational number between $0.101001000\cdots$ and $0.102002000\cdots$.

52. Find a rational number between $0.303003000\cdots$ and $0.304004000\cdots$.

53. Find an irrational number between $0.101001000\cdots$ and $0.102002000\cdots$.

54. Find an irrational number between $0.303003000\cdots$ and $0.304004000\cdots$.

55. Find a rational number between $\frac{3}{11}$ and $\frac{4}{11}$.

56. Find a rational number between $\frac{7}{9}$ and $\frac{9}{11}$.

57. Find an irrational number between $\frac{4}{9}$ and $\frac{5}{9}$.

58. Find an irrational number between $\frac{2}{11}$ and $\frac{3}{11}$.

59. Find a rational number between $0.\overline{5}$ and $\frac{2}{3}$.

60. Find a rational number between 0.1 and $0.\overline{1}$.

In problems 61 and 62, list the given numbers in order from smallest to largest.

61. 0.21, $0.212112111\cdots$, 0.21211, 0.2121, 0.21212

62. 3.14, 3.1414, $3.141411411\cdots$, 3.141, 3.1

The Number π

In problems 63–68, use the approximate value 3.14 for π.

63. The largest circular crater in northern Arizona is 5200 ft across. If you were to walk around this crater, how many miles would you walk? (1 mi = 5280 ft)

64. The Fermi National Accelerator Laboratory has a circular atom smasher that is 6562 ft in diameter. Find the distance in miles that a particle travels in going once around in this accelerator. (1 mi = 5280 ft)

65. The U.S. Department of Energy is studying the possibility of building a circular superconductivity collider that will be 52 mi in diameter. How far would a particle travel in going once around this collider? Give your answer to the nearest mile.

66. The smallest functional phonograph record, a rendition of "God Save the King," is $1\frac{3}{8}$ in. in diameter. To the nearest hundredth of an inch, what is the circumference of this record?

67. The diameter of a circular running track is increased from 90 to 100 yd. By how many yards is the length of the track increased? Give your answer to the nearest yard.

68. You are asked to build a circular running track of length $\frac{1}{4}$ km. To the nearest meter, what would be the radius of the track? (Recall that 1 km is 1000 m.)

 In Other Words

69. Explain in your own words how you can tell a rational from an irrational number.

70. If the rational numbers were used to mark the points on a straight line, what points would be missed? Give three examples.

Using Your Knowledge

The diagram below shows an interesting spiral made up of successive right triangles. This spiral can be used to construct lengths corresponding to the square roots of the integers on the number line. For example, the second triangle has sides 1 and $\sqrt{2}$, so if the hypotenuse is of length x, then

$$x^2 = 1^2 + (\sqrt{2})^2$$
$$= 1 + 2 = 3$$

Therefore, $x = \sqrt{3}$.

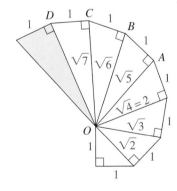

71. Verify that the ray OB is of length $\sqrt{6}$.

72. Verify that the ray OD is of length $\sqrt{8}$.

73. Find the length of the hypotenuse h of the triangle shown in blue in the diagram.

74. If the triangles are numbered 1, 2, 3, . . . , starting with the lowest triangle in the diagram, what would be the number of the triangle whose hypotenuse is of length $\sqrt{17}$?

 Calculator Corner

If your calculator has a pi (π) key, then calculations involving π can easily be done on it. For example, to find the radius of a circle with a circumference of 96 ft, you need to find $96/2\pi$. All you need do is key in

Most calculators will show the answer 15.278875, which, to the nearest 0.01 ft, rounds to 15.28 ft.

1. Use the π key on your calculator to check the answers to problems 63–68, Exercises 5.6.

 Collaborative Learning

The real numbers are an example of a *mathematical system:* a set R of elements with one or more operations ($+$, $-$, $\times$, and $\div$), one or more relations to compare the elements in the set ($<$ and $>$), and some rules, axioms, and laws that the elements in the set satisfy. Form two or more groups of students. Each group is to develop a new mathematical system, and each group can call its system anything it wishes! Here are some guidelines.

1. Use the set $\{0, 1\}$.

2. The operations will be binary addition, subtraction, multiplication, and division.

3. The relations will be $>$ and $<$.

4. Each group will develop the rules of the system but will try to end up with a system similar to the real numbers.

You already know how to write whole numbers in the binary system. What about the *integers*? Half the work is done because all the whole numbers are integers, but what about negative integers? First, use 8 columns, or 8 bits, to describe an integer in base 2. The leftmost column indicates the sign of the number, a 0 for positive and a 1 for negative, and the 7 columns to the right are used to display the number.

1. Write 7, -7, 80, and -80 using this new system.

2. To write *rational numbers,* write the numbers as two groups of 8 columns, or 8 bits. First, write the rational number in scientific notation, that is, in the form $m \times 10^n$. The first group of 8 numbers (on the left) represents m (remember, the leftmost column in this group indicates the *sign* of the number). The second group of numbers indicates the exponent n (again, the leftmost column in this group indicates the *sign* of the number). Each group writes 1.19, 3/4, $-5/2$, and -11.2 using this idea. Do their answers agree?

3. Now, here is the fun part. Go back to Section 5.2 and look at the properties for addition and multiplication listed there. How many of the properties does this new system have?

4. Can you find at least one rule for the real numbers that does not work in the new system? Can you find a way to define irrational numbers in the new system?

 **Research Questions**

In this section we proved that $\sqrt{2}$ is irrational. Ours was a proof by contradiction.

 Online Study Center 1. Go to this textbook's Online Study Center and click on link 5.6.2 to construct your own proof that $\sqrt{5}$ is irrational.

2. On this same site there is a proof that $\sqrt{2}$ is irrational. How does it differ from the one given in the text? Explain!

3. Can you prove that $\sqrt{2}$ is irrational using a different method from the ones we have used?

Online Study Center

4. In Example 4 we found a rational number between 0.121 and 0.122. How do we know such a number even exists? Go to link 5.6.3 on this textbook's Online Study Center. Write your own proof of the first part of Lemma 1:

Between any two different rational numbers a and b there is at least one other rational number.

5.7 Radicals

A plane can break the sound barrier.

The "Radical" Notion of Supersonic Speed

How fast can this plane travel? The answer is classified information, but it exceeds twice the speed of sound (747 mph). It is then said that the plane's speed is more than Mach 2. The formula for calculating the Mach number is

$$M = \sqrt{\frac{2}{\gamma}} \sqrt{\frac{P_2 - P_1}{P_1}}$$

where P_1 and P_2 are air pressures. This expression can be simplified by multiplying both radical expressions and *rationalizing* the denominator.

In this section you will add, subtract, multiply, and divide radical expressions, that is, expressions containing radicals. ▶

In the preceding section we studied irrational numbers of the form $\sqrt{n}$, where n is a positive number but not a perfect square. We did this with the help of the Pythagoreans, who solved the equation $x^2 = 2$. Irrational numbers of the form $\sqrt{n}$ are called **radicals.** In general, $\sqrt[n]{m}$ is a **radical expression** with index n and radicand m. The index 2 is usually omitted, so we write $\sqrt{n}$ instead of $\sqrt[2]{n}$. We study next some simple operations on radicals.

A. Simplifying Radicals

The velocity v (in feet per second) of an object in free fall depends on the distance d that it has fallen. The formula is

$$v = \sqrt{32d}$$

Online Study Center

To further explore square roots, access link 5.7.1 on this textbook's Online Study Center.

Thus, after an object has fallen 1 ft ($d = 1$), its velocity is $\sqrt{32}$ ft/sec, and after 2 ft ($d = 2$), it is $\sqrt{64}$ ft/sec. The number $\sqrt{32}$ is an irrational number, because 32 is not a perfect square, but 64 is a perfect square ($8^2 = 64$), so $\sqrt{64} = 8$ is a rational number. Note that $\sqrt{32}$ is positive. We interpret $\sqrt{n}$ to mean the positive square root of n. Can we simplify $\sqrt{32}$? We say that $\sqrt{n}$ is in **simplest form** if n has no factor (other than 1) that is a perfect square. Using this definition, we can see that $\sqrt{32}$ is not in simplest form because the perfect square 16 is a factor of 32. The simplification can be done using the following property:

> **Multiplication of Radicals**
>
> If a and b are nonnegative real numbers, then
> $$\sqrt{a \cdot b} = \sqrt{a} \cdot \sqrt{b}$$

Thus,

$$\sqrt{32} = \sqrt{16 \cdot 2} = \sqrt{16} \cdot \sqrt{2} = 4\sqrt{2}$$

In general, the simplest form of a number involving the radical sign $\sqrt{}$ is obtained by using the perfect squares 1, 4, 9, 16, 25, 36, 49, 64, 81, 100, and so on, as factors under the radical and then using the multiplication property, as in the next example.

EXAMPLE 1 ▶ Simplifying Radicals

Simplify the following if possible:

(a) $\sqrt{75}$ (b) $\sqrt{70}$

Solution

(a) The largest perfect square dividing 75 is 25. Thus, we write

$$\sqrt{75} = \sqrt{25 \cdot 3} = \sqrt{25} \cdot \sqrt{3} = 5\sqrt{3}$$

(b) There is no perfect square (except 1) that divides 70. (Try dividing by 4, 9, 16, 25, and 36.) Thus, $\sqrt{70}$ cannot be simplified any further. ■

The property $\sqrt{a \cdot b} = \sqrt{a} \cdot \sqrt{b}$ can be used to **rationalize** the denominator of certain expressions, that is, to free the denominator of radicals. Thus, if we wish to rationalize the denominator in the expression $6/\sqrt{3}$, we use the fundamental principle of fractions and multiply the numerator and the denominator of the fraction by $\sqrt{3}$ (because we know that $\sqrt{3} \cdot \sqrt{3} = 3$), as follows:

$$\frac{6}{\sqrt{3}} = \frac{6 \cdot \sqrt{3}}{\sqrt{3} \cdot \sqrt{3}} = \frac{6 \cdot \sqrt{3}}{\sqrt{9}} = \frac{6 \cdot \sqrt{3}}{3} = 2\sqrt{3}$$

EXAMPLE 2 ▶ Rationalizing the Denominator

Rationalize the denominator in the expression $5/\sqrt{10}$.

Solution

We multiply the numerator and the denominator by $\sqrt{10}$ and then simplify.

$$\frac{5}{\sqrt{10}} = \frac{5 \cdot \sqrt{10}}{\sqrt{10} \cdot \sqrt{10}} = \frac{5 \cdot \sqrt{10}}{\sqrt{100}} = \frac{5 \cdot \sqrt{10}}{10} = \frac{\sqrt{10}}{2}$$

■

B. Multiplication and Division of Radicals

Can we simplify $\sqrt{\frac{3}{4}}$? (This is one of the two answers you will get if you solve the equation $x^2 = \frac{3}{4}$.) This time, the perfect square 4 appears in the denominator, so to simplify the expression, we use the following property:

Division of Radicals

If a and b are positive numbers, then

$$\sqrt{\frac{a}{b}} = \frac{\sqrt{a}}{\sqrt{b}}$$

Thus,

$$\sqrt{\frac{3}{4}} = \frac{\sqrt{3}}{\sqrt{4}} = \frac{\sqrt{3}}{2}$$

EXAMPLE 3 ▶ Simplifying Radicals Involving Quotients

Simplify the following:

(a) $\sqrt{\dfrac{32}{25}}$ (b) $\sqrt{\dfrac{36}{7}}$

Solution

(a) $\sqrt{\dfrac{32}{25}} = \dfrac{\sqrt{32}}{\sqrt{25}} = \dfrac{\sqrt{32}}{5}$

$$= \frac{\sqrt{16 \cdot 2}}{5}$$

$$= \frac{4 \cdot \sqrt{2}}{5}$$

(b) $\sqrt{\dfrac{36}{7}} = \dfrac{\sqrt{36}}{\sqrt{7}} = \dfrac{6}{\sqrt{7}}$

But now we must rationalize the denominator by multiplying the numerator and the denominator of the fraction $6/\sqrt{7}$ by $\sqrt{7}$ to obtain

$$\frac{6 \cdot \sqrt{7}}{7}$$

as our final answer. An easier way to get this result would be to multiply the numerator and the denominator of the original fraction $\frac{36}{7}$ by 7 first to obtain

$$\sqrt{\frac{36}{7}} = \sqrt{\frac{36 \cdot 7}{7 \cdot 7}} = \frac{\sqrt{36 \cdot 7}}{\sqrt{7 \cdot 7}} = \frac{6\sqrt{7}}{7}$$

Keep this in mind when doing the exercises! ■

The two properties we have presented are used to do multiplication and division of radicals. Thus,

$$\sqrt{6} \cdot \sqrt{2} = \sqrt{12} \qquad \text{Using multiplication of radicals}$$
$$= \sqrt{4 \cdot 3} = 2\sqrt{3} \qquad \text{Using multiplication of radicals}$$

Similarly,

$$\frac{\sqrt{32}}{\sqrt{2}} = \sqrt{\frac{32}{2}} = \sqrt{16} = 4 \qquad \text{Using division of radicals}$$

EXAMPLE 4 ▶ Multiplication and Division of Radicals

Perform the indicated operations and simplify.

(a) $\sqrt{6} \cdot \sqrt{3}$ (b) $\sqrt{40}/\sqrt{5}$

Solution

(a) $\sqrt{6} \cdot \sqrt{3} = \sqrt{18} = \sqrt{9 \cdot 2} = 3\sqrt{2}$

(b) $\dfrac{\sqrt{40}}{\sqrt{5}} = \sqrt{\dfrac{40}{5}} = \sqrt{8} = \sqrt{4 \cdot 2} = 2\sqrt{2}$ ■

C. Addition and Subtraction of Radicals

The addition and subtraction of radicals can be accomplished using the distributive property. Note that radicals can be combined only when their radicands (the quantities under the radical signs) and indexes are the same. Thus, to add $5\sqrt{2} + 3\sqrt{2}$ or subtract $5\sqrt{2} - 3\sqrt{2}$, we write

$$5\sqrt{2} + 3\sqrt{2} = (5 + 3)\sqrt{2} = 8\sqrt{2}$$

or

$$5\sqrt{2} - 3\sqrt{2} = (5 - 3)\sqrt{2} = 2\sqrt{2}$$

Sometimes we may have to use the multiplication and division of radicals mentioned earlier before the additions or subtractions can be accomplished. Thus, to add $\sqrt{48} + \sqrt{27}$, we use the first property to write $\sqrt{48} = \sqrt{16 \cdot 3} = 4\sqrt{3}$ and $\sqrt{27} = \sqrt{9 \cdot 3} = 3\sqrt{3}$. We then have

$$\sqrt{48} + \sqrt{27} = 4\sqrt{3} + 3\sqrt{3} = 7\sqrt{3}$$

EXAMPLE 5 ▶ Addition and Subtraction of Radicals

Perform the indicated operations.

(a) $\sqrt{50} - \sqrt{8}$ (b) $\sqrt{75} + \sqrt{48} - \sqrt{147}$

Solution

(a) $\sqrt{50} - \sqrt{8} = \sqrt{25 \cdot 2} - \sqrt{4 \cdot 2}$
$$= 5\sqrt{2} - 2\sqrt{2}$$
$$= 3\sqrt{2}$$

(b) $\sqrt{75} + \sqrt{48} - \sqrt{147} = \sqrt{25 \cdot 3} + \sqrt{16 \cdot 3} - \sqrt{49 \cdot 3}$
$$= 5\sqrt{3} + 4\sqrt{3} - 7\sqrt{3}$$
$$= 2\sqrt{3}$$ ■

D. Applications

EXAMPLE 6 ▶ Applications Involving Radicals

The greatest speed s (in miles per hour) at which a bicyclist can safely turn a corner of radius r ft is $s = 4\sqrt{r}$. Find the greatest speed at which a bicyclist can safely turn a corner with a 20-ft radius, and write the answer in simplest form.

Solution

$$s = 4\sqrt{r} = 4\sqrt{20} = 4\sqrt{4 \cdot 5} = 4 \cdot 2\sqrt{5} = 8\sqrt{5} \text{ mph}$$

This is slightly less than 18 mph.   ■

This completes our discussion of the relationship of the various sets of numbers that we have studied. In particular, we see that

$$N \subset W \subset I \subset Q \subset R$$

as shown in Figure 5.10.

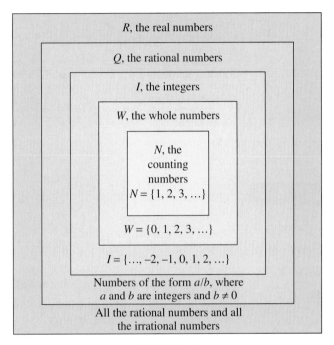

FIGURE 5.10

EXAMPLE 7 ▶ Classifying Real Numbers

Classify the following numbers by making a check mark in the appropriate row:

Set	$\sqrt{4}$	$0.\overline{3}$	$\sqrt{2}$	$-\frac{5}{7}$	0
Natural numbers					
Whole numbers					
Integers					
Rational numbers					
Irrational numbers					
Real numbers					

Solution
The correct check marks are shown below.

Set	$\sqrt{4}$	$0.\overline{3}$	$\sqrt{2}$	$-\frac{5}{7}$	0
Natural numbers	✔				
Whole numbers	✔				✔
Integers	✔				✔
Rational numbers	✔	✔		✔	✔
Irrational numbers			✔		
Real numbers	✔	✔	✔	✔	✔

■

EXERCISES 5.7

A Simplifying Radicals

In problems 1–12, simplify as much as possible.

1. $\sqrt{90}$ 2. $\sqrt{72}$ 3. $\sqrt{122}$

4. $\sqrt{175}$ 5. $\sqrt{180}$ 6. $\sqrt{162}$

7. $\sqrt{200}$ 8. $\sqrt{191}$ 9. $\sqrt{384}$

10. $\sqrt{486}$ 11. $\sqrt{588}$ 12. $\sqrt{5000}$

In problems 13–18, rationalize each denominator.

13. $\dfrac{3}{\sqrt{7}}$ 14. $\dfrac{6}{\sqrt{5}}$ 15. $-\dfrac{\sqrt{2}}{\sqrt{5}}$

16. $-\dfrac{\sqrt{3}}{\sqrt{7}}$ 17. $\dfrac{4}{\sqrt{8}}$ 18. $\dfrac{3}{\sqrt{27}}$

B Multiplication and Division of Radicals

In problems 19–27, simplify each expression.

19. $\sqrt{\frac{3}{49}}$ 20. $\sqrt{\frac{7}{16}}$ 21. $\sqrt{\frac{4}{3}}$

22. $\sqrt{\frac{25}{11}}$ 23. $\sqrt{\frac{8}{49}}$ 24. $\sqrt{\frac{18}{25}}$

25. $\sqrt{\frac{18}{50}}$ 26. $\sqrt{\frac{24}{75}}$ 27. $\sqrt{\frac{32}{125}}$

In problems 28–36, perform the indicated operations and simplify.

28. $\sqrt{7} \cdot \sqrt{8}$ 29. $\sqrt{5} \cdot \sqrt{50}$ 30. $\sqrt{10} \cdot \sqrt{5}$

31. $\dfrac{\sqrt{28}}{\sqrt{2}}$ 32. $\dfrac{\sqrt{22}}{\sqrt{2}}$ 33. $\dfrac{\sqrt{10}}{\sqrt{250}}$

34. $\dfrac{\sqrt{10}}{\sqrt{490}}$ 35. $\dfrac{\sqrt{33}}{\sqrt{22}}$ 36. $\dfrac{\sqrt{18}}{\sqrt{12}}$

C Addition and Subtraction of Radicals

In problems 37–50, perform the indicated operations and simplify.

37. $\sqrt{3} + \sqrt{12}$ 38. $\sqrt{32} - \sqrt{8}$

39. $\sqrt{125} + \sqrt{80}$ 40. $\sqrt{24} - \sqrt{150}$

41. $\sqrt{3^2 + 4^2}$ 42. $\sqrt{5^2 + (12)^2}$

43. $\sqrt{(13)^2 - (12)^2}$ 44. $\sqrt{(25)^2 - (24)^2}$

45. $6\sqrt{7} + \sqrt{7} - 2\sqrt{7}$

46. $\sqrt{3} + 11\sqrt{3} - 3\sqrt{3}$

47. $5\sqrt{7} - 3\sqrt{28} - 2\sqrt{63}$

48. $3\sqrt{28} - 6\sqrt{7} - 2\sqrt{175}$

49. $-3\sqrt{45} + \sqrt{20} - \sqrt{5}$

50. $-5\sqrt{27} + \sqrt{12} - 5\sqrt{48}$

D Applications

51. A playing field is 80 m wide and 100 m long. If Marcie ran diagonally across this field from one corner to the opposite corner, how far did she run? Leave your answer in simplest radical form.

52. The hypotenuse of a right triangle is 10 in. long and one of the sides is 7 in. long. Is the other side longer or shorter than the 7-in. side? Justify your answer.

53. The time t (in seconds) it takes an object dropped from a certain distance (in feet) to hit the ground is

$$t = \sqrt{\frac{\text{distance in feet}}{16}}$$

Find the time it takes an object dropped from a height of 50 ft to hit the ground, and write the answer in simplified form.

54. The time t (in seconds) it takes an object dropped from a certain distance (in meters) to hit the ground is

$$t = \sqrt{\frac{\text{distance in meters}}{5}}$$

Find the time it takes an object dropped from a height of 160 m to hit the ground, and write the answer in simplified form.

55. The compound interest rate r that is paid when you borrow $\$P$ and pay $\$A$ at the end of 2 years is

$$r = \sqrt{\frac{A}{P}} - 1$$

Find the rate when $100 is borrowed and the amount paid at the end of the 2 years is $144.

56. When you are at an altitude of a ft above the Earth, your view V_m (in miles) extends as far as a circle called the *horizon* and is given by

$$V_m = \sqrt{\tfrac{3}{2}a}$$

The greatest altitude reached in a manned balloon is 123,800 ft and was attained by Nicholas Piantanida.
   a. In simplified form, what was the view in miles from this balloon?
   b. If $\sqrt{1857} \approx 43$, what was the view in miles?

If air resistance is neglected, the terminal velocity v of a falling body in meters per second is given by

$$v = \sqrt{20h + v_0}$$

57. Find v if $h = 10$ and $v_0 = 25$ m/sec.

58. Find v if a body is dropped ($v_0 = 0$) from a height of 45 m.

59. If the velocity as measured in feet per second is

$$v = \sqrt{64h + v_0}$$

$h = 12$ ft, and $v_0 = 16$ ft/sec, find v.

60. Find v if a body is dropped ($v_0 = 0$) from a height of 25 ft.

In problems 61–66, evaluate $\sqrt{b^2 - 4ac}$.

61. $a = 1, b = 5, c = 4$

62. $a = 1, b = 3, c = 2$

63. $a = 2, b = -3, c = -20$

64. $a = \dfrac{1}{2}, b = -\dfrac{1}{12}, c = -1$

65. $a = \dfrac{1}{12}, b = \dfrac{1}{3}, c = -1$

66. $a = \dfrac{1}{12}, b = \dfrac{1}{2}, c = \dfrac{2}{3}$

In problems 67–76, classify each number by making a check mark in the appropriate row.

	67.	68.	69.	70.	71.	72.	73.	74.	75.	76.
Set	4.2	$-\frac{3}{8}$	0	$\sqrt{3}$	$\sqrt{9}$	5	$0.\overline{66}$	$1\frac{5}{8}$	$\sqrt{20}$	π
Natural numbers										
Whole numbers										
Integers										
Rational numbers										
Irrational numbers										
Real numbers										

In Other Words

77. Explain why we cannot use the definition $\sqrt{a} \cdot \sqrt{b} = \sqrt{ab}$ when a and b are negative numbers.

78. Explain why every integer is a rational number.

Using Your Knowledge

At the beginning of the next section we shall see that after the man in the picture had fallen 1 ft, his velocity was $\sqrt{32} = 4\sqrt{2}$ ft/sec. Can you estimate $\sqrt{32}$?

Mathematicians use a method called **interpolation** to approximate this answer. Since we know that $\sqrt{25} = 5$ and $\sqrt{36} = 6$, $\sqrt{32}$ should be between 5 and 6. If we place $\sqrt{25}$, $\sqrt{32}$, and $\sqrt{36}$ in a column, the interpolation is done as shown in the diagram.

$$32 - 25 = 7$$
$$\sqrt{25} = 5$$
$$\sqrt{32} = 5 \quad \frac{7}{11}$$
$$\sqrt{36} = 6$$
$$36 - 25 = 11$$

Thus, $\sqrt{32}$ is approximately $5\frac{7}{11}$. If we wish, we can write this answer as a decimal by dividing 7 by 11, obtaining $0.\overline{63}$ and writing the approximation as $5.\overline{63}$. (If you use a calculator to find the square root of 32, the answer is 5.6569.)

Use this knowledge to approximate the following square roots. Give each answer as a mixed number and then as a decimal to two places.

79. $\sqrt{40}$ **80.** $\sqrt{68}$

81. $\sqrt{85}$ **82.** $\sqrt{108}$

Online Study Center

To further explore how Babylonians used their averaging method to find the square root of a number, access link 5.7.2 on this textbook's Online Study Center.

Collaborative Learning

It is said that the Babylonians used an averaging method to find the square root of a number. Here is how the method worked. Suppose you want to find the square root of 4, that is, $\sqrt{4}$. Start with the first guess. Say the guess is 1. Now take the average of 4 and 1, which is 2.5. Now the second guess is the average of 2.5 and $\frac{4}{2.5}$, which is 2.05. Your next guess will be the average of 2.05 and $\frac{4}{2.05}$, that is, 2.0006098. If you keep this procedure going, you will reach the conclusion that $\sqrt{4}$ is 2.

Now form two or more groups, and each group must find $\sqrt{10}$, $\sqrt{867}$, and $\sqrt{900}$.

Now use the Babylonian method to approximate the numbers approximated in problems 79–82. Which is the better approximation? Which method do you prefer?

5.8 Number Sequences

"The Twelve Days of Christmas" and Progressions

Do you know the song "The Twelve Days of Christmas"? How many gifts do you get each day? What is the total number of gifts you receive? If you sing along, you will recall that the song goes like this:

Day 1	A partridge in a pear tree	1 gift
Day 2	A partridge in a pear tree and two turtle doves	1 + 2 gifts
Day 3	A partridge in a pear tree, two turtle doves, and three French hens	1 + 2 + 3 gifts

On the twelfth day, you get $1 + 2 + 3 + \cdots + 11 + 12$ gifts. This sum is an **arithmetic progression.** Can you add it quickly?

According to a story told by E. T. Bell, in late-eighteenth-century Germany, a precocious boy of 10 was admitted to the class in arithmetic in which none of the children were expected to know about progressions. It was easy for the heroic Buttner (the teacher) to give out a long problem in addition whose answer he could find by a formula in a few seconds. The problem was to add

$$1 + 2 + 3 + 4 + \cdots + 98 + 99 + 100$$

The student who first got the answer was to lay his slate on the table, the next to lay his slate on top of the first's, and so on. Buttner had barely finished stating the problem when a boy flung his slate on the table. "*Ligget se* [There it lies]," he said in his peasant dialect. The rest of the hour, while the class worked on the problem, the boy sat with his hands folded, favored now and then by a sarcastic glance from Buttner, who imagined the boy to be just another blockhead. At the end of the period, Buttner looked over the answers. On the boy's slate there appeared but a single number. Do you know what that number was?

In later years, the boy confessed to having recognized the pattern

$$1 + 2 + 3 + 4 + \cdots + 97 + 98 + 99 + 100$$

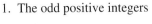

101

in which the sum of each pair of numbers is 101. Since there are 50 pairs of numbers, the total sum would be $50 \times 101 = 5050$, the number on the boy's slate.

In this section you will see the magic formula so you too can add arithmetic progressions! By the way, you will get 6×13 gifts on the twelfth day of Christmas (if you have been good). What about the boy? Carl Friedrich Gauss became one of the most renowned European mathematicians of his time. ▶

The photograph shows a skydiver plunging toward the ground. Do you know how far he will fall in the first 5 sec? A free-falling body travels about 16 ft in the first second, 48 ft in the next second, 80 ft in the third second, and so on. The number of feet traveled in each successive second is

16,   48,   80,   112,   144,   . . .

This list of numbers is an example of a **number sequence.** In general, a list of numbers having a first number, a second number, a third number, and so on is called a **sequence;** the numbers in the sequence are called the **terms.** The following are examples of sequences:

1. The odd positive integers   1, 3, 5, 7, . . .

2. The positive multiples of 3   3, 6, 9, 12, . . .

3. The powers of 10 $10^1, 10^2, 10^3, . . .$

4. The interest on the first three payments on a   \$8.33, \$8.11, \$7.89, . . .
   \$10,000 car being paid over 3 years at 12%
   annual interest

How far will the skydiver fall in the first 5 sec?

A. Arithmetic Sequences

The sequences (1), (2), and (4) on the preceding page are *arithmetic sequences.* An **arithmetic sequence,** or **arithmetic progression,** is a sequence in which each term after the first is obtained by *adding* a quantity called the **common difference** to the preceding term. Thus,

$$16, \quad 48, \quad 80, \quad 112, \quad 144, \quad \ldots$$

is an arithmetic sequence in which each term is obtained by adding the common difference 32 to the preceding term. This means that the common difference for an arithmetic sequence is just the difference between any two consecutive terms.

EXAMPLE 1 ▶ **Finding Common Differences**

Find the common difference in each sequence.

(a) $7, 37, 67, 97, \ldots$ (b) $10, 5, 0, -5, \ldots$

Solution

(a) The common difference is $37 - 7 = 30$ (or $67 - 37$, or $97 - 67$).

(b) The common difference is $5 - 10 = -5$ (or $0 - 5$, or $-5 - 0$).   ▪

It is customary to denote the first term of an arithmetic sequence by a_1 (read "a sub 1"), the common difference by d, and the nth term by a_n. Thus, in the sequence $16, 48, 80, 112, 144, \ldots$, we have $a_1 = 16$ and $d = 32$. The second term of the sequence, a_2, is

$$a_2 = a_1 + 32 = 16 + 32 = 48$$

Since each term is obtained from the preceding one by adding 32,

$$a_3 = a_2 + 32 = (a_1 + 1 \cdot 32) + 32 = a_1 + 2 \cdot 32 = 80$$
$$a_4 = a_3 + 32 = (a_1 + 2 \cdot 32) + 32 = a_1 + 3 \cdot 32 = 112$$
$$a_5 = a_4 + 32 = (a_1 + 3 \cdot 32) + 32 = a_1 + 4 \cdot 32 = 144$$

By following this pattern, we find the **general term** a_n to be

$$a_n = a_1 + (n - 1) \cdot d$$

EXAMPLE 2 ▶ **Finding Terms and Differences**

Consider the sequence $7, 10, 13, 16, \ldots$. Find the following:

(a) a_1, the first term   (b) d, the common difference

(c) a_{11}, the eleventh term   (d) a_n, the nth term

Solution

(a) The first term a_1 is 7.

(b) The common difference d is $10 - 7 = 3$.

(c) The eleventh term is $a_{11} = 7 + (11 - 1) \cdot 3 = 7 + 10 \cdot 3 = 37$.

(d) $a_n = a_1 + (n - 1) \cdot d = 7 + (n - 1) \cdot 3 = 4 + 3n$ ▪

B. Sum of an Arithmetic Sequence

Let us go back to our original problem of finding how far the skydiver falls in 5 sec. The first five terms of the sequence are 16, 48, 80, 112, and 144; thus, we need to find the sum

$$16 + 48 + 80 + 112 + 144$$

Since successive terms of an arithmetic sequence are obtained by adding the common difference d, the sum S_n of the first n terms is

$$S_n = a_1 + (a_1 + d) + (a_1 + 2d) + (a_1 + 3d) + \cdots + a_n \qquad (1)$$

We can also start with a_n and obtain successive terms by subtracting the common difference d. Thus, with the terms written in reverse order,

$$S_n = a_n + (a_n - d) + (a_n - 2d) + \cdots + a_1 \qquad (2)$$

Adding equations (1) and (2), we find that the d's drop out, and we obtain

$$2S_n = (a_1 + a_n) + (a_1 + a_n) + \cdots + (a_1 + a_n)$$
$$= n(a_1 + a_n)$$

Thus,

THEOREM 5.6 Sum of a Finite Arithmetic Sequence

The sum of a finite arithmetic sequence with n terms is

$$S_n = \frac{n(a_1 + a_n)}{2}$$

We are now able to determine the sum S_5, the distance the skydiver dropped in 5 sec. The answer is

$$S_5 = \frac{5(16 + 144)}{2} = 400 \text{ ft}$$

C. Geometric Sequences

The sequence 10, 100, 1000, and so on is *not* an arithmetic sequence because there is no common difference. This sequence is obtained by *multiplying* each term by 10 to get the next term. Such sequences are called *geometric sequences*. A **geometric sequence,** or **geometric progression,** is a sequence in which each term after the first is obtained by multiplying the preceding term by a number r, called the **common ratio.** Thus, the common ratio r can be found by taking the ratio of two successive terms. For example, in the sequence 8, 16, 32, . . . , the first term a_1 is 8, and the common ratio is $\frac{16}{8} = 2$ $\left(\text{or } \frac{32}{16}\right)$. Thus, the first n terms in a geometric sequence are

$$a_1, \qquad a_1 r, \qquad a_1 r^2, \qquad a_1 r^3, \qquad \ldots, \qquad a_1 r^{n-1}$$

EXAMPLE 3 ▶ More Terms and Ratios

Consider the sequence $1, \frac{1}{10}, \frac{1}{100}, \frac{1}{1000}, \ldots$. Find the following:

(a) a_1 (b) r (c) a_n

Solution

(a) a_1 is the first term, 1.

(b) r is the common ratio of any two successive terms. Thus,

$$r = \frac{\frac{1}{10}}{1} = \frac{1}{10}$$

(c) $a_n = a_1 r^{n-1} = 1 \cdot \left(\frac{1}{10}\right)^{n-1} = \frac{1}{10^{n-1}}$ ■

D. Sum of a Geometric Sequence

Can we find the sum S_n of the first n terms in a geometric sequence?

By definition	$S_n = a_1 + a_1 r + a_1 r^2 + \cdots + a_1 r^{n-1}$
Multiplying by r	$rS_n = a_1 r + a_1 r^2 + a_1 r^3 + \cdots + a_1 r^n$
Subtracting	$S_n - rS_n = a_1 - a_1 r^n = a_1(1 - r^n)$
By the distributive property	$S_n(1 - r) = a_1(1 - r^n)$
Dividing by $1 - r$	$S_n = \dfrac{a_1(1 - r^n)}{1 - r}$

THEOREM 5.7 Sum of a Finite Geometric Sequence

The sum of a geometric sequence $a_1, a_1 r, a_1 r^2, a_1 r^3, \ldots, a_1 r^{n-1}$ with a common ratio $r \neq 1$ is

$$S_n = \frac{a_1(1 - r^n)}{1 - r}$$

Thus, the sum of the first three powers of 10—that is, $10 + 10^2 + 10^3$—can be found by noting that $a_1 = 10$, $r = 10^2/10 = 10$, and

$$S_3 = \frac{10(1 - 10^3)}{1 - 10} = \frac{(10)(-999)}{-9} = 1110$$

as expected.

EXAMPLE 4 ▶ Finding Terms and Sums

The first term of a geometric sequence is $\frac{1}{5}$, and $r = \frac{1}{2}$. Find the following:

(a) a_5, the fifth term   (b) S_5, the sum of the first five terms

Solution

(a) The nth term in a geometric sequence is $a_1 r^{n-1}$; thus,

$$a_5 = \left(\frac{1}{5}\right)\left(\frac{1}{2}\right)^{5-1} = \left(\frac{1}{5}\right)\left(\frac{1}{2}\right)^4 = \left(\frac{1}{5}\right)\left(\frac{1}{16}\right) = \frac{1}{80}$$

(b) The sum of the first n terms of a geometric sequence is

$$S_n = \frac{a_1(1 - r^n)}{(1 - r)}$$

GRAPH IT

To find the *sum* of a *sequence* tell your grapher you are doing some math involving sums (press ⎡2nd⎤ ⎡LIST⎤ ⎡▶⎤ ⎡▶⎤ ⎡5⎤) of sequences (press ⎡2nd⎤ ⎡LIST⎤ ⎡▶⎤ ⎡5⎤). To do Example 4(b), you have to enter the expression for the sequence, the name of the variable, and where you want to begin and end the sum, so we enter (.2 × .5 ^ (N − 1), N, 1, 5). Note that to enter the variable N, you have to press ⎡ALPHA⎤ N. Now press ⎡ENTER⎤ and make the answer a fraction by pressing ⎡MATH⎤ ⎡1⎤ ⎡ENTER⎤ obtaining $\frac{31}{80}$ as before.

```
sum(seq(.2*.5^(N-1),N,1,
5)
                    .3875
Ans▶Frac
                    31/80
```

so

$$S_5 = \frac{\frac{1}{5}\left[1 - \left(\frac{1}{2}\right)^5\right]}{1 - \frac{1}{2}} = \frac{\frac{1}{5}\left(\frac{31}{32}\right)}{\frac{1}{2}} = \frac{62}{160} = \frac{31}{80}$$ ∎

E. Infinite Geometric Sequences

Let us now return to the repeating decimals we discussed in Section 5.5. These decimals can be written using an **infinite geometric sequence.** Thus, the decimal $0.333 \cdots$ can be written as

$$0.333 \cdots = \frac{3}{10} + \frac{3}{100} + \frac{3}{1000} + \cdots$$

where the common ratio is $\frac{1}{10}$. The sum of the first n terms of this sequence is

$$S_n = \frac{a_1(1 - r^n)}{1 - r} = \frac{a_1}{1 - r} \cdot (1 - r^n)$$

where $a_1 = \frac{3}{10}$ and $r = \frac{1}{10}$.

If we want to find the sum of *all* the terms, we note that as n increases, $\left(\frac{1}{10}\right)^n$ becomes smaller and smaller. Thus, S_n approaches (becomes closer and closer to)

$$\frac{a_1}{1 - r} = \frac{\frac{3}{10}}{1 - \frac{1}{10}} = \frac{\frac{3}{10}}{\frac{9}{10}} = \frac{1}{3}$$

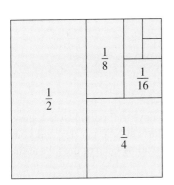

FIGURE 5.11
What do you think is the sum of the sequence $\frac{1}{2} + \frac{1}{4} + \frac{1}{8} + \cdots$?

that is, $0.333 \cdots = \frac{1}{3}$. Figure 5.11 gives a good graphic representation of the sum of an infinite geometric progression approaching a limit. We can generalize this discussion to obtain the following result:

THEOREM 5.8 Sum of an Infinite Geometric Sequence

If r is a number between -1 and 1, the sum of the infinite geometric sequence $a_1, a_1 r, a_1 r^2, \ldots$ is

$$S = \frac{a_1}{1 - r}$$

EXAMPLE 5 ▶ **Writing Decimals as Sequences**

Use the sum of an infinite geometric sequence to write the following repeating decimals as fractions:

(a) $0.666 \cdots$ (b) $0.121212 \cdots$ (c) $3.222 \cdots$

Solution

(a) $0.666 \cdots = \frac{6}{10} + \frac{6}{100} + \frac{6}{1000} + \cdots$. This is a geometric sequence with first term $a_1 = \frac{6}{10}$ and ratio $r = \frac{1}{10}$. The sum of this sequence is

$$\frac{a_1}{1 - r} = \frac{\frac{6}{10}}{1 - \frac{1}{10}} = \frac{\frac{6}{10}}{\frac{9}{10}} = \frac{6}{9} = \frac{2}{3}$$

Thus, $0.666 \cdots = \frac{2}{3}$.

(b) $0.121212\cdots = \frac{12}{100} + \frac{12}{10,000} + \cdots$. This is a geometric sequence with $a_1 = \frac{12}{100}$, $r = \frac{1}{100}$, and sum

$$\frac{a_1}{1-r} = \frac{\frac{12}{100}}{1 - \frac{1}{100}} = \frac{\frac{12}{100}}{\frac{99}{100}} = \frac{12}{99} = \frac{4}{33}$$

Thus, $0.121212\cdots = \frac{4}{33}$.

(c) $3.222\cdots = 3 + \frac{2}{10} + \frac{2}{100} + \frac{2}{1000} + \cdots$. The repeating part, $0.222\cdots$, is a geometric sequence with $a_1 = \frac{2}{10}$, $r = \frac{1}{10}$, and sum

$$\frac{a_1}{1-r} = \frac{\frac{2}{10}}{1 - \frac{1}{10}} = \frac{\frac{2}{10}}{\frac{9}{10}} = \frac{2}{9}$$

Thus, $3.222\cdots = 3\frac{2}{9} = \frac{29}{9}$. ∎ ■

PROBLEM SOLVING

Number Sequences

Suppose you have two job offers for a 2-week (14-day) trial period. Job A starts at $50 per day with a $50 raise each day. Job B starts at $.50 per day and your salary is doubled every day. Find the total amount paid by each of the jobs at the end of the 14 days.

❶ Read the problem.

❷ Select the unknown.

You need to find the total amount each job pays.

❸ Think of a plan.
Find the amount job A pays at the end of 14 days. Then find the amount job B pays at the end of 14 days.

The pay for job A starts at $50 ($a_1 = 50$) and increases by $50 each day ($d = 50$). The salary for the fourteenth day is $a_{14} = 50 + 13 \cdot 50 = 700$. The pay for job B starts at $.50 ($a_1 = 0.50$) and doubles every day ($r = 2$).

❹ Use the formulas for the sum of an arithmetic and for the sum of a geometric progression to find the amount each job pays for 14 days.

For job A, the sum of the arithmetic progression for 14 days is

$$S_{14} = \frac{n(a_1 + a_n)}{2} = \frac{14 \cdot 750}{2} = \$5250$$

For job B, the sum of the geometric progression for 14 days is

$$S_{14} = \frac{a_1(1 - r^n)}{1 - r} = \frac{0.50(1 - 2^{14})}{1 - 2}$$

❺ Use a calculator to find 2^{14}.

$$= \frac{0.50(1 - 2^{14})}{-1}$$
$$= 0.50(2^{14} - 1)$$
$$= 0.50(16,383)$$
$$= \$8191.50$$

Job B pays much more!

continued

⑥ **Verify** your answer.   The verification is left for you.

　　　　　　TRY EXAMPLE 6 NOW.   Cover the solution, write your own solution, and then check your work.

EXAMPLE 6 ▶ Chess and Sequences

The game of chess is said to have originated in Persia. Legend has it that the shah (or king) was so happy that he offered the inventor of the game anything he wanted. The inventor asked that one grain of wheat be placed on the first square of the chessboard, two grains on the second, four on the third, and so on. There are 64 squares on a chessboard.

(a) How many grains were to be placed on the 64th square?

(b) What is the total number of grains the inventor would receive?

Solution

(a) List the number of grains in each square.

Square 1	Square 2	Square 3	Square 4	$\cdots$	Square 64
1	2	$4 = 2^{3-1}$	$8 = 2^{4-1}$	$\cdots$	$2^{63} = 2^{64-1}$

(b) The sum of the geometric progression $1, 2, 4, \ldots, 2^{63}$, where $a_1 = 1$ and $r = 2$ (because the number of grains is doubled for each successive square), is

$$S_{64} = \frac{1(1 - 2^{64})}{1 - 2} = \frac{1 - 2^{64}}{-1}$$
$$= 2^{64} - 1$$

By the way, since 2^{10} is about 1000, $2^{60} = (2^{10})^6$ is about $(1000)^6$, or 1,000,000,000,000,000,000 (1 quintillion). ■

EXERCISES 5.8

Ⓐ **Arithmetic Sequences**

In problems 1–10, find the following:
 a. The first term
 b. The common difference d
 c. The tenth term
 d. The nth term

1. 7, 13, 19, 25, . . .　　**2.** 3, 6, 9, 12, . . .

3. 43, 34, 25, 16, . . .　　**4.** 3, −1, −5, −9, . . .

5. 2, −3, −8, −13, . . .　　**6.** $\frac{2}{3}, \frac{5}{6}, 1, \frac{7}{6}, \ldots$

7. $\frac{-5}{6}, \frac{-1}{3}, \frac{1}{6}, \frac{2}{3}, \ldots$　　**8.** $\frac{-1}{4}, \frac{1}{4}, \frac{3}{4}, \frac{5}{4}, \ldots$

9. 0.6, 0.2, −0.2, −0.6, . . .

10. 0.7, 0.2, −0.3, −0.8, . . .

Ⓑ **Sum of an Arithmetic Sequence**

In problems 11–20, find S_{10} and S_n for the sequences given in problems 1–10.

Ⓒ **Geometric Sequences**

In problems 21–26, find the following:
 a. The first term
 b. The common ratio r
 c. The tenth term
 d. The nth term

21. 3, 6, 12, 24, . . .　　**22.** 5, 15, 45, 135, . . .

23. $\frac{1}{3}$, 1, 3, 9, . . .　　**24.** $\frac{1}{5}$, 1, 5, 25, . . .

25. 16, −4, 1, $-\frac{1}{4}$, . . .　　**26.** 3, −1, $\frac{1}{3}$, $-\frac{1}{9}$, . . .

D Sum of a Geometric Sequence

In problems 27–32, find S_{10} and S_n for the sequences given in problems 21–26. Give answers in simplified exponential form.

E Infinite Geometric Sequences

In problems 33–36, find the sum of each infinite geometric sequence.

33. $6, 3, \frac{3}{2}, \frac{3}{4}, \ldots$

34. $12, 4, \frac{4}{3}, \frac{4}{9}, \ldots$

35. $-8, -4, -2, -1, \ldots$

36. $9, -3, 1, -\frac{1}{3}, \ldots$

In problems 37–40, use sequences to write each repeating decimal as a fraction.

37. $0.777 \cdots$

38. $1.555 \cdots$

39. $2.101010 \cdots$

40. $1.272727 \cdots$

41. A property valued at $30,000 will depreciate (decline in value) $1380 the first year, $1340 the second year, $1300 the third year, and so on.
  a. What will be the depreciation the tenth year?
  b. What will be the value of the property at the end of the tenth year?

42. Strikers at a plant were ordered to return to work and were told they would be fined $100 the first day they failed to do so, $150 the second day, $200 the third day, and so on. If the strikers stayed out for 10 days, what was their total fine?

43. A well driller charges $50 for the first foot; for each succeeding foot, the charge is $5 more than that for the preceding foot. Find the following:
  a. The charge for the tenth foot
  b. The total charge for a 50-ft well

44. When dropped on a hard surface, a Super Ball takes a series of bounces, each one about $\frac{9}{10}$ as high as the preceding one. If a Super Ball is dropped from a height of 10 ft, find the following:
  a. How high it will bounce on the tenth bounce
  b. The approximate distance the ball travels before coming to rest (*Hint:* Draw a picture.)

45. If $100 is deposited at the end of each year in a savings account paying 10% compounded annually, at the end of 5 years the compound amount of each deposit is

$$100, \quad 100(1.10), \quad 100(1.10)^2,$$
$$100(1.10)^3, \quad \text{and} \quad 100(1.10)^4$$

How much money is in the account right after the last deposit? [*Hint:* $(1.10)^5 = 1.61051$.]

46. Sally's father told her that if she behaved well, he would put a nickel in her piggy bank at the end of 1 week, two nickels at the end of 2 weeks, four nickels at the end of 3 weeks, and so on, doubling the number of nickels each successive week. At this rate, in how many weeks would a single deposit amount to over $6?

47. In the Getting Started section for this part we discussed the pattern

$$1 + 2 + 3 + 4 + \cdots + 97 + 98 + 99 + 100$$

$$101$$

where the sum of each pair is 101 and there are $\frac{100}{2} = 50$ pairs. Thus, the total sum is 50×101. Generalize this idea to find the sum of the following sequence:

$$1 + 2 + 3 + \cdots + (n - 1) + n$$

48. a. Use the ideas of problem 47 to find the sum of the following sequence:

$$2 + 4 + 6 + \cdots + (2n - 2) + 2n$$

  b. You can check the answer to part (**a**) by doubling the answer you get for problem 47. Why? Did you get the same answer?

In Other Words

49. What is the difference between an arithmetic sequence and a geometric sequence?

50. Explain why the Fibonacci sequence 1, 1, 2, 3, 5, . . . is neither an arithmetic sequence nor a geometric sequence.

Using Your Knowledge

According to the Health Insurance Association of America and the American Hospital Association, the average daily hospital room charge has been as follows:

1980	1985	1990
$127	$212	$297

Consider the sequence 127, 212, 297, ...

51. Is this sequence an arithmetic sequence, a geometric sequence, or neither?

52. What is the first term of the sequence? What is the fourth term of the sequence and what does it represent?

53. Write the *n*th term of the sequence.

54. On the basis of the pattern found in problems 52 and 53, estimate what the average daily room cost would be in the year 2000 and in the year 2010.

Chapter **5** Summary

Section	*Item*	*Meaning*	*Example*
5.1	$N = \{1, 2, 3, \ldots\}$	The natural numbers	All counting numbers such as 10, 27, 38, and so on
5.1A	$n(A)$	The cardinal number of A	If $A = \{a, b\}$, then $n(A) = 2$.
5.1A	1st, 2nd, 3rd, . . .	Ordinal numbers	This is the *first* one.
5.1A	123-45-6789	Number used for identification	A Social Security number
5.1B	Prime number	A number with exactly two divisors, 1 and itself	2, 3, 5, 7, 11, . . .
5.1B	Composite number	A number with more than two divisors	4, 33, 50, . . .
5.1C	$12 = 2^2 \cdot 3$	Prime factorization of 12	
5.1E	GCF	Greatest common factor	18 is the GCF of 216 and 234.
5.1E	LCM	Least common multiple	252 is the LCM of 18, 21, and 28.
5.2A	$a \cdot 1 = 1 \cdot a = a$	Identity property for multiplication	$1 \cdot 97 = 97$ and $83 \cdot 1 = 83$
5.2A	$W = \{0, 1, 2, \ldots\}$	The set of whole numbers	
5.2A	$0 + a = a + 0 = a$	Identity property for addition	$0 + 13 = 13$ and $84 + 0 = 84$
5.2B	$I = \{\ldots, -1, 0, 1, \ldots\}$	The set of integers	
5.2B	$\xleftrightarrow{\;+\;+\;+\;+\;+\;}$ $\quad -2 -1 \; 0 \; 1 \; 2$	The number line	
5.2B	$n + (-n) = 0$	Additive inverse property	$3 + (-3) = 0$
5.2B	$a - b = a + (-b)$	Definition of subtraction	$3 - 7 = 3 + (-7)$
5.2B	Closed set	A set with an operation defined on it such that when the operation is performed on elements of the set, the result is also an element of the set	The natural numbers are closed under multiplication. The integers are closed under subtraction, but the natural numbers are *not*. ($3 - 5 = -2$ is *not* a natural number.)

Section	Item	Meaning	Example
5.2B	$a + b = b + a$	Commutative property of addition	$3 + 5 = 5 + 3$
5.2B	$a \cdot b = b \cdot a$	Commutative property of multiplication	$6 \cdot 7 = 7 \cdot 6$
5.2B	$a + (b + c) =$ $(a + b) + c$	Associative property of addition	$-4 + (2 + 5) = (-4 + 2) + 5$
5.2B	$a \cdot (b \cdot c) = (a \cdot b) \cdot c$	Associative property of multiplication	$-2 \cdot (4 \cdot 7) = (-2 \cdot 4) \cdot 7$
5.2B	$a \cdot (b + c) =$ $a \cdot b + a \cdot c$	Distributive property	$3 \cdot (4 + 7) = 3 \cdot 4 + 3 \cdot 7$
5.3A	$Q = \left\{ r \middle\| r = \dfrac{a}{b}, \right.$ $\left. a, b \in I, b \neq 0 \right\}$	The set of rational numbers	$\frac{3}{5}, -\frac{7}{3}, 3\frac{1}{2}$ are rational numbers.
5.3B	$\dfrac{b}{a}, a \neq 0$	The reciprocal, or multiplicative inverse, of a/b	$\frac{3}{4}$ and $\frac{4}{3}$ are reciprocals.
5.4A	$a^{-n} = \dfrac{1}{a^n}, a \neq 0$	Definition of negative exponents	$4^{-2} = \dfrac{1}{4^2}$
5.4B	$m \times 10^n$, where m is greater than or equal to 1 and less than 10, and n is an integer	Scientific notation	7.4×10^{-6}
5.5A	$0.\overline{142857}$	A nonterminating, repeating decimal	$0.\overline{142857} =$ $0.142857142857 \cdots$
5.5C	%	Percent sign	3%
5.6	Irrational number	A number that is not rational	$\sqrt{2}, 3\sqrt{5}, \pi$
5.6B	Irrational number	A number that has a nonterminating, nonrepeating decimal representation	$0.101001000 \cdots$
5.6B	R	The set of real numbers	$3, -8, 0, \frac{1}{5}, 0.5$ and $\sqrt{2}$ are real numbers.
5.6C	$C = \pi d$	Circumference of a circle	
5.7A	$\sqrt{}$	Radical sign	$\sqrt{2}$
5.7A	$\sqrt{a \cdot b} = \sqrt{a} \cdot \sqrt{b}$	Multiplication of radicals	$\sqrt{32} = \sqrt{16 \cdot 2} =$ $\sqrt{16} \cdot \sqrt{2} = 4\sqrt{2}$
5.7B	$\sqrt{\dfrac{a}{b}} = \dfrac{\sqrt{a}}{\sqrt{b}}$	Division of radicals	$\sqrt{\dfrac{2}{7}} = \dfrac{\sqrt{2}}{\sqrt{7}}$

Section	Item	Meaning	Example
5.8	d	The common difference of an arithmetic sequence	In the sequence, 4, 9, 14, . . . , $d = 5$.
5.8A	$a_n = a_1 + (n - 1) \cdot d$	The nth term of an arithmetic sequence	In the above sequence, $a_n = 4 + (n - 1) \cdot 5 = 5n - 1$
5.8B	$S_n = \dfrac{n(a_1 + a_n)}{2}$	The sum of the first n terms of an arithmetic sequence	In the above sequence, $S_5 = \dfrac{5(4 + 24)}{2} = 70$
5.8C	r	The common ratio of a geometric sequence	The common ratio of 5, 10, 20, . . . is $r = 2$.
5.8C	$a_n = a_1 r^{n-1}$	The nth term of a geometric sequence	In the sequence 5, 10, 20, . . . , $a_n = 5 \cdot 2^{n-1}$.
5.8D	$S_n = \dfrac{a_1(1 - r^n)}{1 - r}$	The sum of the first n terms of a geometric sequence	For the sequence 5, 10, 20, . . . , $S_4 = \dfrac{5(1 - 2^4)}{1 - 2} = 75$.
5.8E	$S = \dfrac{a_1}{1 - r}$	The sum of the infinite geometric sequence a_1, $a_1 r, a_1 r^2, . . . $, where r is between -1 and 1	The sum of the sequence 2, 1, $\frac{1}{2}, \frac{1}{4}, . . .$ is $S = \dfrac{2}{1 - \frac{1}{2}} = 4$.

 Research Questions

Sources of information for these questions can be found in the Bibliography at the end of the book.

1. Write a report about the introduction of the terms *googol* and *googolplex* in mathematics.

2. Write a report about the discovery of the largest known prime number (see the *Guinness Book of World Records*).

3. Write a report about the discovery of negative numbers and the popularization of 0.

4. The first perfect number is 6 and the next is 28. Write a report about the religious implications of perfect numbers and find out the next two perfect numbers.

5. Write a report on Euclid's formula for perfect numbers.

6. Trace the evolution of different approximations for π including the early Hebrew approximations and the approximation of π using the perimeter of a polygon.

7. In the Discovery section in Exercises 5.3 we have a diagonalization process that shows that there are as many natural numbers as there are rational numbers. This development is due to Georg F. L. P. Cantor. Write a report on Cantor's diagonalization process.

9. In Section 5.1 we stated several divisibility rules. Find the rules for divisibility by 7 and by 11.

10. Write a report on the Golden Ratio and its mathematical and aesthetic significance.

Chapter 5 Practice Test

1. Tell whether the underlined item is used as a cardinal number, an ordinal number, or for identification.
   a. Sally came in <u>third</u> in the 100-yd dash.
   b. Bill's lottery ticket won <u>two</u> dollars.
   c. Jane's auto license number was <u>270–891</u>.

2. a. A prime number is _____.
   b. A composite number is _____.

3. Write 1220 as a product of primes.

4. Write the prime numbers between 50 and 70.

5. Is 143 prime or composite?

6. Of the numbers 2345, 436, 387, and 1530, identify those divisible by
   a. 2.   b. 3.   c. 5.

7. Find the GCF of 216 and 254 and reduce the fraction $\frac{216}{254}$ to lowest terms.

8. Find the LCM of 18, 54, and 60 and evaluate $\frac{1}{18} + \frac{1}{54} - \frac{1}{60}$.

9. A father left $\frac{1}{4}$ of his estate to his daughter, $\frac{1}{2}$ to his wife, and $\frac{1}{8}$ to his son. If the rest went for taxes, what fraction of the estate was that?

10. Change the following to equivalent addition problems and give the results:
    a. $8 - 19$ b. $8 - (-19)$ c. $-8 - 19$ d. $-8 - (-19)$

11. Evaluate a. $4 \times 12 \div 3 \times 10^3 - 2(-6 + 4) \times 10^4$.
    b. $\left(\frac{1}{2}\right)^3 \div \frac{1}{4} \cdot \frac{1}{2} + \frac{1}{3}\left(\frac{7}{2} - \frac{3}{2}\right) - \frac{1}{2} \cdot \frac{1}{3}$.

12. Find a rational number with a denominator of 16 and equal to $\frac{3}{4}$.

13. Find the reciprocals of the following:
    a. $\frac{2}{3}$ b. $-\frac{4}{7}$ c. $2\frac{5}{8}$ d. -8

14. Perform the indicated operations.
    a. $\frac{7}{8} \times \left(-\frac{5}{16}\right)$ b. $-\frac{7}{8} \div \left(-\frac{5}{16}\right)$

15. a. Write 23.508 in expanded form.
    b. Write $(8 \times 10^2) + (3 \times 10^0) + (4 \times 10^{-2})$ in decimal form.

16. Calculate $(6 \times 10^4) \times (8 \times 10^{-6})$ and write the answer in scientific notation.

17. Perform the indicated operations.
    a. $6.73 + 2.8$ b. $9.34 - 4.71$ c. 0.29×6.7 d. $17.36 \div 3.1$

18. The three sides of a triangle are measured to be 18.7, 6.25, and 19.63 cm, respectively. What is the perimeter of the triangle?

19. The dimensions of a college basketball court are 93.5 by 50.6 feet. What is the area of this court?

20. Write the following as decimals:
   a. $\frac{3}{4}$ **b.** $\frac{1}{15}$

21. Write the following as quotients of two integers:
   a. $0.\overline{12}$ **b.** $2.6555\cdots$

22. Write the following as decimals:
   a. 21% **b.** 9.35% **c.** 0.26%

23. Write the following as percents:
   a. 0.52 **b.** 2.765 **c.** $\frac{3}{5}$ **d.** $\frac{2}{11}$ (to one decimal place)

24. A 2-L bottle of soda sells for 86¢ and costs the store 48¢. Find, to two decimal places, the percent of profit on the cost.

25. About 4 million women make between \$25,000 and \$50,000 annually. This represents 15% of all working women. What is the total number of working women?

26. Classify the following as rational or irrational:
   a. $\sqrt{49}$ **b.** $\sqrt{45}$ **c.** $\sqrt{121}$
   d. $0.41252525\cdots$ **e.** $0.212112111\cdots$ **f.** $0.246810\cdots$

27. Find the following:
   a. A rational number between $0.\overline{2}$ and 0.25
   b. An irrational number between $0.\overline{2}$ and 0.25

28. The diameter of a circular hamburger is 4 in. To the nearest tenth of an inch, what is the circumference of this hamburger? (Use $\pi = 3.14$.)

29. The circumference of a basketball is 29.5 in. Find the diameter to the nearest hundredth. (Use $\pi = 3.14$.)

30. Simplify the following if possible:
   a. $\sqrt{96}$ **b.** $\sqrt{58}$

31. Simplify the following:
   a. $\dfrac{4}{\sqrt{20}}$ **b.** $\sqrt{\dfrac{48}{49}}$

32. Perform the indicated operations and simplify.
   a. $\sqrt{8} \cdot \sqrt{6}$ **b.** $\dfrac{\sqrt{56}}{\sqrt{7}}$

33. Perform the indicated operations.
   a. $\sqrt{90} - \sqrt{40}$ **b.** $\sqrt{32} + \sqrt{18} - \sqrt{50}$

34. The length c of the hypotenuse of a right triangle is given by $c = \sqrt{a^2 + b^2}$, where a and b are the lengths of the other two sides. If a right triangle has sides measuring 4 in. and 6 in. respectively, what is the length of the hypotenuse? Simplify the answer.

35. Classify the given numbers by making check marks in the appropriate rows.

Set	**a.** $\sqrt{16}$	**b.** $-\frac{5}{4}$	**c.** $\sqrt{5}$	**d.** $0.\overline{8}$	**e.** $-2\frac{3}{4}$
Natural numbers					
Whole numbers					
Integers					
Rational numbers					
Irrational numbers					
Real numbers					

36. Classify the following as arithmetic or geometric sequences:
   a. 2, 4, 8, 16, . . .   **b.** 5, 8, 11, 14, . . .

37. Find the sum of the first ten terms of 9, 13, 17, 21, ...

38. Find the sum of the first five terms of the sequence $1, \frac{1}{2}, \frac{1}{4}, \frac{1}{8}, \ldots$.

39. The first term of a geometric sequence is $\frac{1}{3}$ and $r = \frac{1}{2}$. Find the following:
   a. a_5, the fifth term   **b.** S_5, the sum of the first five terms

40. Use the sum of an infinite geometric sequence to write the following repeating decimals as fractions:
   a. $0.444 \cdots$ **b.** $0.212121 \cdots$ **c.** $2.555 \cdots$

Answers to Practice Test

ANSWER	IF YOU MISSED Question	REVIEW Section	REVIEW Example(s)	REVIEW Page(s)
1. a. Ordinal number **b.** Cardinal number **c.** Identification	1	5.1	1	225
2. a. A natural number with exactly two distinct divisors **b.** A natural number with more than two distinct divisors	2	5.1	Def. of prime and composite numbers	226
3. $2^2 \times 5 \times 61$ **4.** 53, 59, 61, and 67	3	5.1	2	228
5. Composite ($143 = 11 \times 13$)	4	5.1	Table 5.1	229
6. a. 436 and 1530 are divisible by 2. **b.** 387 and 1530 are divisible by 3. **c.** 2345 and 1530 are divisible by 5.	5	5.1	3	228
	6	5.1	4	229
7. $\text{GCF}(216, 254) = 2; \frac{216}{254} = \frac{108}{127}$	7	5.1	6	231
8. $\text{LCM}(18, 54, 60) = 540; \frac{1}{18} + \frac{1}{54} - \frac{1}{60} = \frac{31}{540}$	8	5.1	7	232–233
9. $\frac{1}{8}$	9	5.1	8	233

ANSWER	IF YOU MISSED	REVIEW		
	Question	Section	Example(s)	Page(s)
10. a. $8 - 19 = 8 + (-19) = -11$	10	5.2	2	244
b. $8 - (-19) = 8 + (+19) = 27$				
c. $-8 - 19 = -8 + (-19) = -27$				
d. $-8 - (-19) = -8 + (+19) = 11$	11	5.2	5, 6	248, 249
11. a. 56,000 **b.** $\frac{3}{4}$ **12.** $\frac{12}{16}$	12	5.3	2	257
13. a. $\frac{3}{2}$ **b.** $-\frac{7}{4}$ **c.** $\frac{8}{21}$ **d.** $-\frac{1}{8}$	13	5.3	4	259
14. a. $-\frac{35}{128}$ **b.** $\frac{14}{5}$	14	5.3	5	261–262
15. a. $2 \times 10 + 3 + 5 \times 10^{-1} + 8 \times 10^{-3}$	15	5.4	1	268
b. 803.04	16	5.4	2–4	269, 270
16. 4.8×10^{-1}	17	5.4	5–7	270–272
17. a. 9.53 **b.** 4.63 **c.** 1.943 **d.** 5.6	18	5.4	9	274
18. 44.58 cm **19.** 4731.1 ft^2	19	5.4	11	276
20. a. 0.75 **b.** $0.0666\cdots$	20	5.5	1	281
21. a. $\frac{4}{33}$ **b.** $\frac{239}{90}$	21	5.5	2, 3	283–284
22. a. 0.21 **b.** 0.0935 **c.** 0.0026	22	5.5	4	285
23. a. 52% **b.** 276.5%	23	5.5	5, 6	285
c. 60% **d.** 18.2%	24	5.5	7	285–286
24. 79.17% **25.** about 26.7 million	25	5.5	8	286
26. a. Rational **b.** Irrational	26	5.6	1, 2	293, 294
c. Rational **d.** Rational				
e. Irrational **f.** Irrational				
27. a. 0.24 (Other answers are possible.)	27	5.6	4, 5	295–296
b. $0.23456\cdots$ (Other answers are possible.)	28	5.6	6	297
28. 12.6 in. **29.** 9.39 in.	29	5.6	7	298
30. a. $4\sqrt{6}$	30	5.7	1	302
b. $\sqrt{58}$ is in simplest form.				
31. a. $\frac{2\sqrt{5}}{5}$ **b.** $\frac{4\sqrt{3}}{7}$	31	5.7	2, 3	302, 303
32. a. $4\sqrt{3}$ **b.** $2\sqrt{2}$	32	5.7	4	303–304
33. a. $\sqrt{10}$ **b.** $2\sqrt{2}$	33	5.7	5	304
34. $2\sqrt{13}$	34	5.7	6	304
35. a. Check natural numbers, whole numbers, integers, rational numbers, and real numbers.	35	5.7	7	305–306
b. Check rational numbers and real numbers.				
c. Check irrational numbers and real numbers.				
d. Check rational numbers and real numbers.				
e. Check rational numbers and real numbers.				
36. a. A geometric sequence	36	5.8	Def., part (C)	310–311
b. An arithmetic sequence				
37. 270	37	5.8	Theorem 5.6	311
38. $\frac{31}{16}$	38	5.8	4	312–313
39. a. $\frac{1}{48}$ **b.** $\frac{31}{48}$	39	5.8	4	312–313
40. a. $\frac{4}{9}$ **b.** $\frac{7}{33}$ **c.** $\frac{23}{9}$	40	5.8	5	313–314

Maintaining a nutritional diet is a key element of a healthy lifestyle. In Section 6.6, you will use ratios to find the nutritional value of certain foods.

Equations, Inequalities, and Problem Solving

We are now ready to begin the study of algebra. The word *algebra* is the European derivation of *al-jabr,* part of the title of al-Khowarizmi's treatise *Hisab al-jabr w'al muqabalah,* "The Science of Reunion and Reduction." The study of algebra starts with its foundations: open sentences, statements, equations, and inequalities.

We will consider how to solve first-degree equations and inequalities (Section 6.1) and how to graph the solutions of the inequalities (Sections 6.2 and 6.3). Next, we study quadratic equations and their methods of solution, factoring and the quadratic formula, and then use this information to solve different types of word problems. We end the chapter by examining ratio, proportion, and variation, emphasizing applications to consumer problems such as unit pricing.

Now, remember that precocious boy of 10 who amazed his teacher by adding $1 + 2 + 3 + \cdots + 100$ with lightning speed? In his doctoral dissertation he provided the first proof of the fundamental theorem of algebra. You can read more about Carl Friedrich Gauss in the Human Side of Math.

Online Study Center

For links to Internet sites related to Chapter 6, please access **college.hmco.com/PIC/bello9e** and click on the Online Study Center icon.

6.1 Solutions of First-Degree Sentences

Crickets, Ants, and Temperatures

Does temperature affect animal behavior? You must know about bears hibernating in the winter and the languid nature of students in the spring. But what about the behavior of crickets and ants? Can you tell whether crickets will stop chirping before ants stop crawling? In the Discovery section, you will find that the number N of chirps a cricket makes per minute *satisfies* the *equation* $N = 4(F - 40)$, where F is the temperature in degrees Fahrenheit. What happens as the temperature increases? In problem 113, Exercises 6.1, you will find that the speed S (in cm/sec) for certain types of ants is $S = \frac{1}{6}(C - 4)$, where C is the temperature in degrees Celsius (see photo on page 326). What happens as the temperature decreases? The relationship between Fahrenheit and Celsius temperature is given by $F = \frac{9}{5}C + 32$. Armed with this information, can you tell whether crickets stop chirping before ants stop crawling? ▶

Elementary algebra was first treated in a systematic fashion by the Arabs during the period before the Renaissance, when Europe was almost at a standstill intellectually. By the early 1600s, algebra had become a fairly well developed branch of mathematics, and mathematicians were beginning to discover that a marriage of algebra and geometry could be highly beneficial to both subjects.

It has been said that algebra is arithmetic made simple, and it is true that a small amount of elementary algebra enables us to solve many problems that would be quite difficult by purely arithmetic means. In this chapter we shall consider some of the simpler algebraic techniques that are used in problem solving.

We have already made frequent use of various symbols, usually letters of the alphabet, as placeholders for the elements of a set of numbers. For example, we wrote

$$a + b = b + a, \qquad a, b \text{ real numbers}$$

as a symbolic way of stating the commutative property of addition. Of course, we mean that a and b can each be replaced by any real number. In this case the set of real numbers is the **replacement set** for a and b. A symbol that can be replaced by any one of a set of numbers is called a **variable.**

Letters of the alphabet as well as symbols such as $\square$ are often used to indicate variables in arithmetic. The study of sentences and expressions involving variables is, however, a part of algebra.

In algebra, as in arithmetic, the commonly used **verb phrases** are

$=$	is equal to	$\neq$	is not equal to
$>$	is greater than	$\geq$	is greater than or equal to
$<$	is less than	$\leq$	is less than or equal to

By using these verb phrases along with specific numbers and variables joined by the usual operations of arithmetic, we can form many types of sentences. Some examples of simple algebraic sentences are

$$x - 1 = 3, \qquad 3x - 2 \neq 4, \qquad x - 1 \geq 3, \text{ and} \qquad x + 7 < -9x$$

The parts that are added or subtracted in these algebraic sentences are called **terms.** Thus, the terms in $3x - 2$ are $3x$ and -2. The numerical part of the term

Looking Ahead
Much of Gauss's work in pure mathematics dealt with number theory, the concept of complex numbers, and the solutions to algebraic equations, which is the focus of this chapter.

For some types of ants, the speed at which they move varies directly with changes in temperature.

$3x$, 3, is its **numerical coefficient,** or simply its **coefficient.** Terms that differ only in their numerical coefficients are called **like terms.** For example, $-7x$ and $3x$ are like terms; $2y^2$ and $-5y^2$ are like terms, but $2y^2$ and $2y$ are not.

A. Equations and Inequalities

In the four preceding sentences, x is a variable, that is, a placeholder for the numbers by which it can be replaced. Until x is replaced by a number, none of these sentences is a statement because it is neither true nor false. For this reason, we call such sentences **open sentences.** Because only one variable is involved, we refer to the sentences as **open sentences in one variable.** Sentences in which the verb phrase is "=" are called **equations;** if the verb phrase is any of the others we have listed, then the sentence is called an **inequality.**

In order to study an open sentence in one variable, we obviously must know what the replacement set for that variable is. We are interested in knowing for which of the possible replacements the sentence is a true statement. To **solve** an open sentence is to find its solution set.

Definition of Solution Set

The set of elements of the replacement set that make the open sentence a true statement is called the **solution set** for the given replacement set.

EXAMPLE 1 ▶ Finding Solution Sets

Suppose the replacement set for x is $\{2, 4, 6\}$. For each of the following open sentences, find the solution set:

(a) $x - 1 = 3$ (b) $3x - 2 \neq 4$

(c) $x - 1 \geq 3$ (d) $x + 7 < -9x$

Solution

(a) We substitute the elements of the replacement set into the open sentence $x - 1 = 3$.

For $x = 2$, we get $2 - 1 = 1$, not 3.
For $x = 4$, we get $4 - 1 = 3$, which makes the sentence a true statement.
For $x = 6$, we get $6 - 1 = 5$, not 3.

Thus, $x = 4$ is the only replacement that makes the sentence $x - 1 = 3$ a true statement, so the solution set is $\{4\}$.

(b) We make the permissible replacements into the open sentence $3x - 2 \neq 4$.

For $x = 2$, we get $3(2) - 2 = 4$, which *is* equal to 4, so the sentence is a false statement.
For $x = 4$, we get $3(4) - 2 = 10$, which is not equal to 4, so the sentence is a true statement.
For $x = 6$, we get $3(6) - 2 = 16$, which also satisfies the " $\neq 4$," so the sentence is a true statement.

Thus, the solution set is $\{4, 6\}$.

(c) We make the permissible replacements into $x - 1 \geq 3$.

> For $x = 2$, we get $2 - 1 = 1$, which is less than 3, not greater than or equal to 3. Hence, the sentence is a false statement.
> For $x = 4$, we get $4 - 1 = 3$, which satisfies the "≥ 3," so the sentence is a true statement.
> For $x = 6$, we get $6 - 1 = 5$, which satisfies the "≥ 3," so the sentence is a true statement.

The solution set is thus $\{4, 6\}$.

(d) We make the permissible replacements into $x + 7 < -9x$.

> For $x = 2$, we get $2 + 7 = 9$, which is not less than $-9(2) = -18$, so the sentence is a false statement.
> For $x = 4$, we get $4 + 7 = 11$, which is not less than $-9(4) = -36$, so the sentence is a false statement.
> For $x = 6$, we get $6 + 7 = 13$, which is not less than $-9(6) = -54$, so the sentence is a false statement.

Since none of the replacements makes the sentence $x + 7 < -9x$ a true statement, the solution set is $\varnothing$. ∎

Now that we have examined how to determine whether a certain number satisfies an equation or inequality, we consider how to find these numbers, that is, how to *solve* equations or inequalities by finding equivalent equations or inequalities whose solutions are obvious.

B. Solving Equations

First, we consider which operations can be performed on a sentence to obtain an **equivalent** sentence, that is, one with exactly the same solution set as the original sentence. Such operations are called **elementary operations.**

Elementary Operations
For the equation
$$a = b$$
the following elementary operations yield equations **equivalent** to the original equation:
1. **Addition** $\qquad a + c = b + c$
2. **Subtraction** $\qquad a - c = b - c$
3. **Multiplication** $\qquad a \times c = b \times c, \qquad c \neq 0$
4. **Division** $\qquad a \div c = b \div c, \qquad c \neq 0$

Briefly stated, we can add or subtract the same number on both sides, or multiply or divide both sides by the same nonzero number. To solve an equation, we use the elementary operations as needed to obtain an equivalent equation of the form

$$x = n \qquad \text{or} \qquad n = x$$

where the number n is the desired solution.

GRAPH IT

Your grapher can solve equations by using the EQUATION SOLVER. Press [MATH] [0]. If there is nothing entered, the display will show

```
EQUATION SOLVER
eqn:0=
```

If an equation has been entered, press [▲] [CLEAR]. There are many optional steps, but we will only cover the basics. To solve $x + 2 = 5$, subtract 5 from both sides and enter $x - 3$ [ENTER] [ALPHA] [ENTER] to obtain the answer 3 as shown.

```
X-3=0
■X=3
  bound={■-1ε99,1...
■left-rt=0
```

For example, to solve the equation

$$55 = 0.20m + 25$$

we must *isolate m*, that is, get m all by itself on one side of the equation. Hence, we proceed as follows:

Subtract 25.	$55 - 25 = 0.20m + 25 - 25$
Simplify.	$30 = 0.20m$
Divide by 0.20.	$\dfrac{30}{0.20} = \dfrac{0.20m}{0.20}$
or	$150 = m$

Thus, the solution set of $55 = 0.20m + 25$ is $\{150\}$.

EXAMPLE 2 ▶ Solving Equations

Solve the equation $x + 2 = 5$.

Solution

To solve the equation, we first want to get the variable x by itself on one side. Therefore, we subtract 2 from both sides to eliminate the $+2$ on the left-hand side.

$$x + 2 - 2 = 5 - 2$$

This simplifies to

$$x = 3$$

The solution set is $\{3\}$. We used elementary operation 2.   ∎

EXAMPLE 3 ▶ Solving More Equations

Solve the equation $2x - 8 = 5x - 6$.

Solution

To have a positive number of x's on the right, we subtract $2x$ on both sides and get

$$2x - 8 - 2x = 5x - 6 - 2x \qquad \text{or} \qquad -8 = 3x - 6$$

To eliminate the -6 on the right side, we add 6 to both sides and obtain

$$-8 + 6 = 3x - 6 + 6 \qquad \text{or} \qquad -2 = 3x$$

Since the 3 on the right multiplies the x, we divide both sides by 3.

$$\frac{-2}{3} = \frac{3x}{3}$$

That is,

$$-\frac{2}{3} = x \qquad \text{or} \qquad x = -\frac{2}{3}$$

The solution set is $\{-\frac{2}{3}\}$, and we used elementary operations 1, 2, and 4.   ∎

GRAPH IT

To do Example 3, subtract $(5x - 6)$ from both sides of the equation to obtain $-3x - 2 = 0$. Press MATH 0 ▲ CLEAR. Enter $-3x - 2$ to the right of the 0. (Recall that the $-$ in $-3x$ is entered as $(-)$.) To obtain the answer, press ENTER ALPHA ENTER and get $-0.666 \cdots$, or $-\frac{2}{3}$.

```
-3X-2=0
■X=-.666666666...
   bound=■-1E99,1...
■left-rt=0
```

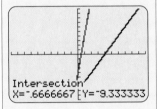
The equations in Examples 2 and 3 are called **linear** or **first-degree** equations in one variable because the highest exponent of the variable is 1. In general, we have the following:

Definition of a First-Degree Expression

An expression of the form $ax + b$, where a and b are real numbers and $a \neq 0$, is called a **first-degree** (or **linear**) **expression in x,** and an equation of the form

$$ax + b = 0 \qquad a \neq 0$$

is called a **first-degree** (or **linear**) **equation in x.**

To help in solving linear equations, we suggest the following procedure:

Procedure to Solve Linear Equations

1. If there are fractions, multiply each term on both sides of the equation by the lowest common denominator (LCD) of the fractions.

2. Simplify both sides of the equation if necessary (remove parentheses and combine like terms).

3. Add or subtract the same expression (terms) on both sides so that the variable is isolated on one side.

4. Add or subtract the same numbers (constants) on both sides so that the variable is isolated on one side.

5. If the coefficient of the variable is not 1, divide both sides by this coefficient.

6. The resulting equation is in the form $x = a$ (or $a = x$), where the number a is the solution of the equation.

7. Check the answer by substituting it into the original equation. Both sides must simplify to the same number.

EXAMPLE 4 ▶ **Solving Equations Involving Fractions**

Solve the equation

$$\frac{(x + 1)}{5} = \frac{(x - 2)}{2} + \frac{9}{5}$$

Solution
We use the suggested procedure for the given equation.

1. Multiply each term by the LCD 10. (You obtain an equivalent equation.)

$$10\frac{(x + 1)}{5} = 10\frac{(x - 2)}{2} + 10 \cdot \frac{9}{5}$$

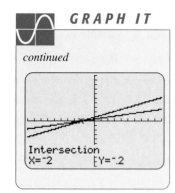

GRAPH IT

continued

Intersection
X=-2   Y=-.2

2. Simplify by removing parentheses and combining like terms.

$$2(x + 1) = 5(x - 2) + 18$$
$$2x + 2 = 5x - 10 + 18 \qquad \text{or} \qquad 2x + 2 = 5x + 8$$

3. Since the coefficient of x is greater on the right side than on the left, we subtract $2x$ on both sides and get

$$2x + 2 - 2x = 5x + 8 - 2x \qquad \text{or} \qquad 2 = 3x + 8$$

4. To eliminate the 8 on the right side, we subtract 8 on both sides.

$$2 - 8 = 3x + 8 - 8 \qquad \text{or} \qquad -6 = 3x$$

5. To make the coefficient of x equal 1, we divide both sides by 3.

$$\frac{-6}{3} = \frac{3x}{3}$$

6. We get

$$-2 = x$$

Thus, the solution is $x = -2$.

7. *Check in the original equation:* For $x = -2$, the left side of the given equation becomes

$$\frac{-2 + 1}{5} = \frac{-1}{5} = -\frac{1}{5}$$

and the right side becomes

$$\frac{-2 - 2}{2} + \frac{9}{5} = -2 + \frac{9}{5}$$
$$= -\frac{10}{5} + \frac{9}{5} = -\frac{1}{5}$$

Since the two sides agree, the solution checks.   ∎

EXAMPLE 5 ▶ Solving for a Variable

Solve for p.

$$1 + 3 \times 2p = 7 + \frac{3(2p)^2}{3p}$$

Solution

First, we simplify the given equation by using the correct order of operations in doing the indicated multiplications and divisions.

$$1 + 3 \times 2p = 7 + \frac{3(2p)^2}{3p} \qquad \text{Given.}$$

$$1 + 6p = 7 + \frac{4p^2}{p} \qquad \text{Simplify, since } \frac{\cancel{3}(2p)^2}{\cancel{3}p} = \frac{4p^2}{p}.$$

$$1 + 6p = 7 + 4p$$

$$1 + 2p = 7 \qquad \text{Subtract } 4p \text{ from both sides.}$$

$$2p = 6 \qquad \text{Subtract 1 from both sides.}$$

$$p = 3 \qquad \text{Divide both sides by 2.}$$

Check: For $p = 3$, the left side becomes

$$1 + 3 \times 6 = 19$$

and the right side becomes

$$7 + \frac{3(6^2)}{9} = 7 + \frac{36}{3} = 7 + 12 = 19$$

Since the two sides agree, the answer $p = 3$ is correct. ■

C. Solving Formulas for a Variable

Sometimes we are asked to solve for a variable in a *literal equation* containing several variables. To solve the literal equation $H = 2.89h + 70.64$ relating a man's height H and the length of his humerus bone h, for h, we use a procedure similar to the one used for solving linear equations (page 329). Here are some suggestions to follow when solving for a *specified variable:*

> **Procedure to Solve for a Specified Variable**
>
> **1.** Add or subtract the same expression on both sides of the equation so that the terms containing the *specified variable* (the one for which you are solving) are isolated on one side. The terms that *do not* contain the specified variable are thus on the other side.
>
> **2.** If necessary, use the distributive property to write the side containing the specified variable as a product of the variable and a sum (or difference) of terms.
>
> **3.** Then use the rules for solving linear equations.

EXAMPLE 6 ▶ **Solving Formulas for a Variable**

Anthropologists know how to estimate the height of a man (in centimeters) using only a bone. They use the formula

$$H = 2.89h + 70.64$$

Height of   Length of
the man   the humerus

a. Solve for h.   **b.** A man is 157.34 cm tall. How long is his humerus?

Solution

a. We have to solve for h, so we isolate h on one side of the equation. Given

$$\begin{aligned}
H &= 2.89h + 70.64 \\
H - 70.64 &= 2.89h + 70.64 - 70.64 &&\text{Subtract 70.64.} \\
H - 70.64 &= 2.89h &&\text{Simplify.} \\
\frac{H - 70.64}{2.89} &= h &&\text{Divide by 2.89.}
\end{aligned}$$

Now we can find h for any given value of H.

b. Substitute 157.34 for H to get

$$h = \frac{157.34 - 70.64}{2.89} = \frac{86.7}{2.89} = 30$$

The man's humerus is 30 cm long. ■

D. Solving Inequalities

Definition of First-Degree Inequality

An **inequality of the first degree in x** is an inequality of the form

$$ax + b < 0 \qquad a \neq 0$$

or of the form $ax + b > 0$.

Note: If $<$ is replaced by $\leq$ and $>$ is replaced by $\geq$, the results are still inequalities of the first degree in x.

We can solve such inequalities by means of elementary operations that produce equivalent inequalities whose solution sets are obvious, such as $x = 3$ or $x = -5$, or, in general, $x = \square$ or $\square = x$. These operations are as follows:

Elementary Operations for Inequalities

For the inequality

$$a < b$$

the following elementary operations yield inequalities **equivalent** to the original inequality:

1. **Addition** $\qquad a + c < b + c$

2. **Subtraction** $\qquad a - c < b - c$

3. **Multiplication** $\qquad ac < bc \qquad$ for $c > 0$

    $\qquad\qquad\qquad\qquad\quad ac > bc \qquad$ for $c < 0$

4. **Division** $\qquad\qquad \dfrac{a}{c} < \dfrac{b}{c} \qquad$ for $c > 0$

    $\qquad\qquad\qquad\qquad\quad \dfrac{a}{c} > \dfrac{b}{c} \qquad$ for $c < 0$

Briefly stated, we can add or subtract the same number on both sides. The sense (direction) of the inequality is **unchanged** if both sides are multiplied or divided by the same *positive* number. The sense (direction) of the inequality is **reversed** if both sides are multiplied or divided by the same *negative* number. For instance, if both sides of $-2 < 1$ are multiplied by -3, we get $6 > -3$. Similarly, if both sides of $-9 < -6$ are divided by -3, the result is $3 > 2$.

The preceding operations have been stated for the inequality $a < b$, but the same operations are valid for $a > b$. You can convince yourself of the validity of these operations by noting that the geometric equivalent of $a < b$ is **a precedes b on the number line.** (Figure 6.1 clarifies this idea.) *Note:* Since $a \leq b$ means

$a < b$ **or** $a = b$, the elementary operations listed above may also be used for inequalities of the type $a \leq b$ and $a \geq b$.

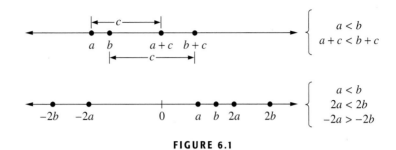

FIGURE 6.1

Linear (first-degree) inequalities can be solved by a procedure similar to that used for equations. This is illustrated in the next examples, where x represents a real number.

EXAMPLE 7 ▶ Solving Inequalities Involving >

Solve the inequality $3x + 2 > x + 6$.

Solution

1. The two sides are already in simplified form.

2. Subtract x from both sides.

$$3x + 2 - x > x + 6 - x$$
$$2x + 2 > 6$$

3. Subtract 2 from both sides.

$$2x + 2 - 2 > 6 - 2$$
$$2x > 4$$

4. Divide both sides by 2.

$$\frac{2x}{2} > \frac{4}{2}$$
$$x > 2$$

5. The solution set is $\{x \mid x > 2\}$.

6. A partial check can be made by substituting a number from the proposed solution set into the original inequality. For instance, 3 is in the set $\{x \mid x > 2\}$. For $x = 3$, the left side becomes

$$3(3) + 2 = 11$$

and the right side becomes

$$3 + 6 = 9$$

Since $11 > 9$, $x = 3$ does satisfy the inequality. Because the solution set contains infinitely many numbers, we cannot check by substituting one number at a time. However, if the number selected did *not* check, then something would be wrong, and we could check the work to find the error. ■

GRAPH IT

To solve $2x - 3 < 5x + 7$, let $Y_1 = 2x - 3$ and $Y_2 = 5x + 7$. Press ZOOM 6. Y_1 is **below** Y_2 to the right of the point at which the lines intersect. This intersection is found by pressing 2nd CALC 5 and ENTER three times. Thus, $Y_1 < Y_2$ to the right of $-3.333\cdots$, that is, when $x > -3\frac{1}{3} = -\frac{10}{3}$.

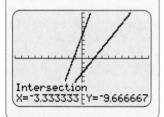

Intersection
X=-3.333333 Y=-9.666667

EXAMPLE 8 ▶ **Solving Inequalities Involving** $<$

Solve the inequality $2x - 3 < 5x + 7$.

Solution

1. The two sides are already in simplified form.

2. Subtract $5x$ from both sides.

$$2x - 3 - 5x < 5x + 7 - 5x$$
$$-3x - 3 < 7$$

3. Add 3 to both sides.

$$-3x - 3 + 3 < 7 + 3$$
$$-3x < 10$$

4. Divide both sides by -3.

$$x > -\frac{10}{3} \qquad \text{Do not forget to change } < \text{ to } >.$$

5. The solution set is $\{x \mid x > -\frac{10}{3}\}$. Be sure to notice that **division by -3 reversed the sense (direction) of the inequality.**

6. The check is left for you to do. An easy number to use is 0.

   If, in step 2, you were to subtract $2x$ from both sides of the inequality, you would avoid dividing by a negative number later. However, in this case, the answer would be $-\frac{10}{3} < x$. If you are asked what x is, then you must write the equivalent answer $x > -\frac{10}{3}$. Which way should you do it? Whichever way you understand best! ■

EXAMPLE 9 ▶ **Solving Inequalities Involving Fractions**

Solve the inequality $-\frac{1}{2}x < \frac{1}{2} + x$.

Solution

This time we want to eliminate the fractions and the negative sign on the left. This is done by multiplying both sides by -2. Since this multiplier is negative, the inequality sign is reversed. Here are the steps.

$$\overset{\text{Reverse the inequality sign!}}{(-2)\left(-\tfrac{1}{2}\right)x > (-2)\left(\tfrac{1}{2} + x\right)}$$

or

$$x > -1 - 2x$$

Now we add $2x$ to both sides to get

$$3x > -1$$

Then we divide both sides by 3 (do not reverse the inequality because 3 is positive) and obtain the answer

$$x > -\frac{1}{3}$$

The solution set is $\{x \mid x \text{ is a real number and } x > -\frac{1}{3}\}$. ■

E. Applications

One of the most important ideas in elementary mathematics is the use of **percent.** Basically, there are three types of problems involving percent. These may be illustrated as follows:

Statement	*Translation*
1. 40% of 60 is what number?	$0.40 \times 60 = n$
2. What percent of 50 is 10?	$r \times 50 = 10$
3. 20 is 40% of what number?	$20 = 0.40 \times n$

All these problems (which, incidentally, can be stated in different ways) can be solved easily by using what we have studied in this section. The basic idea is that "$r\%$ of n is p" translates into the equation

$$0.01rn = p \qquad \text{Recall that } r\% = \frac{r}{100} = 0.01r. \tag{1}$$

Each type of percent problem can be solved by substituting the known data into equation (1) or by **translating** the problem into an equation and then solving for the unknown.

EXAMPLE 10 ▶ Solving Percent Problems Algebraically

(a) 40% of 60 is what number?   (b) What percent of 50 is 10?

(c) 20 is 40% of what number?

Solution

(a) Here, $r = 40$, $n = 60$, and p is unknown. Equation (1) becomes

$$0.01 \times 40 \times 60 = p \qquad \text{or} \qquad 0.40 \times 60 = p$$

This gives $p = 24$, so 40% of 60 is 24.

(b) Here, r is unknown, $n = 50$, and $p = 10$. Equation (1) becomes

$$0.01 \times 50r = 10 \qquad \text{or} \qquad 0.50r = 10$$

To find r, divide both sides by 0.50 and get $r = 20$, so 10 is 20% of 50.

(c) Here, $r = 40$, $p = 20$, and n is unknown. Equation (1) becomes

$$0.01 \times 40n = 20 \qquad \text{or} \qquad 0.40n = 20$$

Dividing by 0.40, we find that $n = 50$, so 20 is 40% of 50. ■

EXAMPLE 11 ▶ Beverages or Insurance?

According to the Bureau of Labor Statistics, the amount of money I (in billions) spent annually on vehicle insurance by persons under 25 years of age is as shown in the graph in **blue** and can be approximated by

$$I = 22.2x + 390 \text{ (billion)}$$

where x is the year number (1, 2, 3, 4 or 5). The amount A spent annually on alcoholic beverages is in **red** and can be approximated by

$$A = 28.2x + 324 \text{ (billion)}$$

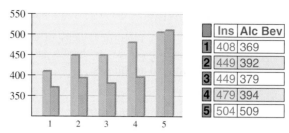

	Ins	Alc Bev
1	408	369
2	449	392
3	449	379
4	479	394
5	504	509

Data sources: http://tinyurl.com/a5tz7 and http://tinyurl.com/ct8c8.

In how many years will the amount **A** of money spent on alcoholic beverages equal the amount I spent on insurance?

Solution

For the amount of money spent on alcoholic beverages to be equal to the amount spent on insurance, A must equal I; that is,

$$28.2x + 324 = 22.2x + 390$$

To find out when this occurs, we must solve for the year number x.

Given:	$28.2x + 324$	$= 22.2x + 390$
Subtract **324**	$28.2x + 324 - \mathbf{324}$	$= 22.2x + 390 - \mathbf{324}$
	$28.2x$	$= 22.2x + \mathbf{66}$
Subtract **22.2x**	$28.2x - \mathbf{22.2x}$	$= 22.2x - \mathbf{22.2x} + \mathbf{66}$
	$6x$	$= 66$
Divide by **6**	x	$= \mathbf{11}$

Thus, $I = A$ when $x = \mathbf{11}$, that is, in **11** years.

Algebra is used in many ways in our daily lives. For example, do you know the exact relationship between your shoe size S and the length L of your foot? Here are the formulas used in the United States.

$$S = 3L - 22 \quad \text{for men} \qquad S = 3L - 21 \quad \text{for women}$$

We use these ideas next.

EXAMPLE 12 ▶ Big Foot and Equations

(a) If a man wears a size 12 shoe, what is the length L of his foot?

(b) If you know that a woman's shoe size is not bigger than a size 6, what can you say about the length of her foot?

Solution

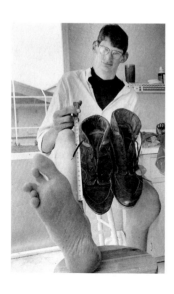

(a) The formula for the length of a man's foot is $S = 3L - 22$

Since the man wears a size 12, $12 = 3L - 22$
We would like to solve for L.

Add 22. $22 + 12 = 3L - 22 + 22$

Simplify. $34 = 3L$

Divide both sides by 3. $\frac{34}{3} = L$

Thus, the man's foot is $\frac{34}{3} = 11\frac{1}{3}$ in. long. By the way, the man in the photo has an 18-in.-long foot.

(b) If the woman's shoe size S is not bigger than 6, then

$$S = 3L - 21 \leq 6$$

Add 21. $$3L - 21 + 21 \leq 6 + 21$$

Simplify. $$3L \leq 27$$

Divide both sides by 3. $$L \leq 9$$

Thus, the woman's foot is less than 9 in. long.   ■

EXERCISES 6.1

A Equations and Inequalities

In problems 1–4, determine which of the given numbers are solutions of the given inequality.

1. $2 + x \leq 2 - x$
   a. 2   b. -2 c. 0   d. 5

2. $3x + 1 < 2x + 4$
   a. $\frac{1}{3}$ b. 4   c. 0   d. 3

3. $3x - 2 \geq 2x - 1$
   a. 0   b. 3   c. -2 d. 1

4. $x > 3 - 2x$
   a. 1   b. 2   c. 0   d. -2

5. Determine which of the following are solutions of $2x - 1 = 3$.
   a. -1 b. 1   c. 2   d. 0

6. Determine which of the following are solutions of $\frac{1}{5}(3x - 2) = 2$.
   a. -1 b. 0   c. 2   d. 4

B Solving Equations

In problems 7–32, solve each equation.

7. $x + 10 = 15$ 8. $x - 5 = 8$

9. $2x - 1 = 5$ 10. $3x + 1 = 4$

11. $2x + 2 = x + 4$ 12. $3x + 1 = x - 3$

13. $3x + 1 = 4x - 8$ 14. $2x + 3 = 3x - 1$

15. $7 = 3x + 4$ 16. $22 = 4x + 2$

17. $4 = 3x - 2$ 18. $1 = 5x - 8$

19. $7n + 10 - 2n = 4n - 2 + 3n$

20. $13a - 6 + a = 5a + 3 + 3a$

21. $2(x + 5) = 13$ 22. $6(x + 2) = 17$

23. $\frac{1}{2}(x - 2) = 5$ 24. $\frac{2}{3}(5x - 4) = 1$

25. $3(x + 1) - x = 2(9 - x)$

26. $14y - 14 - 3y = 10y - 6(1 - y)$

27. $8(x - 1) = x + 2$ 28. $x + 6 = 6(x - 1)$

29. $\frac{1}{4}(x - 2) = \frac{1}{3}(x - 4)$

30. $\frac{1}{2}(3x - 1) = \frac{2}{5}(3x + 1)$

31. $3 \times 2p + 5 = 37 - \dfrac{4p^2}{2p}$

32. $15 + \dfrac{12t^2}{3t} = 3 \times 3t - 5$

C Solving Formulas for a Variable

In problems 33–38, solve the given formula for the indicated letter.

33. $V = \pi r^2 h$ for h 34. $V = \frac{1}{3}\pi r^2 h$ for h

35. $V = LWH$ for W 36. $V = LWH$ for H

37. $P = s_1 + s_2 + b$ for b

38. $P = s_1 + s_2 + b$ for s_2

39. The distance D traveled in time T by an object moving at rate R is given by $D = RT$.
    a. Solve for T.
    b. The distance between two cities A and B is 220 mi. How long would it take a driver traveling at 55 mph to go from A to B?

40. The ideal height H (in inches) of a man is related to his weight W (in pounds) by the formula $W = 5H - 190$.
    a. Solve for H.
    b. If a man weighs 160 lb, what is his height?

41. The number of hours H a growing child should sleep is $H = 17 - \frac{A}{2}$, where A is the age of the child in years.
   a. Solve for A.
   b. At what age would you expect a child to sleep 8 hours?

42. The area A of a trapezoid is given by $A = \frac{1}{2}h(a + b)$.
   a. Solve for b.
   b. If the area of a trapezoid is 60 square units, its height h is 10 units, and side a is 7 units, what is the length of side b?

D **Solving Inequalities**

Find the solution set for each of the following if the replacement set is the set of real numbers.

43. $x - 3 < 1$

44. $x - 2 < 2$

45. $x - 4 > -1$

46. $x + 3 > -2$

47. $2x - 1 > x + 2$

48. $3x - 3 > 2x + 1$

49. $2x + 3 \le 9 + 5x$

50. $x + 8 \ge 2x - 1$

51. $x + 1 > \frac{1}{2}x - 1$

52. $x - 1 < \frac{1}{2}x + 2$

53. $x \ge 4 + 3x$

54. $x - 1 \le 5 + 3x$

55. $\frac{1}{3}x - 2 \ge \frac{2}{3}x + 1$

56. $\frac{1}{4}x + 1 \le \frac{3}{4}x - 1$

57. $2x - 2 > x + 1$

58. $3x - 2 > 2x + 2$

59. $x + 3 > \frac{1}{2}x + 1$

60. $x + 1 < \frac{1}{2}x + 4$

61. $x \ge 2 + 4x$

62. $x - 2 \le 6 + 3x$

63. $2x + 1 < 2x$

64. $5x \le 5x + 4$

65. $8x + 2 \le 3(x + 4)$

66. $9x + 3 \le 4(x + 2)$

67. $3(x + 4) > -5x - 4$

68. $5(x + 2) < -3x + 2$

69. $-2(x + 1) \ge 3x - 4$

70. $-3(2 - x) \ge 5x - 7$

71. $a(x - 1) \le a(2x + 3)$, with $a < 0$

72. $b(1 - 2x) > 5b - 4bx$, with $b < 0$

E **Applications**

73. 40% of 80 is what number?

74. Find 15% of 60.

75. 315 is what percent of 3150?

76. 8 is what percent of 4?

77. What percent of 40 is 5?

78. 20 is what percent of 30?

79. 30% of what number is 60?

80. 10 is 40% of what number?

81. North America has approximately 5% of the world's oil reserves. If the North American reserves represent 50 billion barrels of oil, what are the world's oil reserves? (Round your answer to the nearest billion.)

82. In a recent year, about 280 million tons of pollutants were released into the air in the United States. If 47% of this amount was carbon monoxide, how many tons of carbon monoxide was that?

83. On a 60-item test, a student got 40 items correct. What percent is that? (Round to the nearest percent.)

84. The price of an article on sale was 90% of the regular price. If the sale price was $18, what was the regular price?

85. Two stores sell an item that they normally price at $140. Store A advertises a sale price of 25% off the regular price, and store B advertises a sale price of $100. Which is the lower price?

86. ABC Savings & Loan loans the Adams family $40,000 toward the purchase of a $48,000 house. What percent of the purchase price is the loan?

87. The table shows the amount of money spent annually over a 5-year period on books and maps,

$$B = 1.73x + 30 \text{ (in billions)}$$

and the amount spent on magazines, newspapers, and sheet music,

$$M = 0.53x + 33 \text{ (billion)}$$

where x is the year number (1, 2, 3, 4, or 5). In how many years will the amount B spent on books and maps be the same as the amount M spent on magazines, newspapers, and sheet music?

	Books, Maps	Mag, News
1	28.8	32.1
2	31.6	33.5
3	33.7	35.0
4	34.6	34.5
5	35.8	34.2

Source: U.S. Bureau of Economic Analysis, *Survey of Current Business,* January 2004, www.census.gov/prod/2004pubs/04statab/arts.pdf.

88. According to the Bureau of Labor Statistics, the amount of money I spent annually on vehicle insurance by persons between 25 years and 34 years of age is as shown in the table and can be approximated by

$I = 50.8x + 664.2$ **(billion)**

where x is the year number (1, 2, 3, 4, or 5). The annual amount A spent on alcoholic beverages can be approximated by

$A = 12.6x + 368.2$ **(billion)**

a. Find the value of x so that $I = A$.

b. Based on your answer to part (a), can the amount of money spent on alcoholic beverages ever equal the amount spent on insurance?

	Ins	Alc Bev
1	705	365
2	774	431
3	822	393
4	872	395
5	910	446

89. The graph shows the total enrollment of 18–24-year-old students in degree-granting institutions and can be approximated by

$E = 0.15N + 7.77$ **(millions)**

where N represents the number of years after 1989. Use the equation to find the projected number E of students enrolled in

a. **2002.** How close is your answer to the value 9.9 given in the graph?

b. **2014.** How close is your answer to the value 11.5 given in the graph?

c. The graph does not show a projected value for the year 2010. Use the equation to predict the enrollment for the year 2010.

Enrollment, by age of student

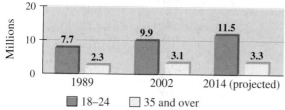

Source: U.S. Dept. of Education, NCES: Integrated Post-secondary Education Data System (IPEDS), "Fall Enrollment Survey." http://nces.ed.gov/pubs2005/2005074_1.pdf.

90. The graph shows the total student enrollment in degree-granting institutions and can be approximated by

$E = 0.24N + 13.5$

where N represents the number of years after 1989. Use the equation to find the projected number E of students enrolled in

a. **2002.** How close is your answer to the value 16.7 given in the graph?

b. **2014.** How close is your answer to the value 19.5 given in the graph?

c. The graph does not project a value for the year 2010. Use the equation to predict the enrollment for 2010.

Enrollment in millions

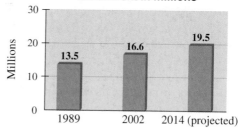

As you recall from Example 12, the relationship between your shoe size S and the length of your foot L (in inches) is given by

$S = 3L - 22$ for men
$S = 3L - 21$ for women

91. If Tyrone wears a size 11 shoe, what is the length L of his foot?

92. If Maria wears a size 7 shoe, what is the length L of her foot?

93. Sam's size 7 tennis shoes fit Sue perfectly! What size women's tennis shoe does Sue wear?

94. The largest shoes ever sold was a pair of size 42 shoes made for the giant Harley Davidson of Avon Park, Florida. How long is Mr. Davidson's foot?

95. How long a foot requires a size 14 shoe, the largest standard shoe size for men?

96. How long is your foot when your shoe size is the same as the length of your foot and you are

a. a man?   **b.** a woman?

97. In 1951, Eric Shipton photographed a 23-in. footprint believed to be that of the Abominable Snowman.

   a. What size shoe would the Abominable Snowman need?

   b. If the Abominable Snowman turned out to be the Abominable Snowwoman, what size shoe would she need?

98. When the variable cost per unit is $12 and the fixed cost is $160,000, the total cost for a certain product is $C = 12n + 160,000$ (n is the number of units sold). If the unit price is $20, the revenue R is $20n$. What is the minimum number of units that must be sold to make a profit? (You need $R > C$ to make a profit.)

99. The cost of first-class mail is 39¢ for the first ounce and 24¢ for each additional ounce. A delivery company will charge $5 for delivering a package weighing up to 2 lb (32 oz). When would the U.S. Post Office price, represented by $P = 0.39 + 0.24(x - 1)$ (where x is the weight of the package in ounces), be cheaper than the delivery company's price?

100. The parking cost at a citywide garage is $C = 1 + 0.75(h - 1)$, where h is the number of hours you park and C is the cost in dollars. When is the cost C less than $10?

101. Do you follow major league baseball? When do you think the average number of runs per game was highest? It was in 1996. The average number of runs scored per game for the National League can be approximated by $N = 0.165x + 4.68$. For the American League the approximation is $A = -0.185x + 5.38$, where x is the number of years after 1996. When will $N > A$—that is, when will the National League run production exceed that of the American League?

 In Other Words

102. In your own words define the *replacement set* for an equation.

103. In your own words define the *solution set* for an equation.

104. If a real number a is in the replacement set of an equation, will it always be in the solution set? Explain.

105. If a particular number s is in the solution set of an equation, will it always be in the replacement set? Explain.

106. Can you think of an equation in which the replacement set and the solution set are the same?

 Using Your Knowledge

The source for problems 107–108 is *The Completely Revised and Updated Fast-Food Guide* or http://chowbaby.com. You can use your knowledge and the facts you have learned in this section to solve problems like this one. Of the calories in a McDonald's biscuit with sausage and eggs, 60% are fat calories (calories derived from the fat in the food). If there are 351 fat calories in this product, what is the total number of calories in a McDonald's biscuit with sausage and eggs? Let this number be c. Since 60% of c is 351, we have

$$60\% \cdot c = 351$$
$$0.60c = 351$$
$$60c = 35,100 \qquad \text{Multiplying both sides by 100}$$
$$c = \frac{35,100}{60} = 585 \qquad \text{Dividing by 60}$$

Thus, there are 585 total calories in a McDonald's biscuit with sausage and eggs. (See the *Fast-Food Guide* for a wealth of nutritional information on fast foods.)

107. Of the total calories in a McDonald's apple pie, about 50% are fat calories. If there are 125 fat calories in a McDonald's apple pie, how many total calories are there in the apple pie?

108. Of the total calories in a Big Mac, 55% are fat calories. If 270 of the calories in a Big Mac are fat calories, how many total calories are there in a Big Mac? (Source: www.mcdonalds.com/app_controller.nutrition.index1.html.)

109. The Burger King Whopper also contains 50% fat calories. If 355 of the calories in a Whopper are fat calories, how many total calories are there in a Whopper? (Source: www.bk.com/Food/Nutrition/NutritionWizard/index.aspx.)

 **Discovery**

110. The number of chirps N that a cricket makes per minute satisfies the equation

$$N = 4(F - 40)$$

where F is the temperature in degrees Fahrenheit. A farmer claimed that a cricket chirped 150 times a minute when the temperature was 80°F. Is this possible?

111. Referring to problem 110, if the temperature were 77.5°F, how many chirps per minute would the cricket make?

112. At what temperature will the cricket of problem 110 stop chirping?

113. The speed S (in centimeters per second) at which a certain type of ant crawls is $S = \frac{1}{6}(C - 4)$, where C is the temperature in degrees Celsius. Find the ant's speed when the temperature is 10°C.

114. At what temperature will the ant in problem 113 stop crawling? Now, does the cricket stop chirping before the ant stops crawling?

 Collaborative Learning

Form two groups of students: the Ants and the Crickets.

1. The Ants group has to find some ants! Record the temperature C in degrees Celsius. Place two meter sticks parallel on the floor about 3 in. apart and let ants walk in between the two meter sticks. Record the distance a single ant travels (in centimeters) and the time it takes it to travel that distance (in seconds). Do this several times as temperatures vary (maybe morning and noon).

Do your results satisfy the equation $S = \frac{1}{6}(C - 4)$, where S is the speed of the ant in centimeters per second and C is the Celsius temperature? How close are your results to the ones in the formula? Explain why results may differ.

2. The Cricket group has to find some crickets! Record the temperature F in degrees Fahrenheit. Time the number of chirps c a cricket makes in 1 min. (You have to try to isolate the chirps of a single cricket!) Do this several times at different temperatures (early after sunset and later in the evening). Do your results satisfy the equation $c = 4(F - 40)$, where F is the temperature in degrees Fahrenheit? How close are your results to the ones in the formula? Explain why results may differ.

 Research Questions

1. When was the first time that the word *equation* was used in a book written in English, and what was the name of the book?

2. We have mentioned that al-Khowarizmi used *al-jabr,* or "restoring," in his work. Find out what inspired al-Khowarizmi to write his book.

3. In his work, al-Khowarizmi noted six types of equations that can be written using squares, square roots, and constants. What were the six types of equations?

4. The work of al-Khowarizmi was continued in the Islamic world. In the first decade of the eleventh century, a major work dealing with algebra, entitled *The Marvelous,* was written. Who was the author of this book, what was its original title, and what topics were covered in the book?

6.2 Graphs of Algebraic Sentences

In King Solomon's Garden

It is said that in ancient times King Solomon was the wisest of men. There is a legend that goes like this: One day when he was resting in his palace garden, Solomon, who was so wise that he could even understand the language of all animals and plants, heard gentle voices close to him. On further inspection he discovered two snails in a solemn meeting.

The first snail said, "Brother, see'st thou yon straight pole that riseth upright from the ground 30 cubits high?" And the second snail answered, "Yea, even so do I."

"It is my desire," said the first snail, "to climb to the very top of it. How long thinketh thou it will take me?"

"That certainly shall depend on the speed with which thou climbest."

"It is not as simple as that," said the would-be climber. "I can ascend but 3 cubits during the day, but in the evening I fall asleep and slip back 2 cubits so that, in effect, I move up but 1 cubit every 24 hours."

"'Tis plain then," said the second snail, "that thou will take 30 days to reach the top of the pole. Why dost thou plague me with such a simple problem? Prithee be silent and allow me to sleep."

King Solomon smiled. He alone knew whether the second snail was right.

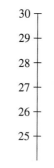

FIGURE 6.2

What do you think? (*Hint:* It will obviously take the snail 25 days to reach 25 cubits of height. From then on, draw a graph on the number line in Figure 6.2 and find the number of days it took the snail to get to the top.)

In this section you learn how to graph algebraic sentences (equations and inequalities) on the number line. ▶

A. Graphs of Algebraic Sentences

The solution set of an open sentence in one variable can always be represented by a set of points on the number line. This set of points is often called the **graph** of the equation or the inequality, as the case may be. We illustrate various types of graphs in the following examples:

EXAMPLE 1 ▶ **Graphing Equations with the Integers as Replacements**

Graph the solution set of the equation $x + 1 = 0$, where the replacement set is the set of integers.

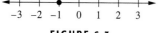

FIGURE 6.3
The singleton set $\{-1\}$.

Solution

Subtracting 1 from both sides, we see that the solution set is the singleton set $\{-1\}$. The graph consists of the single point -1 on the number line. We draw a solid dot to indicate this graph (see Figure 6.3). ■

EXAMPLE 2 ▶ **More Equations with the Integers as Replacements**

Graph the solution set of the inequality $x + 1 \leq 0$, where the replacement set is the set of integers.

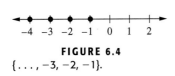

FIGURE 6.4
$\{\ldots, -3, -2, -1\}$.

Solution
Subtracting 1 from both sides yields the equivalent inequality $x \le -1$. Thus, the solution set is the set of all integers that are less than or equal to -1—that is, the set $\{\ldots, -3, -2, -1\}$. To show this graph, we draw dots at the corresponding points on the number line (see Figure 6.4). Start at -1 and, as suggested by the notation $x \le -1$, draw solid dots at $-1, -2, -3$, and so on. ∎

EXAMPLE 3 ▶ Graphing Inequalities Involving ≤

Graph the solution set of the inequality $x + 1 \le 0$, where x is a real number.

Solution
Proceeding as in Example 2, we see that the solution set is the set of all real numbers less than or equal to -1—that is, $\{x \mid x \le -1, x \text{ real}\}$. We display this set by drawing a heavy line starting at -1 on the number line and going to the left. The point at -1 is marked with a solid dot to show that it is included in the set (see Figure 6.5). ∎

FIGURE 6.5
The set $\{x \mid x \le -1\}$.

EXAMPLE 4 ▶ Graphing Inequalities Involving ≠

Graph the solution set of the inequality $x + 2 \ne 5$, where x is a real number.

Solution
The number 3 is the only replacement for x such that $x + 2 = 5$, so the solution set is all real numbers except 3—that is, $\{x \mid x \ne 3, x \text{ real}\}$. The graph consists of the entire number line except for the point 3. In Figure 6.6, the graph is shown as a heavy line, and the point 3 is marked with an open circle to indicate its exclusion from the solution set. ∎

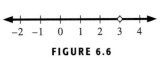

FIGURE 6.6
The set $\{x \mid x \ne 3\}$.

EXAMPLE 5 ▶ More Inequalities with Real Numbers

Graph the solution set of the inequality $-2 \le x < 1$, where the replacement set is the set of real numbers.

Solution
The solution set consists of all the real numbers between -2 and 1, with the -2 included and the 1 excluded. The graph is shown in Figure 6.7. ∎

FIGURE 6.7
The set $\{x \mid -2 \le x < 1\}$.

A piece of the number line such as that in Figure 6.7 is called a **finite interval** (or a **line segment**). The endpoints in Figure 6.7 are -2 and 1. We call the interval **closed** if both endpoints are included, **open** if both endpoints are excluded, and **half-open** if only one of the endpoints is included. The interval in Figure 6.7 is half-open.

If a and b are real numbers with $a < b$, the various types of finite intervals and how they are written in interval notation are as shown in Figure 6.8. The notation for **infinite** intervals is discussed in the Using Your Knowledge section of Exercises 6.2.

$a \le x \le b$ Closed interval $[a, b]$

$a < x < b$ Open interval (a, b)

$a \le x < b$ ⎱
                    Half-open intervals $[a, b)$
$a < x \le b$ ⎰
                                            $(a, b]$

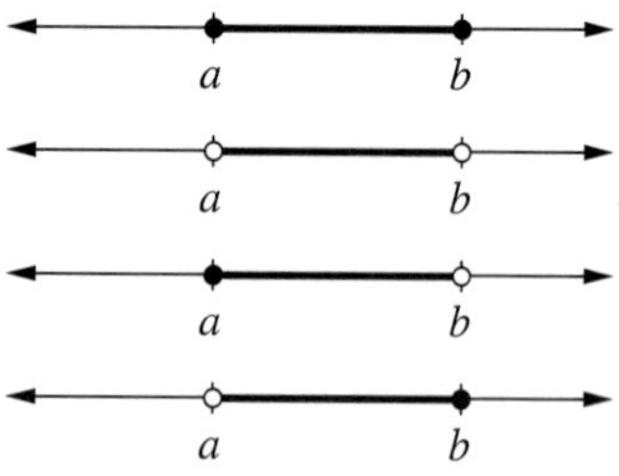

FIGURE 6.8

Finite intervals.

In this section we consider compound algebraic sentences consisting of two or more simple sentences of the type that occurred in the preceding sections. We are concerned with the connectives *and* and *or* used in exactly the same sense as in Chapter 3 and with how to graph the compound algebraic sentences.

B. Sentences with *and*

As we work with finding the solution sets of compound sentences, note that if no replacement set is specified, we assume that the replacement set consists of all real numbers for which the members of the inequalities are defined.

EXAMPLE 6 ▶ Graphing Inequalities with *and*

Consider the sentence $x + 1 < 3$ and $x - 1 > -1$. Find its solution set.

Solution

The sentence given here is a compound sentence of type $p \wedge q$ (p and q), where p is $x + 1 < 3$ and q is $x - 1 > -1$. Such a sentence as $p \wedge q$ is true only when both p and q are true. Consequently, the solution set of the compound sentence is the **intersection** of the solution sets of the two components.

We have

$$x + 1 < 3 \qquad \text{and} \qquad x - 1 > -1$$
$$x + 1 - 1 < 3 - 1 \qquad \text{and} \qquad x - 1 + 1 > -1 + 1$$
$$x < 2 \qquad \text{and} \qquad x > 0$$

Rewriting the second inequality, we see that x must satisfy the conditions

$$0 < x \qquad \text{and} \qquad x < 2$$

Thus, the solution set can be written in set notation as

$$\{x \mid 0 < x\} \cap \{x \mid x < 2\}$$

or, more efficiently, $\{x \mid 0 < x < 2\}$.

This result is most easily seen in Figure 6.9.   ■

We can also obtain this result by first graphing $x + 1 < 3$, or equivalently, $x < 2$ (see Figure 6.9), and then graphing $x - 1 > -1$, or equivalently, $x > 0$. The intersection of these two graphs consists of all numbers between 0 and 2, as shown in Figure 6.9.

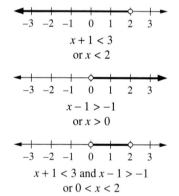

$x + 1 < 3$
or $x < 2$

$x - 1 > -1$
or $x > 0$

$x + 1 < 3$ and $x - 1 > -1$
or $0 < x < 2$

FIGURE 6.9

Note that the open circles indicate that the points are *not* included in the solution set.

EXAMPLE 7 ▶ More Inequalities with *and*

Find the solution set of $x - 1 > 4$ and $x + 2 < 5$.

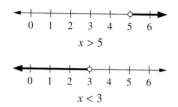

FIGURE 6.10

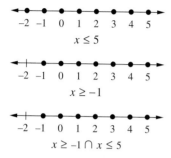

FIGURE 6.11

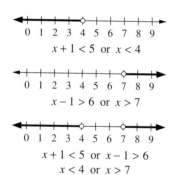

FIGURE 6.12

Solution
We have

$$x - 1 > 4 \qquad \text{and} \qquad x + 2 < 5$$
$$x - 1 + 1 > 4 + 1 \qquad \text{and} \qquad x + 2 - 2 < 5 - 2$$
$$x > 5 \qquad \text{and} \qquad x < 3$$

Since there are no numbers satisfying both of these conditions (see Figure 6.10), there are no solutions; the solution set is empty. ∎

EXAMPLE 8 ▶ Inequalities with *and* Where *x* Is an Integer

Find the solution set of the sentence $x \leq 5$ and $x + 1 \geq 0$ if x is an integer.

Solution
If x is an integer, then the solution set of $x \leq 5$ is the set of all integers less than or equal to 5. Similarly, the solution set of $x + 1 \geq 0$ is the set of all integers greater than or equal to -1. The intersection of these two sets is the set $\{-1, 0, 1, 2, 3, 4, 5\}$, which is the desired solution set (see Figure 6.11). ∎

C. Sentences with *or*

EXAMPLE 9 ▶ Inequalities with *or*

Find the solution set of the sentence $x + 1 < 5$ or $x - 1 > 6$.

Solution
The replacement set is the set of all real numbers. (Why?) Because this is a sentence of the type $p \vee q$ (p or q), we know that the solution set is the **union** of the solution sets of the two components. We have

$$x + 1 < 5 \qquad \text{or} \qquad x - 1 > 6$$
$$x < 4 \qquad \text{or} \qquad x > 7$$

Thus, the required solution set is

$$\{x \mid x < 4\} \cup \{x \mid x > 7\}$$

or stated in another way,

$$\{x \mid x < 4 \text{ or } x > 7\}$$

The graph in Figure 6.12 illustrates the solution. ∎

D. Applications

EXAMPLE 10 ▶ Solving Inequalities Involving Decreasing Cassettes

The number of music cassettes C sold each year is decreasing and can be approximated by $C = -33t + 272$ (in millions), where t is the number of years after 1995. There is enough shelf space in music stores for the cassettes as long as their number does not exceed 150 million. (Source: Recording Industry Association of America.)

(a) Write an inequality that models these conditions. (Do not solve it!)

(b) On the other hand, the number of single compact discs S sold each year is increasing and can be approximated by $S = 15t + 22$ (in millions). If the number of single compact discs that can be produced is at most 175 million, write an inequality that models these conditions. (Do not solve it!)

(c) Write and solve a compound inequality describing the years in which the conditions in (a) and (b) are met.

Solution

(a) Since C cannot exceed 150 million, $C \leq 150$, that is,

$$-33t + 272 \leq 150$$

(b) Since S can be at most 175 million, $S \leq 175$; that is,

$$15t + 22 \leq 175$$

(c) Since both inequalities must be satisfied, we want

$-33t + 272 \leq 150$	and	$15t + 22 \leq 175$
$272 \leq 150 + 33t$	and	$15t \leq 175 - 22$
$272 - 150 \leq 33t$	and	$15t \leq 153$
$122 \leq 33t$	and	$t \leq \dfrac{153}{15}$
$\dfrac{122}{33} \leq t$	and	$t \leq \dfrac{153}{15}$
$3.7 \leq t$		≤ 10.2

Online Study Center

To futher explore inequalities, access links 6.2.1 and 6.2.2 at this textbook's Online Study Center.

This means that when the number of years t is between $1995 + 3.7 = 1998.7 \approx 1999$ **and** $1985 + 10.2 = 2005.2 \approx 2005$, the conditions in parts (a) and (b) will be met. ∎

EXERCISES 6.2

A Graphs of Algebraic Sentences

In problems 1–8, take x to be an integer and graph the solution set of each sentence.

1. $x + 2 = 4$

2. $x - 1 = 3$

3. $x + 1 \geq 2$

4. $x + 2 < 5$

5. $x - 3 \neq 1$

6. $-3 < x \leq 2$

7. $-2 \leq x \leq 4$

8. $-2 < x < 4$

In problems 9–26, take x to be a real number and graph the solution set of each sentence.

9. $x < 4$

10. $x \geq 2$

11. $x - 2 \leq 0$

12. $x - 2 \geq 4$

13. $-2 \leq x \leq 4$

14. $x - 3 \neq 1$

15. $x + 2 > 5$

16. $x - 2 = 1$

17. $-1 < x < 2$

18. $x \geq 3$

19. $x + 4 < 5$

20. $x + 5 < 4$

21. $x + 1 < x$

22. $x + 1 > x$

23. $2x + 3 < x + 1$

24. $-x + 5 \leq 2x + 2$

25. $3x - 7 \geq -7$

26. $2x + 5 < 5$

B Sentences with *and*
C Sentences with *or*

In problems 27–34, let each replacement set be the set of integers. Give the solution set by listing its elements.

27. $x \leq 4$ and $x - 1 \geq -2$

28. $x > 0$ and $x \leq 5$

29. $x + 1 \leq 7$ and $x > 2$

30. $x > -5$ and $x - 1 < 0$

31. $2x - 1 > 1$ and $x + 1 < 4$

32. $x - 1 < 1$ or $3x - 1 > 11$

33. $x < -5$ or $x > 5$ **34.** $x \geq 0$ or $x < -2$

In problems 35–52, let each replacement set be the set of real numbers. Graph the solution set unless it is the empty set.

35. $x + 1 \geq 2$ and $x \leq 4$

36. $x \leq 5$ and $x > -1$

37. $x > 2$ and $x < -2$

38. $x + 2 \leq 4$ or $x + 2 \geq 6$

39. $x - 1 > 0$ and $x + 1 < 5$

40. $x \leq 0$ or $x > 3$

41. $x \leq x + 1$ and $x \geq 2$

42. $x + 2 \geq -2$ or $x < 0$

43. $x - 2 \geq 2$ and $x < 0$

44. $x + 3 \leq 0$ or $x - 1 > 0$

45. $x \geq 0$ and $x - 1 \geq 2$

46. $x \geq 0$ and $x - 1 \leq 2$

47. $x < 0$ or $x - 1 < 2$

48. $x < 0$ and $x - 1 > 2$

49. $x + 1 > 2$ and $x - 2 < 3$

50. $x - 1 > 3$ and $x + 1 > 2$

51. $x - 1 > 0$ or $x + 2 < 4$

52. $x + 1 > 2$ or $x - 1 < 2$

D Applications

53. The average annual cost C of owning a midsized car is \$4953 plus \$0.12 per mile. (Source: www.vtpi.org/tdm/tdm82.htm.)
   a. Write an expression that models the average annual cost of operating a midsized car for m mi.
   b. A company budgets between \$6000 and \$6500 for each car in the company pool. Write an inequality to describe the number of miles m that a midsized car in the company pool can be driven.
   c. Solve the inequality in part (**b**) and interpret the solution.

54. The average annual cost C of operating a full-size car is \$6689 plus \$0.15 per mile. (Source: www.wtvpi.org/tdm/tdm82.htm.)
   a. Write an expression that models the average annual cost of operating a full-size car for m mi.

   b. A company budgets between \$7000 and \$7500 for each car in the company pool. Write an inequality to describe the number of miles m that a full-size car in the company pool can be driven.
   c. Solve the inequality in part (**b**) and interpret the solution.

55. A computer manufacturer finds that when 1000 chips are produced, the average number of defective P_1 chips is 1.2 times the number of defective P_2 chips. On a given day, the total number of defective chips is between 22 and 66. Write an inequality indicating how many P_2 chips are defective that day.

56. A plumber charges a \$35 flat fee for a service call plus \$25 per hour for his time. If you get a repair estimate between \$85 and \$110, write an inequality that estimates the plumber's time.

In Other Words

57. Use the word *between* to indicate which numbers are represented in the following graphs:

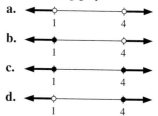

58. Write in words the sets of numbers represented in the following graphs:

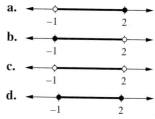

59. Write in words the sets of numbers represented in the following graphs:

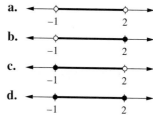

Using Your Knowledge

The graph of the inequality $x \leq -1$ in Example 3 is an **infinite interval** on the number line and can be written as $(-\infty, -1]$ using interval notation. The symbol $-\infty$ (read "negative infinity") does not represent a number; it simply means that the interval includes all numbers less than or equal to -1. The square bracket on the right indicates that -1 is part of the interval. The interval $(-\infty, -1]$ is called a *half-open interval*. If -1 were not included, we would write the *open interval* as follows: $(-\infty, -1)$. Note that the interval notation for $\{x \mid x > -2\}$ is $(-2, \infty)$, where the ∞ (read "infinity") symbol indicates that the interval includes all numbers greater than -2 (-2 itself is *not* included in the interval). The interval is an open interval. The following are some types of infinite intervals, their notation, and their graphs:

Set Notation	Interval Notation	Graph
$\{x \mid x > a\}$	$(a, +\infty)$	$\xleftarrow{\qquad\overset{\circ}{\underset{a}{}}\qquad}$
$\{x \mid x < b\}$	$(-\infty, b)$	$\xleftarrow{\qquad\overset{\circ}{\underset{b}{}}\qquad}$
$\{x \mid x \geq a\}$	$[a, +\infty)$	$\xleftarrow{\qquad\overset{\bullet}{\underset{a}{}}\qquad}$
$\{x \mid x \leq b\}$	$(-\infty, b]$	$\xleftarrow{\qquad\overset{\bullet}{\underset{b}{}}\qquad}$

Use interval notation to write the following:

60. $\{x \mid x \geq -4\}$

61. $\{x \mid x < 5\}$

62. $\{x \mid x \leq -6\}$

63. $\{x \mid x > 9\}$

64. $\{x \mid 3 < x < 7\}$

65. $\{x \mid -4 \leq x < -1\}$

66. $\{x \mid 0 < x \leq 8\}$

67. $\{x \mid -1 \leq x \leq 10\}$

Discovery

Let J be Joe's height.
Let B be Bill's height.
Let F be Frank's height.
Let S be Sam's height.

What can you conclude from the information in the cartoon? To find the answer, you have to know how to translate the given information into symbols.

68. The second panel says that Bill is taller than Frank *and* Frank is taller than Joe. Write a compound inequality indicating these relationships.

Now translate each of the following statements into an equation or an inequality.

69. Joe is 5 ft (60 in.) tall.

70. Bill is taller than Frank.

71. Frank is 3 in. shorter than Sam.

72. Frank is taller than Joe.

73. Sam is 6 ft 5 in. (77 in.) tall.

74. According to the statement in problem 70, Bill is taller than Frank, and according to the statement in problem 72, Frank is taller than Joe. Write these two statements as an inequality of the form $a > b > c$.

75. On the basis of the answer to problem 74 and the fact that you can obtain Frank's height by using the results of problems 70, 71, and 73, what can you say about Bill's height?

6.3 Sentences Involving Absolute Values

Budgeting expenses is an important family activity.

Budget Variance and Absolute Value

Have you ever been on a budget? Businesses and individuals usually try to predict how much money will be spent on certain items over a given period of time, but it is almost impossible to know exactly the final expenditures in different categories. For example, suppose you budget $120 for a month's utilities. Any heat wave or cold snap could make your actual expenses jump to $150. The $30 difference, representing a 25% increase $\left(\frac{30}{120} = 25\%\right)$, might be an acceptable *variance*. To keep your budget "on target," you can make several variance checks during the year, possibly at the end of each month. Now, suppose b represents the budgeted amount for a certain item, a represents the actual expense, and you want to be within $10 of your estimate. The item will **pass** the variance test if the actual expenses a are within $10 of the budgeted amount b, that is, if

$$-10 \leq b - a \leq 10$$

Is there a way to write this information using a single inequality? You can if you use *absolute values*. The **absolute value** of a number x, denoted by $|x|$ (read "the absolute value of x") is its numerical value with the sign disregarded. For example, $|-3| = 3$ and $|+7| = 7$. Thus, $-10 \leq b - a \leq 10$ is equivalent to $|b - a| \leq 10$. Do you see why?

In general, if a and b are as before, a certain item will pass the variance test if $|b - a| \leq c$, where c is the variance. The quantity c can be a definite amount or a percent of the budget. Now suppose you budget $50 for gas and you want to be within 10% of your budget. How much gas money can you spend and still be within your variance? Intuitively, you can see that if you spend between $45 and $55, you will be within your 10% variance.

The amount of variance is given by

$$0.10 \cdot 50 = 5$$

and $|b - a| \leq c$ becomes $|50 - a| \leq 5$, or

$$-5 \leq 50 - a \leq 5$$

Subtracting 50 from each member,

$$-55 \leq -a \leq -45$$

Multiplying each term by -1,

$$55 \geq a \geq 45$$

or

$$45 \leq a \leq 55$$

The answer that you expected!

In problems 37–39 of Exercises 6.3, you will solve some more problems dealing with variance. ▶

A. Absolute Value

Sometimes we need to solve equations or inequalities that involve absolute values. The **absolute value** of a number x is defined to be the distance on the number line from 0 (the origin) to x and is denoted by $|x|$ (read "absolute value of x"). For example, the number 2 is 2 units away from 0, so $|2| = 2$. The number -2 is also 2 units away from 0, so $|-2| = 2$ (see Figure 6.13). Similarly, we see that $|8| = 8$, $|\frac{1}{2}| = \frac{1}{2}$, $|-5| = 5$, and $|-\frac{1}{4}| = \frac{1}{4}$. In general, if x is any real nonnegative number ($x \geq 0$), then $|x|$ is simply x itself. But if x is a negative number ($x < 0$), then $|x|$ is the corresponding positive number obtained by reversing the sign of x.

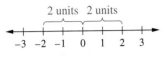

2 units   2 units

FIGURE 6.13

Definition of Absolute Value
The **absolute value** of x is given by $$|x| = \begin{cases} x & \text{if } x \geq 0 \\ -x & \text{if } x < 0 \end{cases}$$

B. Equations and Inequalities with Absolute Values

EXAMPLE 1 ▶ Solving Equations of the Form $|x| = a$

Find and graph the solution set of $|x| = 2$.

Solution

We look for all numbers that are 2 units away from the origin (the 0 point). Since 2 and -2 are the only two numbers that satisfy this condition, the solution set of the equation $|x| = 2$ is $\{2, -2\}$. The graph of this set is shown in Figure 6.14. Note that this solution set can be described by the compound sentence $x = 2$ or $x = -2$. ∎

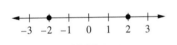

FIGURE 6.14

EXAMPLE 2 ▶ Solving Inequalities of the Form $|x| < a$

Find and graph the solution set of the inequality $|x| < 2$.

Solution

Here we look for all real numbers x that are less than 2 units away from 0. Since 2 and -2 are each exactly 2 units away from 0, we need all points between -2 and 2, that is, all numbers that satisfy the inequality

$$-2 < x < 2$$

The solution set is thus $\{x \mid -2 < x < 2\}$. The graph appears in Figure 6.15. ∎

FIGURE 6.15

Equivalency for $\lvert x \rvert < a$
In general, if a is any positive number, then $$|x| < a \qquad \text{is equivalent to} \qquad -a < x < a$$

EXAMPLE 3 ▶ Graphing Inequalities of the Form $|x| \leq a$

Graph the solution set of $|x| \leq 4$.

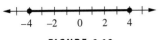

FIGURE 6.16

Solution

Since $|x| \leq 4$ is equivalent to $-4 \leq x \leq 4$, the graph of the solution set is as shown in Figure 6.16. ∎

EXAMPLE 4 ► **Graphing Inequalities of the Form $|x| < a$**

Find the solution set of the inequality $|x - 1| < 4$.

Solution

Since $|x| < a$ is equivalent to $-a < x < a$,

$$|x - 1| < 4 \quad \text{is equivalent to} \quad -4 < x - 1 < 4$$

As in the case of equations, we still want to isolate x, so in the middle member of the inequality we need to add 1. Of course, we must do the same to all the members. Thus, we get

$$-4 + 1 < x - 1 + 1 < 4 + 1$$

or

$$-3 < x < 5$$

Thus, the solution set is $\{x \mid -3 < x < 5\}$, or $(-3, 5)$ in interval notation. (See Using Your Knowledge in Section 6.2 for more information about interval notation.) ∎

GRAPH IT

To graph $|x - 1| < 4$, let Y_1 be the absolute value of $x - 1$ by pressing [Y =] [MATH] [▶] [1] and $x - 1$. Let Y_2 be 4 by pressing [▼] and 4. Press [ZOOM] [6]. When is $|x - 4|$, the V-shaped curve, below 4? Between -3 and 5!

EXAMPLE 5 ► **Graphing Inequalities of the Form $|x| \geq a$**

Find and graph the solution set of $|x| \geq 3$.

Solution

The solution set of $|x| = 3$ consists of all points that are exactly 3 units away from 0. Hence, the solution set of $|x| \geq 3$ consists of all points that are 3 or more units away from 0. As you can see from Figure 6.17, these points can be described by the compound sentence

$$x \geq 3 \quad \text{or} \quad x \leq -3$$

We can therefore write the solution set of $|x| \geq 3$ in the form

$$\{x \mid x \geq 3\} \cup \{x \mid x \leq -3\}$$

Note that the answer can also be written in the form $\{x \mid x \geq 3 \text{ or } x \leq -3\}$ or in interval notation $(-\infty, -3] \cup [3, \infty)$. ∎

FIGURE 6.17

The idea in Example 5 can be generalized as follows:

Equivalency for $|x| > a$
If $a \geq 0$, then $|x| > a$ is equivalent to the compound sentence
$\qquad x > a \qquad \text{or} \qquad x < -a$

GRAPH IT

To get a graph similar to the one in Example 5 using your grapher, enter [Y =] [MATH] [▶] [1] to get the absolute value; then enter x [)]. Now press [2nd] [MATH] [4] 3 to enter ≥ 3, and finally press [ZOOM] [6]. Same graph!

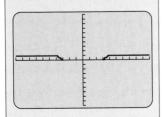

Note: Be sure to remember to reverse the inequality sign for $-a$, as in the next examples.

FIGURE 6.18

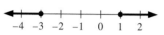

FIGURE 6.19
The dots at -3 and 1 indicate that -3 and 1 are part of the solution set.

EXAMPLE 6 ▶ Graphing Inequalities of the Form $|x| > a$

Graph the solution set of $|x| > 2$.

Solution

$|x| > 2$ is equivalent to the compound sentence $x > 2$ or $x < -2$. The graph of the solution set appears in Figure 6.18. ∎

EXAMPLE 7 ▶ Graphing Inequalities of the Form $|x + a| \geq b$

Graph the solution set of $|x + 1| \geq 2$.

Solution

$|x| \geq a$ is equivalent to $x \geq a$ or $x \leq -a$. Therefore, $|x + 1| \geq 2$ is equivalent to $x + 1 \geq 2$ or $x + 1 \leq -2$; that is, $x \geq 1$ or $x \leq -3$. The graph of the solution set is shown in Figure 6.19. ∎

C. Applications

Absolute value inequalities are used when the difference between two quantities is less or greater than a fixed amount. For example, an average male 5 ft 10 in. tall should weigh between 144 and 154 lb, inclusive. Ideally, 149 lb is the desired weight for a small-framed person 25–59 years of age. (Source: Metropolitan Life Insurance Co.) This means a person in this category should be within 5 lb (above or below) of 149 lb; that is, $|w - 149| \leq 5$, where w is the weight of the person. Do you see why?

$$|w - 149| \leq 5 \quad \text{is equivalent to} \quad -5 \leq w - 149 \leq 5$$

Adding 149,
$$149 - 5 \leq w \leq 5 + 149$$
$$144 \leq w \leq 154$$

This still means that the person should weigh between 144 and 154 lb, inclusive.

EXAMPLE 8 ▶ Inequalities and Margin of Error

When surveys or polls are taken, the margin of error is usually stated. For example, a Roper Starch worldwide poll indicated that 62% of Americans picked out a "letdown in moral values" as one of the "major causes of our problems today." The margin of error e was given as ± 4 points.

(a) Write an absolute value inequality that represents the percentage p of people who believe the statement.

(b) Write an inequality giving a range for the percentage p of people who believe the statement.

Solution

(a) Since the margin of error is 4 points, $|p - 62| \leq 4$.

(b) The inequality in (a) is equivalent to

$$-4 \leq p - 62 \leq 4$$

Online Study Center

To further explore inequalities, access links 6.3.1 and 6.3.2 on this textbook's Online Study Center.

Adding 62,

$$58 \le p \le 66$$

This means that between 58 and 66% of the people, inclusive, believe the statement. ∎

EXERCISES 6.3

Ⓐ Absolute Value

In problems 1–10, evaluate each expression.

1. $|-10|$ **2.** $|15|$ **3.** $\left|-\frac{1}{8}\right|$

4. $\left|\frac{3}{4}\right|$ **5.** $|5-8|$ **6.** $|8-5|$

7. $|0|+|-2|$ **8.** $|-2|-|-3|$ **9.** $-|8|$

10. $-|3|+|-4|$

11. Determine which of the following are solutions of $|1-3x| > 3$.
   a. 2 **b.** $-\frac{1}{2}$ **c.** $\frac{5}{3}$ **d.** 0

12. Determine which of the following are solutions of $|x-2| < 2$.
   a. 0 **b.** 1 **c.** -1 **d.** -2

Ⓑ Equations and Inequalities with Absolute Values

In problems 13–18, find the set of integers for which each sentence is true.

13. $|x| < 1$ **14.** $|x| > -2$ **15.** $|x| = 5$

16. $|x| \le 3$ **17.** $|x| \ge 1$ **18.** $|x| < 4$

In problems 19–36, graph the solution set of each sentence and write the solution set (if possible) in interval notation.

19. $|x| = 1$ **20.** $|x| = 2.5$ **21.** $|x| \le 4$

22. $|x| > 1$ **23.** $|x+1| < 3$ **24.** $|x-2| < 1$

25. $|x| \ge 1$ **26.** $|x| > -1$ **27.** $|x-1| > 2$

28. $|x-3| \ge 1$ **29.** $|2x| < 4$

30. $|3x| \le 9$ **31.** $|3x| \ge 6$

32. $|2x| > 5$ **33.** $|2x-3| \le 3$

34. $|3x+1| \le 8$ **35.** $|2x-3| > 3$

36. $|3x+1| > 8$

Ⓒ Applications

In problems 37–39, use $|b-a| \le c$, where b is the budgeted amount, a is the actual expense, and c is the variance.

37. A company budgets $500 for office supplies. How much money can it spend if its variance is $50?

38. A company budgets $800 for maintenance. How much money can it spend if an acceptable variance is 5% of its budgeted amount?

39. George budgets $300 for miscellaneous monthly expenses. His actual expenses for 1 month amounted to $290. Was he within a 5% budget variance?

40. Write the absolute value inequalities that correspond to the following graphs:

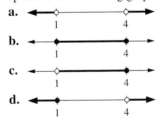

41. An average female, medium frame, 5 ft 6 in. tall should weigh between 130 and 144 lb, inclusive. Using w for the weight,
   a. write an absolute value inequality representing this situation.
   b. simplify the inequality.

42. An average male, large frame, 5 ft 8 in. tall should weigh between 152 and 172 lb, inclusive. Using w for the weight,
   a. write an absolute value inequality representing this situation.
   b. simplify the inequality.

43. A plumber wants to cut a 12-ft length of pipe with no more than a 2% error. If L is the length of the actual cut,
   a. write an absolute value inequality representing the situation.
   b. simplify the inequality.

44. A machinist has to build a 1-in.-long bushing with no more than a 1% error in the length L.
   a. Write an absolute value inequality representing the situation.
   b. Simplify the inequality.

45. In a recent year, the score s on the verbal portion of the SAT for males was 505. If the margin of error on this test is ± 4 points,
   a. write an absolute value inequality representing the average score for males on the SAT.
   b. simplify the inequality.

46. In a recent year, the average cholesterol level c for 20- to 34-year-olds was 186. According to Mosby's *Diagnostic and Laboratory Test Reference* (© 2004), this value can vary by as much as 15%.
   a. Write an absolute value inequality representing the situation.
   b. Simplify the inequality.

In Other Words

47. Write in words: $|x| < a$ is equivalent to $-a < x < a$.

48. Write in words: $|x| > a$ is equivalent to $x > a$ or $x < -a$.

49. What is the solution of $|x - 2| > -5$? Explain. Can you generalize these results?

50. What is the solution of $|x - 1| < 0$? Explain. Can you generalize these results?

Skill Checker

Next section: Quadratic equations! To solve them using the quadratic formula, you need to review Sections 5.2 (Order of Operations) and 5.7 (Simplifying Radicals). Do that first and then try these practice problems.

1. $\sqrt{(-2)^2 - 4(1)(-1)}$ **2.** $\sqrt{1^2 - 4(3)(-5)}$

3. $\sqrt{(-2)^2 - 4(2)(-1)}$ **4.** $\sqrt{(-3)^2 - 4(1)(-2)}$

5. $\sqrt{3^2 - 4(2)(-5)}$ **6.** $\sqrt{5^2 - 4(2)(-7)}$

7. $\sqrt{(-6)^2 - 4(9)(-2)}$ **8.** $\sqrt{(-8^2) - 4(2)(5)}$

9. $\sqrt{8^2 - 4(4)(-5)}$ **10.** $\sqrt{2^2 - 4(2)(-5)}$

6.4   Quadratic Equations

Firefighting and Factoring Quadratics

How much water is a fire engine pumping if the friction loss is 36 lb/in.^{2}? You can find out by solving the equation

$$2g^2 + g - 36 = 0 \qquad (g \text{ in hundreds of gallons per minute})$$

This equation is a *quadratic equation in standard form,* which can be solved by *factoring.* We do this by undoing the multiplication that yielded $2g^2 + g - 36$, writing $(2g + 9)(g - 4) = 0$, and then reasoning that if the product of $(2g + 9)$ and $(g - 4)$ is 0, at least one of the factors must be 0; that is, $2g + 9 = 0$ or $g - 4 = 0$. The first equation gives $g = -\frac{9}{2}$, an impossible answer because g represents hundreds of gallons per minute. The second equation gives $g = 4$ (hundred gallons per minute), an acceptable answer. But wouldn't it be nice if you had a formula that you could use to give you the answer? Fortunately, there is such a formula. It is called the *quadratic formula,* and you will use it in this section to solve quadratic equations. ▶

Firefighters use high-pressure hoses to bring a fire under control.

In the preceding sections we considered the solution of first-degree sentences. We now turn our attention to a certain type of *second-degree* sentence that is called a *quadratic equation*.

Definition of a Quadratic Equation

A **quadratic equation** is a second-degree sentence with standard form

$$ax^2 + bx + c = 0$$

where a, b, and c are real numbers and $a \neq 0$.

We first consider quadratic equations where the second-degree expression $ax^2 + bx + c$ can be written as a product of two first-degree expressions. Suppose that we have the product $(x + p)(x + q)$. Then, by the distributive property $a(b + c) = ab + ac$, with $a = x + p$ and $b + c = x + q$, we get

$$
\begin{aligned}
(x + p)(x + q) &= (x + p)x + (x + p)q \\
&= x^2 + px + qx + pq \\
&= x^2 + (p + q)x + pq
\end{aligned}
$$

A good way to remember how to multiply two binomials is to remember the **FOIL** method. To multiply $(x - 1)(x + 2)$, we write the following:

1. Multiply First terms.

$$\overset{x \cdot x}{(x - 1)(x + 2)} = x^2 \cdots$$

2. Multiply Outside terms.

$$\overset{x \cdot 2}{(x - 1)(x + 2)} = x^2 + 2x \cdots$$

3. Multiply Inside terms.

$$(x - 1)(x + 2) = x^2 + 2x - x \cdots$$
$$\underset{-1 \cdot x}{}$$

4. Multiply Last terms.

$$(x - 1)(x + 2) = x^2 + 2x - x - 2$$
$$\underset{-1 \cdot 2}{}$$

5. Combine like terms.

$$(x - 1)(x + 2) = x^2 + x - 2$$

A. Factoring Quadratic Expressions

Possible Factors	Sum 5
$\pm 1, \pm 6$	
$\pm 2, \pm 3$	$+2, +3$

These ideas can sometimes be "reversed" to write a quadratic expression in the product (factored) form. For example, in order to write $x^2 + 5x + 6$ in factored form, we try to find two integers p and q such that $pq = 6$ and $p + q = 5$. By inspection (looking in the table at left), we see that $p = 2$ and $q = 3$ will work. Therefore,

$$x^2 + 5x + 6 = (x + 2)(x + 3)$$

To **factor** an expression means to write it as a product of lower-degree expressions using integers as coefficients, as in the above illustration. We will use the fact that $x^2 + 5x + 6 = (x + 2)(x + 3)$ in Example 3. Note that $x^2 - 2$ is *not* factorable using integers since we cannot find integers a and b such that $x^2 - 2 = (x + a)(x + b)$. However, $x^2 - 2 = (x + \sqrt{2})(x - \sqrt{2})$!

EXAMPLE 1 ▶ Factoring Quadratic Expressions

Factor the following:

(a) $x^2 + x - 2$ (b) $x^2 - 2x - 8$

Solution

Possible Factors	Sum 1
$\pm 1, \pm 2$	$+2, -1$

(a) We must find two numbers whose product is -2 and whose sum is 1. By inspection or from the table, we see that these numbers are 2 and -1. Thus,

$$x^2 + x - 2 = (x + 2)(x - 1)$$

Possible Factors	Sum -2
$\pm 1, \pm 8$	
$\pm 2, \pm 4$	$-4, +2$

(b) Here we need two numbers whose product is -8 and whose sum is -2. By inspection or from the table, we see that these numbers are -4 and 2. Thus,

$$x^2 - 2x - 8 = (x - 4)(x + 2)$$ ∎

To factor $2x^2 - 5x - 3$, we need to find integers a, b, c, and d such that

$$2x^2 - 5x - 3 = (ax + b)(cx + d)$$

with $ac = 2$ and $bd = -3$.

Using **FOIL** to multiply the right side, ac must be 2 (the coefficient of the first term) and bd must be -3 (the last term in $2x^2 - 5x - 3$). The positive factors of $ac = 2$ are 2 and 1. The possible factors of $bd = -3$ are -1, 3 or 3, -1, or -3, 1 or 1, -3. We try different arrangements of these factors until we obtain the correct middle term, $-5x$, as follows:

$$(ax + b)(cx + d)$$

$b = -1, d = 3$	$(2x - 1)(x + 3) = 2x^2 + 6x - 1x - 3$	Incorrect
$b = 3, d = -1$	$(2x + 3)(x - 1) = 2x^2 - 2x + 3x - 3$	Incorrect
$b = -3, d = 1$	$(2x - 3)(x + 1) = 2x^2 + 2x - 3x - 3$	Incorrect
$b = 1, d = -3$	$(2x + 1)(x - 3) = 2x^2 - 6x + 1x - 3$	Correct!

$$-5x$$

Thus, $2x^2 - 5x - 3 = (2x + 1)(x - 3)$.

EXAMPLE 2 ▶ Factoring More Quadratic Expressions

Factor $6x^2 - 7x - 3$.

Solution

We need to find integers a, b, c, and d such that

$$6x^2 - 7x - 3 = (ax + b)(cx + d)$$

The positive factors of $ac = 6$ are 6, 1 or 3, 2. The factors of $bd = -3$ are -3, 1 or 1, -3. After trying different possibilities, $a = 3$, $c = 2$, $b = 1$, and $d = -3$ will yield $6x^2 - 7x - 3 = (3x + 1)(2x - 3)$. ∎

What if the expression is not factorable? We use the ac rule to determine whether an expression factors.

Factorable Quadratic Expression

$ax^2 + bx + c$ is factorable if there are two integers whose product is ac and whose sum is b. If $ax^2 + bx + c$ is *not* factorable, it is prime.

Using this idea, $6x^2 - 7x - 1$ and $3x^2 + 2x - 4$ are prime (not factorable). Do you see why?

B. Solving Quadratic Equations by Factoring

The next examples show how factoring can sometimes be used to solve quadratic equations.

EXAMPLE 3 ▶ Solving Quadratics by Factoring

Solve the following equations:

(a) $x^2 + 5x + 6 = 0$ (b) $5x^2 - 14x = 3$

Solution

(a) We have already shown that $x^2 + 5x + 6 = (x + 2)(x + 3)$. Thus,

$$(x + 2)(x + 3) = 0$$

With 0 on one side of the equation, we can make use of the property of the real number system that says that **a product of two real numbers is 0 if and only if at least one of them is 0.** Thus, the preceding equation is true if and only if

$$x + 2 = 0 \quad \text{or} \quad x + 3 = 0$$
$$x = -2 \quad \text{or} \quad x = -3$$

The solution set of the given equation is $\{-3, -2\}$.

Check: By substitution in the left side of the given equation, we find for $x = -3$, $x^2 + 5x + 6 = (-3)^2 + 5(-3) + 6 = 9 - 15 + 6 = 0$. This checks the solution $x = -3$. For $x = -2$, we get $x^2 + 5x + 6 = (-2)^2 + 5(-2) + 6 = 4 - 10 + 6 = 0$, which checks the solution $x = -2$.

(b) Subtract 3 from both sides of the equation to obtain

$$5x^2 - 14x - 3 = 0$$

The positive factors of 5 are 5 and 1, whereas the factors of -3 are $-3, 1$ or $1, -3$. Thus,

$$5x^2 - 14x - 3 = (5x + 1)(x - 3) = 0$$

Hence,

$$5x + 1 = 0 \quad \text{or} \quad x - 3 = 0$$

That is,

$$x = -\tfrac{1}{5} \quad \text{or} \quad x = 3$$

Therefore, the solution set of $5x^2 - 14x = 3$ is $\{3, -\tfrac{1}{5}\}$. Make sure that you check that the solution is correct!   ∎

GRAPH IT

To solve $x^2 + 5x + 6 = 0$, enter $Y_1 = x^2 + 5x + 6$ and ZOOM 6 . Where is $Y_1 = 0$? Press 2nd CALC 2 . When the grapher asks for the left bound, use ◀ to move the cursor to the left side of the curve. Press ENTER . Use ▶ to move below the curve to find the right bound. Press ENTER twice. One of the roots is -3 as shown.

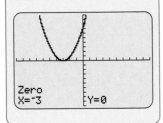

EXAMPLE 4 ▶ Solving Quadratics That Are Differences of Squares

Solve the equation $x^2 - 9 = 0$.

Solution

The equation $x^2 - 9 = 0$ can be solved by factoring because the difference of two squares, $x^2 - p^2$, can always be written as the product $(x - p)(x + p)$. Checking this multiplication, we get

$$(x - p)(x + p) = x^2 \underbrace{- px + px}_{0} - p^2 = x^2 - p^2$$

Thus, $x^2 - 9 = 0$ becomes

$$(x - 3)(x + 3) = 0$$

which gives the solution set $\{-3, 3\}$.

We could have avoided factoring by adding 9 to both sides to obtain

$$x^2 = 9$$

Taking the square roots of both sides, we have

$$x = \pm 3$$

so the solution set is $\{-3, 3\}$, as before.   ∎

EXAMPLE 5 ▶ Solving Quadratics by Root Extraction

Solve the equation $3x^2 + 2 = 50$.

Solution

We first subtract 2 from both sides to obtain

$$3x^2 = 48$$

We then divide both sides by 3 to get

$$x^2 = 16$$

As in Example 4, we find the solution set by taking square roots of both sides. This gives the solution set $\{-4, 4\}$.   ∎

C. The Quadratic Formula

Quadratic Formula

The solutions of a quadratic equation in standard form

$$ax^2 + bx + c = 0 \qquad a \neq 0$$

are given by the **quadratic formula,**

$$x = \frac{-b \pm \sqrt{b^2 - 4ac}}{2a}$$

We use this formula when expressions are *not* factorable. For example, $5x^2 - 7x - 3 = 0$ and $3x^2 + 2x - 4 = 0$ involve the expressions $5x^2 - 7x - 3$

and $3x^2 + 2x - 4$, which are not factorable. To solve these equations we would use the quadratic formula.

The derivation of this formula is given in Exercises 6.4, problem 55. The symbol $\pm$ in the formula means that there are two solutions, one with the plus sign and the other with the minus sign. If the quantity under the radical sign, $b^2 - 4ac$ (called the *discriminant*), is positive, there are two real number solutions. If $b^2 - 4ac$ is 0, the two solutions are the same, so there is actually just one solution, $-b/2a$. If the quantity $b^2 - 4ac$ is negative, there are no real number solutions.

P R O B L E M   S O L V I N G

Quadratic Equations

❶ **Read** the problem.

Solve the equation $x^2 - 2x = 1$.

❷ **Select** the unknown.

Find the values of the variable x such that $x^2 - 2x = 1$.

❸ **Think** of a plan.
Is the equation a quadratic equation? If it is, write it in standard form. Can you factor it? If not, use the quadratic formula to solve it.

The equation $x^2 - 2x = 1$ is a quadratic equation. To write it in standard form, subtract 1 from both sides to obtain $x^2 - 2x - 1 = 0$. Since you cannot factor this equation, you write $1x^2 - 2x - 1 = 0$ and compare it with $ax^2 + bx + c = 0$. Thus, $a = 1$, $b = -2$, and $c = -1$.

❹ **Use** the quadratic formula to carry out the plan.
Find a, b, and c and substitute their values in

$$x = \frac{-b \pm \sqrt{b^2 - 4ac}}{2a}$$

How many solutions should you get?

Substituting these values in

$$\frac{-b \pm \sqrt{b^2 - 4ac}}{2a}$$

you obtain

$$x = \frac{-(-2) \pm \sqrt{(-2)^2 - 4(1)(-1)}}{2(1)}$$

$$= \frac{2 \pm \sqrt{4 + 4}}{2} = \frac{2 \pm \sqrt{8}}{2} = \frac{2 \pm \sqrt{4 \cdot 2}}{2} = \frac{2 \pm 2\sqrt{2}}{2}$$

Make sure the final answer is simplified.

This answer is *not* simplified. Divide each number in the numerator and denominator by 2 to obtain $1 \pm \sqrt{2}$. Thus,

$$x = 1 + \sqrt{2} \quad \text{or} \quad x = 1 - \sqrt{2}$$

❺ **Verify** the answer.

The verification is left to you. Note that $(1 \pm \sqrt{2})^2 = 1 \pm 2\sqrt{2} + 2$.

TRY EXAMPLE 6 NOW.

Cover the solution, write your own solution, and then check your work.

EXAMPLE 6 ▸ Solving Quadratics by Using the Quadratic Formula

Use the quadratic formula to solve $x^2 - 4x - 12 = 0$.

Solution

To obtain the correct values of a, b, and c, we rewrite the equation in standard quadratic form as which we compare with

$$1x^2 + (-4)x + (-12) = 0$$

$$ax^2 + bx + c = 0$$

Now we see that $a = 1$, $b = -4$, and $c = -12$. Thus,

$$x = \frac{-(-4) \pm \sqrt{(-4)^2 - (4)(1)(-12)}}{(2)(1)}$$

$$= \frac{4 \pm \sqrt{16 + 48}}{2} = \frac{4 \pm \sqrt{64}}{2} = \frac{4 \pm 8}{2}$$

so that

$$x = \frac{4 + 8}{2} = \frac{12}{2} = 6$$

or

$$x = \frac{4 - 8}{2} = \frac{-4}{2} = -2$$

Hence, the solution set is $\{-2, 6\}$. ∎

Note: The instructions in the preceding problem required the use of the quadratic formula. In general, however, try to *factor* the equation *first*.

EXAMPLE 7 ▶ Solving Nonfactorable Equations

Solve the equation $3x^2 + x - 5 = 0$.

Solution

This equation is *not* factorable because we are unable to find two integers whose product is $3(-5) = -15$ and whose sum is 1, so we use the quadratic formula. We compare

$$ax^2 + bx + c = 0 \quad \text{and} \quad 3x^2 + 1x - 5 = 0$$

and see that $a = 3$, $b = 1$, and $c = -5$. Hence,

$$x = \frac{-1 \pm \sqrt{1^2 - 4(3)(-5)}}{2(3)} = \frac{-1 \pm \sqrt{61}}{6}$$

The solution set is

$$\left\{ \frac{-1 - \sqrt{61}}{6}, \frac{-1 + \sqrt{61}}{6} \right\}$$

These numbers cannot be expressed exactly in any simpler form. By using the table of square roots in the back of the book (or a calculator), we find that $\sqrt{61} \approx 7.81$. This gives the approximate solutions

$$\frac{-1 - 7.81}{6} \approx -1.47 \quad \text{and} \quad \frac{-1 + 7.81}{6} \approx 1.14$$ ∎

EXAMPLE 8 ▶ Solving Using the Quadratic Formula

Solve the equation $2x^2 - 2x = 1$.

Solution

In order to use the quadratic formula, we must first write the equation in standard quadratic form. We can do this by subtracting 1 from both sides of the given equation to obtain $2x^2 - 2x - 1 = 0$. This equation is *not* factorable because there are no two integers whose product is $2(-1) = -2$ and whose sum is -2, so we use the quadratic formula. To find a, b, and c, we write

$$2x^2 - 2x - 1 = 0$$

which we compare with

$$ax^2 + bx + c = 0$$

Thus, $a = 2$, $b = -2$, and $c = -1$. Now, we can substitute into the quadratic formula to find

$$x = \frac{-(-2) \pm \sqrt{(-2)^2 - 4(2)(-1)}}{2(2)} = \frac{2 \pm \sqrt{4 + 8}}{4} = \frac{2 \pm \sqrt{12}}{4}$$

$$= \frac{2 \pm 2\sqrt{3}}{4} = \frac{1 \pm \sqrt{3}}{2}$$

so the solutions are

$$x = \frac{1 \pm \sqrt{3}}{2}$$

The solution set is

$$\left\{ \frac{1 + \sqrt{3}}{2}, \frac{1 - \sqrt{3}}{2} \right\}$$

or, using decimal approximations, $\{1.37, -0.37\}$ ∎

Examples 6–8 display the tremendous advantage of the quadratic formula over other methods of solving quadratic equations. It is not necessary to attempt to write the left side as a product of first-degree expressions. You need only recognize the values of a, b, and c and make direct substitutions into the quadratic formula. Note that if the equation is easily factorable, it is easier and faster to solve it by factoring.

D. The Pythagorean Theorem

Quadratic equations can be used to find the lengths of the sides of right triangles using the *Pythagorean theorem.*

THEOREM 6.1 Pythagorean Theorem

In any right triangle (a triangle with one $90°$ angle, as shown in Figure 6.20), the square of the longest side (the hypotenuse) is equal to the sum of the squares of the other two sides (the legs). In symbols,

$$c^2 = a^2 + b^2$$

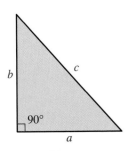

FIGURE 6.20

It is interesting to note that there are infinitely many triples of whole numbers (a, b, c) that satisfy the equation $c^2 = a^2 + b^2$. Such triples are called **Pythagorean triples.**

EXAMPLE 9 ▶ Using the Pythagorean Theorem to Find Lengths

The lengths of the three sides of a right triangle are consecutive integers. What are these lengths?

Solution

Let the length of the shortest side be x. Since the lengths of the sides are consecutive integers, we have the following:

Length of the shortest side	x
Length of the next side	$x + 1$
Length of the hypotenuse	$x + 2$

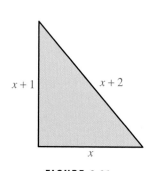

See Figure 6.21. By the Pythagorean theorem,

$$
\begin{aligned}
(x + 2)^2 &= (x + 1)^2 + x^2 && \\
x^2 + 4x + 4 &= x^2 + 2x + 1 + x^2 && \text{Multiply.} \\
x^2 + 4x + 4 &= 2x^2 + 2x + 1 && \text{Simplify.} \\
0 &= x^2 - 2x - 3 && \text{Subtract } x^2, 4x, \text{ and } 4 \text{ from both sides.} \\
x^2 - 2x - 3 &= 0 && \text{Write in standard form.} \\
(x - 3)(x + 1) &= 0 && \text{Factor.} \\
x - 3 = 0 \quad &\text{or} \quad x + 1 = 0 && \text{Use the zero-factor property.} \\
x = 3 \quad &\text{or} \quad x = -1 && \text{Solve } x - 3 = 0 \text{ and } x + 1 = 0.
\end{aligned}
$$

FIGURE 6.21

Since the lengths of the sides must be positive, we discard the negative answer, -1. Thus, the shortest side is 3 units, so the other two sides are 4 and 5 units. ■

Note that if we have one Pythagorean triple of the form $(a, a + 1, c)$, we can find another by substituting the a and c of the original triple into

$$(3a + 2c + 1, 3a + 2c + 2, 4a + 3c + 2)$$

This is so because using $c^2 = a^2 + (a + 1)^2$, we can show that

$$(3a + 2c + 1)^2 + (3a + 2c + 2)^2 = (4a + 3c + 2)^2$$

This proves that there are infinitely many Pythagorean triples of the form $(a, a + 1, c)$. For instance, if we let $a = 3$, $b = 4$, and $c = 5$, we find

$$(3 \cdot 3 + 2 \cdot 5 + 1, 3 \cdot 3 + 2 \cdot 5 + 2, 4 \cdot 3 + 3 \cdot 5 + 2) = (20, 21, 29)$$

as another such triple. If we now let $a = 20$, $b = 21$, and $c = 29$, we get the next such triple: $(119, 120, 169)$, and so on.

E. Applications

Many problems can be studied by using quadratic equations. For example, do you use hair spray containing chlorofluorocarbons (CFCs) for propellants? A U.N.-sponsored conference negotiated an agreement to stop producing CFCs by the year 2000 because they harm the ozone layer. Can we check to see whether the goal has been reached if current levels of production continue? We shall see in Example 10. (Source: www. ejil.org/journal/Vol1/No1/art14.html.)

EXAMPLE 10 ▶ **Applications with CFCs and Quadratics**

The production of CFCs for use as aerosol propellants (in thousands of tons) can be represented by $P(t) = -0.4t^2 + 22t + 120$, where t is the number of years after 1960. When will production be stopped?

Solution

Production will be stopped when $P(t) = 0$. We need to solve the equation

$$P(t) = -0.4t^2 + 22t + 120 = 0$$
$$-4t^2 + 220t + 1200 = 0 \qquad \text{Multiply by 10 (to clear decimals).}$$
$$t^2 - 55t - 300 = 0 \qquad \text{Divide by } -4 \text{ (to obtain } t^2\text{).}$$
$$(t - 60)(t + 5) = 0 \qquad \begin{array}{l}\text{We need two numbers whose}\\ \text{product is } -300 \text{ and whose sum}\\ \text{is } -55 \text{ } (-60 \text{ and } 5).\end{array}$$

$$t - 60 = 0 \qquad \text{or} \qquad t + 5 = 0 \qquad \text{By the zero-factor property}$$
$$t = 60 \qquad \text{or} \qquad t = -5 \qquad \text{Solve each equation.}$$

Online Study Center

To explore further finding solutions as decimals, solving by factoring, or using the quadratic equation, access links 6.4.1, 6.4.2, and 6.4.3 on this textbook's Online Study Center.

Since t represents the number of years *after* 1960, production will be zero (stopped) 60 years after 1960, or in 2020 (not in the year 2000 as promised!). The answer $t = -5$ has to be discarded because it represents 5 years *before* 1960, but the equation applies only to years *after* 1960. ∎

EXERCISES 6.4

A **Factoring Quadratic Expressions**

In problems 1–14, factor each expression.

1. $x^2 + 6x + 8$ **2.** $x^2 + 7x + 10$

3. $x^2 - x - 12$ **4.** $x^2 - 3x - 10$

5. $x^2 + 7x - 18$ **6.** $x^2 - 12x + 11$

7. $x^2 - 10x + 25$ **8.** $x^2 - 8x + 16$

9. $x^2 + 10x + 25$ **10.** $x^2 + 16x + 64$

11. $2x^2 + x - 3$ **12.** $3x^2 + 10x + 3$

13. $6x^2 - 5x + 1$ **14.** $6x^2 - 11x + 3$

B **Solving Quadratic Equations by Factoring**

In problems 15–38, solve each equation.

15. $(x - 2)(x - 4) = 0$ **16.** $(x + 2)(x + 3) = 0$

17. $(x + 2)(x - 3) = 0$ **18.** $(x + 5)(x - 6) = 0$

19. $x(x - 1)(x + 1) = 0$

20. $(x + 1)(x + 2)(x - 3) = 0$

21. $(2x - 1)(x + 2) = 0$

22. $(3x + 5)(4x + 7) = 0$

23. $x^2 - 16 = 0$ **24.** $3x^2 - 27 = 0$

25. $5x^2 = 125$ **26.** $4x^2 + 1 = 65$

27. $(3x - 6)(2x + 3)(5x - 8) = 0$

28. $2x(x + 7)(2x - 3) = 0$

29. $6x^2 - 1 = 215$ **30.** $4x^2 + 1 = 50$

31. $x^2 - 12x + 27 = 0$ **32.** $x^2 - 6x + 8 = 0$

33. $x^2 - 8x = 20$ **34.** $x^2 - 9x = 36$

35. $10x^2 + 7x + 1 = 0$ **36.** $6x^2 + 17x + 5 = 0$

37. $3x^2 + 2x = 5$ **38.** $2x^2 - x = 6$

C **The Quadratic Formula**

In problems 39–54, solve each equation by using the quadratic formula.

39. $2x^2 + 3x - 5 = 0$ **40.** $3x^2 - 7x + 2 = 0$

41. $2x^2 + 5x - 7 = 0$ **42.** $4x^2 - 7x - 15 = 0$

43. $x^2 + 5x + 3 = 0$ **44.** $2x^2 + 7x - 4 = 0$

45. $5x^2 - 8x + 2 = 0$ **46.** $3x^2 + 5x + 1 = 0$

47. $7x^2 - 6x = -1$ **48.** $7x^2 - 12x = -5$

49. $9x^2 - 6x - 2 = 0$ **50.** $2x^2 - 8x + 5 = 0$

51. $2x^2 + 2x = 1$ **52.** $2x^2 - 6x = 5$

53. $4x^2 = -8x + 5$ **54.** $2x^2 = 2x + 5$

55. The following procedure can be used to obtain the quadratic formula. Suppose that the quadratic equation is given in standard form.

$$ax^2 + bx + c = 0, \qquad a \neq 0$$

Then

$$ax^2 + bx = -c \qquad \text{Why?}$$

Now we multiply both sides by $4a$ to get

$$4a^2x^2 + 4abx = -4ac$$

By adding b^2 to both sides, we get

$$4a^2x^2 + 4abx + b^2 = b^2 - 4ac$$

The left side of this equation is the square of $(2ax + b)$. (Verify this!) Thus, we have

$$(2ax + b)^2 = b^2 - 4ac$$

Next, we take the square roots of both sides to get

$$2ax + b = \pm \sqrt{b^2 - 4ac}$$

From this equation, we get

$$2ax = -b \pm \sqrt{b^2 - 4ac} \qquad \text{Explain.}$$

and

$$x = \frac{-b \pm \sqrt{b^2 - 4ac}}{2a} \qquad \text{Explain.}$$

(These solutions can be checked in the original equation.)

D **The Pythagorean Theorem**

Solve problems 56–60 using the Pythagorean theorem.

56. The sides of a right triangle are consecutive even integers. Find their lengths.

57. The hypotenuse of a right triangle is 4 cm longer than the shortest side and 2 cm longer than the remaining side. Find the dimensions of this triangle.

58. The hypotenuse of a right triangle is 16 in. longer than the shortest side and 2 in. longer than the remaining side. Find the dimensions of this triangle.

59. The hypotenuse of a right triangle is 8 in. longer than the shortest side and 1 in. longer than the remaining side. Find the dimensions of this triangle.

60. People have been interested in right triangles for thousands of years. The right triangle relationship $a^2 + b^2 = c^2$ seems to have been known to the Babylonians and the ancient Egyptians. Among the interesting problems about right triangles is this one: Find the right triangles with integer sides such that the hypotenuse is 1 unit longer than one of the legs. You can solve this problem by letting the legs be x and y units long, so the hypotenuse is $y + 1$ units long. Then, you have

$$x^2 + y^2 = (y + 1)^2$$

a. Solve this equation for x^2. You should find $x^2 = 2y + 1$. Since x and y are to be positive integers, $2y + 1$ is an odd integer. This means that x^2 is an odd integer, so x also must be an odd integer. You can see that $x \neq 1$ because this would make $y = 0$ (no triangle!). However, if x is any odd integer greater than 1, there is a right triangle with the desired relationship between the sides. If you choose $x = 3$ and solve $x^2 = 2y + 1$ for y, you find $y = 4$. This gives you the well-known 3-4-5 right triangle.

b. Make a table of the next four triangles of this type.

E **Applications**

In Problems 61–64, use

$$h = 5t^2 + V_0 t$$

where h is the distance (in meters) traveled in t sec by an object thrown downward with an initial velocity V_0 (in meters per second).

61. An object is thrown downward with an initial velocity of 5 m/sec from a height of 10 m. How long does it take the object to hit the ground?

62. An object is thrown downward from a height of 28 m with an initial velocity of 4 m/sec. How long does it take the object to reach the ground?

63. An object is thrown downward from a building 15 m high with an initial velocity of 10 m/sec. How long does it take the object to hit the ground?

64. How long would it take a package thrown downward from a plane with an initial velocity of 10 m/sec to hit the ground 175 m below?

65. It costs a business $(0.1x^2 + x + 50)$ dollars to serve x customers. How many customers can be served if $250 is the cost?

66. The cost of serving x customers is given by $(x^2 + 10x + 100)$ dollars. If $1300 is spent serving customers, how many customers are served?

67. A manufacturer will produce x units of a product when its price is $(x^2 + 25x)$ dollars per unit. How many units will be produced when the price is $350 per unit?

68. When the price of a ton of raw materials is $(0.01x^2 + 5x)$ dollars, a supplier will produce x tons of it. How many tons will be produced when the price is $5000 per ton?

69. To attract more students, the campus theater decides to reduce ticket prices by x dollars from the current $5.50 price.
   a. What is the expression representing the new price after the x dollar reduction?
   b. If the number of tickets sold is $100 + 100x$ and the revenue is the number of tickets sold times the price of each ticket, what is the expression for the revenue?
   c. If the theater wishes to have $750 in revenue, how much is the price reduction?
   d. If the reduction must be less than $1, what is the reduction?

70. An apartment owner wants to increase the monthly rent from the current $250 in n increases of $10.
   a. What is the expression for the new price after the n increases of $10?
   b. If the number of apartments rented is $70 - 2n$ and the revenue is the number of apartments rented times the rent, what is the expression for the revenue?
   c. If the owner wants to receive $17,980 per month, how many $10 increases can the owner make?
   d. What will be the monthly rent?

 In Other Words

71. Write in your own words the procedure you would use to multiply two binomials using the **FOIL** method

72. Write the steps you would use to factor $x^2 + 8x + 7$.

73. Write the steps you would use to solve a quadratic equation by factoring.

74. Write the steps you would use to solve a quadratic equation with the quadratic formula.

The solutions of the quadratic equation $ax^2 + bx + c = 0$ $(a \neq 0)$ are

$$x = \frac{-b \pm \sqrt{b^2 - 4ac}}{2a}$$

75. Which type of solution do you get if $b^2 - 4ac = 0$?

76. Which type of solution do you get if $b^2 - 4ac > 0$?

77. Which type of solution do you get if $b^2 - 4ac < 0$?

 Using Your Knowledge

Have you been to a baseball game lately? Did anybody hit a home run? The trajectory of a baseball is usually very complicated, but we can get help from *The Physics of Baseball,* by Robert Adair. According to Mr. Adair, after t sec, starting 1 sec after the ball leaves the bat, the height of a ball hit at a 35° angle, rotating with an initial backspin of 2000 revolutions per minute (rpm), and hit at about 110 mph, is given by

$$H(t) = -80t^2 + 340t - 260 \text{ (in feet)}$$

78. How many seconds will pass before the ball hits the ground?

The distance traveled by the ball is given by

$$D(t) = -5t^2 + 115t - 110 \text{ (in feet)}$$

79. How far will the ball travel before it hits the ground?

80. Using $D(t)$, how far will a ball travel in 6 sec, the time it takes a high fly ball to hit the ground?

 Discovery

Charlie Brown received a chain letter. Several days later, after receiving more letters, he found that the number he had received was a *perfect square*. (The numbers 1^2, 2^2, 3^2, and so on are perfect squares.) Charlie decided to throw away some of the letters, and being very superstitious, he threw away 13^2 of them.

To his surprise, he found that the number he had left was still a perfect square.

81. What is the maximum number of letters Charlie could have received before throwing any away? (*Hint:* Let x^2 be the initial number of letters, and let y^2 be the number he had left. Then $x^2 = y^2 + 13^2$. Remember that x and y are integers, and you will want to make y as large as possible relative to x because you want x to be as large as possible.)

Peanuts. Reprinted by permission of United Feature Syndicate, Inc.

Calculator Corner

Your calculator can be extremely helpful in finding the roots of a quadratic equation by using the quadratic formula. Of course, the roots you obtain are being approximated by decimals. It is most convenient to start with the radical part in the solution of the quadratic equation and then store this value so that you can evaluate both roots without having to backtrack or copy down any intermediate steps. Look at the following equation:

$$2x^2 + 7x - 4 = 0$$

Using the quadratic formula, the solution is obtained by following these key strokes using a scientific calculator.

$$\boxed{7}\ \boxed{x^2}\ \boxed{-}\ \boxed{4}\ \boxed{\times}\ \boxed{2}\ \boxed{\times}\ \boxed{4}\ \boxed{+/-}\ \boxed{=}$$

$$\boxed{\sqrt{x}}\ \boxed{STO}\ \boxed{7}\ \boxed{+/-}\ \boxed{+}\ \boxed{RCL}\ \boxed{=}\ \boxed{\div}\ \boxed{2}\ \boxed{\div}\ \boxed{2}\ \boxed{=}$$

The display will show 0.5. To obtain the other root, key in

$$\boxed{7}\ \boxed{+/-}\ \boxed{-}\ \boxed{RCL}\ \boxed{=}\ \boxed{\div}\ \boxed{2}\ \boxed{\div}\ \boxed{2}\ \boxed{=}$$

which yields -4. In general, to solve the equation $ax^2 + bx + c = 0$ using your calculator, key in the following:

$$\boxed{b}\ \boxed{x^2}\ \boxed{-}\ \boxed{4}\ \boxed{\times}\ \boxed{a}\ \boxed{\times}\ \boxed{c}\ \boxed{=}$$

$$\boxed{\sqrt{x}}\ \boxed{STO}\ \boxed{b}\ \boxed{+/-}\ \boxed{+}\ \boxed{RCL}\ \boxed{=}\ \boxed{\div}\ \boxed{2}\ \boxed{\div}\ \boxed{a}\ \boxed{=}$$

and

$$\boxed{b}\ \boxed{+/-}\ \boxed{-}\ \boxed{RCL}\ \boxed{=}\ \boxed{\div}\ \boxed{2}\ \boxed{\div}\ \boxed{a}\ \boxed{=}$$

1. If your instructor allows it, do problems 39–54 using a calculator.

Collaborative Learning

Wouldn't it be nice if we looked at a factorable quadratic equation of the form $x^2 + bx + c = 0$ and we knew what the solution was? We are going to see how to do just that!

1. Form three groups. The first group does the first three problems, the second group the next three, and the third group the last three.

	Factors	*Solutions*
1. $x^2 + 3x + 2 = 0$	_____	_____
2. $x^2 + 5x + 6 = 0$	_____	_____
3. $x^2 + 7x + 12 = 0$	_____	_____
4. $x^2 - 3x + 2 = 0$	_____	_____
5. $x^2 - 5x + 6 = 0$	_____	_____
6. $x^2 - 7x + 12 = 0$	_____	_____
7. $x^2 - x - 2 = 0$	_____	_____
8. $x^2 - x - 6 = 0$	_____	_____
9. $x^2 - x - 12 = 0$	_____	_____

Each group should determine the relationship between the factors and the solutions and then make a conjecture regarding the solution of factorable quadratics of the form

$$x^2 + bx + c = 0$$

Do all groups reach the same conjecture?

2. What about $ax^2 + bx + c = 0$? Here is a hint. Look at the ac rule. Find the two integers whose product is ac and whose sum is b. In Example 3(b), $5x^2 - 14x - 3 = 0$, $ac = 5(-3) = -15$, and the two integers whose product is -15 and whose sum is -14 are 1 and -15. The solutions for the equation [as shown in Example 3(b)] are $-\frac{1}{5}$ and 3. Do you see any pattern linking 1 and -15 to $-\frac{1}{5}$ and 3? If you do not, look at problems 35–38. Do you see any pattern now?

Research Questions

1. Find out what techniques the Babylonians used to solve quadratic equations.

2. Discuss the method used by Euclid to solve quadratics.

3. How did the Hindu mathematician Brahmagupta solve quadratics?

We have shown several Pythagorean triples in the text.

Online Study Center 4. To find how to generate Pythagorean triples, access link 6.4.4 on this textbook's Online Study Center. Does the formula generate the ones given in the text?

Online Study Center 5. To see proof of the Pythagorean theorem, access link 6.4.5 on this textbook's Online Study Center. Then write your own proof based on the one you have been shown.

Online Study Center 6. To find more proofs of the Pythagorean theorem ("39 proofs of the Pythagorean theorem"), access link 6.4.6 on this textbook's Online Study Center. Then write out three different proofs of the Pythagorean theorem.

6.5 Modeling and Problem Solving

Putting on the Brakes

Do you remember the **RSTUV** procedure that you studied in Chapter 1? You are certainly going to need it in this section, but with a small modification. The most difficult task in solving word problems is *translating* the problem into the language of algebra so that you obtain an accurate mathematical model representing all the conditions of the problem. Because of this, the third step of the procedure will emphasize this idea of translation. What types of problems will you be working? There are many of them, including some that may save your life! For example, do you know what the stopping distance d is for a car traveling v mph if the driver has a reaction time of t sec? Now suppose you are driving a car at 20 mph and you are 42 ft away from an intersection. If your reaction time is 0.6 sec, can you stop the car in time? See Example 3 and problems 27–34 in Exercises 6.5, and you will be able to tell! ▶

The *Discovery,* one of the active space shuttles in the NASA program, lifts off from Cape Canaveral, Florida.

A. Problem Solving

In the preceding sections we considered how to solve certain kinds of equations. Now we are ready to apply this knowledge to solve problems. These problems will be stated in words and are consequently called **word** or **story problems.** Word problems frighten many students, but you should not panic. We have the **RSTUV** procedure, a surefire method for tackling such problems.

Let us start with a problem that you might have heard about. It has to do with space exploration.

PROBLEM SOLVING

Word Problems

When fully loaded, a space shuttle and its payload (cargo) weigh about 215,000 lb. The shuttle itself weighs 85,000 lb more than the payload. What is the weight of each?

① **Read** the problem carefully and decide what it asks for.

The problem asks for the weight of each, that is, the weight of the shuttle and the weight of the payload.

② **Select** a variable to represent the unknown.

Let p represent the weight of the payload in pounds. Since the shuttle weighs 85,000 lb more than the payload, the shuttle weighs $p + 85,000$ lb.

③ **Think** of a plan. Can you translate the information into an equation or inequality?

Translate the first sentence in the problem into an algebraic statement.

④ **Use** algebra to solve for the unknown.

$$\underbrace{\text{The shuttle and its payload}}\ \underbrace{\text{weigh}}\ \underbrace{\text{215,000 lb.}}$$ ← This is a verbal model for the problem.

$$(p + 85,000) + p = 215,000$$ ← This is an algebraic model for the problem.
$$p + 85,000 + p = 215,000$$
$$2p + 85,000 - 85,000 = 215,000 - 85,000$$
$$2p = 130,000$$
$$p = 65,000$$

Thus, the payload weighs 65,000 lb, and the shuttle weighs $65,000 + 85,000 = 150,000$ lb.

⑤ **Verify** the answer.

To verify the answer, note that the combined weight of the shuttle and its payload is $150,000 + 65,000 = 215,000$ lb, as stated in the problem.

TRY EXAMPLE 1 NOW.

Cover the solution, write your own solution, and then check your work.

EXAMPLE 1 ▶ Modeling and Interest

Angie bought a 6-month, $10,000 certificate of deposit. At the end of the 6 months, she received $650 simple interest. What rate of interest did the certificate pay?

Solution

1. Read the problem. It asks for the rate of simple interest.

2. Select the variable r to represent this rate.

3. Translate the problem. Here, we need to know that the formula for simple interest is

   $$I = Prt$$ ← This is an algebraic model for the problem.

   where I is the amount of interest, P is the principal, r is the interest rate, and t is the time in years. For our problem, $I = \$650$, $P = \$10,000$, r is unknown, and $t = \frac{1}{2}$ year. Thus, we have

   $$650 = (10,000)(r)\left(\tfrac{1}{2}\right) \qquad \text{or} \qquad 650 = 5000r$$

4. Use algebra to solve the equation

$$650 = 5000r$$

Divide by 5000. $\dfrac{650}{5000} = r$

Express decimally. $r = 0.13$

Hence, the certificate paid 13% simple interest.

5. Verify the answer. Is the interest earned on a $10,000, six-month certificate at a 13% rate $650? Evaluating Prt, we have

$$(10{,}000)(0.13)\left(\tfrac{1}{2}\right) = 650$$

Since the answer is yes, 13% is correct.   ■

    The next problem may help you to save some money. When you rent a car, you may be able to choose either a mileage rate or a flat rate. For example, if you wish to rent an intermediate sedan for 1 day, you can pay $25 plus $.20 for each mile traveled or a $50 flat rate. Which is the better deal? That depends on how far you plan to drive. To be more specific, how many miles could you travel for $50 if you used the mileage rate? You will find the answer in the next example.

EXAMPLE 2 ▶ Renting Cars and Solving Equations

Jim Smith rented a car at $25 per day plus $.20 per mile. How many miles can Jim travel for $50?

Solution

Again, we proceed by steps as follows:

1. Read the problem carefully. We are looking for the number of miles Jim can travel for $50.

2. Select m to represent this number of miles.

3. Translate the problem into an equation; that is, make a model that represents the conditions of the problem. To do this, we must realize that Jim is paying $.20 for each mile plus $25 for the day. Thus, we have the following information for Jim's travels:

1 mi	The cost is $0.20(1) + 25$.
2 mi	The cost is $0.20(2) + 25$.
.	
.	
.	
m mi	The cost is $0.20m + 25$. ← Algebraic model

Because we want to know how many miles Jim can drive for $50, we must put the cost for m miles equal to $50; this gives the equation

$$0.20m + 25 = 50$$

4. Use algebra to solve the equation

$$0.20m + 25 = 50$$

Subtract 25. $0.20m = 25$

Multiply by 100. $20m = 2500$ This gets rid of the decimal.

Divide by 20. $m = \dfrac{2500}{20} = 125$

Thus, Jim can travel 125 mi for $50.

5. Verify that $(0.20)(125) + 25 = 50$. (This is left for you to do.)   ■

From this information, you can deduce that if you want to rent an intermediate sedan for 1 day and plan to drive over 125 mi, then the flat rate is the better of the two options.

Talking about distances, the next example could even save some lives. Have you seen the Highway Patrol booklet that indicates the **braking distance** b (in feet) that it takes to stop a car after the brakes are applied? This information is usually given in a chart, but there is a formula that gives close estimates under normal driving conditions. The braking distance formula is

$$b = 0.06v^2 \qquad \leftarrow \text{Algebraic model}$$

where v is the speed of the car (in miles per hour) when the brakes are applied. Thus, if you are traveling 20 mph, you will travel

$$b = 0.06(20^2) = 0.06(400) = 24 \text{ ft}$$

after you apply the brakes.

EXAMPLE 3 ▶ Putting on the Brakes with Quadratics

A car traveled 150 ft *after* the brakes were applied. (It might have left a skid mark that long.) How fast was the car going when the brakes were applied?

Solution

1. Read the problem carefully.

2. Select the variable v to represent the velocity.

3. Translate: The braking distance $b = 150$, and the braking distance formula reads $b = 0.06v^2$. Thus, we have the equation

$$0.06v^2 = 150 \qquad \leftarrow \text{Algebraic model}$$

4. Use algebra to solve the last equation:

Multiply by 100. $6v^2 = 15{,}000$ This gets rid of the decimal.

Divide by 6. $v^2 = \dfrac{15{,}000}{6} = 2500$

Take square roots. $v = 50 \qquad \text{or} \qquad -50$

Since the -50 makes no sense in this problem, we discard it. Thus, the car was going 50 mph when the brakes were applied.

5. Verify the answer by substituting 50 for v in the braking distance formula.   ■

You have probably noticed the frequent occurrence of certain words in the statements of word problems. Because these words are used frequently, Table 6.1 presents a brief mathematics dictionary to help you translate them properly.

The next example shows how some of these words are used in a word problem.

EXAMPLE 4 ▶ Applications: Finding Integers

If 7 is added to twice the square of a number n, the result is 9 times the number. Find n.

Solution

1. Read the problem.

2. Select the unknown, n in this case.

3. Translate the problem.

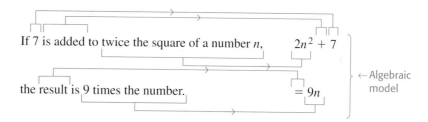

TABLE 6.1 Mathematics Dictionary

Words	Translation	Example	Translation
Add to More than Sum of Increased by Added to	+	Add n to 7. 7 more than n The sum of n and 7 n increased by 7 7 added to n	$n + 7$
Subtract from Less than Minus Difference of Decreased by Subtracted from	−	Subtract 9 from x. 9 less than x x minus 9 Difference of x and 9 x decreased by 9 9 subtracted from x	$x - 9$
Of The product of Times Multiply by	×	$\frac{1}{2}$ of a number x The product of $\frac{1}{2}$ and x $\frac{1}{2}$ times a number x Multiply $\frac{1}{2}$ by x.	$\frac{1}{2}x$
Divide by Divided by The quotient of	÷	Divide 10 by x. 10 divided by x The quotient of 10 and x	$\frac{10}{x}$

The same,
  yields, gives,
  is, equals
	=	x equals 12.
		x is 12.
		x yields 12.
			$x = 12$

4. Use algebra to solve the equation

$$2n^2 + 7 = 9n$$

Since this equation has both a squared term and a first-degree term, we solve it as a quadratic equation, first putting it in the standard form $ax^2 + bx + c = 0$. Hence, we subtract $9n$ from both sides to get

$$2n^2 - 9n + 7 = 0$$

Then we use the quadratic formula with $a = 2$, $b = -9$, and $c = 7$ to get the values of n.

$$n = \frac{-(-9) \pm \sqrt{(-9)^2 - 4(2)(7)}}{2(2)}$$

$$= \frac{9 \pm \sqrt{81 - 56}}{4} = \frac{9 \pm \sqrt{25}}{4}$$

$$= \frac{9 \pm 5}{4} = \frac{9 + 5}{4} \quad \text{or} \quad \frac{9 - 5}{4}$$

$$= \frac{14}{4} \quad \text{or} \quad \frac{4}{4}$$

$$= \frac{7}{2} \quad \text{or} \quad 1$$

5. Verify $n = \frac{7}{2}$.

$$2\left(\frac{7}{2}\right)^2 + 7 = 2 \times \frac{7}{2} \times \frac{7}{2} + 7 = \frac{49}{2} + 7 = \frac{63}{2}$$

and

$$9 \times \frac{7}{2} = \frac{63}{2}$$

Thus, we see that if 7 is added to twice the square of $\frac{7}{2}$, the result is 9 times $\frac{7}{2}$. The verification for $n = 1$ is left for you to calculate. We now have shown that the number n can be either $\frac{7}{2}$ or 1. ■

Sometimes we do not have to find a numerical answer to a problem, but only an algebraic model that the answer will satisfy, as shown in the next example.

EXAMPLE 5 ▶ Finding Equations That Model a Digit

A certain two-digit number is equal to 7 times the sum of its digits. If the tens digit is x and the units digit is y, what equation must x and y satisfy?

Solution

1. Read the problem.

2. The unknowns are x (the tens digit) and y (the units digit).

3. Translate the problem. Since the tens digit is x and the units digit is y, the two-digit number must be $10x + y$. The sum of the digits is $x + y$, and 7 times the sum of the digits is $7(x + y)$, so the required equation is

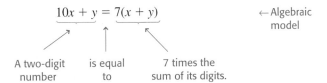

$$10x + y = 7(x + y)$$ ← Algebraic model

A two-digit number · is equal to · 7 times the sum of its digits.

Note: There are many solutions to this problem. Can you find one?

EXAMPLE 6 ▶ Checking for the Best Deal

The Better Business Bank has two types of checking accounts, A and B. Type A has a monthly service charge of $3 plus $.25 for each check written. Type B has a monthly service charge of $5 plus $.10 for each check written. What is the greatest number of checks that can be written before type A becomes the more expensive of the two?

Solution

Let x be the number of checks written. The cost of each account is as follows:

Type A $3 + 0.25x$ dollars
Type B $5 + 0.10x$ dollars

Hence, we need to find the greatest value of x such that

$$\underbrace{3 + 0.25x}_{\text{Type A}} \leq \underbrace{5 + 0.10x}_{\text{Type B}}$$ ← Algebraic model

This inequality can be solved as follows:

Given.	$3 + 0.25x \leq 5 + 0.10x$
Subtract 3 from both sides.	$0.25x \leq 2 + 0.10x$
Subtract 0.10x from both sides.	$0.15x \leq 2$
Divide both sides by 0.15.	$x \leq \dfrac{2}{0.15}$
	$x \leq 13\frac{1}{3}$

Since x must be a whole number, it follows that 13 is the required answer. Note that if you write more than 13 checks, a type A account will be more expensive.

B. Mathematical Modeling

In the preceding problems we have translated verbal sentences into algebraic equations that include all the conditions of the problem. This procedure is called **mathematical modeling.** Intuitively, we say that a model is a replica or copy of the situation at hand. Thus, a **mathematical model** is a structure that displays the features of the problem being solved. Some structures that lend themselves to modeling are equations or formulas, inequalities, and graphs. Now we shall explore a situation that leads to a more sophisticated mathematical model than those studied previously.

EXAMPLE 7 ▶ Distance Learning and Modeling

One of the most radical changes in the way educational material is presented involves what is called *distance learning.* One of the models for distance learning uses television broadcasts (cable or regular channels) of lessons that students can view at home. From the point of view of an educational institution, is this a cost-effective method of presenting (teaching) a 3-credit-hour course? The following information is needed to solve this problem:

The primary *costs* to the institution are

S	Salary of the instructor, $500 for each credit hour
L	Licensing fee for using copyrighted materials on TV, $500
F	Fee per student paid to the copyright holder, $15
N	Number of students

The *revenues* to the institution are

T	Tuition, $40 per credit hour
C	Contribution by state [In Florida, the state provides about $2800 per FTE (1 full-time equivalent $=$ 40 credit hours) at the community-college level and more at the university level.]

Solution

To solve this problem, we use the **RSTUV** procedure. This time, however, there is no formula or "back of the book" to tell us when we are correct.

1. ***R****ead the problem.* Notice that we are referring to only one type of distance learning: television instruction.

2. ***S****elect the unknown.* The question is: From the institution's standpoint, is this a cost-effective method of presenting a mathematics course? A first step might be to define what is meant by *cost-effective.* We can start by looking at the *break-even point,* the point at which the revenue to the school equals the expense for the course.

3. ***T****hink of a plan.* Find the cost and the revenue for the institution based on N students taking the course. Equate the two expressions and solve for N. This will tell us the number of students needed to break even.

4. ***U****se the model to carry out the plan.* Let us figure out the costs. How much is the instructor's salary? $500 per credit hour. For a 3-credit-hour course, we have

Instructor's salary (S)	$3 \cdot 500 = \$1500$
Licensing fee (L)	$\$\ 500$
Copyright holder fee (F) for N students	$15 \cdot N = 15N$
Cost to the institution	$1500 + 500 + 15N = 2000 + 15N$

The revenues to the institution are

Tuition (T) for N students	$40 \cdot N = 40N$
State contribution (C) for N students taking a 3-credit-hour course (Multiply the number of students N by 3 and divide by 40 to find the number of FTEs. Then, multiply the result by $2800.)	$\left(\dfrac{3 \cdot N}{40}\right) \cdot 2800 = 210N$
Revenue to the institution	$40N + 210N = 250N$

The break-even equation

$$Cost = revenue$$

$$2000 + 15N = 250N$$

Solving for N,

$$2000 = 235N$$

$$\frac{2000}{235} = N$$

$$8.51064 = N$$

To the nearest whole number,

$$9 \approx N$$

Note that 9 is the smallest number of students for which the institution will incur no loss. Thus, *using this model,* if 9 students register for the course, the institution makes a small amount of money. Try this problem using the pertinent information ($S, L, F, T, C,$ and FTE) for your state!

How would the break-even point be affected if there were more expenses (for example, utilities, building construction, maintenance, equipment) than the ones detailed here?

5. *Verify the answer.* Try the problem for 9 students and see whether the institution does make a small amount of money above the break-even point. ∎

Online Study Center

To further explore problem solving, access links 6.5.1 and 6.5.2 on this textbook's Online Study Center.

EXERCISES 6.5

A Problem Solving

1. Write an equation that is equivalent to this description: The product of 4 and a number m is the number increased by 18.

2. Given three consecutive, positive, even integers, write an inequality that is equivalent to the statement: The product of the smallest integer—say, $2n$—and the largest integer is always less than the square of the middle integer.

3. The tens digit of a two-digit number is 3 more than the units digit. If the number itself is 26 times the units digit, write an equation to find x, the tens digit.

4. The square of a number x, decreased by twice the number itself, is 10 more than the number. Write an equation that represents this statement.

In problems 5–14, write each statement as an equation, and then solve it.

5. If 4 times a number is increased by 5, the result is 29. Find the number.

6. Eleven more than twice a number is 19. Find the number.

7. The sum of 3 times a number and 8 is 29. Find the number.

8. If 6 is added to 7 times a number, the result is 69. Find the number.

9. If the product of 3 and a number is decreased by 2, the result is 16. Find the number.

10. Five times a certain number is 9 less than twice the number. What is the number?

11. Two times the square of a certain number is the same as twice the number increased by 12. What is the number?

12. If 5 is subtracted from half the square of a number, the result is 1 less than the number itself. Find the number.

13. One-third the square of a number decreased by 2 yields 10. Find the number.

14. One-fifth the square of a certain number, plus 2 times the number, is 15. What is the number?

In problems 15–37, use the **RSTUV** method to obtain the solution.

15. The space shuttle consists of the orbiter, the external tank, two solid-fuel boosters, and fuel. At the time of liftoff, the weight of all these components was 4.16 million lb. The tank and the boosters weighed 1.26 million lb less than the orbiter and fuel. What was the weight of the orbiter and fuel?

16. The external tank and the boosters of the space shuttle weighed 2.9 million lb, and the boosters weighed 0.34 million lb more than the external tank. Find the weight of the tank and of the boosters.

17. Russia and Japan have the greatest number of merchant ships in the world. The combined total is 15,426 ships. If Japan has 2276 ships more than Russia, how many ships does each have?

18. Mary is 12 years old and her brother Joey is 2 years old. In how many years will Mary be just twice as old as Joey?

19. The cost of renting a car is $18 per day plus $0.20 per mile traveled. Margie rented a car and paid $44 at the end of the day. How many miles did Margie travel?

20. In baseball, the slugging average of a player is obtained by dividing his or her total bases (1 for a single, 2 for a double, 3 for a triple, and 4 for a home run) by his or her official number of times at bat. José Cataña has 2 home runs, 1 triple, 2 doubles, and 9 singles. His slugging average is 1.2. How many times has he been at bat?

21. Petěr buys a $10,000, 3-month certificate of deposit. At the end of the 3 months, he receives interest of $350. What was the rate of interest on the certificate?

22. A loan company charges $588 for a 2-year loan of $1400. What simple interest rate is the loan company charging?

23. The cost of renting a sedan for 1 day is $(0.25m + 20)$ dollars, where m is the number of miles traveled.
   a. How many miles could you drive for $71?
   b. If you plan to make a 60-mi round trip, should you use a $40 flat rate or the mileage rate?

24. The Greens need to rent a station wagon to move some furniture. If the round trip distance is 50 mi, should the Greens use a $50 flat rate or a rate of $40 per day plus $0.22 per mile?

25. If P dollars are invested at r percent compounded annually, then at the end of 2 years, the amount will have grown to

$$A = P(1 + r)^2$$

At what rate of interest will $1000 grow to $1210 in 2 years?

26. Use the formula in problem 25 to find at which rate of interest $1000 will grow to $1440 in 2 years.

27. As given in the text, the braking distance b (in feet) for a car traveling v mph is

$$b = 0.06v^2$$

After the driver applied the brakes, a car traveled 54 ft. How fast was the car going when the brakes were applied?

28. The **reaction distance** r (in feet) is the distance a car travels while the driver is moving his or her foot (reacting) to apply the brakes. The formula for this distance is

$$r = 1.5tv$$

where t is the driver's reaction time (in seconds) and v is the speed of the car (in miles per hour). A car going 30 mph travels 22.5 ft while the driver is reacting to apply the brakes. What is the driver's reaction time?

Use the following information in problems 29–34. The stopping distance d (in feet) for a car traveling v mph when the driver has a reaction time of t is given by

$$d = 1.5tv + 0.06v^2$$

29. A car is traveling 30 mph. If the driver's reaction time is 0.5 sec, what is the stopping distance?

30. The reaction time of a driver is 0.5 sec. If the stopping distance is 187.5 ft, how fast is the car going?

31. An automobile going 20 mph is 42 ft away from an intersection when the traffic light turns red. If the automobile stops right at the intersection, what is the driver's reaction time?

32. You may have heard about the reflecting collars that can save the life of a dog or a cat. If your car's headlights will illuminate objects up to 200 ft away and you are driving 50 mph when you see a dog on the road at the edge of the illuminated distance, what must be your reaction time if you can stop just short of hitting the dog?

33. Loren, who has a reaction time of $\frac{2}{3}$ sec, was taking a driving test. When the examiner signaled for a stop, she stopped her car in 44 ft. How fast was she going at the instant of the stop signal?

34. Pedro reacts very quickly. In fact, his reaction time is 0.4 sec. When driving on a highway, Pedro saw a danger signal ahead and tried to stop. If his car traveled 120 ft before stopping, how fast was he going when he saw the signal?

35. Two consecutive integers are such that 6 times the smaller is less than 5 times the larger. Find the largest integers for which this is true.

36. A wallet contains 20 bills, all $1 and $5 bills. If the total value is less than $80, what is the greatest possible number of $5 bills in the wallet? What is the largest total value of money in the wallet?

37. Americans use about 60 billion aluminum cans each year. Approximately two-thirds as many of these cans are either thrown away or left to litter the landscape as are recycled. About how many billion aluminum cans will not be recycled this year?

B **Mathematical Modeling**

In problems 38–48, make models to solve the problems.

38. Referring to Example 7, the office of Institutional Research claims that the break-even point for a 3-credit-hour course is 15 students. This means that other costs O, perhaps utilities, maintenance, and so on, have been added to the costs to the institution.
  a. Find O.
  b. Repeat Example 7 using the costs for your college or university. Assume that the licensing fee and the fee paid to the copyright holder are still $500 and $15 per student, respectively.
  c. Find the amount of other expenses O that your institution must pay.

39. Another and more expensive type of distance learning involves transmitting the contents of a lesson to other sites via satellite. The primary costs to the institution in this case are

  s Studio time, $100 per hour
  U Uplink satellite service, $150 per hour
  G GTE connection fee and satellite rental time, $1000 per hour
  S Salary of instructors, $1000 per credit hour

The revenues for the institution are

  T Tuition (cost per credit hour × number of credit hours × number of students)
  C Contribution of state, $2800 per FTE (1 FTE = 40 credit hours)

If the class carries 3 credit hours, meets 40 hours during the term, and projected enrollment is N students, do or answer the following:
  a. Write an expression for the revenue to the institution.
  b. What is the cost to the institution?
  c. Write the break-even equation for this model.
  d. Solve for N in terms of T.
  e. If the tuition per credit hour is $40, what is the least number of students needed to incur no loss?

40. Should you consider fuel efficiency, measured in miles per gallon, when buying a new car? Suppose that car A costs $20,000 and gets 20 mi/gal, whereas car B costs $25,000 but gets 25 mi/gal. Further, assume that the price of gasoline is $1/gal and that you plan to drive the car for m mi. Taking into account price and fuel efficiency *only,*
  a. write an expression C_A for the cost of car A.
  b. write an expression C_B for the cost of car B.
  c. how many total miles m would you have to drive so that the cost is the same for both cars?
  d. on the basis of this information, how would you make your decision on which car to buy?

41. A paint contractor needs to find short-term (2-month) storage for 5-gal containers of Evermore Stucco Paint at minimum expense. Each container is a right circular cylinder with a diameter of 1 ft and a height of 15 in. All 350 containers must be stored in an upright position. It is not advisable to have more than five containers in each layer (stack). Storage units are available in three sizes as follows:

Size A: 5 × 5 × 4 ft high for $25 a month
Size B: 5 × 10 × 8 ft high for $90 a month
Size C: 10 × 10 × 8 ft high for $128 a month

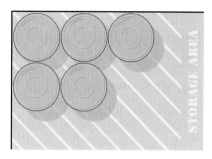

Storage area: Identical row alignment

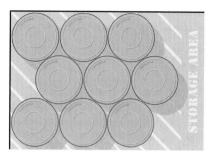

Storage area: Staggered row alignment

Identical Row Alignment

Unit Size	Cost per Month	Number per Row	Rows	Layers	Total	Units Needed	Cost for 2 Months
A							
B							
C							

Staggered Row Alignment

Unit Size	Cost per Month	Number per Row	Rows	Layers	Total	Units Needed	Cost for 2 Months
A							
B							
C							

a. How many containers will fit into each storage area? [*Hint:* How many containers can you put in each row; how many rows and how many layers can you have? Note that you can store the containers in identical rows as shown in the diagram above on the left, or staggered as shown in the diagram above on the right. To answer (**a**), complete the tables above.]

b. What is the least expensive way to store the containers?

c. What is the next-thriftiest choice?

d. If the containers could be layered (stacked) as high as you wished, what would be the least expensive way to store the containers?

42. Is there a relationship between the amount of time you study and your grade-point average (GPA)? Of course there is! This problem will show you how to maximize GPA on the basis of study time.

Suppose that you are taking mathematics, science, and English. Each of the classes is worth 3 credit hours. At the present time you have C's in all classes, but you know the following:

Your mathematics grade can be improved by one letter for each additional 3 hours per week of studying mathematics.

Your science grade can be improved by one letter for each additional 2 hours per week of studying science.

Your English grade can be improved by one letter for each additional 1 hours per week of studying English.

If you decide to spend 10 extra hours per week studying mathematics, science, and English, how many hours should you spend studying each of the three subjects to maximize your GPA? [*Hint:* Organize the possibilities using a table like the following; then find the resulting GPA for each strategy. The GPA for each course is calculated by multiplying the number of points for each grade (A = 4, B = 3, C = 2, D = 1, F = 0) times the number of credits the course carries. For example, a C in a 3-credit course is worth 6 grade points ($2 \cdot 3 = 6$), whereas a B in a 4-credit course is worth 12 grade points ($3 \cdot 4 = 12$). The GPA is then found by adding the total grade-point values for all courses and dividing by the total number of credit hours.]

Hours Spent and Grade Earned			
Mathematics	*Science*	*English*	*GPA*
6(A)	4(A)	0(C)	$\frac{30}{9} = 3.33$

43. Referring to problem 42, suppose you have D's in all your classes, but your mathematics and science courses are 4-credit-hour courses. If you decide to spend 12 extra hours per week studying mathematics, science, and English, how many hours should you spend studying each of the three subjects to maximize your GPA?

44. How often do you rent movies? How much is the charge for returning them late? Rock-Busters charges $3 to rent a movie as long as you return it by 2 P.M. the second day after you check it out. If you return it late, however, you are charged an additional $2.50 per overdue day for new releases and $1.50 per overdue day for other movies. Video Renters charges $2.50 per movie as long as the movie is returned by 3 P.M. of the second day after you check it out. If you return it late, you are charged $2 for each day the movie is overdue, regardless of whether the movie is a new release.
   a. Make a table for new and old movies overdue 1, 2, and 3 days at each store.
   b. Write an algebraic model representing the total cost of having both a new release and an old movie overdue for *n* days at Rock-Busters; then do the same for Video Renters.
   c. How much are the total charges when you return a new release and an old movie 5 days overdue?

45. At Blockbuster in Tampa, it costs $3 to rent a movie. New releases are due the next day at midnight, but you can keep old movies for 2 days without incurring additional charges. Overdue charges are $2 per movie per day. At Red Rabbit, new releases are $3 per day and they must be returned by 7 P.M. the next day. Overdue charges are $3 per movie per additional day. Old movies are $1.60 for 2 days, and overdue charges are $1.50 per movie per day.
   a. Make a table for new and old movies returned after 2, 3, and 4 days at each store.
   b. Construct an algebraic model representing the total cost of having a new movie and an old movie returned after *n* days at Blockbuster; then do the same for Red Rabbit.
   c. At each of these stores, what are the total charges when you have a new movie and an old movie returned after 5 days?

46. On Saturdays, the cost of parking at Park and Shop is $5 for the first hour and $.50 for each additional hour. Safe Park charges $4.75 for the first hour and $.60 for each additional hour.
   a. Make a table to compare the costs for 1, 2, 3, 4, and 5 hours of parking at each location.
   b. Write an algebraic expression representing the cost of parking *n* hours at each location.

47. When traffic lights are installed, planning must include the duration of the green and the yellow lights. How can we determine the duration of the yellow light assuming that the intersection is 80 ft wide and the speed limit is 45 mph?
   a. The stopping distance *d* (in feet) for a car traveling *v* mph when the driver has a reaction time of *t* sec is

   $$d = 1.5tv + 0.06v^2$$

   Reaction times vary from 0.3 to 1 sec, but the average is about 0.6 sec. With this additional information, what is *d*?
   b. On the basis of the answer to part (**a**), what is the nearest a car traveling 45 mph can be to the intersection and still stop safely?
   c. The total maximum distance a car must travel through the intersection is the width of the intersection plus the distance from the intersection. What is this distance?

d. On the basis of the previous information, what should the minimum duration for the yellow light be? Answer to the nearest hundredth of a second. (*Hint: d = rt*, where *r* is rate, but watch your units!)

48. If the reaction time of the driver is 1 sec, what should the minimum duration for the yellow light be?

In Other Words

49. When reading a word problem, what is the first thing you should try to determine?

50. How can you verify a word-problem answer?

Using Your Knowledge

Have you ever heard of *Chamberlain's formula,* which claims to be a model that tells you how many years you should drive your present car before you buy a new one? If *y* is this number of years, then Chamberlain's formula reads

$$y = \frac{GMC}{(G - M)DP}$$

where *G* is the new car's gas mileage, *M* is your present car's gas mileage, *C* is the cost in dollars of the new car, *D* is the number of miles you drive in a year, and *P* is the dollar price of gasoline per gallon.

For instance, suppose that the new car's gas mileage is 24 mi/gal, the old car's mileage is 12 mi/gal, the price of the new car is $12,800, you drive 24,000 mi per year, and the cost of gasoline is $1.40/gal. This means that $G = 24, M = 12, C = 12,800, D = 24,000$, and $P = 1.40$, so

$$y = \frac{(24)(12)(12,800)}{(12)(24,000)(1.4)} = \frac{64}{7} \approx 9.1$$

Thus, according to the formula, your old car should have been driven about 9.1 years before the purchase is justified.

Now try these problems.

51. Suppose the new car's gas mileage is 32 mi/gal, the old car's gas mileage is 10 mi/gal, the price of the new car is $14,000, you drive 30,000 mi/year, and the cost of gasoline is $1.30/gal. How many years should your old car have been driven to justify buying the new one?

52. Suppose that *G, M, D,* and *P* have the same values as in problem 51 and you have driven your present car for 5 years. What price (to the nearest $10) would you be justified in paying for the new car?

53. Disregarding such factors as depreciation, the decrease in value of the car over time; maintenance; replacement cost; and so on, under which conditions does Chamberlain's formula yield unrealistic values for *y*?

6.6   Ratio, Proportion, and Variation

Junk Food and Ratios

You have probably heard the term *junk food,* but do you know what foods fall into this category? The U.S. Department of Agriculture suggests looking at the *ratio* of nutrients to calories and multiplying the result by 100. The procedure adds the RDAs (recommended daily allowances) for the first eight nutrients listed on the label, divides by the number of calories in one serving, and multiplies by 100.

For instance, the sum of the first eight nutrients for the cereal label shown in Figure 6.22 is 139, and the number of calories in one serving is 100. Thus, the ratio of nutrients to calories is

$$\frac{139}{100} \times 100 = 139$$

If the answer is less than 32, the nutritional value of the food is questionable.

NUTRITION INFORMATION		
SERVING SIZE: 1 OZ (28.4 g, ABOUT 4 BISCUITS)		
SERVINGS PER PACKAGE:		20
	CEREAL	WITH $\frac{1}{2}$ CUP VITAMINS A & D SKIM MILK
CALORIES	**100**	140*
PROTEIN	3 g	7 g
CARBOHYDRATE	24 g	30 g
FAT	0 g	0 g*
CHOLESTEROL	0 mg	0 mg*
SODIUM	5 mg	65 mg
POTASSIUM	80 mg	280 mg

*WHOLE MILK SUPPLIES AN ADDITIONAL 30 CALORIES, 4 g FAT, AND 15 mg CHOLESTEROL.

PERCENTAGE OF U.S. RECOMMENDED DAILY ALLOWANCES (U.S. RDA)		
PROTEIN	4	15
VITAMIN A	25	30
VITAMIN C	25	25
THIAMIN	25	30
RIBOFLAVIN	25	35
NIACIN	25	25
CALCIUM	**	15
IRON	10	10
VITAMIN D	10	25
VITAMIN B_6	25	25
FOLIC ACID	25	25
PHOSPHORUS	8	20
MAGNESIUM	8	10
ZINC	10	15
COPPER	4	6

**CONTAINS LESS THAN 2% OF THE U.S. RDA OF THIS NUTRIENT.

FIGURE 6.22

In this section you will study many kinds of ratios and their uses. For example, you will study unit pricing (see Example 2) to make you more aware as a consumer. Then you will learn how to compare ratios by studying proportions, and you will end the section by looking at different types of variation. ▶

A. Ratios

Many quantities can be compared by using ratios. You probably have heard of the student-teacher ratio or the gear ratio in your automobile. Below is the definition of a ratio.

What is the ratio of paper towels to ice cream pints you can buy with the same amount of money?

Definition of a Ratio

A **ratio** is a quotient of two numbers. The ratio of a number a to another number b can be written as

$$a \text{ to } b \quad \textbf{or} \quad a{:}b \quad \textbf{or} \quad \frac{a}{b}$$

The last form is used most often, but is frequently written as a/b.

EXAMPLE 1 ▶ Ratios of Nutrients

Write the ratio of nutrients (139) to calories (100) in three different ways.

Solution

$$139 \text{ to } 100 \qquad 139{:}100 \qquad \frac{139}{100}$$

Of course, if the ratio in Example 1 had been 140 to 100, you could write it in reduced form as

$$7 \text{ to } 5 \quad \text{or} \quad 7{:}5 \quad \text{or} \quad \frac{7}{5}$$

You encounter many ratios in your everyday life. For example, the expression *miles per gallon* is a ratio, the ratio of the number of miles traveled to the number of gallons of gas used. Thus, if your car travels 294 mi on 12 gal of gas, your miles per gallon ratio is $\frac{294}{12} = \frac{49}{2} = 24.5$.

Ratios can also be used to compare prices. For example, most people have the misconception that the more you buy of an item, the cheaper it is. Is this always true? Not necessarily. The photo shows two cans of Hunt's tomato sauce bought in the same store. The 15-oz can costs 68¢, whereas the 8-oz can costs 32¢. Which can is the better buy? To compare these prices, we need to find the price of 1 oz of tomato sauce; that is, we need the **unit price** (the price per ounce). This unit price is given by the ratio

Unit pricing allows shoppers to quickly compare relative costs of the same or similar products when they are sold in different weights or volumes.

$$\frac{\text{Price}}{\text{Number of ounces}}$$

For the 15-oz can   For the 8-oz can

$\frac{68}{15} = 4.5\overline{3}$ $\frac{32}{8} = 4$

Thus, the 15-oz can is more expensive. (Note that you could buy 16 oz, two 8-oz cans, for only 64¢, instead of the 15-oz can for 68¢!)

EXAMPLE 2 ▶ Unit Costs and Ratios

A 4-oz can of mushrooms costs $.74, whereas an 8-oz can of the same brand costs $1.44. Find the cost per ounce for

(a) the 4-oz can.   (b) the 8-oz can.

(Round answers to the nearest tenth of a cent.)

Solution

(a) For the 4-oz can, the cost per ounce is

$$\frac{74}{4} = 18.5¢$$

(b) For the 8-oz can, the cost per ounce is

$$\frac{144}{8} = 18¢ \qquad\qquad ■$$

Note that if we wish to find which of the two cans in Example 2 is the better buy, we do not have to look at the unit prices. Two 4-oz cans would cost us $1.48 for 8 oz, whereas the 8-oz can costs only $1.44.

B. Proportions

Let us return to the problem of the car that traveled 294 mi on 12 gal of gas. How many miles would this car travel on 10 gal of gas? The ratio of miles per gallon for the car is $\frac{294}{12}$, or $\frac{49}{2}$. If we let m be the number of miles the car would travel on

10 gal of gas, the ratio of miles per gallon would be $m/10$. Since the two ratios must be equal,

$$\frac{49}{2} = \frac{m}{10}$$

This equation is an equality between two ratios; it is called a **proportion.**

Definition of Proportion

A **proportion** is an equality between two ratios that can be written as

$$\frac{a}{b} = \frac{c}{d}$$

To solve the proportion

$$\frac{49}{2} = \frac{m}{10}$$

which is simply an equation involving fractions, we proceed as before.

1. Multiply both sides by 10 so that the m is by itself on the right.

$$\frac{49}{2} \cdot 10 = \frac{m}{10} \cdot 10$$

2. Simplify: $245 = m$

3. Thus, the car can go 245 mi on 10 gal of gas.

Proportions often can be solved by a shortcut method that depends on the following definition of equality of fractions:

Equality of Fractions

If $\quad \dfrac{a}{b} = \dfrac{c}{d},\quad$ then $\quad ad = bc$

Thus, to solve the proportion

$$\frac{3}{4} = \frac{15}{x}$$

we use the cross-products and write

$$3 \cdot x = 15 \cdot 4$$
$$3x = 60$$
$$x = 20$$

EXAMPLE 3 ▶ Proportions and Foot Lengths

The ratio of your foot length to your height (in inches) is 1 to 6.7. In 1951, Eric Shipton published photographs of what he thought were the Abominable Snowman's footprints. Each footprint was 23 in. long. How tall is the Abominable Snowman?

Solution

We use the five-step procedure outlined in Section 6.5 as follows:

1. Read the problem carefully.

2. Select a variable to represent the unknown. Here, we let h be the height of the Abominable Snowman.

3. Translate the problem. Since the original ratio is 1/6.7, the ratio for the Snowman is 23/h. But these ratios must be equal, so

$$\frac{1}{6.7} = \frac{23}{h}$$

4. Use cross-products to solve the equation.

$$1 \cdot h = 6.7 \cdot 23$$
$$h = 154.1 \text{ in.} = 12 \text{ ft } 10.1 \text{ in.}$$

5. Verify your answer. If we substitute $h = 154.1$ in the original proportion, we obtain

$$\frac{1}{6.7} = \frac{23}{154.1}$$

which is a true statement because $1 \cdot 154.1 = 6.7 \cdot 23$.   ■

C. Variation

Sometimes we say that a variable is **proportional to** or **varies directly as** another variable x. For example, the number m of miles you drive a car is proportional to, or varies directly as, the number g of gallons of gas used. This means that the ratio m/g is a constant, the miles per gallon the car attains. In general, we have the following definition:

Direct Variation

y **varies directly as** x if there is a constant k such that $y = kx$. The constant k is the constant of variation or proportionality.

In real life, we can find k by experimenting. Suppose you are dieting and want to eat McDonald's fries, but only 100 calories' worth. The fries come in small, medium, or large sizes, with 210, 450, or 540 calories, respectively. (Source: www.shapefit.com/mcdonalds.html.) You can use *estimation* (Chapter 1) and eat approximately 1/2 of a small order ($210/2 \approx 105$ calories), but you can estimate better! The number of fries in the small, medium, and large sizes are 45, 97, and 121—thus, the calories per fry are

$$\frac{210}{45} \approx 4.7 \text{ (small)}, \qquad \frac{450}{97} \approx 4.6 \text{ (medium)}, \qquad \frac{540}{121} \approx 4.5 \text{ (large)}$$

EXAMPLE 4 ▶ **You Want Fries with That?**

The calories C in a bag of McDonald's French fries vary directly as the number n of fries in the bag. Suppose you buy a large bag of fries (121 fries, 540 calories). (Source: www.shapefit.com/mcdonalds.html.)

(a) Write an equation of variation between C and n.

(b) Find k.

(c) If you want to eat 100 calories' worth, how many fries would you eat?

Solution

(a) If C varies directly as n, $C = kn$.

(b) A large bag of fries has $C = 540$ calories and $n = 121$ fries, thus

$$C = kn \qquad \text{becomes} \qquad 540 = k(121)$$

or

$$k = \frac{540}{121} \approx 4.5$$

(c) If you want to eat $C = 100$ calories' worth, then

$$C = kn = 4.5n \qquad \text{becomes} \qquad 100 = 4.5n$$

Solving for n by dividing both sides by 4.5, $n = \frac{100}{4.5} \approx 22$.

So, if you eat about 22 fries, you will have eaten about 100 calories. Note that this is indeed about one half of a small bag! ∎

Now that we have eaten some calories, let us see how we can spend (burn) some calories.

EXAMPLE 5 ▶ **Variation and Bicycling**

If you weigh about 160 lb and you jog (5 mph) or ride a bicycle (12 mph), you use C calories, which are proportional to the time t (in minutes) that you jog or ride.

(a) Find an equation of variation between C and t.

(b) If jogging for 15 min burns 150 calories, find k.

(c) How many calories would you burn if you jogged for 20 min?

(d) To lose 1 lb, you have to burn about 3500 calories. How many minutes do you have to bicycle in order to lose 1 lb?

Solution

(a) Since the calories used C are proportional to the time t (in minutes), an equation of variation is $C = kt$.

(b) If jogging for 15 min ($t = 15$) uses $C = 150$ calories,

$C = kt$ becomes $\qquad\qquad 150 = k(15)$.
Dividing both sides by 15, $\qquad\qquad 10 = k$
Note that now $C = kt$ becomes $\qquad\qquad C = 10t$

(c) We want to know how many calories C you would use if you jogged for $t = 20$. Substitute 20 for t in $C = 10t$, obtaining

$$C = 10(20) = 200$$

Thus, you will use 200 calories if you jog for 20 min.

(d) To lose 1 lb, you need to use $C = 3500$ calories.

$C = 10t$ becomes $\qquad\qquad 3500 = 10t$
Dividing both sides by 10, $\qquad\quad 350 = t$

Thus, you need 350 min of bicycling to lose 1 lb! By the way, if you bicycle for 350 min at 12 mph, you will be 70 mi away! ∎

Sometimes, as a quantity increases, a related quantity decreases proportionately. For example, the more time we spend practicing a task, the less time it will take us to do the task. In this case, we say that the quantities **vary inversely as** each other.

Inverse Variation

y **varies inversely as** x if there is a constant k such that

$$y = \frac{k}{x}$$

EXAMPLE 6 ▶ Applications with Speed and Distance

The rate of speed r at which a car travels is inversely proportional to the time t it takes to travel a given distance.

(a) Write an equation of variation.

(b) If a car travels at 60 mph for 3 hr, what is k, and what does k represent?

Solution

(a) The equation is

$$r = \frac{k}{t}$$

(b) We know that $r = 60$ when $t = 3$. Thus,

$$60 = \frac{k}{3}$$
$$k = 180$$

In this case, k represents the distance traveled, and the new equation of variation is $r = \dfrac{180}{t}$. ∎

EXAMPLE 7 ▶ Boom Boxes and Variation

Have you ever heard one of those loud "boom" boxes or a car sound system that makes your stomach tremble? The loudness L of sound is inversely proportional to the square of the distance d that you are from the source.

(a) Write an equation of variation.

(b) The loudness of rap music coming from a boom box 5 ft away is 100 dB (decibels). Find k.

(c) If you move to 10 ft away from the boom box, how loud is the sound?

Solution

(a) The equation is

$$L = \frac{k}{d^2}$$

(b) We know that $L = 100$ for $d = 5$, so

$$100 = \frac{k}{5^2} = \frac{k}{25}$$

Multiplying both sides by 25, we find that $k = 2500$, and the new equation of variation is

$$L = \frac{2500}{d^2}$$

(c) Since $k = 2500$,

$$L = \frac{2500}{d^2}$$

When $d = 10$,

$$L = \frac{2500}{10^2} = 25 \text{ dB}$$ ∎

D. Application

EXAMPLE 8 ▶ Melting Snow

Figure 6.23 shows the number of gallons of water g (in millions) produced by an inch of snow in different cities. Note that the larger the area of the city, the more gallons of water are produced, so g is directly proportional to A, the area of the city (in square miles).

FIGURE 6.23
Source: Adapted from *USA Today*. Copyright 1993.

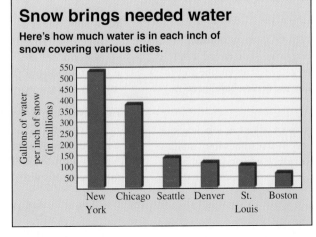

Snow brings needed water

Here's how much water is in each inch of snow covering various cities.

(a) Write an equation of variation.

(b) If the area of St. Louis is about 62 mi², what is k?

(c) Find the amount of water produced by 1 in. of snow falling in Anchorage, Alaska, with an area of 1700 mi².

Solution

a. Since g is directly proportional to A, $g = kA$.

b. From Figure 6.23, we can see that $g = 100$ (million) is the number of gallons of water produced by 1 in. of snow in St. Louis. Since it is given that $A = 62$, $g = kA$ becomes

$$100 = k \cdot 62 \quad \text{or} \quad k = \frac{100}{62} = \frac{50}{31}$$

c. For Anchorage, $A = 1700$; thus $g = \frac{50}{31} \cdot 1700 \approx 2742$ million gal of water. ∎

Online Study Center

To futher explore ratio, proportion, and variation, access link 6.6.1 on this textbook's Online Study Center.

EXERCISES 6.6

Ⓐ Ratios

1. *Voyager* was the first plane to fly nonstop around the world without refueling. At the beginning of the trip, the fuel in the 2000-lb plane weighed 7000 lb. Write the ratio of fuel to plane weight in three different ways.

2. During the last six years of his life, Vincent Van Gogh produced 700 drawings and 800 oil paintings. Write the ratio of drawings to oil paintings in three different ways.

3. The first suspension bridge built in England has a 70-ft span. The world's longest suspension bridge, the Verrazano Narrows Bridge in New York, is 4260 ft long. Write the ratio of the span of the first suspension bridge to that of the Verrazano Narrows Bridge in three different ways.

4. A woman has a 28-in. waist and 34-in. hips. Write the waist-to-hip ratio in three different ways. (If the waist-to-hip ratio is over 1.0 for men or over 0.8 for women, the risk of heart attack or stroke is five to ten times greater than for persons with a lower ratio.)

5. The reduced transmission ratio in an automobile is the ratio of engine speed to drive-shaft speed. Find the reduced transmission ratio for an engine running at 2000 rev/min when the drive-shaft speed is 600 rev/min.

6. Most job seekers can expect about 6 job leads and/or interviews for every 100 resumes they mail out. What is the ratio of job leads and/or interviews to resumes?

7. The average U.S. car is driven 12,000 mi per year and burns 700 gal of gas. How many miles per gallon does the average U.S. car get? (Give the answer to the nearest whole number.)

8. Are generic products always cheaper than name brands? Not necessarily! It depends on where you buy the products. When comparing two cans of mushrooms, the name brand 8-oz can costs $1.09 and the generic 4-oz can costs 53¢.
   a. To the nearest tenth of a cent, what is the cost per ounce of the name brand 8-oz can?
   b. To the nearest tenth of a cent, what is the cost per ounce of the generic 4-oz can?
   c. Which can is the better buy?
   d. In Example 2, the 4-oz can cost 18.5¢ per ounce. Which is the better buy, the generic can here or the 4-oz can of Example 2?

9. As a consumer, you probably believe that cheaper is always better, but be careful. Here is a situation that should give you food for thought!
   a. Dermassage dishwashing liquid costs $1.31 for 22 oz. To the nearest cent, what is the cost per ounce?
   b. White Magic dishwashing liquid costs $1.75 for 32 oz. To the nearest cent, what is the cost per ounce?
   c. On the basis of price alone, which is the better buy, Dermassage or White Magic?

But how much do you use per wash? *Consumer Reports* estimated that it costs 10¢ for 10 washes with Dermassage and 18¢ for the same number of washes with White Magic. Thus, Dermassage is more economical.

10. Is cheaper still better? Here is another problem. A&P Wool Washing Liquid costs $.79 for 16 oz. Ivory Liquid is $1.25 for 22 oz.
   a. To the nearest cent, what is the price per ounce of the A&P Wool Washing Liquid?
   b. To the nearest cent, what is the price per ounce of the Ivory Liquid?
   c. On the basis of price alone, which is the better buy?
   But wait. How much do you have to use? According to *Consumer Reports*, it costs 17¢ for 10 washes with the A&P Wool Washing Liquid, but Ivory Liquid is only 12¢ for 10 washes!

B **Proportions**

In problems 11–16, solve each proportion.

11. $\dfrac{x}{9} = \dfrac{4}{3}$ **12.** $\dfrac{x}{6} = \dfrac{5}{12}$ **13.** $\dfrac{8}{x} = \dfrac{4}{3}$

14. $\dfrac{6}{x} = \dfrac{18}{7}$ **15.** $\dfrac{3}{8} = \dfrac{9}{x}$ **16.** $\dfrac{3}{5} = \dfrac{9}{x}$

17. A machine manufactures 9 toys every 2 hours. Let n be the number of toys it produces in 40 hours. Write a proportion using this information and the ratio of n to 40.

18. A woman has several rectangular flower beds that she wants to be of equal area. If L ft is the length of a bed that is 3 ft wide, and the length is 10 ft when the width is 2 ft, write a proportion using the ratio of L to 10.

19. When flying a hot-air balloon, you get $\frac{1}{2}$ hour of flight time for each 20-lb tank of propane gas. How many tanks of gas do you need for a 3-hour flight?

20. When serving shrimp, you need $\frac{1}{2}$ lb of cooked shrimp without the shell to make 3 servings. How many pounds of cooked shrimp do you need for 90 servings?

21. The official ratio of width to length for the U.S. flag is 10 to 19. If a flag is 35 in. wide, how long should it be?

22. Do you like tortillas? Tom Nall does. As a matter of fact, he ate 74 tortillas in 30 min! How many could he eat in 45 min at that rate?

23. A certain pitcher allowed 10 runs in the last 32 innings he pitched. At this rate, how many runs would he allow in 9 innings? (The answer is called the *ERA*, or *earned run average*, for the pitcher and is usually rounded to two decimal places.)

24. Do you know what a xerus is? It is a small rodent that looks like a cross between a squirrel and a chipmunk. The ratio of tail to body length in one of these animals is 4 to 5. If the body of a xerus is 10 in. long, how long is its tail?

25. A zoologist tagged and released 250 fish into a lake. A few days later, 53 fish were taken at random locations from the lake and 5 of them were found to be tagged. Approximately how many fish are there in the lake?

C **Variation**

26. The amount of annual interest I received on a savings account is directly proportional to the amount of money m you have in the account.
   a. Write an equation of variation.
   b. If $480 produced $26.40 in interest, what is k?
   c. How much annual interest would you receive if the account had $750?

27. The number of revolutions R a record makes as it is being played varies directly with the time t that it is on the turntable.
   a. Write an equation of variation.
   b. A record that lasted $2\frac{1}{2}$ min made 112.5 revolutions. What is k?
   c. If a record makes 108 revolutions, how long does it take to play it?

28. The distance d an automobile travels after the brakes have been applied varies directly as the square of its speed s.
   a. Write an equation of variation.
   b. If the stopping distance for a car going 30 mph is 54 ft, what is k?
   c. What is the stopping distance for a car going 60 mph?

29. The weight of a person varies directly as the cube of the person's height h (in inches). The **threshold weight** T (in pounds) for a person is defined as the "crucial weight, above which the mortality risk for the patient rises astronomically."
   a. Write an equation of variation relating T and h.
   b. If $T = 196$ when $h = 70$, find k.
   c. To the nearest pound, what is the threshold weight for a person 75 in. tall?

30. To remain popular, the number S of new songs a rock band needs to produce each year is inversely proportional to the number y of years the band has been in the business.
   a. Write an equation of variation.

b. If, after 3 years in the business, the band needs 50 new songs, how many songs will it need after 5 years?

31. When a camera lens is focused at infinity, the *f*-stop on the lens varies inversely with the diameter *d* of the aperture (opening).
  a. Write an equation of variation.
  b. If the *f*-stop on a camera is 8 when the aperture is $\frac{1}{2}$ in., what is *k*?
  c. Find the *f*-stop when the aperture is $\frac{1}{4}$ in.

32. Boyle's law states that if the temperature is held constant, then the pressure *P* of an enclosed gas varies inversely as the volume *V*. If the pressure of the gas is 24 lb/in.^{2} when the volume is 18 in.^{3}, what is the pressure if the gas is compressed to 12 in.^{3}?

33. For the gas of problem 32, if the pressure is 24 lb/in.^{2} when the volume is 18 in.^{3}, what is the volume if the pressure is increased to 40 lb/in.^{2}?

34. The weight *W* of an object varies inversely as the square of its distance *d* from the center of the Earth.
  a. Write an equation of variation.
  b. An astronaut weighs 121 lb on the surface of the Earth. If the radius of the Earth is 3960 mi, find the value of *k* for this astronaut. (Do not multiply out your answer.)
  c. What will this astronaut weigh when she is 880 mi above the surface of the Earth?

35. One of the manuscript pages of this book had about 600 words and was typed using a 12-point font. If the average number of words *w* that can be printed on a manuscript page is inversely proportional to the font size *s*,
  a. write an equation of variation.
  b. what is *k*?
  c. how many words could be typed on the page if a 10-point font were used?

36. The price *P* of oil varies inversely with the supply *S* (in million barrels per day). In the year 2000, the price of one barrel of oil was $26.00 and OPEC production was 24 million barrels per day.
  a. Write an equation of variation.
  b. What is *k*?
  c. In 2005, OPEC plans to increase production to 28 millions barrels per day. What would the price of one barrel be? Answer to the nearest cent.

37. According to the National Center for Health Statistics, the number of births *b* (per 1000 women)

is inversely proportional to the age *a* of the woman. The number *b* of births (per 1000 women) for 27-year-olds is 110.
  a. Write an equation of variation.
  b. What is *k*?
  c. What would you expect the number *b* of births (per 1000 women) to be for 33-year-old women?

38. The number of miles *m* you can drive in your car is directly proportional to the amount of fuel *g* in your gas tank.
  a. Write an equation of variation.
  b. The greatest distance yet driven without refueling on a single fill in a standard vehicle is 1691.6 miles. If the twin tanks used to do this carried a total of 38.2 gal of fuel, what is *k*?
  c. How many miles per gallon is this?

39. The distance *d* (in miles) traveled by a car is directly proportional to the average speed *s* (in miles per hour) of the car, even when driving in reverse!
  a. Write an equation of variation.
  b. The highest average speed attained in any non-stop reverse drive of more than 500 mi is 28.41 mph. If the distance traveled was 501 mi, find *k*.
  c. What does *k* represent?

40. Have you called in on a radio contest lately? According to Don Burley, a radio talk-show host in Kansas City, the listener response to a radio call-in contest is directly proportional to the size of the prize.
  a. If 40 listeners call when the prize is $100, write an equation of variation using *N* for the number of listeners and *P* for the prize in dollars.
  b. How many calls would you expect for a $5000 prize?

41. You would think that your blood alcohol content (BAC) would be directly proportional to how many beers you drink in an hour. Strangely enough, for both males and females of a specific weight, the BAC is directly proportional to the number $(N - 1)$ of beers they drank during the last hour.
  a. Write an equation of variation between BAC and the number of beers he drank during the last hour.
  b. For a 150-lb average man, his BAC after 3 beers is 0.052. Find *k*.
  c. What is his BAC after 5 beers?
  d. How many beers can the man drink before going over the 0.08 BAC limit?

42. For a 130-lb average female, her BAC is still directly proportional to the number $(N - 1)$ of beers she had during the last hour.
  a. Write an equation of variation between BAC and the number of beers she drank during the last hour.
  b. For a 130-lb average female, her BAC after 3 beers is 0.06. Find k.
  c. What is her BAC after 5 beers?
  d. How many beers can the woman drink before going over the 0.08 BAC limit?

43. The number of chirps C a cricket makes each minute is directly proportional to 37 less than the temperature F in degrees Fahrenheit.
  a. If a cricket chirps 80 times when the temperature is 57°F, what is the equation of variation?
  b. How many chirps per minute would the cricket make when the temperature is 90°F?

44. According to George Flick, the ship's surgeon of the SS *Constitution*, the number of hours H your life is shortened by smoking cigarettes varies jointly as N and $t + 10$, where N is the number of cigarettes you smoke and t is the time in minutes it takes you to smoke each cigarette. If it takes 5 min to smoke a cigarette and smoking 100 of them shortens your life span by 25 hours, how long would smoking 2 packs a day for a year (360 days) shorten your life span? (*Note:* There are 20 cigarettes in a pack.)

45. The simple interest I in an account varies jointly as the time t (in years) and the principal P (in dollars). After one quarter (3 months), an $8000 principal earned $100 in interest. How much would a $10,000 principal earn in 5 months?

46. At depths of more than 1000 m (a kilometer), water temperature T (in degrees Celsius) in the Pacific Ocean varies inversely as the water depth d (in meters). If the water temperature at 4000 m is 1°C, what would it be at 8000 m?

47. Anthropologists use the cephalic index C in the study of human races and groupings. This index is directly proportional to the width w and inversely proportional to the length L of the head. The width of the head in a skull found in 1921 and named Rhodesian man was 15 cm, and its length was 21 cm. If the cephalic index of Rhodesian man was 98, what would the cephalic index of Cro-Magnon man be, whose head was 20 cm long and 15 cm wide?

48. BAC is inversely proportional to weight W. For a 130-lb male, his BAC after drinking 3 beers in the last hour is 0.06.
  a. Write an equation of variation between BAC and the weight of the person drinking 3 beers in the last hour.
  b. What is the BAC of a 260-lb male after drinking 3 beers in the last hour?
  c. In most states, you are legally drunk if your BAC is 0.08. What is the weight of a male whose BAC is exactly 0.08 after drinking 3 beers in the last hour?
  d. What would your BAC be if you weighed more than the male in part (**c**)?

49. BAC is inversely proportional to weight W. For a 130-lb female, her BAC after drinking 3 beers in the last hour is 0.066.
  a. Write an equation of variation between BAC and the weight of a person drinking 3 beers in the last hour.
  b. What is the BAC of a 260-lb female after drinking 3 beers in the last hour?
  c. In most states, you are legally drunk if your BAC is 0.08. What is the weight of a female whose BAC is exactly 0.08 after drinking 3 beers in the last hour?
  d. What would your BAC be if you weighed more than the female in part (**c**)?

In Other Words

50. Explain the difference between a ratio and a proportion.

51. Explain the difference between direct variation and inverse variation.

52. Write the steps you use to solve a proportion.

53. What does it mean when we say that the proportion $a/b = c/d$ can be solved using cross-products?

54. Can you solve $x/2 + x/4 = 3$ using cross-products? Explain.

55. What does it mean when we say that "the number of miles you drive a car is proportional to the number of gallons of gas the car uses"?

56. What does it mean when we say that "the number of gallons of gas your car uses is proportional to the speed at which you drive the car"? Is it directly proportional or inversely proportional?

Collaborative Learning

Health professionals use many ratios to detect abnormalities. Form three different groups: Library, Internet, and Other. What is the leading cause of death in the United States? Heart disease! Let each of the groups find the answers to the following questions:

1. How many deaths a year are attributed to heart disease? What is the total number of annual deaths? What is the ratio of deaths attributed to heart disease to the total number of deaths? Do the answers of the groups agree? Why or why not?

There are several ratios associated with heart disease. Let us look at some of them.

2. Let each of the groups find out what the HDL/cholesterol ratio measures. What is the recommended value of this ratio?

3. There is another noninvasive way to measure your cardiac health (see problem 4 of Exercises 6.6). This time, divide the groups into males and females. Complete the following table:

	Waist (w)	Hips (h)	Ratio w/h
Women			
Men			

As you recall, for women, the risk of heart disease increases when $w/h > 0.8$, for men, when $w/h > 1$.

Chapter 6 Summary

Section	Item	Meaning	Example
6.1	Variable	A symbol that may be replaced by any one of a set of numbers	x, y, z
6.1A	Open sentence	A sentence in which the variable can be replaced by a number	$x + 3 = 5; x - 1 < 7$
6.1A	Equation	A sentence in which the verb is $=$	$x + 7 = 9$
6.1A	Inequality	A sentence in which the verb is $>, <, \neq, \geq,$ or $\leq$	$x + 7 < 9; x > 8; x \neq 9$
6.1A	Solution set	The set of elements of the replacement set that make the open sentence a true statement	$\{3\}$ is the solution set of $x + 2 = 5$ when the replacement set is the set of whole numbers.
6.1B	First-degree equation	An equation that can be written in the form $ax + b = 0; a \neq 0$. The unknown quantity x (or m) has an exponent of 1 only.	$x + 7 = 8 - 2x; 20 = 3m$
6.1B	Elementary operations	Operations that can be performed on a sentence to obtain an equivalent sentence	Addition (or subtraction) of the same number on both sides of an equation
6.1C	First-degree inequality	An inequality that can be written in the form $ax + b < 0$ or $ax + b > 0$ ($>$ can be replaced by $\geq$ and $<$ by $\leq$ and we still have a first-degree inequality.)	$3x + 5 < 8; 5x - 2 \geq 3x + 1$

Section	Item	Meaning	Example
6.2A	Finite intervals: Closed interval	$a \le x \le b$	$[a, b]$
	Open interval	$a < x < b$	(a, b)
	Half-open intervals	$\begin{cases} a \le x < b \\ \\ a < x \le b \end{cases}$	$[a, b)$ $(a, b]$
6.3A	Absolute value	The distance on the number line from 0 to the number	$\|3\| = 3; \|-7\| = 7; \left\|\frac{-2}{3}\right\| = \frac{2}{3}$
6.4	Quadratic equation	A second-degree sentence that can be written in the form $ax^2 + bx + c = 0, a \ne 0$	$x^2 - 7x = 6; 8x^2 - 3x - 4 = 0$
6.4C	Quadratic formula	For the equation $ax^2 + bx + c = 0,$ $$x = \frac{-b \pm \sqrt{b^2 - 4ac}}{2a}$$	For the equation $x^2 - x - 1 = 0,$ $$x = \frac{1 \pm \sqrt{1 - 4(1)(-1)}}{2}$$ $$= \frac{1 \pm \sqrt{5}}{2}$$
6.4D	Pythagorean theorem	In any right triangle, the square of the hypotenuse c is equal to the sum of the squares of the other two sides.	If a and b are the sides of a right triangle and c is the hypotenuse, $a^2 + b^2 = c^2.$

6.5	**Procedure for Solving Word Problems**

1. **R**ead the problem carefully, and decide what it asks for (the unknown).
2. **S**elect a variable to represent this unknown.
3. **T**hink of a plan.
4. **U**se the rules of algebra to solve for the unknown.
5. **V**erify the solution.

Section	Item	Meaning	Example
6.6A	Ratio	A quotient of two numbers	$\frac{5}{7}$
6.6B	Proportion	An equality between two ratios	$\frac{5}{7} = \frac{10}{14}$
6.6C	Varies directly	y varies directly as x if $y = kx.$	
6.6C	Varies inversely	y varies inversely as x if $y = k/x.$	

📖📖 *Research Questions*

Sources of information for these questions can be found in the Bibliography at the end of the book.

1. Write a short essay about Gauss's childhood.

2. Write a report about Gauss's proof regarding regular polygons in his *Disquisitiones Arithmeticae.*

3. Write a short paragraph about Gauss's inventions.

4. Aside from being a superb mathematician, Gauss did some work in astronomy. Report on some of Gauss's discoveries in this field.

5. In 1807, a famous French mathematician paid Gauss's involuntary 2000-franc contribution to the French government. Find out who this famous mathematician was and the circumstances of the payment.

6. Another French mathematician asked the general commanding the French troops to send an officer to see how Gauss was faring during the war. This mathematician had submitted some results in number theory to Gauss under a pen name. Write a report on this incident; include the circumstances and the pen name and real name of the mathematician.

Chapter 6 Practice Test

1. If the replacement set is the set of integers, solve the following equations:
   a. $x + 7 = 2$ **b.** $x - 4 = 9$

2. If the replacement set is the set of integers, find the solution set for each of the following inequalities:
   a. $x + 5 > 4$ **b.** $2 + x \geq -x - 1$

3. Solve
   a. $2x + 2 = 3x - 2.$ **b.** $\dfrac{x + 3}{5} = \dfrac{x - 1}{3} + \dfrac{6}{5}.$

4. For women, the relationship between shoe size S and length of foot L is $S = 3L - 21.$
   a. Solve for $L.$
   b. If a woman wears a size 6 shoe, what is the length of her foot?

5. Solve
   $$\frac{x}{2} + 2 \geq \frac{-(x + 1)}{4}$$

6. Graph the solution set of
   a. $x - 3 \leq 0.$ **b.** $-2x + 4 > x + 1.$

7. Graph the solution set (if it is not empty) of
   a. $x + 2 \geq 3$ and $x \leq 4.$ **b.** $x - 3 \geq 1$ and $x \leq 0.$

8. Graph the solution set of
   a. $x < 0$ or $x - 2 < 1.$ **b.** $x + 2 < 3$ or $x - 1 > 2.$

9. Solve the equation $|x| = 3$.

10. Graph the solution set of $|x| < 2$.

11. Graph the solution set of $|x - 2| < 2$.

12. Graph the solution set of $|x| > 2$.

13. Graph the solution set of $|x - 2| > 3$.

14. Factor
   a. $x^2 + 3x + 2$.   **b.** $x^2 - 3x - 4$.

15. Solve
   a. $(x - 1)(x + 2) = 0$.   **b.** $x(x - 1) = 0$.

16. Solve, by factoring, $x^2 + 7x + 10 = 0$.

17. Solve, by factoring, $x^2 - 3x = 10$.

18. Solve
   a. $9x^2 - 16 = 0$.   **b.** $25x^2 - 4 = 0$.

19. Use the quadratic formula to solve $2x^2 + 3x - 5 = 0$.

20. Use the quadratic formula to solve $x^2 - 2x = 2$.

21. The hypotenuse of a right triangle is 9 cm longer than one of the legs and 2 cm longer than the other leg. Find the dimensions of the triangle.

22. Suppose that you rent a car for 1 day at the rate of $21 per day plus $.21 per mile. How many miles could you drive for a rental charge of $63?

23. Three times the sum of two consecutive integers is 45. What are the integers?

24. A pair of consecutive integers is such that 12 times the smaller is more than 9 times the larger. What is the least pair of integers for which this is true?

25. A certain two-digit number is equal to 5 times the sum of its digits. If the tens digit is x and the units digit is y, write an equation that x and y must satisfy.

26. On a certain day, the New York Stock Exchange reported that 688 stocks went up, 801 went down, and 501 were unchanged.
   a. Write the ratio of losers to gainers in three different ways.
   b. What is the ratio of losers to the total number of stocks?

27. A supermarket is selling a certain kind of cracker for 50¢ for an 8-oz box and 76¢ for a 12-oz box.
   a. Find the unit price for each box.
   b. Which is the better buy?

28. We know that corresponding sides of similar rectangles are proportional. One rectangle is 5 by 8 ft, and the short side of a similar rectangle is 9 ft long.
   a. Write a proportion for the length x of the long side of the second rectangle.
   b. Find the missing length.

29. The cost C of fuel per hour for running an airplane is directly proportional to the square of the speed x.
   a. Write an equation of variation.
   b. If the cost is \$100 per hour for a speed of 150 mph, find the value of k.
   c. Find the cost per hour for a speed of 180 mph.

30. The time t of exposure to photograph an object at a fixed distance from the camera is inversely proportional to the intensity I of the illumination.
   a. Write an equation of variation.
   b. If the correct exposure is $\frac{1}{30}$ sec when I is 300 units, find the value of k.
   c. If I is increased to 600 units, what is the correct exposure time?

Answers to Practice Test

ANSWER	IF YOU MISSED Question	REVIEW Section	REVIEW Example(s)	REVIEW Page(s)
1. a. $x = -5$ **b.** $x = 13$	1	6.1	1	326–327
2. a. $\{0, 1, 2, 3, \ldots\}$ **b.** $\{-1, 0, 1, 2, 3, \ldots\}$	2	6.1	1	326–327
3. a. $x = 4$ **b.** $x = -2$	3	6.1	2–4	328–330
4. a. $\dfrac{S + 21}{3}$ **b.** 9 in.	4	6.1	6	331–332
5. $\{x \mid x \geq -3\}$	5	6.1	7–9	333–334
6. a. **b.**	6	6.2	1–5	342–343
7. a. **b.** The solution set is empty.	7	6.2	6–8	344–345
8. a. **b.**	8	6.2	9	345
9. $x = \pm 3$	9	6.3	1	350
10.	10	6.3	2, 3	350–351
11.	11	6.3	4	351
12.	12	6.3	5, 6	351, 352
13.	13	6.3	7	352
14. a. $(x + 1)(x + 2)$ **b.** $(x + 1)(x - 4)$	14	6.4	1, 2	356
15. a. $x = -2$ or $x = 1$ **b.** $x = 0$ or $x = 1$	15	6.4	3	357
16. $x = -5$ or $x = -2$	16	6.4	3	357
17. $x = -2$ or $x = 5$	17	6.4	3	357
18. a. $x = \pm \frac{4}{3}$ **b.** $x = \pm \frac{2}{5}$	18	6.4	4, 5	358
19. $x = -\frac{5}{2}$ or $x = 1$	19	6.4	6, 7	359–360
20. $x = 1 + \sqrt{3}$ or $x = 1 - \sqrt{3}$	20	6.4	8	360–361
21. 8 cm, 15 cm, 17 cm	21	6.4	9	362
22. 200 mi	22	6.5	2	369–370

ANSWER	IF YOU MISSED	REVIEW		
	Question	Section	Example(s)	Page(s)
23. 7 and 8	23	6.5	4	371–372
24. 4 and 5	24	6.5	4	371–372
25. $10x + y = 5(x + y)$	25	6.5	5	372–373
26. a. 801 to 688; 801:688; $\frac{801}{688}$ **b.** $\frac{801}{1990}$	26	6.6	1	381
27. a. $6\frac{1}{4}$¢ per oz for the 8-oz box; $6\frac{1}{3}$¢ per oz for the 12-oz box	27	6.6	2	382
b. The 8-oz box				
28. a. $\dfrac{x}{8} = \dfrac{9}{5}$ **b.** 14.4 ft	28	6.6	3	383–384
29. a. $C = kx^2$ **b.** $k = \frac{1}{225}$	29	6.6	4–8	385–388
c. \$144 per hour				
30. a. $t = k/I$ **b.** $k = 10$ **c.** $\frac{1}{60}$ sec	30	6.6	4–8	385–388

To maximize efficiency in activities such as running and cycling and to achieve the greatest health benefits, individuals need to identify the target pulse rate for their age group. In Section 7.2, you will utilize linear functions to explore target pulse rates and relations to represent real-life data such as maximum target pulse rates.

Functions and Graphs

In Chapter 6 we studied first- and second-degree equations with *one* variable. In this chapter we shall study similar equations with *two* variables. The main feature here, however, is an introduction to some simple ideas that belong to the area that is called **analytic geometry,** a blend of algebra and geometry in which algebra is used to study geometry and geometry is used to study algebra. The key to this combination is a workable system of associating **points in the plane** with **ordered pairs of numbers.**

We start the chapter by studying sets of ordered pairs called **relations,** concentrating on a special type of relation called a **function.** We examine function notation and how to represent relations and functions by means of **graphs.** We also study the formula giving the distance between any two points in the Cartesian plane and the slope (inclination) of a line passing through these two points. We explore the ways to write the equations of a line depending on the information that is given and the ways to solve systems of linear equations with two unknowns by using algebraic or graphical methods.

Online Study Center

For links to Internet sites related to Chapter 7, please access **college.hmco.com/PIC/bello9e** and click on the Online Study Center icon.

7.1 Graphing Relations and Functions

(1596–1650)

Looking Ahead
The Cartesian coordinate system, a legacy of the work of René Descartes, is introduced in Section 7.1 and used throughout this chapter in studying functions and graphs.

Functions for Fashions

Did you know that women's clothing sizes are getting smaller? According to a J. C. Penney catalog, "Simply put, you will wear one size smaller than before." Is there a relationship between the new sizes and waist size? Look at Table 7.1 (page 400). It gives a $1\frac{1}{2}$-in. leeway for waist sizes. In Table 7.1, consider the first numbers in the waist sizes corresponding to different dress sizes, that is, 30, 32, 34, and 36. For dress sizes 14 to 22, the waist size is 16 in. more than the dress size. To formalize this relationship, you can write

$$w(s) = s + 16 \quad \text{(read "}w \text{ of } s \text{ equals } s \text{ plus 16")}$$

What would the waist size of a woman who wears size 14 be? It would be

$$w(14) = 14 + 16 = 30$$

For size 16,

$$w(16) = 16 + 16 = 32$$

and so on. It works! Can you do the same for hip sizes? If you get

$$h(s) = s + 26.5$$

you are on the right track. Examples 5 and 6 and problems 28 and 29 in Exercises 7.1 will give you more practice with relationships between number pairs. ▶

The word *relation* might remind you of members of your family—parents, brothers, sisters, cousins, and so on. You have already studied mathematical relations such as "is a subset of," "is less than," and "is equivalent to." Relations can be expressed by using ordered pairs. In this section you first examine relations in general and then concentrate on a very important type of relation called a **function** and its applications. To help understand these concepts, you end the section by learning how to make pictures, or graphs, of these relations and functions.

A. Relations and Functions

> **Definition of a Relation**
>
> A **relation** is a set of ordered pairs of the form (x, y). The set of all possible x values is the **domain** of the relation, and the set of all possible y values is the **range** of the relation.

For instance, the set $R = \{(4, -3), (2, -5), (-3, 4)\}$ is a relation in which all the pairs have been specifically listed. Notice that the set of first members, the **domain**, is $\{-3, 2, 4\}$, and the set of second members, the **range**, is $\{-5, -3, 4\}$. The listing in the relation R shows how the elements of the first set are associated with the elements of the second set to form the ordered pairs of the given

TABLE 7.1 Women's Sizes

	Women's Petite: 4'11"–5'3"					Women's: 5'3½"–5'7½"				
Women's petite size	14WP	16WP	18WP	20WP	22WP	24WP	26WP	28WP	30WP	32WP
Women's size	—	16W	18W	20W	22W	24W	26W	28W	30W	32W
Bust	38–$39\frac{1}{2}$	40–$41\frac{1}{2}$	42–$43\frac{1}{2}$	44–$45\frac{1}{2}$	46–$47\frac{1}{2}$	48–$49\frac{1}{2}$	50–$51\frac{1}{2}$	52–$53\frac{1}{2}$	54–$55\frac{1}{2}$	56–$57\frac{1}{2}$
Waist	30–$31\frac{1}{2}$	32–$33\frac{1}{2}$	34–$35\frac{1}{2}$	36–$37\frac{1}{2}$	38–40	$40\frac{1}{2}$–$42\frac{1}{2}$	43–45	$45\frac{1}{2}$–$47\frac{1}{2}$	48–50	$50\frac{1}{2}$–$52\frac{1}{2}$
Hips	$40\frac{1}{2}$–42	$42\frac{1}{2}$–44	$44\frac{1}{2}$–46	$46\frac{1}{2}$–48	$48\frac{1}{2}$–50	$50\frac{1}{2}$–52	$52\frac{1}{2}$–54	$54\frac{1}{2}$–56	$56\frac{1}{2}$–58	$58\frac{1}{2}$–60

The first J. C. Penney store opened in 1902 in Kemmerer, Wyoming.

relation. Different ways of associating the elements of the two sets will, of course, result in different relations. Thus,

$$S = \{(-3, -5), (2, -3), (4, 4)\}$$

is an example of a relation different from R but formed from the same two sets of first and second numbers.

Rule for Finding Domain and Range

Unless otherwise specified, the **domain** of a relation is taken to be the largest set of real numbers that can be substituted for x and that result in real numbers for y. The **range** is then determined by the rule of the relation.

For example, if

$$Q = \{(x, y) \mid y = \sqrt{x}\}$$

which means that Q is the set consisting of all ordered pairs (x, y) such that $y = \sqrt{x}$, then we may substitute any *nonnegative* real number for x and obtain a real number for y. But x cannot be replaced by a negative number. (Why?) Thus, the domain is the set of all nonnegative real numbers, and the range, in this case, is the same set. Why?

EXAMPLE 1 ▶ Finding Domain and Range

Find the domain and the range of

(a) $R = \{(x, y) \mid y = 2x\}$.   (b) $S = \{(x, y) \mid y = x^2 + 1\}$.

Solution

(a) The variable x can be replaced by any real value because 2 times any real number is a real number. Hence, the domain is $\{x \mid x$ a real number$\}$. The range is also the set of real numbers because every real number is 2 times another real number. Thus, the range is $\{y \mid y$ a real number$\}$.

(b) The variable x can be replaced by any real value because the result of squaring a real number and adding 1 is again a real number. Thus, the domain of S is $\{x \mid x \text{ a real number}\}$. The square of a real number is never negative, $x^2 \geq 0$, so $x^2 + 1 \geq 1$. Hence, the rule $y = x^2 + 1$ implies $y \geq 1$. Consequently, the range of S is $\{y \mid y \geq 1\}$. ∎

In many areas of mathematics and its applications, the most important kind of relation is one for which to each element in the domain there corresponds one and only one element in the range. Example 1 illustrates such a relation with $R = \{(x, y) \mid y = 2x\}$. It is clear here that for each x value there corresponds exactly one y value because the rule is $y = 2x$. On the other hand, the relation $x = y^2$ is not this type of relation, because to each of the x values 1, 4, and 9 there corresponds more than one y value, as the following shows:

For $x = 1$, $y = +1$ or $y = -1$, denoted by $y = \pm 1$, since $1 = (\pm 1)^2$
For $x = 4$, $y = \pm 2$, since $4 = (\pm 2)^2$
For $x = 9$, $y = \pm 3$, since $9 = (\pm 3)^2$

Definition of a Function

A **function** is a relation for which to each domain value x there corresponds exactly one range value y; that is, for every x there is only one y.

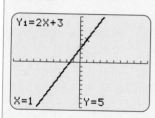

GRAPH IT

Your grapher evaluates functions! First, enter $\boxed{Y=}$ $2x + 3$ $\boxed{\text{ZOOM}}$ $\boxed{6}$ and then press $\boxed{\text{2nd}}$ $\boxed{\text{CALC}}$ $\boxed{1}$. Next, enter the desired x value, say, 1, and $\boxed{\text{ENTER}}$. The y value 5 is shown.

```
Y₁=2X+3




X=1          Y=5
```

If you press $\boxed{\text{TRACE}}$ $\boxed{\blacktriangleleft}$, you can evaluate at other points. If your points (x's) are integers, press $\boxed{\text{ZOOM}}$ $\boxed{8}$ $\boxed{\text{ENTER}}$ $\boxed{\text{TRACE}}$ and $\boxed{\blacktriangleleft}$ or $\boxed{\blacktriangleright}$ to evaluate the function when x is an integer. Try when x is 0, -6, or 4.

You bump into functions every day: the correspondence between the weight of a letter and the amount of postage you pay, the correspondence between the cost of a piece of meat and the number of pounds it weighs, and the correspondence between the number of miles per gallon that you get and the speed at which you drive. These are all simple examples of functions. You can undoubtedly think of many more.

EXAMPLE 2 ▶ Determining If a Relation Is a Function

(a) Is the relation $\{(x, y) \mid y = 2x + 3, x \text{ a real number}\}$ a function? Explain.

(b) Is the relation $\{(x, y) \mid x = y^2 + 1\}$ a function? Explain.

Solution

(a) If x is a real number, then y is the unique real number $2x + 3$. For instance, if $x = 2$, then $y = 2(2) + 3 = 7$; if $x = -\frac{1}{2}$, then $y = 2(-\frac{1}{2}) + 3 = 2$. Clearly, for each real x value, the expression $2x + 3$ gives one and only one y value. Thus, the given relation has exactly one range value corresponding to each domain value and is therefore a function.

(b) For this relation the domain is $\{x \mid x \geq 1\}$. Why? If $x = 5$, then the rule $x = y^2 + 1$ gives $5 = y^2 + 1$, or $y^2 = 4$. Thus, $y = 2$ or -2 because $2^2 = (-2)^2 = 4$. So the pairs $(5, 2)$ and $(5, -2)$ are both elements of this relation. The fact that there are two range values, 2 and -2, for the domain value 5 shows that the given relation is not a function. ∎

How do we denote functions? We often use letters such as f, F, g, G, h, and H to do this. Thus, for the relation in Example 2(a), we use set notation to write

$$f = \{(x, y) \mid y = 2x + 3\}$$

because we know this relation to be a function. Another very commonly used notation to denote the range value that corresponds to a given domain value x is $y = f(x)$. (This is usually read "f of x.")

The $f(x)$ notation, called **function notation,** is quite convenient because it denotes the value of the function for the given value of x. For example, if

$$f(x) = 2x + 3$$

then

$$f(1) = 2(1) + 3 = 5$$
$$f(0) = 2(0) + 3 = 3$$
$$f(-6) = 2(-6) + 3 = -9$$
$$f(4) = 2(4) + 3 = 11$$
$$f(a) = 2(a) + 3 = 2a + 3$$
$$f(w + 2) = 2(w + 2) + 3 = 2w + 7$$

and so on. Whatever appears between the parentheses in $f(\ )$ is to be substituted for x in the rule that defines $f(x)$. The result is y; that is, $f(x) = y$.

Instead of describing a function in set notation, we frequently say "the function defined by $f(x) = \cdots$," where the three dots are to be replaced by the expression for the value of the function. For instance, "the function defined by $f(x) = 2x + 3$" has the same meaning as "the function $f = \{(x, y) \mid y = 2x + 3\}$."

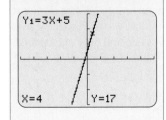

GRAPH IT

To do Example 3, enter $\boxed{Y=}$ and $3x + 5$. Since we want to evaluate at integral values, enter $\boxed{\text{ZOOM}}$ $\boxed{8}$ $\boxed{\text{ENTER}}$ $\boxed{\text{TRACE}}$. Now use the $\boxed{\blacktriangleright}$ or $\boxed{\blacktriangleleft}$ to see different values for x and y. Find $f(4)$ and $f(2)$.

Y₁=3X+5

X=4   Y=17

EXAMPLE 3 ▶ **Evaluating a Function**

Let $f(x) = 3x + 5$. Find

(a) $f(4)$.   (b) $f(2)$.   (c) $f(2) + f(4)$.   (d) $f(x + 1)$.

Solution

(a) Since $f(x) = 3x + 5$,

$$f(4) = 3 \cdot 4 + 5 = 12 + 5 = 17$$

(b) $f(2) = 3 \cdot 2 + 5 = 6 + 5 = 11$

(c) Since $f(2) = 11$ and $f(4) = 17$,

$$f(2) + f(4) = 11 + 17 = 28$$

(d) $f(x + 1) = 3(x + 1) + 5 = 3x + 8$ ∎

EXAMPLE 4 ▶ **Evaluating More Functions**

A function g is defined by $g(x) = x^3 - 2x^2 + 3x - 4$. Find

(a) $g(2)$.   (b) $g(-3)$.   (c) $g(2) - g(-3)$.

Solution

(a) $g(2) = 2^3 - 2(2^2) + 3(2) - 4 = 8 - 8 + 6 - 4 = 2$

(b) $g(-3) = (-3)^3 - 2(-3)^2 + 3(-3) - 4$
$$= -27 - 18 - 9 - 4 = -58$$

(c) $g(2) - g(-3) = 2 - (-58) = 60$ ∎

In the preceding problems we evaluated a specified function. Sometimes, as was the case in Getting Started, we must *find* the function, as shown next.

EXAMPLE 5 ► Finding Functional Relationships from Ordered Pairs

Consider the ordered pairs $(2, 6)$, $(3, 9)$, $(1.2, 3.6)$, and $\left(\frac{2}{5}, \frac{6}{5}\right)$. There is a linear functional relationship $y = f(x)$ between the numbers in each pair. Find $f(x)$ and use it to fill in the missing numbers in the pairs (_____, 12), (_____, 3.3), and (5, _____).

Solution

The given pairs are of the form (x, y). A close examination reveals that each of the y's in the pairs is 3 times the corresponding x; that is, $y = 3x$, or $f(x) = 3x$. Now, in each of the ordered pairs (_____, 12), (_____, 3.3), and (5, _____), the y value must be 3 times the x value. Thus,

$$(\underline{\quad}, 12) = (4, 12)$$
$$(\underline{\quad}, 3.3) = (1.1, 3.3)$$
$$(5, \underline{\quad}) = (5, 15)$$

∎

B. Applications of Functions

In recent years, aerobic exercises such as jogging, swimming, bicycling, and roller blading have been taken up by millions of Americans. To see if you are exercising too hard (or not hard enough), you should stop from time to time and take your pulse to determine your heart rate. The idea is to keep your rate within a range known as the **target zone**, which is determined by your age. The next example explains how to find the **lower limit** of your target zone by using a function.

In-line skaters benefit from a low-impact, highly aerobic workout.

EXAMPLE 6 ▶ Finding Heartbeats per Minute

The lower limit L (heartbeats per minute) of your target zone is a *function* of your age a (in years) and is given by

$$L(a) = -\tfrac{2}{3}a + 150$$

Find the value of L for people who are

(a) 30 years old.   (b) 45 years old.   (c) What age a corresponds to a lower limit $L(a)$ of 112?

Solution

(a) We need to find $L(30)$, and because

$$L(a) = -\tfrac{2}{3}a + 150$$
$$L(30) = -\tfrac{2}{3}(30) + 150$$
$$= -20 + 150 = 130$$

This result means that a 30-year-old person should try to attain at least 130 heartbeats per minute while exercising.

(b) Here, we want to find $L(45)$. Proceeding as before, we obtain

$$L(45) = -\tfrac{2}{3}(45) + 150$$
$$= -30 + 150 = 120$$

(c) If $L(a) = 112$, we have

$-\tfrac{2}{3}a + 150 = 112$	
$-2a + 450 = 336$	Multiply by 3.
$-2a = -114$	Subtract 450.
$a = 57$	Divide by -2.

Thus, if you are 57 years old, your lower limit is 112.

(Find the value of L for your own age.) ∎

C. Graphing Relations and Functions

FIGURE 7.1
Cartesian coordinate system.

We shall now study a method for drawing pictures of relations and functions. Figure 7.1 shows two number lines drawn perpendicular to each other. The horizontal line is labeled x and is called the **x axis**. The vertical line is labeled y and is called the **y axis**. The intersection of the two axes is the **origin.** We mark a number scale with the 0 point at the origin on each of the axes. The four regions into which the plane is divided by these axes are called **quadrants** and are numbered I, II, III, and IV, as shown in Figure 7.1. This diagram forms a **Cartesian coordinate system** (named after René Descartes). We can make pictures of relations on such a coordinate system.

Figure 7.1 shows the usual way in which the positive directions along the axes are chosen. On the x axis, to the right of the origin is positive and to the left is negative. On the y axis, up from the origin is positive and down is negative. In order to locate a point (x, y), we move x units horizontally along the x axis and then y units vertically. For instance, to locate $(-2, 3)$, we move 2 units horizontally in the negative direction along the x axis and then 3 units up, parallel to the y axis. Several points are plotted in Figure 7.1.

The Cartesian coordinate system furnishes us with a one-to-one correspondence between the points in the plane and the set of all ordered pairs of numbers; that is, corresponding to a given ordered pair there is exactly one point, and corresponding to a given point there is exactly one ordered pair. The **graph** of a relation is the set of points corresponding to the ordered pairs of the relation.

EXAMPLE 7 ▶ Graphing Relations

Graph the following relations:

(a) $R = \{(x, y) \mid y = 2x, x \text{ an integer between } -1 \text{ and } 2, \text{ inclusive}\}$

(b) $T = \{(x, y) \mid y^2 = x, y \text{ an integer between } -3 \text{ and } 3, \text{ inclusive}\}$

Solution

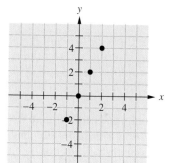

FIGURE 7.2

(a) The domain of R is $\{-1, 0, 1, 2\}$. By using the rule of R, $y = 2x$, we can find the ordered pairs $(-1, -2)$, $(0, 0)$, $(1, 2)$, and $(2, 4)$ that belong to R. Note that since $y = 2x$, the y coordinate is always *twice* the x coordinate. The graph of the relation is shown in Figure 7.2.

(b) We are given the range $\{y \mid y \text{ an integer between } -3 \text{ and } 3, \text{ inclusive}\}$, and by squaring each of these integers, we find the domain, $\{0, 1, 4, 9\}$. The set of pairs $\{(0, 0), (1, 1), (1, -1), (4, 2), (4, -2), (9, 3), (9, -3)\}$ is thus the relation T. The graph of the relation consists of the seven dots shown in Figure 7.3. If the domain of this relation had been all nonnegative real numbers $(x \geq 0)$, the graph would have been a curve called a **parabola** (shown dashed in Figure 7.3). ∎

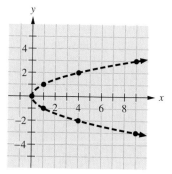

FIGURE 7.3

Since a function is simply a special kind of relation, a function can also be graphed on a Cartesian coordinate system. It is customary to represent the x values along the horizontal axis and the $f(x)$, or y, values along the vertical axis and graph the resulting ordered pairs (x, y).

Because a function has just one value for each value of x in the domain, the graph of the function cannot be cut in more than one point by any vertical line. Thus, we have a simple **vertical line test** for a function.

Vertical Line Test
If a vertical line can be drawn so that it intersects the graph of a relation at more than one point, then the relation is *not* a function.

If the graph of a relation is known, we can tell by inspection whether the relation is a function. For example, the parabola in Figure 7.3 is the graph of a relation, but that relation is *not* a function because any vertical line to the right of the y axis cuts the graph in two points. On the other hand, if we had defined the range of the relation so that its graph were only the lower (or only the upper) portion of the parabola, then the relation would be a function.

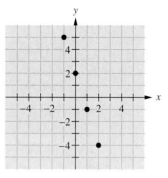

FIGURE 7.4

EXAMPLE 8 ▶ Graphing Functions

Graph the function f defined by $f(x) = 2 - 3x$ if the domain is the set $\{-1, 0, 1, 2\}$.

Solution

We calculate the values of the function using the rule $f(x) = 2 - 3x$ to obtain the values in Table 7.2. Figure 7.4 shows the graph.

TABLE 7.2

x	$y = f(x)$	Ordered Pair
-1	5	$(-1, 5)$
0	2	$(0, 2)$
1	-1	$(1, -1)$
2	-4	$(2, -4)$

There is something remarkable about the set of points graphed in Figure 7.4. Can you see what it is? The points all lie on a straight line! We shall study functions whose graphs are straight lines (linear functions) in the next section.

EXAMPLE 9 ▶ Graphing Functions

Graph the function g defined by $g(x) = 2 - x^2$ if the domain is the set of real numbers.

Solution

The rule $g(x) = 2 - x^2$ tells us that to find the range value corresponding to an x value, we must square the x value and subtract the result from 2. Using this procedure for integral values of x between -3 and 3, inclusive, we obtain the values in Table 7.3. We then graph the ordered pairs and join them with a smooth curve, as shown in Figure 7.5. Notice that the graph is a parabola, as in Example 7.

TABLE 7.3

x	$y = g(x)$	Ordered Pair
-3	-7	$(-3, -7)$
-2	-2	$(-2, -2)$
-1	1	$(-1, 1)$
0	2	$(0, 2)$
1	1	$(1, 1)$
2	-2	$(2, -2)$
3	-7	$(3, -7)$

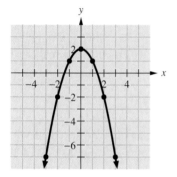

FIGURE 7.5

D. Applications of Graphs

TABLE 7.4

Mon.–Fri.	8 A.M.–5 P.M.
Initial minute $.36	Additional minute $.28

EXAMPLE 10 ▶ Graphing Long-Distance Functions

Long-distance telephone costs are functions of time. The dial-direct rate for a Tampa to Los Angeles call made during business hours is given in Table 7.4. Of course, if you talk for a fraction of a minute, you will be charged for the whole minute, so the words "or fraction thereof" should be understood to be included in each column. Find the costs of calls lasting the following amounts of time:

(a) 1 min or less

(b) More than 1 min but not more than 2 min

(c) More than 2 min but not more than 3 min

(d) More than 3 min but not more than 4 min

(e) Draw a graph showing the results of parts (a), (b), (c), and (d).

Solution

Make a table of values (Table 7.5), as in Examples 8 and 9. Table 7.5 gives the answers for parts (a), (b), (c), and (d).

TABLE 7.5

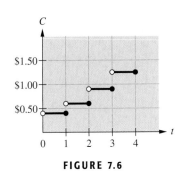

FIGURE 7.6

	Time, t	Cost, C
(a)	$0 < t \leq 1$	\$.36
(b)	$1 < t \leq 2$	\$.36 + (1)(\$.28) = \$0.64
(c)	$2 < t \leq 3$	\$.36 + (2)(\$.28) = \$0.92
(d)	$3 < t \leq 4$	\$.36 + (3)(\$.28) = \$1.20

(e) The graph is shown in Figure 7.6. The open circles in Figure 7.6 indicate that the left-hand endpoints of the line segments are not included in the graph. The solid dots mean that the right-hand endpoints are included. ∎

EXAMPLE 11 ▶ Finding Domain and Range When Dining

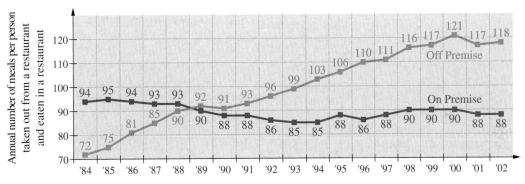

Source: The NPD Group's CREST Service, www.asfsa.org/meetingsandevents/archive/itc2003/16.

FIGURE 7.7

Referring to Figure 7.7,

(a) what is the domain for both graphs?

(b) what is the range for on-premise dining?

(c) what is the *exact* year in which the number of off-premise meals (take-out food) surpassed the number of on-premise meals purchased?

Solution

(a) The domain is the set of all possible x values {84, 85, 86, . . . , 02}.

(b) The range is {85, 86, 88, 90, 93, 94, 95}, the set of all possible y values for the on-premise dining line.

(c) 1989 (In 88, on-premise beat off-premise 93 to 90, but in 89, on-premise lost 90 to 92, so we can conclude that off-premise surpassed on-premise *during 1989.*) ∎

Online Study Center

To further explore graphing, access links 7.1.1, 7.1.2, and 7.1.3 on this textbook's Online Study Center.

EXERCISES 7.1

Ⓐ Relations and Functions

In problems 1–14, find the domain and the range of each relation. (*Hint:* Remember that you cannot divide by 0 or take the square root of a negative number.)

1. $\{(1, 2), (2, 3), (3, 4)\}$ **2.** $\{(3, 1), (2, 1), (1, 1)\}$

3. $\{(1, 1), (2, 2), (3, 3)\}$ **4.** $\{(4, 1), (5, 2), (6, 1)\}$

5. $\{(x, y) \mid y = 3x\}$ **6.** $\{(x, y) \mid y = 2x + 1\}$

7. $\{(x, y) \mid y = x + 1\}$ **8.** $\{(x, y) \mid y = 1 - 2x\}$

9. $\{(x, y) \mid y = x^2\}$ **10.** $\{(x, y) \mid y = 2 + x^2\}$

11. $\{(x, y) \mid y^2 = x\}$ **12.** $\{(x, y) \mid x = 1 + y^2\}$

13. $\left\{(x, y) \mid y = \dfrac{1}{x}\right\}$ **14.** $\left\{(x, y) \mid y = \dfrac{1}{x - 2}\right\}$

In problems 15–22, decide whether each relation is a function. State the reason for your answer in each case.

15. $\{(x, y) \mid y = 5x + 6\}$ **16.** $\{(x, y) \mid y = 3 - 2x\}$

17. $\{(x, y) \mid x = y^2\}$ **18.** $\{(x, y) \mid x + 1 = y^2\}$

19. $\{(x, y) \mid y = \sqrt{x}, x \geq 0\}$

20. $\{(x, y) \mid x = \sqrt{y}, y \geq 0\}$

21. $\{(x, y) \mid x = y^3\}$ **22.** $\{(x, y) \mid y = x^3\}$

23. A function f is defined by $f(x) = 3x + 1$. Find
   a. $f(0)$.   **b.** $f(2)$.   **c.** $f(-2)$.

24. A function g is defined by $g(x) = -2x + 1$. Find
   a. $g(0)$.   **b.** $g(1)$.   **c.** $g(-1)$.

25. A function F is defined by $F(x) = \sqrt{x - 1}$. Find
   a. $F(1)$.   **b.** $F(5)$.   **c.** $F(26)$.

26. A function G is defined by $G(x) = x^2 + 2x - 1$. Find
   a. $G(0)$.   **b.** $G(2)$.   **c.** $G(-2)$.

27. A function f is defined by $f(x) = 3x + 1$. Find
   a. $f(x + h)$.   **b.** $f(x + h) - f(x)$.
   c. $\dfrac{f(x + h) - f(x)}{h}, h \neq 0$.

28. Given are the ordered pairs (2, 1), (6, 3), (9, 4.5), and (1.6, 0.8). There is a simple functional relationship, $y = f(x)$, between the numbers in each pair. What is $f(x)$? Use this to fill in the missing numbers in the pairs (_____, 7.5), (_____, 2.4), and $\left(\_\_\_\_\_, \frac{1}{7}\right)$.

29. Given are the ordered pairs $\left(\frac{1}{2}, \frac{1}{4}\right)$, (1.2, 1.44), (5, 25), and (7, 49). There is a simple functional relationship, $y = g(x)$, between the numbers in each pair. What is $g(x)$? Use this to fill in the missing numbers in the pairs $\left(\frac{1}{4}, \_\_\_\_\_\right)$, (2.1, _____), and (_____, 64).

30. Given that $f(x) = x^3 - x^2 + 2x$, find
   a. $f(-1)$.   **b.** $f(-3)$.   **c.** $f(2)$.

31. If $g(x) = 2x^3 + x^2 - 3x + 1$, find
   a. $g(0)$.   **b.** $g(-2)$.   **c.** $g(2)$.

Ⓑ Applications of Functions

32. The Fahrenheit temperature reading F is a function of the Celsius temperature reading C. This function is given by

$$F(C) = \tfrac{9}{5}C + 32$$

   a. If the temperature is 15°C, what is the Fahrenheit temperature?
   b. Water boils at 100°C. What is the corresponding Fahrenheit temperature?
   c. The freezing point of water is 0°C or 32°F. How many Fahrenheit degrees below freezing is a temperature of −10°C?
   d. The lowest temperature attainable is −273°C; this is the zero point on the absolute temperature scale. What is the corresponding Fahrenheit temperature?

33. Refer to Example 6. The *upper limit U* of your target zone when exercising is also a function of your age a (in years) and is given by

$$U(a) = -a + 190$$

Find the highest and safest heart rate for people who are
   a. 50 years old.   **b.** 60 years old.

34. Refer to Example 6 and problem 33. The target zone for a person a years old consists of all the heart rates between $L(a)$ and $U(a)$, inclusive. Thus, if a person's heart rate is R, that person's target zone is described by $L(a) \leq R \leq U(a)$. Find the target zone for people who are
   a. 30 years old.   **b.** 45 years old.

35. The ideal weight w (in pounds) of a man is a function of his height h (in inches). This function is defined by

$$w(h) = 5h - 190$$

  a. What should the weight be for a man 70 in. tall?
  b. What should the height be for a man who weighs 200 lb?

36. The cost C in dollars of renting a car for 1 day is a function of the number m of miles traveled. For a car renting for $20 per day and $.20 per mile, this function is given by

$$C(m) = 0.20m + 20$$

  a. Find the cost of renting a car for 1 day and driving 290 mi.
  b. If an executive paid $60.60 after renting a car for 1 day, how many miles did she drive?

37. The pressure P (in pounds per square foot) at a depth d ft below the surface of the ocean is a function of the depth. This function is given by

$$P(d) = 63.9d$$

Find the pressure on a submarine at a depth of
  a. 10 ft.   **b.** 100 ft.

38. If a ball is dropped from a point above the surface of the Earth, the distance s (in meters) that the ball falls in t sec is a function of t. This function is given by

$$s(t) = 4.9t^2$$

Find the distance that the ball falls in
  a. 2 sec.   **b.** 5 sec.

39. The function $S(t) = \frac{1}{2}gt^2$ gives the distance that an object falls from rest in t sec. If S is measured in feet, then the gravitational constant g is approximately 32 ft/sec^2. Find the distance that the object falls in
  a. 3 sec.   **b.** 5 sec.

40. An experiment, carefully carried out, showed that a ball dropped from rest fell 64.4 ft in 2 sec. What is a more accurate value of g than that given in problem 39?

C **Graphing Relations and Functions**

In problems 41–50, graph each relation.

41. $\{(x, y) \mid y = x, x \text{ an integer between } -1 \text{ and } 4, \text{ inclusive}\}$

42. $\{(x, y) \mid y = -x, x \text{ an integer between } -1 \text{ and } 4, \text{ inclusive}\}$

43. $\{(x, y) \mid y = 2x + 1, x \text{ an integer between } 0 \text{ and } 5, \text{ inclusive}\}$

44. $\{(x, y) \mid x + 2y = 3, x \text{ an odd integer between } 0 \text{ and } 10\}$

45. $\{(x, y) \mid 2x - y = 4, x \text{ an integer between } -2 \text{ and } 2, \text{ inclusive}\}$

46. $\{(x, y) \mid y = x^2, x \text{ an integer between } -3 \text{ and } 3, \text{ inclusive}\}$

47. $\{(x, y) \mid y = \sqrt{x}, x = 0, 1, 4, 9, 16, 25, \text{ or } 36\}$

48. $\{(x, y) \mid x = \sqrt{y}, x \text{ an integer between } 0 \text{ and } 3, \text{ inclusive}\}$

49. $\{(x, y) \mid x + y < 5, x, y \text{ nonnegative integers}\}$

50. $\{(x, y) \mid y > x, x \text{ and } y \text{ positive integers less than } 4\}$

In problems 51–54, graph the given function for each domain (replacement set for x).

51. $f(x) = x + 1, x \text{ an integer between } -3 \text{ and } 3, \text{ inclusive}$

52. $f(x) = 3x - 1, x \text{ an integer between } -1 \text{ and } 3, \text{ inclusive}$

53. $g(x) = x^2 + 1, x \text{ an integer between } -3 \text{ and } 3, \text{ inclusive}$

54. $h(x) = -x^2, x \text{ an integer between } -3 \text{ and } 3, \text{ inclusive}$

D **Applications of Graphs**

55. Anthropologists can determine a person's height in life by using the person's skeletal remains as a clue. For example, the height (in centimeters) of a man with a humerus bone of length x cm can be obtained by multiplying 2.89 by x and adding 70.64 to the result.
  a. Find a function $h(x)$ that gives the height of a man whose humerus bone is x cm long.
  b. Use your function h from part (**a**) to predict the height of a man whose humerus bone is 34 cm long.

56. A plumber charges $25 per hour plus $30 for the service call. If $c(h)$ is the function representing the total charges, and h is the number of hours worked, find the following:

a. $c(h)$

b. The total charges when the plumber works for 2, 3, or 4 hours

c. Graph the points obtained in part (**b**).

57. A finance company will lend $1000 for a finance charge of $20 plus simple interest at 1% per month. This means that you will pay interest of 0.01 of $1000, that is, $10 per month, in addition to the $20 finance charge. The total cost of the loan can be expressed as a function $F(x)$, where x is the number of months before you repay the loan.

a. Find $F(x)$.

b. Find the total cost of borrowing the $1000 for 8 months.

58. A bank charges 2% per month on a high-risk loan. If this is simple interest, describe the function that gives the cost $C(x)$ of borrowing $5000 for x months. (Compare problem 57.)

59. One of the depreciation (loss-of-value) methods approved by the IRS is the **straight-line method.** Under this method, if a $10,000 truck is to be fully depreciated in 5 years, the yearly depreciation will be $\frac{1}{5}$ of $10,000, that is, $2000.

a. Write an equation that defines the depreciated value V of the truck as a function of the time t in years.

b. Draw a graph of this function.

60. The rates for a Tampa to Atlanta dial-direct telephone call during business hours are 33¢ for the first minute or fraction thereof and 27¢ for each additional minute or fraction thereof. Find the cost C of calls lasting

a. 1 min or less.

b. more than 1 min but not more than 2 min.

c. more than 2 min but not more than 3 min.

d. Draw a graph of the results of parts (**a**), (**b**), and (**c**).

61. The cost of a Boston-to-Nantucket dial-direct telephone call during business hours is given for positive integer values of t (in minutes) by

$$C(t) = 0.55 + 0.23(t - 1)$$

a. Draw a graph of $C(t)$ for $0 < t \le 5$.

b. Santiago made a call from Boston to his friend Jessica in Nantucket. If the call cost Santiago $3.31, how long did they talk?

In Other Words

62. Consider the function $f(x) = \dfrac{1}{x^2 - 1}$. What numbers are excluded from the domain and why?

63. Consider the function $g(x) = \sqrt{x - 1}$. What numbers are excluded from the domain and why?

64. Consider the function $h(x) = \dfrac{x}{\sqrt{x + 1}}$. What numbers are excluded from the domain and why?

65. What is the graph of a function?

66. Is every relation a function? Explain why or why not.

67. Is every function a relation? Explain why or why not.

68. Why does the vertical line test work?

Using Your Knowledge

69. There are many interesting functions that can be defined using the ideas of this section. Let's return to the cricket from Getting Started 6.1 whose chirping frequency is a function of the temperature. The table below shows the number of chirps per minute and the temperature in degrees Fahrenheit. Find a function that relates the number c of chirps per minute and the temperature x.

Temperature (°F)	Chirps per Minute
40	0
41	4
42	8
43	12
44	16

70. The function relating the number of chirps per minute of the cricket and the temperature is given by $f(x) = 4(x - 40)$. If the temperature is 80°F, how many chirps per minute will you hear from your friendly house cricket?

71. An interesting function in physics was discovered by the Italian scientist Galileo Galilei. This function relates the distance an object (dropped from a given height) travels and the time elapsed. The following table shows the time (in seconds) and the distance (in feet) traveled by a rock dropped from a tall building. Find the relationship between the number t of seconds elapsed and the distance $f(t)$ traveled.

Time Elapsed (sec)	Distance (ft)
1	$16 = 16 \times 1$
2	$64 = 16 \times 4$
3	$144 = 16 \times 9$
4	$256 = 16 \times 16$
5	$400 = 16 \times 25$
6	$576 = 16 \times 36$

72. Assume that a rock took 10 sec to reach the ground when dropped from a helicopter. Using the results of problem 71, find the height of the helicopter.

Discovery

Have you ever seen a pendulum clock? Galileo Galilei also made various discoveries about swinging weights, and these discoveries led to the invention of the pendulum clock. Galileo discovered that there was a relationship between the time of the swing of a pendulum and its length. The table below shows corresponding values of these two quantities. (The unit length is about 25 cm.)

Time of Swing $f(x)$	Length of Pendulum (x)
1 sec	1 unit
2 sec	4 units
3 sec	9 units
4 sec	16 units
5 sec	25 units

73. Judging from the table, what is the rule connecting the time $f(x)$ of the swing and the length x of the pendulum?

74. From the pattern given in the table, find the length of a pendulum that takes 6 sec for a swing.

75. Find the length of a pendulum that takes 100 sec for a swing.

76. The University of South Florida has a Foucault pendulum in its physics building. This pendulum takes 7 sec for a swing. Find the length of the pendulum.

This Foucault pendulum is on display in the visitors' entrance lobby of the United Nations General Assembly building in New York.

Collaborative Learning

Many decisions in business and other areas are made on the basis of information provided in the form of a graph. For example, suppose you have a restaurant and you have examined the graph in Example 11. Form two teams and answer the following questions within your team, and then discuss them with the entire group.

1. On the basis of the graph, would you expand dining facilities or kitchen facilities?

2. How can you predict the number of annual on-premise and off-premise meals per person in the year 2010? Can you use the same technique to predict the number for any year?

3. Select a local restaurant with both take-out and on-premise facilities. One team makes a graph of the number of persons purchasing take-out, and the other team makes a graph of the number of persons purchasing on-premise meals during a 3-hour period. Label your x axis 1, 2, and 3 and your y axis with the number of meals purchased. Are the graphs similar? Would the graphs differ if a different 3-hour period were chosen? Explain. (If you have enough students to form several teams, the teams can select different 3-hour periods and see if there is a difference between graphs.)

4. On the basis of your graphs, can you predict how many take-out and how many on-premise meals are sold annually at this particular restaurant?

What about the number of take-out and on-premise meals per person? If you have a prediction, is it close to the one given in the graph of Example 11? Why or why not?

 Research Questions

1. Nicole Oresme is credited with developing a "graphical representation akin to our analytical geometry." Find out what Oresme did with "latitudes," "longitudes," and the law for falling bodies.

2. The idea of a function was discussed by Euler in his book *Introduction to Analysis of the Infinite*. Find out how Euler defined a function and how many basic classes of functions he considered.

3. Who invented the notation $f(x)$ to represent a function?

7.2 Linear Functions, Relations, and Applications

TABLE 7.6

Age	Pulse
0	190
70	120

Fitness and Graphing Functions

Have you been to a fitness center lately? Some of them display a graph showing your desirable heart rate **target zone** based on your age. The idea is to elevate your heart rate so that it is within a prescribed range. If your heart races during exercise, the consequences may be fatal. Thus, it is recommended that your heart rate (pulse) not exceed 190 beats per minute regardless of age. For a 70 year old, the recommended upper limit is 120 beats per minute. This information is shown in Table 7.6.

Now, suppose $U(a)$ is the upper limit in heartbeats per minute for a person whose age is a; can you find $U(a)$? Assume that $U(a)$ is linear; it must be defined by an equation of the form $U(a) = ma + b$. Now find m and b. According to Table 7.6, when $a = 0$, $U(a) = 190$. Thus,

$$U(0) = m \cdot 0 + b = 190$$

and

$$b = 190$$

Also from Table 7.6,

$$U(70) = 120$$

Thus,

$$U(70) = m \cdot 70 + 190 = 120$$

Subtracting 190,

$$70m = -70$$

Solving for m,

$$m = -1$$

Since $m = -1$ and $b = 190$, $U(a) = -1a + 190$; that is, $U(a) = -a + 190$. To graph $U(a)$ between 10 and 70, let $a = 10$; then $U(10) = -10 + 190 = 180$.

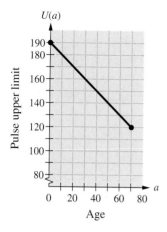

FIGURE 7.8

Graph (10, 180). The point (70, 120) is in Figure 7.8, so plot it and draw a straight line through (70, 120) and (10, 180). The result is the graph for $U(a)$. Note that all points on the graph are solutions of (satisfy) $U(a) = -a + 190$. Why do you think the domain is between 10 and 70?

In Section 7.7C, Example 6, you will also graph $L(a)$, the lower limit in heartbeats per minute. The region between the two graphs is the **target zone** for the heart rate while exercising. You will explore more problems like this one in problems 29–31 of Exercises 7.2. ▶

A relation of the form

$$\{(x, y) \mid y = ax + b\}$$

where a and b are real numbers, always defines the function $f(x) = ax + b$. (Why?) A function of this special form is called a **linear function** because its graph is a straight line. All the points (x, y) on the graph of $y = ax + b$ are solutions of the equation $y = ax + b$. In this section you will learn how to draw the graph of a linear function.

A. Vertical and Horizontal Lines

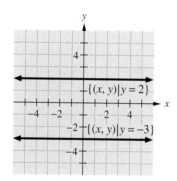

FIGURE 7.9

Let us look first at the case in which $a = 0$ in $y = ax + b$. Then the rule of the function is $y = b$ for *all real values of x*. Note that x is unrestricted (any real number) but for any value of x, $y = b$. This means that the graph consists of all points such as $(0, b), (-1, b), (\sqrt{2}, b), (10.26, b)$, and so on. Thus, the graph is a straight line parallel to the x axis and b units from this axis. If $b > 0$, the line will be above the x axis, and if $b < 0$, the line will be below the x axis. (What if $b = 0$?) Figure 7.9 shows the graphs of $\{(x, y) \mid y = b\}$ for $b = -3$ and for $b = 2$.

A relation such as $\{(x, y) \mid x = c\}$, where c is a real number, is not a function because y can have any real value. For instance, if $R = \{(x, y) \mid x = 2\}$, then $(2, -1), (2, 0), (2, 1.75)$, and so on are all ordered pairs belonging to R. Because x has the fixed value 2, all these points are on the line parallel to and 2 units to the right of the y axis. In general, the equation $x = c$ (or the relation $\{(x, y) \mid x = c\}$) has for its graph a vertical line, that is, a line parallel to and c units from the y axis.

Horizontal and Vertical Lines

The graph of the function $\{(x, y) \mid y = b\}$ or, equivalently, of the equation $y = b$ is a **horizontal** line parallel to the x axis and b units from it.

The graph of the relation $\{(x, y) \mid x = c\}$ or, equivalently, of the equation $x = c$ is a **vertical** line parallel to the y axis and c units from it.

EXAMPLE 1 ▶ **Graphing Horizontal and Vertical Lines**

Graph

(a) $y = -2$.   (b) $x = 2$.

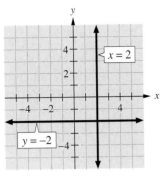

FIGURE 7.10

Solution

(a) The graph of $y = -2$ is a horizontal line parallel to the x axis and two units below it. Note that for any x you choose, y is always -2. The graph is shown in Figure 7.10.

(b) The graph of $x = 2$ is a vertical line parallel to the y axis as shown in Figure 7.10. ∎

B. Graphs of Linear Functions and Equations

We will now examine the general procedure that can be used to draw the graph of a linear function in the form $f = \{(x, y) \mid y = ax + b\}$ or, equivalently, $f(x) = ax + b$.

PROBLEM SOLVING

❶ **Read** the problem.

❷ **Select** the unknown. What are we looking for?

❸ **Think** of a plan. Is the graph a straight line? If it is, we know that two points will determine the line.

❹ **Use** the preceding idea to carry out the plan. Find two points and then join them with a line. (Try to use points that are easy to graph.)

❺ **Verify** the answer. How can we do this?

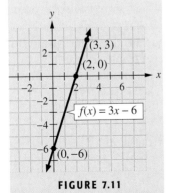

FIGURE 7.11

TRY EXAMPLE 2 NOW.

Graphing Linear Functions

Graph the function defined by $f(x) = 3x - 6$.

We want to draw the graph of $f(x) = 3x - 6$. To do this, we need to find ordered pairs (x, y) that satisfy the relation $f = \{(x, y) \mid y = 3x - 6\}$.

The function is of the form $y = ax + b$, so its graph, consisting of the ordered pairs (x, y) satisfying the equation $y = ax + b$, is a straight line. To graph this function, we find two points on the line and draw the line through them. The result will be the graph of the function.

For ease of computation, we let $x = 0$, to obtain $y = 3 \cdot 0 - 6$, or $y = -6$. Thus, $(0, -6)$ is one of the points. For $y = 0$, we have $0 = 3x - 6$, or $6 = 3x$; that is, $x = 2$. This makes $(2, 0)$ another point on the line. Now we graph $(0, -6)$ and $(2, 0)$ as in Figure 7.11 and draw a line through them. The result is the graph of $f(x) = 3x - 6$.

Select a point on the line, say $(3, 3)$, and verify that it satisfies the equation $y = 3x - 6$. If $x = 3$ and $y = 3$, $3 = 3 \cdot 3 - 6$ is a true statement.

The x coordinate of the point where the line crosses the x axis is called the **x intercept** of the line. Similarly, the y coordinate of the point where the line crosses the y axis is called the **y intercept.** For $f(x) = 3x - 6$, the x intercept is 2 and the y intercept is -6. Note that, in general, we may say either "graph the function f, where $f(x) = ax + b$," or "graph the equation $y = ax + b$."

Cover the solution, write your own solution, and then check your work.

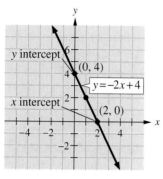

FIGURE 7.12

EXAMPLE 2 ▶ Graphing Equations Using Intercepts

Graph the equation $y = -2x + 4$ using the intercepts.

Solution

Since the equation is of the form $y = ax + b$, its graph is a straight line. We find the y intercept by letting $x = 0$ to obtain $y = -2 \cdot 0 + 4$, or $y = 4$. We then graph the point $(0, 4)$. For $y = 0$, we have $0 = -2x + 4$, or $x = 2$. We graph the point $(2, 0)$ as in Figure 7.12. We then draw a straight line through $(0, 4)$ and $(2, 0)$ to get the graph of $y = -2x + 4$. For verification, we select a point on the line, say, $(1, 2)$. Substituting $x = 1$ and $y = 2$ into $y = -2x + 4$, we have $2 = -2(1) + 4$, a true statement. ■

Now that we have graphed some particular straight lines, we will examine other relations whose graphs are straight lines. We can show that the relation described by $\{(x, y) \mid ax + by = c\}$, where a and b are not both 0, is always a linear relation. If $b \neq 0$, we can solve the equation $ax + by = c$ for y and obtain

$$by = -ax + c$$

$$y = -\frac{a}{b}x + \frac{c}{b}$$

Because a, b, and c are all real numbers, this equation is the rule for a linear function. If $b = 0$ in $ax + by = c$, then

$$ax = c \qquad \text{so that} \qquad x = +\frac{c}{a}$$

which is the rule for a linear relation (not a function) corresponding to a vertical line c/a units from the y axis. Because of these facts, the equation $ax + by = c$, with a and b not both 0, is called a **linear equation;** its graph is always a straight line, which can be drawn using our familiar intercept procedure, as shown next.

EXAMPLE 3 ▶ Graphing More Equations Using Intercepts

Graph the equation $2x + 3y = 6$.

Solution

Because the equation $2x + 3y = 6$ is a linear equation, we know that its graph is a straight line. Thus, any two points on the line will determine the line. For $x = 0$, we have $3y = 6$, or $y = 2$, so the point $(0, 2)$ is on the line. For $y = 0$, we get $x = 3$, so $(3, 0)$ is a second point on the line. We graph these two points and draw a straight line through them to get the graph shown in Figure 7.13.

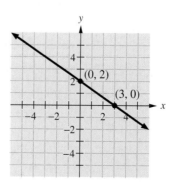

FIGURE 7.13 ■

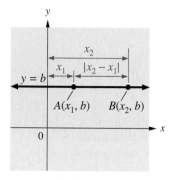

FIGURE 7.14

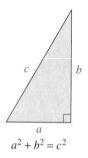

$a^2 + b^2 = c^2$

FIGURE 7.15

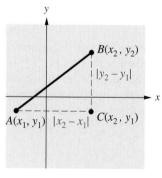

FIGURE 7.16

$|AB| = \sqrt{(x_2 - x_1)^2 + (y_2 - y_1)^2}$

C. Distance Between Two Points

In this section we have noted that the line $y = b$ is a line parallel to the x axis and b units from it (see Figure 7.9). Similarly, the line $x = a$ is a line parallel to the y axis and a units from it (see Example 1). It is not difficult to find the distance between two points on a horizontal or on a vertical straight line, that is, on a line $y = b$ or on a line $x = a$.

Suppose that we have two points—say, A and B—on the same horizontal line, the line $y = b$ in Figure 7.14. Then their coordinates are (x_1, b) and (x_2, b), as in Figure 7.14. Because x_1 and x_2 are the **directed distances** of the respective points from the y axis, the length of AB is $|x_2 - x_1|$. Denoting this length by $|AB|$, we have the formula

$$|AB| = |x_2 - x_1| \tag{1}$$

Similarly, if the two points C and D are on the same vertical line, their coordinates are (a, y_1) and (a, y_2), and the length of CD is

$$|CD| = |y_2 - y_1| \tag{2}$$

To obtain a formula for the distance between *any* two points, we use the *Pythagorean theorem.*

The Pythagorean Theorem

If a triangle is a right triangle (has one 90° angle), then the square of the hypotenuse c equals the sum of the squares of the other two sides a and b; that is,

$$a^2 + b^2 = c^2$$

(See Figure 7.15.)

The Pythagorean theorem will be discussed in greater detail in Section 8.4.

In Figure 7.16, $A(x_1, y_1)$ and $B(x_2, y_2)$ represent two general points. The line BC is drawn parallel to the y axis and the line AC is drawn parallel to the x axis, so ABC is a right triangle with right angle at C. Because A and C are on the same horizontal line, they must have the same y coordinates. Likewise, B and C are on the same vertical line, so they have the same x coordinates. Thus, the coordinates of C must be (x_2, y_1). By the Pythagorean theorem, the length of AB is given by $|AB| = \sqrt{|AC|^2 + |BC|^2}$. By equations (1) and (2), $|AC| = |x_2 - x_1|$ and $|BC| = |y_2 - y_1|$. Thus, we have the following:

The Distance Between Two Points

The distance between two points $A(x_1, y_1)$ and $B(x_2, y_2)$ is given by

$$|AB| = \sqrt{(x_2 - x_1)^2 + (y_2 - y_1)^2} \tag{3}$$

EXAMPLE 4 ▶ **Finding the Distance Between Two Points**

Find the distance between the points $A(2, -3)$ and $B(8, 5)$.

Solution

By equation (3),

$$|AB| = \sqrt{(8-2)^2 + [5-(-3)]^2}$$
$$= \sqrt{6^2 + 8^2} = \sqrt{36 + 64} = \sqrt{100} = 10$$

The distance between A and B is 10 units. ∎

EXAMPLE 5 ▶ **Finding More Distances**

Find the distance between the points $A(-1, -4)$ and $B(-3, 5)$.

Solution

As in the preceding example, we find

$$|AB| = \sqrt{[-3-(-1)]^2 + [5-(-4)]^2}$$
$$= \sqrt{(-2)^2 + 9^2} = \sqrt{4 + 81} = \sqrt{85}$$

With the aid of a table of square roots (or a calculator), we could express this result in decimal form. However, the indicated root form, $\sqrt{85}$ units, is adequate for our purposes. ∎

EXAMPLE 6 ▶ **Distances and Hurricanes**

One of the most destructive forces in nature is a hurricane. In the following grid, each unit represents 100 mi. At 3 P.M. the center of the hurricane was at $(2, -1)$. Hurricane-force winds extend 15 mi out from the center, and it is predicted that the center will be at $(1, -1)$ in 5 hours. Computer models indicate that after that, the hurricane will move directly NW toward Tampa at the same speed.

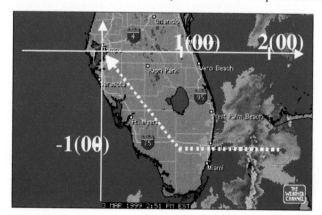

(a) How fast is the hurricane moving?

(b) How far is the hurricane from Tampa at 8 P.M.?

(c) Will Tampa feel hurricane force winds at 8 P.M.?

Solution

(a) The hurricane moved from $(2, -1)$ to $(1, -1)$, a distance of 100 mi (see the grid in the figure) in 5 hours. Thus, the hurricane is moving at 20 mph.

(b) At 8 P.M. the center of the hurricane is at $(1, -1)$, so we have to find the distance from $(1, -1)$ to $(0, 0)$. Let $(x_2, y_2) = (0, 0)$ and $(x_1, y_1) = (1, -1)$. Substituting in the distance formula, $d = \sqrt{(0 - 1)^2 + [0 - (-1)]^2} = \sqrt{2}$. Since each unit represents 100 mi, the actual distance from Tampa is $100\sqrt{2} \approx 141$ mi.

(c) Yes. Theoretically at least, hurricane-force winds extend 150 mi from the center, and the hurricane center is approximately 141 mi from Tampa.   ■

D. Applications

In order to make sensible decisions in problem solving and in certain consumer problems, it is helpful to be able to make a quick sketch of a linear function. Such a situation is illustrated in the next example.

EXAMPLE 7 ▶ Comparing Cellular Costs

Suppose the cost of cellular telephone service from company A is $200 for the telephone plus $.25 per minute of air time (the time spent talking on the telephone). If m is the number of minutes of air time and $C_1(m)$ is the corresponding cost in dollars, then

$$C_1(m) = 200 + 0.25m$$

For company B the cost is $100 for the telephone plus $.50 per minute of air time. If $C_2(m)$ is the corresponding cost, then

$$C_2(m) = 100 + 0.50m$$

(a) Graph C_1 and C_2 on the same set of axes.

(b) When will the cost be the same for these two companies?

(c) If cost were the only consideration, from which company would you get the service if you are planning to have more than 400 minutes of air time per month?

Solution

(a) Since C_1 and C_2 are both linear functions of m, their graphs are straight lines. For $m = 0$, $C_1(0) = 200$, so $(0, 200)$ is on the graph of C_1. For $m = 100$, $C_1(200) = 200 + 0.25(200) = 250$, so $(200, 250)$ is also on this graph. We draw a line through these two points as shown in Figure 7.17. Similarly, for $m = 0$, $C_2(0) = 100$, and for $m = 200$, $C_2(200) = 200$. Thus, the two points $(0, 100)$ and $(200, 200)$ are on the graph of C_2. Again, we draw a line through these two points as shown in Figure 7.17.

(b) Since both lines pass through the point $(400, 300)$, it is clear that the cost is the same for both companies when 400 minutes of air time are used.

(c) You can see from the graph that company A will cost less if you are planning to use more than 400 minutes of air time per month.

Note: Most companies give you a set number of free minutes before they charge for additional minutes.   ■

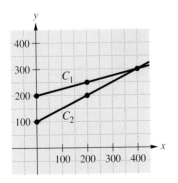

FIGURE 7.17

EXERCISES 7.2

A **Vertical and Horizontal Lines**
B **Graphs of Linear Functions and Equations**

1. In the table, the entries in the two columns have a common, simple linear relationship, $y = ax$. Find the missing entry.

x	3	2	$\frac{1}{2}$	
y	$2\frac{1}{4}$	$1\frac{1}{2}$	$\frac{3}{8}$	3

2. In the table, the entries in the two columns have a common, simple quadratic relationship, $y = ax^2$. Find the missing entry.

x	$\frac{1}{2}$	1	2	
y	$\frac{1}{2}$	2	8	50

In problems 3–18, graph each linear function or equation.

3. $f(x) = 3x + 6$

4. $f(x) = 2x + 5$

5. $f(x) = 3$

6. $f(x) = -2$

7. $x = -1$

8. $x = 4$

9. $f(x) = -x + 2$

10. $f(x) = -2x - 4$

11. $g(x) = -3x - 6$

12. $g(x) = -2x + 6$

13. $3x + 2y = 6$

14. $4x + 3y = 12$

15. $-2x + 3y = 6$

16. $-3x + 2y = 12$

17. $4x - 3y = 12$

18. $3x - 5y = 15$

C **Distance Between Two Points**

In problems 19–28, find the distance between the given points.

19. $(2, 4)$ and $(-1, 0)$

20. $(3, -2)$ and $(8, 10)$

21. $(-4, -5)$ and $(-1, 3)$

22. $(5, 7)$ and $(-2, 3)$

23. $(4, -8)$ and $(1, -1)$

24. $(-2, -2)$ and $(6, -4)$

25. $(3, 0)$ and $(3, -2)$

26. $(4, -1)$ and $(6, -1)$

27. $(-2, 3)$ and $(-2, 7)$

28. $(1, -5)$ and $(8, -5)$

D **Applications**

29. Elite Catering charges $500 to plan a banquet plus $25 per plate. Society Catering's charges are $1000 for the planning and $20 per plate. Let $E(x)$ and $S(x)$ represent the cost of a banquet for x persons catered by Elite and Society, respectively.
    a. Find $E(x)$ and $S(x)$.
    b. Graph $E(x)$ and $S(x)$ on the same set of axes.
    c. Find the number of persons for which the cost is the same with either company.

30. The costs (in dollars) of a repair call taking x hours or fraction thereof from companies A and B are, respectively,

$$C_1(x) = 20x + 30 \quad \text{and} \quad C_2(x) = 10x + 50$$

    a. Graph C_1 and C_2 on the same set of axes.
    b. For how many hours is the cost the same for either company?

31. Economy telephone calls to Spain from Tampa, Florida, have to be made between 6 P.M. and 7 A.M. The rates for calls dialed direct during those hours are $1.16 for the first minute or fraction thereof and $.65 for each additional minute or fraction thereof.
    a. Let t (in minutes) be the duration of a call, and make a table of the cost $C(t)$ for $0 < t \leq 5$.
    b. Draw a graph of $C(t)$ for $0 < t \leq 5$.

32. Use the rates described in problem 31.

    a. José wanted to call his dad in Spain during the economy hours, but José had only $5. How long a call could he make without exceeding his $5?

    b. Maria called her mother in Spain during the economy hours and was charged $7.01. How long was Maria's call?

33. Does your area have an evacuation map? In the case of floods, for example, persons in low-lying areas are ordered to evacuate. Suppose an evacuation order is issued for all people living within a radius of 10 mi from an area with coordinates $(2, -3)$. If your house is located at $(-2, 3)$, do you need to evacuate? (Coordinate units are in miles.)

34. When a train carrying poisonous gas derails, authorities order all people living within a 5-mi radius to evacuate the area. The derailment location has coordinates $(-3, 5)$, and your location is at $(2, 7)$. How far are you from the derailment, and are you in the evacuation zone? (Coordinate units are in miles.)

35. The percent of 12- to 17-year-olds who consumed alcohol in the past month is shown in the figure (top, right column) and can be approximated by

$$D(t) = -\tfrac{3}{2}t + 50$$

using the points $(0, 50)$ and $(20, 20)$, where t is the number of years after 1975. Using this model,

    a. what was that percent in 1975?

    b. what would it be in the year 2005?

    c. in which year would there be no 12- to 17-year-olds who consumed alcohol in the past month?

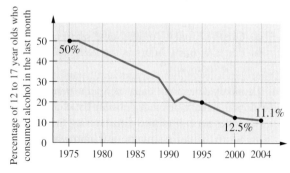

Source: U.S. Center for Disease Control (updated by author.); SAMHSA, National Survey of Drug and Health, 2003 and 2004, Table 2.2B.

Teenage drinking.

36. Do you do text messaging with your cell phone? The graph shows that the percent of teens sending text messages increases with age. We can approximate the percent $P(a)$ by $P(a) = 7a - 63$, where a is the age of the person. Assuming that the trend continues,

    a. use $P(a)$ to estimate the percentage of 18-year-olds that use text messaging.

    b. use $P(a)$ to estimate the percentage of 19-year-olds that use text messaging.

    c. According to this model, by what percent does the amount of text messages sent increase every year?

    d. Based on the graph, what was the percent *increase* in text messages sent between 12- and 13-year-olds? Does yor answer fit the pattern given by $P(a)$?

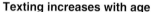
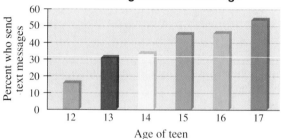

Texting increases with age

Source: tinyuri.com/bza61 or
www.pewinternet.org/pdfs/PIP_Teens_Tech_July2005web.pdg.

37. The table shows the mean verbal and math SAT scores from 2001 to 2005.

Year	Verbal	Math
2001	506	514
2002	504	516
2003	507	519
2004	508	518
2005	508	520

Source:
www.collegeboard.com/prod_downloads/about/news_
info/cbsenior/yr2005/table2_mean_SAT_scores.pdf.

    a. Graph the points representing the data for both the verbal and the math portions.

    b. Calculate the average verbal score V and the average math score M for 2001 to 2005.

    c. Graph $y = V$ and $y = M$ in the same coordinate system. How well do $y = V$ and $y = M$ model the data you graphed?

    d. Are $y = V$ and $y = M$ functions? How do you know?

38. The table shows the mean English and math ACT scores from 2001 to 2005.

Year	English	Math
2000–01	20.5	20.7
2001–02	20.2	20.6
2002–03	20.3	20.6
2003–04	20.4	20.7
2004–05	20.4	20.7

Source: www.act.org/news/data/05/pdf/tsum.pdf.

a. Graph the points representing the data for both the English and the math portions.
b. Calculate the average English score E and the average math score M for 2001 to 2005.
c. Graph $y = E$ and $y = M$ in the same coordinate system. How well do $y = E$ and $y = M$ model the data you graphed?
d. Are $y = E$ and $y = M$ functions? How do you know?

 In Other Words

39. Suppose the three vertices of a triangle are $A(a_1, a_2)$, $B(b_1, b_2)$, and $C(c_1, c_2)$. Explain in detail how you can determine whether the triangle is a right triangle.

40. Explain in detail how you can use the Problem Solving procedure given in this section to graph $y = b$.

41. Explain in detail how you can use the Problem Solving procedure given in this section to graph $x = c$.

42. Explain in detail the steps you would use in graphing $f(x) = ax + b$; then look at the Discovery problem and see if your steps are similar.

 Using Your Knowledge

There are many applications that require the knowledge you have obtained to save you some money! The graph shows the amount owed on a $1000 debt at an 18% interest rate when the minimum $25 payment is made (blue) or when a new monthly payment of $92 is made. Thus, at the current monthly payment of $25, it will take 60 months to pay the $1000 balance (you got to $0!). With a new $92 monthly payment you pay off the $1000 balance in 12 months. To know the balance, you have to know how to read the graph!

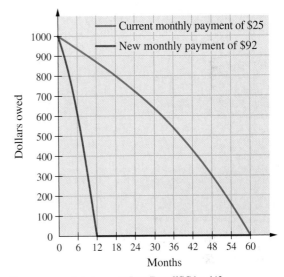

Source: www.dinkytown.net/java/PayoffCC.html43.

43. What is your balance after 6 months if you are paying $25 a month?

44. What is your balance after 6 months if you are paying $92 per month?

45. What is your balance after 18 months if you are paying $25 a month?

46. What is your balance after 48 months if you are paying $25 a month?

 Discovery

The ideas developed in this section for graphing lines can be summarized by means of a flowchart. A **flowchart** is a pictorial representation showing the logical steps that have to be taken in order to perform a task.

The basic component of a flowchart is a box that contains a command. For example, $\boxed{x \to 0}$ says to take x, a previously given quantity, and let it be equal to 0. In mathematics, this instruction is given as

Let $x = 0$.

The figure shows a flowchart that can be used to find the graph of any line not passing through the origin.

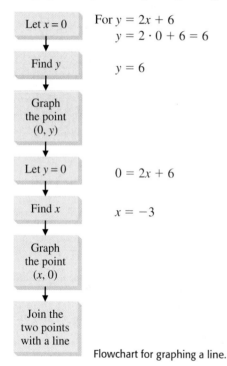

For $y = 2x + 6$
$\quad y = 2 \cdot 0 + 6 = 6$

$y = 6$

$0 = 2x + 6$

$x = -3$

Flowchart for graphing a line.

An example of how it works for the line $y = 2x + 6$ follows:

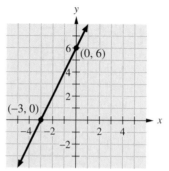

Use the flowchart technique to graph the linear relations you graphed in the following problems of Exercises 7.2.

47. Problem 3   **48.** Problem 4   **49.** Problem 9

50. Problem 10   **51.** Problem 11   **52.** Problem 12

53. Problem 13   **54.** Problem 14   **55.** Problem 15

56. Problem 16   **57.** Problem 17   **58.** Problem 18

Calculator Corner

You can use a calculator to find the distance between two points. For example, to find the distance between $(2, -3)$ and $(8, 5)$, you must use equation (3). Here are the steps you need. Press

$\boxed{(}\;\boxed{8}\;\boxed{-}\;\boxed{2}\;\boxed{)}\;\boxed{x^2}\;\boxed{+}\;\boxed{(}\;\boxed{5}\;\boxed{-}\;\boxed{3}\;\boxed{\pm}\;\boxed{)}\;\boxed{x^2}$
$\boxed{=}\;\boxed{\sqrt{}}$ (or $\boxed{\text{2nd}}\;\boxed{\sqrt{x}}$)

Note that in this case the answer appears as 10. If you work Example 5 using a calculator, your answer will be 9.219544457, an approximation for $\sqrt{85}$.

1. For problems 19, 21, 23, 25, and 27, estimate the distance between the two points with your calculator.

7.3   Slopes and Equations of a Line

A Graphic Look at Engine Deposits

Study the advertisement in Figure 7.18 (page 423). It shows that engine deposits increase if you do *not* use Texaco gasoline and decrease if you do. How fast do the deposits increase or decrease? To find out, you can look at the ratio of deposit

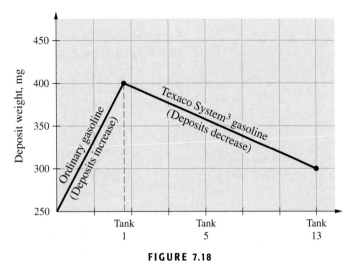

FIGURE 7.18
In the BMW test, Texaco's new System³ gasoline removed performance-robbing deposits left by ordinary gasoline.

weight (in milligrams) to tanks of gas. Without Texaco gas, the deposits increased from 250 to 400 mg after using approximately one tank of gas. Thus,

$$\frac{\text{Mg of deposit}}{\text{Tanks of gas}} = \frac{400 - 250}{1} = 150 \frac{\text{mg}}{\text{tank}}$$

How fast did the deposits decrease? They decreased from 400 to 300 mg after using about 12 tanks of Texaco. The ratio of decrease is given by

$$\frac{\text{Mg of deposit}}{\text{Tanks of gas}} = \frac{300 - 400}{13 - 1} = -\frac{100}{12} = -8\frac{1}{3} \frac{\text{mg}}{\text{tank}}$$

The ratio of milligrams of deposit to tanks of gas, or in general, the ratio of rise (difference in y values) to run (difference in x values) for a line, is called the **slope** of the line. The slope of the increasing line is 150 mg per tank, whereas the slope of the decreasing line is $-8\frac{1}{3}$ mg per tank. Note that the slope of an *increasing* line (rising from left to right) is *positive* and the slope of a *decreasing* line (falling from left to right) is *negative*. Also, the two lines look almost perpendicular to each other, so if one has slope m, the other one must have slope $-1/m$ (see problem 54 in Exercises 7.3). Is this the case? If not, what is wrong? Think of how you can redraw the graph more accurately to reflect the actual situation. ▶

The best investment in the graph in Figure 7.19 is the tax-exempt account. Can you tell when the account produces the most money? Since the line is steeper during the last 5 years, that is, when the account will produce the most. The annual amount produced in the last 5 years is approximately

$$\frac{200{,}000 - 140{,}000}{30 - 25} = \frac{60{,}000}{5} = \$12{,}000$$

That is, the rate of change in the account over the last 5 years is about \$12,000 per year. In general, when you look at a line graph and speak of its "rate of change," you are referring to the **slope** of the line.

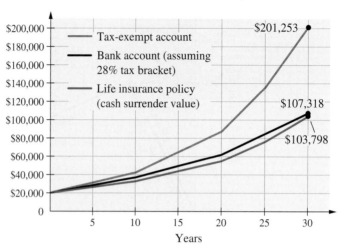

Note: Life insurance policy is single-premium, variable-life coverage for a 35-year-old male. Assumes interest on other accounts is reinvested.

FIGURE 7.19
$20,000 invested at 8% per year in three ways.

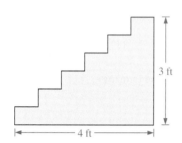

FIGURE 7.20
Slope $= \frac{3}{4}$

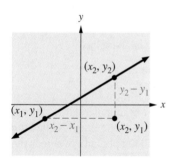

FIGURE 7.21
The slope of a line:
$m = \dfrac{y_2 - y_1}{x_2 - x_1}, x_1 \ne x_2$

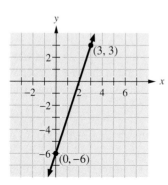

FIGURE 7.22
Line with positive slope.

A. Slope

In mathematics, an important feature of a straight line is its *steepness*. We can measure the steepness of a nonvertical line by means of the ratio of the **vertical rise** (or **fall**) to the corresponding **horizontal run.** We call this ratio the **slope.** For example, a staircase that rises 3 ft in a horizontal distance of 4 ft is said to have a slope of $\frac{3}{4}$ (see Figure 7.20). By the way, most building codes specify that the maximum safe slope for stairs is 0.83. Is this staircase safe?

In general, we use the following definition:

Definition of Slope

A line going through two points (x_1, y_1) and (x_2, y_2), where $x_1 \ne x_2$, has **slope** m, where

$$m = \frac{y_2 - y_1}{x_2 - x_1} = \frac{\text{rise}}{\text{run}} = \frac{\text{change in } y}{\text{change in } x}$$

Figure 7.21 shows the horizontal run $x_2 - x_1$ and the vertical rise $y_2 - y_1$ used in calculating the slope. We do not define slope for a vertical line because $x_2 - x_1 = 0$, and division by 0 is not defined. The slope of a horizontal line is obviously 0 because all points on such a line have the same y values. (See an illustrated summary of these ideas at the end of this section, page 425.)

EXAMPLE 1 ▶ **Finding Slopes Given Two Points**

Find the slope of the line that passes through the points $(0, -6)$ and $(3, 3)$.

Solution

The two given points are shown in Figure 7.22. Suppose that you choose $(x_1, y_1) = (0, -6)$ and $(x_2, y_2) = (3, 3)$. Then you get

$$m = \frac{3 - (-6)}{3 - 0} = \frac{9}{3} = 3$$

If you choose $(x_1, y_1) = (3, 3)$ and $(x_2, y_2) = (0, -6)$, then

$$m = \frac{-6 - 3}{0 - 3} = \frac{-9}{-3} = 3$$

As you can see, it makes no difference which point is labeled (x_1, y_1) and which is labeled (x_2, y_2). Since an interchange of the two points simply changes the sign of both the numerator and the denominator in the slope formula, the result is the same in both cases. ■

EXAMPLE 2 ▶ Finding More Slopes Given Two Points

Find the slope of the line that passes through the two points $(3, -4)$ and $(-2, 3)$. See Figure 7.23.

Solution

We take $(x_1, y_1) = (-2, 3)$, so that $(x_2, y_2) = (3, -4)$. Then

$$m = \frac{-4 - 3}{3 - (-2)} = -\frac{7}{5}$$

■

Examples 1 and 2 are illustrations of the fact that a line that rises from left to right has a **positive slope** and one that falls from left to right has a **negative slope**. Note that the slope of a line gives **the change in y per unit change in x**. A summary for slopes follows:

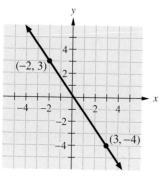

FIGURE 7.23
Line with negative slope.

Slope Summary

A line that *falls* from left to right has a *negative* slope.	The slope of a *vertical* line is *undefined*. Since $x_2 - x_1 = 0$, $$m = \frac{y_2 - y_1}{x_2 - x_1} = \frac{y_2 - y_1}{0}$$ so m is *undefined*.	A line that *rises* from left to right has a *positive* slope.	A *horizontal* line has zero slope. Since $y_2 - y_1 = 0$, $$m = \frac{y_2 - y_1}{x_2 - x_1} = \frac{0}{x_2 - x_1}$$ so $m = 0$.

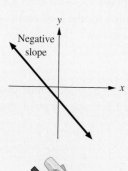
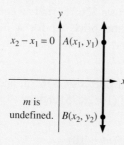
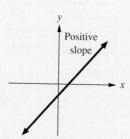
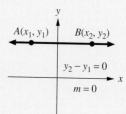

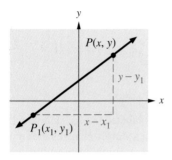

FIGURE 7.24

B. Equations of Lines

The slope of a line can be used to obtain an equation of the line. For example, suppose the line passes through a point $P_1(x_1, y_1)$ and has slope m. See Figure 7.24. We let $P(x, y)$ be any second point (distinct from P_1) on the line. Then, by definition, the slope of the line in terms of these two points is

$$\frac{y - y_1}{x - x_1} = m$$

Multiplying both sides by $(x - x_1)$, we get $y - y_1 = m(x - x_1)$.

Point-Slope Equation of a Line

The **point-slope form** of the equation of a line with slope m and passing through the point (x_1, y_1) is

$$y - y_1 = m(x - x_1) \tag{1}$$

This equation must be satisfied by the coordinates of every point on the line.

EXAMPLE 3 ▶ Finding Equations Given a Point and the Slope

Find an equation of the line that passes through the point $(2, -3)$ and has slope $m = -4$.

Solution

Using the point-slope equation (1), we get

$$y - (-3) = -4(x - 2)$$
$$y + 3 = -4x + 8$$
$$y = -4x + 5$$

∎

EXAMPLE 4 ▶ Finding an Equation Given a Point and the Slope

Taxi fares in Key West are \$2.25 for the first $\frac{1}{5}$ mi and then \$2.50 for each additional mile. If a 10-mi ride costs \$26.75, find an equation for the total cost C of an m-mi ride. What will the price be for a 30-mi ride? (Source: www.keywesttaxi.com/Rates.htm.)

Solution

Since we know that a 10-mi ride costs \$26.75, and each additional mile costs \$2.50, we are given the point $(10, 26.75)$ and the slope 2.5. Using the point-slope form, we have

$$C - 26.75 = 2.5(m - 10)$$

Or $C = 2.5m - 25 + 26.75$

That is, $C = 2.5m + 1.75$

For a 30-mi ride, $m = 30$ and $C = 2.5(30) + 1.75 = \$76.75$.

∎

An important special case of equation (1) is that in which the given point is the point where the line intersects the y axis. Let this point be denoted by $(0, b)$; b is the y intercept of the line. Using equation (1), we obtain

$$y - b = m(x - 0)$$

and by adding b to both sides, we get $y = mx + b$.

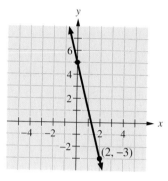

FIGURE 7.25

$y = -4x + 5$

> ### Slope-Intercept Form of a Line
>
> The **slope-intercept form** of the equation of a line with slope m and y intercept b is
>
> $$y = mx + b \qquad (2)$$

Notice that the answer to Example 3 was given in the slope-intercept form. This form is convenient for reading off the slope and the y intercept of the line. Thus, the answer to Example 3 immediately tells us that the slope of the line is -4 and the y intercept is 5 (see Figure 7.25).

EXAMPLE 5 ▶ Finding the Slope and the y Intercept Given an Equation

Find the slope and the y intercept of the line with equation $6x + 3y = 5$.

Solution

By equation (2), the slope-intercept form of the equation of a line is $y = mx + b$, where m is the slope and b is the y intercept. We can solve the given equation for y by subtracting $6x$ from both sides and then dividing by 3 and obtaining

$$y = -2x + \frac{5}{3}$$

an equation in the slope-intercept form, $y = mx + b$, with $m = -2$ and $b = \frac{5}{3}$. Thus, the slope m of the given line is -2, and the y intercept b is $\frac{5}{3}$. ∎

The procedure of Example 5 can be followed for any equation of the form $Ax + By = C$, where $B \neq 0$, to obtain an equation in the slope-intercept form. If $B = 0$, then we can divide by A to get $x = C/A$, an equation of a line parallel to the y axis.

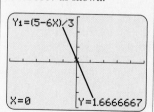

GRAPH IT

To find the x or y intercept of a line, solve for y, press $\boxed{Y=}$, and enter $y = (5 - 6x)/3$. To get integral values, press $\boxed{ZOOM}$ $\boxed{8}$ $\boxed{ENTER}$ $\boxed{TRACE}$. Use the $\boxed{\blacktriangleright}$ or $\boxed{\blacktriangleleft}$ buttons to find the y intercept 1.6666667 as shown.

> ### Standard Equation of a Straight Line
>
> Every equation of the form
>
> $$Ax + By = C$$
>
> where A, B, and C are real numbers and A and B are not both 0, is an equation of a straight line.

On the other hand, we can show that every straight line has an equation of the form $Ax + By = C$. Any line that is not parallel to the y axis has a slope-intercept equation $y = mx + b$, which can be written in the form $mx - y = -b$ by subtracting b and y from both sides. This last equation is of the form

$Ax + By = C$, with $A = m$, $B = -1$, and $C = -b$. For example, the equation $y = 2x + 5$ can be written in the form $2x - y = -5$ by subtracting 5 and y from both sides. This equation is the special case of $Ax + By = C$, with $A = 2$, $B = -1$, and $C = -5$. A line parallel to the y axis has an equation $x = a$, which is already of the form $Ax + By = C$, with $A = 1$, $B = 0$, and $C = a$.

General Equation of a Line

The **general form** of the equation of a straight line is

$$Ax + By = C \tag{3}$$

Every straight line can be described by the equation $Ax + By = C$.

EXAMPLE 6 ▶ **Finding the General Form Given Two Points**

Find the general form of the equation of the line that passes through $(6, 2)$ and $(3, -2)$.

Solution
We first find the slope of the line.

$$m = \frac{-2 - 2}{3 - 6} = \frac{-4}{-3} = \frac{4}{3}$$

Then, using the point-slope form, with $P(6, 2)$ and $m = \frac{4}{3}$, we find the equation

$$y - 2 = \tfrac{4}{3}(x - 6)$$
$$3y - 6 = 4x - 24$$
$$4x - 3y = 18$$

We can check this answer by verifying that $(6, 2)$ and $(3, -2)$ are both solutions of the equation.   ■

The ideas of Example 6 can often be used to obtain a simple formula that summarizes a group of data in a convenient form. For example, Table 7.7 shows the desirable weight range corresponding to a given height for men and for women.

TABLE 7.7

Height (in.)	Men's Weight (lb)	Women's Weight (lb)	Height (in.)	Men's Weight (lb)	Women's Weight (lb)
62	108–134	98–123	70	140–174	130–163
63	112–139	102–128	71	144–179	134–168
64	116–144	106–133	72	148–184	138–173
65	120–149	110–138	73	152–189	
66	124–154	114–143	74	156–194	
67	128–159	118–148	75	160–199	
68	132–164	122–153	76	164–204	
69	136–169	126–158			

Can we find an equation that illustrates the relationship between height and weight? First, we must realize the following two things:

1. Men are heavier, so there will be one equation for men and another equation for women.

2. The table gives only weight *ranges*. For instance, a 76-in. (6-ft 4-in.) man should weigh between 164 and 204 lb.

Hence, to write an equation requires ideally that we be more specific, as in the next example.

EXAMPLE 7 ▶ Relationship Between Height and Weight

Find an equation that gives the relationship between a man's height h (in inches) and his weight w (in pounds) using the lower weights in Table 7.7 as the desirable weights.

Solution

If we examine the heights and the corresponding weights in Table 7.7, we see that the heights increase by 1 in. from one entry to the next and the corresponding weights increase by 4 lb. This means that all the points $(62, 108)$, $(63, 112), \ldots, (76, 164)$ lie on one straight line because the slope of a line connecting any consecutive pair of these points is the same as the slope of the line connecting any other consecutive pair. Thus, we want to find the equation of this line. The slope of the line is easily obtained by using the first pair of points $(62, 108)$ and $(63, 112)$.

$$m = \frac{112 - 108}{63 - 62} = 4$$

Then we can use the point-slope form of the equation to obtain

$$w - 108 = 4(h - 62)$$
$$w - 108 = 4h - 248$$
$$w = 4h - 140$$

We can check this against entries in Table 7.7. For example, for a 72-in. (6-ft) man, the equation gives

$$w = 4(72) - 140$$
$$= 288 - 140 = 148$$

which agrees with the table entry. ∎

C. Parallel Lines

Since the slope of a line determines its direction, you can see that **two lines with the same slope and different *y* intercepts are parallel lines.** The next example makes use of this idea.

EXAMPLE 8 ▶ **Finding Equations of Lines Parallel to a Given Line**

Show that $3y = x + 2$ and $2x - 6y = 7$ describe parallel lines.

Solution

Solve each equation for y and obtain

$$y = \tfrac{1}{3}x + \tfrac{2}{3} \qquad \text{and} \qquad y = \tfrac{1}{3}x - \tfrac{7}{6}$$

These equations show that both lines have slope $\tfrac{1}{3}$. The y intercepts are different, so the lines are parallel. ■

TABLE 7.8

To Find the Equation of a Line, Given the Following:	Use
Two points (x_1, y_1) and (x_2, y_2), $x_1 \neq x_2$	Two-point form: $y - y_1 = m(x - x_1)$, where $m = (y_2 - y_1)/(x_2 - x_1)$
A point (x_1, y_1) and the slope m	Point-slope form: $y - y_1 = m(x - x_1)$
The slope m and the y intercept b	Slope-intercept form: $y = mx + b$

You have learned how to identify and work with many of the properties of lines. At this point, many students ask, "Which formula should we use in the problems?" Table 7.8 tells you which formula to use, depending on what information is given. Study Table 7.8 before you attempt the problems in Exercises 7.3. The resulting equation can always be written in the general form $Ax + By = C$.

PROBLEM SOLVING

Finding the Equation of a Line

❶ Read the problem.

Find the equation of the line parallel to $2y = 6x + 5$ and passing through $(1, 2)$.

❷ Select the unknown.

You are asked to find the line parallel to $2y = 6x + 5$ and passing through $(1, 2)$.

❸ Think of a plan.
Find the slope of the given line and use it in finding the equation of the new line.

If the new line is to be parallel to $2y = 6x + 5$ or, equivalently, $y = 3x + \tfrac{5}{2}$, the new line must have slope $m = 3$. Since the line passes through the point $(1, 2)$ and has slope $m = 3$, use the point-slope formula

❹ Use one of the formulas to find the equation.
Which formula can you use?

$$y - y_1 = m(x - x_1)$$

with $m = 3$ and $(x_1, y_1) = (1, 2)$ to obtain

$$y - 2 = 3(x - 1)$$

❺ Verify the answer.

To verify your answer, solve for y in $y - 2 = 3(x - 1)$ to obtain $y = 3x - 1$. This equation has slope 3, thus is parallel to the given line, and passes through $(1, 2)$ because $2 - 2 = 3(1 - 1)$.

TRY EXAMPLE 9 NOW.

Cover the solution, write your own solution, and then check your work.

EXAMPLE 9 ▶ Finding the Equation of a Line Parallel to Another

Find the equation of a line parallel to $3y = -6x + 8$ and with y intercept -3.

Solution

The line $3y = -6x + 8$ has slope -2. (Why?) Since we want to find a line parallel to $3y = -6x + 8$, the new line also must have slope $m = -2$. We now have the slope and y intercept of the line we want. Using the slope-intercept form $y = mx + b$, with $m = -2$ and $b = -3$, we obtain *the desired equation* $y = -2x - 3$.   ■

D. Applications

In problem 35 in Exercises 7.2 we approximated the percent of 12- to 17-year-olds who consumed alcohol in the past month by $D(t) = -\frac{3}{2}t + 50$. Now we are ready to show you how we did it so that you can do it too!

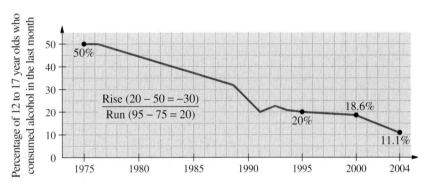

Sources: U.S. Center for Disease Control (updated by author); SAMHSA, National Survey of Drug and Health, 2003 and 2004, Table 2.2B.

FIGURE 7.26

Teenage drinking.

EXAMPLE 10 ▶ Finding the Equation of a Line

(a) Find the equation of the line shown in Figure 7.26. Use the points $(0, 50)$ and $(20, 20)$.

(b) How can you interpret the slope of the line?

Solution

(a) We can find the equation of a line if we know the slope m and the y intercept b. The slope m of the line in the graph is

$$m = \frac{\text{rise}}{\text{run}} \approx \frac{-30}{20} = -\frac{3}{2}$$

and the y intercept $b = 50$ (see Figure 7.26). Thus, the equation of the line is $y = -\frac{3}{2}t + 50$ or, in function notation, $D(t) = -\frac{3}{2}t + 50$, where t is the number of years after 1975.

(b) The slope of the line indicates that the percent of 12- to 17-year-olds who consumed alcohol in the past month is *decreasing* at an annual rate of $\frac{3}{2}\% = 1.5\%$.   ■

Online Study Center

To further explore lines and slopes, access links 7.3.1 and 7.3.2 on this textbook's Online Study Center.

GRAPH IT

We can use a grapher to find the equation of a line given two points. To do Example 1, press [STAT] [1] and enter 0 and 3 under L1 and −6 and 3 under L2; then press [STAT] [▶] [4] [ENTER] to get the equation $y = ax + b$, with $a = 3$ and $b = -6$, that is, the line $y = 3x - 6$. Graph this line. Now graph the two points we entered in the table by using [2ND] [STAT PLOT] [ENTER] [ENTER] [GRAPH]. Does the line pass through the two points? (See Window 1.)

In Example 3 we are given the point $(2, -3)$. Since the slope is $-4 = \frac{-4}{1}$, move 1 unit right and 4 units down to reach the point $(3, -7)$. We now have two points. So we use the same procedure and find the equation of the line $y = ax + b$, with $a = -4$ and $b = 5$, that is, the line $y = -4x + 5$. (See Window 2.) In the Problem Solving section we are given the point $(1, 2)$, and we want a line parallel to $2y = 6x + 5$, that is, $y = \frac{6x + 5}{2}$, which has slope $3 = \frac{3}{1}$. This means that the change in x is 1 and the change in y is 3. Starting at the point $(1, 2)$, we go 1 unit right and 3 units up, ending at $(2, 5)$. We now have two points, $(1, 2)$ and $(2, 5)$, so we can follow our previous steps to make two lists and get an equation of the form $y = ax + b$. For $(1, 2)$ and $(2, 5)$, $a = 3$ and $b = -1$; that is, $y = 3x - 1$ as before. Now check Example 9.

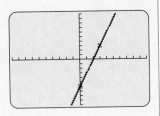

Window 1

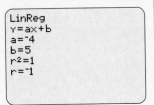

Window 2
$y = -4x + 5$

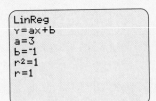

Window 3
$y = 3x - 1$

EXERCISES 7.3

A Slope

In problems 1–10, find the slope m of the line that passes through the two given points.

1. $(1, 2)$ and $(3, 4)$
2. $(1, -2)$ and $(-3, -4)$
3. $(0, 5)$ and $(5, 0)$
4. $(3, -6)$ and $(5, -6)$
5. $(-1, -3)$ and $(7, -4)$
6. $(-2, -5)$ and $(-1, -6)$
7. $(0, 0)$ and $(12, 3)$
8. $(-1, -1)$ and $(-10, -10)$
9. $(3, 5)$ and $(-2, 5)$
10. $(4, -3)$ and $(2, -3)$

B Equations of Lines

In problems 11–16, find the slope-intercept form (if possible) of the equation of the line that has the given properties (m is the slope).

11. Passes through $(1, 2)$; $m = \frac{1}{2}$
12. Passes through $(-1, -2)$; $m = -2$
13. Passes through $(2, 4)$; $m = -1$
14. Passes through $(-3, 1)$; $m = \frac{3}{2}$
15. Passes through $(4, 5)$; $m = 0$
16. Passes through $(3, 2)$; slope is not defined (does not exist)

In problems 17–26, find (**a**) the slope and (**b**) the y intercept for the graph of the given equation.

17. $y = x + 2$

18. $2x + y = 3$

19. $3y = 4x$

20. $2y = x + 4$

21. $x + y = 14$

22. $y - 4x = 8$

23. $y = 6$

24. $2y = 16$

25. $x = 3$

26. $3x = -6y + 9$

In problems 27–32, find the general form of the equation of the line that passes through the two given points.

27. $(1, -1)$ and $(2, 2)$

28. $(-3, -4)$ and $(-2, 0)$

29. $(3, 2)$ and $(2, 3)$

30. $(3, 0)$ and $(0, 5)$

31. $(0, 0)$ and $(1, 10)$

32. $(-4, -1)$ and $(-4, 3)$

33. Use Table 7.7 to find a formula relating the height h of a man and his ideal weight w given by the *second* number in the weight column. See Example 7.

34. Use Table 7.7 to find a formula relating the height h of a woman and her weight w as given by the *first* number in the weight column. See Example 7.

35. Repeat problem 34 using the *second* number in the weight column.

36. On the basis of your answer to problem 35, if a woman weighs 183 lb, how tall should she be? (This is not given in Table 7.7.)

37. Taxi fares in San Francisco are $2 for the first mile and $1.70 for each additional mile. If a 10-mi ride costs $17.30, find an equation for the total cost C of an m-mi ride. What would the price be for a 30-mi ride?

38. A different taxicab company in San Francisco charges $3 for the first mile and $1.50 for each additional mile. If a 20-mi ride costs $31.50, find an equation for the total cost C of an m-mi ride. What would the price of a 10-mi ride be? If you are taking a 50-mi ride, which company is cheaper, this one or the company in problem 37?

39. Pedro took a cab in San Francisco and paid the fare quoted in problem 37. Tyrone paid the fare quoted in problem 38. Amazingly, they paid the same amount! How far did they ride?

40. In New York, a 20-mi cab ride is $32 and consists of an initial charge and $1.50 per mile thereafter.

Find an equation for the total cost C of an m-mi ride. How much would you have to pay for a 30-mi ride?

41. The cost C for San Francisco fares (problem 37) is $2 for the first mile and $1.70 for each mile thereafter. We can find C by following these steps.
  a. What is the cost of the first mile?
  b. If the whole trip is m mi, how many miles do you travel after the first mile?
  c. How much do you pay per mile after the first mile?
  d. What is the cost of each mile after the first mile?
  e. The total cost C is the sum of the cost of the first mile and the cost of the miles after the first. What is that cost? Is your answer the same as your answer to problem 37?

42. New York fares are easier to compute. They are simply $2 for the initial fare and $1.50 for each mile thereafter. Find the total cost C for an m-mi trip. Do you get the same answer as you did in problem 40?

43. Did you know that you could rent cell phones for your overseas travel? If you are in Paris, the cost for a 1-week rental, including 60 min of long-distance calls to New York, the rental fee, and $2 for each airtime minute, was $175.
  a. Find a formula for the total cost C of a rental phone that includes m min of long-distance calls to New York.
  b. What is the weekly rental fee for the phone?

44. Long-distance calls from the Hilton Hotel in Paris, France, to New York cost $7.80 per minute.
  a. Find a formula for the cost C of m min of long-distance calls to New York.
  b. How many minutes can you use so that the charges are identical to those you would pay when renting the telephone of problem 43? Answer to the nearest minute.

C Parallel Lines

In problems 45–50, determine whether the given lines are parallel.

45. $y = 2x + 5$; $4x - 2y = 7$

46. $y = 4 - 5x$; $15x + 3y = 3$

47. $2x + 5y = 8$; $5x - 2y = -9$

48. $3x + 4y = 4$; $2x - 6y = 7$

49. $x + 7y = 7; 2x + 14y = 21$

50. $y = 5x - 12; y = 3x - 8$

51. Find an equation of the line that passes through the point $(1, -2)$ and is parallel to the line $4x - y = 7$.

52. Find an equation of the line that passes through the point $(2, 0)$ and is parallel to the line $3x + 2y = 5$.

53. It is shown in analytic geometry that two lines $a_1x + b_1y = c_1$ and $a_2x + b_2y = c_2$ are perpendicular if and only if

$$a_1a_2 + b_1b_2 = 0$$

that is, **two lines are perpendicular if and only if the sum of the products of the corresponding coefficients of x and y in the general form of the equations is 0.** This leads to a very easy way of writing an equation of a line that is perpendicular to a given line. For instance, if the given line is $3x + 5y = 8$, then for each real number c, $5x - 3y = c$ is a line perpendicular to the given line. This is obvious because $(3)(5) + (5)(-3) = 0$. Notice that all we need to do to form the left side of the second equation is interchange the coefficients of x and y in the first equation and change the sign of one of these coefficients. If we wish to have the second line pass through a specified point, we select the value of c so that this happens. Thus, if the second line is to pass through $(3, 2)$, then c has to be selected so that $(3, 2)$ is in the solution set of the equation. Hence,

$$5(3) - 3(2) = c \quad \text{or} \quad c = 9$$

The line $5x - 3y = 9$ passes through $(3, 2)$ and is perpendicular to the line $3x + 5y = 8$. In each of the following problems, find an equation of the line that is perpendicular to the given line and passes through the given point:

a. $2x + 5y = 7; (2, 0)$ **b.** $y = 2x - 3; (1, 1)$
c. $x - 2y = 3; (2, -2)$ **d.** $4x + 5y = 9; (1, 1)$

54. If the lines $y = m_1x + b_1$, $m_1 \neq 0$, and $y = m_2x + b_2$ are perpendicular, then $m_2 = -1/m_1$. Show this by referring to problem 53. This leads to the simple statement that **the slopes of perpendicular lines are negative reciprocals of each other.** Why do we need the condition $m_1 \neq 0$?

55. A line passes through the two points $(0, 0)$ and $(100, 200)$. A second line passes through the two points $(0, 10)$ and $(790, -405)$. Can you see how to determine whether these are perpendicular lines? (*Hint:* Look at problem 54.)

D Applications

56. The graph below shows the number of U.S. wireless subscriptions (in millions) from 2000 to 2010. Assume that the number of wireless subscriptions (red line) is a straight line.
  a. The slope of the line is rise/run. Find the slope from 2000 to 2008.
  b. What is the slope-intercept equation of the line?
  c. If $P(t)$ represents a function indicating the number of millions of person-trips, where t is the number of years after 2000, what is $P(t)$?
  d. The slope of the line is simply a ratio. What does the ratio represent?
  e. How many millions of person-trips would you predict for the year 2020?

Wireless vs. wireline consumer subscriptions

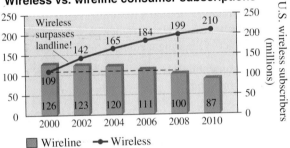

Source: Tom Hamby, President, Bell South Alabama, http://tinyurl.com/atkgj.

57. Suppose that the graph of net sales (in billions of dollars) is a line.

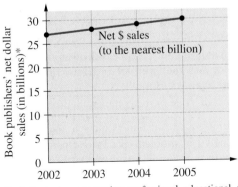

*Includes trade, mass market, professional, educational, and university press publishers.
Source: Book Industry Study Group (updated by author).

a. What is the slope of the line, and what does it represent?

b. What is the y intercept?

c. If $U(t)$ represents the number of sales (in billions) after 2002, what is $U(t)$?

d. What are the projected sales (in billions) for 2010?

58. *Daily fat consumption.* According to the U.S. Department of Agriculture, the daily fat consumption F per person can be approximated by

$$F = 190 + 0.8t \text{ (grams)}$$

where t is the number of years after 2000.

a. What is the slope of the line?

b. What does the slope represent?

59. *Daily seafood consumption.* According to the U.S. Department of Agriculture, the daily consumption T of tuna can be approximated by

$$T = 3.87 - 0.13t \text{ (pounds)}$$

and the daily consumption F of fish and shellfish can be approximated by

$$F = -0.29t + 15.46 \text{ (pounds)}, \text{ where } t \text{ is the}$$
number of years after 2000.

a. Is the consumption of tuna increasing or decreasing?

b. Is the consumption of fish and shellfish increasing or decreasing?

c. Which consumption (tuna or fish and shellfish) is decreasing faster?

60. *Life expectancy of women.* The average life span (life expectancy) y of an American woman is given by $y = 0.15t + 80$, where t is the number of years after 2000. (Source: U.S. National Center for Health Statistics, *Statistical Abstract of the United States.*)

a. What is the slope of this line?

b. Is the life span of American women increasing or decreasing?

c. What does the slope represent?

61. *Life expectancy of men.* The average life span y of an American man is given by $y = 0.15t + 74$, where t is the number of years after 2000. (Source: U.S. National Center for Health Statistics, *Statistical Abstract of the United States.*)

a. What is the slope of this line?

b. Is the life span of American men increasing or decreasing?

c. What does the slope represent?

62. *Velocity of a thrown ball.* The speed v of a ball thrown up with an initial velocity of 15 m/sec is given by $v = 15 - 5t$, where v is the velocity (in meters per second) and t is the number of seconds after the ball is thrown.

a. What is the slope of this line?

b. Is the velocity of the ball increasing or decreasing?

c. What does the slope represent?

63. *Daily fat consumption.* The number of fat grams f consumed daily by the average American can be approximated by $f = 165 + 0.4t$, where t is the number of years after 2000. (Source: U.S. Department of Agriculture, *Statistical Abstract of the United States.*)

a. What is the slope of this line?

b. Is the consumption of fat increasing or decreasing?

c. What does the slope represent?

64. *Milk products consumption.* The number of gallons of milk products g consumed annually by the average American can be approximated by $g = 24 - 0.2t$, where t is the number of years after 2000.

a. What is the slope of this line?

b. Is the consumption of milk products increasing or decreasing?

c. What does the slope represent?

65. *Daily fat intake.* According to the U.S. Department of Agriculture, the total daily fat intake g (in grams) per person can be approximated by $g = 140 + t$, where t is the number of years after 1950.

a. What was the daily fat intake per person in 1950?

b. What was the daily fat intake per person in 1990?

c. What was the daily fat intake per person in the year 2000?

66. *Death rates.* According to the U.S. Health and Human Services, the number D of deaths from heart disease per 100,000 population can be approximated by $D = -5t + 290$, where t is the number of years after 1998.

a. How many deaths per 100,000 population were there in 1998?

b. How many deaths per 100,000 population would you expect in 2008?

c. What is the slope of the line D and what does it represent?

67. The number N (in millions) of recreational boats in the United States can be approximated by $N = 10 + 0.4t$, where t is the number of years after 1975.
  a. How many recreational boats were there in 1975?
  b. How many would you expect in 2010?
  c. What is the annual increase of boats in the United States?

In Other Words

68. Explain why it is impossible to find the slope of a vertical line.

69. Explain why the slope of any horizontal line is 0.

70. Refer to problem 54. If $m_1 = 0$, then the first equation becomes $y = b_1$. What lines are perpendicular to this line? Explain. What are the equations of these perpendicular lines?

Using Your Knowledge

The accompanying table will be used in problems 71–75.

Building Codes Standards for Slope and Safety

Standards	Maximum Slope
Ramps—wheelchair	0.125
Ramps—walking	0.3
Driveway or street parking	0.22

71. The steepest street in the world is Baldwin Street in Dunedin, New Zealand, with a rise of 1 m for every 1.266 m.
  a. What is the slope of this street? (Answer to two decimal places.)
  b. Is it safe to park on this street?

72. Filbert Street, in San Francisco, has a rise of 10 ft for every 31.7 ft.
  a. What is the slope of this street? (Answer to two decimal places.)
  b. Is it safe to park on this street?

73. A walking ramp to the library has a run of 1.6 ft and a rise of 0.4 ft.
  a. What is the slope of this ramp?
  b. According to the standards, is this a safe ramp?

74. A wheelchair ramp connecting the street to the ticket office at the Sun Dome ends 8 ft above street level. What is the shortest run this ramp can have and still meet safety standards?

75. An architect is designing a driveway. When a grid is placed over the plans, the top of the driveway has coordinates (124, 20), and the bottom of the driveway has coordinates (108, 16). Will this driveway meet safety specifications? If the bottom of the driveway has coordinates $(x, 16)$, what should x be so that the slope is the maximum allowable?

Discovery

We've mentioned that graphs could be used to make an algebraic model when solving problems. Make a model to solve problems 76–78.

76. The Home Show at the Fairgrounds charged a flat parking fee and an additional amount for admission to the show. Study the chart and discover the parking fee and the admission price.

Total Paid	$25	$13	$22
Number of Persons	7	3	6

77. Garcia's Plumbing charges a fixed fee for making a service call plus an hourly rate for each hour the plumber spends making the call. Study the table and discover the fixed fee and the hourly rate for each service call.

Total Paid	$110	$60	$160
Number of Hours	3	1	5

78. A cellular calling plan charges a monthly fee plus an additional amount for each minute of air time (the time spent talking on the phone). Study the table and discover the monthly fee and the charge per minute of air time.

Total Paid	Air Time (minutes)
$ 87.50	230
$ 52.50	90
$133.75	415

Skill Checker

Do you remember how to graph relations and functions? We will need this concept in the next section! Try these practice problems.

1. If $f(x) = x^2$, find and graph $f(0)$, $f(1)$, $f(-1)$, $f(2)$, and $f(-2)$.

2. If $f(x) = (x - 1)^2 - 2$, find and graph $f(0)$, $f(1)$, $f(-1)$, $f(2)$, and $f(-2)$.

If you need further practice, go to Section 7.1.

Collaborative Learning

Divide the class into teams of males and females. Place a ruler on the floor, take off your shoes, step on the ruler, and measure the size L of your foot (in inches). Now measure your height H (in inches).

1. Record the results as ordered pairs of the form (L, H).

2. Graph the ordered pairs in a coordinate system. Each unit should be 1 for L and 10 for H.

3. Average all the L's and all the H's in each group and form the ratio

$$\frac{\text{Average of } H\text{'s}}{\text{Average of } L\text{'s}} = \frac{\text{rise}}{\text{run}} = m$$

4. Graph the equation $y = mx$ on the same coordinate system you used in problem 2. Is it a close fit?

5. What would be the formula for finding the height of a male or a female based on foot size?

6. How would you state your result in terms of variation?

7. What would the constant of variation be for males?

8. What would the constant of variation be for females?

7.4 Quadratic Functions and Their Graphs

The Fountain of Parabolas

Have you seen any parabolas lately? They are as near as your fountain: The streams of water follow the path of a quadratic function called a *parabola*. Parabolas, ellipses, circles, and hyperbolas are called **conic sections** because they can be obtained by intersecting (slicing) a cone with a plane, as shown in Figure 7.27. As you will see later, these conic sections occur in many practical applications. For example, your satellite dish, your flashlight lens, and your telescope lens have a parabolic shape. We study parabolas next.

FIGURE 7.27
Parabola.

►

A. Graphing the Parabola $y = f(x) = ax^2 + k$

In Section 7.2 we studied *linear functions* such as $f(x) = 2x + 5$ and $g(x) = -3x - 4$ whose graphs were straight lines. In this section we shall study equations (functions) defined by a quadratic (second-degree) function of the form

$$f(x) = ax^2 + bx + c$$

These functions are called **quadratic functions,** and their graphs are called **parabolas.**

Just as the simplest *line* to graph is $y = f(x) = x$, the simplest *parabola* to graph is $y = f(x) = x^2$. To draw this graph, we select values for x and find the corresponding values of y.

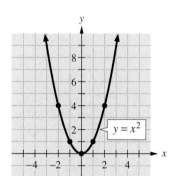

FIGURE 7.28

x Value	$f(x) = y$ Value
$x = -2$	$f(-2) = (-2)^2 = 4$
$x = -1$	$f(-1) = (-1)^2 = 1$
$x = 0$	$f(0) = (0)^2 = 0$
$x = 1$	$f(1) = (1)^2 = 1$
$x = 2$	$f(2) = (2)^2 = 4$

x	y	(x, y)
-2	4	$(-2, 4)$
-1	1	$(-1, 1)$
0	0	$(0, 0)$
1	1	$(1, 1)$
2	4	$(2, 4)$

Then we make a table of ordered pairs, plot the ordered pairs on a coordinate system, and draw a smooth curve through the plotted points as in Figure 7.28.

A very important feature of this parabola is its **symmetry** to the y axis. This means that if you folded the graph of $y = x^2$ along the y axis, the two halves of the graph would coincide because the same value of y is obtained for any value of x and its opposite $-x$. For instance, $x = 2$ and $x = -2$ both give $y = 4$ (see the preceding tables). Because of this symmetry, the y axis is called the **axis of symmetry** or, simply, the **axis** of the parabola. The point $(0, 0)$, where the parabola crosses its axis, is called the **vertex** of the curve. Note that the arrows on the curve in Figure 7.28 mean that the parabola goes on without end.

GRAPH IT

To graph $y = x^2$ and $y = -x^2$, press $\boxed{Y=}$ and enter x^2 for Y_1 and $-x^2$ for Y_2. Press $\boxed{\text{ZOOM}}$ $\boxed{6}$. Note that $y = x^2$ and $y = -x^2$ are reflections of each other across the x axis as shown.

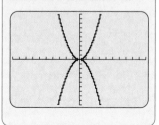

EXAMPLE 1 ▶ Graphing Parabolas Opening Downward

Graph $y = -x^2$.

Solution

We could make a table of x and y values as before. However, note that for any x value, the y value will be the *negative* of the y value on the parabola $y = x^2$. (If you don't believe this, go ahead and make the table and check it, but it's easier to copy the table for $y = x^2$ with the negatives of the y values entered as shown.) Thus, the parabola $y = -x^2$ has the same shape as $y = x^2$, but it is turned in the *opposite* direction (opens *downward*). The graph of $y = -x^2$ is shown in Figure 7.29.

x	y	(x, y)
-2	-4	$(-2, -4)$
-1	-1	$(-1, -1)$
0	0	$(0, 0)$
1	-1	$(1, -1)$
2	-4	$(2, -4)$

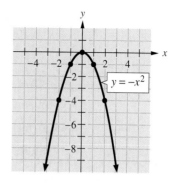

FIGURE 7.29

As you can see from the two preceding examples, when the coefficient of x^2 is positive (as in $y = x^2 = 1x^2$), the parabola opens upward (is **concave up**), but when the coefficient of x^2 is negative (as in $y = -x^2 = -1x^2$), the parabola opens downward (is **concave down**). In either case, the vertex is at $(0, 0)$. To determine the effect of a in $f(x) = ax^2$ in general, let's plot some points and see how the graphs of $g(x) = 2x^2$ and $h(x) = \frac{1}{2}x^2$ compare to the graph of $f(x) = x^2$. All three graphs are shown in Figure 7.30.

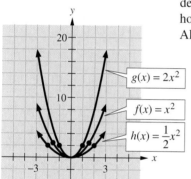

FIGURE 7.30

x	$g(x) = 2x^2$	x	$h(x) = \frac{1}{2}x^2$
-2	$2(-2)^2 = 8$	-2	$\frac{1}{2}(-2)^2 = 2$
-1	$2(-1)^2 = 2$	-1	$\frac{1}{2}(-1)^2 = \frac{1}{2}$
0	$2(0)^2 = 0$	0	$\frac{1}{2}(0)^2 = 0$
1	$2(1)^2 = 2$	1	$\frac{1}{2}(1)^2 = \frac{1}{2}$
2	$2(2)^2 = 8$	2	$\frac{1}{2}(-2)^2 = 2$

Note that the graph of $g(x) = 2x^2$ is narrower than that of $f(x) = x^2$, whereas the graph of $h(x) = \frac{1}{2}x^2$ is wider. The vertex and line of symmetry are the same for the three curves. In general, we have the following:

Properties of the Parabola $g(x) = ax^2$

The graph of $g(x) = ax^2$ is a parabola with the vertex at the origin and the y axis as its line of symmetry.

If a is *positive*, the parabola opens *upward;* if a is *negative*, the parabola opens *downward*.

If $|a|$ is greater than 1 ($|a| > 1$), the parabola is narrower than the parabola $f(x) = x^2$.

If $|a|$ is between 0 and 1 ($0 < |a| < 1$), the parabola is wider than the parabola $f(x) = x^2$.

Using this information, you can draw the graph of any parabola of the form $g(x) = ax^2$, as we illustrate in Example 2.

EXAMPLE 2 ► **Graphing More Parabolas**

Graph

(a) $f(x) = 3x^2$.   (b) $g(x) = -3x^2$.   (c) $h(x) = \frac{1}{3}x^2$.

Solution

(a) By looking at the properties of the parabola $y = ax^2$, we know that the vertex of the parabola $f(x) = 3x^2$ is at the origin and that the y axis is its line of symmetry. Since $3 > 0$, we also know that the parabola $f(x) = 3x^2$ opens upward and is narrower than the parabola $y = x^2$. We pick three easy points to complete our graph, which is shown in Figure 7.31.

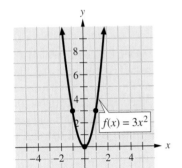

FIGURE 7.31

x	$y = 3x^2$
-1	$y = 3(-1)^2 = 3$
0	$y = 3(0)^2 \quad = 0$
1	$y = 3(1)^2 \quad = 3$

(b) The parabola $g(x) = -3x^2$ opens downward but is still narrower than the parabola $y = x^2$. As a matter of fact, the parabola $g(x) = -3x^2$ is the reflection of the parabola $f(x) = 3x^2$ across the x axis. Again, we pick three points to complete the graph, which is shown in Figure 7.32.

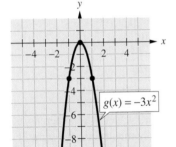

FIGURE 7.32

x	$y = -3x^2$
-1	$y = -3(-1)^2 = -3$
0	$y = -3(0)^2 \quad = 0$
1	$y = -3(1)^2 \quad = -3$

(c) The parabola $h(x) = \frac{1}{3}x^2$ opens upward because $\frac{1}{3} > 0$ but is wider than the parabola $y = x^2$. This time, instead of selecting $x = -1, 0,$ and 1, we select $x = -3, 0,$ and 3 for ease of computation. The completed graph is shown in Figure 7.33.

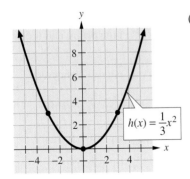

FIGURE 7.33

x	$y = \frac{1}{3}x^2$
-3	$y = \frac{1}{3}(-3)^2 = 3$
0	$y = \frac{1}{3}(0)^2 \quad = 0$
3	$y = \frac{1}{3}(3)^2 \quad = 3$

■

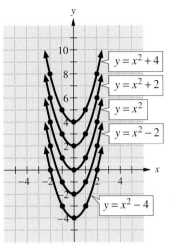

FIGURE 7.34

What do you think will happen if we graph the parabola $y = x^2 + 2$? Two things: First, the parabola opens upward because the coefficient of x^2 is understood to be $+1$. Second, all the points will be 2 units higher than those for the same value of x on the parabola $y = x^2$. Thus we can make the graph of $y = x^2 + 2$ by following the pattern of $y = x^2$. The graphs of $y = x^2 + 2$, $y = x^2 + 4$, $y = x^2 - 2$, and $y = x^2 - 4$ are shown in Figure 7.34. The points used to make the graphs are listed in the following table:

For $y = x^2 + 2$		For $y = x^2 + 4$		For $y = x^2 - 2$		For $y = x^2 - 4$	
x	y	x	y	x	y	x	y
0	2	0	4	0	-2	0	-4
± 1	3	± 1	5	± 1	-1	± 1	-3
± 2	6	± 2	8	± 2	2	± 2	0

Vertex → (pointing to the 0, 2 row)

Note that adding or subtracting a positive number k on the right-hand side of equation $y = x^2$ raises or lowers the graph (and the vertex) by k units.

EXAMPLE 3 ► **Graphing Parabolas Opening Downward**

Graph $y = -x^2 - 2$.

Solution

Since the coefficient of x^2 (which is understood to be -1) is negative, the parabola opens downward. It is also 2 units lower than the graph of $y = -x^2$. Thus the graph of $y = -x^2 - 2$ is a parabola opening downward with its vertex at $(0, -2)$. Letting $x = 1$, we get $y = -3$, and for $x = 2$, $y = -6$. Graph the two points $(1, -3)$ and $(2, -6)$ and, by symmetry, the points $(-1, -3)$ and $(-2, -6)$. The parabola passing through all these points is shown in Figure 7.35.   ■

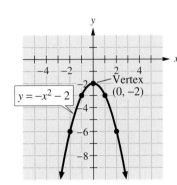

FIGURE 7.35

B. Graphing a Parabola of the Form
$y = f(x) = a(x - h)^2 + k$

So far we have graphed only parabolas of the form $y = ax^2 + k$. What do you think the graph of $y = (x - 1)^2$ looks like? As before, we make a table of values.

x	$y = (x - 1)^2$		x	y
$x = -1$	$y = (-1 - 1)^2 = (-2)^2 = 4$		-1	4
$x = 0$	$y = (0 - 1)^2 = (-1)^2 = 1$ ← y intercept	0	1
$x = 1$	$y = (1 - 1)^2 = (0)^2 = 0$ ← Vertex	1	0
$x = 2$	$y = (2 - 1)^2 = 1^2 = 1$		2	1
$x = 3$	$y = (3 - 1)^2 = 2^2 = 4$		3	4

or

The graph appears in Figure 7.36.

Note that the shape of the graph is identical to that of $y = x^2$, but it is shifted 1 unit to the *right*. Thus, the vertex is at $(1, 0)$, and the line of symmetry is as shown in Figure 7.36. Similarly, the graph of $y = -(x + 1)^2$ is identical to that of $y = -x^2$ but shifted 1 unit to the *left*. Thus the vertex is at $(-1, 0)$, and the line of symmetry is as shown in Figure 7.37. Some easy points to plot are $x = 0$, $y = -(1)^2 = -1$ and $x = 1$, $y = -(1 + 1)^2 = -2^2 = -4$. When we plot the points $(0, -1)$ and $(1, -4)$, by symmetry, the points $(-2, -1)$ and $(-3, -4)$ are also on the graph.

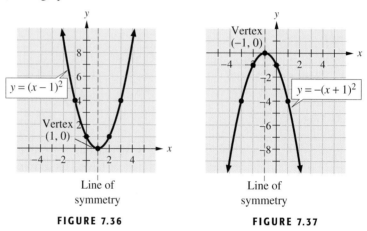

FIGURE 7.36   **FIGURE 7.37**

EXAMPLE 4 ▶ Graphing Parabolas Using Shifts

Graph $y = (x - 1)^2 - 2$.

Solution

The graph of this equation is identical to the graph of $y = x^2$ except for its position. The new parabola is shifted 1 unit to the right (because of the -1) and 2 units down (because of the -2). Thus the vertex is $(1, -2)$. Note the line of symmetry. Figure 7.38 indicates these two facts and shows the finished graph of $y = (x - 1)^2 - 2$.

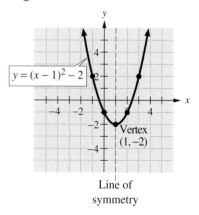

FIGURE 7.38

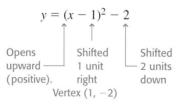

From these examples we can see that

1. The graph of $y = -x^2 - 2$ (Example 3) is exactly the same as the graph of $y = -x^2$ (Example 1) but moved 2 units *down*. In general, the graph of $y = ax^2 + k$ is the same as the graph of $y = ax^2$ but moved vertically k units. The vertex is at $(0, k)$.

2. The graph of $y = (x - 1)^2$ is the same as that of $y = x^2$ but moved 1 unit *right*. The vertex is at $(1, 0)$.

3. The graph of $y = (x - 1)^2 - 2$ (Example 4) is exactly the same as the graph of $y = (x - 1)^2$ but moved 2 units *down*. The vertex is at $(1, -2)$.

Here is the summary of this discussion.

Properties of the Parabola $y = a(x - h)^2 + k$

The graph of the parabola $y = a(x - h)^2 + k$ is the same as that of $y = ax^2$ but moved h units horizontally and k units vertically. The *vertex* is at the point (h, k), and the line of symmetry is $x = h$.

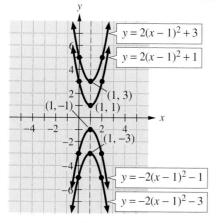

FIGURE 7.39

In conclusion, follow the given directions to graph an equation of the form

$$y = a(x - h)^2 + k \qquad \text{Vertex } (h, k)$$

Opens upward for $a > 0$, downward for $a < 0$.

Shifts the graph right or left.

Moves the graph up or down.

The graphs of $y = 2(x - 1)^2 + 1$, $y = 2(x - 1)^2 + 3$, $y = -2(x - 1)^2 - 1$, and $y = -2(x - 1)^2 - 3$ are shown in Figure 7.39.

C. Graphing the Parabola $y = f(x) = ax^2 + bx + c$

How can we graph $f(x) = ax^2 + bx + c$? If we learn to write $f(x) = ax^2 + bx + c$ as $f(x) = a(x - h)^2 + k$, we can do it by using the techniques we just learned. We do this by completing the square. Here's how.

$f(x) = ax^2 + bx + c$	Given.
$= (ax^2 + bx) + c$	Group.
$= a\left(x^2 + \dfrac{b}{a}x + \quad\right) + c$	Factor a.
$= a\left[x^2 + \dfrac{b}{a}x + \left(\dfrac{b}{2a}\right)^2 - \left(\dfrac{b}{2a}\right)^2\right] + c$	To complete the square, add and subtract $\left(\dfrac{1}{2} \cdot \dfrac{b}{a}\right)^2 = \left(\dfrac{b}{2a}\right)^2$ inside the brackets.
$= a\left(x + \dfrac{b}{2a}\right)^2 - a\left(\dfrac{b}{2a}\right)^2 + c$	Use the distributive property and factor inside the brackets.
$= a\left(x + \dfrac{b}{2a}\right)^2 - a \cdot \dfrac{b^2}{4a^2} + c$	Square $\dfrac{b}{2a}$.
$= a\left(x + \dfrac{b}{2a}\right)^2 - \dfrac{b^2}{4a} + c$	Multiply $-a \cdot \dfrac{b^2}{4a^2} = -\dfrac{b^2}{4a}$.
$= a\left(x + \dfrac{b}{2a}\right)^2 + \dfrac{4ac - b^2}{4a}$	Find the LCD of $-b^2/4a$ and c.

Thus to write

$$f(x) = a\left(x + \frac{b}{2a}\right)^2 + \frac{4ac - b^2}{4a}$$

as

$$f(x) = a(x - h)^2 + k$$

we must have

$$h = -\frac{b}{2a} \quad \text{and} \quad k = \frac{4ac - b^2}{4a}$$

the coordinates of the vertex. Note that you *do not* have to memorize the y coordinate of the vertex. After you find the x coordinate, substitute in the equation and find y.

Here is a summary of our discussion.

Graphing the Parabola $y = f(x) = ax^2 + bx + c$

1. To find the vertex use one of the following methods:

   Method 1   Let $x = -b/2a$ in the equation and solve for y.

   Method 2   Complete the square and compare with $y = a(x - h)^2 + k$.

2. Let $x = 0$. The result c is the y intercept.

3. Since the parabola is symmetric to its axis, use this symmetry to find additional points.

4. Let $y = 0$. Find x by solving $ax^2 + bx + c = 0$. If the solutions are real numbers, they are the x intercepts. If not, the parabola does not intersect the x axis.

5. Draw a smooth curve through the points found in steps 1–4. Remember that if $a > 0$, the parabola opens *upward;* if $a < 0$, the parabola opens *downward.*

We demonstrate this procedure in Example 5.

EXAMPLE 5 ▶ Graphing Parabolas Using Two Methods

Graph $y = x^2 + 3x + 2$.

Solution

1. We first find the vertex using either of the two methods.

Method 1

Use the vertex formula for the x coordinate. Since $a = 1$, $b = 3$, and $c = 2$,

$$x = -\frac{b}{2a} = -\frac{3}{2}$$

Substituting for x in the equation gives

$$y = x^2 + 3x + 2$$
$$= \left(-\frac{3}{2}\right)^2 + 3\left(-\frac{3}{2}\right) + 2$$
$$= \frac{9}{4} - \frac{9}{2} + 2$$
$$= \frac{9}{4} - \frac{18}{4} + \frac{8}{4} = -\frac{1}{4}$$

The vertex is at $\left(-\frac{3}{2}, -\frac{1}{4}\right)$.

Method 2

Complete the square.

$$y = (x^2 + 3x + \quad) + 2$$
$$= \left[x^2 + 3x + \left(\frac{3}{2}\right)^2\right] + 2 - \left(\frac{3}{2}\right)^2$$
$$= \left(x + \frac{3}{2}\right)^2 + 2 - \frac{9}{4}$$
$$= \left(x + \frac{3}{2}\right)^2 - \frac{1}{4}$$

The vertex is at $\left(-\frac{3}{2}, -\frac{1}{4}\right)$.

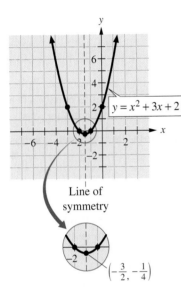

Line of symmetry

$\left(-\frac{3}{2}, -\frac{1}{4}\right)$

FIGURE 7.40

2. Let $x = 0$; then $y = x^2 + 3x + 2$ becomes $y = 2$. The y intercept is 2.

3. By symmetry, the point $(-3, 2)$ is also on the graph.

4. Let $y = 0$; $y = x^2 + 3x + 2$ becomes

$$0 = x^2 + 3x + 2$$
$$= (x + 2)(x + 1)$$

Thus $x = -2$ or $x = -1$. The graph intersects the x axis at $(-2, 0)$ and $(-1, 0)$.

5. Since the coefficient of x^2 is 1, $a > 0$, and the parabola opens upward.

We draw a smooth curve through these points to obtain the graph of the parabola, as shown in Figure 7.40. ∎

Online Study Center

To further explore graphing polynomials and quadratics, access links 7.4.1 and 7.4.2 on this textbook's Online Study Center.

EXAMPLE 6 ▶ Parabolas Opening Down

Graph $y = -2x^2 + 4x - 3$.

Solution

1. To find the vertex, we can use either of these methods.

Method 1

Use the vertex formula. Here $a = -2$ and $b = 4$, so

$$x = -\frac{b}{2a}$$
$$= \frac{-4}{2(-2)}$$
$$= 1$$

Method 2

Complete the square.

$$y = -2x^2 + 4x - 3$$
$$= -2(x^2 - 2x + \quad) - 3$$
$$= -2(x^2 - 2x + 1) - 3 + 2$$
$$= -2(x - 1)^2 - 1$$

The vertex is at $(1, -1)$.

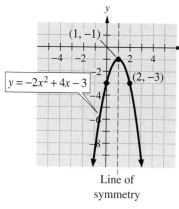

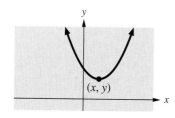

FIGURE 7.41

If we substitute $x = 1$ in $y = -2x^2 + 4x - 3$,

$$y = -2(1)^2 + 4(1) - 3$$
$$= -2 + 4 - 3$$
$$= -1$$

Thus the vertex is at $(1, -1)$.

2. If $x = 0$, $y = -2x^2 + 4x - 3 = -3$, the y intercept.

3. We graph the vertex $(1, -1)$ and the y intercept -3. To make a more accurate graph, we need some more points. Since the parabola is symmetric, we can find a point across from the y intercept by letting $x = 2$. Then

$$y = -2(2)^2 + 4(2) - 3$$
$$= -8 + 8 - 3 = -3$$

as expected.

4. For $y = 0$, $0 = -2x^2 + 4x - 3$. However, the right-hand side is not factorable, and the vertex $(1, -1)$ is below the x axis. This means that this equation has no solution, and there are no x intercepts; the graph does not cross the x axis.

5. Since $a = -2 < 0$, the parabola opens downward. The completed graph is shown in Figure 7.41. ∎

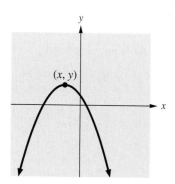

FIGURE 7.42
$f(x)$ has a minimum at the vertex (x, y).

D. Solving Applications Involving Parabolas

Every parabola of the form $y = ax^2 + bx + c$ that we have graphed has its vertex at either its maximum (highest) or minimum (lowest) point on the graph. If the graph opens upward, the vertex is the minimum (Figure 7.42), and if the graph opens downward, the vertex is the maximum (Figure 7.43).

Thus, if we are dealing with a quadratic function, we can find its maximum or minimum by finding the vertex of the corresponding parabola. This idea can be used to solve many real-world applications. For example, suppose that a CD company manufactures and sells x CDs per week. If the revenue is given by $R = 10x - 0.01x^2$, we can use the techniques we've just studied to maximize the revenue. We do this next.

EXAMPLE 7 ▶ Finding Maximum Revenue Using Parabolas

If $R = 10x - 0.01x^2$, how many CDs does the company have to sell in order to obtain maximum revenue?

Solution

We first write the equation as $R = -0.01x^2 + 10x$. Since the coefficient of x^2 is negative, the parabola opens downward (is concave down), and the vertex is its highest point. Letting

$$x = -\frac{b}{2a} = -\frac{10}{-0.02} = 500$$

$R = 10(500) - 0.01(500)^2 = 5000 - 2500 = 2500$. Thus, when the company sells $x = 500$ CDs a week, the revenue is a maximum: $2500. ∎

FIGURE 7.43
$f(x)$ has a maximum at the vertex (x, y).

We close this section by presenting a summary of the material we have studied in Table 7.9.

TABLE 7.9 Summary of Parabolas

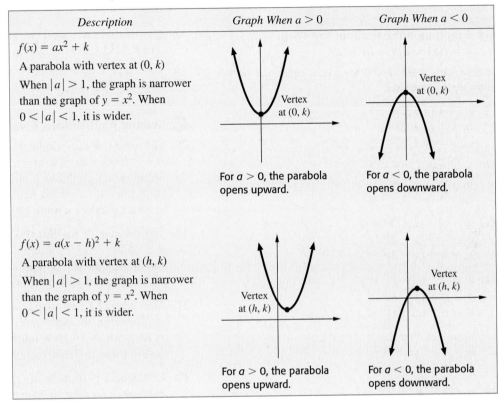

Description	Graph When $a > 0$	Graph When $a < 0$
$f(x) = ax^2 + k$ A parabola with vertex at $(0, k)$ When $|a| > 1$, the graph is narrower than the graph of $y = x^2$. When $0 < |a| < 1$, it is wider.	Vertex at $(0, k)$ For $a > 0$, the parabola opens upward.	Vertex at $(0, k)$ For $a < 0$, the parabola opens downward.
$f(x) = a(x - h)^2 + k$ A parabola with vertex at (h, k) When $|a| > 1$, the graph is narrower than the graph of $y = x^2$. When $0 < |a| < 1$, it is wider.	Vertex at (h, k) For $a > 0$, the parabola opens upward.	Vertex at (h, k) For $a < 0$, the parabola opens downward.

EXERCISES 7.4

A Graphing the Parabola $y = f(x) = ax^2 + k$

In problems 1–8, graph the given equations on the same coordinate axes.

1. a. $y = 2x^2$
   b. $y = 2x^2 + 2$
   c. $y = 2x^2 - 2$

2. a. $y = 3x^2 + 1$
   b. $y = 3x^2 + 3$
   c. $y = 3x^2 - 2$

3. a. $y = -2x^2$
   b. $y = -2x^2 + 1$
   c. $y = -2x^2 - 1$

4. a. $y = -4x^2$
   b. $y = -4x^2 + 1$
   c. $y = -4x^2 - 1$

5. a. $y = \frac{1}{4}x^2$

   b. $y = -\frac{1}{4}x^2$

6. a. $y = \frac{1}{5}x^2$

   b. $y = -\frac{1}{5}x^2$

7. a. $y = \frac{1}{3}x^2 + 1$

   b. $y = -\frac{1}{3}x^2 + 1$

8. a. $y = \frac{1}{4}x^2 + 1$

   b. $y = -\frac{1}{4}x^2 + 1$

B Graphing a Parabola of the Form
$y = f(x) = a(x - h)^2 + k$

In problems 9–16, graph the given equations on the same coordinate axes.

9. a. $y = (x + 2)^2 + 3$
   b. $y = (x + 2)^2$
   c. $y = (x + 2)^2 - 2$

10. a. $y = (x - 2)^2 + 2$
    b. $y = (x - 2)^2$
    c. $y = (x - 2)^2 - 2$

11. a. $y = -(x + 2)^2 - 2$
    b. $y = -(x + 2)^2$
    c. $y = -(x + 2)^2 - 4$

12. a. $y = -(x - 1)^2 + 1$
    b. $y = -(x - 1)^2$
    c. $y = -(x - 1)^2 + 2$

13. a. $y = -2(x + 2)^2 - 2$
    b. $y = -2(x + 2)^2$
    c. $y = -2(x + 2)^2 - 4$

14. a. $y = -2(x - 1)^2 + 1$
    b. $y = -2(x - 1)^2$
    c. $y = -2(x - 1)^2 + 2$

15. a. $y = 2(x + 1)^2 + \frac{1}{2}$
    b. $y = 2(x + 1)^2$

16. a. $y = 2(x + 1)^2 - \frac{1}{2}$
    b. $y = 2(x + 1)^2$

C Graphing the Parabola
$y = f(x) = ax^2 + bx + c$

In problems 17–28, use the five-step procedure in the text to sketch the graph. Label the vertex and the intercepts.

17. $y = x^2 + 2x + 1$ **18.** $y = x^2 + 4x + 4$

19. $y = -x^2 + 2x + 1$ **20.** $y = -x^2 + 4x - 2$

21. $y = -x^2 + 4x - 5$ **22.** $y = -x^2 + 4x - 3$

23. $y = 3 - 5x + 2x^2$

24. $y = 3 + 5x + 2x^2$

25. $y = 5 - 4x - 2x^2$
    (*Hint:* $\sqrt{56} \approx 7.5$)

26. $y = 3 - 4x - 2x^2$
    (*Hint:* $\sqrt{40} \approx 6.3$)

27. $y = -3x^2 + 3x + 2$
    (*Hint:* $\sqrt{33} \approx 5.7$)

28. $y = -3x^2 + 3x + 1$
    (*Hint:* $\sqrt{21} \approx 4.6$)

D Solving Applications Involving Parabolas

29. The profit P (in dollars) for a company is $P = -5000 + 8x - 0.001x^2$, where x is the number of items produced each month. How many items does the company have to produce in order to obtain maximum profit? What is this profit?

30. The revenue R for Shady Glasses is given by $R = 1500p - 75p^2$, where p is the price of each pair of sunglasses (R and p in dollars). What should the price be to maximize revenue?

31. After spending x thousand dollars in an advertising campaign, the number N of units sold is given by $N = 50x - x^2$. How much should be spent in the campaign to obtain maximum sales?

32. The number N of units of a product sold after a television commercial blitz is $N = 40x - x^2$, where x is the amount spent in thousands of dollars. How much should be spent on television commercials to obtain maximum sales?

33. If a ball is batted up at 160 ft/sec, its height h ft after t sec is given by $h = -16t^2 + 160t$. Find the maximum height reached by the ball.

34. If a ball is thrown upward at 20 ft/sec, its height h ft after t sec is given by $h = -16t^2 + 20t$. How many seconds does it take for the ball to reach its maximum height, and what is this height?

35. If a farmer digs potatoes today, she will have 600 bu (bushels) worth $1 per bu. Every week she waits, the crop increases by 100 bu, but the price decreases $.10 a bushel. Show that she should dig and sell her potatoes at the end of 2 weeks.

36. A man has a large piece of property along Washington Street. He wants to fence the sides and back of a rectangular plot. If he has 400 ft of fencing, what dimensions will give him the maximum area?

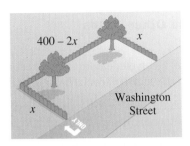

37. Have you read the story "The Celebrated Jumping Frog of Calaveras County"? According to the *Guinness Book of Records,* the second greatest distance covered by a frog in a triple jump is 21 ft $5\frac{3}{4}$ in. at the annual Calaveras Jumping Jubilee; this occurred on May 18, 1986.

　a. If Rosie the Ribiter's (the winner) path in her first jump is approximated by $R = -\frac{1}{98}x^2 + \frac{6}{7}x$ (where x is the horizontal distance covered in inches), what are the coordinates of the vertex of Rosie's path?

　b. Find the maximum height attained by Rosie in her first jump.

　c. Use symmetry to find the horizontal length of Rosie's first jump.

　d. Make a sketch for R showing the initial position (0, 0), the vertex, and Rosie's ending position after her first jump.

38. Amazingly, Rosie's jump is not the best triple jump on record. That distinction belongs to Santjie, a South African frog who jumped 33 ft $5\frac{1}{2}$ in. on May 21, 1977.

　a. If Santjie's path in his first jump is approximated by $S = -\frac{1}{200}x^2 + \frac{7}{10}x$ (where x is the distance covered in inches), what are the coordinates of the vertex of Santjie's path?

　b. Find Santjie's maximum height in his first jump.

　c. Use symmetry to find the horizontal length of Santjie's first jump.

　d. Make a sketch for S showing the initial position (0, 0), the vertex, and Santjie's ending position after his first jump.

39. A baseball hit at an angle of $35°$ has a velocity of 130 mph. Its trajectory can be approximated by the equation $d = -\frac{1}{400}x^2 + x$, where x is the distance the ball travels in feet.

　a. What are the coordinates of the vertex of the trajectory?

　b. Find the maximum height attained by the ball.

　c. Use symmetry to find how far the ball travels horizontally.

　d. Make a sketch for d showing the initial position (0, 0), the vertex, and the ending position of the baseball.

40. Is there a relationship between mothers who smoke and the percent of low-birth-weight babies?

　a. What is the (approximate) lowest percent of low-birth-weight babies for mothers who smoke? At what (approximate) age does it occur?

　b. What is the (approximate) highest percent of low-birth-weight babies for mothers who smoke? At what (approximate) age does it occur?

　c. If the function representing the graph is of the form $f(x) = ax^2 + bx + c$, what can you say about a?

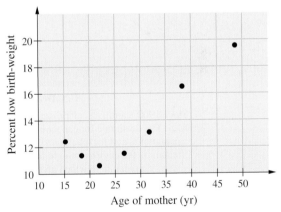

Source: www.uvm.edu/~dhowell/gradstat/psych340/
Lectures/CorrelReg/correl1.html.

Mothers who smoke.

41. The relationship between the percent of low birth weight and age for mothers who *do not* smoke is shown on the next page.

　a. What is the (approximate) lowest percent of low-birth-weight babies for mothers who do not smoke? At what (approximate) age does it occur?

b. What is the (approximate) highest percent of low-birth-weight babies for mothers who do not smoke?

c. At what (approximate) age do mothers who smoke have the lowest percent of low-birth-weight babies?

d. At what (approximate) age do mothers who *do not* smoke have the highest percent of low-birth-weight babies?

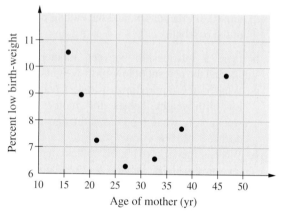

Source: www.uvm.edu/~dhowell/gradstat/psych340/Lectures/CorrelReg/correl1.html.

Mothers who do not smoke.

In Other Words

42. Explain how you determine whether the graph of a quadratic function opens up or down.

43. What causes the graph of the function $f(x) = ax^2$ to be wider or narrower than the graph of $f(x) = x^2$?

44. What is the effect of the constant k on the graph of the function $f(x) = ax^2 + k$?

45. How does a parabola that has two x intercepts and vertex at $(1, 1)$ open; that is, does it open up or down? Why?

46. Why does the graph of a function never have two y intercepts?

In the graph below, the death rate for cardiovascular disease from 1920 to 2001 is nearly a parabola and can be approximated by $f(x) = a(x - h)^2 + k$.

47. Is a positive or negative?

48. What is the maximum point on the curve?

49. What are the approximate coordinates for the vertex of the curve?

50. On the basis of the graph, what would you predict the number of deaths (per 100,000 population) due to cardiovascular disease to be in the year 2010?

MILLENNIUM FACT BOX

U.S. Death Rates for Selected Causes

Since 1900, cardiovascular disease has generally remained the number-one killer in the U.S. The death rate for cancer has slowly climbed, while death rates for influenza and pneumonia and for tuberculosis have declined and approached zero. One prominent feature is the extremely high death rate for influenza in 1918, reporting an influenza epidemic in which about 500,000 Americans died from the disease.

— Cardiovascular disease
— Influenza and pneumonia
- - - Tuberculosis
— Cancer

Sources: National Center for Health Statistics. Centers for Disease Control and Prevention. (Updated by author.)

Discovery

A **parabola** is the set of all points equidistant from a fixed point $F(0, p)$ (called the **focus**) and a fixed line $y = -p$ (called the **directrix**). If $P(x, y)$ is a point on the parabola, this definition says that $FP = DP$; that is, the distance from F to P is the same as the distance from D to P.

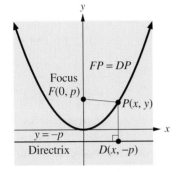

51. Find FP.   **52.** Find DP.

53. Set $FP = DP$ and solve for x^2.

54. For the parabola $x^2 = 4y$,
  a. locate the focus.
  b. write the equation of the directrix.

Many applications of the parabola depend on an important focal property of the curve. If the parabola is a mirror, a ray of light parallel to the axis reflects to the focus,

and a ray originating at the focus reflects parallel to the axis. (This can be proved by using methods of calculus.)

If the parabola is revolved about its axis, a surface called a *paraboloid of revolution* is formed. This is the shape used for automobile headlights and searchlights that throw a parallel beam of light when the light source is placed at the focus; it is also the shape of a radar dish or a reflecting telescope mirror that collects parallel rays of energy (light) and reflects them to the focus.

We can find the equation of the parabola needed to generate a paraboloid of revolution by using the equation $x^2 = 4py$ as follows: Suppose a parabolic mirror has a diameter of 6 ft and a depth of 1 ft. Then, we find the value of p that makes the parabola pass through the point $(3, 1)$. This means that we substitute into the equation and solve for p. Thus we have

$$3^2 = 4p(1)$$

so that $4p = 9$ and $p = 2.25$. The equation of the parabola is $x^2 = 9y$, and the focus is at $(0, 2.25)$.

55. A radar dish has a diameter of 10 ft and a depth of 2 ft. The dish is in the shape of a paraboloid of revolution. Find an equation for a parabola that would generate this dish and locate the focus.

56. The cables of a suspension bridge hang very nearly in the shape of a parabola. A cable on such a bridge spans a distance of 1000 ft and sags 50 ft in the middle. Find an equation for this parabola.

7.5 Exponential and Logarithmic Functions

What Do Cells Know About Exponents!

Are you taking biology? Have you studied cell reproduction? The photographs below show a cell reproducing by a process called *mitosis*. In mitosis, a single cell or bacterium divides and forms two identical daughter cells. Each daughter cell then doubles in size and divides. As you can see, the number of bacteria present is a function of time. If we start with one cell and assume that each cell

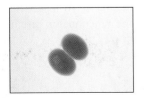 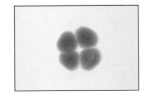 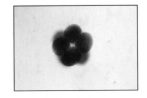 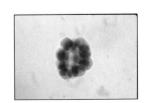

Mitosis.

divides after 10 min, then the number of bacteria present at the end of the first 10-min period ($t = 10$) is

$$2 = 2^1 = 2^{10/10}$$

At the end of the second 10-min period ($t = 20$), the two cells divide, and the number of bacteria present is

$$4 = 2^2 = 2^{20/10}$$

Similarly, at the end of the third 10-min period ($t = 30$), the number is

$$8 = 2^3 = 2^{30/10}$$

Thus we can see that the number of bacteria present at the end of t minutes is given by the function

$$f(t) = 2^{t/10}$$

Note that this also gives the correct result for $t = 0$, because $2^0 = 1$.

The function $f(t) = 2^{t/10}$ is called an *exponential function* because the variable t is in the exponent.

In this section we shall graph exponential functions and see how such functions can be used to solve real-world problems. ▶

A. Graphing Exponential Functions

Exponential functions take many forms. For example, the following functions are exponential functions because the variable (or unknown) is an exponent:

$$f(x) = 3^x \qquad F(y) = \left(\frac{1}{2}\right)^y \qquad H(z) = (1.02)^{z/2}$$

In general, we have the following definition:

Definition of an Exponential Function

An **exponential function** is a function defined for all real values of x by

$$f(x) = b^x, \quad b > 0, b \neq 1$$

The variable b in the definition of an exponential function *must not* equal 1 because $f(x) = 1^x = 1$ is a constant function, *not* an exponential function.

In this definition, b is a constant called the **base,** and the **exponent** x is the variable.

The exponential function defined by $f(t) = 2^{t/10}$ can be graphed and used to predict the number of bacteria present after a period of time t. To make this graph, we first construct a table giving the value of the function for certain convenient times.

t	0	10	20	30
$f(t) = 2^{t/10}$	1	2^1	2^2	2^3

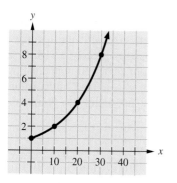

FIGURE 7.44

The corresponding points can then be graphed and joined with a smooth curve, as shown in Figure 7.44. In general, we graph an exponential function by plotting several points calculated from the function and then drawing a smooth curve through these points.

EXAMPLE 1 ▶ Graphing Exponential Fuctions

Graph on the same coordinate system

(a) $f(x) = 2^x$.   (b) $g(x) = \left(\frac{1}{2}\right)^x$.

Solution

(a) We first make a table with convenient values for x and find the corresponding values for $f(x)$. We then graph the points and connect them with the smooth curve $y = 2^x$, as shown in Figure 7.45.

x	-2	-1	0	1	2
$f(x) = 2^x$	$2^{-2} = \frac{1}{4}$	$2^{-1} = \frac{1}{2}$	$2^0 = 1$	$2^1 = 2$	$2^2 = 4$

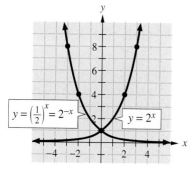

FIGURE 7.45

(b) If we let $x = -2$,

$$g(-2) = \left(\frac{1}{2}\right)^{-2} = \frac{1}{\left(\frac{1}{2}\right)^2} = 4$$

Similarly, for $x = -1$,

$$g(-1) = \left(\frac{1}{2}\right)^{-1} = \frac{1}{\left(\frac{1}{2}\right)^1} = 2$$

For $x = 0$, 1, and 2, the function values are $\left(\frac{1}{2}\right)^0 = 1, \left(\frac{1}{2}\right)^1 = \frac{1}{2}$, and $\left(\frac{1}{2}\right)^2 = \frac{1}{4}$, as shown in the table.

x	-2	-1	0	1	2
$g(x) = \left(\frac{1}{2}\right)^x$	4	2	1	$\frac{1}{2}$	$\frac{1}{4}$

GRAPH IT

To do Example 1, enter $Y_1 = 2^x$. Enter $Y_2 = \left(\frac{1}{2}\right)^x$ by pressing `(` `1` `÷` `2` `)` `^` `χ, τ, θ, η` and press `ZOOM` `6`. The result is shown in the window.

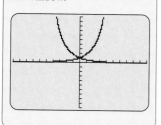

We can save time if we realize that $\left(\frac{1}{2}\right)^x = (2^{-1})^x$, whose values are shown in the table for $f(x)$ in part (a). The graph of $g(x) = \left(\frac{1}{2}\right)^x = 2^{-x}$ is shown in Figure 7.45. Note that the graphs approach the x axis but do not intersect it. In these examples, the x axis is called the *horizontal asymptote* of the graphs. ■

Also, the two graphs in Figure 7.45 are symmetric to each other with respect to the y axis. In general, we have the following fact:

Symmetric Graphs

The graphs of $y = b^x$ and $y = b^{-x}$ are **symmetric** with respect to the y axis.

GRAPH IT

The Graph It at the end of this section shows how to do this numerical work with a grapher.

In our definition of the function $y = b^x$, it was required only that $b > 0$ and $b \neq 1$. For many practical applications, however, there is a particularly important base, the irrational number e. The value of e is approximately 2.7182818. The reasons for using this base are made clear in more advanced mathematics courses, but for our purposes, we need only note that e is defined as the value that the quantity $(1 + 1/n)^n$ approaches as n increases indefinitely. In symbols,

$$\left(1 + \frac{1}{n}\right)^n \to e \approx 2.7182819 \quad \text{as} \quad n \to \infty$$

To show this, we use increasing values of n (1000, 10,000, 100,000, 1,000,000) and evaluate the expression $(1 + 1/n)^n$ using a calculator with a $\boxed{x^y}$ key.

For $n = 1000$, $(1 + 1/n)^n \quad = (1.001)^{1000} \quad = 2.7169239$ Enter 1.001
For $n = 10,000$, $(1 + 1/n)^n \quad = (1.0001)^{10,000} \quad = 2.7181459$ $\boxed{x^y}$ 1000 $\boxed{=}$.
For $n = 100,000$, $(1 + 1/n)^n \quad = (1.00001)^{100,000} \quad = 2.7182682$
For $n = 1,000,000$, $(1 + 1/n)^n = (1.000001)^{1,000,000} = 2.7182805$

As you can see, the value of $(1 + 1/n)^n$ is indeed getting closer to $e \approx 2.7182819$.

To graph the functions $f(x) = e^x$ and $g(x) = e^{-x}$, we make a table giving x different values (say, $-2, -1, 0, 1, 2$) and find the corresponding $y = e^x$ and $y = e^{-x}$ values. This can be done with a calculator with an $\boxed{e^x}$ key. On such calculators, you usually have to enter $\boxed{\text{INV}}$ or $\boxed{\text{2nd}}$ to find the value of e^x. [Enter 1 $\boxed{\text{2nd}}$ (or $\boxed{\text{INV}}$) $\boxed{e^x}$, and the calculator will give the value 2.7182818.] We use these ideas next.

EXAMPLE 2 ▶ Graphing e^x and e^{-x}

Use the values in the table to graph $f(x) = e^x$ and $g(x) = e^{-x}$. (Use the same coordinate system.)

x	-2	-1	0	1	2
e^x	0.1353	0.3679	1	2.7183	7.3891

x	-2	-1	0	1	2
e^{-x}	7.3891	2.7183	1	0.3679	0.1353

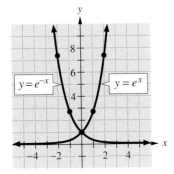

FIGURE 7.46

Solution

Plotting the given values, we obtain the graphs of $f(x) = e^x$ and $g(x) = e^{-x}$ shown in Figure 7.46. Note that e^x and e^{-x} are symmetric with respect to the y axis.

B. Applications Involving Exponential Functions

Do you have some money invested? Is it earning interest compounded annually, quarterly, monthly, or daily? Does the frequency of compounding make a difference? Some banks have instituted what is called *continuous interest compounding*.

We can compare continuous compounding and n compoundings per year by examining their formulas.

Continuous compounding $A = Pe^{rt}$

n compoundings per year $A = P(1 + r/n)^{nt}$

where A = compound amount
 P = principal
 r = interest rate
 t = time in years
 n = periods per year (second formula only)

Note that as the number of times the interest is compounded increases, n increases, and it can be shown that

$$\left(1 + \frac{r}{n}\right)^{nt}$$

gets closer to e^{rt}. Let's see the difference between the two formulas.

EXAMPLE 3 ▶ Finding the Compound Amount

Find the compound amount

(a) for $100 compounded continuously for 18 months at 6%.

(b) for $100 compounded quarterly for 18 months at 6%.

Solution

(a) Here, $P = 100$, $r = 0.06$, and $t = 1.5$, so

$$A = 100e^{(0.06)(1.5)}$$
$$= 100e^{0.09}$$

A calculator gives the value

$$e^{0.09} \approx 1.0942$$

Thus

$$A = (100)(1.0942) \approx 109.42$$

and the compound amount is $109.42.

(b) As before, $P = 100$, $r = 0.06$, $t = 1.5$, and $n = 4$, so

$$1 + \frac{r}{n} = 1 + \frac{0.06}{4} = 1.015 \quad \text{and} \quad nt = 4 \cdot 1.5 = 6$$

Thus the compound amount for $100 at the same rate, compounded quarterly, is given by $A = 100(1.015)^6 \approx 109.34$. Note that at 18 months, the difference between continuous and quarterly compounding is only $.08! For more comparisons, see Using Your Knowledge in Exercises 7.5. ■

EXAMPLE 4 ▶ Finding the Amount

A radioactive substance decays so that G, the number of grams present, is given by

$$G = 1000e^{-1.2t}$$

GRAPH IT

In Example 4, let $Y_1 = 1000e^{-1.2t}$. To use a $[-1, 10]$ by $[-100, 1000]$ window and $\text{Yscl} = 100$, press WINDOW and enter -1 for Xmin, 10 for Xmax, -100 for Ymin, and 1000 for Ymax, with $\text{Yscl} = 100$. Do you now see that the substance is *decaying*? Now let's calculate the value of the function when $x = 2$ as required in the example. Press 2nd CALC 1. Answer the grapher question by entering $X = 2$ and press ENTER. The result is $Y = 90.717953$ as shown in the window. Rounding the answer to the nearest gram, the answer is 91, as before.

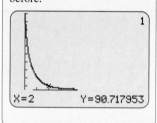

where t is the time in years. Find, to the nearest gram, the amount of the substance present

(a) at the start.   (b) in 2 years.

Solution

(a) Here, $t = 0$, so $G = 1000e^0 = 1000(1) = 1000$; that is, 1000 g of the substance are present at the start.

(b) Since $t = 2$, $G = 1000e^{-2.4}$. To evaluate G, we use a calculator and obtain

$$e^{-2.4} \approx 0.090718$$

so that

$$G \approx (1000)(0.090718)$$
$$= 90.718$$

There are about 91 g present in 2 years. ∎

EXAMPLE 5 ▶ Finding Population

Thomas Robert Malthus invented a model for predicting population based on the idea that when the birth rate (B) and the death rate (D) are constant and no other factors are considered, the population P is given by

$$P = P_0 e^{kt}$$

where P = population at any time t
   P_0 = initial population
   k = annual growth rate ($B - D$)
   t = time in years after 1980

According to the *Statistical Abstract of the United States,* the population in 2000 was 281,421,906, the birth rate was 0.0147, and the death rate was 0.0087. Use this information to predict the number of people in the United States in the year 2010.

Solution

The initial population is $P_0 = 281,421,906$, $k = 0.0147 - 0.0087 = 0.006$, and the number of years from 2000 to 2010 is $t = 10$.

$$P = 281,421,906e^{0.006 \times 10}$$
$$= 281,421,906e^{0.06}$$
$$\approx 298,824,065$$

Thus, the predicted population is 298,824,065. The *Statistical Abstract* predicts 298,710,000. Both predictions are wrong; we passed 300 million in October 2006! ∎

C. Logarithmic Functions

Most of the functions and equations we have discussed in this section did not have variables used as *exponents*. The equation

$$3^x = 9$$

is an *exponential* equation with solution 2 because $3^2 = 9$. The 2 is the exponent (power) to which the base 3 must be raised to obtain 9. The exponent 2 is called the **logarithm** to the base 3 of 9, and we write

$2 = \log_3 9$ (read "2 equals the log base 3 of 9")

Please note that the logarithm is simply an exponent, the exponent to which the base must be raised to get a certain number. In general,

---

Definition of a Logarithm

$y = \log_b x$ is equivalent to $b^y = x$ $b > 0, b \neq 1$

Remember that this definition simply means that the logarithm of a number is an exponent. For example,

$4 = \log_2 16$	is equivalent to	$2^4 = 16$	The logarithm is the exponent.
$2 = \log_5 25$	is equivalent to	$5^2 = 25$	The logarithm is the exponent.
$-3 = \log_{10} 0.001$	is equivalent to	$10^{-3} = 0.001$	The logarithm is the exponent.

The graph of the function $f(x) = \log_b x$ is found by interchanging the roles of x and y in the function $y = g(x) = b^x$ and graphing $x = b^y$. Geometrically, this is done by reflecting the graph of $g(x) = b^x$ about the line $y = x$, as shown in Figures 7.47 and 7.48.

GRAPH IT

Note that you cannot graph $f(x) = \log_2 x$ automatically using a grapher. The button $\boxed{\text{LOG}}$ on your grapher means $\log_{10}$, that is, the log to the base 10.

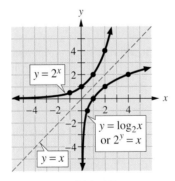

FIGURE 7.47
The graph of $y = f(x) = \log_2 x$ is obtained by reflecting the graph of $y = 2^x$ along the line $y = x$. Thus the points $\left(-1, \frac{1}{2}\right)$, $(0, 1)$, $(1, 2)$, and $(2, 4)$ become $\left(\frac{1}{2}, -1\right)$, $(1, 0)$, $(2, 1)$, and $(4, 2)$.

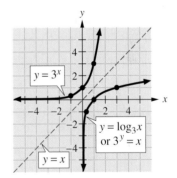

FIGURE 7.48
The graph of $y = f(x) = \log_3 x$ is obtained by reflecting the graph of $y = 3^x$ along the line $y = x$. Thus the points $\left(-1, \frac{1}{3}\right)$, $(0, 1)$, and $(1, 3)$ become $\left(\frac{1}{3}, -1\right)$, $(1, 0)$, and $(3, 1)$.

EXAMPLE 6 ▶ Graphing Logarithmic Functions

Graph $y = f(x) = \log_4 x$.

Solution
By the definition of logarithm, $y = \log_4 x$ is equivalent to $4^y = x$. We can graph $4^y = x$ by first assigning values to y and calculating the corresponding x values.

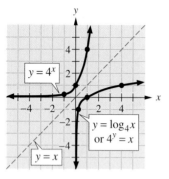

y	$x = 4^y$	Ordered Pair
0	$x = 4^0 = 1$	$(1, 0)$
1	$x = 4^1 = 4$	$(4, 1)$
2	$x = 4^2 = 16$	$(16, 2)$
-1	$x = 4^{-1} = \frac{1}{4}$	$\left(\frac{1}{4}, -1\right)$
-2	$x = 4^{-2} = \frac{1}{16}$	$\left(\frac{1}{16}, -2\right)$

FIGURE 7.49

We then graph the ordered pairs and connect them with a smooth curve, the graph of $y = \log_4 x$. To confirm that our graph is correct, graph $y = 4^x$ on the same coordinate axes (see Figure 7.49). As expected, the graphs are reflections of each other across the line $y = x$. ◾

Here are some properties of the graphs of $y = f(x) = \log_b x$.

Properties of Logarithmic Graphs

1. The point $(1, 0)$ is always on the graph because $\log_b 1 = 0$, since $b^0 = 1$.

2. For $b > 1$, the graph will rise from left to right (increase).

3. For $0 < b < 1$, the graph will fall from left to right (decrease).

4. The y axis is the vertical asymptote for $f(x)$.

5. The domain of $f(x) = \log_b x$ is $(0, \infty)$ and the range is $(-\infty, \infty)$.

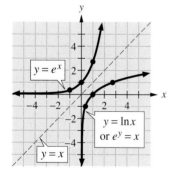

FIGURE 7.50

We have mentioned that e is an important base used in applications of exponential functions. The corresponding logarithmic function $y = \log_e x$ is so important that there is a special symbol for $\log_e x$; it is **ln x** (read "el-en of x"). Thus

$$\ln x = \log_e x$$

As before, to graph $y = \ln x$, we interchange the roles of x and y in $y = e^x$ and graph $x = e^y$ (see Example 2) as shown in Figure 7.50. The graph of $y = \ln x$ is a reflection $y = e^x$ along the line $y = x$. We interchanged the ordered pairs $(-1, 0.3679)$, $(0, 1)$, and $(1, e)$ and graphed $(0.3679, -1)$, $(1, 0)$, and $(e, 1)$.

Since a logarithm is simply an exponent, the expression $\ln e^k$ is the exponent to which the base e must be raised in order to obtain e^k. There is only one number that will do this, and it is k itself. Here is the idea.

Inverse Property of Logarithms

For all real numbers k, $\ln e^k = k$.

We use this idea in the following examples.

D. Applications Involving Exponential or Logarithmic Equations

Exponential and logarithmic equations have many applications in such areas as business, engineering, social science, psychology, and science. The following examples will give you an idea of the variety and range of their use.

EXAMPLE 7 ▶ Finding the Time It Takes Money to Double

With continuous compounding, a principal of P dollars accumulates to an amount A given by the equation

$$A = Pe^{rt}$$

where r is the interest rate and t is the time in years. If the interest rate is 6%, how long would it take for the money in your bank account to double?

Solution
With $A = 2P$ and $r = 0.06$, the equation becomes

$$2P = Pe^{0.06t}$$

or

$$2 = e^{0.06t} \qquad \text{Divide by } P.$$

We want to solve this equation for t, so we take natural logarithms of both sides.

$$\ln 2 = \ln e^{0.06t} = 0.06t \ln e = 0.06t \qquad \text{Recall that } \ln e^k = k.$$

Thus, $\ln 2 = 0.06t$ and

$$t = \frac{\ln 2}{0.06}$$

Using a calculator, we find $\ln 2 \approx 0.69315$ so that

$$t \approx \frac{0.69315}{0.06} \approx 11.6$$

This means that it would take about 11.6 years for your money to double. The verification is left to you. ■

EXAMPLE 8 ▶ Estimating World Population

In 1984, the population of the world was about 4.8 billion and the yearly growth rate was 2%. The equation giving the population P in terms of the time t is

$$P = 4.8e^{0.02t}$$

Estimate the world population P in the year 2050.

Solution
We are looking for the population P in the year 2050. Since $P = 4.8$ for $t = 0$, the equation shows that t is measured from the year 1984. To estimate the population in 2050, we use $t = 66$ in the equation.

$$P = 4.8e^{(0.02)(66)} = 4.8e^{1.32}$$

The value $e^{1.32} \approx 3.7434$ can be found with a calculator. Hence

$$P \approx (4.8)(3.7434) \approx 17.97$$

Thus, our estimate for the population in 2050 is about 18 billion. What would it be today? The verification is left to you.   ∎

EXAMPLE 9 ▶ Finding the Time to Reach a Given Level

If B is the number of bacteria present in a laboratory culture after t min, then, under ideal conditions,

$$B = Ke^{0.05t}$$

where K is a constant. If the initial number of bacteria is 1000, how long would it take for there to be 50,000 bacteria present?

Solution
Since $B = 1000$ for $t = 0$, we have

$$1000 = Ke^0 = K$$

The equation for B is then

$$B = 1000e^{0.05t} \quad \text{where } t \text{ is the unknown}$$

Now let $B = 50,000$.

$$50,000 = 1000e^{0.05t}$$
$$50 = e^{0.05t}$$

To solve for t, take natural logarithms of both sides.

$$\ln 50 = \ln e^{0.05t}$$
$$= 0.05t \qquad \ln e = 1$$

Thus

$$t = \frac{\ln 50}{0.05} \approx \frac{3.9120}{0.05} \approx 78.2 \text{ min}$$

It will take about 78.2 min for there to be 50,000 bacteria present. The verification is left to you.   ∎

EXAMPLE 10 ▶ Finding the Half-Life of an Element

The element cesium-137 decays at the rate of 2.3% per year. Find the half-life of this element.

Solution
The half-life of a substance is found by using the equation

$$A(t) = A_0 e^{-kt}$$

where $A(t)$ is the amount present at time t (years), k is the decay rate, and A_0 is the initial amount of the substance present. In this problem,

$$k = 2.3\% = 0.023 \quad \text{and} \quad A(t) = \tfrac{1}{2}A_0$$

and we want to find t.

With this information, the basic equation becomes

$$\frac{1}{2}A_0 = A_0 e^{-0.023t}$$
$$\frac{1}{2} = e^{-0.023t} \qquad \text{Divide by } A_0.$$

Now take natural logarithms of both sides.

$$\ln \tfrac{1}{2} = \ln e^{-0.023t}$$
$$\ln 1 - \ln 2 = -0.023t$$
$$t = \frac{\ln 2}{0.023} \qquad \text{Since } \ln 1 = 0$$
$$\approx \frac{0.69315}{0.023} \approx 30.14$$

Thus, the half-life of cesium-137 is about 30.14 years. The verification is left to you.   ∎

GRAPH IT

The numerical work preceding Example 2 can be done with a grapher. To find $(1.001)^{1000}$, press [2nd] [QUIT] [CLEAR] (to clear your home screen) and then 1 [·] 0 0 1 [^] 1 0 0 0 [ENTER]. For a more dramatic approximation for $e = 2.718281828$, tell the grapher to graph the sequence of numbers $(1 + 1/n)^n$ as n increases. Press [MODE], move the cursor three lines down to the line starting with "FUNC," and select "SEQ." Go down to the next line and select "DOT"; then press [ENTER]. (You have told the grapher you are about to graph a sequence.) Now press [Y=]. The symbols u(n) and v(n) are on your screen. Let's define the sequence $u(n)$ by entering $(1 + 1/n)^n$. (The n is entered by pressing [X, τ, θ, η]). To adjust the window, press [WINDOW] and select Xmin = −1, Xmax = 10, Xscl = 1, Ymin = −1, Ymax = 3, and Yscl = 1. Now press [GRAPH]. The result is shown in Window 1.

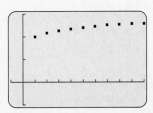

Window 1

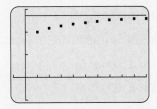

Window 2

To show that the sequence of dots is getting closer to e, use the draw feature to graph $y = e$. [Since you have entered a sequence, you can't enter the function $f(x) = e$.] Press [2nd] [DRAW] [6] and then enter [2ND] [e^x] 1 [ENTER]. The result is shown in Window 2. Do you now see how the sequence of dots representing $(1 + 1/n)^n$ approaches e?

━━━ **EXERCISES 7.5** ━━━━━━━━━━━━━━━━━━━━━━━━

Ⓐ Graphing Exponential Functions

In problems 1–6, find the value of the given exponential for the indicated values of the variable.

1. 5^x
   a. $x = -1$
   b. $x = 0$
   c. $x = 1$

2. 5^{-x}
   a. $x = -1$
   b. $x = 0$
   c. $x = 1$

3. 3^t
   a. $t = -2$
   b. $t = 0$
   c. $t = 2$

4. 3^{-t}
   a. $t = -2$
   b. $t = 0$
   c. $t = 2$

5. $10^{t/2}$
   a. $t = -2$
   b. $t = 0$
   c. $t = 2$

6. $10^{-t/2}$
   a. $t = -2$
   b. $t = 0$
   c. $t = 2$

In problems 7–10, graph the functions given in parts (**a**) and (**b**) on the same coordinate system.

7. a. $f(x) = 5^x$
   b. $g(x) = 5^{-x}$

8. a. $f(t) = 3^t$
   b. $g(t) = 3^{-t}$

9. a. $f(x) = 10^x$
   b. $g(x) = 10^{-x}$

10. a. $f(t) = 10^{t/2}$
   b. $g(t) = 10^{-t/2}$

Ⓑ Applications Involving Exponential Functions

In problems 11–18, find the compound amount if the compounding is (**a**) continuous or (**b**) quarterly.

11. $1000 at 9% for 10 years

12. $1000 at 9% for 20 years

13. $1000 at 6% for 10 years

14. $1000 at 6% for 20 years

15. The population of a town is given by the equation $P = 2000(2^{0.2t})$, where t is the time in years from 2000. Find the population in
   a. 2005.   **b.** 2010.   **c.** 2015.

16. A colony of bacteria grows so that its number B is given by the equation $B = 1200(2^t)$, where t is in days. Find the number of bacteria
   a. at the start ($t = 0$).   **b.** in 5 days.
   c. in 10 days.

17. A radioactive substance decays so that G, the number of grams present, is given by $G = 2000e^{-1.05t}$, where t is the time in years. Find the amount of the substance present
   a. at the start.   **b.** in 1 year.
   c. in 2 years.

18. Solve problem 17 when the equation for the number of grams present in a decaying radioactive substance is $G = 2000e^{-1.1t}$.

In problems 19–26, you may use a calculator with $\boxed{e^x}$ and $\boxed{x^y}$ keys, or you can use your grapher.

19. In 2000, the number of persons of Hispanic origin living in the United States was 35,305,818. If their birth rate is 0.0251 and their death rate is 0.006, predict the number of persons of Hispanic origin living in the United States for the year 2010. (Use a calculator.) (*Hint:* See Example 5.) (Source: U.S. Census 2000.)

20. In 2000, the number of African Americans living in the United States was 34,658,190. If their birth rate is 0.0181 and their death rate is 0.0011, predict the number of African Americans living in the United States in the year 2010. (*Hint:* See Example 5.) (Source: U.S. Census 2000.)

21. The number of compact discs (CDs) sold in the United States (in millions) since 1985 can be approximated by the exponential function $S(t) = 32(10)^{0.19t}$, where t is the number of years after 1985. To the nearest million,
   a. what was the number of CDs sold in 1990?
   b. what was the number sold in the year 2000?
   c. what would be the prediction for the number sold in 2010?

22. The number of cell phones sold in the United States (in thousands) since 1989 can be approximated by the exponential function $S(t) = 900(10)^{0.27t}$, where t is the number of years after 1989.
   a. What was the number of cell phones sold in 1990?
   b. Predict the number that will be sold in the year 2010.

23. According to the *Statistical Abstract of the United States,* about $\frac{2}{3}$ of all aluminum cans distributed are recycled. If a company distributes 500,000 cans, the number still in use after t years is given by the exponential function $N(t) = 500{,}000\left(\frac{2}{3}\right)^t$. How many cans are still in use after
   a. 1 year?   **b.** 2 years?   **c.** 10 years?

24. If the value of an item each year is about 60% of its value the year before, after t years the salvage value of an item costing C dollars is given by $S(t) = C(0.6)^t$. Find the salvage value of a computer costing $10,000

a. 1 year after it was purchased.

b. 10 years after it was purchased.

25. The atmospheric pressure A (in pounds per square inch) can be approximated by the exponential function $A(a) = 14.7(10)^{-0.000018a}$, where a is the altitude in feet.

a. The highest mountain in the world, Mount Everest, is about 29,000 ft high. Find the atmospheric pressure at the top of Mount Everest.

b. In the United States, the highest mountain is Mount McKinley in Alaska, whose highest point is about 20,000 ft. Find the atmospheric pressure at the top of Mount McKinley.

26. The atmospheric pressure A (in pounds per square inch) can also be approximated by $A(a) = 14.7e^{-0.21a}$, where a is the altitude in miles.

a. If we assume that the altitude of Mount Everest is about 6 mi, what is the atmospheric pressure at the top of Mount Everest?

b. If we assume that the altitude of Mount McKinley is about 4 mi, what is the atmospheric pressure at the top of Mount McKinley?

Ⓒ Logarithmic Functions

In problems 27 and 28 graph the function.

27. $f(x) = \log_5 x$ **28.** $f(x) = \log_6 x$

Ⓓ Applications Involving Exponential or Logarithmic Equations

In problems 29–32, assume continuous compounding and follow the procedure in Example 7 to find how long it takes a given amount to double at the given interest rate.

29. $r = 5\%$ **30.** $r = 7\%$

31. $r = 6.5\%$ **32.** $r = 7.5\%$

33. Suppose that the population of the world grows at the rate of 1.2% and that the population in 2000 was about 6.1 billion. Follow the procedure of Example 8 to estimate the population in the year 2010. (Source: www.un.org/esa/population/publications/wpp2000/highlights.pdf.)

34. Repeat problem 33 for a growth rate of 1.75%.

In problems 35–38, assume that the number of bacteria present in a culture after t min is given by $B = 1000e^{0.04t}$. Find the time it takes for the number of bacteria present to be

35. 2000.   **36.** 5000.   **37.** 25,000.   **38.** 50,000.

39. When a bacteria-killing solution is introduced into a certain culture, the number of live bacteria is given by the equation $B = 100,000e^{-0.2t}$, where t is the time in hours. Find the number of live bacteria present at the following times:

a. $t = 0$ **b.** $t = 2$

c. $t = 10$ **d.** $t = 20$

40. The number of honey bees in a hive is growing according to the equation $N = N_0e^{0.015t}$, where t is the time in days. If the bees swarm when their number is tripled, find how many days until this hive swarms.

In problems 41–44, follow the procedure of Example 10 to find the half-life of the substance.

41. Plutonium, whose decay rate is 0.003% per year

42. Krypton, whose decay rate is 6.3% per year

43. A radioactive substance whose decay rate is 5.2% per year

44. A radioactive substance whose decay rate is 0.2% per year

45. The atmospheric pressure P in pounds per square inch at an altitude of h feet above the Earth is given by the equation $P = 14.7e^{-0.00005h}$. Find the pressure at an altitude of

a. 0 ft.   **b.** 5000 ft.   **c.** 10,000 ft.

46. If the atmospheric pressure in problem 45 is measured in inches of mercury, then $P = 30e^{-0.207h}$, where h is the altitude in miles. Find the pressure

a. at sea level.   **b.** at 5 mi above sea level.

47. Has free agency affected salaries? According to the National Football League Players Association (NFLPA), average NFL salaries (in thousands of dollars) are as shown in the graph (on the next page) and can be approximated by

$$S = 250e^{0.11t}$$

where t is the number of years after 1986.

a. In how many years will average salaries be $800,000? (Answer to the nearest year.)

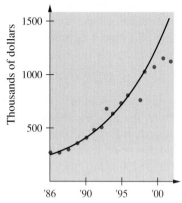

How close is your estimate? To see the current figures, visit the NFL players' association Web site at www.nflpa.org/main/main.asp. To see the figures for a recent year, try http://tinyurl.com/dtpka.

b. On the basis of this graph, in how many years will salaries reach the 1 million dollar mark? (Answer to the nearest year.)

48. According to the *Statistical Abstract of the United States,* about $\frac{2}{3}$ of all aluminum cans distributed are recycled. If a company distributes 500,000 cans, the number in use after t years is

$$N(t) = 500,000 \left(\frac{2}{3}\right)^t$$

How many years will it take for the number of cans to reach 100,000? (Answer to the nearest year.)

49. Do you have a fear of flying? The U.S. Department of Transportation has good news for nervous fliers: The number of general aviation accidents A has gone down significantly in the last 30 years! It can be approximated by

$$A = 5000e^{-0.04t}$$

where t is the number of years after 1970.
a. How many accidents were there in 1970?
b. How many accidents were there in 1990?
c. In what year do you predict the number of accidents to be 1000? (Answer to the nearest year.)

50. After exercise the diastolic blood pressure of normal adults is a function of time and can be approximated by

$$P = 90e^{-0.5t}$$

where t is the time in minutes. How long would it be before the diastolic pressure decreased to 80? (Answer to two decimal places.)

51. The number of bacteria present in a certain culture can be approximated by

$$B = 50,000e^{0.2t}$$

where t is measured in hours and $t = 0$ corresponds to 12 noon. Find the number of bacteria present at
a. noon.
b. 2 P.M.
c. 6 P.M.

52. If a bactericide (a bacteria killer) is introduced into a bacteria culture, the number of bacteria can be approximated by

$$B = 50,000e^{-0.1t}$$

where t is measured in hours. Find the number of bacteria present
a. when $t = 0$.
b. when $t = 1$.
c. when $t = 10$.

53. Sales begin to decline d days after the end of an advertising campaign and can be approximated by
$$S = 1000e^{-0.1d}$$
a. How many sales will be made on the last day of the campaign—when $d = 0$?
b. How many sales will be made 10 days after the end of the campaign?

54. The demand function for a certain commodity is approximated by

$$p = 100e^{-q/2}$$

where q is the number of units demanded at a price of p dollars per unit.
a. If there is a 100-unit demand for the product, what will its price be?
b. If there is no demand for the product, what will its price be?

55. The concentration C of a drug in the bloodstream at time t (in hours) can be approximated by

$$C = 100\,(1 - e^{-0.5t})$$

a. What will the concentration be when $t = 0$?
b. What will the concentration be after 1 hour? (Round to 1 decimal place.)

56. The number of people $N(t)$ reached by a particular rumor at time t is approximated by

$$N(t) = \frac{5050}{1 + 100e^{-0.06t}}$$

a. Find $N(0)$.
b. Find $N(0)$.

57. The stellar magnitude M of a star is defined by

$$M = -2.5 \log \left(\frac{B}{B_0}\right)$$

where B is the brightness of the star and B_0 the minimum of brightness.

a. Find the stellar magnitude of the North Star, 2.1 times as bright as B_0. (Round to 4 decimal places.)

b. Find the stellar magnitude of Venus, 36.2 times as bright as B_0. (Round to 4 decimal places.)

58. The percent P of adult height a male has reached at age A ($13 \leq A \leq 18$) is

$$P = 16.7 \log (A - 12) + 87$$

a. What percent of adult height has a 13-year-old male reached? (Round to the nearest percent.)

b. What percent of adult height has an 18-year-old male reached? (Round to the nearest thousandth of a percent.)

 Using Your Knowledge

In this section you learned that for continuous compounding, the compound amount is given by $A = Pe^{rt}$. For ordinary compound interest, the compound amount is given by $A = P(1 + r/n)^{nt}$, where r is the annual interest rate and n is the number of periods per year. Suppose you have \$1000 to put into an account where the interest rate is 6%. How much more would you have at the end of 2 years for continuous compounding than for monthly compounding?

For continuous compounding, the amount is given by

$$A = 1000e^{(0.06)(2)} = 1000e^{0.12}$$
$$= (1000)(1.1275) \quad \text{Use a calculator.}$$
$$= 1127.50$$

Thus the amount is \$1127.50. For monthly compounding,

$$\frac{r}{n} = \frac{0.06}{12} = 0.005 \quad \text{and} \quad nt = 24$$

Thus the amount is given by $A = 1000(1 + 0.005)^{24} = 1000(1.005)^{24}$. Compound interest tables or your calculator gives the value

$$(1.005)^{24} = 1.1271598$$

so that $A = 1127.16$ (to the nearest hundredth). Thus the amount is \$1127.16. Continuous compounding earns you only \$.34 more! But, see what happens when the period t is extended in problems 59 and 60!

59. Make the same comparison where the time is 10 years.

60. Make the same comparison where the time is 20 years.

 In Other Words

61. The definition of the exponential function $f(x) = b^x$ does not allow $b = 1$.
a. What type of graph will $f(x)$ have when $b = 1$?
b. Is $f(x) = b^x$ a function when $b = 1$? Explain.

62. List some reasons to justify the condition $b > 0$ in the definition of the exponential function $f(x) = b^x$.

63. Discuss the relationship between the graphs of $f(x) = b^x$ and $g(x) = b^{-x}$.

64. In Example 5, we predicted the U.S. population for the year 2010 to be 298,824,065. The U.S. Census Bureau predicted 299,862,000. Can you give some reasons for this discrepancy? To check the latest projections, try www.census.gov/ipc/www. usinterimproj/natprojtab01a.pdf.

 Skill Checker

In the next section you will solve a system of simultaneous equations by graphing. To graph equations simultaneously, you better practice graphing just one equation at a time. Here are some practice problems for you. Graph them!

1. $x + y = 4$
2. $2x - y = 4$
3. $2x - y = 5$
4. $x - 2y = 4$

If you are unsure of how to graph these problems, review Sections 7.1 and 7.2.

Research Questions

1. Who was the first person to publish a book describing the rules of logarithms, what was the name of the book, and what does *logarithm* mean?

2. Describe how Jobst Burgi used "red" and "black" numbers in his logarithmic table.

3. The symbol e was first used in 1731. Name the circumstances under which the symbol e was first used.

7.6   Two Linear Equations in Two Variables

Women and Men in the Work Force

Figure 7.51 shows the percentage of women (W) and men (M) in the work force since 1955. Can you tell from the graph in which year the percentage will be the same? It will happen after 2005! If t is the number of years after 1955, W the percentage of women in the work force, and M the percentage of men, the graphs

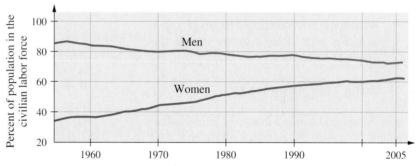

Source: U.S. Bureau of Labor Statistics.

FIGURE 7.51

Women and men in the work force.

will *intersect* at a point (a, b). The coordinates of the point of intersection (if there is one) are the common solution of both equations. If the percents W and M are approximated by

$$W = 0.055t + 34.4 \text{ (percent)}$$
$$M = -0.20t + 83.9 \text{ (percent)}$$

you can find the year after 1955 when the same percentage of women and men were in the work force by graphing M and W and locating the point of intersection. We will do that later! ▶

There are many applications of algebra that require the solution of a **system of linear equations.** We shall consider only the simple case of two equations in two variables.

As we have seen, a linear equation $Ax + By = C$ has a straight line for its graph. Hence, two such equations will graph into two straight lines in the plane. Two distinct lines in the plane can either *intersect* at a point or else be *parallel* (have no intersection). If the lines intersect, then the coordinates of the point of intersection are called the **solution of the system of equations.** If the lines are parallel, then, of course, the system has no solution.

A. Solution by Graphing

One way to solve a system of two equations in two variables is to graph the two lines and read the coordinates of the point of intersection from the graph. The next two examples illustrate this graphical method.

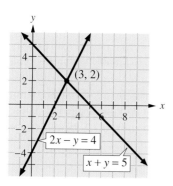

FIGURE 7.52

 GRAPH IT

To do Example 1, make sure you are in the function mode. Solve each equation for y and enter them as $Y_1 = 2x - 4$ and $Y_2 = 5 - x$. Press ZOOM 6. To find the intersection, press 2nd CALC 5 and ENTER three times. The intersection is $x = 3$ and $y = 2$ as shown.

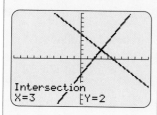

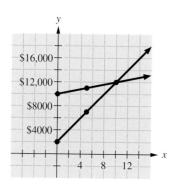

FIGURE 7.53

EXAMPLE 1 ▶ **Solving Systems by the Graphical Method**

Use the graphical method to find the solution of the system

$$2x - y = 4$$
$$x + y = 5$$

Solution

We graph the two equations as shown in Figure 7.52. The point of intersection of the two lines appears to be (3, 2). We check this set of coordinates in the given system. For $x = 3$ and $y = 2$,

$$2x - y = 2 \cdot 3 - 2 = 4$$
$$x + y = 3 + 2 = 5$$

Thus, the two equations are both satisfied, and the desired solution is $x = 3$, $y = 2$, or (3, 2). ∎

EXAMPLE 2 ▶ **Using Systems of Equations to Analyze Costs**

If you are about to build a house, you might consider the following two types of heating:

1. Solar heating, which requires a $10,000 initial investment and a $200 annual cost

2. Oil heating, with a $2000 initial investment and an annual cost of $1000

(a) Write an equation giving the total cost y (in dollars) of solar heating over x years.

(b) Repeat part (a) for oil heating.

(c) Graph the equations obtained in parts (a) and (b) for a period of 12 years.

(d) When will the total costs for the two systems be equal?

Solution

(a) The cost y of solar heating is given by $y = 10,000 + 200x$.

(b) The cost y of oil heating is given by $y = 2000 + 1000x$.

(c) Note that the x coordinate represents the number of years and the y coordinate represents the number of dollars in the corresponding total cost. For the solar heating equation, we find the two points (0, 10,000) and (5, 11,000). These two points and the line through them are shown in Figure 7.53. For the oil heating equation, we find the two points (0, 2000) and (5, 7000). These two points and the line through them are also shown in Figure 7.53.

(d) As you can see from Figure 7.53, the lines intersect at the point where $x = 10$ and $y = 12,000$. Thus, the total costs are equal at the end of 10 years. (After 10 years, the total cost for solar heating is much less than that for oil heating.) ∎

One word of warning: You should always check apparent solutions by substituting the respective values of x and y back into the original equations. **Picture (graph) solutions often give only approximations of actual solutions.** Remember what happened in Getting Started?

B. Solution by Algebraic Methods

If the solution of the system is a pair of simple numbers that can be read exactly from the graph, then the graphical method is quite satisfactory. Naturally, you should always check a proposed solution in the given equations. Unfortunately, most systems have solutions that are not easy to read exactly from graphs, and for such systems an algebraic method of solving is needed. There is a quite simple method that is best explained by means of examples, as follows:

EXAMPLE 3 ▶ Solving Systems by Algebraic Methods

Solve the system of Example 1 by algebraic means.

Solution

The point (x, y) where the lines $2x - y = 4$ and $x + y = 5$ intersect is the intersection of the solution sets of the two equations. This means that the number pair consisting of the x value and the y value of this point must satisfy both equations. Hence, we can use the following procedure: Since the y terms in the two equations are the same except for sign, we add the two equations term by term to eliminate the y terms and get an equation in one variable.

$$
\begin{array}{r}
2x - y = 4 \\
\underline{x + y = 5} \\
3x \quad\;\; = 9
\end{array}
$$

Thus, $x = 3$. By substituting this x value into the second equation, we get

$$3 + y = 5 \quad \text{or} \quad y = 2$$

We have now found the same solution as before, $(3, 2)$.   ■

If both variables occur in both equations, we can proceed as follows:

Solving a System by Elimination

1. Eliminate one of the variables by

    a. multiplying one or both equations by nonzero constants, if necessary, so that the resulting coefficients for one of the variables are of equal magnitude and opposite in sign in the two respective equations and

    b. adding corresponding terms of the two equations.

2. Solve the resulting linear equation for the remaining variable.

3. Substitute the solution back into one of the original equations to find the value of the second variable.

This procedure is illustrated in the next example.

EXAMPLE 4 ▶ Solving Systems Algebraically

Find the point of intersection of the two lines $3x - 2y = 19$ and $2x + 5y = -19$.

Solution

Step 1a We multiply the first equation by 5 and the second by 2 to get

$$15x - 10y = 95$$
$$4x + 10y = -38$$

Opposites

Step 1b We then add the two equations to get

$$19x = 57$$

Step 2 Solving for x gives $x = 3$.

Step 3 We substitute $x = 3$ into the first of the given equations to obtain

$$3 \cdot 3 - 2y = 19$$
$$9 - 2y = 19$$
$$-2y = 10$$
$$y = -5$$

The point of intersection is thus $(3, -5)$.   ■

EXAMPLE 5 ▶ Systems with No Solution

Solve the following system (if possible):

$$2x - y = 5$$
$$4x - 2y = 7$$

Solution

Step 1a We multiply the first equation by -2 to obtain the system

$$-4x + 2y = -10$$
$$4x - 2y = 7$$

Step 1b Adding these two equations, we get the *impossible* result

$$0 = -3$$

Thus, the system has no solution. The lines represented by the two equations are parallel, and there is no point of intersection.   ■

Notice that the method of addition of the equations (used in Examples 4 and 5) will detect parallel lines as in Example 5 by arriving at an *impossible* result: $0 =$ a nonzero number. The next example shows that different equations can represent the same line.

EXAMPLE 6 ▶ Solving Systems with Many Solutions

Solve the following system (if possible):

$$2x - y = 5$$
$$4x - 2y = 10$$

Solution

Step 1a We multiply the first equation by 2 to get

$$4x - 2y = 10$$

which is exactly the second equation. Thus, we see that the two equations represent the same line, and every solution of one of these equations satisfies the other. If we put $x = a$ in the first equation, and then solve for y to get $y = 2a - 5$, we can write the general solution of the system as $(a, 2a - 5)$, where a is any real number. For example, $(0, -5)$, $(3, 1)$, and $(-1, -7)$ are solutions of the system. The system has infinitely many solutions. ∎

$24.95 $19.95

FIGURE 7.54

C. Solving an Application

Most of the problems that we've discussed use x and y values ranging from -10 to 10. This is not the case when working real-life applications! For example, the price for connecting to the Internet using America Online (AOL) or CompuServe (CS) used to be identical: $9.95 per month plus $2.95 for every hour over 5 hours. If you are planning on using more than 5 hours, however, both companies had a special plan (see Figure 7.54).

> AOL: $19.95 for 20 hours plus $2.95 per hour after 20 hours*
> CS: $24.95 for 20 hours plus $1.95 per hour after 20 hours

For convenience, let's round the numbers as follows:

> AOL: $20 for 20 hours plus $3 per hour after 20 hours
> CS: $25 for 20 hours plus $2 per hour after 20 hours

EXAMPLE 7 ▶ Comparing Internet Prices

Make a graph of both prices to see which is a better deal.

Solution

Let's start by making a table where h represents the number of hours used and p the monthly price. Keep in mind that when the number of hours is 20 or less the price is fixed: $20 for AOL and $25 for CS.

h	Price for AOL, p	
0	20	
5	20	
10	20	
15	20	
20	20	
25	35	$20 + (25 - 20)3 = 35$
30	50	$20 + (30 - 20)3 = 50$
35	65	$20 + (35 - 20)3 = 65$

h	Price for CS, p	
0	25	
5	25	
10	25	
15	25	
20	25	
25	35	$25 + (25 - 20)2 = 35$
30	45	$25 + (30 - 20)2 = 45$
35	55	$25 + (35 - 20)2 = 55$

In order to graph these two functions, we let h run from 0 to 50 and p run from 0 to 100 in increments of 5. After 20 hours, the price p for h hours for AOL is

$$p = 20 + (h - 20) \cdot 3 = 20 + 3h - 60 = 3h - 40$$

The price for CS is

$$p = 25 + (h - 20) \cdot 2 = 25 + 2h - 40 = 2h - 15$$

The graph is shown in Figure 7.55. As you can see, if you plan on using 20 hours or less, the price is fixed for both AOL and CS. From 20 to 25 hours, AOL has a lower price, but if you are using more than 25 hours, the price for CS is lower. At exactly 25 hours, the price is the same for both, $35. Note that the complete graph is in quadrant I, since neither the number of hours h nor the price p is negative. You can use these ideas to compare different Internet plans, but in 1998, CompuServe became a wholly owned subsidiary of AOL! ∎

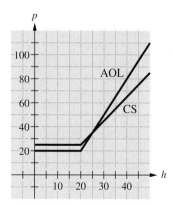

FIGURE 7.55

*As of this writing, AOL has a new plan costing $25.90 per month with no limit on the number of hours used.

D. Solving by Substitution

To avoid graphing and to obtain a more precise answer, we can use the **substitution** method shown next.

EXAMPLE 8 ▶ Finding When Internet Services Cost the Same

When is the price for both services the same?

Solution

As you recall, for $h > 20$,

$$\text{AOL: } p = \underbrace{20}_{\$20} + \underbrace{3(h - 20)}_{\substack{\$3 \text{ for each} \\ \text{hour after 20}}}$$

$$\text{CS: } p = \underbrace{25}_{\$25} + \underbrace{2(h - 20)}_{\substack{\$2 \text{ for each} \\ \text{hour after 20}}}$$

To find when the price p is the same for both services, we *substitute* $p = 20 + 3(h - 20)$ into the second equation to obtain

$20 + 3(h - 20) = 25 + 2(h - 20)$	
$20 + 3h - 60 = 25 + 2h - 40$	Use the distributive property.
$3h - 40 = 2h - 15$	Simplify.
$3h = 2h + 25$	Add 40 to both sides.
$h = 25$	Subtract h from both sides.

Thus, if you use 25 hours, the price is the same for AOL and CS.

Check: The price for 25 hours of AOL service is

$$p = 20 + 3(25 - 20) = 20 + 15 = \$35$$

The price for 25 hours of CS service is

$$p = 25 + 2(25 - 20) = 25 + 10 = \$35$$

Thus when using 25 hours, the price is the same for both services, $35. ■

Before you try the exercises, here is a summary that will help you decide what method to use.

Online Study Center

To further explore linear equations in two variables and to solve a system with two equations and two unknowns, access links 7.6.1 and 7.6.2 on this textbook's Online Study Center.

Solving Systems of Equations: A Summary

Method	Suggested Use	Disadvantages
Graphical	When the coefficients of the variables and the solutions are integers. You get a picture of the situation.	If the solutions are not integers, they are hard to read on the graph.
Substitution	When one of the variables is isolated (alone) on one side of the equation	If fractions are involved, you may have much computation.
Elimination	When fractions, decimals, or variables with coefficients that are the same or additive inverses of each other ($2x$ and $-2x$, for example) are present	You may have lots of computations involving signed numbers.

EXERCISES 7.6

Ⓐ Solution by Graphing

In problems 1–6, find the solution of the given system by the graphical method.

1. $x + y = 3; 2x - y = 0$

2. $x + y = 5; x - 4y = 0$

3. $2x - y = 10; 3x + 2y = 1$

4. $2x - 3y = 1; x + 2y = 4$

5. $3x + 4y = 4; 2x - 6y = 7$

6. $y = 5x - 12; y = 3x - 8$

Ⓑ Solution by Algebraic Methods

In problems 7–30, find the solution of the given system by the method of addition that was used in Examples 4 and 5. If there is no solution or infinitely many solutions, state so.

7. $x + y = 3; 2x - y = 0$

8. $x + y = 5; x - 4y = 0$

9. $x + y = 6; 3x - 2y = 8$

10. $2x - y = 5; 5x + 3y = 18$

11. $2x - y = 10; 3x + 2y = 1$

12. $2x - 3y = 1; x + 2y = 4$

13. $5x + y = 4; 15x + 3y = 8$

14. $2x - y = -5; 4x - 2y = -10$

15. $2x + 5y = 12; 5x - 3y = -1$

16. $2x + 3y = 9; 11x + 7y = 2$

17. $3x + 4y = 4; 2x - 6y = 7$

18. $y = 5x - 12; y = 3x - 8$

19. $11x + 3 = -3y; 5x + 2y = 5$

20. $10x + 6y = 1; 5x = 9 - 3y$

21. $x = 2y - 3; x = -2y - 1$

22. $y = 4x - 2; 4x = 2y + 3$

23. $2x + 3y + 11 = 0; 5x + 6y + 20 = 0$

24. $3x + y = 4; 2x = 4y - 9$

25. $3x - 12y = -8; 2x + 2y = 3$

26. $4x + 8y = 7; 3x + 4y = 6$

27. $r - 4s = -10; 2r - 8s = 13$

28. $3r + 4s = 15; 4r - s = 20$

29. $6u - 2v = -27; 4u + 3v = 8$

30. $8w - 13z = 3z + 4; 12w - 3 = 18z$

31. Company A will cater a banquet for x guests at a cost of y dollars, where

$$y = 8x + 1000$$

For company B the cost is

$$y = 10x + 800$$

   a. Use the method of addition to find the number of guests for which the two costs are equal.

   b. Graph the two equations on the same set of axes from $x = 0$ to $x = 200$.

   c. For which company is the cost less if there are more than the number of guests you found in part (**a**)?

32. A solar hot-water heating system has an initial investment of $5400 and an annual cost of $50. An electric hot-water heating system has an initial investment of $1000 and an annual cost of $600. Let y dollars be the total cost at the end of x years.

   a. Write an equation giving the total cost of the solar heating system.

   b. Write an equation giving the total cost of the electric heating system.

   c. Graph the two equations you found in parts (**a**) and (**b**) on the same set of axes from $x = 0$ to $x = 10$.

   d. Use the method of addition to solve the system of equations.

   e. How long will it take for the total cost of the solar heat to become less than the total cost of the electric heat?

33. The supply y of a certain item is given by the equation $y = 2x + 8$, where x is the number of days elapsed. If the demand is given by $y = 4x$, in how many days will the supply equal the demand?

34. The supply of a certain item is $y = 3x + 8$, where x is the number of days elapsed. If the demand is given by $y = 4x$, in how many days will the supply equal the demand?

35. A company has 10 units of a certain item and can manufacture 5 units each day. If the demand for

the item is $y = 7x$, where x is the number of days elapsed, in how many days will the demand equal the supply?

36. Clonker Manufacturing has 12 clonkers in stock. It manufactures 3 other clonkers each day. If the clonker demand is 7 each day, in how many days will the supply equal the demand?

C Solving an Application

In problems 37–46, use the following information: You want to watch 10 movies at home each month. You have two options.

> Option 1: Get cable service. The cost is $20 for the installation fee and $35 per month.
> Option 2: Buy a VCR and spend $25 each month renting movies. The cost is $200 for a VCR and $25 a month for movie rental fees.

37. a. If C is the cost of installing cable service plus the monthly fee for m months, write an equation for C in terms of m.
   b. Complete the following table where C is the cost of cable service for m months:

m	6	12	18
C			

   c. Graph the information obtained in parts (**a**) and (**b**). (*Hint:* Let m run from 1 to 24 and C run from 0 to 1000 in increments of 100.)

38. a. If C is the total cost of buying the VCR plus renting the movies for m months, write an equation for C in terms of m.
   b. Complete the following table, where C is the cost of buying a VCR and renting movies for m months:

m	6	12	18
C			

   c. Graph the information obtained in parts (**a**) and (**b**).

39. Make a graph of the information obtained in problems 37 and 38 on the same coordinate axes.

40. On the basis of the graph for problem 39, when is the cable service cheaper?

41. On the basis of the graph for problem 39, when is the VCR and rental option cheaper?

42. At Grady's restaurant, servers earn $80 a week plus tips, which amount to $5 per table.
   a. Write an equation for the weekly wages W based on serving t tables.
   b. Complete the following table, where W is the wages and t is the number of tables served:

t	5	10	15	20
W				

   c. Graph the information obtained in parts (**a**) and (**b**).

43. At El Centro restaurant, servers earn $100 a week, but the average tip per table is only $3.
   a. Write an equation for the weekly wages W based on serving t tables.
   b. Complete the following table, where W is the wages and t is the number of tables served:

t	5	10	15	20
W				

   c. Graph the information obtained in parts (**a**) and (**b**).

44. Graph the information from problems 42 and 43 on the same coordinate axes. On the basis of the graph, answer the following questions:
   a. When does a server at Grady's make more money than a server at El Centro?
   b. When does a server at El Centro make more money than a server at Grady's?

45. Do you have a cell phone? How much do you pay a month? At the present time, GTE and AT&T have cell phone plans that cost $19.95 per month. However, GTE costs $.60 per minute of airtime during peak hours while AT&T costs $.45 per minute of airtime during peak hours. GTE offers a free phone with their plan while AT&T's phone costs $45. For comparison purposes, since the monthly cost is the same for both plans, the cost C is based on the price of the phone plus the number of minutes of airtime m used.
   a. Write an equation for the cost C of the GTE plan.
   b. Write an equation for the cost C of the AT&T plan.
   c. Using the same coordinate axes, make a graph for the cost of the GTE and the AT&T plans. (*Hint:* Let m and C run from 0 to 500.)

46. On the basis of the graphs obtained in problem 45, which plan would you buy, GTE or AT&T? Explain.

 Using Your Knowledge

According to the *Statistical Abstract of the United States,* the prices for art books and the number of books *supplied* in 5-year intervals are as shown in the following table:

Demand Function		Supply Function	
Price ($)	Quantity (hundreds)	Price ($)	Quantity (hundreds)
$28	17	$26	12
$35	15	$35	15
$42	13	$44	18

Note that *fewer* books are *demanded* by consumers as the price of the books goes up. (Consumers are not willing to pay that much for the book.)

47. Graph the points corresponding to the demand function in the coordinate system shown.

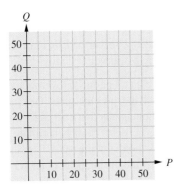

48. Draw a line passing through the points. The line is an approximation to the demand function.

On the other hand, book sellers *supply fewer* books when prices decrease and more books when prices increase.

49. Graph the points corresponding to the supply function on the same coordinate system.

50. Draw a line passing through the points. The line is an approximation to the supply function.

51. What is the point of intersection of the two lines?

52. At the point of intersection, what is the price?

53. At the point of intersection, what is the quantity of books sold?

 In Other Words

54. What does the solution of a system of linear equations represent?

55. How can you tell graphically whether a system has *no* solution?

56. How can you tell graphically whether a system has one solution?

Suppose you have a system of equations and you solve both equations for y and obtain

$$y = m_1 x + b_1$$
$$y = m_2 x + b_2$$

What can you say about the graph of the system when

57. $m_1 = m_2$ and $b_1 \neq b_2$? How many solutions do you have? Explain.

58. $m_1 = m_2$ and $b_1 = b_2$? How many solutions do you have? Explain.

59. $m_1 \neq m_2$? How many solutions do you have? Explain.

 Collaborative Learning

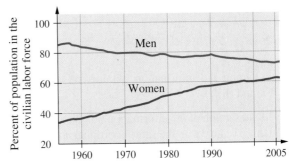

Source: U.S. Bureau of Labor Statistics.

Women and men in the work force.

We promised in Getting Started that we would get back to this problem: finding the year in which the percent of men and women in the work force are equal. Divide your group into three teams. Here are the options.

Team 1 Extend the lines for the percent of men (M), the percent of women (W), and the horizontal axis.

1. In what year do the lines seem to intersect?

2. What would be the percent of men and women in the labor force in that year?

Team 2 Get some graph paper with a 20-by-20 grid, each grid representing 1 unit. Graph the lines

$$M = -0.20t + 83.9 \text{ (percent)}$$
$$W = 0.55t + 34.4 \text{ (percent)}$$

where t is the number of years after 1955.

1. Where do the lines intersect?

2. What does the point of intersection represent?

Team 3 The point at which the lines intersect will satisfy both equations; consequently at that point $M = W$. Substitute $-0.20t + 83.9$ for M and $0.55t + 34.4$ for W and solve for t.

1. What value did you get?

2. What is the point of intersection for the two lines and what does it represent?

Compare the results of teams 1, 2, and 3. Do they agree? Why or why not? Which is the most accurate method for obtaining an answer for this problem?

7.7 Linear Inequalities

GETTING STARTED

Rhyme and Reason

Do you like poetry? Perhaps you will understand the graphing of inequalities if we do it in verse.

How to Graph a Linear Inequality, by Julie Ashmore*

FIGURE 7.56

Look here, my children, and you shall see
How to graph an inequality.
Here's a simple inequality to try:
x plus 2 is less than y. $x + 2 < y$
First, make the "less than" "equal to";
So now y equals x plus 2. $y = x + 2$
Then pick a point for x: say, 10;
Now plug that single constant in.
Add 10 plus 2 and you'll get y; $x = 10, y = 10 + 2$
See if this pair will satisfy!
x: 10, y: 12; you'll find it's right,   (10, 12)
So graph this point to expedite.
Now find a second ordered pair   See Figure 7.56.
That fits in your equation there.
x: 3, y: 5 will do quite well.   (3, 5)
And it's correct, as you can tell.
Plot this point, and then you've got
To draw a line from dot to dot.   See Figure 7.56.
Make it neat and make it straight;
A ruler's edge I'd advocate.
The next step's hard! You've got to choose

---

*Julie Ashmore was a young student in the John Burroughs School in St. Louis, Missouri, when she wrote this verse.

Which side of this line you must use.	
Change "equal to" back to "less than,"	$x + 2 < y$
Just as it was when you began.	
Pick a point on one side! I	
Use 3 for x and 1 for y.	$(3, 1)$
Is 1 greater than 3 plus 2?	
No! This side will never do!	$3 + 2 \not< 1$
On the other side, let's try	
1 for x and 4 for y.	$(1, 4)$
1 plus 2 (which equals 3)	
Is less than 4, as you can see.	$1 + 2 = 3 < 4$
Shade in the side that dot is on;	
We've one more step to come upon.	See Figure 7.56.
Do the points upon your line	
Fit the equation I assigned?	
Use 1 and 3 for this last test;	$(1, 3)$
They're "equal to," but they're not "less."	
Make your line dotted to show this is true.	$1 + 2 = 3$
And that is *all* you have to do!	See Figure 7.56. ▶

In this section we will see how to graph a relation in which the rule is a linear inequality, that is, an inequality of the type $ax + by + c \geq 0$. Let us use these ideas in a more practical way.

A. Graphing Linear Inequalities

Suppose that you want to rent a car for a few days. Here are some prices for an intermediate car rented in Florida in a recent year:

Rental A $36 per day, $.15 per mile
Rental B $49 per day, $.33 per mile

The total cost C for the rental A car is

$$C = \underbrace{36d}_{\substack{\text{Cost for} \\ d \text{ days}}} + \underbrace{0.15m}_{\substack{\text{Cost for} \\ m \text{ miles}}}$$

Now suppose that you want the cost C to be $180. Then $180 = 36d + 0.15m$. Graph this equation by finding the intercepts. When $d = 0$, $180 = 0.15m$, or

$$m = \frac{180}{0.15} = 1200$$

When $m = 0$, $180 = 36d$, or

$$d = \frac{180}{36} = 5$$

Join $(0, 1200)$ and $(5, 0)$ with a line, and then graph the discrete points corresponding to 1, 2, 3, or 4 days.

But what if you want the cost to be less than $180? Then you have

$$36d + 0.15m < 180$$

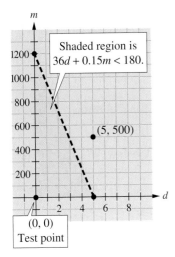

FIGURE 7.57

You have graphed the points on the line $36d + 0.15m = 180$. Where are the points for which $36d + 0.15m < 180$? As the graph in Figure 7.57 shows, the line $36d + 0.15m = 180$ divides the plane into three parts.

1. The points (half-plane) **below** the line

2. The points **on** the line

3. The points (half-plane) **above** the line

It can be shown that if any point on one side of the line $Ax + By = C$ satisfies the inequality $Ax + By < C$, then all points on that side satisfy the inequality and no point on the other side of the line does. Let's select $(0, 0)$ as a test point. Since

$$36 \cdot 0 + 0.15 \cdot 0 < 180$$
$$0 < 180$$

is true, all points below the line (shown shaded in Figure 7.57) satisfy the inequality. [As a check, note that $(5, 500)$, which is above the line, doesn't satisfy the inequality because $36 \cdot 5 + 0.15 \cdot 500 < 180$ is a false statement.] The line $36d + 0.15m = 180$ is *not* part of the graph and is shown dashed. To apply this result to our rental car problem, note that d must be an integer. The solution to our problem consists of the points on the heavy line segments at $d = 1, 2, 3$, and 4. The graph shows, for instance, that you can rent a car for 2 days and go about 700 mi at a cost less than \$180. What we have just set up is called a *linear inequality*. Here are the definitions we need in order to proceed.

Definition of a Linear Inequality

A **linear inequality** is a statement that can be written in the form

$$Ax + By \leq C \quad \text{or} \quad Ax + By \geq C$$

where A and B are not both zero.

Note: If $<$ and $>$ are substituted for $\leq$ and $\geq$, we still have a linear inequality. Here is a summary of the procedure we used to graph the linear inequality above.

Graphing a Linear Inequality

1. Graph the line associated with the inequality. If the inequality involves $\leq$ or $\geq$, draw a solid line; this means the line is included in the solution. If the inequality involves $<$ or $>$, draw the line dashed, which means the line is not part of the solution.

2. Choose a test point [$(0, 0)$ if possible] not on the line.

3. If the test point satisfies the inequality, shade the region containing the test point; otherwise, shade the region on the other side of the line.

❶ Read the problem.

❷ Select the unknown.

❸ Think of a plan.
  (a) First, graph
     $2x = 3y - 6$.
  (b) Select a test point.
  (c) Check to see whether it
      satisfies the inequality.

❹ Use the test point to find the
    solution set.
    What do you have to do after
    you check to see whether the
    test point satisfies the
    inequality $2x < 3y - 6$?

❺ Verify the solution.

TRY EXAMPLE 1 NOW.

Graphing Linear Inequalities

Graph the linear inequality $2x < 3y - 6$.

Find the solution set for the linear inequality $2x < 3y - 6$; that is, find its
graph.

(a) Start by graphing $2x = 3y - 6$, as in Figure 7.58 below. When $x = 0$, then
    $y = 2$, so graph $(0, 2)$. When $y = 0$, then $x = -3$, so graph $(-3, 0)$. Join
    $(0, 2)$ and $(-3, 0)$ with a *dashed* line, since the graph of $2x = 3y - 6$ is not
    part of the solution set.

(b) To find the points satisfying $2x < 3y - 6$, find out if $(0, 0)$ satisfies
    $2x < 3y - 6$.

(c) When $x = 0$ and $y = 0$, $2x < 3y - 6$ becomes

$$2 \cdot 0 < 3 \cdot 0 - 6$$

or

$$0 < -6$$

which is false. Thus, $(0, 0)$ and all points below the line $2x = 3y - 6$ are *not* in
the solution set of $2x < 3y - 6$. The solution set must consist of all points *above*
the line $2x = 3y - 6$.

Remember that to emphasize the fact that points on the line $2x = 3y - 6$ are *not*
part of the solution set, the line itself is shown dashed in Figure 7.58.

The verification that the points in the shaded region satisfy the inequality
$2x < 3y - 6$ is left to you.

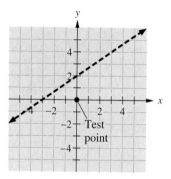

FIGURE 7.58

Cover the solution, write your own solution, and then check your work.

Example 1 illustrates the solution of an inequality that involves the less than or equal to sign. Notice carefully that the line corresponding to the equals sign is a part of the solution and is drawn solid in the graph.

EXAMPLE 1 ▶ Graphing Linear Inequalities

Graph the linear inequality $3x - 4y \leq 12$.

Solution

(a) We first draw the graph of the line $3x - 4y = 12$ (see Figure 7.59). We know that our answer requires one of the half-planes determined by this line.

(b) To find which of the half-planes is required, we can select any point not on the line and check whether it satisfies the original inequality. If the origin is not on the line, then $(0, 0)$ is a good choice, because it is so easy to check.

(c) Does $(0, 0)$ satisfy the inequality $3x - 4y \leq 12$? Yes, it does, because $3(0) - 4(0) = 0 < 12$. Thus, the half-plane containing $(0, 0)$ is the one we need. This half-plane is shaded in Figure 7.59. ■

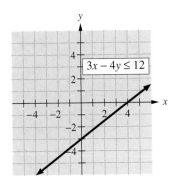

FIGURE 7.59

B. Solving Systems of Inequalities by Graphing

The solution set of a system of linear inequalities in two variables can often be found as in the next example, where the individual inequalities are first solved separately and then the final shading shows the intersection of these solution sets. This intersection is the solution set of the system.

Keep in mind that the lines (boundaries) are included when the inequalities are $\leq$ or $\geq$. In this case the line is drawn solid. They are *not* included for $<$ or $>$. In that case the line is drawn dashed.

EXAMPLE 2 ▶ Graphing Systems of Inequalities

Graph the solution set of the system of inequalities $-2 \leq x \leq 3$.

Solution

Our system consists of the inequalities $-2 \leq x$ and $x \leq 3$. We first graph the lines $x = -2$ and $x = 3$ as shown in Figure 7.60. The inequality $-2 \leq x$ is satisfied by all the points on or to the right of the line $x = -2$, as indicated by the arrows. Similarly, the inequality $x \leq 3$ is satisfied by all the points on or to the left of the line $x = 3$, as indicated by the arrows. The given system is satisfied by all the points common to these two regions, that is, all the points on either line or between the two lines. The solution set is shown shaded in Figure 7.60. ■

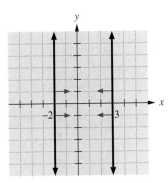

FIGURE 7.60

EXAMPLE 3 ▶ Graphing Systems with Two Inequalities

Graph the solution set of the system of inequalities

$$x + 2y \leq 5$$
$$x - y < 2$$

Solution

We first graph the lines $x + 2y = 5$ and $x - y = 2$, as shown in Figure 7.61. The inequality $x + 2y \leq 5$ is satisfied by the points *on or below* the line $x + 2y = 5$,

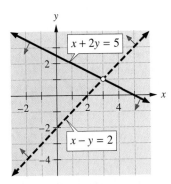

FIGURE 7.61

as indicated by the arrows attached to the line. The inequality $x - y < 2$ is satisfied by the points *above* the line $x - y = 2$, as indicated by the arrows attached to the line. This line is drawn dashed to indicate that the points on it do *not* satisfy the inequality $x - y < 2$. The solution set of the system is shown in Figure 7.61 by the shaded region and the portion of the solid line forming one boundary of this region. The point of intersection of the two lines is *not* in the solution set. ■

We can also graph the solution set of a system consisting of more than two inequalities, as illustrated by the next example.

EXAMPLE 4 ▶ Graphing Systems with Four Inequalities

Graph the solution set of the system of inequalities

$$x + 2y \le 6$$
$$3x + 2y < 10$$
$$x \ge 0$$
$$y \ge 0$$

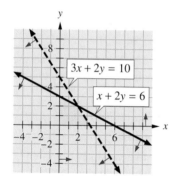

FIGURE 7.62

Solution

We first graph the lines $x + 2y = 6$, $3x + 2y = 10$, $x = 0$, and $y = 0$, as in Figure 7.62. The inequality $x + 2y \le 6$ is satisfied by the set of points on or below the line $x + 2y = 6$, as indicated by the arrows attached to the line. The inequality $3x + 2y < 10$ is satisfied by the set of points below the line $3x + 2y = 10$. Notice that this line is drawn dashed to show that it does not satisfy the given inequality. The set of inequalities $x \ge 0$ and $y \ge 0$ is satisfied by points in the first quadrant or points on the portions of the axes bounding the first quadrant. The solution set of the system can easily be identified; it is the shaded region plus the solid portions of the boundary lines. ■

EXAMPLE 5 ▶ Graphing Systems Involving "Or"

Graph the solution set of the system of inequalities $x + 1 \ge 0$ or $y - 2 \le 0$.

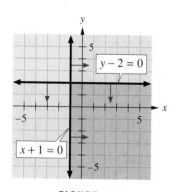

FIGURE 7.63

Solution

We first graph the lines $x + 1 = 0$ and $y - 2 = 0$, as shown in Figure 7.63. The inequality $x + 1 \ge 0$ is satisfied by the points on or to the right of the line $x + 1 = 0$, as indicated by the arrows attached to this line. The inequality $y - 2 \le 0$ is satisfied by the points on or below the line $y - 2 = 0$, as indicated by the arrows attached to this line. Since this is an "or" set of inequalities, the solution consists of all the points that are on or to the right of the line $x + 1 = 0$ and all the points that are on or below the line $y - 2 = 0$. The shaded region and the two solid lines in Figure 7.63 show this set of points. ■

C. Applications

In Section 7.1 the target zone used to gauge your effort when performing aerobic exercises was discussed. This target zone is determined by your pulse rate p and your age a and is found in the next example.

EXAMPLE 6 ▶ Finding the Target Zone with Inequalities

The target zone for aerobic exercise is defined by the following inequalities, in which a is the age in years and p is the pulse rate:

$$10 \leq a \leq 70$$
$$p \geq -\tfrac{2}{3}a + 150 \qquad \text{Lower limit}$$
$$p \leq -a + 190 \qquad \text{Upper limit}$$

Graph these inequalities and label the resulting target zone.

Solution

Since a is between 10 and 70, inclusive, it is convenient to label the horizontal axis starting at 10, using 10-unit intervals, and ending at 70. The vertical axis is used for the pulse rate p, and we start at 70 (the normal pulse rate) and go up to 200, as shown in Figure 7.64. We then graph the line $p = -\tfrac{2}{3}a + 150$ after finding the two points (30, 130) and (60, 110). Notice that because of the $-\tfrac{2}{3}a$, values of a that are divisible by 3 are most convenient to use. The inequality $p \geq -\tfrac{2}{3}a + 150$ is satisfied for all points that are *on or above* this line. Next, we graph the line $p = -a + 190$. Two points easy to find on this line are (10, 180) and (70, 120). The inequality $p \leq -a + 190$ is satisfied by all points that are *on or below* this line. The target zone is shown in Figure 7.64. ∎

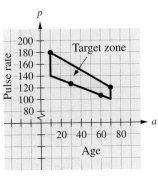

FIGURE 7.64

EXERCISES 7.7

Ⓐ Graphing Linear Inequalities

In problems 1–10, graph the given inequality.

1. $x + 2y \geq 2$

2. $x - 2y > 0$

3. $x \leq 4$

4. $y \leq 3$

5. $3x - y < 6$

6. $3x + 4y \geq 12$

7. $2x + y \leq 4$

8. $2x - 3y < 0$

9. $4x + y > 8$

10. $x - 4y \leq 4$

Ⓑ Solving Systems of Inequalities by Graphing

In problems 11–26, graph the solution set of the given system of inequalities.

11. $-4 \leq x < 3$

12. $2 \leq y \leq 5$

13. $x \leq -1$ or $x > 1$

14. $y < 0$ or $y > 3$

15. $x - y \geq 2; x + y \leq 6$

16. $x + 2y \leq 3; x \leq y$

17. $2x - 3y \leq 6; 4x - 3y \geq 12$

18. $2x - 5y \leq 10; 3x + 2y \leq 6$

19. $2x - 3y \leq 5; x \geq y; y \geq 0$

20. $x \leq 2y; 2x \geq y; x + y < 4$

21. $x + 3y \leq 6; x \geq 0; y \geq 0$

22. $2x - y \leq 2; y \geq 1; x \geq \tfrac{1}{2}$

23. $x \geq 1; y \geq 1; x - y \leq 1; 3y - x < 3$

24. $x - y \geq -2; x + y \leq 6; x \geq 1; y \geq 1$

25. $x + y \geq 1; x \leq 2; y \geq 0; y \leq 1$

26. $1 < x + y < 8; x < 5; y < 5$

In problems 27–30, select the set of conditions (**a**) or (**b**) that corresponds to the shaded region in each figure.

27.

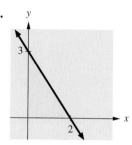

a. $3x + 2y < 6$, $x \le 2$ and $y \le 3$
b. $3x + 2y \le 6$, $x \ge 0$ and $y \ge 0$

28.

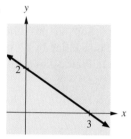

a. $2x + 3y \ge 6$, $x \ge 3$ and $y \ge 2$
b. $2x + 3y \ge 6$, $x \ge 0$ and $y \ge 0$

29.

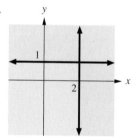

a. $x \ge 2$ or $y \ge 1$ **b.** $x \ge 2$ and $y \ge 1$

30.

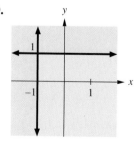

a. $x \le -1$ and $y \ge 1$ **b.** $x \le -1$ or $y \ge 1$

31. Which shaded region corresponds to the set of inequalities $x + y \ge 1$ and $x - y \le 1$?

a.
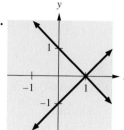

b.
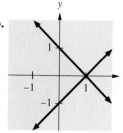

c.
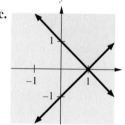

d.
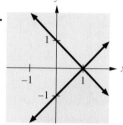

32. Write the set of inequalities that corresponds to the shaded region in the figure. Explain.

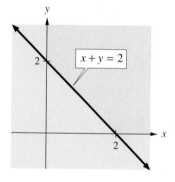

$x + y = 2$

C **Applications**

In problems 33–38, translate the statement into an inequality.

33. The height h (in feet) of any mountain does not exceed that of Mt. Everest, 29,035 ft.

34. The number e of possible eclipses in a year is at most 7.

35. The number e of possible eclipses in a year is at least 2.

36. The altitude h (in feet) attained by the first liquid-fueled rocket was no more than 41 ft.

37. There are no less than 4×10^{25} nematode sea worms in the world. (Let n be the number of nematodes.)

38. There are at least 713 million people p that speak Mandarin Chinese.

39. When the variable cost per unit is \$12 and the fixed cost is \$160,000, the total cost for a certain product is $C = 12n + 160,000$ (n is the number of units sold). If the unit price is \$20 and the revenue R is $20n$, write an inequality for the situation and find the minimum number of units that must be sold to make a profit. (*Hint:* You need $R > C$ to make a profit.)

40. The cost of first-class mail is 39¢ for the first ounce and 24¢ for each additional ounce. A delivery company will charge \$4.00 for delivering a package weighing up to 2 lb (32 oz). Write an inequality indicating when the U.S. Postal service price $P = 0.39 + 0.24(x - 1)$ (where x is the weight of the package in ounces) would be cheaper than the delivery company's price, and then solve the inequality.

41. The cost to park at a garage is $C = 1 + 0.75(h - 1)$, where h is the number of hours you park and C is the cost in dollars. Write an inequality to indicate when the cost C is less than \$10, and then solve the inequality.

42. Cigarette smoking produces air pollution. The annual number of cigarettes per person consumed in the United States is given by $N = 4200 + 70x$, where x is the number of years after 1960. How many years after 1960 will consumption be less than 1000 cigarettes per person annually? (Answer to the nearest whole year.) What year will that take place?

43. The desirable weight range corresponding to a given height for a man is shown in Table 7.7. We found the equation for the lower weights in terms of the height to be $w = 4h - 140$. The equation for the upper weights is $w = 5h - 176$. Thus, the desirable weights for men from 62 to 76 in. in height satisfy the system of inequalities

$$62 \leq \ h \leq 76$$
$$w \geq 4h - 140$$
$$w \leq 5h - 176$$

Graph this system of inequalities and show the region corresponding to the range of desirable weights. Be sure to take the horizontal axis as the h axis.

44. The desirable weight range corresponding to a given height for a woman is also shown in Table 7.7. The equation for the lower weights in terms of the height h is $w = 4h - 150$ and for the upper weights is $w = 5h - 187$. Thus, the desirable weights for women from 62 to 72 in. in height satisfy the system of inequalities

$$62 \leq h \leq 72$$
$$w \geq 4h - 150$$
$$w \leq 5h - 187$$

Graph this system of inequalities and show the region corresponding to the range of desirable weights. Be sure to take the horizontal axis as the h axis.

 In Other Words

45. Describe the steps you would take to graph the inequality $ax + by > c$.

46. Describe the graph of the inequality $y < mx + b$.

47. Describe the graph of the inequality $x \geq k$.

48. Describe the graph of the inequality $y < k$.

 Using Your Knowledge

Suppose you want to find two integers x and y with $y > 1$ such that the sum of the two integers is less than 10 and the difference $3x - 2y$ is greater than 6. Here is an easy way to find the possible pairs (x, y).

49. Graph the corresponding system of inequalities

$$y > 1$$
$$x + y < 10$$
$$3x - 2y > 6$$

You should end up with a shaded region in the plane. Mark each integer point (point with integer coordinates) within this region, and then just read the coordinates. These are the possible pairs of integers.

50. Find all the pairs of integers (x, y) such that x is greater than 1, the sum of the two integers is less than 12, and the difference $2y - x$ is at least 8.

You can use what you know about linear inequalities to save money when you rent a car. Here are the rental prices for an intermediate car obtained from a telephone survey conducted in Tampa, Florida, in a recent year.

Rental A $41 a day, unlimited mileage
Rental B $36 a day, $0.15 per mile

51. If you compare the rental A price ($41) with the rental B price, you see that rental B appears to be cheaper.
   a. How far can you drive a rental B car in 1 day if you wish to spend exactly $41? (Answer to the nearest mile.)
   b. If you are planning on driving 100 mi in 1 day, which car would you rent, rental B or rental A?

52. How far can you drive a rental B car in 1 day if you wish to spend exactly $42? (Answer to the nearest mile.)

53. On the basis of your answers to problems 51 and 52, which is the cheaper rental price? (*Hint:* It has to do with the miles you drive.)

54. Will the rental prices ever be identical?

Discovery

55. Here is a problem just for fun. A square is 2 in. on a side. Five points are marked inside or on the perimeter of the square. Show that the distance between at least two of the five points is less than or equal to $\sqrt{2}$. (*Hint:* Cut the square up into four equal squares by connecting the midpoints of the opposite sides.)

56. The figure in the next column is a flowchart for graphing the inequality $ax + by > c$, $abc \neq 0$. Explain why this works. (*Hint:* Try it for some of the problems.)

57. Can you discover what changes have to be made in the flowchart if the inequality is $ax + by < c, c \neq 0$?

58. Can you discover what changes have to be made in the flowchart if $c = 0$?

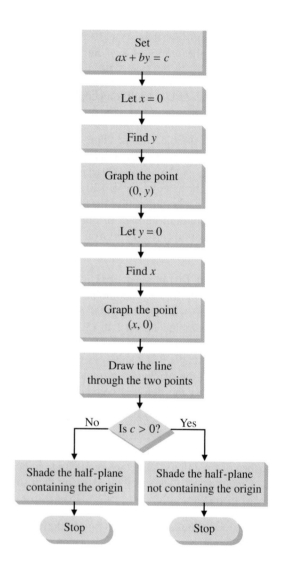

 Collaborative Learning

As you can see in the chart (page 485), many fatal accidents involving drivers and pedestrians were alcohol related. The two graphs show the effect of alcohol on males and females over a 2-hr period. Form three teams. The assignments follow.

Team 1 For the "Effects of Alcohol" graph on the left (males), use the points (6, 0.1) and (2, 0.01) to find the slope of the line. Then use (6, 0.1) to find the equation of the line relating B the blood alcohol level and D the number of drinks. Finally, write the inequality corresponding to the shaded region on the graph.

Team 2 For the "Effects of Alcohol" graph on the right (females), let the slope be rise/run. What is this value? Use the point (4, 0.1) and the slope to find the equation of the line relating B the blood alcohol level and D the number of drinks. Finally, write the inequality corresponding to the shaded region on the graph.

Team 3 The two graphs shown apply to 170-lb males and 137-lb females. A more general formula relating the number of drinks D and the weight W that will produce a blood alcohol level of 0.10 in 2 hr is $0.00125W - 0.01875D = 0.1$. Using W as the vertical axis and D as the horizontal axis, graph the equation and then write an inequality indicating that the blood alcohol level exceeds 0.1.

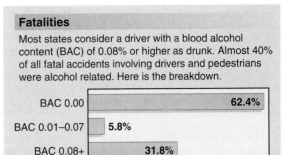

Fatalities

Most states consider a driver with a blood alcohol content (BAC) of 0.08% or higher as drunk. Almost 40% of all fatal accidents involving drivers and pedestrians were alcohol related. Here is the breakdown.

BAC 0.00 — 62.4%
BAC 0.01–0.07 — 5.8%
BAC 0.08+ — 31.8%

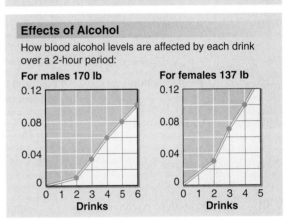

Effects of Alcohol

How blood alcohol levels are affected by each drink over a 2-hour period:

For males 170 lb

For females 137 lb

Sources: (*top*) http://www.alcoholstats.com/documentsVirtual/BACpiechart.pdf; (*bottom*) http://www.madd.org/stats/0,1056,1183,00.html.

Blood alcohol levels.

Chapter 7 Summary

Section	*Item*	*Meaning*	*Example*
7.1A	Relation	A set of ordered pairs	$S = \{(3, 2), (5, 2), (7, 4)\}$
7.1A	Domain	The set of all possible x values of a relation	The domain of S is $\{3, 5, 7\}$.
7.1A	Range	The set of all possible y values of a relation	The range of S is $\{2, 4\}$.
7.1A	Function	A relation such that to each domain value there corresponds exactly one range value	$\{(x, y) \mid y = 2x\}$
7.1A	$f(x)$	Function notation	$f(x) = 3x + 2$

Section	Item	Meaning	Example
7.1C	Graph of a relation	The set of points corresponding to the ordered pairs of a relation	
7.1C	Vertical line test	The graph of a function cannot be cut in more than one point by any vertical line.	Function Function Not a function
7.2B	x intercept	x coordinate of the point where the line crosses the x axis ($y = 0$)	The x intercept of $y = 2x - 4$ is 2.
7.2B	y intercept	y coordinate of the point where the line crosses the y axis ($x = 0$)	The y intercept of $y = 2x - 4$ is -4.
7.2B	Linear equation	An equation that can be written in the form $ax + by = c$	$3x + 5y = -2$; $3y = 2x - 1$
7.2C	Distance formula	$d = \sqrt{(x_2 - x_1)^2 + (y_2 - y_1)^2}$	The distance between $(3, 5)$ and $(5, 12)$ is $\sqrt{53}$.
7.3A	Slope of a line	$m = \dfrac{y_2 - y_1}{x_2 - x_1}$	The slope of the line through $(3, 5)$ and $(5, 12)$ is $\frac{7}{2}$.
7.3B	Point-slope equation	$y - y_1 = m(x - x_1)$	$y - 5 = \frac{7}{2}(x - 3)$ is the point-slope equation of the line described above.
7.3B	Slope-intercept equation	$y = mx + b$ (m is the slope, b is the y intercept)	$y = \frac{7}{2}x - \frac{11}{2}$ is the slope-intercept equation of the line described above.
7.3B	General equation of a line	$Ax + By = C$	
7.3C	Parallel lines	Two lines with the same slope and different y intercepts	$y = 2x + 5$ and $y = 2x - 3$ are equations of parallel lines.

 Research Questions

Sources of information for most of these questions can be found in the Bibliography at the end of the book.

1. Some historians claim that the official birthday of analytic geometry is November 10, 1619. Write a report on why this is so and the events that led Descartes to the discovery of analytic geometry.

2. Find out what led Descartes to make his famous pronouncement *"Je pense, donc je suis"* (I think, therefore I am) and write a report about the contents of one of his works, *La Geometrie.*

3. She was nineteen, a capable ruler, a good classicist, a remarkable athlete, and an expert hunter and horsewoman. Find out who this queen was and what connections she had with Descartes.

4. On her arrival at the University of Stockholm, one newspaper reporter wrote "Today we do not herald the arrival of some vulgar, insignificant prince of noble blood. No, the Princess of Science has honored our city with her arrival." Write a report identifying this woman and discussing the circumstances leading to her arrival in Sweden.

5. When she was 6 years old, the "princess's" room was decorated with a unique type of wallpaper. Write a paragraph about this wallpaper and its influence on her career.

6. In 1888, the "princess" won the Prix Bordin offered by the French Academy of Sciences. Write a report about the contents of her prize-winning essay and the motto that accompanied it.

Chapter 7 Practice Test

1. Find the domain and range of the relation

   $$R = \{(5, 3), (3, -1), (2, 2), (0, 4)\}$$

2. Find the domain and range of the relation

   $$R = \{(x, y) \mid y = -3x\}$$

3. Which of the following relations are functions?
   a. $\{(x, y) \mid y^2 = x\}$ **b.** $\{(3, 1), (4, 1), (6, 1)\}$ **c.** $\{(x, y) \mid y = x^2\}$

4. A function is defined by $f(x) = x^2 - x$. Find the following:
   a. $f(0)$ **b.** $f(1)$ **c.** $f(-2)$

5. For a car renting for $15 per day plus $.10 per mile, the cost for 1 day is

   $$C(m) = 15 + 0.10m \text{ (dollars)}$$

   where m is the number of miles driven. If a person paid $35.30 for 1 day's rental, how far did the person drive?

6. Graph the relation

   $$R = \{(x, y) \mid y = 3x, x \text{ is an integer between } -1 \text{ and } 3, \text{ inclusive}\}$$

7. Graph the relation

   $$Q = \{(x, y) \mid x + y < 3, x \text{ and } y \text{ are nonnegative integers}\}$$

8. Graph the function defined by $g(x) = 2x^2 - 1$, where x is an integer and $-2 \le x \le 2$.

9. Graph the function defined by $f(x) = 2x - 6$.

10. Graph the equation $3x - 2y = 5$.

11. Find the distance between the two given points.
  a. $(4, 7)$ and $(7, 3)$ **b.** $(-3, 8)$ and $(-3, -2)$

12. Find the slope of the line that passes through the two points $(-1, -3)$ and $(9, -2)$.

13. a. Find the slope-intercept form of the equation of the line that passes through the point $(3, -1)$ and has slope -2.
  b. Find the slope-intercept form of the equation $2y = 4 - 8x$. What is the slope and what is the y intercept of the line?

14. Find the general equation of the line passing through the points $(-1, -3)$ and $(9, -2)$.

15. Determine whether the two given lines are parallel. If they are not parallel, find the coordinates of the point of intersection.
  a. $2x + y = 1$; $12x + 3y = 4$ **b.** $y = 2x - 5$; $4x - 2y = 7$

16. Find the general equation of the line that passes through the point $(1, -2)$ and is parallel to the line $2x - 3y = -5$.

17. Graph $y = -x^2 - 3$.

18. Graph $y = (x + 2)^2 - 1$ and give the coordinates of the vertex.

19. Graph $y = x^2 + 6x + 8$ and give the coordinates of the vertex.

20. Graph $f(x) = 3^x$ and $g(x) = \left(\frac{1}{3}\right)^x$ on the same coordinate system.

21. Graph $f(x) = e^{x/2}$ and $g(x) = \ln(x/2)$ on the same coordinate system.

22. A principal of P dollars accumulates to an amount $A = Pe^{rt}$, where r is the interest rate and t is the time in years. If the interest rate is 8%, how long would it take for the money to double?

23. Find the point of intersection of the lines

$$3x + 2y = 9 \quad \text{and} \quad 2x - 3y = 19$$

24. Solve the following system if possible. If not possible, explain why.

$$y = 2x - 3$$
$$6x - 3y = 9$$

25. Graph the solution set of the inequality $4x - 3y \le 12$.

26. Graph the solution set of the system of inequalities

$$x + 3y \le 6 \quad \text{and} \quad x - y \ge 2$$

27. Graph the solution set of the system of inequalities

$$x + 2y \le 3$$
$$x \le y$$
$$x \ge 0$$

Answers to Practice Test

ANSWER	IF YOU MISSED	REVIEW		
	Question	Section	Example(s)	Page(s)
1. Domain, $\{0, 2, 3, 5\}$; range, $\{-1, 2, 3, 4\}$	1	7.1	1	400
2. Domain, the set of all real numbers; range, the set of all real numbers	2	7.1	1	400
3. **b.** and **c.**	3	7.1	2	401
4. **a.** 0 **b.** 0 **c.** 6	4	7.1	3–5	402–403
5. 203 mi	5	7.1	6	404
6.	6	7.1	7	405
7.	7	7.1	7, 8	405–406
8.	8	7.1	9	406
9.	9	7.2	2	415

ANSWER	IF YOU MISSED	REVIEW		
	Question	Section	Example(s)	Page(s)
10.	10	7.2	3	415
11. a. 5 units **b.** 10 units	11	7.2	4, 5	417
12. $\frac{1}{10}$	12	7.3	1, 2	424–425
13. a. $y = -2x + 5$ **b.** $y = -4x + 2; m = -4, b = 2$	13	7.3	3–5	426–427
14. $x - 10y = 29$	14	7.3	6	428
15. a. Not parallel; intersect at $\left(\frac{1}{6}, \frac{2}{3}\right)$ **b.** Parallel	15	7.3	8	430
16. $2x - 3y = 8$	16	7.3	9	431
17.	17	7.4	1–3	438, 440, 441
18. Vertex $= (-2, -1)$	18	7.4	4	442

ANSWER	IF YOU MISSED	REVIEW		
	Question	Section	Example(s)	Page(s)
19. Vertex $= (-3, -1)$	19	7.4	5, 6	444–446
20. $g(x) = \left(\frac{1}{3}\right)^x$　　$f(x) = 3^x$	20	7.5	1	453
21.	21	7.5	2	454
22. $\dfrac{\ln 2}{0.08} \approx 8.66$ years	22	7.5	7–10	459–461
23. $(5, -3)$	23	7.6	1, 2	467
24. The two equations represent the same line. The solution set is $\{(a, 2a - 3) \mid a \text{ is a real number}\}$.	24	7.6	3–6	468–470
25.	25	7.7	1	479

ANSWER	IF YOU MISSED	REVIEW		
	Question	Section	Example(s)	Page(s)
26.	26	7.7	2, 3	479–480
27.	27	7.7	4	480

Satellites play an important role in advanced communication technology. Angles are critical in setting up equipment to receive signals from satellites. In Section 8.1, you will explore the use of angles in applications such as architecture and space exploration.

Geometry

This chapter is devoted to the study of **geometry.** The word *geometry* literally means earth (*geo-*) measurements (*-metry*). You might have studied geometry in the past as a collection of theorems and proofs. We will not do that here. Instead, we will concentrate on measurements involving *linear, square,* and *cubic* units.

We start with some undefined terms (*points, lines, planes*), see how they relate to each other, and use linear measures to measure distances and perimeters of many-sided figures called **polygons.** We then discuss how to measure and classify *angles* by their measures, and we classify *triangles* by the number of equal sides they contain. We will relate these ideas to many other topics, including motorcycle riding, satellite trajectories, angles formed by veins in leaves, and traffic signs.

We study *similar* triangles, *similar* figures, and *circumferences* of circles and their many applications: hat sizes, ring sizes, and so on. We then study *areas* of polygons (squares, rectangles, parallelograms, and triangles) and circles and the Pythagorean theorem. Next, we look at *surface areas* and *volumes* of three-dimensional objects such as cubes, cylinders, cones, pyramids, and spheres. We then look at a classic problem, called the *Bridges of Königsberg,* solved by Leonhard Euler in 1736. Finally, we discuss non-Euclidean geometry and topology.

Online Study Center

For links to Internet sites related to Chapter 8, please access **college.hmco.com/PIC/bello9e** and click on the Online Study Center icon.

8.1   Points, Lines, Planes, and Angles

HUMAN SIDE OF MATH

One of the most famous mathematicians of all time is Euclid, who taught in about 300 B.C. at the university in Alexandria, the main Egyptian seaport.

(Flourished 300 B.C.)

Geometry evolved from the more or less rudimentary ideas of the ancient Egyptians (about 1500 B.C.), who were concerned with

(continued)

Construction of the Tower of Pisa began in 1174, and settling was already noticeable after only three of its eight stories were built. Still leaning, the tower was completed in the fourteenth century.

Getting the Right Angle

Have you heard the expression "I am looking at it from a new angle"? In algebra, you measured the inclination of a line by using the *slope* of a line. In geometry, you measure the inclination of a line by measuring the *angle* the line makes with the horizontal using an instrument called a **protractor.** (See Figures 8.23 and 8.24 on page 499.) Angles are everywhere. You can see angles in things as large as the cables on a suspension bridge and as small as the veins in a plant leaf. Figure 8.1 shows the angles in the flight path of the *Ranger 9* lunar probe.

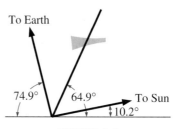

FIGURE 8.1
Ranger 9 terminal trajectory orientation and glide path (NASA).

The Leaning Tower of Pisa has an angle of inclination of about 5° (read "5 degrees") from the vertical in the photo below. Thus the tower makes an 85° *acute* angle, an angle that is less than 90°, with the ground. How do we know that? Because if the tower were completely vertical, it would make a 90° angle, that is, a *right* angle with the ground. Try to find the angle that the left side of the tower makes with the ground. Your answer will be an *obtuse* angle, an angle that measures more than 90°.

But we are getting ahead of the story! We shall see that angles are formed by *rays, rays* are part of *lines,* and *lines* are sets of *points,* and we have not discussed any of these terms yet! We shall do so in more detail in this section. ▶

The basic elements of Euclidean geometry are **points, lines,** and **planes.** These three words cannot be precisely defined, because this would require the use of other words that are also undefined. For example, we can say that a line is a set of points. But what is a point? A point is that which has no dimension. But again, what is dimension? However, since all other geometric terms are defined on the basis of these three words, we must develop an intuitive idea of their meaning.

A. Points and Lines

A **point** may be regarded as a location in space. A point has no breadth, no width, and no length. We can picture a point as a dot, such as *A* in Figure 8.2. (The sharper the pencil, the better the picture.)

We can think of a line as the path of a point moving in a fixed direction in space. A line extends without end in both its directions. If we have two points *A* and *B*, there is only *one* line that can be drawn through the two points: the practical problems involving measurement of areas and volumes. Greek geometers, however, tried to apply the principles of Greek logic to the study of geometry and to prove theorems by a sequence of logical steps that proceeded from certain basic assumptions to a conclusion.

Euclid's greatest contribution was his collection and systematization of most of the Greek mathematics of his time. His reputation rests mainly on his work titled *The Elements,* which contains geometry, number theory, and some algebra. Most U.S. textbooks on plane and solid geometry contain the material in the geometry portions of *The Elements.* Over a thousand editions of *The Elements* have been published since the first printed edition appeared in 1482, and for more than 2000 years this work has dominated the teaching of geometry.

Looking Ahead
Many of the geometric concepts that were the focus of Euclid's studies and writings also constitute the topics in this chapter.

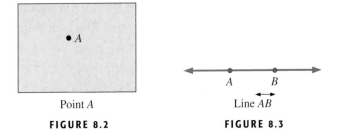

Point *A*

FIGURE 8.2

Line $\overleftrightarrow{AB}$

FIGURE 8.3

line *AB*, denoted by $\overleftrightarrow{AB}$ (see Figure 8.3) and extending without end in both directions.

Selecting any point *A* on a line divides the line into *three* sets: the point *A* itself and two **half-lines,** one on each side of *A,* as shown in Figure 8.4. The half-lines *do not* include the point *A,* as indicated by the open circle at one end of the arrow. When the endpoint is included, the result is a **ray.** In Figure 8.5, **ray** $\overrightarrow{AB}$ has initial point *A* and extends in the direction of *B,* whereas **ray** $\overrightarrow{BA}$ has initial point *B* and extends in the direction of *A.* The **line segment** *AB,* shown in Figure 8.5, consists of the points *A* and *B* and all the points between *A* and *B.*

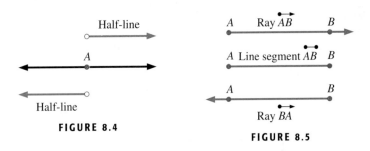

FIGURE 8.4

FIGURE 8.5

Figure 8.6 shows the figures and notations we have described.

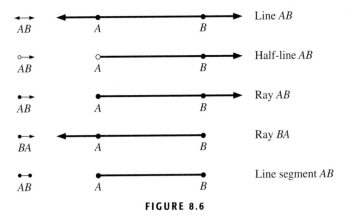

FIGURE 8.6

We can use the sets of points we have described and the set of operations of intersection ($\cap$) and union ($\cup$) to describe new sets, as shown next.

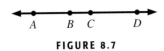

FIGURE 8.7

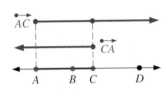

FIGURE 8.8

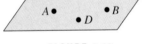

FIGURE 8.9

EXAMPLE 1 ▶ What Do You Get with the Descriptions?

Refer to Figure 8.7 and state what each of the following describes:

(a) $\overrightarrow{AC} \cap \overrightarrow{CA}$ (b) $\overline{AB} \cap \overrightarrow{BD}$

(c) $\overline{AC} \cup \overrightarrow{BD}$ (d) $\overline{AB} \cap \overline{CD}$

(e) $\overleftrightarrow{BC} \cap \overleftrightarrow{BA}$ (f) $\overline{AB} \cup \overline{BD}$

Solution

(a) Figure 8.8 shows the two rays $\overrightarrow{AC}$ and $\overrightarrow{CA}$. The set of points they have in common is the segment $\overline{AC}$. Thus, $\overrightarrow{AC} \cap \overrightarrow{CA} = \overline{AC}$.

(b) We can see from Figure 8.9 that the segment $\overline{AB}$ and the ray $\overrightarrow{BD}$ have only the point B in common. Therefore, $\overline{AB} \cap \overrightarrow{BD} = \{B\}$.

(c) Figure 8.10 shows that the union of the segment $\overline{AC}$ and the ray $\overrightarrow{BD}$ is the ray $\overrightarrow{AD}$. Thus, $\overline{AC} \cup \overrightarrow{BD} = \overrightarrow{AD}$.

(d) Figure 8.11 shows that the segments $\overline{AB}$ and $\overline{CD}$ have no points in common. Hence, $\overline{AB} \cap \overline{CD} = \varnothing$.

(e) In Figure 8.8, $\overleftrightarrow{BC}$, $\overleftrightarrow{BA}$, and $\overleftrightarrow{AD}$ are all symbols for the same line, so we have $\overleftrightarrow{BC} \cap \overleftrightarrow{BA} = \overleftrightarrow{AD}$.

(f) As we can see in Figure 8.7, the union of the segment $\overline{AB}$ and the segment $\overline{BD}$ is the segment $\overline{AD}$; thus, $\overline{AB} \cup \overline{BD} = \overline{AD}$.

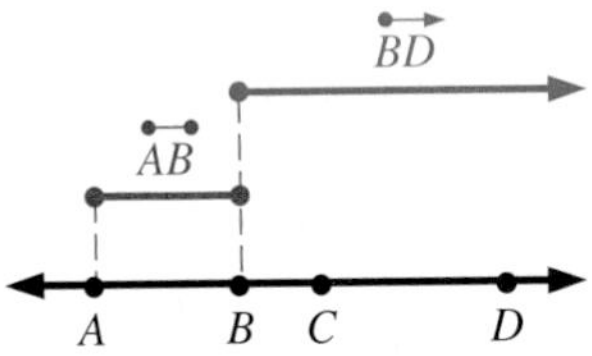

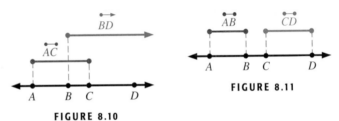

FIGURE 8.10

FIGURE 8.11

B. Planes

As with the terms *point* and *line*, we give no formal definition of a **plane.** To help visualize a plane, we can think of the surface of a very large flat floor or of a straight wall extending indefinitely in all directions. Here are some basic properties of planes.

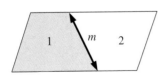

FIGURE 8.12
The plane *ABC.*

1. If we have three noncollinear (not on the same line) points *A*, *B*, and *C*, there is only *one* plane containing the three points, the plane *ABC* (see Figure 8.12).

2. Any line *m* in a plane separates the plane into *three* parts: the line *m* itself and two **half-planes.** The points on the line *m* do not belong to either half-plane, although the line is often called the *edge* of both half-planes. As indicated by the shading in Figure 8.13, we regard the points of the plane that are on one side of line *m* as forming one of the half-planes and the points of the plane on the other side of line *m* as forming the other half-plane.

FIGURE 8.13
Two half-planes, 1 and 2.

3. The *intersection* of two planes is a **line.** For example, the intersection of planes 1 and 2 (see Figure 8.14) is the line ℓ. Two planes that have no common point are **parallel** (see Figure 8.15).

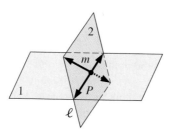

FIGURE 8.14
Intersecting planes.

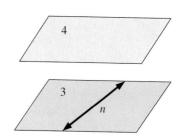

FIGURE 8.15
Parallel planes.

If a line is not in a given plane, then there are two possibilities.

1. The line **intersects** the plane in exactly one point. For instance, line m in plane 2 of Figure 8.14 intersects plane 1 in the point P.

2. The line is **parallel** to the plane. In Figure 8.15, any line such as n in plane 3 will be parallel to plane 4.

If two distinct lines in space are given, then there may be a plane that contains both lines. If the lines are parallel or if they intersect, then there is exactly one plane that contains both lines. If the lines are neither parallel nor intersecting so that no plane can contain both lines, then they are called **skew** lines. A simple example of skew lines is the line of intersection of the ceiling and the front wall of an ordinary rectangular classroom and the line of intersection of the floor and one of the side walls. Figure 8.16 shows a rectangular box. The edges determine various straight lines. For instance, lines m and n intersect at a vertex (corner) of the box; m and p are skew lines, and m and q are parallel lines, that is, lines in the same plane (coplanar) that *never* intersect.

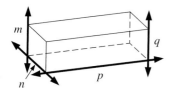

FIGURE 8.16
A rectangular box. Lines m and p are skew lines. Lines m and q are parallel lines.

C. Angles

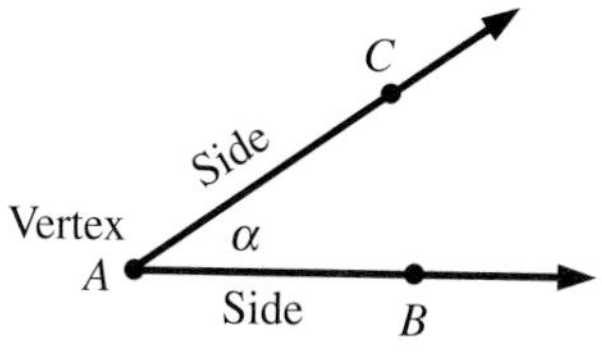

A very important concept in mathematics is that of a *plane angle*. In elementary geometry, we think of a **plane angle** as the figure formed by two rays with a common endpoint, as in Figure 8.17. The common endpoint (A in Figure 8.17) is called the **vertex** of the angle, and the two rays ($\overrightarrow{AB}$ and $\overrightarrow{AC}$ in Figure 8.17) are called the **sides** of the angle. We often use the symbol $\angle$ (read "angle") in naming angles. The angle in Figure 8.17 can be named in the following three ways:

1. By using a letter or a number inside the angle. Thus, we would name the angle in Figure 8.17 $\angle \alpha$ (read "angle alpha").

2. By using the vertex letter only, such as $\angle A$ in Figure 8.17.

3. By using three letters, one from each ray, with the vertex letter in the middle. The angle in Figure 8.17 would be named $\angle BAC$ or $\angle CAB$.

FIGURE 8.17

EXAMPLE 2 ▶ Naming Angles, Vertices, and Sides

Consider the angle in Figure 8.18.

(a) Name the angle in three different ways.

(b) Name the vertex of the angle.

(c) Name the sides of the angle.

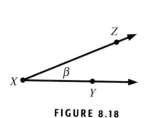

FIGURE 8.18

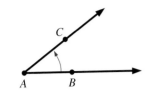

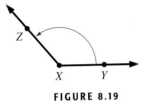

FIGURE 8.19

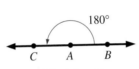

360°

FIGURE 8.20

A complete revolution.

Solution

(a) The angle can be named $\angle\beta$ (Greek letter beta), $\angle X$, or $\angle YXZ$ (or $\angle ZXY$).

(b) The vertex is the point X.   (c) The sides are the rays $\overrightarrow{XZ}$ and $\overrightarrow{XY}$. ■

For most practical purposes, we need to have a way of measuring angles. We first consider the amount of **rotation** needed to turn one side of an angle so that it *coincides with* (falls exactly on top of) the other side. Figure 8.19 shows two angles, $\angle CAB$ and $\angle ZXY$, with curved arrows to indicate the rotation needed to turn the rays $\overrightarrow{AB}$ and $\overrightarrow{XY}$ so that they coincide with the rays $\overrightarrow{AC}$ and $\overrightarrow{XZ}$, respectively. Clearly, the amount needed for $\angle ZXY$ is greater than that for $\angle CAB$. To find how much greater, we have to measure the amounts of rotation.

The most common unit of measure for an angle is the *degree*. We can trace the degree system back to the ancient Babylonians, who were responsible for the base 60 system of numeration. The Babylonians considered a *complete revolution* of a ray as indicated in Figure 8.20 and divided that into 360 equal parts. Each part is **1 degree,** denoted by **1°**. Thus, a complete revolution is equal to 360°. One-half of a complete revolution is 180° and gives us an angle that is called a **straight angle** (see Figure 8.21). One-quarter of a complete revolution is 90° and gives a **right angle** (see Figure 8.22). Notice the small square at Y to denote that it is a right angle.

FIGURE 8.21
The straight angle *CAB*.

FIGURE 8.22
The right angle *XYZ*.

Definition of Perpendicular Lines

Two lines that intersect at right angles are said to be *perpendicular* to each other. The lines are called **perpendicular lines.**

For example, two adjacent outside edges of a page of this book are perpendicular to each other. You can learn some more about the different relationships that can exist between two lines in the Research Questions.

EXAMPLE 3 ▶ Time on a Clock and Degrees

Through how many degrees does the hour hand of a clock move in going through the following time intervals?

(a) 1 to 2 o'clock   (b) 1 to 4 o'clock

(c) 12 to 5 o'clock   (d) 12 to 9 o'clock

Solution

(a) One complete revolution is 360°, and the face of the clock is divided into 12 equal parts. Thus, the hour hand moves through

$$\frac{360°}{12} = 30°$$

in going from 1 to 2 o'clock.

(b) From 1 to 4 o'clock is 3 hours. Since a 1-hour move corresponds to 30°, a 3-hours move corresponds to 3(30°) = 90°. (Thus, the hour hand moves through one right angle.)

(c) From 12 to 5 o'clock is 5 hours. Hence, the hour hand moves through 5(30°) = 150°.

(d) From 12 to 9 o'clock is 9 hours, so the hour hand moves through 9(30°) = 270°. ■

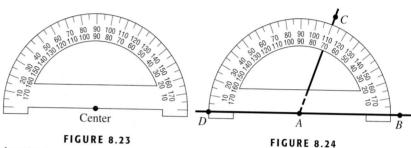

FIGURE 8.23

A protractor.

FIGURE 8.24

Measuring an angle.

In practice, the size of an angle is measured with a protractor (see Figure 8.23). The protractor is placed with its center at a vertex of the angle and the straight side of the protractor along one side of the angle, as in Figure 8.24. The measure of ∠*BAC* is then read as 70° (because it is obviously less than 90°), and the measure of ∠*DAC* is read as 110°. Surveying and navigational instruments, such as the sextant, use the idea of a protractor to measure angles very precisely.

We have already named two angles: a *straight angle* (180°) and a *right angle* (90°). Certain other angles are classified as follows:

> **Definition of Acute and Obtuse Angles**
>
> An **acute angle** is an angle of measure *greater* than 0° and *less* than 90°.
> An **obtuse angle** is an angle of measure *greater* than 90° and *less* than 180°

In Figure 8.24, ∠*BAC* is an acute angle, and ∠*DAC* is an obtuse angle.

Geometric figures frequently appear in highway signs. Do you know what the sign in the margin means? It is an advance warning for a railroad crossing. The angles *B* and *D* that are marked in Figure 8.25 are called *vertical angles.*

FIGURE 8.25

> **Definition of Vertical Angles**
>
> When two lines intersect, the opposite angles so formed are called **vertical angles.**

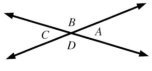

FIGURE 8.26

Two pairs of vertical angles are shown in Figure 8.26. Since the sides of angle *C* are just extensions of the sides of angle *A*, these two angles are of equal size. Similarly, angles *B* and *D* are of equal size. You complete a mathematical proof of this fact in problem 67, Exercises 8.1. Thus, in Figure 8.26, the measure of angle A, denoted by $m\angle A$, is the same as that of angle C. This fact is simply written as $m\angle A = m\angle C$. Similarly, $m\angle B = m\angle D$.

In Figure 8.26, angles A and B together form a straight angle, so the sum of their measures must be 180°. For this reason, A and B are called *supplementary angles*.

Definition of Supplementary Angles

Any two angles whose measures add up to 180° are called **supplementary angles.**

Other pairs of supplementary angles in Figure 8.26 are B and C, C and D, and A and D. Figure 8.26 illustrates the fact that *supplements of the same angle are equal.* For example, angles A and C are both supplements of $\angle B$, so $m\angle A = m\angle C$.

EXAMPLE 4 ▶ Finding the Measure of Angles

Refer to Figure 8.26.

(a) If the measure of $\angle A$ is 25°, what are the measures of the other three angles?

(b) If the two lines are to be drawn so that $\angle B$ is twice the size of $\angle A$, what should the measure of $\angle A$ be?

Solution

(a) Angles A and B are supplementary, so their measures add to 180°. Hence, the measure of $\angle B$ is 180° minus the measure of $\angle A$; that is, $180° - 25° = 155°$. Since $m\angle D = m\angle B$, the measure of $\angle D$ is also 155°. Also, $m\angle C = m\angle A$, so the measure of $\angle C$ is 25°.

(b) We let the measure of angle A be $x°$. Then the measure of angle B is $2x°$. Because angles A and B are supplementary, we must have

$$x + 2x = 180$$
$$3x = 180$$
$$x = 60$$

Thus, if we make $\angle A$ a 60° angle, $\angle B$ will be a 120° angle, twice the size of $\angle A$. ∎

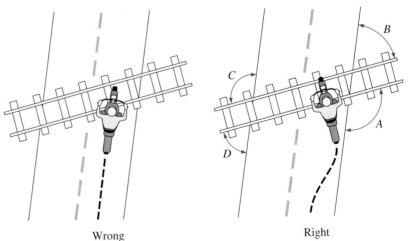

Wrong   Right

FIGURE 8.27

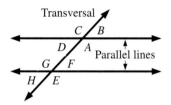

FIGURE 8.28
Corresponding angles:
A and *E*, *B* and *F*
C and *G*, *D* and *H*
Alternate interior angles:
A and *G*, *D* and *F*
Alternate exterior angles:
B and *H*, *C* and *E*
Hint: All the acute angles in the diagram have the same measure. All the obtuse angles have the same measure.

If you drive a motorcycle, you should look closely at Figure 8.27; it tells you to cross the railroad tracks at right angles (because there is less danger of a wheel catching in the tracks). Now look at angles *A*, *B*, *C*, and *D* marked on the right side of Figure 8.27. What can you say about angles *B* and *D*? Since they are vertical angles, they are of equal measure. (Similarly, angles *A* and *C* are of equal measure.) As you can see, the railroad track crosses the two parallel black lines in the figure. In geometry, a line that crosses two or more other lines is called a **transversal.** Thus, each railroad track is a transversal of the pair of parallel black lines. If a transversal crosses a pair of parallel lines, some of the resulting angles are of equal measure (see Figure 8.28). The exact relationships are as follows:

Corresponding angles are of equal measure.

$$m\angle A = m\angle E \qquad m\angle B = m\angle F$$
$$m\angle C = m\angle G \qquad m\angle D = m\angle H$$

Alternate interior angles are of equal measure.

$$m\angle A = m\angle G \qquad m\angle D = m\angle F$$

Alternate exterior angles are of equal measure.

$$m\angle B = m\angle H \qquad m\angle C = m\angle E$$

In Figure 8.28, angles *A* and *B* form a straight angle and are thus supplementary. Because $m\angle B = m\angle F$, angles *A* and *F* are also supplementary. The same idea applies to angles *D* and *G* as well as to angles *B* and *E* and angles *C* and *H*. We can summarize these facts by saying:

Parallel Lines Cut by a Transversal

Interior angles on the same side of the transversal are supplementary, and exterior angles on the same side of the transversal are also supplementary. In summary, when parallel lines are cut by a transversal,

1. All the acute angles are equal.

2. All obtuse angles are equal.

3. The sum of the measures of an acute angle and an obtuse angle is 180°.

The next example will help to make the concept clearer.

EXAMPLE 5 ▶ Finding More Measures of Angles

In Figure 8.29, find the measures of the following angles:

(a) *Y*   (b) *Z*   (c) *X*   (d) *R*   (e) *S*   (f) *T*   (g) *U*

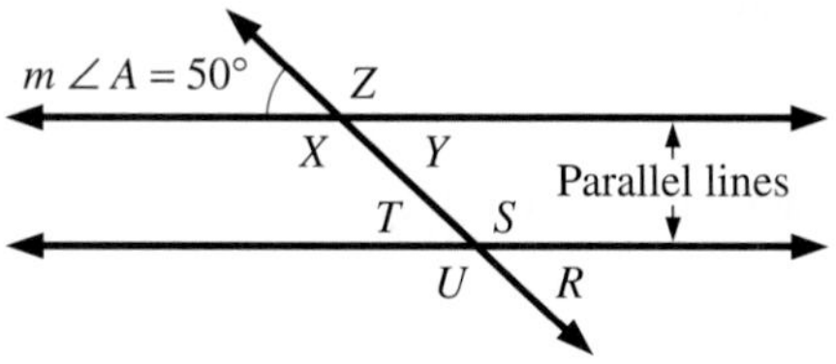

FIGURE 8.29

Solution

First, note that the measure of each acute angle ($\angle A$, $\angle Y$, $\angle T$, and $\angle R$) is 50° and the measure of each obtuse angle ($\angle Z$, $\angle X$, $\angle S$, and $\angle U$) is 130°.

(a) Since Y and A are vertical angles, $m\angle Y = m\angle A = 50°$.

(b) Angles A and Z are supplementary, so $m\angle Z = 180° - 50° = 130°$.

(c) Angles X and Z are vertical angles. Thus, $m\angle X = m\angle Z = 130°$.

(d) Angles R and A are alternate exterior angles, so $m\angle R = m\angle A = 50°$.

(e) Angles S and Y are interior angles on the same side of the transversal and so are supplementary. Therefore, $m\angle S = 180° - m\angle Y = 130°$.

(f) Angles T and A are corresponding angles. Thus, $m\angle T = m\angle A = 50°$.

(g) Angles U and A are exterior angles on the same side of the transversal. Thus, $m\angle U = 180° - m\angle A = 130°$. ∎

Parallel lines and the associated angles allow us to obtain one of the most important results in the geometry of triangles. In Figure 8.30, ABC represents any triangle. The line $\overleftrightarrow{XY}$ has been drawn through the point C parallel to the side $\overleftrightarrow{AB}$ of the triangle. Note that $m\angle 1 = m\angle 2$ and $m\angle 3 = m\angle 4$ because they

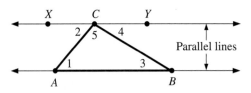

FIGURE 8.30

are respective pairs of alternate interior angles. Furthermore, angles 2, 5, and 4 form a straight angle, so

$$m\angle 2 + m\angle 5 + m\angle 4 = 180°$$

By substituting $\angle 1$ for $\angle 2$ and $\angle 3$ for $\angle 4$, we obtain

$$m\angle 1 + m\angle 5 + m\angle 3 = 180°$$

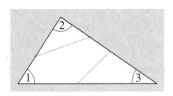

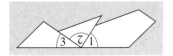

FIGURE 8.31

Here is a way to show this. Cut a triangle out of a sheet of paper as shown in Figure 8.31. Label the angles 1, 2, and 3 and cut them off the triangle. Place the vertices of angles 1, 2, and 3 together. Now, two of the sides form a straight line! Thus, we have shown that *the sum of the measures of the three interior angles of any triangle is 180°.*

Sum of the Measure of the Interior Angles of a Triangle

The sum of the measures of the three **interior angles** of any triangle is **180°**.

EXAMPLE 6 ▶ Finding Measures of Angles in a Triangle

(a) In a triangle ABC, $m\angle A = 47°$ and $m\angle B = 59°$. Find the measure of $\angle C$.
(b) Is it possible for a triangle ABC to be such that $\angle A$ is 2 times the size of $\angle B$ and $\angle C$ is 3 times the size of $\angle B$?

Solution

(a) Because $m\angle A + m\angle B + m\angle C = 180°$, we have

$$47° + 59° + m\angle C = 180°$$
$$m\angle C = 180° - 47° - 59°$$
$$= 180° - 106° = 74°$$

(b) To answer this question, let $m\angle B = x°$, so $m\angle A = 2x°$ and $m\angle C = 3x°$. Then, since the sum of the angles is $180°$,

$$x + 2x + 3x = 180$$
$$6x = 180$$
$$x = 30 \qquad 2x = 60 \qquad 3x = 90$$

This means that there is such a triangle, and $m\angle A = 60°$, $m\angle B = 30°$, and $m\angle C = 90°$. (Note that this is a right triangle because one of the angles is a right angle.) ■

In part (b) of Example 6 we found angles A and B to be of measure $60°$ and $30°$, respectively, so $m\angle A + m\angle B = 90°$.

Definition of Complementary Angles

Two angles whose sum is $90°$ are called **complementary angles,** and each angle is called the **complement** of the other.

EXAMPLE 7 ▶ Finding the Complement of an Angle

(a) Find the complement of a $38°$ angle.

(b) Can two complementary angles be such that one is three times the size of the other?

Solution

(a) For an angle to be the complement of a $38°$ angle, its measure must be $90° - 38° = 52°$.

(b) Let the smaller angle be of measure $x°$, so the larger is of measure $3x°$. Since the angles are to be complementary,

$$x + 3x = 90$$
$$4x = 90$$
$$x = \frac{90}{4} = 22\tfrac{1}{2} \qquad 3x = 67\tfrac{1}{2}.$$

The answer is yes, and the angles measure $22\tfrac{1}{2}°$ and $67\tfrac{1}{2}°$. ■

Online Study Center

To further explore points, lines, and planes, access links 8.1.1 and 8.1.2 on this textbook's Online Study Center.

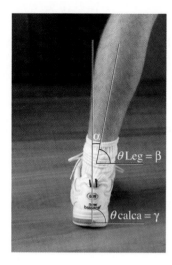

Unsupported   Supported

D. Applications

EXAMPLE 8 ▶ Applications to Kinesiology

How far is the leg in the photo from being vertical? In physical education and kinesiology, they measure the rear foot angle α. This angle can be adjusted by using an insole as shown in the illustrations below the photo. You can find the rear foot angle if you measure angle α or angle β.

(a) What type of angles are $\angle\alpha$ and $\angle\beta$?

(b) Find $m\angle\alpha$.

(c) If $m\angle\alpha = 15°$, find $m\angle\beta$.

Solution

(a) $\angle\alpha$ and $\angle\beta$ are complementary angles.

(b) $m\angle\alpha = 90° - m\angle\beta$.

(c) Since $\angle\alpha$ and $\angle\beta$ are complementary angles,

$$m\angle\beta = 90° - m\angle\alpha$$
$$= 90° - 15° \qquad \text{Remember } m\angle\alpha = 15°.$$
$$= 75°$$

Thus, $m\angle\beta = 75°$. ■

EXERCISES 8.1

Ⓐ Points and Lines

In problems 1 and 2, draw a line or a portion of a line that corresponds to each symbol.

1. a. $\overrightarrow{PQ}$ **b.** $\overrightarrow{QP}$ **c.** $\overset{\bullet}{Q}\overset{\bullet}{P}$

2. a. $\overrightarrow{PQ}$ **b.** $\overrightarrow{QP}$ **c.** $\overleftrightarrow{PQ}$

In problems 3–14, use the figure below and determine what each union or intersection describes.

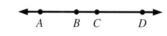

3. $\overrightarrow{AB} \cap \overleftrightarrow{BC}$ **4.** $\overleftrightarrow{AC} \cap \overleftrightarrow{BC}$

5. $\overleftrightarrow{AC} \cup \overrightarrow{BC}$ **6.** $\overrightarrow{AD} \cup \overleftrightarrow{CB}$

7. $\overrightarrow{AC} \cap \overrightarrow{DA}$ **8.** $\overleftrightarrow{BD} \cap \overrightarrow{DC}$

9. $\overrightarrow{AC} \cup \overrightarrow{DC}$ **10.** $\overleftrightarrow{AC} \cup \overrightarrow{DB}$

11. $\overleftrightarrow{BA} \cap \overset{\bullet}{C}\overset{\bullet}{D}$ **12.** $\overset{\bullet}{C}\overset{\bullet}{B} \cap \overset{\bullet}{C}\overset{\bullet}{D}$

13. $AC \cap DC$ **14.** $AB \cap DB$

15. The following figure shows a triangular pyramid.
   a. Name all the edges.
   b. Which pairs of edges determine skew lines?
   c. Do any of the edges determine parallel lines?

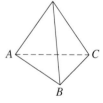

16. The figure below represents a pyramid with a square base $ABCD$.
   a. Name all the edges of the pyramid.
   b. Which pairs of edges determine parallel lines?
   c. Which determine skew lines?
   d. Which lines are intersecting?

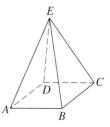

B Planes

Determine whether each of the following statements seems to be true or false in Euclidean geometry. You will find it helpful to use pencils, pieces of cardboard, walls, floors, and so on to represent lines and planes.

17. Given any plane *ABC* and any point *P* not on *ABC*, there is exactly one plane that contains *P* and is parallel to plane *ABC*.

18. Given any line *m* and any point *P* not on *m*, there is exactly one plane that contains *P* and is parallel to *m*.

19. Given any line *m* and any point *P* not on *m*, there is exactly one line that contains *P* and is parallel to *m*.

20. Given any plane *ABC* and any point *P* not on *ABC*, there are any number of lines containing *P* and parallel to *ABC*.

21. Given any line *m* and any point *P* not on *m*, there are any number of lines that contain *P* and are skew to *m*.

22. Given any plane *ABC* and any line *m* parallel to *ABC*, there is exactly one plane that contains *m* and intersects *ABC*.

23. Two nonparallel lines always determine a plane.

24. Given any line *m* and any point *P* not on *m*, there is exactly one plane that contains both *m* and *P*.

25. Given a plane *ABC* and a line *m* that intersects *ABC*, there is a plane that contains *m* and that does not intersect *ABC*.

26. If a plane intersects two parallel planes, the lines of intersection with each of the parallel planes are parallel.

C Angles

Problems 27–40 refer to the figure below. Note that perpendicular lines are indicated by small red boxes.

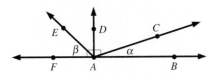

27. Name the following angles in another way:
  a. ∠α   **b.** ∠EAF

28. Name the following angles in another way:
  a. ∠CAB   **b.** ∠β

29. List all the acute angles in the figure.

30. List all the right angles in the figure.

31. List all the obtuse angles in the figure.

32. Name an angle that is the complement of
  a. ∠α.   **b.** ∠β.

33. Name an angle that is the complement of
  a. ∠EAF.   **b.** ∠BAC.

34. Name an angle that is supplementary to
  a. ∠α.   **b.** ∠β.

35. Name an angle that is supplementary to
  a. ∠BAD.   **b.** ∠EAF.

36. If $m\angle\alpha = 15°$, find $m\angle CAD$.

37. If $m\angle\beta = 55°$, find $m\angle DAE$.

38. If $m\angle DAE = 35°$, find $m\angle\beta$.

39. If $m\angle CAD = 75°$, find $m\angle\alpha$.

40. If $m\angle\alpha = 15°$ and $m\angle\beta = 55°$, find $m\angle CAE$.

Problems 41–48 refer to the two intersecting lines shown in the figure below.

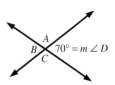

41. Name the angle that is vertical to the 70° angle.

42. Name the two angles that are each supplementary to the 70° angle.

43. Find $m\angle A$.   **44.** Find $m\angle B$.

45. What is the measure of an angle complementary to the 70° angle?

46. Find the sum of the measures of angles *A*, *B*, and *C*.

47. Find the sum of the measures of angles *A* and *C*.

48. If $m\angle D = x°$ (instead of 70°), write an expression for the measure of ∠*A*.

Problems 49–51 refer to the two parallel lines and the transversal shown in the figure below.

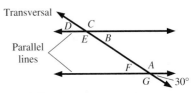

49. Find the following:
  a. $m\angle A$ **b.** $m\angle B$ **c.** $m\angle C$

50. Find the following:
  a. $m\angle D$ **b.** $m\angle E$ **c.** $m\angle F$

51. Name all the angles that are supplementary to $\angle B$.

52. Refer to the angles shown in the figure below.
  a. If $m\angle AOB = 30°$ and $m\angle AOC = 70°$, find $m\angle BOC$.
  b. If $m\angle AOB = m\angle COD$, $m\angle AOD = 100°$, and $m\angle BOC = 2x°$, find $m\angle COD$ in terms of x.

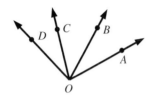

53. If $m\angle A = 41°$, find $m\angle B$ given that
  a. the two angles are complementary.
  b. the two angles are supplementary.

54. If $m\angle A = 19°$, find $m\angle B$ given that
  a. the two angles are complementary.
  b. the two angles are supplementary.

55. Given that $m\angle A = (3x + 15)°$, $m\angle B = (2x - 5)°$, and the two angles are complementary, find x.

56. Rework problem 55 if the two angles are supplementary.

In problems 57–60, the figures show the number of degrees in each angle in terms of x. Use algebra to find x and the measure of each angle.

57.

$10x - 5$ $2x + 5$

58.

$4x$
$2x$

59.

$5x + 10$
$2x + 10$

60.

$8x - 14$ $6x - 2$

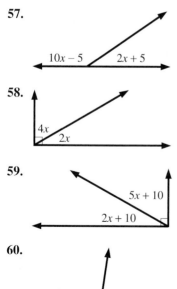

61. Through how many degrees does the hour hand of a clock move during the following time intervals?
  a. 11 to 12 o'clock   **b.** 11 to 5 o'clock

62. Through how many degrees does the hour hand of a clock move during the following time intervals?
  a. 12 to 7 o'clock   **b.** 12 back to 12 o'clock

63. In a triangle ABC, $m\angle A = 37°$ and $m\angle C = 53°$. Find $m\angle B$.

64. In a triangle ABC, $m\angle B = 67°$ and $m\angle C = 105°$. Find $m\angle A$.

65. In a triangle ABC, $m\angle A = (x + 10)°$, $m\angle B = (2x + 10)°$, and $m\angle C = (3x + 10)°$. Find x.

66. In a triangle ABC, $m\angle A$ is 10° less than $m\angle B$, and $m\angle C$ is 40° greater than $m\angle B$. Find the measure of each angle.

67. Refer to Figure 8.26. Complete the first two equations and give the reason for each step in the following proof that vertical angles are of equal measure:
  a. $m\angle A + m\angle B = $ _____
  b. $m\angle C + m\angle B = $ _____
  c. $m\angle A + m\angle B = m\angle C + m\angle B$
  d. $m\angle A = m\angle C$

D **Applications**

In Problems 68–72, refer to the photo and classify the given angle.

68. $\angle\alpha$

69. $\angle\beta$

70. $\angle\gamma$

71. $\angle \delta$

72. Classify and name the two types of angles shown in the photo.

73. Name the type of angle you need when you want to use the phone shown in the photo below.

74. About how many degrees is the angle shown?

75. How many degrees would the angle be when the phone is not being used and is closed?

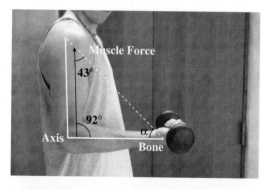

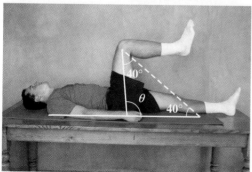

Joint range of motion (ROM) is a common measure for students in physical education and kinesiology, and it involves the measurement of angles.

76. Find $m\angle\alpha$.

77. Find $m\angle\theta$.

In Other Words

78. You have heard the saying "The shortest distance between two points is a straight line." Explain why this is technically incorrect.

79. Name three undefined geometric terms.

80. Find the word *point* in a dictionary. How many definitions does the word have? In your own words, give three definitions that can apply to geometry.

81. Describe the different ways in which a ray and a plane can intersect.

82. Describe the different ways in which a line and a ray can intersect.

83. We have already defined an acute and an obtuse angle. Look at the definitions of *acute* and *obtuse* in a dictionary and explain the following:
   a. What an *acute pain* means
   b. What *obtuse intelligence* means

84. Describe and show sketches of the following:
   a. How a line can intersect an angle
   b. How two angles can intersect

Using Your Knowledge

Angles are extremely important in surveying and in navigation. With the knowledge you have gained in this section, you should be able to do the following problems:

85. A surveyor measured a triangular plot of ground and reported the three angles as 48.2°, 75.9°, and 56.1°. How much of an error did the surveyor make?

In land surveying, angles are measured with respect to due north and due south. For example, a direction of N 30° W means a direction that is 30° west of due north, and S 60° E means a direction that is 60° east of due south. Surveyors use acute angles only. In aerial navigation, angles are always measured clockwise from due north. A navigator's bearing of 90° corresponds to due east, 180° to due south, and 270° to due west. Thus, a navigator's bearing of 225° would correspond to a surveyor's bearing of S 45° W.

86. Write a navigator's bearing of 135° in a surveyor's terminology.

87. Write a navigator's bearing of 310° in a surveyor's terminology.

88. Write the direction S 40° W as a navigator's bearing.

89. Write the direction N 40° W as a navigator's bearing.

 Research Questions

1. There are at least four definitions or descriptions of *point*. Call them 1, 2, 3, and 4. Find out what they are and how they differ, and then fill in the blanks in the chart with A for always, S for sometimes, or N for never.

	Definition of Point			
	1	2	3	4
Through two points there is exactly one line.				
A line is an infinite set of points.				
A point has no size.				

2. There are four different relationships that two lines can have. For example, two lines can be *identical*. What are the other three relationships?

3. Go to the library or to this textbook's Online Study Center and create a list of several applications of geometry. A suggested link is 8.1.3.

4. Was algebra invented before geometry? When was Euclid born? Where? What did his book *The Elements* contain? Find the answers to these questions and write a short sentence about each topic. You can start by accessing links 8.1.4 and 8.1.5 on this textbook's Online Study Center.

5. Book I of *The Elements* contains five important postulates used in what is now called *Euclidean geometry*. What are these five postulates?

8.2 Triangles and Other Polygons

Soap Bubble Polygons

Look at the soap bubbles in the photograph. What do you notice about the shapes shown? First, all of them are made up of line segments, and second, each shape can be traced by starting and ending at one point. You will study such shapes, called **polygons,** in this section.

Plane polygons can be named according to the number of sides. Thus, a *three-sided* polygon is a *tri*angle and a *four-sided* polygon is a *quad*rilateral. Later, you will learn how to name polygons with five, six, seven, and eight sides.

How many polygons do you see in the picture?

Can you find the sum of the angles in the polygons you see?

As you recall, the sum of the measures of the angles in a triangle is $180°$. Because a quadrilateral can always be divided into two triangles, the sum of the measures of the angles is $360°$. Can you find the sum of the measures of the angles in polygons with five, six, seven, and in general, n sides? See if you can discover a pattern for the sum of the angles in Table 8.1 and fill in the blank.

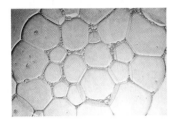

Soap bubbles exhibit various polygonal shapes when closely packed.

TABLE 8.1

Number of Sides	Sum of the Angles
Three-sided polygon	$1 \cdot 180°$
Four-sided polygon	$2 \cdot 180°$
Five-sided polygon	————

The answer is $3 \cdot 180°$. Why? Read part C of this section on page 518 and you will see! ▶

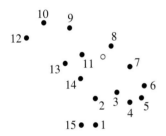

A popular children's puzzle consists of joining in order a set of numbered dots by straight line segments to form a path from the first to the last point. An example of such a puzzle and its solution is shown in Figure 8.32. (What is pictured, an antelope or a bird?)

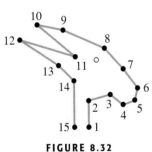

FIGURE 8.32

A. Broken Lines and Polygons

A **path** (such as the one shown in the puzzle) consisting of a sequence of connected straight line segments is called a **broken line.** Such a path can be traced without lifting the pencil from the paper. Certain broken line paths have the following two characteristics that are of interest to us in this section:

1. A **simple path** is a path that does *not* cross itself.

2. A **closed path** is a path that starts and ends at the same point.

Not all geometric figures are constructed using straight line segments. The following are more general concepts:

1. A **simple curve** is a curve that does not cross itself.

2. A **closed curve** is a curve that starts and ends at the same point.

Most of our discussion, however, will deal with simple and closed *paths* rather than simple and closed *curves,* because simple, closed paths are used to define polygons.

The path in the solution of the puzzle in Figure 8.32 is simple but not closed. Figure 8.33 shows four broken lines. The paths in panels A and C are simple, but the paths in panels B and D are not simple. Moreover, the paths in panels B and C are closed, but the paths in panels A and D are not closed. Can you draw a path that is not simple and not closed? Can you draw a curve that is not simple and not closed? Figure 8.34 gives examples of curved paths.

Any path that consists of a simple, closed broken line is called a **polygon.** In Figures 8.35 and 8.36, paths *ABCDE* are examples of polygons. The line segments of the path are the **sides** of a polygon, and the endpoints of the sides are the **vertices.**

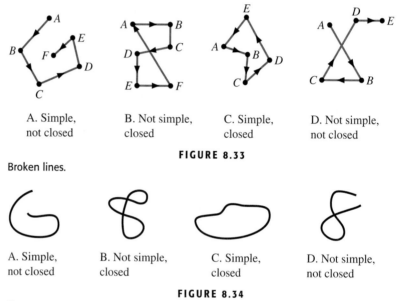

A. Simple,
not closed

B. Not simple,
closed

C. Simple,
closed

D. Not simple,
not closed

FIGURE 8.33

Broken lines.

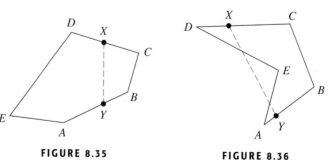

A. Simple,
not closed

B. Not simple,
closed

C. Simple,
closed

D. Not simple,
not closed

FIGURE 8.34

Curves.

A polygon is said to be **convex** if no line segment $\overleftrightarrow{XY}$ joining any two points on the path (boundary) ever extends outside the polygon (see Figure 8.35). The points *X* and *Y* may be any two points not on the same side of the polygon. Except for its endpoints, the line segment $\overleftrightarrow{XY}$ lies entirely inside the polygon. (Of course, if *X* and *Y* were on the same side, then the segment would lie on that side.)

A polygon that is not convex is called a **concave,** or **reentrant,** polygon (see Figure 8.36). Here a portion of the line segment $\overleftrightarrow{XY}$ lies outside the polygon—something that never occurs in a convex polygon.

FIGURE 8.35
Convex polygon.

FIGURE 8.36
Concave polygon.

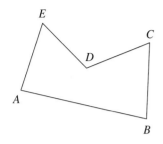

FIGURE 8.37

EXAMPLE 1 ▶ Naming Sides and Vertices of Polygons

Consider the polygon $ABCDE$ in Figure 8.37.

(a) Name its sides.   (b) Name its vertices.

(c) Is this a concave or a convex polygon?

Solution

(a) The sides are $\overleftrightarrow{AB}$, $\overleftrightarrow{BC}$, $\overleftrightarrow{CD}$, $\overleftrightarrow{DE}$, and $\overleftrightarrow{EA}$.

(b) The vertices are A, B, C, D, and E.

(c) This is a concave, or reentrant, polygon.   ▪

Plane polygons are customarily named according to the number of sides. Table 8.2 states some of the usual names.

TABLE 8.2

Sides	Name	Sides	Name
3	Triangle	8	Octagon
4	Quadrilateral	9	Nonagon
5	Pentagon	10	Decagon
6	Hexagon	12	Dodecagon
7	Heptagon		

As you probably know, traffic signs are most often in the shapes of polygons. For example, the stop sign is in the shape of an octagon. In fact, it is in the shape of a *regular* octagon. A **regular polygon** is a polygon with all its sides of equal length and all its angles of equal size. A regular triangle has three equal sides and three 60° angles; it is called an **equilateral triangle.** A regular quadrilateral has four equal sides and four 90° angles. You undoubtedly know that it is called a **square.** No special names are given to other regular polygons.

EXAMPLE 2 ▶ Naming Shapes of Traffic Signs

Some standard traffic signs are shown below.

(a) Which ones are regular polygons?

(b) Name the shape of the school sign.

(c) Name the shape of the yield sign.

(d) Name the shape of the warning sign.

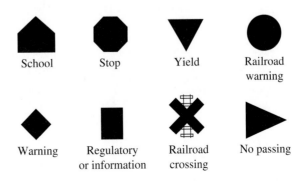

School   Stop   Yield   Railroad warning

Warning   Regulatory or information   Railroad crossing   No passing

Solution

(a) The stop sign, the yield sign, and the warning sign are all in the shape of regular polygons.

(b) The school sign has five sides, so it is in the shape of a pentagon (but not a regular pentagon).

(c) The yield sign has three equal sides; it is in the shape of an equilateral triangle.

(d) The warning sign has four equal sides and four equal angles; it is in the shape of a regular quadrilateral, a square. ■

B. Triangles and Quadrilaterals

Triangles are often classified according to their angles, as shown in Table 8.3. Triangles are also classified according to the number of equal sides, as shown in Table 8.4.

Note that the angles opposite the equal sides of an **isosceles triangle** are of equal measure. (See the Using Your Knowledge section of Exercises 8.2.) Moreover, an **equilateral triangle** is also **equiangular;** it has three 60° angles. The triangle that you see in the optical illusion in Figure 8.38 is an equilateral triangle.

Certain quadrilaterals (four-sided polygons) also have special names and properties, as shown in Table 8.5 on page 513.

FIGURE 8.38
Can you see the triangular region? It is an optical illusion. The triangle exists only in your mind.

TABLE 8.3 Classification of Triangles by Angles

Definition	Illustration
Right triangle: A triangle containing a right angle	55° 35° 90°
Acute triangle: A triangle in which all the angles are acute	60° 45° 75°
Obtuse triangle: A triangle containing an obtuse angle	30° 135° 15°

TABLE 8.4 Classification of Triangles by Number of Equal Sides

Definition	Illustration
Scalene triangle: A triangle with *no* equal sides (Note that the sides are labeled |, ||, and ||| to show that the lengths of the sides are different.)	
Isosceles triangle: A triangle with two equal sides	
Equilateral triangle: A triangle with **all** three sides equal	

TABLE 8.5 Quadrilaterals

Definition	Illustration
Trapezoid: A quadrilateral with exactly one pair of parallel sides	
Kite: A quadrilateral with two distinct pairs of consecutive equal sides	
Parallelogram: A quadrilateral in which the opposite sides are parallel	
Rectangle: A parallelogram with a right angle	
Rhombus: A parallelogram with all sides equal	
Square: A rectangle with all sides equal	

EXAMPLE 3 ▶ Classifying Triangles by Angles and Sides

Classify the given triangles according to their angles and their sides.

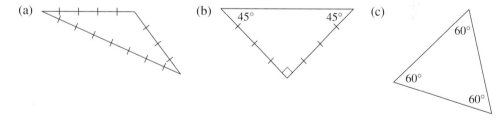

Solution

(a) The triangle has an obtuse angle and no equal sides; it is an obtuse, scalene triangle.

(b) The triangle has two equal sides and a right angle; it is an isosceles, right triangle.

(c) The triangle has three $60°$ angles; it is an equilateral triangle, which is also equiangular. ∎

In Tables 8.3 and 8.4, we classified triangles according to their angles and their sides, respectively. Angles and sides are used to compare triangles. If two triangles have exactly the same shape (their corresponding angles are equal) but

The Smiley on the right is an enlargement of the Smiley on the left and exactly the same shape but not the same size. The Smileys are similar.

not necessarily the same size (their sides are *proportional*), then the triangles are called *similar triangles.*

Definition of Similar Triangles

Triangles $\triangle ABC$ and $\triangle DEF$ are **similar** (denoted by $\triangle ABC \sim \triangle DEF$) if and only if they have **equal** corresponding angles and their corresponding sides are **proportional.**

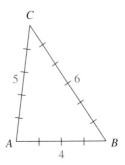

FIGURE 8.39

Look at the two similar triangles $\triangle ABC$ and $\triangle DEF$ in Figure 8.39. Because they have the same shape, the corresponding angles are equal.

Corresponding Angles of Similar Triangles

In Figure 8.39, $m\angle A = m\angle D$, $m\angle B = m\angle E$, and $m\angle C = m\angle F$. The corresponding sides of the two triangles are $\overline{AB}$ and $\overline{DE}$, $\overline{BC}$ and $\overline{EF}$, and $\overline{AC}$ and $\overline{DF}$. The length of each side of $\triangle DEF$ is one-half the length of the corresponding side of $\triangle ABC$. Recall that the ratio of two numbers a and b is the fraction a/b. For Figure 8.39, we have

$$\frac{DE}{AB} = \frac{EF}{BC} = \frac{DF}{AC} = \frac{1}{2}$$

as the ratios of corresponding sides.

We use these ideas in the next example.

EXAMPLE 4 ▶ Finding Dimensions for Similar Triangles

Two similar triangles are shown in Figure 8.40. Find d and f for $\triangle DEF$.

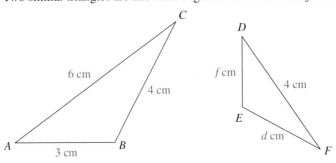

FIGURE 8.40

Solution

Since $\triangle ABC \sim \triangle DEF$, the corresponding sides must be proportional. Thus,

$$\frac{DE}{AB} = \frac{EF}{BC} = \frac{DF}{AC}$$

so that

$$\frac{f}{3} = \frac{d}{4} = \frac{4}{6}$$

Since $\frac{4}{6} = \frac{2}{3}$, we have

$$\frac{d}{4} = \frac{2}{3} \quad \text{and} \quad \frac{f}{3} = \frac{2}{3}$$

Solving these equations for d and f, we get

$$d = \frac{8}{3} = 2\frac{2}{3} \quad \text{and} \quad f = \frac{6}{3} = 2 \qquad \blacksquare$$

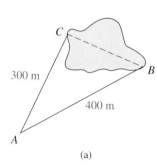

(a)

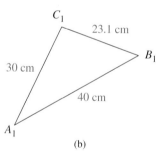

(b)

FIGURE 8.41

EXAMPLE 5 ▶ Applications of Similar Triangles

Wally wanted to find the distance across a lake, so he located a point A from which he could sight points B and C on opposite ends of the lake (see Figure 8.41a). With a surveyor's sextant, Wally found that $m\angle A = 35°$. He was able to measure $\overset{\bullet}{AB}$ and $\overset{\bullet}{AC}$ as 400 and 300 m, respectively. He took these data to his office and drew a triangle which he labeled $A_1B_1C_1$ with $m\angle A_1 = 35°$, $A_1B_1 = 40$ cm, and $A_1C_1 = 30$ cm. He then measured $\overset{\bullet}{B_1C_1}$ and found it to be 23.1 cm long. How could Wally use all this information to find the distance across the lake?

Solution

Since Wally made $m\angle A_1 = m\angle A$ and used the same scale (10 m/cm) for $\overset{\bullet}{A_1B_1}$ and $\overset{\bullet}{A_1C_1}$, he knows that $\triangle A_1B_1C_1 \sim \triangle ABC$ (see Figure 8.41b). Therefore, if the length of $\overset{\bullet}{BC}$ is x m, then

$$\frac{400}{40} = \frac{300}{30} = \frac{x}{23.1}$$

so that

$$\frac{x}{23.1} = 10 \quad \text{and} \quad x = 231$$

Thus, Wally found the distance across the lake to be 231 m. $\qquad \blacksquare$

The definition of similar triangles indicates that they have the same corresponding angles and shape but their size may be different. Triangles that have both the same shape, and size are called **congruent triangles.** This actually means that the two triangles look *identical.* Note that if two triangles are congruent, they are also similar.

Warning: The converse of the last sentence is *not* true. In other words, two similar triangles are *not* necessarily congruent. (See the photograph in the margin.)

Minime and Dr. Evil are similar but not congruent.

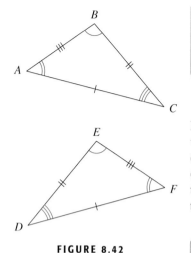

FIGURE 8.42
$\triangle ABC \cong \triangle DEF$

Definition of Congruent Triangles

Triangles $\triangle ABC$ and $\triangle DEF$ are **congruent** (denoted by $\triangle ABC \cong \triangle DEF$) if and only if they have **equal** corresponding **angles** and **sides.**

Objects that are *exactly* the same shape and size are said to be **congruent.** Note that the symbol to denote congruency ($\cong$) is made up of two parts: "$\sim$," which means the same shape (similar), and "$=$," which means the same size (equal). When looking at congruent triangles, be sure that the sides are of equal (length) and corresponding angles are of equal measure (in degrees). When two triangles such as the ones in Figure 8.42 are congruent, there are six facts that are true about the triangles:

The length of side $\overleftrightarrow{AB}$ is the same as the length of side $\overleftrightarrow{EF}$.	$AB = EF$
The length of side $\overleftrightarrow{BC}$ is the same as the length of side $\overleftrightarrow{DE}$.	$BC = DE$
The length of side $\overleftrightarrow{AC}$ is the same as the length of side $\overleftrightarrow{DF}$.	$AC = DF$
The measure of angle A is the same as the measure of angle F.	$m\angle A = m\angle F$
The measure of angle B is the same as the measure of angle E.	$m\angle B = m\angle E$
The measure of angle C is the same as the measure of angle D.	$m\angle C = m\angle D$

To show graphically that two triangles are congruent, draw two triangles on a piece of paper and then carefully cut them out along their outlines. If you can place one of the triangles on top of the other so that the three vertices of the top triangle match up perfectly with the three vertices of the bottom triangle (you may have to flip and/or rotate the triangles first), then the triangles are congruent. Practice this idea by tracing out the congruent triangles in Figure 8.42.

Note: Corresponding equal sides are marked with small straight line segments and corresponding congruent angles are marked with arcs.

You can determine whether two triangles are congruent even if you don't have all the information about the measures of their sides and angles. First, we need some definitions.

Included Side of Angles

Given any triangle, a side that lies between two angles is called the **included side** of the angles.

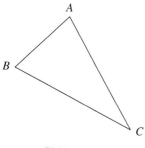

FIGURE 8.43

Similarly,

> **Included Angle of Side**
>
> The angle formed by two sides is the **included angle** of the side.

For example, $\overset{\bullet\;\;\bullet}{AC}$ is the included side of $\angle A$ and $\angle C$ in Figure 8.43, whereas $\angle A$ is the included angle of $\overset{\bullet\;\;\bullet}{AC}$ and $\overset{\bullet\;\;\bullet}{AB}$ in the same figure. Here are the theorems you need to prove that two triangles are congruent.

> **The Congruent Triangle Theorems**
>
> 1. Side-Angle-Side (SAS). If two sides and the included angle of one triangle are equal to the corresponding sides and the included angle of a second triangle, then the two triangles are congruent.
>
> 2. Angle-Side-Angle (ASA). If two angles and the included side of one triangle are equal to the corresponding two angles and the included side of a second triangle, then the two triangles are congruent.
>
> 3. Side-Side-Side (SSS). If the three sides of one triangle are equal to the corresponding sides of a second triangle, then the two triangles are congruent.

✦Online Study Center

To further explore congruent triangles, access link 8.2.1 and 8.2.2 on this textbook's Online Study Center.

EXAMPLE 6 ▶ Exploring the Congruent Triangle Theorems

(a) Show that $\triangle ABE$ and $\triangle DCE$ pictured in Figure 8.44 are congruent.

(b) Show that $\triangle ABD$ and $\triangle ACD$ pictured in Figure 8.45 are congruent.

(c) Why is there not an Angle-Angle-Angle congruent triangle theorem?

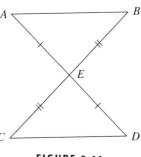

FIGURE 8.44

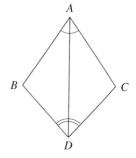

FIGURE 8.45

Solution

(a) Let $\angle X$ be the included angle of $\overset{\bullet\;\;\bullet}{AE}$ and $\overset{\bullet\;\;\bullet}{BE}$, and let $\angle Y$ be the included angle of $\overset{\bullet\;\;\bullet}{CE}$ and $\overset{\bullet\;\;\bullet}{DE}$. Since $\angle X$ and $\angle Y$ are vertical angles, $m\angle X = m\angle Y$. According to the diagram, $AE = DE$ and $BE = CE$. Thus, $\triangle ABE \cong \triangle DCE$ by SAS.

(b) Viewing $\overset{\bullet\;\;\bullet}{AD}$ as the included side in each triangle, ASA tells us that $\triangle ABD \cong \triangle ACD$.

(c) The two triangles pictured in Figure 8.46 are equilateral triangles and so all the angles are 60°. These triangles are *similar* but not congruent.

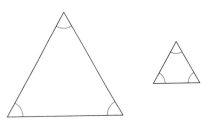

FIGURE 8.46
Triangles with all angles equal.

C. Angles of a Polygon

In Example 5, Wally found the lengths of the sides of a triangle. Next, we shall examine how to find the measure of the angles of any given polygon.

PROBLEM SOLVING

Finding the Measure of an Angle

① Read the problem.

Find the measure of angle β in Figure 8.47.

② Select the unknown.

The measure of angle β is the unknown.

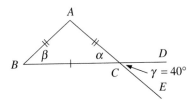

FIGURE 8.47

③ Think of a plan.
Find the equal sides. What angles are given? Are there any vertical angles?

You want to find the measure of angle β. According to Figure 8.47, $AB = AC$ and $\gamma = 40°$.

④ Use the fact that α and γ are vertical angles to find α. Look at the angles opposite the equal sides.

Since α and γ are vertical angles, they must be equal; that is, $\alpha = \gamma = 40°$. Angle α and angle β are opposite the equal sides of $\triangle ABC$ (triangle ABC); hence, they must also be of equal measure. Thus, the measure of angle β is 40°.

⑤ Verify the answer.

The verification is left to you.

TRY EXAMPLE 7 NOW.

Cover the solution, write your own solution, and then check your work.

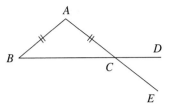

FIGURE 8.48

EXAMPLE 7 ▶ Finding Measures in a Triangle with Two Equal Sides

In $\triangle ABC$ in Figure 8.48, $AB = AC$ and $m\angle DCE = 37°$. Find $m\angle A$.

Solution

$\angle ACB$ and $\angle DCE$ are vertical angles, so $m\angle ACB = m\angle DCE = 37°$. $\angle ACB$ and $\angle B$ are opposite the equal sides of $\triangle ABC$ and are thus of equal measure. So $m\angle B = 37°$. Then,

$$m\angle A = 180° - m\angle ACB - m\angle ABC$$
$$m\angle A = 180° - 37° - 37°$$
$$= 106°$$ ∎

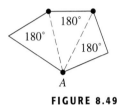

FIGURE 8.49

We have already noted that the sum of the measures of the three angles of any triangle is 180°. We can use this idea to find the corresponding sum S for any convex polygon. For example, suppose the polygon is a pentagon, as shown in Figure 8.49. Select any vertex—say, A—and draw lines from A to each nonadjacent vertex, as shown by the dashed lines in Figure 8.49. The resulting number of triangles is 3 (2 fewer than the 5 sides of the polygon), and the number of degrees in the angles of these three triangles is $3 \times 180° = 540°$.

We can generalize this result to any convex polygon; if the polygon has n sides, there will be $n - 2$ triangles. We have the following result:

The Sum of the Measures of the Interior Angles of a Convex Polygon

The sum S of the measures of the interior angles of any convex polygon with n sides is given by

$$S = (n - 2)180°$$

Online Study Center

To see a proof of the conjecture that the sum S of the measures of any convex polygon with n sides is $S = (n - 2)180$, access link 8.2.3 on this textbook's Online Study Center.

Furthermore, if the polygon is a regular polygon, then the angles are all equal. In this case, the measure of a single angle of the polygon is just the preceding result divided by n, the total number of angles. This idea is illustrated in the next example.

EXAMPLE 8 ▶ Finding Measures of Angles in a Heptagon

Find the measure of an angle of a regular heptagon.

Solution

Since a heptagon has 7 sides, the formula gives

$$S = (7 - 2)180° = 5 \times 180° = 900°$$

Because the polygon is regular, each angle has measure

$$\frac{900°}{7} = 128\frac{4}{7}°$$ ∎

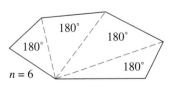

FIGURE 8.50
This hexagon breaks into four triangles, each containing 180° of angles.

That the sum S of the measures of the interior angles of any convex polygon with n sides equals $(n - 2)180°$ for the case $n = 6$ is illustrated in Figure 8.50.

What if you want to know the measure of *each* of the angles in the polygon? Look at Figure 8.51 and then fill in the blank:

The measure of each of the *n* angles in a **regular** convex polygon with *n* sides is <u>Your Answer</u>.

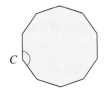

$(n = 6)$	$(n = 8)$	$(n = 10)$
$A = (6 - 2) \cdot 180/6 = 120°$	$B = (8 - 2) \cdot 180/8 = 135°$	$C = (10 - 2) \cdot 180/10 = 144°$

FIGURE 8.51

EXERCISES 8.2

Ⓐ Broken Lines and Polygons

In problems 1 and 2, sketch a broken line path as described.

1. a. Closed but not simple
   b. Simple but not closed

2. a. Both simple and closed
   b. Neither simple nor closed

In problems 3–7, use the alphabet,

ABCDEFGHIJKLMNOPQRSTUVWXYZ

3. State which letters form a path or curve that is
   a. simple.   **b.** closed.

4. State which letters form a path or curve that is
   a. closed but not simple.
   b. simple but not closed.

5. State which letters form a path or curve that is
   a. simple and closed.
   b. neither simple nor closed.

6. If the lowest points of the legs of the letter **M** are joined by a straight line segment, will the resulting polygon be concave or convex?

7. If the highest points of the legs of the letter **V** are joined by a straight line segment, will the resulting polygon be concave or convex?

8. Refer to the traffic signs in Example 2, and name the shape of the
   a. information sign.   **b.** no passing sign.

Ⓑ Triangles and Quadrilaterals

In problems 9–16, name each quadrilateral.

9.

10.

11.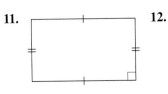

12.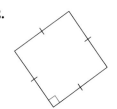

13.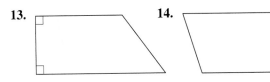

14.

15.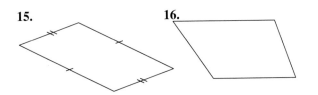

16.

In problems 17–24, classify each triangle as scalene, isosceles, or equilateral and as acute, right, or obtuse.

17.

18.
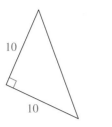

19.
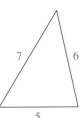

20.
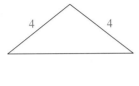

21.

22.

23.
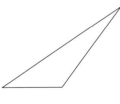

24.
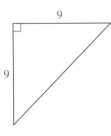

25. Which (if any) of the following triangles are similar?

a.   **b.**   **c.**
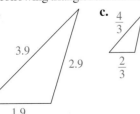

26. The marked angles in the following triangles are all equal. Which (if any) of the triangles are similar?

a.   **b.**   **c.**
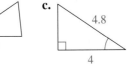

27. The following triangles are similar. Find the lengths marked x and y.

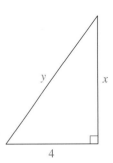

28. The following parallelograms are similar. Find the length of the diagonal $\overset{\bullet\bullet}{PR}$.

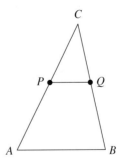

Problems 29–32 refer to the following figure. In this figure, $\overset{\bullet\bullet}{PQ}$ is parallel to $\overset{\bullet\bullet}{AB}$. In each problem, certain lengths are given. Find the missing length, indicated by a blank.

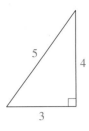

	AP	*PC*	*BQ*	*BC*
29.	3	4	6	___
30.	5	4	___	6
31.	2	___	3	5
32.	___	4	4	8

33. The sides of a triangle measure 6, 9, and 12 cm, respectively. The longest side of a similar triangle measures 7 cm. Find the lengths of the other two sides of the smaller triangle.

34. In problem 33, if the length of the shortest side of the smaller triangle is 5 cm, find the lengths of the other two sides.

35. The sides of a triangle measure 2, 3, and 4 in., respectively. The perimeter of a similar triangle is 36 in. Find the length of each side of the second triangle.

36. Jackie has a piece of wire 18 in. long. She wants to bend this into a triangle similar to a triangle whose sides are 3, 4, and 5 in., respectively. What must be the dimensions of her triangle?

37. A telephone pole casts a shadow 30 ft long at the same time that a 5-ft fence post casts a shadow 8 ft long. How tall is the telephone pole?

38. Betty wants to measure the height of a flagpole. She puts up a vertical post 8 ft tall and moves back in line with the flagpole and the post until her line of sight hits the tops of both post and pole (see the following figure). If the known distances are as shown in the figure, what is the height of the flagpole?

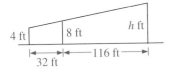

39. Ronny wants to find the height of a tree standing on level ground. He places a 5-ft stake vertically so that the sun throws a shadow of the tree and the stake as shown in the figure. He then measures the lengths of the shadows, with the results shown in the figure below. Why are the two triangles similar? What is the height of the tree?

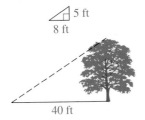

40. Gail wants to measure the distance *AB* across a small lake (see the figure in the next column). She walks 240 m away in a direction perpendicular to the line $\overleftrightarrow{AB}$ to a point *C* from which she can sight the point *B*. She then walks back 80 m along $\overleftrightarrow{CA}$ to a point *P*, and then walks in a direction perpendicular to $\overleftrightarrow{CA}$ to a point *Q* in line with *B* and *C*. She

finds that *PQ* = 60 m. With this information, what does Gail find for the distance *AB*?

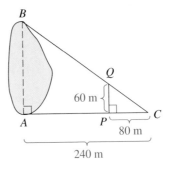

41. Sonny, the surveyor, needs to find the length of a tunnel to be bored through a small hilly area. He makes the marked angles equal and finds the measurements shown in the figure below. What length should Sonny find for the tunnel?

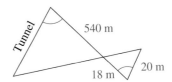

42. To find the distance between two points *P* and *Q* separated by an inaccessible area, Andy makes the measurements shown in the figure below. He returns to his office and draws △*ABC* with $m\angle A = m\angle O$, $m\angle B = m\angle P$, and *AB* = 10 cm. He then measures $\overleftrightarrow{BC}$ and finds it to be 18 cm. What length does Andy find for $\overleftrightarrow{PQ}$?

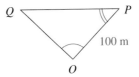

43. Use the congruent triangle theorems to show that △*ABC* is an isosceles triangle.

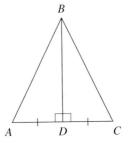

44. Show that $\triangle ABD \cong \triangle BCD$.

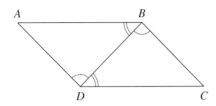

45. If $AD = BC$ and $AC = BD$, show that $\triangle ACD \cong \triangle BCD$.

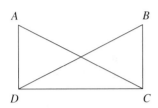

46. Old McBello has finally decided to sell his farm, which consists of two triangular plots of land (call them plot *ABC* and plot *DEF*). As an interested buyer, you inquire about the areas of each of the plots. He tells you that he has measured the lengths of the sides of each plot and that $AB = DF$, $BC = DE$, and $AC = EF$. He then declares that plot *ABC* has exactly the same area as plot *DEF*. Is Old McBello telling the truth, or is he trying to pull the wool over your eyes? (*Hint:* Draw the picture.)

C Angles of a Polygon

47. In the given figures, $AC = BC$. Find the measures of the three angles of $\triangle ABC$.

a.

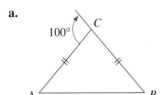

b.

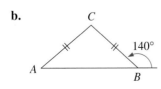

48. *ABCD* is a parallelogram. Find the measure of $\angle BAD$.

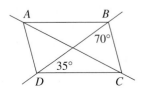

49. In the given figure, $\overleftrightarrow{AE}$ and $\overleftrightarrow{BD}$ are parallel and $\overleftrightarrow{CE}$ is perpendicular to $\overleftrightarrow{BD}$. Find $m\angle BCD$.

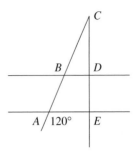

50. For the given figure, find $m\angle C$. What type of triangle is $\triangle ABC$?

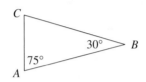

51. Find the sum of the measures of the angles of a regular polygon with 14 sides.

52. Find the sum of the measures of the angles of a regular polygon with 20 sides.

In problems 53–58, find the measure of one angle for each polygon.

53. A regular pentagon   **54.** A regular hexagon

55. A regular octagon   **56.** A regular nonagon

57. A regular decagon   **58.** A regular dodecagon

59. The information in Table 8.5, page 513, can be represented as the **family tree** or **hierarchy** shown below, where each name includes all the shapes below it to which it is connected.
  a. Use set notation to show the relationship between the set *P* of all polygons, the set *T* of all triangles, and so on.
  b. Draw a Venn diagram to show the relationship between *P*, *T*, *I*, and *E*.

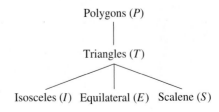

60. Show a family tree or hierarchy for the quadrilaterals in Table 8.5. Here are the first two steps.

Quadrilateral (*Q*)

Kite (*K*)   Trapezoid (*T*)

61. Draw a Venn diagram to represent the hierarchy shown in problem 60.

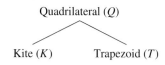

In Other Words

62. When a boat, car, or plane is manufactured, a model of it is constructed first. Explain what the model and the actual structure have in common.

63. Some distances are not easy to measure directly (the distance across a lake, as for Wally and in problem 40). Describe in your own words a situation in which indirect measurements using similar triangles are used.

64. Johnny claims that he has drawn two similar triangles. He says the sides of the first triangle are 4, 5, and 7 cm long and that the corresponding sides of the second triangle are 8, 10, and 12 cm long. Betty says that Johnny is making a mistake. Write out a complete explanation and justification for Betty's claim.

Using Your Knowledge

In problems 65–68, use the following table, which lists all possible triangle classifications. For example, IR stands for an isosceles, right triangle, SO stands for a scalene, obtuse triangle, and so on. The problems ask you to draw certain types of triangles. In any case in which you think no such triangle exists, write "impossible."

65. Draw an example of each of the following types:
 a. SA   **b.** IA   **c.** EA

	Acute	*Right*	*Obtuse*
Scalene	SA	SR	SO
Isosceles	IA	IR	IO
Equilateral	EA	ER	EO

66. Draw an example of each of the following types:
 a. SR   **b.** IR   **c.** ER

67. Draw an example of each of the following types:
 a. SO   **b.** IO   **c.** EO

68. Which of the triangles in the table are impossible?

Collaborative Learning

In Example 5 we saw that similar triangles and ratios can be used to find distances that cannot be measured directly because of their inaccessibility. Suppose you want to measure the height of the flagpole at your school.

1. Form several groups and find different procedures to measure the height of the flagpole.

2. The Greek mathematician Thales of Miletus was able to find the height of the Great Pyramid in Egypt by using a method called "shadow reckoning." Find out how this was done, and calculate the height of the Great Pyramid.

3. Discuss your findings with the other groups. Did they get the same answer?

8.3   Perimeter and Circumference

Hats, Rings, and Circumferences

Do you know your hat size? If you don't, you will be glad to know that it is supposed to be about the same as your ring size! (see *Rules of Thumb* by Tom Parker). How can that be? The ancient Greeks discovered that if they divided the circumference *C* of a circle by the length of its diameter *d*, they always obtained approximately the same number, π, regardless of the size of the circle. You can verify this by measuring the circumference of a soda can (the distance around) and dividing by its diameter. The answer should be about 3.14. The formula used to do this is the same one used by the Greeks, $C/d = \pi$. Now, back to rings and hats.

The average person's head is about 23 in. in circumference. If you divide 23 by π (3.14)—that is, $23/\pi$—the answer is the diameter $d \approx 7.32$. Unfortunately, hats are sized in increments of $\frac{1}{8}$ in. What size, to the nearest $\frac{1}{8}$ in., is the closest to 7.32 in.? You know that $7\frac{2}{8} = 7\frac{1}{4} = 7.25$ and can easily find $7\frac{3}{8} = 7.375$. Thus, $7\frac{3}{8}$ is the closest to the measure of $d \approx 7.32$ in.

Now you know that your hat size is the average diameter of your head in inches. How can this be the same as your ring size? Assume that your hat size is 7 in. in diameter. What would a 7 ring size measure? According to a jewelry handbook, the diameter d of a ring (in inches) is given by

$$d = 0.458 + 0.032s$$

where s is the ring size. When $s = 7$,

$$d = 0.458 + 0.032 \cdot 7 = 0.682 \text{ in.}$$

Since the diameter of the ring, and hence of the finger, is 0.682 in., the circumference of the finger should be $C = 3.14 \cdot 0.682 = 2.14$ in. Thus, if your hat size is 7, the circumference of your ring finger should be about 2.14 in., and of course, your ring size should be 7.

You can repeat this little experiment by measuring your own head, dividing by π, and obtaining your hat size. Now substitute your hat size for s in the formula and multiply the result by π. The answer you get should be the same as the circumference of your finger! What if it is not even close? Keep in mind that hat sizes and ring sizes should be the same for an *average* person. If you lose or gain weight, the circumference of your finger decreases or increases, respectively, but your head size does not.

In this section you study circles and the perimeters of circles, called **circumferences,** and the different applications you can solve using these ideas. ▶

In geometry, units of length are used to measure distances along lines and are consequently called **linear measures.** Table 8.6 shows the standard units of length in the U.S. customary system and in the metric system.

TABLE 8.6 Standard Units of Length

U.S. System	Metric System
Inch (in.) Foot (ft) = 12 in. Yard (yd) = 3 ft Mile (mi) = 1760 yd = 5280 ft	Millimeter (mm) = 0.001 m Centimeter (cm) = 0.01 m Meter (m) = base unit Kilometer (km) = 1000 m

Note that the metric system is a simple decimal system in contrast to the awkward conversions in the U.S. system. Although you will not be asked to convert from one of these systems to the other in this chapter, you should know that the

U.S. inch is legally defined to be exactly 2.54 cm long. Table 8.7 gives a few conversions to help you see the relationship between metric and U.S. lengths.

TABLE 8.7 Metric/U.S. Conversions

1 in. = 2.54 cm	1 cm = 0.39 in.
1 yd = 0.91 m	1 m = 1.09 yd
1 mi = 1.61 km	1 km = 0.62 mi

So a centimeter is about $\frac{4}{10}$ of an inch, a meter is just a bit longer than a yard, and a kilometer is about $\frac{2}{3}$ of a mile.

A. Polygonal Paths and Perimeters

Many of the applications of geometry involve finding the length of a polygonal path. For instance, fencing a field, laying tile around a rectangular pool, or finding the amount of baseboard needed for a room all involve measuring around polygons. The distance around a plane figure is generally called the **perimeter** of the figure. In the case of a polygon, the perimeter is just the sum of the lengths of the sides. Table 8.8 gives the formulas for the perimeters in terms of the sides for some of the polygons we have discussed. Note that perimeters are always expressed in linear measures.

How many kinds of polygons can you see in this view looking up through the center of a tower that supports power lines?

TABLE 8.8

Name	Geometric Shape	Perimeter
Triangle	s_1, h, s_2, b	$P = s_1 + s_2 + b$
Trapezoid	s_1, h, b_1, s_2, b_2	$P = s_1 + s_2 + b_1 + b_2$
Parallelogram	h, W, L	$P = 2L + 2W$
Rectangle	W, L	$P = 2L + 2W$
Square	S	$P = 4S$

EXAMPLE 1 ▶ Perimeter for the Mona Lisa

The *Mona Lisa* by Leonardo da Vinci, assessed at $100 million for insurance purposes, measures 30.5 by 20.9 in. Find the perimeter of this picture.

Solution

Because the picture is rectangular, its perimeter is given by

$$P = 2L + 2W$$

where $L = 30.5$ and $W = 20.9$. Thus,

$$P = 2(30.5) + 2(20.9)$$
$$= 102.8 \text{ in.}$$

The ideas we have been studying here can be combined with the algebra you know to solve certain kinds of problems. Here is an interesting problem.

EXAMPLE 2 ▶ Width of the Largest Poster

The largest recorded poster was a rectangular greeting card 166 ft long and with a perimeter of 458.50 ft. How wide was this poster?

Solution

The perimeter of a rectangle is $P = 2L + 2W$, and we know that $L = 166$ and $P = 458.50$. Thus, we can write

$$2(166) + 2W = 458.50$$
$$332 + 2W = 458.50$$
$$2W = 126.50$$
$$W = 63.25$$

So the poster was 63.25 ft wide.

EXAMPLE 3 ▶ Finding the Dimensions of Emily's Rectangle

John and Emily Gardener want to fence in a small, rectangular plot for Emily's kitchen garden. John has 20 m of fencing and decides that the length of the plot should be $1\frac{1}{2}$ times its width. What will be the dimensions of Emily's garden plot?

Solution

We let x meters be the width of the plot. Then the length must be $\frac{3}{2}x$ m. We know the perimeter is to be 20 m, so

$$2L + 2W = P$$
$$2\left(\tfrac{3}{2}x\right) + 2x = 20$$
$$3x + 2x = 20$$
$$5x = 20$$
$$x = 4 \qquad \text{and} \qquad \tfrac{3}{2}x = 6$$

Emily's plot will be 6 m long and 4 m wide.

B. Circles

You are probably familiar with the geometric figure called a *circle*. The circle has been of great interest ever since prehistoric times; it has always been used for many decorative and practical purposes. Here is a modern definition.

> **Definition of a Circle**
>
> A **circle** is the set of all coplanar (on the same plane) points at a given fixed distance (the **radius**) from a given point (the **center**).

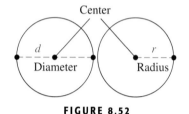

Center

d Diameter

r Radius

FIGURE 8.52

Figure 8.52 shows two circles with a radius and diameter indicated. Note that the **diameter** consists of two collinear radii, so that $d = 2r$; that is, the length of the diameter is twice the length of the radius, or the radius is half of the diameter. The perimeter of a circle is known as the **circumference.**

> **Circumference of a Circle**
>
> The **circumference** C of a circle of diameter d and radius r is given by $C = \pi d$ or $C = 2\pi r$.

The irrational number $\pi \approx 3.14159$ was discussed in an earlier chapter. Unless otherwise noted, you may use the approximate value 3.14 for the problems in this book.

EXAMPLE 4 ▶ Fencing Your Pool for Safety

Andy has a circular swimming pool with a diameter of 25 ft. To keep his little boy from falling into the water, Andy wants to put a low wire fence around the circumference of the pool. How much fencing does he need?

Solution
We use $C = \pi d$ and get

$$C \approx (3.14)(25) = 78.5 \text{ ft}$$

Since 3.14 is a little less than π, Andy should play it safe and get 79 ft of fencing. ∎

EXAMPLE 5 ▶ Diameters of Hamburgers

According to the *Guinness Book of World Records,* one of the largest beef hamburgers had a circumference of 27.50 ft. Find the diameter of this hamburger to the nearest hundredth of a foot. Use $\pi \approx 3.14$.

Solution
Since the circumference of a circle is $C = \pi d$, we have

$$27.50 \approx 3.14d$$

Dividing by 3.14, we find

$$d = 8.76 \qquad \text{to the nearest hundredth}$$

Thus, the diameter of this mammoth beef hamburger was about 8.76 ft. (It also weighed 2859 lb.) (*Note:* If we used a more accurate value of π, we would find the diameter to be 8.75 ft, to the nearest hundredth of a foot.) ■

Before we leave this section, let us shed some light on diameters and circumferences. To start, do you know how light bulbs are measured? An obvious answer is by wattage (how many watts). But there is more. Bulb shapes are identified by the letters A (typical house bulb), B (candelabra base Christmas tree bulb), and C (miniature night-light bulb). The bulb size is then given by a letter for shape followed by the bulb *maximum diameter* (in eighths of an inch). Thus, a T-8 bulb is a tubular bulb $\frac{8}{8} = 1$ in. in diameter. (Source: John Lord, *Sizes*: Harper Perennial, 1995.)

EXAMPLE 6 ▶ Circumference and Diameter of a Bulb

(a) What is the circumference and diameter of a 75-watt A-19 bulb?

(b) A bulb burns out at home. You measure its circumference (it is hard to measure the diameter!), and it is 8.25 in. Which type of bulb do you need?

Solution

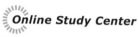
Online Study Center

To further explore circles and their applications to graphs, wheels, and rainbows, access link 8.3.1 on this textbook's Online Study Center.

(a) The diameter is $\frac{19}{8} = 2.375$ in. and $C = \pi d = \pi(2.375)$, or about 7.46 in.

(b) Since $C = \pi d$, we have $8.25 = \pi d$ and $d = 8.25/\pi \approx 2.63$. The question is, how many eighths is 0.63? We know that $\frac{1}{8} = 0.125$, so $\frac{5}{8} = 0.625$. This means that $d \approx 2.63 \approx 2\frac{5}{8} = \frac{21}{8}$. But you remember that bulbs are measured in eighths of an inch, so we need an A-21 bulb (the wattage is up to you!). ■

C. Applications

EXAMPLE 7 ▶ Distance Traveled by Venus

The planet Venus revolves around the sun in a nearly circular orbit whose diameter is about 100 million km. Find the distance traveled by Venus in one revolution around the Sun. Use 3.14 for π.

Solution

The distance traveled is the circumference C.

$$C = \pi d$$
$$= (3.14)(100) \text{ million km}$$
$$= 314 \text{ million km}$$

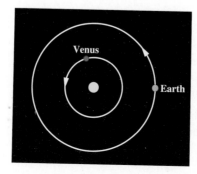
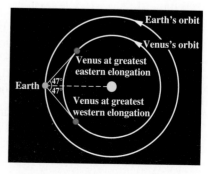

■

EXERCISES 8.3

A Polygonal Paths and Perimeters

In problems 1–8, find the perimeter of each polygon.

1.
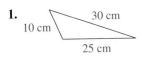

2.
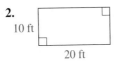

3.
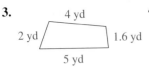

4.

5.
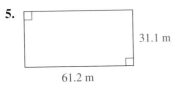

6.
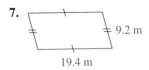

7.
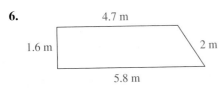

8.
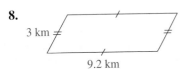

9. If one side of a regular pentagon is 6 cm long, find the perimeter of the pentagon.

10. If one side of an octagonal stop sign is 6 in. long, what is the perimeter of the stop sign?

11. The largest rectangular omelet ever cooked was 30 ft long and had an 80-ft perimeter. How wide was it?

12. Do you have a large swimming pool? If you were to walk around the largest pool in the world, in Casablanca, Morocco, you would walk more than 1 km. To be exact, you would walk 1110 m. If the pool is 480 m long, how wide is the pool?

13. A baseball diamond is actually a square. A batter who hits a home run must run 360 ft around the bases. What is the distance to first base?

14. The playing surface of a football field is 120 yd long. A player jogging around the perimeter of this surface covers 346 yd. How wide is the playing surface?

15. Have you seen the largest scientific building in the world? It is in Cape Canaveral, Florida. If you were to walk around the perimeter of this building, you would cover 2468 ft. If this rectangular building is 198 ft longer than it is wide, what are its dimensions?

16. The largest regular hexagon that can be cut from a circular sheet of cardboard has each side equal to the radius of the circle. If you cut such a hexagon from a sheet of radius 5 in., how much shorter is the perimeter of the hexagon than the circumference of the circle?

B Circles

In problems 17–24, find the circumference. First, give the answer in terms of π; then calculate the approximate answer using $\pi \approx 3.14$.

17.
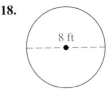

18.
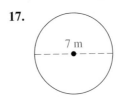

19.
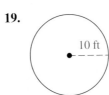

20.
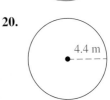

21.
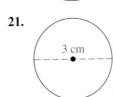

22.
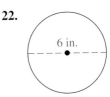

23.
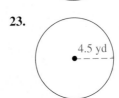

24.
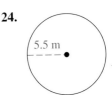

In the following problems, first give an exact answer in terms of π and then calculate the required approximate answer using $\pi \approx 3.14$:

25. The diameter of a bicycle tire is 61 cm. Through what distance does the bicycle go when the wheel makes one complete turn? Give your answer to the nearest centimeter.

26. The lid on a garbage can has a diameter of 17 in. Find the length of the circumference of this lid. Give your answer to the nearest tenth of an inch.

27. The minute hand of a clock is 8 cm long. How far does the tip of the hand move in 1 hr? Give your answer to the nearest tenth of a centimeter.

28. The diameter of a CD is 4 in. How far does a point on the rim of this CD travel when the CD makes three-fourths of a complete revolution?

29. Shirley wants to put a low, decorative wooden border around her circular flower bed. If the radius of the bed is 4 ft, how long must the border be? Give your answer to the nearest tenth of a foot.

30. Have you seen a long-playing record? A long-playing record has a radius of 6 in. How far does a point on the rim move when the record goes around once? Give your answer to the nearest tenth of an inch.

31. The circumference of a circle is $15\,\pi$ cm. Find the diameter and the radius of this circle.

32. A thin metal rod 8 ft long is to be bent into a circular hoop. Find the radius of this hoop to the nearest tenth of an inch.

33. To make a wedding band for a man who wears a size 12 ring, a strip of gold 7 cm long is needed. Find the diameter of this ring.

34. One of the largest pizzas ever made had a circumference of 251.2 ft. What was its diameter?

35. You already know from problem 30 that a long-playing record has a 6-in. radius. However, do you know the circumference of the smallest functional record? It is an amazing $4\frac{1}{8}$ in.! Find the diameter of this tiny record to the nearest hundredth of an inch.

36. The outside diameter of a motorcycle tire is 26 in. How many revolutions does the tire make in one *mile?* Round the answer to the nearest whole number. (*Hint:* 1 mi = 5280 ft.)

37. A motorcycle tire is guaranteed for 20,000 mi. Use the information from problem 36 to find out for how many revolutions the tire is guaranteed.

38. The largest doughnut ever made weighed 2099 lb and had a 70-ft circumference. To the nearest whole number, what is the diameter of the smallest pan in which the doughnut could fit?

39. A circular pool is enclosed by a square deck 20 yd on each side. If the distance between the edge of the deck and the pool is 2 yd as shown, what is the distance around the pool?

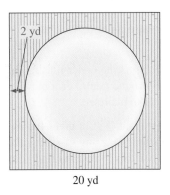

20 yd

40. It takes about 15 min to walk around the Colosseum in Rome. At this rate, how long would it take you to walk straight through along a diameter? (Assume the Colosseum is perfectly round and give the answer to the nearest minute.)

C Applications

If you are traveling by car, the distance between two points is shown on travel maps.

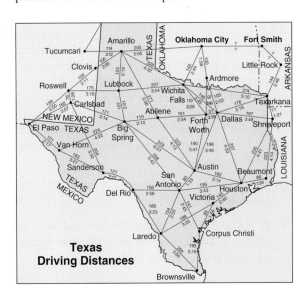

Texas Driving Distances

41. Find the distance traveled when going from Laredo to San Antonio to Del Rio and back to Laredo.

42. Find the distance traveled when going from Dallas to Houston to Shreveport and back to Dallas.

43. Find the distance traveled when starting at Austin, driving to Ft. Worth, then to Abilene, and then back to Austin.

44. The distance from Tucumcari (New Mexico) to Fort Smith (Arkansas) is about 555 mi. How far is it from Oklahoma City to Fort Smith?

 In Other Words

45. Which do you think is the better illustration of a circle: a perfectly round penny or a bicycle tire? (See the definition of a circle.)

46. The formula $C = 1.44 + 0.1s$ gives the circumference C (in inches) of a ring of size s. Explain how you would use this formula to find your ring size.

47. If two tires, one new and one worn, are installed on a car, which one will turn more times per mile? Explain your answer.

48. If your calculator has a π key, use it to calculate the answer to problem 30. In your own words, explain the discrepancy between the answer you obtained following the instructions for problem 30 and the new answer on your calculator.

 Using Your Knowledge

Suppose that circle A has a 3-in. circumference and circle B has a 1-in. circumference, as shown in the left diagram in the figure in the next column. If circle B rolls around the perimeter of circle A without slipping and returns to its original position, how many revolutions will it have made? (*Hint:* Look at the diagram below! Using your knowledge of circumferences, can you prove your case?)

49. In the diagram on the left, what is the length of the arc shown in red?

50. If circle B makes one revolution and ends as shown in the diagram on the right, how far has point b traveled relative to circle B?

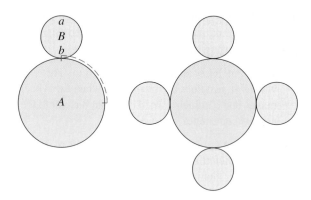

51. How far does point b travel relative to circle B when point a makes one complete revolution relative to circle B?

52. On the basis of your previous answers, _____ in. = 1 revolution of circle B. Thus, how many revolutions does circle B make around circle A, whose circumference is 3 in., to return to its original position?

 Calculator Corner

Many calculators come equipped with a $\boxed{\pi}$ key (to access it, you may have to press $\boxed{\text{2nd}}$ $\boxed{\pi}$) that gives a nine-decimal-place approximation for π. If you use this approximation to solve Example 5, the answer obtained when dividing 27.50 by π comes out to be 8.75352187, or 8.75 to the nearest hundredth of a foot, as mentioned in the text.

1. Rework problems 25, 26, 30, 32, and 34 of Exercises 8.3 on your calculator.

 Collaborative Learning

At the beginning of this section we claimed that your hat size and your ring size are identical. By the way, there is a little flaw here. This is true only of men's hat sizes. Women's hat sizes are measured differently and more logically! Women's hat sizes are simply the circumference of the inner band of the hat. Nevertheless, we would like to see if there is a relationship between men's hat sizes—for men and women—and ring sizes. Form several groups of males and females.

1. Measure the circumference of the head $C(H)$ of the participants.

2. Measure the circumference of their ring finger $C(f)$.

3. Complete the table to the right, where $C(H)$ is the circumference of the head, $C(f)$ the circumference of the finger, d the diameter of the finger, H the men's hat size, and s the ring size.

$C(H)$	$\dfrac{C(H)}{\pi} = H$	$C(f)$	$\dfrac{C(f)}{\pi} = d$	$\dfrac{d - 0.458}{0.032} = s$

4. Is there a relationship between H and s?

5. Is there a relationship between hat size and ring size for women? If so, what is it?

8.4 Area Measure and the Pythagorean Theorem

If you know areas, you can find the best pizza deal.

Pizza Area

What is your telephone area code? What areas were severely affected by storms last year? The idea of **area** occurs in many contexts, but in geometry, **area** is measured by defining a **unit region.** For example, a square whose side is **1 unit in length** has an area of **1 square unit.** To measure larger areas, you measure how many **unit squares** are contained in the given region. This means that areas are always measured in square units. In this section you encounter formulas that give the area of several polygons as well as the area of a circle.

Now suppose you want to find the area covered by the giant 111-ft-diameter pizza made by Pizza Hut in Singapore on June 9, 1990. You will learn later that the formula for the area A of a circle of radius r is $A = \pi r^2$. Since the diameter of the pizza is 111 ft, its radius is 55.5 ft, and its area is $A = \pi \cdot 55.5^2 = 3080.25\pi$, or about 9677 square ft, usually written as 9677 ft^2.

In the same way, you can use areas to do some comparison shopping for pizzas. Suppose Pizza Hut sells medium pizzas (12 in. in diameter) with one topping for \$7.99. If you assume that the ingredients are of equal quality and the pizzas of equal thickness, is it better to buy one of these pizzas or get two 10-in. pizzas for \$4.99 each? You can compare by finding out how much you pay per square inch. The area of the \$7.99 pizza is $A = \pi(6 \text{ in.})^2 = 36\pi$, or about 113 in.^{2}. Thus, the price per square inch is $\frac{7.99}{113} \approx 0.07$, that is, about 7¢/in.^{2}. On the other hand, the area of one 10-in. pizza is $A = \pi(5)^2 = 25\pi$, or about 78.5 in.^{2}. Thus, the area of two pizzas is about 157 in.^{2}, and the price per square inch is $\frac{2 \cdot 4.99}{157}$, or about 6.36¢/in.^{2}. In this case, smaller is better; you pay less per square inch if you buy the two small pizzas.

By the way, at 7¢/in.^{2}, how much would the 111-ft-diameter pizza be worth? Close to \$100,000! Check this out by keeping in mind that 1 square foot (1 ft^2) is exactly 144 in.^{2}. Here are two more questions for you. How many 12-in. pizzas are there in the Singapore pizza, and if you assume that a 12-in. pizza serves eight people, how many people can be served with the Singapore pizza? Think of areas and use the techniques you used here to find the answers. ▶

Now consider the question of how much glass might be used in the construction of a skyscraper. If we think a moment, we realize that the meaning of the question is not clear. Are we looking for the number of pieces of glass, the number of pounds of glass, or what? Let us be more specific and say that we want to know the total *area* covered by the glass in this building. In order to answer questions like this, we must have a good understanding of what we mean by *area*.

FIGURE 8.53
The unit of area measure.

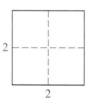

FIGURE 8.54
Area = 4 square units.

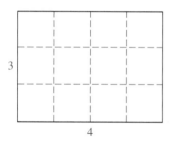

FIGURE 8.55
Area = 12 square units.

FIGURE 8.56
Area = *bh*.

A. Area of a Polygon

We start by choosing the **unit region** to be that of a **square,** each of whose sides is **1 unit in length,** and we say that this region has an **area of 1 square unit** (see Figure 8.53). The side of the unit square may be 1 in., 1 ft, 1 mi, 1 cm, 1 m, and so on. The corresponding units of area are the square inch (in.^{2}), the square foot (ft^2), the square mile (mi^2), the square centimeter (cm^2), the square meter (m^2), and so on.

Some of the commonly used units of area are given in Table 8.9. Note that **area** is always measured in **square** units.

TABLE 8.9 Standard Units of Area

U.S. System	Metric System
Square inch (in.^{2})	Square millimeter (mm^2)
Square foot (ft^2)	Square centimeter (cm^2)
Square yard (yd^2)	Square meter (m^2)
Square mile (mi^2)	Square kilometer (km^2)

In everyday language, the area measure of a plane region is the *number of unit regions contained in the given region.* Thus, suppose that we have a square with side 2 units long. Then, as shown in Figure 8.54, we can draw lines joining the midpoints of the opposite sides and dividing the square into $2 \times 2 = 4$ unit squares. We say that the area of the square is 4 square units. Similarly, for a rectangle that is 3 units by 4 units, we can draw lines parallel to the sides that divide the rectangle into $3 \times 4 = 12$ unit squares (see Figure 8.55). We say that the area of the rectangle is 12 square units.

The previous illustrations show how the area measure can be found if the sides of the rectangle are whole numbers. In order to avoid a complicated mathematical argument, we define the area of a rectangle of sides *b* units long and *h* units long to be *bh* square units (see Figure 8.56).

Area of a Rectangle

The area of a rectangle with base length *b* and height *h* is given by $A = bh$.

Notice that we are following the usual custom of saying "area of a rectangle" to mean "area of a rectangular region." Similarly, we shall say "area of a polygon" to mean "area of a polygonal region."

Knowing the area of a rectangle, we can find the area of a parallelogram. The idea is to construct a rectangle with the same area as the parallelogram. As Figure 8.57 shows, we simply cut the right triangle *ADE* from one end of the parallelogram and attach it to the other end. This forms a rectangle *CDEF* with the same area as the parallelogram. Since the base and height are the same for both figures, the area of a parallelogram of base *b* and height *h* is *bh*.

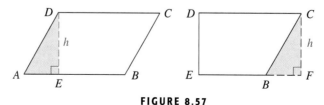

FIGURE 8.57

> **Area of a Parallelogram**
>
> The area A of a parallelogram with base b and height h is given by $A = bh$.

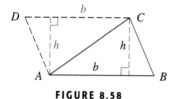

FIGURE 8.58

Be sure to note that h is the perpendicular height, not the length of a side of the parallelogram.

We can now find the area of any triangle. Suppose that triangle ABC in Figure 8.58 is given. We draw a line through C parallel to $\overleftrightarrow{AB}$ and a line through A parallel to $\overleftrightarrow{BC}$. These lines meet at a point D, and the quadrilateral $ABCD$ is a parallelogram. Clearly, the segment $\overline{AC}$ is a diagonal of the parallelogram and divides the parallelogram into two equal pieces. Because the area of parallelogram $ABCD$ is bh, the area of the triangle ABC is $\frac{1}{2}bh$. Thus, we have a formula for the area of any triangle with base b and height h.

Note: If the angle B in Figure 8.58 is a right angle, the parallelogram obtained is a rectangle. This does not change the formula for the area of a triangle.

EXAMPLE 1 ▸ Finding the Area of a Triangle

Find the area of the triangle shown in Figure 8.59.

> **Area of a Triangle**
>
> The area A of a triangle with base b and height h is given by $A = \frac{1}{2}bh$.

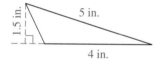

FIGURE 8.59

Solution

This triangle has a base of 4 in. and a height of 1.5 in. Thus,

$$A = \tfrac{1}{2}(4)(1.5) = 3 \text{ in.}^2$$ ∎

EXAMPLE 2 ▸ Finding the Area of a Subdivided Region

Find the area of the region given in Figure 8.60.

> **Area of Polygonal Regions**
>
> The formulas we have developed can be used for finding the areas of many polygonal regions. This is done by subdividing these regions into nonoverlapping rectangular and/or triangular regions, finding the areas of these subdivisions, and adding the results. This procedure is illustrated in the next example.

FIGURE 8.60

Solution

We subdivide the region as shown in Figure 8.60. The area of the lower rectangle is $6 \times 3 = 18$ m^2. The area of the shaded triangle is $\frac{1}{2}(2 \times 3) = 3$ m^2. The upper two triangles each have area 3 m^2, and the upper rectangle has area $2 \times 3 = 6$ m^2. Thus, the required area is

$$18 - 3 + 3 + 3 + 6 = 27 \text{ m}^2$$ ∎

Thus far we have been concerned entirely with the areas of polygonal regions. How about the circle? Can we find the area of a circle by using one of the preceding formulas? Interestingly enough, the answer is yes. The required area can be found by using the formula for the area of a rectangle. Here is how we go about it. Look at Figure 8.61. Cut the lower half of the circular region into small equal slices called **sectors,** and arrange them as shown in Figure 8.61. Then cut the remaining half of the circle into the same number of slices, and arrange them along with the others as shown in Figure 8.62. The result is approximately a parallelogram whose longer side is of length πr (half the circumference of the circle) and whose shorter side is r (the radius of the circle).

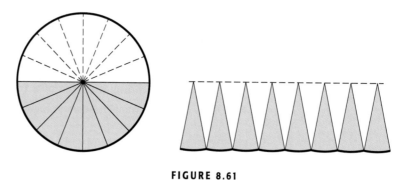

FIGURE 8.61

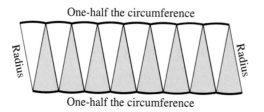

FIGURE 8.62

The more pieces we cut the circle into, the more accurate this approximation becomes. We should also observe that the more pieces we use, the more nearly the parallelogram becomes a rectangle of length πr and height r. Mathematicians have proved that the area of the circle is the same as the area of this rectangle, that is, $(\pi r)(r)$, or πr^2. Thus, we arrive at the formula for the area of a circle of radius r.

Area of a Circle

The area A of a circle with radius r is given by

$$A = \pi r^2$$

EXAMPLE 3 ▶ Area for the Arecibo Radiotelescope

The circular dish for the Arecibo radiotelescope has a radius of 500 ft. What area does the dish cover? (Use $\pi \approx 3.14$.)

Solution
Here, $r = 500$, so

$$A = \pi(500)^2$$
$$= \pi(250,000) \approx 785,000 \text{ ft}^2$$

Your knowledge of algebra can be used in conjunction with the geometric formulas you have just studied. This is illustrated in the next example.

EXAMPLE 4 ▶ Finding the Area of a Subdivided Region

The Fermi National Accelerator Laboratory has the atom smasher shown in the photograph. The smasher covers an area of 1.1304 mi^2. What is the diameter of this atom smasher? Use $\pi \approx 3.14$, and give the answer to the nearest hundredth of a mile.

The large circle is the main accelerator at the Fermi National Accelerator Laboratory at Batavia, Illinois. A proton beam travels around this circle 50,000 times a second, producing highly energized particles used in nuclear research.

Solution
The formula for the area of a circle of radius r is

$$A = \pi r^2$$

Here, $A = 1.1304$ and $\pi \approx 3.14$. Thus,

$$1.1304 \approx 3.14 r^2$$
$$r^2 \approx \frac{1.1304}{3.14} = 0.36$$
$$r \approx 0.60 \text{ mi}$$

So the diameter is about 1.20 mi. If your calculator has a square root key, you can do this calculation by keying in $\boxed{1.1304}$ $\boxed{\div}$ $\boxed{3.14}$ $\boxed{=}$ $\boxed{\sqrt{x}}$.

Here is the summary of some of the area formulas for polygons.

TABLE 8.10 Area Formulas for Polygons

Name	Shape	Area
Rectangle		$A = bh$
Parallelogram		$A = bh$
Triangle		$A = \frac{1}{2}bh$
Trapezoid		$A = \frac{1}{2}(b_1 + b_2)h$

B. The Pythagorean Theorem

The solution of Example 4 required taking the square root of both sides of an equation. The same technique is used in many problems involving the sides of a right triangle. One of the most famous and important theorems of all time is the **Pythagorean theorem,** which says **the square of the hypotenuse (the longest side) of a right triangle is equal to the sum of the squares of the other two sides.** Figure 8.63 illustrates the theorem for a 3-4-5 right triangle, a triangle with sides of length 3 and 4 units and a 5-unit hypotenuse.

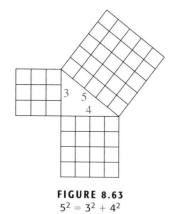

FIGURE 8.63
$5^2 = 3^2 + 4^2$

THEOREM 8.1 Pythagorean Theorem

If the two legs of a right triangle have lengths a and b and the hypotenuse has length c, then

$$c^2 = a^2 + b^2 \qquad \text{or equivalently} \qquad a^2 + b^2 = c^2$$

It is important to note that the converse of the Pythagorean theorem is also true. (The proof, which can be constructed with a little trigonometry, is omitted.) We can include both theorems in the statement that follows:

Generalized Pythagorean Theorem

A triangle is a right triangle if and only if the square of one of the sides is the sum of the squares of the other two sides.

Online Study Center

To further explore geometric proofs of the Pythagorean theorem, access links 8.4.1 and 8.4.2 on this textbook's Online Study Center.

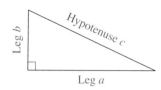

FIGURE 8.64

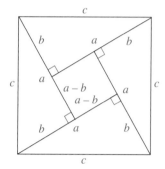

FIGURE 8.65

The perpendicular sides of a right triangle are often called **legs.** In Figure 8.64, the legs are labeled a and b, and the hypotenuse is labeled c.

The early Egyptians, and even the Babylonians, knew some special cases of this result; in particular, they knew the result illustrated in Figure 8.63. However, the ancient Greeks seem to have been the first to prove the general theorem. There are many different proofs of the theorem, some of which depend on areas in a simple fashion. Here is one of these proofs.

The area of the large square in Figure 8.65 is c^2. The area of the small square is $(a - b)^2 = a^2 - 2ab + b^2$, and the area of each of the four right triangles is $\frac{1}{2}ab$. Thus,

$$c^2 = a^2 - 2ab + b^2 + 4(\tfrac{1}{2}ab)$$
$$= a^2 - 2ab + b^2 + 2ab$$
$$= a^2 + b^2$$

EXAMPLE 5 ▶ Finding the Hypotenuse Given Two Sides

Find the length of the hypotenuse of a right triangle whose legs are 5 and 12 units long.

Solution

We use the Pythagorean theorem to get

$$c^2 = a^2 + b^2$$
$$= 5^2 + 12^2 = 25 + 144 = 169$$

Therefore, $c = \sqrt{169} = 13$ units. ∎

C. Applications

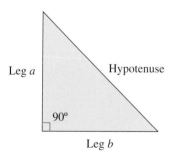

FIGURE 8.66

EXAMPLE 6 ▶ Finding Monitor Dimensions with Pythagoras

The screen in a rectangular computer monitor shown in Figure 8.66 is 3 in. longer than it is high and its diagonal is 6 in. longer than its height. What are the dimensions of the monitor?

Solution

We are asked to find the dimensions, which involves finding the length, height, and diagonal of the screen.

Since all measurements are given in terms of the height, let H be the height.

It's a good idea to start with a picture so that we can see the relationship among the measurements. Note that the length is $3 + H$ (3 in. longer than the height), and the diagonal is $6 + H$ (6 in. longer than the height). We enter this information in Figure 8.67.

According to the Pythagorean theorem,

$$H^2 + (3 + H)^2 = (6 + H)^2$$

$$H^2 + 9 + 6H + H^2 = 36 + 12H + H^2 \qquad \text{Expand } (3 + H)^2 \text{ and } (6 + H)^2.$$

$$H^2 + 9 + 6H + H^2 - 36 - 12H - H^2 = 0 \qquad \text{Subtract } 36 + 12H + H^2.$$

$$H^2 - 6H - 27 = 0 \qquad \text{Simplify.}$$

$$(H - 9)(H + 3) = 0 \qquad \text{Factor } (-9 \cdot 3 = -27; \ -9 + 3 = -6).$$

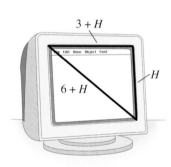

FIGURE 8.67

$H - 9 = 0$	or	$H + 3 = 0$	Use the principle of zero product.
$H = 9$		$H = -3$	Solve each equation.

Since H is the height, we discard -3, so the height of the monitor is 9 in., the length is 3 more inches, or 12 in., and the diagonal is 6 more inches than the height, or $9 + 6 = 15$ in.

Looking at the diagram and using the Pythagorean theorem, we see that $9^2 + 12^2$ must be 15^2; that is,

$$9^2 + 12^2 = 15^2$$

Since $81 + 144 = 225$ is a true statement, our dimensions are correct. ∎

By the way, television sets claiming to be 25 in. or 27 in. (meaning that the length of the screen measured diagonally is 25 or 27 in.) hardly ever measure 25 or 27 in. This is easy to confirm using the Pythagorean theorem. For example, a 27-in. Panasonic has a screen that is 16 in. high and 21 in. long. Can it really be 27 in. diagonally? If this were the case, $16^2 + 21^2$ would equal 27^2. Is this true? Measure a couple of TV or computer screens and see whether the manufacturers' claims are true!

EXERCISES 8.4

Ⓐ Area of a Polygon

In problems 1–12, find the area of each region. Use 3.14 for π.

1.

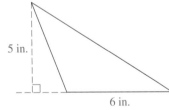

2.

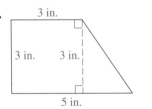

3.

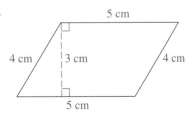

4.

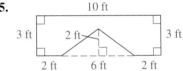

5.

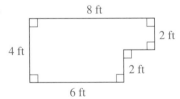

6.

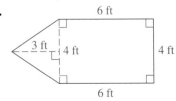

7.

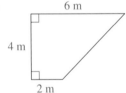

8.

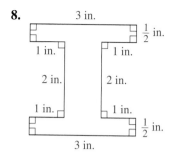

9.

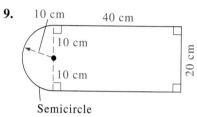

10.

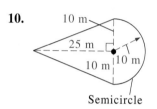

11.

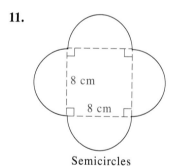

Semicircles

12.

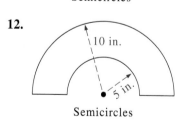

Semicircles

In problems 13–16, find the shaded area. Use 3.14 for π.

13.

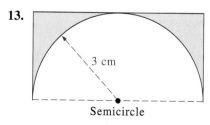

Semicircle

14.

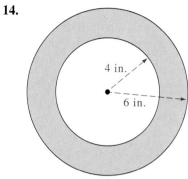

15.

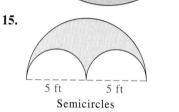

5 ft   5 ft
Semicircles

16.

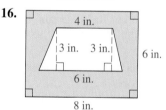

B **The Pythagorean Theorem**

17. A rectangle has a diagonal 17 ft long and a length of 15 ft. How wide is it?

18. A rectangle is 24 cm long and 7 cm wide. How long is its diagonal?

19. The base of an isosceles triangle is 16 in. long and the equal sides are each 17 in. long. Find the height h of the triangle, that is, the perpendicular distance from the base to the opposite vertex.

20. The two diagonals of a rhombus are 4 and 8 cm long, respectively. Find the length s of the sides of the rhombus. Diagonals are perpendicular.

21. A concrete wall is to be constructed on a triangular lot as shown in the diagram. If the estimated cost of the wall is $12 per linear foot, what is the estimated total cost of the wall?

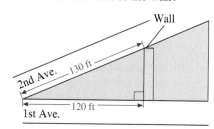

22. A telephone pole is supported by two cables attached to points on the pole 15 ft above ground and to points on the level ground 8 ft from the foot of the pole. What is the total length of the two cables?

23. The base of a parallelogram is 12 ft long, the height is 8 ft, and the shorter diagonal is 10 ft. How long are the other sides of the parallelogram?

24. The base of a parallelogram is 10 cm long, the height is 5 cm, and the longer diagonal is 13 cm long. How long is the shorter diagonal?

C Applications

25. The playing surface of a football field is 120 yd long and $53\frac{1}{3}$ yd wide. How many square yards of artificial turf are needed to cover this surface?

26. The floors of three rooms in a certain house measure 9 by 10 ft, 12 by 12 ft, and 15 by 15 ft, respectively.
   a. How many square yards of carpet are needed to cover these three floors?
   b. If the price of the carpet is $14/yd^2, how much would it cost to cover these floors?

27. A rectangular room is to have 288 ft^2 of floor space. If the room is 16 ft long, how wide must it be?

28. The Louisiana Superdome covers an area of 363,000 ft^2. Find the diameter of this round arena to the nearest foot. Use $\pi \approx 3.14$.

29. The largest cinema screen in the world is in the Pictorium Theater in Santa Clara, California; it covers 6720 ft^2. If this rectangular screen is 70 ft tall, how wide is it?

30. The area of the biggest pizza was about 5024 ft^2. What was its diameter?

31. What is the area of a circular region whose diameter is 8 cm? (Leave your answer in terms of π.)

32. Find the height of a triangle of area 70 in.^{2} if its base is 20 in. long.

33. Glass for picture frames costs $3/ft^2. If the cost of the glass for a rectangular frame is $4.50 and the frame is $1\frac{1}{2}$ times as long as it is wide, find its dimensions in inches.

34. Diazinon, a toxic chemical used for insect control in grass, is banned for commercial use after 2004. Each ounce of this chemical, diluted in 3 gal of water, covers 125 ft^2. The transportation department used 32 oz of Diazinon to spray the grass in the median strip of a highway. If the strip was 16 ft wide, how long was it?

35. John Carpenter has a tabletop that is 4 ft wide by 5 ft long. He wants to cut down the length and the width both by the same amount so as to decrease the area by $4\frac{1}{4}$ ft^2. What would be the new dimensions?

36. In problem 35, suppose that John wants to decrease the 4-ft side and increase the 5-ft side by the same amount so as to decrease the area by $\frac{3}{4}$ ft^2. What would be the new dimensions?

37. Show that the area of an equilateral (all sides equal) triangle of side length s is $s^2\sqrt{3}/4$. (*Hint:* A perpendicular from one vertex to the opposite side bisects that side.)

38. State what happens to the area of a triangle if
   a. the length of its base is doubled.
   b. both its base and height are doubled in length.

39. State what happens to the area of a rectangle if both its dimensions are
   a. doubled.
   b. tripled.
   c. multiplied by a constant k.

40. Use the result of problem 37 to find the area of a regular hexagon whose side is 4 in. long.

41. The circumference of a circle and the perimeter of a square are both 20 cm long. Which has the greater area, and by approximately how much? Use $\pi = 3.14$.

42. In the figure below, the two circles have their centers on the diagonal $\overset{\leftrightarrow}{AC}$ of the square $ABCD$. The circles just touch each other and the sides of the square, as shown in the figure. If the side of the square is of length s, find the total area of the two circles in terms of s.

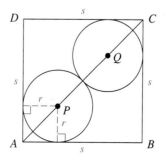

43. A regular octagon is to be formed by cutting equal isosceles right triangles from the four corners of a square. If the side of the square is of length a and the equal sides of the isosceles triangles are of length s, show that

$$s = \frac{2 - \sqrt{2}}{2}a$$

 In Other Words

44. Which of the following triangles are right triangles? Explain your reasoning.
  a. Sides 3, 5, 6   **b.** Sides 5, 12, 13
  c. Sides 7, 24, 25   **d.** Sides 9, 10, 15

45. What would be the most appropriate U.S. units (in.^{2}, yd^2, mi^2) and metric units (cm^2, m^2, and km^2) for measuring the area of
  a. a nickel.   **b.** your state.
  c. your classroom.   **d.** a house.

46. Explain in your own words the difference between a 2-in. square and 2 in.^{2}.

 Using Your Knowledge

You can use what you learned in this section to help you solve some commonly occurring problems. Do the following problems to see how:

47. A gallon of Lucite wall paint costs $14 and covers 450 ft^2. Three rooms in a house measure 10 by 12 ft, 14 by 15 ft, and 12 by 12 ft, and the ceiling is 8 ft high.
  a. How many gallons of paint are needed to cover the walls of these rooms if you make no allowance for doors and windows?
  b. What will be the cost of the paint? (The paint is sold by the gallon only.)

48. The figure below shows the front of a house. House paint costs $17/gal and each gallon covers 400 ft^2.

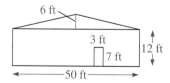

  a. What is the minimum number of gallons of paint needed to cover the front of the house? (The paint is sold by the gallon only.)
  b. How much will the paint for the front of the house cost?

49. A house and lot are shown in the figure that follows. The entire lot, except for the buildings and the drive, is lawn. A bag of lawn fertilizer costs $4 and covers 1200 ft^2 of grass.
  a. What is the minimum number of bags of fertilizer needed for this lawn?
  b. What will be the cost of the fertilizer?

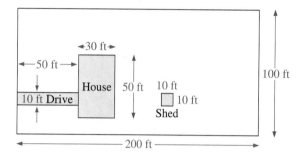

50. A small pizza (11-in. diameter) costs $5 and a large pizza (15-in. diameter) costs $8. Use $\pi \approx 3.14$ and find to the nearest square inch
  a. the area of the small pizza.
  b. the area of the large pizza.
  c. Which is the better deal, two small pizzas or one large pizza?

51. A frozen apple pie of 8-in. diameter sells for $1.25. The 10-in.-diameter size sells for $1.85.
  a. What is the unit price (price per square inch), to the nearest hundredth of a cent, of the 8-in. pie?
  b. What is the unit price, to the nearest hundredth of a cent, of the 10-in. pie?
  c. Which pie gives you the most for your money?

In Problems 52–61, use 3.14 for π, and give your answer to two decimal places.

52. *Area of the largest crop circle.* The largest recorded crop circle appeared in Wiltshire, England, and is 240 m in diameter. What is the area of the circle?

Complex wheat pattern at least 500 ft in diameter reported on July 15, 2005, at Garsington, Oxfordshire, England.

53. *More crop circles.* The photo shows several of the mysterious wheat circles appearing in Rockville, California. The biggest circle is claimed to be 140 ft in diameter. What is the area of this circle?

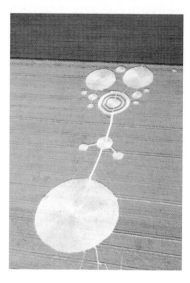

54. *Area of a Dutch crop circle.* The diameter of the largest reported Dutch crop circle is 12 m. What is its area?

55. *Area of a hurricane's eye.* The photo shows Hurricane Katrina striking Florida the night of August 25, 2005. The diameter of the eye at landfall was 32 mi. What was the area of the eye?

56. *More winds.* At Katrina's landfall, winds extended outward 120 mi from the center of the storm. What total area did Katrina cover?

57. *Comparing hurricanes.* In comparison, Hurricane Camille (1969) had a 10-mi-diameter eye at landfall. What was the difference in the area of the eye of Katrina and the eye of Camille?

58. *Mowing automatically.* A self-propelled lawnmower is tied to a pole in the backyard with a 20-ft rope. If the mower goes around in decreasing circles (because the rope is getting wrapped around the pole), what is the area the mower can mow?

59. *Area of sectors in wheel of fortune.* A wheel of fortune has six equal sectors, half of which are red and half of which are yellow. If the radius of the wheel is 3 ft, what area is covered by the red sectors?

60. *Area of Disneyland carousel.* The magical carousel in Disneyland Paris has a radius of about 27 ft. What area does it cover?

61. The recordable area on a CD is the difference between the area of the complete CD and the area covered by the opening in the middle. If the opening has a diameter of $1\frac{1}{2}$ in. and the diameter of the CD is 4.5 in., what is the recordable area for the CD?

⌕ Discovery

Artists and architects, past and present, have used the **Golden Ratio** in their art and their architecture. Do you know what the Golden Ratio is? Begin with a square (shaded in the following figure). Let E be the midpoint of the base $\overleftrightarrow{AD}$. Put the point of your pencil compass at E and the pencil point at C, and draw the circular arc as shown. Extend $\overleftrightarrow{AD}$ to meet this arc and call the point of intersection F. Now draw a line perpendicular to $\overleftrightarrow{AF}$ at F and extend $\overleftrightarrow{BC}$ to meet this perpendicular at G. The rectangle $ABGF$ is known as the **Golden Rectangle.** Its proportions are supposed to be particularly pleasing to the eye. The ratio of the longer to the shorter side of this rectangle is the Golden Ratio.

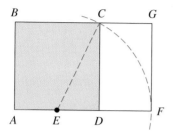

62. Can you discover the numerical value of the Golden Ratio? (*Hint:* Let the side of the square $ABCD$ be $2a$. Then $\overset{\cdot}{E}\overset{\cdot}{C}$ is the hypotenuse of a right triangle with legs of length a and $2a$, respectively.)

Artists are interested in the areas of their paintings and sometimes meet with the problem of drawing a rectangle of height h that will have the same area as that of a given rectangle. The figure below shows the problem.

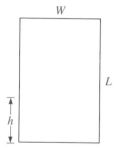

The given rectangle is of length L and width W, and the artist wants a rectangle of height h that will have the same area. Of course, you could find the area LW and divide by h to get the second dimension of the desired rectangle.

But the artist can do the job very quickly without any arithmetic at all! Here is how. Look at the following figure. Draw a line across the given rectangle at height h (line $\overleftrightarrow{BD}$ in the figure). Next, draw a line through A and $\overleftrightarrow{C}$ and extend the top line of the rectangle to meet line $\overleftrightarrow{AC}$ at point F. Then drop a perpendicular from F to the extended base of the given rectangle. The rectangle $ABDE$ is the desired rectangle. Can you discover why? Look at the figure again. The triangles ABC and FDC are similar (have exactly the same shape). Corresponding sides of similar triangles are always in the same ratio. Thus,

$$\frac{y}{W} = \frac{x}{h}$$

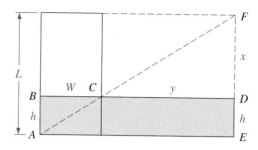

63. Use this result to show that the area of rectangle $ABDE$ is equal to the area of the original rectangle. (*Hint:* What is the area of the rectangle taken away from the given rectangle? What is the area of the rectangle added on?)

64. Draw a careful diagram of a rectangle 4 cm wide and 6 cm high. Use the construction described above to find a rectangle 5 cm high with the same area. If you do this carefully, the result will help convince you that this is a neat construction.

Collaborative Learning

Have you heard of "squaring the circle"? The expression means trying to find the square with the same area as a circle of a given radius. The Greek mathematician Archimedes (287–212 B.C.) used this method to try to approximate the area of a circle.

Form 3 groups. Each of the groups will approximate the area of a circle according to the suggested procedure.

Group 1 *Approximating with a square*

The diagram below shows a square inscribed in a circle. Triangle ABC is a right triangle with hypotenuse $2r$.

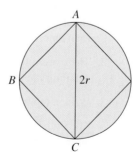

1. Find the length of the line segment AB.

2. The area of the square is the length of the line segment $\overset{\cdot}{A}\overset{\cdot}{B}$ squared. What is this area? How close is the area of the square to the area of a circle?

Group 2 *Approximating with a hexagon*

The diagram below shows a hexagon inscribed in a circle. The area A of the hexagon is the area of the six triangles shown; that is, $A = 6\left(\frac{1}{2}rh\right)$.

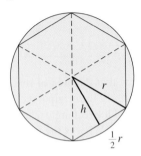

1. Find the height h in terms of r.

2. Find A.

Group 3 *Approximating with an n-sided polygon*

The following diagram shows an n-sided polygon inscribed in a circle. The area of the polygon is the area of the n triangles shown; that is, $A = n$(area of triangle) $= n\left(\frac{1}{2}hb\right) = \frac{1}{2}nbh$.

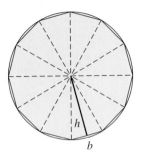

1. The polygon has n sides each of length b. What is nb?

2. When n increases, nb is the distance around the circle, the circumference. What is nb then?

3. As n increases, what is A?

☀️ Online Study Center

To further explore this process, access link 8.4.3 on this textbook's Online Study Center.

📖 **Research Questions**

Pythagoras is such an important part of mathematical lore and history that many pages and effort have been devoted to the study of his life and achievements. Here are some topics to ponder.

1. Many scholars ascribe the Pythagorean theorem to, surprise, Pythagoras! However, other versions of the theorem like the following exist:

   The ancient Chinese proof   Bhaskara's proof   Euclid's proof
   Garfield's proof   Pappus's generalization

   Select three of these versions and write a paper giving details, if possible, about where they appeared, who authored them, and what was stated in each theorem.

2. There are different versions of Pythagoras's death. Write a short paper detailing the circumstances of his death, where it occurred, and how.

3. Write a report about Pythagorean triples.

4. The Pythagoreans studied arithmetic, music, geometry, and astronomy. Write a report about the Pythagoreans' theory of music.

5. Write a report about the Pythagoreans' theory of astronomy.

8.5 Volume and Surface Area

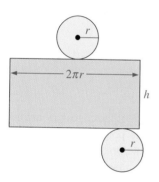

FIGURE 8.68

From the Grocer's Shelf

Which of the two bottles on the right side of the picture has more ketchup? Which has more sauce, the bottle on the left or the can? As it turns out, both ketchup bottles hold the same amount. They have the same **volume.** The volume of sauce in the can and in the bottle are also the same. One of the ketchup bottles has greater **surface area,** which in this case gives the false impression that it has greater volume and holds more. We have already studied perimeter (the length of the **boundary** of a polygon) and area (the space **enclosed** by the polygon). The counterparts of perimeter and area in three dimensions are **surface area** and **volume.** Like perimeter, **surface area** measures a **boundary,** the surface of a three-dimensional figure, whereas **volume,** like area, measures the space **enclosed** by the figure. Keep in mind, however, that perimeter is measured in **linear** units, surface area in **square** units, and volume in **cubic** units.

How can we find the surface area of the can or, in general, the surface area of a **cylinder?** If a vertical cut is made down the side of the can and the metal flattened, it forms a rectangle whose length is the same as the circumference of the can ($2\pi r$) and whose width is the height h of the can (see Figure 8.68).

The area of this rectangle is $2\pi r \cdot h$. Then we find the area of the top and bottom lids, πr^2 for each, to obtain the total surface area for the can or, in general, for a cylinder of radius r and height h. Thus, the surface area S is given by

$$S = 2\pi rh + 2\pi r^2$$

Now, how could you determine whether a can and a bottle hold the same volume? (No fair reading the label!) One way is to fill the can with water and then pour it into the bottle. If the bottle is exactly filled, they have the same volume. We will study volumes and surface areas in this section. ▶

The photograph below shows the Transamerica pyramid in San Francisco. At 853 ft in height (48 stories) it is the tallest building on the city's skyline. The pyramid portion of the structure is built on a square base and is thus an example of a *square pyramid*. Pyramids are one example of solids bounded by polygons.

A. Three-Dimensional Figures

A solid bounded by plane polygons is called a **polyhedron.** The polygons are the **faces** of the polyhedron, the sides of the polygons are the **edges,** and the vertices of the polygons are the **vertices** of the polyhedron. The Egyptian pyramids are polyhedrons with five faces, one of which is a square (the base) and the others are triangles. A square pyramid has eight edges and five vertices. The Transamerica pyramid is a striking example of the use of a square pyramid in the design of a modern building.

A **convex polyhedron** is one that lies entirely to one side of the plane of each of its faces. A polyhedron that is not convex is called **concave,** or **reentrant.** The polyhedrons shown in Figure 8.69 are a **cube,** a **rectangular parallelepiped,** a six-sided polyhedron with triangular faces (panel C), and a seven-sided polyhedron (panel D). The first three polyhedrons are convex, and the fourth is concave (reentrant).

The Transamerica Corporation building in San Francisco, more commonly called the Transamerica pyramid.

A. Cube

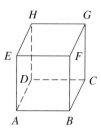

B. Rectangular parallelepiped

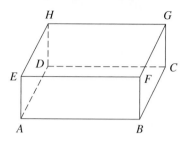

C.

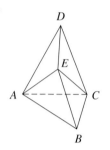

D.

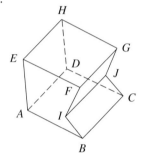

FIGURE 8.69   Polyhedrons.

If two faces of a polyhedron lie in parallel planes and if the edges that are not in these planes are all parallel to each other, then the polyhedron is called a **prism.** In Figure 8.69, panels A, B, and D all illustrate prisms. The faces of a prism that are in the two parallel planes are called the **bases.** The parallel lines joining the bases are the **lateral edges.** The two bases of a prism are congruent polygons. Figure 8.70 shows a **triangular prism.**

If all but one of the vertices of a polyhedron lie in one plane, then the polyhedron is a **pyramid.** The face that lies in this one plane is the **base,** and the remaining vertex is the **vertex of the pyramid.** Figure 8.71 shows a pentagonal pyramid; the base is a pentagon. Prisms and pyramids are named by the shapes of their bases.

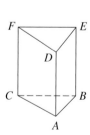

FIGURE 8.70
Triangular prism.

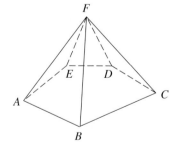

FIGURE 8.71
Pentagonal pyramid.

EXAMPLE 1 ▶ Naming the Edges and Vertices of a Pyramid

Name the edges and the vertices of the pyramid in Figure 8.71.

Solution

The edges are the line segments $\overleftrightarrow{AB}$, $\overleftrightarrow{BC}$, $\overleftrightarrow{CD}$, $\overleftrightarrow{DE}$, $\overleftrightarrow{EA}$, $\overleftrightarrow{EA}$, $\overleftrightarrow{FB}$, $\overleftrightarrow{FC}$, $\overleftrightarrow{FD}$, and $\overleftrightarrow{FE}$. The vertices are the points A, B, C, D, E, and F. ∎

B. Formulas for Volume and Surface Area

The *volume* of a three-dimensional region is measured in terms of a *unit volume*, just as area is measured in terms of a unit area. For the unit volume, we choose the region enclosed by a *unit cube*, as shown in Figure 8.72. Some of the commonly used units of volume are given in Table 8.11. Note that *volume* is always measured in *cubic* units.

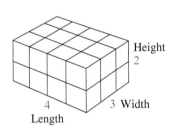

FIGURE 8.72
Unit cube;
volume = 1 cubic unit.

TABLE 8.11 Standard Units of Volume

U.S. System	Metric System
Cubic inch (in.^{3})	Cubic millimeter (mm^3)
Cubic foot (ft^3)	Cubic centimeter (cm^3)
Cubic yard (yd^3)	Cubic meter (m^3)

Volumes can be considered in a manner similar to that used for areas. If a rectangular parallelepiped (box) is such that the lengths of its edges are all whole numbers, then the region can be cut up by planes parallel to the faces, as in Figure 8.73. In general, we define the volume of a rectangular box of length l, width w, and height h to be lwh. Thus,

FIGURE 8.73
$V = 4 \times 3 \times 2 = 24$ cubic units.

Volume of a Rectangular Box

The volume V of a rectangular box with length l, width w, and height h is given by

$$V = lwh = \text{length} \times \text{width} \times \text{height}$$

The surface area of a rectangular box is the sum of the areas of the six faces.

Surface Area of a Rectangular Box

The surface area S of a rectangular box with length l, width w, and height h is given by

$$S = 2lw + 2lh + 2wh$$

For the rectangular box in Figure 8.73, we have

$$S = 2(4 \times 3) + 2(4 \times 2) + 2(3 \times 2)$$
$$= 24 + 16 + 12 = 52 \text{ square units}$$

Table 8.12 gives the formulas for the volume and the surface area of some commonly occurring three-dimensional figures.

TABLE 8.12 Volumes and Surface Areas

Name	Figure	Volume (V)	Surface Area (S)
Cube		$V = a^3$	$S = 6a^2$
Rectangular box		$V = lwh$	$S = 2(lw + lh + wh)$
Cylinder		$V = \pi r^2 h$	$S = 2\pi rh + 2\pi r^2$
Cone		$V = \frac{1}{3}\pi r^2 h$*	$S = \pi r^2 + \pi rs$
Sphere		$V = \frac{4}{3}\pi r^3$	$S = 4\pi r^2$

*The formula for the volume of a circular cone can be verified by pouring water or sand from a hollow cone into a cylinder of the same base and height. This will check the coefficient $\frac{1}{3}$.

Which formulas do you need to find the capacity of the tank?

EXAMPLE 2 ▶ Finding the Volume and Surface Area of Rubik's Cube

You are probably familiar with the Rubik's Cube puzzle like the one shown in the photograph on the next page. Each of the little cubes is $\frac{3}{4}$ in. on a side.

(a) Find the volume of Rubik's Cube.

(b) Find the surface area if no allowance is made for the dips between the small cubes.

Solution

(a) Since there are three little cubes in a row, the edge of the large cube is

$$(3)\left(\tfrac{3}{4}\right) = \tfrac{9}{4} \text{ in.}$$

Using the formula for the volume of a cube, we get

$$V = a^3 = \left(\tfrac{9}{4}\right)^3 = \tfrac{729}{64} \text{ in.}^3$$

or expressed as a decimal,

$$V = 11.390625 \qquad \text{or} \qquad \text{about } 11.4 \text{ in.}^3$$

(b) The formula for the surface area of a cube gives

$$S = 6a^2 = 6\left(\tfrac{9}{4}\right)^2 = \tfrac{243}{8} \qquad \text{or} \qquad \text{about } 30.4 \text{ in.}^2 \qquad ■$$

Let us denote the area measure of the base of a prism or a pyramid by B and the height by h. It is shown in solid geometry that the formulas for the volumes of these figures are as follows:

Volume of a Prism and Pyramid	
$V = Bh$	Volume of a prism
$V = \tfrac{1}{3}Bh$	Volume of a pyramid

EXAMPLE 3 ▶ Finding the Volume of a Crocoite Crystal

If a crocoite crystal is in the form of a prism with a triangular base of height 3 cm and base length 2.6 cm, and if the crystal is 30 cm long, what is its volume?

Solution

First we find B, the area of the base of the prism. Since the base is a triangle of base length 2.6 cm and height 3 cm,

$$B = \tfrac{1}{2}(2.6)(3)$$
$$= 3.9 \text{ cm}^2$$

Then we use the formula for the volume of a prism to obtain

$$V = Bh$$
$$= (3.9)(30)$$
$$= 117 \text{ cm}^3 \qquad ■$$

EXAMPLE 4 ▶ Finding the Volume and Surface Area of a Polyhedron

Figure 8.74 (page 552) shows a polyhedron that consists of a rectangular box surmounted by a square pyramid with the top of the box for its base. The dimensions of the polyhedron are shown in Figure 8.74.

(a) Find the volume of the polyhedron.   (b) Find the total surface area.

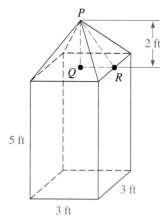

FIGURE 8.74

Solution

(a) The volume of the rectangular box portion in Figure 8.74 is

$$V = lwh = (3)(3)(5) = 45 \text{ ft}^3$$

The volume of the pyramid is

$$V = \tfrac{1}{3}Bh = \tfrac{1}{3}(3 \times 3)(2) = 6 \text{ ft}^3$$

Thus, the entire volume is 45 ft³ + 6 ft³, or 51 ft³.

(b) The area of the base of the figure is

$$B = 3 \times 3 = 9 \text{ ft}^2$$

The area of the four sides of the rectangular box portion is

$$A_1 = 4(3 \times 5) = 60 \text{ ft}^2$$

To find the area of the four triangular faces of the pyramid, we must first find the altitude of the triangles. To do this, we draw a line $\overleftrightarrow{PQ}$ from the top vertex and perpendicular to the base of the pyramid. This line will meet the base at its midpoint Q. Then we draw a second line $\overleftrightarrow{QR}$ from the foot of the perpendicular to the midpoint of an edge of the base. Finally, we draw the line $\overleftrightarrow{PR}$. The triangle PQR is a right triangle, whose hypotenuse $\overrightarrow{PR}$ is the desired altitude. Thus, we have

$$PR = \sqrt{2^2 + \left(\tfrac{3}{2}\right)^2} = \sqrt{4 + \tfrac{9}{4}} = \sqrt{\tfrac{25}{4}} = \tfrac{5}{2} \text{ ft}$$

Consequently, the area of the four triangular faces of the pyramid is

$$A_2 = 4 \times \tfrac{1}{2}bh = 4 \times \tfrac{1}{2}(3)\left(\tfrac{5}{2}\right) = 15 \text{ ft}^2$$

The total surface area of the polyhedron is

$$B + A_1 + A_2 = 9 + 60 + 15 = 84 \text{ ft}^2 \qquad \blacksquare$$

EXAMPLE 5 ▶ Diameters of Spheres Given Surface Area and Volume

Natasha is inflating a toy globe. Suppose that the globe is a perfect sphere. What is the diameter of the globe when the number of square inches of its surface area is equal to the number of cubic inches of its volume?

Solution

Let r in. be the radius of the globe. Then the surface area is $4\pi r^2$ in.², and the volume is $\tfrac{4}{3}\pi r^3$ in.³. The number of units of surface area is to be equal to the number of units of volume, so

$$\tfrac{4}{3}\pi r^3 = 4\pi r^2$$
$$\tfrac{1}{3}r = 1 \qquad \text{Divide both sides by } 4\pi r^2.$$
$$r = 3$$

Since the diameter is twice the radius, the required diameter is 6 in.   ■

EXAMPLE 6 ▶ Finding the Volume of a Tank

The water tank for a small town is in the shape of a cone (vertex down) surmounted by a cylinder, as shown in Figure 8.75. If a cubic foot of water is about 7.5 gal, what is the capacity of the tank in gallons? (Use $\pi \approx 3.14$.)

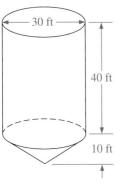

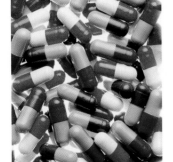

FIGURE 8.75

Solution

For the cylindrical portion of the tank, $r = 15$ and $h = 40$, so

$$V = \pi(15)^2(40) = 9000\pi \text{ ft}^3$$

For the conical portion, $r = 15$ and $h = 10$. Thus,

$$V = \tfrac{1}{3}\pi(15)^2(10) = 750\pi \text{ ft}^3$$

The total volume is the sum of these, that is, 9750π ft^3. Using 7.5 gal/ft^3 and 3.14 for π, we get

$$9750\pi \text{ ft}^3 \approx (9750)(3.14)(7.5)$$
$$\approx 230,000 \text{ gal} \qquad\blacksquare$$

EXAMPLE 7 ▶ How Much Water Is in the Tank?

A solid metal sphere of radius 3 m just fits inside a cubical tank. If the tank is full of water and the sphere is slowly lowered into the tank until it touches bottom, how much water is left in the tank?

Solution

The amount of water left in the tank is the difference between the volume of the tank and the volume of the sphere. Thus, since the length of any edge of the cube equals the diameter of the sphere, the required volume is

$$V = 6^3 - \tfrac{4}{3}\pi(3^3)$$
$$= 216 - 36\pi$$
$$\approx 103 \text{ m}^3 \qquad\blacksquare$$

EXAMPLE 8 ▶ Volume of a Capsule

The capsules in the photo are in the shape of a cylinder 12 mm long with two half spheres with a diameter of 2 mm at each end. What is the volume of each capsule? Use 3.14 for π, and round the answer to the nearest hundredth of a cubic millimeter.

Solution

We have to find the volume V of the cylinder plus the volume of the two half spheres (which make up one whole sphere).

The volume of the cylinder is $V = \pi r^2 h$. Here, $r = 1$ mm and $h = 12$ mm, so

$$V = (3.14)(1 \text{ mm})^2(12 \text{ mm})$$
$$\approx 37.68 \text{ mm}^3$$

The volume of the sphere is

$$V = \tfrac{4}{3}\pi r^3$$
$$= \tfrac{4}{3}(3.14)(1 \text{ mm})^3$$
$$= \frac{4(3.14)(1 \text{ mm})^3}{3}$$
$$\approx 4.19 \text{ mm}^3$$

The volume of the whole capsule is $37.68 + 4.19 = 41.87$ mm^3.

$r = 1$ mm

$h = 12$ mm

$r = 1$ mm

"The Burger King 2.8 oz. flame-broiled hamburger has 75% more beef than McDonald's hamburger."

FIGURE 8.76

One question students always ask is, Do we have to memorize all these formulas? Judge for yourself after reading this actual case.

Here's an example in which an incorrect formula has been used: The ad claims that the Burger King 2.8-oz. hamburger has 75% more beef than the McDonald's hamburger. Is that the idea you get from Figure 8.76? We can compare the two hamburgers by comparing the *volume* of the beef in each burger. The volume V of a burger is $V = Ah$, where A is the area of the top of the burger and h is the height. The top area of the McDonald's burger is $M = \pi r^2$, where r is the radius (half the distance across the middle) of the burger. For the McDonald's burger, $r = 1$ in., so its area is

$$M = \pi(1 \text{ in.})^2 = \pi \text{ in.}^2 \ (\pi \text{ square inches})$$

For the Burger King burger, $r = 1.75$ in., so

$$B = \pi(1.75 \text{ in.})^2 \approx 3\pi \text{ in.}^2 \qquad \text{This means approximately equal.}$$

For the McDonald's burger, the volume is $V = \pi h$, and for the Burger King burger, the volume is $V = 3\pi h$. The difference in volumes is $3\pi h - \pi h = 2\pi h$, and the *percent* increase for the Burger King burger is given by

$$\text{Percent increase} = \frac{\text{increase}}{\text{base}} = \frac{2\pi h}{\pi h} = 2$$

Thus on the basis of this discussion, Burger King burgers have 200% more beef. What other assumptions must you make for this to be so? (You'll have an opportunity in In Other Words to give your opinion!)

EXERCISES 8.5

Ⓐ Three-Dimensional Figures

1. Refer to Figure 8.69C on page 548, and name
   a. the vertices.
   b. the edges.

2. Refer to Figure 8.70 on page 548 and repeat problem 1.

3. Refer to Figure 8.69D on page 548, and name the bottom face.

4. Refer to Figure 8.69D on page 548, and name the left-hand back face.

In each of problems 5–8, make a sketch of the figure.

5. A triangular pyramid

6. A triangular prism surmounted by a triangular pyramid with the top base of the prism as the base of the pyramid

7. A six-sided polyhedron that is convex and is not a parallelepiped

8. An eight-sided polyhedron with triangular faces

9. **a.** If the edges of a cube are doubled in length, what happens to the volume?
   b. What if the lengths are tripled?

B **Formulas for Volume and Surface Area**

10. A pyramid has a rectangular base. Suppose that the edges of the base and the height of the pyramid are all doubled in length. What happens to the volume?

11. For a rectangular solid that is 20 in. long, 10 in. wide, and 8 in. high, find
    a. the volume V.
    b. the total surface area S.

12. A rectangular solid is constructed by putting together two cubes each of edge x ft. If the number of cubic feet in the total volume is equal to the number of square feet in the total external surface area of this solid, what is the value of x?

13. A solid consists of a cube of side $6x$ surmounted by a regular square pyramid whose base is the top face of the cube and whose height is $4x$. Find formulas for
    a. the volume of this solid.
    b. the total external surface area of this solid.

14. A cylindrical can is to have the area of its curved surface equal to the sum of the areas of its base and its top. What must be the relation between the height and the diameter of this can?

15. A pentagonal prism has a base whose area is 10 in.². If the prism is 5 in. high, what is its volume?

16. The base of a prism is a triangle whose base is 3 ft and whose height is 4 ft. If the prism is 5 ft high, what is its volume?

17. The edge of the base of a square pyramid is 4 in. long, and the pyramid is 6 in. high. Find the volume of the pyramid.

18. A convex polyhedron consists of two pyramids with a common base that is an equilateral triangle 3 in. on a side. The height of one of the pyramids is 2 in. and the height of the other is 4 in. What is the volume of the polyhedron?

19. A container consists of a cube 10 cm on an edge surmounted by a pyramid of height 15 cm and with the top face of the cube as its base. How many liters does this container hold? (*Hint:* 1 L = 1000 cm³.)

20. The Great Pyramid of Egypt (the Pyramid of Cheops) is huge. It is 148 m high, and its square base has a perimeter of 930 m. What is the volume of this pyramid?

In problems 21–24, use the approximate value 3.14 for π to find the volume and the total surface area of
    a. the circular cylinder of given radius and height.
    b. the circular cone of given radius and height.

21. Radius 5 in., height 9 in.

22. Radius 10 cm, height 6 cm

23. Radius 3 ft, height 4 ft

24. Radius 6 cm, height 12 cm

25. Find the volume of a sphere of radius 6 in.

26. Find the volume of a sphere of radius 12 cm.

27. The Peachtree Plaza Hotel tower in Atlanta, Georgia, is 70 stories high. If the height of this cylinder is 754 ft and its diameter is 116 ft, what is the volume?

28. A grain silo is in the shape of a cone, vertex down, surmounted by a cylinder. If the diameter of the cylinder is 10 ft, the cylinder is 30 ft high, and the cone is 10 ft high, how many cubic yards of grain does the silo hold?

29. A pile of salt is in the shape of a cone 12 m high and 32 m in diameter. How many cubic meters of salt are in the pile?

30. The fuel tanks on some ships are spheres of which only the top halves are above deck. If one of these tanks is 120 ft in diameter, how many gallons of fuel does it hold? Use 1 ft³ ≈ 7.5 gal.

31. A popular-sized can in U.S. supermarkets is 3 in. in diameter and 4 in. high (inside dimensions). About how many grams of water will one of these cans hold? (Recall that 1 cm³ of water weighs 1 g, and 1 in. = 2.54 cm.)

32. An ice cream cone is 7 cm in diameter and 10 cm deep (see the figure below). The inside of the cone is packed with ice cream and a hemisphere of ice cream is put on top. If ice cream weighs $\frac{1}{2}$ g/cm^3, how many grams of ice cream are there in all?

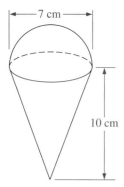

33. The circumferences of a baseball, soccer ball, and basketball are 9, 27, and 30 in., respectively.
  a. Find the surface area and volume of each.
  b. How does the surface area of a baseball compare to that of a basketball?
  c. How does the volume of a baseball compare to that of a soccer ball?

34. A cylindrical storage tank for gasoline has a 300-ft diameter and is 200 ft high.
  a. Find the capacity of this tank if one cubic foot of gasoline equals 7.5 gal.
  b. How many gallons of paint are needed to paint the exterior and top of this tank with two coats of paint if a gallon of paint covers 400 ft^2?

35. A bowling ball has a 27-in. circumference and weighs 16 lb.
  a. Find the volume of the ball before the holes are drilled.
  b. If three cylindrical holes with diameters 1.25, 1, and 1 in., respectively, each 2.5 in. deep, are drilled into the ball, what is the weight of the ball now?

36. A basketball with a 30-in. circumference is tightly packed into a cubical box for shipment.
  a. What is the volume of the ball?
  b. What is the smallest possible volume for the box?
  c. What percent of the space in the box is occupied by the basketball?

37. A grocery bag has a 7-by-12-in. base and is 17 in. high.
  a. What is the volume of the bag?
  b. What is the minimum outer surface area for the bag?
  c. A ton of recycled paper will save 17 trees. If one bag weighs 2 oz, how many bags are needed to save 34 trees?

38. The Pet Kennel measures 26 in. long, 19 in. high, and 16 in. wide. What is the volume of the pet kennel?

39. The Pet Taxi is 16 in. long, 10 in. high, and 11 in. wide. What is the volume of the Pet Taxi?

40. The inside dimensions of the toaster are 11 in. wide, 8 in. deep, and 5 in. high. What is the volume of the inside of the toaster?

41. The inside dimensions of the microwave are 16 in. by 11 in. by 13 in. What is the volume of the inside of the microwave?

42. Find the volume of the large 4-ft-high, 1-ft-diameter cylinder. Use 3.14 for π, and give the answer to the nearest hundredth.

43. Each sugar container is 5 in. high and 3 in. in diameter. Find the volume of one container. Use 3.14 for π, and give the answer to the nearest hundredth.

44. Find the volume of the small 2-ft-high, 1-ft-diameter cylinder in the photo below on the left. Use 3.14 for π, and give the answer to the nearest hundredth.

45. Using the information in problem 43, find the volume of the sugar in the three containers. Note that one of the containers is half-full. Answer to the nearest whole number.

46. *Moving on out.* Have you moved to the dorm or to an apartment lately? You probably needed some boxes with some of the dimensions shown. Give the answer in both cubic inches and cubic feet. (1 cubic foot = 1728 cubic inches.)
   a. Find the volume of the small box.
   b. Find the volume of the medium box.
   c. Find the volume of the large box.
   d. Which of the three volumes shown for the boxes agrees exactly with your answer, the box in **a**, **b**, or **c**?

- **Small Box**
  $16'' \times 12'' \times 12''$ 1.5 cu/ft
- **Medium Box**
  $18'' \times 18'' \times 16''$ 3.0 cu/ft
- **Large Box**
  $18'' \times 18'' \times 24''$ 4.5 cu/ft
- **Extra-Large Box**
  $24'' \times 18'' \times 24''$ 6.0 cu/ft

47. *Does it all fit?* You probably needed a truck to move. The U-Haul truck has the dimensions shown.

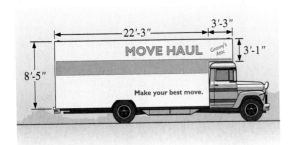

Inside dimensions: 22'-3" × 7'-7" × 8'-5" (*L* × *W* × *H*)
Granny's Attic: 3'-3" × 7'-7" × 3'-1" (*L* × *W* × *H*)

a. Approximate the inside dimensions of the truck to 22 ft by 7 ft by 8 ft. What is the volume?

b. Approximate the dimensions of Granny's Attic to 3 ft by 7 ft by 3 ft. What is the volume?

c. You estimate that you have a 1300 ft³ of stuff to be moved. Does your stuff theoretically fit in the truck? Explain. *Hint* Don't forget the space in Granny's Attic (over the truck cab).

48. *U-Haul recommendations.* If you have a two- to three-room apartment, U-Haul recommends a 1200 to 1600 ft³ truck.

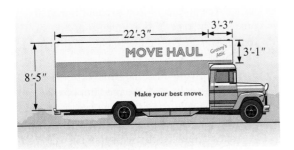

Inside dimensions: 20'-10" × 7'-6" × 8'-1" (*L* × *W* × *H*)
Granny's Attic: 2'-10" × 7'-6" × 3'-1" (*L* × *W* × *H*)

a. Approximate the inside dimensions of the truck to 20 ft by 7 ft by 8 ft. What is the volume?

b. Approximate the dimensions of Granny's Attic to 3 ft by 7 ft by 3 ft. What is the volume?

c. If you have a two-room apartment, U-Haul recommends a truck with a minimum of 1200 ft³ of space. Does this truck meet that recommendation?

ⓒ Applications

What is the volume (capacity) of the garbage cans that are used at your school? The garbage can in the photograph consists of two parts: the cylindrical bottom (2.5 ft high, 1.5 ft in diameter) and the hemisphere on top (half of a sphere).

49. What is the volume of the cylindrical part?

50. What is the volume of the hemispherical top?

51. What is the volume of the entire trash can?

52. The chemistry building has 3 floors and there are 4 trash cans on each floor. What is the total volume of the trash cans on the 3 floors of the chemistry building?

53. A garbage truck carries about 500 ft³ of trash. How many full cans of trash can it carry?

54. If we assume that every building has 3 floors and 4 trash cans on each floor, how many buildings can the truck service? (*Hint:* How many 12-can loads fit in the truck?)

✏ In Other Words

55. Remember the hamburgers in Figure 8.76? Write two explanations of how the Burger King burger can have 75% more beef than the McDonald's burger and still look like the one in the picture.

56. To make a fair comparison of the amount of beef in two hamburgers, should you compare the circumferences, areas, or volumes? Explain.

57. The McDonald burger has a 2-in. diameter, whereas the Burger King burger has a 3.50-in. diameter. How much bigger (in percent) is the circumference of the Burger King burger?

58. Can you now explain the claim in the ad? Is the claim correct? Explain.

59. Explain in your own words why lengths are measured in linear units, areas in square units, and volume in cubic units.

60. Explain in your own words what happens to the volume and surface area of a sphere if the diameter is doubled.

61. Is the number of cubic units in the volume of a sphere always larger than the number of square units in its surface area? Explain in your own words why or why not.

62. Suppose the radius of a circular cone is doubled and the height is halved. Explain in your own words what happens to the volume.

 **Discovery**

63. A **regular polyhedron** is one whose faces are all congruent regular polygons, that is, regular polygons of exactly the same shape and size. The appearance of such a polyhedron at any vertex is identical with its appearance at any other vertex; the same is true at the edges. The early Greeks discovered that only five regular polyhedrons are possible. Repeat this discovery and explain why it is so.

Consider the regular polygons one at a time: equilateral triangles, squares, regular pentagons, and so on. In the case you use squares, how many squares can you put together at a vertex to form a polyhedron? Look at a cube. It is the only regular polyhedron with squares for its faces.

The regular polyhedrons are as follows:

A **tetrahedron** with 4 equilateral triangles for faces

A **cube** with 6 squares for faces

An **octahedron** with 8 equilateral triangles for faces

A **dodecahedron** with 12 pentagons for faces

An **icosahedron** with 20 equilateral triangles for faces

64. Copy the pattern of equilateral triangles in the following figure onto a piece of stiff cardboard. Cut around the outside edges and fold on the heavy lines. You can build an icosahedron by holding the cut edges together with transparent tape. There are five triangles at each vertex.

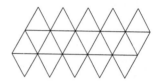

65. Count the number of faces (F), vertices (V), and edges (E) for each of the figures listed in the table below. Compare the value of E with the value of $F + V$ and discover Euler's famous formula for polyhedrons.

Figure	F	V	E	F + V
8.69B				
8.69C				
8.69D				
8.70				
8.71				

66. Can you check the formula you got in problem 65 by using the diagram in problem 64?

 Calculator Corner

When higher exponents are present in an expression, calculators with a $\boxed{y^x}$ key are especially helpful. As the notation indicates, this key raises a number y to a power x. For instance, to perform the calculations in Example 7, we enter

The result is given as 102.9026645, or about 103 m^3.

1. Rework problem 25 using your calculator.

Collaborative Learning

We have studied cubes, prisms, pyramids, cylinders, and cones. What can you build using all of these geometric figures? Your own castle, of course! Form several groups. Here is the challenge.

1. Let each group find pictures of several castles and count the number of geometric figures seen in them (prisms, pyramids, cylinders, and so on). The team that identifies the most figures wins.

2. Let each team create a drawing of a castle. The team that uses the most geometric figures in the design wins. Good luck!

Online Study Center

To further explore geometric figures, access links 8.5.1 and 8.5.2 on this textbook's Online Study Center.

8.6 Networks, Non-Euclidean Geometry, and Topology

Crossing That Bridge

What do you think the following activities have in common: urban engineers mapping traffic patterns, chemists modeling complex molecules, managers creating the organizational charts for a large corporation, and drawing your family tree? Each of the models consists of points (locations, atoms, job positions, and people) connected by lines. The study of these graphs is now called **graph** or **network theory.**

A famous puzzle known as the **Bridges of Königsberg** probably started the study of the traversability of networks. There was a river flowing through the city, and in the river were two islands (*A* and *D*) connected to each other and to the city by seven bridges (*a*, *b*, *c*, *d*, *e*, *f*, and *g*), as shown in Figure 8.77. The people of the city loved a Sunday walk and thought it would be fun to follow a route that would take them across each of the seven bridges exactly once. But they found that no matter where they started or what path they took, they could not cross each bridge exactly once.

The great Swiss mathematician Leonhard Euler solved the problem in 1736. First, he redrew the map by making islands *A* and *D* smaller and lengthening the bridges (Figure 8.78). He also made shores *B* and *C* smaller (Figure 8.79).

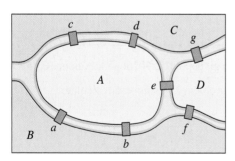

FIGURE 8.77

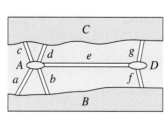

FIGURE 8.78

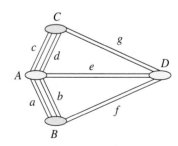

FIGURE 8.79

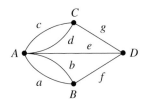

FIGURE 8.80

These changes distort the picture but *do not* change the problem. Then, in a stroke of genius, Euler thought of land areas *A*, *B*, *C*, and *D* as **points** and bridges *a* through *g* as **arcs** connecting them. The resulting **network** is diagrammed in Figure 8.80. Euler's question became: *Can **all** the arcs be traced with a pencil only once without lifting the pencil off the paper?* If the answer to a problem like this is yes, the network is called **traversable.**

In a network, the points are the endpoints of arcs and are called **vertices** (singular **vertex**). Euler noticed that if a path goes through a vertex, the vertex must have two arcs: one in and one out. Thus, if a network has an odd vertex, it *must* be the starting or finishing point for a traversable path. But all four vertices in the Königsberg network are odd! Thus, the network is *not* traversable. We will study more about networks, traversability, and the bridges of New York and then end the section by considering other geometries and topology. ▶

A. Networks

Any connected set of line segments or arcs is called a **network.** (For the purposes of this section, your intuitive notion of what an arc is is sufficient.) If the network can be drawn by tracing each line segment or arc exactly once without lifting the pencil from the paper, the network is said to be **traversable.** Any simple network (one that does not cross itself) is traversable. If the network is both simple and closed, then you can choose any point of the network as the starting point, and this point will also be the terminal point of the drawing. If the network is simple but not closed, then you must start at one of the endpoints and finish at the other. Figure 8.81 shows several examples of simple networks that are either closed or not closed.

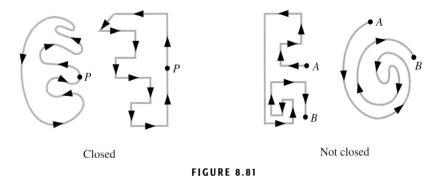

Closed   Not closed

FIGURE 8.81

Simple networks.

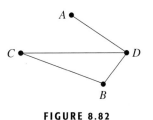

FIGURE 8.82

Let us examine the network in Figure 8.80 more closely. The number of edges for vertex *A* is 5; it is an odd vertex. As a matter of fact, all vertices in Figure 8.80 are odd. Figure 8.82 shows another network in which *A* and *D* are odd vertices but *B* and *C* are even. Since every arc has two endpoints, it is impossible for a network to have an odd number of odd vertices. Thus, a network can have 0, 2, 4, 6, . . . , any even number of odd vertices. Notice that there are four odd vertices in the Königsberg problem and two odd vertices in Figure 8.82.

EXAMPLE 1 ▶ Finding Which Network Is Traversable

Which of the networks below are traversable? If a network is traversable, indicate your beginning and ending points.

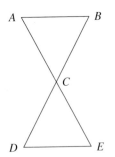

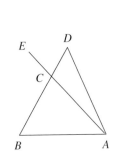

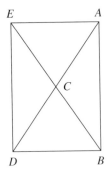

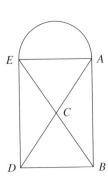

a. No odd vertices b. Two odd vertices c. Two odd vertices d. Four odd vertices e. Two odd vertices

Solution

(a) The network is traversable. You can start at any point and you will end at that same point.

(b) The figure is traversable, but you must start at one of the odd vertices, B or D. If you start at B, you end at D, and if you start at D, you end at B.

(c) The figure is traversable. This time you must start at A or E (the odd vertices).

(d) The figure is not traversable. (Do you know why?)

(e) The figure is traversable. You can start at one of the odd vertices, D or B, and end at the other. ∎

From what you have studied so far, can you make any generalizations about the traversability of a network? Here are the conclusions Euler reached.

Traversability Rules

1. A network with *no* odd vertices is traversable. You can start at any point and you will end at that same point.

2. A network with exactly *two* odd vertices is traversable. You must start at one of the odd vertices and end at the other.

3. A network with *more than two* odd vertices is not traversable.

EXAMPLE 2 ▶ A New York Tour Using Networks

The city of New York is composed of five boroughs, as shown in Figure 8.83: Bronx (*B*), Brooklyn and Queens (*B-Q*), Manhattan (*M*), and Staten Island (*SI*), connected by a network of bridges and tunnels, some of them passing through Randall's Island (*RI*) and some leading to New Jersey (*NJ*). Determine whether you could take a tour going over each bridge and through each tunnel exactly once. (Since there is no water separating Brooklyn and Queens, we call this land mass *B-Q*.)

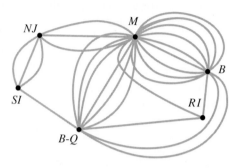

FIGURE 8.83

Solution
The network is shown in Figure 8.83. Note that the only *two* odd vertices are Manhattan (19) and Randall's Island (3), so the network is traversable. Your path must start at one of these two odd vertices and end at the other. ∎

EXAMPLE 3 ▶ A House Tour Using Networks
Figure 8.84 shows a house floor plan with six rooms (*A, B, C, D, E,* and *F*) and with the openings representing doors. Is it possible to take a walk through the house and pass through each door exactly once?

Solution
Think of each room and the outdoors as a vertex and the corresponding number of doors as the number of paths to that vertex. Since there are only two odd vertices, *B* and *E*, the network is traversable, but you must start at one of the odd vertices and end at the other, as shown in Figure 8.85. ∎

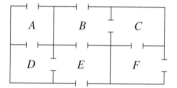

FIGURE 8.84

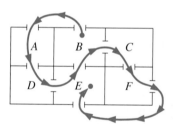

FIGURE 8.85

Network theory has many practical applications. It is of great importance in computer science and technology; it is also used to solve problems in the design of city streets, to analyze traffic patterns, to find the most efficient routes for garbage collection, and so on. Networks are further used in connection with PERT (Program Evaluation and Review Technique) diagrams in planning complicated projects. These diagrams help to determine how long a project will take and when to schedule different phases of the project.

B. Non-Euclidean Geometry

The theory of networks we have studied is part of a branch of modern geometry called **graph theory.** Most of the geometry we studied earlier is based on axioms and *postulates* (statements assumed to be true) derived from Euclid's famous book, *The Elements,* written in about 300 B.C. In *The Elements,* five postulates were assumed and then used to prove many propositions about geometric figures. The fifth postulate, the parallel postulate, was quite different from the others and harder to understand until John Playfair (1748–1819), a Scottish physicist and mathematician, stated it in a logically equivalent form.

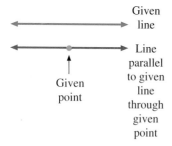

Given
line

Line
parallel
to given
line
through
given
point

Given
point

FIGURE 8.86

The postulate is illustrated in Figure 8.86.

One of the most revealing but unsuccessful attempts to prove the postulate was made by the Italian Jesuit Girolamo Saccheri (1667–1733) in a book called *Euclid Freed of Every Flaw*. The "proof" used the method of *contradiction;* that is, it assumed the postulate to be false and showed that this led to a contradiction. The negation (denial) of the fifth postulate consisted of two parts.

1. Given a line and a point not on the line, there is *no* line through the point parallel to the original line.

2. Given a line and a point not on the line, there are *at least two* lines through the point parallel to the original line.

But, to Saccheri's surprise, no contradiction could be found. Less than one hundred years later, three mathematicians, Carl Gauss, Janos Bolyai, and Nikolai Lobachevsky, used the assumptions in the negation of the parallel postulate to develop what is now called **non-Euclidean geometry.**

Carl Friedrich Gauss became interested in the parallel postulate at the age of 15. He formulated an axiom contradicting Euclid's by assuming that *more* than one parallel line could be drawn through a point not on a given line. Rather than attempting to reach a contradiction, Gauss began to see a geometry different from Euclid's but internally consistent. Unfortunately, Gauss never published his research, perhaps to avoid the inevitable public controversy.

Janos Bolyai (1802–1860) and Nikolai Lobachevsky (1792–1856) separately decided to include an *alternative* postulate. Janos Bolyai was a Hungarian army officer and son of a respected mathematician who had studied under Gauss. As Gauss had done earlier, Bolyai assumed that more than one parallel line through a point not on a given line existed, an assumption that led him to exclaim: "Out of nothing I have created a strange new world." Bolyai's work was dated 1829 but published in 1832 as an appendix to his father's two-volume book. When his father sent a copy of the work to Gauss, the reply was less than encouraging. Janos Bolyai never published again. On the other hand, Nikolai Lobachevsky published the first complete text on non-Euclidean geometry in 1829 containing one of the alternative postulates.

The use of this postulate led to an entirely new type of geometry called *hyperbolic* **geometry.**

In hyperbolic geometry lines are represented by *geodesics* on a *pseudosphere.* A **pseudosphere** is formed by revolving a curve called a *tractrix* about a line *AB,* as shown in Figure 8.87. A **geodesic** is the shortest and least-curved arc between two points on a surface. Can you see why there is more than one line through a given point parallel to the given line?

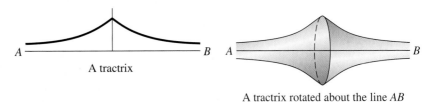

A tractrix

A tractrix rotated about the line *AB*

FIGURE 8.87
Lobachevskian hyperbolic model.

A generation after the discovery of hyperbolic geometry, Bernhard Riemann (1826–1866), who was a disciple of Gauss, developed ***elliptic* geometry.** In this geometry, the fifth postulate was stated as follows:

> **Riemann's Alternative to the Parallel Postulate**
>
> Given a line and a point not on that line, there is *no* line parallel to the original line.

Of course, if there are *no* lines parallel to a given line in elliptic, or Riemannian, geometry, every two lines must intersect! How can this be? Because the term *line* is undefined and can be interpreted differently in different geometries.

We have already considered lines in the Euclidean plane. In elliptic, or Riemannian, geometry, lines are great circles on a sphere, as shown in Figure 8.88. A **great circle** is a circle that has its center at the center of the sphere and that divides the sphere into two equal parts. Now, can you see why any two lines (great circles) on a sphere must intersect?

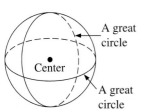

FIGURE 8.88
Riemannian spherical model.

EXAMPLE 4 ▶ **Angle Measures of a Quadrilateral**

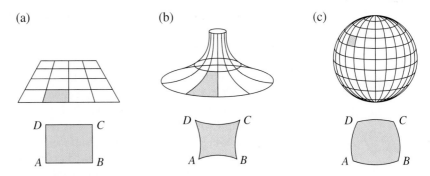

Look at the illustrations showing quadrilaterals on a plane, on a pseudosphere, and on a sphere. What can you say about the measure of the angles *D*?

Solution

(a) In the Euclidean plane, the measure of angle *D* is 90°.

(b) In the pseudosphere the measure of angle *D* is less than 90°.

(c) On the sphere, the measure of angle *D* is more than 90°. ■

Online Study Center

To further explore Euclidean, spherical, and hyperbolic geometry and the pseudosphere, access link 8.6.1 on this textbook's Online Study Center.

C. Topology

Another branch of modern geometry is **topology.** We have already studied properties of geometric figures that were assumed to be rigid (unchanging). If we allow the figures to be transformed by stretching or compressing, the study of properties that remain unchanged is called *geometric topology.* One of the first persons to introduce topology was the German mathematician Augustus Ferdinand Möbius (1790–1896), so topology is a relatively new branch of mathematics.

Let us look at a very simple example. Consider an ordinary rubber band as shown in Figure 8.89. Take a point on this band and call it *P*. If we move along the band in a fixed direction from *P*, we eventually return to the point *P*. This is a property of the band that is unaltered even if the shape of the band is changed, twisted, or tied in knots, as long as the band is not broken or cut. If the band is cut in order to get a new shape, then the cut ends must be rejoined in order to keep this property unchanged.

The three figures formed by the rubber bands in Figure 8.89 are said to be *topologically equivalent* because they can be obtained one from another by just changing the shape of the band as restricted above. Any two figures are topologically equivalent if one can be formed from the other by continuous deformation without cutting or tearing the figure. For example, a baseball and a flat solid plate are equivalent because if the ball were made of soft clay, it could be compressed into the flat plate without cutting or tearing the clay.

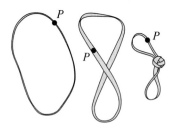

FIGURE 8.89

EXAMPLE 5 ▶ Finding Figures That Are Topologically Equivalent

Are the figures in each of the following pairs topologically equivalent?

(a) A straight line segment and the letter C

(b) A doughnut and a solid steel cylinder

Solution

(a) These are equivalent because the line would just have to be bent in the shape of the C.

(b) These are not equivalent because the doughnut would have to be cut before it could be shaped into a cylinder. ■

One of the things that Möbius is noted for is his study of one-sided surfaces. These surfaces include the one that is called a Möbius strip, shown in Figure 8.90(a). You can make such a strip quite easily. Take a strip of paper, give it a half-twist, and join the ends together. This connects the *opposite* sides of the original strip so that the new figure has only one side. You can check this by making an X, as in Figure 8.90(b), on the surface of the strip and then running a pencil line from the X along the strip until you come back to the starting point. Also note that the Möbius strip has only one edge. Try cutting the strip along the line you have drawn and see what you get.

The number of holes that occur in a figure are of interest to topologists. This number is called the *genus* of the figure; it gives the largest number of complete cuts that can be made without cutting the figure apart. For example, if you take a coffee cup with the usual small handle [Figure 8.91(a)], the only cut through the figure that leaves it in one piece is a cut through the handle, as in

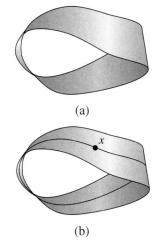

(a)

(b)

FIGURE 8.90
Möbius strip.

Figure 8.91(b). Note that the coffee cup is a figure with just one hole (in the handle), so its genus is 1.

(a)   (b)

FIGURE 8.91

EXAMPLE 6 ▶ Finding the Genus of Objects

Give the genus of each of the following objects:

(a) A 25-cent coin (a quarter)   (b) This button

(c) This nail file

Solution

(a) There is no hole in the quarter, so the genus is zero.

(b) There are four holes in the button, so the genus is 4.

(c) There is one hole in the nail file, so the genus is 1.   ■

D. Applications

EXAMPLE 7 ▶ Applications of Geometry to McDonald's

A software program called Vertical Mapper for Surface Analysis created a Natural Neighborhood (Voroni) diagram for McDonald's restaurants (marked with stars) in San Francisco, as shown in Figure 8.92. The edges of each area (also known as a *Thiessen polygon*) are equidistant between adjacent outlets. This technique can be used to select locations that maximize distance from sister stores or competing ones. Thiessen polygons create a sense of a chain's location strategy and provide a tool to visualize the effect of potential changes in strategy.

(a) Do the edges of the areas shown (in red) form a traversable network?

(b) If you draw a line segment between two adjacent outlets (stars), where does the line segment intersect the (red) edge?

FIGURE 8.92

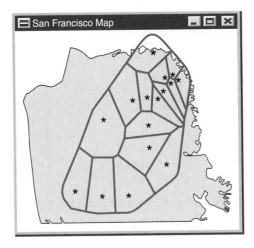

Solution

(a) There are more than two odd vertices in the network. This means that the network is not traversable.

(b) The point of intersection of the line segment and the edge is the midpoint of the line segment.   ■

Note that the distance between most adjacent outlets is the same. Can you generalize this?

EXERCISES 8.6

A Networks

In problems 1–10, find

    a. the number of even vertices.

    b. the number of odd vertices.

    c. whether the network is traversable and which vertices are possible starting points if the network is traversable.

1.

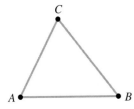

2.

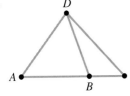

3.

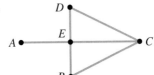

4.

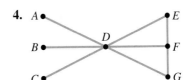

5.

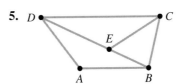

6.

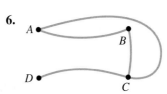

7.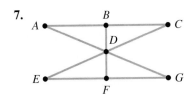

8.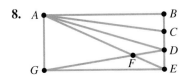

9. The network formed by the edges of a square pyramid

10. The network formed by the edges of a rectangular box

11. Use a network to find whether it is possible to draw a simple connected broken line that crosses each line segment of the figure below exactly once. The line segments are those that join successive dots.

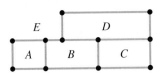

12. Repeat problem 11 for the figure below.

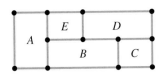

In problems 13–20, use the traversability rules to determine whether it is possible to take a walk through the house and pass through each door exactly once. Which of the paths must start and end outside?

13.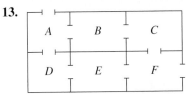

14.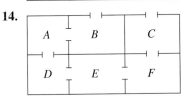

15.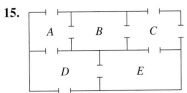

16.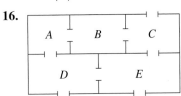

17.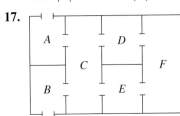

18.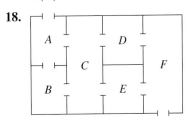

19.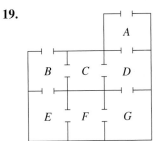

20.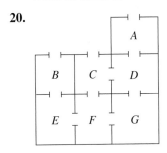

21. Use a network to find whether it is possible to take a walk through the building with the floor plan in the figure below and pass through each doorway exactly once. Is it possible if you must start and end outside?

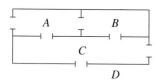

22. Repeat problem 21 for the floor plan in the figure below.

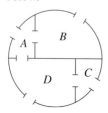

B **Non-Euclidean Geometry**

23. State the parallel postulate in Euclidean geometry.

24. State the parallel postulate in hyperbolic geometry.

25. State the parallel postulate in elliptic geometry.

26. In which geometry is the sum of the measures of the angles of a triangle 180°? (*Hint:* See the illustrations in Example 4.)

27. In which geometry is the sum of the measures of the angles of a triangle less than 180°? (*Hint:* See the illustrations in Example 4.)

28. In which geometry is the sum of the measures of the angles of a triangle more than 180°? (*Hint:* See the illustrations in Example 4.)

29. Which model is often used to describe a surface in Euclidean geometry?

30. Which model is often used to describe a surface in hyperbolic geometry?

31. Which model is often used to describe a surface in elliptic geometry?

32. In which geometry is a globe representing the surface of the Earth most appropriate?

C **Topology**

33. State which of the following saber saw and knife blades are topologically equivalent:

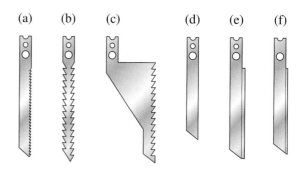

34. Which of the following statements are true and which are false?
  a. An ordinary lead pencil and a golf ball are topologically equivalent.
  b. A screw bolt and the nut that goes on it are topologically equivalent.
  c. A teacup with one handle and a teacup with two handles are topologically equivalent.
  d. An ordinary washer that goes on a bolt and the nut that screws it down are topologically equivalent.
  e. Two uncut fruits, a plum and a peach, are topologically equivalent.

35. State the genus of each of the blades shown in problem 33.

36. State the genus of each of the following:
  a. A comb

  b. A drawing aid

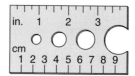

37. Assume that both the following are made of solid cord. Student Topo claims that the two are topologically equivalent, while student Geo claims they are not equivalent. Who do you think is correct? Explain. (*Hint:* Suppose that you cut through a loop of the left-hand figure and try to unwind it.)

38. Are the two clips shown here topologically equivalent? Explain.

 **In Other Words**

39. Explain in your own words why it is impossible for a network to have an odd number of odd vertices. (Try to construct one.)

40. Explain in your own words how a network can be used to plan a bus route that can be traversed without traveling any path twice.

Discovery

A plane curve that does not cross itself is called **simple,** just as in the case of a broken line. A **closed curve** is one that starts and ends at the same point. **A simple closed curve divides the plane into two parts, the region interior and the region exterior to the curve.** This important statement is the **Jordan curve theorem,** a very deep theorem and one that is very difficult to prove in spite of the fact that it seems so obvious.

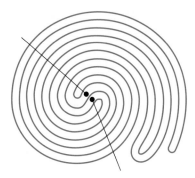

Which dot is inside?

A closed curve that is not simple divides the plane into three or more parts. We call the points where the curve crosses itself **vertices.** Any other points on the curve may also be designated as vertices. We mark the vertices with black dots as in the figure, where five vertices are indicated. The regions into which the plane is divided are numbered 1, 2, 3, and 4 in the figure. The simple curves with vertices as endpoints and containing no other endpoints are called **arcs.** In the figure, there are seven arcs. Can you count them?

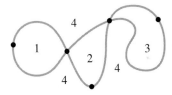

A nonsimple closed curve.

For the following networks (41 and 42), fill in the table on page 572 with the number of vertices (V), regions (R) into which the plane is divided, and arcs (A) for each figure. (Line segments connecting vertices are also called arcs.) See if you can discover a formula for A in terms of V and R. This formula is one form of the **Euler formula for networks.**

41.   **42.**

43. **44.**

45. **46.**

47. **48.**

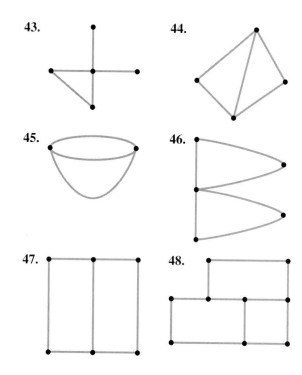

Figure	Vertices (V)	Regions (R)	Arcs (A)
41			
42			
43			
44			
45			
46			
47			
48			

 Web It Exercises

All the topics discussed in this section are thoroughly discussed on the Web. In some cases you need additional software to view the animation, but usually you can download the necessary tools for free. For example, go to link 8.6.1 on this textbook's Online Study Center and click on "NonEuclid Home." You can make a triangle in non-Euclidean geometry! Click on "Using NonEuclid—My First Triangle."

1. Find the definition of a parallel line in the text. According to that definition, can parallel lines intersect?

2. Now, suppose you have two lines L_1 and L_2 that are parallel to a third line L_3. What can you say about L_1 and L_2 in Euclidean geometry? What about them in hyperbolic geometry?

Do not answer yet; look at the diagram! For more explanation, click on "Parallel lines" at link 8.6.2 on this textbook's Online Study Center.

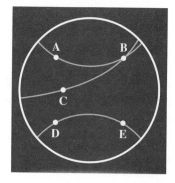

3. Remember the Cartesian coordinate system discussed in Section 7.1? Go to the site mentioned above and learn how to set up an xy coordinate system and then answer the question, How does the graph of the equation $y = x$ look in hyperbolic geometry?

Chapter **8** Summary

Section	Item	Meaning	Example
8.1A	$\overleftrightarrow{AB}$	The line AB	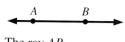
8.1A	Ray $\overrightarrow{AB}$	A half-line and its endpoint A	The ray AB 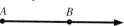

Section	Item	Meaning	Example
8.1A	Line segment $\overset{\bullet\quad\bullet}{AB}$	The points A and B and the part of the line between A and B	The line segment AB
8.1B	Skew lines	Lines that are not on the same plane	
8.1C	Plane angle	The figure formed by two rays with a common endpoint	
8.1C	Vertex of an angle	The common point of the two rays	
8.1C	Sides of the angles	The rays forming the angle	
8.1C	Degree	$\frac{1}{360}$th of a complete revolution	
8.1C	Straight angle	One-half of a complete revolution	
8.1C	Right angle	One-quarter of a complete revolution	
8.1C	Vertical angles	The opposite angles formed by two intersecting lines	
8.1C	Supplementary angles	Angles whose measures add to 180°	
8.1C	Transversal	A line that crosses two or more other lines	
8.1C	Complementary angles	Angles whose measures add to 90°	
8.1C	Perpendicular lines	Lines that intersect at right angles	
8.2A	Broken line	A sequence of connected straight line segments	
8.2A	Simple path	A path that does not cross itself	

Section	*Item*	*Meaning*	*Example*
8.2A	Closed path	A path that starts and ends at the same point	
8.2A	Polygon	A simple, closed broken line	
8.2A	Sides (of polygon)	The line segments of the path	
8.2A	Vertices (of polygon)	The endpoints of the sides of the polygon	
8.2A	Convex polygon	A polygon in which no line segment joining any two points on the boundary ever extends outside the polygon	
8.2A	Concave polygon	A polygon that is not convex	
8.2A	Regular polygon	A polygon with all sides of equal length and all angles of equal size	
8.2B	Equilateral triangle	A triangle with three equal sides	
8.2B	Right triangle	A triangle with one angle that is $90°$	
8.2B	Acute triangle	A triangle with three angles, each of which is less than $90°$	
8.2B	Obtuse triangle	A triangle with one angle that is greater than $90°$	
8.2B	Scalene triangle	A triangle with no equal sides	
8.2B	Isosceles triangle	A triangle with two equal sides	
8.2B	Trapezoid	A quadrilateral with two parallel and two nonparallel sides	

Section	Item	Meaning	Example
8.2B	Parallelogram	A quadrilateral with both pairs of opposite sides parallel	
8.2B	Rectangle	A parallelogram whose angles are right angles	
8.2B	Rhombus	A parallelogram with four equal sides	
8.2B	Square	A rectangle with four equal sides	
8.2B	Similar triangles	Triangles with exactly the same shape but not necessarily the same size	
8.2B	Congruent triangles	Triangles that have both the same shape and size	
8.2C	$S = (n - 2) \cdot 180°$	Sum of the measures of the angles of a polygon of n sides	
8.3A	Perimeter	Distance around a polygon	
8.3B	Circle	The set of all coplanar (on the same plane) points at a given fixed distance (the radius) from a point (the center)	
8.3B	Circumference $C = \pi d = 2\pi r$	The perimeter of a circle of diameter d (radius r)	
8.4A	$A = bh$	The area of a rectangle of base b and height h	
8.4A	$A = bh$	The area of a parallelogram of base b and height h	
8.4A	$A = \frac{1}{2}bh$	The area of a triangle of base b and height h	
8.4A	$A = \pi r^2$	The area of a circle of radius r	

Section	*Item*	*Meaning*	*Example*
8.4B	Pythagorean theorem $c^2 = a^2 + b^2$	The square of the hypotenuse c of a right triangle equals the sum of the squares of the other two sides, a and b.	
8.5A	Polyhedron	A solid bounded by plane polygons	
8.5A	Convex polyhedron	A polyhedron that lies entirely on one side of the plane of each of its faces	
8.5A	Concave polyhedron	A polyhedron that is not convex	
8.5A	Prism $V = Bh$	A polyhedron two of whose faces are parallel and whose edges that are not on these faces are all parallel	
8.5A	Pyramid $V = \frac{1}{3}Bh$	A polyhedron with all but one of its vertices in one plane The volume (V) of a pyramid or prism of height h and base B	
8.5B	$V = a^3$ $S = 6a^2$	The volume (V) and surface area (S) of a cube of edge a	
8.5B	$V = lwh$ $S = 2(lw + lh + wh)$	The volume (V) and surface area (S) of a rectangular box of length l, width w, and height h	
8.5B	$V = \pi r^2 h$ $S = 2\pi rh + 2\pi r^2$	The volume (V) and surface area (S) of a circular cylinder of radius r and height h	
8.5B	$V = \frac{1}{3}\pi r^2 h$ $S = \pi r^2 + \pi rs$	The volume (V) and surface area (S) of a circular cone of radius r, height h, and slant height s	

Section	*Item*	*Meaning*	*Example*
8.5B	$V = \frac{4}{3}\pi r^3$ $S = 4\pi r^2$	The volume (V) and surface area (S) of a sphere of radius r	
8.6A	Network	A connected set of line segments or curves	
8.6A	Traversable network	A network that can be drawn by tracing each line segment or curve exactly once without lifting the pencil from the paper	
8.6B	Euclid's fifth postulate	Given a line and a point not on that line, there is one and only one line through the point parallel to the original line.	
8.6B	Hyperbolic geometry	A type of geometry developed by Gauss, Bolyai, and Lobachevsky	Given a line and a point not on that line, there is *more* than one line through that point parallel to the original line.
8.6B	Elliptic geometry	A type of geometry developed by Bernhard Riemann	Given a line and a point not on that line, there is *no* line parallel to the original line.
8.6C	Topology	A branch of modern geometry	
8.6C	Topologically equivalent	The relation between two figures such that each can be obtained from the other by just changing the other's shape	□ and ⬭ are topologically equivalent.

Research Questions

Sources of information for these questions can be found in the Bibliography at the end of the book.

1. Write a report about the author of the phrase "There is no royal road to geometry" and that person's other contributions to geometry.

2. Write a paragraph about Thales of Miletus (640–546 B.C.) and his contributions to geometry.

3. Write a report on how surveyors and navigators use geometry. Read one of the following articles from *The World of Mathematics* by James Newman and write a report on its relevance to the geometry we have studied:
   a. "Commentary on Descartes and Analytical Geometry"
   b. "Commentary on a Famous Problem"
   c. "Projective Geometry"

4. Write a report on Camille Jordan (1838–1922), his discoveries in topology, and how they relate to geometry.

5. The Pythagorean theorem reveals three numbers (3, 4, 5) such that $a^2 + b^2 = c^2$. Write a report on the evolution of this theorem through different ages and civilizations.

6. In 1637, Pierre de Fermat scribbled a note proposing that there are no positive integers a, b, and c such that $a^n + b^n = c^n$, when $n > 2$. Write a report on Fermat's last theorem and the false proof of the theorem offered in March 1988. Has the theorem been proved?

7. Charlotte Angas Scott (1858–1931) wrote *An Introductory Account of Certain Modern Ideas and Methods in Plane Analytical Geometry.* Write a report on her work and her life.

8. Grace Chisholm Young (1868–1944) developed some interesting material in the field of solid geometry. In her *First Book of Geometry,* she included many diagrams that could be cut and folded to make three-dimensional figures. Try to make some of the figures that Young described.

9. Compare Euclidean geometry and non-Euclidean geometry.

10. Write a report about topology. Explain how you can show that two geometric figures are topologically equivalent by drawing them on a deflated balloon.

Chapter **8** Practice Test

1. Refer to the line shown below and state what each of the following describes:

   a. $\overleftrightarrow{WY} \cap \overleftrightarrow{XZ}$ **b.** $\overleftrightarrow{WY} \cap \overleftrightarrow{YZ}$ **c.** $\overleftrightarrow{WX} \cup \overrightarrow{XZ}$

   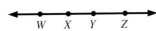

2. Sketch a triangular prism. Label the vertices; then name the edges that are
   a. parallel lines.   **b.** skew lines.   **c.** intersecting lines.

3. Through how many degrees does the hour hand of a clock turn in the following time intervals?
   a. 12 to 4 o'clock   **b.** 3 to 5 o'clock

4. If $\angle A$ and $\angle B$ are supplementary angles, and $m\angle A = 3 \times m\angle B$, find the measures of the two angles.

5. In the figure below, lines $\overleftrightarrow{PQ}$ and $\overleftrightarrow{RS}$ are parallel. Find
   a. $m\angle C$.   **b.** $m\angle E$.   **c.** $m\angle D$.

   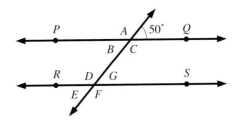

6. **a.** In a triangle ABC, $m\angle A = 38°$ and $m\angle B = 43°$. Find $m\angle C$.
   b. In a triangle ABC, $m\angle A = m\angle B = 2 \times m\angle C$. Find the measures of the three angles.

7. Sketch a broken line that is
   a. simple but not closed.   **b.** closed but not simple.

8. Triangles ABC and XYZ are similar, with $m\angle A = m\angle X$ and $m\angle B = m\angle Y$. If $\overline{AB}$, $\overline{BC}$, and $\overline{AC}$ are 2, 3, and 4 in. long, respectively, and $\overline{XY}$ is 3 in. long, find the lengths of $\overline{YZ}$ and $\overline{XZ}$.

9. What is the measure of one of the interior angles of a regular polygon with
   a. nine sides?   **b.** ten sides?

10. A rectangular plot of ground is to be enclosed with 120 yd of fencing. If the plot is to be twice as long as it is wide, what must be its dimensions?

11. The center of a circle is also the center of a square of side 2 cm. The circle passes through the four vertices of the square. Find the circumference of the circle. (Leave your answer in terms of π.)

12. If you rolled up an $8\frac{1}{2}$-by-11-in. sheet of paper into the largest possible cylinder $8\frac{1}{2}$ in. high, what would be the diameter of the cylinder? (Leave your answer in terms of π.)

13. In problem 11, find the area of the region that is inside the circle and outside the square. (Leave your answer in terms of π.)

14. A window is in the shape of a rectangle surmounted by a semicircle. The width of the window is 3 ft, which is also the diameter of the circle, and the height of the rectangular part is 4 ft. Find the total area of the window.

15. A circle of diameter 2 in. has its center at the center of a square of side 2 in. Find the area of the region that is inside the square and outside the circle.

16. A rectangle is 84 ft long and 13 ft wide. Find the length of its diagonal.

17. Find the total surface area of a rectangular box whose base is 3 ft wide and 5 ft long and whose height is 2 ft.

18. The triangle shown at the right is the base of a pyramid that is 4 ft high. What is the volume of this pyramid?

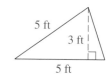

19. A solid consists of a cone and a hemisphere mounted base to base. If the radius of the common base is 2 in. and the volume of the cone is equal to the volume of the hemisphere, what is the height of the cone?

20. A sphere and a cylinder have the same radius, 10 in. If the total surface area of the cylinder equals the surface area of the sphere, what is the height of the cylinder?

21. Draw two networks, each with five vertices, such that one of them is traversable and the other is not traversable.

22. State Euclid's fifth postulate, the parallel postulate.

23. a. Which object is *not* topologically equivalent to the rest?
  b. What is the genus of the golf ball?
  c. What is the genus of the button?
  d. What is the genus of the paper clip?

Answers to Practice Test

ANSWER	IF YOU MISSED Question	REVIEW Section	Example(s)	Page(s)
1. a. $\overleftrightarrow{XY}$ **b.** Point Y **c.** $\overrightarrow{WZ}$	1	8.1	1	496
2.	2	8.1	Fig. 8.16, 2	497–498

a. $\overleftrightarrow{AD}$, $\overleftrightarrow{BE}$, and $\overleftrightarrow{CF}$; $\overleftrightarrow{EF}$ and $\overleftrightarrow{BC}$; $\overleftrightarrow{DE}$ and $\overleftrightarrow{AB}$; $\overleftrightarrow{DF}$ and $\overleftrightarrow{AC}$

b. $\overleftrightarrow{AB}$ and $\overleftrightarrow{CF}$; $\overleftrightarrow{AB}$ and $\overleftrightarrow{DF}$; $\overleftrightarrow{AB}$ and $\overleftrightarrow{EF}$; $\overleftrightarrow{BC}$ and $\overleftrightarrow{AD}$; $\overleftrightarrow{BC}$ and $\overleftrightarrow{DE}$; $\overleftrightarrow{BC}$ and $\overleftrightarrow{DF}$; $\overleftrightarrow{AC}$ and $\overleftrightarrow{BE}$; $\overleftrightarrow{AC}$ and $\overleftrightarrow{DE}$; $\overleftrightarrow{AC}$ and $\overleftrightarrow{EF}$; $\overleftrightarrow{DE}$ and $\overleftrightarrow{CF}$; $\overleftrightarrow{EF}$ and $\overleftrightarrow{AD}$; $\overleftrightarrow{DF}$ and $\overleftrightarrow{BE}$

c. $\overleftrightarrow{AB}$, $\overleftrightarrow{AC}$, and $\overleftrightarrow{AD}$; $\overleftrightarrow{AD}$, $\overleftrightarrow{DE}$, and $\overleftrightarrow{DF}$; $\overleftrightarrow{BE}$, $\overleftrightarrow{AB}$, and $\overleftrightarrow{BC}$; $\overleftrightarrow{BE}$, $\overleftrightarrow{EF}$, and $\overleftrightarrow{ED}$; $\overleftrightarrow{CF}$, $\overleftrightarrow{AC}$, and $\overleftrightarrow{BC}$; $\overleftrightarrow{CF}$, $\overleftrightarrow{EF}$, and $\overleftrightarrow{DF}$

ANSWER	IF YOU MISSED Question	REVIEW Section	Example(s)	Page(s)
3. a. $120°$ **b.** $60°$	3	8.1	3	498–499
4. $m\angle A = 135°$, $m\angle B = 45°$	4	8.1	4	500
5. a. $m\angle C = 130°$ **b.** $m\angle E = 50°$ **c.** $m\angle D = 130°$	5	8.1	5	501–502
6. a. $m\angle C = 99°$ **b.** $m\angle A = m\angle B = 72°$, $m\angle C = 36°$	6	8.1	6	502–503
7. a. **b.**	7	8.2	Figs. 8.33, 8.34	510

ANSWER	IF YOU MISSED Question	REVIEW Section	Example(s)	Page(s)
8. $\overline{XZ} = 6$ in., $\overline{YZ} = 4\frac{1}{2}$ in.	8	8.2	4, 5	514–515
9. a. $140°$ **b.** $144°$	9	8.2	8	519
10. 20 yd by 40 yd	10	8.3	1–3	527
11. $2\sqrt{2}\pi$ cm	11	8.3	4, 5	528–529
12. $\frac{11}{\pi}$ in.	12	8.3	5	529
13. $(2\pi - 4)$ cm^2	13	8.4	1–3	535–537
14. $12 + 9\ \pi/8 \approx 15.5$ ft^2	14	8.4	1–3	535–537
15. $4 - \pi \approx 0.86$ in.^{2}	15	8.4	1–3	535–537
16. 85 ft	16	8.4	5, 6	539–540
17. 62 ft^2	17	8.5	2	550–551
18. 10 ft^3	18	8.5	3	551
19. 4 in.	19	8.5	4	551–552
20. 10 in.	20	8.5	5	552

ANSWER	IF YOU MISSED		REVIEW		
		Question	Section	Example(s)	Page(s)
21. Traversable Not traversable		21	8.6	1	562
22. Given a line and any point not on that line, there is one and only one line through that point that is parallel to the given line.		22	8.6	Fig. 8.86	564
23. **a.** The button **b.** 0 **c.** 4 **d.** 0		23	8.6	5, 6	566, 567

As advancements in technology occur, keeping information secure becomes more important. Check digits are used in order to secure information on many items such as credit cards, money orders, and airline tickets. In Section 9.3, you will study modular arithmetic, its operations, and applications such as the use of check digits for security purposes.

Mathematical Systems and Matrices

In this chapter we will study different mathematical systems. First, we consider rectangular arrays of numbers called **matrices** and the operations that can be performed with these matrices, as well as their applications. As it turns out, matrices can be used to solve systems of linear equations by using *elementary row operations*. They can also be used to code and decode messages, a method employed by intelligence services around the world.

We then start the study of mathematical systems by considering *clock arithmetic* on a 12- or 5-hour clock and generalize the idea to *modular* arithmetic. We continue with a discussion of abstract mathematical systems, including two mathematical structures called **groups** and **fields,** and end the chapter with a study of game theory, a subject with numerous applications to business, games of chance, and military science.

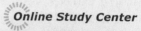

Online Study Center

For links to Internet sites related to Chapter 9, please access college.hmco.com/PIC/bello9e and click on the Online Study Center icon.

9.1 Matrix Operations

Applications of Matrices

The information in Table 9.1 is presented as a rectangular array of numbers. Thus, the entry in the third row, fourth column tells us that a 176-lb person would burn off 250 calories by gardening for 1 hour. In mathematics, when such arrays are enclosed in brackets, the result is called a **matrix.** In this section we shall study matrices and their addition, subtraction, and multiplication.

TABLE 9.1 How Fast You Burn Off Calories*

| Moderate Activity | Weight (in pounds) | | | | | |
	110	132	154	176	187	209
	Calories Burned per Hour					
Housework	175	210	245	285	300	320
Making bed	165	185	210	230	245	255
Gardening	155	190	215	250	265	280
Mowing lawn	195	225	250	280	295	305
Watching TV	60	70	80	85	90	95
Golf (foursome)	210	240	270	295	310	325
Tennis	335	380	425	470	495	520
Bowling	210	240	270	300	310	325
Jogging	515	585	655	725	760	795
Walking briskly	355	400	450	500	525	550
Dancing	350	395	445	490	575	540
Cycling	190	215	245	270	280	295

*Read down the column closest to the given weight for the approximate use of calories per hour.

Suppose that we wish to know how many calories are burned in 3 hours using the same activities and weights listed in the matrix. We simply multiply each entry by 3. For example, the entry in the fourth row, third column should be 750, telling us that a 154-lb person would burn 750 calories by mowing the lawn for 3 hours.

Matrices have a wide range of applications, including spreadsheets, inventory analysis, and communications. In this section we shall study the operations we can perform using matrices. ▶

Since the late 1940s, matrices have become an important tool in business and the social sciences, just as they have been in the physical sciences and technology from decades earlier. Matrices can be used to store information. For example, suppose the ABC Sports Company manufactures two types of billiard tables, regular and deluxe. Each of these models requires bolts, clamps, and screws according to Table 9.2. If we remember that the rows are labeled bolts, clamps, and screws, in that order, then the rectangular array of numbers on the right.

TABLE 9.2

Hardware	Regular	Deluxe
Bolts	5	3
Clamps	2	4
Screws	7	10

$$R = \begin{bmatrix} 5 & 3 \\ 2 & 4 \\ 7 & 10 \end{bmatrix}$$

Looking Ahead
Some of Germain's communication with Gauss involved her work in modular arithmetic, which is the focus in Section 9.3.

 GRAPH IT

Your grapher does matrices, but you have to specify the dimension and name of the matrix first. Thus, to enter the matrix A, press MATRX ◄ 1 2 ENTER 2 ENTER. Now enter the elements of A by pressing 1 ENTER 2 ENTER (−) 3 ENTER and 0 ENTER as shown.

```
MATRIX (A) 2 ×2
[1       2      ]
[-3      0      ]



2,2=0
```

Practice some more by entering B and C!

gives all the necessary information regarding the assembly requirements of each billiard table.

Definition of a Matrix

A rectangular array of numbers enclosed in square brackets (or sometimes parentheses) is called a **matrix.** (The plural is **matrices.**) The entries in a matrix are called its **elements.**

The matrix

$$\begin{bmatrix} a_{11} & a_{12} \\ a_{21} & a_{22} \\ a_{31} & a_{32} \end{bmatrix}$$

where the subscripts give the **row** and **column** of each element, has 3 rows and 2 columns. We speak of it as a 3 by 2 (3×2) matrix. The number of rows and the number of columns in a matrix are called the **dimensions** of the matrix. Thus, a matrix with m rows and n columns is an $m \times n$ matrix.

For convenience, we often denote matrices by single capital letters, such as A, B, and C. We might write

$$A = \begin{bmatrix} 1 & 2 \\ -3 & 0 \end{bmatrix}, \qquad B = \begin{bmatrix} 1 \\ 3 \\ 5 \end{bmatrix}, \qquad \text{and} \qquad C = \begin{bmatrix} 1 & 0 & -2 & 4 \end{bmatrix}$$

Of these, A is a 2×2 matrix, B is a 3×1 matrix, and C is a 1×4 matrix. A matrix such as B, consisting of a single column, is called a **column matrix.** Likewise, a matrix such as C, consisting of a single row, is called a **row matrix.** A matrix such as A, having the same number of columns as rows, is called a **square matrix.**

Definition of Equality of Matrices

Two matrices A and B are said to be **equal** (denoted $A = B$) if and only if the corresponding elements are equal. (This means that they have the same dimensions and that each element of A equals the element of B in the same row and column.)

For example, if

$$A = \begin{bmatrix} 1 & 2 \\ -3 & 0 \end{bmatrix} \overset{\neq}{\qquad} B = \begin{bmatrix} 1 & 4 \\ -3 & 0 \end{bmatrix} \qquad C = \begin{bmatrix} 1 & 2 \\ -3 & 0 \end{bmatrix}$$

then $A \neq B$ because the element in the first row, second column of A is 2 and that in B is 4. On the other hand, $A = C$ because each element of A is the same as the corresponding element of C.

A. Addition, Subtraction, and Multiplication by a Scalar

Suppose that the ABC Sports Company makes 5 of each of its 2 models of billiard tables. Of course, it would need 5 times as many bolts, clamps, and screws as for just one of each. As you can see, the matrix of parts requirements for 5 of each table is obtained by multiplying each element of the original matrix by 5.

$$5 \times \begin{bmatrix} 5 & 3 \\ 2 & 4 \\ 7 & 10 \end{bmatrix} = \begin{bmatrix} 5 \times 5 & 5 \times 3 \\ 5 \times 2 & 5 \times 4 \\ 5 \times 7 & 5 \times 10 \end{bmatrix} = \begin{bmatrix} 25 & 15 \\ 10 & 20 \\ 35 & 50 \end{bmatrix}$$

If you write R for the original parts requirement matrix, then it is natural to write $5R$ for the new matrix. Thus,

$$5R = 5\begin{bmatrix} 5 & 3 \\ 2 & 4 \\ 7 & 10 \end{bmatrix} = \begin{bmatrix} 5 \cdot 5 & 5 \cdot 3 \\ 5 \cdot 2 & 5 \cdot 4 \\ 5 \cdot 7 & 5 \cdot 10 \end{bmatrix} = \begin{bmatrix} 25 & 15 \\ 10 & 20 \\ 35 & 50 \end{bmatrix}$$

This idea motivates the definition of multiplication of a matrix by a real number.

> **Definition of Scalar Multiplication of a Matrix**
>
> If A is a matrix and k is a real number, then the **scalar multiplication of A by k,** symbolized by kA or Ak, is the operation that multiplies every element of A by k. The number k is called the **scalar.**

Thus, if

$$A = \begin{bmatrix} 1 & 2 \\ -3 & 0 \end{bmatrix}$$

then

$$-2A = \begin{bmatrix} (-2) \times 1 & (-2) \times 2 \\ (-2) \times (-3) & (-2) \times 0 \end{bmatrix} = \begin{bmatrix} -2 & -4 \\ 6 & 0 \end{bmatrix}$$

EXAMPLE 1 ▶ Finding Components in Scalar Multiplication

If

$$A = \begin{bmatrix} 1 & x \\ 0 & y \end{bmatrix} \quad \text{and} \quad B = \begin{bmatrix} 2 & 6 \\ 0 & -10 \end{bmatrix}$$

find x and y to satisfy the equation $2A = B$.

Solution

Substituting for A and B in the equation $2A = B$, we get

$$2\begin{bmatrix} 1 & x \\ 0 & y \end{bmatrix} = \begin{bmatrix} 2 & 6 \\ 0 & -10 \end{bmatrix}$$

Thus,

$$\begin{bmatrix} 2 & 2x \\ 0 & 2y \end{bmatrix} = \begin{bmatrix} 2 & 6 \\ 0 & -10 \end{bmatrix}$$

By the definition of equality of matrices,

$$2x = 6 \quad \text{and} \quad 2y = -10$$

Hence, $x = 3$ and $y = -5$. ∎

Suppose that a work crew at the ABC Sports Company assembles 6 of each of their billiard tables during the first half of the day and 5 of each during the second half. (It gets a little bit tired of regulars and deluxes before the day is over.) What is the matrix of parts requirements for the day's work?

We see that the matrix $6R$ will be the appropriate matrix for the first half-day and that $5R$ will be the matrix for the second half-day. Thus,

$$6R = \begin{bmatrix} 30 & 18 \\ 12 & 24 \\ 42 & 60 \end{bmatrix} \begin{matrix} \text{Bolts} \\ \text{Clamps} \\ \text{Screws} \end{matrix} \quad \text{and} \quad 5R = \begin{bmatrix} 25 & 15 \\ 10 & 20 \\ 35 & 50 \end{bmatrix} \begin{matrix} \text{Bolts} \\ \text{Clamps} \\ \text{Screws} \end{matrix}$$

are the matrices for the parts requirements for the two half-days. We can see that the total requirements can be found by adding the corresponding elements as follows:

$$\begin{bmatrix} 30 + 25 & 18 + 15 \\ 12 + 10 & 24 + 20 \\ 42 + 35 & 60 + 50 \end{bmatrix} = \begin{bmatrix} 55 & 33 \\ 22 & 44 \\ 77 & 110 \end{bmatrix}$$

It seems perfectly natural to regard this as $6R + 5R$, the sum of the two matrices. These and similar considerations motivate the following definition of addition of matrices:

Definition of the Sum of Two Matrices

The **sum** of two matrices A and B *of the same dimensions* is written $A + B$ and is the matrix obtained by adding corresponding elements of A and B.

GRAPH IT

To do Example 2(b) on this page, first enter A and B as matrices. Now go to the home screen by pressing ⌈2nd⌉ ⌈MODE⌋. Finally, enter $2A + 3B$ by pressing 2 ⌈MATRX⌉ 1 ⌊+⌋ 3 ⌈MATRX⌉ 2 ⌈ENTER⌋. The result is shown.

```
2[A]+3[B]
          [[7    17]
           [18   14]]
```

EXAMPLE 2 ▶ Addition of Matrices

If

$$A = \begin{bmatrix} 2 & 1 \\ 3 & 4 \end{bmatrix} \quad \text{and} \quad B = \begin{bmatrix} 1 & 5 \\ 4 & 2 \end{bmatrix}$$

find

(a) $A + B$. (b) $2A + 3B$.

Solution

(a) $A + B = \begin{bmatrix} 2 + 1 & 1 + 5 \\ 3 + 4 & 4 + 2 \end{bmatrix} = \begin{bmatrix} 3 & 6 \\ 7 & 6 \end{bmatrix}$

(b) $2A + 3B = \begin{bmatrix} 4 & 2 \\ 6 & 8 \end{bmatrix} + \begin{bmatrix} 3 & 15 \\ 12 & 6 \end{bmatrix} = \begin{bmatrix} 7 & 17 \\ 18 & 14 \end{bmatrix}$ ∎

Note that, in general, the addition of matrices of the same dimension is commutative, that is, $A + B = B + A$.

Definition of the Difference of Two Matrices

If A and B are two matrices of the same dimensions, then the **difference** $A - B$ is given by $A - B = A + (-1)B$, which is the same as the matrix formed by subtracting each element of B from the corresponding element of A.

EXAMPLE 3 ▶ **Finding the Difference of Two Matrices**

If A and B are as in Example 2, find $2A - 3B$.

Solution
With $2A$ and $3B$ as in Example 2, we have

$$2A - 3B = \begin{bmatrix} 4 & 2 \\ 6 & 8 \end{bmatrix} - \begin{bmatrix} 3 & 15 \\ 12 & 6 \end{bmatrix}$$

$$= \begin{bmatrix} 4 - 3 & 2 - 15 \\ 6 - 12 & 8 - 6 \end{bmatrix} = \begin{bmatrix} 1 & -13 \\ -6 & 2 \end{bmatrix} \qquad ■$$

B. Matrix Multiplication

Let us return to the ABC Sports Company and recall that the assembly requirement matrix was

$$
R = \begin{array}{c} \\ \text{Bolts} \\ \text{Clamps} \\ \text{Screws} \end{array}
\begin{array}{cc} \text{Regular} & \text{Deluxe} \\ \begin{bmatrix} 5 & 3 \\ 2 & 4 \\ 7 & 10 \end{bmatrix} \end{array}
$$

Suppose now that Jane's work crew assembles regulars and Jim's assembles deluxes. Jane turns out 18 regulars and Jim turns out 16 deluxes per day. We can specify the total output by means of the column matrix

$$
C = \begin{bmatrix} 18 \\ 16 \end{bmatrix} \quad \begin{array}{l} \text{Regular} \\ \text{Deluxe} \end{array}
$$

Can we use these matrices to calculate how many bolts, clamps, and screws are needed for the daily output of Jane's and Jim's worker crews? Yes! We proceed as follows: To calculate the number of bolts, we use a row-column type of multiplication, as indicated by the following scheme:

$$[\,⑤\quad ③\,]\begin{bmatrix} ⑱ \\ ⑯ \end{bmatrix} = 5 \times 18 + 3 \times 16 = 90 + 48 = 138$$

Similarly, we find the number of clamps to be

$$[\,②\quad ④\,]\begin{bmatrix} ⑱ \\ ⑯ \end{bmatrix} = 2 \times 18 + 4 \times 16 = 36 + 64 = 100$$

and the number of screws to be

$$[\textcircled{7}\quad \textcircled{10}]\begin{bmatrix} \textcircled{18} \\ \textcircled{16} \end{bmatrix} = 7 \times 18 + 10 \times 16 = 126 + 160 = 286.$$

Next, let us write the result of our computation in the briefer form

Bolts $\qquad$ $\begin{bmatrix} 5 & 3 \\ 2 & 4 \\ 7 & 10 \end{bmatrix}\begin{bmatrix} 18 \\ 16 \end{bmatrix} = \begin{bmatrix} 138 \\ 100 \\ 286 \end{bmatrix}$ $\qquad$ Bolts
Clamps $\qquad\qquad\qquad\qquad\qquad\qquad\qquad$ Clamps
Screws $\qquad\qquad\qquad\qquad\qquad\qquad\qquad$ Screws

If we call the daily parts requirement matrix D, then we can write

$$RC = D$$

and regard RC as a type of product. Notice that this product is a column matrix in which the first element is obtained from a row-column multiplication using the first row of R. Similarly, the second element is obtained in the same way using the second row of R, and likewise for the third element.

Now suppose the ABC Sports Company needs to produce 18 regular and 16 deluxe billiard tables today and 10 regular and 20 deluxe tables tomorrow. To represent the production matrix P for the 2 days, we write

Today $\quad$ Tomorrow
$$P = \begin{bmatrix} 18 & 10 \\ 16 & 20 \end{bmatrix}\quad \begin{matrix} \text{Regular} \\ \text{Deluxe} \end{matrix}$$

How many bolts, clamps, and screws are needed in the next 2 days? We need to make the computation for tomorrow only because we have already done it for today. Thus,

$$\begin{bmatrix} 5 & 3 \\ 2 & 4 \\ 7 & 10 \end{bmatrix}\begin{bmatrix} 10 \\ 20 \end{bmatrix} = \begin{bmatrix} 5 \times 10 + 3 \times 20 \\ 2 \times 10 + 4 \times 20 \\ 7 \times 10 + 10 \times 20 \end{bmatrix} = \begin{bmatrix} 110 \\ 100 \\ 270 \end{bmatrix}$$

If we denote the final requirement matrix by W, then it is natural to write

Today $\quad$ Tomorrow
Bolts $\qquad$ $\begin{bmatrix} 5 & 3 \\ 2 & 4 \\ 7 & 10 \end{bmatrix}\begin{bmatrix} 18 & 10 \\ 16 & 20 \end{bmatrix} = \begin{bmatrix} 138 & 110 \\ 100 & 100 \\ 286 & 270 \end{bmatrix}$
Clamps
Screws

or

$$RP = W$$

where each element in W is calculated by the following rule: **The element of W in the ith row and jth column is the row-column product of the ith row of R and the jth column of P.**

For example, the element of W in the first row, second column position is the row-column product of the first row of R and second column of P; that is,

$$[5\quad 3]\begin{bmatrix} 10 \\ 20 \end{bmatrix} = 5 \times 10 + 3 \times 20 = 110$$

Check all the elements of W to be sure that you understand the rule.

> **Definition of Matrix Multiplication**
>
> The matrix $W = RP$ is called the **product** of the matrices R and P in that order, and the calculation of W according to the rule given on the previous page is called **matrix multiplication.**

The preceding discussion can easily be generalized to the multiplication of two matrices A and B. Suppose that A has as many columns as B has rows. Then we can do a row-column multiplication with any row of A and any column of B. Thus, we have the following definition:

> **Definition of Conformable Matrices**
>
> The matrices A and B are said to be **conformable** with respect to multiplication in the order AB if the number of columns in A is the same as the number of rows in B. In this case, each element of the product AB is formed as follows: The element in row i and column j of AB is the row-column product of the ith row of A and the jth column of B.

If two matrices are not conformable, then we do not attempt to define their product. Notice that A and B might be conformable for multiplication in the order AB but not in the order BA. Thus, if A is $m \times n$ and B is $n \times k$, then A and B are conformable for multiplication in the order AB but not in the order BA unless $m = k$. Note that AB will be an $m \times k$ matrix. For instance, if A is 2×3 and B is 3×4, then we can form AB, and it will be a 2×4 matrix, but we cannot form BA.

Note: Even when the matrices are conformable for both orders of multiplication, there is no reason why the results need to be equal. In fact, it is generally true that $AB \neq BA$.

This failure of matrix multiplication to be commutative was one of the exciting properties of matrices to the mathematicians who first studied them.

EXAMPLE 4 ▶ Finding Products of Matrices

Let

$$A = \begin{bmatrix} 2 & 1 & 0 \\ 3 & 2 & 4 \end{bmatrix} \quad \text{and} \quad B = \begin{bmatrix} 1 & 4 \\ 3 & 1 \\ 2 & 3 \end{bmatrix}$$

Calculate the product AB.

Solution

Because A is 2×3 and B is 3×2, they are conformable. We can proceed, according to the definition of conformable matrices, to get the 2×2 product matrix.

$$
\begin{aligned}
AB &= \begin{bmatrix} 2 & 1 & 0 \\ 3 & 2 & 4 \end{bmatrix}\begin{bmatrix} 1 & 4 \\ 3 & 1 \\ 2 & 3 \end{bmatrix} \\
&= \begin{bmatrix} 2 \times 1 + 1 \times 3 + 0 \times 2 & 2 \times 4 + 1 \times 1 + 0 \times 3 \\ 3 \times 1 + 2 \times 3 + 4 \times 2 & 3 \times 4 + 2 \times 1 + 4 \times 3 \end{bmatrix} \\
&= \begin{bmatrix} 5 & 9 \\ 17 & 26 \end{bmatrix}
\end{aligned}
$$

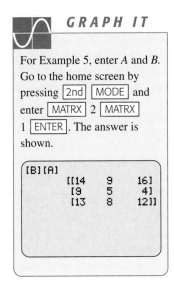

EXAMPLE 5 ▶ **Finding More Products of Matrices**

Using the same matrices as in Example 4, form the product BA (if possible).

Solution

We first check that B has as many columns as A has rows (two in each case). This shows that the matrices are conformable for multiplication in the order BA. Thus, by the definition of conformable matrices, we get the 3×3 product matrix.

$$BA = \begin{bmatrix} 1 & 4 \\ 3 & 1 \\ 2 & 3 \end{bmatrix} \begin{bmatrix} 2 & 1 & 0 \\ 3 & 2 & 4 \end{bmatrix}$$

$$= \begin{bmatrix} 1 \times 2 + 4 \times 3 & 1 \times 1 + 4 \times 2 & 1 \times 0 + 4 \times 4 \\ 3 \times 2 + 1 \times 3 & 3 \times 1 + 1 \times 2 & 3 \times 0 + 1 \times 4 \\ 2 \times 2 + 3 \times 3 & 2 \times 1 + 3 \times 2 & 2 \times 0 + 3 \times 4 \end{bmatrix}$$

$$= \begin{bmatrix} 14 & 9 & 16 \\ 9 & 5 & 4 \\ 13 & 8 & 12 \end{bmatrix}$$

■

Examples 4 and 5 offer unassailable evidence that in general $AB \neq BA$. If A and B were both square matrices of the same size, do you think that $AB = BA$ always? Try it for some simple 2×2 matrices!

C. The Identity Matrix

Is there an identity for matrix multiplication? The answer is yes. For example, the multiplication of any 2×2 matrix A by

$$I = \begin{bmatrix} 1 & 0 \\ 0 & 1 \end{bmatrix}$$

gives the matrix A back again. We find that

$$IA = AI = A \qquad \text{Try it for } \begin{bmatrix} a & b \\ c & d \end{bmatrix}.$$

It can be shown that I is the only matrix that has this property for *all* 2×2 matrices. Because of this uniqueness, I is called the **multiplicative identity** for 2×2 matrices.

Similarly for 3×3 matrices, the multiplicative identity is the matrix

$$I = \begin{bmatrix} 1 & 0 & 0 \\ 0 & 1 & 0 \\ 0 & 0 & 1 \end{bmatrix}$$

The **main diagonal** of a square matrix is the diagonal set of numbers running from the upper left corner to the lower right corner of the matrix. The two identity matrices, one for 2×2 and the other for 3×3 matrices, are particular examples of identity matrices which, in general, have 1s on the main diagonal and 0s everywhere else.

> **The Identity Matrix**
>
> The $n \times n$ **identity** matrix is the $n \times n$ matrix with 1s on the main diagonal and 0s for all other elements.

Additional properties of matrices are discussed in the Discovery section of Exercises 9.1.

D. Applications

Matrices can be used to simplify the display of information in real-life applications. For example, according to the U.S. Division of Labor Statistics, the monthly expenditures (in dollars) for entertainment (E), food (F), and housing (H) per person under 25 and 25–34 years old are as shown.

$$
\begin{array}{c}
 \\
< \textbf{25} \\
\textbf{25–34}
\end{array}
\begin{array}{ccc}
E & F & H
\end{array} \\
\begin{pmatrix}
90 & 220 & 490 \\
140 & 370 & 880
\end{pmatrix}
$$

EXAMPLE 6 ▶ Applications to Entertainment, Food, and Housing

(a) What would be the annual expenditures for persons in the categories under 25 and 25–34?

(b) If it is estimated that entertainment, food, and housing will increase by 10%, 15%, and 20%, respectively, how much more would each person in the under 25 and 25–34 categories have to spend each month?

Solution

(a) To find out the annual expenditures, we multiply by 12.

$$
12 \begin{array}{c}
\begin{array}{ccc} E & F & H \end{array} \\
\begin{pmatrix} 90 & 220 & 490 \\ 140 & 370 & 880 \end{pmatrix}
\end{array}
=
\begin{array}{c}
\begin{array}{ccc} E & F & H \end{array} \\
\begin{pmatrix} 1080 & 2640 & 5880 \\ 1680 & 4440 & 10560 \end{pmatrix}
\end{array}
$$

(b) To find the increase, we multiply the original matrix by $\begin{pmatrix} 0.10 \\ 0.15 \\ 0.20 \end{pmatrix}$, obtaining

$$
\begin{pmatrix} 90 & 220 & 490 \\ 140 & 370 & 880 \end{pmatrix}
\begin{pmatrix} 0.10 \\ 0.15 \\ 0.20 \end{pmatrix}
=
\begin{pmatrix} 90 \cdot 0.10 + 220 \cdot 0.15 + 490 \cdot 0.20 \\ 140 \cdot 0.10 + 370 \cdot 0.15 + 880 \cdot 0.20 \end{pmatrix}
$$

$$
= \begin{pmatrix} 140 \\ 245.50 \end{pmatrix}
$$

Thus, persons under 25 will pay $140 more, and persons 25–34 will pay $245.50 more.

Online Study Center

To further explore matrix operations, access links 9.1.1 and 9.1.2 on this textbook's Online Study Center.

EXERCISES 9.1

(A) **Addition, Subtraction, and Multiplication by a Scalar**

For problems 1–6, suppose that

$$A = \begin{bmatrix} 2 & 1 \\ 0 & -1 \end{bmatrix} \quad B = \begin{bmatrix} -2 & 4 \\ 3 & 1 \end{bmatrix} \quad C = \begin{bmatrix} 3 & 5 \\ 2 & 0 \end{bmatrix}$$

1. Find the following:
 a. $4A$ **b.** $-3B$ **c.** $C - B$

2. Find the following:
 a. $2A + B$ **b.** $4A + 3B$ **c.** $2A - C$

3. Find the following:
 a. $A + B + C$ **b.** $A + B - C$

4. Find the following:
 a. $A - B + 2C$ **b.** $2A + 2B - C$

5. Find the following:
 a. $7A + 4B - 2C$ **b.** $2A - 2B - 3C$

6. Find the following:
 a. $5A + 4B - 4C$ **b.** $4A - 2B - 4C$

For problems 7–12, suppose that

$$A = \begin{bmatrix} 1 & -1 & 2 \\ 3 & 0 & -2 \\ 4 & 2 & 1 \end{bmatrix} \quad B = \begin{bmatrix} -1 & 2 & 1 \\ 4 & 3 & -1 \\ 0 & 1 & -1 \end{bmatrix}$$

$$C = \begin{bmatrix} 0 & -1 & 3 \\ 1 & -2 & 4 \\ 3 & -3 & 0 \end{bmatrix}$$

7. Find the following:
 a. $A + B$ **b.** $A - B$ **c.** $A - C$

8. Find the following:
 a. $2A$ **b.** $-2B$ **c.** $-3C$

9. Find the following:
 a. $2A + 3B$ **b.** $-2A + 3B$

10. Find the following:
 a. $4A + 4B$ **b.** $4A - 4B$

11. Find the following:
 a. $3A - 2C$ **b.** $B + C$

12. Find the following:
 a. $3A + B - 4C$ **b.** $5A + 2B + 3C$

(B) **Matrix Multiplication**

In problems 13–20, use

$$A = \begin{bmatrix} 1 & -2 & 1 \\ 2 & 0 & 2 \\ -1 & 1 & 3 \end{bmatrix} \quad B = \begin{bmatrix} 3 & 2 & 0 \\ 1 & 1 & -1 \\ 2 & 0 & 1 \end{bmatrix}$$

$$C = \begin{bmatrix} 1 & 0 & 2 \\ 3 & 2 & 1 \\ 2 & 0 & 1 \end{bmatrix}$$

and evaluate the given expression.

13. AB **14.** AC

15. BA **16.** CA

17. $(A - B)(A + B)$ **18.** $(A - C)(A + C)$

19. $A^2 - B^2$ **20.** $A^2 - C^2$

(C) **The Identity Matrix**

21. If I is the 2×2 identity matrix, find I^2.

22. If n is a positive integer, is it true that $I^n = I$? Why or why not?

In problems 23–27, verify that $AB = I$ and calculate BA.

23. $A = \begin{bmatrix} 2 & 1 \\ 1 & 1 \end{bmatrix}$; $B = \begin{bmatrix} 1 & -1 \\ -1 & 2 \end{bmatrix}$

24. $A = \begin{bmatrix} -2 & 3 \\ 1 & -1 \end{bmatrix}$; $B = \begin{bmatrix} 1 & 3 \\ 1 & 2 \end{bmatrix}$

25. $A = \begin{bmatrix} 2 & 5 \\ 1 & 3 \end{bmatrix}$; $B = \begin{bmatrix} 3 & -5 \\ -1 & 2 \end{bmatrix}$

26. $A = \begin{bmatrix} 1 & 0 & 1 \\ 0 & 1 & 1 \\ 1 & 0 & 0 \end{bmatrix}$; $B = \begin{bmatrix} 0 & 0 & 1 \\ -1 & 1 & 1 \\ 1 & 0 & -1 \end{bmatrix}$

27. $A = \begin{bmatrix} 1 & 0 & 1 \\ 0 & 2 & -1 \\ 2 & 1 & 2 \end{bmatrix}$; $B = \begin{bmatrix} 5 & 1 & -2 \\ -2 & 0 & 1 \\ -4 & -1 & 2 \end{bmatrix}$

28. Let R be the row matrix $[2 \quad 1]$. Is it possible for there to be a 2×2 matrix, say, J, such that $JR = RJ = R$? Explain.

29. If A is a nonsquare matrix of dimensions $m \times n$, do you think there could be a square matrix J such that $AJ = JA = A$? Explain.

30. Find all possible 2×2 matrices A such that $A = 3A$.

31. A square matrix A is called **idempotent** if $A^2 = A$. All identity matrices and all zero matrices (elements all 0s) are idempotent. Show that the following matrices are also idempotent:

a. $\begin{bmatrix} 1 & 2 \\ 0 & 0 \end{bmatrix}$ **b.** $\begin{bmatrix} \frac{1}{2} & \frac{1}{2} \\ \frac{1}{2} & \frac{1}{2} \end{bmatrix}$

32. A square matrix A is called **nilpotent** if there is a positive integer n such that A^n is a zero matrix (elements all 0s). Show that the following matrices are nilpotent:

a. $\begin{bmatrix} 2 & 1 \\ -4 & -2 \end{bmatrix}$ **b.** $\begin{bmatrix} 0 & 0 & 1 \\ 2 & 0 & 3 \\ 0 & 0 & 0 \end{bmatrix}$

D Applications

33. The E-Z Rest Furniture Company makes armchairs and rocking chairs in three models: E, an economy model; M, a medium-priced model; and L, a luxury model. Each month the company turns out 20 model E armchairs, 15 model M armchairs, 10 model L armchairs, 12 model E rockers, 8 model M rockers, and 5 model L rockers.
a. Write this information as a 2×3 matrix.
b. Use your answer to part (**a**) to obtain a matrix showing the total production for 6 months.

34. Suppose the costs of materials for E-Z Rest's armchairs (see problem 33) are $30 for model E, $35 for model M, and $45 for model L, and for its rockers, $35 for model E, $40 for model M, and $60 for model L.
a. Write this information as a 3×2 matrix.
b. Suppose that costs increase by 20%. Write the new matrix by multiplying your answer to part (**a**) by 1.2.

35. At the end of 6 months, the E-Z Rest Furniture Company (see problem 33) has sold armchairs as follows: 90 model E, 75 model M, and 50 model L. They also sold rockers as follows: 60 model E, 20 model M, and 30 model L. Use matrix methods to obtain a matrix showing how many of each model are left at the end of the 6 months. Use your answer to problem 33 and assume that there was no unsold stock at the beginning of the period.

36. The ABC Sports Company figures that the bolts, clamps, and screws used in assembling its billiard tables cost, respectively, 5¢, 10¢, and 2¢ each. Find the cost of these assembly materials for each table by carrying out the multiplication CR, where C is the **cost matrix** $[5 \quad 10 \quad 2]$ and R is the assembly requirement matrix given at the beginning of section B (page 584).

37. The ABC Sports Company has the projected production schedule shown in the table below for the 5 months preceding December.

Model	July	Aug.	Sept.	Oct.	Nov.
Regular	100	200	300	400	300
Deluxe	50	100	200	200	300

Let M be the matrix with these numbers as the elements, and let R be the same as in problem 36. Use these matrices to find the schedule of assembly requirements for the 5 months and fill in the missing items in the following table:

Hardware	July	Aug.	Sept.	Oct.	Nov.
Bolts	650				
Clamps		800			
Screws			4100		

38. Let A be the matrix of assembly requirements you calculated in problem 37, and let C be the cost matrix of problem 36. Calculate CA to find the cost of assembly materials for each of the 5 months.

39. Tom, Dick, and Harry are in a computer network whose communication matrix C is as shown. The 1s in the first row indicate that Tom can communicate with Dick and Harry directly.

$$\begin{array}{c} \\ T \\ D \\ H \end{array} \begin{array}{c} T \ \ D \ \ H \\ \begin{bmatrix} 0 & 1 & 1 \\ 1 & 0 & 0 \\ 1 & 0 & 0 \end{bmatrix} \end{array} = C$$

a. Find C^2, the matrix of possible two-step communications in the network.
b. What does the 2 in the first row, first column mean?
c. How many two-step communications between two different people are shown in C^2?

40. Suppose you weigh 132 lb. Referring to Table 9.1 in Getting Started, how many hours of TV watching do you need to equal 1 hour of housework?

41. Let the point (a, b) in the XY plane be represented by the column matrix

$$\begin{bmatrix} a \\ b \end{bmatrix}$$

Then the right triangle with vertices at $(1, 0)$, $(1, 2)$, and $(0, 2)$ can be represented by the matrix

$$\begin{bmatrix} 1 & 1 & 0 \\ 0 & 2 & 2 \end{bmatrix}$$

Draw this triangle. Then multiply

$$\begin{bmatrix} -2 & 0 \\ 0 & 2 \end{bmatrix} \begin{bmatrix} 1 & 1 & 0 \\ 0 & 2 & 2 \end{bmatrix}$$

to get the vertices of another triangle. Draw the new triangle and explain what the multiplication did to the original triangle.

 In Other Words

42. The system of matrices with the operations of addition, subtraction, and multiplication is a mathematical system. Name and discuss all the *differences* you see between the system of matrices and the real numbers. (For example, if a and b are any real numbers, $a + b = b + a$. Is this true for any two matrices A and B?)

 Using Your Knowledge

Here are six special 2×2 matrices.

$$r_1 = \begin{bmatrix} 1 & 0 \\ 0 & -1 \end{bmatrix} \quad r_2 = \begin{bmatrix} -1 & 0 \\ 0 & 1 \end{bmatrix} \quad r_3 = \begin{bmatrix} 0 & 1 \\ 1 & 0 \end{bmatrix}$$

$$R_1 = \begin{bmatrix} 0 & -1 \\ 1 & 0 \end{bmatrix} \quad R_2 = \begin{bmatrix} -1 & 0 \\ 0 & -1 \end{bmatrix} \quad R_3 = \begin{bmatrix} 0 & 1 \\ -1 & 0 \end{bmatrix}$$

It is interesting to see what effect these matrices have when they multiply a column matrix

$$P = \begin{bmatrix} a \\ b \end{bmatrix}$$

that represents a point (a, b) in the xy plane. For example,

$$r_1 \times P = \begin{bmatrix} 1 & 0 \\ 0 & -1 \end{bmatrix} \begin{bmatrix} a \\ b \end{bmatrix} = \begin{bmatrix} a \\ -b \end{bmatrix}$$

which represents the point $(a, -b)$, the reflection of (a, b) across the x axis (see the following figure). Because the matrices r_1, r_2, and r_3 all correspond to reflections across certain lines in the plane, we can think of these three matrices as **reflectors.**

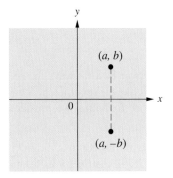

Similarly,

$$R_1 \times P = \begin{bmatrix} 0 & -1 \\ 1 & 0 \end{bmatrix} \begin{bmatrix} a \\ b \end{bmatrix} = \begin{bmatrix} -b \\ a \end{bmatrix}$$

represents the point $(-b, a)$, which can be obtained from the point (a, b) by a rotation of $90°$ around the origin (see the following figure). Because all three of the matrices R_1, R_2, and R_3 correspond to rotations in the plane, we can think of them as **rotators.**

What is accomplished by each of the following?

43. $r_2 \times P$ **44.** $r_3 \times P$

45. $R_2 \times P$ **46.** $R_3 \times P$

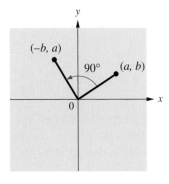

47. A triangle with vertices at $(2, 1)$, $(3, 1)$, and $(3, 2)$ can be represented by the matrix

$$T = \begin{bmatrix} 2 & 3 & 3 \\ 1 & 1 & 2 \end{bmatrix}$$

Form the products $r_1 \times T$, $r_2 \times T$, and so on. Then draw graphs to show the effect of each multiplication.

48. Consider another special matrix

$$K = \begin{bmatrix} k & 0 \\ 0 & k \end{bmatrix}$$

What happens to the triangle in problem 47 if you form the product $K \times T$? Try this for $k = 2$.

 **Discovery**

We want to discover in what ways the set S of 2×2 matrices behaves like the set of integers under addition and multiplication. To do this, we consider the following questions:

49. Is S closed with respect to addition? (*Hint:* Is the sum of two 2×2 matrices a 2×2 matrix?)

50. Is S associative with respect to addition?

51. Can you find an additive identity for the set S? *Hint:* If

$$A = \begin{bmatrix} 1 & 3 \\ 4 & 6 \end{bmatrix}$$

find a matrix B such that

$$\begin{bmatrix} 1 & 3 \\ 4 & 6 \end{bmatrix} + \begin{bmatrix} & \\ & \end{bmatrix} = \begin{bmatrix} 1 & 3 \\ 4 & 6 \end{bmatrix}$$

52. Does every matrix in S have an additive inverse? For example, if

$$A = \begin{bmatrix} 1 & 2 \\ -3 & 4 \end{bmatrix}$$

try to find a matrix B such that

$$\begin{bmatrix} 1 & 2 \\ -3 & 4 \end{bmatrix} + \begin{bmatrix} & \\ & \end{bmatrix} = \begin{bmatrix} 0 & 0 \\ 0 & 0 \end{bmatrix}$$

53. Does $A + B = B + A$ for any two 2×2 matrices in S?

54. Consider the set S of 2×2 matrices.
  a. Is S closed with respect to multiplication? (*Hint:* Is the product of two 2×2 matrices a 2×2 matrix?)
  b. Is S associative with respect to multiplication?
  c. Can you find a multiplicative identity for the set S?

55. The inverse of a 2×2 matrix A is defined to be a matrix B such that $AB = BA = I$, where I is the 2×2 identity matrix. Can you show that the matrix given below has no inverse?

$$A = \begin{bmatrix} 1 & 2 \\ 0 & 0 \end{bmatrix}$$

56. Can you discover whether the matrix given below has an inverse?

$$A = \begin{bmatrix} 2 & 4 \\ 1 & 2 \end{bmatrix}$$

 Collaborative Learning

Example 6 gives the average expenditures for entertainment, food, and housing for persons under 25 and 25–34. Form three groups.

1. Have each of the groups construct a matrix similar to the one in Example 6 showing the expenses for a particular month. To construct each of the entries in the matrix, take the *average* expenses for the members in the group. Are the results close to the national averages given in Example 6? Discuss why or why not.

2. To find the average *monthly* expenses in each category for *all* three groups, which operation would you have to use?

3. Find the matrix representing the average monthly expenses for entertainment, food, and housing for *all* three groups.

4. How would you estimate the *annual* expenses for entertainment, food, and housing for persons in your group?

 Skill Checker

In the next section we are going to discuss a method of solving systems of equations using matrices. To be successful, you have to remember how to solve a system of two equations in two unknowns. Here are some practice problems. If you do not remember how to do these problems, review the material in Section 7.6. Solve the following systems:

1. $3x + y = 4$
   $2x + y = 3$

2. $2x - y = 3$
   $4x - 3y = 7$

3. $-2x + 3y = -10$
   $5x - 2y = 14$

4. $4x - 3y = -2$
   $-3x + 2y = 0$

 *Online Study Center*

For more practice, access link 9.1.3 on this textbook's Online Study Center.

9.2 Solving Systems of Linear Equations with Matrices

Once upon a Matrix

Have you ever read *Alice in Wonderland*? Do you know who wrote this book? Lewis Carroll did. Although better known as the author of this popular tale, Carroll was a mathematician and logician. He wrote another book about Alice, called *Through the Looking Glass*. In this book, a conversation between the characters Tweedledee and Tweedledum included the following exchange:

Tweedledee: The sum of your weight and twice mine is 361 pounds.
Tweedledum: Contrariwise, the sum of your weight and twice mine is 360 pounds.

Alice meets Tweedledee and Tweedledum in
Through the Looking Glass.

If Tweedledee weighs x pounds and Tweedledum weighs y pounds, recall from Section 7.6 that the two sentences can be translated as

$$2x + y = 361$$
$$x + 2y = 360$$

Do you remember how to solve this system? You are going to solve it again, but this time you will write the equivalent operations using matrices. Follow each of the steps very carefully.

Given.	$\begin{aligned} 2x + y &= 361 \\ x + 2y &= 360 \end{aligned}$	$\begin{bmatrix} 2 & 1 & \vert & 361 \\ 1 & 2 & \vert & 360 \end{bmatrix}$
Multiply the second equation by -2.	$\begin{aligned} 2x + y &= 361 \\ -2x - 4y &= -720 \end{aligned}$	$\begin{bmatrix} 2 & 1 & \vert & 361 \\ -2 & -4 & \vert & -720 \end{bmatrix}$
Add the two equations.	$\begin{aligned} 2x + y &= 361 \\ -3y &= -359 \end{aligned}$	$\begin{bmatrix} 2 & 1 & \vert & 361 \\ 0 & -3 & \vert & -359 \end{bmatrix}$
Divide the second equation by -3.	$\begin{aligned} 2x + y &= 361 \\ y &= 119\tfrac{2}{3} \end{aligned}$	$\begin{bmatrix} 2 & 1 & \vert & 361 \\ 0 & 1 & \vert & 119\tfrac{2}{3} \end{bmatrix}$

Now you can substitute $119\frac{2}{3}$ for y in the equation $2x + y = 361$ and solve for x. Did you notice that the operations performed on the equations were identical to those performed on the matrices? The matrices used are called the **augmented matrices,** and they can be written more quickly because the variables are omitted. In this section you will solve systems of three equations using matrices. ▶

We have considered systems of two linear equations in two unknowns and discovered how to solve such systems. There are many practical applications that require the consideration of m linear equations in n unknowns. We shall not try to be so general here; instead, we shall look in detail only into the case of three equations in three unknowns. The techniques we shall use are applicable to all the more general cases.

A. Solution by Matrices

As you can see from Getting Started, the first step in solving a system of equations is to write the **augmented matrix,** a matrix consisting of the coefficients of the variables and the constants in the equation. The system of equations and the augmented matrix in Getting Started and another system with three variables and its augmented matrix are shown.

$$
\begin{array}{cc}
\textit{System of Equations} & \textit{Augmented Matrix} \\
\begin{aligned}
2x + y &= 361 \\
x + 2y &= 360
\end{aligned} \quad \rightarrow &
\begin{bmatrix} 2 & 1 & | & 361 \\ 1 & 2 & | & 360 \end{bmatrix}
\end{array}
$$

$$
\begin{aligned}
2x - y + z &= 3 \\
x + y &= -1 \quad \rightarrow \\
3x - y - 2z &= 7
\end{aligned}
\qquad
\begin{bmatrix} 2 & -1 & 1 & | & 3 \\ 1 & 1 & 0 & | & -1 \\ 3 & -1 & -2 & | & 7 \end{bmatrix}
$$

Note that the bar separates the coefficients of the variables from the constants and that a 0 is entered as the coefficient of the variable z in the second equation because z was not present.

We can solve the system of equations

$$
\begin{aligned}
2x - y + z &= 3 \\
x + y &= -1 \\
3x - y - 2z &= 7
\end{aligned}
\tag{1}
$$

by means of matrices using operations similar to the elementary operations used in Chapter 7. These new operations are called *row operations* and produce **row-equivalent** matrices having the same solutions as the original system. When two matrices A and B are **row equivalent,** we shall write $A \sim B$. Here is the procedure we will use to solve system (1).

$$
\boxed{\text{Start with}} \Rightarrow
\begin{bmatrix} 2 & -1 & 1 & | & 3 \\ 1 & 1 & 0 & | & -1 \\ 3 & -1 & -2 & | & 7 \end{bmatrix}
\quad \boxed{\text{end with}} \Rightarrow
\begin{bmatrix} 1 & 0 & 0 & | & c_1 \\ 0 & 1 & 0 & | & c_2 \\ 0 & 0 & 1 & | & c_3 \end{bmatrix}
$$

Note: The second augmented matrix tells us that $x = c_1$, $y = c_2$, and $z = c_3$.

The operations we can use are as follows:

Elementary Row Operations on Matrices

1. Interchanging the order of the rows

2. Multiplying all the elements of a row by any nonzero number

3. Multiplying all the elements of a row by any nonzero number and adding the corresponding numbers to any other row of numbers

---

To symbolize the steps, we use the notation R_1, R_2, and R_3 for the respective rows of the matrix, along with the following typical abbreviations:

Notation	*Meaning*
1. $R_1 \leftrightarrow R_2$	Interchange R_1 and R_2.
2. $2 \times R_1$	Multiply each element of R_1 by 2.
3. $2 \times R_1 + R_2 \to R_2$	Replace R_2 by $2 \times R_1 + R_2$.

Thus, we write

$$\begin{bmatrix} 2 & -1 & 1 & | & 3 \\ 1 & 1 & 0 & | & -1 \\ 3 & -1 & -2 & | & 7 \end{bmatrix} \sim \begin{bmatrix} 1 & 1 & 0 & | & -1 \\ 2 & -1 & 1 & | & 3 \\ 3 & -1 & -2 & | & 7 \end{bmatrix}$$

$$R_1 \leftrightarrow R_2$$

Next, we proceed to get 0s in the second and third rows of the first column.

$$\begin{bmatrix} 1 & 1 & 0 & | & -1 \\ 2 & -1 & 1 & | & 3 \\ 3 & -1 & -2 & | & 7 \end{bmatrix} \sim \begin{bmatrix} 1 & 1 & 0 & | & -1 \\ 0 & -3 & 1 & | & 5 \\ 0 & -4 & -2 & | & 10 \end{bmatrix}$$

$$-2 \times R_1 + R_2 \to R_2$$
$$-3 \times R_1 + R_3 \to R_3$$

Then we get a 0 in the third row of the second column.

$$\begin{bmatrix} 1 & 1 & 0 & | & -1 \\ 0 & -3 & 1 & | & 5 \\ 0 & -4 & -2 & | & 10 \end{bmatrix} \sim \begin{bmatrix} 1 & 1 & 0 & | & -1 \\ 0 & -3 & 1 & | & 5 \\ 0 & 0 & -10 & | & 10 \end{bmatrix}$$

$$-4 \times R_2 + 3 \times R_3 \to R_3$$

Now we multiply each member of the third row by $-\frac{1}{10}$ and obtain

$$\begin{bmatrix} 1 & 1 & 0 & | & -1 \\ 0 & -3 & 1 & | & 5 \\ 0 & 0 & 1 & | & -1 \end{bmatrix} \qquad \text{This means that } z = -1.$$

$$-\frac{1}{10} \times R_3 \to R_3$$

To get a 0 in the second row, third column, we multiply the third row by -1 and add it to the second row.

$$\begin{bmatrix} 1 & 1 & 0 & | & -1 \\ 0 & -3 & 0 & | & 6 \\ 0 & 0 & 1 & | & -1 \end{bmatrix}$$

$$-1 \times R_3 + R_2 \to R_2$$

Now we multiply the second row by $-\frac{1}{3}$.

$$\begin{bmatrix} 1 & 1 & 0 & | & -1 \\ 0 & 1 & 0 & | & -2 \\ 0 & 0 & 1 & | & -1 \end{bmatrix} \qquad \text{This means that } y = -2.$$

$$-\frac{1}{3} \times R_3 \to R_2$$

Finally, we multiply the second row by -1 and add it to the first row.

$$\begin{bmatrix} 1 & 0 & 0 & | & 1 \\ 0 & 1 & 0 & | & -2 \\ 0 & 0 & 1 & | & -1 \end{bmatrix}$$ This means that $x = 1$.

$-1 \times R_2 + R_1 \rightarrow R_1$

Thus, system (1) has the unique solution $x = 1$, $y = -2$, and $z = -1$. We can verify this by substituting 1, -2, and -1 for x, y, and z, respectively, in all three equations in system (1). For example, the second equation says that $x + y = -1$. For $x = 1$ and $y = -2$, we get $1 + (-2) = -1$, which is true. Verify this solution in the other two equations.

We shall now look at additional illustrations of this procedure, but these will help only if we take a pencil and paper and carry out the detailed row operations as they are indicated.

GRAPH IT

Suppose you want to solve the equation $4x = 8$. Multiply both sides of the equation by 4^{-1}, the inverse of the coefficient of the unknown, to obtain $4^{-1} \times 4x = 4^{-1} \times 8$, or $x = 2$. Now enter A, the coefficient matrix, and B, the constant matrix, as shown.

$$A = \begin{bmatrix} 2 & -1 & 2 \\ 2 & 2 & -1 \\ -1 & 2 & 2 \end{bmatrix}$$

$$B = \begin{bmatrix} 3 \\ 0 \\ -12 \end{bmatrix}$$

Let X be the unknown. If $AX = B$, $X = A^{-1}B$ as before. To find $A^{-1}B$, go to the home screen and enter MATRX 1 x^{-1} MATRX 2 ENTER . The result is read as $x = 2$, $y = -3$, and $z = -2$.

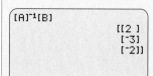

EXAMPLE 1 ▶ **Solving Systems with Three Variables Using Matrices**

Solve the system

$$\begin{aligned} 2x - y + 2z &= 3 \\ 2x + 2y - z &= 0 \\ -x + 2y + 2z &= -12 \end{aligned}$$

Solution

The augmented matrix is

$$\begin{bmatrix} 2 & -1 & 2 & | & 3 \\ 2 & 2 & -1 & | & 0 \\ -1 & 2 & 2 & | & -12 \end{bmatrix}$$

To get a 0 in row 2, column 1, multiply R_1 by -1 and add the result to R_2. To get a 0 in row 3, column 1, multiply R_3 by 2 and add the result to R_1.

$$\begin{bmatrix} 2 & -1 & 2 & | & 3 \\ 0 & 3 & -3 & | & -3 \\ 0 & 3 & 6 & | & -21 \end{bmatrix}$$

$-R_1 + R_2 \rightarrow R_2$
$R_1 + 2 \times R_3 \rightarrow R_3$

To get a 0 in row 3, column 2, multiply R_2 by -1 and add the result to R_3. To get a 1 in row 2, column 2, multiply R_2 by $\frac{1}{3}$.

$$\begin{bmatrix} 2 & -1 & 2 & | & 3 \\ 0 & 1 & -1 & | & -1 \\ 0 & 0 & 9 & | & -18 \end{bmatrix}$$

$-R_2 + R_3 \rightarrow R_3$
$\frac{1}{3} \times R_2 \rightarrow R_2$

To get a 1 in row 3, column 3, multiply R_3 by $\frac{1}{9}$. Then add row 3 to row 2 to get a 0 in row 2, column 3.

$$\begin{bmatrix} 2 & -1 & 2 & | & 3 \\ 0 & 1 & 0 & | & -3 \\ 0 & 0 & 1 & | & -2 \end{bmatrix}$$

This means that $y = -3$.
This means that $z = -2$.

$$\frac{1}{9} \times R_3 \to R_3$$
$$R_3 + R_2 \to R_2$$

Add row 2 and row 1 to get a 0 in row 1, column 2. Multiply row 3 by -2 and add the result to row 1 to get a 0 in row 1, column 3.

$$\begin{bmatrix} 2 & 0 & 0 & | & 4 \\ 0 & 1 & 0 & | & -3 \\ 0 & 0 & 1 & | & -2 \end{bmatrix}$$

$$R_2 + R_1 \to R_1$$
$$-2 \times R_3 + R_1 \to R_1$$

Finally, multiply row 1 by $\frac{1}{2}$.

$$\begin{bmatrix} 1 & 0 & 0 & | & 2 \\ 0 & 1 & 0 & | & -3 \\ 0 & 0 & 1 & | & -2 \end{bmatrix}$$

This means that $x = 2$.

$$\frac{1}{2} \times R_1 \to R_1$$

The final answer, $x = 2$, $y = -3$, and $z = -2$, can be checked in the given system. ∎

EXAMPLE 2 ▶ Systems with No Solution

Solve the system

$$2x - y + 2z = 3$$
$$2x + 2y - z = 0$$
$$4x + y + z = 5$$

Solution

To get a 0 in row 2, column 1, multiply R_1 by -1 and add the result to R_2. To get a 0 in row 3, column 1, multiply R_1 by -2 and add the result to R_3.

$$\begin{bmatrix} 2 & -1 & 2 & | & 3 \\ 2 & 2 & -1 & | & 0 \\ 4 & 1 & 1 & | & 5 \end{bmatrix} \sim \begin{bmatrix} 2 & -1 & 2 & | & 3 \\ 0 & 3 & -3 & | & -3 \\ 0 & 3 & -3 & | & -1 \end{bmatrix}$$

$$-R_1 + R_2 \to R_2$$
$$-2 \times R_1 + R_3 \to R_3$$

To get a 0 in row 3, column 2, multiply R_2 by -1 and add the result to R_3.

$$\begin{bmatrix} 2 & -1 & 2 & | & 3 \\ 0 & 3 & -3 & | & -3 \\ 0 & 0 & 0 & | & 2 \end{bmatrix}$$

$$-R_2 + R_3 \to R_3$$

The last row corresponds to the equation

$$0x + 0y + 0z = 2$$

which is false for all values of x, y, and z. Hence, the given system has *no solution.*   ∎

Note: It is important to notice that if the procedure introduces any row with all 0s to the left and a nonzero number to the right of the vertical line, then the system has *no* solution.

EXAMPLE 3 ▶ **Systems with Infinitely Many Solutions**

Solve the system

$$\begin{aligned} 2x - y + 2z &= 3 \\ 2x + 2y - z &= 0 \\ 4x + y + z &= 3 \end{aligned}$$

Solution

To get a 0 in row 2, column 1, multiply R_1 by -1 and add the result to R_2. To get a 0 in row 3, column 1, multiply R_1 by -2 and add the result to R_3.

$$\begin{bmatrix} 2 & -1 & 2 & | & 3 \\ 2 & 2 & -1 & | & 0 \\ 4 & 1 & 1 & | & 3 \end{bmatrix} \sim \begin{bmatrix} 2 & -1 & 2 & | & 3 \\ 0 & 3 & -3 & | & -3 \\ 0 & 3 & -3 & | & -3 \end{bmatrix}$$

$$-R_1 + R_2 \rightarrow R_2$$
$$-2 \times R_1 + R_3 \rightarrow R_3$$

To get a 0 in row 3, column 2, multiply R_2 by -1 and add the result to R_3.

$$\begin{bmatrix} 2 & -1 & 2 & | & 3 \\ 0 & 3 & -3 & | & -3 \\ 0 & 0 & 0 & | & 0 \end{bmatrix}$$

$$-R_2 + R_3 \rightarrow R_3$$

The last row corresponds to the equation

$$0x + 0y + 0z = 0$$

which is true for all values of x, y, and z. We now show that the system has infinitely many solutions and any solution of the first two equations will be a solution of the system.

The first two equations are

$$\begin{aligned} 2x - y + 2z &= 3 \\ 3y - 3z &= -3 \end{aligned}$$

This system is equivalent to

$$\begin{aligned} 2x - y &= 3 - 2z \\ y &= -1 + z \end{aligned} \qquad \text{Solve the second equation for } y.$$

Let $z = k$, where k is any real number. Then the last equation gives $y = -1 + k$. Substitution into the first equation results in

$$2x + 1 - k = 3 - 2k$$

so that

$$2x = 2 - k$$
$$x = 1 - \tfrac{1}{2}k$$

Thus, if k is any real number, then $x = 1 - \tfrac{1}{2}k$, $y = k - 1$, and $z = k$ is a solution of the system. You should verify this by substituting into the original system. The system in this example has infinitely many solutions because the value of k can be quite arbitrarily chosen. For instance, if $k = 2$, then the solution is $x = 0$, $y = 1$, and $z = 2$; if $k = 5$, then $x = -\tfrac{3}{2}$, $y = 4$, and $z = 5$; if $k = -4$, then $x = 3$, $y = -5$, and $z = -4$; and so on. ∎

Note: If in the final form of the matrix there is no row with all 0s to the left and a nonzero number to the right of the vertical line but there is a row with all 0s both to the left and to the right, then the system has *infinitely many solutions.*

Examples 1, 2, and 3 illustrate the three possibilities for three linear equations in three unknowns. The system may have a **a unique solution** as in Example 1, the system may have **no solution** as in Example 2, and the system may have **infinitely many solutions** as in Example 3. The final form of the matrix always shows which case is at hand.

B. Applications

PROBLEM SOLVING

Matrices and Nutrition

❶ Read the problem.

A dietitian wants to arrange a diet composed of three basic foods A, B, and C. The diet must include 170 units of calcium, 90 units of iron, and 110 units of vitamin B. Table 9.3 gives the number of units per ounce of each of the needed ingredients contained in each of the basic foods.

TABLE 9.3

Ingredients	Units per Ounce		
	Food A	Food B	Food C
Calcium	15	5	20
Iron	5	5	10
Vitamin B	10	15	10

If a, b, and c are the number of ounces of basic foods A, B, and C taken by an individual, find the number of ounces of each of the basic foods needed to meet the diet requirements.

❷ Select the unknown.

You want to find the values of a, b, and c, the number of ounces of basic foods A, B, and C taken by an individual.

❸ Think of a plan.
What is the amount of calcium needed?

Since the individual gets 15 units of calcium from A, 5 from B, and 20 from C, the amount of calcium is

$$15a + 5b + 20c = 170$$

continued

What is the amount of iron needed?

The amount of iron is

$$5a + 5b + 10c = 90$$

What is the amount of vitamin B needed?

The amount of vitamin B is

$$10a + 15b + 10c = 110$$

Write the equations obtained using matrices. How can you simplify each of the three original equations?

The simplified system of three equations and three unknowns, obtained by dividing each term in each of the equations by 5, is written as

$$\begin{bmatrix} 3 & 1 & 4 & | & 34 \\ 1 & 1 & 2 & | & 18 \\ 2 & 3 & 2 & | & 22 \end{bmatrix}$$

❹ **Use** matrices to solve the system.

Interchange R_1 and R_2.

$$\begin{bmatrix} 3 & 1 & 4 & | & 34 \\ 1 & 1 & 2 & | & 18 \\ 2 & 3 & 2 & | & 22 \end{bmatrix} \sim \begin{bmatrix} 1 & 1 & 2 & | & 18 \\ 3 & 1 & 4 & | & 34 \\ 2 & 3 & 2 & | & 22 \end{bmatrix}$$

$$R_1 \leftrightarrow R_2$$

$$\sim \begin{bmatrix} 1 & 1 & 2 & | & 18 \\ 0 & -2 & -2 & | & -20 \\ 0 & 1 & -2 & | & -14 \end{bmatrix} \sim \begin{bmatrix} 1 & 1 & 2 & | & 18 \\ 0 & 1 & 1 & | & 10 \\ 0 & 1 & -2 & | & -14 \end{bmatrix}$$

$$\begin{array}{ll} R_2 - 3R_1 \rightarrow R_2 & -\frac{1}{2} \times R_2 \rightarrow R_2 \\ R_3 - 2R_1 \rightarrow R_3 & \end{array}$$

$$\sim \begin{bmatrix} 1 & 1 & 2 & | & 18 \\ 0 & 1 & 1 & | & 10 \\ 0 & 0 & -3 & | & -24 \end{bmatrix} \sim \begin{bmatrix} 1 & 1 & 2 & | & 18 \\ 0 & 1 & 1 & | & 10 \\ 0 & 0 & 1 & | & 8 \end{bmatrix}$$

$$\begin{array}{ll} R_3 - R_2 \rightarrow R_3 & -\frac{1}{3} \times R_3 \rightarrow R_3 \end{array}$$

$$\sim \begin{bmatrix} 1 & 1 & 0 & | & 2 \\ 0 & 1 & 0 & | & 2 \\ 0 & 0 & 1 & | & 8 \end{bmatrix} \sim \begin{bmatrix} 1 & 0 & 0 & | & 0 \\ 0 & 1 & 0 & | & 2 \\ 0 & 0 & 1 & | & 8 \end{bmatrix}$$

$$\begin{array}{ll} -R_3 + R_2 \rightarrow R_2 & -R_2 + R_1 \rightarrow R_1 \\ -2 \times R_3 + R_1 \rightarrow R_1 & \end{array}$$

This means that $a = 0$, $b = 2$, and $c = 8$. Thus, we need 0 oz of A, 2 oz of B, and 8 oz of C to meet the requirements.

❺ **Verify** the solution.

Substitute $a = 0$, $b = 2$, and $c = 8$ into the first equation; you obtain $15 \cdot 0 + 5 \cdot 2 + 20 \cdot 8 = 10 + 160 = 170$. Use the same procedure to check the second and third equations.

TRY EXAMPLE 4 NOW.

Cover the solution, write your own solution, and then check your work.

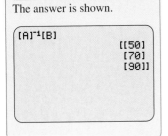
TABLE 9.4

Size	Price per Pound
Small	50¢
Medium	70¢
Large	90¢

EXAMPLE 4 ▶ Applications Tough As Nails

Tom Jones, who was building a workshop, went to the hardware store and bought 1 lb each of three kinds of nails: small, medium, and large. After completing part of the work, Tom found that he had underestimated the number of small and large nails he needed for the project. So he bought another pound of the small nails and 2 lb more of the large nails. After completing some more of the construction, he again ran short of nails and had to buy another pound of each of the small and the medium nails. While looking over his bills, he found that the hardware store had charged him $2.10 for nails the first time, $2.30 the second time, and $1.20 the third time. The prices for the various sizes of nails were not listed. Find these prices.

Solution

We let x¢/lb, y¢/lb, and z¢/lb be the prices for the small, medium, and large nails, respectively. Then, we know that

$$x + y + \ z = 210$$
$$x \quad\ \ + 2z = 230$$
$$x + y \quad\ \ = 120$$

We solve this system as follows:

$$\begin{bmatrix} 1 & 1 & 1 & | & 210 \\ 1 & 0 & 2 & | & 230 \\ 1 & 1 & 0 & | & 120 \end{bmatrix} \sim \begin{bmatrix} 1 & 1 & 1 & | & 210 \\ 0 & 1 & -1 & | & -20 \\ 0 & 0 & 1 & | & 90 \end{bmatrix}$$

$$R_1 - R_2 \rightarrow R_2$$
$$R_1 - R_3 \rightarrow R_3$$

$$\sim \begin{bmatrix} 1 & 1 & 1 & | & 210 \\ 0 & 1 & 0 & | & 70 \\ 0 & 0 & 1 & | & 90 \end{bmatrix} \sim \begin{bmatrix} 1 & 0 & 0 & | & 50 \\ 0 & 1 & 0 & | & 70 \\ 0 & 0 & 1 & | & 90 \end{bmatrix}$$

$$R_3 + R_2 \rightarrow R_2 \qquad -R_3 + R_1 \rightarrow R_1$$
$$-R_2 + R_1 \rightarrow R_1$$

This means that $x = 50$, $y = 70$, and $z = 90$, giving the schedule of prices in Table 9.4. ∎

EXAMPLE 5 ▶ Finding the Number of Calls

Do you have a cell phone? Do you know what a LATA is? LATA stands for Local Access Transport Area; the areas are unique to the telecommunication industry. Telephone calls between two parties in different LATAs must, by law, be transported by a long-distance carrier. The map shows the state of New York divided into LATA regions. Suppose long-distance calls between Kingston (region 133), Albany (region 134), and Syracuse (region 136) and the percentage of calls from one LATA to another are as described in Figure 9.1. A survey reveals that the number of calls per minute to Kingston was 115, to Albany 205, and to Syracuse 80. How many calls came from Kingston, Albany, and Syracuse, respectively?

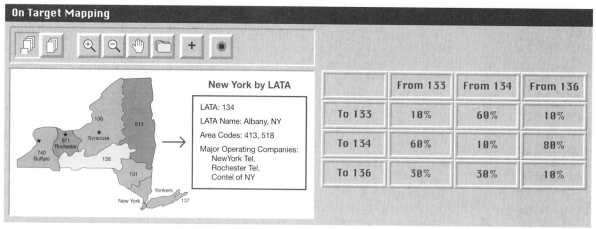

Source: www.tetrad.com.

FIGURE 9.1

Solution

We solve the system

$$\begin{bmatrix} 0.10 & 0.60 & 0.10 & | & 115 \\ 0.60 & 0.10 & 0.80 & | & 205 \\ 0.30 & 0.30 & 0.10 & | & 80 \end{bmatrix}$$

First, multiply all entries by 10.

$$\begin{bmatrix} 1 & 6 & 1 & | & 1150 \\ 6 & 1 & 8 & | & 2050 \\ 3 & 3 & 1 & | & 800 \end{bmatrix} \sim \begin{bmatrix} 1 & 6 & 1 & | & 1150 \\ 0 & -35 & 2 & | & -4850 \\ 0 & -15 & -2 & | & -2650 \end{bmatrix}$$

$$\begin{array}{c} -6R_1 + R_2 \to R_2 \\ -3R_1 + R_3 \to R_3 \end{array}$$

$$\sim \begin{bmatrix} 1 & 6 & 1 & | & 1150 \\ 0 & -50 & 0 & | & -7500 \\ 0 & -15 & -2 & | & -2650 \end{bmatrix} \sim \begin{bmatrix} 1 & 6 & 1 & | & 1150 \\ 0 & 1 & 0 & | & 150 \\ 0 & -15 & -2 & | & -2650 \end{bmatrix}$$

$$R_3 + R_2 \to R_2 \qquad\qquad R_2/-50 \to R_2$$

$$\sim \begin{bmatrix} 1 & 0 & 1 & | & 250 \\ 0 & 1 & 0 & | & 150 \\ 0 & -15 & -2 & | & -2650 \end{bmatrix} \sim \begin{bmatrix} 1 & 0 & 1 & | & 250 \\ 0 & 1 & 0 & | & 150 \\ 0 & 0 & -2 & | & -400 \end{bmatrix}$$

$$-6R_2 + R_1 \to R_1 \qquad\qquad 15R_2 + R_3 \to R_3$$

$$\sim \begin{bmatrix} 1 & 0 & 1 & | & 250 \\ 0 & 1 & 0 & | & 150 \\ 0 & 0 & 1 & | & 200 \end{bmatrix} \sim \begin{bmatrix} 1 & 0 & 0 & | & 50 \\ 0 & 1 & 0 & | & 150 \\ 0 & 0 & 1 & | & 200 \end{bmatrix}$$

$$R_3/-2 \to R_3 \qquad\qquad R_1 - R_3 \to R_1$$

This means that 50 calls came from Kingston, 150 from Albany, and 200 from Syracuse. ■

EXERCISES 9.2

A Solution by Matrices

In problems 1–10, find all the solutions (if possible).

1.
$x + y - z = 3$
$x - 2y + z = -3$
$2x + y + z = 4$

2.
$x + 2y - z = 5$
$2x + y + z = 1$
$x - y + z = -1$

3.
$2x - y + 2z = 5$
$2x + y - z = -6$
$3x + 2z = 3$

4.
$x + 2y - z = 0$
$2x + 3y = 3$
$2y + z = -1$

5.
$3x + 2y + z = -5$
$2x - y - z = -6$
$2x + y + 3z = 4$

6.
$4x + 3y - z = 12$
$2x - 3y - z = -10$
$x + y - 2z = -5$

7.
$x + y + z = 3$
$x - 2y + z = -3$
$3x + 3z = 5$

8.
$x + y + z = 3$
$x - 2y + 3z = 5$
$5x - 4y + 11z = 20$

9.
$x + y + z = 3$
$x - 2y + z = -3$
$x + z = 1$

10.
$x - y - 2z = -1$
$x + 2y + z = 5$
$5x + 4y - z = 13$

11. Show that elementary operation 3 yields an equivalent system. [*Hint:* Consider the first two equations of system (1) in the text. Show that if (m, n, p) satisfies both these equations, then it satisfies the system consisting of the first equation and the sum of the first two equations, and conversely.]

B Applications

12. The sum of $8.50 is made up of nickels, dimes, and quarters. The number of dimes is equal to the number of quarters plus twice the number of nickels. The value of the dimes exceeds the combined value of the nickels and the quarters by $1.50. How many of each coin are there?

13. The Mechano Distributing Company has three types of vending machines that dispense snacks as listed in the table below. Mechano fills all the machines once a day and finds them all sold out before the next day. The total daily sales are candy, 760; peanuts, 380; and sandwiches, 660. How many of each type of machine does Mechano have?

Snack	Vending Machine Type		
	I	*II*	*III*
Candy	20	24	30
Peanuts	10	18	10
Sandwiches	0	30	30

14. Suppose the total daily income from the various types of machines in problem 13 is type I, $32.00; type II, $159.00; and type III, $192.00. What is the selling price for each type of snack? (Use the answers from problem 13.)

15. Gro-Kwik Garden Supply has three types of fertilizer that contain chemicals A, B, and C in the percentages shown in the table that follows. In what proportions must Gro-Kwik mix these three types to get an 8-8-8 fertilizer (one that has 8% of each of the three chemicals)?

Chemical	Type of Fertilizer		
	I	*II*	*III*
A	6%	8%	12%
B	6%	12%	8%
C	8%	4%	12%

16. Three water supply valves A, B, and C are connected to a tank. If all three valves are opened, the tank is filled in 8 hours. The tank also can be filled by opening A for 8 hours and B for 12 hours while keeping C closed or by opening B for 10 hours and C for 28 hours, while keeping A closed. Find the time needed by each valve to fill the tank by itself. (*Hint:* Let x, y, and z, respectively, be the fractions of the tank that valves A, B, and C can fill alone in 1 hour.)

17. A 2×2 matrix $\begin{bmatrix} a & b \\ c & d \end{bmatrix}$ is said to be **nonsingular** if $ad - bc \neq 0$. Determine whether the following matrices are nonsingular:

a. $\begin{bmatrix} 1 & 2 \\ 2 & 4 \end{bmatrix}$ **b.** $\begin{bmatrix} 2 & -3 \\ 3 & 5 \end{bmatrix}$ **c.** $\begin{bmatrix} 0 & 2 \\ 2 & 4 \end{bmatrix}$

18. The definition in problem 17 is important because every nonsingular matrix has a unique multiplicative inverse. To find the inverse of the matrix $\begin{bmatrix} a & b \\ c & d \end{bmatrix}$, you have to find a matrix $\begin{bmatrix} x & y \\ z & w \end{bmatrix}$ such that $\begin{bmatrix} a & b \\ c & d \end{bmatrix}\begin{bmatrix} x & y \\ z & w \end{bmatrix} = \begin{bmatrix} 1 & 0 \\ 0 & 1 \end{bmatrix}$. This means that you must solve the systems

$ax + bz = 1$ and $ay + bw = 0$
$cx + dz = 0$ $cy + dw = 1$

To solve the first system, you can write

$$\begin{bmatrix} a & b & | & 1 \\ c & d & | & 0 \end{bmatrix} \sim \begin{bmatrix} ac & bc & | & c \\ ac & ad & | & 0 \end{bmatrix}$$

$$\sim \begin{bmatrix} ac & bc & | & c \\ 0 & ad - bc & | & -c \end{bmatrix}$$

Now you see that the second equation has a unique solution if and only if $ad - bc \neq 0$. Use these ideas to find the inverses of the following:

a. $\begin{bmatrix} 3 & 2 \\ 2 & 1 \end{bmatrix}$ **b.** $\begin{bmatrix} 1 & -2 \\ 2 & 1 \end{bmatrix}$

 In Other Words

19. A system of three equations with three unknowns x, y, and z has the final augmented matrix

$$\begin{bmatrix} 1 & 0 & 0 & | & a \\ 0 & 1 & 0 & | & b \\ 0 & 0 & 1 & | & c \end{bmatrix}$$

What is the solution to the system?

20. A system of three equations with three unknowns x, y, and z has the final augmented matrix

$$\begin{bmatrix} 1 & 0 & 0 & | & a \\ 0 & 1 & 0 & | & b \\ 0 & 0 & 0 & | & c \end{bmatrix}$$

If c is not 0, what does that mean about the solution of the system?

21. If in problem 20, $c = 0$, what does that mean about the solution of the system?

 Using Your Knowledge

Suppose you are solving the system

$$a_1 x + b_1 y + c_1 z = d_1$$
$$a_2 x + b_2 y + c_2 z = d_2$$
$$a_3 x + b_3 y + c_3 z = d_3$$

There are three possible outcomes for the solutions to this system.
1. There is a unique solution $x = a$, $y = b$, and $z = c$ for the system.

2. There is no solution for the system.

3. There are infinitely many solutions for the system.

22. If the solution to the system is $x = a$, $y = b$, and $z = c$, write the final augmented matrix for the system.

23. If the system has no solution, write a possible final augmented matrix for the system. Then find two others.

24. If the system has infinitely many solutions, write a possible final augmented matrix for the system. Then find two others if you can.

 Collaborative Learning

Matrices are used in banking to convey large quantities of information using limited space. The table shown at the top of page 609 was prepared by PCCensus to determine mortgage potential.
    Form two or more teams.

Team 1
1. Construct a 3×3 matrix where the rows indicate "Tenure" and the columns the *number* of houses in each tract falling into the three tenure categories.

2. Which tract has the smallest number of vacant homes?

3. Which tract has the largest number of owner-occupied homes?

4. On the basis of your answers, which tract has the most mortgage potential; that is, in which tract do you think financial institutions would invest?

Team 2
1. Construct a 3×3 matrix where the rows indicate "Tenure" and the columns the *percent* of houses in each tract falling into the three tenure categories.

2. Which tract has the highest percent of vacant homes?

3. Which tract has the highest percent of owner-occupied homes?

4. On the basis of your answers, which tract has the most mortgage potential; that is, in which tract do you think financial institutions would invest?

Group discussion: How would you prefer to analyze the information given, as *numbers* or as *percents*? Discuss the advantages of each. In your opinion, which would be the best way to present the information in the table?

Housing Tenure	Tract 1302		Tract 1303		Tract 1304	
Total Housing Units	776		1,357		2,571	
Tenure						
Owner occupied	296	38%	686	51%	1,379	54%
Renter occupied	304	39%	439	32%	815	32%
Vacant	176	23%	232	17%	377	15%

Source: www.tetrad.com/new/banking.html.

5. Suppose tracts 1502, 1503, and 1504 are demographically comparable with tracts 1302, 1303, and 1304 and are represented by the matrix below. Complete the second matrix using percents instead of numbers.

$$\begin{bmatrix} 290 & 700 & 1500 \\ 310 & 390 & 850 \\ 200 & 210 & 450 \end{bmatrix} \quad \begin{bmatrix} 36\% & \_\_\_ & \_\_\_ \\ \_\_\_ & \_\_\_ & \_\_\_ \\ \_\_\_ & \_\_\_ & \_\_\_ \end{bmatrix}$$

6. If you want to find the total number of vacant homes in the "02 tracts," which matrix operation would you use? Do it!

7. If you wish to construct a third demographically comparable neighborhood with tracts 1602, 1603, and 1604, but you are planning to have twice as many homes as those in the "1500" neighborhood, which matrix operation would you use to plan the number of houses in the "1600" neighborhood? Do it!

8. How many vacant homes would you expect in the "1600" neighborhood? How many owner-occupied? If you plan to loan $10,000 to each of the owner-occupied homes so the owner can renovate the properties in the "1600" neighborhood, how much money would you need for this task?

9.3 Clock and Modular Arithmetic

Airbills, Money Orders, Credit Cards, and Checking Digits

Look at the Federal Express airbill in Figure 9.2 on the following page with tracking number 9048724285. Divide 904872428 (the number to the left of 5) by 7. The remainder is 5. Now try it with airbill 0054604362 from Figure 9.2. When you divide 5460436 by 7, the remainder is 2. Next, consider the postal money order number 42888671414 in Figure 9.2. Divide the number to the left of the last 4 by 9 this time. The remainder is 4. In mathematics, when you divide a by b and the remainder is c, a is said to be **congruent** to c modulo b, written

$$a \equiv c \pmod{b}$$

The relationships in the Federal Express airbills can be written as

$$904872428 \equiv 5 \pmod 7 \quad \text{and} \quad 005460436 \equiv 2 \pmod 7$$

and that in the money order as $4288867141 \equiv 4 \pmod 9$.

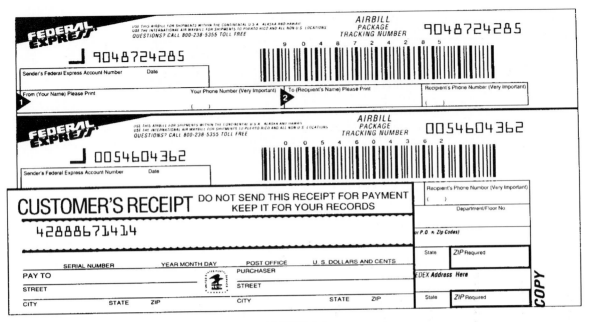

FIGURE 9.2

In this section you will study modular arithmetic, its operations and relationships, as well as some applications (see the Using Your Knowledge section of Exercises 9.3).

The last number in the airbill (5) or the money order (4) is sometimes called the **checking digit,** or **CD** for short. Checking digits are used in a variety of situations. Airline tickets, UPS packages, and drivers' licenses, for example, use checking digits to encode information. Why? For security reasons. Suppose you try to change the postal money order amount from $5 to $500. The post office *knows* that money order number 42888671414 is for $5, so if the amount was altered, the checking number would have to be altered as well. Do you know what other numbers would work? Mathematically, you have to solve the congruence $x \equiv 4 \pmod 9$. Practically, x must somewhat resemble 42888671414. [For example, $22 \equiv 4 \pmod 9$ because when 22 is divided by 9, the remainder is 4, but money order 22 was probably used a long time ago.] Can you find "better" numbers that will work without changing the CD number? The CD number 4 is usually written under three bars $|||$ so that an erasure or alteration of the CD will disturb the bars and expose the fraud. ▶

In earlier chapters of this book we made frequent reference to *mathematical systems*. For example, we defined the set of natural numbers together with the operations of addition and multiplication. We then discussed certain properties of the set of natural numbers with respect to these operations. The set of natural numbers, together with the operations of addition and multiplication, constitutes a mathematical system.

In general, a *mathematical system* consists of the following items:

1. A *set of elements*

2. One or more *operations*

3. One or more *relations* that enable us to compare the elements in the set

4. Some *rules, axioms,* or *laws* that the elements in the set satisfy

For example, when we refer to the system of integers, we have

1. *Elements.* The elements of this system are the integers in the set
$I = \{\ldots, -2, -1, 0, 1, 2, \ldots\}$.

2. *Operations.* Within the system of integers, we can always perform the operations of addition, subtraction, and multiplication. We can sometimes divide, but most often the result of dividing one integer by another is not an integer. The set of integers is not closed under division.

3. *Relations.* We have three possible relations between any two integers *a* and *b*. They are

$$a < b, \qquad a > b, \qquad \text{and} \qquad a = b$$

4. *Rules or laws.* Addition and multiplication in the system of integers satisfy the commutative and the associative laws, as well as the distributive law.

Most of the mathematical systems we have discussed involve an *infinite* number of elements. In this section we shall study mathematical systems that contain only a *finite* number of elements. If we are thoroughly familiar with the workings of a finite system, we can generalize this knowledge and apply it to other systems.

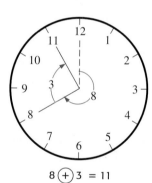

$8 \oplus 3 = 11$

FIGURE 9.3

A. Clock Arithmetic

As you are aware, the numbers on the face of a clock are used to tell the time of the day or night. The set of numbers used for this purpose is

$$S = \{1, 2, 3, 4, 5, 6, 7, 8, 9, 10, 11, 12\}$$

We shall define addition on this set by means of an addition table, but we first present some examples to justify the entries in the table. If it is now 11 A.M., and we have to be in class in 3 hours, it is obvious that we have to go to class at 2 P.M. For this reason, we define $11 \oplus 3 = 2$, where $\oplus$ (read "circle plus") is the operation of clock addition. If our class was to meet in 5 hours, and it was now 10 A.M., we would have to be in class at 3 P.M., so $10 \oplus 5 = 3$.

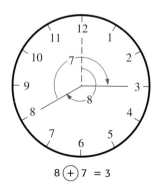

$8 \oplus 7 = 3$

FIGURE 9.4

EXAMPLE 1 ▶ Addition in Clock Arithmetic

Find the following sums in clock arithmetic:

(a) $8 \oplus 3$ \qquad (b) $8 \oplus 7$ \qquad (c) $11 \oplus 12$

Solution

(a) $8 \oplus 3$ means 3 hours after 8 o'clock, so $8 \oplus 3 = 11$ (see Figure 9.3).

(b) $8 \oplus 7$ means 7 hours after 8 o'clock. Thus, $8 \oplus 7 = 3$ (see Figure 9.4).

(c) $11 \oplus 12$ means 12 hours after 11 o'clock. Thus, $11 \oplus 12 = 11$. ∎

We can now construct Table 9.5, which shows the addition facts in clock arithmetic. Verify the entries in this table before proceeding further.

TABLE 9.5 Addition in Clock Arithmetic

$+$	1	2	3	4	5	6	7	8	9	10	11	12
1	2	3	4	5	6	7	8	9	10	11	12	1
2	3	4	5	6	7	8	9	10	11	12	1	2
3	4	5	6	7	8	9	10	11	12	1	2	3
4	5	6	7	8	9	10	11	12	1	2	3	4
5	6	7	8	9	10	11	12	1	2	3	4	5
6	7	8	9	10	11	12	1	2	3	4	5	6
7	8	9	10	11	12	1	2	3	4	5	6	7
8	9	10	11	12	1	2	3	4	5	6	7	8
9	10	11	12	1	2	3	4	5	6	7	8	9
10	11	12	1	2	3	4	5	6	7	8	9	10
11	12	1	2	3	4	5	6	7	8	9	10	11
12	1	2	3	4	5	6	7	8	9	10	11	12

Of course, other operations can be defined on this set. For example, the clock difference $3 \ominus 4$ (read "3 circle minus 4") is a number n with the property that $3 = 4 \oplus n$. By looking at Table 9.5 we can see that the number 11 satisfies this equation because $3 = 4 \oplus 11$. Accordingly, we state the following definition:

Definition of Subtraction Using Clock Arithmetic

$$a \ominus b = n \qquad \text{if and only if} \qquad a = b \oplus n$$

With this definition, $4 \ominus 6 = 10$ because $4 = 6 \oplus 10$.

If a is a positive number, we can define $-a$ in clock arithmetic as the number obtained by going counterclockwise a hours. With this convention, it follows that

$$-a = 12 - a$$

For example, $-4 = 12 - 4 = 8$; this agrees with Figure 9.5.

With this definition of $-a$, it also follows that

$$a \ominus b = a \oplus (-b)$$

which may be regarded as an alternative way to define circle minus.

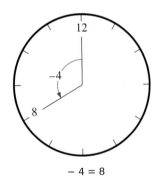

$-4 = 8$

FIGURE 9.5

EXAMPLE 2 ▶ Solving Equations in Clock Arithmetic

Find numbers n such that the following hold:

(a) $n \oplus 5 = 3$ (b) $2 \ominus 3 = n$

Solution

(a) We inspect the fifth row in Table 9.5. We have to find a number n such that when we add n to 5, we obtain 3. The answer is 10, because $10 \oplus 5 = 3$.

(b) $2 \ominus 3 = 2 \oplus (-3) = 2 \oplus (12 - 3) = 2 \oplus 9 = 11$. (Check in the table that $3 \oplus 11 = 2$.) ∎

As usual, multiplication can be defined in terms of addition. Thus, $3 \otimes 5 = (5 \oplus 5) \oplus 5 = 10 \oplus 5 = 3$, so $3 \otimes 5 = 3$.

Definition of Multiplication Using Clock Arithmetic

$$a \otimes b = \underbrace{b \oplus b \oplus b \oplus \cdots \oplus b}_{a \text{ times}}$$

(The symbol $\otimes$ is read "circle times.")

With this definition, $4 \otimes 3 = 3 \oplus 3 \oplus 3 \oplus 3 = 12$ and $2 \otimes 8 = 8 \oplus 8 = 4$. *Note:* A quicker way to obtain the answer is to multiply the two numbers by ordinary arithmetic, divide by 12, and take the remainder as the answer. (If the remainder is 0, the answer is 12, the zero point on a clock.) For example, $6 \otimes 5 = 6$ because $6 \times 5 = 30$ and 30 divided by 12 equals 2 with a remainder of 6. Similarly, $6 \otimes 4 = 12$ because $6 \times 4 = 24$, and 24 divided by 12 is 2 with a 0 remainder.

EXAMPLE 3 ▶ Finding Products in Clock Arithmetic

Find the following products in clock arithmetic:

(a) $3 \otimes 5$ (b) $4 \otimes 8$ (c) $9 \otimes 9$

Solution

(a) $3 \otimes 5 = 3$ because $3 \times 5 = 15$, which divided by 12 is 1 with a remainder of 3.

(b) $4 \otimes 8 = 8$ because $4 \times 8 = 32$, and 32 divided by 12 is 2 with a remainder of 8.

(c) $9 \otimes 9 = 9$ because $9 \times 9 = 81$, which divided by 12 is 6 with a remainder of 9. ∎

Division is defined in clock arithmetic in terms of multiplication. For example, $\frac{4}{8} = 2$ because $4 = 8 \otimes 2$, and $\frac{3}{5} = 3$ because $3 = 5 \otimes 3$.

Definition of Division Using Clock Arithmetic

$$\frac{a}{b} = n \qquad \text{if and only if} \qquad a = b \otimes n$$

In discussing multiplication in clock arithmetic, we noted that 12 in this arithmetic corresponds to 0 in ordinary arithmetic. To show that 12 behaves further like 0 in ordinary arithmetic, let us try to find a number n such that

$$\tfrac{1}{12} = n$$

This equation is true only if $1 = 12 \otimes n$. But there is no n in clock arithmetic such that $12 \otimes n = 1$. So, in clock arithmetic, we cannot divide 1 by 12, just as in ordinary arithmetic we cannot divide 1 by 0.

EXAMPLE 4 ▶ Finding Quotients in Clock Arithmetic

Find the following quotients in clock arithmetic:

(a) $\frac{2}{7}$ (b) $\frac{8}{8}$ (c) $\frac{3}{4}$

Solution

(a) $\frac{2}{7} = n$ if and only if $2 = 7 \otimes n$. Thus, we wish to find an n such that $7 \otimes n = 2$ because $n = 2$ satisfies the given equation, $\frac{2}{7} = 2$. Note that the answer $n = 2$ can be found by trial and error; that is, we let $n = 1$ and see whether $7 \otimes 1 = 2$ (it doesn't). We then let $n = 2$ and check to see whether $7 \otimes 2 = 2$. Because $7 \otimes 2 = 2$, the desired number has been obtained.

(b) $\frac{8}{8} = n$ if and only if $8 = 8 \otimes n$, an equation that is satisfied by $n = 1$, $n = 4$, $n = 7$, and $n = 10$. Note that unlike ordinary division problems, in which the answer is unique, in clock arithmetic a division problem may have many solutions.

(c) $\frac{3}{4} = n$ if and only if $3 = 4 \otimes n$. There is no n such that $3 = 4 \otimes n$, so the problem $\frac{3}{4} = n$ has no solution.   ■

B. Modular Arithmetic

The four fundamental operations were defined on the 12-hours clock. These operations can be generalized to a type of mathematical system called **modular arithmetic.** This system can be defined on the set of integers as follows:

Definition of Congruent Modulo m

Two integers a and b are said to be **congruent modulo m,** denoted by $a \equiv b \pmod{m}$, if $a - b$ (or $b - a$) is a multiple of m (m an integer).

Thus, $14 \equiv 4 \pmod 5$ because $14 - 4 = 10$, which is a multiple of 5, and $2 \equiv 18 \pmod 4$ because $18 - 2 = 16$, which is a multiple of 4.

EXAMPLE 5 ▶ Truth or Falsity in Modular Arithmetic

Are the following statements true or false?

(a) $4 \equiv 1 \pmod 3$ (b) $7 \equiv 2 \pmod 4$ (c) $1 \equiv 6 \pmod 5$

Solution

(a) True because $4 - 1 = 3$, a multiple of 3

(b) False because $7 - 2 = 5$ is *not* a multiple of 4

(c) True because $6 - 1 = 5$, a multiple of 5   ■

The addition of two numbers in modular arithmetic is simple. For example, in arithmetic modulo 5, we have $3 + 3 \equiv 1$ because $3 + 3 = 6 \equiv 1 \pmod 5$. Thus, to add nonnegative numbers in a system modulo m, we proceed as follows:

> **Addition Modulo _m_**
>
> 1. Add the numbers in the ordinary way.
>
> 2. If the sum is less than _m_, the answer is the sum obtained.
>
> 3. If the sum is greater than or equal to _m_, the answer is the remainder obtained by dividing the sum by _m_.

For example, in modulo 8

$3 + 4 \equiv 7 \pmod 8$	Because $3 + 4 = 7$ is less than 8
$3 + 5 \equiv 0 \pmod 8$	Because $3 + 5 = 8$, which divided by 8 leaves a 0 remainder
$3 + 7 \equiv 2 \pmod 8$	Because $3 + 7 = 10$, which divided by 8 is 1 with a remainder of 2

We now define addition in modular arithmetic for modulo 5, as shown in Table 9.6.

TABLE 9.6 Addition Modulo 5

+	0	1	2	3	4
0	0	1	2	3	4
1	1	2	3	4	0
2	2	3	4	0	1
3	3	4	0	1	2
4	4	0	1	2	3

When adding numbers in modular arithmetic, we may want to use the following theorem:

> **Addition Theorem for Congruence**
>
> If $a \equiv b \pmod m$ and $c \equiv d \pmod m$, then
>
> $$a + c \equiv b + d \pmod m$$

For example, to add 18 and 31 (mod 5), we can use the facts that

$$18 \equiv 3 \pmod 5 \qquad \text{and} \qquad 31 \equiv 1 \pmod 5$$

to obtain

$$18 + 31 \equiv 4 \pmod 5$$

Proof

The theorem can be proved by using the meanings of the two congruences. Thus,

$a \equiv b \pmod m$	means	$a - b = km,$	k an integer
$c \equiv d \pmod m$	means	$c - d = pm,$	p an integer

Hence,

$$(a - b) + (c - d) = km + pm = (k + p)m$$

Also,

$$(a - b) + (c - d) = (a + c) - (b + d)$$

so that

$$(a + c) - (b + d) = (k + p)m$$

That is, $(a + c) - (b + d)$ is a multiple of m; this means that

$$a + c \equiv b + d \ (\text{mod } m)$$

To multiply two numbers in arithmetic modulo 5, we multiply the numbers in the ordinary way. If the product is less than 5, the answer is this product. If the product is greater than 5, the answer is the remainder when the product is divided by 5. For example, $4 \times 4 \equiv 1$ because the product $4 \times 4 = 16$, which leaves a remainder of 1 when divided by 5. Multiplication modulo 5 is shown in Table 9.7.

TABLE 9.7 Multiplication Modulo 5

$\times$	0	1	2	3	4
0	0	0	0	0	0
1	0	1	2	3	4
2	0	2	4	1	3
3	0	3	1	4	2
4	0	4	3	2	1

Subtraction in modular arithmetic is defined as follows:

Definition of Subtraction in Modular Arithmetic

$$a - b \equiv c \ (\text{mod } m) \qquad \text{if and only if} \qquad a \equiv b + c \ (\text{mod } m)$$

Thus, $3 - 4 \equiv 4 \ (\text{mod } 5)$ because $3 \equiv (4 + 4) \ (\text{mod } 5)$, and $2 - 4 \equiv 3$ because $2 \equiv (4 + 3) \ (\text{mod } 5)$.

EXAMPLE 6 ▶ Subtraction (Mod 5)

Find the following:

(a) $(4 - 1) \ (\text{mod } 5)$ (b) $(2 - 3) \ (\text{mod } 5)$ (c) $(1 - 4) \ (\text{mod } 5)$

Solution

(a) $4 - 1 \equiv 3 \ (\text{mod } 5)$

(b) $2 - 3 \equiv n \ (\text{mod } 5)$ if and only if $2 \equiv (3 + n) \ (\text{mod } 5)$. From Table 9.6, $3 + 4 \equiv 2 \ (\text{mod } 5)$. Thus, $2 - 3 \equiv 4 \ (\text{mod } 5)$.

(c) $1 - 4 \equiv n \ (\text{mod } 5)$ if and only if $1 \equiv 4 + n \ (\text{mod } 5)$. From Table 9.6, $1 \equiv 4 + 2 \ (\text{mod } 5)$. Thus, $1 - 4 \equiv 2 \ (\text{mod } 5)$.   ■

We examine Table 9.8 to find which properties the set $S = \{0, 1, 2, 3, 4\}$ has under the operation of addition modulo 5.

TABLE 9.8 Addition Modulo 5

+	0	1	2	3	4
0	0	1	2	3	4
1	1	2	3	4	0
2	2	3	4	0	1
3	3	4	0	1	2
4	4	0	1	2	3

1. All entries in Table 9.8 are elements of S, so S is *closed* with respect to addition (mod 5).

2. The operation of addition (mod 5) is *associative*. (This fact is a consequence of the properties of ordinary addition because division by 5 and taking the remainder can be done after the ordinary addition is done.)

3. The operation of addition (mod 5) is *commutative;* that is, if x and y are any elements of S, then $x + y \equiv y + x$ (mod 5). This can easily be checked by inspecting Table 9.8. As we can see, the results of the operations $2 + 4$ and $4 + 2$ appear in positions that are *symmetric* with respect to a diagonal line drawn from top left to bottom right of the table. If this type of symmetry is present for any operation table; that is, if the top half of the table is the reflection of the bottom half across the diagonal, then the operation is commutative.

4. The *identity* for addition modulo 5 is 0 because for any x in S, $x + 0 = x$; that is, $0 + 0 = 0$, $1 + 0 = 1$, $2 + 0 = 2$, $3 + 0 = 3$, and $4 + 0 = 4$.

5. Every element has an additive *inverse* (mod 5). The inverse of 0 is 0 because $0 + 0 \equiv 0$ (mod 5). The inverse of 1 is 4 because $1 + 4 \equiv 0$ (mod 5). The inverse of 2 is 3 because $2 + 3 \equiv 0$ (mod 5). The inverse of 3 is 2 because $3 + 2 \equiv 0$ (mod 5). The inverse of 4 is 1 because $4 + 1 \equiv 0$ (mod 5).

As we have seen in the preceding discussion, modular arithmetic is a mathematical system in which two operations are defined on the same set. For the set $S = \{0, 1, 2, 3, 4\}$, we defined addition and multiplication modulo 5 as shown in Tables 9.6 and 9.7. Are there any properties that involve both operations? The answer is yes. For example, to find $3 \times (2 + 4)$ (mod 5), we can proceed in one of the two following ways:

1. $3 \times (2 + 4) \equiv 3 \times (1) \equiv 3$ (mod 5)

2. $3 \times (2 + 4) \equiv (3 \times 2) + (3 \times 4) \equiv 1 + 2 \equiv 3$ (mod 5)

To get the answer in the second way, we used a property that involves both addition and multiplication. This property is called the *distributive property;* it is true in general that if a, b, and c are elements of S, then

$$a \times (b + c) \equiv (a \times b) + (a \times c) \,(\text{mod } m)$$

because $a \times (b + c) - (a \times b) - (a \times c) = 0$ in ordinary arithmetic.

EXAMPLE 7 ▶ Solving Equations (Mod 5)

Find a replacement for n that will make the given sentence true in arithmetic modulo 5.

(a) $4 \times (3 + 1) \equiv (4 \times n) + (4 \times 1)$

(b) $n \times (1 + 3) \equiv (2 \times 1) + (2 \times 3)$

Solution

(a) By the distributive property, $4 \times (3 + 1) \equiv (4 \times 3) + (4 \times 1)$. Thus, $n = 3$ will make the sentence $4 \times (3 + 1) \equiv (4 \times n) + (4 \times 1)$ true.

(b) By the distributive property, $2 \times (1 + 3) \equiv (2 \times 1) + (2 \times 3)$. Thus, $n = 2$ will make the sentence $n \times (1 + 3) \equiv (2 \times 1) + (2 \times 3)$ true. ■

EXAMPLE 8 ▶ Solving More Equations (Mod 5)

Find a replacement for n that will make the given sentence true.

(a) $3 + n \equiv 2 \ (\text{mod } 5)$ (b) $\frac{3}{4} \equiv n \ (\text{mod } 5)$

Solution

(a) By inspection of Table 9.6 (or 9.8), we see that $3 + 4 \equiv 2 \ (\text{mod } 5)$. Thus, $n = 4$ will make the sentence $3 + n \equiv 2 \ (\text{mod } 5)$ true.

(b) $\frac{3}{4} \equiv n \ (\text{mod } 5)$ is true if and only if $3 \equiv 4 \times n \ (\text{mod } 5)$. By inspection of Table 9.7, we see that $4 \times 2 \equiv 3 \ (\text{mod } 5)$. Hence, $n = 2$ will make the sentence $\frac{3}{4} \equiv n \ (\text{mod } 5)$ true. ■

In Examples 7 and 8 we have given answers in the set $\{0, 1, 2, 3, 4\}$. If we allow n to be any integer that makes the given congruence true, then each answer can be modified by adding or subtracting any desired multiple of 5. This is justified by the addition theorem for congruency proved earlier. For instance, in Example 8(a), all the possible integer answers are given by $n \equiv 4 \ (\text{mod } 5)$. This means that n can have any of the values $\dots, -6, -1, 4, 9, 14, \dots$; each of these makes the congruence $3 + n \equiv 2 \ (\text{mod } 5)$ true.

Online Study Center

To further explore clock arithmetic, access links 9.3.1 and 9.3.2 on this textbook's Online Study Center.

EXAMPLE 9 ▶ Finding Your Sign with Congruences

Suppose you were born in year Y. Your Chinese zodiac sign is R, where R is the remainder obtained from the formula $Y/12 = N + R$ and corresponds to monkey (0), cock (1), dog (2), boar (3), rat (4), ox (5), tiger (6), rabbit (7), dragon (8), snake (9), horse (10), and goat (11). The cock, boar, and ox are sometimes known as the rooster, pig, and buffalo.

(a) What is your sign if you were born in 1982?

(b) How can you find your Chinese zodiac sign using congruences?

Solution

(a) $\frac{1982}{12} = 165$ with a remainder R of 2, corresponding to the dog.

(b) In modular arithmetic, the remainder is the answer when a number is divided by the modulus. Thus, if Y is your birth year, to find the sign, use $Y \equiv R \ (\text{mod } 12)$. Note that in part (a) $1982 \equiv 2 \ (\text{mod } 12)$. Your sign is the number corresponding to R. ■

Are you a monkey or a tiger? A ram or a dragon? Please, do not be offended! Traditional Chinese belief is that a person's year of birth holds the key to their long-life character and well-being. The Chinese zodiac is based on a 12-year cycle, and your sign depends on your year of birth. We show you how to find your sign in the Example 9.

Now that you have learned about airbills, money orders, and even your horoscope, it is time for a break! Grill some spicy burgers with Bello Pepper Sauce. Horrors! The bottle is empty; there is no more. You go to the Web, get the manufacturer's phone number, and call. They want to know the UPC (Universal Product Code, or **bar code** for short) on the side of the bottle. You calmly read

0   1   2   0   6   1   1   0   1   8   5   8

The operator politely replies, "That is not a valid number." What are these Bello people, psychics? You look at the bottle again for reassurance. The operator was correct. In your hastiness, you have misread a number. The correct number is

0   1   2   0   6   1   0   0   1   8   5   8

How did the operator know that the number you gave him at first was incorrect? The Bello employees know about bar codes and check digits. The first six numbers on the bar code encode information about the manufacturer, the next five numbers have information about the product, and the last number, the 8, is the check digit. In general, how do we get the check digit c? We will tell you now.

Suppose the first 11 numbers in the bar code are $n_1, n_2, n_3, \ldots, n_{11}$.
Write:

Compute	$n_1 + n_2 + n_3 + n_4 + n_5 + n_6 + n_7 + n_8 + n_9 + n_{10} + n_{11} + c \equiv 0 \pmod{10}$
In our case,	$3n_1 + n_2 + 3n_3 + n_4 + 3n_5 + n_6 + 3n_7 + n_8 + 3n_9 + n_{10} + 3n_{11} + c \equiv 0 \pmod{10}$
Or	$3 \cdot 0 + 1 + 3 \cdot 2 + 0 + 3 \cdot 6 + 1 + 3 \cdot 0 + 0 + 3 \cdot 1 + 8 + 3 \cdot 5 + c \equiv 0 \pmod{10}$
That is,	$0 + 1 + 6 + 0 + 18 + 1 + 0 + 0 + 3 + 8 + 15 + c \equiv 0 \pmod{10}$
	$52 + c \equiv 0 \pmod{10}$

Since $52 + 8 \equiv 0 \pmod{10}$, the check digit c is 8.

EXAMPLE 10 ▶ Finding a Check Digit

Find the check number c for Quaker Quick grits if the first 11 numbers are

$$0 \quad 3 \quad 0 \quad 0 \quad 0 \quad 0 \quad 0 \quad 4 \quad 1 \quad 5 \quad 0 \quad c$$

Solution
Using the scheme above, we compute the number

	$3 \cdot 0 + 3 + 3 \cdot 0 + 0 + 3 \cdot 0 + 0 + 3 \cdot 0 + 4 + 3 \cdot 1 + 5 + 3 \cdot 0 + c \equiv 0 \pmod{10}$
Or	$0 + 3 + 0 + 0 + 0 + 0 + 0 + 4 + 3 + 5 + 0 + c \equiv 0 \pmod{10}$
That is,	$15 + c \equiv 0 \pmod{10}$

Since $15 + 5 \equiv 0 \pmod{10}$, the check digit c is 5. Remember, $15 + 5 = 20$, which leaves a remainder of 0 when divided by 10, so the check digit is correct! ■

Have you heard of identity theft? Identity theft occurs when someone uses your personal information, such as your credit card number, without your permission to commit fraud or other crimes. We are ready to learn more about these credit card numbers.

Check digit (8)

Look at the credit card number. The very first digit (**4**) represents the type of entity that issued the credit card. For example, a **4** means that the card was issued by a bank or financial institution. The different categories (from **0** to **9**) are shown in the accompanying table. The first *six* digits in the credit card (**499977**) identify the issuer of the card. For example, the first *six* numbers for a MasterCard may be 51xxxx, for a Visa 4xxxxx, and 34xxxx for an American Express.

Issuer: 499977

4: Banking/Financial

(0–9)	Card Issuer Categories
0	Special industry assignment
1, 2	Airlines
3	Entertainment
4, 5	Banking and financial
6	Merchandising/banking
7	Petroleum
8	Telecommunications
9	National assignment

CARD TYPE	Prefix	Length
MASTERCARD	51xxxx–55xxxx	16
VISA	4xxxxx	13, 16
AMEX	34xxxx–37xxxx	15
Diners Club/	300xxx–305xxx	14
Carte Blanche	36xxxx	
	38xxxx	
Discover	6011xx	16

The *final digit* of the credit card number (8) is called the **check digit.** There is a formula, called the *Luhn algorithm,* that is used to decide if a credit card is valid. This formula was named after the IBM scientist Hans Peter Luhn (1896–1964), who was awarded U.S. Patent 2950048 ("Computer for Verifying Numbers") for inventing the technique in 1960. The Luhn algorithm is used by banks and other credit-issuing entities to combat credit-card fraud. We can decide whether a credit card is valid or not by following these three steps.

Using the Luhn Algorithm to Find Valid Credit Card Numbers

1. Starting with the second-to-last digit (the first digit to the *left* of the check digit) and moving left, **double** the value of all the alternating digits. If the result of doubling a number has two digits, add the two numbers. For example, when you double 9, you get 18, which yields $1 + 8 = 9$. Now add all these numbers together.

2. Starting from the left, take all the unchanged digits (i.e., the numbers that you did *not* double in step 1) and add them together.

3. Now add the numbers obtained in steps 1 and 2. If this number ends with a zero, then the credit card number is valid. If not, then the credit card number is not valid.

EXAMPLE 11 ▶ Checking Credit Card Numbers

A credit card is numbered 4417 1234 5678 9112.

(a) What type of institution issued the card?

(b) What is the check digit of the card?

(c) Is the credit card valid? If not, what should the check digit be to make it valid?

Solution

(a) The card starts with the number 4, which denotes a banking or financial institution.

(b) The check digit is the last number in 4417 1234 5678 9112, which is 2.

(c) We check the number by following the three-step procedure. The card number is

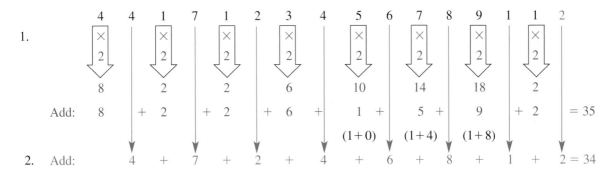

3. The sum of the results from steps 1 and 2 is $35 + 34 = 69$, which does not end in zero, so the credit card number is not valid. If we change the check digit from a 2 to a 3, the sum will be 70, and the number will be valid. ∎

EXERCISES 9.3

A Clock Arithmetic

In problems 1–8, find the sum in clock arithmetic.

1. 9 ⊕ 7   **2.** 2 ⊕ 8

3. 8 ⊕ 3   **4.** 5 ⊕ 7

5. 7 ⊕ 8   **6.** 9 ⊕ 9

7. 8 ⊕ 11   **8.** 12 ⊕ 3

In problems 9–14, find the difference in clock arithmetic.

9. 8 ⊖ 3   **10.** 5 ⊖ 8   **11.** 9 ⊖ 12

12. 6 ⊖ 9   **13.** 8 ⊖ 7   **14.** 1 ⊖ 12

In problems 15–22, find all n satisfying each equation in clock arithmetic.

15. $n ⊕ 7 = 9$ **16.** $n ⊕ 8 = 2$

17. $2 ⊕ n = 1$ **18.** $7 ⊕ n = 3$

19. $3 ⊖ 5 = n$ **20.** $2 ⊖ 4 = n$

21. $1 ⊖ n = 12$ **22.** $3 ⊖ 7 = n$

In problems 23–28, find the indicated products in clock arithmetic.

23. 4 ⊗ 3   **24.** 3 ⊗ 8   **25.** 9 ⊗ 2

26. 3 ⊗ 9   **27.** 2 ⊗ 8   **28.** 12 ⊗ 3

29. Make a table of multiplication facts in clock arithmetic.

In problems 30–34, find the indicated quotients in clock arithmetic.

30. $\frac{9}{7}$ **31.** $\frac{3}{5}$

32. $\frac{3}{9} = n$ (*Hint:* There are three answers.)

33. $\frac{1}{11}$ **34.** $\frac{1}{12}$

In problems 35–40, find all n satisfying the given equation in clock arithmetic.

35. $\frac{n}{5} = 8$ **36.** $\frac{n}{2} = 4$ **37.** $\frac{n}{2} ⊕ 4 = 8$

38. $\frac{n}{7} = 9$ **39.** $\frac{2}{n} = 3$ **40.** $\frac{12}{12} = n$

B Modular Arithmetic

In problems 41–46, classify each statement as true or false.

41. $2 \equiv 4 \pmod 3$ **42.** $5 \equiv 2 \pmod 3$

43. $6 \equiv 7 \pmod 5$ **44.** $5 \equiv 3 \pmod 2$

45. $8 \equiv 9 \pmod{10}$ **46.** $12 \equiv 8 \pmod 4$

In problems 47–50, find the indicated sums.

47. $(3 + 4) \pmod 5$ **48.** $(2 + 9) \pmod{10}$

49. $(3 + 1) \pmod 5$ **50.** $(3 + 6) \pmod 7$

In problems 51–54, find the indicated products.

51. $(4 \times 2) \pmod 5$ **52.** $(4 \times 3) \pmod 5$

53. $(2 \times 3) \pmod 5$ **54.** $(3 \times 3) \pmod 5$

In problems 55–58, find the indicated differences.

55. $(2 - 4) \pmod 5$ **56.** $(3 - 4) \pmod 5$

57. $(1 - 3) \pmod 5$ **58.** $(0 - 2) \pmod 5$

In problems 59–74, find a value of n in the set $\{0, 1, 2, 3, 4\}$ that will make the given congruence true.

59. $4 \times (3 + 0) \equiv (4 \times 3) + (4 \times n) \pmod 5$

60. $2 \times (1 + 3) \equiv (2 \times 1) + (n \times 3) \pmod 5$

61. $2 \times (0 + 3) \equiv (2 \times n) + (2 \times 3) \pmod 5$

62. $4 \times (1 + n) \equiv (4 \times 1) + (4 \times 2) \pmod 5$

63. $2 + n \equiv 3 \pmod 5$ **64.** $n + 3 \equiv 1 \pmod 5$

65. $2 \times n \equiv 4 \pmod 5$ **66.** $3 \equiv 2 \times n \pmod 5$

67. $n - 3 \equiv 4 \pmod 5$ **68.** $2 \equiv n - 1 \pmod 5$

69. $3 \equiv n - 4 \pmod 5$ **70.** $n - 2 \equiv 1 \pmod 5$

71. $\frac{n}{2} \equiv 4 \pmod 5$ **72.** $\frac{n}{3} \equiv 2 \pmod 5$

73. $\frac{3}{4} \equiv n \pmod 5$ **74.** $\frac{1}{2} \equiv n \pmod 5$

In problems 75–78, use Table 9.7, page 616, and multiplication modulo 5.

75. Is the set $S = \{0, 1, 2, 3, 4\}$ closed with respect to $\times$?

76. Is the set operation $\times$ commutative?

77. Is there an identity for the operation $\times$? If so, what is this identity?

78. Find the inverse (if possible) of the following:
a. 0   **b.** 1   **c.** 2   **d.** 3   **e.** 4

Problems 79–86 refer to the Chinese zodiac signs discussed in Example 9.

79. Find the sign for a person born in 1985.

80. Find the sign of a baby born in 2000.

81. A person tells you that his sign is the ox. Can you tell in what exact year the person was born? Can you find out if the person tells you that he is in his 20s?

82. A goat sign person tells you that she is 30 something. How old is the person?

83. A person tells you that his sign is the dragon. What additional information do you need to find his exact age?

84. A person tells you that her sign is the snake and that she was born in the 60s. Can you determine her exact age?

85. List all the years of the rat, starting in the 1940s.

86. List all the years of the tiger, starting in the 1950s. When is the next year of the tiger?

 In Other Words

87. Find out about military, or international, or airline time. How many hours does a military clock have? Discuss the advantages of military time over regular time.

88. In algebra, if a and b are real numbers and $a \cdot b = 0$, then $a = 0$ or $b = 0$.
a. In your own words, write an equivalent theorem for clock arithmetic.
b. How many numbers a and b can you find such that $a \otimes b = 12$?

89. In your own words, write all the similarities and differences between clock arithmetic and regular arithmetic.

90. Suppose that you are tracking a package for Federal Express. The customer claims that the package number is 0005234981 or 0005234983. Which is correct? Explain your answer.

 Using Your Knowledge

Look at the back cover of this book. What did you find? The ISBN (International Standard Book Number) is a 10-digit number that encodes certain information about the book. (Some books show the ISBN only on the copyright page.) Say that a book's ISBN is 0-669-28957-4. The following diagram shows the reference of each part of the ISBN:

The book

English-speaking country   0-669-28957-4

Publisher

But what about the 4 at the right end? This digit is a check digit that is used to verify orders. The check number is obtained from the other digits as follows: Write the numbers 10, 9, 8, 7, 6, 5, 4, 3, and 2 above the first nine digits of the ISBN, and then multiply each of these digits by the number above it and add the results (see the diagram at the bottom of this page).

To get the check number (4 in our ISBN), divide the sum 282 by 11 and take the remainder; $r = 7$. The check number c is the solution of $c + r \equiv 0 \pmod{11}$; that is, $c + 7 \equiv 0 \pmod{11}$ for the replacement set $\{0, 1, 2, \ldots, 10\}$. Since $4 + 7 = 11 \equiv 0 \pmod{11}$, you see that the check number is $c = 4$. Note that the easy way to get the check number is to subtract the remainder r from 11.

Use these ideas to solve the following problems. Note that if the check number is 10, it is written as the Roman numeral X.

10	9	8	7	6	5	4	3	2
↓	↓	↓	↓	↓	↓	↓	↓	↓
0	6	6	9	2	8	9	5	7
↓	↓	↓	↓	↓	↓	↓	↓	↓

$$(10 \cdot 0) + (9 \cdot 6) + (8 \cdot 6) + (7 \cdot 9) + (6 \cdot 2) + (5 \cdot 8) + (4 \cdot 9) + (3 \cdot 5) + (2 \cdot 7)$$
$$= \quad 0 \; + \; 54 \; + \; 48 \; + \; 63 \; + \; 12 \; + \; 40 \; + \; 36 \; + \; 15 \; + \; 14$$
$$= 282$$

91. If the first nine digits of a book's ISBN are 0-06-040613, find the check number.

92. If the first nine digits of a book's ISBN are 0-517-53052, find the check number.

93. Find the check number for ISBN 0-312-87867.

94. The last digit of the book number in the ISBN 0-060-4098■3 was blurred as indicated. What must this digit be?

95. Find what the blurred digit must be in the ISBN 0-03-0589■4-2.

96. Find what the blurred digit must be in the ISBN 0-716■-0456-0.

In problems 97–102, find the check number.

97. G.E. light bulb

    0   43168   38540   *c*

98. Campbell's Pork and Beans

    0   51000   02952   *c*

99. Del Monte Fruit Cocktail

    0   24000   02140   *c*

100. Celestial Seasonings Green Tea

    0   70734   00501   *c*

101. Fuji Film CR-R Disk

    0   74101   74010   *c*

102. Tylenol PM

    3   00450   17615   *c*

In problems 103–105, find the missing number.

103. Goya Tembleque (A dessert)

    0   4133☐   03183   7

104. Peanut M and M

    0   40000   1213☐   9

105. Pompeian Red Vinegar

    0   ☐0404   00100   2

In problems 106–110, decide if the given credit card number is valid. If not, find the check digit that will make it valid. You can check your work at http://elliottback.com/tools/validate-credit-card/.

106. 5123 4567 8901 2345

107. 5500 0100 1479 0003

108. 4111 1111 1234 0023

109. 3465 3421 7902 4679

110. 3728 5290 7342 8237

111. 3002 4456 2940 41

112. 3000 0030 0100 04

113. 3663 9023 3938 49

114. 3823 4467 2390 38

115. 6011 0030 0507 1031

 **Discovery**

Look at the lower left-hand corner of the check. The nine-digit number in the lower left is the bank identification number, sometimes called the *routing number*, and the last digit, 6, is the check digit. The scheme uses the numbers 7, 3, and 9 as multipliers to find the check digit.

$$0 \quad 6 \quad 3 \quad 1 \quad 0 \quad 4 \quad 0 \quad 6 \quad 6$$
$$7 \cdot 0 + 3 \cdot 6 + 9 \cdot 3 + 7 \cdot 1 + 3 \cdot 0 + 9 \cdot 4 + 7 \cdot 0 + 3 \cdot 6 + 9 \cdot 6$$
$$= 0 + 18 + 27 + 7 + 0 + 36 + 0 + 18 + 54$$
$$= 160 \equiv 0 \pmod{10}$$

If the sum is not equivalent to 0 mod 10, there is an error! In problems 116–120, discover if there is an error in the given routing number.

116. 001234567   **117.** 031313562

118. 063107513   **119.** 063107514

120. 111036002

After you finish, you can verify these numbers at http://yourfavorite.com/free/verification.htm.

 Collaborative Learning

Do you remember Pascal's triangle? The triangle is formed by starting and ending each row with the number 1. In the *third* row, the number 2 is obtained by adding $1 + 1$ in the preceding row. Now, in the *fourth* row, the 3 is the sum of $1 + 2$ in the preceding row. And so on.

```
      1
     1 1
    1 2 1
   1 3 3 1
```

1. Construct the next 3 rows of the triangle.

2. Now let us convert the number in each of the rows to arithmetic modulo 2. The first four rows are shown.

```
      1               1
     1 1             1 1
    1 2 1           1 0 1
   1 3 3 1         1 1 1 1
```

Complete the first 7 rows.

3. Now let a black square represent a 1 and a white square a 0. The first four rows are shown.

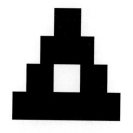

```
   1
  1 1
 1 0 1
1 1 1 1
```

Complete the next 3 rows.

4. Form three or more groups of students. Each group constructs a diagram using modulo 3, 4, 5, and so on.

 **Online Study Center**

5. Access link 9.3.3 on this textbook's student Online Study Center. Write a report on your findings. Discuss the findings with the other groups.

 Research Questions

1. What is a lunometer, and how is it connected to Hans Peter Luhn?

2. Write a report about Hans Peter Luhn. Your report should include his date of birth, where he went to school, and two of his greatest achievements.

3. What is KWIC, and how does it work?

4. Write a report on the Luhn algorithm, how it works, and its uses.

9.4 Abstract Mathematical Systems: Groups and Fields

TABLE 9.9

+	E	O
E	E	O
O	O	E

Evens, Odds, and Mathematical Systems

Suppose you encounter the mysterious Table 9.9. What do you think this table represents? If you think for a moment, you will see that the table exhibits the fact that if you add two even numbers, the answer is even; an even and an odd number yields an odd; an odd and an even number yields an odd, and an odd and an odd number yields an even. As you recall from Section 9.3, a mathematical system consists of a set of elements, {E, O} in this case, one or more operations (+ here), one or more relations (=), and some rules, axioms, or laws that the elements in the set satisfy (E + O = O, for example). In this section you will study mathematical systems in general. To prepare for this study, can you finish defining the operation $\times$ in Table 9.10 at the top of page 626? If you do it successfully, you are ready to start studying abstract mathematical systems.

TABLE 9.10

×	E	O
E	E	____
O	____	____

▶

A. Abstract Mathematical Systems

In this section we shall concentrate on the rules and laws that are obeyed by abstract mathematical systems. To illustrate the ideas involved, we shall define a set $A = \{a, b, c, d, e\}$ and an operation $*$, which we shall call *star*. The operations on the set A can be defined by a table similar to the ones used to define the addition of natural numbers or the addition in modulo 5. Suppose that the operation $*$ is defined by Table 9.11.

To perform the operation star on any pair of elements—say, b and c—we find the element that is in the b row and the c column (rows are horizontal and columns are vertical), as shown in Table 9.12. From this table we can see that $b * c = e$. Similarly, $a * b = c$ and $c * d = b$.

TABLE 9.11

*	a	b	c	d	e
a	b	c	d	e	a
b	c	d	e	a	b
c	d	e	a	b	c
d	e	a	b	c	d
e	a	b	c	d	e

TABLE 9.12

*	a	b	c	d	e
a	b	c	d	e	a
b	...c	...d	...e	a	b
c	d	e	a	b	c
d	e	a	b	c	d
e	a	b	c	d	e

EXAMPLE 1 ▶ Finding Results Under the * Operation

Use Table 9.11 to find the result of each of the following operations:

(a) $b * e$ (b) $d * b$ (c) $(a * b) * c$

Solution

(a) $b * e$ corresponds to the entry in the b row, e column; that is, $b * e = b$.

(b) $d * b$ corresponds to the entry in the d row, b column; that is, $d * b = a$.

(c) $a * b$ is in parentheses, so we must find $a * b$ first. Thus, $(a * b) * c = c * c = a$. ◼

By proceeding as in Example 1, we can find the result $x * y$ for all elements x and y that are in A, and it is evident that the result is also in A. The name that we give to this property is **closure.**

Definition of a Closed Set

The set A is **closed** under the operation $*$ if, for all a and b in A, $a * b$ is also in A.

Intuitively, we see that the set A is closed under the operation $*$ if the operation is always possible and if no new elements are introduced in the table defining the operation. For example, the set of natural numbers is closed under the operation of addition because for any two natural numbers a and b, $a + b$ is also a natural number. On the other hand, the set of natural numbers is not closed under subtraction because, for example, $5 - 7 = -2$, which is not a natural number.

EXAMPLE 2 ▶ Determining the Closure of Sets

Consider the sets $A = \{0, 1\}$ and $B = \{1, 2\}$. Are these sets closed under ordinary multiplication?

Solution

The set A is closed under multiplication because all possible products $0 \times 0 = 0$, $0 \times 1 = 0$, $1 \times 0 = 0$, and $1 \times 1 = 1$ are in A. On the other hand, the set B is not closed under multiplication because $2 \times 2 = 4$, which is not in B. ∎

Another important property previously discussed in connection with the natural numbers is *associativity*.

Definition of an Associative Operation

An operation $*$ defined on a set A is **associative** if, for all a, b, and c in A,

$$(a * b) * c = a * (b * c)$$

For example, the intersection of sets that we studied in Chapter 2 is associative because for any three sets A, B, and C, we have

$$A \cap (B \cap C) = (A \cap B) \cap C$$

EXAMPLE 3 ▶ Checking Associativity Under $*$

In Table 9.11, check to see whether each of the following is true:

(a) $(a * b) * d = a * (b * d)$ 　　　 (b) $(c * a) * e = c * (a * e)$

Solution

(a) Table 9.11 gives $a * b = c$, so $a * b$ can be replaced by c to obtain $(a * b) * d = c * d$. Then, because the figure gives $c * d = b$, we see that $(a * b) * d = b$. Similarly, we find that $a * (b * d) = a * a = b$. The result in both cases is b, so $(a * b) * d = a * (b * d)$.

(b) Again, by using Table 9.11, we find that $(c * a) * e = d * e = d$ and $c * (a * e) = c * a = d$. Because the result is d in both cases, we have $(c * a) * e = c * (a * e)$. ∎

Can we conclude from Example 3 that the operation $*$ is associative? The answer is no because we have not checked all the possibilities. Try some other possibilities and state whether $*$ is associative.

The next property we shall discuss is the *commutative property*.

> **Definition of a Commutative Operation**
>
> An operation $*$ defined on a set A is **commutative** if, for every a and b in A,
>
> $$a * b = b * a$$

For example, the intersection of sets that we studied in Chapter 2 is commutative because for any two sets A and B, $A \cap B = B \cap A$.

EXAMPLE 4 ▶ Checking Commutativity Under $*$

In Table 9.11, check to see whether each of the following is true:

(a) $b * d = d * b$ (b) $e * c = c * e$

Solution

(a) $b * d = a$ and $d * b = a$; thus, $b * d = d * b$.

(b) $e * c = c$ and $c * e = c$; thus, $e * c = c * e$. ◼

Can we conclude from Example 4 that the operation $*$ is commutative? Again, the answer is no because we have not checked all the possibilities. However, there is a simple check that we can make. Since the top half of the table is the reflection of the bottom half across the diagonal going from the upper left to the lower right, the operation is commutative.

As we noted in Chapter 5, the set of integers has an additive identity (0) such that for any integer a, $a + 0 = a = 0 + a$. Similarly, 1 is the identity for multiplication because $a \cdot 1 = a = 1 \cdot a$. The idea of an *identity* can be generalized by means of the following definition:

> **Definition of an Identity for an Operation**
>
> An element e in a set A is said to be an **identity** for the operation $*$ if, for each element x in A,
>
> $$x * e = x = e * x$$

For example, for the operator $*$ defined on the set $A = \{a, b, c, d, e\}$ in Table 9.11, the identity element is e, as we can easily check. Notice that the column directly under the identity element e is identical with the column at the far left, and the row opposite the element e is identical with the row across the top of the table. This appearance of the operation table is characteristic for any set that has an identity element under the operation.

EXAMPLE 5 ▶ Completing the Table for the $*$ Operation

Let $A = \{a, b, c\}$, and let $*$ be an operation defined on A. If c is the identity element, what do you know about the table that defines the operation $*$?

Solution

From the preceding discussion, we can complete a partial table like Table 9.13. The column at the far right is identical to the column at the far left; the row across the bottom is identical to the row across the top. ◼

TABLE 9.13

$*$	a	b	c
a	___	___	a
b	___	___	b
c	a	b	c

Closely related to the idea of an identity is the idea of an *inverse*. For example, the additive inverse of 3 is -3 because $3 + (-3) = 0$ (the additive identity). Similarly, the multiplicative inverse of 3 is $\frac{1}{3}$ because $3 \times \frac{1}{3} = 1$ (the multiplicative identity). In order to find the inverse of a number a under addition, we need a number b such that $a + b = 0$; similarly, to find the inverse of a number a under multiplication, we need a number b such that $a \times b = 1$. These ideas can be summarized by the following definition:

Definition of Inverse

If a and b are in A, we say that **a is the inverse of b** under the operation $*$ if $a * b = e = b * a$, where e is the identity.

EXAMPLE 6 ▶ Finding Identities and Inverses

Consider Table 9.14, which defines the operation #. Find the following:

(a) The identity   (b) The inverse of a (c) The inverse of b

TABLE 9.14

#	a	b
a	b	a
b	a	b

Solution

(a) The identity is b because the column under b is identical to the column at the far left and the row opposite b is identical to the row across the top.

(b) We have to find an element to fill the blank in a # _____ $= b$. From the table we can see that this element is a because a # $a = b$. Thus, a is its own inverse.

(c) We have to find an element to fill the blank in b # _____ $= b$. From the table we can see that this element is b because b # $b = b$. So, b is its own inverse. ◼

EXAMPLE 7 ▶ Checking Identities, Inverses, and Commutativity

Consider the operation # defined by Table 9.15.

(a) Is there an identity element?

(b) Do any of the elements have inverses?

(c) Is the operation commutative?

TABLE 9.15

#	a	b	c
a	c	a	b
b	b	c	a
c	a	b	c

Note that this table of operations corresponds to rotations in the plane as shown in Figure 9.6 at the top of page 630.

Solution

(a) There is no column identical to the column under #, so there is no identity element.

(b) Since there is no identity element, there are no inverses.

(c) The operation is not commutative. For example,

$$a \, \# \, b = a \qquad \text{but} \qquad b \, \# \, a = b$$

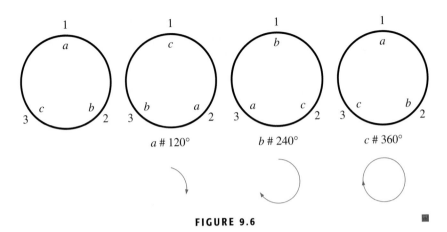

FIGURE 9.6

As in the case of modular arithmetic, there is a property that involves two operations. This property is called the *distributive property* and is defined as follows:

Definition of the Distributive Property

The operation ∗ defined on a set A is said to be **distributive** over the operation # if, for all a, b, and c in A, we have

$$a * (b \# c) = (a * b) \# (a * c)$$

For example, the operation of multiplication defined on the set of real numbers is distributive over addition because, for any real numbers a, b, and c,

$$a \times (b + c) = (a \times b) + (a \times c)$$

On the other hand, addition is *not* distributive over multiplication because

$$a + (b \times c) \neq (a + b) \times (a + c)$$

as shown by the example

$$3 + (4 \times 5) = 3 + 20 = 23$$
$$(3 + 4) \times (3 + 5) = 7 \times 8 = 56$$

Thus,

$$3 + (4 \times 5) = 23 \neq (3 + 4) \times (3 + 5) = 56$$

Now let us consider two operations, F and S, defined for any two natural numbers a and b as follows:

1. a F b means to select the *first* of the two numbers.

2. a S b means to select the *smaller* of the two numbers. (If $a = b$, then a S a is defined to be a.)

Thus, 3 F 5 = 3, 5 F 3 = 5, and 4 F 1 = 4. Similarly, 3 S 5 = 3, 5 S 3 = 3, 4 S 1 = 1, and 2 S 2 = 2.

EXAMPLE 8 ▶ Establishing Distributivity

Is the operation F distributive over S?

Solution

Recall that a F b means to select the first of the numbers and that a S b means to select the smaller of the two numbers a and b (if the numbers are the same, select the number). To show that F distributes over S, we have to show that

$$a \text{ F } (b \text{ S } c) = (a \text{ F } b) \text{ S } (a \text{ F } c)$$

We know that a F $(b$ S $c) = a$ (because a is the first number). Also, a F $b = a$ and a F $c = a$, so

$$(a \text{ F } b) \text{ S } (a \text{ F } c) = a \text{ S } a = a$$

Thus,

$$a \text{ F } (b \text{ S } c) = (a \text{ F } b) \text{ S } (a \text{ F } c)$$

This shows that F is distributive over S. ∎

B. Groups and Fields

Interest in mathematical systems such as those we just studied started in the early nineteenth century with the study of groups.

Definition of a Group

A **group** is a mathematical system consisting of a set S and an operation $*$ with the following four properties:

1. **Closure** property

2. **Associative** property

3. **Identity** property. There is an identity element in S.

4. **Inverse** property. Each element of S has an inverse in S.

If the system also has the **commutative** property, then the group is called a **commutative** (or **abelian**) **group.**

For example, the set of integers with the operation of addition has the following properties:

Closure	The sum of two integers is always an integer.
Associative	If a, b, and c are integers, then $a + (b + c) = (a + b) + c$.
Identity	The integer 0 is the identity element such that $0 + a = a + 0 = a$ for every integer a.
Inverse	If a is an integer, then $-a$ is an integer such that $a + (-a) = 0$.
Commutative	If a and b are integers, then $a + b = b + a$.

Thus, the set of integers is a commutative group under addition.

Although group theory was developed by many mathematicians, the French mathematician Evariste Galois is usually considered the pioneer in this field. However, the origin of group theory can be traced back to the efforts of the Babylonians in solving equations of degree greater than 2. At the time of the Renaissance, the Italian mathematicians Girolamo Cardano and Nicolo de Brescia, commonly referred to as *Tartaglia* ("the Stammerer"), made the first successful attempts to solve equations of third and fourth degree. After their discoveries, it seemed natural to pursue methods to solve equations of degree 5 and higher. Galois investigated the general properties of the equations involved and the properties of their solutions. These studies led to the theory of groups.

EXAMPLE 9 ▶ **Checking for a Commutative Group**

Does the set $S = \{0, 1, 2, 3, 4\}$, together with the operation of multiplication modulo 5, form a commutative group?

Solution
The multiplication table for this system is shown in Table 9.16. We check the five properties as follows:

TABLE 9.16 Multiplication Modulo 5

×	0	1	2	3	4
0	0	0	0	0	0
1	0	1	2	3	4
2	0	2	4	1	3
3	0	3	1	4	2
4	0	4	3	2	1

1. The set S is *closed* because all the entries in the table are elements of S.

2. The operation is *commutative* because the reflection of the bottom half of the table along the diagonal is identical to the top half.

3. The operation is *associative*. [You have to check that $(a \times b) \times c = a \times (b \times c)$, where a, b, and c are in $\{0, 1, 2, 3, 4\}$.]

4. The *identity* for this system is 1.

5. All the elements, except 0, have *inverses*. Zero does not have an inverse because there is no number a in S such that $0 \times a \equiv 1$. The inverses are as follows:

    The inverse of 1 is 1 ($1 \times 1 \equiv 1$).
    The inverse of 2 is 3 ($2 \times 3 \equiv 1$).
    The inverse of 3 is 2 ($3 \times 2 \equiv 1$).
    The inverse of 4 is 4 ($4 \times 4 \equiv 1$).

Since the system does not have an inverse for 0, the system is not a group under the operation of multiplication modulo 5.   ■

EXAMPLE 10 ▶ **Checking for Another Commutative Group**

If the number 0 is omitted from the set in Example 9, is the resulting system a commutative group?

Solution
The multiplication table for this system is Table 9.16 with the 0 row and the 0 column omitted. From the discussion in Example 9, we see that this system has all the required properties and is thus a commutative group.   ■

The distributive property is the only property we have studied that involves two operations. We now turn our attention to a kind of mathematical system, called a *field,* which consists of a set S and two operations.

Definition of a Field

A **field** is a mathematical system consisting of a set S and two operations—say, $*$ and $\#$—defined on S and having the following properties:

1. **Closure** property

2. **Associative** property

3. **Identity** property

4. **Inverse** property (except that there is no inverse for the identity with respect to $*$)

5. **Commutative** property

6. **Distributive** property of $\#$ with respect to $*$

PROBLEM SOLVING

Groups and Fields

① Read the problem.

Show that the system consisting of the set $S = \{0, 1, 2, 3, 4\}$ and the operations of addition and multiplication modulo 5 is a field.

② Select the unknown.

You have to show that the set S together with the operations $+$ and $\times$ modulo 5 form a field.

③ Think of a plan. What do we have to show?

Thus, you have to show that $+$ and $\times$ have the closure, associative, identity, inverse, commutative, and distributive properties.

④ Use the results of Example 9 to verify that the operations of $+$ and $\times$ modulo 5 satisfy the properties stated in the Definition of a Field.

The operation $+$ modulo 5 has the closure, associative, identity, inverse, and commutative properties, as shown in Table 9.8, and the operation $\times$ modulo 5 has the closure, commutative, associative, identity, and inverse properties (except that 0 has no inverse) as shown in Example 9 of this section. Finally, $\times$ is distributive over $+$ in the set of real numbers. Thus, $\times$ is distributive over $+$ in the set S.

⑤ Verify the answer.

You can examine Tables 9.6 and 9.16 and verify the properties mentioned. In general, it can be shown that if p is any prime number, then the integers modulo p form a field under $+$ and $\times$.

TRY EXAMPLE 11 NOW.

Cover the solution, write your own solution, and then check your work.

EXAMPLE 11 ▶ Checking for a Field

Let $S = \{$Odd, Even$\}$, and let $+$ and $\times$ be two operations defined by Tables 9.17 and 9.18 on page 634. You may recognize that these operations correspond to adding or multiplying odd and even numbers. For instance, an odd number added

TABLE 9.17

+	*Odd*	*Even*
Odd	Even	Odd
Even	Odd	Even

TABLE 9.18

×	*Odd*	*Even*
Odd	Odd	Even
Even	Even	Even

to an even number gives an odd number; thus, Odd + Even = Odd. Similarly, Even × Even = Even. Does the set $S = \{Odd, Even\}$ form a field under the + and × operations?

Solution

1. *Closure:* The tables show that S is closed under the two operations.

2. *Associative:* Both operations are associative. For example,
Odd + (Odd + Even) = Odd + Odd = Even, and
(Odd + Odd) + Even = Even + Even = Even, so that
Odd + (Odd + Even) = (Odd + Odd) + Even. (You can check the other cases in the same way.)

3. *Commutative:* The two tables show that both operations are commutative.

4. *Identity:* The identity for + is Even, because the column under Even in Table 9.17 is identical to the column at the far left, and the row adjacent to Even is identical to the top row. You can check in the same way that the identity for × is Odd.

5. *Inverse:* The inverse of Even under + is Even because Even + Even = Even. The inverse of Odd under + is Odd, because Odd + Odd = Even. The inverse of Odd under × is Odd because Odd × Odd = Odd. There is no inverse of Even under × (see the definition of a field).

6. *Distributive:* The distributive property of × over + holds because it holds for multiplication over addition for the real numbers.

Thus, the set $S = \{Odd, Even\}$ with the operations + and × satisfies all the requirements of the definition of a field, so the system is a field.　∎

Online Study Center

To further explore group theory, access link 9.4.1 on this textbook's Online Study Center.

EXERCISES 9.4

A Abstract Mathematical Systems

In problems 1–7, consider the set $S = \{a, b, c\}$ and the operation @ defined by the following table.

@	a	b	c
a	c	a	b
b	a	b	c
c	b	c	a

1. Find the following:
a. $a @ b$　**b.** $b @ c$　**c.** $c @ a$

2. Find the following:
a. $a @ (b @ c)$　**b.** $(a @ b) @ c$
c. Are the results in parts (**a**) and (**b**) identical?

3. Find the following:
a. $b @ (a @ b)$　**b.** $(b @ a) @ b$
c. Are the results in parts (**a**) and (**b**) identical?

4. Find the following:
a. $(a @ b) @ a$　**b.** $a @ (b @ a)$
c. Are the results in parts (**a**) and (**b**) identical?

5. Find the following:
a. $b @ c$　**b.** $c @ b$
c. Are the results in parts (**a**) and (**b**) identical?

6. Is the operation @ a commutative operation? Explain.

7. Is the set S closed with respect to the operation @? Explain.

8. Suppose a F b means to select the first of two numbers a and b, as in Example 8 of this section. Let $A = \{1, 2, 3\}$.
a. Make a table that will define the operation F on the set A.
b. Is A closed under the operation F?
c. Is the operation F commutative?

d. For any three natural numbers a, b, and c, show that

$$a \text{ F } (b \text{ F } c) = (a \text{ F } b) \text{ F } c$$

9. Let the operation F be defined as in problem 8, and let N be the set of natural numbers.
   a. Is the set N closed under the operation F? Explain.
   b. Is the operation F associative? Explain.
   c. Is the operation F commutative? Explain.

10. Let S be the set of all non-negative multiples of 5 (0, 5, 10, and so on).
   a. Is the set S closed with respect to ordinary multiplication?
   b. Is the operation of ordinary multiplication commutative on S? Explain.
   c. Is the operation of ordinary multiplication associative on S? Explain.

11. Are the following sets closed under the given operation?
   a. The odd numbers under addition
   b. The odd numbers under multiplication
   c. The even numbers under addition
   d. The even numbers under multiplication

12. Give an example of an operation under which the set $\{0, 1\}$ is
   a. not closed.
   b. not associative.
   c. not commutative.

Let $S = \{\varnothing, \{a\}, \{b\}, \{a, b\}\}$. The following table will be used in problems 13–18. The entries in the table represent the set intersection of the elements in the corresponding rows and columns.

∩	$\varnothing$	$\{a\}$	$\{b\}$	$\{a, b\}$
$\varnothing$	$\varnothing$	$\varnothing$	$\varnothing$	$\varnothing$
$\{a\}$	$\varnothing$	$\{a\}$	$\varnothing$	$\{a\}$
$\{b\}$	$\varnothing$			
$\{a, b\}$	$\varnothing$			

13. Supply the missing entries in the table.

14. Find the following:
   a. $(\{a\} \cap \{b\}) \cap \{a, b\}$
   b. $\{a\} \cap (\{b\} \cap \{a, b\})$
   c. Are the results in parts (**a**) and (**b**) identical?

15. Find the following:
   a. $(\{b\} \cap \{a, b\}) \cap \{a\}$
   b. $\{b\} \cap (\{a, b\} \cap \{a\})$
   c. Are the results in parts (**a**) and (**b**) identical?

16. In Chapter 2 you learned that for any three sets A, B, and C, $A \cap (B \cap C) = (A \cap B) \cap C$. On the basis of this result, would you say that the operation $\cap$ defined in the table above is associative?

17. Is the set S closed with respect to the operation $\cap$? Explain.

18. Is the operation $\cap$ commutative? How can you tell?

19. Suppose that $a \text{ L } b$ means to select the larger of the two numbers a and b (if the numbers are the same, select the number).
   a. Complete the following table.
   b. If there is an identity element, what is it?

L	1	2	3	4
1				4
2			3	
3		3		
4	4			4

The table below will be used in problems 20 and 21.

S	1	2	3	4
1	1	1	1	1
2	1	2	2	2
3	1	2	3	3
4	1	2	3	4

20. Does the set $A = \{1, 2, 3, 4\}$ have an identity? If so, what is this identity?

21. Find the inverses of the following:
   a. 1   **b.** 2   **c.** 3   **d.** 4

22. Consider the set $S = \{-1, 0, 1\}$ and the operation of ordinary multiplication. Complete the table.

×	−1	0	1
−1			
0			
1			

23. Does the set $S = \{-1, 0, 1\}$ of problem 22 have an identity under multiplication? If so, what is this identity?

24. For the set $S = \{-1, 0, 1\}$ of problem 22 under the operation of multiplication, find the inverses (if possible) of the following:
  a. 1   **b.** -1 **c.** 0

25. If $\mathcal{U}$ is the set of all subsets of any nonempty set A,
  a. find the identity element for the operation of set intersection ($\cap$).
  b. can you find more than one identity?

The table below will be used in problems 26–28.

@	0	1	2	3
0	0	1	2	3
1	1	2	3	0
2	2	3	0	1
3	3	0	1	2

26. Is the set $S = \{0, 1, 2, 3\}$ closed under the operation @?

27. Does the set S have an identity with respect to the operation @? If so, what is this identity?

28. Find the inverses (if they exist) of the following:
  a. 0   **b.** 1   **c.** 2   **d.** 3

In problems 29–32, let F be defined as in problem 8 and let a L b mean to select the larger of the two numbers a and b (if $a = b$, assume a L $b = a$) as in problem 19.

29. Find the following:
  a. 3 F (4 L 5)   **b.** 4 F (5 L 6)

30. Find the following:
  a. 4 L (4 F 5)   **b.** 5 L (6 F 7)

31. Does the distributive property a F (b L c) = (a F b) L (a F c) hold for all real numbers a, b, and c? Explain.

32. Does the distributive property a L (b F c) = (a L b) F (a L c) hold for all real numbers a, b, and c? Explain.

33. In ordinary arithmetic, is multiplication distributive over subtraction?

34. In ordinary arithmetic, is division distributive over subtraction? [*Hint:* Look at the two forms $a \div (b - c) = (a \div b) - (a \div c)$ and $(a - b) \div c = (a \div c) - (b \div c)$.]

35. In the arithmetic of fractions, is multiplication distributive over addition?

36. In the arithmetic of fractions, is addition distributive over multiplication?

B **Groups and Fields**

37. Let $S = \{a, b, c\}$, and let $*$ be defined by the following table. Is S a group with respect to the operation $*$?

$*$	a	b	c
a	b	c	a
b	c	a	b
c	a	b	c

In problems 38–49, determine whether the given set forms a group under the given operation. For each set that is not a group, give one specific example of a condition that is not satisfied.

38. The odd integers under the operation of addition

39. The odd integers under the operation of multiplication

40. The even integers under the operation of addition

41. The even integers under the operation of multiplication

42. The positive integers under the operation of addition

43. The positive integers under the operation of multiplication

44. The integers under the operation of addition

45. The integers under the operation of multiplication

46. The real numbers under the operation of multiplication

47. The real numbers under the operation of addition

48. The set $\{-1, 0, 1\}$ under the operation of addition

49. The set $\{-1, 0, 1\}$ under the operation of multiplication

50. Complete the following table so that the result will be a group under the given operation:

#	a	b	c
a			b
b		b	c
c			

51. Let $S = \{a, b, c, d, e\}$, and let # be defined on S by the following table. Is the set S under the operation # a group? If not, give one specific example of a condition that is not satisfied.

#	a	b	c	d	e
a	a	b	c	d	e
b	b	e	a	c	d
c	d	c	e	a	b
d	c	d	b	e	a
e	e	a	d	b	c

In problems 52–56, determine whether the given sets form a field under the operations of addition and multiplication.

52. The set of positive odd integers

53. The set of positive even integers

54. The set of integral multiples of 5

55. The set of integral multiples of 2

56. The set of all real numbers

 In Other Words

Suppose you are given a set S and a table defining the operation ◆. For problems 57–60, explain in your own words the procedure for determining whether

57. the set S is closed under the operation ◆.

58. the operation ◆ is commutative.

59. there is an identity for the operation ◆.

60. two elements—say, a and b—are inverses of each other.

61. Explain a procedure you can use to verify that the operation ∗ of problem 37 is associative. How many cases do you have to check to be convinced that ∗ is associative?

62. Suppose that an associative operation · is defined for every pair of elements of a set S and that S is closed under this operation. There is a theorem that can save work when trying to show that this system is a group. This theorem says that if there is an identity element e in S such that $e \cdot x = x$ for every element x of S and if every element x of S has an inverse $\hat{x}$ in S such that $\hat{x} \cdot x = e$, then the system is a group. Explain the procedure you would follow to use this theorem to solve problem 37. How many cases do you have to check now?

 Using Your Knowledge

The distributive property can be used to shorten the labor in certain multiplication problems. For instance, the product of 6 and 999 can easily be found by writing

$$6 \times 999 = 6 \times (1000 - 1) = 6000 - 6 = 5994$$

Use this idea to calculate the following products.

63. 6×9999 **64.** 8×99

65. 7×59 **66.** 8×999

67. 4×9995 **68.** 3×9998

 **Discovery**

The distributive property can be used in an interesting way in number puzzles. Have you ever seen a magician ask a person in the audience to think of a number and perform several calculations with it? Then, without knowing the original number, the magician knows the person's final number! The following is one of these puzzles:

Think of a number.
Add 3 to it.
Triple the result.
Subtract 9.
Divide by the number with which you started.
The result is 3.

In the following table are the calculations of four persons who selected different numbers:

	First	Second	Third	Fourth
Think of a number.	4	6	8	10
Add 3 to it.	7	9	11	13
Triple the result.	21	27	33	39
Subtract 9.	12	18	24	30
Divide by the number with which you started.	3	3	3	3
The result is always 3.				

69. Why does the puzzle work? (*Hint:* Let x be the number you select, and work through the puzzle.)

70. What is the result of the following puzzle?

Think of a number.
Add 2 to it.
Double the result.
Subtract 4.
Divide by the number with which you started.
The result is _____.

 Collaborative Learning

Here is an actual mathematical group that provides aerobic benefits. Form several groups each with a copy of the table that follows. Select a leader for each group, and let the rest of the members of the group stand at attention. Consider the set $C = \{At, R, L, A\}$, where the

elements *At*, *R*, *L*, and *A* stand for the commands "Attention!" "Right face!" "Left face!" and "About face!" Let the operation [FB] mean *followed by. L* followed by *A* has the same effect in terms of the final orientation of the group members as the single command *R*. So we enter *R* in row *L* column *A*. With this information, each group leader should give the proper commands to fill in the blanks in the table.

[FB]	At	R	L	A
At				
R				
L				R
A				

1. Is the set *C* closed under the operation [FB]? Explain.

2. What is the identity element?

3. What is the inverse of *At*?

4. What is the inverse of *R*?

5. What is the inverse of *L*?

6. What is the inverse of *A*?

7. Is the operation associative? Explain.

8. Is the set consisting of the set *C* and the operation [FB] a group? Explain.

9. Is the operation [FB] commutative? Explain.

9.5   Game Theory

Have It Your Way

Have you noticed that when a fast-food chain such as McDonald's opens a restaurant, another fast-food chain, maybe Burger King, starts a competing restaurant within a couple of blocks? Why do you think this happens? Either both chains conduct extensive fast-food demand, demographic, and location surveys, reach the same conclusion, and target the same population, or one of the chains does most of the research and the other just follows. Now suppose McDonald's and Burger King wish to open restaurants. They can decide to open on opposite sides of town (*O*) or in a central location (*C*) near each other. If they both open on opposite sides of town (*O*) or in a central location (*C*) near each other, each will get 50% of the targeted business. If McDonald's opens in the central location (*C*) and Burger King on the other side of town (*O*), McDonald's gets 70%

It is not unusual to see competing fast-food franchises in close proximity to one another.

of the business (and Burger King 30%). However, if McDonald's opens on the other side of town (O) and Burger King in the central location (C), McDonald's gets 40% of the business (and Burger King 60%). What should each of the franchisers do to ensure their greatest incomes? The percentage gains or losses for McDonald's can be the entries in a **payoff matrix** as shown below. The optimal strategies can be obtained by examining the rows and columns in this matrix.

$$
\begin{array}{c}
\textbf{Burger King} \\
\begin{array}{cc}
C & O
\end{array} \\
\textbf{McD} \quad
\begin{array}{c}
C \\
O
\end{array}
\begin{bmatrix}
50 & 70 \\
40 & 50
\end{bmatrix}
\end{array}
$$

To see how another simple situation can be represented by a matrix, look at the simple game of matching pennies. In this game, two players R and C each toss a penny. If they match (HH or TT), R keeps both pennies, but if they do not, C keeps both pennies. In the matrix representation for this game, the column headings H and T represent the outcomes (Heads or Tails) for C, the row headings represent the outcomes for R, and the matrix entries give the payoffs for R. Note that row 1, column 1 corresponds to the outcome HH, and since there is a match, R wins one penny ($+1$). On the other hand, in row 2, column 1, the outcomes are TH and since there is no match, R loses a penny (-1) as shown in the matrix.

$$
\begin{array}{c}
\textbf{Column } (C) \\
\begin{array}{cc}
H & T
\end{array} \\
\textbf{Row } (R) \quad
\begin{array}{c}
H \\
T
\end{array}
\begin{bmatrix}
+1 & -1 \\
-1 & +1
\end{bmatrix}
\end{array}
$$

Now, for which of these situations (choosing a location or matching pennies) do you think it is easier to develop a winning strategy? Surprisingly enough, you will see in this section that it is easier to develop a winning strategy for the restaurant situation. ▶

In business, economics, the sciences, and the military, decisions have to be made in competitive situations that are similar to games played according to formal rules. The branch of mathematics that deals with the analysis of competition and conflict is called **game theory** and was started because of the efforts of John von Neumann, Emil Borel, and Oskar Morgenstern. **Two-person games** (or **matrix games**) are those that have only two adversaries or players always making intelligent choices. Matching pennies and the selection of a restaurant location by two different franchises are examples of two-person games. Let us construct the **payoff matrix** denoting the percentage gains for McDonald's as described in Getting Started. The following is the information we need:

1. If McDonald's selects O or C, and Burger King chooses the same location, each will get 50% of the business.

2. If McDonald's picks C and Burger King O, McDonald's gets 70% (and Burger King 30%).

3. If McDonald's selects O, and Burger King C, McDonald's gets 40% (and Burger King 60%).

From (1), we can enter 50 in row 1, column 1 and in row 2, column 2.
From (2), we can enter 70 in row 1, column 2.

From (3), we can enter 40 in row 2, column 1.

Burger King

$$
\text{McD} \quad
\begin{array}{c}
 \\
C \\
O
\end{array}
\begin{array}{cc}
C & O \\
\left[\begin{array}{cc} 50 & 70 \\ 40 & 50 \end{array}\right]
\end{array}
$$

This game and the penny-matching game are **constant-sum** games, because the sum of the payoffs is a constant (in the McDonald's–Burger King game, the McDonald's percentage plus the Burger King percentage is a constant 100%). In the penny-matching game, the sum of the payoffs is 0. Thus, matching pennies is called a **zero-sum** game.

A. Saddle Points

In game theory, a player seeks an **optimal strategy,** a strategy that maximizes the player's gain or minimizes his or her loss. What should the strategy for McDonald's (the row player) be? To select row 1 or row 2? If it selects row 1, it has a chance of getting 70% of the business! But Burger King will then pick column 1, and each will end up with 50%. Certainly, row 1 is better than row 2 for McDonald's because the best it can do in row 2 is 50%, but it is also possible to end up with only 40% if Burger King picks column 1. Thus, the best strategy for McDonald's is to **select row 1.** If we think of McDonald's as player R, the row player, this discussion suggests the following method for developing an **optimal pure strategy** for player R:

Optimal Pure Strategy for Player R
1. Circle the smallest element in each row.
2. Choose the row that has the largest of the circled values.

Note that this strategy guarantees the greatest gain to R, independent of the other player's choice.

What strategy should Burger King follow? If it selects column 2, McDonald's will certainly select row 1, giving McDonald's 70% of the business. Thus, the best strategy for Burger King is to **select column 1.**

Optimal Pure Strategy for Player C
1. Box the largest element in each column.
2. Choose the column with the smallest boxed number.

In this game, the **optimal pure strategy** for McDonald's is to **select row 1** and the **optimal pure strategy** for Burger King is to **select column 1.** Note that 50 is both the **smallest** element in its row and the **largest** element in its column. 50 is called a **saddle point,** and it is the **value** of this game.

> **Definition of Strictly Determined**
>
> A matrix game is **strictly determined** if there is an entry in the payoff matrix that is both the *smallest* element in its row and the *largest* element in its column. Such an entry is called a **saddle point** for the game, and the **value** of the game is the value of this saddle point.

EXAMPLE 1 ▶ Finding Values of a Game

If the entries represent payoff values for the row player, determine which of the following matrices define strictly determined games and find the values of the games:

(a) $\begin{bmatrix} 1 & 3 & -2 \\ -1 & 0 & 4 \\ 2 & -3 & 1 \end{bmatrix}$ (b) $\begin{bmatrix} -2 & 1 & -3 \\ 3 & 4 & 1 \\ -2 & -4 & -1 \end{bmatrix}$

(c) $\begin{bmatrix} 2 & 3 & 2 & 4 \\ 1 & 4 & -3 & 3 \\ 1 & -1 & 0 & -1 \\ 2 & 4 & 2 & 5 \end{bmatrix}$

Solution

(a) For the row player, do the following:

1. Circle the smallest element in each row (-2, -1, and -3).

2. The row player's optimal strategy is to choose the row that has the largest circled value, that is, the row containing -1 (*row 2*).

$\begin{bmatrix} 1 & 3 & ⊖2 \\ ⊖1 & 0 & 4 \\ 2 & ⊖3 & 1 \end{bmatrix}$ ← Choose row 2.

For the column player, do the following:

1. Box the largest element in each column (2, 3, and 4).

2. The column player's optimal strategy is to choose the column with the smallest boxed number, that is, the column containing 2 (*column 1*).

$\begin{bmatrix} 1 & \boxed{3} & -2 \\ -1 & 0 & \boxed{4} \\ \boxed{2} & -3 & 1 \end{bmatrix}$

↑
Choose
column 1.

Since there is no value that is both the smallest element in its row and the largest in its column, there is no saddle point. This game is *not strictly determined.*

(b) The row player circles -3, 1, and -4. Since 1 is the *largest* of the three numbers, the row player selects *row 2*. The row player's optimal strategy is to select the row containing 1 (*row 2*).

$$\begin{bmatrix} -2 & 1 & \boxed{-3}\!\!\bigcirc \\ 3 & 4 & \bigcirc\!\!1 \\ -2 & \bigcirc\!\!-4 & -1 \end{bmatrix}$$ $\leftarrow$ Choose row 2.

The column player boxes 3, 4, and 1. Since 1 is the *smallest* of the three numbers, the optimal strategy for the column player is to select *column 3*, the column containing 1.

$$\begin{bmatrix} -2 & 1 & -3 \\ \boxed{3} & \boxed{4} & \boxed{\bigcirc\!1} \\ -2 & -4 & -1 \end{bmatrix}$$
$\uparrow$
Choose
column 3.

Since the 1 in the second row, third column is both the *smallest* element in its row and the *largest* element in its column (it is *circled and boxed*), this is a saddle point. The value of this strictly determined game is 1.

(c) The row player circles the *smallest* numbers in each row, 2 (twice) in row 1, -3 in row 2, -1 (twice) in row 3, and 2 (twice) in row 4. Since 2 is the largest of 2, -3, -1, and 2, the row player should select row 1 or row 4 (each containing 2 twice).

  The column player boxes the *largest* numbers in each column, the *4*s in column 1, the *4*s in column 2, the *2*s in column 3, and *5* in column 4. The column player should select column 1 or 3.

  Since the number 2 (columns 1 and 3) is both the smallest element in rows 1 and 4 and the largest element in columns 1 and 3, the game is *strictly determined* and its value is 2, occurring at four different saddle points. Note that 2 is both circled and boxed each time.

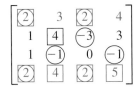

As you can see from Example 1(c), a game can have more than one saddle point (four in this case), but they must all have the same value.

B. Mixed Strategies

The penny-matching game discussed in Getting Started does not have a saddle point. How can we develop a strategy to play this game a large number of times?

Clearly, player R should not always play the same strategy (say, H), for if she did, player C might always play T and win! So R must play H some fraction p of the time and win $(+1)$ and T the rest, or $1 - p$, of the time and lose (-1). What can C do? Let us look at the payoff matrix.

Expected Payoff

If C always chooses H, R's expected payoff is

$$E_H = (p)(1) + (1 - p)(-1) = 2p - 1$$

If C always chooses T, R's expected payoff is

$$E_T = (p)(-1) + (1 - p)(1) = -2p + 1$$

$$
\begin{array}{c}
\textbf{Column} \\
\begin{array}{cc}
H & T
\end{array} \\
\textbf{Row}\;
\begin{array}{c}
H \\
T
\end{array}
\left[
\begin{array}{cc}
+1 & -1 \\
-1 & +1
\end{array}
\right]
\end{array}
$$

Note that these two equations can be obtained by multiplying $[p \quad 1 - p]$ by the first and second columns, respectively, in the payoff matrix. Since the expected payoff for R should be the same, regardless of C's choice,

$$E_H = E_T$$

or

$$2p - 1 = -2p + 1$$

Solving for p,

$$p = \tfrac{1}{2}$$

This means that if C chooses H, R's expected payoff will be

$$E_H = 2p - 1 = 2\left(\tfrac{1}{2}\right) - 1 = 0$$

and if C chooses T, R's expected payoff will be

$$E_T = -2p + 1 = -2\left(\tfrac{1}{2}\right) + 1 = 0$$

EXAMPLE 2 ▶ Finding Payoff Matrices and Saddle Points

At the present time, most restaurants have a smoking section (S) and a nonsmoking section (N). Suppose that the average smoker spends \$15 per meal, and the average nonsmoker \$10. A customer enters a restaurant, and the manager has designated smoking and nonsmoking sections. If the person is a smoker and the smoking section is open, the restaurant makes \$15, but if only the nonsmoking section is open, the smoker leaves, and the restaurant makes nothing. If the customer is a nonsmoker and only the smoking section is available, the nonsmoker will leave, and the restaurant will make nothing, whereas if the person is a nonsmoker and the nonsmoking section is open, the restaurant will make \$10.

(a) Construct the payoff matrix for the manager.

(b) Is there a saddle point for this game?

(c) What proportion of the customers should the manager expect to be smokers and what proportion nonsmokers?

Solution

(a) If we use the rows for the manager's choice and the columns for the customer's choice, the payoff matrix is

Customer

$$\begin{array}{c} & S \quad\ \ N \\ \textbf{Manager} \begin{array}{c} S \\ N \end{array} \left[\begin{array}{cc} \$15 & 0 \\ 0 & \$10 \end{array}\right] \end{array}$$

(b)   **Customer**

$$\begin{array}{c} & S \quad\ \ N \\ \textbf{Manager} \begin{array}{c} S \\ N \end{array} \left[\begin{array}{cc} \boxed{\$15} & ⓪ \\ ⓪ & \boxed{\$10} \end{array}\right] \end{array}$$ There is no saddle point for this matrix.

(c) Let p be the proportion of customers that are smokers and $1 - p$ the proportion that are nonsmokers. The manager's expectation for a smoking customer is

$$E_S = p(15) + (1 - p)(0) = 15p$$

The manager's expectation for a nonsmoking customer is

$$E_N = p(0) + (1 - p)(10) = 10 - 10p$$

To make each customer's payoff be the same, we must have

$$E_S = E_N$$
$$15p = 10 - 10p$$
$$25p = 10$$
$$p = \tfrac{2}{5}$$

Thus, the manager should expect $\tfrac{2}{5}$ of his customers to be smokers and $\tfrac{3}{5}$ to be nonsmokers.   ■

EXAMPLE 3 ▶ Finding the Expected Test Using Matrices

How hard do you study for your tests? It probably depends on the subject and your instructor's reputation. The payoff matrix gives the average grade expectations for students when a test that can be classified as Hard (H), Medium (M), or Easy (E) is given.

Instructor

$$\begin{array}{c} & \ H \quad\ M \quad\ E \\ \textbf{Student} \begin{array}{c} H \\ M \\ E \end{array} \left[\begin{array}{ccc} 70 & 80 & 90 \\ 75 & 70 & 80 \\ 50 & 60 & 70 \end{array}\right] \end{array}$$

On the basis of this information, what type of test should students expect?

Solution

First, we note that there is no saddle point. Think of the students and the instructor as competitors in a two-person game. Certainly the students would never choose row 3 because all the grades there are lower than the corresponding grades in row 1. Thus, the students eliminate row 3. (We say that row 1 **dominates** row 3). The result is

Instructor

$$\begin{array}{c} & H \quad M \quad E \\ \textbf{Student} \quad \begin{array}{c} H \\ M \end{array} & \left[\begin{array}{ccc} 70 & 80 & 90 \\ 75 & 70 & 80 \end{array} \right] \end{array}$$

On the other hand, the instructor would never let the students make 90s and 80s, that is, would never choose column 3. (Note that both columns 1 and 2 **dominate** column 3 because their corresponding values are smaller than those in column 3). Thus, we can eliminate column 3 from further consideration; this leaves the 2×2 matrix

Instructor

$$\begin{array}{c} & H \quad M \\ \textbf{Student} \quad \begin{array}{c} H \\ M \end{array} & \left[\begin{array}{cc} 70 & 80 \\ 75 & 70 \end{array} \right] \end{array}$$

Now let h be the proportion of times a hard test is given. Then $1 - h$ of the time the test will not be hard. The student's expectation for a hard test is

$$E_h = h(70) + (1 - h)(75)$$

The student's expectation for not having a hard test is

$$E_n = h(80) + (1 - h)(70)$$

Thus,

$$h(70) + (1 - h)(75) = h(80) + (1 - h)(70)$$
$$-5h + 75 = 10h + 70$$
$$h = \tfrac{1}{3}$$

Thus, on the basis of the information given, the students should expect a hard test $\frac{1}{3}$ of the time and a medium test $\frac{2}{3}$ of the time. No easy tests are given! ∎

Online Study Center

To further explore game theory, saddle points, and strategies, access links 9.5.1 and 9.5.2 on this textbook's Online Student Study Center

---

EXERCISES 9.5

A Saddle Points

In problems 1–8, decide whether each payoff matrix represents a strictly determined game. If the game is strictly determined, find the optimal pure strategy for each player and the value of the game.

1. $\begin{bmatrix} 4 & 5 \\ 2 & 3 \end{bmatrix}$

2. $\begin{bmatrix} 3 & 1 \\ 2 & 0 \end{bmatrix}$

3. $\begin{bmatrix} 5 & 2 \\ 3 & 4 \end{bmatrix}$

4. $\begin{bmatrix} -3 & 2 \\ 0 & -3 \end{bmatrix}$

5. $\begin{bmatrix} 6 & 6 & 4 \\ 2 & -5 & -10 \\ 8 & 5 & -15 \end{bmatrix}$

6. $\begin{bmatrix} 10 & -3 & -8 \\ 12 & -2 & 5 \\ 8 & 0 & 4 \end{bmatrix}$

7. $\begin{bmatrix} 1 & 6 & -2 \\ 3 & 5 & 2 \\ 4 & 5 & 4 \end{bmatrix}$

8. $\begin{bmatrix} 5 & 1 & 2 \\ 8 & 0 & -4 \\ 4 & 1 & 3 \end{bmatrix}$

B Mixed Strategies

In problems 9–12, do the following:
  a. Determine whether there is a saddle point.
  b. Find the mixed strategy for the row player.

9. $\begin{bmatrix} 2 & 1 \\ -1 & 4 \end{bmatrix}$

10. $\begin{bmatrix} 3 & 2 \\ 2 & 4 \end{bmatrix}$

11. $\begin{bmatrix} 3 & -2 \\ -1 & 0 \end{bmatrix}$

12. $\begin{bmatrix} 10 & 5 \\ 6 & 12 \end{bmatrix}$

13. Use the given payoff matrix to find the optimal mixed strategy and the value of the game for the row player.

$$\begin{bmatrix} 6 & 0 \\ 2 & 4 \\ 0 & 3 \end{bmatrix}$$

14. Use the given payoff matrix to find the optimal mixed strategy and the value of the game for the row player.

$$\begin{bmatrix} 5 & 30 \\ 0 & 20 \\ 25 & 15 \end{bmatrix}$$

15. You are preparing for a test, but you don't know whether it is an essay test or a multiple choice test. You can spend 2 or 4 hours studying for this test, and the table below shows the probable scores that you would make. On the basis of this table, what should you do?

		Essay	Multiple Choice
Study	2 hours	70	80
	4 hours	85	75

16. John must prepare for an English test and for a mathematics test. He decides to study 3 hours for one of the tests and 2 hours for the other. The table shows the probable scores he would make on each test. What should he do?

Studying Time	English	Math
English 2 hours Mathematics 3 hours	80	85
English 3 hours Mathematics 2 hours	90	75

17. Ann is planning to invest some of her savings in bonds, stocks, and money market funds. Her expected return will depend on whether interest rates rise or fall during the year. The table shows the expected returns. Find Ann's optimum investment strategy and the corresponding expected return.

	Interest Rates	
Investments	Rise (%)	Fall (%)
Bonds	6	11
Stocks	10	12
Money market	15	10

18. When stocks increase in price on a certain day, their prices tend to decrease the next day, and vice versa. The payoff matrix shows the pertinent percents. In the long run, what fraction of the time will the stock increase in price?

		Tomorrow's Price	
		Increase	Decrease
Today's Price	Increase	0.2	0.8
	Decrease	0.6	0.4

19. Two gas stations R and C are competing with each other at a certain location. The percentage of the business captured by each station is dependent on the price and is given in the payoff matrix. How should station R price its gasoline?

		Station C	
		$1.00	$1.10
Station R	$1.00	60%	50%
	$1.10	30%	70%

20. Manhattan is divided into three sectors: Uptown (*U*), Midtown (*M*), and Lower (*L*). The proportions of the time that a taxicab operating in Manhattan picks up a passenger in any sector and drops the passenger off in any sector are given by the payoff matrix. What is the expected percent of fares going from upper Manhattan to a different sector?

		Drop Off		
		U	M	L
Pick Up	U	0.5	0.4	0.1
	M	0.3	0.6	0.1
	L	0.2	0.3	0.1

21. Car manufacturer's ads rely on the performance or safety of the cars being advertised. Market studies show that ads based on performance are effective for 70% of the younger buyers and 20% of the older buyers, whereas ads based on safety are effective for 80% of the older buyers and 40% of the younger buyers. If this process is viewed as a game between the manufacturer and buyers, do or answer the following:
   a. Give the payoff matrix for the game.
   b. In what proportion should the two kinds of ads be mixed for maximum effectiveness?

22. You and your friend play a game with pennies and nickels. Each of you puts down a coin. If the coins match, you take both, and if they do not

match, your friend takes both. Write the matrix for this game. What is your optimal mixed strategy and what is your corresponding expected payoff for this game?

23. A farmer has a large acreage planted in strawberries. A freeze is predicted and the farmer must decide whether to turn on his water sprays to protect the crop. He estimates that if there is a freeze and he turns on the sprays, he will save the crop and be able to sell it for $6000. If there is no freeze, he will be out the $400 in water and pumping costs. If he does not turn on the sprays and there is a freeze, he will lose $4000, but if there is no freeze, he can sell the crop for $4000. Make a matrix to show these figures and find the farmer's optimal mixed strategy and his corresponding expected payoff.

 In Other Words

24. Why do you think the value of a strictly determined game is called a saddle point?

25. Explain why, when a row i dominates another row j in a payoff matrix, j can be eliminated.

26. Explain why, when a column i dominates a column j in a payoff matrix, j can be eliminated.

 Using Your Knowledge

We are now ready to use your knowledge in love games. A young student is planning to send flowers, poems, or candy to his sweetheart, but there are several things to take into account. She may be allergic to flowers or on a very strict diet. The payoff matrix for the situation is

		Sweetheart's Response	
		Allergic	On a diet
	Flowers	−3	4
He Sends	Poems	2	5
	Candy	3	−2

27. Which of the rows is dominated by row 2?

28. Which row can be eliminated?

29. What is the student's optimal mixed strategy?

30. What is the student's expected value for this love game?

 Collaborative Learning

We would be remiss if we failed to give you the opportunity to solve a classic problem: The Prisoner's Dilemma. Suppose two suspects, Bonnie and Clyde, are taken into custody and questioned separately by police. Each has the choice of confessing to the crime or not. If each confesses, both will be prosecuted and go to prison for 10 years. If neither confesses, both will serve a 1-year sentence on minor charges. However, if only one confesses and agrees to testify against the other, the one who confesses will go free and the other will go to prison for 20 years, the maximum charge. Let us write their prison sentences as ordered pairs of the form (B, C). The matrix for this situation will look like this:

1. Fill in the entries in the matrix with the four ordered pairs corresponding to the situations in which both confess, neither confesses, and one confesses and the other doesn't.

2. If Bonnie wants to minimize her time in jail, what is her best strategy: confess or not?

3. If Clyde wants to minimize his time in jail, what is his best strategy: confess or not?

4. What happens if they both reason the same way, so that both confess?

5. What happens if they both keep quiet and don't confess?

6. What do you think the "dilemma" is?

Online Study Center

To further explore the Prisoner's Dilemma, access links 9.5.3 and 9.5.4 on this textbook's Online Study Center.

Research Questions

1. John von Neumann, Emil Borel, and Oskar Morgenstern are contributors to the development of game theory. Access link 9.5.5 on this textbook's Online Study Center and find out the contributions of each mathematician to the theory.

2. Write a small history of game theory stating the main contributors as well as their contribution.

3. What is the Prisoner's Dilemma, who discovered it, and what is its importance? State a different version and solution to the problem.

Chapter 9 Summary

Section	Item	Meaning	Example
9.1	Matrix	A rectangular array of numbers enclosed in square brackets (or parentheses)	$\begin{bmatrix} 3 & 4 \\ 5 & 6 \end{bmatrix}$
9.1	$A = B$	The corresponding elements of matrices A and B are equal.	$\begin{bmatrix} 2 & 4 \\ 5 & 6 \end{bmatrix} = \begin{bmatrix} \frac{6}{3} & 2 \cdot 2 \\ 3 + 2 & 6 \end{bmatrix}$
9.1A	kA	Multiplication of each element of matrix A by the number k	$3\begin{bmatrix} 1 & 2 \\ 3 & 4 \end{bmatrix} = \begin{bmatrix} 3 & 6 \\ 9 & 12 \end{bmatrix}$
9.1A	$A + B$	The sum of matrices A and B	$\begin{bmatrix} 1 & 2 \\ 3 & 4 \end{bmatrix} + \begin{bmatrix} 4 & 5 \\ 3 & 2 \end{bmatrix} = \begin{bmatrix} 5 & 7 \\ 6 & 6 \end{bmatrix}$
9.1B	AB	The product of the matrices A and B	$[3 \quad 4]\begin{bmatrix} 1 & 2 \\ 0 & 5 \end{bmatrix} = [3 \quad 26]$
9.1C	I	The identity matrix	For a 2×2 matrix, $I = \begin{bmatrix} 1 & 0 \\ 0 & 1 \end{bmatrix}$
9.2A	Augmented matrix	A matrix consisting of the coefficients of the variables and the constants in a system of equations	The augmented matrix for the system $\begin{aligned} x + y + z &= 3 \\ x + y \quad\;\; &= 1 \\ y + z &= 3 \end{aligned}$ is $\left[\begin{array}{ccc|c} 1 & 1 & 1 & 3 \\ 1 & 1 & 0 & 1 \\ 0 & 1 & 1 & 3 \end{array}\right]$

9.3	Mathematical system	A set of elements with one or more operations; relations; and rules, axioms, or laws	The set of integers with the four fundamental operations
9.3	$a \equiv b \pmod{m}$	$a - b$ is a multiple of m.	$3 \equiv 8 \pmod{5}$
9.4	Closure property	A set S is closed under an operation $*$ if, for all a and b in A, $a * b$ is in A.	The set I of integers is closed under the operation $+$.
9.4	Associative property	An operation $*$ is associative if, for all a, b, and c in A, $(a * b) * c = a * (b * c)$.	The operation $+$ is associative.
9.4	Commutative property	An operation $*$ is commutative if, for all a and b in A, $a * b = b * a$.	The operation $+$ is commutative.
9.4	Identity	An element e is an identity for $*$ if, for every a in A, $a * e = a = e * a$.	0 is the identity for addition.
9.4	Inverse	An element b is the inverse of a if $a * b = e = b * a$, where e is the identity for $*$.	-3 and 3 are additive inverses.
9.4	Distributive property	The operation $*$ is distributive over the operation $\#$ if, for every a, b, and c in A, $a * (b \# c) = (a * b) \# (a * c)$.	$\times$ is distributive over $+$.
9.4	Group	A mathematical system consisting of a set S and an operation $*$ that has the closure, associative, identity, and inverse properties	The set of integers under addition is a group.
9.4	Commutative group	A group with the commutative property	The set of integers is a commutative group under $+$.
9.4	Field	A set S and the operations $*$ and $\#$ with the closure, associative, commutative, and identity properties; the distributive property of $\#$ with respect to $*$; and the inverse property (except that there is no inverse for the identity with respect to $*$)	The set R of real numbers and the operations $+$ and $\times$ is a field.
9.5	Zero-sum game	A game in which the payoff to one player is the opposite of the payoff to the opposing player	
9.5A	Optimal strategy	A strategy that maximizes a player's gain or minimizes his or her loss	

9.5A	Strictly determined game	A game in which the payoff game matrix has a saddle point
9.5A	Saddle point	A point in the payoff matrix that is both the smallest element in its row and the largest element in its column

Research Questions

Sources of information for these questions can be found in the Bibliography at the end of the book.

1. Write a report on the uses of matrices in science with particular emphasis on the work of Werner Heisenberg, W. J. Duncan, and A. R. Collar.

2. Find out which mathematician first introduced the term *matrix*.

3. Who received the 1973 Nobel Prize for economics and how did he use matrices?

4. Write a report on how matrices are used in game theory and the contribution of John von Neumann to this field.

5. In 1858, Arthur Cayley, an English mathematician, published *A Memoir on the Theory of Matrices*. What was discussed in this book, and what was the name of the only theorem appearing in the book?

6. In the Human Side of Math it was stated that Sophie Germain won a French Academy prize for one of her research papers. Write a short report giving the name of this paper and detailing its content.

7. Emmy Noether (1882–1935) was a creative mathematician who made significant contributions to the area of abstract algebra. Write a short report about the life of this famous woman.

Chapter 9 Practice Test

1. Given the three matrices below, find x and y such that $2A + B = C$.

$$A = \begin{bmatrix} 2 & x \\ 3 & y \end{bmatrix} \quad B = \begin{bmatrix} 2 & -1 \\ 3 & 2 \end{bmatrix} \quad C = \begin{bmatrix} 6 & 5 \\ 9 & 10 \end{bmatrix}$$

For problems 2–3, suppose that

$$A = \begin{bmatrix} 2 & 0 & 1 \\ 2 & -1 & 3 \\ 4 & 1 & 2 \end{bmatrix} \quad \text{and} \quad B = \begin{bmatrix} -2 & 1 & 0 \\ 4 & 3 & -2 \\ 1 & 2 & -1 \end{bmatrix}$$

2. Find $A + B$.

3. Find $2A - 3B$.

4. To build three types of zig-zag toys, the Zee-Zee Toy Company requires materials as listed below. Suppose that Zee-Zee decides to build 20 type I, 25 type II, and 10 type III zig-zag toys. Use matrices to find the number of units of frames, wheels, chains, and paint needed for the job.

Type	Frames	Wheels	Chains	Paint
I	1	2	1	1
II	1	3	2	1
III	4	4	2	2

5. Zee-Zee finds that frames cost $2 each, wheels $1 each, chains $.75 each, and paint $.50 per unit. Use matrices to find the total cost of these items for each type of zig-zag toy in problem 4.

6. If

$$A = \begin{bmatrix} 4 & -5 \\ 1 & -1 \end{bmatrix} \quad \text{and} \quad B = \begin{bmatrix} -1 & 5 \\ -1 & 4 \end{bmatrix}$$

calculate AB and BA.

In problems 7 and 8 let

$$A = \begin{bmatrix} 2 & 0 & 1 \\ 2 & -1 & 3 \\ 4 & 1 & 2 \end{bmatrix} \quad \text{and} \quad B = \begin{bmatrix} -2 & 1 & 0 \\ 4 & 3 & -2 \\ 1 & 2 & -1 \end{bmatrix}$$

7. Find AB and BA.

8. Find $(A + B)^2$.

9. Use matrices to solve the system

$$x + 3y = 19$$
$$y + 3z = 10$$
$$3x + z = -5$$

10. On counting the money in her piggy bank, Sally found that she had 122 coins, all nickels, dimes, and quarters. If the total value of the coins was $15, and the total value of the quarters was 4 times the total value of the nickels, how many of each kind of coin did Sally have?

11. The augmented matrix of a system of three equations in the three unknowns x, y, and z is reduced to

$$\begin{bmatrix} 2 & 1 & 1 & | & 1 \\ 0 & 3 & 2 & | & 4 \\ 0 & 0 & 1 & | & 2 \end{bmatrix}$$

Find the solution of the system.

12. Suppose the reduced matrix in problem 11 were

$$\begin{bmatrix} 2 & 1 & 1 & | & 1 \\ 0 & 3 & 2 & | & 4 \\ 0 & 0 & 0 & | & 2 \end{bmatrix}$$

What could be said about the system of equations?

13. Suppose the reduced matrix in problem 11 were

$$\left[\begin{array}{ccc|c} 2 & 1 & 1 & 1 \\ 0 & 3 & 2 & 4 \\ 0 & 0 & 0 & 0 \end{array}\right]$$

Find the solution of the system.

In problems 14 and 15, find the answers in 12-hour clock arithmetic.

14. a. $3 \oplus 11$ **b.** $8 \oplus 9$ **c.** $3 \ominus 9$ **d.** $5 \ominus 12$

15. a. $3 \otimes 5$ **b.** $6 \otimes 8$ **c.** $\frac{3}{11}$ **d.** $\frac{5}{6}$

16. Are the following statements true or false?
   a. $5 \equiv 2 \pmod 3$ **b.** $9 \equiv 5 \pmod 4$
   c. $9 \equiv 2 \pmod 6$

17. Find the values of n in the following:
   a. $3 + 2 \equiv n \pmod 5$ **b.** $4 \times 3 \equiv n \pmod 5$
   c. $2 - 4 \equiv n \pmod 5$ **d.** $\frac{3}{2} \equiv n \pmod 5$

In problems 18 and 19, find all possible replacements for n for which each congruence is true.

18. a. $6 + n \equiv 1 \pmod 7$ **b.** $3 - n \equiv 4 \pmod 7$

19. a. $2n \equiv 1 \pmod 3$ **b.** $\frac{n}{2} \equiv 2 \pmod 3$

Problems 20–24 refer to the operation $*$ defined by the table.

$*$	$\#$	$\$$	$\%$	$\cent$
$\#$	$\#$	$\$$	$\%$	$\cent$
$\$$	$\$$	$\%$	$\cent$	$\#$
$\%$	$\%$	$\cent$	$\#$	$\$$
$\cent$	$\cent$	$\#$	$\$$	$\%$

20. Use the table to find the result of each of the following operations:
   a. $(\$ * \cent) * \#$
   b. $\$ * (\cent * \#)$
   c. $(\$ * \#) * (\% * \cent)$

21. Is the set $S = \{\#, \$, \%, \cent\}$ closed with respect to the operation $*$? Explain.

22. Is the operation $*$ commutative? Explain.

23. Find the identity element for the operation $*$.

24. Find the inverse of each of the following:
   a. $\#$
   b. $\$$
   c. $\%$
   d. $\cent$

25. The following table defines the operation @.
    a. Find the identity element if possible.
    b. Does any element have an inverse?
    c. Is the operation commutative?

@	$	#	&
$	&	$	#
#	#	&	$
&	$	#	&

26. Suppose that a S b means to select the second of the two numbers a and b and that a L b means to select the lesser of the two numbers (if the numbers are equal, select the number).
    a. Is S distributive over L? Explain.
    b. Is L distributive over S? Explain.

27. Is the set $\{0, 1, 2\}$ together with addition modulo 3 a commutative group? Explain.

28. Is the set $\{1, 2\}$ together with multiplication modulo 3 a commutative group? Explain.

29. Is the set of all rational numbers along with the ordinary operations of addition and multiplication a field? Explain.

Problems 30–32 refer to the set $\{0, 1\}$ and the operations $\oplus$ and $\otimes$ as defined by the following tables:

$\oplus$	0	1
0	0	1
1	1	0

$\otimes$	0	1
0	0	0
1	0	1

30. Is the set $\{0, 1\}$ together with the operation $\oplus$ a group?

31. Is the set $\{0, 1\}$ together with the operation $\otimes$ a group?

32. Is the set $\{0, 1\}$ together with the two operations $\oplus$ and $\otimes$ a field?

33. Determine which of the following payoff matrices represent strictly determined games and find their values:

a. $\begin{bmatrix} -2 & 0 & 8 \\ 1 & 3 & -2 \\ 2 & -3 & 1 \end{bmatrix}$ **b.** $\begin{bmatrix} -2 & -4 & -1 \\ -4 & 2 & -6 \\ 3 & 4 & 1 \end{bmatrix}$ **c.** $\begin{bmatrix} 1 & 2 & 1 & 4 \\ 1 & -1 & 0 & 1 \\ 2 & 8 & -6 & 6 \\ 2 & 3 & 2 & 4 \end{bmatrix}$

34. In the following strictly determined game, find the optimal pure strategy for the row player and give the payoff when both players use their optimal strategies:

$\begin{bmatrix} 4 & -6 \\ 2 & -7 \end{bmatrix}$

35. The Hills Area Rapid Transit (HART) is planning a new rapid transit system. Planners estimate that 60% of the commuters presently using HART will continue to do so next year but that 40% of them will switch to using their own cars. On the other hand, the planners predict that 30% of commuters presently using their own cars this year will switch to HART but 70% will continue to use cars.

a. Give the payoff matrix for the game.

b. What is the expected percentage of commuters using HART next year?

Answers to Practice Test

ANSWER	IF YOU MISSED	REVIEW		
	Question	Section	Example(s)	Page(s)
1. $x = 3, y = 4$	1	9.1	1	586–587
2. $\begin{bmatrix} 0 & 1 & 1 \\ 6 & 2 & 1 \\ 5 & 3 & 1 \end{bmatrix}$	2	9.1	2	587
3. $\begin{bmatrix} 10 & -3 & 2 \\ -8 & -11 & 12 \\ 5 & -4 & 7 \end{bmatrix}$	3	9.1	3	588
4. $[20 \quad 25 \quad 10] \begin{bmatrix} 1 & 2 & 1 & 1 \\ 1 & 3 & 2 & 1 \\ 4 & 4 & 2 & 2 \end{bmatrix}$ Frames Wheels Chains Paint $= [\ 85 \qquad 155 \qquad 90 \qquad 65\ ]$	4	9.1	4, 5	590, 591
5. Type I, $5.25; type II, $7.00; type III, $14.50	5	9.1	6	592
6. $AB = BA = \begin{bmatrix} 1 & 0 \\ 0 & 1 \end{bmatrix}$	6	9.1	4, 5	590, 591
7. $AB = \begin{bmatrix} -3 & 4 & -1 \\ -5 & 5 & -1 \\ -2 & 11 & -4 \end{bmatrix}$; $BA = \begin{bmatrix} -2 & -1 & 1 \\ 6 & -5 & 9 \\ 2 & -3 & 5 \end{bmatrix}$	7	9.1	4, 5	590, 591
8. $\begin{bmatrix} 11 & 5 & 2 \\ 17 & 13 & 9 \\ 23 & 14 & 9 \end{bmatrix}$	8	9.1	4, 5	590, 591
9. $x = -2, y = 7, z = 1$	9	9.2	1–3	600–603
10. 40 nickels, 50 dimes, 32 quarters	10	9.2	4	605
11. $x = -\frac{1}{2}, y = 0, z = 2$	11	9.2	1	600–601
12. The system has no solution.	12	9.2	2	601–602
13. $x = -\dfrac{(k+1)}{6}, y = \dfrac{(4-2k)}{3}$, and $z = k$, where k is any real number	13	9.2	3	602–603
14. a. 2 **b.** 5 **c.** 6 **d.** 5	14	9.3	1, 2	611, 612
15. a. 3 **b.** 12 **c.** 9 **d.** No solution	15	9.3	3, 4	613, 614
16. a. True **b.** True **c.** False	16	9.3	5	614
17. a. $n = 0$ **b.** $n = 2$ **c.** $n = 3$ **d.** $n = 4$	17	9.3	6	616

ANSWER	IF YOU MISSED	REVIEW		
	Question	Section	Example(s)	Page(s)
18. a. $n = 2 + 7k$, k any integer **b.** $n = 6 + 7k$, k any integer	18	9.3	8	618
19. a. $n = 2 + 3k/2$, k any integer **b.** $n = 1 + 3k$, k any integer	19	9.3	8	618
20. a. # **b.** # **c.** %	20	9.4	1, 3	626, 627
21. Yes. All the entries in the table are elements of S.	21	9.4	2	627
22. Yes. The table is symmetrical across the diagonal from upper left to lower right.	22	9.4	4	628
23. #	23	9.4	5, 6	628, 629
24. a. # **b.** ¢ **c.** % **d.** $	24	9.4	6	629
25. a. No identity element **b.** No **c.** No	25	9.4	7	629
26. a. Yes. a S $(b$ L $c) = b$ L c and $(a$ S $b)$ L $(a$ S $c) = b$ L c. Therefore, S is distributive over L. **b.** Yes. a L $(b$ S $c) = a$ L c and $(a$ L $b)$ S $(a$ L $c) = a$ L c. Thus, L is distributive over S.	26	9.4	8	631
27. Yes. The system has the five properties (closure, associative, identity, inverse, and commutative), so the system is a commutative group.	27	9.4	9	632
28. Yes. The same explanation as for problem 27.	28	9.4	10	632
29. Yes. The system has the six properties (closure, associative, commutative, identity, distributive of multiplication over addition, and inverses, except there is no inverse for 0 with respect to multiplication), so the system is a field.	29	9.4	11	633–634
30. Yes. All the requirements of the definition of a group are satisfied.	30	9.4	9–11	632–634
31. No. The element 0 has no inverse.	31	9.4	9–11	632–634
32. Yes. All the requirements of the definition of a field are satisfied.	32	9.4	11	633–634
33. a. Not strictly determined **b.** Strictly determined. Value = 1. **c.** Strictly determined. Value = 2.	33	9.5	1	641–642
34. Row player's optimal strategy is play row 1. (Column player's optimal strategy is play column 2.) Row player's payoff when both players use their optimal strategies is −6.	34	9.5	2, 3	643–645
35. a. $\begin{array}{c} & H \quad C \\ H & \\ \text{Next Year } C & \begin{bmatrix} 0.6 & 0.4 \\ 0.3 & 0.7 \end{bmatrix} \end{array}$ **b.** 50%	35	9.5	2, 3	643–645

Millions of people purchase lottery tickets daily. In Section 10.4, you will study how the Sequential Counting Principle, permutations, and combinations can be used to calculate odds of events such as winning the lottery.

Counting Techniques

You use counting every day. "How many shopping days are left until Christmas?" or "How many students are in your class?" In this chapter, you will learn how to determine the number of ways in which more complicated events can occur without actually counting them. For example, do you know how many ways 6 numbers can be randomly picked from a set of 49 numbers? The answer is 13,983,816. If there are 53 numbers to pick from, it gets worse. Then you have 22,957,480 ways of selecting the 6 winning numbers! You can see that your chances of picking the right numbers to win a lottery jackpot are pretty slim. The counting methods discussed in this chapter are the following:

1. **Tree diagrams,** a useful technique when the number of outcomes is small

2. The **sequential counting principle (SCP),** a method based on generalizations made about tree diagrams

3. **Permutations,** a procedure that counts the number of ordered arrangements that can be made with r objects selected from a set of n objects when the order is important

4. **Combinations,** a method that counts the number of arrangements that can be made with r objects selected from a set of n objects when order is not important

All these counting techniques will be extremely useful when you examine probability in the next chapter. Make sure that you master the techniques in this chapter before you go on to the next.

Online Study Center

For links to Internet sites related to Chapter 10, please access **college.hmco.com/PIC/bello9e** and click on the Online Study Center icon.

Trees and Breakfast Possibilities

Why does it seem nearly impossible to win the lottery or open a combination lock by just guessing at the numbers? Because the number of possibilities is so large! Consider another problem. At breakfast, the server asks, "How do you want your eggs? Fried (f), poached (p), or scrambled (s)? Rye (r) or white toast (w)? Juice (j) or coffee (c)?" How many choices do you have? To answer this type question, you can use a tree diagram, a picture that details the possibilities at each step. In this case, the diagram will be a *three*-step (three-event) process (pick eggs, toast, and beverage). The first step will have 3 branches indicating the 3 different ways in which you can order your eggs, as shown in Figure 10.1. Each of the 3 branches will have 2 branches (rye or white), and each of these branches will have 2 other branches (coffee or juice). The total number of possibilities corresponds to the number of branches, $3 \times 2 \times 2 = 12$. By tracing each path from left to right you can see all 12 possibilities. For example, the third choice in the list is fwc (fried egg, white toast, and coffee).

Now look at a lottery game called Cash 3, played with three identical urns, each containing 10 balls numbered from 0 to 9. A ball is chosen from the first urn (10 choices), then another one from the second urn (10 choices), and a third ball from the last urn (10 choices). How many possibilities are there in all? If you drew the tree diagram, it would have $10 \times 10 \times 10 = 1000$ branches, the number of possibilities for Cash 3.

HUMAN SIDE OF MATH

Gottfried Wilhelm Leibniz was born in Leipzig, Germany, in 1646. By the (1646–1716) time he was 20, he had already begun to have ideas for a kind of *universal mathematics,* which later developed into the symbolic logic of George Boole.

In 1673, he became acquainted with some British mathematicians and exhibited his calculating machine (the first mechanical device that could do multiplication).

He made contributions to law, religion, history, literature, and logic. In 1682, he helped establish a journal, the *Acta Eruditorum.*

Leibniz's outstanding achievements in mathematics were his discovery of *calculus* (independent of Isaac Newton) and his work on *combinatorial analysis.*

Looking Ahead

Leibniz's combinatorial analysis, which involves counting techniques, is the subject of this chapter.

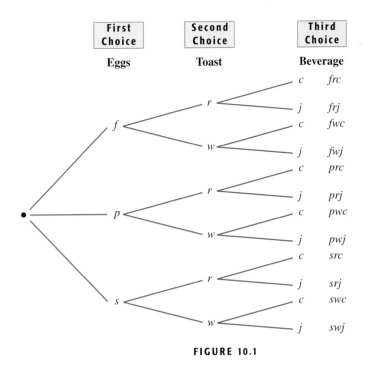

FIGURE 10.1

What about the combination lock? Since there are 40 possible numbers to choose as your first number, 40 for the second, and 40 for the third, the total number of possibilities is $40 \times 40 \times 40 = 64{,}000$. This is why it is nearly impossible to open the lock just by picking numbers at random!

In this section you will study tree diagrams and then develop a formula for counting the number of possibilities associated with different events. ▶

Counting is not always as easy as 1, 2, 3. For example, can we tell in how many different ways Funky Winkerbean can answer the questions on a true/false test? There are many situations in which the answer to the question "How many?" is the first step in the solution of a problem. Earlier in the book we counted the number of subsets of a given set, and we used Venn diagrams to count the number of elements of various sets. In this chapter we shall consider a few counting techniques that are important in many applications and that we shall use when we study probability.

FUNKY WINKERBEAN by Tom Batiuk by permission of North America Syndicate, Inc.

A. Tree Diagrams

| 1st | 2nd | Possible |
| ? | ? | Responses |

	T	TT
T <		
	F	TF
	T	FT
F <		
	F	FF

FIGURE 10.2

Let us return to Funky Winkerbean. He is still taking a true/false test and guessing at the answers. If we assume that there are just two questions and that Funky answers both, in how many different ways can he respond?

In order to answer this question, we have to find all the possible ways in which the two questions can be answered. We do this by constructing a **tree diagram,** as shown in Figure 10.2. In the figure, the first set of branches of the tree shows the 2 ways in which the first question can be answered (T for true, F for false), whereas the second set of branches shows the ways in which the second question can be answered. By tracing each path from left to right, we find that there are 4 end results, which correspond to the $2 \times 2 = 4$ ways in which the two questions can be answered. The 4 possibilities are TT, TF, FT, and FF.

The tree-diagram technique is used again in the next example.

EXAMPLE 1 ▶ Using Tree Diagrams to Count Choices

A woman wants to purchase a car. She has a choice of 2 body styles (convertible or hardtop) and 3 colors (red, blue, or green). Make a tree diagram and find how many choices she has.

Solution

We make a tree diagram as in the Funky Winkerbean problem. As shown in Figure 10.3, for each of the body styles, convertible (c) or hardtop (h), the woman has 3 choices of color, red (r), blue (b), or green (g). Thus, she has $2 \times 3 = 6$ choices.

FIGURE 10.3

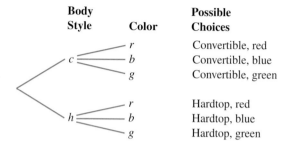

B. Sequential Counting Principle (SCP)

In Example 1 we found that if there are 2 ways to do some event (select a body style) and 3 ways to do a second event (select a color), then there are $2 \times 3 = 6$ ways of doing the two events in succession in the stated order. This example illustrates a basic principle.

> **Sequential Counting Principle (SCP)**
>
> If an event can occur in m ways and a second event can then occur in n ways, then the sequence of two events can occur in $m \times n$ ways.

The SCP is sometimes called the **Fundamental Counting Principle (FCP).** Note that it is assumed that the second event can occur in n ways for *each* of the ways in which the first event can occur.

For example, the number of ways in which two cards can be drawn in succession and without replacement from a pack of 52 cards is $52 \times 51 = 2652$ because the first card can be drawn in 52 ways, whereas the second card can be drawn in only 51 ways (one card has been withdrawn). If the first card is returned to the deck (before the second card is drawn), then the answer is 52×52.

EXAMPLE 2 ▶ Using the SCP to Find the Number of Meals

Johnny's Homestyle Restaurant has 12 different meals and 5 different desserts on the menu. How many meal choices followed by a dessert choice does a customer have?

Solution
There are 12 choices for the meal and 5 choices for the dessert. Thus, by the SCP, there are $12 \times 5 = 60$ choices in all. ∎

The SCP can be extended to cases in which 3, 4, or more things occur in succession. Thus, if the customer in Example 2 also has the choice of selecting a cookie (vanilla, chocolate, or almond) to go with the dessert, then the number of choices the customer has is

$$\boxed{\text{Meals}} \quad \boxed{\text{Dessert}} \quad \boxed{\text{Cookie}} \leftarrow 3 \text{ choices (vanilla, chocolate, almond)}$$
$$12 \quad \times \quad 5 \quad \times \quad 3 \quad = 180 \leftarrow \text{Total}$$

We now state a more general sequential counting principle.

> **Sequential Counting Principle (SCP)**
>
> If an event can occur in m ways, and then a second event can occur in n ways, and then a third event can occur in r ways, and so on, then the sequence of events can occur in $m \times n \times r \times \cdots$ ways.

EXAMPLE 3 ▶ **Using the SCP to Find the Number of Letters**

In a Peanuts cartoon, Charlie Brown has just received a chain letter. If he sends this letter to 6 of his friends, these 6 friends send letters to 6 of their friends, and all these people send letters to 6 of their friends, how many letters will be sent starting with the ones sent by Charlie Brown?

Solution

Charlie Brown sends 6 letters, and each of the people receiving one of these sends 6 letters. Thus, by the SCP, these 6 people send a total of $6 \times 6 = 36$ letters. Then, each of the 36 people receiving one of these sends 6 letters. Again, by the SCP, these 36 people send a total of $36 \times 6 = 216$ letters. Therefore, the total number of letters is the sum

$$6 + (6 \times 6) + (36 \times 6) = 6 + 36 + 216 = 258 \qquad ■$$

EXAMPLE 4 ▶ **Hit or Miss?**

Alice and Betsy agreed to meet at 2 P.M. and go shopping in one of the 3 clothing stores in their hometown. However, they forgot to specify which store. If Alice and Betsy were both on time and each went to 1 of the 3 stores, find

(a) the number of ways in which they could miss each other.

(b) the number of ways in which they could meet.

Solution

(a) Alice could go to any one of the 3 stores (first event), and Betsy could go to either of the remaining 2 stores (second event). Thus, by the SCP, there would be

$$3 \times 2 = 6$$

ways in which they could *miss* each other.

(b) In order to meet, they would both have to go to the same one of the 3 stores. Thus, there are just 3 ways in which they could *meet*.   ■

Sometimes it is advantageous to use a diagram to represent the individual events in a sequence of events. For example, suppose we want to find out how many different telephone numbers are possible. Telephone numbers consist of a three-digit area code, followed by a seven-digit number (ten digits in all); we must find the number of ways in which the blanks in the diagram can be filled.

The answer is $10 \times 10 \times 10 \times \cdots \times 10 = 10^{10}$, or 10 billion numbers. But wait, the following are some restrictions (see note on following page):

1. The area code cannot begin with 0 or 1 (why?).

2. The second digit in the area code must be 0 or 1.

3. The seven-digit number after the area code cannot begin with 0 or 1.

Now, how many telephone numbers are possible? To solve this new problem, we will again use ten blanks to represent the individual numbers to be picked.

PROBLEM SOLVING

The Sequential Counting Principle

❶ Read the problem.

How many telephone numbers can be made using the three given restrictions?

❷ Select the unknown.

We want to find the number of telephone numbers that are possible under the given restrictions.

❸ Think of a plan.
We have to find the number of choices we have to fill each of the ten blanks.

The idea is to find the number of ways in which each of the ten blanks can be filled and then use the SCP to find the total number of possibilities. Note that special restrictions apply to blank 1 (no 0s or 1s), blank 2 (must be 0 or 1), and blank 4 (no 0s or 1s).

❹ Use the SCP to carry out the plan. How many numbers can we place in the following?

The first blank
The second blank
The third blank
The fourth blank
Each of the remaining six blanks

The first restriction permits us to use 8 numbers (2, 3, 4, 5, 6, 7, 8, 9) in blank 1; the second restriction lets us use 2 numbers (0, 1) in blank 2; there are no restrictions for the third blank, so we can use 10 numbers in blank 3; and the third restriction allows 8 numbers in blank 4 (no 0s or 1s). We can use 10 numbers in each of the remaining six blanks. Using the SCP, the number of different telephone numbers is

$$\underline{(8}\ \underline{2}\ \underline{10)}\ \underline{8}\ \underline{10}\ \underline{10}\ \text{-}\ \underline{10}\ \underline{10}\ \underline{10}\ \underline{10}$$

$$128 \times 10^7 = 1{,}280{,}000{,}000$$

❺ Verify the solution.

TRY EXAMPLE 5 NOW.

Cover the solution, write your own solution, and then check your work.

Online Study Center

To further explore how telephone numbers are formed and assigned, go to links 10.1.1 and 10.1.2 on this textbook's Online Study Center.

Note: Because of a shortage of area codes and telephone numbers, these restrictions were changed in January 1995. For example, Mobile's area code was changed to 334 and Cape Canaveral's to 321. Note that if the second digit in the area code can be any digit, there will be 10 choices for that second digit. In order to have more phone numbers available, the third restriction now stipulates that the three numbers after the area code can't be the same as the area code.

EXAMPLE 5 ▶ Counting the Number of Possible Phrases

A game consists of 4 cubes, each with 6 different words or phrases inscribed, one on each face. If it is assumed that the first cube has pronouns on its faces, the second auxiliary verbs, the third verbs, and the fourth adverbs, find how many different phrases can be formed.

Solution

We draw 4 blanks representing the 4 events.

_____ _____ _____ _____

There are 6 choices for each of the blanks (each cube has 6 sides), so we enter a 6 on each line.

$$\underline{6}\ \underline{6}\ \underline{6}\ \underline{6}$$

The number of possibilities, by the SCP, is $6 \times 6 \times 6 \times 6 = 6^4 = 1296$. ∎

EXAMPLE 6 ▶ Counting Possibilities in a Slot Machine

A slot machine has 3 dials, each having 20 symbols, as listed in Table 10.1. The 6 symbols (bar, bell, cherry, lemon, orange, plum) are all *different*.

(a) How many symbol arrangements are possible on the 3 dials?

(b) In how many ways can we get 3 bars (the biggest payoff)?

TABLE 10.1

Symbol	Dial 1	Dial 2	Dial 3
Bar	1	3	1
Bell	1	3	3
Cherry	7	7	0
Lemon	3	0	4
Orange	3	6	7
Plum	5	1	5

↑ ↑ ↑
Each dial has 20 symbols.

Solution

(a) We draw 3 blanks representing the 3 dials.

___ ___ ___

There are 20 choices for each of the blanks (each dial has 20 symbols), so we enter a 20 on each line.

<u>20</u> <u>20</u> <u>20</u>

The number of possibilities is $20 \times 20 \times 20 = 8000$.

(b) The number of different ways we can get 3 bars is $1 \times 3 \times 1 = 3$, because we have 1 bar on the first dial, 3 on the second, and 1 on the third. ■

The Clearwater Hilton Inn has a free-chance coupon, as shown in Figure 10.4, that works like this: When you check out, ask the desk clerk to hand you the 3 dice. Pick your lucky number, and give the dice a toss. Look at the sum of the numbers on the 3 dice. If your number comes up, the room charges are canceled. Does this sound easy? Look at the next example.

FREE CHANCE

COUPON

This coupon is better than all the rest! When you check out, ask the clerk to hand you the three dice. Pick your lucky number and give them a toss. If your number comes up, the room charges for your entire stay are on us!!!

(Validated Coupon Issued At Check In)

FIGURE 10.4

EXAMPLE 7 ▶ Finding Outcomes when 3 Dice Are Thrown

Three dice are thrown.

(a) How many different outcomes are possible?

(b) If you picked 4 as your lucky number, in how many ways could you get a *sum* of 4?

Solution

(a) We draw 3 blanks, one for each die (singular of dice), to represent the possible outcomes.

___ ___ ___

There are 6 choices for each blank because a die can come up with any number from 1 to 6. So we enter a 6 on each line.

$$\underline{6}\ \underline{6}\ \underline{6}$$

By the SCP, there are $6 \times 6 \times 6 = 216$ outcomes possible.

(b) One way to solve this part of the problem is to reason that in order to get a sum of 4, one of the 3 dice must come up 2 and the other 2 dice must come up 1. Thus, the only choice we have is which die is to come up 2. This means that there are only *3* ways to get a sum of 4 out of the 216 possible outcomes (see the tree diagram in Figure 10.5). ∎

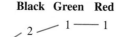

Black Green Red

2 ⟍ 1 —— 1   (Sum $2 + 1 + 1 = 4$)

1 ⟨ 2 —— 1   (Sum $1 + 2 + 1 = 4$)
      1 —— 2   (Sum $1 + 1 + 2 = 4$)

FIGURE 10.5

EXAMPLE 8 ▶ Counting Possible Zip Codes

Zip codes start with a 5-digit number.

(a) How many different 5-digit Zip codes are possible?

(b) How many are possible if 0 is not to be used as the first digit?

Solution

(a) We draw 5 blanks, each to be filled with the number of choices for that digit.

$$\underline{\quad}\ \underline{\quad}\ \underline{\quad}\ \underline{\quad}\ \underline{\quad}$$

Since there are 10 digits (0, 1, 2, 3, 4, 5, 6, 7, 8, 9), there are 10 choices for each blank.

$$\underline{10}\ \underline{10}\ \underline{10}\ \underline{10}\ \underline{10}$$

Thus, by the SCP, there are $10 \times 10 \times 10 \times 10 \times 10$, that is,

$$10^5 = 100{,}000$$

different 5-digit Zip codes possible. This result gives a good estimate of the number of cities and towns in the United States. (Towns too small to have their own Zip code make up for cities and towns with more than one Zip code.)

(b) If we cannot use a 0 for the first digit, we will have only 9 choices for the first box, but the others will still have 10 choices. Again, by the SCP, there are $9 \times 10 \times 10 \times 10 \times 10$, that is,

$$9 \times 10^4 = 90{,}000$$

different possible 5-digit Zip codes that do not start with a 0. ∎

EXAMPLE 9 ▶ Counting Choices when 2 Cards Are Drawn

Two cards are drawn in succession and *without* replacement from a deck of 52 cards. Find the following:

(a) The number of ways in which we can obtain the ace of spades and the king of hearts, *in that order*.

(b) The total number of ways in which 2 cards can be *dealt*.

(c) The total number of ways in which 2 cards can be dealt *with* replacement; that is, the first card is drawn, recorded, and placed back in the deck, and then the second card is drawn and recorded.

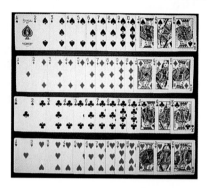

Solution

(a) There is 1 way of selecting the ace of spades and 1 way of selecting the king of hearts. Thus, there is $1 \times 1 = 1$ way of selecting the ace of spades and the king of hearts, in that order.

(b) There are 52 ways of selecting the first card and 51 ways of selecting the second card. Thus, there are $52 \times 51 = 2652$ ways in which the two cards can be dealt.

(c) There are 52 ways of selecting the first card and 52 ways of selecting the second card. Thus, there are $52 \times 52 = 2704$ ways in which the 2 cards can be dealt. ∎

C. Applications

EXAMPLE 10 ▶ Treatment Choices and Costs

The techniques we have studied can be used to determine the cost effectiveness of alternative courses of action. Suppose your doctor tells you that you can be treated with drug A or B. You may then need a second visit (or not) depending on your tolerance for the drug. The costs for drugs A and B are $80 and $50, respectively. Your doctor charges $50 per visit.

(a) How many choices are possible?

(b) What are the highest and lowest possible costs for your treatment?

Solution

(a) We can use the SCP to find the number of choices. Draw two blanks representing the choice of drugs (A or B) and whether you need a second visit (Yes or No). Since there are 2 choices for each blank, there is a total of 4 choices.

$$\underset{\text{Drug}}{\underline{2}} \times \underset{\text{2nd visit}}{\underline{2}} = 4$$

(b) Let us draw a tree diagram as in Figure 10.6 and label the possible costs.

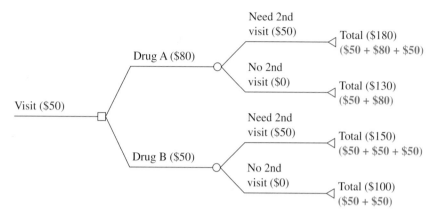

FIGURE 10.6
The highest possible cost is $180, the lowest $100.

Online Study Center

Tree diagrams are so important in the decision-making process that software companies publish, sell, and support decision analysis software. Problems 33 and 40 and Example 10 are adapted from one such company. To further explore these topics, access link 10.1.3 on this textbook's Online Study Center.

EXERCISES 10.1

A Tree Diagrams

1. A man has 2 suits and 4 shirts. Use a tree diagram to find how many different outfits consisting of a suit and a shirt he can wear.

2. At the end of a meal in a restaurant, Elsie wants to have pie à la mode (pie topped with ice cream) for dessert. There are 5 flavors of ice cream—chocolate, vanilla, strawberry, peach, and coffee—and there are 2 kinds of pie—apple and cherry. Make a tree diagram to find how many choices Elsie has for dessert.

3. Research Associates selects 4 people and asks their preferences regarding 2 different styles of blue jeans. If it is important to know which person prefers which style, how many different outcomes are possible? Use a tree diagram.

B Sequential Counting Principle (SCP)

4. In 1935, a chain letter fad started in Denver, Colorado. The scheme worked like this. You would receive a letter with a list of 5 names, send a dime to the person at the top of the list, cross that name out, and add your name to the bottom of the list. Suppose you received one of these letters today and sent it to 5 other persons, each of whom sent it to 5 other persons, each of whom sent it to 5 others, and so on.
   a. How many letters would have your name on the list?

   b. How much money would you receive if the chain were not broken?
   Note: Don't do this! Chain letters are illegal!

5. Refer to the slot machine in Example 6, and determine in how many ways you can get
   a. 3 bells.   b. 3 oranges.   c. 3 plums.

6. An ordinary deck of playing cards contains 52 cards, 26 red and 26 black. If a card is dealt to each of 2 players, find in how many different ways this can be done if the following occur:
   a. Both cards are red.
   b. Both cards are black.
   c. One card is black and the other is red.

7. In poker, a pair of aces with any other pair is a good hand. (*Hint:* See Example 9.)
   a. In how many ways can you get a pair of aces when 2 cards are dealt from the deck?
   b. If the two pairs are aces and eights, the hand is considered to be unlucky! In how many ways can you get a pair of aces and then a pair of eights, in that order, when 4 cards are dealt from the deck? (This superstition dates back to 1876 when Wild Bill Hickok was shot dead by Jack McCall during a poker game. What hand was Wild Bill holding when he died? A pair of aces and a pair of eights!)

8. Mr. C. Nile and Mr. D. Mented agreed to meet at 8 P.M. in one of the Spanish restaurants in Ybor City. They were both punctual, and they both remembered the date agreed on. Unfortunately, they forgot to specify the name of the restaurant. If there are 5 Spanish restaurants in Ybor City and each man goes to one of these, find
   a. the number of ways they could miss each other.
   b. the number of ways they could meet.

9. In how many ways can 1 man and 1 woman, in that order, be selected from 5 men and 6 women?

10. How many different sets of 2 initials can be made from the letters of the English alphabet?

11. How many different sets of 3 initials can be made from the letters of the English alphabet? (*Hint:* AAA is different from BBB.)

12. A man wants to buy a ring. Suppose that he has 2 choices of metals (gold and silver) and 3 choices of stones (diamond, emerald, and ruby). How many choices does he have?

13. The Good Taste Restaurant has 7 entrees, 6 vegetables, and 9 desserts on its menu. If you want to order 1 entree, 1 vegetable, and 1 dessert, how many choices do you have?

14. An airline recently introduced its newest hub. There are 2 flights from Tampa to Dayton and 2 flights from Dayton to Lansing. In how many ways can you fly from Tampa to Lansing via Dayton?

15. An airline has 8 flights from Miami to Washington and 2 flights from Washington to Dayton. In how many ways can you fly from Miami to Dayton via Washington?

16. In Connecticut, auto license plates carry 3 digits followed by 3 letters.
   a. How many arrangements are possible for the 3 letters?
   b. How many arrangements are possible for the 3 numbers?
   c. How many different license plates can be made using 3 numbers followed by 3 letters?

17. How many two-digit numbers are there in the set of natural numbers? (*Hint:* 10 is a two-digit number, but 01 is not.)

18. How many three-digit numbers are there in the set of natural numbers?

19. Social Security numbers consist of nine digits.
   a. If the first digit cannot be 0, how many Social Security numbers are possible?
   b. How many Social Security numbers are possible if there are *no* restrictions?

20. Telephone numbers within the same area code consist of seven digits. For local calls, the first digit cannot be a 0 or a 1. How many local telephone numbers are possible?

21. A combination lock has 40 numbers on its face. To open this lock, you move clockwise to a certain number, then counterclockwise to another number, and finally clockwise again to a third number.
   a. If no number is used twice, what is the total number of combinations?
   b. What is the total number of combinations if repetition of numbers is allowed?

22. Romano's Restaurant has 6 items you can add to your pizza. The dessert menu lists 5 different desserts. If you want a pizza with 1 of the 6 items added and a dessert, how many choices do you have for your dinner?

Problems 23–30 refer to the menu at the top of page 667.

23. How many choices does Billy have if he decides to get the following?
   a. An item from group A
   b. A soup
   c. A dessert
   d. An item from group A, a soup, and a dessert

24. How many choices does Sue have if she decides to get the following?
   a. An item from group B
   b. An item from group B, a soup, and a dessert

25. How many choices does Pedro have if he decides to get the following?
   a. An item from group A or B
   b. An item from group A or B, a soup, and a dessert

26. How many choices do Bob and Sue have if they decide to get the following?
   a. One item from group A and one item from group B (no soup or dessert)
   b. The family dinner, which includes soup and dessert, one item from group A, and one item from group B

Family Dinners

(Choice of Soup and Dessert)

FOR 2 PERSONS 13.00 —— Select 1 from Group A and 1 from Group B
FOR 3 PERSONS 18.00 —— Select 1 from Group A and 2 from Group B
FOR 4 PERSONS 24.00 —— Select 2 from Group A and 2 from Group B
FOR 5 PERSONS 30.00 —— Select 2 from Group A and 3 from Group B
FOR 6 PERSONS 36.00 —— Select 3 from Group A and 3 from Group B

Entree

EGG ROLL (One Per Person)
Soup: ROAST PORK WONTON or EGG DROP
(Served Individually)

A	**B**
SHRIMP WITH LOBSTER SAUCE	CHICKEN CHOW MEIN
CHOW HAR KEW	SHRIMP CHOW MEIN
BUTTERFLY SHRIMP	CHICKEN CHOP SUEY
CHOW GAI KEW	SHRIMP CHOP SUEY
WOR SUE GAI	BEEF CHOP SUEY
MOO GOO GAI PAN	BEEF WITH BEAN SPROUT
LEMON CHICKEN	ROAST PORK EGG FOO YOUNG
GREEN PEPPER STEAK	ROAST PORK LO MEIN
WOR SUE OPP (PRESSED DUCK)	BARBECUED SPARERIBS
SWEET & SOUR PORK	

Served with fried rice and hot tea
Group A in Exchange for Group B——$1.50 extra

CHOICE OF DESSERTS: PINEAPPLE CHUNKS,
ALMOND COOKIES, KUMQUATS, FORTUNE COOKIES

EXTRA SERVICE ON ANY FAMILY DINNER INCLUDING SOUP $1.00

---

27. Sam and Sally are having a family dinner, which includes soup and dessert, one item from group A, and one item from group B. If Sally decides to get an item from group A, which forces Sam to have an item from group B, how many choices do they have for dinner?

28. In problem 27, if Sam does not want to eat shrimp, how many choices do they have for dinner?

29. In problem 27, if Sam wants to avoid all the varieties of chop suey, how many choices do they have?

30. In problem 27, if Sam does not want the roast pork, how many choices do they have?

31. Write out all the different ways in which the elements of the set {a, b, c} can be matched in pairs with the elements of the set {@, &, %} if the order in each pair is important. For example, the pair (a, @) is different from the pair (@, a).

32. Determine how many different couples Escort Dating Service can select if it has 120 men and 210 women registered with the service.

C Applications

Problems 33–36 are adapted from David C. Skinner, *Introduction to Decision Analysis,* 3rd ed. (Florida: Probabilistic Publishing, 2006).

33. Many investors buy stocks during a bull market (rising prices); others do so during a bear market (falling prices). You can decide to buy a risky stock or a conservative mutual fund. What is your total number of choices?

34. When you buy insurance you can select a low premium (high deductible) or a high premium (low deductible). Then you may or may not have an accident. How many possibilities are there?

35. You inherit some money and want to invest it in mutual funds, stocks, or futures. Each of these investments can go higher or lower. How many investment possibilities are there?

36. You are in charge of a company that is about to bring its new product to market. Here are your choices: select a large, medium, or small market; then launch, license out, or sell rights to the product. How many choices do you have?

Problems 37–38 are adapted from Robert Clemen, *Making Hard Decisions: An Introduction to Decision Analysis* (Belmont, Calif.: Wadsworth Publishing, 1996).

37. The research and development department of a company can develop (*D*) or abandon (*A*) a certain product. If it decides to develop the product, it may obtain a patent (*P*) for it or not (*N*). If the patent is obtained, it can license (*Li*) the product or sell the product (*S*) directly. If it sells the product directly, sales can be high (*H*), low (*L*), or medium (*M*). Draw a tree diagram showing all the possibilities.

Problems 38 and 39 are adapted from Ann Haddix et al., *Prevention Effectiveness: A Guide to Decision Analysis and Economic Evaluation* (New York: Oxford University Press, 2002).

38. What decisions have to be made in case of a hurricane? First, you should listen to the weather forecast. The possibilities are "Will make landfall" or "Will not make landfall." Next, you have to make a decision: evacuate or not. Finally, the hurricane makes landfall or not. Draw a tree diagram showing all the possibilities.

39. A health care worker gets accidentally stuck with a needle. What are the possible courses of action? Here are some. Treat with AZT to try to prevent HIV (or not treat). If the person is treated with AZT, there may be some side effects (or no side effects) and the person may get the HIV virus (or not). Those who do not get the AZT treatment may or may not get the HIV virus. Draw a tree diagram showing all possibilities.

40. A company is considering two projects (P_1 and P_2) that must be approved (*A*) or rejected (*R*) by the Research and Development (R&D) department. If the product is approved by R&D, it can be a market success (*S*) or a failure (*F*). Draw a tree diagram showing all possibilities.

 In Other Words

41. The SCP indicates that if one event can occur in *m* ways *and* another event in *n* ways, then the sequence of events can occur in $m \times n$ ways. State a similar principle when a single event can occur in *m* ways *or* in *n* ways.

42. Many states are changing the configuration of their license plates. Florida, for example, changed from 3 digits followed by 3 letters to 3 letters followed by 2 digits and 1 letter. Explain why you think this change was made.

 Using Your Knowledge

43. If a "word" is any arrangement of 4 different letters, how many 4-letter "words" can be formed from the letters B, O, N, and K?

44. How many 3-letter "words" can be formed from the letters B, O, N, and K, where each letter is to be used once?

45. In problem 11 we found the number of different sets of 3 initials that are possible. If a town has 27,000 inhabitants, each with exactly 3 initials, can you show that at least two of the inhabitants have the same initials?

 Discovery

In a trial in Sweden, the owner of a car was charged with overtime parking. The police officer who accused the man had noted the position of the air valves on the front and rear tires on the curb side of the car and ascertained that one valve pointed to the place occupied by 12 o'clock on a clock (directly upward), and the other one to 3 o'clock. (In both cases the closest hour was selected.) After the allowed time had elapsed, the car was still there, with the valves pointing to 12 and 3 o'clock. In court, however, the man claimed that he had moved the car and returned later. The valves just happened to come to rest in the same position as before! At this time an expert was called to compute the probability of such an event's happening.

If you were this expert and you assumed that the two wheels move independently of each other, use the SCP to find

46. the number of ways in which the front air valve could come to rest.

47. the number of ways in which the front *and* rear air valves could come to rest.

The defendant, by the way, was acquitted! The judge remarked that if all four wheels had been checked (assuming that they moved independently) and found in the same position as before, he would have rejected the claim as too improbable and convicted the man.

48. Again, assume you are the expert and find the number of positions in which the four air valves could come to rest.

49. The claim that the two wheels on an automobile move independently is not completely warranted. For example, if the front air valve points to 12 and the rear (on the same side) points to 3, after a complete revolution of the front wheel, where will the rear air valve be positioned? (Assume no slippage of the wheels.)

50. On the basis of your answer to problem 49, how many positions were possible for the two wheels on the curb side of the car if the owner did move it and return later?

Collaborative Learning

A tree diagram serving as a model for a certain situation can be analyzed on the basis of expected costs, effectiveness, or a combination of both. Examine the diagram at the top of the next column, where the notation $500/10 means that the cost of the treatment is $500 and the quality of the treatment 10.

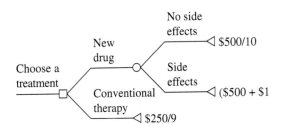

1. Which is the most inexpensive treatment? How much is it?

2. Which is the most expensive treatment? How much is it?

3. According to the diagram, which treatment offers the best quality? Discuss why you think that is.

4. According to the diagram, which treatment offers the lowest quality? Discuss why you think that is.

A decision-making model for a rental property is shown below.

5. Discuss the conditions that have to be met to maximize the amount of money made. Are the conditions realistic? Explain.

6. Discuss the conditions that have to be met to minimize the amount of money made. Are the conditions realistic? Explain.

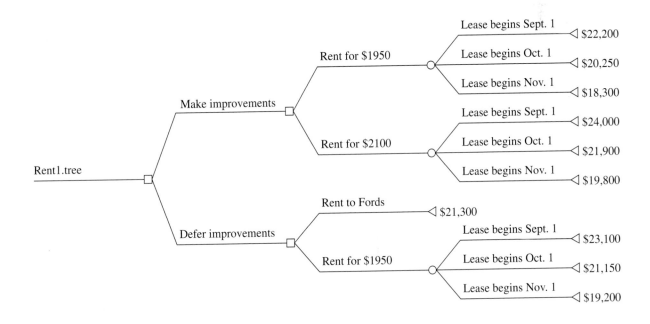

10.2   Permutations

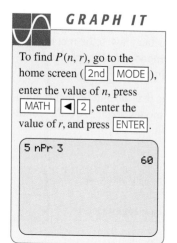

Permutations in the Medical Lab

Have you heard of animal research in medical labs? The possibility of adverse reactions and side effects makes experimenting with medicines risky to test on humans, so lab animals are used for these tests. For example, some patients take blood-pressure medicines, blood thinners, and cholesterol-lowering medicines. To test for adverse reactions or differences in the way individual medicines work, a lab may want to experiment with animals by giving them 3 different medicines chosen from a group of 5 medicines and studying the results. How many lab animals are needed to perform the experiment?

In mathematics, an ordered arrangement of r objects selected from n objects without repetition is called a **permutation.** The number of such permutations is denoted by

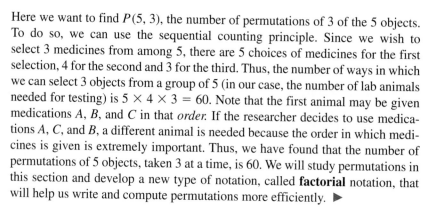

$$P(n, r) \quad \text{or} \quad {}_nP_r \quad \text{Read as "permutations of } n \text{ objects selected } r \text{ at a time"}$$

Available Chosen

Here we want to find $P(5, 3)$, the number of permutations of 3 of the 5 objects. To do so, we can use the sequential counting principle. Since we wish to select 3 medicines from among 5, there are 5 choices of medicines for the first selection, 4 for the second and 3 for the third. Thus, the number of ways in which we can select 3 objects from a group of 5 (in our case, the number of lab animals needed for testing) is $5 \times 4 \times 3 = 60$. Note that the first animal may be given medications A, B, and C in that *order.* If the researcher decides to use medications A, C, and B, a different animal is needed because the order in which medicines is given is extremely important. Thus, we have found that the number of permutations of 5 objects, taken 3 at a time, is 60. We will study permutations in this section and develop a new type of notation, called **factorial** notation, that will help us write and compute permutations more efficiently. ▶

GRAPH IT

To find $P(n, r)$, go to the home screen ([2nd] [MODE]), enter the value of n, press [MATH] [◄] [2], enter the value of r, and press [ENTER].

```
5 nPr 3
              60
```

In Section 10.1 we used the sequential counting principle (SCP) to determine the number of ways in which a sequence of events could happen. A special case of this principle occurs when we want to count the possible *arrangements* of a given set of elements.

EXAMPLE 1 ▶ Counting Arrangements with No Repetitions

In how many different orders can we write the letters in the set $\{a, b, c\}$ if no letter is repeated in any one arrangement?

Solution

We have 3 choices for the first letter, 2 for the second, and 1 for the third. By the SCP, the number of arrangements is $3 \times 2 \times 1 = 6$. ■

A. Permutations

If we are asked to display the arrangements in Example 1, we can draw the tree diagram shown in Figure 10.7, in which each path corresponds to one such arrangement. There are 6 paths, so the total number of arrangements is 6. Notice

| 1st | 2nd | 3rd |
| Letter | Letter | Letter |

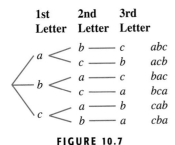

FIGURE 10.7

that in this example *abc* and *acb* are treated as different arrangements, because the **order** of the letters is not the same in the two arrangements. This type of arrangement, in which the *order is important,* is called a *permutation.*

Definition of Permutation

An ordered arrangement of *n* distinguishable objects, taken *r* at a time and with no repetitions, is called a **permutation** of the objects. The **number of permutations** or arrangements of *n* distinct objects taken *r* at a time is denoted by **$P(n, r)$**.

The notation $_nP_r$ is sometimes used instead of $P(n, r)$.

In Example 1 we saw that the number of permutations of the letters in the set $\{a, b, c\}$ is $3 \times 2 \times 1 = 6$. Thus, $P(3, 3) = 6$. A notation that is convenient to represent $3 \times 2 \times 1$ is 3! (read "3 factorial").

Definition of Factorial

The symbol *n*! (read **"n factorial"**), where *n* is a positive integer, represents the product of *n* and every positive integer less than *n*.

$$n! = n \times (n - 1) \times (n - 2) \times \cdots \times 3 \times 2 \times 1$$

Note:
0! = 1 (See page 673.)
1! = 1
2! = 2
3! = 6
4! = 24
5! = 120
6! = 720
7! = 5040
8! = 40,320
9! = 362,880
10! = 3,628,800

Thus,

$$4! = 4 \times 3 \times 2 \times 1 = 24$$

and

$$5! = 5 \times 4 \times 3 \times 2 \times 1 = 120$$

EXAMPLE 2 ▶ Computations Involving Factorials

Compute the following:

(a) 6!　　(b) 7!　　(c) $\dfrac{6!}{3!}$

Solution
By the definition of a factorial,

(a) $6! = 6 \times 5 \times 4 \times 3 \times 2 \times 1 = 720$

(b) $7! = 7 \times 6 \times 5 \times 4 \times 3 \times 2 \times 1 = 5040$

(c) $\dfrac{6!}{3!} = \dfrac{6 \times 5 \times 4 \times \cancel{3} \times \cancel{2} \times \cancel{1}}{\cancel{3} \times \cancel{2} \times \cancel{1}} = 120$　　Note that $\dfrac{6!}{3!} \neq 2!$.　■

EXAMPLE 3 ▶ Using Permutations to Count Outcomes

Wreck-U Car Club organizes a race in which 5 automobiles, A, B, C, D, and E, are entered. There are no ties.

(a) In how many ways can the race finish?

(b) In how many ways can the first 3 finishers finish the race?

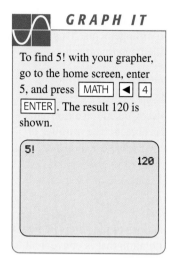

GRAPH IT

To find 5! with your grapher, go to the home screen, enter 5, and press [MATH] [◄] [4] [ENTER]. The result 120 is shown.

```
5!
                    120
```

First   Second   Third   ...   nth
blank   blank   blank   blank

FIGURE 10.8

Note that the permutation formula is used when the *order* is important.

Solution

(a) The number of ways in which the race can finish if there are no ties is the number of permutations of 5 things taken 5 at a time. By the definition of a permutation, this number is $P(5, 5)$, so, by the SCP,

$$P(5, 5) = 5 \cdot 4 \cdot 3 \cdot 2 \cdot 1 = 5! = 120$$

(b) Here we need the number of ordered arrangements of 3 out of the 5 cars—that is, the number of permutations of 5 things taken 3 at a time. Again, by the definition of permutation, this number is $P(5, 3)$, and by the SCP,

$$P(5, 3) = 5 \cdot 4 \cdot 3 = 60$$ ■

By the definition of permutation, $P(n, r)$ is the number of permutations of n objects using r of these objects at a time. We can now obtain formulas to compute these numbers.

1. If we use all n of the objects, we must find $P(n, n)$. We can think of n blanks to be filled with n objects, as shown in Figure 10.8. We have n choices for the first blank, $(n - 1)$ choices for the second blank, $(n - 2)$ choices for the third blank, and so on, until we come to the last blank, for which there is only 1 object left. By the SCP, we have

> **Permutations Formula (*n* at a Time)**
>
> The number of permutations of n objects, taken n at a time, is
>
> $$P(n, n) = n(n - 1)(n - 2) \cdots (3)(2)(1) = n!$$

2. The procedure used to evaluate $P(n, n)$ can be applied to $P(n, r)$. We think of r blanks to be filled by r of the n objects. There are n choices for the first blank, $n - 1$ choices for the second blank, $n - 2$ choices for the third blank, and so on, until there are $n - r + 1$ choices for the rth blank. Thus, by the SCP, we have

> **Permutations Formula (*r* at a Time)**
>
> The number of permutations of n objects, taken r at a time, is
>
> $$P(n, r) = n(n - 1)(n - 2) \cdots (n - r + 1)$$

You should keep in mind that the n in $P(n, r)$ is the number of objects available and the r is the number of spaces to be filled. (Notice that if $r = n$, then the preceding two formulas agree exactly.)

The symbols $_nP_r$, $P_{n,r}$, and P_r^n are sometimes used to represent the number of permutations of n things, r at a time.

Some calculators have a factorial ($n!$) key and a $P(n, r)$ or $_nP_r$ key. By following the instructions for the calculator, you can get answers to many of the following examples and problems without having to do all the detailed arithmetic.

EXAMPLE 4 ▶ **Using the Permutations Formulas**

Compute the following:

(a) $P(6, 6)$ (b) $P(7, 3)$ (c) $P(6, 2)$

Solution

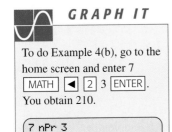

GRAPH IT

To do Example 4(b), go to the home screen and enter 7 MATH ◀ 2 3 ENTER. You obtain 210.

```
7 nPr 3
                210
```

(a) Here we use the formula for $P(n, n)$ with $n = 6$.

$$P(6, 6) = 6 \cdot 5 \cdot 4 \cdot 3 \cdot 2 \cdot 1 = 6! = 720$$

(b) We wish to find the number of permutations of 7 objects, taken 3 at a time. We use the formula for $P(n, r)$ with $n = 7$ and $r = 3$ (n choices, r blanks).

$$P(7, 3) = 7 \cdot 6 \cdot 5 = 210$$

(c) We proceed as in part (b), but with $n = 6$ and $r = 2$.

$$P(6, 2) = 6 \cdot 5 = 30$$ ■

By using the definitions of $n!$ and $(n - r)!$, we can obtain the useful formula

A Factorial Formula for Permutations

The number of permutations of n objects taken r at a time is given by

$$P(n, r) = \frac{n!}{(n - r)!}, \, r \leq n$$

We can verify this formula as follows:

$$\frac{n!}{(n - r)!} = \frac{n(n - 1)(n - 2) \cdots (n - r + 1)(n - r) \cdots (3)(2)(1)}{(n - r) \cdots (3)(2)(1)}$$

$$= n(n - 1)(n - 2) \cdots (n - r + 1) = P(n, r)$$

as the formula states. Notice that for $r \leq n$,

$$P(n, n) = n(n - 1)(n - 2) \cdots (n - r + 1)[(n - r)!]$$
$$= P(n, r)[(n - r)!]$$

We want this formula to hold also in the case $n = r$; that is, we want

$$P(n, n) = P(n, n)(0!)$$

For this reason, $0!$ is defined to be 1.

Zero Factorial

$$0! = 1$$

EXAMPLE 5 ► Using the Formula for *P(n, r)*

Use the factorial formula for $P(n, r)$ to compute the answer to part (b) of Example 4.

Solution

$$P(7, 3) = \frac{7!}{(7-3)!} = \frac{7!}{4!}$$

$$= \frac{7 \times 6 \times 5 \times 4 \times 3 \times 2 \times 1}{4 \times 3 \times 2 \times 1}$$

$$= 7 \times 6 \times 5$$

$$= 210 \qquad \blacksquare$$

Note that it is easier to consider $P(7, 3)$ as in Example 4(b); that is, $7 \times 6 \times 5$. Similarly, $P(8, 2) = 8 \times 7 = 56$ and $P(5, 3) = 5 \times 4 \times 3 = 60$.

B. The Complementary Counting Principle

The number of elements in a set A can sometimes be calculated more easily by an indirect rather than a direct method. If $\mathcal{U}$ is the universal set, the number of elements in A can be obtained by subtracting the number of elements in A' from the number in $\mathcal{U}$. This gives us the **complementary counting principle.**

Complementary Counting Principle

$$n(A) = n(\mathcal{U}) - n(A')$$

EXAMPLE 6 ► Selecting Dogs

How many ways are there to select *at least* 1 male dog if 4 dogs are available?

Solution

The only alternative to selecting at least 1 male dog is selecting no male dogs; that is, the 4 dogs are all females. This is just one of all the possible cases. There are 4 places to fill, with 2 possible choices for each place (male or female). The total number of possible arrangements is

$$n(\mathcal{U}) = 2 \times 2 \times 2 \times 2 = 2^4 = 16$$

Thus,

$$n(\text{at least one male}) = 16 - n(\text{no males})$$
$$= 16 - 1 = 15 \qquad \blacksquare$$

C. The Additive Counting Principle

Another useful counting principle is the **additive counting principle,** giving the number of elements in the union of two sets, which we obtained earlier in the book.

The use of this formula is illustrated in the next example.

The additive counting principle is sometimes called the principle of inclusion-exclusion (PIE for short).

> **Additive Counting Principle**
>
> If A and B are two sets, then
>
> $$n(A \cup B) = n(A) + n(B) - n(A \cap B)$$

EXAMPLE 7 ▶ Divisibility Involving *Or*

How many 2-digit numbers are divisible by 2 or by 5?

Solution

Let A be the set of 2-digit numbers divisible by 2, and let B be the set of 2-digit numbers divisible by 5. Using the SCP, for 2-digit numbers divisible by 2, the first digit can be any digit from 1 to 9 (9 choices), and the second digit can be 0, 2, 4, 6, or 8 (5 choices). Thus,

$$n(A) = \underline{9} \times \underline{5} = 45$$

For 2-digit numbers divisible by 5, the first digit can be any digit from 1 to 9 (9 choices), and the second digit can be 0 or 5 (2 choices). Thus,

$$n(B) = 9 \times 2 = 18$$

Since $A \cap B$ is the set of numbers divisible by both 2 and 5, the first digit can still be any digit from 1 to 9 (9 choices), and the second digit can only be 0 (1 choice). Thus,

$$n(A \cap B) = 9 \times 1 = 9$$

and the desired number is

$$n(A \cup B) = n(A) + n(B) - n(A \cap B)$$
$$= 45 + 18 - 9 = 54$$

That is, the number of 2-digit numbers divisible by either 2 or 5 is 54.   ■

D. Applications

Now that you have learned about permutations, you have to be careful using the formulas because there are teachers watching. As a matter of fact, Bob Swain, a Souderton High School mathematics teacher, found a mistake in a Boston Market (a restaurant chain) ad. The ad, featuring a famous quarterback, claimed that there are 3360 combinations of 3-item meals available to customers. (Actually, it was determined later that what the ad intended to convey was the fact that you could have 3360 three-item *side dishes* to accompany your main meal.) Anyway, Mr. Swain proved this statement wrong and in the process got a free lunch for himself and 30 of his students! (Who says there is no free lunch anymore?)

EXAMPLE 8 ▶ Counting Outcomes at Boston Market

Suppose Boston Market has 16 side dishes.

(a) In how many ways can you *select* 3 different dishes?

(b) How many permutations of 16 objects taken 3 at a time are there?

(c) Suppose you select carrots, potatoes, and broccoli. If you had selected broccoli, potatoes, and carrots, would the *end result* be different?

(d) Is the order in which you select your side dishes important?

Solution

(a) You can use the SCP by filling in the 3 blanks below. Since you have 16 choices for the first vegetable, 15 for the next, and 14 for the last, the number of ways in which you can *select* the 3 side dishes (no repetitions) is

$$\underline{16} \times \underline{15} \times \underline{14} = 3360$$

(b) The number of permutations of 16 items taken 3 at a time is

$$P(16, 3) = \frac{16!}{(16 - 3)!} = \frac{16 \cdot 15 \cdot 14 \cdot 13!}{13!} = \underline{16} \cdot \underline{15} \cdot \underline{14} = 3360$$

(c) You end up with the same 3 side dishes, so the end result is the same.

(d) The order is not important, so this is *not* a permutation.   ■

Online Study Center

To further explore counting, generating, and listing permutations, access links 10.2.1, 10.2.2, and 10.2.3 on this textbook's Online Study Center.

EXERCISES 10.2

Ⓐ Permutations

1. In how many different orders can the letters in the set $\{a, b, c, d\}$ be written?

2. In how many different ways can 4 people be seated in a row?

3. If 6 horses are in a race and they all finish with no ties, in how many ways can the horses finish the race?

4. In how many different ways can 7 people be lined up at the checkout counter in a supermarket?

5. An insurance agent has a list of 5 prospects. In how many different orders can the agent telephone these 5 prospects?

6. If the agent in problem 5 decides to telephone 3 of the prospects today and the other 2 prospects tomorrow, in how many ways can the agent telephone the prospects?

In problems 7–20, compute the given number.

7. 8!

8. 10!

9. 9!

10. $\frac{10!}{7!}$

11. $\frac{11!}{8!}$

12. $\frac{8!}{2!6!}$

13. $\frac{9!}{5!4!}$

14. $P(9, 4)$

15. $P(10, 2)$

16. $\frac{P(6, 3)}{4!}$

17. $\frac{P(5, 2)}{2!}$

18. $2 \cdot P(8, 3)$

19. $3 \cdot P(8, 5)$

20. $4 \cdot P(3, 3)$

21. A student is taking 5 classes, each of which requires 1 book. In how many ways can she stack the 5 books she must carry?

22. Suppose 10 people are entered in a race. If there are no ties, in how many ways can the first 3 places come out?

23. A basketball coach must choose 4 players to play in a particular game. (The team already has a center.) In how many ways can the remaining 4 positions be filled if the coach has 10 players who can play any position?

24. Rework problem 23 if the coach does not have a center already and must fill all 5 positions from the 10 players.

In problems 25–28, assume the cards are drawn without replacement. (*Hint:* See Example 9, Section 10.1)

25. In how many ways can 3 hearts be drawn from a standard deck of 52 cards?

26. In how many ways can 2 kings be drawn from a standard deck of 52 cards?

27. In how many ways can 2 red cards be drawn from a standard deck of 52 cards?

28. In how many ways can 4 diamonds be drawn from a standard deck of 52 cards?

29. How many three-digit numbers can be formed from the digits 1, 3, 5, 7, and 9 with no repetitions allowed?

30. How many even three-digit numbers can be formed from the digits 2, 4, 5, 7, and 9 with no repetitions allowed? (*Hint:* Try filling the units place first.)

31. A red die and a green die are tossed. In how many ways is it possible for both dice to come up even numbers? (Distinguish between the two dice.)

32. In problem 31, in how many ways is it possible for one of the dice to come up an odd number and the other to come up an even number?

B The Complementary Counting Principle

33. Out of 5 children, in how many ways can a family have *at least* 1 boy?

34. If two dice are tossed, in how many ways can at least one of the dice come up a 6? (*Hint:* There are 5 ways a single die can come up not a 6.)

Six coins are tossed; in how many ways can you have

35. no tails.

36. at least one tail.

37. at least two heads.

38. at most one head.

C The Additive Counting Principle

39. How many of the first 100 natural numbers are multiples of 2 or multiples of 5?

40. How many of the first 100 natural numbers are multiples of 2 or multiples of 3?

D Applications

41. How many license plates using six digits can the state of Vermont issue if repetition of digits is
  a. permitted?
  b. *not* permitted?
  c. In the 2000 census the population of Vermont was 608,827. Why do you think Vermont allows repetition of digits in its license plates? (It now uses three letters and three numerals.)

42. How many license plates using one digit followed by three letters and then three digits can the state of California issue if
  a. repetition of letters *and* numbers is permitted?
  b. no repetition of letters *or* numbers is permitted?

Online Study Center

To see U.S. and Canadian license plates, access links 10.2.4 and 10.2.5 on this textbook's Online Study Center.

43. Most radio stations licensed after 1927 use four call letters starting with K or W—for example, WFLA in Tampa or KROW in Huntsville, MO. Assuming no repetitions, how many four-letter sets are possible?

44. Your nine-digit Social Security number is divided into three parts: area (XXX), group (XX), and serial (XXXX). The area indicates the state on the original application, the group has no special significance, and the serial represent a straight numerical progression.
  a. How many Social Security numbers are possible if repetitions of digits are permitted?
  b. How many if repetitions of digits are not permitted?
  c. How many if repetitions of digits are not permitted and groups under 10 have never been assigned? You can go to link 10.2.5 to read more about Social Security numbers.

In Other Words

45. Give at least two reasons why 0! had to be defined as 1.

46. How would you define $P(n, 0)$? Verify your answer by finding $P(n, 0)$ using the formula for $P(n, r)$.

47. In your own words, what is the additive counting principle?

48. Explain under what circumstances it is advantageous to use the additive counting principle.

Using Your Knowledge

If you are interested in horse racing, here are some problems for you.

49. Five horses are entered in a race. If there are no ties, in how many ways can the race end?

50. In problem 49, if we know that 2 horses (A and B) are going to be tied for first place, in how many ways can the race end?

51. It seems unlikely that if 5 horses are entered in a race, 3 of them will be tied for first place. However, this actually happened! In the Astley Stakes, at Lewes, England, in August 1880, Mazurka, Wandering Nun, and Scobell triple dead-heated for first place. If it is known that these 3 horses tied for first place, in how many ways could the rest of the horses finish?

52. You probably answered 2 in response to problem 51, because it is unlikely that there will be a tie for fourth place. However, the other 2 horses, Cumberland and Thora, *did* tie for fourth place. If ties are allowed, in how many different ways could Cumberland and Thora have finished the race in the preceding problem?

Discovery

In this section you learned that the number of permutations of n distinct objects is $n!$. Thus, if you wish to seat 3 people across the table from you, the number of possible arrangements is 3!. However, if 3 persons are to be seated at a circular table, the number of possible

arrangements is only $2! = 2$. If the persons are labeled A, B, and C, the two arrangements look like those in the figure.

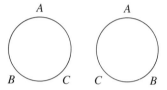

At first glance it may seem that there should be $3! = 6$ different arrangements, like those in the next figure.

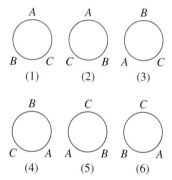

However, a closer look will reveal that arrangements (1), (4), and (5) are identical; in all of them B is to the right and C to the left of A. Similarly, (2), (3), and (6) are identical, because in every case C is to the right and B is to the left of A. To avoid this difficulty, if you have, say, 4 persons to be seated at a circular table, you seat one of them and use this person as a reference. The rest of the people can be seated in 3! ways. You have the following relationship:

Number of Persons	Number of Different Ways They Can Be Seated Around a Circular Table
3	2!
4	3!

53. From this discussion, can you discover in how many ways n persons can be seated around a circular table?

54. In how many ways can 4 people (including A and B) be seated at a circular table so that A and B are facing each other?

55. In problem 54, find the number of ways in which the people can be seated so that *A* and *B* are *not* facing each other.

56. In problem 54, find the number of ways in which the people can be seated so that *A* and *B* are next to each other.

 Calculator Corner

Many calculators have a factorial $\boxed{x!}$ or $\boxed{n!}$ key. Thus, to find 6!, you first enter the number 6 and then activate the factorial key. The steps are

$$\boxed{6}\ \boxed{\text{2nd}}\ \boxed{x!} \quad \text{or} \quad \boxed{6}\ \boxed{\text{inv}}\ \boxed{x!}$$

In addition, some calculators even have a key that will calculate $P(n, r)$, the $\boxed{_nP_r}$ key. To enter the *n* and the *r*, you must use other special keys. If the $\boxed{a}$ and the $\boxed{b}$ keys are those special keys on your calculator and you wish to find $P(7, 3)$, as in Example 4(b), you enter

$$\boxed{7}\ \boxed{a}\ \boxed{3}\ \boxed{b}\ \boxed{\text{2nd}}\ \boxed{_nP_r} \quad \text{The answer is 210.}$$

1. Use your calculator to check the answers to problems 7 and 14.

 Collaborative Learning

In Example 8 we discussed the number of ways in which the three side dishes could be *selected,* the number of *permutations* of the three side dishes possible, and the number of *different choices* you had for the three side dishes. The three important words here are *selected, permutations,* and *different choices.* Let us discuss this further.

Form three groups and consider the set

$$A = \{a, b, c, d, e\}.$$

Group 1

1. In how many ways can you *select* three letters from the set *A*?

2. Make a list to confirm your answer.

3. Is the order in which you *select* the letters important? Explain why or why not.

Group 2

4. How many *permutations* of three letters can be made using the letters in set *A*?

5. Make a list to confirm your answer.

6. Is the order in which you *select* the letters important? Explain why or why not.

Group 3

7. How many *different choices* consisting of three letters can be made using the letters in set *A*?

8. Make a list to confirm your answer.

9. Is the order in which you *select* the letters important? Explain why or why not.

Now, discuss your findings among the members of all three groups and establish a procedure to solve problems involving the words *select, permutations,* and *different choices.*

 Research Questions

Many notations for *n* factorial, or factorial *n*, have been used by mathematicians. Find the name of the person and the year in which the given notation for *n* factorial, or factorial *n,* was used.

Notation	Person	Year
1. M		
2. $n*$		
3. $[p]^n$		
4. $a^{m\,\vert\,r}$		
5. $\lfloor n$		
6. $n!$		

10.3 Combinations

Planetary Conjunctions and Combinations

Have you heard of planetary conjunctions? When two or more planets are in line with Earth and the Sun, as shown in Figure 10.9, you have what is known as a **planetary conjunction.** In certain cultures, planetary conjunctions were believed to exert special influences on events. According to Hindu tradition, a special dreaded conjunction was that of the seven planetary bodies known to man at that time (Sun, Moon, Mercury, Venus, Mars, Jupiter, and Saturn), an event that was supposed to occur in 26,000 years and result in the end of the world.

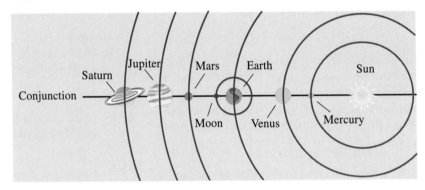

FIGURE 10.9

Rabbi Ben Ezra, a Jewish scholar, used an original computational method to show that the number of possible conjunctions of two or more of these planetary bodies was 120. How did he do it? Using *permutations* and *combinations,* Ben Ezra reasoned that the number of conjunctions of 2 planets was $P(7, 2) = 42$, but a conjunction of the Sun and Moon, for example, was the same as a conjunction of the Moon and Sun. Since there were 2! ways of arranging 2 planets, he divided $P(7, 2)$ by 2! to obtain the correct result, 21. For 3 planets, the number of conjunctions was $P(7, 3) = 210$, but this number had to be divided by $3! = 6$ to account for repetitions; thus the number of possible conjunctions involving 3 planets was 35.

In mathematics, a set of objects that can be selected disregarding their order is called a **combination** of the objects. The number of combinations of n objects taken r at a time is denoted by $C(n, r)$, read "the number of combinations of n objects taken r at a time" or "n choose r." In our example,

$$C(7, 2) = \frac{P(7, 2)}{2!} \quad \text{and} \quad C(7, 3) = \frac{P(7, 3)}{3!}$$

In general,

$$C(n, r) = \frac{P(n, r)}{r!}$$

Now, can you help Ben Ezra find how many different planetary conjunctions are possible? You need to find $C(7, 2) + C(7, 3) + C(7, 4) + \cdots + C(7, 7)$. If you arrive at a total of 120, you are on the way to understanding the formula for combinations.

Here is another way of thinking about the concept itself. $C(n, r)$ counts the number of subsets of r objects that can be made from a set of n objects. Thus, $C(4, 2)$ counts all the subsets of 2 objects that can be formed from a set of 4 objects. (If the set is $\{a, b, c, d\}$, the subsets are $\{a, b\}$, $\{a, c\}$, $\{a, d\}$, $\{b, c\}$, $\{b, d\}$, and $\{c, d\}$, a total of 6.) By the formula,

$$C(4, 2) = \frac{P(4, 2)}{2!} = \frac{12}{2} = 6$$

Now, you have both the concept and the formula for combinations! ▶

A. Combinations

In Section 10.2, we found the number of ordered arrangements that are possible with n distinguishable objects. Sometimes we may wish to count the number of subsets of these objects that can be selected if we *disregard the order* in which the objects are selected. Such subsets are called **combinations.** We use the symbol $C(n, r)$ to denote the number of combinations of r objects that can be formed from a set of n objects. The symbols $_nC_r$, $C_{n,r}$, C_r^n, and $\binom{n}{r}$ are also used to represent $C(n, r)$.

EXAMPLE 1 ▶ Using Combinations to Find Sums of Money

How many different sums of money can be made from a set of coins consisting of a penny, a nickel, and a dime if exactly 2 coins are selected?

Solution

Because the *order* in which we select the coins is *not* important (selecting a nickel and then a dime gives the same sum as selecting a dime and then a nickel), the question asked is equivalent to finding $C(3, 2)$, the number of combinations of 2 things that can be formed using a set of 3 things. One of the sums is 6¢ (it makes no difference whether the penny is selected first and then the nickel, or vice versa), the second sum is 11¢, and the third sum is 15¢. Hence, $C(3, 2)$, the number of combinations of 3 objects taken 2 at a time, is 3. ■

EXAMPLE 2 ▶ Using Combinations of Different Set Elements

Consider the set $S = \{a, b, c, d\}$.

(a) How many combinations of 2 elements are possible using elements of the set S?

(b) How many permutations of 2 elements are possible using elements of the set S?

(c) How many subsets of 2 elements does the set S have?

Solution

(a) The 6 possible combinations are shown in Table 10.2 on page 682. Hence, $C(4, 2) = 6$.

(b) $P(4, 2) = 4 \times 3 = 12$. The 12 permutations are shown in Table 10.2.

(c) This problem is equivalent to finding the number of combinations that can be made from 4 objects using 2 at a time; hence, the answer is 6, as in part (a).

TABLE 10.2

Combinations	Permutations
ab	ab, ba
ac	ac, ca
ad	ad, da
bc	bc, cb
bd	bd, db
cd	cd, dc

We can see from Table 10.2 that every combination determines 2! permutations, so $P(4, 2) = 2! \cdot C(4, 2)$. We use a similar argument to solve Example 3.

GRAPH IT

To find $C(n, r)$ with your grapher, go to the home screen (2nd MODE) enter the n (26) MATH ◄ 3 and the r (3). When you press enter, the answer 2600 will appear.

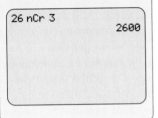

```
26 nCr 3
                2600
```

EXAMPLE 3 ► **Combinations and Sets of Three Letters**

How many sets of three letters can be made from the English alphabet?

Solution

Here we want to find $C(26, 3)$. One of the possible combinations is, for example, $\{A, B, C\}$. This choice determines $3! = 6$ permutations (ABC, ACB, BAC, BCA, CAB, and CBA). If we were to make a table similar to Table 10.2, we would see that to each combination there corresponds $3! = 6$ permutations. Hence, there are 6 times as many permutations as there are combinations. That is, $P(26, 3) = 6 \cdot C(26, 3)$; but $P(26, 3) = 26 \times 25 \times 24 = 15{,}600$, so $15{,}600 = 6 \times C(26, 3)$, or $C(26, 3) = 2600$. ■

The number of ways in which we can select a combination of r objects from a set of n objects is $C(n, r)$. The r objects in any one of these combinations can be arranged (permuted) in $r!$ ways. By the SCP, the total number of permutations is $r! \times C(n, r)$; but this number is $P(n, r)$. Hence,

$$P(n, r) = (r!)C(n, r)$$

so dividing both sides by $r!$, we have

Combinations Formula

The number of combinations of n objects, taken r at a time, is

$$C(n, r) = \frac{P(n, r)}{r!} \qquad 0 \le r \le n$$

The following useful form of the formula for $C(n, r)$ is obtained in the next example:

Factorial Formula for Combinations

The number of combinations of n objects, taken r at a time, is

$$C(n, r) = \frac{n!}{r!(n - r)!}$$

EXAMPLE 4 ▶ Using the Combinations Formula

Show that $C(n, r)$ has the form given on page 682: $\dfrac{n!}{r!(n-r)!}$

Solution

We know that

$$C(n, r) = \frac{P(n, r)}{r!} \quad \text{and} \quad P(n, r) = \frac{n!}{(n-r)!}$$

Thus, by substituting for $P(n, r)$, we obtain

$$C(n, r) = \frac{P(n, r)}{r!} = \frac{n!}{(n-r)!} \div r! = \frac{n!}{r!(n-r)!}$$

as given. ■

The meaning of $C(n, r)$ can also be stated in terms of a set of n elements.

Combinations as Subsets

$C(n, r)$ is the number of subsets of r elements each that can be formed from a set of n elements.

Note that this is just a repetition of the statement made at the beginning of this section. In general, we have the following:

Interpretations of $C(n, r)$

1. The number of ways to select r different objects from n different objects when the order is **not** important

2. The number of combinations of r different objects from n different objects

3. The number of subsets with r elements of a set with n elements

4. $C(n, r) = \dfrac{n!}{r!(n - r)}$

EXAMPLE 5 ▶ Finding the Number of Subsets

How many subsets of *at least* 3 elements can be formed from a set of 4 elements?

Solution

If we wish to have *at least* 3 elements in the subset, we can have either 3 or 4 elements. Using the preceding statement, the number of subsets of 3 elements that can be formed from a set of 4 elements is

$$C(4, 3) = \frac{4!}{3!1!} = 4$$

and the number of subsets of 4 elements that can be formed from a set of 4 elements is

$$C(4, 4) = \frac{4!}{4!0!} = 1$$

Thus, the number of subsets of at least 3 elements that can be formed from a set of 4 elements is $4 + 1 = 5$. (Try it with the set $\{a, b, c, d\}$.) ■

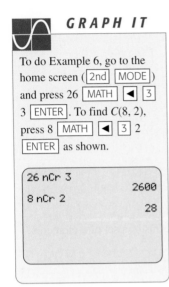

GRAPH IT

To do Example 6, go to the home screen ([2nd] [MODE]) and press 26 [MATH] [◀] [3] 3 [ENTER]. To find $C(8, 2)$, press 8 [MATH] [◀] [3] 2 [ENTER] as shown.

26 nCr 3
                    2600
8 nCr 2
                      28

EXAMPLE 6 ▶ Computations Involving the Combinations Formula

Compute the following: (a) $C(26, 3)$ (b) $C(8, 2)$

Solution

(a) From the second formula for $C(n, r)$, we get

$$C(26, 3) = \frac{26!}{3!23!} = \frac{26 \times 25 \times 24 \times 23!}{3!23!} = 2600$$

(b) Similarly,

$$C(8, 2) = \frac{8!}{2!6!} = \frac{8 \times 7 \times 6!}{2!6!} = 28$$ ∎

You can make your work easier by noting possible cancellations, as done in the solutions for Example 6.

EXAMPLE 7 ▶ Finding the Number of Different 2-Card Hands

How many different 2-card hands are possible if we use an ordinary deck of 52 cards?

Solution

Since the order in which you receive the cards is not important, the question asked is equivalent to "How many combinations are there of 52 elements, 2 at a time?" Using the first formula for $C(n, r)$, we find

$$C(52, 2) = \frac{P(52, 2)}{2!} = \frac{52 \cdot 51}{2 \cdot 1} = 1326$$ ∎

Suppose you are asked to find the number of combinations of 10 objects, 8 at a time. You can see that if you take away any combination of 8 of the objects, a combination of 2 of the objects is left. This shows that $C(10, 8) = C(10, 2)$. This result can be verified directly as follows:

$$C(10, 8) = \frac{P(10, 8)}{8!} = \frac{10 \cdot 9 \cdot 8 \cdot 7 \cdot 6 \cdot 5 \cdot 4 \cdot 3}{8 \cdot 7 \cdot 6 \cdot 5 \cdot 4 \cdot 3 \cdot 2 \cdot 1} = \frac{10 \cdot 9}{2 \cdot 1}$$

and

$$C(10, 2) = \frac{P(10, 2)}{2!} = \frac{10 \cdot 9}{2 \cdot 1}$$

Therefore,

$$C(10, 8) = C(10, 2)$$

In general,

Alternative Formula for Combinations

The alternative formula for finding combinations is

$$C(n, r) = C(n, n - r)$$

The second formula for $C(n, r)$, which was verified in Example 4, makes this possible because

$$C(n, n - r) = \frac{n!}{(n - r)![n - (n - r)]!}$$

$$= \frac{n!}{(n - r)!(n - n + r)!} = \frac{n!}{(n - r)!r!} = C(n, r)$$

ROMANO'S
Greek - Italian Restaurant

Menu For Lunch & Take Out

555-6666

PIZZA

OUR SPECIAL DOUGH LIGHT AND CRISPY
TO YOUR EXPECTATION ONE-SIZE ONLY
10"-6 PIECES

PLAIN CHEESE..................5.95
ANY 1 ITEM.....................6.95
ANY 2 COMBINATIONS............7.95
ANY 3 COMBINATIONS............8.95
SPECIAL (All Items)...............9.95

ITEMS

PEPPERONI, ONION, PEPPERS, MUSHROOMS,
SAUSAGE, AND MEATBALL

SOFT DRINKS

Sm.   Lg.

COKE, 7-UP.....................75....95
TAB, ROOT BEER...............75....95
COFFEE or ICE TEA..................95

DESSERTS

Try Our Delicious Homemade Desserts

RICE PUDDING...................1.50
GALACTOBURICO.................1.50
Greek Custard with Fillo
BAKLAVA.........................2.50
Walnuts, Honey, and Fillo
BOUGATZA.......................2.50
Walnuts, Honey, Cinnamon, and Fillo
SPUMONI.........................2.50
Italian-style Ice Cream

FIGURE 10.10

EXAMPLE 8 ▶ Counting Pizzas

Romano's Restaurant offers the pizza menu in Figure 10.10. Find how many different pizzas you can order with the following:

(a) 1 item   (b) 2 items   (c) 3 items   (d) 4 items

Solution

(a) Because there are exactly 6 different items available, there are 6 different pizzas with 1 item. Note if you use the formula for $C(6, 1)$, it gives $\frac{6}{1} = 6$.

(b) The *order* in which the items are added is *not* important. (If you order pepperoni and onion, you get the same pizza as if you order onion and pepperoni.) Thus, you need to find the number of combinations of 6 things, taken 2 at a time.

$$C(6, 2) = \frac{P(6, 2)}{2!} = \frac{6 \cdot 5}{1 \cdot 2} = 15$$

(c) Here, you need $C(6, 3)$.

$$C(6, 3) = \frac{P(6, 3)}{3!} = \frac{6 \cdot 5 \cdot 4}{3 \cdot 2 \cdot 1} = 20$$

(d) The answer is $C(6, 4)$, which is the same as $C(6, 2) = 15$ ∎

EXAMPLE 9 ▶ Counting Possibilities Selecting 3 Side Dishes

Let us go back to the problem of selecting 3 side dishes from 16 available dishes (Section 10.2, Example 8) and count the number of choices we have for our 3 side dishes

(a) without any repetitions (we have to pick 3 different dishes).

(b) with one repetition (say *aba* or *ccd* or *eff*).

(c) with 3 repetitions (say *aaa*, *bbb*, or *ccc*).

Solution

(a) Without repetition, there are

$$C(16, 3) = \frac{16!}{3!13!} = \frac{16 \cdot 15 \cdot 14}{6} = 560 \text{ choices}$$

(b) With one repetition there are 16 choices for the first dish, 1 choice for the second (it has to be the same as the first), and $C(15, 1)$ for the last dish, a total of

$$\underbrace{16}_{\text{1st dish}} \cdot \underbrace{1}_{\text{2nd}} \cdot \underbrace{C(15, 1)}_{\text{3rd}} = 16 \cdot \frac{15!}{1 \cdot 14!} = 16 \cdot 15 = 240 \text{ choices}$$

(c) With three repetitions, we have a total of 16 choices (say, 3 macaroni or 3 broccoli or 3 whatever!).

Note that the total number of choices is $560 + 240 + 16 = 816$ and that is the total number of choices we have to select our 3 side dishes, which is what Mr. Swain said! ∎

Online Study Center

To further explore combinations and permutations, access links 10.3.2 and 10.3.3 on this textbook's Online Study Center.

Before you go to the exercises, you should know that some calculators have a $C(n, r)$ or a $_nC_r$ key so you can work most of the problems without doing all the detailed arithmetic. If you do not have one of those calculators, go to link 10.3.1 and use the one provided there.

EXERCISES 10.3

Ⓐ Combinations

In problems 1–6, evaluate each of the following:

1. $C(5, 2)$ and $P(5, 2)$ **2.** $C(6, 4)$ and $P(6, 4)$

3. $C(7, 3)$ and $P(7, 3)$ **4.** $C(5, 0)$ and $P(5, 0)$

5. $C(9, 6)$ and $P(9, 6)$ **6.** $C(7, 0)$ and $P(7, 0)$

In problems 7–10, find the number of combinations that can be made from each of the following:

7. 5 objects taken 4 at a time

8. 9 objects taken 3 at a time

9. 10 objects taken 2 at a time

10. 12 objects taken 3 at a time

11. How many subsets of 2 elements can be made from a set of 8 elements?

12. How many subsets of 5 elements can be made from a set of 7 elements?

13. How many different 8-element subsets can be made from a set of 12 elements?

14. How many different 10-element subsets can be made from a set of 15 elements?

15. Let T be a set of 10 elements.
   a. How many subsets of 3 elements does T have?
   b. How many subsets of less than 3 elements does T have?
   c. How many subsets of no elements does T have?
   d. How many subsets of more than 9 elements does T have?

16. How many different sums of money can be formed from a penny, a nickel, a dime, a quarter, and a half-dollar if exactly 3 coins are used?

17. Rework problem 16 using 4 coins.

18. Rework problem 16 using at least 2 coins.

19. Let $A = \{1, 2, 3, 4, 5\}$.
   a. How many subsets of 3 elements does the set A have?
   b. How many subsets of A have no more than 3 elements?

20. If 20 people all shake hands with each other, how many handshakes are there?

21. The Greek alphabet has 24 letters. In how many ways can 3 different Greek letters be selected if the order does not matter?

22. The mathematics department is sending 5 of its 10 members to a meeting. How many different sets of 5 members can be selected?

23. A committee is to consist of 3 members. If there are 4 men and 6 women available to serve on this committee, how many different committees can be formed?

24. The Book of the Month Club offers a choice of 3 books from a list of 40. How many different selections of 3 books each can be made from this list?

25. How many different 5-card poker hands are possible using a deck of 52 cards?

26. A restaurant offers 8 different kinds of sandwiches. How many different sets of 2 sandwiches could you select?

27. The U.S. Senate has 100 members. How many different 5-member committees can be formed from the Senate?

28. In how many ways can a committee of 7 be formed from a group of 12 eligible people?

29. Johnny has a $1 bill, a $5 bill, a $10 bill, and a $20 bill in his pocket. How many different sums of money can Johnny make with these bills if he uses at least 1 bill each time?

30. Desi has 6 coins: a penny, a nickel, a dime, a quarter, a half-dollar, and a dollar. How many different sums of money can Desi form using just 2 of these coins for each sum?

31. Refer to problem 30. How many different sums of money can Desi form if she uses at least 1 coin each time?

32. How many different committees can be formed from 8 people if each committee must consist of at least 3 people?

33. In how many ways can 8 people be divided into 2 equal groups?

34. A diagonal of a polygon is a line segment joining two nonadjacent vertices. How many diagonals does a polygon of
   a. 8 sides have?
   b. *n* sides have?
   (*Hint:* Think of *all* the lines joining the vertices two at a time. How many of these lines are sides and not diagonals?)

 In Other Words

35. Write in your own words the difference between a permutation of 3 objects and a combination of 3 objects.

36. You know that a *combination* for your locker uses the numbers 1, 2, 3.
   a. Will the combination 1, 2, 3 necessarily open the locker?
   b. What are the permutations of 1, 2, and 3? Will one of these open the locker? Explain.

37. Discuss why a *combination lock* should really be called a *permutation lock*.

38. Consider $P(n, r)$ and $C(n, r)$.
   a. Discuss the conditions under which $P(n, r) = C(n, r)$ and explain why.
   b. Is $P(n, r)$ greater than or less than $C(n, r)$? Explain.

 Using Your Knowledge

The following figure shows the famous Pascal's triangle. The triangle counts the number of subsets of *k* elements that can be made from a set of *n* elements, that is, $C(n, k)$. If you consider *n* to be the row number and *k* the diagonal number, you can find $C(5, 2)$ by going to the fifth row, second diagonal. The answer is 10. Note that the value 10 is obtained by adding the 4 and

6 above 10 in the triangle. Similarly, the 5 in the fifth row, first diagonal is found by adding the 1 and 4 above 5 in the preceding row.

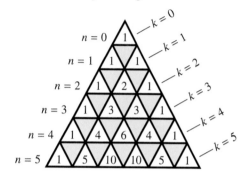

39. Construct the next two rows of Pascal's triangle.

40. Use Pascal's triangle to find the following:
   a. $C(6, 4)$ **b.** $C(7, 3)$

41. You might have learned in algebra that

$$(a + b)^0 = \qquad\qquad 1$$
$$(a + b)^1 = \qquad\qquad 1a + 1b$$
$$(a + b)^2 = \qquad 1a^2 + 2ab + 1b^2$$
$$(a + b)^3 = 1a^3 + 3a^2b + 3ab^2 + 1b^3$$

Compare Pascal's triangle with the given expressions and then find the following:
   a. $(a + b)^4$ **b.** $(a + b)^5$

42. Use Pascal's triangle to find $(a + b)^6$.

 **Discovery**

Suppose a fair coin is flipped 5 times in succession. How many different outcomes are possible? Since the coin can fall in either of 2 ways (heads or tails), the number of different outcomes is $2^5 = 32$.

In how many different ways can the outcome be 2 heads and 3 tails? If you think a moment, you will realize that the answer is the number of combinations of 5 things, taken 2 at a time. (Look at the 5 tosses and determine in how many ways you can select 2 of them when the order is unimportant.) Thus, the correct answer is

$$C(5, 2) = \frac{5 \cdot 4}{2 \cdot 1} = 10$$

See if you can discover the answers to the following questions:

43. In how many different ways can the outcome be 0 heads and 5 tails? Call this number $C(5, 0)$.

44. Rework problem 43 for 1 head and 4 tails.

45. Rework problem 43 for 3 heads and 2 tails.

46. Rework problem 43 for 4 heads and 1 tail.

47. Rework problem 43 for 5 heads and 0 tails.

48. Add the answer you found for 2 heads and 3 tails and your answers for problems 43–47. You should come out with

$$C(5, 0) + C(5, 1) + C(5, 2) + C(5, 3) + C(5, 4) + C(5, 5) = 32$$

Explain why.

49. The result of problem 48 is a special case of the general result

$$C(n, 0) + C(n, 1) + C(n, 2) + \cdots + C(n, n) = 2^n$$

Can you tell why this must be a correct result? (*Hint:* Think of the coin's being flipped n times.)

 Calculator Corner

Some calculators can evaluate $C(n, r)$. To do this, you must enter n and r using special keys, say $\boxed{a}$ and $\boxed{b}$, on your calculator. Thus, to evaluate $C(26, 3)$, as in Example 6(a), press $\boxed{26}$ $\boxed{a}$ $\boxed{3}$ $\boxed{b}$ $\boxed{2nd}$ $\boxed{_nC_r}$. As before, the answer is 2600.

1. Check the answers to Examples 6, 7, and 8 using your calculator.

 **Collaborative Learning**

Suppose that your group is in charge of designing license plates for the Department of Motor Vehicles.

1. If you use the numbers 0–9 to create your plates, as Washington, D.C., did, how many license plates are possible?

2. **Online Study Center**

Go to an almanac or the *Statistical Abstract of the United States* or access link 10.3.4 on this textbook's Online Study Center and determine the states in which your answer to the first question might provide enough license plates.

3. The population of Wyoming is about 493,782. If you used the scheme described in the first question, do you think there would be enough license plates for all motorists? (*Hint:* Suppose you have two cars!)

4. If you were in charge of designing license plates for the state of California, which method would you use to make sure that there were enough license plates available? (Here's how they actually do it.)

5. How can you make sure that there are enough license plates available?

6. At the present time there are two types of license plates in California: one digit, three letters and three digits; and three letters followed by three digits.
  a. How many license plates are possible using each of the two schemes?
  b. Which one do you think is more likely to be used? Explain why.

10.4   Miscellaneous Counting Methods

"Counting" on Winning the Lottery

In Section 10.1 you were asked, "Why does it seem so difficult to win the lottery?" The answer, you found, is because there are so many possible *permutations* or so many *combinations,* depending on the game you play. For example, suppose you play a lottery in which you pick four digits and you win if the digits are drawn in the *exact* order you have chosen. To find your chance of winning, you must find in how many ways you can select the four digits. The number of ways in which you can fill the 4 blanks

____ ____ ____ ____

using the sequential counting principle and the 10 digits (0 to 9) to fill each blank, is

$$\underline{10} \times \underline{10} \times \underline{10} \times \underline{10}$$

or 10,000 ways. Only one number will win, so your chance is 1 in 10,000. (If you pick 1345 and the number selected is 5431, you are out of luck.) Here the Sequential Counting Principle (SCP) was used as the counting method.

You can also play this game by selecting the 24-way box. In this case, you select a set of 4 numbers and win if the numbers come out *in any order.* How many ways can you win now? Since there are $P(4, 4)$, or 24, ways of permuting the 4 numbers you choose, your chances are increased to 24 in 10,000. Here permutations were used as the counting method.

Other lottery games are even more challenging. For example, in many state lotteries you pick 6 different numbers from a set of 49 numbers. You win the jackpot if you match *(in any order)* the 6 winning lottery numbers. How many chances of winning do you have now? To find out, you must find the number of ways in which 6 numbers can be selected from a set of 49 numbers when the order is not important; that is, you must find $C(49, 6)$. If you have a calculator or if you are patient, you can compute $C(49, 6)$ and obtain 13,983,816. Your chances are only 1 in almost 14 million. Here combinations were used as the counting method.

The examples and problems in this section will use the SCP, permutations, and combinations. The question you must consider is when to use which method. You begin this section by considering a diagram that may help you to make this choice. ▶

In the preceding sections we discussed the use of the sequential counting principle (SCP), permutations, and combinations in simple counting problems. Very often, the most difficult step in dealing with a counting problem is to decide which method or formula to use. We shall try to make this decision easier in the following discussion.

A. Permutations and Combinations

As you recall, if the problem involves two or more events that are to occur in succession, you must use the SCP. For problems that involve choosing r items from a set of n different items, with no repetitions allowed, remember the diagram shown in Figure 10.11.

FIGURE 10.11

The next example will help to clarify this idea.

EXAMPLE 1 ▶ Combinations or Permutations?

An employment agency has 5 highly skilled workers listed. Find in how many ways 2 of these workers can be selected given the following:

(a) The first one is to be in charge and the second one is to help.

(b) They are simply to do a job.

Solution

In both parts of this problem, 2 workers are to be selected from 5.

(a) If the first worker is to be in charge and the second to help, then the *order* in which they are picked is important. Hence, we must use *permutations*. The answer is

$$P(5, 2) = 5 \cdot 4 = 20$$

(b) Here the order is *not* important. (It makes no difference if Joe and Sally, or Sally and Joe are picked; both will do the job.) Thus, we must use *combinations,* and the answer is

$$C(5, 2) = \frac{5 \cdot 4}{2 \cdot 1} = 10$$ ∎

Sometimes we must combine more than one principle in solving a counting problem. We illustrate this in the next example.

EXAMPLE 2 ▶ Choices to Select 3 Programs from 6

A television network has 6 different half-hour programs during prime time (7 to 10 P.M.). You want to watch 3 programs in one evening.

(a) How many choices do you have?

(b) If exactly 1 of the programs must be after 9 P.M., how many choices do you have?

Solution

In this problem you can choose the programs, but you must watch them at the times when they are presented. No permutations are allowed. Thus, to answer the questions, you must use combinations.

(a) Here, you simply need to select 3 of the 6 programs. This means

$$C(6, 3) = \frac{6 \cdot 5 \cdot 4}{3 \cdot 2 \cdot 1} = 20 \text{ choices}$$

(b) Divide the problem into two parts as follows:

1. Select 1 program *after 9 P.M.* There are 2 choices.

2. Select 2 other programs from 4 *before 9 P.M.* There are

$$C(4, 2) = \frac{4 \cdot 3}{2 \cdot 1} = 6 \text{ choices}$$

Now use the SCP to combine the two sets of choices. This gives

$$6 \cdot 2 = 12 \text{ choices}$$

Before 9 ↗ ↖ After 9 ∎

EXAMPLE 3 ▶ **More Television Choices**

Here is another view of the television problem. A local station manager has 10 different half-hour programs available and needs to schedule 6 of them in the hours from 7 to 10 P.M. The station manager feels that 4 of the programs are unsuitable for showing before 9 P.M. but is obligated to show 2 of these sometime during the evening. How many choices does this leave for the evening's schedule?

Solution

We divide this problem into two parts, as we did in Example 2(b). Here, however, the station manager can control the order as well as the choice of programs. Therefore, this problem requires permutations.

1. For the hours *from 7 to 9 P.M.*, there are 6 (10 − 4) programs available and the manager must select 4 of them. Since the order of showing has to be considered, the number of choices is $P(6, 4) = 6 \cdot 5 \cdot 4 \cdot 3 = 360$.

2. For the hour *from 9 to 10 P.M.*, there are 4 programs, of which 2 must be selected. The number of choices is $P(4, 2) = 4 \cdot 3 = 12$.

Now, we can use the SCP to give us the total number of choices, which is $360 \cdot 12 = 4320$. (Pity the poor station manager!)

9 to 10 P.M. ↑   ↖ 7 to 9 P.M.   ■

EXAMPLE 4 ▶ **Number of Choices Selecting from a Menu**

Roy and Rosie are eating out at an Asian restaurant (see the menu on page 667). They select a special family dinner that allows an individual choice of 1 of 2 soups, 1 entree from 10 items in group A, 1 entree from 9 items in group B, and an individual choice of 1 of 4 desserts. How many different possibilities are there for dinner?

Solution

We consider the following two cases:

1. Roy picks an entree from group A and Rosie picks one from group B. Thus, Roy has a choice of 2 soups, 10 entrees, and 4 desserts, so by the SCP, he has $2 \cdot 10 \cdot 4 = 80$ choices. At the same time, Rosie has a choice of 2 soups, 9 entrees, and 4 desserts, so she has $2 \cdot 9 \cdot 4 = 72$ choices. Hence, by the SCP, together they have $80 \cdot 72 = 5760$ different choices available.

Roy ─↑   ↑─ Rosie

2. Roy picks an entree from group B and Rosie picks one from group A. This simply exchanges the choices we found in case 1, so the number of choices for both is again 5760.

Thus, the total number of possibilities is $2 \cdot 5760 = 11{,}520$.   ■

B. Permutations of Nondistinct Objects

In the preceding examples, all the objects considered were distinct (you could tell them apart). Here is a different type of problem. If you go to Madison, Wisconsin, and look at the white pages of the telephone book, you might find that the last name listed is Hero Zzyzzx (pronounced "Ziz-icks"). Can you find in how many different ways the letters in Mr. Zzyzzx's last name can be arranged? Before tackling this problem, look at a simpler one. It is conceivable that no one calls Mr. Zzyzzx by his proper last name; perhaps he is named Zzx (Zicks) for short. In how many different ways can the letters in the name Zzx be arranged? To do this problem, you ignore the capitalization and rewrite the name as $z_1 z_2 x$ so that you now have three distinct things. Then look at all the possible arrangements of z_1, z_2, and x. After this step, erase the subscripts and look at the arrangements again. Table 10.3 shows the two sets of arrangements. Notice that with the subscripts you have 3 distinct objects, which can be ordered in $P(3, 3) = 3! = 6$ ways. The second half of the table, with subscripts erased, shows that two permutations of z_1, z_2, and x, in which the 2 z's are simply interchanged, become identical. Hence, to find the number of distinct arrangements without subscripts, you must divide the number with subscripts by the number of ways in which the identical letters can be permuted. Because there are 2 z's, they can be permuted in 2! ways, so the number of arrangements of zzx is

TABLE 10.3

With Subscripts		Without Subscripts
$z_1 z_2 x$	$z_2 z_1 x$	zzx zzx
$z_1 x z_2$	$z_2 x z_1$	zxz zxz
$x z_1 z_2$	$x z_2 z_1$	xzz xzz

$$\frac{3!}{2!} = 3$$

A similar argument leads to the general result.

Distinct Arrangements

Suppose that a set of n objects consists of r different types, objects of the same type being indistinguishable. If there are n_1 objects of type 1, n_2 objects of type 2, ... , and n_r objects of type r, then the total number of *distinct* arrangements of the n objects is

$$\frac{n!}{n_1! n_2! \cdots n_r!}$$

With this formula, you can find the number of distinct arrangements of the letters in the name Zzyzzx. Now regard the Z and the z as distinct, so there are 6 letters, 1 Z, 3 z's, 1 y, and 1 x. The formula gives

$$\frac{6!}{1! 3! 1! 1!} = 6 \cdot 5 \cdot 4 = 120$$

EXAMPLE 5 ▶ Arranging Minnie's Letters

In how many different ways can the letters in the name Minnie be arranged?

Solution

There is a total of $n = 6$ letters in the name: 1 M, 2 i's, 2 n's, and 1 e. Thus, $n_1 = 1, n_2 = 2, n_3 = 2$, and $n_4 = 1$. Hence, the number of distinct arrangements is

$$\frac{6!}{1! 2! 2! 1!} = \frac{6 \cdot 5 \cdot 4 \cdot 3 \cdot 2 \cdot 1}{1 \cdot 1 \cdot 2 \cdot 1 \cdot 2 \cdot 1} = 180 \qquad ■$$

EXAMPLE 6 ▶ Que Zzzzzzzzzra Zzzzzzzzzra

The last name in the San Francisco phone book used to be (are you ready?) Zachary Zzzzzzzzzra. (Please, don't ask how to pronounce it!) In how many distinguishable ways can the letters in Zzzzzzzzzra be arranged?

Solution

Here, $n = 11$, $n_1 = 1$ (1 Z), $n_2 = 8$ (8 z's), $n_3 = 1$ (1 r), and $n_4 = 1$ (1 a). Thus, the number of distinct arrangements is

$$\frac{11!}{1!8!1!1!} = \frac{11 \cdot 10 \cdot 9 \cdot 8!}{8!} = 11 \cdot 10 \cdot 9 = 990 \qquad \blacksquare$$

C. Applications

There is one counting technique that has not been discussed in this section: tree diagrams. These diagrams can be used to "help make better decisions in litigation management." Let us see how. (Source: TreeAge Software.)

EXAMPLE 7 ▶ An Application to Law

A lawyer handling legal cases "first identifies the factual and legal uncertainties in a case and then decides: Should we litigate or settle?" If we litigate, we can lose or win a summary judgment. If we lose, the jury finds liability (high, medium, or low) or there may not be any liability. Draw a tree diagram and show all the possibilities for the case.

Solution

We draw the tree shown in Figure 10.12 and label the branches.

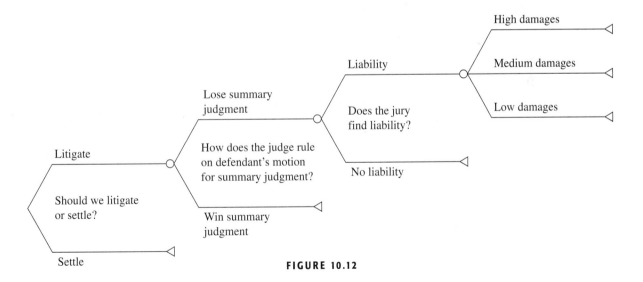

FIGURE 10.12

Online Study Center

To further explore applications to law, access link 10.4.1 on this textbook's Online Study Center.

As you can see, there are 6 distinct possibilities identified by the ◁ symbol. In the next chapter we will be able to assign probabilities to each of the events involved. For example, "The reality is that more than 90% of all cases end up being resolved through settlement, rather than trial." ∎

EXERCISES 10.4

A Permutations and Combinations

1. Three cards are dealt in succession and without replacement from a standard deck of 52 cards.
   a. In how many different orders can the cards be dealt?
   b. How many different 3-card hands are possible?

2. An employment agency has 6 temporary workers.
   a. In how many ways could 4 of them be assigned to the research department?
   b. In how many ways could 3 of them be assigned to 3 different companies?

3. The playbook for the quarterback of the Dallas Cowboys contains 50 plays.
   a. In how many different ways could the quarterback select 3 plays to use in succession in the next 3 downs?
   b. In how many different ways could he select a set of 3 plays to study?

4. A student must take 3 different courses on Mondays. In how many different ways can the student do this given the following?
   a. There are 6 different courses, all available at each of the 3 hours 8, 9, and 10 A.M.
   b. Only 1 of these courses is available each hour between 8 A.M. and 2 P.M. (6 hours).

5. Rework problem 4(b) if the student wants to keep the hour from 12 noon to 1 P.M. free for lunch.

6. A student wishes to schedule mathematics, English, and science. These classes are available every hour between 9 A.M. and noon (3 hours).
   a. How many different schedules are possible?
   b. How many schedules are possible if this student wants to take mathematics at 11 A.M. with her favorite instructor, Mr. Eldridge?

7. Peter must select 3 electives from a group of 7 courses.
   a. In how many ways can Peter do this?
   b. If all 7 of these courses are available each of the 4 hours from 8 A.M. to noon, from how many different schedules (hours and what course at each hour) can Peter choose?

8. At the University of South Florida, a student must take at least 2 courses from each of 5 different areas in order to satisfy the general distribution requirement. Each of the areas has the number of courses indicated in the table.

Area	I	II	III	IV	V
Courses	2	50	20	40	100

   a. If Sandy has satisfied all the requirements except for area V, and she wishes to take 3 courses in this area, how many choices does she have?
   b. Bill has already satisfied his requirements in areas I, II, and III. Now he wishes to take the minimum number of courses in areas IV and V. How many choices does he have?

9. A class consists of 14 boys and 10 girls. They want to elect officers so that the president and secretary are girls, and the vice president and treasurer are boys.
   a. How many possibilities are there?
   b. How many are there if 2 of the boys refuse to participate?

10. A company has 6 officers and 4 directors (10 different people). In how many ways can a committee of 4 be selected from these 10 people so that the following hold?
    a. 2 members are officers and 2 are directors.
    b. 3 members are officers and 1 is a director.
    c. All the members are officers.
    d. There are no restrictions.

11. There are 4 vacancies on the scholarship committee at a certain university. In order to balance the men and women on the committee, 1 woman and 3 men are to be appointed. In how many ways can this be done if the following are available to serve?
    a. 7 men and 8 women
    b. 5 men and 2 women

12. Romano's Restaurant has the menu shown in Example 8 of Section 10.3. In how many ways can a meal consisting of a pizza with 3 toppings, 2 beverages, and a dessert be chosen? The menu shows that there are 6 toppings for the pizza, 6 beverages, and 5 desserts offered.

B Permutations of Nondistinct Objects

13. How many distinct arrangements can be made with the letters in the word TALLAHASSEE?

14. How many distinct arrangements can be made with the letters in the word MISSISSIPPI?

15. Do you know what a *palindrome* is? It is a word or phrase with the same spelling when written forward or backward. The longest single-word palindrome in the English language is the word REDIVIDER. How many distinct arrangements can be made with the letters in this word?

16. There is a place in Morocco with a name that has 8 vowels in a row in its spelling! Do you know what place this is? It is spelled IJOUAOUOUENE. How many distinct arrangements can be made with the letters in this name?

17. A contractor needs to buy 7 electronic components from 3 different subcontractors. The contractor wants to buy 2 of the components from the first subcontractor, 3 from the second, and 2 from the third. In how many ways can this be accomplished?

18. An advertiser has a contract for 20 weeks that provides 3 different ads each week. If it is decided that in no 2 weeks will the same 3 ads be shown, how many different ads are necessary? [*Hint:* You need to find the least n such that $C(n, 3) \geq 20$.]

19. A cable television network wishes to show 5 movies every day for 3 weeks (21 days) without having to show the same 5 movies any 2 days in the 3 weeks. What is the least number of movies the network must have in order to do this? (See the hint in problem 18.)

20. Repeat problem 19 but assume the network wants to show the movies for 8 weeks.

21. Polly needs to take biology, English, and history. All these are available every hour between 9 A.M. and 3 P.M. (6 hours). If Polly must schedule 2 of these courses between 9 A.M. and 1 P.M. and 1 course between 1 and 3 P.M., how many schedules (hours and course each hour) are available to her?

22. Roy must elect 3 courses from among 4 courses in group I and 3 courses in group II. If he must take at least 1 of his 3 electives from each group, how many choices does he have? (*Hint:* First find how many choices he has if he elects only 1 course from group I. Then find how many choices he has if he elects 2 courses from group I. Since he must do one or the other of these, the final answer is the sum of the two answers.)

C Applications

Problems 23–25 are adapted from David C. Skinner, *Introduction to Decision Analysis* (Florida: Probabilistic Publishing, 1996).

23. Suppose you wish to invest $1000 for a year. You have *three* choices: a mutual fund, a management company, or a CD. Each of the investments may have a *high* or a *low* yield. Mutual funds return 6–10%, management companies 5–10%, and CDs 3–4%.
   a. Make a tree diagram showing all possibilities. At the end of each branch write the amount of money expected from the investment.
   b. Based on your tree diagram, what is the highest value for the investment?
   c. Based on your tree diagram, what is the lowest value for the investment?

24. Do you bet at all? Let us look at a hypothetical problem and perhaps you won't. Suppose you are considering betting on the horses. You can either place the bet or not. If you place a $200 bet, you can win $50,000 or lose your $200 (−$200).
   a. Make a tree diagram showing all possibilities. At the end of each branch write the amount associated with that branch.
   b. Based on your diagram, what is the highest (best) amount you can get?
   c. Based on your diagram, what is the lowest (worst) amount you can get?

25. A company is considering the introduction of a new product. The market size for the product can be extremely high or low. If the market size is extremely high, sales can be high ($2000), medium ($1000), or low ($750). If the market size is low, sales can be high ($750), medium ($500), or low ($300).
   a. Make a tree diagram showing all possibilities. At the end of each branch, write the amount expected from that outcome.

b. Based on your tree diagram, what is the highest amount of money you can get and under what conditions?

c. Based on your tree diagram, what is the lowest amount of money you can get and under what conditions?

26. Here is a personal investment decision. You can buy high, low, or preferred stocks. For each of these types of stocks, the market can go up, stay the same, or go down. In the case of the high risk stocks, if the markets go up, you get $1500; if it stays the same, you get $100; but if it goes down, you lose $1000 (−$1000). With the low-risk stock, you get $1000 when the market goes up and $200 if it stays the same, but you lose $100 (−$100) if the market goes down. The preferred stock will always pay $500 regardless of market conditions.

a. Make a tree diagram showing all possibilities. At the end of each branch, write the amount associated with the outcome corresponding to that branch.

b. Based on your tree diagram, what is the highest amount of money you can expect and under what conditions?

c. Based on your tree diagram, what is the lowest amount of money you can expect and under what conditions?

 In Other Words

27. Describe in your own words how you would decide which of the formulas to use (SCP, permutations, combinations) in a counting problem.

28. Which formula would you use in a counting problem involving indistinguishable objects? Explain.

 Using Your Knowledge

Do you know an easy way of finding how many positive integers are exact divisors of a given positive integer? For example, how many exact divisors does 4500 have? The easy way to answer this question is to write 4500 first as a product of its prime divisors.

$$4500 = 2^2 3^2 5^3$$

Now you can see that every exact divisor of 4500 must be of the form $2^a 3^b 5^c$, where a is 0, 1, or 2; b is 0, 1, or 2; and c is 0, 1, 2, or 3. Because there are 3 choices for a, 3 choices for b, and 4 choices for c, the SCP tells you that the number of exact divisors of 4500 is $3 \cdot 3 \cdot 4 = 36$. Notice that the exponents in the prime factorization of 4500 are 2, 2, and 3 and the number of exact divisors is the product $(2 + 1)(2 + 1)(3 + 1)$. Try this out for a small number, say 12, for which you can check the answer by writing out all the exact divisors.

29. How many exact divisors does 144 have?

30. How many exact divisors does 2520 have?

31. If the integer $N = 2^a 3^b 5^c 7^d$, how many exact divisors does N have?

32. How many exact divisors does the number $2^4 3^2 7^3$ have?

 Collaborative Learning

The tree on page 697 shows a model of a decision whether to vaccinate for a specific disease. The numbers appearing on individual branches indicate the probability that the event associated with that branch will occur. For example, if the decision is made to vaccinate, the probability that there are no complications is 0.99, or 99%. In addition, different numbers are given at the end of each branch. Those numbers represent the *quality* of that particular course of action. Thus, if the decision is made not to vaccinate and there is no disease, the quality number is 10 (see the bottom branch).

1. What is the best quality number in the diagram and what course of action has to be taken to obtain that number?

2. Discuss which one of the two courses of action associated with the highest quality number is, in your opinion, best for the patient.

3. What is the worst quality number in the diagram and under which conditions does it occur?

4. Why do you think there are two courses of action that merit a 5 for their quality number? What actions could be taken to get a 5?

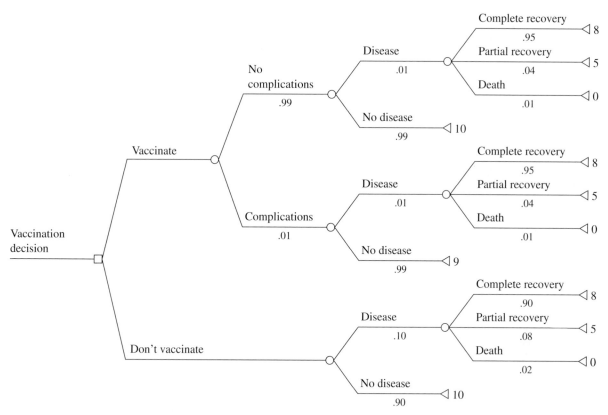

This tree models a decision of whether to vaccinate for a specific disease and is to be used in the Collaborative Learning exercises.

Chapter **10** Summary

Section	Item	Meaning	Example
10.1B	SCP	Sequential counting principle: If one event can occur in m ways, a second event can then occur in n ways, a third event can occur in r ways, and so on, then the sequence of events can occur in $m \times n \times r \times \cdots$ ways.	If there are 3 roads to go to the beach and 2 dates are available, then you have $3 \times 2 = 6$ different choices.
10.2A	Permutation	An ordered arrangement of n distinguishable objects, taken r at a time with no repetitions	$P(3, 2) = 3 \times 2 = 6$
10.2A	$n!$ (n factorial)	$n(n - 1)(n - 2) \cdots (3)(2)(1)$	$3! = 3 \cdot 2 \cdot 1 = 6$
10.2A	$P(n, r)$	$n(n - 1)(n - 2) \cdots (n - r + 1)$ or $\dfrac{n!}{(n - r)!}$	$P(6, 2) = \dfrac{6!}{(6 - 2)!}$ or $6 \times 5 = 30$
10.2A	$0!$	1	

10.2B	Complementary counting principle	$n(A) = n(\mathcal{U}) - n(A')$	If $\mathcal{U} = \{a, b, c, d\}$ and $A = \{a, c, d\}$, then $A' = \{b\}$ and $n(A) = n(\mathcal{U}) - n(A')$; or $3 = 4 - 1$.
10.2C	Additive counting principle	$n(A \cup B) = n(A) + n(B) - n(A \cap B)$	If $A = \{a, b, c\}$ and $B = \{b, c, d\}$, then $A \cup B = \{a, b, c, d\}$, $A \cap B = \{b, c\}$, and $n(A \cup B) = n(A) + n(B) - n(A \cap B)$; or $4 = 3 + 3 - 2$.
10.3	Combination	A selection of r objects without regard to order, taken from a set of n distinguishable objects	The combination of 2 letters that can be made from the letters in $\{a, b, c\}$ are ab, ac, and bc.
10.3	$C(n, r)$	$\dfrac{n!}{r!(n-r)!} = \dfrac{P(n, r)}{r!}$	$C(6, 2) = \dfrac{6!}{2!4!}$
10.4	Permutations of a set of n objects, not all different	$\dfrac{n!}{n_2! n_2! \cdots n_r!}$	The number of arrangements of the letters aabbbc is $\dfrac{6!}{2!3!1!}$.

Research Questions

Sources of information for these questions can be found in the Bibliography at the end of the book.

1. Who introduced the symbol $n!$, in what work, and why?

2. Write a report about the journal that Leibniz helped establish, the position he attained with the journal, and his mathematical achievements.

3. Research and write a paper on Leibniz's life, giving particular emphasis to *De Ars Combinatoria,* one of his works.

4. Write a report about Ben Ezra and his contributions to combinatorics.

5. This man rediscovered Euclid's 32nd proposition, invented the first calculating machine at the age of 18, and inspired by a persistent toothache, became a successful researcher of the properties of the cycloid (a geometric curve). Find out who this mathematician was and write a paper detailing the events mentioned.

Chapter 10 Practice Test

1. A student wants to take two courses, A and B, both of which are available at 9, 10, and 11 A.M. Make a tree diagram to show all the possible schedules for that student. Use a notation like (A, 9) to mean course A at 9 A.M., (B, 11) to mean course B at 11 A.M., and so on.

2. A restaurant offers a choice of 2 soups, 3 entrees, and 5 desserts. How many different meals consisting of a soup, an entree, and a dessert are possible?

3. **a.** Two dice are thrown. How many different outcomes are possible?
   b. In how many ways could you get a sum of 5?

4. Two cards are drawn in succession and without replacement from a standard deck of 52 cards. In how many ways could these be a black jack and a red card, in that order?

5. An airline has 3 flights from city A to city B and 5 flights from city B to city C. In how many ways could you fly from city A to city C using this airline?

6. Compute the following: **a.** 7!   **b.** $\dfrac{7!}{4!}$

7. Compute the following: **a.** $3! \times 4!$ **b.** $3! + 4!$

8. Find the following: **a.** $P(5, 5)$ **b.** $P(6, 6)$

9. Find the following: **a.** $P(8, 2)$ **b.** $P(7, 3)$

10. In how many ways can 4 people be arranged in a row for a group picture?

11. Three married couples are posing for a group picture. They are to be seated in a row of 6 chairs, with each husband and wife together. In how many ways can this be done? (*Hint:* First count the number of ways in which the couples can be arranged.)

12. Bobby has 6 pigeons: 2 white, 2 gray, and 2 gray and white. In how many ways can Bobby select 3 of his pigeons and include exactly 1 white bird?

13. How many counting numbers less than 50 are divisible by 2 or 5?

14. How many different sums of money can be made from a set of coins consisting of a penny, a nickel, a dime, and a quarter if exactly 2 coins are selected?

15. How many subsets of 3 elements does the set $\{a, b, c, d, e, f\}$ have?

16. Find the following: **a.** $C(5, 2)$ **b.** $C(6, 4)$

17. Find the following: **a.** $C(6, 0)$ **b.** $\dfrac{C(5, 4)}{C(5, 3)}$

18. Two cards are drawn in succession and without replacement from a standard deck of 52 cards. How many different sets of 2 cards are possible?

19. A student wants to schedule mathematics, English, science, and economics. These 4 classes are available every hour between 8 A.M. and noon (4 hours). How many different schedules are possible?

20. Billy has 5 coins: a penny, a nickel, a dime, a quarter, and a half-dollar. How many different sums of money can Billy form by using 1, 2, or 3 of these coins?

21. The A-1 Company needs 3 skilled employees, 1 to be in charge and 2 to help. If the company has 5 competent applicants, in how many ways can the employees be selected?

22. On a certain night, there are 8 half-hour programs scheduled on a television station. How many choices do you have if you want to watch 4 of these programs?

23. How many distinct arrangements can you make with the letters in the word *boogaboo?*

24. How many distinct arrangements can be made with the letters in the palindrome "Madam I'm Adam"? (Disregard the apostrophe and capitalization.)

25. On a particular day, a total of 1977 stocks were traded on the New York Stock Exchange for a volume of 137,350,000 shares. Of the 1977 stocks traded, 1189 advanced (a), 460 declined (d), and 328 were unchanged (n). Suppose at the end of the day, you marked a, d, or n after each stock traded. How many distinct arrangements of all the a's, d's, and n's are possible? (Do not try to simplify your answer.)

Answers to Practice Test

ANSWER	IF YOU MISSED	REVIEW		
	Question	Section	Example(s)	Page(s)
1. (tree diagram: (A, 9) → (B, 10), (B, 11); (A, 10) → (B, 9), (B, 11); (A, 11) → (B, 9), (B, 10))	1	10.1	1	658
2. 30	2	10.1	2	659
3. a. 36 **b.** 4	3	10.1	3, 4	660
4. 52	4	10.1	5–9	661–664
5. 15	5	10.1	10	664–665
6. a. 5040 **b.** 210	6	10.2	2	671
7. a. 144 **b.** 30	7	10.2	2	671
8. a. 120 **b.** 720	8	10.2	3, 4	671–672, 673
9. a. 56 **b.** 210	9	10.2	4	673
10. 24	10	10.2	5	674
11. 48	11	10.2	5	674
12. 72	12	10.2	6	674
13. 29	13	10.2	7	675
14. 6	14	10.3	1	681
15. 20	15	10.3	2, 3, 5	681, 682, 683
16. a. 10 **b.** 15	16	10.3	6	684
17. a. 1 **b.** $\frac{1}{2}$	17	10.3	6	684
18. 1326	18	10.3	7	684
19. 24	19	10.4	1	690
20. 25	20	10.4	1	690
21. 30	21	10.4	1	690
22. 70	22	10.4	2–4	690–691
23. 840	23	10.4	5, 6	692–693
24. 34,650	24	10.4	5, 6	692–693
25. $\frac{1977!}{1189!460!328!}$	25	10.4	5, 6	692–693

Advertising new products is a key marketing strategy that increases the probability of sales. In Section 11.4, you will study conditional probability, which can be used to determine the conditions under which sales may be made.

Probability

In this chapter we will examine probability, the science of determining the likelihood or chance that an event will occur.

The study of probability dates back to the Assyrians and Sumerians who had games similar to dice. The Egyptians even had dicelike objects called *tali* made from the heel bones of animals, polished and engraved so that when thrown, they could land on any of four different sides with different probabilities because the *talis* were not uniformly shaped.

In Section 11.1 we shall study how to assign probabilities to events. We will use the ideas of Chapter 10 (tree diagrams, the sequential counting principle, permutations, and combinations) to handle more complex problems and consider some of the theories and formulas that facilitate the computations in probability. Because probability is heavily dependent on arithmetic, the absence of these formulas and methods long hampered the development of the theory of probability. It was not until the contributions of Blaise Pascal, Pierre Fermat, and Pierre-Simon Laplace that the theory was fully developed.

Sections 11.4 and 11.5 are devoted to the study of conditional probability and independent events, ideas that are used in medicine, law, and insurance.

We end the chapter by studying odds and mathematical expectation. If a state has a lottery, we can verify the odds of the game and the probability of obtaining different prizes. But our applications of mathematical expectation are not limited to gambling. We also use the idea as a management tool to determine which way to proceed when confronted with different business decisions involving outcomes with given probabilities and to consider the results that can be expected when these outcomes occur.

Online Study Center

For links to Internet sites related to Chapter 11, please access college.hmco.com/PIC/bello9e and click on the Online Study Center icon.

11.1   Sample Spaces and Probability

HUMAN SIDE OF MATH

Blaise Pascal
was born in
the French
province of
Auvergne in
June of 1623.
His most
original

(1623–1662)

contribution was to the
theory of probability, but
this notoriety was shared
by the famous French
mathematician Pierre
Fermat.

By the age of 17, Pascal
had written an amazing
essay on conic sections
(parabolas, circles, ellipses,
and hyperbolas), including
new and deep theorems
on the properties of these

(continued)

The odds of appearing on the
Tonight Show with Jay Leno are
better than the odds of winning
the lottery in your state (if there
is one).

Probability and the *Tonight Show with Jay Leno*

Which is more probable, appearing on the *Tonight Show with Jay Leno* or winning the jackpot in your state lottery with a single ticket? (If there is no lottery in your state, the probability of winning it is, of course, 0.) According to a book called *What Are the Chances,* the probability of appearing on the *Tonight Show with Jay Leno* is 1 in 490,000, that is, $\frac{1}{490,000}$. What is the probability of winning the jackpot in your state lottery? In most lotteries (Georgia, Kentucky, New Jersey, and Virginia) you buy a $1 ticket and pick 6 numbers from 1 to 49. (In Florida you pick numbers from 1 to 53.) If you match all 6 numbers, you win the jackpot. How probable is that? Since there is only 1 set of winning numbers and $C(49, 6)$ possible number combinations, your probability is 1 in $C(49, 6)$, that is, 1 in 13,983,816. It is much more probable that you will appear on the *Tonight Show with Jay Leno!* If you are playing the Florida lottery, the probability that you hold a winning ticket is even smaller because there are more possible number combinations, namely, $C(53, 6)$ instead of $C(49, 6)$. This makes the probability of winning the Florida jackpot 1 in 22,957,480.

Probability theory was developed by mathematicians studying gambling games. In 1654, Antoine Gombaud, better known as the Chevalier de Méré, offered even money that in 4 rolls of a die, at least one 6 would be rolled. He reasoned that since the chance that a 6 will be rolled when a die is tossed is $\frac{1}{6}$, in 4 rolls, the chances of getting at least one 6 should equal $\frac{4}{6} = \frac{2}{3}$. Do you think that he was right? You will be able to give the answer after you read this section! ▶

The theory of probability is an important tool in the solution of many problems in the modern world. Although the most obvious applications are in gambling games, important applications occur in many situations involving an element of uncertainty. Probability theory is used to estimate whether a missile will hit its target, to determine premiums on insurance policies, and to make important business decisions such as where to locate a supermarket or how many checkout clerks to employ so that customers will not be kept waiting in line too long. Various sampling techniques, which are used in opinion polls and in the quality control of mass-produced items, are based on the theory of probability.

We want the **probability** of a given event to be a mathematical estimate of the likelihood that this event will occur. The following examples show how a probability can be assigned to a given event.

A. Probabilities

EXAMPLE 1 ▶ Finding the Probability of Heads

A fair coin is tossed; find the probability of getting heads.

Solution

At this time, we are unable to solve this problem because we have not even defined the term *probability.* However, our intuition tells us the following:

1. When a fair coin is tossed, it can turn up in either of 2 ways. Assuming that the coin will not stand on edge, heads and tails are the only 2 possible **outcomes.**

2. If the coin is balanced (and this is what we mean by saying "the coin is fair"), the 2 outcomes are considered **equally likely.**

curves. At the age of 18, Pascal had invented the world's first calculating machine and had begun to work in physics and mechanics. But he continued his scientific work for only a few years and quit at the age of 27 to devote himself to religious contemplation.

At the age of 31, a problem was proposed to him on the division of the pot in an unfinished gambling game. Pascal wrote to Fermat about the problem, and in the ensuing correspondence these two men shared equally in establishing basic results in the theory of probability.

Looking Ahead

Mathematicians' interest in probability theory grew because of Pascal's and Fermat's writings and prompted the thorough development of the theories contained in this chapter.

3. The probability of obtaining heads when a fair coin is tossed, denoted by $P(H)$, is 1 out of 2. That is, $P(H) = \frac{1}{2}$. ∎

Activities such as tossing a coin (as in Example 1), drawing a card from a deck, or rolling a pair of dice are called **experiments.** The set $\mathcal{U}$ of all possible outcomes for an experiment is called the **sample space** for the experiment. These terms are illustrated in Table 11.1 below.

Returning to Example 1, we see that the set of all possible outcomes for the experiment is $\mathcal{U} = \{H, T\}$. But there are only two subsets of $\mathcal{U}$ that can occur, namely, $\{H\}$ and $\{T\}$, and each of these is called an **event.** If we get heads—that is, if the event $E = \{H\}$ occurs—we say that we have a **favorable outcome** or a **success.** Since there are 2 equally likely events in $\mathcal{U}$ and 1 of these is E, we assign the value $\frac{1}{2}$ to the event E. But what if the coin is *not* fair?

Here is a quote from an article appearing in the English newspaper *The Guardian* on Friday, January 4, 2002: "When spun on edge 250 times, a Belgian one-euro coin came up heads 140 times and tails 110. 'It looks very suspicious to

TABLE 11.1 Experiments and Sample Spaces

Experiment	Possible Outcomes	Sample Space $\mathcal{U}$
A penny is tossed.	Heads or tails are equally likely outcomes.	$\{H, T\}$
There are 3 beige and 3 red balls in a box; 1 ball is drawn at random.	A beige or a red ball is equally likely to be drawn.	$\{b_1, b_2, b_3, r_1, r_2, r_3\}$
A penny and a nickel are tossed.	*Penny* / *Nickel*: H H, H T, T H, T T	$\{(H, H), (H, T), (T, H), (T, T)\}$
One die is rolled.	The numbers from 1 to 6 are all equally likely outcomes.	$\{1, 2, 3, 4, 5, 6\}$
The pointer is spun, as shown in Figure 11.1.	The pointer is equally likely to point to 1, 2, 3, or 4.	$\{1, 2, 3, 4\}$
An integer between 1 and 50 (inclusive) is selected at random.	The integers from 1 to 50 are all equally likely to be selected.	$\{1, 2, 3, \ldots, 50\}$

FIGURE 11.1

me,' said Barry Blight, a statistics lecturer at the London School of Economics." Does it look suspicious to you?

On the basis of that experiment, what would the probability of heads be for *that* coin? This time, the number of favorable outcomes for $E = \{H\}$ is 140 and the total number of outcomes is 250. Thus, $P(H) = \frac{140}{250} = \frac{14}{25}$, not $\frac{1}{2}$! Can you consider this coin fair?

We now expand on the problem discussed in Example 1. Suppose that a fair coin is tossed 3 times. Can we find the probability that 3 heads come up? As before, we proceed in three steps as follows:

1. The set of all possible outcomes for this experiment can be found by drawing a tree diagram, as shown in Figure 11.2. As we can see, the possibilities for the first toss are labeled H and T, and likewise for the other two tosses. The number of outcomes is 8.

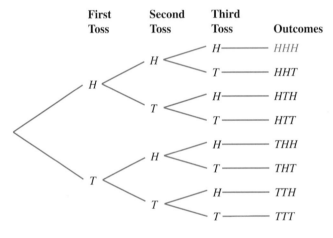

FIGURE 11.2

Tree diagram.

2. The 8 outcomes are equally likely.

3. We conclude that the probability of getting 3 heads, denoted by $P(HHH)$, is 1 out of 8; that is, $P(HHH) = \frac{1}{8}$.*

If we want to know the probability of getting *at least* 2 heads, the 4 outcomes *HHH*, *HHT*, *HTH*, and *THH* are favorable out of the 8 outcomes shown in Figure 11.2, so the probability of getting at least 2 heads is $\frac{4}{8} = \frac{1}{2}$.

In examples such as these, in which all the possible outcomes are equally likely, the task of finding the probability of any event E can be simplified by using the following definition:

Definition of the Probability of an Event

Suppose an experiment has $n(\mathcal{U})$ possible outcomes, **all equally likely.** Suppose further that the event E occurs in $n(E)$ of these outcomes. Then, the **probability** of event E is given by

$$P(E) = \frac{\text{number of ways } E \text{ can occur}}{\text{number of possible outcomes}} = \frac{n(E)}{n(\mathcal{U})} \tag{1}$$

---

*Technically, we should write $P(\{HHH\})$ instead of $P(HHH)$. However, we shall write $P(HHH)$ whenever the meaning is clear.

In the previous example, $n(\mathcal{U}) = 8$ and $n(E) = 4$, so

$$P(E) = \frac{n(E)}{n(\mathcal{U})} = \frac{4}{8} = \frac{1}{2}$$

We illustrate the use of equation (1) in the following examples. Note that $0 \le P(E) \le 1$ because $0 \le n(E) \le n(\mathcal{U})$.

PROBLEM SOLVING

Computing Probability

❶ **Read** the problem.

There are 75 possible numbers in Bingo. Find the probability that the first number selected is the following:

(a) 25　　(b) 80　　(c) An odd number　　(d) A number less than 80

❷ **Select** the unknown.

We want to find the probability of four events: (a), (b), (c), and (d).

❸ **Think** of a plan.
We have to find out how many outcomes are possible when the first number is selected and in how many ways each of the four given events (a), (b), (c), and (d) can occur.

If we select a number from a group of 75 numbers, there are 75 equally likely outcomes; that is, the universal set is

$$\mathcal{U} = \{1, 2, 3, \ldots, 75\} \quad \text{and} \quad n(\mathcal{U}) = 75$$

　Let T be the event that number 25 is selected.
　Let E be the event that number 80 is selected.
　Let O be the event that an odd number is selected.
　Let L be the event that a number less than 80 is selected.

❹ **Use** equation (1) on page 704 to find the probability of each event. In how many ways can we do the following?

(a) Select number 25.

(a) There is only one way of selecting number 25; thus,

$$P(T) = \frac{n(T)}{n(\mathcal{U})} = \frac{1}{75}$$

(b) Select number 80.

(b) There is no way number 80 can be selected; thus, $P(E) = \frac{0}{75} = 0$. This event is *impossible*.

(c) Select an odd number.

(c) There are 38 odd numbers $\{1, 3, 5, 2 \cdot 38 - 1 = 75\}$ that are 75 or less; thus, $P(O) = \frac{38}{75}$.

(d) Select a number that is less than 80.

(d) All 75 numbers in Bingo are less than 80; thus, $P(L) = \frac{75}{75} = 1$. This event is *certain* (a "sure thing"!).

❺ **Verify** the solution.

Are all the probabilities we have computed between 0 and 1, inclusive?

TRY EXAMPLE 2 NOW.

Cover the solution, write your own solution, and then check your work.

EXAMPLE 2 ▶ Probabilities with 1 Die

A single die is rolled. Find the probabilities of obtaining the following:

(a) A number greater than 4   (b) An odd number

Solution

(a) Let E be the event in which a number greater than 4 appears. When a die is rolled, there are 6 equally likely outcomes, so $n(\mathcal{U}) = 6$. There are two outcomes (5 and 6) in E; that is, $n(E) = 2$. Hence,

$$P(E) = \frac{n(E)}{n(\mathcal{U})} = \frac{2}{6} = \frac{1}{3}$$

(b) Let O be the event in which an odd number appears. Three outcomes (1, 3, 5) are in O. Thus, $P(O) = \frac{3}{6} = \frac{1}{2}$. ∎

EXAMPLE 3 ▶ Urn Probabilities

Ten balls numbered from 1 to 10 are placed in an urn. If 1 ball is selected at random, find the probabilities of the following:

(a) An even-numbered ball is selected (event E).

(b) Ball number 3 is chosen (event T).

(c) Ball number 3 is *not* chosen (event T').

Solution

(a) There are 5 outcomes (2, 4, 6, 8, 10) in E out of 10 equally likely outcomes. Hence, $P(E) = \frac{5}{10} = \frac{1}{2}$.

(b) There is only 1 outcome (3) in the event T out of 10 equally likely outcomes. Thus, $P(T) = \frac{1}{10}$.

(c) There are 9 outcomes (all except the 3) in T' out of the 10 possible outcomes. Hence, $P(T') = \frac{9}{10}$. ∎

In Example 3 we found $P(T) = \frac{1}{10}$ and $P(T') = \frac{9}{10}$, so $P(T') = 1 - P(T)$. This is a general result because $T \cup T' = \mathcal{U}$ and $T \cap T' = \varnothing$. Thus,

$$n(T \cup T') = n(T) + n(T') = n(\mathcal{U})$$

Therefore,

$$\frac{n(T)}{n(\mathcal{U})} + \frac{n(T')}{n(\mathcal{U})} = \frac{n(\mathcal{U})}{n(\mathcal{U})}$$

or, by the definition of the probability of an event,

$$P(T) + P(T') = 1$$

Thus, the probability $P(T')$ of an event not occurring is $1 - P(T)$.

Probability of an Event Not Occurring

The probability $P(T')$ of an event not occurring is

$$P(T') = 1 - P(T)$$

The next example illustrates the use of this idea.

EXAMPLE 4 ▶ Probability of at Least 1 Head

A coin is thrown 3 times. Find the probability of obtaining at least 1 head.

Solution

Let E be the event that we obtain at least 1 head. Then E' is the event that we obtain 0 heads; that is, that we obtain 3 tails. From the preceding discussion, $P(E) = 1 - P(E')$. Here, $P(E')$ is the same as $P(TTT) = \frac{1}{8}$; hence, $P(E) = 1 - P(TTT) = 1 - \frac{1}{8} = \frac{7}{8}$. ∎

EXAMPLE 5 ▶ Heredity and Probability

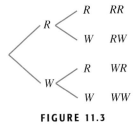

FIGURE 11.3

The science of heredity uses the theory of probability to determine the likelihood of obtaining flowers of a specified color when crossbreeding. Suppose we represent with letters the genes that determine the color of an offspring flower. For example, a white offspring has genes *WW*, a red offspring has genes *RR*, and a pink offspring has genes *RW* or *WR*. When we crossbreed 2 pink flowers, each plant contributes one of its color genes to each of its offspring. The tree diagram in Figure 11.3 shows the 4 possibilities. Assuming that these possibilities are all equally likely, what are the probabilities of obtaining the following?

(a) A white flower (b) A pink flower (c) A red flower

Solution

(a) We see from the tree diagram that the probability of obtaining a white flower (*WW*) is $\frac{1}{4}$.

(b) The probability of obtaining a pink flower (*RW* or *WR*) is $\frac{2}{4} = \frac{1}{2}$.

(c) The probability of obtaining a red flower (*RR*) is $\frac{1}{4}$. ∎

EXAMPLE 6 ▶ Probabilities with 2 Dice

Two dice are rolled. Find the following:

(a) The sample space for this experiment

(b) The probability that the sum of the two numbers facing up is 12

(c) The probability that the sum of the two numbers facing up is 7

Solution

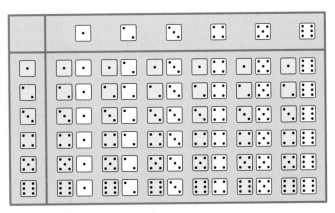

FIGURE 11.4

(a) Figure 11.4 shows the 36 possible outcomes. The sample space for this experiment follows on page 708.

$$\left\{\begin{array}{cccccc} (1,1) & (1,2) & (1,3) & (1,4) & (1,5) & (1,6) \\ (2,1) & (2,2) & (2,3) & (2,4) & (2,5) & (2,6) \\ (3,1) & (3,2) & (3,3) & (3,4) & (3,5) & (3,6) \\ (4,1) & (4,2) & (4,3) & (4,4) & (4,5) & (4,6) \\ (5,1) & (5,2) & (5,3) & (5,4) & (5,5) & (5,6) \\ (6,1) & (6,2) & (6,3) & (6,4) & (6,5) & (6,6) \end{array}\right\}$$

Note: Since we want the sum of the two numbers, we can think of the sample space as

$$\left\{\begin{array}{cccccc} 2 & 3 & 4 & 5 & 6 & 7 \\ 3 & 4 & 5 & 6 & 7 & 8 \\ \vdots & \vdots & \vdots & \vdots & \vdots & \vdots \\ 7 & 8 & 9 & 10 & 11 & 12 \end{array}\right\}$$

(b) The probability that the sum of the two numbers facing up is 12 is $\frac{1}{36}$ because there is only 1 favorable case, (6, 6), and there are 36 possible outcomes, all equally likely.

(c) There are 6 favorable cases to obtain a sum of 7 [(6, 1), (5, 2), (4, 3), (3, 4), (2, 5), and (1, 6)] out of 36 possible. Thus, the probability is $\frac{6}{36} = \frac{1}{6}$. ■

EXAMPLE 7 ▶ **Finding the Probability of 1 King**

Find the probability of getting a king when drawing 1 card at random from a standard deck of 52 playing cards. (See the photo in Section 10.1, Example 9.)

Solution

A standard deck of playing cards consists of 4 suits (clubs, diamonds, hearts, and spades) of 13 cards each. The clubs and spades are printed in black; the diamonds and hearts are printed in red. Each suit contains 9 cards numbered from 2 to 10, an ace, and 3 face (picture) cards: a jack, a queen, and a king. Since 4 of the 52 cards are kings, the probability of drawing a king is $\frac{4}{52} = \frac{1}{13}$. ■

B. Empirical Probability

Because the probabilities in the preceding examples are based on the **theory** that the outcomes are equally likely (a balanced coin, fair dice), they are called **theoretical probabilities.** What about the Belgian euro coin (see page 704)? After performing an **experiment** in which the coin was tossed 250 times and the number of heads observed was 140, we concluded that the probability of heads for this coin was

$$P(H) = \frac{140 \ \blacktriangleleft \quad \text{number of times heads occurred}}{250 \ \blacktriangleleft \quad \text{number of times the experiment is performed}}$$

$P(H)$ is the called the **empirical (expected** or **experimental) probability** of heads. In general,

---

Empirical Probability of an Event E

The empirical probability of E is

$$P(E) = \frac{\text{number of times } E \text{ has occurred}}{\text{total number of times the experiment is performed}}$$

EXAMPLE 8 ▶ **Empirical Probability**

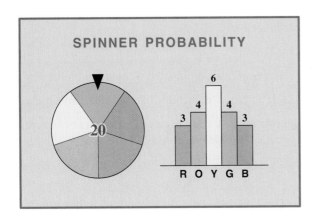

A spinner has 5 equal-sized sectors colored red, orange, yellow, green, and blue. If in 20 spins we get 3 red, 4 orange, 6 yellow, 4 green, and 3 blue outcomes, what is the empirical probability of

(a) getting orange (O) in 20 spins?

(b) getting yellow (Y) in 20 spins?

(c) getting green (G) in 20 spins?

(d) getting blue (B) in 20 spins?

Solution

(a) $P(O) = \dfrac{4}{20} = \dfrac{1}{5} = \dfrac{\text{orange outcomes}}{\text{total number of spins}}$

(b) $P(Y) = \dfrac{6}{20} = \dfrac{3}{10} = \dfrac{\text{yellow outcomes}}{\text{total number of spins}}$

(c) $P(G) = \dfrac{4}{20} = \dfrac{1}{5} = \dfrac{\text{green outcomes}}{\text{total number of spins}}$

(d) $P(B) = \dfrac{3}{20} = \dfrac{\text{blue outcomes}}{\text{total number of spins}}$ ■

EXAMPLE 9 ▶ **Applications of Empirical Probability to Credit Card Choices**

An online survey of 324 people conducted by Insight Express asked the question, "What is your primary credit card?" The results are shown in the bar graph. (Source: http://tinyurl.com/9cvvg.)

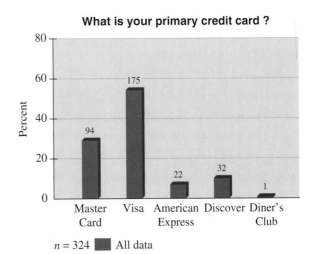

If a person is selected at random from the 324 surveyed, what is the empirical probability that

(a) the person's primary card is MasterCard?

(b) the person's primary card is Visa?

(c) Which event has the highest empirical probability? What is that probability?

(d) Which event has the lowest empirical probability? What is that probability?

(e) If you were the manager of a retail store and you can only accept two types of credit cards, which two cards would you accept?

Solution

(a) According to the graph, 94 out of 324 people use MasterCard as their primary card; thus, the empirical probability of selecting a person whose primary card is a MasterCard is

$$P(MC) = \frac{94}{324} = \frac{47}{162}$$

(b) Similarly, 175 people use Visa as their primary card; thus,

$$P(V) = \frac{175}{324}$$

(c) The event with the highest empirical probability (longest bar in the graph) corresponds to the selection of a person who uses Visa as his or her primary card. The probability is

$$P(V) = \frac{175}{324}$$

(d) The event with the lowest empirical probability (shortest bar in the graph) corresponds to the selection of a person who that uses Diner's Club as his or her primary card. The probability of that is

$$P(DC) = \frac{1}{324}$$

(e) Visa and MasterCard (those are the cards that most people in the survey use as their primary card)

The formulas to find the theoretical or empirical probability of an event are very similar. Is there a relationship between the numerical results obtained when using the formulas? Suppose somebody claims to have a "fair" coin. Can we determine if the coin is indeed fair? (Remember the Belgian euro coin!) We can do it by performing an experiment in which we toss the coin 1, 10, 100, 1000, and 10,000 times. But who has the time to do this? We used a "coin-toss simulator" (see Figure 11.5). You can get similar results using the one at http://argyll.epsb.ca/jreed/math7/strand4/4201.htm#13.

NUMBER OF TOSSES	NUMBER OF HEADS	PROBABILITY	
		EXPERIMENTAL $P(H)$	THEORETICAL $P(H)$
1	1	1	$\frac{1}{2} = 0.5$
10	7	$\frac{7}{10} = 0.7$	$\frac{1}{2} = 0.5$
100	54	$\frac{54}{100} = 0.54$	$\frac{1}{2} = 0.5$
1000	493	$\frac{493}{1000} = 0.493$	$\frac{1}{2} = 0.5$
10,000	5022	$\frac{5022}{10000} = 0.5022$	$\frac{1}{2} = 0.5$
100,000	49,875	$\frac{49875}{100000} = 0.49875$	$\frac{1}{2} = 0.5$

FIGURE 11.5

Online Study Center

To further explore probability (theoretical and experimental) and the Law of Large Numbers; access links 11.1.1, 11.1.2, and 11.1.3 on this textbook's Online Study Center.

What do you notice about the decimal value of the **empirical** probabilities (0.7, 0.54, 0.493, 0.5022, and 0.49875)? As the number of tosses gets bigger, the value gets near 0.5, the theoretical probability. This is a good indication (though not a proof) that the coin is "fair." More important it illustrates the fact that as an experiment is repeated a "large" number of times, the empirical probability of an event tends to get closer to the theoretical probability of the event. This principle is appropriately known as the **Law of Large Numbers.**

EXERCISES 11.1

A Probabilities

On a single toss of a die, what is the probability of obtaining the following?

1. The number 5

2. An even number

3. A number greater than 4

4. A number less than 5

A single ball is taken at random from an urn containing 10 balls numbered 1 through 10. What is the probability of obtaining the following?

5. Ball number 8

6. An even-numbered ball

7. A ball different from 5

8. A ball whose number is less than 10

9. A ball numbered 12

10. A ball that is either less than 5 or odd

In problems 11–16, assume that a single card is drawn from a well-shuffled deck of 52 cards. Find the probability of the following:

11. An ace is drawn.

12. The king of spades is drawn.

13. A spade is drawn.

14. One of the face (picture) cards (jack, queen, or king) is drawn.

15. A face (picture) card or a spade is drawn.

[*Hint:* $P(F \text{ or } S) \neq P(F) + P(S)$]

16. A red card or a face (picture) card is drawn.

[*Hint:* $P(R$ or $F) \neq P(R) + P(F)$]

17. An executive has to visit 1 of his 5 plants for an inspection. If these plants are numbered 1, 2, 3, 4, and 5, and if he selects the plant he will visit at random, find the probability that he will
  a. visit plant number 1.
  b. visit an odd-numbered plant.
  c. not visit plant number 4.

18. Four fair coins are tossed.
  a. Draw a tree diagram to show all the possible outcomes.
  b. Find the probability that 2 or more heads come up.
  c. Find the probability that exactly 1 head comes up.

19. A disk is divided into 3 equal parts numbered 1, 2, and 3, respectively (see the figure below). After the disk is spun and comes to a stop (assuming it will not stop on a line), a fixed pointer points to 1 of the 3 numbers. Suppose that the disk is spun once. Find the probabilities that the disk stops on
  a. the number 3.
  b. an even number.

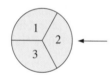

Many companies send packages from one location to another using delivery services such as Airborne and Federal Express. Assume that a person can select from 8 such delivery services A, B, C, D, E, F, G, and H. Assuming one of the services is to be selected at random to mail a package, do problems 20–23.

20. Find the probability that company A is selected.

21. Find the probability that company G is selected.

22. Find the probability that company C is not used to mail the package.

23. Find the probability that 1 of the first 3 companies is selected.

24. The genetic code of an organism is the self-reproducing record of the protein pattern in that organism. This code is formed by groups of small molecules that can be of 4 kinds: adenine (A), cytosine (C), guanine (G), and thymine (T).

  a. Draw a tree diagram to find all possible groups of 2 molecules. *Note:* It is possible for both molecules to be of the same kind.
  b. Assume that all the outcomes in part (**a**) are equally likely. Find the probability of obtaining 2 adenine molecules in a row.
  c. Find the probability of obtaining a guanine molecule and a cytosine molecule, in that order.
  d. Find the probability of obtaining two cytosine molecules in a row.

In problems 25–30, find the probability on a single toss of a pair of dice of obtaining (see Example 6)

25. a sum of 7.

26. a sum of 2.

27. the same number on both dice.

28. different numbers on the two dice.

29. an even number for the sum.

30. an odd number for the sum.

B **Empirical Probability**

In problems 31–44, write the answers as a fraction in reduced form.

31. A spinner has 4 equal sectors colored yellow, blue, green, and red. If in 100 spins we get 25 blue, 28 red, 24 green, and 23 yellow outcomes, find the empirical probability of

  a. getting blue $P(B)$.
  b. getting red $P(R)$.
  c. getting green $P(G)$.
  d. getting yellow $P(Y)$.
  e. What are the theoretical probabilities of $P(B)$, $P(R)$, $P(G)$, and $P(Y)$?
  f. Which outcome has the same empirical and theoretical probability?

32. A spinner has 6 equal sectors colored yellow, blue, green, white, orange, and red. If in 100 spins we get 15 yellow, 18 blue, 20 green, 13 white,

15 black, and 19 red outcomes, find the empirical probability of

a. getting yellow $P(Y)$.
b. getting blue $P(B)$.
c. getting green $P(G)$.
d. getting white $P(W)$.
d. getting orange $P(O)$.
e. getting red $P(R)$.

33. A spinner has 5 unequal sectors: $\frac{1}{6}$ red, $\frac{1}{9}$ blue, $\frac{5}{18}$ green, $\frac{2}{9}$ white and $\frac{2}{9}$ yellow. In 100 spins we get these results.

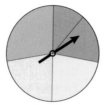

Spins 100	Red 24	Blue 9	Green 29	White 20	Yellow 18

Fill in the blanks with the appropriate reduced fractions.

Outcome	Empirical Probability	Theoretical Probability
a. Red	$P(R) =$	$P(R) =$
b. Blue	$P(B) =$	$P(B) =$
c. Green	$P(G) =$	$P(G) =$
d. White	$P(W) =$	$P(W) =$
e. Yellow	$P(Y) =$	$P(Y) =$

34. When 2 dice are thrown, the 36 possible outcomes are as shown (see Example 6). Now, suppose that 2 dice are thrown 100 times, and the following sums occur the indicated number of times. (A sum of 2 occurs 3 times, a sum of 3 occurs 4 times, and so on.)

Sum 2	Sum 3	Sum 4	Sum 5	Sum 6	Sum 7	Sum 8	Sum 9	Sum 10	Sum 11	Sum 12
3	4	9	12	14	17	15	13	5	6	2

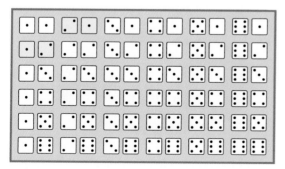

Note that there are 2 ways of getting a sum of 3, (1, 2) and (2, 1), as shown.

Fill in the blanks with the appropriate reduced fractions.

Outcome	Empirical Probability	Theoretical Probability
a. The sum is 2.	$P(\text{sum } 2) =$	$P(\text{sum } 2) =$
b. The sum is 5.	$P(\text{sum } 5) =$	$P(\text{sum } 5) =$
c. The sum is 7.	$P(\text{sum } 7) =$	$P(\text{sum } 7) =$
d. The sum is 9.	$P(\text{sum } 9) =$	$P(\text{sum } 9) =$
e. The sum is 12.	$P(\text{sum } 12) =$	$P(\text{sum } 12) =$
f. The sum is greater than 10.	$P(\text{sum } G\ 10) =$	$P(\text{sum } G\ 10) =$
g. The sum is less than 4.	$P(\text{sum } L\ 4) =$	$P(\text{sum } L\ 4) =$
h. The sum is greater than 2.	$P(\text{sum } G\ 2) =$	$P(\text{sum } G\ 2) =$
i. The sum is less than 12.	$P(L\ 12) =$	$P(L\ 12) =$

35. Suppose 2 dice are rolled 50 times and that the results are as shown in the bar graph. Find the empirical probability of

a. $P(11)$, the sum is 11.
b. $P(7)$, the sum is 7.
c. $P(O)$, the sum is odd.
d. $P(\sim O)$, the sum is not odd.
e. Which outcome has the highest empirical probability?
f. Which outcome has the lowest empirical probability?
g. Which three outcomes have the same empirical probability?
h. Do these three outcomes have the same empirical and theoretical probability?

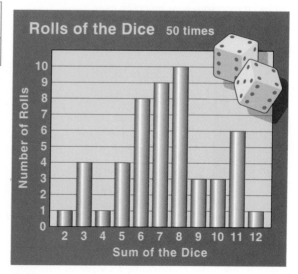

Rolls of the Dice 50 times

36. This time 2 dice are rolled 36 times, and the results are shown in the bar graph. What is the empirical probability of obtaining
  a. a sum of 2 or 12?
  b. a sum that is odd and greater than 8?
  c. a sum that is even and less than 5?
  d. a sum less than 12?
  e. a sum of 1?
  f. Which of the sums have the same empirical and theoretical probability? (See the sample space for the theoretical probability in Example 6.)

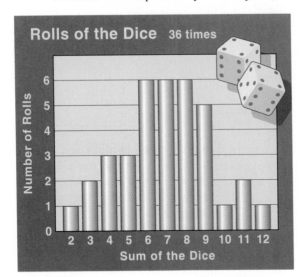

37. An online survey of 502 persons conducted by Insight Express asked the question, "How many credit cards do you own (credit cards only, no debit cards)?" The results are shown in the bar graph. If a person is selected at random from the 502 surveyed, what is the empirical probability that
  a. the person owns no credit cards?
  b. the person owns exactly one credit cards?

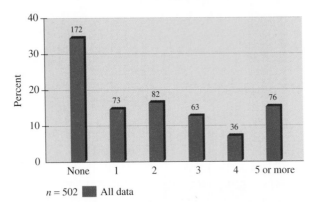

$n = 502$ ▇ All data

Source: http://tinyurl.com/cu2ju.

c. What is the outcome with the lowest empirical probability?
  d. What is the outcome with the highest empirical probability?
  e. What is the empirical probability that the person owns at least one credit card?

38. In a recent study conducted by Impulse Research for Oreida, half of 1510 moms with kids at home say that "When they serve French fries, they also cook hamburgers." Find the empirical probability that a mom selected at random from the 1510 French fry–serving moms surveyed also will cook
  a. hog dogs.
  b. sandwiches.
  c. chicken.
  d. fish sticks or sandwiches.

What goes with French fries?

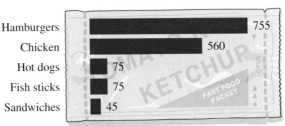

Source: www.theallineed.com/family/05062502.htm.

39. In a survey conducted by Insight Express, 500 Americans were asked if they have been a victim of identity theft. The results were as shown in the graph. If a person is chosen at random from the group of 500 Americans, what is the empirical probability that
  a. the person was a victim of identity theft?
  b. the person was not a victim of identity theft?
  c. the person was not sure if he or she was a victim of identity theft?

Source: Insight Express,
www.insightexpress.com/presentation/
IdentityTheft/frame.htm.

40. In a statistical survey among adults 18–49 years old, the percent of respondents unwilling to give up certain activities is as shown. If one of the respondents is selected at random, what is the empirical probability that the person would not give up

a. exercise?

b. drinking coffee?

c. Which activity are the respondents least likely to give up?

Activities We Won't Give Up

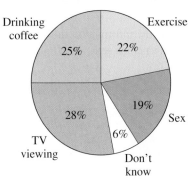

Source: Opinion Research for Dunkin Donuts.

41. Which activity is your favorite thing to do on a flight? A survey of 239 respondents planning to

Things to do at 35,000 feet

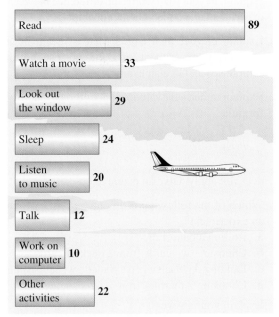

Source: Survey conducted by ORC's CARAVAN Omnibus Services, http://tinyurl.com/9ey32.

fly revealed their favorite in-flight activity was as shown in the graph. If you are flying with these 239 people and select one of them at random, what is the empirical probability that

a. he or she will be reading?

b. he or she will be watching a movie?

c. he or she will not be watching a movie?

d. he or she will be sleeping?

42. A group of 1594 respondents was asked the question, "Should intelligent design be taught in high school biology class." The responses were as shown. What is the empirical probability that if one of the respondents is selected at random, the respondent would answer

a. "Yes."

b. "No."

c. "Not sure."

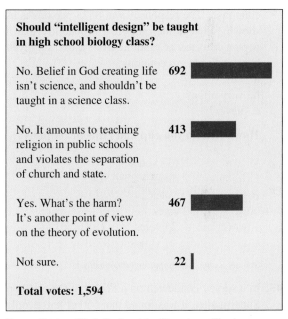

Source: BusinessWeek On-line, http://tinyurl.com/7byvh.

43. Do you know what *regifting* is? It is loosely defined as "giving someone a gift that was originally given to you without telling the recipient the origin of the gift." Now that you know, how often do you engage in regifting? In a survey of 5543 persons, the answers to that question were as shown on page 716. If one of the persons is selected at random, what is the empirical probability that the person

a. very often engages in regifting?

b. never engages in regifting?

c. If you get a gift and you assume that the results of the survey apply to that gift, what is the probability that you are getting a regift? Assume that the people who give regifts are the ones that did not answer "Never."

Response to the question: How often do you engage in regifting?

Very often → 140 votes
Every now and then → 817 votes
Not very often → 1755 votes
Never → 2804 votes
I'm not sure → 27 votes
Current number of voters: 5543

Source: Excite.com, http://tinyurl.com/knjux.

44. Do you clip coupons? What do you do with them? A survey conducted by Excite.com asked the question, "How often do you clip and redeem coupons?" The response from 8523 persons is as shown. What is the empirical probability that a person selected at random from those surveyed clips and redeems coupons
  a. rarely?
  b. very often?
  c. never?

How often do you clip and redeem coupons?

Very often → 2137 votes
Every now and then → 2386 votes
Rarely → 2443 votes
Never → 1532 votes
I'm not sure → 25 votes
Current number of voters: 8523

Source: Excite.com, http://tinyurl.com/a9lwr.

45. In a survey conducted on a Friday at Quick Shop Supermarket, it was found that 650 of 850 people who entered the supermarket bought at least 1 item. Find the probability that a person entering the supermarket on a Friday will purchase
  a. at least 1 item.   **b.** no item.

46. Two common sources of nicotine are cigarettes and cigars. Suppose that 35% of the adults in the United States smoke cigarettes but not cigars and 10% smoke both cigarettes and cigars. Find the probability that a randomly selected adult does not smoke cigarettes. (*Hint:* Draw a Venn diagram.)

Use the following survey table in problems 47–50. The table gives the numbers of males and females falling into various salary classifications.

Salary	Sex		Totals
	M	F	
Low	40	200	240
Average	300	160	460
High	500	300	800
Totals	840	660	1500

On the basis of the information in the table, find the probability that a person selected at random from those surveyed

47. a. is a female.   **b.** is a male.

48. a. has a high income.   **b.** has a low income.

49. is a female with
  a. an average income.
  b. a high income.

50. is a male with
  a. a low income.
  b. a high income.

 In Other Words

51 Explain, in your own words the difference between experimental and theoretical probability.

52. Explain in your own words what is meant by a probability experiment and the corresponding sample space. Give an example.

53. Explain under what circumstances the probability formula

$$P(E) = \frac{n(E)}{n(\mathcal{U})}$$

does *not* apply. Give an example.

54. Which of the following is an outcome and which an experiment?
  a. Rolling a pair of dice
  b. Choosing two cards from a deck of cards
  c. Landing on black when a roulette wheel is spun
  d. Choosing 2 marbles from a jar
Explain your answers.

55. Which of the following experiments do not have equally likely outcomes?
   a. Choose a number at random from 1 to 7.
   b. Toss a coin.
   c. Choose a letter at random from the word *MISSISSIPPI*.
   d. Choose a number at random from 1 to 10.
   Explain your answers.

56. Two dice are rolled. One student claims that there are only 3 ways in which a sum of 7 can occur. Explain what is wrong.

 Using Your Knowledge

Do you have to have surgery soon? The chances are that you will have no trouble at all! The following table gives the statistics for the numbers of certain operations and the numbers of successes in a recent year. Using these statistics, estimate the probability of

57. a gallbladder operation's being successful.

58. an appendectomy's being successful.

59. a hernia operation's being successful.

Type	Number of Operations	Number of Successes
Gallbladder	472,000	465,300
Appendectomy	784,000	781,000
Hernia	508,000	506,000

 Collaborative Learning

Have you read the column "Ask Marilyn" in *Parade* magazine? Sometimes a great deal of controversy is generated by the answers given. Here are a couple of instances you can discuss.

1. You're at a party with 199 other guests when robbers break in and announce they're going to rob one of you. They put 199 blank pieces of paper in a hat, plus one marked "you lose." Each guest must draw a piece, and the person who draws "you lose" gets robbed. The robbers think you're cute, so they offer you the option of drawing first, last, or any time in between. (Source: *Parade*, 3 January 1999, p. 16, and *Chance*.)

   a. Marilyn said she would choose to draw first, explaining, "It would make no difference to my chances of losing—any turn is the same—but at least I'd get to leave this party as soon as possible." What is the probability that you pick the paper marked "you lose" if you are the first person who draws? Is the probability different if you are the 100th person who draws? Explain why or why not.

   b. One letter argues for drawing first: "You said any turn is the same, but I believe that would be true only if the partygoers all had to replace the papers they drew before another selection was made. But if they keep the papers (the scenario intended by the question), wouldn't the odds of losing increase as more blanks were drawn? If so, drawing first is best." If 100 blanks have been drawn, the chance that the next slip says "you lose" is indeed 1 in 100. Why doesn't this mean you should draw early if you have the choice?

   c. The first letter mentions the distinction between sampling with replacement and without replacement, which Marilyn does not directly address. Is the answer really the same in both scenarios? Discuss your answer.

   d. When would you choose to make your draw? Why?

2. Here is another problem that appeared in Marilyn's column. Given that a family has 2 children and at least 1 is a boy, what is the probability that the family has 2 boys?

   a. Marilyn said the answer was $\frac{1}{3}$. Can you prove this?

   b. Readers said that it was easy to give stories for which the answer would be $\frac{1}{2}$. Can you make up one such story?

11.2   Counting Techniques and Probability

Poker Improbabilities

Have you ever played poker? John Scarne's *Complete Guide to Gambling* claims that 95 out of 100 adults have played the game! In his book *Poker Stories* (1896) John F. B. Lillard tells the story of a professional gambler stopping for a beer at a saloon in Butte, Montana. As luck would have it, a poker game was in progress, so the hustler decided to join in and make some money. After playing for a while, the hustler dealt himself four aces.

GRAPH IT

To verify the answer (getting four aces), go to the home screen ([2nd] [MODE]) and press 48 [MATH] [◄] [3] 1 [ENTER] [÷] 52 [MATH] [◄] [3] 5 [ENTER]. The answer is a decimal. How do you know the answer is equivalent to $\frac{1}{54,145}$? Press 1 [÷] 54,145 [ENTER]. You get the same answer, so you are correct!

```
48 nCr 1
                      48
Ans/52 nCr 5
          1.846892603E-5
1/54145
          1.846892603E-5
```

Do you know the probability of getting four aces in a 5-card poker hand? Since there are four aces in a deck, there is only 1 way of getting four aces if the fifth card can be any of the 48 remaining; hence, there are $1 \cdot C(48, 1) = 48$ ways of getting four aces of the $C(52, 5) = 2,598,960$ possible poker hands (see Example 6 for the computation). Thus, the probability of four aces in poker is only

$$\frac{48}{2,598,960} = \frac{1}{54,145}$$

Assured of winning, the hustler made a fair-sized bet that forced every player to drop out except for one old stalwart with gray whiskers and a deadpan poker face. The old cowboy didn't blink; he merely shoved all his chips into the pot and called. The hustler showed his four aces and reached for the pot.

"Not so fast, sonny," said the cowboy, laying down three clubs and two diamonds. Can you find the probability of getting three clubs and two diamonds? It is

$$\frac{C(13, 3) \cdot C(13, 2)}{C(52, 5)} = \frac{22,308}{2,598,960} = \frac{143}{16,660}$$

"What do you mean, not so fast?" the hustler said. "My four aces have a lower probability, and they should win."

"Of course they should—ord'narily," the cowboy said, "But in this town a Lollapalooza beats anythin', and that's what I've got, three clubs and two diamonds, a Lollapalooza."

The hustler knew he had just been out-hustled, but he figured he could still change his "luck." On the next deal, the hustler dealt himself a Lollapalooza and gave four aces to the old cowboy with the gray whiskers. Again, he made a fair-sized bet, and again, the old cowboy stayed while the rest dropped out. The hustler pushed all his chips to the center of the table. The cowboy called again.

"Well," the hustler said grinning, "This time I can't lose. Seems I've got the Lollapalooza!"

But the old cowboy was already bellied-up to the table, raking in the pot. "Sorry, pardner," he said, as the hustler looked on, "Only one Lollapalooza per night!"

In this section you will use the sequential counting principle (SCP), permutations, and combinations to find the probabilities of many events, including different poker hands. ▶

The counting techniques that we studied in Chapter 10 play a key role in many probability problems. We now see how these techniques are used in such problems.

A. Using Tree Diagrams

EXAMPLE 1 ▶ Stock Probabilities Using a Tree

Have you heard of the witches of Wall Street? These are people who use astrology, tarot cards, or other supernatural means to predict whether a given stock will go up, go down, or stay unchanged. Not being witches, we assume that a stock is equally likely to go up (U), go down (D), or stay unchanged (S). A broker selects two stocks at random from the New York Stock Exchange list.

(a) What is the probability that both stocks go up?

(b) What is the probability that both stocks go down?

(c) What is the probability that one stock goes up and one goes down?

Solution

In order to find the total number of equally likely possibilities for selecting the two stocks, we draw the tree diagram shown in Figure 11.6 at the top of page 720.

(a) There is only 1 outcome (UU) out of 9 in which both stocks go up. Thus, the probability that both stocks go up is $\frac{1}{9}$.

(b) There is only 1 outcome (DD) in which both stocks go down, so the probability that both go down is $\frac{1}{9}$.

(c) There are 2 outcomes (UD, DU) in which one stock goes up and one down. Hence, the probability of this event is $\frac{2}{9}$.

(Notice that the tree diagram shows that there are 4 outcomes in which one stock stays unchanged and the other goes either up or down. The probability of this event is thus $\frac{4}{9}$.)

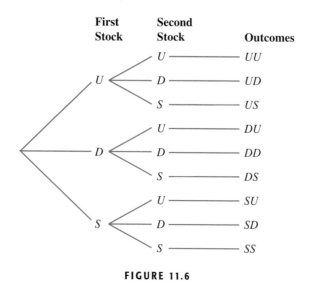

First Stock **Second Stock** **Outcomes**

U — UU
U — D — UD
— S — US

U — DU
D — D — DD
— S — DS

U — SU
S — D — SD
— S — SS

FIGURE 11.6

B. Using Permutations and Combinations

In many games of chance, probability is used to determine payoffs. For example, a slot machine has three dials with 20 symbols on each dial, as listed in Table 11.2. In the next example we shall find the probability of getting certain arrangements of these symbols on the 3 dials.

EXAMPLE 2 ▶ Slot Machine Probabilities

Refer to the slot machine described in Table 11.2 to answer the following questions.

(a) What is the probability of getting 3 bars?

(b) What is the probability of getting 3 bells?

(c) What is the probability of getting 3 oranges?

(d) What is the probability of getting 3 plums?

(e) On the basis of your answers to these questions, which outcome should have the greatest payoff and which the least?

Solution

(a) We draw 3 blanks representing the 3 dials.

———— ———— ————

There are 20 choices for each of the blanks (each dial has 20 symbols), so we enter a 20 on each blank.

$\underline{20}\ \underline{20}\ \underline{20}$

The total number of possibilities is $20 \times 20 \times 20 = 8000$. Now, the number of ways to get 3 bars is $1 \times 3 \times 1 = 3$ because the first dial has 1 bar, the second has 3 bars, and the third has 1 bar. Thus,

$$P(3 \text{ bars}) = \frac{\text{number of favorable cases}}{\text{number of possible outcomes}} = \frac{3}{8000}$$

TABLE 11.2

Symbol	Dial		
	1	*2*	*3*
Bar	1	3	1
Bell	1	3	3
Cherry	7	7	0
Lemon	3	0	4
Orange	3	6	7
Plum	5	1	5

(b) The number of ways of getting 3 bells is $1 \times 3 \times 3 = 9$. Thus,

$$P(3 \text{ bells}) = \frac{9}{8000}$$

(c) The number of ways of getting 3 oranges is $3 \times 6 \times 7 = 126$. Thus,

$$P(3 \text{ oranges}) = \frac{126}{8000} = \frac{63}{4000}$$

(d) The number of ways of getting 3 plums is $5 \times 1 \times 5 = 25$. Thus,

$$P(3 \text{ plums}) = \frac{25}{8000} = \frac{1}{320}$$

(e) Since 3 bars is the outcome with the lowest probability and 3 oranges is the outcome with the highest probability, the greatest payoff should be for 3 bars, and the least for 3 oranges. (This is how payoffs are actually determined.) ∎

EXAMPLE 3 ▶ Probability of Being in a Committee

Suppose you are 1 of a committee of 10 people, of whom 2 are to be chosen for a particular task. If these are selected by drawing names out of a hat, what is the probability that you will be 1 of the 2 selected?

Solution

It is easier to first calculate the probability that you will *not* be selected. Since there are 9 people not including you, there are $P(9, 2)$ ways of selecting 2 *not* including you. Also there are $P(10, 2)$ ways of selecting 2 people from the entire 10. Hence, the probability P' that you will *not* be selected is

$$P' = \frac{P(9, 2)}{P(10, 2)} = \frac{9 \times 8}{10 \times 9} = \frac{4}{5}$$

So the probability that you *will* be selected is

$$P = 1 - P' = 1 - \frac{4}{5} = \frac{1}{5}$$ ∎

The next example deals with a problem involving ordinary playing cards. Note that in solving part (a), you can use combinations, permutations, or the SCP. The important thing is to be **consistent** in the computation.

EXAMPLE 4 ▶ Probabilities When 2 Cards Are Chosen

Two cards are drawn in succession and without replacement from an ordinary deck of 52 cards. Find the probability that

(a) the cards are both aces. (b) an ace and a king, in that order, are drawn.

Solution

(a) Here, the **order is not important** because we are simply interested in getting 2 aces. We can find this probability by using **combinations.** The number of ways to draw 2 aces is $C(4, 2)$ because there are 4 aces and we want a combination of any 2 of them. The number of combinations of 2 cards picked from the deck of 52 cards is $C(52, 2)$. Thus, the probability of both cards being aces is

$$\frac{C(4, 2)}{C(52, 2)} = \frac{P(4, 2)}{2!} \div \frac{P(52, 2)}{2!} = \frac{P(4, 2)}{P(52, 2)} = \frac{4 \cdot 3}{52 \cdot 51} = \frac{1}{221}$$

We can also find the probability using permutations. The number of ways to draw 2 aces is $P(4, 2)$ and the number of ways of picking 2 cards from 52 is $P(52, 2)$. The final answer is the same!

By the SCP, there are 4 ways of drawing the first ace and 3 for the second, out of 52 choices for the first card and 51 for the second. The final answer is the same!

For illustrative purposes, the problem is done three ways.

1. Using combinations

$$\frac{C(4, 2)}{C(52, 2)}$$

2. Using permutations

$$\frac{P(4, 2)}{P(52, 2)}$$

3. Using the SCP

$$\frac{4 \cdot 3}{52 \cdot 51}$$

Note that you can find the final answer several ways. Just be consistent!

(b) In this part of the problem, we want to consider the **order** in which the 2 cards are drawn, so we use **permutations.** The number of ways of selecting an ace is $P(4, 1)$ and the number of ways of selecting a king is $P(4, 1)$. By the SCP, the number of ways of doing these two things in succession is $P(4, 1)P(4, 1)$. The total number of ways of drawing 2 cards is $P(52, 2)$, so the probability of drawing an ace and a king, in that order, is

$$\frac{P(4, 1)P(4, 1)}{P(52, 2)} = \frac{4 \cdot 4}{52 \cdot 51} = \frac{4}{663}$$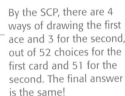

In part (a) of Example 4, we found that

$$\frac{C(4, 2)}{C(52, 2)} = \frac{P(4, 2)}{P(52, 2)}$$

This equation is a special case of a general result that can be obtained as follows:

$$\frac{C(m, r)}{C(n, r)} = C(m, r) \div C(n, r)$$
$$= \frac{P(m, r)}{r!} \div \frac{P(n, r)}{r!}$$
$$= \frac{P(m, r)}{r!} \times \frac{r!}{P(n, r)} = \frac{P(m, r)}{P(n, r)}$$

EXAMPLE 5 ▶ Not Picking Aces, Kings, Queens, or Jacks

Suppose we take all the cards from one suit—say, hearts—out of a standard deck of 52 cards. Shuffle the 13 hearts and then draw 3 of them. What is the probability that none of the 3 will be an ace, king, queen, or jack?

Solution

Here the **order does not matter,** so we use **combinations.** The number of combinations of 13 things taken 3 at a time is $C(13, 3)$. There are 9 cards not including the ace, king, queen, or jack, and the number of combinations of these taken 3 at a time is $C(9, 3)$. Thus, the required probability is

$$\frac{C(9, 3)}{C(13, 3)} = \frac{P(9, 3)}{P(13, 3)} = \frac{9 \cdot 8 \cdot 7}{13 \cdot 12 \cdot 11} = \frac{42}{143}$$

Using permutations ⬑   ⬐ Using the SCP
(Same answer!)   (Same answer!)   ■

In Example 5, because we were taking 3 cards from both the 13 hearts and the 9 cards, we could use either combinations or permutations. The next example is one that requires the use of combinations.

EXAMPLE 6 ▶ Probability of 4 Aces and 1 King

A poker hand consists of 5 cards. What is the probability of getting a hand of 4 aces and 1 king?

Solution

Here, the **order is not to be considered** because any order of getting the aces and the king will result in a hand that consists of 4 aces and 1 king. Now, the number of ways in which 4 aces can be selected is $C(4, 4)$, and the number of ways in which 1 king can be selected is $C(4, 1)$. Hence, by the SCP, the number of ways of getting 4 aces and 1 king is $C(4, 4)C(4, 1)$. Furthermore, the total number of 5-card hands is $C(52, 5)$, so the required probability is

$$\frac{C(4, 4)C(4, 1)}{C(52, 5)} = \frac{1 \cdot 4}{C(52, 5)}$$

Since

$$C(52, 5) = \frac{52 \cdot 51 \cdot 50 \cdot 49 \cdot 48}{5 \cdot 4 \cdot 3 \cdot 2 \cdot 1} = 2,598,960$$

the probability of getting 4 aces and 1 king is

$$\frac{C(4, 4)C(4, 1)}{C(52, 5)} = \frac{4}{2,598,960} = \frac{1}{649,740}$$

which is very small indeed!   ■

EXAMPLE 7 ▶ Probability of 1 Ace and No Face Card

Five cards are drawn from a standard deck. What is the probability of getting exactly 1 ace and no face cards?

Solution

Since there are 4 aces, the number of ways of getting exactly 1 ace is $C(4, 1)$. There are 36 cards that are not aces or face cards. The number of ways of getting

4 of these is $C(36, 4)$. Thus, the number of ways of getting 1 ace and 4 of the 36 cards is, by the SCP,

$C(4, 1)C(36, 4)$

The total number of ways that 5 cards can be drawn from the entire deck is $C(52, 5)$, so the required probability is

$$\frac{C(4, 1)C(36, 4)}{C(52, 5)} = \frac{4 \cdot 36 \cdot 35 \cdot 34 \cdot 33}{4 \cdot 3 \cdot 2 \cdot 1 \cdot C(52, 5)}$$

$$= \frac{235,620}{2,598,960} \quad \text{(See Example 6 to get the denominator.)}$$

$$= \frac{33}{364} \qquad \blacksquare$$

Do we always use the SCP and/or permutations and/or combinations in solving a probability problem? Not necessarily; sometimes it is easier just to look at the possible outcomes or to reason the problem out directly. This is illustrated in the next example.

EXAMPLE 8 ▶ Stuffing Envelopes at Random

A careless clerk was supposed to mail 3 bills to 3 customers. He addressed 3 envelopes but absentmindedly paid no attention to which bill he put in which envelope.

(a) What is the probability that exactly 1 of the customers received the proper bill?

(b) What is the probability that exactly 2 of the customers received the proper bills?

Solution
In Table 11.3, the headings C_1, C_2, and C_3 represent the customers, and the numbers 1, 2, and 3 below represent the bills for the customers with the corresponding subscripts. Thus, the rows represent the possible outcomes.

TABLE 11.3

C_1	C_2	C_3	
1	2	3	
①	3	2	← Favorable
2	1	③	← Favorable
2	3	1	
3	1	2	
3	②	1	← Favorable

(a) Table 11.3 shows that there are 6 possibilities and 3 cases in which exactly 1 (that is, 1 and only 1) of the customers received the proper bill. Therefore, the required probability is $P = \frac{3}{6} = \frac{1}{2}$.

(b) Here the probability is 0 because if 2 customers received their proper bills, then the third one did also. This means that there is no case in which 2 and only 2 received the proper bills. ∎

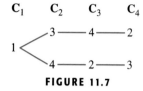

FIGURE 11.7

In Example 8, if there were 4 customers and 4 bills, then there would be a total of $P(4, 4) = 24$ cases in all. (This is just the number of ways in which the bills could be permuted.) You can see that it would be quite cumbersome to list all these cases. Instead, draw a tree showing the possible favorable cases if 1 customer, C_1, receives the proper bill (see Figure 11.7). This shows the 1 (representing the bill for C_1) under C_1; then C_2 can have only the 3 or the 4. If C_2 has the 3, then C_3 must have the 4 and C_4 the 2. (Otherwise, more than 1 customer would receive the proper bill.) If C_2 has the 4, then C_3 must have the 2 and C_4 the 3. These are the only 2 favorable cases possible if C_1 gets bill 1. The same argument holds if one of the other customers gets the proper bill; there are just 2 ways in which none of the other customers gets a proper bill. Since there are 4 customers, the SCP shows that there are only $4 \times 2 = 8$ favorable cases. Thus, the probability that exactly 1 of the customers gets the proper bill is $\frac{8}{24} = \frac{1}{3}$.

C. Applications

EXAMPLE 9 ▶ An Application to Oil Drilling

An oil company is considering drilling an exploratory oil well. If the rocks under the drilling site are characterized by what geologists call a "dome" structure, the chances of finding oil are 60%. The well can be **dry,** a **low** producer, or a **high** producer of oil. The probabilities for these outcomes are given in the table.

Production	Dome (60%)	No Dome (40%)
Dry	0.6	0.85
Low producer	0.25	0.125
High producer	0.15	0.025

Source: Robert T. Clemen and Terry Reilly, *Making Hard Decisions with Decisions Tools Suite.* (Pacific Grove, CA: Brooks Cole Publishing, 2004).

(a) Draw a tree diagram for the data given in the table.

(b) What is the probability that the well is dry?

Solution

(a) Since the probability of finding oil when there is a dome structure is 60%, the probability of finding oil when there is **no** dome structure is $100\% - 60\% = 40\%$. We draw the tree diagram in Figure 11.8 on page 726 and label the first two branches "Dome (0.6)" and "No dome (0.4)." We then label three branches starting from the end of the "Dome" outcome and three branches starting from the end of the "No dome" outcome with the probabilities for a dry, low-producing, and high-producing well.

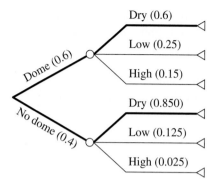

(b) To find the probability that the well is dry, we start at the branch in Figure 11.8 labeled "Dome (0.6)" and continue through the branch labeled "Dry (0.6)." The probability of that path is $0.6 \times 0.6 = 0.36$. The other possibility for a dry well is to start at the branch labeled "No dome (0.4)" and continue through the branch labeled Dry (0.850). The probability for that branch is $0.4 \times 0.850 = 0.34$. The total probability for a dry well is the sum of the two probabilities, $0.36 + 0.34 = 0.70$.   ■

EXERCISES 11.2

Ⓐ Using Tree Diagrams

1. A man has 3 pairs of shoes, 2 suits, and 3 shirts. If he picks a pair of shoes, a suit, and a shirt at random, what is the probability that he picks his favorite shoes, suit, and shirt?

2. At the end of a meal in a restaurant, a person wants to have pie à la mode (pie topped with ice cream) for dessert. There are 5 flavors of ice cream—chocolate, vanilla, strawberry, peach, and coffee—and there are 2 kinds of pie—apple and cherry. If the waiter picks the pie and ice cream at random, what is the probability that the person will get apple pie with vanilla ice cream?

3. A fair die is rolled 3 times in succession. What is the probability that even numbers are rolled all 3 times?

Ⓑ Using Permutations and Combinations

4. Jim belongs to a club of 40 members. A committee of 3 is to be selected at random from the 40 members. Find the probability that Jim is 1 of the 3 selected members.

5. Helen and Patty both belong to a club of 25 members. A committee of 4 is to be selected at random from the 25 members. Find the probability that both Helen and Patty will be selected.

6. Two cards are drawn at random, in succession and without replacement, from a deck of 52 cards.
  a. Find the number of ways in which the ace of spades and a king can be selected, in that order.
  b. What is the probability of drawing the ace of spades and a king, in that order?

7. Mr. C. Nile and Mr. D. Mented agreed to meet at 8 P.M. in one of the Spanish restaurants in Ybor City. They were both punctual, and they both remembered the date agreed on. Unfortunately, they forgot to specify the name of the restaurant. If there are 5 Spanish restaurants in Ybor City, and the 2 men each go to 1 of these, find the probability that they
  a. meet each other.   **b.** miss each other.

8. P.U. University offers 100 courses, 25 of which are mathematics. All these courses are available each hour, and a counselor randomly selects 4 different courses for a student. Find the probability that the selection will not include a mathematics course. (Do not simplify your answer.)

9. A piggy bank contains 2 quarters, 3 nickels, and 2 dimes. A person takes 2 coins at random from this bank. Label the coins $Q_1, Q_2, N_1, N_2, N_3, D_1,$ and D_2 so that they can all be regarded as different. Then find the probabilities that the values of the 2 coins selected are the following:

a. 35¢   **b.** 50¢

10. A committee of 2 is chosen at random from a population of 5 men and 6 women. What is the probability that the committee will consist of 1 man and 1 woman?

In problems 11–15, assume that 2 cards are drawn in succession and without replacement from an ordinary deck of 52 cards. Find the probability that

11. 2 kings are drawn.

12. 2 spades are drawn.

13. 1 spade and 1 king other than the king of spades (in that order) are drawn.

14. 1 spade and 1 king other than the king of spades (not necessarily in that order) are drawn.

15. 2 red cards are drawn.

In problems 16 and 17, assume that there is an urn containing five $50 bills, four $20 bills, three $10 bills, two $5 bills, and one $1 bill and that the bills all have different serial numbers so that they can be distinguished from each other. A person reaches into the urn and withdraws one bill and then another.

16. a. In how many ways can two $20 bills be withdrawn?
   b. How many different outcomes are possible?
   c. What is the probability of selecting two $20 bills?

17. a. In how many ways can a $50 bill and a $10 bill be selected in that order?
   b. What is the probability of selecting a $50 bill and a $10 bill in that order?
   c. What is the probability of selecting two bills, one of which is a $50 bill and the other a $10 bill?

18. If 2% of the auto tires manufactured by a company are defective and 2 tires are randomly selected from an entire week's production, find the probability that neither is defective.

19. In problem 18, find the probability that at least 1 of the 2 selected tires is defective.

20. A box contains 10 computer disks and 2 are defective. If 2 disks are randomly selected from the box, find the probability that both are defective.

21. A survey showed that 10% of high school football players later played football in college. Of these, 5% went on to play professional football. Find the probability that a randomly selected high school football player will play both collegiate and professional football.

22. An urn contains 5 white balls and 3 black balls. Two balls are drawn at random from this urn. Find the probability that
   a. both balls are white.
   b. both balls are black.
   c. 1 ball is white and the other is black.

23. In this problem, do not simplify your answers. What is the probability that a 5-card poker hand will contain the following?
   a. 2 kings, 2 aces, and 1 other card
   b. 3 kings and 2 aces

24. A box of light bulbs contains 95 good bulbs and 5 bad ones. If 3 bulbs are selected at random from the box, what is the probability that 2 are good and 1 is bad?

25. A plumbing company needs to hire 2 plumbers. Five people (4 men and 1 woman) apply for the job. Since they are all equally qualified, the selection is made at random (2 names are pulled out of a hat). What is the probability that the woman is hired?

26. Low-calorie food is required to contain no more than 40 calories per serving. The Food and Drug Administration (FDA) suspects that a company is marketing illegally labeled low-calorie food. If an inspector selects 3 cans at random from a shelf holding 10 cans (3 legally labeled and 7 illegally labeled), what are the probabilities that the following are legally labeled?
   a. All 3 cans selected
   b. Only 2 of the 3 cans selected

In problems 27–32 a poker hand consisting of 5 cards is drawn. Find the probability of obtaining

27. a royal flush (ten, jack, queen, king, ace of the same suit).

28. a straight flush (5 consecutive cards of the same suit).

29. 4 of a kind (4 cards of the same face value).

30. a full house (one pair and one triple of the same face value).

31. a flush (5 cards of the same suit but not a straight or royal flush).

32. a straight (5 consecutive cards, not all of the same suit).

33. Referring to Example 9, what is the probability of drilling a low-producing well?

34. Referring to Example 9, what is the probability of drilling a high-producing well?

35. On the basis of the answers to Example 9 and problems 33 and 34, what is the outcome with the highest probability? Explain.

Applications

Problems 36–40 refer to the following table showing the death penalty and racial status distribution for 326 convicted murderers.

Death Penalty	White	Black
Yes	19	17
No	141	149
Total	160	166

Source: M. Radelet, "Racial Characteristics of the Death Penalty," *American Sociological Review* 46: 918–927.

36. a. Start a tree diagram similar to that in Example 9 with two branches labeled white and black.
  b. What is the probability that a person selected at random from the 326 convicts is white? Write the answer on the appropriate branch.
  c. What is the probability that a person selected at random from the 326 convicts is black? Write the answer on the appropriate branch.

37. The first set of branches for the tree corresponding to the table were labeled "White" and "Black." The second set of branches should be labeled "Yes" and "No" for whites and "Yes" and "No" for blacks.
  a. What is the probability that a white convict received the death penalty? Write the answer on the appropriate branch.
  b. What is the probability that a black convict received the death penalty? Write the answer on the appropriate branch.

38. Using a tree diagram, find the probability that a convict (either race) did not receive the death penalty.

39. Using the table, find the probability that a convict (either race) did not receive the death penalty.

40. Is your answer to problem 39 the same as your answer to problem 38? Explain why or why not.

In Other Words

Three people, a mathematician, a statistician, and a fool, observed 10 tosses of a coin. Heads came up 10 times. Do you agree with statements 41–43? Explain why or why not.

41. Tails are "due." Bet on tails.

42. Heads are "hot." Bet on heads.

43. It is a random fluke. Don't bet.

44. Explain which strategy you think
  a. the statistician will pick.
  b. the mathematician will pick.
  c. the fool will pick.

If you do this correctly, you will answer the classic riddle, "How do you tell the difference between a mathematician, a statistician, and a fool?"

Using Your Knowledge

Dr. Benjamin Spock, a famous pediatrician, was accused of violating the Selective Service Act by encouraging resistance to the Vietnam War. In his trial, the defense challenged the legality of the method used to select the jury. In the Boston District Court, jurors are selected in three stages, as follows:

45. The clerk of the court selects 300 names at random from the Boston City Directory. If the directory lists 76,000 names (40,000 women and 36,000 men), what is the probability of selecting 150 men and 150 women? (Do not simplify.)

46. The 300 names are placed in a box, and the names of 30 potential jurors are drawn. If the names in the box correspond to 160 women and 140 men, find the probability that 15 men and 15 women are selected. (Do not simplify.)

47. The subgroup of 30 is called a *venire*. From the venire, 12 jurors are selected. If the venire consists of 16 women and 14 men, what is the probability that the final jury consists of 6 men and 6 women? (Do not simplify.)

By the way, it was shown that the Spock trial judge selected only about 14.6% women, whereas his colleagues selected about 29% women. This showed that the trial judge systematically reduced the proportion of women and had not selected a jury legally.

 Calculator Corner

You can use a calculator to compute expressions such as $C(52, 5)$ in Example 6. To do this, enter

$$\boxed{52}\ \boxed{a}\ \boxed{5}\ \boxed{b}\ \boxed{\text{2nd}}\ \boxed{_nC_r}$$

 Collaborative Learning

Have you heard of the birthday problem? Here it is. Discuss or prove why this is so: In a group of 23 people, *at least 2* have the same birthday with probability higher than $\frac{1}{2}$. Divide into groups and discuss this problem. (*Hint:* It is easier to find the answer to the related question: In a group of 23 people, what is the probability that all of them have *different* birthdays?)

1. Suppose your group consists of 5 students. What is the probability that all 5 students have different birthdays? Assume there are 365 days in a year. *Hint:* The birthday for the first student falls on any one of the 365 days in the year, with probability $\frac{365}{365}$, so the probability that the second student has a different birthday is $\frac{364}{365}$. For the third student, the probability is $\frac{363}{365}$, for the fourth $\frac{362}{365}$, and for the fifth $\frac{361}{365}$. Thus, the probability that all 5 students have different birthdays is

$$\frac{365}{365} \times \frac{364}{365} \times \frac{363}{365} \times \frac{362}{365} \times \frac{361}{365} \approx 0.97$$

Check this!

2. If you follow the pattern in problem 1, the probability that in a group consisting of 10 students all have different birthdays is

$$\frac{365}{365} \times \frac{364}{365} \times \frac{363}{365} \times \frac{362}{365} \times \cdots \times \frac{365 - 10 + 1}{365}$$

What is the probability that in a group of 23 students all have different birthdays?

3. If the probability that all 23 students in the group of problem 2 have different birthdays is q, the probability that at least 2 have the same birthday is $1 - q$. Find this probability.

4. Is the answer to problem 3 more than 50%?

 *Online Study Center*

To further explore the birthday problem, access link 11.2.1 on this textbook's Online Study Center.

11.3   Computation of Probabilities

Advertising and Probability

Does advertising influence a consumer's decision when buying a car? A car dealership conducted a survey of people inquiring about a new model car. Of the people surveyed, 45% had seen an advertisement for the car in the paper, 50% eventually bought one of these cars, and 25% had neither seen the ad nor bought a car. What is the probability that a person selected at random from the survey read the ad *and* bought a car?

At this time, we are unable to answer this question. However, let us assume that 100 persons were surveyed. The information from the survey is as follows:

45 saw the ad (S).
50 bought a car (B).
25 neither saw the ad nor bought a car ($S' \cap B'$).

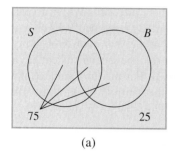

(a)

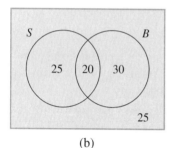

(b)

FIGURE 11.9

We can draw a Venn diagram, as in Figure 11.9(a), using the sets S and B. Since 25 persons neither saw the ad nor bought a car, we place 25 persons outside both circles. This means that we must have 75 persons in $S \cup B$. Since S has 45 persons and B has 50 $(45 + 50 = 95)$, there must be 20 persons in $S \cap B$, as shown in Figure 11.9(b). Thus, the number of persons that saw the ad *and* bought a car is 20.

If we generalize this idea to probability (see Example 3), the probability that a person selected at random read the ad and bought a car is $\frac{20}{100} = \frac{1}{5}$. Compare this with the probability that a person bought a car without even seeing the ad. Did seeing the ad make a lot of difference?

In this section we will formalize some of the ideas we have used. What do we think the probability for an impossible event should be? What about the probability of an event that is certain to occur? The answers to these questions are given in properties 1 and 2 that follow. We will use these properties to solve many of the examples in this section. ▶

A. Using Properties of Probabilities

In this section we give four properties that are useful in the computation of probabilities. The letters E, A, and B stand for events in a sample space $\mathscr{U}$.

Property 1 (Impossible Events)
An impossible event E has a probability of 0.
$\qquad P(E) = 0 \quad$ if and only if $\quad E = \varnothing \qquad\qquad$ (1)

Property 1 states that an **impossible event** has a probability 0. The next example illustrates this idea.

EXAMPLE 1 ▶ Impossible Probabilities

A die is rolled. What is the probability that a 7 comes up?

Solution
The sample space for this experiment is

$$\mathscr{U} = \{1, 2, 3, 4, 5, 6\}$$

so it is impossible for a 7 to be rolled. Thus, $P(7) = 0$. ∎

Property 2 (Range of Probability)
The probability of any event E is a number between 0 and 1, inclusive.
$\qquad 0 \leq P(E) \leq 1 \qquad\qquad$ (2)

Property 2 says that the probability of any event is a number between 0 and 1, inclusive. This follows because the number of favorable cases cannot be less than 0 or more than the total number of possible cases. Thus, $P(E) = 1$ means that the event E is *certain* to occur (a "sure thing").

> **Property 3 (Probabilities with *Or*)**
>
> The probability of *A* or *B* is given by
> $$P(A \cup B) = P(A) + P(B) - P(A \cap B) \qquad (3)$$

Property 3 says that the probability of event *A or B* is the probability of event *A* plus the probability of event *B*, decreased by the probability of events *A and B*. Note the key words *or* and *and*. The subtraction of $P(A \cap B)$ is to ensure that events belonging to both *A and B* are not counted twice.

EXAMPLE 2 ▶ Tossing Two Coins

A penny and a nickel are tossed. What is the probability that one *or* the other of the coins will turn up heads?

Solution

If we use subscripts p and n for penny and nickel, respectively, we can list the possible cases as follows: (H_p, H_n), (H_p, T_n), (T_p, H_n), and (T_p, T_n). Since there are 3 favorable cases out of the 4 possible,

$$P(H_p \cup H_n) = \tfrac{3}{4}$$

We can check that equation (3) gives

$$P(H_p \cup H_n) = P(H_p) + P(H_n) - P(H_p \cap H_n)$$
$$= \tfrac{1}{2} + \tfrac{1}{2} - \tfrac{1}{4} = \tfrac{3}{4}$$

as before.

EXAMPLE 3 ▶ Probability of an Ace or a Red Card

A card is drawn from a deck of 52 playing cards. Find the probability that the card is either an ace or a red card.

Solution

Let *A* be the event in which the card drawn is an ace, and let *R* be the event in which the card drawn is red. Then by equation (3),

$$P(A \cup R) = P(A) + P(R) - P(A \cap R)$$

Now, $P(A) = \tfrac{4}{52}$, $P(R) = \tfrac{26}{52}$, and $P(A \cap R) = \tfrac{2}{52}$, so

$$P(A \cup R) = \tfrac{4}{52} + \tfrac{26}{52} - \tfrac{2}{52} = \tfrac{7}{13}$$

Equation (3), in case all outcomes in the sample space are equally likely, is derived from the fact that

$$P(E) = \frac{n(E)}{n(\mathcal{U})}$$

Hence,

$$P(A \cup B) = \frac{n(A \cup B)}{n(\mathcal{U})} = \frac{n(A) + n(B) - n(A \cap B)}{n(\mathcal{U})}$$

$$= \frac{n(A)}{n(\mathcal{U})} + \frac{n(B)}{n(\mathcal{U})} - \frac{n(A \cap B)}{n(\mathcal{U})}$$

$$= P(A) + P(B) - P(A \cap B)$$

If the outcomes are not equally likely, then the same result follows by replacing $n(E)$ and $n(\mathcal{U})$ by $w(E)$ and $w(\mathcal{U})$, where these mean the sum of the weights of the outcomes in E and $\mathcal{U}$, respectively.

EXAMPLE 4 ▶ Probability of a Red or Yellow Ball

An urn contains 5 red, 2 black, and 3 yellow balls. Find the probability that a ball selected at random from the urn will be red or yellow.

Solution
By equation (3),

$$P(R \cup Y) = P(R) + P(Y) - P(R \cap Y)$$
$$= \tfrac{5}{10} + \tfrac{3}{10} - 0 = \tfrac{4}{5} \qquad \blacksquare$$

In Example 4, notice that $P(R \cap Y) = 0$. This means the events of selecting a red ball and selecting a yellow ball cannot occur simultaneously, that is, $R \cap Y = \varnothing$. We say that A and B are **mutually exclusive** if $A \cap B = \varnothing$. For any two mutually exclusive events A and B, it follows that $P(A \cap B) = 0$, and property 3 becomes

---

Property 4 (Mutually Exclusive Events)

For any two mutually exclusive events A and B,

$$P(A \cup B) = P(A) + P(B); \quad \text{that is,} \quad P(A \cap B) = 0 \qquad (4)$$

EXAMPLE 5 ▶ Determining Mutually Exclusive Events

Show that the events R and Y of Example 4 are mutually exclusive.

Solution
Because $P(R \cap Y) = 0$, $R \cap Y = \varnothing$ (property 1), so R and Y are mutually exclusive. $\qquad \blacksquare$

EXAMPLE 6 ▶ Determining Mutually Exclusive Events

In a game of blackjack (also called twenty-one), a player and the dealer each get 2 cards. Let A and B be the events defined as follows:

  A The player gets an ace and a face card for 21 points.
  B The dealer gets an ace and a 10 for 21 points.

Are A and B mutually exclusive events?

Solution
No. Both player and dealer can get 21 points. (In the game of blackjack, 21 points wins, and in most casinos, the dealer would be the winner with this tie score.) $\qquad \blacksquare$

EXAMPLE 7 ▶ Video Games and Probability

The video games that you can attach to your television set have both audio and video. It is estimated that the probability of the audio being defective is 0.03 and the probability of at least one or the other (audio or video) being defective is 0.04, but the probability of both being defective is only 0.01. What is the probability that the video is defective?

Solution

Let A stand for the event that the audio is defective and V for the event that the video is defective. Then, using formula (3), with $P(A) = 0.03$, $P(A \cup V) = 0.04$, and $P(A \cap V) = 0.01$, we get

$$P(A \cup V) = P(A) + P(V) - P(A \cap V)$$

so that

$$0.04 = 0.03 + P(V) - 0.01$$

Thus,

$$P(V) = 0.02 \qquad \blacksquare$$

We have used the formula $P(T') = 1 - P(T)$ to calculate the probability of the complement of an event. For example, if the probability that it will rain today is $\frac{1}{4}$, the probability that it will not rain today is $1 - \frac{1}{4} = \frac{3}{4}$, and if the probability of a stock going up in price is $\frac{3}{8}$, the probability that the stock will not go up in price is $1 - \frac{3}{8} = \frac{5}{8}$. We now see how this property is used in the field of life insurance.

TABLE 11.4 Table of Mortality for 100,000 Americans

Exact Age	Males		Females	
	Number of Lives*	Life Expectancy	Number of Lives*	Life Expectancy
0	100,000	74.14	100,000	79.45
10	99,023	64.86	99,204	70.08
20	98,447	55.20	98,937	60.25
30	97,100	45.90	98,432	50.53
40	95,348	36.64	97,487	40.97
50	91,782	27.85	95,387	31.75
60	84,671	19.72	90,802	23.06
70	69,719	12.75	80,252	15.35
80	43,126	7.31	58,721	8.95
90	12,095	3.70	23,791	4.47
100	486	2.00	1,811	2.29
110	1	1.10	6	1.16

*Number of survivors out of 100,000 born alive.
Source: www.ssa.gov/OACT/STATS/table4c6.html.

According to Table 11.4, of 98,447 males alive at age 20 (column 2, row 3), 486 were still alive at age 100, and only 1 was left at age 110! Females get a better deal. Of 98,937 females alive at age 20 (column 4, row 3), 1811 were still alive at age 100, and 6 were still around at age 110!

A table similar to this one is used to calculate a portion of the premium on life insurance policies. Use Table 11.4 to do the next example.

EXAMPLE 8 ▶ **Longevity Probabilities from a Table**

Find the probability that a person who is alive at age 20 is

(a) still alive at age 70 if the person is a female.

(b) not alive at age 70 if the person is a female.

(c) still alive at age 70 if the person is a male.

(d) not alive at age 70 if the person is a male.

Solution

(a) Based on Table 11.4, the probability that a female alive at age 20 is still alive at age 70 is given by

$$P(\text{female alive at 70}) = \frac{\text{number of females alive at 70}}{\text{number of females alive at 20}} = \frac{80{,}252}{98{,}937}$$

(b) Using formula (4), we find that the probability of a female not being alive at age 70 is

$$1 - \frac{80{,}252}{98{,}937} = \frac{18{,}685}{98{,}937}$$

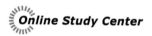
Online Study Center

To further explore topics in probability, including the ones in this section, access links 11.3.1 and 11.3.2 on this textbook's Online Study Center.

(c) $P(\text{male alive at 70}) = \dfrac{\text{number of males alive at 70}}{\text{number of males alive at 20}} = \dfrac{69{,}719}{98{,}447}$

(d) $1 - \dfrac{69{,}319}{98{,}447} = \dfrac{29{,}128}{98{,}447}$ ■

EXERCISES 11.3

A Using Properties of Probabilities

In problems 1–4, find the answer to each question and indicate which of the four properties presented in this section you used to solve the problem.

1. A die is rolled. Find the probability that the number that comes up is a 0.

2. A die is rolled. Find the probability that an odd or an even number occurs.

3. Two dice are rolled. Find the probability that the sum of the two faces that turn up is between 0 and 13.

4. An absentminded professor wished to mail 3 report cards to 3 of her students. She addressed 3 envelopes but, unfortunately, did not pay any attention to which card she put in which envelope. What is the probability that exactly 2 students receive their own report cards? (Assume that all 3 envelopes were delivered.)

A single ball is drawn from an urn containing 10 balls numbered 1 through 10. In each of problems 5–8, find the probability that the ball chosen is

5. an even-numbered ball or a ball with a number greater than 7.

6. an odd-numbered ball or a ball with a number less than 5.

7. an even-numbered ball or an odd-numbered ball.

8. a ball with a number that is greater than 7 or less than 5.

In problems 9–13, a single card is drawn from a deck of 52 cards. Find the probability the card chosen is

9. the king of hearts or a spade.

10. the ace of hearts or an ace.

11. the ace of diamonds or a diamond.

12. the ace of clubs or a black card.

13. the king of hearts or a picture card (jack, queen, or king).

14. The U.S. Weather Service reports that in a certain northern city it rains 40 days and snows 50 days in the winter. However, it rains and snows on only 10 of those days. Based on this information, what is the probability that it will rain or snow in that city on a particular winter day? (Assume that there are 90 days of winter.)

15. Among the first 50 stocks listed in the New York Stock Exchange transactions on a certain day (as reported in the *Wall Street Journal*), there were 26 stocks that went down, 15 that went up, and 9 that remained unchanged. On the basis of this information, find the probability that a stock selected at random from this list would not have remained unchanged.

The following table shows the probability that there is a given number of people waiting in line at a checkout register at Dear's Department Store.

Number of Persons in Line	Probability
0	0.10
1	0.15
2	0.20
3	0.35
4 or more	0.20

In problems 16–20, find the probability of having

16. exactly 2 persons in line.

17. more than 3 persons in line.

18. at least 1 person in line.

19. more than 3 persons or fewer than 2 persons in line.

20. more than 2 persons or fewer than 3 persons in line.

In solving problems 21–25, refer to Table 11.4 on page 733.

21. What is the probability that a person who is alive at age 20 will not be alive at age 60
  a. if the person is a male?
  b. if the person is a female?

22. What is the probability that a person who is alive at age 30 will be alive at age 70
  a. if the person is a male?
  b. if the person is a female?

23. What is the probability that a person who is alive at age 30 will not be alive at age 70
  a. if the person is a male?
  b. if the person is a female?

24. What is the probability that a person who is alive at age 50 will live 80 more years
  a. if the person is a male?
  b. if the person is a female?

(Assume that none of the persons in Table 11.4 attained 130 years of age.)

25. What is the probability that a person who is alive at age 50 will live less than 80 more years? (See problem 24.)
  (a) if the person is a male?
  (b) if the person is a female?

Problems 26–30 refer to the following table. This table shows the number of correctly and incorrectly filled out tax forms obtained from a random sample of 100 returns examined by the Internal Revenue Service (IRS) in a recent year.

	Short form (1040A)	Long Form (1040)		
	No Itemized Deductions	No Itemized Deductions	Itemized Deductions	Totals
Correct	15	40	10	65
Incorrect	5	20	10	35
Totals	**20**	**60**	**20**	**100**

26. Find the probability that a form was a long form (1040) or an incorrectly filled out form.

27. Find the probability that a form had no itemized deductions and was correctly filled out.

28. Find the probability that a form was not filled out incorrectly.

29. Find the probability that a form was not a short form (1040A).

30. Find the probability that a form was a long form (1040) with no itemized deductions and filled out incorrectly.

31. A traffic light follows the pattern green, yellow, red for 60, 5, and 20 seconds, respectively. What is the probability that a driver approaching this light will find it green or yellow?

32. A driver approaching the green light decides to go ahead through the intersection whether the light changes or not. If it takes the driver 6 seconds to get through the intersection, what is the probability that the driver makes it through the intersection before the light turns red? See problem 31.

In Other Words

33. Explain, in your own words, what it means to say "The probability of an event is 0."

34. Explain, in your own words, what it means to say "The probability of an event is 1."

35. Explain, in your own words, the circumstances under which you can use the formula

$$P(A \cup B) = P(A) + P(B)$$

36. Explain why the probability of an event cannot be negative.

Using Your Knowledge

In this section you used a mortality table to calculate the probability that a person alive at a certain age will be alive at a later age. There are other tables that give the probabilities of different events. For example, many mortgage companies use a credit-scoring table to estimate the likelihood that an applicant will repay a loan. One such table appears below.

A Hypothetical Credit-Scoring Table

Age	Under 25 (12 pts)	25–29 (5 pts)	30–34 (0 pts)	35–39 (1 pt)	40–44 (18 pts)	45–49 (22 pts)	50 + (31 pts)
Time at Address	1 yr or less (9 pts)	1–2 yr (0 pts)	2–3 yr (5 pts)	3–5 yr (0 pts)	5–9 yr (5 pts)	10 yr + (21 pts)	
Age of Auto	None (0 pts)	0–1 yr (12 pts)	2 yr (16 pts)	3–4 yr (13 pts)	5–7 yr (3 pts)	8 yr + (0 pts)	
Monthly Auto Payment	None (18 pts)	$1–$80 (6 pts)	$81–$99 (1 pt)	$100–$139 (4 pts)	$140 + (0 pts)		
Housing Cost	$1–$125 (0 pts)	$126–$274 (10 pts)	$275 + (12 pts)	Owns clear (12 pts)	Lives with relatives (24 pts)		
Checking and Savings Accounts	Both (15 pts)	Checking only (2 pts)	Savings only (2 pts)	Neither (0 pts)			
Finance Company Reference	Yes (0 pts)	No (15 pts)					
Major Credit Cards	None (0 pts)	1 (5 pts)	2 + (15 pts)				
Ratio of Debt to Income	No debts (41 pts)	1–5% (16 pts)	6–15% (20 pts)	16% + (0 pts)			

In the table on page 736 your score depends on the number of points you get on the nine tabulated items. To obtain your score, you add the scores (shown in color) on the individual items. For example, if your age is 21, you get 12 points. If you have lived at your present address for less than a year, you get 9 more points. Moreover, if your car is 2 years old, you get another 16 points. So far your score is $12 + 9 + 16$. This should give you the idea.

A lender using the scoring table selects a cutoff point from a table, such as the following table that gauges the probability that an applicant will repay a loan:

Total Score	Probability of Repayment	Total Score	Probability of Repayment
60	0.70	100	0.92
65	0.74	105	0.93
70	0.78	110	0.94
75	0.81	115	0.95
80	0.84	120	0.955
85	0.87	125	0.96
90	0.89	130	0.9625
95	0.91		

37. John Dough, 27 years old, living for 3 years at his present address, has a 2-year-old automobile on which he pays $200 monthly. He pays $130 per month for his apartment and has no savings account, but he does have a checking account. He has no finance company reference. He has one major credit card, and his debt-to-income ratio is 12%. On the basis of the credit-scoring table, what is the probability that Mr. Dough will repay a loan?

38. What is the probability in problem 37 if Mr. Dough pays off his car, sells it, and rides the bus to work?

39. Find the probability that you will repay a loan, based on the information in the table.

Discovery

The Venn diagrams we studied in Chapter 2 can often be used to find the probability of an event by showing the number of elements in the universal set and the number of elements corresponding to the event under consideration. For example, if there are 100 employees in a certain firm and it is known that 82 are males (M), 9 are clerk typists (C), and 2 of these clerk typists are male, we can draw a diagram corresponding to this situation.

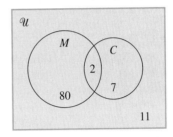

From this diagram, we can conclude that

$$P(M) = \frac{82}{100}, \quad P(C) = \frac{9}{100}, \quad P(M \cap C) = \frac{2}{100},$$

$$\text{and } P(M \cup C) = \frac{82}{100} + \frac{9}{100} - \frac{2}{100} = \frac{89}{100}$$

Use this technique to solve the following problems. In problems 40–42, assume that of the 100 persons in a company, 70 are married, 80 are college graduates, and 60 are both married and college graduates. Find the probability that if a person is selected at random from this group, the person will be

40. married and a college graduate.

41. married or a college graduate.

42. not married and not a college graduate.

In a recent election, voters were asked to vote on two issues, A and B. A Gallup poll indicated that of 1000 eligible voters, 600 persons voted in favor of A, 500 persons voted in favor of B, 200 persons voted in favor of both A and B, and 50 persons voted against both issues. If an eligible voter is selected at random, find the probability that he or she voted

43. for A but not B.   **44.** for B but not A.

45. for both A and B.   **46.** against both A and B.

47. not at all.

Collaborative Learning

In order to see how credit ratings are done at Fair, Isaac and Co. (FICO), access links 11.3.3 and 11.3.4 on this textbook's Web site. The scores there are between 300 and 800. Higher than 660 it is OK, between 620 and 660 is fair, and below 620 you may be in trouble. To reach these scores, they consider past delinquency, how you used credit in the past, the age of the credit file, the number of times you ask for credit, and your mix of credit (cards, installment, and revolving). Discuss how these five factors can affect your credit-worthiness.

11.4   Conditional Probability

Surveys and Conditional Probability

Have you ever been to a car dealership and taken a look at consumer magazines that rank different automobiles? If you are the manager of a dealership, you want to know whether people who read the reports in *Consumer Reports* or *Car and Driver* are more likely to buy a car from you. The first step is to conduct a survey of potential buyers. Suppose the results of such a survey are as follows:

> 70% of the people read the report (R).
> 45% bought a car from you, the dealer (B).
> 20% neither read the report nor bought a car from you, the dealer.

You want to find the effect of reading the report (R) on buying a car from you (B). Thus, you must compare the probability that the person bought a car $P(B)$ with the probability that the person bought a car given that the person read the report, denoted by $P(B|R)$ and read as "the probability of B given R."

You first make a Venn diagram of the situation. Label two circles R and B and place 20%, the percentage of people who neither read the report nor bought a car, outside these two circles. This means that 80% of the people must be inside the two circles. But 70% + 45% = 115%, so 35% (115% − 80%) of the people must be in $B \cap R$, as shown in Figure 11.10.

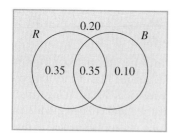

FIGURE 11.10

You can see from Figure 11.10 that $P(B) = 0.35 + 0.10 = 0.45$. To find $P(B|R)$, you have to look at all the people who bought cars *given* that they read the report; that is, you must look inside the circle labeled R. Inside this circle, 35% of the people out of the 70% who read the report bought a car; that is,

$$P(B|R) = \frac{P(B \cap R)}{P(R)} = \frac{0.35}{0.35 + 0.35} = \frac{0.35}{0.70} = 0.50$$

Thus, $P(B) = 0.45$ and $P(B|R) = 0.50$; this means that people are more likely to buy a car from you, the dealer, if they have read the report. For this reason many dealers give copies of consumer magazine articles to potential customers. In this section you will study *conditional* probability, that is, probability computed using a subset of the sample space. ▶

A.   Conditional Probability

As the "Wizard of Id" cartoon on page 739 shows, it is sometimes the case that in considering the probability of an event A, we obtain additional information that may suggest a revision of the probability of A. For example, assume that in

The WIZARD OF ID by permission of Johnny Hart and Creator's Syndicate, Inc.

Getwell Hospital, 70 of the patients have lung cancer (C), 60 of the patients smoke (S), and 50 have cancer and smoke. If there are 100 patients in the hospital, and 1 is selected at random, then $P(C) = \frac{70}{100}$ and $P(S) = \frac{60}{100}$. But suppose a patient selected at random tells us that he or she smokes. What is the probability that this patient has cancer? In other words, what is the probability that a patient has cancer, given that the patient smokes? The expression "given that the patient smokes" means that we must restrict our attention to those patients who smoke. We have thus added a **restrictive condition** to the problem. Essentially, the condition that the person smokes requires that we use S as our sample space.

To compute $P(C \mid S)$ (read "the probability of C, given S"), we recall that there are 50 favorable outcomes (people who have lung cancer and smoke) and 60 elements in the new sample space (people who smoke). Hence, $P(C \mid S) = \frac{50}{60} = \frac{5}{6}$.

We note that

$$P(C|S) = \frac{n(C \cap S)}{n(S)}$$

$$= \frac{n(C \cap S)/n(\mathcal{U})}{n(S)/n(\mathcal{U})} = \frac{P(C \cap S)}{P(S)}$$

This discussion suggests the following definition:

Definition of Conditional Probability

If A and B are events in a sample space $\mathcal{U}$ and $P(B) \neq 0$, the **conditional probability of A, given B,** is denoted by $P(A|B)$ and is defined by

$$P(A|B) = \frac{P(A \cap B)}{P(B)} \qquad (1)$$

Notice that the conditional probability of A, given B, results in a new sample space consisting of the elements in $\mathcal{U}$ for which B occurs. This gives rise to a second method of handling conditional probability, as illustrated in the following five examples:

EXAMPLE 1 ▶ Probability of a 3 When an Odd Number Turns Up

A die is rolled. Find the probability that a 3 comes up if it is known that an odd number turns up.

Solution

Method 1 Let T be the event in which a 3 turns up and Q be the event in which an odd number turns up. By equation (1),

$$P(T|Q) = \frac{P(T \cap Q)}{P(Q)} = \frac{\frac{1}{6}}{\frac{3}{6}} = \frac{1}{3}$$

Method 2 We know that an odd number turns up, so our new sample space is $\mathcal{U} = \{1, 3, 5\}$. Only one outcome (3) is favorable, so

$$P(T|Q) = \tfrac{1}{3}$$ ∎

EXAMPLE 2 ▶ Probability of a 6 After Heads

A coin is tossed; then a die is rolled. Find the probability of obtaining a 6, given that heads comes up.

Solution

Method 1 Let S be the event in which a 6 is rolled, and let H be the event in which heads comes up.

$$P(S|H) = \frac{P(S \cap H)}{P(H)} = \frac{\frac{1}{12}}{\frac{1}{2}} = \frac{1}{12} \cdot 2 = \frac{1}{6}$$

Method 2 We know that heads comes up, so our new sample space is $\mathcal{U} = \{(H, 1), (H, 2), (H, 3), (H, 4), (H, 5), (H, 6)\}$. Only one outcome is favorable, $(H, 6)$, so $P(S|H) = \tfrac{1}{6}$.   ∎

EXAMPLE 3 ▶ Probability of 4 After Different Outcomes

Two dice were thrown, and a friend tells us that the numbers that came up were different. Find the probability that the sum of the two numbers was 4.

Solution

Method 1 Let D be the event in which the two dice show different numbers, and let F be the event in which the sum is 4. By equation (1),

$$P(F|D) = \frac{P(F \cap D)}{P(D)}$$

Now, $P(F \cap D) = \frac{2}{36}$ because there are two outcomes, $(3, 1)$ and $(1, 3)$, in which the sum is 4 and the numbers are different, and there are 36 possible outcomes. Furthermore,

$$P(D) = \frac{36 - 6}{36} = \frac{30}{36} \qquad \text{so} \qquad P(F|D) = \frac{P(F \cap D)}{P(D)} = \frac{\frac{2}{36}}{\frac{30}{36}} = \frac{1}{15}$$

Method 2 We know that the numbers on the two dice were different, so we have $36 - 6 = 30$ (36 minus 6 outcomes that show the same number on both dice) elements in our sample space. Of these, only two, $(3, 1)$ and $(1, 3)$, have a sum of 4. Hence, $P(F|D) = \frac{2}{30} = \frac{1}{15}$.   ∎

EXAMPLE 4 ▶ Probability of 7 After 6

Two dice are rolled, and a friend tells you that the first die shows a 6. Find the probability that the sum of the numbers showing on the two dice is 7.

Solution

Method 1 Let S_1 be the event in which the first die shows a 6, and let S_2 be the event in which the sum is 7. Then

$$P(S_2|S_1) = \frac{P(S_2 \cap S_1)}{P(S_1)} = \frac{\frac{1}{36}}{\frac{6}{36}} = \frac{1}{6}$$

Method 2 We know that a 6 comes up on the first die, so our new sample space is $\mathcal{U} = \{(6, 1), (6, 2), (6, 3), (6, 4), (6, 5), (6, 6)\}$. Hence, $P(S_1|S_2) = \frac{1}{6}$ because there is only one favorable outcome, $(6, 1)$. ■

B. Applications

EXAMPLE 5 ▶ Genes and Probability

Suppose we represent with the letters B and b the genes that determine the color of a person's eyes. If the person has two b genes, the person has blue eyes; otherwise, the person has brown eyes. If it is known that a man has brown eyes, what is the probability that he has two B genes? (Assume that both genes are equally likely to occur.)

Solution

The tree diagram for the four possibilities appears in Figure 11.11.

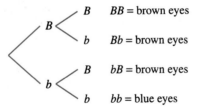

FIGURE 11.11

Method 1 Let T be the event in which the man has two B genes, and let B be the event in which the man has brown eyes. By the definition of conditional probability,

$$P(T|B) = \frac{P(T \cap B)}{P(B)} = \frac{\frac{1}{4}}{\frac{3}{4}} = \frac{1}{3}$$

Method 2 It is known that the man has brown eyes, so we consider the 3 outcomes corresponding to these cases (BB, Bb, and bB). Because only 1 of these equally likely outcomes (BB) is favorable, the probability that a man has two B genes if it is known that he has brown eyes is $\frac{1}{3}$. ■

Other important applications also make use of conditional probability. For example, the Framingham Heart Disease Study focused on strokes and heart failure. Table 11.5 on page 742 is based on this study and shows the number of

**TABLE 11.5 Strokes, per
1000 People Aged 45–74**

Blood Pressure	*Number*	*Strokes*
Normal	390	8
Borderline	315	14
High	295	31

adults (per 1000) aged 45–74 with certain blood pressure types and the number of strokes in each category. As we can see, the incidence of stroke for people aged 45–74 increases almost fourfold as blood pressure goes from normal to high (from 8 per 1000 to 31 per 1000). Note that the numbers in the body of Table 11.5 are all per 1000. This means that Table 11.5 gives approximate conditional probabilities. The number 31 in the last line of Table 11.5 means that the probability that a person will have a stroke and has high blood pressure is about $\frac{31}{1000}$. We look at some other aspects of this study in the next example.

EXAMPLE 6 ▶ Blood Pressure and Probability

Assume that numbers in Table 11.5 are accurate, and find the probability that

(a) a person in the 45–74 age group has a stroke (S), given that the person has normal blood pressure (N).

(b) a person in the 45–74 age group has a stroke (S), given that the person has borderline blood pressure (B).

(c) a person in the 45–74 age group has a stroke (S).

(d) a person has normal blood pressure (N), given that the person had a stroke (S).

Solution

(a) There were 8 strokes among the 390 who have normal blood pressure, so

$$P(S|N) = \frac{8}{390} = \frac{4}{195}$$

(b) The idea is similar to that of part (a); in the second row there are 14 people that had a stroke out of 315; thus,

$$P(S|B) = \frac{14}{315}$$

(c) There are three mutually exclusive sets, N, B, and H, so the required probability is the sum of the probabilities for the three sets, that is,

$$\frac{8}{1000} + \frac{14}{1000} + \frac{31}{1000} = \frac{53}{1000}$$

(d) Here, we know the person has had a stroke, so we can use the idea of conditional probability. The population for this condition (having a stroke) consists of the $8 + 14 + 31 = 53$ people. Of these, 8 have normal blood pressure. Thus, the required probability is $\frac{8}{53}$. You also can use the formula

$$P(N|S) = \frac{P(N \cap S)}{P(S)}$$

$$= \frac{\dfrac{8}{1000}}{\dfrac{53}{1000}} = \frac{8}{53}$$

Same answer!

EXAMPLE 7 ▶ Monty Hall Probabilities

Have you heard of the Monty Hall Problem? It goes like this: In the game show *Let's Make a Deal*, contestants are presented with 3 doors (A, B, C), only one of which has a big prize behind it (the others are empty). You do not know what is behind any of the doors. You choose a door. Monty then counters by showing you what is behind one of the other doors (which is empty) and asks you if you would like to stick with the door you have or switch to the other unknown door.

(a) What is the probability that you win given that you decide to **stay?**

(b) What is the probability that you win given that you decide to **switch?**

Solution

(a) Since there are 3 doors, the probability that you win by selecting the door with the prize is $\frac{1}{3}$.

(b) The winning prize (W) can be behind door A, B, or C. Two of the doors are empty (E). Here are the three possibilities.

	A	B	C
1.	W	E	E
2.	E	W	E
3.	E	E	W

Now, suppose that you select door A (it really does not matter which door you select), and you decide to **switch.** (Remember, Monty has to show you the empty door!)

In case 1, you will certainly lose.
In case 2, Monty will have to show you door C and you pick B and win.
In case 3, Monty will have to show you door B and you pick C and win.

Thus, you win 2 out of 3 times by switching! ■

Online Study Center

To further explore the Monty Hall problem, access links 11.4.1 and 11.4.2 on this textbook's Online Study Center.

EXERCISES 11.4

A Conditional Probability

1. A die was rolled. Find the probability that a 5 came up, given that an even number turned up.

2. A coin was tossed; then a die was rolled. Find the probability of obtaining a 7, given that tails came up.

3. Two dice were rolled, and a friend tells us that the numbers that came up were identical. Find the probability that the sum of the numbers was
   a. 8.   **b.** 9.
   c. an even number.   **d.** an odd number.

4. Referring to Example 5 of this section, find the probability that a person has two *b* genes, given that the person has
   a. brown eyes.   **b.** blue eyes.

5. For a family with 2 children, the sample space indicating boy (*B*) or girl (*G*) is *BB*, *BG*, *GB*, and *GG*. If each of the outcomes is equally likely, find the probability that the family has 2 boys, given that the first child is a boy.

6. A family has 3 children. If each of the outcomes in the sample space is equally likely, find the probability that the family has 3 girls given that
   a. the first child is a girl.
   b. the first child is a boy.

7. Referring to problem 6, find the probability that the family has exactly 2 girls, given that the first child is a girl.

8. The following table gives the approximate number of suicides per 100,000 persons, classified according to country and age for 1 year:

Age	Country		
	United States	Canada	Germany
15–24	10	10	20
25–44	20	15	30
45–64	30	13	50
65 or over	40	14	50

Based on the table, find the probability that

a. a person between 25 and 44 years of age committed suicide, given that the person lived in the United States. (For national suicide rates per 100,000 by gender and age, see www5.who.int/mental_health/main.cfm?p=0000000515.)

b. a person between 25 and 44 years of age committed suicide, given that the person lived in Canada.

c. a person committed suicide, given that the person lived in Germany.

9. The personnel director of Gadget Manufacturing Company has compiled the following table, which shows the percent of men and women employees who were absent the indicated number of days. Suppose there are as many women as men employees.

Absences (Days)	Sex	
	Men	Women
0	20%	20%
1–5	40%	20%
6–10	40%	20%
11 or more	0%	40%
Total	100%	100%

a. Find the probability that an employee missed 6–10 days, given that the employee is a woman.

b. Find the probability that an employee is a woman, given that the employee missed 6–10 days.

10. The following table describes the student population in a large college:

Class	Male	Female
Freshman (%)	25	15
Sophomore (%)	13	10
Junior (%)	12	8
Senior (%)	10	7

a. Find the probability that a randomly selected student is female.

b. Find the probability that a randomly selected student is a junior.

c. If the selected student is a junior, find the probability that the student is female.

11. An examination of Professor Larin's records for the last 10 years shows the following distribution of grades in his courses:

Grade	Percentage of All Grades
A	15
B	30
C	40
D	10
F	5

a. If one of Professor Larin's students is randomly selected, what is the probability that the student received neither an A nor a B?

b. If 200 of Professor Larin's students were selected, how many would be expected to have received a B or a D?

c. If it is known that one of Professor Larin's students did not get a C or a D, what is the probability that the student received a B?

In problems 12–14, assume that 2 cards are drawn in succession and without replacement from a standard deck of 52 cards. Find the probability that

12. the second card is the ace of hearts, given that the first card was the ace of spades.

13. the second card is a king, given that the first card was a king.

14. the second card is a 7, given that the first card was a 6.

B Applications

The following information will be used in problems 15 and 16.

The Merrilee Brokerage House studied two groups of industries (computers and petroleum) and rated them as low risks or high risks, as shown.

Industry	Low	High
Computers	5	10
Petroleum	20	15

15. If a person selected one of these stocks at random (that is, each stock had probability $\frac{1}{50}$ of being selected), find the probability that the person selected a computer stock, given that the person selected a low-risk stock.

16. If a person selected one of the stocks at random, find the probability that the person selected a petroleum stock, given that the person selected a high-risk stock.

17. A stock market analyst figures the probabilities that two related stocks, A and B, will go up in price. She finds the probability that A will go up to be 0.6 and the probability that both stocks will go up to be 0.4. What should be her estimate of the probability that stock B goes up, given that stock A goes up?

18. The Florida Tourist Commission estimates that a person visiting Florida will visit Disney World, Busch Gardens, or both with probabilities 0.5, 0.3, and 0.2, respectively. Find the probability that a person visiting Florida will visit Busch Gardens, given that the person did visit Disney World.

19. A recent survey of 400 instructors at a major university revealed the data shown in the following table. Based on the data, what are the probabilities of the following?
    a. An instructor received a good evaluation, given that the instructor was tenured.
    b. An instructor received a good evaluation.

Status	Good Evaluations	Poor Evaluations
Tenured	72	168
Nontenured	84	76

20. Referring to the data in problem 19, find the probability that an instructor received
    a. a poor evaluation, given that the instructor was tenured.
    b. a poor evaluation.

21. Billy was taking a history test, and his memory started playing tricks on him. He needed the date when Columbus reached America, and he remembered that it was 1492 or 1294 or 1249 or 1429 but was not sure which. Then he remembered that the number formed by the first three digits was not divisible by 3. What is the probability that he guessed the right date? (*Hint:* Recall that a number is divisible by 3 if the sum of its digits is divisible by 3. Use this information to find which dates this leaves Billy to choose from.)

22. Nancy was asked to guess a preselected number between 1 and 50 (inclusive). By asking questions first, Nancy learned that the number was divisible by 2 and/or by 3. What is the probability that Nancy guessed the right number after correctly using her information? (*Hint:* Eliminate the numbers that are not divisible by 2 or by 3. This eliminates all the odd numbers that are not multiples of 3.)

23. A doctor for a pharmaceutical company treats 100 patients with an experimental drug and another 100 patients with a conventional drug. The results of the experiment are given in the following table. What is the probability that
    a. a patient chosen at random from the group of 200 patients has improved, $P(I)$?
    b. a patient taking the experimental drug has improved, $P(I|E)$?

Type	Improved (I)	Same (S)
Experimental (E)	70	30
Conventional (C)	65	35

24. The University Apartments has 1000 units classified by size and location as shown in the following table. What is the probability of selecting at random
    a. a first-floor apartment?
    b. a first-floor, three-bedroom apartment?

c. a second-floor apartment, given that it is a one-bedroom?

d. a two- or three-bedroom apartment, given that it is located on the first floor?

Floor	Bedrooms		
	One	Two	Three
First floor	20%	30%	10%
Second floor	15%	20%	5%

Problems 25–27 refer to the following table showing the death penalty and racial distribution for 326 convicted murderers.

Death Penalty	White	Black
Yes	19	17
No	141	149
Total	**160**	**166**

Source: M. Radelet, "Racial Characteristics of the Death Penalty," *American Sociological Review* 46: 918–927.

Let *D* be the person gets the death penalty, *W* the person is white, and *B* the person is black.

25. Find $P(D|W)$.   **26.** Find $P(D|B)$.

27. On the basis of your answers to problems 25 and 26, is there much difference between outcomes for whites and blacks? Explain.

Problems 28–32 refer to the following table showing the death penalty, racial status, and race of the victim distribution for the same 326 murderers.

Race of Victim	Death Penalty	White	Black
White	Yes	19	11
	No	132	52
	Total	**151**	**63**

Race of Victim	Death Penalty	White	Black
Black	Yes	0	6
	No	9	97
	Total	**9**	**103**

28. Find $P(D|WW)$, where *WW* means that the defendant is white and the victim is white.

29. Find $P(D|BW)$.   **30.** Find $P(D|WB)$.

31. Find $P(D|BB)$.

32. On the basis of your answers to problems 28–31, is there much difference between outcomes for whites and blacks? Explain and compare your answer with problem 27.

 In Other Words

33. Of the two methods of solving conditional probability problems, which do you prefer? Why?

34. Can you find two events *A* and *B* such that $P(A|B) = P(A)$? What is the relationship between *A* and *B*? Explain.

 Using Your Knowledge

The *Statistical Abstract of the United States* gives the number of crime victims per 1000 persons, 12 years old and over, as shown in the following table:

Sex	Robbery	Assault	Personal Larceny
Male	5	18	52
Female	2	9	42

Use the information in this table to do the following problems:

35. a. Find the probability that the victim of one of the three types of crime was a male.

b. Find the probability that the victim of one of the three types of crime was a female.

c. Considering your answers to parts (**a**) and (**b**), which sex would you say is more likely to be the victim of one of these three types of crime?

36. If it is known that an assault was committed, what is the probability that the victim was a
a. male?   **b.** female?

37. If it is known that the victim was a female, what is the probability that the crime was assault?

38. If it is known that the victim was a male, what is the probability that the crime was robbery?

 Collaborative Learning

The study cited in problems 25–32 was done several years ago. Go to the *Statistical Abstract of the United States* or to the Web and find more recent and similar statistics to the ones given in the tables of problems 25–32. Discuss any changes in the outcomes.

11.5 Independent Events

Probabilities in Bingo and Birthdays

Have you played Bingo lately? The world's biggest Bingo contest was held in Cherokee, North Carolina, and offered a $200,000 prize to any player who could fill a 24-number card by the 48th number called (there are 75 possible numbers in Bingo). What is the probability that you would win this game? The probability that any given number on your 24-number card is drawn is $\frac{48}{75}$, the probability of drawing a second number on your card is $\frac{47}{74}$, and so on. To win, you must get *all* 24 numbers on your card in 48 draws. The probability is

$$\frac{48}{75} \cdot \frac{47}{74} \cdot \frac{46}{73} \cdot \ldots \cdot \frac{25}{52} = \frac{1}{799,399}$$

Note that the individual probabilities have been multiplied to find the final answer. In this section, you will study **independent** events. If two events A and B are independent, $P(A \cap B) = P(A) \cdot (B)$.

A classic use of this formula is the birthday problem. Given a group of people, what is $P(L)$, the probability that *at least* two people have the same birthday? It is easier to find $P(L')$, the probability that *no* two people have the same birthday, and then to compute $P(L) = 1 - P(L')$. Assuming that all birthdays are equally likely, the probability that a second person has a different birthday than a first is $\frac{364}{365}$, the probability that a third person has a different birthday than the other two is $\frac{363}{365}$, and the probability that an nth person has a different birthday than all the others is

$$\frac{365 - n + 1}{365}$$

Thus,

$$P(L) = 1 - \frac{364}{365} \cdot \frac{363}{365} \cdot \frac{362}{365} \cdot \ldots \cdot \frac{365 - n + 1}{365}$$

Now, compute some of these probabilities and note some others. When there are two people in the room, $n = 2$,

$$P(L) = 1 - \frac{364}{365} \approx 1 - 0.997 = 0.003$$

When there are three people in the room, $n = 3$,

$$P(L) = 1 - \frac{364}{365} \cdot \frac{363}{365} \approx 1 - 0.992 = 0.008$$

When $n = 10$, $P(L) = 0.117$; when $n = 22$, $P(L) = 0.476$; and when $n = 23$, $P(L) = 0.507$. Thus, with 22 people in a room, the probability that at least 2 have the same birthday is slightly under $\frac{1}{2}$; add one more person and it becomes slightly better than $\frac{1}{2}$. ▶

One of the more important concepts in probability is that of **independence.** In this section we shall define what we mean when we say that two events are independent. For example, the probability of obtaining a sum of 7 when two dice are

A water tower advertises bingo on a Florida Seminole Native American reservation.

rolled *and* it is known that the first die shows a 6 is $\frac{1}{6}$; that is $P(S|6) = \frac{1}{6}$. It is of interest that the probability of obtaining a 7 when two dice are rolled is also $\frac{1}{6}$, so $P(S|6) = P(S)$. This means that the additional information that a 6 came up on the first die does not affect the probability of the sum's being 7. It can happen, in general, that the probability of an event A is not affected by the occurrence of a second event B. Hence, we state the following definition:

Definition of Independent Events

Two events A and B are said to be **independent** if and only if

$$P(A|B) = P(A) \tag{1}$$

If A and B are independent, we can substitute $P(A)$ for $P(A|B)$ in the equation

$$P(A|B) = \frac{P(A \cap B)}{P(B)} \qquad \text{See Section 11.4, equation (1).}$$

to obtain

$$P(A) = \frac{P(A \cap B)}{P(B)}$$

Then, multiplying by $P(B)$, we get

$$P(A \cap B) = P(A) \cdot P(B)$$

Consequently, we see that an equivalent definition of independence is as follows:

Alternate Definition of Independent Events

Two events A and B are **independent** if and only if

$$P(A \cap B) = P(A) \cdot P(B) \tag{2}$$

A. Independent Events

The preceding ideas can be applied to experiments involving more than two events. We define independent events to be such that the occurence of any one of these events does not affect the probability of any other. The most important result for applications is that if n events, $E_1, E_2, \ldots, E_n$, are known to be independent, then the following multiplication rule holds:

Multiplication Rule for Independent Events

$$P(E_1 \cap E_2 \cap E_3 \cap \cdots \cap E_n) = P(E_1) \cdot P(E_2) \cdot \cdots \cdot P(E_n) \tag{3}$$

The next examples illustrate these ideas.

EXAMPLE 1 ▶ Determining If Events Are Independent

Two coins are tossed. Let E_1 be the event the first coin comes up tails, and let E_2 be the event the second coin comes up heads. Are E_1 and E_2 independent?

Solution

Because $P(E_1 \cap E_2) = \frac{1}{4}$, $P(E_1) = \frac{1}{2}$, $P(E_2) = \frac{1}{2}$, and $\frac{1}{2} \cdot \frac{1}{2} = \frac{1}{4}$, we see that $P(E_1 \cap E_2) = P(E_1) \cdot P(E_2)$. Hence, E_1 and E_2 are independent. ■

EXAMPLE 2 ▶ Probability of 2 Black Balls

We have two urns, I and II. Urn I contains 2 red and 3 black balls, whereas urn II contains 3 red and 2 black balls. A ball is drawn at random from each urn. What is the probability that both balls are black?

Solution

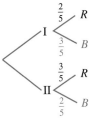

FIGURE 11.12

Let $P(B_1)$ be the probability of drawing a black ball from urn I, and let $P(B_2)$ be the probability of drawing a black ball from urn II. Clearly, B_1 and B_2 are independent events. Thus, $P(B_1) = \frac{3}{5}$ and $P(B_2) = \frac{2}{5}$, so $P(B_1 \cap B_2) = \frac{3}{5} \cdot \frac{2}{5} = \frac{6}{25}$ (see Figure 11.12). ■

EXAMPLE 3 ▶ Grades and Probability

Bob is taking math, Spanish, and English. He estimates that his probabilities of receiving A's in these courses are $\frac{1}{10}$, $\frac{3}{10}$, and $\frac{7}{10}$, respectively. If he assumes that the grades can be regarded as independent events, find the probability that Bob makes

(a) all A's (event A).

(b) no A's (event N).

(c) exactly two A's (event T).

Solution

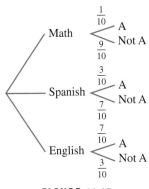

FIGURE 11.13

(a) $P(A) = P(M) \cdot P(S) \cdot P(E) = \frac{1}{10} \cdot \frac{3}{10} \cdot \frac{7}{10} = \frac{21}{1000}$, where M is the event in which he makes an A in math, S is the event in which he makes an A in Spanish, and E is the event in which he makes an A in English (see the tree diagram in Figure 11.13).

(b) $P(N) = P(M') \cdot P(S') \cdot P(E') = \frac{9}{10} \cdot \frac{7}{10} \cdot \frac{3}{10} = \frac{189}{1000}$

(c) There are the following three ways of getting exactly two A's:

1. Getting A's in math and Spanish and not in English. The probability of this event is
$$P(M) \cdot P(S) \cdot P(E') = \frac{1}{10} \cdot \frac{3}{10} \cdot \frac{3}{10} = \frac{9}{1000}$$

2. Getting A's in math and English and not in Spanish. The probability of this event is
$$P(M) \cdot P(S') \cdot P(E) = \frac{1}{10} \cdot \frac{7}{10} \cdot \frac{7}{10} = \frac{49}{1000}$$

3. Getting A's in Spanish and English and not in math. The probability of this event is
$$P(M') \cdot P(S) \cdot P(E) = \frac{9}{10} \cdot \frac{3}{10} \cdot \frac{7}{10} = \frac{189}{1000}$$

Since the three events we have just considered are mutually exclusive, the probability of getting exactly two A's is the sum of the probabilities we calculated. Thus,

$$P(T) = \frac{9}{1000} + \frac{49}{1000} + \frac{189}{1000} = \frac{247}{1000}$$

where T is the event of getting exactly two A's. ■

EXAMPLE 4 ▶ Witches, Probability, and the Stock Market

Do you recall the witches of Wall Street? (See Section 11.2, Example 1.) The witches in this case are persons who claim that they use occult powers to predict the behavior of stocks on the stock market. One of the most famous of the witches claims to have a 70% accuracy record. A stockbroker selects three stocks, A, B, and C, at random from the New York Stock Exchange listing and asks this witch to predict their behavior. Assuming that the 70% accuracy claim is valid, find the probability that the witch will

(a) correctly predict the behavior of all three stocks.

(b) incorrectly predict the behavior of all three stocks.

(c) correctly predict the behavior of exactly two of the three stocks.

The witches of Wall Street claim they can predict the behavior of stocks on the stock market. Do you think the witches read the *Wall Street Journal?*

Solution

(a) The probability of the witch's correctly predicting the behavior of all three stocks is the product $P(A) \cdot P(B) \cdot P(C)$ or

$$(0.70)(0.70)(0.70) = 0.343$$

(b) The probability of the witch's incorrectly predicting the behavior of all three stocks is the product $P(A') \cdot P(B') \cdot P(C')$ or

$$(0.30)(0.30)(0.30) = 0.027$$

(c) The probability of the witch's correctly predicting the behavior of two specific stocks and incorrectly predicting the behavior of the third stock is the product $P(A) \cdot P(B) \cdot P(C')$ or

$$(0.70)(0.70)(0.30) = 0.147$$

Because there are 3 ways of selecting the two specific stocks, we use the SCP and multiply the last result by 3. Thus, the probability of correctly predicting the behavior of exactly two of the stocks is

$$3(0.147) = 0.441$$

You can visualize the calculations in Example 4(c) by looking at the tree diagram in Figure 11.14, where C represents a correct prediction and I represents an incorrect prediction. Each branch is labeled with the probability of the event it represents. Note that if you find and add the probabilities at the ends of all the branches, the sum will be 1.

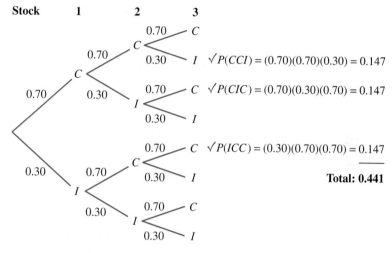

FIGURE 11.14

B. Stochastic Processes

A **stochastic process** is a sequence of experiments in which the outcome of each experiment depends on chance. For example, the repeated tossing of a coin or of a die is a stochastic process. Tossing a coin and then rolling a die is also a stochastic process.

In the case of repeated tosses of a coin, we assume that on each toss there are two possible outcomes, each with probability $\frac{1}{2}$. If the coin is tossed twice, we can construct a tree diagram corresponding to this sequence of experiments (see Figure 11.15 on page 752).

In Figure 11.15 we have put on each branch the probability of the event corresponding to that branch. To obtain the probability of, say, a tail and then a head, $P(T \cap H)$, we multiply the probabilities on each of the branches going along the path that leads from the start to the final outcome, as indicated in Figure 11.15. This multiplication gives $\left(\frac{1}{2}\right)\left(\frac{1}{2}\right) = \frac{1}{4}$, in agreement with the results we have previously obtained.

It is possible to show, by means of the SCP, that the terminal probabilities are always correctly obtained by using this multiplication technique. Notice that Figure 11.15 illustrates that

$$P(H \cap T) = P(H|T) \cdot P(T) \quad \text{and} \quad P(T \cap T) = P(T|T) \cdot P(T)$$

	First Toss	Second Toss	Final Outcome	Probability

$$P(H|H) \atop \frac{1}{2}$$ — H (H, H) $P(H \cap H) = \frac{1}{2} \times \frac{1}{2} = \frac{1}{4}$

$P(H) \atop \frac{1}{2}$ H

$$\frac{1}{2} \atop P(T|H)$$ — T (H, T) $P(H \cap T) = \frac{1}{2} \times \frac{1}{2} = \frac{1}{4}$

$$P(H|T) \atop \frac{1}{2}$$ — H (T, H) $P(T \cap H) = \frac{1}{2} \times \frac{1}{2} = \frac{1}{4}$

$\frac{1}{2} \atop P(T) \atop$ T

$$\frac{1}{2} \atop P(T|T)$$ — T (T, T) $P(T \cap T) = \frac{1}{2} \times \frac{1}{2} = \frac{1}{4}$

FIGURE 11.15

EXAMPLE 5 ▶ Probability of a Head and a 5

A coin and a die are tossed. What is the probability of getting a head and a 5?

Solution

Since the outcomes depend only on chance, this is a stochastic process for which the multiplication procedure can be used. Since the probability of getting a head on the coin is $\frac{1}{2}$, and the probability of getting a 5 on the die is $\frac{1}{6}$, the probability of getting a head and a 5 is

$$P(H, 5) = \frac{1}{2} \times \frac{1}{6} = \frac{1}{12}$$

EXAMPLE 6 ▶ Outcomes from an Unbalanced Coin

Jim has two coins, one fair (F) and the other unbalanced (U) so that the probability of its coming up heads is $\frac{2}{3}$. He picks up one of the coins and tosses it, and it comes up heads. What is the probability that the outcome came from the unbalanced coin?

Solution

We draw a tree diagram as shown in Figure 11.16. The probabilities at the ends of the branches are obtained by the multiplication technique. The asterisked probabilities (*) may be taken as weights for the corresponding events. Thus, the required probability is $\dfrac{\frac{2}{6}}{\frac{1}{4} + \frac{2}{6}} = \dfrac{4}{7}$

FIGURE 11.16

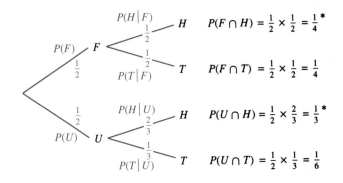

$$P(H|F) \atop \frac{1}{2}$$ — H $P(F \cap H) = \frac{1}{2} \times \frac{1}{2} = \frac{1}{4}$ *

$P(F) \atop \frac{1}{2}$ F

$$\frac{1}{2} \atop P(T|F)$$ — T $P(F \cap T) = \frac{1}{2} \times \frac{1}{2} = \frac{1}{4}$

$$P(H|U) \atop \frac{2}{3}$$ — H $P(U \cap H) = \frac{1}{2} \times \frac{2}{3} = \frac{1}{3}$ *

$\frac{1}{2} \atop P(U) \atop$ U

$$\frac{1}{3} \atop P(T|U)$$ — T $P(U \cap T) = \frac{1}{2} \times \frac{1}{3} = \frac{1}{6}$

The required probability also can be found by using the equation

$$P(U|H) = \frac{P(U \cap H)}{P(H)} = \frac{\frac{1}{3}}{\frac{1}{4} + \frac{1}{3}} = \frac{4}{7}$$

■

EXAMPLE 7 ▶ Longevity Probabilities from a Table

Referring to Table 11.4 (page 733), find the probability that two people who are 50 and 55 years old, respectively, will both be alive at age 70.

Solution

The probability that a 50-year-old person lives to 70 is $\frac{386}{698} = \frac{193}{349}$, and the probability that a 55-year-old person lives to 70 is $\frac{386}{646} = \frac{193}{323}$. Thus, the probability that both persons live to 70 is

$$\frac{193}{349} \times \frac{193}{323} = \frac{37,249}{112,727} \approx 0.33$$

■

EXAMPLE 8 ▶ Pancakes, Hats, and Marilyn

The following problem appeared in the "Ask Marilyn" question-and-answer column in *Parade* magazine:

> You have a hat in which there are three pancakes: One is golden on both sides, one is brown on both sides, and one is golden on one side and brown on the other. You withdraw one pancake, look at one side, and see that it is brown. What is the probability that the other side is brown?

Robert Batts
Acton, Massachusetts

So, what do you think? There are several ways of solving the problem; we shall use a tree diagram.

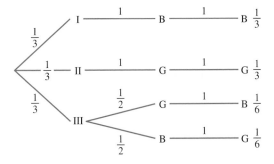

FIGURE 11.17

Solution

Let us make a tree diagram as in Figure 11.17 and label the pancakes and their sides I(B, B), II(G, G), and III(G, B). Let S be the second side is brown, and F the first side is brown. We want

$$P(S|F) = \frac{P(S \cap F)}{P(F)}$$

For both sides to be brown, we must select I(B, B) with probability $\frac{1}{3} = P(S \cap F)$. To find $P(F)$, let us study the tree.

If we select I(B, B), both sides are certainly brown, with probability 1.

If we select II(G, G), both sides are certainly golden with probability 1.

Finally, for III(G, B), the first side can be G or B each with probability $\frac{1}{2}$. If you picked G first, the second side is B with probability 1. If you picked B first, the second side is G with probability 1. Thus, the required probability is

Online Study Center

To further explore the three-pancake problem, access links 11.5.1 and 11.5.2 on this textbook's Online Study Center.

$$\frac{\frac{1}{3}}{\frac{1}{3} + \frac{1}{6}} = \frac{\frac{2}{6}}{\frac{3}{6}} = \frac{2}{3}$$ ∎

EXERCISES 11.5

Ⓐ Independent Events

1. Two coins are tossed. Let E_1 be the event in which the first coin comes up heads, and let E_2 be the event in which the second coin comes up tails. Are E_1 and E_2 independent?

2. A bubble gum machine has 50 cherry-flavored gums, 20 grape-flavored, and 30 licorice-flavored; a second machine has 40 cherry, 50 grape, and 10 licorice. A gum is drawn at random from each machine. Find the probability that
  a. both gums are cherry flavored.
  b. both gums are licorice flavored.
  c. the gum from the first machine is cherry flavored, and the one from the second machine is grape flavored.

3. A computer repair shop has estimated the probability that a computer sent to the shop has a bad modem is $\frac{1}{4}$, the probability that the computer has a bad CPU is $\frac{1}{8}$, and the probability that it has a bad drive is $\frac{1}{3}$. If we assume that modems, CPUs, and drives are independent, find the probability that
  a. a modem, CPU, and a drive in a computer sent to the shop are bad.
  b. only a modem and a CPU in a computer sent to the shop are bad.
  c. none of the three parts (modem, CPU, or drive) is bad.

4. The following table gives the kinds of stocks available in three brokerage houses, H_1, H_2, and H_3. A brokerage house is selected at random, and one type stock is selected. Find the probability that the stock is
  a. a petroleum stock.   **b.** a computer stock.

Houses	Petroleum	Computer
H_1	3	2
H_2	2	3
H_3	2	2

5. A coin is tossed 3 times. Find the probability of obtaining
  a. heads on the first and last toss, and tails on the second toss.
  b. at least 2 heads.   **c.** at most 2 heads.

6. A die is rolled 3 times. Find the probability of obtaining
  a. an odd number each time.
  b. 2 odd numbers first and an even one on the last roll.
  c. at least 2 odd numbers.

7. A card is drawn from an ordinary deck of 52 cards, and the result is recorded on paper. The card is then returned to the deck and another card is drawn and recorded. Find the probability that
  a. the first card is a spade.
  b. the second card is a spade.
  c. both cards are spades.
  d. neither card is a spade.

8. Rework problem 7, assuming that the 2 cards are drawn in succession and without replacement. (*Hint:* Make a tree diagram and assign probabilities to each of the branches.)

9. A family has 3 children. Let M be the event "the family has at most 1 girl," and let B be the event "the family has children of both sexes."
  a. Find $P(M)$.   **b.** Find $P(B)$.
  c. Find $P(B \cap M)$.
  d. Determine whether B and M are independent.

10. Two cards are drawn in succession and without replacement from an ordinary deck of 52 cards. What is the probability that
    a. the first card is a king and the second card is an ace?
    b. both cards are aces?
    c. neither card is an ace?
    d. exactly 1 card is an ace?

In Problems 11–15, assume the spinner has 5 *unequal* sectors $\frac{1}{6}$ red, $\frac{1}{9}$ blue, $\frac{5}{18}$ green, $\frac{2}{9}$ white, and $\frac{2}{9}$ yellow as shown. If the spinner is spun twice, find the probability (as a reduced fraction) that the spinner will

11. land in the green and then in the red sector.

12. land in the white and then in the blue sector.

13. land twice on green.

14. not land on the white sector.

15. land in a sector other than red.

Do you know what the message below says? If the 20 marbles used to spell the word are placed in a glass jar and two marbles are chosen at random in succession and without replacement, what is the probability (as a reduced fraction) of choosing

16. 2 red marbles?

17. a red and then a blue marble?

18. a marble other than red each time?

19. a green and then a yellow marble?

20. 2 marbles of the same color?

Think braille!

A jar contains 7 red, 6 green, 8 blue, and 4 yellow marbles. A marble is chosen at random from the jar. After replacing it, a second marble is chosen. What is the probability of choosing

21. a red and then a yellow marble?

22. 2 yellow marbles?

23. no blue marbles?

24. 2 marbles of the same color?

25. at least one red marble?

The circle graph below will be used in problems 26–30. In a survey of different spam categories conducted by Clearswift, the percent of each type of spam was as shown in the graph. Suppose 3 spam messages from the ones included in the survey are selected at random. Find the probability (written as a decimal) that

26. all of them were about healthcare.

27. none of them was about healthcare.

28. the first one was about healthcare, the second one about finance, and the third regarded direct products.

29. there was no direct product spam among the 3.

30. If you were to regulate spamming, which category would you investigate first?

Spam Categorization Breakdown, August 2005

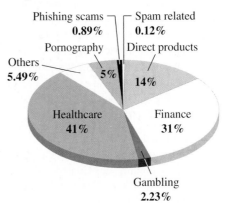

Note: Results do not add to 100% because of rounding.
Source: www.clickz.com/stats/sectors/email/article.php/3549111 #table1.

31. A company has estimated that the probabilities of success for three products introduced in the market are $\frac{1}{4}$, $\frac{2}{3}$, and $\frac{1}{2}$, respectively. Assuming independence, find the probability that
  a. the three products are successful.
  b. none of the products is successful.

32. In problem 31, find the probability that exactly 1 product is successful.

33. A coin is tossed. If heads comes up, a die is rolled; but if tails comes up, the coin is thrown again. Find the probability of obtaining
  a. 2 tails.
  b. heads and the number 6.
  c. heads and an even number.

34. In a survey of 100 persons, the data in the following table were obtained:

Type of Person	Lung Cancer (L)	No Lung Cancer (L')
Smoker (S)	42	28
Nonsmoker (S')	18	12

  a. Are S and L independent?
  b. Are S' and L' independent?

35. Referring to Table 11.4 (page 733), find the probability that two persons, one 30 years old and the other 40 years old, will live to be 60.

36. In problem 35, find the probability that both persons will live to be 70.

37. The Apollo module has five components: the main engine, the propulsion system, the command service module, the lunar excursion module (LEM), and the LEM engine. If each of the systems is considered independent of the others and the probability that each of the systems performs satisfactorily is 0.90, what is the probability all the systems will perform satisfactorily?

38. A die is loaded so that 1, 2, 3, and 4 each has probability $\frac{1}{8}$ of coming up, while 5 and 6 each has probability $\frac{1}{4}$ of coming up. Consider the events $A = \{1, 3, 5\}$ and $B = \{2, 4, 5\}$. Determine whether A and B are independent.

39. On one of the experimental flights of a space shuttle, the mission was cut short due to a malfunction of a battery aboard the ship. The batteries in the shuttle are guaranteed to have a failure rate of only 1 in 20. The system of three batteries is designed to operate as long as any one of the batteries functions properly. Find the probability that
  a. all three batteries fail.
  b. exactly two fail.

40. In a certain city, the probability of catching a burglar is 0.30, and the probability of convicting a caught burglar is 0.60. Find the probability that a burglar will be caught and convicted.

41. In Example 4, what is the probability of a Wall Street witch predicting correctly the behavior of one of the stocks and predicting incorrectly the behavior of the other two?

42. In Example 4, suppose the broker had selected four stocks. What is the probability that the witch would give a correct prediction for two of the stocks and an incorrect prediction for the other two stocks?

B Stochastic Processes

43. Three boxes, labeled A, B, and C, contain 1 red and 2 black balls, 2 red and 1 black ball, and 1 red and 1 black ball, respectively. First a box is selected at random, and then a ball is drawn at random from that box. Find the probability that the ball is red. (*Hint:* Draw a tree diagram, assign the probabilities to the separate branches, and compute the terminal probabilities by using the multiplication technique. Then add the terminal probabilities for all the outcomes in which the ball is red.)

44. There are three filing cabinets, each with two drawers. All the drawers contain letters. In one cabinet, both drawers contain airmail letters; in a second cabinet, both drawers contain ordinary letters; and in the third cabinet, one drawer contains airmail and the other contains ordinary letters. A cabinet is selected at random, and then a drawer is picked at random from this cabinet. When the drawer is opened, it is found to contain airmail letters. What is the probability that the other drawer of this cabinet also contains airmail letters? (*Hint:* Use the same procedure as in problem 43.)

45. John has two coins, one fair and the other unbalanced so that the probability of its coming up heads is $\frac{3}{4}$. He picks one of the coins at random, tosses it, and it comes up heads. What is the probability that he picked the unbalanced coin?

46. A box contains 3 green balls and 2 yellow balls. Two balls are drawn at random in succession and without replacement. If the second ball is yellow, what is the probability that the first one is green?

In Other Words

47. a. Explain, in your own words, what is meant by the statement "Two events A and B are independent."
  b. If A and B are independent events and you know $P(A)$ and $P(B)$, how can you calculate $P(A \cap B)$?

48. In Getting Started you found the probability that at least two people have the same birthday. Is this the same as finding the probability that another person has the same birthday as you do? Explain.

Using Your Knowledge

Suppose that a fair coin is flipped 10 times in succession. What is the probability that exactly 4 of the flips turn up heads? This is a problem in which repeated trials of the same experiment are made, and the probability of success is the same for each of the trials. This type of procedure is often called a **Bernoulli trial,** and the final probability is known as a **binomial probability.**

Let us see if we can discover how to calculate such a probability. We represent the 10 flips and one possible success like this:

H	T	T	H	H	T	T	T	H	T
1	2	3	4	5	6	7	8	9	10

Because each flip is independent of the others, the probability of getting the particular sequence shown is $\left(\frac{1}{2}\right)^{10}$. All we need do now is find in how many ways we can succeed, that is, in how many ways we can get exactly 4 heads. This is the same as the number of ways we can select 4 of the 10 flips, that is, $C(10, 4)$. The successful

ways of getting 4 heads are all mutually exclusive, so the probability of getting exactly 4 heads is

$$\frac{C(10, 4)}{2^{10}}$$

Let us suppose now that the coin is biased so that the probability of heads on any one toss is p and the probability of tails is $q = 1 - p$. The probability of getting the arrangement we have shown is now $p^4 q^6$. (Why?) Hence, the probability of getting exactly 4 heads is

$$C(10, 4)p^4 q^6$$

You should be able to convince yourself that if n is the number of trials, p is the probability of success in each trial, and $q = 1 - p$ is the probability of failure, then the probability of exactly x successes is

$$C(n, x)p^x q^{n-x}$$

49. Suppose that a fair coin is tossed 50 times in succession. What is the probability of getting exactly 25 heads? (Do not multiply out your answer.)

50. If a fair coin is tossed 6 times in succession, what is the probability of getting at least 3 heads?

51. Suppose that the coin in problem 50 is biased 2 to 1 in favor of heads. Can you calculate the probability of getting at least 3 heads?

52. Suppose that a fair coin is tossed an even number of times, 2, 4, 6, ... What happens to the probability of getting heads in exactly half the tosses as the number of tosses increases?

53. A fair die is tossed 5 times in succession. What is the probability of getting exactly two 3s?

54. In problem 53, what is the probability of obtaining at least two 3s?

Discovery

Suppose you have two switches, S_1 and S_2, installed in series in an electric circuit and these switches have probabilities $P(S_1) = \frac{9}{10}$ and $P(S_2) = \frac{8}{10}$ of working. As you can see from the figure, the probability that the circuit works is the probability that S_1 and S_2 work; that is,

$$P(S_1) \cdot P(S_2) = \frac{9}{10} \times \frac{8}{10} = \frac{72}{100}$$

$$A \text{———} S_1 \text{———} S_2 \text{———} B$$

If the same two switches are installed in parallel (see the figure below), then we can calculate the probability that the circuit works by first calculating the probability that it does not work, as follows:

The probability that S_1 fails is $1 - \frac{9}{10} = \frac{1}{10}$.
The probability that S_2 fails is $1 - \frac{8}{10} = \frac{2}{10}$.
The probability that both S_1 and S_2 fail is
$\frac{1}{10} \times \frac{2}{10} = \frac{2}{100}$.

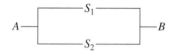

Thus, the probability that the circuit works is $1 - \frac{2}{100} = \frac{98}{100}$. By comparing the probability that a series circuit works $\left(\frac{72}{100}\right)$ with the probability that a parallel circuit works $\left(\frac{98}{100}\right)$, we can see that it is better to install switches in parallel.

55. What is the probability that a series circuit with three switches, S_1, S_2, and S_3, with probabilities $\frac{1}{3}$, $\frac{1}{2}$, and $\frac{3}{4}$ of working, will work?

56. What is the probability if the switches are installed in parallel?

We have just seen that, under certain circumstances, it is better to install parallel rather than series circuits to obtain maximum reliability. However, in the case of security systems, independent components in series are the most reliable. For example, consider a triple-threat security system that uses voice patterns, finger-prints, and handwriting to screen persons entering a maximum security area. Here is how the system operates. To enter a secure area, a person must pass through a room that has a door at each end and contains three small booths. In the first booth, theperson punches in his or her four-digit identification number. This causes the machine inside the booth to intone four words, which the person must repeat. If the voice pattern matches the pattern that goes with the identification number, the machine says "Thank you" and the person goes to the next booth. After entering his or her number there, the person signs his or her name on a Mylar sheet. If the signature is acceptable, the machine flashes a green light and the person goes to the third booth. There, he or she punches in the identification number once more and then pokes a finger into a slot, fingerprint down. If a yellow light flashes "IDENTITY VERIFIED," the door opens and the person can enter the high-security area.

57. Assume each of the machines is 98% reliable. What is the probability that a person fools the first machine?

58. What is the probability that a person fools the first two machines?

59. What is the probability that a person fools all three machines?

60. On the basis of your answer to problem 39, how would you rate the reliability of this security system?

11.6   Odds and Mathematical Expectation

Lottery Odds

Look at the information in Table 11.6. The odds of winning the first prize by picking 6 out of 6 numbers (there are 49 numbers to pick from) are said to be 1 in 13,983,816. But in Section 11.1 we found that the *probability* of winning the first prize is $\frac{1}{13,983,816}$. Isn't there a difference between odds and probability? Of course there is! What most state lotteries mistakenly report as *odds* are actually the *probabilities* of winning.

To explain further, the **probability** of an event is a fraction whose numerator is the number of times the event can occur and whose denominator is the total number of possibilities in the sample space. Thus, if we throw a die, the probability of getting a number greater than 4 is $\frac{2}{6}$ because there are two favorable outcomes of numbers greater than 4 (5 and 6) out of 6 total possibilities (1, 2, 3, 4, 5, and 6.) The **odds** in favor of an event are defined as the **ratio** of favorable to

TABLE 11.6 State Lottery

| Prize Divisions and Odds of Winning ||
Matching Numbers	Odds
6 of 6 numbers	1 in 13,983,816
5 of 6 numbers	1 in 54,200.84
4 of 6 numbers	1 in 1032.4
3 of 6 numbers	1 in 56.66
Overall odds	1 in 53.66

unfavorable occurrences for the event. Thus, the odds for getting a number greater than 4 are *2 to 4* because there are 2 favorable outcomes (5 and 6) and 4 unfavorable ones (1, 2, 3, and 4). These odds are sometimes written as 2 : 4 (read "2 to 4"). Now, back to the lottery ticket. Since the probability of winning the first prize is 1 to 13,983,816, the odds of winning the first prize are 1 to 13,983,815, not 1 to 13,983,816.

Some probabilities are given in Table 11.7. What are the corresponding odds?

TABLE 11.7

Situations	Probability	Odds in Favor
Getting married if you are 18 or older	0.64	64 : 36
Having 3 or more children	0.11	
Developing high blood pressure	$\frac{2}{5}$	
Getting accepted to medical school	$\frac{6}{10}$	
Never eating candy	$\frac{1}{33}$	

Note that if the odds for the event are 64 to 36, the probability of the event should be

$$\frac{64}{64 + 36} = \frac{64}{100} = 0.64$$

We shall study more about the relationship between odds and probability in this section. ▶

In this chapter we have several times used games of chance to illustrate the concepts of probability. In connection with these games, we often encounter such statements as "The odds are 1 to 5 for rolling a 1 with a die" or "The odds are 12 to 1 against picking an ace from a deck of cards." When a person gives us 1 to 5 odds for rolling a 1 with a die, it usually means that if a 1 does occur, we pay $5 and, if a 1 does not occur, the person pays $1. These statements simply compare the number of favorable outcomes to the number of unfavorable outcomes. Thus, odds of 1 to 5 mean that there are 5 times as many unfavorable as favorable outcomes.

A. Odds

> **Odds in Favor of an Event**
>
> If an event E is such that the total number of favorable outcomes is f and the total number of unfavorable outcomes is u, the **odds in favor of E are f to u.**

For instance, there are 4 aces in a standard deck of 52 cards. Thus, if a single card is drawn from the deck, there are 4 ways of getting an ace (favorable) and 48 ways of not getting an ace (unfavorable). Thus, the odds in favor of drawing an ace are 4 to 48, or 1 to 12.

Favorable   Unfavorable

EXAMPLE 1 ▶ Odds When Rolling 1 Die

A fair die is rolled. What odds should a person give

(a) in favor of a 1 turning up?   (b) against a 1 turning up?

Solution

(a) In this case, there is 1 favorable outcome, so $f = 1$, and there are 5 unfavorable outcomes, so $u = 5$. Thus, the odds are 1 to 5.

(b) There are 5 ways in which 1 does not turn up (favorable) and 1 way in which 1 turns up (unfavorable), so the odds against a 1 turning up are 5 to 1.   ■

EXAMPLE 2 ▶ Probabilities Based on a Record

A horse named Camarero has a record of 73 wins and 4 losses. Based on this record, what is the probability of a win for this horse?

Solution
Here, $f = 73$ and $u = 4$, since the probability of an event is

$$\frac{\text{Number of favorable outcomes}}{\text{Number of possible outcomes}}$$

Favorable →
Possible → $\dfrac{f}{f+u} = \dfrac{73}{73+4} = \dfrac{73}{77}$ ■

If n is the total number of possible outcomes, and f and u are as before, then we know that

$$P(E) = \frac{f}{n} \qquad \text{and} \qquad P(\text{not } E) = \frac{u}{n}$$

Therefore,

$$\frac{P(E)}{P(\text{not } E)} = \frac{f/n}{u/n} = \frac{f}{u}$$

Thus, an equivalent definition of odds in favor of the event E is

$$P(E) \quad \text{to} \quad P(\text{not } E)$$

or, since $P(\text{not } E) = 1 - P(E)$,

$$P(E) \quad \text{to} \quad 1 - P(E)$$

In the case of the die in Example 1(a), the odds are $\frac{1}{6}$ to $\frac{5}{6}$, which is the same as 1 to 5. Note that if $P(E)$ and $P(\text{not } E)$ are expressed as fractions with the same denominator, we can compare just the numerators. For example, if $P(E) = \frac{2}{7}$, then $P(\text{not } E) = 1 - \frac{2}{7} = \frac{5}{7}$, so the odds in favor of E are 2 to 5.

EXAMPLE 3 ▶ From Probability to Odds

A horse named Blue Bonnet has won 5 of her last 8 races and is thus assigned a probability of $\frac{5}{8}$ of winning her ninth race. Assuming this probability is correct, what are the odds *against* Blue Bonnet's winning that race?

Solution

Since $P(\text{winning}) = \frac{5}{8}$, we know $P(\text{not winning}) = 1 - \frac{5}{8} = \frac{3}{8}$. Thus, the odds in favor of Blue Bonnet are 5 to 3, and the odds *against* her are 3 to 5. ∎

EXAMPLE 4 ▶ Odds for the Florida Lottery

To play Lotto in Florida you select 6 numbers from 1 to 53 and pay \$1. To win, you must match (in any order) 3, 4, 5, or 6 of the winning numbers drawn in the official drawing for the date played.

(a) What is the probability that you match 6 out of the 6 numbers?

(b) What are the odds in favor of matching 6 out of 6 numbers?

(c) What is the probability that you match exactly 5 out of the 6 numbers?

(d) What are the odds in favor of matching exactly 5 out of 6 numbers?

Solution

Recall that the probability of an event is

$$\frac{\text{Favorable outcomes}}{\text{Total outcomes}}$$

(a) Since there are 53 numbers to choose from and we have to select 6 of them, the total number of outcomes is $C(53, 6)$. Of these, only 1 is favorable, thus

$$\frac{\text{Favorable outcomes}}{\text{Total outcomes}} = \frac{1}{C(53, 6)} = \frac{1}{22{,}957{,}480}$$

(b) The odds in favor are 1 to 22,957,479.

(c) To match 5 out of the 6 numbers, we have to select 5 numbers from the 6 winning numbers. This can be done in $C(6, 5)$ ways. We then have to select one more number from the 47 $(53 - 6 = 47)$ remaining numbers. This can be done in 47 different ways. Thus, we have $C(6, 5) \cdot 47$ favorable outcomes out of $C(53, 6)$. The probability is

$$\frac{C(6, 5) \cdot 47}{C(53, 6)} = \frac{6 \cdot 47}{22{,}957{,}480} = \frac{282}{22{,}957{,}480}$$

(d) The odds in favor of this are 282 to 22,957,198.

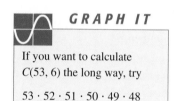

GRAPH IT

If you want to calculate $C(53, 6)$ the long way, try

$$\frac{53 \cdot 52 \cdot 51 \cdot 50 \cdot 49 \cdot 48}{6 \cdot 5 \cdot 4 \cdot 3 \cdot 2 \cdot 1}$$

Simplify!
With a grapher, enter 5 3 MATH ◀ 3 6. Then press ENTER. The answer is shown.

```
53 nCr 6
              22957480
```

TABLE 11.8 Florida Lotto Odds and Estimated Prizes

Prize Level	Estimated Prize*	Odds of Winning
6-of-6	Jackpot†	1:22,957,480
5-of-6	$5,000	1:81,409.50
4-of-6	$70	1:1,415.82
3-of-6	$5	1:70.79
Overall		1:67.36

*The estimated prize payouts to each winner are based on averages
of previous payouts. FLORIDA LOTTO is a parimutual game.
Actual prize amounts vary depending on sales and the number of
winners in each prize category.
†If no FLORIDA LOTTO ticket matches all six winning numbers,
the jackpot rolls over, and the cash in the jackpot prize pool is carried over to the next drawing.
Source: Florida Lottery.

The official Florida Lotto table (Table 11.8) confuses *odds* with probability, erro-
neously stating that the *odds* of winning the jackpot are 1:22,957,480 when, as
we have shown, the *probability* of winning the jackpot is $\frac{1}{22,957,480}$. Can you figure
out how they obtained the odds for getting 5 of 6 numbers?   ■

B. Expected Value

In many games of chance we are concerned with betting. Suppose that a given
event E has probability $P(E) = f/n$ of occurring and $P(\text{not } E) = u/n$ of not occurring. If we now agree to pay f dollars if E does not occur in exchange for receiving u dollars if E does occur, then we can calculate our "expected average winnings" by multiplying $P(E)$ (the approximate proportion of the times we win) by
u (the amount we win each time). Similarly, our losses will be $P(\text{not } E) \times f$,
because we lose f dollars approximately $P(\text{not } E)$ of the times. If the bet is to be
fair, the average net winnings should be 0. Let us see if this is the case. Our net
winnings will be

$$\underbrace{P(E) \times u}_{\text{Our gain}} - \underbrace{P(\text{not } E) \times f}_{\text{Our loss}} = \frac{f}{n} \times u - \frac{u}{n} \times f$$

$$= \frac{fu - uf}{n} = 0$$

as they should be. Since the odds in favor of E are f to u, we state the following
definition:

Definition of a Fair Bet

If the probability that event E will occur is f/n and the probability that E
will not occur is u/n, where n is the total number of possible outcomes,
then the odds of f to u in favor of E occurring constitute a **fair bet.**

EXAMPLE 5 ▶ Odds for a Fair Bet

A woman bets that she can roll a 7 in one roll of a pair of dice. What odds should she give for the bet to be fair?

Solution

$P(7) = \frac{6}{36} = \frac{1}{6}$. Here $1 = f$ and $f + u = 6$, so $u = 5$. Hence, the odds should be 1 to 5. ∎

Sometimes we wish to compute the *expected value*, or *mathematical expectation*, of a game. For example, if a woman wins $6 when she obtains a 1 in a single roll of a die and loses $12 for any other number, we can see that if she plays the game many times, she will win $6 one-sixth of the time and she will lose $12 five-sixths of the time. We then expect her to gain $(\$6)(\frac{1}{6}) - (\$12)(\frac{5}{6}) = -\$9$, that is, to lose $9 per try on the average.

For another example, if a fair die is rolled 600 times, we would expect $(\frac{1}{6})(600) = 100$ ones to appear. This does not mean that exactly 100 ones *will appear* but that this is the *expected average* number of ones for this experiment. In fact, if the number of ones were far away from 100, we would have good reason to doubt the fairness of the die.

Definition of Expected Value

If the k possible outcomes of an experiment are assigned the values a_1, $a_2, \ldots, a_k$ and they occur with probabilities $p_1, p_2, \ldots, p_k$, respectively, then the **expected value** of the experiment is given by

$$E = a_1 p_1 + a_2 p_2 + a_3 p_3 + \cdots + a_k p_k$$

A casino game called Keno is played with 80 balls numbered 1 through 80. Twenty winning balls are chosen at random. A popular bet is the $2, 10-number bet in which you select 10 numbers and the casino will pay you, say, $4 if exactly 5 of the numbers you picked match 5 of the 20 that were selected. What is the probability of that happening?

There are 80 numbers altogether, 10 you pick and 70 you do not. There are $C(10, 5)$ ways to pick the 5 matching numbers from your 10. That leaves $C(70, 15)$ for the numbers you do not pick.

Thus, the number of ways of matching 5 numbers from the 10 is

$$C(10, 5) \cdot C(70, 15)$$

The number of combinations when 20 balls are picked from 80 is $C(80, 20)$. Thus,

$$P(\text{match 5}) = \frac{C(10, 5) \cdot C(70, 15)}{C(80, 20)}$$

$$\approx 0.0514277$$

Table 11.9 gives the payoffs and probabilities for a $2, 10-number bet.

TABLE 11.9

Match	Pays	Probability
5	$ 4	0.0514277
6	$ 40	0.0114794
7	$ 280	0.0016111
8	$ 1800	0.0001354
9	$ 8000	0.0000061
10	$50,000	0.0000001

PROBLEM SOLVING

Mathematical Expectation

❶ Read the problem.

Find the mathematical expectation for a $2, 10-number bet in Keno.

❷ Select the unknown.

You want to find the mathematical expectation of a $2 bet.

❸ Think of a plan.
Find how much you can expect to get if you match 5, 6, 7, 8, 9, or 10 numbers. Subtract the $2 cost of the ticket.

To find how much you can expect to get for matching 5, 6, 7, 8, 9, or 10 numbers, multiply the payoffs by their probabilities and add to get your winnings. Then, subtract $2.

❹ Use Table 11.9 to find the expected value of the winnings for matching the following:

Here are the payoffs times the probabilities rounded to two decimal places.

5 numbers	
6 numbers	
7 numbers	
8 numbers	
9 numbers	
10 numbers	

$$\$4 \cdot 0.0514277 = \$0.21$$
$$\$40 \cdot 0.0114794 = \$0.46$$
$$\$280 \cdot 0.0016111 = \$0.45$$
$$\$1800 \cdot 0.0001354 = \$0.24$$
$$\$8000 \cdot 0.0000061 = \$0.05$$
$$\$50{,}000 \cdot 0.0000001 = \underline{\$0.01}$$
$$\$1.42$$

Subtract the cost of the $2 ticket to find E.

$$E = \$1.42 - \$2.00 = -\$.58$$

Thus, your expected value is $-\$.58$. (You lose 58 cents!)

❺ Verify the solution.

Do this with a calculator!

TRY EXAMPLE 6 NOW.

Cover the solution, write your own solution, and then check your work.

EXAMPLE 6 ▶ Finding Expected Value

A die is rolled. If an even number comes up, a person receives $10; otherwise, the person loses $20. Find the expected value of this game.

Solution
We let $a_1 = \$10$ and $a_2 = -\$20$. Now $p_1 = \frac{3}{6} = \frac{1}{2}$ and $p_2 = \frac{1}{2}$, so $E = (\$10)(\frac{1}{2}) - (\$20)(\frac{1}{2}) = -\$5$. ∎

In Example 6, the player is expected to lose \$5 per game in the long run, so we say that this game is not fair.

> **Definition of a Fair Game**
>
> A game is **fair** if its expected value is 0.

EXAMPLE 7 ▶ Deciding If a Game Is Fair

A coin is tossed. If heads comes up, we win \$1; if tails comes up, we lose \$1. Is this a fair game?

Solution
Here, $a_1 = \$1$, $a_2 = -\$1$, and $p_1 = p_2 = \frac{1}{2}$, so $E = (1)(\frac{1}{2}) - 1(\frac{1}{2}) = 0$. Thus, by the definition of a fair game, the game is fair. ∎

EXAMPLE 8 ▶ Payment for a Fair Game

A die is rolled. A person receives double the number of dollars corresponding to the dots on the face that turns up. How much should a player pay for playing in order to make this a fair game?

Solution
The player can win \$2, \$4, \$6, \$8, \$10, and \$12, each with probability $\frac{1}{6}$, so expected winnings (the player does not lose) are

$$E = 2(\tfrac{1}{6}) + 4(\tfrac{1}{6}) + 6(\tfrac{1}{6}) + 8(\tfrac{1}{6}) + 10(\tfrac{1}{6}) + 12(\tfrac{1}{6}) = \tfrac{42}{6} = \$7$$

A person paying \$7 can expect winnings of 0. Thus, \$7 is a fair price to pay for playing this game. ∎

EXAMPLE 9 ▶ Maximizing Business Expectations

Dear's Department Store wishes to open a new store in one of two locations. It is estimated that if the first location is chosen, the store will make a profit of \$100,000 per year if successful and will lose \$50,000 per year otherwise. For the second location, it is estimated that the annual profit will be \$150,000 if successful; otherwise, the annual loss will be \$80,000. If the probability of success at each location is $\frac{3}{4}$, which location should be chosen in order to maximize the expected profit?

Solution
For the first location, $a_1 = \$100,000$, $p_1 = \frac{3}{4}$, $a_2 = -\$50,000$, and $p_2 = \frac{1}{4}$. Thus, the expected profit is

$$E_1 = \$100,000(\tfrac{3}{4}) - \$50,000(\tfrac{1}{4}) = \$75,000 - \$12,500 = \$62,500$$

For the second location, $a_1 = \$150,000$, $p_1 = \frac{3}{4}$, $a_2 = -\$80,000$, and $p_2 = \frac{1}{4}$. Thus, the expected profit is

$$E_2 = \$150,000(\tfrac{3}{4}) - \$80,000(\tfrac{1}{4}) = \$112,500 - \$20,000 = \$92,500$$

The expected profit from the second location ($92,500) is greater than that for the first location ($62,500), so the second location should be chosen.   ■

Decision problems that depend on mathematical expectation require three things for their solutions: *options, values,* and *probabilities*. From Example 9 we have the information given in Table 11.10. With this information we can find the expected value for each option and hence make the desired decision.

TABLE 11.10

	Options			
	Site 1		Site 2	
Values	$100,000	−$50,000	$150,000	−$80,000
Probabilities	$\frac{3}{4}$	$\frac{1}{4}$	$\frac{3}{4}$	$\frac{1}{4}$

However, sometimes it is easier to write all the information using a tree diagram as we shall show next.

EXAMPLE 10 ▶ Maximizing Personal Decisions

Suppose you have two choices for a personal decision. Let us call these choices *A* and *B*. With choice *A*, you can make $20 with probability 0.24, $35 with probability 0.47, and $50 with probability 0.29. With choice *B* you can **lose** $9 with probability 0.25, make nothing ($0) with probability 0.35, and make $95 with probability 0.40. Make a tree diagram and determine what your decision should be if you want to maximize your profit.

Solution

We first draw the tree diagram for this situation as in Figure 11.18 with the branches labeled with their respective probabilities and the monetary outcomes indicated at the end of the corresponding branches.

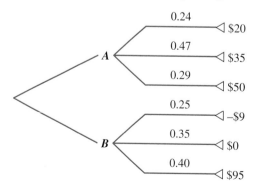

FIGURE 11.18

The expected value E_A for A is

$$E_A = (0.24)(\$20) + (0.47)(\$35) + (0.29)(\$50)$$
$$= \quad \$4.80 \quad + \quad \$16.45 \quad + \quad \$14.50$$
$$= \$35.75$$

Online Study Center

To further explore expected value examples, access link 11.6.1 on this textbook's Online Study Center. To explore odds and probabilities regarding lotto games, visit link 11.6.2.

The expected value E_B for B is

$$E_B = (0.25)(-\$9) + (0.35)(\$0) + (0.40)(\$95)$$
$$= \quad -\$2.25 \quad + \quad \$0 \quad + \quad \$38$$
$$= \$35.75$$

Both decisions have the same expected value! We will discuss possible courses of action in Collaborative Learning in the exercises. ■

EXERCISES 11.6

A Odds

In problems 1–7, find the odds in favor of obtaining

1. a 2 in one roll of a single die.

2. an even number in one roll of a single die.

3. an ace when drawing 1 card from an ordinary deck of 52 cards.

4. a red card when drawing 1 card from an ordinary deck of 52 cards.

5. 2 tails when an ordinary coin is tossed twice.

6. at least 1 tail when an ordinary coin is tossed twice.

7. a vowel when 1 letter is chosen at random from among the 26 letters of the English alphabet.

In problems 8–12, find the odds against obtaining

8. a 4 in one roll of a single die.

9. an odd number in one roll of a single die.

10. the king of spades when drawing 1 card from an ordinary deck of 52 cards.

11. one of the face (picture) cards (jack, queen, king) when drawing 1 card from an ordinary deck of 52 cards.

12. at most 1 tail when an ordinary coin is tossed twice.

13. Using the information in Table 11.8 on page 762, what is the probability of winning $5 when you buy one Florida Lottery ticket? Write the answer

as a proper fraction. What are the odds in favor of winning $5? (*Hint:* The table is wrong!)

14. Using the information in Table 11.8, what is the probability of winning $70 when you buy one Florida Lottery ticket? Write the answer as a proper fraction. What are the odds in favor of winning $70? (*Hint:* The table is wrong!)

15. If the correct odds in favor of Johnny's winning a race are 3 to 2, what is the probability that Johnny wins?

The information for problems 16–25 was taken from the book *What Are the Chances?* by Bernard Siskin, Jerome Staller, and David Rorvik. Find the missing numbers.

Event	Probability	Odds
16. Being the victim of a serious crime in your lifetime	$\frac{1}{20}$	_____
17. Being the victim of a serious crime in San Antonio, where there are 630 violent crimes per 100,000 population	_____	_____
18. Having complications during surgery in June	_____	1 to 4
19. Having complications during surgery in July when new interns and residents are brought in	$\frac{1}{2}$	_____

Event	Probability	Odds
20. Having high cholesterol levels	$\frac{1}{4}$	_____
21. Publishing 1 of the 10 best-selling novels of the year	_____	10 to 4867
22. Getting rich by hard work	_____	41 to 9
23. Being a top executive of a major company without going to college	_____	3 to 21
24. Completing four years of college	0.19	_____
25. Growing up being incompetent in math	0.33	_____

B Expected Value

26. A coin is tossed twice. If heads comes up either time, we get $2; but if heads does not occur, we lose $4. What is the expected value of this game?

27. Two dice are rolled. If the sum of the dots showing is even, we get $10; otherwise, we lose $20. What is the expected value of this game?

28. A die is rolled. A person receives the number of dollars corresponding to the dots on the face that turns up. How much should a player pay in order to make this game fair?

29. In a recent charity raffle, there were 10,000 tickets in all. If the grand prize was a used Lincoln Continental (priced at $21,500), what is a fair price to pay for a ticket?

30. If in problem 29 the charity paid $20,000 for the Lincoln and wished to make a profit of $10,000 from the raffle, for how much should each ticket sell?

31. A man offers to bet $3 against $5 that he can roll a 7 on one throw of his pair of dice. If he wins fairly consistently, are his dice fair? Explain.

32. Louie gets a raffle ticket with a $1000 prize and offers to sell it to you for $10. If 200 tickets were sold and one is selected at random to win, determine whether $10 is a fair price. Explain.

33. If in Example 9 of this section the probabilities of success in the first and second locations are $\frac{2}{3}$ and $\frac{2}{5}$, respectively, what location should be chosen in order to have a maximum expected profit? See Table 11.10 on page 766.

34. Gadget Manufacturing Company is debating whether to continue an advertising campaign for a new product. Its research department has predicted the gain or loss to be derived from the decision to continue or discontinue the campaign, as summarized in the following table. The president of the firm assigns odds of 4 to 1 in favor of the success of the advertising campaign. Find
  a. the expected value for the company if the advertising campaign is continued.
  b. the expected value for the company if the advertising campaign is discontinued.
  c. the best decision based on the answers to parts (**a**) and (**b**).

Advertising Campaign	Successful	Unsuccessful
Continue	$20,000	−$10,000
Discontinue	$30,000	$ 5000

35. Repeat problem 34 assuming the president of the firm assigns odds of 4 to 1 against the success of the advertising campaign.

36. An oil drilling company is considering two sites for its well. The probabilities for getting a dry, a low-producing, or a high-producing well at site A are 0.6, 0.25, and 0.15, respectively. The costs for the three eventualities are −$100,000, $150,000, and $500,000. For site B, the probability of finding a dry well, resulting in a $200,000 loss, is 0.2. The company estimates that the probability of a low-producing well is 0.8, and in that case it would make $50,000. Make a tree diagram for this situation and find the expected value for
  a. site A.   **b.** site B.
  c. On the basis of your answers to (**a**) and (**b**), which site should the company select?

37. Suppose you have the choice of selling hot dogs at two stadium locations. At location A you can sell 100 hot dogs for $4 each, or if you lower the price and move to location B, you can sell 300 hot dogs for $3 each. The probability of being assigned to

A or *B* is equally likely. Make a tree diagram for this situation and find the expected value for

a. *A.*   **b.** *B.*

c. Which location has the better expected value?

In Other Words

38. Explain the difference between the probability of an event and the odds in favor of an event.

39. Explain why betting in Keno is not a fair bet.

40. Explain in your own words what information is needed to solve a decision problem that depends on mathematical expectation.

Using Your Knowledge

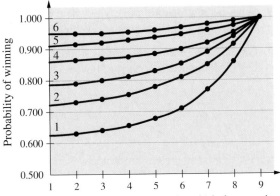

Curves represent the number of runs by which the team is ahead.

The graph shows the probabilities that a baseball team that is ahead by 1, 2, 3, 4, 5, or 6 runs after a certain number of innings goes on to win the game. As you can see, if a team is leading by 1 run at the end of the first inning, the probability that this team wins is about 0.62. If a team is ahead by 2 runs at the end of the first inning, then the probability of this team's winning the game is about 0.72.

Suppose a team is ahead by 1 run at the end of the first inning. The probability that this team wins the game is about 0.62, which can be written

$$\frac{62}{100} = \frac{31}{50}$$

Therefore, the odds in favor of this team's winning the game should be 31 to 19 (50 − 31). Use the graph shown above to solve the following problems:

41. Find the probability that a team leading by 2 runs at the end of the sixth inning
a. wins the game.   **b.** loses the game.

42. For the same situation as in problem 41, find the odds
a. in favor of this team's winning the game.
b. against this team's winning the game.

43. At the end of the sixth inning, a team is ahead by 4 runs. A man offers to bet $10 on this team. How much money should be put up against his $10 to make a fair bet?

Discovery

In American roulette, the wheel has 38 compartments, 2 of which, the 0 and 00, are colored green. The rest of the compartments are numbered from 1 through 36, and half of them are red and the other half black. The wheel is spun in one direction, and a small ivory ball is spun in the other direction. If the wheel is fair, all the compartments are equally likely and hence the ball has probability $\frac{1}{38}$ of landing in any one of them. If a player bets, say, $1 on a given number and the ball comes to rest on that number, the player receives from the croupier 36 times his or her stake, that is, $36. In this case, the player wins $35 with probability $\frac{1}{38}$ and loses $1 with probability $\frac{37}{38}$. The expected value of this game is

$$E = \$35(\tfrac{1}{38}) - \$1(\tfrac{37}{38}) = -5\tfrac{5}{19}¢$$

This may be interpreted to mean that in the long run, for every dollar that you bet in roulette, you are expected to lose $5\frac{5}{19}¢$.

A second way to play roulette is to bet on red or black. Suppose a player bets $1 on red. If the ball stops on a red number (there are 18 of them), the player receives twice his or her stake; in this case the player wins $1. If a black number comes up, the player loses his or her stake. If a 0 or 00 turns up, then the wheel is spun again until it stops on a number different from 0 and 00. If this is black, the player loses the $1, but if it is red, the player receives only his or her original stake (gaining nothing).

44. What is the expected value of this game?

45. If you place 50¢ on red and 50¢ on black, what is the expected value of the game? (*Hint:* The answer is not 0.)

 Collaborative Learning

In Example 10 the expected values for choices A and B were the same. We may rely on conditional probability to make our decision. Form two groups. One will study choice A and the other choice B.

Group A

1. What is the probability that you lose money given that you choose A?

2. What is the probability that you make more than $50 given that you choose A?

3. Based on your answers to 1 and 2, make an argument in favor of choosing A.

Group B

4. What is the probability that you do not make any money given that you choose B?

5. What is the probability that you make more than $50 given that you chose B?

6. Based on your answers to 4 and 5, make an argument in favor of choosing B.

Now, one more debate. Which do you think is best: to play $50 one week in the lottery or to play $1 for 50 weeks? (The possible number of combinations for the lottery is $C(n, r)$, where n is how many numbers you can select and r is how many correct numbers you need to win.) Assume tickets cost $1.

7. What is the probability that you win your state lottery with one ticket? What about with 50 tickets?

8. What is the probability that you lose in the state lottery? What about losing 50 times in a row?

9. What is the probability that you win at least once when playing the lottery 50 weeks in a row?

10. On the basis of your answers to 7 and 9, which is best: $50 one week or $1 for 50 weeks?

Chapter 11 Summary

Section	Item	Meaning	Example
11.1	Experiment	An activity that generates well-defined outcomes	Tossing a coin, drawing a card from a deck
11.1	Sample space	The set of all possible outcomes for an experiment	The sample space for tossing a coin is $\{H, T\}$.
11.1	$P(E)$	The probability of event E, $$P(E) = \frac{n(E)}{n(\mathcal{U})}$$	When tossing a coin, the probability of tails is $P(T) = \frac{1}{2}$.
11.1	$P(T')$	$1 - P(T)$	The probability of a 3 when rolling a die is $\frac{1}{6}$. The probability of not rolling a 3 is $\frac{5}{6}$.
11.1	Empirical probability of an event E	$$P(E) = \frac{\text{number of times E occurred}}{\text{number of times experiment is performed}}$$	If a coin is thrown 890 times and heads come up 451 times, $P(H) = \frac{451}{890}$.

Section	*Item*	*Meaning*	*Example*
11.3	$P(E) = 0$	E is an impossible event.	Rolling a 7 on a die
11.3	$P(E) = 1$	E is a certain event.	Rolling less than 7 on a die
11.3	$P(A \cup B)$	$P(A) + P(B) - P(A \cap B)$	
11.3	Mutually exclusive events	Two events that cannot occur simultaneously; $P(A \cap B) = 0$	Getting a 5 and a 6 on one roll of a die
11.4	$P(A|B)$, the probability of A, given B	$\dfrac{P(A \cap B)}{P(B)}$	The probability that a die turns up 3 (T) when rolled, given that an odd number (O) comes up is $\dfrac{P(T \cap O)}{P(O)} = \dfrac{\frac{1}{6}}{\frac{1}{2}} = \dfrac{1}{3}.$
11.5	Independent events	$P(A \cap B) = P(A) \cdot P(B)$	Rolling a 6 with a die and tossing tails with a coin are independent events.
11.5A	$P(E_1 \cap E_2 \cap \cdots \cap E_n)$	$P(E_1) \cdot P(E_2) \cdot \cdots \cdot P(E_n)$, when the events are independent	The event H: obtaining heads when a coin is tossed, and the event S: obtaining a 6 when a die is rolled, are independent because $P(H \cap S) = P(H) \cdot P(S)$.
11.5B	Stochastic process	A sequence of experiments in which the outcome of each experiment depends on chance	
11.6A	Odds in favor	The ratio of favorable to unfavorable occurrences	The odds for a 3 when rolling a die are 1 to 5.
11.6B	Expected value	$E = a_1 p_1 + a_2 p_2 + \cdots + a_n p_n$, where the a's are the values that occur with probability p_1, p_2, and so on	The expected value of getting heads when a coin is tossed and you are paid \$2 is $E = 2 \cdot \frac{1}{2} = \$1.$

 Research Questions

Sources of information for these questions can be found in the Bibliography at the end of the book.

1. Write a report about Antoine Gombaud, the Chevalier de Méré, and the gambling problem he proposed to Pascal.

2. Write a report about the correspondence between Pascal and Fermat and its influence in the development of the theory of probability.

3. Find the name, author, and year of publication of the first work on the mathematical treatment of probability.

4. The *Arts Conjectandi* was published in 1713 after the death of its author, a brilliant member of a mathematical family. Write a paragraph detailing the circumstances under which the book was published, its contents, and a genealogical table of the family.

5. The theory of probability was studied by the Russian mathematicians P. L. Chebyshev, A. A. Markov, and Andrei Nikolaevich Kolmogorov. Write a paragraph about their contributions to probability.

6. He was the examiner of Napoleon and later his interior minister, a senator, and a count. He was also the author of *Theorie Analytique des Probabilities*. Who was this mathematician and what were his contributions to probability?

7. Probability theory contains several paradoxes, among them the Petersburg paradox. Write a paper explaining this paradox and telling which mathematicians tried to solve it.

8. Find out how insurance companies use mortality tables to establish the cost of life insurance premiums.

9. Investigate and write several paragraphs about three areas that use probability (weather, sports, and genetics, for example).

10. Write a short explanation of the Needle problem.

Chapter 11 Practice Test

1. A single fair die is rolled. Find the probability of obtaining
   a. a number different from 7.   b. a number greater than 2.

2. A box contains 5 balls numbered from 1 to 5. If a ball is taken at random from the box, find the probability that it is
   a. an even-numbered ball.   b. ball number 2.   c. not ball number 2.

3. A box contains 2 red balls, marked R_1 and R_2, and 3 white balls, marked W_1, W_2, and W_3.
   a. If 2 balls are drawn in succession and without replacement from this box, find the number of elements in the sample space for this experiment. (We are interested in which balls are drawn and the order in which they are drawn.)
   b. Do part (**a**) assuming the balls are drawn in succession *with* replacement.

4. Two cards are drawn at random and without replacement from a standard deck of 52 cards. Find the probability that
   a. both cards are red.   b. neither card is an ace.

5. A card is drawn at random from a standard deck of 52 cards and then replaced. Then another card is drawn. Find the probability that
   a. both cards are red.   b. neither card is red.

6. A fair coin is tossed 5 times. What is the probability of obtaining at least 1 head?

7. An urn contains 5 white, 3 black, and 2 red balls. Find the probability of obtaining in a single draw
   a. a white or a black ball.   b. a ball that is not red.

8. Three cards are drawn in succession and without replacement from a standard deck of 52 cards. What is the probability that they are all face cards (jack, queen, king)?

9. A student estimates that the probability of his passing math or English is 0.9, the probability of his passing English is 0.8, but the probability of his passing both is 0.6. What should be his estimate of the probability of his passing math?

10. Two dice are rolled. Find the probability that the sum turning up is 11, given that the first die showed a 5.

11. Two dice are rolled. Find the probability that the sum turning up is 11, given that the second die showed an even number.

12. Two dice are rolled.
  a. Find the probability that they show a sum of 10.
  b. Find the probability that the first die comes up an odd number.
  c. Are these two events independent? Explain.

13. A certain drug used to reduce hypertension (high blood pressure) produces side effects in 4% of the patients. Three patients who have taken the drug are selected at random. Find the probability that
  a. they all had side effects.
  b. none of them had side effects.

14. Roland has to take an English course and a history course, both of which are available at 8 A.M., 9 A.M., and 3 P.M. If Roland picks a schedule at random, what is the probability that he will have English at 8 A.M. and history at 3 P.M.?

15. The probability that a digital tape is defect-free is 0.97. If 2 tapes are selected at random, what is the probability that both are defective?

16. A card is selected at random from a deck of 52 cards. What are the odds in favor of the card's being
  a. a king?   **b.** not a king?

17. The probability of an event is $\frac{3}{7}$. Find
  a. the odds in favor of this event occurring.
  b. the odds against this event occurring.

18. The odds in favor of an event occurring are 3 to 7.
  a. What are the odds against this event occurring?
  b. What is the probability that the event will not occur?

19. A coin is tossed twice. If exactly 1 head comes up, we receive $5, and if 2 tails come up, we receive $5; otherwise, we get nothing. How much should we be willing to pay in order to play this game?

20. The probabilities of being an "instant winner" of $2, $5, $25, or $50 in the Florida lottery are $\frac{1}{10}$, $\frac{1}{50}$, $\frac{1}{600}$, and $\frac{1}{1200}$, respectively. What is the mathematical expectation of being an "instant winner"?

Answers to Practice Test

ANSWER	IF YOU MISSED	REVIEW		
	Question	Section	Example(s)	Page(s)
1. a. 1 **b.** $\frac{2}{3}$	1	11.1	1, 2	702–703, 706
2. a. $\frac{2}{5}$ **b.** $\frac{1}{5}$ **c.** $\frac{4}{5}$	2	11.1	3	706
3. a. 20 **b.** 25	3	11.1	4, 5	707
4. a. $\frac{25}{102}$ **b.** $\frac{188}{221}$	4	11.1	6	707–708
5. a. $\frac{1}{4}$ **b.** $\frac{1}{4}$	5	11.1	6	707–708
6. $\frac{31}{32}$	6	11.1	4	707
7. a. $\frac{4}{5}$ **b.** $\frac{4}{5}$	7	11.2	1–4	719–722
8. $\frac{11}{1105}$	8	11.2	5–8	723–725
9. 0.7	9	11.3	2, 7	731, 732–733
10. $\frac{1}{6}$	10	11.4	1	740
11. $\frac{1}{18}$	11	11.4	2, 3	740
12. a. $\frac{1}{12}$ **b.** $\frac{1}{2}$ **c.** No. $P(A \cap B) = \frac{1}{36} \neq P(A)P(B)$	12	11.4	3, 4	740–741
13. a. 0.000064 **b.** 0.884736	13	11.5	1–4	749–751
14. $\frac{1}{6}$	14	11.5	5	752
15. 0.0009	15	11.5	6, 7	752–753
16. a. 1 to 12 **b.** 12 to 1	16	11.6	1	760
17. a. 3 to 4 **b.** 4 to 3	17	11.6	3, 4	761
18. a. 7 to 3 **b.** $\frac{7}{10}$	18	11.6	3	761
19. \$3.75	19	11.6	8	765
20. $38\frac{1}{3}$¢	20	11.6	6–8	764–765

Statistics play a major role in our everyday lives. Statistics can be used to measure team performances, rainfall, economic trends, and test scores. Throughout this chapter we will analyze data with the use of statistical tools. One such tool is the normal distribution curve, which we use as a problem-solving tool in Section 12.4.

Statistics

Americans are fascinated by numbers. Consider the *Guinness Book of World Records,* almanacs, surveys, and so on. But what are the meanings of all these numbers, and how can we interpret them? In this chapter we discuss different ways of organizing and reporting data. The simplest way is to use the **frequency distribution** of Section 12.1, which is a type of table that tells us how many objects of different types we have in each of several categories. Such a distribution can then be represented by a graph called a **histogram.**

When we want to describe an entire sample or population by a number, we use an **average.** The three most common averages are the **mean,** the **median,** and the **mode.** Each of these averages, which are presented in Section 12.2, uses one number to try to tell us where the "middle" of a set of data is. However, averages cannot tell us how far data values are spread out away from this "middle." For this we use the **range** and the **standard deviation.** Using means, standard deviations, and *z*-scores, we can compute how far a data point is from the "middle." These topics are covered in Sections 12.3 and 12.4.

In real life, newspapers and magazines present data using many varieties of graphs, including line, bar, and circle graphs. We examine these in Section 12.5.

An important aspect of statistics is predicting the likelihood of future outcomes based on data gathered from earlier observations. For example, can we predict winning Olympic times based on athletes' past performances or the incidence of cancer based on exposure to ultraviolet sunlight? We will make predictions and study **scattergrams** and **correlations** in the last two sections of the chapter.

Online Study Center

For links to Internet sites related to Chapter 12, please access **college.hmco.com/PIC/bello9e** and click on the Online Study Center icon.

12.1   Sampling and Frequency Distributions

Comparison Shopping for Jeans

A buyer for a large department store wanted to compare prices of high-quality jeans for men and women. One way to do this is to look in a consumer magazine at the 15 best-rated jeans for men and women and their prices. The prices appear in Table 12.1. How can we organize the information so that we can make meaningful comparisons? Let us look at prices for women's jeans.

Since these prices range from $52 to $19, we will break them into three **classes** (you could as easily make it 4 or 5). To do this, we divide the range of prices by the number of classes to get the **width** of each class.

$$\frac{52 - 19}{3} = \frac{33}{3} = \mathbf{11} \text{ (width)}$$

The **lower limits** for our classes will be 19, 19 + **11** = 30, and 30 + **11** = 41, as shown in Table 12.2. The corresponding **upper limits** are 29, 40, and 51. Notice that there is a gap between the end of one class and the beginning of the next

TABLE 12.1 Jeans' Prices

Women's Jeans		Men's Jeans	
Sears Jeans That Fit	$⑲+	Wrangler Prorodeo	$20
Wrangler Prorodeo	26	Wrangler American Hero	17
Chic Heavenly Blues	48+	Levi's 509	31
P.S. Gitano	21	Wrangler Rustler	15
Gap Straight Leg	30	J. C. Penney Long Haul	23+
Lee Easy Rider	33	Guess/Georges Marciano	㊿
L. L. Bean Stretch	36	Wrangler American Hero L	23
Lands' End Square Rigger	26+	Levi's 550	45
L. L. Bean Double L	27	Lands' End Square Rigger	20+
Levi's 501	35	Levi's 501	34
Lee Relaxed Rider	29	Gap Tapered Leg	34
Calvin Klein	㊼	L. L. Bean Double L	27
Levi's 902	34	Sears Roebucks	⑭+
Bonjour	20	Gap Easy Fit	32
Gap Classic Contour	30	Lee Riders Straightleg	20

TABLE 12.2 Women's Jeans' Prices by Classes

Step 1		Step 2	Step 3
Lower Limit	Upper Limit	Tally	Frequency
19	30	卌 ‖	7
30	41	卌		6
41	52*	‖	2

*Note that 52 *must* go in the third class because there are no classes above it.

The analysis of numerical data is fundamental to so many different fields such as biology, geology, genetics, and evolution. Charles Darwin (1809–1882), Gregor Mendel (1822–1884), and Karl Pearson (1857–1936) contributed greatly to these subjects.

Looking Ahead
We start this chapter by studying what is usually called *descriptive statistics,* which consists of summarizing in a concise way the information collected. We end the chapter by discussing *inferential statistics*—using methods that generalize the results obtained from a sample to the population and measure their reliability.

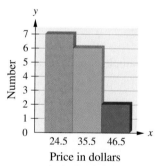

FIGURE 12.1
Women's jeans' prices histogram.

Note: Unless otherwise specified, data values falling on class upper limits are included in the next *higher* class.

(from 29 to 30, for example). We can fix this by making the upper limits of the classes 30, 41, and 52 (instead of 29, 40, and 51). See *step 1* in Table 12.2. Now, in which class will the $30 price go?

With this convention we can tally the prices falling in each class (*step 2*) and note their frequency (*step 3*). We can then make a picture of this information, called a **histogram,** in which each of the classes is represented by a bar whose width is 11 units, the **class width,** and whose height is given by the **frequency** as shown in Figure 12.1. Note that this histogram does not show the lower and upper limits. Rather, each class is described by a single value called its **midpoint.** Sometimes the lower and upper limits are shown as well (see the Problem Solving section).

To see a misuse of statistics when constructing a histogram, look at the Discovery section. ▶

The word *statistics* brings to the minds of most people an image of a mass of numerical data. **Statistics** can be defined as the science of collecting, organizing, and summarizing data (*descriptive statistics*) so that valid conclusions can be drawn from them (*inferential statistics*).

A. Sampling

Who will be the next governor in your state? What about the next president of the United States? A veritable army of statisticians, analysts, and *pollsters* (people who conduct surveys or *polls*) spend a lot of time and money to try to determine what percent of the vote each candidate will receive. How do they do it? Obviously, it is impossible to ask each registered voter (the **target population**) for whom he or she plans to vote, so analysts concentrate on a select, smaller number of people (a **sample population**) to represent the entire group and then project the result to all registered voters. In order for the conclusions reached to be valid, a **simple random sample** must be used.

> **Definition of a Simple Random Sample**
>
> A **simple random sample** of a given population is a sample for which
>
> 1. Each member is selected at random.
>
> 2. Every member of the population has the same chance of being chosen.

Thus, to pick a simple *random* sample of 100 students from a college with 5000 students, number the students from 1 to 5000 and write these numbers on cards. Then mix up the cards and draw 100 numbers. The result will be a *simple random sample* consisting of the 100 students corresponding to the drawn cards. Note that if you decide to select only *even* numbered cards, this would **not** be a random sample because only the students numbered 2, 4, 6, and so on would be chosen. Similarly, if you decide to select the first 100 students arriving at school, this would **not** be a random sample.

EXAMPLE 1 ▶ Finding IBM Chip Population and Sample

Suppose you want to determine whether the chips installed on the latest shipment of computers satisfy quality control standards. Assume the shipment consists of 1000 computers and you want to pick a simple random sample of 50 computers.

(a) What is the population?

(b) What procedure can you use to select the simple random sample?

(c) What is the sample?

Solution

(a) The population consists of the 1000 computers.

(b) Number the computers from 1 to 1000, write their numbers on cards, mix the cards, and draw 50 numbers.

(c) The sample consists of the 50 computers whose numbers were drawn.   ■

How do these ideas relate to each other and to you? Let us look at a simple example that you may have encountered and even avoided! You go to the mall, and there is somebody conducting a survey. To simplify things, suppose that the person conducting the survey selects the first 10 people that walk in and asks each of them if he or she owns a cell phone. Each person falls into one of two categories: yes (Y) or no (N). The responses are

1 Y, 2 N, 3 Y, 4 Y, 5 N, 6 Y, 7 N, 8 Y, 9 Y, 10 Y

1. What do you think the target population and the sample population are? Are they the same?

2. Do we have a simple random sample?

3. How can we summarize the data?

1. The **target** or implied population may be cell phone owners in the United States, the state, or this particular mall. The **sample** population consists of the 10 people surveyed. The target and the sample populations are **not** the same.

2. We do not have a simple random sample. The definition of simple random sample requires that every member of the population have the same chance of being chosen. A mall survey leaves out anyone not visiting the mall that day.

3. There are several ways we can summarize the data. One is using a table or we can even make a picture (graph) of the results, as shown on page 779. The table is called a **frequency distribution;** the graph is a **histogram,** which can be converted easily into a **frequency polygon,** as we will show in the next sections.

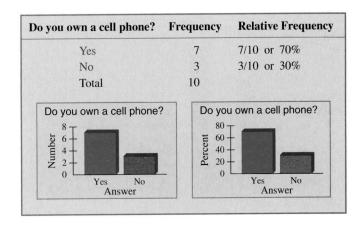

Do you own a cell phone?	Frequency	Relative Frequency
Yes	7	7/10 or 70%
No	3	3/10 or 30%
Total	10	

TABLE 12.3 Frequency Distributions for Class Scores

Score	Tally Marks	Frequency
0	||	2
1	|	1
2	|	1
3		0
4	||	2
5	||||	4
6	|||	3
7	||	2
8	||||	5
9	|||	3
10	||	2
		25 Total

TABLE 12.4 Frequency Distribution with Grouped Data for Class Scores

Score	Frequency
0–1	3
2–3	1
4–5	6
6–7	5
8–9	8
10–11	2

B. Frequency Distributions

Statistical studies start by collecting data. In order to *organize* and *summarize* these data to detect any trends that may be present, we can use three types of tools: *frequency distributions, histograms,* and *frequency polygons.*

Let us look at a statistics problem that should interest a teacher and students, both of whom might wonder how well the students are learning a certain subject. Out of 10 possible points, the class of 25 students made the following scores:

6	5	4	0	9
2	0	8	8	1
10	6	8	5	5
8	7	9	10	9
6	5	8	4	7

This listing shows at once that there were some good scores and some poor ones, but because the scores are not arranged in any particular order, it is difficult to conclude anything else from the list.

A **frequency distribution** is often a suitable way of organizing a list of numbers to show what patterns are present. First, the scores from 0 through 10 are listed in order in a column (see Table 12.3). Then, by going through the original list in the order in which it is given, we can make tally marks on the appropriate lines of our table. Finally, in a third column we can list the number of times that each score occurs; this number is the **frequency** of the score.

It is now easier to see that a score of 8 occurred more times than any other number. This score was made by

$$\frac{5}{25} = \frac{1}{5} = 20\% \text{ of the students}$$

Ten of the students, or 40% of the class, received scores of 8 or better. Only 6, or 24%, received scores less than 5.

If there are very many items in a set of numerical data, then it is usually necessary to shorten the frequency distribution by grouping the data into intervals. For instance, in the preceding distribution, we can group the scores in intervals of 2 to obtain the listing in Table 12.4.

Of course, some of the detailed information in the first table has been lost in the second table, but for some purposes a condensed table may furnish all the information that is required.

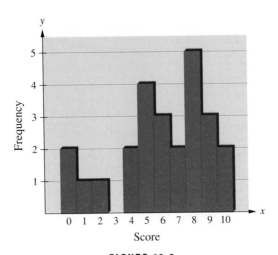

FIGURE 12.2

Histogram for Table 12.3.

FIGURE 12.3

Frequency polygon.

C. Histograms

It is also possible to present the information contained in Table 12.3 by means of a special type of graph called a **histogram,** consisting of vertical bars with no space between bars. In the histogram of Figure 12.2, the units on the y axis represent the frequencies, whereas those on the x axis indicate the scores.

D. Frequency Polygons

From the histogram in Figure 12.2 (above left) we can construct a **frequency polygon** (or line graph) by connecting the midpoints of the tops of the bars, as shown in Figure 12.3. It is customary to extend the graph to the baseline (x axis) using the midpoints of the extended intervals at both ends. This "ties the graph down" but has no predictive significance.

EXAMPLE 2 ▶ Frequency Distribution and Histogram for Wages

Here are the hourly wages of a group of 30 workers who are performing similar tasks but, because of differences in seniority and skill, are paid at different rates:

$8.00	$7.90	$8.00	$8.10	$7.90	$7.90
7.90	7.80	7.90	8.00	7.80	8.00
8.10	7.70	7.90	7.80	8.10	8.00
8.00	8.10	8.20	7.80	8.20	8.10
7.70	8.00	7.80	7.70	7.80	8.00

(a) Make a frequency distribution of these rates.

(b) What is the most frequent rate?

(c) How many workers are being paid less than $8 per hour?

(d) Make a histogram of the wage rate distribution.

(e) Make a frequency polygon of the distribution.

Solution

(a) Table 12.5 lists the wage rates from the lowest ($7.70) to the highest ($8.20). We tally these from the given data and obtain the desired frequency distribution.

TABLE 12.5

Wage	Tally Marks	Frequency
7.70	|||	3
7.80	卌		6
7.90	卌		6
8.00	卌 |||	8
8.10	卌	5
8.20	||	2
		30 Total

GRAPH IT

You can use a grapher to do histograms, but you have to be extremely careful with class widths and endpoints. To produce the histogram in Figure 12.4, enter the values in Example 2 as a list. To do this, press [STAT] [1] and enter the numbers. Tell the grapher you want to plot data by pressing [2nd] [STAT PLOT] [1]. On the next screen, select [ON] and the histogram icon. Now press [Window] and select 7.7 for the minimum, 8.3 for the max (if you select 8.2 for the maximum, you will not see the last bar), and .1 for the Xscl. Finally, select 0 for Ymin, 10 for Ymax, 1 for Yscl, and 1 for Xres. Press [GRAPH]. If you followed these steps faithfully, you will be rewarded with the graph shown below.

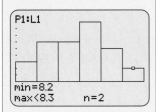

Note that the last bar goes from a min of 8.2 to a max < 8.3 and is $n = 2$ units high.

(b) From the frequency distribution, we read off the most frequent rate to be \$8 per hour.

(c) Again, we read from the frequency distribution that 15 workers are being paid less than \$8 per hour.

(d) The desired histogram appears in Figure 12.4.

(e) Figure 12.4 (below) also shows the frequency polygon. ∎

In making a frequency distribution in which the data are to be grouped, we can use the following procedure:

Procedure for Making a Frequency Distribution

1. Decide on the number of classes into which the data are to be grouped. This depends on the number of items that have to be grouped but is usually between 3 and 15.

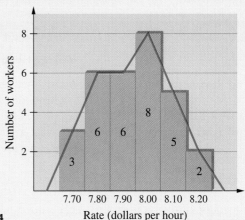

FIGURE 12.4

(a) The *width* of each class is given by

$$\text{Class width} \approx \frac{\text{largest data value} - \text{smallest data value}}{\text{desired number of classes}}$$

The symbol ≈ means "approximately." If we want the class width to be a whole number, we round *up* to the nearest whole number.

(b) The lowest and highest values in each class are called the **lower class limit** and **upper class limit,** respectively. Note that the difference between the lower class limit of one class and that of the next should be the class width.

(c) The center of the class is called the **midpoint.**

(d) To make sure that the bars in a histogram touch, we can do either of the following:

   (i) Use the halfway points between the upper limit of one class and the lower limit of the next class, the **class boundaries,** as the endpoints of the bars.

   (ii) Use the upper class limit of the first class as the lower class limit of the second class and stipulate that values falling on class limits will be included in the next higher class.

2. Sort or tally the data into the appropriate classes.

3. Count the number of items in each class.

4. Display the results in a table.

5. If desired, make a histogram and/or frequency polygon of the distribution.

By following the procedure for making a frequency distribution, we will see that it is really not very difficult to tabulate a frequency distribution and construct a histogram or frequency polygon.

PROBLEM SOLVING

Making Histograms

❶ **Read** the problem.

Make a frequency distribution with three classes, and construct the corresponding histogram for the men's jeans' prices shown in the Getting Started section.

❷ **Select** the unknown.

We want to make a frequency distribution and then a histogram for the men's jeans' prices.

❸ **Think** of a plan.

We need to create three classes and determine their frequencies.

❹ **Use** the procedure we have studied to carry out the plan. What is the class width? What are the class limits? What are the class boundaries? Are these boundaries convenient for this problem?

Since the highest price is 60 and the lowest 14, the class width is

$$\frac{60 - 14}{3} \approx 15.3$$

which is rounded *up* to 16. The lower limits for our classes are 14, 30, and 46, making the upper limits 29, 45, and 61. Thus, the class boundaries are the halfway points between 29 and 30 (29.5), 45 and 46 (45.5), and $45.5 + 16 = 61.5$. However, these boundaries are not convenient or natural, so we choose to make our class limits 14 to 30, 30 to 46, and 46 to 62. The classes can be described by the inequalities shown in Table 12.6, where p represents the price. The tallies and frequencies are shown in Table 12.6 and the histogram in Figure 12.5.

Make the frequency distribution.
Draw the histogram.

TABLE 12.6

Class	Tally	Frequency
$14 \leq p < 30$	𝍤 ||||	9
$30 \leq p < 46$	𝍤	5
$46 \leq p < 62$	|	1

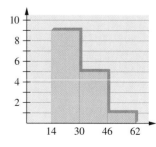

FIGURE 12.5

❺ **Verify** the answer.

TRY EXAMPLE 3 NOW.

Cover the solution, write your own solution, and then check your work.

Note: Sometimes the upper class boundaries are included in the corresponding lower class, as in the next example and in problems 23 and 27 of Exercises 12.1. You will see that the inequalities in the distribution tables show this.

EXAMPLE 3 ▶ Frequency Distribution and Histograms for Voters

In a study of voter turnout in 20 cities with populations of over 100,000 in the United States, the following data were found:

Turnout Rate as a Percent of the Voting-Age Population

85.2	72.4	81.2	62.8	71.6
72.1	87.2	76.6	58.5	70.0
76.5	74.1	70.0	80.3	65.9
74.9	70.8	67.0	72.5	73.1

Inspection of the data shows that the smallest number is 58.5 and the largest is 87.2. This time we go from 55 to 90, with the convenient class width of 5 units.

TABLE 12.7

Voting Rate $r\%$	Tally Marks	Frequency
$55 < r \le 60$	|	1
$60 < r \le 65$	|	1
$65 < r \le 70$	||||	4
$70 < r \le 75$	⧸⧸⧸⧸ |||	8
$75 < r \le 80$	||	2
$80 < r \le 85$	||	2
$85 < r \le 90$	||	2

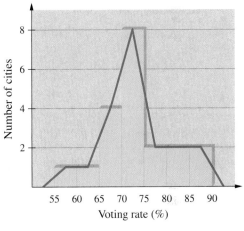

FIGURE 12.6

(a) Make a frequency distribution of the data on voting rate (r) using a class interval of 5% so that the classes will be $55 < r \le 60$, $60 < r \le 65$, ..., $85 < r \le 90$.

(b) Make a histogram and a frequency polygon of this distribution.

(c) In what percent of the cities was the voting rate greater than 80%?

(d) In what percent was the voting rate less than or equal to 70%?

Solution

(a) The required frequency distribution appears in Table 12.7. (You should check this table.)

(b) The histogram and frequency polygon are shown in Figure 12.6. These are constructed from the frequency distribution, just as before with ungrouped data.

(c) In 4 out of 20 cities, the voting rate was greater than 80%. Thus, the required percent is $\frac{4}{20} = 20\%$.

(d) In 6 out of 20 cities, the voting rate was less than or equal to 70%. Thus, in 30% of the cities, no more than 70% of the voting-age population voted. ∎

E. Applications

The data in Table 12.8 represent the number and percent of homes that were owner occupied, renter occupied, or vacant.

TABLE 12.8

Housing Tenure	1302	1303	1304	1305	1404	1405
Owner occupied	524 (77%)	850 (44%)	35 (30%)	879 (39%)	1454 (62%)	594 (44%)
Renter occupied	130 (19%)	981 (51%)	76 (66%)	1240 (55%)	836 (36%)	702 (52%)
Vacant	25 (4%)	89 (5%)	5 (4%)	132 (6%)	56 (2%)	56 (4%)
Total Housing Units	**679**	**1920**	**116**	**2251**	**2346**	**1352**

Source: U.S. Census Bureau, Census 2000, Summary File 1.

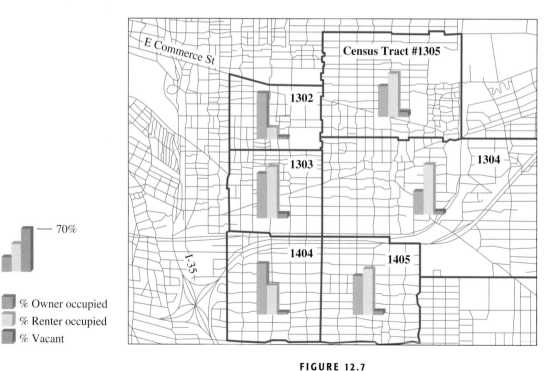

FIGURE 12.7

Housing by tenure.

A better visual representation of this information can be obtained by using histograms. A software program created by Tetrad Computer Applications translates the information into the histograms shown in Figure 12.7.

EXAMPLE 4 ▶ Applications of Histograms to Housing

An investment banker is studying the histograms in Figure 12.7.

(a) Which tract has the most owner-occupied homes?

(b) Which tract has the most vacant homes?

(c) If you are the banker, in which tract would you invest your money?

Solution

(a) Tract 1404 has the most owner-occupied homes (1454).

(b) Tract 1305 has the most vacant homes (132).

(c) The banker wants to find out which tract has the most owner-occupied houses. Unfortunately, the histograms only give the relative frequency (**percent**) of owner-occupied homes, so we have to use Table 12.8. Since tract **1404** has the most owner-occupied homes (1454), a smart banker will select tract 1404, even though the relative frequency (percent) of owner-occupied homes in tract 1302 is higher (77%) than in tract 1404 (62%). ■

Online Study Center

To further explore "Applied Population and Samples," access link 12.1.1 on this textbook's Online Study Center. For an interactive histogram site, access link 12.1.2.

EXERCISES 12.1

Ⓐ Sampling

1. What is descriptive statistics?

2. What is inferential statistics?

3. What is the difference between a target population and a sample population?

4. The shoe sizes of all the members of the U.C. basketball team were recently collected and organized in a frequency distribution. Do these data represent a sample or a population? Explain.

5. A survey in *USA Today* reports that 29% of the 1006 households surveyed stated that their favorite shopping day was Saturday.
   a. What is the implied population?
   b. What is the sample?

6. The students in a statistics class took a random sample of 50 colleges and universities regarding student fees at the 50 institutions and concluded that their own student fees were higher than at most colleges in the country.
   a. What is the population?
   b. What is the sample?

7. A television station asked viewers to respond either yes or no to a certain question by calling a 900 number to respond yes and a different 900 number to respond no. Each call cost $.50.
   a. Is this a simple random sample?
   b. Explain why or why not.

8. In 1936, on the basis of a sample of 2,300,000 voters selected from automobile owners and telephone subscribers, the *Literary Digest* predicted that Alf Landon, the Republican candidate for president, would be elected. Actually, the Democratic candidate, Franklin D. Roosevelt, won.
   a. What was the population?
   b. Was the sample a random sample? Explain.

9. Consider the population of all students in your class.
   a. How could you get a random sample of 10 students from this population?
   b. List three ways of getting samples from this population that are *not* random samples.

10. An important concept in the manufacturing process is quality control. Products are inspected during production, and the equipment is adjusted to correct any defects. Since not every product can be examined, how would you draw a random sample of 10 of the next 50 CD players coming off an assembly line?

11. Do you surf the net? Have you been annoyed by the pop-up ads? An ad from an educational Web site asks you to participate in a survey to determine how people feel about educational television.
    a. What is the target population?
    b. What is the sample population?
    c. Are these two populations the same? Explain.

12. Do you buy your books at the bookstore or online? The finance department of a college conducted a survey at the bookstore, where students were selected at random and asked to participate in the survey.

   a. Is this a random sample?

   b. What is the target population?

   c. What is the sample population?

   d. Are these two populations the same? Explain.

13. Do you buy your vegetables and bakery products at a regular or a specialty store? The owner of El Mirasol wanted to know how many people would buy fresh tortillas at his market, so he surveyed the first 100 people who entered the store and received 50 responses. He then conducted a second survey of 75 people who had actually bought some goods at the store and also received 50 responses. Which survey do you think would produce more accurate results, and why?

14. A politician wants to introduce a bill in Congress and uses the results of a survey of his constituents to help sell the bill to his colleagues. He introduces the bill by saying, "All American patriots really want this bill."

   a. What are the target and sample populations?

   b. Is it reasonable to assume that the target and sample populations are the same? Explain.

B **Frequency Distributions**

15. The athletic department asks 1000 students if they have been to a football game in the last year. The results indicate that 720 students said yes (Y), 160 said no (N), and 120 did not respond (**NR**).

   a. Display the results in a frequency distribution.

   b. Why do you need to include the people who did not respond?

16. Use the information in problem 15 to make a relative frequency distribution showing the percentage of students in each of the three categories.

17. How many hours per week do you spend reading? According to NOP World Co., Americans read an average of 5.7 hours per week. (Source: http://tinyurl.com/dam8m.) The following data represent the number of hours of reading per week for 20 people:

4	3	4	0	4	3	6	3	1	6
6	5	2	7	5	5	6	2	2	7

   a. Make a frequency distribution of the number of hours spent reading per person. Label the columns "Number of Hours," "Tally Marks," and "Frequency."

b. What is the most frequent number of hours spent reading per person?

c. How many people read more hours than the average? What percent is that?

d. How many people read fewer hours than the average? What percent is that?

18. Do you spend more hours watching TV or online? According to the Digital Future Project, the average number of hours spent online (12.5) first surpassed the number of hours watching television (11.6) in 2003. The following data represent the number of hours spent online per week for 30 people:

```
12   13    4    0   14    3   15   13   11   13
 6    5   12    7    5    5    6    2   10   12
12    3   10   13    9   11   14    1   11    8
```

a. Make a frequency distribution for the number of hours spent online per person. Label the columns "Number of Hours," "Tally Marks," and "Frequency."

b. What is the most frequent number of hours spent online per person?

c. How many people are online more hours than average? What percent is that?

d. How many people are online fewer hours than average? What percent is that?

19. The following data show the results of 30 rolls of a die that has had the edges on the 1-face rounded (sometimes called a **loaded die**). In 30 rolls, the 6, which is opposite the 1, showed only once!

```
3   6   3   1   2
1   1   1   1   1
5   5   5   5   3
3   1   2   5   4
5   3   1   5   1
5   1   3   4   1
```

a. Make a frequency distribution for the outcomes. Label the columns "Outcome," "Tally Marks," and "Frequency."

b. Which outcome is the most frequent? How many times did it occur?

c. Which outcome is the least frequent? How many times did it occur?

20. Another loaded die is rolled several times, and the results are as shown in the histogram.

a. Make a relative frequency histogram of the results.

b. What percent of the time does the 1 occur?

c. What percent of the time does the 4 occur?

d. In part (**a**) you made a *relative* frequency histogram from a frequency histogram. Can you make a frequency histogram from a relative frequency histogram? Explain.

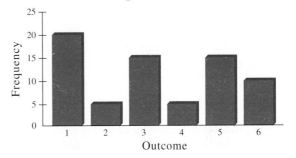

Histogram for loaded die

21. Thirty students were asked to list the television programs each had watched during the preceding week. From this list, the number of hours each had spent watching television during the week was calculated. The results are

```
1    5    4    7   10    8    2    3    9    6
6   12    8   14    3    4    8    7    2    1
0    3    5    8   10   12    0   15    1    4
```

a. Make a frequency distribution of the number of hours of television watched per student. Label the three columns "Number of Hours," "Tally Marks," and "Frequency."

b. What is the most frequent number of hours watched per student?

c. How many students watched television more than 10 hours?

d. How many students watched television 5 hours or less?

e. What percent of the students watched television more than 7 hours?

C Histograms

22. Make a histogram for the data given in problem 21.

23. Have you read in the newspapers or magazines about cases in which individuals became so disgusted with the amount of time they had to wait to see a doctor or a dentist that they sued for lost wages? The waiting times for 50 patients are given in the table below.

Waiting Time t (min)	Number of Patients
0 < t ≤ 3.5	10
3.5 < t ≤ 7.0	8
7.0 < t ≤ 10.5	6
10.5 < t ≤ 14.0	16
14.0 < t ≤ 17.5	6
17.5 < t ≤ 21.0	4

a. Make a histogram for this set of data.
b. What percent of the patients waited 7.0 min or less?
c. What percent of the patients had to wait more than 10.5 min?

24. General Foods, in testing a new product, which it called Solid H, had 50 people (25 men and 25 women selected at random) taste the product and indicate their reactions on the picture ballot shown. The boxes on the ballot were then assigned scores of +3, +2, +1, 0, −1, −2, and −3 in order from left to right, top to bottom.

Please check the box under the picture that expresses how you feel toward the product you have just tasted.

The table below shows the frequencies for this taste test. (Incidentally, no significant difference was found between the men's and women's reactions.)

Score	Frequency
−3	0
−2	1
−1	1
0	13
+1	19
+2	11
+3	5

a. Make a histogram of these data.
b. What percent of the tasters liked Solid H?
c. What percent were undecided?

25. Would you like to be a writer? Look at the following list of authors whose books were published at the ages given:

	Age
Allen Dulles (*The Boer War: A History*)	8
Hilda Conkling (*Poems by a Little Girl*)	9
Betty Thorpe (*Fioretta*)	10
Nathalia Crane (*Janitor's Boys*)	10
David Statler (*Roaring Guns*)	9
Erlin Hogan (*The Four Funny Men*)	8
Minou Drouet (*First Poems*)	8
Dorothy Straight (*How the World Began*)	6
Kali Diana Grosvenor (*Poems by Kali*)	7
Benjamin Friedman (*The Ridiculous Book*)	9

a. Make a frequency distribution showing the number of authors for each age.
b. Make a histogram for the distribution in part (**a**).
c. What percent of the authors were less than 8 years old when they published their first books?

26. How tall are you? The following are 10 famous people and their heights:

	Height (in.)
Honoré de Balzac (Fr. novelist)	62
Napoleon Bonaparte (Fr. emperor)	66
Yuri Gagarin (Soviet cosmonaut)	62
Hirohito (Japanese emperor)	65
Nikita Khrushchev (Soviet leader)	63
James Madison (U.S. president)	64
Margaret Mead (U.S. anthropologist)	62
Pablo Picasso (Spanish painter)	64
Mickey Rooney (U.S. actor)	63
Tutankhamen (Egyptian king)	66

a. Make a frequency distribution showing the number of people for each height.

b. Make a histogram for the distribution in part (**a**).

27. The list below shows the 30 stocks that comprise the Dow Jones Industrial Average and their prices:

DOW JONES INDUS. AVG MEMBERS		
COMPANY ▲	SYMBOL	PRICE
3M CO	▶ MMM	77.42
ALCOA INC	▶ AA	29.17
ALTRIA GROUP INC	▶ M0	76.12
AMER INTL GROUP	▶ AIG	69.06
AMERICAN EXPRESS	▶ AXP	53.16
AT&T INC	▶ T	24.74
BOEING CO	▶ BA	68.97
CATERPILLAR INC	▶ CAT	62.30
CITIGROUP INC	▶ C	48.61
COCA-COLA CO	▶ KO	41.15
DISNEY (WALT) CO	▶ DIS	25.54
DUPONT (EI)	▶ DD	40.00
EXXON MOBIL CORP	▶ XOM	61.45
GENERAL ELECTRIC	▶ GE	34.95
GENERAL MOTORS	▶ GM	20.34
HEWLETT-PACKARD	▶ HPQ	31.36
HOME DEPOT INC	▶ HD	41.73
HONEYWELL INTL	▶ HON	37.03
IBM	▶ IBM	82.94
INTEL CORP	▶ INTC	25.52
JOHNSON&JOHNSON	▶ JNJ	61.65
JPMORGAN CHASE	▶ JPM	39.55
MCDONALDS CORP	▶ MCD	35.00
MERCK & CO	▶ MRK	33.60
MICROSOFT CORP	▶ MSFT	27.02
PFIZER INC	▶ PFE	24.44
PROCTER & GAMBLE	▶ PG	58.72
UNITED TECH CORP	▶ UTX	56.21
VERIZON COMMUNIC	▶ VZ	31.55
WAL-MART STORES	▶ WMT	45.04

Source: http://tinyurl.com/yrgrb.

a. Make a frequency distribution for the price of these stocks, grouped in intervals of $10. The first three lines of your table should look like this.

Price	Tally Marks	Frequency
$0 < P \le 10$		0
$10 < P \le 20$		0
$20 < P \le 30$	||||| ||	7

b. What is the most frequent price interval for these stocks?

c. How many of the stocks sold for more than $40 per share?

d. How many of the stocks sold for $30 or less per share?

e. What percent of the stocks sold for prices between $30 and $50 per share?

f. What percent of the stocks sold for $30 or less per share?

28. Make a histogram for the data obtained in problem 27(**a**).

Ⓓ **Frequency Polygons**

29. Do you know that certain isotopes (different forms) of the elements are used in nuclear reactors and for medical purposes such as the treatment of cancer? At the present time, about 1400 isotopes have been observed, but, of these, only 332 occur naturally. The table below lists the number of elements having 1–10 naturally occurring isotopes. For instance, there are 22 elements having only 1 such isotope but only 1 element having the maximum number, 10. Make a histogram and a frequency polygon for these data.

Number of Naturally Occurring Isotopes	Number of Elements
1	22
2	21
3	9
4	6
5	7
6	9
7	11
8	3
9	1
10	1

30. Twenty apprentices were asked to measure the diameter of a steel rod with a **micrometer** (an

instrument that can measure to thousandths of an inch). Their results (in inches) were

0.254	0.245	0.253	0.251
0.249	0.252	0.251	0.252
0.247	0.251	0.250	0.247
0.251	0.249	0.246	0.249
0.250	0.248	0.249	0.253

a. Make a frequency distribution of these measurements.
b. What single measurement has as many measurements above it as below it?
c. What percent of the measurements are between 0.249 and 0.251 in., inclusive?
d. What would you take as the best estimate of the diameter? Why?

31. The following is a quotation from *Robinson Crusoe*, which many of you have probably read:

> Upon the whole, here was an undoubted testimony that there was scarce any condition in the world so miserable, but was something negative or something positive, to be thankful for in it.

a. There are 151 letters in this quotation. Make a frequency distribution of the 151 letters.
b. Which letter occurs most frequently?
c. What percent of the letters are vowels?

32. Four coins were tossed 32 times, and each time the number of heads occurring was recorded, as follows:

1	2	2	1
2	3	0	3
3	1	3	2
2	3	2	1
2	1	2	3
4	3	3	4
1	2	2	1
2	0	1	2

Label three columns "Number of Heads," "Tally Marks," and "Frequency," and prepare a frequency distribution for these data.

33. a. Make a histogram for the data in problem 32.
b. Now make a frequency polygon for the data in problem 32.

34. A high school class was asked to roll a pair of dice 3000 times. The sums of the top faces of the dice, the frequency of these sums, and the theoretical number of times the sums should have occurred appear in the following table. Make a histogram for these data showing the actual frequency with a solid line and the theoretical frequency with a dotted line (perhaps of a different color).

Sum	Actual Frequency	Theoretical Frequency
2	79	83
3	152	167
4	252	250
5	312	333
6	431	417
7	494	500
8	465	417
9	338	333
10	267	250
11	129	167
12	91	83

35. In a study of air pollution in a certain city, the following concentrations of sulfur dioxide in the air (in parts per million) were obtained for 30 days:

0.04	0.17	0.18	0.13	0.10	0.07
0.09	0.16	0.20	0.22	0.06	0.05
0.08	0.05	0.11	0.07	0.09	0.07
0.08	0.02	0.08	0.08	0.18	0.01
0.03	0.06	0.12	0.01	0.11	0.04

a. Make a frequency distribution for these data grouped in the intervals 0.00–0.04, 0.05–0.09, 0.10–0.14, 0.15–0.19, and 0.20–0.24.
b. For what percent of the time was the concentration of sulfur dioxide more than 0.14 part per million?

36. Make a histogram and a frequency polygon for the data in problem 35.

37. The following are the minimum weekly salaries (rounded to the nearest hundred dollars) for persons engaged in film production:

$7800	$5200	$4600	$1900	$1800	$1600
$1500	$1400	$ 800	$ 700	$ 700	$ 600

a. Make a frequency distribution using four classes with the upper class limit of the first class as the lower class limit of the second.

b. Make a histogram and a frequency polygon from your frequency distribution.

38. The following numbers represent the salaries of the 15 best-paid players in the National Hockey League (in millions of dollars):

$11,000,000	$11,000,000	$10,000,000
$10,000,000	$10,000,000	$10,000,000
$ 9,880,939	$ 9,500,000	$ 9,326,519
$ 9,000,000	$ 9,000,000	$ 9,000,000
$ 8,866,445	$ 8,850,000	$ 8,850,000

Source: http://tinyurl.com/5xcqy.

a. Make a frequency distribution using three classes with the upper class limit of the first class as the lower class limit of the second.

b. Make a histogram and a frequency polygon from your frequency distribution.

39. Here are the salaries of the 25 best-paid baseball players in 1947 (in thousands of dollars):

$90 $75 $65 $60 $44 $30 $30 $28 $26 $25
$23 $23 $20 $20 $20 $20 $20 $20 $20 $20
$20 $20 $20 $20 $20

a. Make a frequency distribution using five classes with the upper class limit of the first class as the lower class limit of the second.

b. Make a histogram and a frequency polygon from your frequency distribution.

40. How do you use the Internet? Is there a difference between the activities you do on the Internet and the activities your professors (academics) do? A survey conducted by BSA-Ipsos indicated the following relative frequencies of Internet use among students and academics:

	Students	Academics
Personal	63%	11%
School	34%	0
Work	3%	89%

Source: www.definetheline.com/resources/BSA-Ipsos-Education-Survey-June2005.pdf.

a. Make a relative frequency histogram for the students' data.

b. Make a relative frequency histogram for the academics' data.

c. Which category has the shortest bar (frequency) in the histogram for the students? What does that mean?

d. Which category has the longest bar (frequency) in the histogram for the academics? What does that mean?

41. Are you looking for a printer for your computer? The following are the list prices (in dollars) of the 15 best-rated color printers:

$500 $300 $475 $280 $550 $265 $300 $310
$250 $240 $300 $250 $190 $290 $180

a. Find the class width using five classes.

b. Make a frequency table with the five classes showing the class boundaries and frequencies.

c. Make a histogram showing the boundaries.

42. One of the top-grossing films of all time is *Titanic*, which grossed more than $600 million by 1999. But which films are the losers? The following are the amounts lost by the ten biggest movie failures (in millions):

$35	$25	$23.3	$20	$20
$18.5	$17	$16.6	$15	$14

a. Find the class width using three classes.

b. Make a frequency table with the three classes showing the class boundaries and frequencies.

c. Make a histogram showing the boundaries.

43. Do you think you're getting old? Still, you're probably not nearly as old as Shigechiyo Izumi of Japan, who almost made it to 121 years. As of this writing, the following were the authenticated ages of the 25 oldest people, rounded to the nearest year:

121 115 115 114 113 113 113 113 112
112 112 112 112 111 111 111 111 111
111 110 110 110 110 110 109

a. Find the class width using five classes.

b. Make a frequency table with the five classes showing the class boundaries and frequencies.

c. Make a histogram showing the boundaries.

E **Applications**

Problems 44–47 refer to the data in Example 4.

44. Which tract has the most renter-occupied homes?

45. Which tract has the fewest renter-occupied homes?

46. Which tract has the fewest vacant homes?

47. On the basis of your previous answers, which tract would you select to sell renter's insurance on?

Problems 48 and 49 refer to the following table. An ad for a home pregnancy test claimed 99.5% accuracy as shown in the table at right.

Status	Actually Pregnant	Actually Not Pregnant
Test says pregnant	197	0
Test says not pregnant	1	2
Total	**198**	**2**

48. a. What was the sample size?
   b. How many times was the test incorrect?
   c. What percent is that?

49. Look at the definition of a simple random sample. Do you think that a simple random sample was used to obtain the data in the table? Why or why not?

50. Midway Airlines published ads in the *New York Times* and the *Wall Street Journal* claiming that "84 percent of frequent business travelers to Chicago prefer Midway Metrolink to American, United, and TWA." If it is known that Midway only has 8% of the traffic between New York and Chicago, can you explain how it may have arrived at this figure?

51. At the bottom of the ad cited in problem 50, the fine print stated that the survey was "conducted among Metrolink passengers between New York and Chicago." Is the sample used a representative random sample? (See the definition of a simple random sample.) Why or why not?

 In Other Words

52. When making a histogram, why is it necessary to make the class boundaries the endpoints of the bar?

53. Explain the difference between class limits and class boundaries.

54. A survey of the weight of 200 persons and a histogram of the last digit of each weight show that 0 occurred 130 times and 5 occurred 123 times. What might be wrong with the survey?

 Using Your Knowledge

Around 1940 it was estimated that it would require approximately 10 years of computation to find the value of the number π (pi) to 1000 decimal places. But in the early 1960s, a computer calculated the value of π to more than 100,000 decimal places in less than 9 hours! Since then, 1.24 trillion decimal places for π have been calculated. Here are the first 40 decimal places for π.

3.14159   26535   89793   23846
26433   83279   50288   41971

55. Make a frequency distribution of the digits after the decimal point. List the digits from 0 to 9 in your first column.

56. What are the most and the least frequently occurring digits?

Mathematicians are interested in knowing whether the digits after the decimal point all occur with the same frequency. This question can hardly be answered with so few decimal places. You should notice that only two of the digits occur with a frequency more than one unit away from what you should expect in 40 decimal places.

 **Discovery**

Misuses of Statistics   In this section we have shown an honest way of depicting statistical data by means of a histogram. But you can lie with statistics! Here is how. In a newspaper ad for a certain magazine, the circulation of the magazine was as shown below. The heights of the bars in the diagram seem to indicate that sales in the first 9 months were tripled by the first quarter of the next year (a whopping 200% rise in sales!).

57. Can you discover what was the approximate jump in sales from the first 9 months to the first quarter of the next year? If so, what was it?

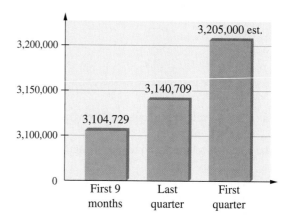

58. Can you discover what was the approximate percent rise in sales? If so, what was it?

59. Can you discover what is wrong with the graph? If so, what was it?

60. A more recent actual misuse of statistics:

CNN.com posted a visually distorted graph showing poll results on Schiavo case.

According to a poll conducted March 18–20, when asked if they "agree[d] with the court's decision to have the feeding tube removed" (from Terry Schiavo), 62 percent of Democratic respondents agreed, compared with 54 percent of Republicans and 54 percent of Independents. The results of the poll are shown in the graph.

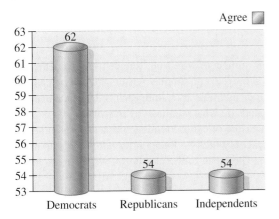

Results by Party

Sample: Interviews conducted by telephone March 18–20, 2005, with 909 adults in the United States.
Sampling error: ± 7% pts.
Source: Adapted from http://mediamatters.org/items/200503220005.

a. The categories are clearly labeled, but the frequencies are not. Should they be "frequencies" or "relative frequencies"? Can you discover how to label the units on the *y* axis?

b. Can you discover the actual percent difference between Republicans and Democrats?

c. Note that the sampling error is ±7% points. Keeping this in mind, can you discover how close the opinion of the Republicans and the Democrats could be? Explain.

d. Can you discover how to redraw the graph to show a more accurate presentation of the poll's findings? CNN has updated its graphic. You can see the updated version at the link provided in the source note.

 Collaborative Learning

Many of the problems in the Discovery section deal with the misuses of statistics. Now, it is your turn to find some of these misuses. Form two or more groups.

1. Discuss how statistics can be misused or misleading. Concentrate on examples involving sampling and histograms.

2. Get the book *How to Lie with Statistics* by Darrell Huff and find some other misuses of statistics involving sampling and histograms.

Online Study Center

For help with Collaborative Learning problem 1 and to further explore problems 47–49, access link 12.1.3 on this textbook's Online Study Center.

The chart below shows cosmetic products that cause allergies.

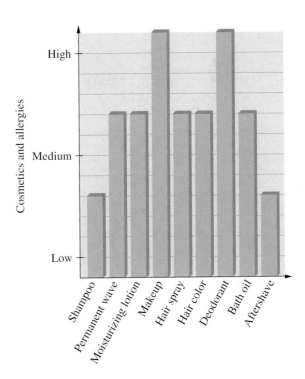

3. Which two products cause the most allergies? Suppose you want to sell hypoallergenic products. On the basis of the chart, which products would you concentrate on manufacturing? Discuss why.

4. Look at the definition of *histogram* given in the text. According to that definition, is the chart a histogram? Discuss why or why not. If not, what type of chart is it? Does the chart convey the idea that some products are more likely to produce allergic reactions, and if so, do you think it makes a difference whether the chart is a histogram or not?

5. A type of chart called a *Pareto* chart is used in quality control programs. The chart is used to improve quality control in production and service industries. In a Pareto chart, the tallest bar is placed on the left and the rest of the bars in descending order by height, so the chart highlights the major causes of problems. Convert the chart to a Pareto chart. Is it more clear now which products produce allergies?

12.2   Measures of Central Tendency: The Mean, Median, and Mode

Tongue Twister Averages

Is there a relationship between the number of words in a tongue twister and the difficulty in reciting it? Table 12.9 shows several tongue twisters and the percentage of successful attempts out of 30 total attempts at reciting each. What is the average number of words in each? It depends on what we mean by *average*.

The most commonly used measure of central tendency of a set of *n* numbers is the mean (the arithmetic average), which is obtained by *adding* all the numbers in the set and *dividing* by *n*. The mean of the number of words in the given tongue twisters is thus

$$\frac{11 + 7 + 5 + 4 + 10 + 6 + 6 + 5 + 13 + 35 + 9 + 8 + 4}{13} = \frac{123}{13} = 9.5$$

TABLE 12.9 Tongue Twisters

Phrase	Success Rate	Words
The seething sea ceaseth and thus the seething sea sufficeth us.	3%	11
The sixth sick sheik's sixth sheep's sick	30%	7
The Leith police dismisseth us.	67%	5
Sixty-six sick chicks	77%	4
Toy boat (said five times fast)	67%	10
Tie twine to three tree twigs	80%	6
She sells seashells by the seashore.	83%	6
Long slim slick sycamore saplings	87%	5
How much wood could a woodchuck chuck if a woodchuck could chuck wood?	80%	13

(continued)

TABLE 12.9 Tongue Twisters *(continued)*

Phrase	*Success Rate*	*Words*
Peter Piper . . . (you know this one)	57%	35
Three new blue beans in a new-blown bladder	90%	9
Twixt six thick thumbs stick six thick sticks.	93%	8
Better baby buggy bumpers	97%	4

Another measure of central tendency is the **median,** the middle value of an ordered set of numbers (there are as many values above as below the median). Let us arrange the number of words in each twister in ascending order.

$$4\ 4\ 5\ 5\ 6\ 6\ \underline{7}\ 8\ 9\ 10\ 11\ 13\ 35$$

In this case, the median number of words is 7.

The easiest average to compute is the **mode,** the value occurring most often. We see that 4, 5, and 6 are modes for these numbers; they occur twice each. If Peter Piper were replaced by "Zack zapped Zeus zinc," the most common number of words (the **mode**) would then be 4.

So what is the **average** number of words in these tongue twisters? Either 4, 5, 6, 7, or 9.5. Which of the numbers is most representative? In this section we shall study these three types of **averages**—the mean, the median, and the mode—and how they can be used and misused (see the Discovery section) in different situations. ▶

Alberto and Barney have just gotten back their test papers. There are 9 questions, and each one counted 10 points. Their scores are given in Table 12.10.

TABLE 12.10 Alberto's and Barney's Scores

Person	*Question*									*Total*
	1	2	3	4	5	6	7	8	9	
Alberto	10	7	10	7	7	10	9	10	2	72
Barney	10	8	10	7	7	7	10	7	7	73

TABLE 12.11 Frequency Table

Score	*Alberto*	*Barney*
2	1	0
7	3	5
8	0	1
9	1	0
10	4	3

Who do you think wrote a better paper? As you can see, Alberto's **average** score is $\frac{72}{9} = 8$, and Barney's **average** score is $\frac{73}{9} = 8.1$. Barney clearly has the higher average and concludes that he wrote the better paper. Do you agree?

Alberto does not agree because he did as well as or better than Barney on 6 of the 9 questions. Alberto thinks that Barney's higher average does not tell the whole story, so he tries something else. First, he makes a frequency distribution of the two sets of scores, as shown in Table 12.11.

On inspecting this list, Alberto says, "I did better than you did, Barney, because I scored 10 more often than any other number, and you scored 7 more often than any other number." Would you agree with Alberto?

Reprinted with special permission of North America Syndicate, Inc.

The first given averages, 8 and 8.1, are the *means*. These are the ones that most of us think of as the averages.

Definition of the Mean

The **mean** of a set of n numbers is the sum of the numbers divided by n. The mean is usually denoted by $\bar{x}$ (read "x bar"). Thus,

$$\bar{x} = \frac{\text{sum}}{n}$$

Alberto used a different kind of measure, called the *mode*.

Definition of the Mode

The **mode** of a set of numbers is that number of the set that occurs most often. If no number in the set occurs more than once, there is **no** mode. If several numbers all occur an equal number of times and more than all the rest, then all these several numbers are *modes*.

Thus, it is possible for a set of numbers to have more than one mode or no mode at all.

The mean and the mode are useful because they give an indication of a sort of center of the set. For this reason, they are called **measures of central tendency.**

EXAMPLE 1 ▶ Mean and Mode of Golf Scores

Ten golf professionals playing a certain course scored 69, 71, 72, 68, 69, 73, 71, 70, 69, and 68. Find the following:

(a) The mean (average) of these scores　　(b) The mode of these scores

Solution

(a) $\bar{x} = \dfrac{69 + 71 + 72 + 68 + 69 + 73 + 71 + 70 + 69 + 68}{10} = \dfrac{700}{10} = 70$

(b) The score that occurred most often—the mode—is 69 (3 times). ■

There is a third commonly used measure of central tendency, called the *median*.

Definition of the Median

The **median** of a set of numbers is the middle number when the numbers are arranged in order of magnitude. If there is no single middle number, then the median is the mean (average) of the two middle numbers.

Let us list Alberto's and Barney's scores in order of magnitude as follows:

Alberto	2	7	7	7	(9)	10	10	10	10
Barney	7	7	7	7	(7)	8	10	10	10

TABLE 12.12 Measures of Central Tendency

Measure	Alberto	Barney
Mean	8	8.1
Mode	10	7
Median	9	7

The median is circled in each case.

Now look in Table 12.12 for the three measures we have found for the scores. The mode and the median in this case would appear to some people to be evidence that perhaps Alberto did write a better paper than Barney.

But Barney is not convinced, and he does not give up easily. He concedes that Alberto had a better **mode** but claims that in many practical situations the mode does not even exist! Look at the prices of gas. "What?" says Alberto. Most gas station signs always show the mean and the median but rarely the mode. The mode is not a good indicator. Go out and check for yourself! Here are some samples.

Mean:
$$\frac{2.29 + 2.39 + 2.49}{3} = 2.39$$
Median: 2.39
No mode

Mean:
$$\frac{2.31 + 2.41 + 2.51}{3} = 2.41$$
Median: 2.41
No mode

Mean:
$$\frac{2.35 + 2.45 + 2.55}{3} = 2.45$$
Median: 2.45
No mode

"Si, si!" Alberto says, "but look at the prices I found" in the next example.

EXAMPLE 2 ▶ Mean, Median, and Mode of Gas Prices

Find the mean, median, and mode of the gas prices. Which is the most representative of the actual prices?

Solution

$$\text{Mean:} \quad \frac{2.31 + 2.39 + 2.49}{3} \approx \mathbf{2.3967}$$

Median: 2.39
No mode

This time probably the median represents the actual prices better. Do you agree?   ∎

EXAMPLE 3 ▶ Mean, Median, and Mode of Calorie Loss

Have you been exercising lately? You must exercise if you want to keep your weight down. The following are 10 different activities with the corresponding hourly energy expenditures (in calories) for a 150-lb person:

Activity	Calories/Hour	Activity	Calories/Hour
Fencing	300	Square dancing	350
Golf	250	Squash	600
Running	900	Swimming	300
Sitting	100	Volleyball	350
Standing	150	Wood chopping	400

(a) Find the mean of these numbers.

(b) Find the median number of calories spent in these activities.

(c) Find the mode of these numbers.

Solution

(a) The mean $\bar{x}$ is obtained by adding all the numbers and dividing the sum by 10. Thus,

$$\bar{x} = \frac{\text{sum}}{n}$$

$$= \frac{300 + 250 + 900 + 100 + 150 + 350 + 600 + 300 + 350 + 400}{10}$$

$$= 370 \text{ calories per hour}$$

GRAPH IT

To do Example 3(a), clear any lists by pressing [2nd] [+] [4] [ENTER], and tell your grapher that you want to do statistics by pressing [STAT] [1]. Enter the values as the list L_1 by pressing 300 [ENTER] 250 [ENTER], and so on. Now you have the list $\{300, 250, \ldots, 400\}$. To find the mean of the numbers in L_1, go to the home screen ([2nd] [MODE]) and press [2nd] [STAT] [◄] [3]. What mean do you want? The mean of the numbers in L_1, so enter [2nd] [L_1] [ENTER]. What if you want the median? Press [2nd] [STAT] [◄] [4] [2nd] [L_1] [ENTER].

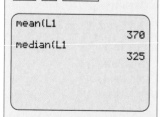

```
mean(L1
                370
median(L1
                325
```

You can close parentheses after L_1 if desired!

(b) To find the median, we must first arrange the numbers in order of magnitude, as follows:

Sitting	100
Standing	150
Golf	250
Fencing	300
Swimming	300
Square dancing	350
Volleyball	350
Wood chopping	400
Squash	600
Running	900

$\leftarrow$ Median $= \dfrac{300 + 350}{2} = 325$

We have an even number of items, and there is no "middle" value. The median is the mean (average) of the two middle items.

(c) The mode is the number with the greatest frequency if there is one such number. In this case, the numbers 300 and 350 both occur twice, whereas all other numbers occur just once. Thus, there are two modes, 300 and 350; the data are *bi*modal. ■

EXAMPLE 4 ▶ Mean, Median, and Mode for Wages

For the frequency distribution of wage rates given in the table at the left, find the following:

(a) The mean rate   (b) The mode   (c) The median rate

Solution

(a) Referring to the table at the left, make the calculation shown in Table 12.13; the mean rate is $7.94 per hour.

Wage	Tally Marks	Frequency
7.70	|||	3
7.80	⑷|	6
7.90	⑷|	6
8.00	⑷|||	8
8.10	⑷	5
8.20	||	2
		30 Total

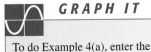 *GRAPH IT*

To do Example 4(a), enter the numbers in the first column as list L_1 and the frequencies as L_2. Go to the home screen (2nd MODE) and press STAT ▶ 1 . This means that you are doing one-variable statistics. But on which variables? It should be on the variables you entered, L_1 and L_2. Press 2nd L_1 , 2nd L_2 ENTER . The mean is 7.94 as before. If you scroll down by pressing ▼, the median is given as 7.95.

TABLE 12.13

Wage Rate	Frequency	Frequency × Rate
7.70	3	23.10
7.80	6	46.80
7.90	6	47.40
8.00	8	64.00
8.10	5	40.50
8.20	2	16.40
	30	$30\overline{)238.20}$
		$7.94 = \bar{x}$

(b) The mode is the most frequent rate, $8 per hour.

(c) By adding down the frequency column, we see that 15 workers get $7.90 or less, and the other 15 get $8.00 or more per hour. The median rate is the mean of $7.90 and $8.00, that is, $7.95 per hour. ■

TABLE 12.14

Score (Number Correct)	Proportion of Students
0	0.05
1	0.05
2	0.10
3	0.35
4	0.25
5	0.20

EXAMPLE 5 ▶ **Median and Mean for Student Scores**

Table 12.14 shows the distribution of scores made by a large number of students taking a five-question true/false test.

(a) Which is the median score?   (b) What is the mean score?

Solution

(a) Table 12.14 gives the proportion of students making each score, so the sum of these proportions must be 1. To find the median (the middle value 0.5), we add the proportions starting from the top until we get a sum of 0.5 or more. This occurs when we add the first four items. Thus the median score is 3.

(b) To find the mean score, we add the products of the scores and their proportions as follows:

$$0 \times 0.05 = 0$$
$$1 \times 0.05 = 0.05$$
$$2 \times 0.10 = 0.20$$
$$3 \times 0.35 = 1.05$$
$$4 \times 0.25 = 1.00$$
$$\underline{5 \times 0.20 = 1.00}$$
$$\text{Sum} = 3.30$$

Since the sum of the proportions is 1, the mean score is the sum 3.30.   ■

EXAMPLE 6 ▶ **Conclusions About a Physics Test**

In a physics class, half the students scored 80 on a midterm exam. Most of the remaining students scored 70, except for a few who scored 20. Which of the following statements is true?

(a) The mean and the median are the same.

(b) The mean and the mode are the same.

(c) The mean is less than the median.

(d) The mean is greater than the median.

Solution

Since half the students scored 80 and the next best score was 70, the median score on the midterm is 75 (midway between 80 and 70). Also, not all the remaining students scored 70, so the mean is less than 75. Thus, statement (c) is correct.   ■

EXAMPLE 7 ▶ **True Statements About the Mean, Median, and Mode**

Figure 12.8 shows the distribution of scores on a placement test for students at South High School. In the chart, x is the score and y is the frequency. Which of the following statements is true?

(a) The mode and the mean are the same.

(b) The mode and the median are the same.

(c) The median is less than the mode.

(d) The median is greater than the mode.

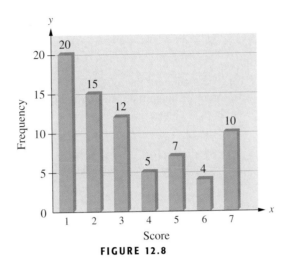

FIGURE 12.8

Solution

The chart shows that the mode is $x = 1$. It also shows that more than 20 students scored higher than 1. Hence, the median must be greater than 1, and therefore, statement (d) is correct. ∎

EXAMPLE 8 ▶ Mean and Median of the Number of Rooms

Remember the housing units we studied in Section 12.1? Averages are used to find the number of rooms per unit. For example, in Tract 1302 (column 1), the average (mean) number of rooms per unit is 4.8 (see Table 12.15).

(a) Find the average (mean) number of one-room houses in the four tracts.

(b) Find the average (mean) number of two-room houses in the four tracts.

(c) Find the median number of two-room houses in the four tracts.

Solution

(a) The average (mean) number of one-room houses is the sum of the numbers in the first row divided by 4, the number of tracts; that is,

$$\frac{43 + 19 + 39 + 20}{4} = \frac{121}{4} = 30.25, \text{ or approximately } 30$$

TABLE 12.15

Housing Tenure (by Number of Rooms)	Tract 1302		Tract 1303		Tract 1304		Tract 1305	
1 room	43	6%	19	1%	39	2%	20	1%
2 rooms	28	4%	67	5%	98	4%	137	7%
3 or 4 rooms	247	32%	481	35%	817	32%	764	37%
5 or 6 rooms	367	47%	611	45%	1315	51%	900	44%
7 or 8 rooms	64	8%	160	12%	246	10%	235	11%
9 or more rooms	27	3%	19	1%	56	2%	4	0%
Total housing units	776		1357		2571		2060	
Rooms per unit	4.8		4.9		4.9		4.7	
Persons per occupied unit	2.1		2.5		2.7		2.4	
Persons per occupied room	0.4		0.5		0.6		0.5	

(b) This time we add the numbers in the second row and divide by 4.

$$\frac{28 + 67 + 98 + 137}{4} = \frac{330}{4} = 82.5, \text{ or approximately } 83$$

(c) The median of 28, 67, 98, and 137 is obtained by arranging the numbers in order of magnitude and finding the mean (average) of the two middle numbers as shown.

$$\left.\begin{array}{c} 28 \\ 67 \\ 98 \\ 137 \end{array}\right\} \quad \frac{67 + 98}{2} = 82.5, \text{ or about } 83$$

∎

In this section we introduced three **measures of central tendency.** The following shows how they compare:

1. The **mean** (arithmetic average) is the most commonly used of the three measures. A set of data always has a unique mean, and this mean takes account of each item of the data. On the negative side, finding the mean takes the most calculation of the three measures. Another bad feature of the mean is its sensitivity to extreme values. For instance, the mean of the data 2, 4, 6, and 8 is $\frac{20}{4} = 5$, but the mean of 2, 4, 6, and 28 is $\frac{40}{4} = 10$, a shift of 5 units toward the extreme value 28.

2. The **mode** has the advantage of requiring no calculation. However, the mode may not exist, as in the case of the data 2, 4, 6, and 8. On the other hand, the mode may be most useful. For example, suppose a shoe manufacturer surveys 100 women to see which of three styles, A, B, or C, of shoes each one prefers and finds style A selected by 30 women, style B by 50, and style C by 20. The mode is 50, and there is not much doubt about which style the manufacturer will feature.

3. The **median** always exists and is unique, as in the case of the mean. However, the median requires very little computation and is not sensitive to extreme values. Of course, in order to find the median, the data must be arranged in order of magnitude, and this may not be practical for large sets of data. But the most important disadvantage of the median is its failure to take account of each item of data. Hence, in many statistical problems, the median is not a reliable measure.

Online Study Center

To further explore the mean, median, mode, and standard deviation, access links 12.2.1 and 12.2.3 on this textbook's Online Study Center. To further explore descriptive statistics, access link 12.2.2.

EXERCISES 12.2

1. Find the mean and the median for each set of numbers.
   a. 1, 5, 9, 13, 17   b. 1, 3, 9, 27, 81
   c. 1, 4, 9, 16, 25
   d. For which of these sets are the mean and the median the same? Which measure is the same for all three sets? Which (if any) of the sets has a mode?

2. Show that the median of the set of numbers 1, 2, 4, 8, 16, 32 is 6. How does this compare with the mean?

3. Out of 10 possible points, a class of 20 students made the following test scores:

   0, 0, 1, 2, 4, 5, 5, 6, 6, 6, 7, 8, 8, 8, 8, 9, 9, 9, 10, 10

Find the mean, the median, and the mode. Which of these three measures do you think is the least representative of the set of scores?

4. Find the mean and the median of the following set of numbers:

0, 3, 26, 43, 45, 60, 72, 75, 79, 82, 83

5. An instructor gave a short test to a class of 25 students. The scores on the basis of 10 were

Score	Number of Students
3	2
4	1
5	3
6	2
7	6
8	4
9	4
10	3

The instructor asked two students, Agnes and Betty, to calculate the mean (average) score. Agnes made the calculation

$$\frac{3 + 4 + 5 + 6 + 7 + 8 + 9 + 10}{8} = \frac{52}{8} = 6.5$$

and said the mean (average) score is 6.5. Betty calculated a *weighted average* by multiplying each score by the number of students attaining that score, adding the results, and dividing by the total number of students as shown.

$$\frac{2 \cdot 3 + 1 \cdot 4 + 3 \cdot 5 + 2 \cdot 6 + 6 \cdot 7 + 4 \cdot 8 + 4 \cdot 9 + 3 \cdot 10}{25}$$

$$= \frac{177}{25} = 7.08$$

She then said the mean (average) was 7.08. Who is correct, Agnes or Betty? Why?

6. An investor bought 150 shares of Fly-Hi Airlines stock. He paid $60 per share for 50 shares, $50 per share for 60 shares, and $75 per share for 40 shares. What was his average cost per share? (Compare with problem 5.)

7. Make a frequency distribution of the number of letters per word in the following quotation: "For

seven days seven priests with seven trumpets invested Jericho, and on the seventh day they encompassed the city seven times."

a. Find the mode(s) of the number of letters per word.

b. Find the median. (You can use your frequency distribution to do this.)

c. Find the mean of the number of letters per word.

d. Do you think your answers would give a good indication of the average length of words in ordinary English writing? Why or why not?

8. The following are the temperatures at 1-hour intervals in Denver, Colorado, from 1 P.M. on a certain day to 9 A.M. the next day:

1 P.M.	90	8 P.M.	81	3 A.M.	66
2 P.M.	91	9 P.M.	79	4 A.M.	65
3 P.M.	92	10 P.M.	76	5 A.M.	66
4 P.M.	92	11 P.M.	74	6 A.M.	64
5 P.M.	91	12 M	71	7 A.M.	64
6 P.M.	89	1 A.M.	71	8 A.M.	71
7 P.M.	86	2 A.M.	69	9 A.M.	75

a. What was the mean temperature? The median temperature?

b. What was the mean temperature from 1 P.M. to 9 P.M.? The median temperature?

c. What was the mean temperature from midnight to 6 A.M.? The median temperature?

9. Suppose that a dime and a nickel are tossed. They can fall in four different ways: (H, H), (H, T), (T, H), and (T, T), where the first letter indicates how the dime falls and the second letter, the nickel. How many tosses do you think it would take, on average, to get all four possibilities at least once? A good way to find out is by experimenting. Take a dime and a nickel and toss them to get your data. For example, on the first trial it took 11 tosses to get all four possibilities, (H, H), (H, T), (T, H), and (T, T). You can keep track of what happens with a frequency distribution like the following:

	Trial 1	Trial 2	Trial 3
(H, H)	|	|	⦀⦀
(H, T)	|	|	⦀
(T, H)	⦀⦀	|	⦀
(T, T)	⦀⦀	⦀⦀	|
	11	7	12 (etc.)

You will need to make tally marks in the trial column until there is at least one mark for each possibility. Then write the total number of tosses at the bottom of the column. A new column will be needed for each trial, of course. Do 20 trials.

a. When you finish the 20 trials, make a frequency distribution of the number of tosses required to give all four possibilities.

b. Use the frequency distribution you obtained in part (a) to find the median number of tosses.

c. Find the mean number of tosses needed to obtain all four possibilities.

10. The mean score on a test taken by 20 students is 75; what is the sum of the 20 test scores?

11. A mathematics professor lost a test paper belonging to one of her students. She remembered that the mean score for the class of 20 was 81 and that the sum of the 19 other scores was 1560. What was the grade on the paper she lost?

12. If in problem 11 the mean was 82, and the sum of the 19 other scores was still 1560, what was the grade on the lost paper?

13. The mean salary for the 20 workers in company A is $90 per week, whereas in company B the mean salary for its 30 workers is $80 per week. If the two companies merge, what is the mean salary for the 50 employees of the new company?

14. A student has a mean score of 88 on five tests taken. What score must she obtain on her next test to have a mean (average) score of 80 on all six tests?

15. The table below shows the distribution of families by income in a particular urban area.

Annual Income ($)	Proportion of Families
0–9999	0.02
10,000–14,999	0.09
15,000–19,999	0.25
20,000–24,999	0.30
25,000–34,999	0.11
35,000–49,999	0.10
50,000–79,999	0.07
80,000–119,999	0.05
120,000+	0.01

a. What proportion of the families have incomes of at least $25,000?

b. What is the median income range?

c. Find the mean of the lower limits for the annual incomes.

d. Find the amount below which 36% of the families have lower incomes.

16. In a history test given to 100 students, 50 earned scores of 80. Of the other students, 40 earned scores between 60 and 75, and the other 10 earned scores between 10 and 40. Which one of the following statements is true about the distribution of scores?

a. The mean and the median are the same.

b. The mean is less than the median.

c. The median is less than the mean.

17. In a mathematics test given to 50 students, 25 earned scores of 90. Most of the other students scored 80, and the remaining students scored 30. Which of the following statements is true about the distribution of scores?

a. The mode is the same as the mean.

b. The median is greater than the mean.

c. The mode is greater than the mean.

d. The mean is greater than the median.

18. The following graph shows the distribution of scores on a placement test given to juniors at West High School. Which of the following statements applies to this distribution?

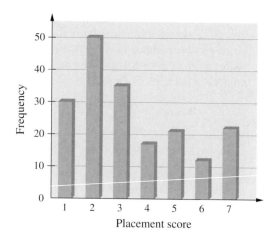

a. The mode and the mean are the same.

b. The mode is greater than the mean.

c. The mode and the median are the same.

d. The mode is less than the median.

19. One hundred students took a mathematics test. None of the students scored over 95 or less than 50. Which of the following is the most reasonable estimate of the mean (average) score?
a. 80   **b.** 60   **c.** 70   **d.** 55

20. Gretchen owns five $5000 municipal bonds. Each quarter she receives the following interest payments: $81.25, $112.50, $118.75, $125.00, and $137.50. Which of the following is the most reasonable estimate of the mean (average) interest payment?
a. $108   **b.** $116   **c.** $120   **d.** $105

Problems 21 and 22 refer to Table 12.15 on page 801.

21. Find the mean and median number of three- or four-room houses in the four tracts.

22. Find the mean and median number of five- or six-room houses in the four tracts.

The *ABC Consumer Comfort Index* rates the economy, the buying climate, and personal finances. The three resulting numbers are added and divided by 3. For a particular week the results were as follows:

General Population

Overall index	25
State of economy	42
Personal finances	34
Buying climate	0

Note that

$$\frac{42 + 34 + 0}{3} = \frac{76}{3} \approx \boxed{25}$$

Find the mean of the following overall indexes classified by

23. *Income*

Under $15K	−18
$15K–$24.9K	8
$25K–$39.9K	5
$40K–$49.9K	46
Over $50K	62

24. *Education*

< High school	−13
High sch. grad.	16
College +	45

The table that follows will be used in problems 25–29.

Type of Spam	July	August
Internet	14	22
Other	28	32
Scams	18	20
Products	40	40
Spiritual	2	2
Financial	30	28
Leisure	16	14
Adult	28	24
Health	24	18

Source: Data from Brightmail Probe Network.

The table shows the number and type of spam (unsolicited "junk" e-mail sent to large numbers of people to promote products or services) received by the same 200 persons in July and August.

25. Find the mean number of spams received in July.

26. Find the median number of spams received in July.

27. Find the mode (if it exists) of the number of spams received in July.

28. Name the categories in which the number of spams received did not change from July to August.

29. Which category increased the most from July to August?

Attendance at Top Five Amusement/Theme Parks in the United States (in millions)*

Name	Year 1	Year 2	Year 3	Year 4
The Magic Kingdom at Walt Disney World Lake Buena Vista, FL	11.2	12.9	13.8	17.0
Disneyland Anaheim, CA	10.3	14.1	15.0	14.3
Epcot at Walt Disney World Lake Buena Vista, FL	9.7	10.7	11.2	11.8
Disney-MGM Studios at Walt Disney World Lake Buena Vista, FL	8.0	9.5	10.0	10.5
Universal Studios Florida Orlando, FL	7.7	8.0	8.4	8.9

*For the latest figures, try http://tinyurl.com/b3nkz.

The table at the bottom of page 805 shows the attendance (in millions) of the top five amusement/theme parks in the United States for four successive years. Find the mean (average) and median attendance at

30. the Magic Kingdom.   **31.** Disneyland.

32. Epcot.   **33.** Disney-MGM.

34. Universal Studios Florida.

Do you surf or shop on the Web? The tables below give the top Web sites and the top education sites.

35. What is the average number of visitors to the ten top Web sites overall?

Top Ten Web Sites Overall

	Top Domains/ Top Web Sites	Unique visitors* (millions)
1	Yahoo! sites	118
2	Time Warner Network	116
3	MSN Microsoft sites	112
4	Google sites	79
5	eBay	64
6	Ask Jeeves	42
7	Amazon sites	39
8	CNET Networks	33
9	Viacom Online	32
10	Monster Worldwide	30

*Unique visitors over 1-month period.
Source: comScore Media Metrix (comScore Media Metrix is a division of comScore Networks, Inc.).

36. What is the average number of visitors to the top ten education sites?

37. The following table shows the average annual salary for five consecutive years for a person with an associate degree:

1	39,468
2	39,276
3	40,827
4	46,778
5	49,733

a. Find the mean, median, and mode of the person's salary for the 5 years.
b. Which is the most representative of the person's salary: the mean, the median, or the mode?

38. The following table shows the average annual salary for five consecutive years for a person with less than a high school diploma:

1	20,484
2	19,935
3	21,611
4	22,679
5	23,845

Source: Bureau of Labor Statistics.

a. Find the mean, median, and mode of the person's salary for the 5 years. Compare with the person in Problem 37!
b. Which is the most representative of the person's salary: the mean, the median, or the mode?

Top Ten: Education Sites

Rank	Education Web Sites	Unique visitors* (millions)	Rank	Education Web Sites	Unique visitors* (millions)
1	Fastweb	4.06	6	UMich.edu	2.53
2	Learning Network Property	3.67	7	Cornell.edu	2.49
3	Berkeley.edu	3.05	8	Harvard.edu	2.44
4	UTexas.edu	2.87	9	MIT.edu	2.39
5	UIUC.edu	2.68	10	Thinkquest site	2.32

*Unique visitors over 1-month period.

In Other Words

39. Explain in your own words what is meant by the median of a set of scores. Is the median a good measure of a set of scores?

40. What is meant by the mode of a set of scores? Is this a good measure? If so, for what purpose? Give an example.

Using Your Knowledge

Have you ever been in the checkout line at a supermarket or department store for so long that you were tempted to walk out? There is a mathematical theory called **queuing** (pronounced "cueing") theory that studies ways in which lines at supermarkets, department stores, and so on can be reduced to a minimum. The following problems show how a store manager can estimate the average number of people waiting at a particular counter:

41. Suppose that in a 5-min interval customers arrive as indicated in the following table. (Arrival time is assumed to be at the beginning of each minute.) In the first minute, A and B arrive. During the second minute, B moves to the head of the line (A was gone because it took 1 min to serve him), and C and D arrive, and so on. From the figure, find
   a. the average (mean) number of people in line.
   b. the mode of the number of people in line.

Time	Customers
1	A, B
2	C, D
3	
4	E, F
5	

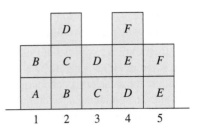

42. Use the ideas of problem 41 and suppose that the list of arrivals is as shown in the table below. (Assume it takes 1 min to serve the first customer in line and that the customer leaves immediately.)
   a. Draw a diagram showing the line during each of the first 5 min.
   b. Find the mean of the number of people in line during the 5 min.
   c. Find the mode of the number of people in line.

Time	Customers
1	A
2	B
3	C, D, E
4	F
5	

Discovery

Misuses of Statistics We have just studied three measures of central tendency: the mean, the median, and the mode. All these measures are frequently called averages. Suppose that the chart below shows the salaries at Scrooge Manufacturing Company.

43. Scrooge claims that the workers should not unionize; after all, he says, the "average" salary is $21,000. Can you discover what "average" this is?

44. Manny Chevitz, the union leader, claims that Scrooge Manufacturing really needs a union. Just

Boss $100,000   Boss's son $50,000   Boss's assistant $25,000   Boss's secretaries $10,000   Workers $6,000 each

look at their salaries! A meager $6000 on the average. Can you discover what "average" he means?

45. B. Crooked, the politician, wants both union and management support. He says that the workers are about average as far as salary is concerned. You can figure it out. The company's "average" salary is $8000. Can you discover what "average" B. Crooked has in mind?

Calculator Corner

If you have a calculator with $\boxed{\Sigma+}$ (read "sigma plus") and $\boxed{\bar{x}}$ keys, you are in luck. The calculation for the mean is done automatically for you. First, place the calculator in the statistics mode (press $\boxed{\text{mode}}$ $\boxed{\text{stat}}$ or $\boxed{\text{2nd}}$ $\boxed{\text{stat}}$). To find the mean of the numbers in Example 2, enter $\boxed{300}$ $\boxed{\Sigma+}$ $\boxed{250}$ $\boxed{\Sigma+}$ $\boxed{900}$ $\boxed{\Sigma+}$ $\boxed{100}$ $\boxed{\Sigma+}$ $\boxed{150}$ $\boxed{\Sigma+}$ $\boxed{350}$ $\boxed{\Sigma+}$ $\boxed{600}$ $\boxed{\Sigma+}$ $\boxed{300}$ $\boxed{\Sigma+}$ $\boxed{350}$ $\boxed{\Sigma+}$ $\boxed{400}$ $\boxed{\Sigma+}$ $\boxed{\text{2nd}}$ $\boxed{\bar{x}}$. The display gives the mean $\bar{x} = 370$.

Collaborative Learning

Form two groups. One group will investigate race discrimination complaints and the other sex discrimination complaints, as shown in the table below. Then answer the questions.

1. What was the average number of complaints for race and for sex in the 5-year period?

2. What was the average number of resolutions for race and for sex in the 5-year period?

3. What were the average monetary benefits for race and for sex per resolved case in the 5-year period?

Discussion Which average monetary benefits per resolved case were greater, race or sex discrimination? Why?

Discrimination Complaints: Number of Complaints Received, Number of Cases Resolved, and Amount of Monetary Benefits for Different Types of Workplace Discrimination in a 5-Year Period

	Year 1	Year 2	Year 3	Year 4	Year 5
Race					
Complaints received	31,695	31,656	29,986	26,287	29,199
Resolutions	27,440	25,253	31,674	35,127	36,419
Monetary benefits (millions)	$ 33.3	$ 39.7	$ 30.1	$ 37.2	$ 41.8
Sex					
Complaints received	23,919	25,860	26,181	23,813	24,728
Resolutions	21,606	21,545	26,726	30,965	32,836
Monetary benefits (millions)	$ 44.0	$ 44.1	$ 23.6	$ 47.1	$ 72.5

Source: U.S. Equal Employment Opportunity Commission.

12.3 Measures of Dispersion: The Range and Standard Deviation

Ratings Deviations for Movies and TV

Which programs get the best television ratings, weekly series or movies? The ratings for the ten best series and ten best movies according to the *Top 10 Almanac* are shown in Table 12.16. If you were a manufacturer selling your product to consumers, would you buy television advertisement time during a series or during a movie? Before you make up your mind, look at Table 12.16

TABLE 12.16

	Ratings	
Rank	Series	Movies
1	24	25
2	23	23
3	23	22
4	22	21
5	21	21
6	20	20
7	20	20
8	20	20
9	19	19
10	18	19
Total	**210**	**210**

TABLE 12.17

Rank	Movie Ratings (x)	$x - \bar{x}$	$(x - \bar{x})^2$
1	25	$25 - 21 = 4$	16
2	23	$23 - 21 = 2$	4
3	22	$22 - 21 = 1$	1
4	21	$21 - 21 = 0$	0
5	21	$21 - 21 = 0$	0
6	20	$20 - 21 = -1$	1
7	20	$20 - 21 = -1$	1
8	20	$20 - 21 = -1$	1
9	19	$19 - 21 = -2$	4
10	19	$19 - 21 = -2$	4
Total	**210**		**32**

and note that the series' ratings **range** from 18 to 24 (6 points), whereas the movies' ratings **range** from 19 to 25 (also 6 points). Since the range is the same in both distributions, you might look next at the mean rating in each category. But the mean is 21 in both cases. What else can you look at to try to determine the best air time for an advertiser? There is a measurement that indicates how the data differ from the mean, and it is called the **standard deviation (s).** The standard deviation, like the range, is a measure of the spread of data. To obtain the standard deviation of a set of numbers, start by computing the difference between each measurement and the mean, that is, $x - \bar{x}$, as shown in Table 12.17 for the movie ratings data.

Unfortunately, if you add the $x - \bar{x}$ values, you get a sum of 0. (This will always be true when finding standard deviations. Can you figure out why?) Therefore, you square each $x - \bar{x}$ value before you do the addition and then arrive at a sum of 32. If you were looking at the entire population of movies on TV, you would then divide this number by the population size to get a type of "average squared difference" of ratings from the mean. However, this sample does not include the entire population, so, as a rule, divide instead by 1 less than the number in the sample (here, that number is $10 - 1 = 9$) to make the final value of the standard deviation a bit larger. You then have $s^2 = \frac{32}{9}$, in units of squared ratings points. What kind of unit is that? To return to ratings points, you need to take the square root of $\frac{32}{9}$.

$$s = \sqrt{\frac{32}{9}} \quad \text{or} \quad \frac{4\sqrt{2}}{3}$$

(Remember that $\sqrt{32} = \sqrt{16 \cdot 2} = 4\sqrt{2}$.) A calculator or square root table gives $s \approx 1.89$. For the series data, $s \approx 1.94$ (try it); this means that series ratings are slightly more dispersed, or spread out. You might like advertising time during movies because their ratings are closer together and slightly less variable.

In this section you will learn how to find the range and calculate the standard deviation of a set of sample data. ▶

Most of the time we want to know more about a set of numbers than we can learn from a measure of central tendency. For instance, the two sets of numbers {3, 5, 7} and {0, 5, 10} both have the same mean and the same median, 5, but the two sets of numbers are quite different. Clearly, some information about how the numbers vary will be useful in describing the set.

A number that describes how the numbers of a set are spread out, or dispersed, is called a **measure of dispersion.** A very simple example of such a measure is the range.

Definition of the Range

The **range** of a set of numbers is the difference between the greatest and the least of the numbers in the set.

The two sets {3, 5, 7} and {0, 5, 10} have ranges $7 - 3 = 4$ and $10 - 0 = 10$, respectively. Because the range is determined by only two numbers of the set, it gives very little information about the other numbers of the set. The range gives only a general notion of the spread of the given data.

As noted in Getting Started, if we add all the deviations from the mean, the $x - \bar{x}$ values, we always get a sum of 0. Accordingly, the most commonly used measure of dispersion, the *standard deviation,* uses the squares of the deviations $x_k - \bar{x}$ in the following definition:

Definition of the Standard Deviation

Let a set of n numbers be denoted by $x_1, x_2, x_3, \ldots, x_n$, and let the mean of these numbers be denoted by $\bar{x}$. Then the **standard deviation s** is given by

$$s = \sqrt{\frac{(x_1 - \bar{x})^2 + (x_2 - \bar{x})^2 + (x_3 - \bar{x})^2 + \cdots + (x_n - \bar{x})^2}{n - 1}}$$

If the standard deviation is to be calculated for an **entire** population, then the sum of the squared differences is divided by N, the size of the entire population. For a **sample** drawn from a population (as in this book), $n - 1$ is used, where n is the sample size.

In order to find the standard deviation, we have to find the following:

1. The mean $\bar{x}$ of the set of numbers
2. The difference (deviation) between each number of the set and the mean
3. The squares of these deviations
4. The sum of the squared deviations divided by $n - 1$
5. The square root s of this quotient

The last four steps motivate the name **root-mean-square deviation,** which is often used for the standard deviation. As we shall see, the number s gives a good indication of how the data are spread about the mean.

GRAPH IT

To do Example 1, erase any lists first (press [2nd] [+] [4] [ENTER]). The grapher says "Done." Press [STAT] [1] and enter 7, 9, 10, 11, and 13 (press [ENTER] after each entry). To find the standard deviation, go to the home screen ([2nd] [MODE]) and press [2nd] [STAT] [◄] [7] [2nd] [L₁] and [ENTER]. The answer is shown.

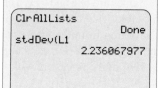

EXAMPLE 1 ▶ Standard Deviation of Children's Ages

The ages of five schoolchildren were found to be 7, 9, 10, 11, and 13. Find the standard deviation s for this set of ages.

Solution

We follow the five steps given at the bottom of page 810, as shown in Table 12.18.

1. The mean of the 5 ages is

$$\bar{x} = \frac{7 + 9 + 10 + 11 + 13}{5} = \frac{50}{5} = 10 \qquad \text{(Column 1)}$$

2. We now find the difference (deviation) between each number and the mean (column 2).

3. We square the numbers in column 2 to get column 3.

TABLE 12.18 Calculation of the Standard Deviation

Age x	Difference from Mean $x - \bar{x}$	Square of Difference $(x - \bar{x})^2$
7	−3	9
9	−1	1
10	0	0
11	1	1
13	3	9
50 Sum of ages $\bar{x} = \frac{50}{5} = 10$ Mean of ages		20 Sum of squares $\frac{20}{5-1} = 5$

4. We find the sum of the squares in column 3 divided by $5 - 1$.

$$\frac{9 + 1 + 0 + 1 + 9}{5 - 1} = \frac{20}{5 - 1} = 5$$

5. The standard deviation is the square root of the number found in step 4. Thus, $s = \sqrt{5} \approx 2.2$. Note that $\sqrt{5}$ can be found from the table at the end of the book or with a calculator. ∎

The number s, although it seems complicated to compute, is a most useful number to know. In many practical applications, about 68% of the data are within 1 standard deviation from the mean. That is, 68% of the numbers lie between $\bar{x} - s$ and $\bar{x} + s$. Also, about 95% of the data are within 2 standard deviations from the mean; that is, 95% of the numbers lie between $\bar{x} - 2s$ and $\bar{x} + 2s$.

For example, if the mean of a set of 1000 numbers is 200 and the standard deviation is 25, then approximately 680 of the numbers lie between 175 and 225, and all but about 50 of the numbers lie between 150 and 250. Thus, even with no further information, the number s gives a fair idea of how the data are spread about the mean. These ideas are discussed more fully in Section 12.4.

EXAMPLE 2 ▶ Mean, Median, Mode, and Standard Deviation

A consumer group checks the price of 1 dozen large eggs at 11 chain stores, with the following results:

Store Number	1	2	3	4	5	6	7	8	9	10	11
Price (cents)	70	68	72	60	63	75	66	65	72	69	68

Find the mean, median, mode, and standard deviation. What percent of the data are within 1 standard deviation from the mean?

Solution

In Table 12.19, the data are arranged in order of magnitude. The mean is found to be 68¢. The median is the middle price, 68¢. The modes are 68¢ and 72¢. The calculation of the standard deviation is shown in the table. The result is $s = 4.3$. To find the percent of the data within 1 standard deviation from the mean, we first find $\bar{x} - s = 63.7$ and $\bar{x} + s = 72.3$. By examining the data, we see that 8 of the prices are between these two numbers. Thus, 73% of the prices are within 1 standard deviation from the mean price.

Note that you are not expected to calculate square roots. Use Table I in the back of the book or a calculator. Both forms of the answer, $\sqrt{18.8}$ and 4.3, are given in the answer section.

TABLE 12.19

x	$x - \bar{x}$	$(x - \bar{x})^2$
60	−8	64
63	−5	25
65	−3	9
66	−2	4
68	0	0
68	0	0
69	1	1
70	2	4
72	4	16
72	4	16
75	7	49
748		**188**

$$\bar{x} = \frac{748}{11} = 68 \qquad \frac{188}{10} = 18.8$$
$$s = \sqrt{18.8} \approx 4.3$$

We have been studying for some time now. I think we need a vacation. Let us go from New York to Honolulu. What do we need to do? Book a flight, a hotel, and a car. Figure 12.9 gives the prices we found.

Flight Price	Hotel Price	Rental	Price
734.20 (US Dollars)	**CORAL REEF** 247 Rooms / 15 Floors **USD 70.00 — USD 150.00***	*Omala*	US $21.99 /Day US $23.99 /Day
816.20 (US Dollars)	**WAIKIKI BEACHCOMBER** 500 Rooms **USD 105.00 — USD 495.00***	**BUCKS**	US $24.99 /Day US $26.99 /Day
837.20 (US Dollars)	**RADISSON ALA MOANA** 1168 Rooms / 36 Floors **USD 215.00 — USD 215.00***	International	US $31.98 /Day US $34.98 /Day
837.20 (US Dollars)	**WAIKIKI PARKSIDE** 247 Rooms / 15 Floors **USD 59.00 — USD 148.00***	Davis	US $33.99 /Day US $37.99 /Day
837.20 (US Dollars)	**PAGODA HOTEL** 360 Rooms / 12 Floors **USD 68.00 — USD 145.00***	*–Widget–*	US $34.99 /Day US $36.99 /Day

* Prices depend on time of travel.

FIGURE 12.9

We can ask a lot of questions about these prices. Some have a simple answer, and others are harder to compute (unless you have a calculator!). We shall answer some questions now and leave some for the exercises.

EXAMPLE 3 ▶ Vacation Price Ranges, Modes, Means, and Deviations

(a) What is the range of prices for the flight? What is the mode?

(b) In the last column, there are two prices for a rental car. The first price is the price of renting an economy model for one day, the second for renting a compact. What is the range of prices for the economy model?

(c) What is the mean rental price for an economy model?

(d) What is the standard deviation of the prices of an economy model?

Solution

(a) The range is the difference between the highest price ($837.20) and the lowest price ($734.20), which is $837.20 − $734.20 = $103. The mode is the price that occurs most often, $837.20.

(b) The prices for the economy model range from $21.99 to $34.99. Thus, the range is $34.99 − $21.99 = $13.00.

(c) To find the mean for the economy model, it is much more practical (and expedient) to use whole numbers. The mean of the numbers is

$$\bar{x} = \frac{22 + 25 + 32 + 34 + 35}{5} = \$29.60 \approx \$30.00$$

(d) To find the standard deviation, use the table on page 814.

x	$x - \bar{x}$	$(x - \bar{x})^2$
22	−7.60	57.76
25	−4.60	21.16
32	2.40	5.76
34	4.40	19.36
35	5.40	29.16
148		**133.20**

$$\bar{x} = \tfrac{148}{5} = 29.60$$

$$\tfrac{133.20}{4} = 33.3$$
$$s = \sqrt{33.3} \approx 5.8$$

Online Study Center

To further explore standard deviations, access links 12.3.1, 12.3.2, and 12.3.3 on this textbook's Online Study Center.

OK, vacation is over; back to work on the exercises! ■

EXERCISES 12.3

In problems 1–10, do the following:
   a. State the range.
   b. Find, to two decimal places, the standard deviation s.

1. 3, 5, 8, 13, 21
2. 1, 4, 9, 16, 25
3. 5, 10, 15, 20, 25
4. 6, 9, 12, 15, 18
5. 5, 6, 7, 8, 9
6. 4, 6, 8, 10, 12
7. 5, 9, 1, 3, 8, 7, 2
8. 2, 0, 4, 6, 8, 10, 8, 2
9. −3, −2, −1, 0, 1, 2, 3
10. −6, −4, −2, 0, 2, 4, 6

11. Out of 10 possible points, a class of 20 students made the following test scores:

0, 0, 1, 2, 4, 4, 5, 6, 6, 6, 7, 8, 8, 8, 8, 9, 9, 9, 10, 10

   a. What is the mode?
   b. What is the median?
   c. What is the mean?
   d. Calculate the standard deviation to the nearest hundreth.
   e. What percent of the scores lie within 1 standard deviation from the mean?
   f. What percent of the scores lie within 2 standard deviations from the mean?

12. Suppose that the 4 students who scored lowest on the test in problem 11 dropped the course. Answer the same questions as in problem 11 for the remaining students.

13. Elmer Duffer plays golf on a par 75 course that is really too tough for him. His scores in his last 10 games are 103, 110, 113, 102, 105, 110, 111, 110, 106, and 110.
   a. What is the mode of Elmer's score?
   b. What is his median score?
   c. What is his mean score?
   d. Calculate the standard deviation of his scores to the nearest hundreth.
   e. Which of his scores are more than 1 standard deviation from his mean score? What percent of the games is this?

14. Answer the same questions as in problem 13 for the lowest eight of Elmer's ten scores.

15. The daily numbers of pounds of garbage for six different households were

6, 2, 17, 3, 5, 9

Find the range, mean, and standard deviation of the weights.

16. Most domestic U.S. airlines have a 40-lb limit on carry-on luggage. The carry-on luggage weights (in pounds) for a random sample of ten passengers during a domestic flight were

30, 30, 32, 35, 37, 40, 40, 40, 42, 44

Find the range, mean, and standard deviation of the weights.

17. The response times of six emergency fire calls were measured to the nearest minute and found to be 6, 7, 9, 12, 3, and 5 min. Find the range, mean, and standard deviation for the calls.

18. From 1918 to 1931, Babe Ruth was the American League home-run champion 12 times (he did not win in 1922 and 1925). The numbers of home runs he hit to earn the titles were

| 11 | 29 | 54 | 59 | 41 | 46 |
| 47 | 60 | 54 | 46 | 49 | 46 |

Find the range, mean, and standard deviation for the number of home runs.

Problems 19–26 refer to Figure 12.9 on page 813.

19. There are two prices (low and high) given for each of the five hotels. Find the range between the high and the low price for each of the hotels.

20. Calculate the average (mean) price between the low and the high price for each of the five hotels.

21. Calculate the average (mean) of the answers in Problem 20.

22. a. Calculate the standard deviation of the prices in problem 20.
   b. What percent of the prices lie within 1 standard deviation of the mean?
   c. What percent of the prices lie within 2 standard deviations of the mean?

23. The second price for each of the rental cars in Figure 12.9 is the price for renting a compact car for 1 day. Find the range of the prices.

24. Calculate the average (mean) of the prices for renting a compact car for 1 day.

25. Calculate the standard deviation of the prices for renting a compact car for 1 day.

26. What percent of the prices for renting a compact car for 1 day lie
   a. within 1 standard deviation of the mean?
   b. within 2 standard deviations of the mean?

In Other Words

27. Suppose the standard deviation of a set of numbers is 0. What does this tell you about the numbers? Explain.

28. In problem 27 we assumed that the standard deviation for a set of numbers was 0. In general, what types of numbers can you obtain as an answer when calculating the standard deviation? Explain.

29. Suppose you are allowed to choose four numbers from 1 to 5. If repetitions are allowed, what is the largest possible result for the standard deviation? Explain.

30. Two classes, each with 100 students, took an examination with a maximum possible score of 100. In the first class, the mean score was 75, and the standard deviation was 5. In the second class, the mean score was 70, and the standard deviation was 15. Which of the two classes do you think had more scores of 90 or better? Why?

Using Your Knowledge

A **binomial experiment** is one that consists of a number of identical trials, each trial having only two possible outcomes (like tossing a coin that must fall heads or tails). Let us consider one of the outcomes as a success and the other as a failure. If p is the probability of success, then $1 - p$ is the probability of failure.

Suppose the experiment consists of n trials; then the theoretical expected number of successes is pn. For instance, if the experiment consists of tossing a fair coin 100 times, then the expected number of heads is $(\frac{1}{2})(100) = 50$. This means that if the experiment of tossing the coin 100 times is repeated many times, then the average number of heads is theoretically 50. In general, if a binomial experiment is repeated many times, then the theoretical mean (average) number of successes is pn, where p is the probability of success in one trial and n is the number of trials in the experiment.

If we let P_k denote the probability of k successes and $n - k$ failures in a binomial experiment with n trials, then the set of numbers $P_0, P_1, P_2, \ldots, P_n$ constitutes a **binomial frequency distribution.** The following simple formula has been obtained for the standard deviation of such a distribution:

$$s = \sqrt{np(1 - p)}$$

For example, if the experiment consists of tossing a fair coin 10,000 times and tallying the number of heads, then $n = 10,000$, $p = \frac{1}{2}$, $\bar{x} = 5000$, and

$$s = \sqrt{10,000(\tfrac{1}{2})(1 - \tfrac{1}{2})} = \sqrt{2500} = 50$$

If this experiment (tossing the coin 10,000 times) were repeated many times, then we would expect the average number of heads to be close to 5000. Although we are not justified in expecting the number of heads in any one experiment to be exactly 5000, we may expect that about 68% of the time the number of heads will be between 4950 and 5050.

31. If a fair die is rolled, the probability that it comes up 2 is $\frac{1}{6}$, and the probability that it comes up not 2 is $\frac{5}{6}$. If we regard 2 as a success and any other number as a failure, what is the standard deviation for the experiment of rolling the die 180 times?

32. Suppose that in rolling a die we regard a 3 or a 4 as a success and any other number as a failure. What is the standard deviation for the experiment of rolling the die 18 times? How far away from the mean would the number of successes have to be before we became suspicious of the die's honesty?

33. Suppose a die is loaded so that the probability that a 6 comes up is $\frac{1}{4}$. If we regard a 6 as a success and any other number as a failure, what is the standard deviation for the experiment of rolling the die 400 times?

 Discovery

34. According to *The Education Reporter,* the mean salary of teachers in Maryland is about $50,000 a year. Suppose all teachers in Maryland get a $1000 raise. What happens to the standard deviation? Explain.

35. This time, suppose all teachers in Maryland get a 10% raise (remember, their average salary is $50,000 annually). What happens to the standard deviation this time? Explain.

 Calculator Corner

If you have a calculator with a $\boxed{\sigma_{n-1}}$ key, it will compute the standard deviation for a set of data at the push of a button. For example, to find the standard deviation

of Example 1, set the calculator in the statistics mode and enter

$\boxed{7}\ \boxed{\Sigma+}\ \boxed{9}\ \boxed{\Sigma+}\ \boxed{10}\ \boxed{\Sigma+}\ \boxed{11}\ \boxed{\Sigma+}\ \boxed{13}\ \boxed{\text{2nd}}\ \boxed{\sigma_{n-1}}$

The result is given as 2.2.

 Collaborative Learning

The table below shows the percent of women in managerial/administrative positions.

Women in Managerial/Administrative Positions

Anglo	African American	Hispanic	Asian/Other
85.7%	6.6%	5.2%	2.5%

Source: "Women of Color in Corporate Management: A Statistical Picture," *American Catalyst.*

1. Calculate the mean, median, and range of the numbers.

Discussion Which of the three measurements gives a better indication of the wide variability of the scores? Explain.

2. Round the percents to the nearest whole number and find the standard deviation of the four numbers.

3. What percent of the numbers obtained in part 2 are within 1 standard deviation of the mean?

4. What percent of the numbers obtained in part 2 are within 2 standard deviations of the mean?

Discussion What does the information obtained in parts 3 and 4 tell you? Does the information indicate that there is a wide variability in the numbers? Explain.

12.4 The Normal Distribution: A Problem-Solving Tool

S.A.T. Deviations

Do you remember your S.A.T. verbal and mathematics scores? In a recent year, the mean score for the verbal portion was μ (read "mu") = 425 with a standard deviation of σ (read "sigma") = 110, whereas the mathematics scores had a mean of 475 and a standard deviation of 125. Suppose you scored 536 on the verbal portion and your friend scored 600 on the mathematics portion. Which is the better score? It might not be the 600! To be able to compare scores, you must learn about the *normal curve* shown in Figure 12.10 on page 817. ▶

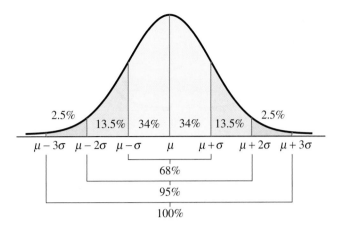

FIGURE 12.10

Area under a normal curve.

Note: Nearly 100% of the scores lie within 3 standard deviations of the mean. Thus, the normal curve is not sketched outside this domain.

(Some books give these values as 34.1%, 13.6%, and 2.3%; we use the approximate values 34%, 13.5%, and 2.5% for convenience.)

A normal curve describes data that have a very large (or infinite) number of values distributed among the population in a **bell shape.** A large number of the values are near the middle with a few values trailing off in either direction. Statisticians call a distribution with a bell-shaped curve a **normal distribution.**

Properties of a Normal Curve

The **normal curve** is a bell-shaped curve with the following four important properties:

1. It is smooth and symmetric (if you fold the graph in half along the center line, the two parts of the curve coincide exactly).

2. Its highest point occurs over the mean μ of the *entire* population.

3. It levels out and approaches the x axis but never touches it.

4. The total area under any normal curve is 1 and the proportion of data values between 1, 2, and 3 standard deviations to either side of the mean is as shown in Figure 12.10.

Now, for the rest of the story. If we assume that scores on the verbal and mathematics portions of the S.A.T. are normally distributed and have the means and standard deviations mentioned, we can label the curves in Figures 12.11 and 12.12 at the top of page 818 with their respective means, $\mu = 425$ and 475. Remember that $\bar{x}$ is the mean of a sample population and s is its standard deviation, but μ is the mean of the entire population and σ is its standard deviation. In Figure 12.11, 1 standard deviation to the right of the mean will be $\mu + \sigma$, or $425 + 110 = 535$, whereas in Figure 12.12, 1 standard deviation to the right of the 475 mean will be 600 ($475 + 125$). Now, a score of 536 on the verbal will be slightly to the right of 1 standard deviation (535), whereas a score of 600 on the mathematics will be **exactly** 1 standard deviation from the mean. Believe it or not, a 536 verbal score is comparatively better than a 600 mathematics score!

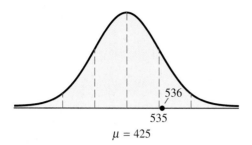

536

535

$\mu = 425$

FIGURE 12.11
Verbal S.A.T. scores.

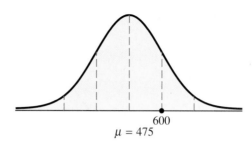

600

$\mu = 475$

FIGURE 12.12
Math S.A.T. scores.

Note: We are using μ and σ, the mean and standard deviation of the *entire* population, instead of $\bar{x}$ and s, the mean and standard deviation of a sample.

The normal distribution we have mentioned is an example of a continuous probability distribution studied by the French mathematician Abraham De Moivre and the German mathematician Carl Gauss (as a matter of fact, normal distributions are sometimes called *Gaussian* distributions in his honor). First, we must learn to recognize normal distributions.

A. The Normal Distribution

Look at the curves in Figure 12.13. They are *not* normal distributions! The curve labeled (a) is not symmetric, (b) is not bell shaped, (c) crosses the x axis, and (d) has tails turning up away from the x axis. We show some normal curves in Figure 12.14.

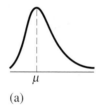 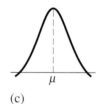 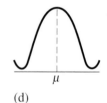

(a)   (b)   (c)   (d)

FIGURE 12.13

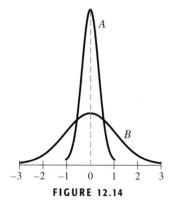

FIGURE 12.14

EXAMPLE 1 ▶ Mean and Standard Deviations in Normal Curve

Consider the normal curves in Figure 12.14.

(a) What is the mean for A?

(b) What is the mean for B?

(c) What is the standard deviation for A?

(d) What is the standard deviation for B?

(e) What percent of the values would you expect to lie between -3 and -1 in B?

(f) What percent of the values would you expect to lie between 0 and 1 in A?

Solution

(a) The mean for A is 0 (under the highest point).

(b) The mean for B is also 0.

(c) The interval from 0 to 1 must have 3 standard deviations, so each of them must be $\frac{1}{3}$ unit.

(d) The standard deviation for B is 1 (there are 3 to the right of 0).

(e) Since there are 2 standard deviations between -3 and -1, $2.5\% + 13.5\% = 16\%$ of the values would be in that region. (Refer to Figure 12.10 on page 817, for the values.)

(f) Half (50%) of the values should be between 0 and 1. ■

PROBLEM SOLVING

The Normal Distribution

❶ **Read** the problem.

Refer to the S.A.T. data given in Getting Started and Figure 12.10 on page 817.

(a) What percent of the scores would you expect to be under 425 on the verbal portion of the S.A.T.?

(b) What percent of the scores would you expect to be between 350 and 475 on the mathematics portion of the S.A.T.?

(c) If 1000 students took the S.A.T., how many students should score more than 600 on the mathematics portion?

❷ **Select** the unknown. Where do we find the different percentages under the normal curve?

We are asked several questions about S.A.T. scores. There are two things that are essential: the information in Figure 12.10 (Memorize it!) and the corresponding information given in Figures 12.11 and 12.12 on page 818.

❸ **Think** of a plan. Look at Figure 12.10. What percentage of the values are in the following locations?

If we examine Figure 12.10, we see the following:

(a) To the left of μ

50% of the values are to the left of μ.

(b) Between $\mu - \sigma$ and μ

34% of the values are between $\mu - \sigma$ and μ.

(c) To the right of $\mu + \sigma$

$13.5\% + 2.5\% = 16\%$ of the values are to the right of $\mu + \sigma$.

❹ **Use** the values shown in Figures 12.11 and 12.12 to answer the questions.

Thus, we have the following:

(a) In Figure 12.11, $\mu = 425$, so we expect 50% of the scores to be to the left of μ and to be less than 425 on the verbal portion.

❺ **Verify** the solution.

(b) In Figure 12.12, $\mu = 475$ and $\mu - \sigma = 350$. Thus, 34% of the scores are between $\mu - \sigma = 350$ and $\mu = 475$.

(c) In Figure 12.12, 16% of the values are to the right of $\mu + \sigma$. Since 16% of $1000 = 160$, 160 students should score higher than 600 on the mathematical portion.

TRY EXAMPLE 2 NOW.

Cover the solution, write your own solution, and then check your work.

EXAMPLE 2 ▶ **Heights and Standard Deviations for Girls**

The heights of 1000 girls are measured and found to be normally distributed, with a mean of 64 in. and a standard deviation of 2 in.

(a) About how many of the girls are over 68 in. tall?

(b) About how many are between 60 and 64 in. tall?

(c) About how many are between 62 and 66 in. tall?

Solution

We refer to Figure 12.10 on page 817 for the required percentages; then draw your own curve and label it.

(a) Because 68 in. is 2 standard deviations above the mean, about 2.5%, or 25, of the girls are over 68 in. tall.

(b) We see that 64 in. is the mean, and 60 in. is exactly 2 standard deviations below the mean; hence, we add 13.5% and 34% and find that 47.5% of the girls are between 60 and 64 in. tall. This represent 47.5% of 1000, or 475, girls.

(c) Because 62 in. is 1 standard deviation below the mean, and 66 in. is 1 standard deviation above the mean, we add 34% and 34% to find that about 68% of 1000, or 680, girls are between these two heights. ∎

EXAMPLE 3 ▶ **Normal Distributions and Reading Tests**

A standardized reading comprehension test is given to 10,000 high school students. The scores are found to be normally distributed, with a mean of 500 and a standard deviation of 60. If a score below 440 is considered to indicate a serious reading deficiency, about how many of the students are rated as seriously deficient in reading comprehension?

Solution

Since 440 is exactly 1 standard deviation below the mean, scores below 440 are more than 1 standard deviation below the mean. By referring to the percentages in Figure 12.10 on page 817, we see that we must add 13.5% and 2.5% to get the total percentage of students who scored more than 1 standard deviation below the mean. Thus, 16% of the 10,000 students, or 1600 students, are rated as seriously deficient in reading comprehension. ∎

In Getting Started we were able to compare two scores on two different tests by referring to the normal curve. Now suppose Rudie earned a score of 80 on her U.S. history test and a score of 80 on her geometry test. Which of these is the better score? Without additional information, we cannot answer this question. However, if we are told that the mean score in the U.S. history test was 60, with a standard deviation of 25.5, and the mean score in the geometry test was 70, with a standard deviation of 14.5, then we can use a technique similar to the one used in Getting Started to compare Rudie's two scores.

B. z-Scores

In order to make a valid comparison, we have to restate the scores on a common scale. A score on this scale is known as a **standardized score** or a **z-score.**

> **Definition of z-Score (Standardized Score)**
>
> If x is a given score and μ and σ are the mean and standard deviation of the entire set of scores, then the **corresponding z-score** is
>
> $$z = \frac{x - \mu}{\sigma}$$

Since the numerator of z is the difference between x and the mean, *the z-score gives the number of standard deviations that x is from the mean.*

EXAMPLE 4 ▶ z-Scores for Rudie

Compare Rudie's scores in U.S. history and geometry, given all the preceding information.

Solution
Rudie's z-scores are

$$\text{U.S. history} \quad z = \frac{80 - 60}{25.5} \approx 0.78$$

$$\text{Geometry} \quad z = \frac{80 - 70}{14.5} \approx 0.69$$

Thus, Rudie did better in U.S. history than in geometry. ∎

C. Distribution of z-Scores

For a normal distribution of scores, if we subtract μ from each score, the resulting numbers will have a mean of 0. If we then divide each number by the standard deviation σ, the resulting numbers will have a standard deviation of 1. Thus, the z-scores are distributed as shown in Figure 12.15. For instance, 34% of the z-scores lie between 0 and 1, 13.5% lie between 1 and 2, and 2.5% are greater than 2. For such a distribution of scores, the probabilities of randomly selecting z-scores between 0 and a given point to the right of 0 have been calculated and appear in tables such as Table II in the back of the book. To read the probability that a score falls between 0 and 0.25 standard deviation above the mean, we go down the column under z to 0.2 and then across to the column under 5; the number there is 0.099, the desired probability. This probability is actually the area under the curve between 0 and 0.25.

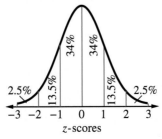

FIGURE 12.15
Distribution of z-scores.

EXAMPLE 5 ▶ Probabilities Under the Normal Curve

For the normally distributed population shown in Figure 12.16 the mean is $\mu = 100$, and the standard deviation is 15. Find the probability that a randomly selected item of the data falls between 100 and 120.

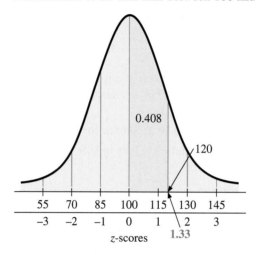

0.408

120

| 55 | 70 | 85 | 100 | 115 | 130 | 145 |
| −3 | −2 | −1 | 0 | 1 | 2 | 3 |

z-scores   1.33

FIGURE 12.16

Solution
We first find the z-score for 120.

$$z = \frac{x - \mu}{\sigma} = \frac{120 - 100}{15} = \frac{4}{3} \approx 1.33$$

We then refer to Table II and read down the column under z to the number 1.3 and then across the column under 3 to the desired probability, 0.408.

EXAMPLE 6 ▶ Reading Tables

Refer to Table II. In the column under 5 and across from 2.0, the entry 0.480 appears. What does this mean?

Solution
This means that if an item is selected at random from a normally distributed set of data, the probability that this item is within 2.05 standard deviations from the mean is 0.480.

EXAMPLE 7 ▶ Finding Probabilities Using Tables

Referring to Example 5, find the probability that a randomly selected item is less than 110.

Solution
The probability that a randomly selected item of the data is less than 100 is 50% because 50% of the scores are to the left of the mean (see Figure 12.17). To find the probability that a randomly selected item is less than 110, we can find the probability that the item is between 100 and 110 and add this probability to 50% = 0.50. The z-score for 110 is

$$z = \frac{x - \mu}{\sigma}$$

$$= \frac{110 - 100}{15} \approx 0.67$$

The value by 0.6 and under 7 in Table II is 0.249. Thus, the probability that a randomly selected item of the data is less than 110 is

$$0.50 + 0.249 = 0.749$$

(See Figure 12.17.)

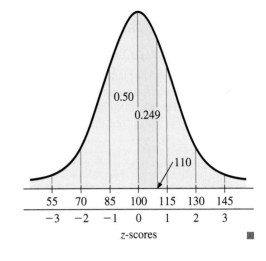

FIGURE 12.17

z-scores

EXAMPLE 8 ▶ We Scream for More Ice Cream

Time for a break: We are headed to the ice cream parlor! The combined weight of the two scoops on a double-dip ice cream cone satisfies a normal distribution with a mean of 8 oz and a standard deviation of $\frac{1}{4}$ oz. But one double-dip looks a little bit larger. As a matter of fact, it weighs 8.5 oz, and you picked it! What is the probability that a randomly selected cone is smaller than yours?

Solution

To find the answer, we find the probability that the amount of ice cream in a randomly selected cone is less than yours, that is, that the z-score for the other cones is less than the z-score of yours. The z-score for your 8.5-oz ice cream cone is

$$z = \frac{x - \mu}{\sigma} = \frac{8.5 - 8}{0.25} = \frac{0.50}{0.25} = +2$$

Now refer to Table II. The value for $z = +2$ is 0.477 (red area), and the total value for all dips with z-scores under 2 is the yellow area plus the red area, that is, $0.50 + 0.477 = 0.977$. This means that the probability that the other dips are smaller than yours is 97.7%. You have a good eye for ice cream!

You also can do this problem by looking at Figure 12.15 on page 821 and observing that the area under the curve to the left of $z = +2$ is approximately $1 - 0.025 = 0.975$, or 97.5%.

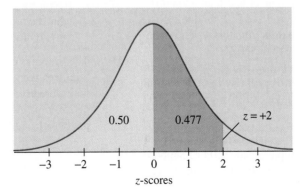

z-scores

EXERCISES 12.4

A The Normal Distribution

1. Farmer Brown has planted a field of experimental corn. By judicious sampling, it is estimated that there are about 20,000 plants and that a graph of their heights looks like that shown in the figure below.

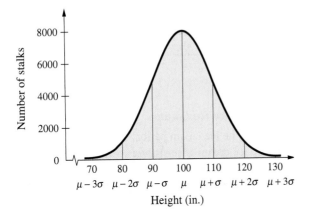

a. What is the mean height of farmer Brown's corn?

b. What is the standard deviation from the mean?

c. What percent of the cornstalks is between 90 and 110 in. tall?

d. About how many stalks are between 80 and 90 in. tall?

2. Suppose you were informed that the annual income of lawyers is normally distributed, with a mean of $40,000 and a standard deviation of $10,000.

a. What would you estimate for the percent of lawyers with incomes over $50,000?

b. What percent of lawyers would you estimate have an annual income of less than $20,000?

c. If a lawyer were selected at random, what would be the probability that his or her annual income were more than $60,000?

d. If the information given here were correct, would you think it very likely that 50% of all lawyers have annual incomes of over $50,000? Why or why not?

3. Part of a test given to young children consists of putting together a simple jigsaw puzzle. Suppose that such a puzzle is given to 1000 children, each child is timed, and a graph of the times is made.

Suppose that the graph is a normal curve with a mean time of 120 seconds and a standard deviation of 15 seconds.

a. About how many of the children finished the puzzle in less than 90 seconds?

b. How many took more than 150 seconds?

c. If you rated as "average" all the children within 1 standard deviation from the mean, how many children would fall into this classification?

4. For a certain standardized placement test, it was found that the scores were normally distributed, with a mean of 200 and a standard deviation of 30. Suppose that this test is given to 1000 students.

a. How many are expected to make scores between 170 and 230?

b. How many are expected to score above 260?

c. What is the expected range of all the scores?

5. A psychology teacher gave an objective-type test to a class of 500 students and, after seeing the results, decided that the scores were normally distributed. The mean score was 50, and the standard deviation was 10. The teacher assigned a grade of A to all scores of 70 or over, B to scores of 60 to 69, C to scores of 40 to 59, D to scores of 30 to 39, and F to scores below 30. About how many of each grade did the teacher assign?

6. In a study of 100 common stocks, it was found that the annual dividend rates were normally distributed, with a mean of 4.0% and a standard deviation of 0.5%.

a. About how many of these stocks do you think paid dividends of over 5%?

b. About how many paid between 3% and 5%?

c. If you picked one of these stocks at random, what do you think would be the probability that it paid at least 4.5%?

7. The lifetimes of a random sample of 200 automobile tires were found to be normally distributed, with a mean of 26,000 mi and a standard deviation of 2500 mi. About how many of these tires gave out before 21,000 mi?

8. Suppose that 100 measurements of the specific gravity of copper gave a mean of 8.8 with a standard deviation of 0.2. Between what limits did about 95% of the measurements fall?

9. Suppose that ten measurements of the length of a wooden beam have a mean of 20 ft and a standard deviation of 0.5 in. Between what limits do almost all the measurements fall?

10. An experiment consists of tossing 100 dimes repeatedly and noting the number of heads each time. The graph of the number of heads turns out to be very nearly a normal curve, with a mean of 50 and a standard deviation of 5.
    a. Within what limits would you expect the number of heads to be 95% of the time?
    b. What percent of the time would you expect the number of heads to be between 45 and 55?
    c. Suppose that a particular dime arouses your suspicion by turning up heads too often. You toss this dime 100 times. How many times will it have to turn up heads in order for you to be almost 100% certain that it is not a fair coin? (*Hint:* Most of the data in a normal distribution fall within 3 standard deviations of the mean.)

11. The purchasing director of Druid Enterprises is considering the purchase of 8000 ball bearings. The purchase is dependent on receiving at least a dozen ball bearings that will last 40 days. If the manufacturer claims that the lifetime of each ball bearing is 30 days, with a standard deviation of 5 days, what will be the decision of the purchasing director and why?

12. The Department of Transportation (DOT) counted the number of vehicles using a certain road for a period of 50 days. The mean number of vehicles using the road was 350 and the standard deviation was 10.
    a. How many days was the road used by more than 360 vehicles?
    b. How many days was the number of vehicles using the road between 340 and 380?
    c. What is the lowest number of vehicles you would expect on this road on any given day?
    d. What is the highest number of vehicles you would expect on this road on any given day?

13. In a recent year, the scores on the mathematics portion of the Scholastic Aptitude Test (S.A.T.) had a 455 mean and a 112 standard deviation. This same year, the mathematics portion of the American College Test (A.C.T.) had a mean of 17.3 and a standard deviation of 7.9. A student scored 570 on the S.A.T., whereas another student scored 25 on the A.C.T. Which student has the higher score relative to the test?

14. Suppose you are the manager of a cereal packing company. Every box must contain at least 16 oz of cereal. Your packing machine has a normal distribution for the weights of the cereal with a standard deviation of 0.05 oz and a mean equal to the setting on the machine. What will you make this setting to ensure that all of the packages contain at least 16 oz of cereal?

B **z-Scores**

15. Examine the normal distributions shown.

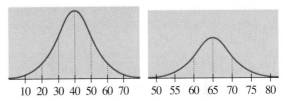

    a. What is the standard deviation for the first?
    b. What is the standard deviation for the second?
    c. What is the mean for the first distribution?
    d. What is the mean for the second distribution?
    e. In the first distribution, what value corresponds to a z-score of +2? What about a z-score of −1?
    f. In the second distribution, what value corresponds to a z-score of +2? What about a z-score of −1?

16. Examine the normal distribution below and determine what percent of the scores you would expect to be
    a. between 40 and 50.
    b. between 20 and 50.
    c. between 10 and 70.

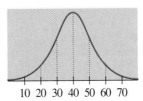

17. You are still managing the cereal packing company of problem 14, and now you know that the weights of the boxes have a normal distribution with a mean of 16.15 oz and a standard deviation of 0.05 oz. What is the weight of a box with a z-score of 0?

18. Now we are going to the dogs, but not literally! According to the St. Bernard club, your 170-lb Bernie is too hefty compared with other dogs his age. (Source: www.acay.com.au/~dissi/sbc/weight.htm.)

   Bernie goes on a diet, and his *z*-score decreases from +2 to −2. If the standard deviation for St. Bernards' weight is 5 lb,

   a. what is the mean weight for Bernie's breed and age?

   b. what is Bernie's weight after the diet?

19. Pedro took an exam in a class in which the mean was 70 with a standard deviation of 10. If his *z*-score was +2, what was his exam score?

20. If you take an exam and your score is the same as the mean score for the class, what is your *z*-score for the exam?

21. In a certain normal distribution of scores, the mean is 5, and the standard deviation is 1.25. Find the *z*-score corresponding to a score of
   a. 6.   **b.** 7.   **c.** 7.5.

22. In a certain normal distribution of scores, the mean is 10, and the standard deviation is 2. Find the *z*-score corresponding to a score of
   a. 11.   **b.** 13.   **c.** 14.2.

23. Gretchen scored 85 on a test in German and also on a test in English. If the mean in the German test was 75, with a standard deviation of 20, and the mean in the English test was 80, with a standard deviation of 15, which of Gretchen's 85s was the better score?

24. Juan scored 88 on a Spanish test and 90 on an algebra test. If the mean on the Spanish test was 78, with a standard deviation of 7.5, and the mean on the algebra test was 82, with a standard deviation of 6.5, which of Juan's scores was the better score?

C Distribution of z-Scores

25. A student's commute to school is normally distributed with a mean of 30 minutes and a standard deviation of 5 minutes. What is the probability that the student gets to school in 20 to 30 minutes?

26. An express workout at the gym is normally distributed with a mean of 30 minutes and a standard deviation of 5 minutes. What is the probability that Latasha completes the workout in 30 to 35 minutes?

27. Elias at La Cubanita restaurant says he will *always* serve you within 5 minutes! If service at the restaurant is normally distributed with a mean of 4.5 minutes and a standard deviation of 1.5 minutes, what is the probability that Elias is right?

28. Referring to problem 27, within what time period will you be certain (100% probability) of being served?

In problems 29–33 assume a normally distributed set of test scores with a mean of $\mu = 100$ and a standard deviation of 15.

29. Find the probability that a person selected at random will have a score between
   a. 100 and 110.   **b.** 100 and 130.

30. Find the probability that a person selected at random will have a score between 80 and 120. [*Hint:* In Example 5 we found the probability that the score is between 100 and 120 to be 0.408. The probability that the score is between 80 and 100 is also 0.408. (Recall the symmetry of the normal curve.)]

31. Find the probability that a person selected at random will have a score
   a. between 55 and 145.   **b.** less than 60.

32. Find the probability that a person selected at random will have a score
   a. between 75 and 100.   **b.** more than 80.

33. Find the probability that a person selected at random will have a score between 110 and 130. (*Hint:* In problem 29 you found the probability that the score will be between 100 and 110 and the probability that the score will be between 100 and 130. You should be able to see how to combine these two results to get the desired probability.)

34. In problem 10 it was noted that the distribution of heads if 100 dimes are tossed repeatedly is approximately a normal distribution, with a mean of 50 and a standard deviation of 5. Find the probability of getting 60 heads if 100 fair coins are tossed. [*Hint:* To use the normal curve, consider 60 to be between 59.5 and 60.5 and proceed as in problem 23. This will give a very good approximation and is much easier to calculate than the exact probability, which is $C(100, 60) \cdot 2^{100}$.]

35. The heights of the male students in a large college were found to be normally distributed, with a mean of 5 ft 7 in. and a standard deviation of 3 in. Suppose these students are to be divided into five equal-sized groups according to height. Approximately what is the height of the shortest student in the tallest group?

36. If 10 of the boys who show up for the basketball team are over 6 ft 3 in. tall, about how many male students are there in the college mentioned in problem 35?

37. What is your major and how much money do you think you will make? The monthly income of computer programmers is normally distributed with a mean μ of \$4200 and standard deviation σ of \$200. (Source: *College Graduates Average Salaries Report.*)

Find the z-value for a monthly income of \$4400. Explain what the z-value you obtained means.

38. Referring to problem 37, find the z-value for a \$3800 monthly income and explain what it means.

39. Remember the ice cream of Example 8? Within what weight will you be certain (100% probability) that the ice cream will be?

40. A professor's commute is normally distributed with a mean of 45 minutes and a standard deviation of 10 minutes.
   a. What is the probability that the professor gets to work in 30 minutes or less?
   b. If the professor has a 9 A.M. class and leaves home at 8 A.M., how often is the professor late for class?

41. According to e-medicine, sleeping 8 to 8.4 hours per night is considered fully restorative for adults. Suppose Caruca sleeps an average of 8 hours per night with a standard deviation of 15 minutes. What is the probability that Caruca will get her restorative sleep (between 8 and 8.4 hours)? (*Hint:* 15 minutes = 0.25 hours.)

42. Suppose you know your test score in this class is above the mean, but you don't know by how much. How many students scored less than you on this test?

The following information will be used in problems 43–48. In Florida, the daily water usage per person is normally distributed with a mean of 110 gal and standard deviation of 10 gal.

43. Between which two values does 68% of the daily water usage per person lie?

44. Between which intervals will 95% and 99% of the daily water usage lie?

45. What is the probability that a person selected at random will use less than 110 gal per day?

46. What is the probability that a person selected at random will use more than 110 gal per day?

47. What percent of the people use between 100 and 110 gal?

48. What percent of the people use between 100 and 120 gal?

49. The amount of daily tips waiters receive at a famous restaurant is normally distributed with a mean of $100 and a standard deviation of $20. A statistically inclined waitress has decided that her service is *poor* if her daily tips are less than $70. On the basis of her theory, what is the probability that she has provided poor service?

50. Referring to problem 49, if the standard deviation is $10, what is the probability that she has provided poor service?

In Other Words

51. Can you have two normal curves with the same mean and different standard deviations? Explain and make a sketch.

52. Can you have two normal curves with the same standard deviation and different means? Explain and make a sketch.

53. If you have two normal curves, is it true that the curve with the larger mean must also have the larger standard deviation? Explain your answer.

54. Explain in your own words the meaning of a *z*-score.

Using Your Knowledge

Standardized, or *z*, scores provide a way of making comparisons among different sets of data. To compare scores within one set of data, we use a measurement called a **percentile**. Percentiles are used extensively in educational measurements and enable us to convert raw scores into meaningful comparative scores. If you take an exam and are told that you scored in the 95th percentile, it does not mean that you scored 95% on the exam but rather that you scored higher than 95% of the persons taking the exam. The formula used to find the percentile corresponding to a particular score is as follows:

$$\text{Percentile} \atop \text{of score } x = \frac{\text{number of scores less than } x}{\text{total number of scores}} \cdot 100$$

Thus, if 80 students take a test and 50 students score less than you do, you will be in the $\frac{50}{80} \cdot 100 = 62.5$ percentile. Use this knowledge to solve the following problems:

55. A student took a test in a class of 50 students, and 40 of the students scored less than she did. What was her percentile?

56. A student took a test in a class of 80, and only 9 students scored better than he did. What was his percentile?

57. The scores in a class were as follows:

$$83, 85, 90, 90, 92, 93, 97, 97, 98, 100$$

   a. What percentile corresponds to a score of 90?
   b. What percentile corresponds to a score of 97?
   c. What percentile corresponds to a score of 100?
   d. What percentile corresponds to a score of 83?

58. The scores on a placement test were scaled so that some of the scores and the corresponding percentiles were as follows:

Score	Percentile
119	98th
150	85th
130	72nd
90	50th
60	26th
30	10th
20	1st

What percent of the scores fell between 30 and 90? (*Hint:* Look at the definition of the percentile. The percentile gives the percent of what?)

Discovery

We have discovered that we can make fairly accurate predictions about the dispersion of the measurements in a normal distribution. For instance, 68% of the measurements fall between $\mu - \sigma$ and $\mu + \sigma$, 95% between $\mu - 2\sigma$ and $\mu + 2\sigma$, and nearly 100% between $\mu - 3\sigma$ and $\mu + 3\sigma$. But what can we say in the case the distribution is not normal?

The great Russian mathematician Pafnuti Lvovich Chebyshev (1821–1894) discovered the following remarkable result:

Chebyshev's Theorem

For any distribution with a finite number N of measurements and for any h such that $h > 1$, the number of measurements within h standard deviations of the mean is **at least** equal to

$$\left(1 - \frac{1}{h^2}\right)N$$

For example, if $h = 2$, then $1 - 1/h^2 = \frac{3}{4}$, so at least $\frac{3}{4}N$, or 75%, of the measurements fall between $\mu - 2\sigma$ and $\mu + 2\sigma$. This is not as large a percentage as for a normal distribution, but the amazing thing is that this result holds for any kind of distribution at all—as long as there is only a finite number of measurements!

Suppose we have 20 numbers with a mean of 8 and a standard deviation of 2. How many of the numbers can we *guarantee* to fall between 2 and 14? Because 2 and 14 are each 3 standard deviations from the mean, we take $h = 3$ in Chebyshev's theorem and obtain

$$\left(1 - \frac{1}{3^2}\right)(20) = \frac{160}{9} \approx 17.8$$

Thus, the theorem guarantees that at least 17 of the 20 numbers fall between 2 and 14.

In the same way, by taking $h = 1.5$, we find that

$$\left(1 - \frac{1}{1.5^2}\right)(20) = \left(\frac{5}{9}\right)(20) \approx 12.1$$

Hence, we can guarantee that at least 11 of the 20 numbers fall within 1.5 standard deviations from the mean, that is, between 5 and 11.

Notice that Chebyshev's theorem makes no claim at all for the case in which $h \leq 1$.

59. If 100 measurements have a mean of 50 and a standard deviation of 5, how many of the measurements must be between
a. 40 and 60?   **b.** 35 and 65?   **c.** 43 and 57?

60. What is the smallest value of h that is large enough to guarantee that of a set of measurements at least the following percentages will be within h standard deviations from the mean?
a. 96%   **b.** 91%   **c.** 64%

61. Find the mean and the standard deviation of the following numbers: 1, 1, 1, 2, 6, 10, 11, 11, 11. How many of these numbers lie within 1 standard deviation from the mean? How many lie within 2 standard deviations from the mean? How do these results compare with those predicted by Chebyshev's theorem?

62. Do you think it is possible for all the items in a population to be *less* than 1 standard deviation from the mean? Justify your answer. *Hint:* The formula for the standard deviation shows that

$$n\sigma^2 = (x_1 - \bar{x})^2 + (x_2 - \bar{x})^2 + \cdots + (x_n - \bar{x})^2$$

63. Can Chebyshev's theorem be used to find the percentage of the measurements that must fall between 2 and 3 standard deviations from the mean? What can you say about this percentage?

Calculator Corner

You can find the probability that a randomly selected item will fall between the mean and a given z-score by using the $\boxed{R(t)}$ key on your calculator. Thus, to find the probability that a randomly selected item falls between the mean and a score of 120, as in Example 5, place the calculator in the statistics mode and press

The result is given as 0.40878. Note that there is a slight difference (due to rounding) between this answer and the one in the text.

1–6. Use a calculator to rework problems 29–34.

Collaborative Learning

The figure on page 830 shows intelligence quotient (IQ) scores under the normal curve. A raging controversy, fueled by a book titled *The Bell Curve*, claiming that intelligence is largely inherited, has sparked new interest in IQ tests. Form several groups.

1. One group's assignment is to get a copy of *The Bell Curve* and find out what the book's actual claims are.

2. A second group can find a copy of the winter '98 issue of *Scientific American* (where the figure below was found) and find out what the *g* (global, general) factor is and how it measures intelligence.

Now, for the math questions.

3. If IQ scores are normally distributed as shown in the diagram, what are the mean, median, and mode for the scores?

Discussion Do you think that IQ scores are normally distributed? For example, the percentages given for the total population distribution apply to "young white adults in the United States."

4. What is the standard deviation suggested in the figure?

5. What is the *z*-score for a person with a score of 120?

6. What is the *z*-score for a person with a score of 80?

7. If scores are normally distributed, for what percent of the people should IQ scores fall between 90 and 110? What percent of the total population distribution is shown in that range? Are the percents different? Why do you think that is?

8. If scores are normally distributed, for what percent of the people should IQ scores fall between 125 and 130? What percent of the total population distribution is shown in that range? Are the percents different? Why do you think that is?

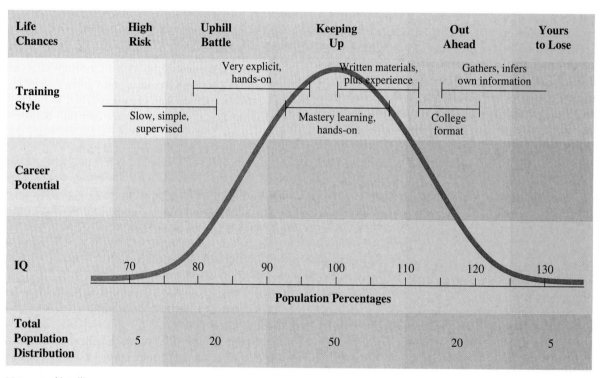

Measure of intelligence.

12.5 Statistical Graphs: A Problem-Solving Tool

Gassing Up and Pie Charts

Americans are often fascinated with numbers. Many of the facts of American life are expressed in books such as *The First Really Important Survey of American Habits, The Top 10 Almanac, The Day America Told the Truth, The Great Divide,* and *On an Average Day.* These books are devoted to surveys of American habits using numbers, tables, and graphs. For example, do you gas up your car when it is $\frac{3}{4}$ full, $\frac{1}{2}$ empty, $\frac{3}{4}$ empty, or almost empty? As you would expect, different age groups of different sexes behave differently. Thus, 2%, 42%, 38%, and 18% of males in the 21–34 age group fill their tanks when they are $\frac{3}{4}$ full, $\frac{1}{2}$ empty, $\frac{3}{4}$ empty, or almost empty, respectively. For females, the corresponding percentages are 1%, 2%, 6%, and 91%. Why do you think this difference exists? Whatever the reason, you can compare the data better if the information is contained in a table with a caption and column headings such as Table 12.20. Table 12.20 enables you to instantly compare differences based on sex.

TABLE 12.20 Percent of People Who Gas Up Car When It Is . . .

| Age | $\frac{3}{4}$ Full | | $\frac{1}{2}$ Empty | | $\frac{3}{4}$ Empty | | Almost Empty | |
	Male	Female	Male	Female	Male	Female	Male	Female
21–34	2	1	42	2	38	6	18	91

You can also show the division of a total quantity (100%) into its component parts by using a **circle graph** or **pie chart.** If you are interested only in a rough sketch, use the four-step procedure shown in Figure 12.18. Now here is a word of caution. Mathematicians and statisticians use the starting point shown in step 4 and move counterclockwise (see Figure 12.19a on page 832). Computer-generated pie charts start at the 12 o'clock point and move clockwise (see Figure 12.19b). Both are correct, and both methods are used in this book.

Step 1. Make a circle.

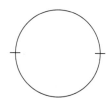

Step 2. Divide it into 2 equal parts.

Step 3. Subdivide each of the 2 parts into 5 equal parts.

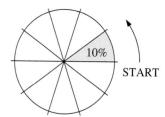

Step 4. Each of the subdivisions represents $\frac{1}{10}$, or 10%.

FIGURE 12.18

(a) Percent of males gassing car
when it is

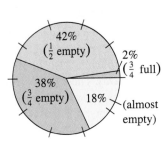

(b) Percent of females
gassing car when it is

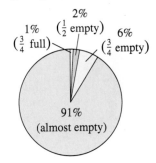

FIGURE 12.19

A graph of a set of data can often provide information at a glance that might be difficult and less impressive when gleaned from a table of numbers. No table of numbers would make the visual impact created by the graph in Figure 12.20, for example.

It is almost always possible to alter the appearance of a graph to make things seem better (or worse) than they are. For instance, Figure 12.21 is a portion of Figure 12.20 but with a compressed vertical scale. Obviously, things look better on this graph! Can you see why an economist might feel it politically advantageous to publish one of these graphs rather than the other?

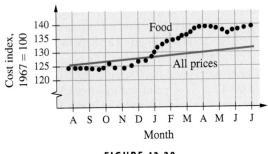

FIGURE 12.20

The high cost of eating.

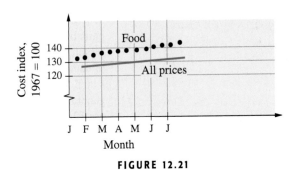

FIGURE 12.21

A. Line Graphs

The construction of line graphs is similar to the type of graphing we did in Chapter 7. As before, we draw two perpendicular lines called *axes*. The horizontal line is still the *x* axis, and the vertical line the *y* axis. Each of these lines has equally spaced points with numbers or other identifying information assigned to it. For example, the numbers on the *x* axis may represent the numbers of hours a person has worked, whereas those on the *y* axis may indicate the earnings for that person. We make our first line graph next.

EXAMPLE 1 ▶ Downloading or CDs

Do you buy CDs in the store or download music from the Internet? A survey by Pricewaterhouse Coopers revealed that Americans spent $10 million downloading music from the Internet in 2003. Here are the projections (in millions of dollars) for the next 4 years. Make a line graph of these data.

Year	Millions
2004	$ 30
2005	$125
2006	$300
2007	$600

Solution

The categories (years 2004, 2005, 2006, and 2007) are on the *horizontal* axis, and the amounts or *frequencies* ($30, $125, $300, and $600 million) are on the *vertical* axis. For convenience, we use a $100 million scale on the vertical axis. To graph the first point corresponding to 2004, we start at 2004, go up 30 units, and graph the point. For 2005, we go to 2005, go up 125 units, and graph the point. We do the same for 2006 and 2007. Finally, we join the points with line segments as shown.

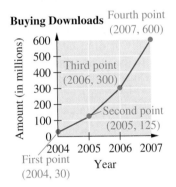

Sometimes, we draw more than one line on the same graph, as shown next.

EXAMPLE 2 ▶ Unemployment Line Graph

Construct a line graph for the data in Table 12.21.

TABLE 12.21

| Month | Unemployment Rate | |
	Women	Men
January	6.4%	5.3%
February	8.0%	7.6%
March	6.0%	5.1%

Solution
In Figure 12.22 we start by making a time scale and labeling the points January, February, and March. (Time is usually shown on the *x* axis.) We then label the *y* axis with the percents from 1 to 9.

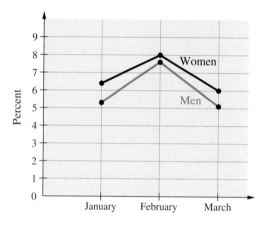

FIGURE 12.22
Unemployment rate.

To graph the point corresponding to the unemployment rate for women in January, go to January on the horizontal scale and move 6.4 units up. Mark the point with a dot. Now go to February on the horizontal axis and move 8.0 units up. Make a dot. Finally, starting at March, go 6 units up and mark the point. Now join the first point to the second and the second to the third with two line segments. Note that we made a grid of horizontal and vertical lines to make the work easier to read. The same result can be obtained by doing the graph on graph paper. We use a similar procedure to add the men (shown in blue).   ■

B. Bar Graphs

Newspapers and magazines often publish **bar graphs** of the type shown in Figure 12.23. These graphs again have the advantage of displaying the data in a form that is easy to understand. It would be difficult for most people to obtain the same information from a table of numbers. As in the case of line graphs, bar graphs can also be made to distort the truth. For example, consider the two graphs in Figures 12.24 and 12.25. The graph in Figure 12.24 does not have the bars starting at 0, so it gives a somewhat exaggerated picture of the proportion of gasoline saved at lower speeds, even though the numerical data are the same for both graphs. The graph in Figure 12.25 gives a correct picture of the proportion of gasoline saved. Why do you think the first bar graph rather than the second would be published?

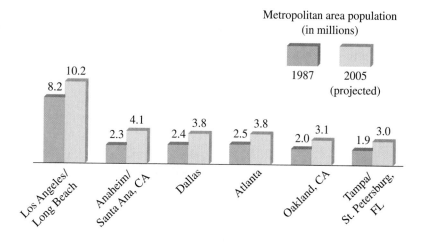

FIGURE 12.23

Bar graph showing metropolitan areas that will gain more than 1 million residents by the year 2005.

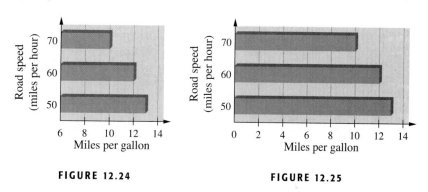

FIGURE 12.24   **FIGURE 12.25**

EXAMPLE 3 ▶ Bar Graphs and Embarrassing Moments

What would embarrass you the most on your first date? See the data below from Listerine! Draw a vertical bar graph using it.

Bad breath (BB)	40%
Acne breakout (AB)	23%
Fly open (FO)	22%
Greasy hair (GH)	14%

Note: Percentages do not sum to 100 due to rounding.
Source: Data from Wirthline Worldwide for Listerine.

Solution

To identify the graph, we label it "First-Date Blunders." We have four categories represented on the *horizontal* axis. The *frequencies* go from 14 to 40, so we make the *vertical* axis go from 0 to 50 at 10-unit intervals as shown. The bars are 40,

23, 22, and 14 units long, corresponding to the percents given in the table. How does it look when done by a graphic artist? The result is shown below.

Worst first-date blunders

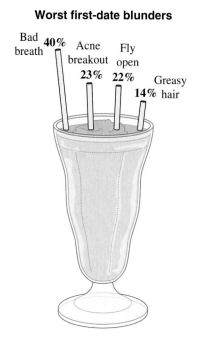

First-Date Blunders

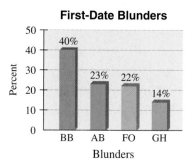

EXAMPLE 4 ► Horizontal Bar Graphs at a Restaurant

A recent poll listed the features patrons liked in a restaurant as follows:

Self-service salad bar	57%	All-you-can-eat specials	37%
Varied portion sizes	47%	Self-service soup bar	30%
More varied menu	42%		

Use horizontal bars to make a bar graph of this information.

Solution

We label the x axis with percents equally spaced and at intervals of 10 as in Figure 12.26. However, since the highest percent used in the problem is 57, we can shorten the x axis and stop at 60%. We find the points on the graph just as we did for the line graphs, but instead of connecting the dots with a line, we draw a bar. The labels can be placed alongside the vertical axis or inside the bars as shown.

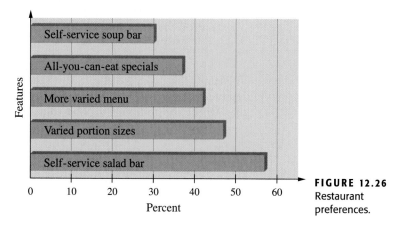

FIGURE 12.26
Restaurant preferences.

C. Circle Graphs

Graphs like those in Figures 12.27 and 12.28 are called **circle graphs** or **pie charts.** Such graphs are a very popular means of displaying data, and they are also susceptible to being drawn to make things look better (or worse) than they

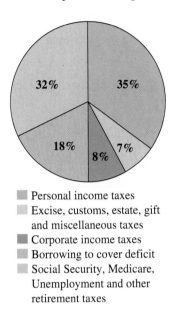

Personal income taxes
Excise, customs, estate, gift
and miscellaneous taxes
Corporate income taxes
Borrowing to cover deficit
Social Security, Medicare,
Unemployment and other
retirement taxes

FIGURE 12.27
Circle graph.

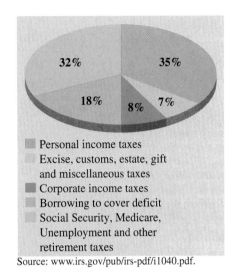

Personal income taxes
Excise, customs, estate, gift
and miscellaneous taxes
Corporate income taxes
Borrowing to cover deficit
Social Security, Medicare,
Unemployment and other
retirement taxes

Source: www.irs.gov/pub/irs-pdf/i1040.pdf.

FIGURE 12.28
Income.

are. For instance, compare the graph in Figure 12.27 with the version of the same data that was published by the Internal Revenue Service. Does the visual impression of Figure 12.28 make you feel that individual income taxes are not quite so large a chunk of federal income as Figure 12.27 indicates?

Circle graphs are quite easy to draw if you know how to use a simple compass and a protractor. For the graph in Figure 12.27, which shows where the typical dollar of federal money comes from, 23¢ is 23% of a dollar. The entire circle corresponds to 360°, so you would use 23% of 360°, or 82.8° ≈ 83°, for the slice that represents 23¢, and likewise for the other slices. This is the most accurate and honest way to present data on a circle graph.

EXAMPLE 5 ▶ Circle Graphs and Budgets

A marketing executive for a food manufacturer wants to show that food is an important part of a family budget. She finds that the typical budget is as follows. Make a circle graph for these data.

Monthly Family Budget

Savings	$ 300
Housing	500
Clothing	200
Food	800
Other	200
Total	$2000

Solution

First determine what *percent* of the total amount each of the items represents.

Savings $\frac{300}{2000} = \frac{3}{20} = 15\%$ Food $\frac{800}{2000} = \frac{2}{5} = 40\%$

Housing $\frac{500}{2000} = \frac{1}{4} = 25\%$ Other $\frac{200}{2000} = \frac{1}{10} = 10\%$

Clothing $\frac{200}{2000} = \frac{1}{10} = 10\%$

Then find out how many degrees each slice covers.

$$15\% \text{ of } 360° = 0.15 \times 360° = 54°$$

$$25\% \text{ of } 360° = 0.25 \times 360° = 90°$$

$$10\% \text{ of } 360° = 0.10 \times 360° = 36°$$

$$40\% \text{ of } 360° = 0.40 \times 360° = 144°$$

$$10\% \text{ of } 360° = 0.10 \times 360° = 36°$$

Now measure the required number of degrees with a protractor, and label each of the slices as shown in Figure 12.29. As a check, make sure the sum of the percentages is 100 and the sum of the degrees for the slices is 360.

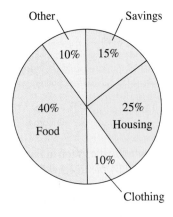

FIGURE 12.29
Monthly family budget.

Online Study Center

To further explore the types of graphs we have studied with real data, access links 12.5.1 and 12.5.2 on this textbook's Online Study Center.

In Example 2 we drew two lines on the same graph. Sometimes we have to draw more than one circle graph to solve a problem, as shown in Example 6.

EXAMPLE 6 ▶ Predictions About El Niño, La Niña and Normal Years

Can you predict the probability of rainfall using a circle graph or pie chart? The Bureau of Meteorology does in Australia. Rainfall predictions can be given using numerical data, either in a table or as a circle graph. The table below gives the probability of dry, wet, or normal weather based on the type of year (El Niño, normal, or La Niña). The information from the table can be summarized by using three circle graphs. Make three different circle graphs corresponding to the three columns in the table.

El Niño Year	*Normal Year*	*La Niña Year*
50% dry	33.3% dry	17% dry
17% wet	33.3% wet	50% wet
33% normal	33.3% normal	33% normal

Solution

The third circle graph below shows the information for a La Niña year. The easiest category to graph is wet because wet represents 50%, or half of the circle. The top half of the circle shows dry (17%) and normal (33%). Because 17 is about one-half of 33, the tan region representing the dry weather is about half the size of the normal region. Can you see how the first and second circle graphs were done?

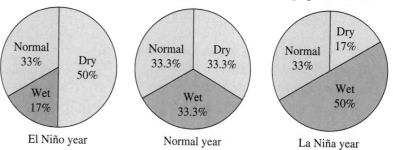

El Niño year   Normal year   La Niña year

Source: Australian Government Bureau of Meteorology.

■

EXERCISES 12.5

A Line Graphs

In problems 1–6, make a line graph for the given data.

1. Rice consumed per capita in the United States

1970	6.7 lb	1985	9.1 lb
1975	7.6 lb	1990	16.6 lb
1980	9.5 lb		

2. Daily calories needed to maintain weight for females in given age intervals

11–14	2200 calories
15–18	2100 calories
19–22	2050 calories
23–50	2000 calories
51–75	1800 calories
75+	1500 calories

3. Percent of married couples with and without children

	1960	1970	1980	1990	2000
With children	44%	40%	31%	27%	24%
Without children	30%	30%	30%	30%	27%

Source: Censusscope.org.

4. Percent of males squeezing the toothpaste tube from top or bottom

Age	Bottom	Top
21–34	37	63
35–44	33	67
45–54	10	90
55+	10	90

5. Percent of females squeezing the toothpaste tube from top or bottom

Age	Bottom	Top
21–34	33	67
35–44	43	57
45–54	30	70
55+	45	55

6. Projected percent of age ranges in the United States

	1950	1990	2010	2030	2050
0–14	27%	22%	18%	19%	19%
15–64	65%	66%	68%	62%	62%
65+	8%	12%	13%	20%	19%

7. How much candy do you eat? According to the Census Bureau, the per capita consumption of candy for five consecutive years is as follows:

Year	1	2	3	4	5
Pounds	25.1	23.6	24	24.6	24.7

Make a line graph for the data using 1–5 as the years and 20–25 as the amount (in pounds).

8. Does your Internet use interfere with your TV watching? A Stanford group survey shows the percent of persons reporting less time watching television based on the number of hours they spent using the Internet.

Internet time	<1 hour	1–5 hours	5–10 hours	>10 hours
Percent watching less TV	27%	43%	56%	65%

Source: www.stanford.edu/group/siqss/itandsociety/v01i02/v01i02a10.pdf.

Make a line graph for the data using the four categories in the horizontal axis and the percents from 0 to 70 at intervals of 10 units.

In Problems 9–20, make a line graph for the given data representing a recent 5-year period. (Source: U.S. Department of Labor. Bureau of Labor Statistics, www.bls.gov/data/home.htm.) How do you compare with these averages?

9. *Clothing* The following data show the average amount of money spent on apparel by men between the ages of 16 and 25:

1	247
2	221
3	209
4	294
5	262

Graph the data using 1–5 as the years and 200–300 as the amounts at $10 intervals.

10. The following data show the average amount of money spent on apparel by women between the ages of 16 and 25:

1	434
2	382
3	377
4	405
5	359

Graph the data using 1–5 as the years and 350–450 as the amounts at $10 intervals.

11. *Food in general* The following data show the average amount of money spent on food by persons under 25:

1	2838
2	3075
3	3354
4	3213
5	3724

Graph the data using 1–5 as the years and 2600–4000 as the amounts at $200 intervals.

12. *Food at home* The following data show the average amount of money spent on food at home by persons under 25:

1	2758
2	2547
3	2890
4	2951
5	2936

Graph the data using 1–5 as the years and 2500–3000 as the amounts at $100 intervals.

13. *Housing* The following data show the average amount of money spent on housing by persons under age 25:

1	5860
2	6151
3	6585
4	7109
5	7585

Graph the data using 1–5 as the years and 5000–8000 as the amounts at $500 intervals.

14. *Housing* The following data show the average amount of money spent on housing by persons between the ages of 25 and 34:

1	11,774
2	12,015
3	12,519
4	13,050
5	13,828

Graph the data using 1–5 as the years and 11,000–14,000 as the amounts at $500 intervals.

15. *Entertainment* The following data show the average amount of money spent on entertainment by persons under age 25:

1	1051
2	974
3	1149
4	1091
5	1152

Graph the data using 1–5 as the years and 900–1200 as the amounts at $50 intervals.

16. *Entertainment* The following data show the average amount of money spent on entertainment by persons between the ages of 25 and 34:

1	1865
2	1757
3	1776
4	1876
5	2001

Graph the data using 1–5 as the years and 1700–2100 as the amounts at $50 intervals.

17. *Healthcare* The following data show the average amount of money spent on health care by persons under age 25:

1	425
2	445
3	551
4	504
5	530

Graph the data using 1–5 as the years and 400 to 600 as the amounts at $50 intervals.

18. *Healthcare* The following data show the average amount of money spent on health care by persons between the ages of 25 and 34:

1	1236
2	1185
3	1170
4	1256
5	1286

Graph the data using 1–5 as the year and 1000–3000 are the amounts at $50 intervals.

19. *Wages-salaries* The following data show the average amount of annual wages-salaries earned by persons between the ages of 25 and 34:

1	37,455
2	38,548
3	39,372
4	42,770
5	46,301

Graph the data using 1–5 as the years and 37,000–47,000 as the amounts at $1000 intervals.

20. *Wages-salaries* The following data show the average amount of annual wages-salaries earned by persons under age 25:

1	13,098
2	14,553
3	16,210
4	16,908
5	17,650

Graph the data using 1–5 as the years and 12,000–18,000 as the amounts at $1000 intervals.

B Bar Graphs

In problems 21–23, draw a bar graph based on the data provided.

21. According to Merrill Lynch Relocation Management, Inc., hundreds of people accepted a transfer from their company. The following are the numbers of employees relocated in various industries:

Industry	Number
Computers	803
Petroleum	324
Transportation	271
Public utilities	265
Retail/wholesale	231

Use horizontal bars to make a graph of this information.

22. A manager in the food industry had to predict food consumption trends for the year 2005. The following are the data:

Food	Increase in Average Consumption by 2005
Fish	44%
Fresh vegetables	35%
Fresh fruits	23%
Beef	30%
Pork	12%
Poultry	7%

a. Use horizontal bars to make a graph of this information.

b. If you wish to be in a growing business, which industry would you choose to own or manage in the year 2005? Which business would you avoid?

23. The personnel department of a major corporation was asked to make a vertical bar graph indicating the percent of salary to be replaced by pension and Social Security income for married persons who worked for 40 years. The following is the information supplied by the Bureau of Labor Statistics:

Salary at Time of Retirement	Percent Received from Pension and Social Security
$20,000	93%
$30,000	74%
$40,000	64%

a. Make the graph.

b. If you retired at $20,000, what amount would you expect from your pension and Social Security?

24. A recent survey of 5000 households showed that more people planned to make major purchases this year. The following are the percents:

	Bought Last Year	Plan to Buy This Year
Home	3.6%	5.0%
Car	8.6%	11.3%
Appliance	27.1%	43.2%

Use side-by-side vertical bars and make a histogram comparing the percent of consumers planning to buy each of the items with the percent who bought each item last year.

25. In a survey conducted by *Chain Store Age*, respondents indicated where their cameras were bought. Some of the responses were

Source of Purchase	This Year	2 Years Ago
Discount stores	18.6%	23.7%
Sears	11.4%	6.2%
Department stores	15.7%	4.1%
Specialty stores	30.0%	29.9%

a. Use side-by-side vertical bars to compare the given percents during the specified time periods.

b. On the basis of the graph, where do most people now buy their 35-mm cameras?

c. On the basis of the graph, which store category lost the most sales in the 2-year interval?

d. On the basis of the graph, which stores would you say had the most consistent camera sales over the 2-year period?

26. In another survey conducted by *Chain Store Age*, respondents indicated which store provided the best value for home electronics. The following are some of the results:

Source for Best Values	Last Year	This Year
Department stores	15%	20%
Electronic stores	31%	20%
Sears	17%	14%
Discount stores	8%	10%

a. Use the same horizontal bars to include each of the four categories for the given time periods.

b. On the basis of the graph, which stores provided the best value for home electronics?

27. Do you go to the movies often? The following survey shows the percent of people who go at least once a month:

Categories Age Bracket	Frequencies Percent
18–24	83%
25–34	54%
35–44	43%
45–54	37%
55–64	27%
65 and up	20%

Source: Data from TELENATION/MarketFacts, Inc.

a. Draw a vertical bar graph for the data.
b. What age bracket goes to the movies the most frequently?
c. What age bracket goes to the movies the least?

28. How many unwanted calls do you get daily? The number of unwanted calls received by the given percent of the people is shown below.

Unwanted Calls	Frequencies
0	15%
1–2	41%
3–5	28%
6 and up	12%

Source: Data from Bruskin/Goldring Research for Sony Electronics.

a. Draw a vertical bar graph for the data.
b. Which is the most common number of calls received?
c. What percent of the people received no unwanted calls?

29. Which branch of the military has the most women? A Defense Department survey shows the following numbers:

Branch	Frequencies
Air Force	19.4%
Army	15.4%
Navy	14.4%
Marines	6%

a. Draw a horizontal bar graph of the data.
b. Which branch has the highest percent of women?
c. Which branch has the lowest percent of women?
d. Can you use the information to find out if there are more women in the Air Force than in the Army? Explain.

30. Who has the most Internet knowledge? A survey of *USA Today* adult respondents answering the question, "Who has the most Internet knowledge?" revealed the following data:

Response	Percent
Kids	72%
Adults	21%
Both the same	2%

a. Draw a horizontal bar graph for the data.
b. According to the survey, who has the most knowledge?
c. What percent of the people think that kids and adults have the same knowledge?

C Circle Graphs

In problems 31–36, make a circle graph for the data.

31. Have you been to a meeting lately? The following are the numbers of hours per week spent in meetings by chief marketing executives:

Hours	Percent	Hours	Percent
Fewer than 5	2	20–24	22
5–9	10	25–29	16
10–14	17	29+	17
15–19	16		

32. Clairol, Inc., reports that the percent of industry sales for favorite hair colorings are

Blond	40%	Black	8%
Brunette	38%	Other	1%
Red	13%		

33. Have you looked in your refrigerator and found some UFOs (unidentified food objects)? The makers of Saran Wrap found the following information from survey respondents:

9% don't have leftovers.
61% have leftovers that are 6 days old or less.
23% have leftovers that are 1–4 weeks old.
5% have leftovers that are more than 4 weeks old.
2% don't know.

34. Where do you think you use the most water in the home? According to *National Wildlife Magazine*, the percents of water used in different parts of the home are as follows:

Toilet	40%	Laundry	12%
Shower/bath	20%	Kitchen	10%
Bathroom sink	15%	Outside	3%

35. What are you recording on your VCR? A survey by the A. C. Nielsen company indicated the percents of taping sources were as follows:

Pay services	10%
Basic cable	7%
PBS	4%
Major networks	66%
Independent networks	13%

36. How often do you wash your car? The following are the results of a survey of 1000 drivers:

    220 wash them weekly.
    180 wash them every 2 weeks.
    230 wash them once a month.
    370 never wash them!

37. The National Restaurant Association surveyed 500 customers at fast-food restaurants serving breakfast and compiled the following figures:

Who Eats Breakfast Out?

Age	Number	Age	Number
18–24	80	50–64	90
25–34	130	65+	50
35–49	150		

a. Make a circle graph for these data.

b. If you were the manager of a fast-food restaurant serving breakfast, which age group would you cater to?

38. Where does money used for advertising go? For every dollar spent in advertising, it goes as follows:

    27¢ to newspapers
    21¢ to television
    16¢ to direct mail
    7¢ to radio
    6¢ to magazines
    3¢ to business publications
    20¢ to other sources

Make a circle graph for these data.

39. A company specializing in leisure products surveyed 500 people and found their favorite activities were as follows:

   75 read.   60 had family fun.
  250 watched TV.   65 had other activities.
   50 watched movies.

Make a circle graph for these data.

40. Here are the top five frozen pizza brands. Make a circle graph for the dollar sales in the data.

Brandname	Dollar Sales (in millions)	Unit Sales (in millions)
DiGiorno	$501.3	98.2
Tombstone	$284.7	84.3
Red Baron	$239.4	74.2
Frechetta	$189.2	37.6
Private Label	$176.7	101.5

Source: Snack & Wholesale Bakery, February 2005, p. 12; www.aibonline.org/resources/statistics/2005pizza.html.

41. The U.S. Department of Labor updated its theoretical budget for a retired couple. The high budget for such a couple is apportioned approximately as follows:

Food	22%
Housing	35%
Transportation	11%
Clothing	6%
Personal care	3%
Medical care	6%
Other family costs	7%
Miscellaneous	7%
Income taxes	3%

Make a circle graph to show this budget.

42. The pie chart shown here appeared side by side with the one in Figure 12.28 on page 837. Make a circle graph to represent the same data. Compare the impression made by your circle graph with that made by the pie chart.

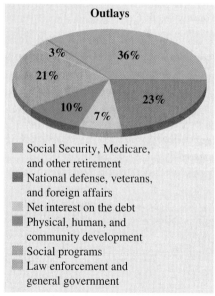

Outlays

Social Security, Medicare, and other retirement
National defense, veterans, and foreign affairs
Net interest on the debt
Physical, human, and community development
Social programs
Law enforcement and general government

Source: www.irs/gov/pub/irs-pdf/il040.pdf.

43. According to an advertisement for a color television, the top six brands of television sets were voted as best by the following percentages of about 2700 people:

Brand	Percent	Brand	Percent
1	50.1	4	8.5
2	21.1	5	5.8
3	8.8	6	5.7

a. Make a circle graph to illustrate this information.
b. Make a bar graph for the same data.
c. Which of these do you think makes the stronger impression? Why?

44. A survey made by the University of Michigan's Institute for Social Research showed that many of the women surveyed enjoy keeping house. The survey found that about 67% of the women who responded had an unqualified liking for housework, while only 4% had an unqualified dislike for housework. Make a bar graph to illustrate these data.

45. The Mighty Midget Canning Company wants to impress the public with the growth of Mighty Midget business, which it claims has doubled over the previous year. It publishes the pictorial graph shown below.

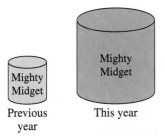

a. Can you see something wrong with this picture? (*Hint:* Your mind compares the volumes pictured here. The volume of a cylinder is $\pi r^2 h$. What happens if you double the radius r and the height h?)
b. Draw a bar graph that correctly represents the situation.

46. The U. B. Wary Company wants to give its stockholders a very strong impression of the rapid rate at which earnings have grown and prints the histogram shown in the following figure in its annual report. Redraw this graph to give a more honest impression.

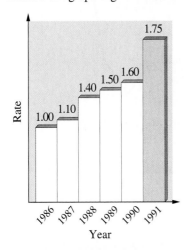

47. Do you know that playing golf is not a particularly good way to lose weight? Here is the calorie consumption per hour for five popular activities. Make a bar graph with horizontal bars, with each bar identified at the left side of your graph.

Activity	Calories Consumed per Hour
Bicycling (15 mph)	730
Running (6 mph)	700
Swimming (40 yd/min)	550
Walking (4 mph)	330
Golf (walking and carrying your clubs)	300

(These figures apply to a person weighing 150 lb; you have to add or subtract 10% for each 15-lb difference in weight.)

48. Women are generally lighter than men, so they require fewer calories per day. Here are three occupations and their energy-per-day requirements. Make a vertical bar graph with the bars for male and female side by side for comparison. On the average, about what percent more calories does the male require than the female for these three occupations?

Occupation	Calories Required per Day
University student, male	2960
University student, female	2300
Laboratory technician, male	2850
Laboratory technician, female	2100
Office worker, male	2500
Office worker, female	1900

49. Although the best a cold remedy can do is ease the discomfort (without curing the cold), the relief seems to be worth plenty to victims. Here is how people in the United States spent money on cold remedies in a recent year. Make a bar graph with the horizontal bars representing the data. Be sure to identify the bars.

Cold Remedy	Millions Spent
Cough drops and sore throat remedies	$130
Nasal sprays, drops, and vaporizers	$160
Aspirin substitutes	$275
Cold and cough syrups	$310
Aspirin	$575

50. The more you learn, the more you earn! The following are the median incomes by educational attainment for persons 25 years or older:

Education	Men	Women
Not high school	$ 22,636	$13,217
High school	32,024	19,156
Some college	39,031	23,015
AA degree	40,608	26,104
Bachelor's degree	56,779	32,816
Master's degree	67,202	41,270
Professional degree	115,931	63,904
Doctoral degree	91,982	56,807

Source: www.cencus.gov/prod/2002pubs/p23-210.pdf

Make a side-by-side vertical bar graph for these data. Find the women's income as a percentage of the men's in each category.

51. The chart below shows the annual sales for the ABC Bookstore for a 10-year period.
  a. Approximately what were the sales in year 4?
  b. When did the sales start to level off?

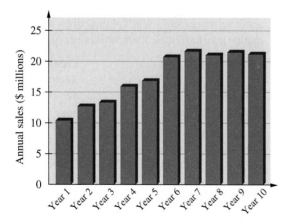

52. The circle graph shows Harry's time allotment for Mondays, Wednesdays, and Fridays. What percent of the time is Harry allowing for classes and study?

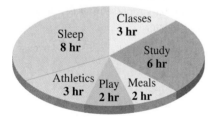

Harry's activities.

In Other Words

You are the manager of a store whose sales (in millions) for January, February, and March were $20, $21, and $23 million, respectively.

53. Explain how to make a line graph that would
  a. make sales look better.
  b. make sales look flat.

54. Explain how to make a bar graph that would
  a. make sales look better.
  b. give the impression that sales are not increasing.

55. Explain in your own words the difference between a bar graph and a histogram.

56. When is a pie chart especially useful?

Discovery

Misleading graphs can be used in statistics to accomplish whatever deception you have in mind. The graphs shown below and on the following page, for example, give exactly the same information. However, the graph below seems to indicate a steep increase in government payrolls, whereas the graph on page 847 shows the stability of the same payrolls!

57. Can you discover what is wrong?

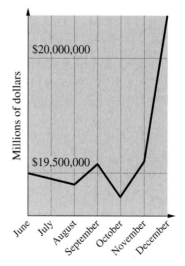

Government payrolls up.

Graph for problem 57.

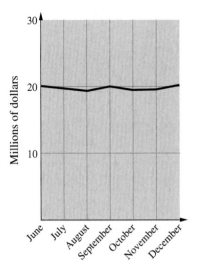

Government payrolls stable.

 Skill Checker

In the next section we will be finding the equations of certain lines called "best fit" or "least squares" lines. Write an equation in the slope-intercept form $y = mx + b$, where m is the slope and b the y intercept. Answer the questions referring to the lines in the figure at the top of the next column.

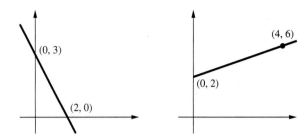

1. What is the slope of the first line in the figure?

2. What is the y intercept of this line?

3. Write the equation of this line in slope-intercept form.

4. What is the slope of the second line in the figure?

5. What is the y intercept of this line?

6. Write the equation of this line in slope-intercept form.

 Collaborative Learning

 *Online Study Center*

This exercise requires access to the Web. Access link 12.5.3 on this textbook's Online Study Center. Answer and discuss the questions provided with each of the five graphs at the site.

12.6 Making Predictions: Linear Regression

Olympic Predictions

Table 12.22 on page 848 gives the winning times (in seconds) for the women's 200-m dash in the Olympic Games from 1948 to 1984. The points in Figure 12.30 (a scattergram) on page 848 are the graphs of the corresponding number pairs (1948, 24.4), (1952, 23.7), and so on. As we can see, the points do not lie on a straight line. However, we can draw a straight line that goes "between" the points and seems to fit the data fairly well. Such a line, called the **line of best fit,** has been drawn in Figure 12.30. Statisticians sometimes use lines like this to make predictions. For example, the winning time for this event in the 1988 Olympics as indicated by the line in the graph is about 21.4 seconds. The actual winning time (made by Florence Griffith Joyner of the United States) was 21.34 seconds. (The time predicted by the graph is less than $\frac{3}{10}$ of 1% in error!) In this section you will see how to make predictions by drawing lines to fit given data.

TABLE 12.22

Year	1948	1952	1956	1960	1964	1968	1972	1976	1980	1984
Time (sec)	24.4	23.7	23.4	23.2	23.0	22.5	22.4	22.37	22.03 (No U.S. participation)	21.81

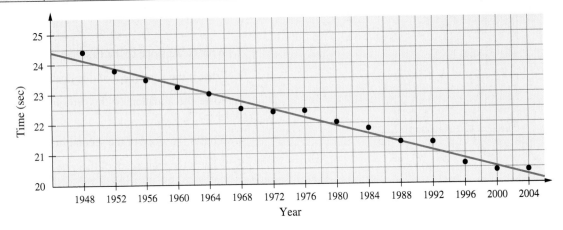

FIGURE 12.30
Olympic women's winning times: 200-m dash. (Graph updated by author.) ▶

Marion Jones won the gold medal in the 200-m dash in the 2000 Summer Olympics.

How do we draw a line to fit data such as those in Table 12.22? One way is just to use our best visual judgment, but a better way is to calculate a *least-squares line*. If we have n points, say $(x_1, y_1), (x_2, y_2), \ldots, (x_n, y_n)$, in the xy plane, a line $y = mx + b$ is called the **least-squares (regression) line** for these points if the sum of the squares of the differences between the actual y values of the points and the corresponding y values on the line is as small as possible. The line in Figure 12.30 is the least-squares line for the data in Table 12.22.

EXAMPLE 1 ▶ Predicting Olympic Times

Table 12.23 gives the winning times in seconds for the men's 100-m freestyle swim in the Olympic Games from 1960 to 1984. Make a graph of these data, and then draw a line "between" the points and predict the winning time for this event in the 1988 Olympics.

TABLE 12.23

Year	1960	1964	1968	1972	1976	1980	1984
Time (sec)	55.2	53.4	52.2	51.22	49.99	50.40	49.80

Solution

The required graph is shown in Figure 12.31. The line in Figure 12.31 is the least-squares line for the given data. From this line we can read the predicted time for the 1988 Olympics as about 48.3 seconds. The actual winning time was 48.63 seconds, so the predicted time is off by less than 0.7%. Do the same for the 1996, 2000, and 2004 Olympics and see how close you come!

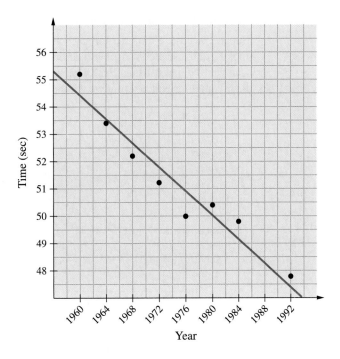

FIGURE 12.31
Winning times: Olympic men's 100-m freestyle swim. (Graph updated by author.) ■

GRAPH IT

To enter the data in Example 1 as a set of ordered pairs (x, y) = (year, time), it is easier to let 1960 = 0, 1964 = 4, and so on. Clear any lists in your grapher (2nd + 4 ENTER), go to the home screen (2nd MODE), press STAT 1, and enter 0, 4, 8, . . . , 24 under L_1 and 55.2, 53.4, . . . , 49.80 under L_2. To find the relationship between the two, press STAT ▶ 4 ENTER.* This means $y = -0.22x + 54.36$. Thus, in 1960 ($x = 0$), the equation predicts a time of 54.36 seconds. For 1988 ($x = 28$), time would be $y = -0.22(28) + 54.36$, or 48.2 seconds, very close to the value in the graph!

```
LinReg
y=ax+b
a=⁻.2179464286
b=54.35964286
```

To see the graph, press Y= VARS 5 ▶ ▶ 1 . This copies the equation and makes it equal to Y_1. Next, make a window like Figure 12.31. Press WINDOW and enter $X\min = 0$, $X\max = 32$, $X\text{scl} = 1$, $Y\min = 47$, $Y\max = 56$, and $Y\text{scl} = 1$. Press GRAPH . Voila!

*You also can enter 2nd L_1 , 2nd L_2 , VARS 5 ▶ ▶ 1 . This will find the regression equation and copy it to Y= .

EXAMPLE 2 ▸ **Best-Fit Line for Sales and Advertising**

Suppose we have the data for sales and advertising shown in Figure 12.32. By looking at the scattergram, it is clear that sales increase as advertising is increased. Can we find a best-fitting line? This time we draw the *two* lines shown. What are their equations and how did we get them?

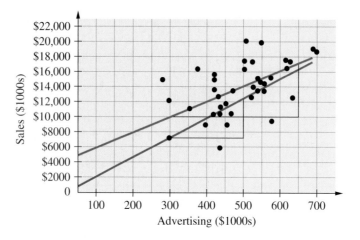

FIGURE 12.32

Sales and advertising.

Solution

As you recall, m = slope = rise/run. For the red line, m is about

$$\frac{12,000 - 7000}{500 - 300} = \frac{5000}{200} = 25$$

The intercept b is about 1000. Thus, $y = 25x + 1000$ (thousands).
 For the blue line,

$$m = \frac{17,000 - 10,000}{650 - 300} = \frac{7000}{350} = 20$$

and $b = 4500$. Hence, $y = 20x + 4500$ (thousands).
 This means that if we spend $400 in advertising and use the red line as our predictor, we can expect $y = 25(400) + 1000 = \$11,000$ (thousands) in sales. On the other hand, if we use the blue line for our estimation, expected sales are $y = 20(400) + 4500 = \$12,500$ (thousands). ■

 Which is the best approximation? The one that is done using the method of "least squares." Look at Figure 12.33, an enlargement of Figure 12.30. In Figure 12.33, d_1 and d_2 represent the difference between the y coordinate of the data point and the corresponding y coordinate on the line itself. Since d_1 is above the line, d_1 is positive. On the other hand, since d_2 is below the line, d_2 is negative. If we simply add d_1 and d_2, the result may be 0. However, the squares of d_1 and d_2 cannot be negative. To construct a line that is as close as possible to the one shown, our job is to "minimize" the sum of the squares of the d's for all data points, thus the name "least squares" line. To facilitate writing the formulas for the calculation of such a line, we shall use the symbol Σ to indicate a sum.

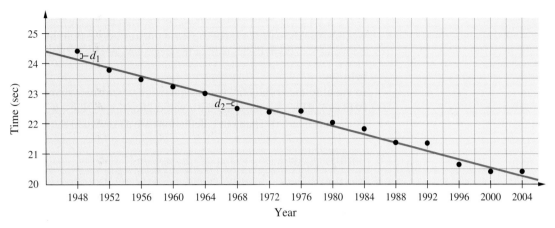

FIGURE 12.33
Winning times: Olympic women's 200-m dash. (Graph projected by author for 1996, 2000, and 2004.)

Although the derivation of the least-squares formula for the best-fit line is too advanced to be given here, the formulas themselves are not difficult to describe, as shown next.

Regression Line (Line of Best Fit)

The best-fit line associated with the n points $(x_1, y_1), (x_2, y_2), \ldots, (x_n, y_n)$ has the form

$$y = mx + b \tag{1}$$

where

$$\text{Slope} = m = \frac{n\Sigma(xy) - (\Sigma x)(\Sigma y)}{n(\Sigma x^2) - (\Sigma x)^2} \tag{2}$$

Same denominator

$$\text{Intercept} = b = \frac{(\Sigma x^2)(\Sigma y) - (\Sigma x)(\Sigma xy)}{n(\Sigma x^2) - (\Sigma x)^2}$$

or

Total number of points →

$$b = \frac{\Sigma y - m(\Sigma x)}{n} \tag{3}$$

$\Sigma xy = $ sum of products $= x_1 y_1 + x_2 y_2 + \cdots + x_n y_n$
$\Sigma x = $ sum of x values $= x_1 + x_2 + \cdots + x_n$
$\Sigma y = $ sum of y values $= y_1 + y_2 + \cdots + y_n$
$\Sigma x^2 = $ sum of squares of x values $= x_1^2 + x_2^2 + \cdots + x_n^2$

Note: You can use either formula for b in equation (3)
The denominator in equations (2) and (3) are identical.

To use these formulas efficiently, we make a table with the headings x, y, x^2, and xy, as shown on page 852. The first two columns simply list the x's and y's, the third column lists the x^2's, and the last column gives the xy products. After filling out the table, we add the four columns to get Σx, Σy, Σx^2, and Σxy, the four sums that are required in equations (2) and (3).

EXAMPLE 3 ▶ Least-Squares (Regression) Line for 200-m Dash

We illustrate the calculation of the least-squares (regression) line for the data for the women's 200-m dash given at the beginning of this section.

Solution

To simplify the arithmetic, we designate the successive Olympics starting with 1948 as 1, 2, 3, . . . ; these are the x values. For the y values, we take the number of seconds over 20. Then our calculations are as shown in the table.

Year	x	y	x^2	xy
1948	1	4.4	1	4.4
1952	2	3.7	4	7.4
1956	3	3.4	9	10.2
1960	4	3.2	16	12.8
1964	5	3.0	25	15.0
1968	6	2.5	36	15.0
1972	7	2.4	49	16.8
1976	8	2.37	64	18.96
1980	No U.S. participation			
1984	10	1.81	100	18.1
Totals	**46**	**26.78**	**304**	**118.66**

We thus have found $\Sigma x = 46$, $\Sigma y = 26.78$, $\Sigma x^2 = 304$, and $\Sigma xy = 118.66$, and we are ready to use equations (2) and (3). Since there are 9 points, $n = 9$, and we get

$$m = \frac{(9)(118.66) - (46)(26.78)}{(9)(304) - (46)^2} = -0.264$$

and

$$b = \frac{(304)(26.78) - (46)(118.66)}{(9)(304) - (46)^2} = 4.327$$

(These answers are rounded to three decimal places.) The required line has the equation

$$y = -0.264x + 4.327$$

If we put $x = 9$ in this equation, we get $y = 1.951 \approx 1.95$. Thus, the predicted winning time for the 1980 Olympics is $20 + 1.95$, or about 22 seconds. (Of course, you must not carry this type of prediction too far because after about 36 Olympics, the winner would reach the finish line before she started the race!) ■

The best way to find m and b in the least-squares line $y = mx + b$ is to organize your work into a table as we have done. Also, your answers to the exercises may differ from ours depending on how you round during the intermediate steps. One last word of warning: Please, do not confuse Σx^2 and $(\Sigma x)^2$. For Σx^2, we *first square* each x value and then find the total sum. For $(\Sigma x)^2$, we *first sum* the

x values and then square the total as in the table. Finally, $n(\Sigma xy)$ and $n(\Sigma x^2)$ mean we multiply n by the appropriate sums. We illustrate this in Example 4.

EXAMPLE 4 ▶ Regression Line for Predicting Birth Weight

What is the relationship between the age of the mother and low birth weight for the child? Of course, they are related (the older the mother, the lower the birth weight of the child), but what happens if, in addition, the mother smokes during pregnancy? The following table shows the age x of seven pregnant women who smoked during their pregnancy and the percent y who had a low-birth-weight baby (defined as infants weighing less than 2500 g, or 5 lb 8 oz). Find the regression line and predict the percent of low-birth-weight babies for 30 year olds. (Source: Adapted from http://www.uvm.edu/~dhowell/gradstat/psych340/Lectures/CorrelReg/correl.html.)

Solution

Women	x	y	x^2	xy
1	16	12	256	192
2	18	11	324	198
3	23	10	529	230
4	27	11	729	297
5	33	13	1089	429
6	38	16	1444	608
7	46	19	2116	874
Totals	**201**	**92**	**6487**	**2828**

From the table, $\Sigma x = 201$, $\Sigma y = 92$, $\Sigma x^2 = 6487$, and $\Sigma xy = 2828$. Since there are $n = 7$ points, using equations (2) and (3),

$$m = \frac{(7)(2828) - (201)(92)}{(7)(6487) - (201)^2} \approx 0.260$$

Using the second formula for b,

$$b = \frac{92 - (0.260)(201)}{7} \approx 5.677$$

Thus,

$$y = 0.260x + 5.677$$

For 30 year olds the percent of low-birth-weight babies is

$$y = 0.260(30) + 5.677 = 13.477\%$$ ■

Can you make weather predictions? Groundhogs are supposed to, but the National Hurricane Center does it better! The graph on page 854 was the first advisory for Hurricane Twenty-Four and shows the National Hurricane Center (NHC) maximum 1-minute wind speed forecast as a broad blue line on a chart of wind speed versus forecast period. Examining the blue line, we can see from its inception (marked NOW in the graph) until 72 hours later, the hurricane was growing in intensity (the wind speed at the beginning was about 30 mph, and 72 hours later it had reached 90 mph). Can we find an equation for the blue line and predict the wind speed? Suppose we concentrate on the period starting at 48 hours and ending at 72 hours marked with the two blue dots. If we let x be the hours and y the wind speed, the first dot has coordinates (48, 80), and the second has coordinates (72, 90). To find the equation of this line, we first find the slope

$$m = \frac{y_2 - y_1}{x_2 - x_1} = \frac{90 - 80}{72 - 48} = \frac{10}{24} = \frac{5}{12}$$

Then we use the point-slope formula $y - y_1 = m(x - x_1)$ to obtain

$$y - 80 = \frac{5}{12}(x - 48)$$

Multiplying by 12,

$$12y - 960 = 5(x - 48) = 5x - 240$$

Adding 960,

$$12y = 5x + 720$$

Dividing by 12,

$$y = \frac{5}{12}x + 60$$

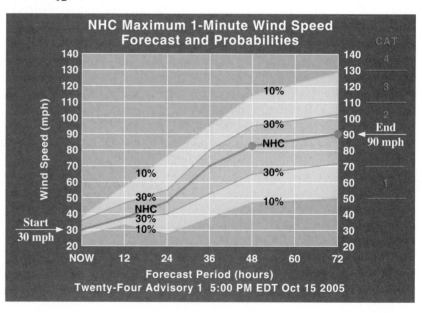

If you want to find the wind velocity y at any point x between 48 and 72 hours, simply substitute for x in the equation! For example, if you want to know the wind velocity 60 hours after the beginning of the forecast, let $x = 60$ in the equation $y = \frac{5}{12}x + 60$, obtaining

$$y = \frac{5}{12}(60) + 60 = 85$$

This means that the wind velocity 60 hours after the beginning of the forecast was 85 mph.

```
LinReg
y=ax+b
a=.4166666667
b=60
```

The work we have done is the same as obtaining the **regression line** for the points (48, 80) and (72, 90). With a graphing calculator (see the Graph It, page 849), press $\boxed{\text{STAT}}$ $\boxed{1}$ and enter 48 and 72 under L_1 and enter 80 and 90 under L_2. Now press $\boxed{\text{STAT}}$ $\boxed{\blacktriangleright}$ $\boxed{4}$ $\boxed{\text{ENTER}}$. The grapher says $y = ax + b$, where $a = 0.4166666667 \approx 0.417$ and $b = 60$. Note that $\frac{5}{12} \approx 0.417$.

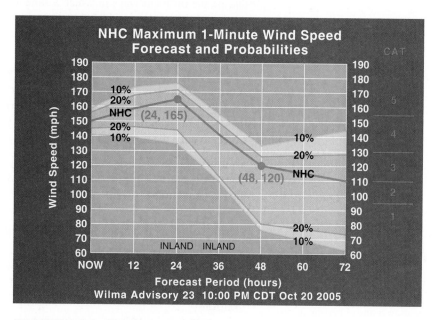

EXAMPLE 5 ▸ Hurricane Predictions

The graph shows the twenty-third advisory for Hurricane Wilma.

(a) What was the wind speed at the beginning of the advisory?

(b) In what interval of time was the hurricane increasing in intensity?

(c) In what interval of time was the hurricane decreasing in intensity?

(d) Find the regression line for the points between (24, 165) and (48, 120).

(e) What was the wind velocity 32 hours after the beginning of the forecast?

Solution

(a) 150 mph

(b) The slope of the line is positive (going up from left to right) from the beginning (NOW) of the forecast to 24 hours later.

(c) From 24 to 72 hours

```
LinReg
y=ax+b
a=-1.875
b=210
```

(d) Use a grapher and follow the procedure we used before. Enter 24 and 48 for L_1 and 165 and 120 for L_2. Then press $\boxed{\text{STAT}}$ $\boxed{\blacktriangleright}$ $\boxed{4}$ $\boxed{\text{ENTER}}$. The grapher says $y = ax + b$, where $a = -1.875$ and $b = 210$. Note that since the wind intensity is *decreasing,* the slope is negative.

(e) To find the wind velocity 32 hours after the beginning of the forecast, let $x = 32$ in $y = -1.875x + 210$, obtaining $y = -1.875(32) + 210 = 150$ mph.

By the way, look at the scale to the right of the graph (under CAT for category); at 150 mph, Wilma was classified as a category 4 hurricane!

Another way in which statisticians make predictions is based on a **sampling procedure.** The idea is quite simple. The statistician takes data from a **random sample** of the population and assumes that the entire population behaves in the same way as the sample. The difficulties lie in making certain that the sample is random and represents the population in a satisfactory manner. We shall not discuss the different ways in which a sample is selected but only mention that one of these ways uses a table of random numbers. (Such tables can be generated by a computer.)

EXAMPLE 6 ▶ Expected Number with Same Favorite Event

A certain school has 3000 students. A random sample of 50 students is selected, and the students are asked to name their favorite track and field events. If 10 of these students name the 100-m dash, how many students in the school would you expect to have the same favorite event?

Solution

You should expect $\frac{10}{50}$, or 20%, of the student body to favor the 100-m dash. Thus, the required number is 20% of 3000, or $(0.20)(3000) = 600$. ∎

EXAMPLE 7 ▶ Expected Number of Failures

In a certain county, 15% of the eleventh-grade students failed a required literacy test. If 60 of the eleventh-grade students in this county were selected at random, how many of these would we expect to have failed the test?

Solution

We would expect 15% of the 60 students to have failed the test; that is, $(0.15)(60) = 9$ students. ∎

Online Study Center

To further explore different ways to draw a line of best fit, access links 12.6.1 and 12.6.2 on this textbook's Online Study Center.

EXERCISES 12.6

In problems 1–3, each table gives the winning results in an Olympic event from 1960 to 1984.

a. Make a graph of the data.

b. Draw the best line that you can "between" the points, and predict the results for the 1988 Olympics.

c. Find the equation for the least-squares (regression) line. Treat 1960 as 0, 1964 as 4, and so on. Use the equation to predict the results for 1988. Compare to (**b**).

d. Use the equation in part (**c**) to predict the results for 2000 and 2004.

1. Men's 400-m hurdles, time in seconds

Year	1960	1964	1968	1972
Time	49.3	49.6	48.1	47.8
Year	1976	1980	1984	
Time	47.6	48.7	47.8	

2. Women's 100-m freestyle swimming, time in seconds

Year	1960	1964	1968	1972
Time	61.2	59.5	60.0	58.6
Year	1976	1980	1984	
Time	55.7	54.8	55.9	

3. Women's running high jump, height in inches

Year	1960	1964	1968	1972
Height	72.75	74.75	71.75	75.63
Year	1976	1980	1984	
Height	76.00	77.50	79.50	

4. The following table shows the world records for the men's mile run since 1965 (1965 = 0). The times are the numbers of seconds over 3 minutes.

Year	1965	1966	1967	1975
Time	53.6	51.3	51.1	49.4
Year	1980	1981	1985	
Time	48.8	47.3	46.3	

a. Make a graph of these data.
b. Draw the best line that you can "between" the points, and predict when the mile will first be run in 3 minutes 45 seconds.
c. Find the equation for the regression line and use it to predict when the mile will be run in 3 minutes 45 seconds. Compare with (**b**).

5. In the Shell Marathon, a Japanese experimental vehicle achieved the equivalent of 6409 mi/gal! Of course, the mileage you get depends on your speed. The table below shows the speed x in miles per hour and the distance y a car ran on 1 gal of gas.

x	30	35	40	45
y	34	31	32	30
x	50	55	60	65
y	29	30	28	27

a. Make a graph of these data.
b. Draw a line between the points and predict how many miles per gallon you would get if your speed were 70 mph.
c. Find the equation for the regression line and repeat part (**b**).

6. The table below shows the number x of television ads Top Flight Auto ran during a certain week and the number (y) of cars it sold during the same week.

x	3	10	0	7	13	8	14
y	7	15	10	8	14	10	20

a. Make a graph of these data.
b. If the management decides that it can afford only 6 ads per week, can you predict how many cars will be sold?
c. Find the equation for the regression line and repeat part (**b**).

7. Does the number of student absences in a course influence the number of failures? The table below shows the average number x of absences per student in a certain course and the number y of students failing the course.

x	5	7	2	4	3
y	16	20	9	12	10

a. Make a graph of these data.
b. Can you predict how many students would fail the course if the average number of absences per student were 6?
c. Find the equation for the regression line and repeat part (**b**).

8. The manager of the concession stand at a baseball stadium is trying to predict the number of hot dogs that must be bought for an upcoming game. The number x (in thousands) of advance tickets sold and the number y (in thousands) of hot dogs sold the day of the game for the last five games are shown in the following table:

x	23	32	19	29	20
y	16	23	13	22	16

a. Find the equation for the regression line.
b. If advance ticket sales are 26,000 tickets, how many hot dogs should the concessionaire buy?

9. The following table shows the serving size (in grams) and number of calories recently given for different sandwiches at Burger King:

Item	Serving Size (in grams)	Calories
Total Bacon Cheeseburger (1)	168 g	450
Total BK 1/4 lb BURGER™ (1)	222 g	590
Total DOUBLE WHOPPER® with Cheese Sandwich (1)	418 g	1120
Total KING SUPREME™ Sandwich (1)	209 g	625
Total Chicken WHOPPER® (1)	231 g	430

Source:www.bk.com/Food/Nutrition/NutritionWizard/index.aspx.

a. Draw the graph for these data.
b. Find the equation of the regression line for the data.
c. Use the equation to predict the caloric intake when you consume 300 grams of a Burger King sandwich.

10. The table below shows the serving size (in grams) and number of calories for various McDonald's sandwiches.

Item	Serving Size (in grams)	Calories
Hamburger	102 g	280
Cheeseburger	116 g	330
Quarter Pounder +	166 g	430
Quarter Pounder with Cheese +	194 g	530
Big Mac	215 g	590

Source:www.mcdonalds.com/app_controller.nutrition.index1.html.

a. Draw the graph for these data.
b. Find the equation of the regression line for the data.
c. Use the equation to predict the caloric intake when you consume 300 grams of McDonald's sandwiches. Compare with problem 9(c).

11. The table below shows the current wind chill estimate when the air temperature is 10°F.

Weather Revisions May Warm Up Cold

The current wind-chill estimate when the temperature of air is 10 degrees above zero, and a proposed revision		
Wind Speed (mph)	Current Wind Chill	Proposed Wind Chill
5	6	10
10	−9	0
15	−18	−8
20	−24	−14

Source: www.usatoday.com/weather/windchil.htm.

a. Draw the graph for these data.
b. Find the equation for the regression line for the data.
c. Use the equation to predict the wind chill when the wind speed is 25 mph.

12. The third column in the table shows the new proposed wind chill when the air temperature is 10°F.
a. Draw the graph for this data.
b. Find the equation of the regression line for the data.
c. Use the equation to predict the new wind chill when the wind speed is 25 mph.
d. Is it lower or higher than the current wind chill?

13. The table below shows the approximate amount of trash produced in the United States from 1980 to 2000. Let x represent the year after 1980 (1980 is year 0) and y represent the amount of trash (millions of tons).

Year	Million Tons
1980	150
1990	200
2000	220

Source: http://www.howstuffworks.com/landfill2.htm.

a. Draw the graph for this data.

b. Find the equation of the regression line for the data.

c. Use the equation to predict the amount of trash y that will be produced in 2005 and 2010.

14. How much of the trash is recycled? The table below shows the amount of paper and paperboard recycled from 1980 to 2000. Let x represent the year after 1990 (1990 is year 0) and y represent the amount of trash (millions of tons).

Year	Million Tons
1980	55
1990	70
2000	90

a. Draw the graph for these data.

b. Find the equation of the regression line for the data.

c. Use the equation to predict the amount of paper and paperboard that will be recycled in 2005 and 2010.

15. The table below shows the Actual Manufacturer's Suggested Retail Price for a Chevrolet Blazer and the invoice price (the price that the manufacturer supposedly charges its franchised dealers). First, round all numbers to the nearest thousand and let x be the MSRP and y be the invoice price in thousands of dollars.

Chevrolet Blazer

Model Trim Name	MSRP	Invoice Price
LS 2-door wagon, 2WD	$20,195	$18,276
LS 2-door wagon, 4WD	$23,195	$20,991
LS 4-door wagon, 2WD	$24,145	$21,851
LS 4-door wagon, 4WD	$26,145	$23,661
Xtreme 2-door wagon, 2WD	$22,295	$20,177

Source: http://auto.consumerguide.com/auto/new/reviews/full/index.cfm/id/23171.htm#prices.

a. Draw the graph for these data.

b. Find the equation of the regression line for the data.

c. Use the equation to predict the invoice price for a car with a $30,000 MSRP.

Note: The invoice price is really not the net price paid by the dealer to the manufacturer because of "holdbacks," moneys that the manufacturer will refund the dealer.

16. Suppose you want to buy a Sony Carousel CD or DVD changer. You can go online and find the MSRP for several models as well as the lowest price for that day.

The table below shows several models with their MSRP and actual price. First, round all numbers to the nearest ten dollars, and let x be the MSRP and y be the invoice price.

Model	MSRP	Actual Price
Sony DVP CX995V	$500.00	Check Prices $399
Sony wx-4500x	$700.00	Check prices $248
Sony 10 Disc-CD Changer	$99.95	Check prices $89
Sony 10 Disc CD Changer Recorder	$170.00	Check prices $199
Sony SCD-CE595	$129.95	Check prices $116
Sony CDP-CX355	$149.95	Check prices $99.88

Source: Adapted from http://auto.consumerguide.com/cp/electronics/browse/index.cfm/type/prod/id/11275.htm.

a. Draw the graph for these data.

b. Find the equation of the regression line for the data.

c. Use the equation to predict the actual price for a changer with a $300 MSRP.

17. Before we end regression problems, let us go and get some gourmet potato chips. The following table shows the brands, the number of chips per serving, and the number of calories in one serving. Let x be the number of chips in a serving and y the number of calories.

Brand	Chips/Serving	Calories
Terra	10	140
Taro	10	140
Sweet Potato	17	160
Glenny's	12	160
Sesame	9	150

a. Draw the graph for these data.
b. Find the equation of the regression line for the data.
c. Use the equation to predict the number of calories in a serving of 15 chips.

18. What about the price you pay per chip? Refer to the table below. Let x be the number of chips in a bag and y the price.

Brand	Chips	Price
Terra	80	$5.00
Taro	60	$3.25
Sweet Potato	100	$3.25
Glenny's	60	$3.00
Sesame	70	$3.00

a. Draw the graph for these data.
b. Find the equation of the regression line for the data.
c. What should the price of a bag with 90 chips be?

19. Of 50 students selected at random at a school, 12 said they preferred their hamburgers plain. How many of the 3000 students enrolled in the school would you expect to prefer their hamburgers plain?

20. A state welfare department selected 150 people at random from its welfare roll of 10,000 people. On investigation, it was found that 9 of the 150 had gotten on the welfare roll through fraud. About how many of the 10,000 people on the roll would you expect to be guilty of fraud?

21. An automobile tire manufacturer selected a random sample of 150 tires from a batch of 10,000 tires. It was found that 3 of the 150 tires were defective. How many of the 10,000 tires should the manufacturer expect to be defective?

22. An automobile manufacturer selected a random sample of 150 of its cars and found a defective steering assembly in 5 of the 150 cars. If the manufacturer had turned out 5000 cars under the same conditions, how many of these should it expect to have defective steering assemblies?

23. The best shooting percent for one basketball season is 72.7% and belongs to Wilt Chamberlain. If he attempted 586 baskets, how many would you expect him to have made?

24. A recent study indicates that 9 out of 50 males and 19 out of 50 females squeeze their toothpaste tubes from the bottom. In a group of 200 males and 200 females, how many males and how many females would you expect to be squeezing their toothpaste tubes from the bottom?

25. Can you predict the weather? Groundhogs are supposed to be able to! On February 2 of every year several famous groundhogs emerge from hibernation. If they see their shadows, that means 6 more weeks of winter. The following are the records of six famous groundhogs for varying numbers of years:

Punxsutawney Phil
   (15 years)   10 right, 5 wrong
Sun Prairie (7 years)   3 right, 4 wrong
West Orange, N.J.
   (13 years)   7 right, 6 wrong
Staten Island, N.Y.
   (7 years)   6 right, 1 wrong
Lilburn, Ga. (7 years)   6 right, 1 wrong
Chicago (6 years)   1 right, 5 wrong

If these six groundhogs predict the weather for the next 20 years, how many times would you expect

a. Punxsutawney Phil to be right?
b. Sun Prairie to be right?
c. West Orange to be wrong?
d. Staten Island to be wrong?
e. Lilburn to be wrong?
f. Chicago to be right?

26. The dropout rate (defined as "neither enrolled in school nor working") for 16–19-year-old African Americans is about 14%. The rate for whites is 9%. (Source: U.S. Bureau of Labor Statistics.)

    A school has 500 African-American and 1500 white students. How many African Americans and how many whites would you expect to drop out?

27. In a recent year 20% of Americans identified themselves as liberal, and 32% as conservative. (Source: CBS News Poll.) If 500 Americans are chosen at random, how many people who identified themselves as liberals and how many as conservatives would you expect?

28. The table below shows the marriage rate y for 1000 Americans from 1980 to 1995.

x	1980	1985	1990	1995	2000
y	10.5	10	9.5	9	8.5

Source: U.S. National Center for Health Statistics.

a. Make a graph for these data.
b. Predict the marriage rate for the year 2005.
c. If you select a representative sample of 5000 Americans in the year 2000, how many would you expect to marry? (*Hint:* Follow the pattern!)

29. The table below shows the percent y of preteens and teens aged 12–17 who "consumed alcohol in the past month" in the given year.

x	1975	1980	1985	1990	1995
y	50	45	35	30	20

Source: U.S. Center for Disease Control.

a. Make a graph for the data.
b. Predict the percent of teens aged 12–17 who would have "consumed alcohol in the past month" in the year 2000.
c. If you select a representative sample of 1000 preteens and teens aged 12–17 in the year 2000, how many would you expect to have "consumed alcohol in the past month"? (*Hint:* Follow the graph!)

30. How much income y do you need to get a 30-year mortgage of x dollars at a 7% rate with a 20% downpayment? The table below gives you an idea.

x	$50,000	$150,000	$200,000
y	$18,200	$54,600	$72,800

Source: Fannie Mae.

a. Make a graph for these data.
b. Predict how much income you need for a $75,000 and a $100,000 loan. (*Hint:* Follow the graph!)

31. Find the least-squares (regression) line for the data in Example 1 on page 848. Let x be the number of the Olympics with 1960 as number 1, 1964 as 2, etc., and let y be the number of seconds.

In Problems 32–35, use 1960 = 1, 1964 = 2, and so on.

32. Find the least-squares line for problem 1.

33. Find the least-squares line for problem 2.

34. Find the least-squares line for problem 3.

35. Find the least-squares line for problem 4. (*Hint:* Let x be the number of years after 1965, and let y be the number of seconds over 45.)

The following graph will be used in problems 36 and 37:

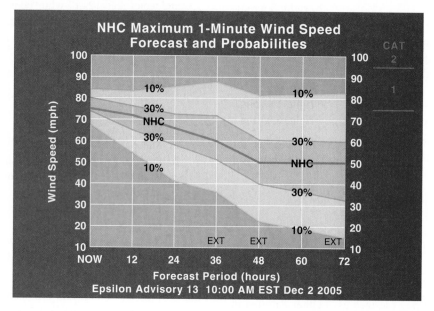

36. a. What was the wind speed at the beginning of the advisory?
b. In what interval of time was there no change in the hurricane intensity?
c. In what interval of time was the hurricane decreasing in intensity?

37. a. Find the regression line for the points between $(0, 75)$ and $(36, 60)$.
b. What was the wind velocity 24 hours after the beginning of the forecast?
c. What was the wind velocity between 48 and 72 hours after the beginning of the forecast?
d. What is the equation of the regression line between 48 and 72 hours?

38. The graph below represents the average monthly temperatures for the first 6 months of the year. What would you predict to be the increase in the average temperature from February to May?

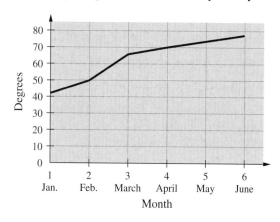

39. The publisher of a science fiction magazine wants to determine which features of the magazine are the most popular with its readers and decides to make a survey. Which of the following procedures would be the most appropriate for obtaining a statistically unbiased sample?
a. Survey the first 100 subscribers from an alphabetical listing of all the subscribers.
b. Survey a random sample of people from the telephone directory.
c. Have the readers voluntarily mail in their preferences.
d. Survey a random sample of readers from a list of all subscribers.

40. A tax committee in a small city wants to estimate the average county tax paid by the citizens of its city and decides to make a survey. Which of the following procedures would be the most appropriate for obtaining a statistically unbiased sample?
a. Survey all the residents of one section of the city.
b. Survey a random sample of people at the largest shopping mall in the city.
c. Survey a random sample of all the people in the city.
d. Survey all the people who work for the largest employer in the city.

In Other Words

41. To make predictions, statisticians take data from random samples. Discuss why the following procedures might not yield a random sample of the population of California:
  a. Every third woman shopper on Rodeo Drive is selected.
  b. Every third man in Berkeley is selected.
  c. Every third person leaving a baseball game in Oakland is selected.

42. If you use a sampling procedure to make a prediction, what do you assume about the sample? How can you make sure that you have a satisfactory sample?

Using Your Knowledge

The least-squares (regression) line in Example 3 is $y = -0.264x + 4.327$. Can we find another approximation for this line without all the calculations? Refer to Figure 12.33 on page 851. Let us concentrate on the first and last points on the line, (1, 24.4) and (11, 21.4).

43. What is the y intercept b of the line?

44. What is the slope m of the line?
  (*Hint:* Slope = rise/run)

45. What is the equation of the line?

46. Use the equation of the line obtained in problem 45 to predict the winning time in the year 2000.

47. What is the predicted winning time in the year 2000 using $y = -0.264x + 4.327$?

48. Compare your answers for problems 46 and 47. What is the difference between the answers?

Collaborative Learning

Online Study Center

In problems 27 and 28 we emphasized that to obtain an unbiased sample we have to make sure that the sample is random and represents the population in a satisfac-

tory manner. Professional "pollsters" give some advice! Form several groups of students and access link 12.6.3 on this textbook's Online Study Center. Questions for each group follow. After the group answers the questions, report to the rest of the members and discuss the results.

Group 1 20 Questions Journalists Should Ask About Polls

1. Name three questions that will help you decide if a poll is "scientific."

2. Name three organizations that conduct "scientific" polls.

3. What is the main difference between "scientific" and "unscientific" polls?

4. What method is used by pollsters to pick interviewees?

5. What kind of mistakes can skew poll results?

Group 2 Margin of Error

1. What does it mean when a report or survey gives a sampling with a margin of error of x percentage points?

2. Suppose 1000 persons are selected at random and 50% favor a certain proposal with a 3% margin of error. Explain what that means.

3. What is the 95% confidence level, and how does it relate to probability?

4. If 2000 people are surveyed and 50% favor a certain proposal, what would be the margin of error? What if only 100 people were surveyed?

What do market research professionals say about sampling and sampling errors? Go to link 12.6.4 on this textbook's Online Study Center. Based on the graph given,

1. if the sample size is 1000, what is the margin of error?

2. if the sample size is 100, what is the margin of error?

3. if you want a margin of error of ± 4, how many persons should you sample?

4. if you want a margin of error of ± 2.5, how many persons should you sample?

12.7 Scattergrams and Correlation

Correlation of Cancer to Solar Radiation

You have probably heard the expression "What does that have to do with the price of tea in China?" In many practical applications of mathematics such as business, medicine, social sciences, and economics, you find pairs of variables that have to be considered simultaneously. (Perhaps tea production and tea prices!) In general, you are looking for certain patterns or co-relations. For example, in Norway, the annual herring catch has dwindled from more than a million tons to fewer than 4000, and the rates of breast and colon cancer have nearly doubled. Is there a connection? Is there a relationship between herring catch and cancer? Yes, if you consider the fact that herring is rich in vitamin D, a nutrient that you can get from sunlight, and there is not too much sunlight in Norway! If this were the case, people living in sunny areas, where they receive greater amounts of sunlight, and hence more vitamin D, should have less breast and colon cancer.

Drs. Frank and Cedric Garland have shown that a population's vitamin D intake can be a predictor of breast and colon cancer. How can they make this claim? They looked at the number of deaths per 1000 women in places where there was not too much sunlight (say, New York, Chicago, and Boston) and the number of deaths per thousand in sunny places (Las Vegas, Honolulu, and Phoenix, for example). The graph of the results, called a **scattergram,** is shown in Figure 12.34. As you can see, the sunnier it is, the fewer deaths from breast cancer there are. It is said that there is a **negative correlation** between the amount of solar radiation (sunshine) received and the number of breast-cancer deaths. On the other hand, there is a **positive correlation** (not shown) between the amount of solar radiation received and the number of skin cancers. In this section you will study positive and negative correlations and how "good" these correlations are by means of a correlation coefficient devised by Karl Pearson.

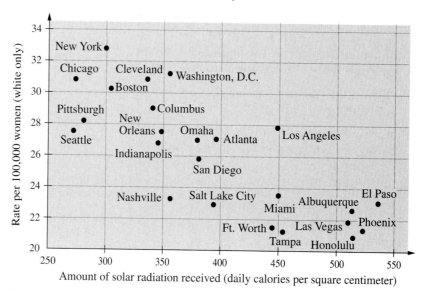

Source: Frank Garland.

FIGURE 12.34
Deaths from breast cancer (negative correlation).

A. Correlation

In general, three kinds of correlations are possible. The scattergrams in Figure 12.35 illustrate typical cases. A good illustration of a negative correlation appears in Section 12.6, Example 1.

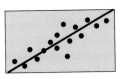
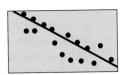
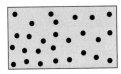

Positive correlation   Negative correlation   No correlation
(Line slopes upward)   (Line slopes downward)   (No particular pattern)

FIGURE 12.35
Different types of correlations.

How can we best represent the data in a scattergram by a line? We use this rule: The slope of the line should show the direction of the data and have approximately the same number of points above as below the line.

EXAMPLE 1 ▶ Determining Accurate Representation of Data

For each of the following graphs, determine if the line is an accurate representation of the data. Explain your reasoning.

(a)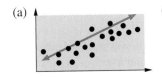

(b)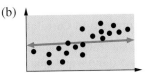

(c)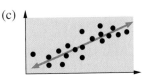

(d)

Solution

(a) No. There are too many points below the line.

(b) No. The slope of the line does not show the direction of the data, even though there are 9 points above and 9 points below the line.

(c) Yes. The slope follows the direction of the line and has about the same number of points above and below the line.

(d) No. There are too many points above the line.

■

GRAPH IT

To do Example 1, clear all lists by pressing $\boxed{\text{2nd}}$ $\boxed{+}$ $\boxed{4}$ $\boxed{\text{ENTER}}$. Press $\boxed{\text{STAT}}$ $\boxed{1}$ and enter the numbers in the left column of Table 12.25 on page 866 as L_1 and the numbers in the right column as L_2. To make a window corresponding to Figure 12.37, press $\boxed{\text{WINDOW}}$ and enter $X\min = 0$, $X\max = 4.4$, $X\text{scl} = 0.4$, $Y\min = 1.6$, $Y\max = 4.4$, and $Y\text{scl} = 0.4$. Now press $\boxed{\text{2nd}}$ $\boxed{\text{STAT PLOT}}$ $\boxed{1}$ $\boxed{\text{ENTER}}$ $\boxed{\blacktriangledown}$ $\boxed{\text{ENTER}}$ and $\boxed{\text{GRAPH}}$.

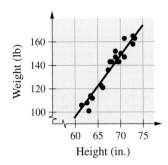

FIGURE 12.36
Height/weight.

Now suppose 20 students tried out for the basketball team at West Side High. The coach listed their heights and weights as in Table 12.24. We have graphed the ordered pairs (height, weight) as shown in Figure 12.36. The scattergram indicates how the height and weight are related. As we might expect, in any group of boys (or girls), the greater height usually corresponds to the greater weight.

The line drawn "between" the points in Figure 12.36 is the least-squares line for the data in Table 12.24. Notice that most of the points lie close to the line and that we have about the same number of points above the line as below. Because this line slopes upward, we say that the scattergram shows a **positive correlation** between the heights and the weights of the 20 students.

TABLE 12.24 Height/Weight for 20 Students

Height (in.)	61.4	62.6	63.0	63.4	63.8	65.7	66.1	67.3	67.7	68.1
Weight (lb)	106	108	101	114	112	123	121	136	143	143
Height (in.)	68.9	68.9	69.3	69.7	70.1	70.9	70.9	72.8	72.8	73.2
Weight (lb)	147	152	143	143	150	147	163	158	165	163

TABLE 12.25

High School GPA	College GPA
2.2	2.0
2.4	2.0
2.5	2.7
2.7	2.3
2.9	3.0
3.0	2.5
3.2	2.8
3.5	3.4
3.9	4.0
4.0	3.9

EXAMPLE 2 ▶ Scattergram and Correlation for GPAs

Ten students were selected at random, and a comparison was made of their high school grade point averages (GPAs) and their grade-point averages at the end of their first year in college (see Table 12.25).

(a) Make a scattergram and decide what kind of correlation is present.

(b) Find the equation for the regression line.

(c) If a student has a 2.5 high school GPA, what would be the predicted college GPA?

Solution

(a) Graph the given ordered pairs as shown in Figure 12.37. This scattergram indicates a *positive* correlation between the high school and college GPAs. Are you surprised?

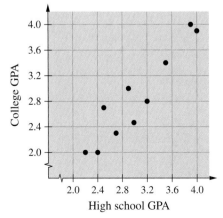

FIGURE 12.37

Online Study Center

To further explore the correlation coefficient, access links 12.7.1, 12.7.2, and 12.7.3 on this textbook's Online Study Center.

GRAPH IT

With a TI-83, you can find all the information in the formula after you enter the values from Table 12.25. Press STAT ▶ 2 ENTER and the values in the formula will be displayed.

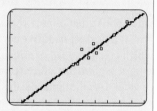

Points in Table 12.25 and the regression line $y = 1.09x - 0.443$.

(b) To find the equation for the regression line, we follow equations (2) and (3) of Section 12.6 and use a TI-83.

$$m = \frac{n\Sigma(xy) - (\Sigma x)(\Sigma y)}{n(\Sigma x^2) - (\Sigma x)^2} \tag{2}$$

$$= \frac{10(90.42) - (30.3)(28.6)}{10(95.25) - (30.3)^2} \approx 1.09$$

$$b = \frac{\Sigma y - m(\Sigma x)}{n} \tag{3}$$

$$= \frac{28.6 - (1.09)(30.3)}{10} \approx -0.443$$

Thus, the regression formula is $y = 1.09x - 0.443$.

(c) When the high school GPA $x = 2.5$, the college GPA would be
$y = 1.09(2.5) - 0.443 = 2.282$. ■

B. Coefficient of Correlation

In Example 2 we found a **positive** correlation between high school and college GPAs and drew their regression line. How accurately does their regression line represent the data? The points in the calculator screen fit the line closely, so the predictions we make should be fairly accurate. If the points are widely scattered, the predictions are not likely to be accurate. The closer the *sample* data points are (like the ones in Table 12.25) to the *regression* line, the more likely the entire *population* of predicted points (x, y) forms a line. The better the fit, the more confident we are that our regression line is a "good" estimator of the entire population line. But can we quantify how good? One common measure of the "goodness" of the linear relationship in the sample is called the **sample coefficient of correlation,** denoted by r. This coefficient is obtained from the sample data as shown.

Coefficient of Correlation

The **coefficient of correlation** for n points $(x_1, y_1), (x_2, y_2), \ldots, (x_n, y_n)$ is

$$r = \frac{n\Sigma(xy) - (\Sigma x)(\Sigma y)}{\sqrt{n(\Sigma x^2) - (\Sigma x)^2} \ \sqrt{n(\Sigma y^2) - (\Sigma y)^2}} \tag{4}$$

The value of r is between -1 (a perfect **negative** correlation) and $+1$ (a perfect **positive** correlation). When r is close or near 0, there is *no linear* correlation.

Note: The numerator for r is the same as that for m in equation (3). The denominator is also very similar to that for m.

TABLE 12.26

High School GPA	College GPA
2.2	2.0
2.4	2.0
2.5	2.7
2.7	2.3
2.9	3.0
3.0	2.5
3.2	2.8
3.5	3.4
3.9	4.0
4.0	3.9

TABLE 12.27

n	Critical Values for r 95%	99%
4	.950	.999
5	.878	.959
6	.811	.917
7	.754	.875
8	.707	.834
9	.666	.798
10	.632	.765
11	.602	.735
12	.576	.708
13	.553	.684
14	.532	.661
15	.514	.641
16	.497	.623
17	.482	.606
18	.468	.590
19	.456	.575
20	.444	.561

EXAMPLE 3 ▶ Finding the Coefficient of Correlation

Find the coefficient of correlation (r) for the data shown in Table 12.26.

Solution

Almost all the values in equation (4) were already computed in Example 2. Thus, $n = 10$, $\Sigma(xy) = 90.42$, $\Sigma x = 30.3$, $(\Sigma y) = 28.6$, $\Sigma x^2 = 95.25$, and $(\Sigma x)^2 = (30.3)^2$. The only missing value is $\Sigma y^2 = 86.44$.

Using the formula, or a calculator, we find $r = 0.94$ (to two decimal places). This value, which is "close" to 1, shows that the GPAs in high school and college are highly correlated. ∎

Is there a way of quantifying *how* highly correlated our variables are? If r is $+1$, we have a **perfect positive** correlation. For $r = -1$, a **perfect negative** correlation. For other values of r we use Table 12.27, which contains a list of critical values that can be used to decide whether there is a *significant* correlation between two variables at the 95% or 99% confidence level. Use the table like this.

Interpreting r

1. Find r for n points of data.

2. Go to line n in Table 12.27.

3. If the absolute value of r is greater than the number in the column labeled 95% on line n, we can be 95% confident that there is a *significant* linear correlation between the variables.*

4. If the absolute value of r is greater than the number in the column labeled 99% on line n, we can be 99% confident that there is a *significant* linear correlation between the variables.

*For the definition of absolute value, see page 350 in Section 6.3A.

Remember, **regression** gives the equation of a line that quantifies the relationship between two variables. The **correlation coefficient** measures the direction and strength of the linear relationship.

Now let us go back to Example 3 and interpret r by following these steps.

1. We found $r = 0.94$ with $n = 10$.

2. See the line corresponding to $n = 10$ in Table 12.27.

3. The absolute value of $r = |0.94| = 0.94$, which exceeds the 0.765 in line 10 and column 99%. Thus, we can be 99% confident that there is a *significant* correlation between the variables.

EXAMPLE 4 ▶ Finding the Correlation Coefficient

Use the data from Section 12.6, Example 4 (see Table 12.28), to find r and then interpret the results.

Solution

Referring to the data in the table, $n = 7$, $\Sigma x = 201$, $\Sigma y = 92$, $\Sigma x^2 = 6487$, and $\Sigma xy = 2828$. The only missing value is $\Sigma y^2 = 1272$.

Equation (4) states

$$r = \frac{n\Sigma(xy) - (\Sigma x)(\Sigma y)}{\sqrt{n(\Sigma x^2) - (\Sigma x)^2}\ \sqrt{n(\Sigma y^2) - (\Sigma y)^2}}$$

TABLE 12.28

Women	x	y	x^2	xy
1	16	12	256	192
2	18	11	324	198
3	23	10	529	230
4	27	11	729	297
5	33	13	1089	429
6	38	16	1444	608
7	46	19	2116	874
Total	**201**	**92**	**6487**	**2828**

```
LinReg
y=ax+b
a=.2603833866
b=5.666134185
r²=.771681673
r=.8784541382
```

Note: If r^2 and r are not showing, press 2ND 0 and ▼ until you get to Diagnostic On; then press ENTER . Repeat the procesure and this time you will see r^2 and r.

So for the data of Example 4,

$$r = \frac{7(2828) - (201)(92)}{\sqrt{7(6487) - (201)^2}\ \sqrt{7(1272) - (92)^2}}$$

$$\approx 0.88$$

(You can calculate this with your grapher. See the grapher screen in the margin.) To interpret the value $r = 0.88$, go to Table 12.27.

1. $n = 7$, $r = 0.88$

2. Go to the line labeled 7.

3. $|0.88| = 0.88$ is larger than 0.875 (barely!), so we can be 99% confident that there is a *significant* correlation between the age of a smoking pregnant woman and low birth-weight for the baby.

What would the relationship be if the mother did not smoke? Figure 12.38 shows that relationship. In order to simplify computations, round to the nearest percent.

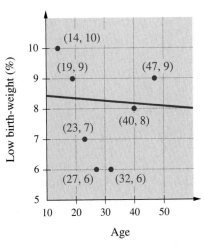

Source: www.uvm.edu/~dhowell/gradstat/
psych340/Lectures/CorrelReg/correl1.html.

FIGURE 12.38 ∎

```
LinReg
Y=ax+b
a=-.0221563154
b=8.496510816
r²=.0270562698
r=-.1644879017
```

```
QuadReg
Y=ax²+bx+c
a=.012868126
b=-.8110631366
c=19.04106455
R²=.8836521692
```

The quadratic regression.

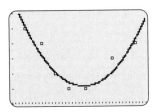

$y = 0.013x^2 - 0.811x + 19.04$

```
EDIT CALC TESTS
4↑LinReg(ax+b)
5:QuadReg
6:CubicReg
7:QuartReg
8:LinReg(a+bx)
9:LnReg
0↓ExpReg
```

EXAMPLE 5 ▶ Finding the Equation for a Regression Line

Use the data in Figure 12.38 on page 869 to

(a) find the equation for the regression line.

(b) find the coefficient of correlation r.

(c) provide an interpretation for the value r found in part (b).

Solution

Enter the ages 14, 19, 23, 27, 32, 40, and 47 under L_1 and the corresponding percent of low birth weight 10, 9, 7, 6, 6, 8, and 9 under L_2. Now press $\boxed{\text{STAT}}$ $\boxed{\blacktriangleright}$ $\boxed{4}$ $\boxed{\text{ENTER}}$. (See the grapher screen in the margin.)

(a) The regression line is $y = -0.022x + 8.5$.

(b) $r = -0.16$

(c) There is a negative correlation between low birth weight and age for non-smoking mothers (as age increases, the percent of low-birth-weight babies decreases). However, Table 12.27 shows that for $n = 7$, r is less (not more) than the numbers under 95% and 99%, so we conclude that there is no *significant* linear correlation between the variables. You would not expect that the percent of low-birth-weight babies would decrease as the mother's age increases! ∎

If we look at the graph of the data for Example 5, you may realize that the quantities are not linearly related. As a matter of fact, the data are closer to a **quadratic** model (parabola). Can we get a regression curve for a parabola? With a TI-83, press $\boxed{\text{STAT}}$ $\boxed{\blacktriangleright}$ $\boxed{5}$ $\boxed{\text{ENTER}}$ and you will get the ordered pairs relating age and percent of low-birth-weight babies, as well as the parabola $y = 0.013x^2 - 0.811x + 19.04$ graphed. A TI-83 grapher can compute best-fit curves of the type shown. (Just press $\boxed{\text{STAT}}$ $\boxed{\blacktriangleright}$ and you will see them!)

General Forms of Equations	
Type of Curve	*Equation*
Linear	$y = ax + b$
Quadratic	$y = ax^2 + bx + c$
Cubic	$y = ax^3 + bx^2 + cx + d$
Quartic	$y = ax^4 + bx^3 + cx^2 + dx + e$
Linear	$y = a + bx$
Logarithmic	$y = a + b \ln x$
Exponential	$y = a \cdot b^x$
Power	$y = a \cdot x^b$

EXAMPLE 6 ▶ Creating and Interpreting Scattergrams

Table 12.29 shows the relationship between the speed of a car and its gas mileage (mpg).

(a) Make a scattergram for the data.

(b) Show that the linear correlation between speed and gas mileage is $r = 0$.

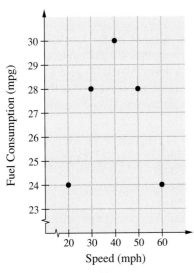

FIGURE 12.39

```
QuadReg
y=ax²+bx+c
a=¯.0142857143
b=1.142857143
c=6.8
R²=.9920634921
```

The size or the mass of a vehicle has the largest effect on the risk of injuries to those traveling in it. The relative risk of driver injury or fatality when cars of similar mass crash head-on declines as the mass of the cars increases (see Table 12.30 below and Table 12.31 on the next page). The five data sets were collected from real crashes on U.S. and German roads. (Sources: Evans and Wasielewski 1987 and Ernst et al. 1993.)

TABLE 12.30

Mass	Risk of Injury
700	2.5
800	2.1
900	2.0
1100	1.7
1200	1.5
1400	1.3
1600	1.0

TABLE 12.29

Speed	20	30	40	50	60
Mpg	24	28	30	28	24

(c) From the shape of the scattergram, what type of curve will give best-fit regression curve?

(d) Find the equation of the curve.

Solution

(a) The scattergram is shown in Figure 12.39.

(b) $r = \dfrac{n\Sigma(xy) - (\Sigma x)(\Sigma y)}{\sqrt{n(\Sigma x^2) - (\Sigma x)^2}\,\sqrt{n(\Sigma y^2) - (\Sigma y)^2}}$

$= \dfrac{5(5360) - 200(134)}{\sqrt{n(\Sigma x^2) - (\Sigma x)^2}\,\sqrt{n(\Sigma y^2) - (\Sigma y)^2}} = 0$

(c) A parabola

(d) Using a TI-83, enter the speeds as list L_1 and the mpgs as list L_2. Then press [STAT] [▶] [5] [ENTER] and you will get the equation of the parabola $y = -0.014x^2 + 1.14x + 6.8$. ∎

EXAMPLE 7 ▶ Using a Grapher to Find a Curve of Best Fit

Use a grapher to find a reasonable curve to fit the data relating the mass of five cars (in kilograms) and the relative risk of injury or fatality when involved in a head-on crash (Figure 12.40). Then predict the risk of injury or fatality if you have a head-on crash and your car weighs 1500 kg.

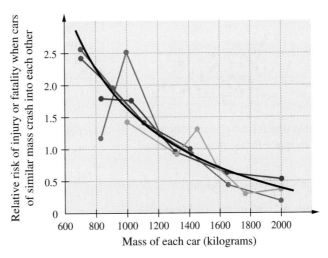

Source: *American Scientist* 90: 248.

FIGURE 12.40

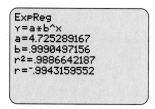

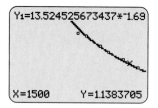

Solution

Start by entering the mass as list L_1 and the risk of injury as list L_2. From the graph, the relationship is not linear. We try three possibilities: logarithmic, exponential, and power. The results are shown in the margin. All curves give a good correlation (close to -1), but the best fit is the exponential curve ($r = -0.994$). Try it! In particular, the best exponential curve is given by $y = 4.725 \cdot (0.999)^x$.

TABLE 12.31

Type of Curve	r	a	b
Logarithmic	-0.994	13.525	-1.694
Exponential	-0.994	4.725	0.999
Power	-0.986	2080.69	-1.024

When your car weighs $x = 1500$ kg, we evaluate the expression by pressing ⟨2nd⟩ ⟨CALC⟩ ⟨1⟩, then entering 1500 and ⟨ENTER⟩. Your relative risk of injury is 1.138. (See the graph in the margin.) ∎

EXERCISES 12.7

Ⓐ Correlation

In problems 1–4, determine if the line is a good representative of the data. Explain why or why not.

1.

2.

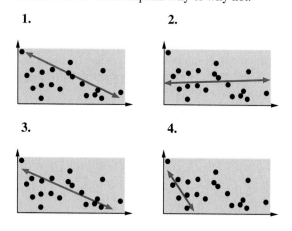

3.

4.

In problems 5–12, state which kind of correlation you would expect in a scattergram for the indicated ordered pairs.

5. (Length of person's leg, person's height)

6. (Outdoor temperature, cost of air conditioning a house)

7. (Student's weight, student's score on mathematics test)

8. (Person's salary, cost of person's home)

9. (Altitude, atmospheric pressure)

10. (Weight of auto, miles per gallon of fuel)

11. (Student's score on college aptitude test, student's GPA)

12. (Speed of auto, miles per gallon)

In problems 13–20, select the phrase that correctly completes the sentence.

13. The value of the linear correlation coefficient r is always
 a. close to 0.
 b. close to 1.
 c. between -1 and 1, inclusive.
 d. positive.

14. The linear correlation coefficient r measures
 a. whether a cause and effect relationship exists between two variables.
 b. whether a scattergram shows an interesting pattern.
 c. whether two variables are related.
 d. the direction and strength of the linear relationship.

15. A linear regression line $y = 10 + 0.9x$ is computed to predict the final exam score y on the basis of the first score x on the first test. Suppose Maria scores a 90 on the first test. What would be the predicted value of her score on the final exam?

a. 81   **b.** 89   **c.** 91

d. Cannot be determined

16. The scattergram below shows the calories and sodium content of several brands of meat hot dogs.

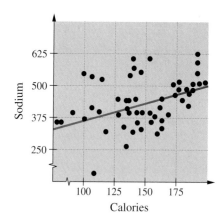

On the basis of the least-squares regression line shown, a hot dog containing 100 calories would have a sodium content of about

a. 70.   **b.** 350.

c. 400.   **d.** 600.

17. The scattergram below shows the amount of carbon monoxide and nitrogen oxide emitted in the exhaust of cars per mile driven.

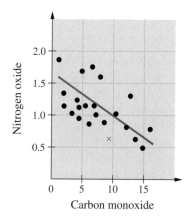

On the basis of the scattergram, the least-squares line would predict that a car emitting 10 g of carbon monoxide per mile driven would emit how many grams of nitrogen oxide per mile driven?

a. 1.1

b. 2.2

c. 10

d. 1.4

18.

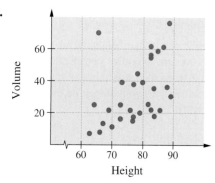

The scattergram above indicates

a. a positive association between height and volume.

b. a negative association between height and volume.

c. neither (**a**) nor (**b**).

d. no association between height and volume.

19. A plausible value r for the correlation between vehicle weight and miles per gallon (mpg) shown in the scattergram below is

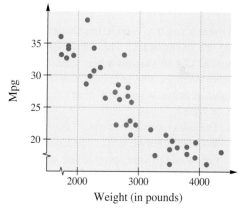

a. -1.0.   **b.** $+0.8$.   **c.** -0.9.   **d.** 1.0.

20. A researcher wishes to determine whether the rate of water flow y (in liters per second) over an experimental soil bed can be used to predict the amount of soil x washed away (in kilograms). If the equation for the regression line is

$$y = 1.3x + 0.4$$

the correlation between amount of eroded soil and flow rate would be

a. $\frac{1}{1.3}$.   **b.** 1.3.

c. positive, but we cannot say what the exact value would be.

d. either positive or negative. It is impossible to say anything about the correlation from the information given.

21. Do people of similar heights tend to date each other? A student measures herself, her roommate, and three women in adjoining dorm rooms. She then measures the height of their dates. Here are the results.

Women	65	63	65	69	64
Men	71	67	69	73	68

Which statement is true?

a. The correlation makes no sense because there is always a correlation between genders.

b. There is a strong positive correlation between the heights of men and women, since the women are always smaller than the men they date.

c. There is a positive correlation between the heights of men and women on a date.

d. There is a strong positive correlation between the heights of men and women on a date, since the men are always taller than the women they date.

22. Is there a positive correlation between beer sales and ice cream sales? The owner of a deli store noticed that in seasons when sales of beer were above average, sales of ice cream also tended to be above average. Which is a valid conclusion from these facts?

a. The sales records are wrong. There should be no association between beer and ice cream sales.

b. For a significant proportion of customers, drinking beer causes a desire for ice cream or eating ice cream causes a thirst for beer.

c. A scattergram of monthly ice cream sales versus monthly beer sales would show that a straight line describes the pattern in the graph, but it would have to be a horizontal line.

d. None of the above.

B Coefficient of Correlation

In problems 23–30,

   a. make a scattergram for the data.

   b. find the equation of the regression line.

   c. find r, the coefficient of correlation.

   d. interpret the results for r (see Table 12.27 on page 868).

23.

x	2	4	7	3	1	5
y	1	3	8	4	2	2
x	8	10	6	9		
y	6	7	6	4		

24.

x	9	10	8	8	6	4
y	4	2	3	6	4	8
x	5	5	3	2		
y	7	5	8	10		

25. The following table gives the weights and the highway miles per gallon for eight automobiles:

Weight (lb)	2800	1900	2000
Miles per gallon	19	34	28
Weight (lb)	3300	3100	2900
Miles per gallon	19	24	23
Weight (lb)	4000	2600	
Miles per gallon	16	24	

26. The following table gives the scores of ten students on an English exam and their corresponding scores on an economics exam:

English	50	95	55	20
Economics	75	95	70	35
English	85	75	45	20
Economics	70	80	40	15
English	80	90		
Economics	60	90		

e. What would the predicted economics score be for a student with a 70 in English?

27. The following are some recent statistics on years of schooling successfully completed and average annual salaries (in thousands of dollars) for men over 25:

Years of schooling	8	12
Average salary	$16.800	$23.300
Years of schooling	15	16
Average salary	$25.800	$33.900

e. A man (over 25) has 10 years of schooling. What is his predicted salary?

28. The following table gives the gain in reading speed for students in a speed-reading program:

Weeks in program	2	3	3
Speed gain (words per minute)	40	60	80
Weeks in program	4	5	6
Speed gain (words per minute)	100	110	150
Weeks in program	8	9	
Speed gain (words per minute)	190	220	

e. If a student spent 7 weeks in the program, what is the expected gain in reading speed?

29. A student was curious about the effect of antifreeze on the freezing point of a water-antifreeze mixture. He went to the chemistry lab, where he made the measurements in the following table:

Percent antifreeze (by volume)	10	20	30
Freezing point (degrees C)	-4	-10	-20
Percent antifreeze (by volume)	40	50	
Freezing point (degrees C)	-24	-36	

e. If the percent of antifreeze by volume is 25%, what is the freezing point of the mixture?

30. The following table gives the heights of students and their scores on an English test.

Height (in.)	62	67	70	64
Test score	85	60	75	70
Height (in.)	72	68	65	61
Test score	95	35	60	80
Height (in.)	73	67		
Test score	45	100		

e. What would the score be for a 66-in. student?

31. The power chart below shows the number of power outages experienced by the Central Power Company during the 12 weeks starting June 1.
  a. During which week was there no outage?
  b. Which week had the most outages?
  c. When did the decline in the number of outages seem to start?
  d. Is any kind of correlation shown by the chart?

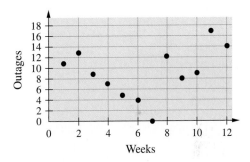

32. The following chart shows the hours studied by 10 students and their grades on an examination.
  a. Which kind of correlation (if any) is shown?
  b. If a student studied for 12 hours, about what grade would you expect her to make?

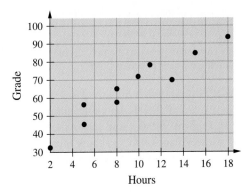

In Other Words

33. During the past 10 years there has been a positive correlation between the number of cars with air bags sold in Florida and the number of traffic accidents.
   a. Do air bags cause traffic accidents?
   b. What other factors may cause traffic accidents and the number of cars with air bags to increase together?

34. Describe what happens to the related items of data if the corresponding scattergram shows
   a. a positive correlation.
   b. a negative correlation.

Using Your Knowledge

You should not assume that there is any cause-and-effect relationship between two variables simply

Is there a correlation between the number of storks and the number of babies born?

because the correlation is high. A classic example of a high positive correlation is that of the number of storks found in English villages and the number of babies born in these villages. Do you think there is a cause-and-effect relationship here? What about a cause-and-effect relationship between smoking and drinking? There are many factors that may *influence* outcomes,

but are not necessarily the *cause* of the outcomes. For example, the British government conducts surveys of household spending. Is there a relationship between spending on tobacco and spending on alcohol? The resulting scattergram shows an overall positive linear relationship between spending on alcohol and spending on tobacco. How strong is the relationship? We will use our knowledge to answer the question.

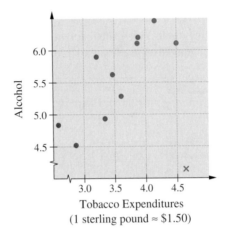

Tobacco Expenditures
(1 sterling pound ≈ $1.50)

The graph above shows a positive linear relationship between tobacco expenditures and alcohol expenditures, with Northern Ireland (blue *x*) as an *outlier*. Can the spending habits of the inhabitants of Northern Ireland alone influence the results?

Region	Alcohol	Tobacco
North	6.47	4.03
Yorkshire	6.13	3.76
Northeast	6.19	3.77
East Midlands	4.89	3.34
West Midlands	5.63	3.47
East Anglia	4.52	2.92
Southeast	5.89	3.20
Southwest	4.79	2.71
Wales	5.27	3.53
Scotland	6.08	4.51
N. Ireland	4.02	4.56

35. Use the table above and a grapher to find the coefficients of correlation r for the 11 regions of Great Britain and interpret the results (use Table 12.27).

36. Do the same as in problem 35, but delete Northern Ireland from the sample.

37. Notice the difference between the coefficient of correlation *r* obtained in problems 35 and 36. Do the different values of *r* mean that Ireland caused the correlation?

38. As we have mentioned, the point corresponding to Northern Ireland and shown with a blue *x* in the graph is called an **outlier.** When do you think you can ignore this point in the calculations? To help you out, access link 12.7.4 on this textbook's Online Study Center; then answer the question.

 Collaborative Learning

Do you know the difference between correlation and causation? Here is an excerpt from an article intimat-

ing that this type of reasoning was used in *The Bell Curve,* a book linking race and IQ:

"There is a direct correlation, mathematicians have found, between children's achievement on mathematics tests and shoe size. A clear signal that big feet make you smarter?" (A letter to the editor commented: "No no no no. It's big shoes that make you smarter.")

Here is a discussion question posed in *Chance News:*

"In *The Bell Curve,* it is shown that having a high IQ is correlated with going to an elite college. As one expert points out, it is probably also true that having parents with a high income is also correlated with going to an elite college. How would you try to determine which is the more important variable in predicting who goes to elite colleges?"

Chapter 12 Summary

Section	Item	Meaning	Example
12.1	Statistics	The science of collecting, organizing, and summarizing data so that valid conclusions can be drawn from them	
12.1A	Population	The entire collection of elements to be studied	In a shipment of 1000 shoes, the set of 1000 shoes is the population.
12.1A	Random sample	A subset of a population	If the 1000 shoes are numbered and 10 are chosen at random, the random sample consists of the 10 shoes chosen.
12.1B	Frequency distribution	A way of organizing a list of numbers	
12.1B	Frequency	Number of times an entry occurs	In the set of numbers {1, 4, 4, 7} the number 4 appears with frequency 2.
12.1C	Histogram	A special type of graph consisting of vertical bars with no space between the bars	
12.1D	Frequency polygon	A line graph connecting the midpoints of the tops of the bars in a histogram	

Section	*Item*	*Meaning*	*Example*
12.2	Mean, $\bar{x}$	The sum of the scores divided by the number of scores	The mean of 3, 7, and 8 is 6.
12.2	Mode	The number that occurs most often	The mode of 1, 2, 2, and 3 is 2.
12.2	Median	If the numbers are arranged in order of magnitude for an odd number of scores, the median is the middle number; for an even number of scores, the median is the average of the two middle numbers.	Median ↓ 1 3 ⑧ 15 19 \ 1 2 5 9 11 18 ↑ ↑ $\frac{5+9}{2} = 7 \leftarrow$ Median
12.3	Range	The difference between the greatest and the least numbers in a set	The range of 2, 8, and 19 is 17.
12.3	Standard deviation, s	$$\sqrt{\frac{(x_1 - \bar{x})^2 + \cdots + (x_n - \bar{x})^2}{n-1}},$$ where $\bar{x}$ is the mean and n is the number of items	
12.5	z-score	$z = \dfrac{x - \mu}{\sigma}$, where x is a score, μ is the mean, and σ is the standard deviation of the population	
12.6	Regression (best-fit) line	The best-fit line associated with the n points $(x_1, y_1), (x_2, y_2), \ldots, (x_n, y_n)$ has the form $y = mx + b$ where $$m = \frac{n\Sigma(xy) - (\Sigma x)(\Sigma y)}{n(\Sigma x^2) - (\Sigma x)^2}$$ and $$b = \frac{(\Sigma x^2)(\Sigma y) - (\Sigma x)(\Sigma xy)}{n(\Sigma x^2) - (\Sigma x)^2}$$ $$= \frac{\Sigma y - m(\Sigma x)}{n}$$	The regression line for the points $(1, 3), (2, 5),$ and $(3, 7)$ is $y = 2x + 1$, where $m = 2$ and $b = 1$.
12.7	Correlation	A measure of the direction and strength of a straight-line relationship between two variables	A $+1.0$ is a perfect positive correlation and a -1.0 is a perfect negative correlation.
12.7	Coefficient of correlation	The coefficient of correlation for n points $(x_1, y_1), (x_2, y_2), \ldots, (x_n, y_n)$ is $$r = \frac{n\Sigma(xy) - (\Sigma x)(\Sigma y)}{\sqrt{n(\Sigma x^2) - (\Sigma x)^2}\,\sqrt{n(\Sigma y^2) - (\Sigma y)}}$$	The coefficient of correlation for the points $(1, 3), (2, 5),$ and $(3, 7)$ is 1.

Research Questions

Sources of information for most of these questions can be found in the Bibliography at the end of the book.

1. Go to the library, look through newspapers and magazines, and find some examples of histograms and bar and circle graphs. Are there any distortions in the drawings? Discuss your findings.

2. Write a brief report about the contents of John Graunt's *Bills of Mortality.*

3. Write a paragraph on how statistics are used in different sports.

4. Write a report on how surveys are used to determine the ratings and rankings of television programs by such organizations as A. C. Nielsen.

5. Discuss the Harris and Gallup polls and the techniques used in their surveys.

6. Prepare a report or an exhibit of how statistics are used in medicine, psychology, and/or business.

7. Discuss the life and work of Adolph Quetelet (1796–1894).

8. Research and write a report on how Gregor Mendel (1822–1884), Sir Francis Galton (1822–1911), and Florence Nightingale (1820–1910) used statistics in their work.

9. Look at the scattergram in Section 12.7, Getting Started. Find out the average daily amount of solar radiation in your area, then make a prediction about the rate of breast-cancer mortality per 100,000 women in your area.

Chapter **12** Practice Test

1. A college president wants to find out which courses are popular with students. The president decides to conduct a survey of a sample of 30 students from the English department. Will these 30 students correspond to a simple random sample of the whole student body? Explain your answer.

The following scores were made on a scholastic aptitude test by a group of 25 high school seniors:

85	65	89	83	98
67	88	87	88	90
95	77	91	73	88
99	67	91	72	86
79	83	61	70	75

Use these data for problems 2 and 3.

2. Group the scores into intervals of $60 < s \leq 65, 65 < s \leq 70, 70 < s \leq 75$, and so on. Then make a frequency distribution with this grouping.

3. **a.** Make a histogram for the frequency distribution in problem 2.
   b. Make a frequency polygon for the distribution in problem 2.

4. During a certain week, the following maximum temperatures (in degrees Fahrenheit) were recorded in a large eastern city: 78, 82, 82, 71, 69, 73, and 70.
   a. Find the mean of these temperatures.
   b. Find the mode of the temperature readings.
   c. Find the median high temperature for the week.

5. a. Find the range of temperatures in problem 4.
   b. Find, to two decimal places, the standard deviation of the temperatures in problem 4.

6. A fair coin is tossed 256 times. If this experiment is repeated many times, the numbers of heads will form an approximately normal distribution, with a mean of 128 and a standard deviation of 8.
   a. Within what limits can we be almost 100% confident that the total number of heads in 256 tosses will lie?
   b. What is the probability that heads will occur fewer than 112 times?

7. A normal distribution consists of 1000 scores, with a mean of 100 and a standard deviation of 20.
   a. About how many of the scores are above 140?
   b. About how many scores are below 80?
   c. About how many scores are between 60 and 80?

8. A testing program shows that the breaking points of fishing lines made from a certain plastic fiber are normally distributed, with a mean of 10 lb and a standard deviation of 1 lb.
   a. What is the probability that one of these lines selected at random has a breaking point of more than 10 lb?
   b. What is the probability that one of these lines selected at random has a breaking point of less than 8 lb?

9. On a multiple choice test taken by 1000 students, the scores were normally distributed, with a mean of 50 and a standard deviation of 5. Find the z-score corresponding to a score of
   a. 58.   b. 62.

10. Agnes scored 88 on a French test and 90 on a psychology test. The mean score in the French test was 76, with a standard deviation of 18, and the mean score in the psychology test was 80, with a standard deviation of 16. If the scores were normally distributed, which of Agnes's scores was the better score?

11. With the data in problem 9, find the probability that a randomly selected student will have a score between 50 and 62. (Use Table II in the back of the book.)

12. The following are the amounts (to the nearest billion) the federal government has spent on education for selected years:

1984	1988	1989	1990
17	21	24	26

   a. Make a line graph for this information.
   b. Use your graph to estimate how much was spent on education in 1986.

13. Here is a list of five of the most active stocks on the New York Stock Exchange in a recent year. Make a bar graph of the yield rates of these stocks.

Stock	Price ($)	Dividend ($)	Yield Rate (%)
Fed DS	60.25	1.48	2.5
Ford M	44.75	2.00	4.5
Noes Ut	20.75	1.76	8.5
Exxon	42.75	2.00	4.7
Gen El	43.25	1.40	3.2

14. In a recent poll, the features that patrons liked in a restaurant were

Low-calorie entrees   67%   All-you-can-eat specials   27%
Varied portion sizes   47%   Self-service soup bar   35%
Cholesterol-free entrees 52%

Use horizontal bars and make a bar graph of this information.

15. The bar graph in the figure below shows the 2000 sales and the projected 2004 sales of the Wesellum Corporation. Read the graph, and estimate the percentage increase that was projected for 2004 over 2000.

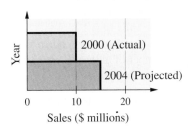

16. The typical family budget is as follows:

Monthly Family Budget

Savings	$ 200
Housing	600
Clothing	300
Food	800
Other	100
Total	$2000

Make a circle graph for these data.

17. Graph the five points given in the table below. Draw the best line you can "between" these points, and estimate the value of y for $x = 6$.

x	1	2	3	4	5
y	10.2	7.6	5.8	4.4	2.0

18. Find the equation for the best fit-line for the points you have just graphed and estimate the value of y for $x = 6$.

19. The testing department of Circle Tire Company checks a random sample of 150 of a certain type of tire that the company makes and finds a defective tread on 3 of these tires. In a batch of 10,000 of these tires, how many are expected to have defective treads?

20. In a large county, 50,000 high school students took a reading comprehension test, and 4000 of these students got a rating of excellent. In a random sample of 100 of these students, how many should be expected to have gotten an excellent rating on this test?

21. Which of the graphs shows the best representation for the data and why?

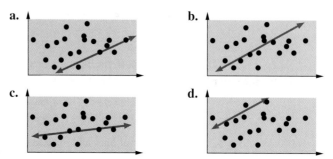

22. Which kind of correlation would you expect for the indicated ordered pairs?
   a. (Value of a family's home, family's annual income)
   b. (Number of hours of training, number of minutes in which runner can do the mile run)
   c. (Person's shoe size, person's salary)
   d. (Number of children getting polio immunization, number of children contracting polio)

23. The coefficient of correlation r for the regression line shown is about

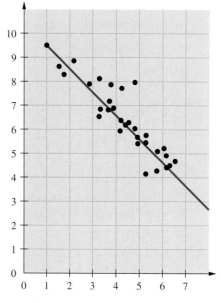

   a. $+0.94$.   **b.** 94.   **c.** -0.94.   **d.** -94.

24. Business intelligence software is one of the fastest-growing segments in the software market. If x represents the year and y the product sales (in billions of dollars), find the line of best fit (regression line) for the data.

Year	x	y
2002	2	4
2003	3	4.2
2004	4	5
2005	5	5.8

25. What is the coefficient of correlation r to the data in problem 24 and what does it represent? (Use the table below.)

n	95%	99%
4	.950	.999
5	.878	.959
6	.811	.917
7	.754	.875

Answers to Practice Test

ANSWER	IF YOU MISSED		REVIEW		
	Question	Section	Example(s)	Page(s)	
1. No. Not every member of the student body had the same chance of being chosen (only the ones in the English department).	1	12.1	**Definition of a simple random sample**	777	
2.	2	12.1	2(a)	780	

2.

Score	Tally Marks	Frequency
$60 < s \le 65$	||	2
$65 < s \le 70$	|||	3
$70 < s \le 75$	|||	3
$75 < s \le 80$	||	2
$80 < s \le 85$	|||	3
$85 < s \le 90$	ℕ ||	7
$90 < s \le 95$	|||	3
$95 < s \le 100$	||	2

ANSWER	IF YOU MISSED	REVIEW		
	Question	Section	Example(s)	Page(s)
3. a.–b.	3	12.1	2(d), (e); 3	780, 783–784
4. a. 75°F **b.** 82°F **c.** 73°F	4	12.2	1–5	796–800
5. a. 13°F **b.** 5.60°F	5	12.3	1–3	811–814
6. a. 104–152 **b.** 0.025	6	12.4	1	818–819
7. a. 25 **b.** 160 **c.** 135	7	12.4	2	820
8. a. 0.5 **b.** 0.025	8	12.4	3	820
9. a. 1.6 **b.** 2.4	9	12.4	4	821
10. The score in French was the better score.	10	12.4	4	821
11. 0.492	11	12.4	5	822
12. a. **b.** About $19 billion	12	12.5	1, 2	833–834
13.	13	12.5	3, 4	835–836
14. Features That Patrons Liked 1 Low-calorie entrees 4 All-you-can-eat specials 2 Varied portion sizes 5 Self-service soup bar 3 Cholesterol-free entrees	14	12.5	3, 4	835–836

<table>
<tr><th rowspan="2">ANSWER</th><th>IF YOU MISSED</th><th colspan="3">REVIEW</th></tr>
<tr><th>Question</th><th>Section</th><th>Example(s)</th><th>Page(s)</th></tr>
<tr><td>15. About 50%</td><td>15</td><td>12.5</td><td>3, 4</td><td>835–836</td></tr>
<tr><td>16.
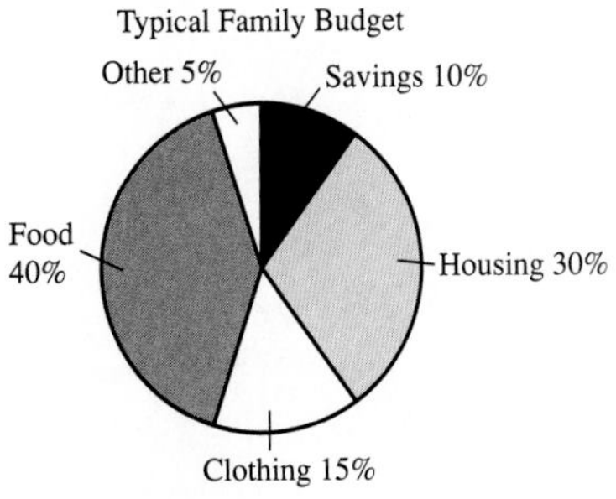

Typical Family Budget
Other 5%
Savings 10%
Food 40%
Housing 30%
Clothing 15%

</td><td>16</td><td>12.5</td><td>5, 6</td><td>837–839</td></tr>
<tr><td>17. a. For $x = 6, y \approx 0$.
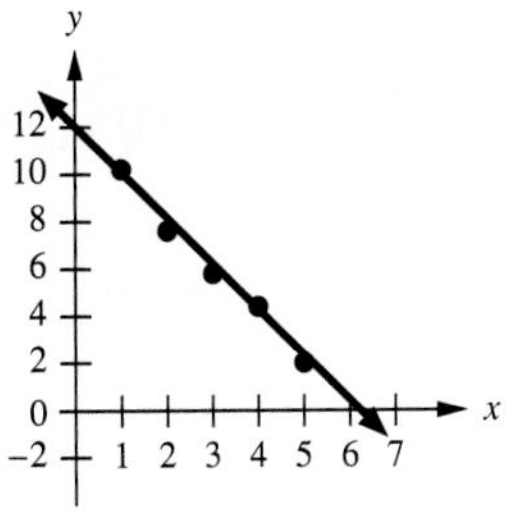

y
x
12
10
8
6
4
2
0
−2
1 2 3 4 5 6 7

</td><td>17</td><td>12.6</td><td>1, 3</td><td>848–849, 852</td></tr>
<tr><td>18. $y = -1.96x + 11.88$; for $x = 6, y = 0.12$</td><td>18</td><td>12.6</td><td>1, 3, 4</td><td>848–849, 852, 853</td></tr>
<tr><td>19. About 200</td><td>19</td><td>12.6</td><td>6</td><td>856</td></tr>
<tr><td>20. About 8</td><td>20</td><td>12.6</td><td>6</td><td>856</td></tr>
<tr><td>21. (b) The slope follows the direction of the line and has about the same number of points above and below the line.</td><td>21</td><td>12.7</td><td>1</td><td>865</td></tr>
<tr><td>22. a. Positive b. Negative c. None d. Negative</td><td>22</td><td>12.7</td><td>2</td><td>866–867</td></tr>
<tr><td>23. (c) −0.94</td><td>23</td><td>12.7</td><td>1, 2</td><td>865–867</td></tr>
<tr><td>24. $y = 0.62x + 2.58$</td><td>24</td><td>12.7</td><td>3</td><td>868</td></tr>
<tr><td>25. $r \approx 0.97$. We can be 95% confident that there is a significant correlation between the year x and sales y.</td><td>25</td><td>12.7</td><td>4, 5</td><td>868–870</td></tr>
</table>

Many people invest money throughout their lives in order to obtain additional income, to save for retirement, or to help save money for their children's education. In Section 13.5, you will study long-term investments, their cost, and their potential for profitability.

Your Money and Your Math

Do you know the difference between **simple** and **compound** interest or how to use percents to figure out sales taxes or discounts? As an informed consumer, you should be aware of these topics; if you are not, you will be particularly interested when they are discussed in Section 13.1.

What about your credit cards? How much interest do you pay on them? What are their annual fees? Can you get some cards *without* an annual fee? Credit cards and other types of credit such as **revolving charge accounts** and **add-on interest** are studied in Section 13.2

As a consumer, you do have certain rights regarding credit purchases. In Section 13.3 you study the **Truth-in-Lending Act,** its provisions, and the benefits afforded you as a consumer under the act.

Section 13.4 discusses the American dream: buying a house. It starts by discussing how much you can afford, how much of a down payment you should make, the different types of loans available to you, and how monthly payments are estimated. It also discusses an expense often over-looked when buying a home: the closing costs. What do they entail and when do they have to be paid? When you study this section, you will gain a wealth of information regarding the mechanics and strategies for buying a home. Finally, after discussing how to manage charge accounts and credit cards and even buying a house, you may have money left for **investing** so you can make even more money. In Section 13.5 we learn how to **invest** money in stocks, bonds, and mutual funds.

Online Study Center

For links to Internet sites related to Chapter 13, please access **college.hmco.com/PIC/bello9e** and click on the Online Study Center icon.

13.1 Interest, Taxes, and Discounts

HUMAN SIDE OF MATH

Have you ever wished for a money tree? The picture shows an East Indian money tree. The tin coins, used by the people of the Malay Peninsula in the nineteenth century, were broken off as needed.

The first coins were probably made about 2500 years ago in Lydia, now part of western Turkey. The coins were of a natural mixture of gold and silver called *electrum,* and they were stamped with a design showing that the king guaranteed them to be of uniform size. These coins were accepted by traders as a convenient medium of exchange and inspired other countries to make their own coins.

The first paper money was used in China about 1400 years ago. Europeans were skeptical about a piece of paper having any value, and it was not until the seventeenth century that paper money was accepted.

(continued)

Simple and Compound Interest

Do you know the difference between simple and compound interest? First, you should know that **interest** is the amount paid for using borrowed money. When you deposit money in a savings account, buy a certificate of deposit, or loan money to someone, the person or institution receiving the money will pay you *interest* for the use of your money. On the other hand, if you borrow money from a bank, you must pay interest for the money you borrow.

How is interest computed? It depends! **Simple interest I** depends on the **principal P** (the amount borrowed or invested), the **interest rate r** (the portion of the principal charged for using the principal and usually expressed as a percent), and the **time** or **term t** (the number of years during which the borrower or investor has the use of all or part of the money). To calculate simple interest, use the formula

$$I = Prt$$

Now suppose you invest \$1000 at 6% simple interest for 1 year. Here, $P = \$1000$, $r = 6\% = 0.06$, and $t = 1$. Thus,

$$I = \$1000 \times 0.06 \times 1 = \$60$$

At the end of the year, you get your \$1000 back plus \$60 in interest; that is, \$1000 + \$60 = \$1060.

But suppose you invest the \$1000 at the same nominal rate of 6% and the interest is **compounded** quarterly (this means the interest is calculated four times a year instead of just once). How much money would you get at the end of the year? This information is shown in Table 13.1.

Wellspring Savings Bank
Your source of financial security since 1875
Certificate of Deposit Accounts

	Interest	Est. Annual Yield
Short-Term		
90–179 days	3.50%	—
180–364 days	3.65%	—
Long-Term		
1–2 years	4.00%	4.07%
2–3 years	4.60%	4.70%
3–8 years	5.00%	5.12%

TABLE 13.1

Quarter	Interest	New Principal*
1st	$\$1000 \times 0.06 \times \frac{1}{4} = \$15$	$\$1000 + \$15.00 = \$1015.00$
2nd	$\$1015 \times 0.06 \times \frac{1}{4} = \$15.23$	$\$1015 + \$15.23 = \$1030.23$
3rd	$\$1030.23 \times 0.06 \times \frac{1}{4} = \$15.45$	$\$1030.23 + \$15.45 = \$1045.68$
4th	$\$1045.68 \times 0.06 \times \frac{1}{4} = \$15.69$	$\$1045.68 + \$15.69 = \$1061.37$

*Banks round *up* when collecting money, *down* when paying it.

Thus, at the end of the year, you get \$1061.37, which is \$1.37 more than the simple interest amount.

In this section you will study simple interest and compound interest and their applications. ▶

Until 1863, most of the currency in the United States consisted of notes issued by various state banks. Then Congress established national banks with the authority to issue bank notes. In 1913, the Federal Reserve System was established, and Federal Reserve notes gradually replaced national bank notes and became the official currency of the United States.

Looking Ahead
In this chapter you will study many financial applications.

Some of these applications involve calculating best buys, checking credit account statements, comparing interest rates, using credit cards, and purchasing a home or a car. In this chapter, you investigate these activities.

Realistic problems dealing with personal finances are usually solved with the aid of a calculator. Hence, the examples will show how a calculator can be used to obtain the answer. Of course, a calculator is not absolutely necessary but is just a tool to help you do the arithmetic faster and more easily.

Let us start with a concept that occurs in most financial transactions: interest. In general, there are two types of interest, simple and compound. We shall study simple interest first.

A. Simple Interest

How much money would you get if you received 12% interest on $1 million for 10 years? The answer depends on how this interest is calculated!

The **simple interest** for 1 year on a principal P at the rate r is just the principal times the rate, that is, Pr. The simple interest for t years is obtained by multiplying by t.

Reprinted with special permission of King Features Syndicate, Inc.

Simple Interest

The formula for calculating simple interest I on a principal P at the rate r for t years is

$$I = Prt$$

Notice that in the calculation of simple interest, the principal is just the original principal; the periodic interest that you earned does *not* earn further interest. Thus, if on $1 million ($P = \$1,000,000$) you receive 12% interest ($r = 12\% = 0.12$) for 10 years ($t = 10$), the simple interest is

$$I = P \times r \times t$$
$$= \$1,000,000 \times 0.12 \times 10$$
$$= \$1,200,000$$

Of course, at the end of the 10 years, you would also get your $1,000,000 back, so you would receive $1,000,000 + $1,200,000 = $2,200,000 at simple interest.

> **Final Amount**
>
> The final amount A is given as $A = P + I$.

However, if your annual interest were calculated on your original principal *plus* all previously earned interest—that is, if your interest were **compounded annually**—you would receive the much greater amount of $3,105,848 (to the nearest dollar). You will examine compound interest later in this section.

As a consumer, you will be interested in three important applications of simple interest: loans and deposits, taxes, and discounts. These applications are considered next.

Of course, you know that borrowing money from (or depositing money in) a bank or lending institution involves interest. Here is an example.

EXAMPLE 1 ▶ Finding Interest and Total Amount

A loan company charges 32% simple interest for a 2-year, $600 loan.

(a) What is the total interest on this loan?

(b) What is the interest for 3 months?

(c) What is the total amount A that must be paid to the loan company at the end of 2 years?

Solution

(a) The interest is given by $I = Prt$, where $P = \$600$, $r = 32\% = 0.32$, and $t = 2$. Thus,

$$I = \$600 \times 0.32 \times 2 = \$384$$

On a calculator with a percent key $\boxed{\%}$, press

$$\boxed{6}\,\boxed{0}\,\boxed{0}\,\boxed{\times}\,\boxed{3}\,\boxed{2}\,\boxed{\%}\,\boxed{\times}\,\boxed{2}\,\boxed{=}$$

The interest for the 2 years is $384.

(b) Here, $P = \$600$, $r = 32\% = 0.32$, and $t = \frac{3}{12} = \frac{1}{4}$ because 3 months is $\frac{3}{12}$ of a year. Thus,

$$I = \$600 \times 0.32 \times \tfrac{1}{4}$$
$$= \$600 \times 0.08 \qquad (0.32 \times \tfrac{1}{4} = 0.08)$$
$$= \$48$$

The interest for 3 months is $48.

(c) At the end of 2 years, the loan company must be paid the original $600 plus the interest of $384; that is,

$$A = \$600 + \$384 = \$984$$

The company must be paid $984.   ∎

B. Taxes

You have probably heard the saying "There is nothing certain but death and taxes." Here is a simple problem "it is certain" you can do.

EXAMPLE 2 ▶ Finding Tax and Total Cost

A state has a 6% sales tax. Mary Rios buys an item priced at $84.

(a) What is the sales tax on this item?

(b) What is Mary's total cost for this item?

Solution

(a) The sales tax S is 6% of $84; that is,

$$S = 0.06 \times \$84 = \$5.04$$

The tax is $5.04.

(b) The total cost is the price, $84, plus the tax.

$$\$84 + \$5.04 = \$89.04$$

A calculator with a percent key $\boxed{\%}$ will give the total cost automatically if you press $\boxed{8}\ \boxed{4}\ \boxed{+}\ \boxed{6}\ \boxed{\%}\ \boxed{=}$. ∎

C. Discounts

In Examples 1 and 2, the consumer had to pay interest or taxes. But there is some hope! Sometimes you can obtain a **discount** on certain purchases. Such a discount is usually stated as a percent. For example, a coupon may entitle you to a 20% discount on certain purchases.

EXAMPLE 3 ▶ Finding Discounts and Cost

Ralph McWaters purchased a $42 item and used his coupon to get 20% off.

(a) How much was his discount?

(b) How much did he have to pay for the item?

Solution

(a) His discount rate d was 20% of $42. So in money his discount was

$$d = 0.20 \times \$42 = \$8.40$$

(b) Since he had a discount of $8.40, he had to pay

$$\$42 - \$8.40 = \$33.60$$

for the item. A calculator with a percent key $\boxed{\%}$ will obtain the final price if you press $\boxed{4}\ \boxed{2}\ \boxed{-}\ \boxed{2}\ \boxed{0}\ \boxed{\%}\ \boxed{=}$. ∎

D. Compound Interest

When interest is **compounded,** the interest is calculated not only on the original principal but also on the earned interest. For example, if you deposit $1000 in a savings account that pays 6% interest compounded annually, then in the first year, the account will earn interest calculated as

$$I = Prt$$
$$= \$1000 \times 0.06 \times 1 = \$60$$

If you make no withdrawals, then at the beginning of the second year the accumulated amount will be

$$\$1000 + \$60 = \$1060$$

which is the new principal. In the second year, this new principal will earn interest

$$I = Prt$$
$$= \$1060 \times 0.06 \times 1 = \$63.60$$

Thus, at the beginning of the third year, the accumulated amount will be

$$\$1060 + \$63.60 = \$1123.60$$

and so on.

You can see that when interest is compounded, the earned interest increases each year ($60, $63.60, and so on, as in the preceding illustration). This is so because the interest at the end of a year is calculated on the accumulated amount (principal plus interest) at the beginning of that year. Piecewise calculation of the accumulated amount is a very time-consuming procedure, but it can be avoided by developing a general formula for the amount A_n accumulated after n interest periods and the use of special tables or a calculator. To develop the formula for A_n, let I be the compound interest, P be the original principal, r be the rate per period, and A_1 be the compound amount at the end of the first period.

$$I = Pr \qquad \text{Interest for the first period}$$
$$A_1 = P + I$$
$$= P + Pr \qquad \text{Substitute } Pr \text{ for } I.$$
$$= P(1 + r) \qquad \text{Use the distributive property.}$$

After the end of the second period, the compound amount A_2 is

$$A_2 = A_1 + A_1 r$$
$$= A_1(1 + r) \qquad \text{Use the distributive property.}$$
$$= P(1 + r)(1 + r) \qquad \text{Substitute } P(1 + r) \text{ for } A_1.$$
$$= P(1 + r)^2 \qquad \text{Substitute } (1 + r)^2 \text{ for } (1 + r)(1 + r).$$

If you continue this procedure, after n periods you will derive the formula for the future value A_n.

Amount After Compounding for n Periods (Future Value A_n)

If P dollars are deposited at an interest rate r and compounded n times, the **future value** A_n is

$$A_n = P(1 + r)^n$$

Thus, if you deposit $1 at 6% compounded annually for 20 years,

$$A_{20} = \$1(1 + 0.06)^{20}$$

Fortunately, there are tables that give the value of the accumulated amount for a $1 initial deposit at compound interest r for n time periods. Table 13.2 on page 892 is such a table. To find the value of the accumulated amount

TABLE 13.2 Amount (in dollars) to Which $1 Will Grow in *n* Periods Under Compound Interest

n	1%	2%	3%	4%	5%	6%	7%	8%	9%	10%	11%	12%
1	1.0100	1.0200	1.0300	1.0400	1.0500	1.0600	1.0700	1.0800	1.0900	1.1000	1.1100	1.1200
2	1.0201	1.0404	1.0609	1.0816	1.1025	1.1236	1.1449	1.1664	1.1881	1.2100	1.2321	1.2544
3	1.0303	1.0612	1.0927	1.1249	1.1576	1.1910	1.2250	1.2597	1.2950	1.3310	1.3676	1.4049
4	1.0406	1.0824	1.1255	1.1699	1.2155	1.2625	1.3108	1.3605	1.4116	1.4641	1.5181	1.5735
5	1.0510	1.1041	1.1593	1.2167	1.2763	1.3382	1.4026	1.4693	1.5386	1.6105	1.6851	1.7623
6	1.0615	1.1262	1.1941	1.2653	1.3401	1.4185	1.5007	1.5869	1.6771	1.7716	1.8704	1.9738
7	1.0721	1.1487	1.2299	1.3159	1.4071	1.5036	1.6058	1.7138	1.8280	1.9487	2.0762	2.2107
8	1.0829	1.1717	1.2668	1.3686	1.4775	1.5938	1.7182	1.8509	1.9926	2.1436	2.3045	2.4760
9	1.0937	1.1951	1.3048	1.4233	1.5513	1.6895	1.8385	1.9990	2.1719	2.3579	2.5580	2.7731
10	1.1046	1.2190	1.3439	1.4802	1.6289	1.7908	1.9672	2.1589	2.3674	2.5937	2.8394	3.1058
11	1.1157	1.2434	1.3842	1.5395	1.7103	1.8983	2.1049	2.3316	2.5804	2.8531	3.1518	3.4785
12	1.1268	1.2682	1.4258	1.6010	1.7959	2.0122	2.2522	2.5182	2.8127	3.1384	3.4985	3.8960
13	1.1381	1.2936	1.4685	1.6651	1.8856	2.1329	2.4098	2.7196	3.0658	3.4523	3.8833	4.3635
14	1.1495	1.3195	1.5126	1.7317	1.9799	2.2609	2.5785	2.9372	3.3417	3.7975	4.3104	4.8871
15	1.1610	1.3459	1.5580	1.8009	2.0789	2.3966	2.7590	3.1722	3.6425	4.1772	4.7846	5.4736
16	1.1726	1.3728	1.6047	1.8730	2.1829	2.5404	2.9522	3.4259	3.9703	4.5950	5.3109	6.1304
17	1.1843	1.4002	1.6528	1.9479	2.2920	2.6928	3.1588	3.7000	4.3276	5.0545	5.8951	6.8660
18	1.1961	1.4282	1.7024	2.0258	2.4066	2.8543	3.3799	3.9960	4.7171	5.5599	6.5436	7.6900
19	1.2081	1.4568	1.7535	2.1068	2.5270	3.0256	3.6165	4.3157	5.1417	6.1159	7.2633	8.6128
20	1.2202	1.4859	1.8061	2.1911	2.6533	3.2071	3.8697	4.6610	5.6044	6.7275	8.0623	9.6403
21	1.2324	1.5157	1.8603	2.2788	2.7860	3.3996	4.1406	5.0338	6.1088	7.4002	8.9492	10.8038
22	1.2447	1.5460	1.9161	2.3699	2.9253	3.6035	4.4304	5.4365	6.6586	8.1403	9.9336	12.1003
23	1.2572	1.5769	1.9736	2.4647	3.0715	3.8197	4.7405	5.8715	7.2579	8.9543	11.0263	13.5523
24	1.2697	1.6084	2.0328	2.5633	3.2251	4.0489	5.0724	6.3412	7.9111	9.8497	12.2392	15.1786
30	1.3478	1.8114	2.4273	3.2434	4.3219	5.7435	7.6123	10.0627	13.2677	17.4494	22.8923	29.9599
36	1.4308	2.0399	2.8983	4.1039	5.7918	8.1473	11.4239	15.9682	22.2512	30.9127	42.8181	59.1356
42	1.5188	2.2972	3.4607	5.1928	7.7616	11.5570	17.1443	25.3395	37.3175	54.7637	80.0876	116.7231
48	1.6122	2.5871	4.1323	6.5705	10.4013	16.3939	25.7289	40.2106	62.5852	97.0172	149.7970	230.3908

$(1 + 0.06)^{20}$ in Table 13.2, go down the column headed *n* until you reach 20, and then go across to the column headed 6%. The accumulated amount given there is $3.2071. If you wish to know the accumulated amount for an original deposit of $1000 instead of $1, multiply the $3.2071 by 1000; you obtain $3207.10.

In using Table 13.2, there is a warning: *The entries in this table have been rounded to four decimal places from more accurate values.* Consequently, you should not expect answers to be accurate to more than the number of digits in the table entry. If more accuracy is needed, you must use a table with more decimal places or a calculator. To find $(1 + 0.06)^{20}$, press ⌷ ⌷1⌷ ⌷.⌷ ⌷0⌷ ⌷6⌷ ⌷)⌷ ⌷yˣ⌷ ⌷2⌷ ⌷0⌷ ⌷=⌷ and obtain 3.207135472, or 3.2071.

Many financial transactions call for interest to be compounded more often than once a year. In such cases, the interest rate is customarily stated as a nominal annual rate, it being understood that **the actual rate per interest period is the nominal rate divided by the number of periods per year.** For instance, if interest is at 18%, compounded monthly, then the actual interest rate is $\frac{18}{12}\% = 1.5\%$ per month, because there are 12 months in a year.

EXAMPLE 4 ▶ Finding the Accumulated Amount and Interest

Find the accumulated amount and the interest earned for the following:

(a) $8000 at 8% compounded annually for 5 years

(b) $3500 at 12% compounded semiannually for 10 years

Solution

(a) Here, $P = 8000$ and $n = 5$. In Table 13.2, we go down the column under n until we come to 5 and then across to the column under 8%. The number there is 1.4693. Hence, the accumulated amount will be

$8000 \times 1.4693 = \$11,754$ to the nearest dollar.

With a calculator, press $\boxed{8}\,\boxed{0}\,\boxed{0}\,\boxed{0}\,\boxed{\times}\,\boxed{(}\,\boxed{(}\,\boxed{1}\,\boxed{+}\,\boxed{.}\,\boxed{0}\,\boxed{8}\,\boxed{)}\,\boxed{y^x}\,\boxed{5}\,\boxed{=}$. The interest earned is the difference between the $11,754.63 and the original deposit; that is, $11,754.63 − $8000 = $3754.63.

(b) Because *semiannually* means twice a year, interest is compounded every 6 months, and the actual rate per interest period is the nominal rate, 12% divided by 2, or 6%. In 10 years, there are $2 \times 10 = 20$ interest periods. Hence, we go down the column under n until we come to 20 and then across to the column headed 6% to find the accumulated amount 3.2071. Since this is the amount for $1, we multiply by $3500 to get

$3500 \times 3.2071 = \$11,225$ to the nearest dollar

The amount of interest earned is $11,225 − $3500 = $7725. ∎

With a calculator, press $\boxed{3}\,\boxed{5}\,\boxed{0}\,\boxed{0}\,\boxed{\times}\,\boxed{(}\,\boxed{(}\,\boxed{1}\,\boxed{+}\,\boxed{.}\,\boxed{0}\,\boxed{6}\,\boxed{)}\,\boxed{y^x}\,\boxed{2}\,\boxed{0}$ and obtain $11,224.97. The interest earned is $11,224.97 − $3500 = $7724.97, which, rounded to the nearest dollar, yields the same answer, $7725. Note that we used $r = 12\%/2 = 6\%$.

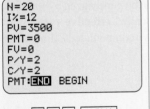

GRAPH IT

To do Example 4(b), press $\boxed{\text{APPS}}\,\boxed{1}\,\boxed{\text{ENTER}}$ and enter the values shown. *Note:* $I\% = 12$, the *annual* rate.

```
N=20
I%=12
PV=3500
PMT=0
FV=0
P/Y=2
C/Y=2
PMT:END BEGIN
```

Press $\boxed{\blacktriangle}\,\boxed{\blacktriangle}\,\boxed{\blacktriangle}\,\boxed{\text{ALPHA}}$ $\boxed{\text{ENTER}}$ to get the *FV* −$11,224.97.

Future Value for Compound Interest

If P dollars are deposited at an annual interest r, compounded m times a year, and the money is left on deposit for n periods, the **future value** (or final amount) A_n is

$$A_n = P\left(1 + \frac{r}{m}\right)^n$$

The interest can also be compounded *continuously* for n years.

> ### Future Value for Continuously Compounded Interest
>
> If P dollars are deposited and earn continuously compounded interest at an annual rate r for n years, then the **future value** A_n is
>
> $$A_n = Pe^{rn}$$

EXAMPLE 5 ▶ Finding Compound Interest

Suppose you invest \$1000 at 8%. How much interest will you earn in 5 years if the money is compounded

(a) quarterly?   (b) continuously?

Solution

(a) Here $P = \$1000$, $r = 8$, $m = 4$, $r/m = 8/4 = 2$, and $n = 4 \times 5 = 20$. From Table 13.2 with $n = 20$ and under the heading 2%, we find 1.4859. Thus, $A_5 = 1000 \times 1.4859 = \$1485.90$. The interest is $\$1485.90 - \$1000 = \$485.90$. With a calculator, press $\boxed{1}\,\boxed{0}\,\boxed{0}\,\boxed{0}\,\boxed{\times}\,\boxed{(}\,\boxed{(}\,\boxed{1}\,\boxed{+}\,\boxed{.}\,\boxed{0}\,\boxed{2}\,\boxed{)}\,\boxed{)}\,\boxed{y^x}\,\boxed{20}\,\boxed{=}$ and obtain \$1485.95.

(b) In this case, $A_n = Pe^{rn} = 1000e^{(0.08)(5)} = 1000e^{0.40}$. With a calculator with an $\boxed{e^x}$ key, press $\boxed{1}\,\boxed{0}\,\boxed{0}\,\boxed{0}\,\boxed{\times}\,\boxed{(}\,\boxed{(}\,\boxed{.}\,\boxed{4}\,\boxed{0}\,\boxed{\text{2nd}}\,\boxed{e^x}\,\boxed{)}\,\boxed{=}$ and get \$1491.82, which is a little bit more than the \$1485.95 you get when compounding quarterly. Caution: When entering quantities involving $\boxed{e^x}$ using a scientific calculator, you must enter the exponent x first and then press $\boxed{\text{2nd}}\,\boxed{e^x}$. ■

The type of comparison made in Example 5 can be generalized by looking at the **APY** (**A**nnual **P**ercentage **Y**ield, sometimes called **effective rate**) of two investments. For example, if the nominal interest is r compounded k times a year, the **future value** of \$1 is given by

$$1\left(1 + \frac{r}{k}\right)^k = \left(1 + \frac{r}{k}\right)^k$$

and the APY is your future value minus the original \$1, that is,

> ### Formula for APY (Effective Rate)
>
> $$\text{APY} = \left(1 + \frac{r}{k}\right)^k - 1 \qquad k \text{ the number of compoundings per year, } r \text{ the nominal rate}$$
>
> $$\text{APY} = e^r - 1 \qquad \text{for continuous compounding}$$

EXAMPLE 6 ▶ Comparing Investments Using APY

Find the APY for an 8% investment when the money is compounded

(a) quarterly.   (b) continuously.

Solution

(a) The nominal rate $r = 8\%$, and the number of compoundings $k = 4$. Thus,

$$\begin{aligned}
\text{APY} &= \left(1 + \frac{0.08}{4}\right)^k - 1 \\
&= (1 + 0.02)^4 - 1 \\
&= (1.02)^4 - 1 \\
&= 1.08243216 - 1 \qquad \text{Using a calculator} \\
&= 0.08243216 \\
&= 8.24\% \qquad\qquad\quad \text{To two decimal places}
\end{aligned}$$

(b) The nominal rate is still $r = 8\%$, but the interest is compounded continuously. Thus,

$$\begin{aligned}
\text{APY} &= e^r - 1 \\
&= e^{0.08} - 1 \\
&= 1.083287068 - 1 \qquad \text{Using a calculator} \\
&= 8.33\% \qquad\qquad\quad\; \text{To two decimal places} \qquad \blacksquare
\end{aligned}$$

EXAMPLE 7 ▶ Purchasing Power and Inflation

Suppose you are earning $30,000 a year. How much salary would you need to earn 10 years from now to maintain your purchasing power if the inflation rate is 2.5%?

Online Study Center

To further explore compound interest, access links 13.1.1 to 13.1.3 on this textbook's Online Study Center.

Solution

We assume that the inflation rate is growing continuously, so your $30,000 salary P in $n = 10$ years at $r = 2.5\%$ will have to amount to $A = \$30,000e^{(0.025)(10)} = 30,000e^{0.25}$. With a calculator, press $\boxed{3}\,\boxed{0}\,\boxed{0}\,\boxed{0}\,\boxed{0}$ $\boxed{\times}\,\boxed{(}\,\boxed{.}\,\boxed{2}\,\boxed{5}\,\boxed{\text{2nd}}\,\boxed{e^x}\,\boxed{)}\,\boxed{=}$ and get $38,520.76. $\blacksquare$

EXERCISES 13.1

Ⓐ Simple Interest

In problems 1–10, find the simple interest.

Principal	Rate	Time
1. $3000	8%	1 year
2. $4500	7%	1 year
3. $2000	9%	3 years
4. $6200	8%	4 years
5. $4000	10%	6 months
6. $6000	12%	4 months
7. $2500	10%	3 months
8. $12,000	9%	1 month
9. $16,000	7%	5 months
10. $30,000	8%	2 months

Ⓑ Taxes

11. The state sales tax in Florida is 6%. Desiree Cole bought $40.20 worth of merchandise.
   a. What was the tax on this purchase?
   b. What was the total price of the purchase?

12. The state sales tax in Alabama is 4%. Beto Frias bought a refrigerator priced at $666.
   a. What was the sales tax on this refrigerator?
   b. What was the total price of the purchase?

13. Have you seen the FICA (Federal Insurance Contribution Act, better known as Social Security) deduction taken from your paycheck? For 1989, the

FICA tax rate was 7.51% of your annual salary. Find the FICA tax for a person earning $24,000 a year.

14. The FICA tax for 1990 and subsequent years is 7.65% of your annual salary. Walter Snyder makes $30,000 a year. What would his FICA tax deduction be based on this rate?

According to the instructions for figuring a single person's estimated federal income tax for a recent year, the tax was 15% of the taxable income if that income was between $0 and $20,350. If the taxable income was between $20,350 and $49,300, the tax was $3052.50 plus 28% of the amount over $20,350. Use this information in problems 15 and 16.

15. Mabelle was single and had a taxable income of $25,850. How much was her estimated income tax?

16. Bob was single and had a taxable income of $9500. What was his estimated income tax?

For current tax information go to www.irs.gov or www.irs.com/.

C Discounts

17. An article selling for $200 was discounted 20%.
   a. What was the amount of the discount?
   b. What was the final cost after the discount?
   c. If the sales tax was 5%, what was the final cost after the discount and including the sales tax?

18. A Sealy mattress sells regularly for $900. It is offered on sale at 50% off.
   a. What is the amount of the discount?
   b. What is the price after the discount?
   c. If the sales tax rate is 6%, what is the total price of the mattress after the discount and including the sales tax?

19. A jewelry store is selling rings at a 25% discount. The original price of a ring was $500.
   a. What is the amount of the discount?
   b. What is the price of the ring after the discount?

20. If you have a Magic Kingdom Club Card from Disneyland, Howard Johnson's offers a 10% discount on double rooms. A family stayed at Howard Johnson's for 4 days. The rate per day was $45.
   a. What was the price of the room for the 4 days?
   b. What was the amount of the discount?
   c. If the sales tax rate was 6%, what was the total bill?

21. Some oil companies are offering a 5% discount on the gasoline you buy if you pay cash. Anita

Gonzalez filled her gas tank, and the pump registered $28.40.
   a. If she paid cash, what was the amount of her discount?
   b. What did she pay after her discount?

22. U-Mart had Comfort bicycles selling regularly for $350. The bicycles were put on sale at 20% off. The manager found that some of the bicycles were dented or scratched, and offered an additional 10% discount.

   a. What was the price of a dented bicycle after the two discounts? (Careful! This is not a 30% discount, but 20%, followed by 10%.)
   b. Would it be better to take the 20% discount followed by the 10% discount or to take a single 28% discount?

D Compound Interest

In problems 23–26, use Table 13.2 or a calculator to find the final accumulated amount and the total interest if interest is compounded annually. (Give answers to five significant digits.) Answers may vary because of rounding or calculator use.

23. $100 at 6% for 8 years

24. $1000 at 9% for 11 years

25. $2580 at 12% for 9 years

26. $6230 at 11% for 12 years

In problems 27–32, use Table 13.2 or a calculator to find the final accumulated amount and the total interest. (Give answers to the same number of digits as the table entry or to the nearest dollar.)

27. $12,000 at 10% compounded semiannually for 8 years

28. $15,000 at 14% compounded semiannually for 10 years

29. $20,000 at 8% compounded quarterly for 3 years

30. $30,000 at 12% compounded quarterly for 4 years

31. $40,000 at 20% compounded semiannually for 24 years

32. $50,000 at 16% compounded semiannually for 15 years

33. When a child is born, grandparents sometimes deposit a certain amount of money that can be used later to send the child to college. Mary and John Glendale deposited $1000 when their grandaughter Anna was born. The account was paying 6% compounded annually.
  a. How much money will there be in the account when Anna becomes 18 years old?
  b. How much money would there be in the account after 18 years if the interest had been compounded semiannually?

34. When Natasha was born, her mother deposited $100 in an account paying 6% compounded annually. After 10 years, the money was transferred into another account paying 10% compounded semiannually.
  a. How much money was in the account after the first 10 years?
  b. How much money was in the account at the end of 18 years? (Give answers to the nearest cent.)

35. Jack loaned Janie $3000. She promised to repay the $3000 plus interest at 10% compounded annually at the end of 3 years.
  a. How much did she have to pay Jack at the end of 3 years?
  b. How much interest did she pay?

36. Bank A pays 8% interest compounded quarterly; bank B pays 10% compounded semiannually. If $1000 were deposited in each bank, how much money would there be at the end of 5 years in
  a. the bank A account.
  b. the bank B account.
  c. In which bank would you deposit your money?

37. How much more would there be at the end of 5 years if $1000 were invested at 12% compounded quarterly rather than semiannually?

38. Suppose you can invest $1000 in a fund that pays 6% compounded annually or into a fund that pays 6.5% annually but does not reinvest the interest. How long would it be before the accumulated interest in the first fund exceeded the total interest paid out by the second fund? Make a table to show what happens and explain.

39. Find the compound amount when $2000 is compounded continuously at 8% for 6 months.

40. Find the compound amount when $3000 is invested at 8% compounded continuously for 5 years.

41. An investor can buy a 2-year, $10,000 certificate of deposit paying 11% compounded semiannually or place the money in an account paying $10\frac{1}{2}$% compounded continuously. Which of the two investments will yield the most interest, and how much extra interest will the investor get by making the correct choice?

42. The president of a bank is considering changing savings account interest to continuous compounding. At the present time, the bank pays 5% interest compounded daily. How much difference in interest will there be in a $1000 deposit left in an account for 180 days? How much difference in interest payments will there be in a 6-month period if the bank has $2 million deposited in savings accounts? (*Hint:* $e^{0.025} = 1.0253151$.) Assume a year has 360 days.

In Other Words

43. Discuss the difference between simple and compound interest.

44. If you increase the price of a product by 10% and then decrease that price by 10%, would the new price be the same as the original price? Explain.

In problems 45–52, find the APY (effective annual rate).

45. 6%, compounded semiannually

46. 6%, compounded monthly

47. 8%, compounded quarterly

48. 12%, compounded monthly

49. 9%, compounded quarterly

50. 15%, compounded semiannually

51. 18%, compounded monthly

52. 15%, compounded monthly

Using Your Knowledge

In problems 53–56, check the APY for the given nominal rate.

High Yield Rates for 1-Year CD

	Institution	Date	Rate	CM	APY	Min. Deposit
53.	Imperial Capital Bank, LaJolla, CA	8/2	3.98	M	4.05	2000
54.	Net Bank Alpharetta, GA	8/2	3.92	W	4.00	1000
55.	ING DIRECT Wilmington, DE	8/2	3.90	S	3.90	1
56.	Interwest Natl Bank New York, NY	8/2	3.88	D	3.95	2500

Source: www.bankrate.com.

Online Study Center

If you want a calculator to figure out your APY, access link 13.1.4 on this textbook's Online Study Center.

 Collaborative Learning

Form three groups. The objective is to answer the question: What is the rule of 70 (or 72)?

Group 1 Call a bank, an investment company, or a financial planner and ask.

Group 2 Go to the Internet and find the answer.

Group 3 Go to the library and use an encyclopedia or other reference materials to answer the question.

Discuss your findings.

If you invest P dollars at r percent compounded continuously, the amount A you will receive after t years is $A = Pe^{rt}$. If you want to double your money, $A = 2P$ and $2P = Pe^{rt}$.

1. Solve for t.

2. If $\ln 2 \approx 0.069315$, what is t?

3. How long would it take P dollars to double at
    a. 4%.   **b.** 8%.   **c.** 12%.

Can you see where the name rule of 70 (or 72) came from? Restate the rule.

13.2   Credit Cards and Consumer Credit

Online Study Center

To further explore the terminology associated with credit cards, access link 13.2.1 on this textbook's Online Study Center. In order to decide on the "lowest interest" or "no annual fee" cards, access link 13.2.2.

Everything You Always Wanted to Know About Credit Cards*

Do you have a credit card or are you planning to get one soon? You can save money if you know the **interest rate** (the percent you pay on the card balance), the **annual fee** (the amount paid for the privilege of having the card), and the **grace period** (the interest-free period between purchases and billing given to consumers who pay off their balances entirely) on your cards. The interest rate can be **fixed** or **variable** (depending on the amount you owe or some standard such as the **prime rate,** the rate banks charge their best customers). How can you save money? First, you can get a card with no annual fee for a savings of $20 to $50. Next, you can eliminate interest payments by paying off your entire balance each month. Note that if a balance of *any amount* is carried over from the previous month, most banks will charge interest from the date of each new purchase, even *before* the monthly statement arrives.

*For more information regarding credit cards, their benefits, and the amount of time or the payments needed to pay off a card, you may search the Internet.

The type of card you should have to maximize savings depends on your monthly balance. If you plan to pay your balance in full each month, the annual fee is the largest expense. Get a card with no annual fee and the longest grace period available. If you plan to have a high monthly balance, choose a card with a low interest rate. (For example, if your average balance is $1000, you will pay $180 annually on an 18% annual percentage rate (APR) card but only $120 on a 12% APR card, a savings of $180 − $120, or $60.) Now, suppose you already have a credit card. How can you decide on your best course of action? Follow these steps.

1. Find the annual fee and interest rate on your card.

2. Look at your past statements and find your average monthly balance.

3. Figure your annual cost by multiplying the interest rate by the monthly balance and add the annual fee.

Here is an example. Suppose you have a card with a $25 annual fee and an 18% APR. If the average monthly balance on your card is $500 and you can get a card with a 14% APR and no annual fee, you will be saving 4% (18% − 4%) of $500, or $20, in interest and the $25 annual fee, a total of $45. On the other hand, if your annual fee is $20 and your APR is 14%, changing to another card with no annual fee and 19% APR makes no sense. (You pay 14% of $500, or $70, plus the $20 fee, $90 in all with the first card and 19% of $500, or $95, with the other.) How do you find the interest rate and annual fee on credit cards? Find a consumer magazine or an organization (such as BankCard Holders of America) that publishes the latest information regarding annual fees and interest rates for credit cards, or look in the World Wide Web (Internet). ▶

To obtain a credit card, you have to meet requirements set by the institution that issues the card. These requirements will vary but are illustrated in the next example. Moreover, banks sometimes have promotions in which credit cards are offered under other special conditions.

Company A will issue a credit card to an applicant who meets the following requirements:

1. The applicant must have a good credit rating.*

2. If single, the applicant must have a gross annual income of at least $30,000.

3. If married, the couple's combined gross annual income must be at least $40,000.

EXAMPLE 1 ▶ Qualifying for a Credit Card

Using the three preceding requirements, determine which of the following applicants qualifies for a credit card from company A:

(a) Annie Jones is single, has a good credit rating, and earns $25,000 per year.

(b) John Smith has an excellent credit rating and earns $35,000 per year. His wife has no paying job.

---

*Your credit rating is usually determined by a credit bureau, an organization that tracks the history of individuals' spending and repayment habits.

(c) Don and Daryl Barnes each earn $24,000 per year and have a good credit rating.

(d) Bill Spender is single, earns $40,000 per year and has only a fair credit rating.

Solution

Only Don and Daryl Barnes meet the requirements of company A. They have a good credit rating, and their combined annual income is $48,000. ■

One of the costs associated with credit cards is the **finance charge** that is collected if you decide to pay for your purchases later than the allowed payment (grace) period. Usually, if the entire balance is paid within a certain period of time (25–30 days), there is no charge. However, if you want more time, then you will have to pay the finance charge computed at the rate printed on the monthly statement you receive from the company issuing the card. Figure 13.1 shows the top portion of such a statement. As you can see, the periodic (monthly) rate is 1.5%. This rate is used to calculate the charge on $600. Where does the $600 come from? The back of the statement indicates that "The Finance Charge is computed on the Average Daily Balance, which is the sum of the Daily Balances divided by the number of days in the Billing Period." Fortunately, the computer calculated this average daily balance and came out with the correct amount. In Example 2, we shall verify only the finance charge.

EXAMPLE 2 ▶ Minimum Payments and Finance Charges

Find the finance charge (interest) to be paid on the statement in Figure 13.1 if the monthly rate is 1.5% computed on the average daily balance of $600.

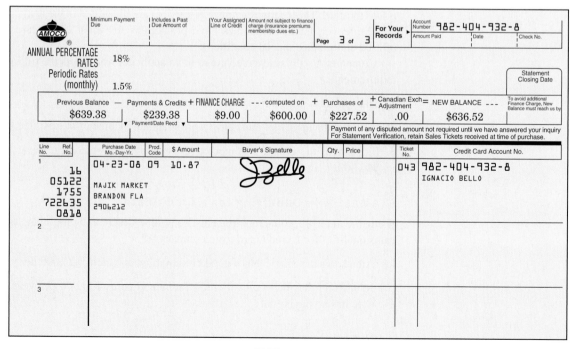

FIGURE 13.1

Solution

The finance charge is 1.5% of $600, that is,

$$0.015 \times \$600 = \$9$$

Next, let us look at a different problem. Suppose you wish to obtain a credit card. First, you have to apply to the issuing bank for such a card (sometimes the bank will preapprove you). If your application is accepted, then you must pay a fee. (Most credit unions and banks issue cards free.) After some time, the card finally arrives in the mail. Now suppose you wish to use your card at a restaurant where these cards are accepted. Instead of collecting cash, the cashier will "swipe" your card in a machine that will print on a receipt certain information that is encoded in the black strip on the back of the card: the card number, your name, and the expiration date of the card; it will also print the name and identification numbers of the restaurant as well as the date of the transaction and an authorization code issued by the bank owning the card. You will sign the receipt and be given a copy for your records. Figure 13.2 shows such a receipt.

At the end of the billing period, a statement is sent to you. If the balance due is $10 or less, you must pay the account *in full*. Otherwise, you must make a minimum payment of $10 or 5% of the balance due, *whichever is greater*. (Terms vary from one bank to another.)

EXAMPLE 3 ▶ **Finding Finance Charges**

The customer who signed the receipt in Figure 13.2 received a statement at the end of the month. The new balance was listed as $37.50. Find the following:

(a) The minimum payment due

(b) The finance charge that will be due the next month if only the minimum payment is made now

FIGURE 13.2

Solution

(a) Because the new balance is $37.50 and 5% of $37.50 is $1.875, the minimum payment is $10. (Remember, you pay $10 or 5% of the new balance, whichever is greater.)

(b) After paying the $10, the customer's new balance is

$37.50 − $10 = $27.50

The finance charge is 1.5% of $27.50, or

$0.015 \times \$27.50 = \$.4125$

Thus, if no additional credit card purchases are made, the finance charge will be $.41.

Note: The finance charge is *not* the only profit made by the credit card company. Credit card companies charge the businesses that accept your credit card 1% to 3% of each sale. (Rates vary and are based on the average charge per transaction.) ■

Many large department stores prefer to handle their own credit card business. This procedure offers the following two main advantages to the stores:

1. The stores save the commission on sales that the national credit card companies charge for their services.

2. The interest (finance charges) collected from their customers is a welcome source of revenue to the stores.

Most charge accounts at department stores are called **revolving charge accounts.** Although the operational procedure for these accounts is similar to that employed by the national credit card companies, there may be some differences between them, as noted in the following list:

1. The interest for revolving charge accounts is $1\frac{1}{2}\%$ to $1\frac{3}{4}\%$ per month on the unpaid balance for balances under $500. If the balance is over $500, some accounts charge only 1% interest per month on the amount over $500.

2. The minimum monthly payment may be established by the department store, and it may or may not be similar to that of the national credit card companies.

EXAMPLE 4 ▶ Finance Charges, Balance, and Payments

Mary Lewis received her statement from Sears, where she has a revolving charge account. Her previous balance was $225.59, and she charged an additional $288.09 to her account. She also had $105.97 in credits. Find the following:

(a) The finance charge for the month (Sears charges interest on the average daily balance of $222.95, as described in the form shown in Figure 13.3.)

(b) The new balance

(c) The minimum monthly payment

Solution

(a) Since the average daily balance was $222.95, the finance charge is $1\frac{3}{4}\%$ (21% annual rate) of the $222.95, that is,

$$0.0175 \times \$222.95 = \$3.91 \qquad \text{(rounded up)}$$

SEARS Card
BOX 34577
Anytown, US 12345

Ms. Mary Lewis
123 Anywhere Street
Hometown, US 55555

Account Number:	12 34567 89012 3
Billing date:	January 13

ACCOUNT SUMMARY

Previous balance		$225.59
Total charges	+	288.09
Total credits	−	105.97
FINANCE CHARGE	+	3.91
New balance		$411.62
Scheduled payment:		35.00
Minimum due:		**35.00**
Due date:		**February 12**

TRANSACTIONS
Dec 15 PANT, 1 PC DRESS, 2PC DRESS, 2PC DRESS........$98.96
Dec 16 SHIRT, SHIRT, 2PC DRESS, 1PC DRESS,
       1PC DRESS, EARS, EARRS 14K........................ $178.02
Dec 16 PARTS, REPAIRS.. 6.88
Dec 16 SEASONAL/OUTDOOR SHOP............................. 4.23
Dec 17 *CREDIT - RETURN, CREDIT - RETURN..................−43.98*
Dec 17 *BLOUSE,*
       CREDIT - CANCELLATION: LADIES DRESSES,
       CREDIT - CANCELLATION: LADIES
       SPORTSWEAR
       CREDIT - CANCELLATION: LADIES
       SPORTSWEAR,
       NET AMOUNT FOR TRANSACTION....................−31.99
Dec 31 *CREDIT - RETURN, CREDIT - RETURN.................−30.00*

HELPFUL INFORMATION
Available Credit: **$5,445**

If the amount of Available Credit is not sufficient, or you have a question, call:
1−800−000−0000
M - S 9-9, SUN 12-5 ET

Mail any billing error notices to:
BOX 35065
Realtown, US 88888
Please include your account number with any correspondence.

SEARS BONUS CLUB
CONGRATULATIONS! You have earned a Bonus Certificate worth $3.33 in savings at Sears. You will find it enclosed with this statement.

FINANCE CHARGE SCHEDULE	Average Daily Balance	ANNUAL PERCENTAGE RATE	Monthly Periodic Rate	Average Daily Balance	FINANCE CHARGE
All	Over $0.00	21.0%	1.75%		
	$222.95			Total	$3.91

NOTE: See other side for important information

FIGURE 13.3

(continued)

On a Sears Revolving Charge Account your monthly payments decrease as your account balance
decreases . . . and, likewise your monthly payments increase as your account balance increases.
Payments are flexible with your balance, as shown on the table below. Any premiums for Group
Insurance, for which you may have contracted, other than Accidental Death and Disability
Insurance, are in addition to your minimum monthly payments.

New Balance	Minimum Payment
$ .01 to $ 10.00	Balance
10.01 to 200.00	$10.00
200.01 to 250.00	15.00
250.01 to 300.00	20.00
300.01 to 350.00	25.00
350.01 to 400.00	30.00
400.01 to 450.00	35.00
450.01 to 500.00	40.00
Over $500.00	1/10 of New Bal.

Source: Courtesy of Sears, Roebuck & Co.

FIGURE 13.3 *continued*

(b) The new balance is calculated as follows:

Previous balance	$225.59
Finance charge	3.91
New purchases	288.09
Total credits	−105.97
New balance	$411.62

(c) The minimum monthly payment is found by using the information given at
the bottom of Figure 13.3. (This is a copy of the table that appears on the
back of the statement.) Because the new balance is between $400.01 and
$450, the minimum payment is $35. ■

Some stores charge interest on the unpaid balance. How do we find the new
balance under this method? The next example will tell you.

EXAMPLE 5 ▶ Finance Charges and New Balance

Ms. Spoto received a statement from a department store where she has a charge
account. Her previous balance was $280. She made a $20 payment and charged
an additional $30.12 to her account. If the store charges 1.5% of the unpaid bal-
ance, find the following:

(a) The finance charge   (b) The new balance

Solution

(a) The unpaid balance is $280 − $20 = $260, so the finance charge is 1.5% of
$260, which is 0.015 × $260 = $3.90.

(b) The new balance is computed as follows:

Unpaid balance	$260.00
Finance charge	3.90
Purchases	30.12
New balance	$294.02

 ■

There is another way of charging interest when consumers buy on credit, the
add-on interest used by furniture stores, appliance stores, and car dealers. For
example, suppose you wish to buy some furniture costing $2500 and you make

a $500 down payment. The amount to be financed is $2000. If the store charges a 10% add-on rate for 5 years (60 monthly payments), the interest will be

$$I = Prt$$
$$= \$2000 \times 0.10 \times 5 = \$1000$$

Thus, the total amount to be paid is $2000 + $1000 = $3000. The monthly payment is found by dividing this total by the number of payments.

$$\text{Monthly payment} = \frac{\$3000}{60} = \$50$$

Note that the add-on interest is charged on the *entire* $2000 for the 5 years, but the customer does *not* have the full use of the entire amount for the 5 years. It would be fairer to charge interest on the *unpaid balance only.*

EXAMPLE 6 ▶ Total Interest and Monthly Payments

A used car costing $8500 can be bought with $2500 down and a 12% add-on interest rate to be paid in 48 monthly installments. Find the following:

(a) The total interest charged   (b) The monthly payment

Solution

(a) The amount to be financed is $8500 − $2500 = $6000. The interest is 12% of $6000 for 4 years. Thus,

$$\text{Interest} = 0.12 \times \$6000 \times 4 = \$2880$$

(b) Total amount owed = $6000 + $2880 = $8880

$$\text{Monthly payment} = \frac{\$8880}{48} = \$185 \qquad ■$$

Now suppose you receive this notice in the mail:

> Limited-time 0% APR
> Your credit line has been increased to $26,200!

with this fine print added at the end:

> Effective on or after the first day following your statement closing date in October, the Daily Periodic Rate (DPR) for new cash advances and for new purchases posting to your account through your April statement closing date is 0% (corresponding **ANNUAL PERCENTAGE RATE (APR)** of 0%). Thereafter, the DPR for these promotional cash advance balances will be .035589% (corresponding **APR** of 12.99%), and the DPR for these promotional purchase balances will be .035589% (corresponding **APR** of 12.99%). **Important Reminder:** The transaction fee for credit card access checks, including the enclosed checks, is 3% of each transaction (Min. $5, Max. $50). See your Credit Card Agreement for any other applicable transaction fees.

If you have two credit cards with a $1000 and a $500 balance, respectively, each charging 9% APR, should you take this deal? Before you do, you should read the fine print. The limited time at 0% is 6 months (from the end of October to April) and after that your interest rate on the new card will be 12.99% (which we round to 13%). Let us compare the total amounts you would pay.

Online Study Center

To further explore how credit card companies calculate your credit card interest each month, access link 13.2.3 on this textbook's Online Study Center. For more topics about credit cards, access link 13.2.4.

PROBLEM SOLVING

Comparing Credit Card Options

❶ **Read** the problem.

We have two credit cards now. We have to compare the total amount these two cards cost with the total amount we would pay for a new card at 0% interest for 6 months and 13% thereafter.

❷ **Select** the unknown.

We have two unknowns: (1) the amount to pay on the two old (9%) credit cards and (2) the amount to pay on the new 13% credit card with the limited 0% interest.

❸ **Think** of a plan.

Let us figure the amount we have to pay for the two 9% credit cards and the amount we have to pay for the new 13% card and compare the results.

❹ **Use** the techniques you are studying to carry out the plan.

Since the balance on the old cards is $1500($1000 + $500), find the amount we pay on these two cards when $n = 2$ and $r = 9\%$. In Table 13.2, go to the 9% column and the $n = 2$ row. The entry there is 1.1881; thus, the amount to be paid is

$$1500 \cdot 1.1881 = \$1782.15$$

```
N=1.5
I%=13
PV=1500
PMT=0
FV=0
P/Y=1
C/Y=1
PMT:END BEGIN
```

Or use the formula for the amount $A = 1500(1 + 0.09)^2 = \$1782.15$.

For the new card, you do not pay interest for 6 months and then pay 13% for 18 months. To find the amount with a grapher, press APPS 1 ENTER and enter the values on the screen in the margin. As usual, press ▲ ▲ ▲ ALPHA ENTER to get FV, the future value of the loan, as $1801.81, or use $A = 1500(1 + .13)^{1.5}$ = **$1801.81**. Thus, this new card with no interest for 6 months and with a 13% interest thereafter is more expensive: $1801.81 versus $1782.15.

But there is more. To pay off the two 9% cards you have to write two checks, one for $1000 and one for $500. The **Important Reminder** tells you that the fee is 3% of each transaction—that is, $30 and $15—for an additional $45 cost. Definitely, stay with the old cards unless you intend to pay off the new card in 6 months at 0% interest.

❺ **Verify** the answer.

The verification (with a calculator if you wish) is left to you.

Cover the solution, write your own solution, and then check your work.

EXERCISES 13.2

1. Chad and Susan Johnson are married and have a good credit rating. Chad earns $25,000 per year. What is the least that Susan must earn for the couple to qualify for a card from company A? (See Example 1.)

2. Jim and Alice Brown are married and have an excellent credit rating. They both work and earn the same salaries. What is the least that each salary must be for the Browns to qualify for a card from company A? (See Example 1.)

In problems 3–7, find the new balance, assuming that the bank charges $1\frac{1}{2}\%$ per month on the unpaid balance.

	Previous Balance	Payment	New Purchases
3.	$100	$10	$50
4.	$300	$190	$25
5.	$134.39	$25	$73.98

	Previous Balance	Payment	New Purchases
6.	$145.96	$ 55	$44.97
7.	$378.93	$ 75	$248.99

In problems 8–17, find the following:

a. The finance charge for the month
b. The new balance
c. The minimum monthly payment

Use the following rates and payments table:

Monthly Rate	$1\frac{1}{2}\%$	1%
Unpaid balance	Up to $500	Over $500
New balance	Under $200	Over $200
Minimum payment	$10	5% of new balance

	Previous Balance	New Purchases
8.	$ 50.40	$173
9.	$ 85	$150
10.	$154	$ 75
11.	$344	$ 60
12.	$666.80	$ 53.49
13.	$ 80.45	$ 98.73
14.	$ 34.97	$ 50
15.	$ 55.90	$ 35.99
16.	$ 98.56	$ 45.01
17.	$ 34.76	$ 87.53

18. Phyllis Phillips has a revolving charge account that charges a finance charge on the unpaid balance using the following schedule:

$1\frac{1}{2}\%$ per month of that portion of the balance up to $300
1% per month on that portion of the balance over $300

If the previous month's balance was $685, find the finance charge.

19. Daisy Rose has a credit card that charges a finance charge on the previous balance according to the following schedule:

2% per month on balances up to $100

$1\frac{1}{2}\%$ per month on balances between $100 and $200
1% per month on balances of $200 or over

If the previous month's balance was $190, find the finance charge.

20. Mr. Dan Dapper received a statement from his clothing store showing a finance charge of $1.50 on a previous balance of $100. Find the monthly finance charge rate.

21. Paul Peters received a statement from the ABC Department Store showing a previous balance of $90. If the ABC store's finance charge is 1.5% on the previous balance, find the finance charge for the month.

22. In problem 21, if the monthly rate were 1.25%, what would be the finance charge for the month?

23. A $9000 used car can be purchased with $1600 down, the balance plus a 9% add-on interest rate to be paid in 36 monthly installments. Find
a. the total interest charged.
b. the monthly payment, rounded to the nearest dollar.

24. The Ortegas move into their first apartment and decide to buy furniture priced at $400 with $40 down, the balance plus 10% add-on interest to be paid in monthly installments in 1 year. Find
a. the total interest charged.
b. the monthly payment.

25. Wayne Pinski wishes to buy a stove and a refrigerator from an appliance dealer. The cost of the two items is $2400, and Wayne pays $400 down and finances the balance at 15% add-on interest to be paid in 18 monthly installments. Find
a. the total interest charged.
b. the monthly payment, to the nearest dollar.

26. Bill Seeker bought a boat costing $8500 with $1500 down, the balance plus add-on interest to be paid in 36 monthly installments. If the add-on interest rate was 18%, find
a. the total interest charged.
b. the monthly payment, to the nearest dollar.

27. Felicia Johnson bought a freezer costing $500 on the following terms: $100 down and the balance plus a 10% add-on interest rate to be paid in 18 monthly installments. Find
a. the total interest to be paid by Ms. Johnson.
b. the amount of her monthly payment, to the nearest dollar.

28. Cissie owes $1000 to a department store that charges a monthly interest rate of 1.5% on the unpaid balance. Cissie considers paying off this debt at the rate of $200 at the end of each month for 5 months and then paying off the balance at the end of the sixth month. She also considers making payments of $200 plus the month's interest at the end of each month for 5 months. Make a table showing the monthly payments and the interest under each scheme. How much would Cissie save by using her second scheme? How do you account for this savings? Explain fully.

In Other Words

29. An article in *Money* magazine states, "If you pay in full each month, you can get the most from the grace period by making big credit purchases just after your statement closing date and paying your bill on time at the last minute." Explain why.

30. The same article says to "buy bigger ticket items toward the end of the billing cycle and pay as much of the bill as possible as soon as you get your bill." Explain why.

Using Your Knowledge

A table shows that a monthly payment of $61 (to the nearest dollar) for 18 months will repay $1000 with interest at 1% per month on the unpaid balance. Can we find the equivalent add-on interest rate? Yes; here is how to do it. Eighteen payments of $61 make a total of $1098 (18 × $61), which shows that the total inter-

est paid is $98. As this is the interest for 18 months ($1\frac{1}{2}$, or $\frac{3}{2}$, years), the equivalent add-on interest rate is

$$\frac{98}{(1000)(\frac{3}{2})} = \frac{98}{1500} \approx 0.065$$

or 6.5%, to the nearest tenth of a percent.

A table shows that the following monthly installment payments will repay $1000 in the stated term and at the stated rate of interest on the unpaid balance. Find the equivalent add-on interest rate to the nearest tenth of a percent.

	Monthly Payment	Term	Rate per Month on Unpaid Balance
31.	$47	2 years	1%
32.	$64	18 months	$1\frac{1}{2}$%
33.	$50	2 years	$1\frac{1}{2}$%
34.	$48	2 years	$1\frac{1}{4}$%

Web It Exercises

The questions in this section require access to the Internet. For help with exercises 1–3, access links 13.2.3 or 13.2.4 on this textbook's Online Study Center.

1. Find and discuss the difference between a "bank card," a "travel and entertainment card," and a "house card."

2. What are secured, unsecured, guaranteed, and debit cards?

3. How do credit companies calculate your credit card interest each month?

13.3 Annual Percentage Rate (APR) and the Rule of 78

Truth-in-Lending: APR to Z

In the preceding section we studied several types of consumer credit: credit cards, revolving charges, and add-on interest. Before 1969, it was almost impossible to compare the different types of credit accounts available to consumers. In an effort to standardize the credit industry, the government enacted the federal Truth-in-Lending Act of 1969. A key feature of this law is the inclusion of the **total payment,** the **amount financed,** and the **finance charges** in credit contracts. In conjunction with this law, the Board of Governors of the Federal

Reserve System issued Regulation Z requiring all lenders that make consumer loans to disclose certain information regarding the cost of consumer credit.

How can we compare loans? To do so, two items are of crucial importance: the **finance charge** and the **annual percentage rate (APR).** A look at annual percentage rates will enable us to compare different credit options.

For example, suppose you can borrow $200 for a year at 8% add-on or get the same $200 by paying $17.95 each month. Which is the better deal? In the first instance, you borrow $200 at 8% add-on, which means that you pay 8% of $200, or $16, in finance charges. The charge you pay per $100 financed is $\frac{16}{200} \times 100 = \$8$. On the other hand, if you pay $17.95 per month for 12 months, you pay a total of $215.40. Here the finance charge per $100 financed is $\frac{15.40}{200} \times 100 = \$7.70$. Obviously, the second loan is a better deal.

Can we find the APR for each loan? To help in doing so, tables have been prepared so that we can translate the finance charge per $100 to the APR (see Table 13.3). In this section we shall discuss the APR and one of the methods that will enable us to get a refund on our interest in case we decide to pay off our loan early. ▶

As you have seen in Sections 13.1 and 13.2, there are many ways of stating the interest rates used to compute credit costs. A few examples are 12% *simple interest,* 12% *compounded annually,* 12% *add-on interest,* and 1% *per month on the unpaid balance.* How can you compare various credit costs? Without some help, it is difficult to do this. In response, Congress enacted the Truth-in-Lending Act on July 1, 1969. This law helps the consumer to know exactly what credit costs. Under this law, all sellers (car dealers, banks, credit card companies, and so on) must disclose to the consumer

1. the finance charge.

2. the *annual percentage rate* (APR).

TABLE 13.3 True Annual Interest Rate (APR)

Number of Payments	14%	$14\frac{1}{2}$%	15%	$15\frac{1}{2}$%	16%	$16\frac{1}{2}$%	17%	$17\frac{1}{2}$%	18%
6	$ 4.12	$ 4.27	$ 4.42	$ 4.57	$ 4.72	$ 4.87	$ 5.02	$ 5.17	$ 5.32
12	7.74	8.03	8.31	8.59	8.88	9.16	9.45	9.73	10.02
18	11.45	11.87	12.29	12.72	13.14	13.57	13.99	14.42	14.85
24	15.23	15.80	16.37	16.94	17.51	18.09	18.66	19.24	19.82
30	19.10	19.81	20.54	21.26	21.99	22.72	23.45	24.18	24.92
36	23.04	23.92	24.80	25.68	26.57	27.46	28.35	29.25	30.15
42	27.06	28.10	29.15	30.19	31.25	32.31	33.37	34.44	35.51
48	31.17	32.37	33.59	34.81	36.03	37.27	38.50	39.75	41.00

Note: Numbers in the body of the table are finance charges per $100 of amount financed.

A. APR

Recall that the finance charge is the total dollar amount you are charged for credit. It includes interest and other charges such as service charges, loan and finder's fees, credit-related insurance, and appraisal fees. The **annual percentage rate (APR)** is the charge for credit stated as a percent.

In general, the lowest APR corresponds to the best credit buy regardless of the amount borrowed or the period of time for repayment. For example, suppose you borrow $100 for a year and pay a finance charge of $8. If you keep the entire $100 for the whole year and then pay $108 all at one time, then you are paying an APR of 8%. On the other hand, if you repay the $100 plus the $8 finance charge in 12 equal monthly payments (8% add-on), you do not have the use of the $100 for the whole year. What, in this case, is your APR? The formulas needed to compute the APR are rather complicated, and as a consequence, tables such as Table 13.3 have been constructed to help you find the APR. These tables are based on the cost per $100 of the amount financed. To use Table 13.3, you must first find the finance charge per $1 of the amount financed and then multiply by 100. Thus, to find the APR on the $100 borrowed at 8% add-on interest and repaid in 12 equal payments of $9, first find the finance charge per $100 as follows:

1. The finance charge is $108 − $100 = $8.

2. The charge per $100 financed is

$$\frac{\text{Finance charge}}{\text{Amount financed}} \times 100 = \frac{\$8}{\$100} \times 100 = \$8$$

Since there are 12 payments, look across the row labeled 12 in Table 13.3 until you find the number closest to $8. This number is $8.03. Then read the heading of the column in which the $8.03 appears to obtain the APR. In this case, the heading is $14\frac{1}{2}\%$. Thus, the 8% add-on rate is equivalent to a $14\frac{1}{2}\%$ APR. (Of course, Table 13.3 gives the APR only to the nearest $\frac{1}{2}\%$.)

EXAMPLE 1 ▶ APR on Furniture

Mary Lewis bought some furniture that cost $1400. She paid $200 down and agreed to pay the balance in 30 monthly installments of $48.80 each. What was the APR for her purchase?

Solution

We first find the finance charge per $100 as follows:

Payments	$30 \times \$48.80 =$	$1464
Amount financed		−1200
Finance charge		$ 264

$$\text{Finance charge per } \$100 \qquad \frac{\$264}{\$1200} \times \$100 = \$22$$

We now turn to Table 13.3 and read across the row labeled 30 (the number of payments) until we find the number closest to $22. This number is $21.99. We then read the column heading to obtain the APR, 16%. ■

Note: There is also a formula to approximate the APR. The formula is discussed in problems 37–46.

GRAPH IT

Want to find the exact interest with your grapher? Go to the TVM solver by pressing
APPS 1 1 . Next, enter the values for *N*(30), *I%*(0), *PV*(1200), *PMT*(−48.80), *FV*(0), *P/Y*(12), *C/Y*(12), and END . Move the cursor up to *I%*. Press ALPHA ENTER . The APR is about 16%, as shown by the *I%*.

```
N=30
I%=16.00944902
PV=1200
PMT=-48.8
FV=0
P/Y=12
C/Y=12
PMT:END BEGIN
```

B. The Rule of 78

In all the preceeding examples it has been assumed that the consumer will faithfully make the payments until the debt is satisfied. But what if you wish to pay in full before the final due date? (Perhaps your rich aunt gave you some money.)

In many cases you are entitled to a partial refund of the finance charge! The problem is to find how much you should get back. One way of calculating the refund is to use the **rule of 78.** This rule assumes that the final payment includes a portion, say, a, of the finance charge, the payment before that includes $2a$ of the finance charge, the second from the final payment includes $3a$ of the finance charge, and so on. If the total number of payments is 12, then the finance charge is paid off by the sum of $a + 2a + 3a + 4a + 5a + 6a + 7a + 8a + 9a + 10a + 11a + 12a = 78a$ dollars. If the finance charge is F dollars, then

$$78a = F$$

so $a = \frac{1}{78}F$. This is the reason for the name "rule of 78." Now suppose you borrow $1000 for 1 year at 8% add-on interest. The interest is $80, and the monthly payment is one-twelfth of $1080, that is, $90. If you wish to pay off the loan at the end of 6 months, are you entitled to a refund of half the $80 interest charge? Not according to the rule of 78. Your remaining finance charge payments, according to this rule, are

$$\tfrac{1}{78}F + \tfrac{2}{78}F + \tfrac{3}{78}F + \tfrac{4}{78}F + \tfrac{5}{78}F + \tfrac{6}{78}F = \tfrac{21}{78}F$$

Since $F = \$80$, you are entitled to a refund of $\frac{21}{78} \times \$80$, or $21.54. There are six payments of $90 each for a total of $540, so you would need to pay $540 − $21.54 = $518.46 to cover the balance of the loan.

Notice that to obtain the numerator of the fraction $\frac{21}{78}$, we had to add $1 + 2 + 3 + 4 + 5 + 6$. If there were n payments remaining, then to find the numerator, we would have to add

$$1 + 2 + 3 + \cdots + (n - 2) + (n - 1) + n$$

There is an easy way to do this. Let us call the sum S. Then we can write the sum S twice, once forward and once backward.

$$S = 1 + \quad 2 \quad + \quad 3 \quad + \cdots + (n - 2) + (n - 1) + n$$
$$S = n + (n - 1) + (n - 2) + \cdots + \quad 3 \quad + \quad 2 \quad + 1$$

If we add these two lines, we get

$$2S = (n + 1) + (n + 1) + (n + 1) + \cdots + (n + 1) + (n + 1) + (n + 1)$$

and because there are n terms on the right,

$$2S = n(n + 1)$$
$$S = \frac{n(n + 1)}{2}$$

Thus, for $n = 6$, we obtain

$$S = \frac{6 \times (6 + 1)}{2} = \frac{6 \times 7}{2} = 21$$

as before. For $n = 12$, we find $S = (12 \times 13)/2 = 78$, which again agrees with our previous result.

Now suppose the loan is for 15 months, and you wish to pay it off in full after 10 payments, so that there are 5 payments remaining. By arguing in the same way as for the 12-month loan, you can see that you are entitled to a refund of a fraction a/b of the finance charge, where the numerator is

$$a = 1 + 2 + 3 + 4 + 5 = \frac{5 \times 6}{2} = 15$$

and the denominator is

$$b = 1 + 2 + 3 + \cdots + 15 = \frac{15 \times 16}{2} = 120$$

Thus, you are entitled to $\frac{15}{120} = \frac{1}{8}$ of the total finance charge.

In general, if the loan calls for a total of n payments and the loan is paid off with r payments remaining, then the unearned interest is a fraction a/b of the total finance charge, with the numerator

$$a = 1 + 2 + 3 + \cdots + r = \frac{r(r + 1)}{2}$$

and the denominator

$$b = 1 + 2 + 3 + \cdots + n = \frac{n(n + 1)}{2}$$

Thus,

$$\frac{a}{b} = \frac{r(r + 1)}{n(n + 1)} \qquad \text{The 2s cancel.}$$

Thus, the unearned interest u is as shown in the box.

Formula for the Unearned Interest u

$$u = \frac{r(r + 1)}{n(n + 1)} \times F$$

where F is the finance charge.

For example, if an 18-month loan is paid off with 6 payments remaining, the amount of unearned interest u is

$$u = \frac{6 \times 7}{18 \times 19} \times F = \frac{7}{57}F$$

Although the denominator is no longer 78 (except for a 12-payment loan), the rule is still called the rule of 78.

EXAMPLE 2 ▶ Refunds and Payoffs

Cal Olleb purchased a television set on a 15-month installment plan that included a $60 finance charge and called for payments of $25 monthly. If Cal decided to pay off the loan at the end of the eighth month, find

(a) the amount of the interest refund u using the rule of 78.

(b) the amount needed to pay off the loan.

Solution

(a) Here, $n = 15$ and $r = 7$. We substitute into the formula to obtain

$$u = \frac{7 \times 8}{15 \times 16} \times \$60 = \frac{7}{30} \times \$60 = \$14$$

(b) There are 7 payments of $25 each left, that is, $175. Thus, Cal needs

$$\$175 - \$14 = \$161$$

to pay off the loan.   ■

There is another way of calculating refunds and payoffs: by using a formula.

Actuary Formulas for the Refund u and the Payoff

The refund u is given by

$$u = \frac{r \cdot \text{PMT} \cdot V}{100 + V}$$

where r is the number of remaining payments, PMT is the payment, and V is the value from the APR table corresponding to r.

The payoff is given by

$$\text{PMT}\left[\frac{1 - (1 + i)^{-r}}{i}\right]$$

where PMT is the payment, r is the number of remaining payments, and $i = \dfrac{\text{APR}}{12}$.

EXAMPLE 3 ▶ Refunds and Payoffs Using Formulas

Refer to Example 1, where Mary Lewis bought furniture costing $1400 with $200 down and 30 payments of $48.80. Assume that Mary wants to pay off the loan after 24 payments. Find

(a) the refund using the rule of 78. (b) the refund using the formula.

(c) the payoff using the rule of 78. (d) the payoff using the formula.

Solution

(a) Using the rule of 78, the refund is

$$u = \frac{r\,(r + 1)}{n\,(n + 1)} \times F$$

where $r = 30 - 24 = 6$, $n = 30$, and $F = \$264$ ($30 \times \$48.60 - \$1200$). Thus, the refund using the rule of 78 is

$$u = \frac{6\,(6 + 1)}{30\,(30 + 1)} \times 264$$
$$= \frac{42}{930} \times 264$$
$$= \$11.92$$

(b) The refund using the formula is

$$u = \frac{r \cdot \text{PMT} \cdot V}{100 + V}$$

where $r = 6$, PMT $= \$48.80$, and V is the value from the APR table corresponding to 6 and an APR of 16% (recall that the APR in Example 1 was 16%). This value is $4.72. Thus, the refund is

$$u = \frac{6 \cdot 48.80 \cdot 4.72}{100 + 4.72} = \$13.20$$

(c) The payoff using the rule of 78 is

$$r \cdot \text{PMT} - u = 6 \cdot \$48.80 - \$11.92 = \$280.88$$

(d) The payoff using the formula is

$$\text{PMT} \cdot \left[\frac{1 - (1 + i)^{-r}}{i} \right]$$

where $i = \dfrac{\text{APR}}{12}$. Since in Example 1 the APR $= 16\%$,

$$i = \frac{\text{APR}}{12} = \frac{0.16}{12}$$

and the payoff is

$$\$48.80 \cdot \left[\frac{1 - \left(1 + \dfrac{0.16}{12}\right)^{-6}}{\dfrac{0.16}{12}} \right] = \$279.60$$

∎

C. Applications

Realistically, there are more factors associated with loans than the APR and the rule of 78. In most cases you need a calculator to do the work! Access link 13.3.1 on this textbook's Online Study Center; then select "Calculators." We illustrate the use of such calculators in Example 4.

EXAMPLE 4 ▶ Monthly Payments from the Net

Suppose that the purchase price of a car is $15,000. There is no cash rebate, your trade-in is $4000, you do not owe any money on your trade-in, the down payment is $2000, and you want to finance the car at 10% for 36 months. What is your monthly payment?

Solution

Select "autos" and under "Auto Calculators" choose "How much will my monthly payment be?" Enter the information as shown in Figure 13.4

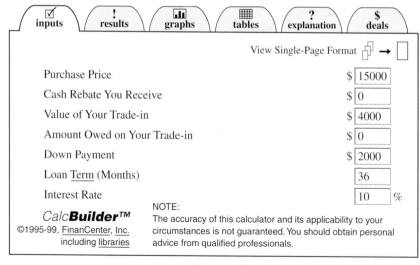

FIGURE 13.4
How much will my monthly payment be?

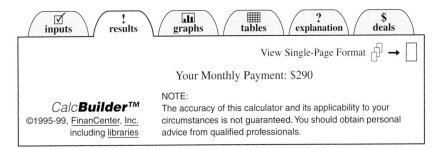

FIGURE 13.5

As you can see in Figure 13.5, your monthly payment will be $290. By the way, you can compare monthly payments and total interest paid if you select "Which term of loan should I choose?" under "autos." You can even figure what car you can afford by selecting the topic! ■

EXERCISES 13.3

Ⓐ APR

In problems 1–10, find the APR.

	Amount Financed	Finance Charge	Number of Payments
1.	$2500	$ 194	12
2.	$2000	$ 166	12
3.	$1500	$ 264	24
4.	$3500	$ 675	24
5.	$1500	$ 210	18
6.	$4500	$1364	36
7.	$4500	$1570	48
8.	$4000	$ 170.80	6
9.	$5000	$1800	48
10.	$4000	$ 908.80	30

Ⓑ The Rule of 78

In problems 11–15, find the following:
  a. The unearned finance charge
  b. The amount needed to pay off the loan

	Finance Charge	Number of Payments	Frequency	Amount	Number of Payments Left
11.	$15.60	12	Monthly	$25	4
12.	$23.40	12	Monthly	$35	5
13.	$31.20	12	Monthly	$45	6
14.	$52.00	18	Weekly	$10	9
15.	$58.50	20	Weekly	$10	5

In problems 16–19, find
  a. the refund using the rule of 78.
  b. the refund using the formula.
  c. the payoff using the rule of 78.
  d. the payoff using the formula.

Amount Financed	Finance Charge	No. of Payments	Payments Left
16. $2500	$194	12	6
17. $1500	$264	24	6
18. $1500	$210	18	12
19. $4500	$1570	48	12

(*Hint:* The APR was found in problems 1, 3, 5, and 7.)

20. Alfreda Brown bought a car costing $6500 with $500 down and the rest to be paid in 48 equal installments of $173.
   a. What was the finance charge?
   b. What was the APR?

21. Gerardo Norega bought a dinette set for $300, which he paid in 12 monthly payments of $27.
   a. What was the finance charge?
   b. What was the APR on this sale?

22. Yu-Feng Liang bought a used car for $6500. He made a down payment of $1000 and paid off the balance in 48 monthly payments of $159 each.
   a. What was the finance charge?
   b. What was the APR?

23. A used sailboat is selling for $1500. The owner wants $500 down and 18 monthly payments of $63.
   a. What finance charge does the owner have in mind?
   b. What is the APR for this transaction?

24. Natasha Gagarin paid $195 interest on a $2000 purchase. If she made 12 equal monthly payments to pay off the account, what was the APR for this purchase?

25. Virginia Osterman bought a television set on a 12-month installment plan that included a $31.20 finance charge and called for payments of $50 per month. If she decided to pay the full balance at the end of the eighth month, find the following:
   a. The interest refund
   b. The amount needed to pay off the loan

26. Marie Siciliano bought a washing machine on a 12-month installment plan that included a finance charge of $46.80 and called for monthly payments of $70. If Marie wanted to pay off the loan after 7 months, find the following:
   a. The interest refund
   b. The amount needed to pay off the loan

27. A couple buys furniture priced at $800 with $80 down and the balance to be paid at 10% add-on interest. If the loan is to be repaid in 12 equal monthly payments, find
   a. the finance charge.
   b. the monthly payment.
   c. the interest refund if the couple decides to pay off the loan after 8 months.
   d. the amount needed to pay off the loan.

28. Dan Leizack is buying a video recorder that costs $1200. He paid $200 down and financed the balance at 15% add-on interest to be repaid in 18 monthly payments. Find the following:
   a. The finance charge
   b. The monthly payment
   c. The interest refund if he pays off the loan after 9 months
   d. The amount needed to pay off the loan

29. Joe Clemente bought a stereo costing $1000 with $200 down and 10% add-on interest to be paid in 18 equal monthly installments. Find
   a. the finance charge.
   b. the monthly payment.
   c. the interest refund if he pays off the loan after 15 months.
   d. the amount needed to pay off the loan.

ⓒ Applications

The following information will be used in problems 30–35. (Internet access is required. See link 13.3.1 on this textbook's Online Study Center for this section.) If you are buying a car, one of the factors to consider (but by no means the only one!) is the monthly payment. What are your options? Assume you are buying a $20,000 car, your trade-in is worth $2000, you have $1000 to give down, the sales tax is 6%, and you want to finance the car for 36 months at 8%.

30. a. What is the total purchase price?
   b. What is the monthly payment?
   c. What down payment do you need so that the monthly payment is about $510?

31. You can lower the monthly payment by getting a lower interest rate. What is the payment if the interest rate is
   a. 7%?   **b.** 6%?

32. On the basis of your answers to problems 30 and 31,
   a. by how much is the payment lowered when you lower the rate by 1%?
   b. if you want a monthly payment of about $536, what rate of interest will you have to negotiate?

33. What is the total cost to purchase the car when your finance rate is
   a. 4%?   **b.** 5%?   **c.** 6%?
   d. By about how much is the total cost to purchase increased when the rate is increased by 1%?

34. An alternative plan is to lease the car. Under the same assumptions as before and assuming the residual value of the car at the end of 3 years is $12,400, what is the monthly payment for leasing the car?

35. What is the total cost of a car if you lease it for three years and buy it at its residual value at the end of the 3 years?

 In Other Words

36. Do you think the rule of 78 gives the debtor a fair break? In your own words, explain why or why not.

 Using Your Knowledge

Approximating the APR by Formula The APR for the loans just discussed can also be approximated (however, not within the $\frac{1}{4}$ of 1% accuracy required by Regulation Z*) by using the following formula:

$$\text{APR} = \frac{2mI}{P(n+1)}$$

where m = the number of payment periods per year
   I = the interest (or finance charge)
   P = the principal (amount financed)
   n = the number of periodic payments to be made

*Regulation Z limits the use of this formula for approximating APRs to the "exceptional instance where circumstances may leave a creditor with no alternative."

Thus, in Example 1, $m = 12$, $I = \$264$, $P = \$1200$, $n = 30$, and

$$\text{APR} = \frac{2mI}{P(n+1)} = \frac{2 \times 12 \times 264}{1200 \times 31} = 17\%$$

In problems 37–46, use the APR formula to find the APR to one decimal place in the specified problem. In each case, state the difference between the two answers.

37. Problem 1   **38.** Problem 2

39. Problem 3   **40.** Problem 4

41. Problem 5   **42.** Problem 6

43. Problem 7   **44.** Problem 8

45. Problem 9   **46.** Problem 10

 Web It Exercises

The exercises in this section require access to the Internet.

1. Look at Example 4 and discuss some expenses that have been omitted from the calculation.

2. Instead of buying a car you can lease (rent) one but you should understand several new terms before you do so. Discuss the meaning and importance of the following terms applicable to leasing a car:
   a. Residual value
   b. Sales tax

3. Suppose you wish to buy a $15,000 car to be financed at 7% for 36 months with a $1000 down payment and a $2000 trade-in. Assume the sales tax is 6% of the car price, the residual value is the price of the $15,000 car less 3 years depreciation, and you will buy the leased car for its residual price at the end of the 3 years.
   a. What is the depreciation? (Access link 13.3.1 on this textbook's Online Study Center.)
   b. What is the total cost to purchase the leased car?
   c. What is the total cost to purchase the other car?
   d. Which is cheaper?

4. On the basis of the results in the exercises above, discuss the benefits of leasing versus purchasing a car.

13.4   Buying a House

Houses: How Much Down? How Much a Month?

As for most people, the single largest credit purchase (and investment) of your life will be buying a house. This purchase will require many decisions in what may be unfamiliar areas. This section will help you make these decisions wisely. The first question is: how much house can you afford? Look at the rules on the next page to figure this out. Next, what type of loans are available? See the discussion after Example 1 and the Using Your Knowledge section of Exercises 13.4. The information there is accurate, except for the rates. What will be the difference in the monthly payment amount if the interest rate is 1% higher? From Table 13.4 on page 920, you will see that for each $1000 borrowed, the payment difference between a 7%, 30-year loan (first row, last column) and an 8% loan (third row, last column) is $7.32 − $6.64 = $.68. Now, this does not seem like much, but if you are borrowing $100,000, the difference in your monthly payment will be $68! Can you figure out how much the payment difference per $1000 willbe on a 30-year loan if the rate is lowered from 13% to 12%? (Current ratesmay be lower than those appearing in the Using Your Knowledge section of Exercises 13.4.)

Of course, your monthly payment should not be the only consideration when buying a house. You must consider the amount of **down payment** required, the **interest rate,** the number of **years taken to pay off** the loan, whether there are any **penalties** for prepayment (paying the loan early), the **loan application fee** covering the cost of appraisal and credit report ($250–$300), and one of the most overlooked items when buying a home: the **closing costs.** These may include, but not be limited to, any or all of the following:

> **Points** are fees lenders charge to increase profits. *Each point is equivalent to 1% of the loan amount*, so if you are borrowing $80,000 and are paying 1 point, you have to pay $800 at closing.
>
> The **loan origination fee** is typically 1% of the loan amount ($700 on a $70,000 loan).
>
> **Private Mortgage Insurance (PMI)** is insurance to protect the lender and is required if your down payment is less than 20% of the purchase price. Fees vary, but typically first year premiums are 1% of the loan amount with a 5% down payment, 0.4% with a 10% down payment, and 0.3% with a 15% down payment, plus a monthly fee. You must prepay the first year's premium at closing, or you can decide to make a lump sum payment of 2.95% of the mortgage amount with a 10% down payment or 2.30% with a 15% down payment.

What else must be paid at closing? First monthly mortgage payment, title search and title insurance, property survey, deed recording, document preparation, homeowners insurance, prorated property taxes, and lawyers' fees. These expenses can add $2000–$4000 to the immediate price of the home. In this section you will learn more of the details! ▶

Many home owners say that buying a house is both a harrowing experience and a rewarding one at the same time.

One of the first, if not the first, decision you must make when buying a house is how much to spend. Certain rules of thumb are used as guidelines in order to help people decide how much they should spend on a home. The following are three such rules:

Guidelines for Purchasing a Home

1. Spend no more than 2 to $2\frac{1}{2}$ times your annual income.

2. Limit housing expenses (the amount going to the mortgage payment) to 1 week's pay out of each month's gross pay (before deductions).

3. Do not let the amount of the monthly payment of principal, interest, taxes, and insurance exceed 28% of your monthly gross pay.

EXAMPLE 1 ▶ Can You Afford It?

John and Pat Harrell are graduate students and earn a total of $35,000 annually. Can they afford an $80,000 home with a $60,000 mortgage that requires monthly payments of $710 including principal, interest, taxes, and insurance (PITI)?

Solution

The following are the maximum amounts they can spend according to the three criteria given above:

1. $2.5 \times \$35,000 = \$87,500$

2. If the Harrells earn $35,000 annually, in 1 week they earn

$$\frac{\$35,000}{52} = \$673.08$$

Online Study Center

To further explore how to calculate the amount you can borrow, access links 13.4.1 to 13.4.3 on this textbook's Online Study Center.

3. The Harrells' gross pay is $35,000. Their monthly gross pay is

$$\frac{\$35,000}{12} = \$2916.67 \text{ per month}$$

and 28% of $2916.67 = $816.67.

Thus, the Harrells qualify under the first and third criteria but not under the second. Of course, they must come up with the $20,000 down payment! ■

Now that you know how much house you can afford, you need to borrow money to buy it. How do you do that? By getting a **mortgage loan,** a contract in which the lender agrees to lend you money to buy a specific house or property. The contract creates a lien (a charge against the property making it security for payment) and you, in turn, agree to repay the money according to the terms of the contract. There are many different types of mortgage loan plans (see the Using Your Knowledge section of Exercises 13.4); two of these will be considered here: **conventional loans** and **Federal Housing Authority (FHA) loans** (Figure 13.6 on page 920).

Conventional loans are arranged between you and a private lender. In these loans, the **amount of the down payment,** the **repayment period,** and the **interest rate** are agreed on by the **borrower** and the **lender.** The lender usually

Current Mortgage Rates		
Conventional Loans ($50k–$322k)		
Program	**Rate**	**APR**
15-year fixed	5.250%	5.628%
30-year fixed	5.875%	6.215%
Jumbo Loans ($323k–$650k)		
Program	**Rate**	**APR**
15-year fixed	5.500%	5.895%
30-year fixed	6.250%	6.765%
FHA Loans ($30k–$261k)		
Program	**Rate**	**APR**
15-year fixed	5.750%	6.345%
30-year fixed	6.000%	6.485%

Source: www.123borrow.com/.

FIGURE 13.6

Online Study Center

Note: Interest rates vary. To get the best current rates in your state, access link 13.4.4 on this textbook's Online Study Center.

Online Study Center

To find the maximum for your area, visit 13.4.5 on this textbook's Online Study Center.

requires taxes and insurance to be paid in advance through a reserve (**escrow**) account. Lenders sometimes require borrowers to pay for private mortgage insurance (PMI) if the down payment is less than 20% of the loan amount. In addition, they may require that the buyer do the following:

1. Be steadily employed and a resident of the state in which the property is located

2. Have enough savings to make one or two mortgage payments

3. Have the necessary down payment in hand (not borrowed)

Maximum amounts for conventional loans are set by individual lenders. Loans up to 80% of the value of the property are quite common, and loans of 90 to 95% can often be obtained.

FHA loans are made by private lenders and are insured by the Federal Housing Administration. The FHA does *not* make loans; it simply insures the lender against loss in case you, the borrower, fail to repay the loan in full. To pay expenses and cover this insurance, the FHA charges (at closing) an insurance premium of 3.8% of the loan for 30-year loans, 2.4% for 15-year loans, or an additional 0.5% is added to the interest rate to pay the insurance. To qualify for an FHA loan, a buyer must have the following:

1. A total housing expense less than 29% of the buyer's gross income

2. Total monthly payments (all debts with 12 or more payments plus total housing expenses) less than 41% of the gross income

Interest rates on these loans are sometimes 1 or 2 percentage points lower than conventional loan rates. The FHA loan maximum for a single-family dwelling depends on its location and range from $172,632 to $312,895.

These limits were established by the FHA Down Payment Simplification Act of 2002, which requires that the borrower makes a cash investment of at least 3% of the purchase price (or appraised value). This 3% can be a down payment or a combination of 2.25% down payment and 0.75% closing costs. The maximum FHA loans are based on the location of the property, as shown in Table 13.4.

TABLE 13.4 FHA Maximum Loan Values by State

Low Closing Cost States	*High Closing Cost States*
Arizona, California, Colorado, Guam, Idaho, Illinois, Indiana, New Mexico, Nevada, Oregon, Utah, Virgin Islands, Washington, Wisconsin, Wyoming	All other states
98.75% if price is ≤$50,000	98.75% if price is ≤$50,000
97.65% if price is $50,001–$125,000	97.75% if price is >$50,000
97.15% if price is >$125,000	

EXAMPLE 2 ▶ Conventional and FHA Loans

A family wishes to buy a $164,000 house in a state with high closing costs.

(a) If a conventional lender is willing to loan the family 90% of the price of the house, what will be the amount of the loan?

(b) What will be the down payment with that loan?

(c) If the family decides to obtain an FHA loan instead, what will be the minimum cash investment?

(d) What will be the maximum FHA loan the family can get?

Solution

(a) 90% of $164,000 = $147,600

(b) $164,000 − $147,600 = $16,400 (which is 10% of $164,000)

(c) With an FHA loan, in a high-closing-cost state, the minimum cash investment the family will have to pay down is 3% of $164,000 = $4920. Why didn't we use 97.75% of $164,000 = $160,310, making the down payment $3690 ($164,000 − $160,310)? Because $3690 is less than the required $4920 minimum cash investment of 3%.

(d) The maximum FHA loan the family can get is

$164,000 − $4920 = $159,080   ■

Here is the procedure we shall use to determine the maximum loan amount and cash investment for an FHA loan.

Finding Down Payment and FHA Loan Amount

Step 1. Calculate the minimum cash investment: 3% of price.
Step 2. Calculate the FHA down payment:

Acquisition price − FHA loan amount

Down payment: Higher value between steps 1 and 2
Loan amount: Acquisition price − down payment

EXAMPLE 3 ▶ Finding Down Payment and FHA Loan Amount

Marcus McWaters is buying a $100,000 home in Florida (a high-closing-cost state) and paying $2500 in closing costs.

(a) What is the minimum cash investment?

(b) What is the maximum FHA loan he can get?

Solution

(a) The minimum cash investment is 3% of $100,000, or $3000.

(b) ***Step 1.*** The minimum cash investment is $3000.
   Step 2. The FHA down payment is

Acquisition price − FHA loan amount
  = ($100,000 + $2500) − (97.75% of $100,000)
  = $102,500 − $97,750
  = $4750

Step 3. The down payment is \$4750 (the higher value in steps 1 and 2). The loan amount is

Acquisition price − down payment
= \$102,500 − \$4750
= \$97,750 ∎

The last item we shall discuss in connection with mortgages is the **actual amount** of the monthly payment. This amount depends on three factors.

1. The amount borrowed

2. The interest rate

3. The number of years taken to pay off the loan

Table 13.5 shows the monthly payments for \$1000 borrowed at various rates and for various times. To figure the actual monthly payment, find the appropriate interest rate and the payment period, and then multiply the amount shown in Table 13.5 by the number of thousands of dollars borrowed. Thus, to figure the monthly payment on an \$80,000 mortgage at 10% for 30 years, look down the column for 30 years until you come to the row labeled 10%. The amount per \$1000 is \$8.78. Multiply this amount by 80 (there are 80 thousands in \$80,000) to obtain \$702.40 for the required monthly payment.

TABLE 13.5 Monthly Payments (Principal and Interest) for Each \$1000 Borrowed

Interest Rate (%)	Payment Period				
	10 Years	*15 Years*	*20 Years*	*25 Years*	*30 Years*
5	10.61	7.91	6.60	5.85	5.37
$5\frac{1}{2}$	10.85	8.17	6.88	6.14	5.68
6	11.10	8.44	7.16	6.44	6.00
$6\frac{1}{2}$	11.36	8.71	7.46	6.75	6.32
7	11.60	8.97	7.75	7.05	6.64
$7\frac{1}{2}$	11.86	9.26	8.04	7.37	6.98
8	12.12	9.54	8.35	7.70	7.32
$8\frac{1}{2}$	12.38	9.83	8.66	8.04	7.67
9	12.67	10.14	9.00	8.39	8.05
$9\frac{1}{2}$	12.94	10.44	9.32	8.74	8.41
10	13.22	10.75	9.65	9.09	8.78
$10\frac{1}{2}$	13.49	11.05	9.98	9.44	9.15
11	13.78	11.37	10.32	9.80	9.52
$11\frac{1}{2}$	14.06	11.68	10.66	10.16	9.90
12	14.35	12.00	11.01	10.53	10.29
$12\frac{1}{2}$	14.64	12.33	11.36	10.90	10.67
13	14.93	12.65	11.72	11.28	11.06
$13\frac{1}{2}$	15.23	12.98	12.07	11.66	11.45
14	15.53	13.32	12.44	12.04	11.85

(continued)

TABLE 13.5 *continued*

Interest Rate (%)	Payment Period				
	10 Years	15 Years	20 Years	25 Years	30 Years
$14\frac{1}{2}$	15.83	13.66	12.80	12.42	12.25
15	16.13	14.00	13.17	12.81	12.64
$15\frac{1}{2}$	16.44	14.34	13.54	13.20	13.05
16	16.75	14.69	13.91	13.59	13.45
$16\frac{1}{2}$	17.06	15.04	14.29	13.98	13.85
17	17.38	15.39	14.67	14.38	14.26

GRAPH IT

To solve part (c), press
[APPS] [1] [1], enter the values for $N(360)$, $I(7\%)$, $PV(-95{,}500)$, $PMT(0)$, $FV(0)$, $P/Y(12)$, and $C/Y(12)$, and enter END. Now press
[▲][▲][▲][▲] [ALPHA] [ENTER].
The payment is given as $632.04. For part (e), enter the values $N = 360$, $I = 9\%$, $PV = -97{,}000$, $PMT = 0$, $FV = 0$, $P/Y = 12$, and $C/Y = 12$ and enter END. Finally, press [▲][▲][▲][▲] [ALPHA] [ENTER]. The payment is given as $780.48.

EXAMPLE 4 ▶ Payments and FHA Loans

Athanassio and Gregoria Pappas wish to obtain a 30-year loan to buy a $100,000 house in Tarpon Springs, Florida, a high-closing-cost state.

(a) If they can get a loan of 95% of the value of the house, what is the amount of the loan?

(b) What will be the down payment with that loan?

(c) If the interest rate is 7%, what will be the monthly payment?

(d) What will be the minimum cash investment with an FHA loan?

(e) If the FHA loan carries 9% interest, what will be the monthly payment?

Solution

(a) 95% of $100,000 = $95,000

(b) $100,000 − $95,000 = $5000

(c) We read from Table 13.5 that the amount per $1000 on a 30-year loan at 7% is $6.64. Thus, the monthly payment will be

$95 \times \$6.64 = \$630.80$ The mortgage loan is for $95,000.

(d) With an FHA loan, the minimum cash investment will be

3% of $100,000 = $3,000

Note that 97.75% of $100,000 = $97,750, leaving a down payment of $2250, less than the 3% required minimum cash investment.

(e) In Table 13.5 we find the amount per $1000 on a 30-year loan at 9% to be $8.05. The amount to be financed is $97,000 ($100,000 − $3000). Thus, the monthly payment will be

$97 \times \$8.05 = \$780.85$

Note: This is not the entire payment because for an FHA loan, interest, insurance, and taxes must be added to this amount. ■

Online Study Center

To further explore Example 4, access link 13.4.6 on this textbook's Online Study Center.

What if we don't have Table 13.5 to find the monthly payment? There is a formula that will do it for us. Here it is, but be warned that you need a calculator to use it!

> **Formula to Find the Monthly Payment for a Loan**
>
> The monthly payment M for a loan of P dollars for n months at monthly rate i is
>
> $$M = \frac{Pi}{1 - (1 + i)^{-n}}$$

For example, the monthly payment M on the 30-year, 9%, $97,000 loan of part (e) is

$$M = \frac{Pi}{1 - (1 + i)^{-n}}$$

Here, $P = 97{,}000$, $i = \frac{9\%}{12} = 0.0075$, and $n = 12 \times 60 = 360$ months. Substituting, we obtain

$$M = \frac{(97{,}000)(0.0075)}{1 - (1 + 0.0075)^{-360}}$$

Using a calculator,

$$M = \$780.48$$

This figure is very close to the $780.85 we obtained in part (e) but be very careful when using your calculator! Make sure that you enter the denominator as $[1 - (1 + 0.0075)^{-360}]$.

EXAMPLE 5 ▶ Finding the Monthly Payment Using the Formula

Use the formula to find the monthly payment M on a $175,000 loan at 6% for 15 years.

Solution
In this case, $P = \$175{,}000$, $i = \frac{6\%}{12} = 0.005$, and $n = 12 \times 15 = 180$ months. Substituting,

$$M = \frac{(175{,}000)(0.005)}{1 - (1 + 0.005)^{-180}}$$

$$= \$1476.75$$

Your screen should look like the one in the margin where the -180 is entered by pressing the gray key $\boxed{(-)}$, indicating the additive inverse, or negative, of 180.

175000*.005/(1-(
1+0.005)^(-180))

                1476.749449

EXERCISES 13.4

1. A family has a $40,000 annual salary. Can they afford an $80,000 house with a $70,000 mortgage requiring payments of $750 per month, including principal, interest, taxes, and insurance?
   a. Use the first criterion given in the text.
   b. Use the second criterion given in the text.
   c. Use the third criterion given in the text.

2. A family earns $36,000 annually. Can they afford a $95,000 house with a $60,000 mortgage requiring monthly payments of $570, including principal, interest, taxes, and insurance?
   a. Use the first criterion given in the text.
   b. Use the second criterion given in the text.
   c. Use the third criterion given in the text.

3. The Browning family of Colorado wants to buy a $77,000 house.
   a. If they can get a loan of 80% of the value of the house, what is the amount of the loan?
   b. What will be the down payment on this loan?
   c. If they decide to obtain an FHA loan, what will be the minimum cash investment? (Do not forget that the maximum FHA loan for this location has to be determined using Table 13.4 on page 920.)

4. The Scotdale family of Arizona wants to buy a $60,000 house.
   a. If they can get a conventional loan of 95% of the purchase price, what will be the amount of the loan?
   b. What will be the down payment with this loan?
   c. If they use an FHA loan, what will be the minimum cash investment?

In problems 5–10, find the total monthly payment, including taxes and insurance, for the given mortgage loan.

	Amount	Rate	Time (Years)	Annual Taxes	Annual Insurance
5.	$60,000	6%	20	$800	$360
6.	$80,000	$6\frac{1}{2}$%	30	$1200	$380
7.	$90,000	9%	25	$1200	$960
8.	$80,000	$9\frac{1}{2}$%	20	$1400	$740
9.	$173,000	$5\frac{1}{2}$%	30	$2400	$1200
10.	$80,000	$10\frac{1}{2}$%	15	$1000	$390

In problems 11–14, find
   a. the minimum cash investment.
   b. the maximum FHA loan amount.

	Sale Price	State
11.	$45,000	Ajo, Arizona
12.	$150,000	Orlando, Florida
13.	$75,000	Eola, Oregon
14.	$95,000	Bath, Maine

15. Assume that the buyer in problem 11 is paying closing costs of $1200.
   a. What is the acquisition cost?
   b. What is the maximum FHA loan amount?
   c. Use the formula to find the monthly payment if the loan is at 6% for 15 years.

16. Assume that the buyer in problem 12 is paying closing costs of $6000.
   a. What is the acquisition cost?
   b. What is the maximum FHA loan amount?
   c. Use the formula to find the monthly payment if the loan is at 6% for 30 years.

17. Assume that the buyer in problem 13 is paying a promotionally reduced $100 in closing costs.
   a. What is the acquisition cost?
   b. What is the maximum FHA loan amount?
   c. Use the formula to find the monthly payment if the loan is at $7\frac{1}{2}$% for 25 years.

18. Assume that the buyer in problem 14 is paying a reduced $500 in closing costs.
   a. What is the acquisition cost?
   b. What is the maximum FHA loan amount?
   c. Use the formula to find the monthly payment if the loan is at 9% for 10 years.

In problems 19–22, use Table 13.4 on page 920.

19. The Aikido family wants to obtain a conventional loan for 30 years at 11%. Suppose it finds a lender that will lend 95% for the $60,000 house it has selected, and its taxes and insurance amount to $1500 per year.
   a. What will be their down payment on the loan?
   b. What will be their total monthly payment, including taxes and insurance?

20. The Perez family is planning to buy a $90,000 house. Suppose the family gets a loan of 80% of the price of the house, and this is a 25-year loan at 10%.

   a. What will be the family's down payment on the loan?

   b. If the family's taxes and insurance amount to $810 annually, what will be the monthly payment, including taxes and insurance?

21. The Green family of Kansas obtained a 30-year FHA loan at $6\frac{1}{2}\%$ to buy a $100,000 house. The family made the minimum required down payment, and taxes and insurance amounted to $2400 annually.

   a. What was the family's down payment?

   b. What was the total monthly payment?

22. A family was planning to buy a $95,000 house with an FHA loan carrying $9\frac{1}{2}\%$ interest over a 20-year period. If the family could get the largest possible loan, $75,000 for this location, and taxes and insurance amounted to $1200 annually, find

   a. the family's down payment.

   b. the total monthly payment.

23. The Bixley family has a $50,000 mortgage loan at 10% for 30 years.

   a. What is the family's monthly mortgage payment?

   b. How many payments will the family have to make in all?

   c. What is the total amount the family will pay for principal and interest?

   d. What is the total interest the family will pay?

   e. If the loan was 80% of the price of the house, is the price more or less than the total interest?

24. The Peminides have a $35,000 mortgage loan at 9% for 30 years.

   a. What is their monthly mortgage payment?

   b. How many payments will they have to make in all?

   c. What is the total amount they will pay for principal and interest?

   d. If their loan was 80% of the price of the house, is the price more or less than the total interest?

25. If you think house prices are high, there is bad news! The costs mentioned so far are not all-inclusive. You also have to pay **closing costs.** These costs include various fees and are usually paid at the time of closing, that is, when the final mortgage contract is signed. They are in addition to the agreed-on down payment. The following are some typical closing costs for a $50,000 house with the buyer making a 20% down payment (prices vary):

Credit report fee	$ 45	
$\frac{3}{12}$ estimated taxes		To escrow
of $600	$150	account
Insurance premium		
for 1 year	$300	To escrow
$\frac{2}{12}$ insurance premium	$ 50	account
Title insurance	$220	
Mortgage recording fee	$ 20	
Loan fee, 1 point		
(1% of loan amount)	$?	
Total closing costs	$?	

   a. What would be the total cash payment, down payment plus closing costs, at the time of closing?

   b. If the buyer had to make escrow account (an account maintained by the mortgage company and used to pay property taxes and insurance) deposits each month for taxes and insurance, what would be the combined monthly payment under mortgage terms of 14% for 30 years?

   c. Suppose the lender agreed to add the closing costs to the loan amount instead of asking for cash. What would be the combined monthly payment with the same terms as in part (**b**)?

Online Study Center

To further explore typical closing costs, access link 13.4.1 on this textbook's Online Study Center.

26. The following are some different closing costs for a $75,000 house with a 10% down payment (prices vary):

Credit report fee	$ 45
Mortgage recording fee	$ 15
Lot survey	$250
Loan fee (1.5 percent	
of loan amount)	$?
Insurance premium for 1 year	$210
Total closing costs	$?

   a. What would be the total cash payment, down payment plus closing costs, at the time of closing?

   b. If no escrow account was required, what would be the monthly payment for a 25-year, 10% loan?

c. If the lender added the closing costs to the loan amount instead of asking for cash, what would be the monthly payment under the same terms as in part (**b**)?

27. The following are some closing costs for a $120,000 home with 20% down (prices vary):

Credit fee	$ 45
Mortgage recording fee	$ 25
Plot plan	$250
1-point loan fee	$?
Title insurance	$350
Total closing costs	$?

a. Find the cash payment, down payment plus closing costs, at the time of closing.
b. If the buyer had to make escrow account deposits each month for $1200 taxes and insurance, what would be the combined monthly payment under mortgage terms of 9% for 30 years?
c. Suppose the lender agrees to add the closing costs to the loan amount instead of asking for cash. What would be the combined monthly payment with the same terms as in part (**b**)?

28. The following are some closing costs for a $150,000 home with 25% down (prices vary):

Credit fee	$ 45
Mortgage recording fee	$ 25
Lot survey	$300
$1\frac{1}{2}$-point loan fee	$?
Title insurance	$420
Total closing costs	$?

a. Find the cash payment, down payment plus closing costs, at the time of closing.
b. If the buyer had to make escrow account deposits each month for $1500 taxes and insurance, what would be the combined monthly payment under mortgage terms of 8% for 20 years?
c. Suppose the lender agrees to add the closing costs to the loan amount instead of asking for cash. What would be the combined monthly payment with the same terms as in part (**b**)?

The following information about closing costs for an $84,000 house with a 10% down payment will be used in problems 29–30:

2 points	2% of loan contract
Appraisal and credit report	$235 conventional loan ($200 FHA)
Recording fee	$ 25
Title insurance	$295

29. Assume you are getting a 15-year conventional loan with a 10% interest rate. Use the given information to find the following:
a. The total cash payment, down payment plus closing costs, at the time of closing
b. The monthly payment (assuming no escrow account)

30. Assume you are getting a 30-year FHA loan with an 8% interest rate. Use the given information to find the following:
a. The total cash payment, down payment plus closing costs, at the time of closing
b. The monthly payment (assuming no escrow account)

 In Other Words

31. Write in your own words the advantages and disadvantages of an FHA loan.

32. Write in your own words the advantages and disadvantages of a conventional loan.

33. Suppose you obtain a $100,000 conventional loan and finance it at 9% for 30 years. Which is greater, the price of the house or the interest you pay on the loan? Answer the same question for a $50,000 loan. Explain your answers.

 Using Your Knowledge

There are so many different types of loans available that we cannot discuss all the possible financing alternatives you may have when you buy a house. The accompanying information derived from *Money* magazine might help you to make some sense out of the existing confusion. Most of the types of loans mentioned are still used, but interest rates are much *lower* in many cases!

A Gallery of Loans   The chart on page 928 will help you shop for the housing loan that's best for you. Your choice should be determined by your income and

Type of Loan	Typical Minimum Down Payment	Initial Interest Rate (Rates Vary)	Interest Rate After 5 Years	Who Should Consider This Type Loan
Fixed rate Conventional	10–20%	5–8%	Unchanged	High-income people who believe interest rates won't drop much
Graduated payment	5%	5–8%	Unchanged	People who feel certain their incomes will rise substantially
Growing equity	10%	5–7%	Unchanged	Borrowers who can afford high payments and want to pay off their loans early
Adjustable rate Typical adjustable rate	10%	$5–6\frac{1}{2}$%	Unknown	People who expect interest rates to drop
Dual rate	10%	5–7%	Unknown	Borrowers who, in return for lower monthly payments, are willing to give some of their equity to lenders if interest rates rise
Balloon payment	20%	5–7%	Loan is usually repaid by then	Borrowers who believe they'll be able to refinance their loans at lower rates in the future
Equity sharing Shared equity	10%	7–8%	Unchanged	People who are willing to give investors part of their houses' tax benefits and future appreciation in return for help in raising down payments and making monthly loan payments
Partnership mortgage	10%	7–8%	Unchanged	Low-income people who are willing to give investors most of the tax deductions a home generates and some of the future appreciation in exchange for down payments and help in monthly payments
Inflation indexed Adjustable balance	10%	$6\frac{1}{2}$%	Unknown	Borrowers who are confident that their incomes will keep pace with inflation

your expectations about inflation and interest rates. As they rise and fall, so will rates for inflation-indexed and adjustable-rate mortgages. Fixed-rate loans don't fluctuate, but they can be expensive. All the loans listed are widely available, except for the adjustable-balance mortgages. They have been offered so far only in Utah, but they may become more common in the future if inflation worsens.

34. Find out the current rates for the type loans in the chart. To find some of these rates, access link 13.4.8 on this textbook's Online Study Center and click on your state.

 Web It Exercises

The questions in this section require access to the Internet.

1. Discuss the following questions:
  a. By how many years would a mortgage be reduced if you make biweekly rather than monthly payments? What does the answer depend on?
  b. Which is better, biweekly payments or extra money in each regular payment?

Now, do the problems and find out some facts. Assume you are considering a $100,000 mortgage at 8% for 30 years. For help with your answers, access link 13.4.9 on this textbook's Online Study Center.

2. For the $100,000 mortgage at 8% for 30 years,
  a. what would the payment be?
  b. what is the total amount of interest you will pay?
  c. if you decide to make biweekly payments, how many years will it take to pay off the entire mortgage?

3. Which do you think is better, to make biweekly payments or to pay $60 extra on each payment? (*Hint:* Find out how many years it will take to pay off the entire mortgage when you add the extra $60 in each payment.)

4. If you want to consider a 15-year mortgage with the same payment as the 8%, 30-year mortgage, what interest rate do you need?

<table>
<tr><td>**13.5**</td><td>**Investing in Stocks, Bonds, and Mutual Funds**</td></tr>
</table>

Types of Investments

Now that we have learned how to handle interest, taxes, credit cards, and mortgages, what can we do if we have some money left? We can invest it! **Investing** is the art of committing money or capital to an endeavor with the expectation of obtaining additional income or profit. Your investments may be **short term** [bank savings accounts, money-market funds, certificates of deposit (CDs)] or **long-term** (stocks, bonds, and mutual funds). We shall concentrate on long-term investments and, in particular, on their cost and profitability (or lack thereof). ▶

A. Becoming an Owner: Investing in Stocks

When you buy stocks (shares) in a company, you become part owner of the company in which you have invested. What do you get then? The right to vote at the shareholder's meeting and receive **dividends** (profits) that the company allocates its owners.

To buy or sell stocks you may use the services of a **broker,** an individual or firm that charges a fee or **commission** for executing (doing) buy and sell orders submitted by an investor like you. Here are three types of brokers you may use:

Full service: Gives investment and financial advice. Charges a commission on the number of shares you buy or sell. The charges are usually a percent of the transaction and typically range from $100 to $300 per trade.

Online Study Center

Who are the world's greatest investors? There are many opinions about this, but you can access link 13.5.1 on this textbook's Online Study Center and read more about it.

Discount: Charges a reduced commission, typically $30 to $55 per transaction, and usually does not provide investment advice. (If they do, they will charge you for it!)

Deep discount: Only executes (does) stock and option trades at a flat rate (say, $25, but as low as $10) regardless of the size of the trade.

We are now ready to learn how to buy and sell stocks. How much will it cost, and how much money (return on your investment) can you make? The cost of one share of stock and related information can be found in newspapers in stock tables similar to Table 13.6.

TABLE 13.6

Online Study Center

For more information about brokers and fees, access links 13.5.2 and 13.5.3 on this textbook's Online Study Center.

52W high	52W low	Stock	Ticker	Div	Yield %	P/E	Vol 00s	High	Low	Close	Net chg
s45.39	19.75	ResMed	RMD			52.5	3831	42.00	39.51	41.50	-1.90
11.63	3.55	Revlon A	REV				162	6.09	5.90	6.09	+0.12
77.25	55.13	RioTinto	RTP	2.30	3.2		168	72.75	71.84	72.74	+0.03
31.31	16.63	RitchieBr	RBA			20.9	15	24.49	24.29	24.49	-0.01
8.44	**1.75**	**RiteAid**	**RAD**				**31028**	**4.50**	**4.20**	**4.31**	**+0.21**
s38.63	18.81	RobtHalf	RHI			26.5	6517	27.15	26.50	26.50	+0.14
51.25	27.69	Rockwell	ROK	1.02	2.1	14.5	6412	47.99	47.00	47.54	+0.24

Highest and lowest levels at which the stock traded the last 52 weeks — Company name — Ticker symbol — Dividend per share — Percent return on dividend — Price to earnings ratio — Total number of shares traded — Price range the stock traded throughout the day — Last trading price recorded when the market closed — Dollar value change in the stock price from the previous day closing price

Source: www.investopedia.com/university/tables/tables1.asp.

EXAMPLE 1 ▶ **Finding the Cost and Return for Stocks**

Mida Lariz bought 100 shares of Coca-Cola at $42.50 per share and paid a $50 commission on the transaction. A year later she sold the 100 shares at $47.75 per share and paid a 2% commission on the sale. If Coca-Cola paid $1.12 per share in dividends, find

(a) the total cost of the 100 shares.

(b) the total dividend.

(c) the commission when selling the 100 shares.

(d) the capital gain (profit) after she sold the 100 shares.

(e) the total return for the year.

(f) the percent of return.

Solution

(a) Total cost = ($42.50 per share) $\times$ (100 shares) + commission

$$= \$4250 + \$50 = \$4300$$

Thus, the total cost was $4300.

(b) Total dividend = ($1.12 per share) $\times$ (100 shares) = $112
Thus, the total dividend was $112.

(c) The commission on the sale was

$$2\% \times (\text{sale price}) = 0.02 \times (\$47.75 \times 100)$$
$$= 0.02 \times \$4775$$
$$= \$95.50$$

Thus, the commission on the sale was $95.50.

(d) Capital gain = (change in price per share) $\times$ 100 − (commissions)
$$= (\$47.75 - \$42.50) \times 100 - \$50 - \$95.50$$
$$= \$525 - \$50 - \$95.50$$
$$= \$379.50$$

The capital gain (profit) was $379.50.

(e) Total return = capital gain + dividends
$$= \$379.50 + \$112$$
$$= \$491.50$$

Thus, the total return was $491.50.

(f) The percent of return PR $= \dfrac{\text{total return}}{\text{total cost of entire transaction}}$

$$= \frac{\$491.50}{\$4250 + \$50 + \$95.50}$$
$$= \frac{\$491.50}{\$4395.50}$$
$$= 11.18\%$$

B. Becoming a Lender: Investing in Bonds

Instead of becoming part owner of a company by becoming a stockholder, you may decide to lend money to the company or even the government. What do you get then? You get **interest** on your money, and eventually, you get all your money back. A **bond** is nothing more than an IOU ("I owe you") from a company or government to you, the lender. When you invest in bonds, the bond you buy shows the amount of money being borrowed (**face value**), the interest rate (**coupon rate** or **yield**) that the borrower has to pay, the interest payments (**coupon payments**), and the deadline for paying the money back (**maturity dates**).

For example, you can buy a bond with a **face value** of $1000, a **coupon rate** of 7%, and a **maturity** of 10 years. This means that you will get $70 (7% of

$1000) each year for the next 10 years. After the 10 years, the bond **matures,** and you get your $1000 back.

Many bonds pay interest semiannually, so you will receive *two* payments of $35 a year for 10 years (the interest is **not** compounded because you are not reinvesting the money). As with stocks, there are **bond tables** in the newspaper similar to Table 13.7 showing the **issuer,** the **coupon** (rate), the **maturity** date, the bid **price** and the **yield** of **bonds.**

TABLE 13.7

	Coupon	Mat. date	Bid $	Yld %
Corporate				
AGT Lt	8.800	Sep 22/25	100.46	8.75
Air Ca	6.750	Feb 02/04	94.00	9.09
AssCap	5.400	Sep 04/01	100.01	5.38
Avco	5.750	Jun 02/03	100.25	5.63
Bell	6.250	Dec 01/03	101.59	5.63
Bell	6.500	May 09/05	102.01	5.95
BMO	7.000	Jan 28/10	106.55	6.04
BNS	5.400	Apr 01/03	100.31	5.24
BNS	6.250	Jul 16/07	101.56	5.95
CardTr	5.510	Jun 21/03	100.52	5.27
Cdn Pa	5.850	Mar 30/09	93.93	6.83
Clearn	0.000	May 15/08	88.50	8.61
CnCrTr	5.625	Mar 24/05	99.78	5.68
Coke	5.650	Mar 17/04	99.59	5.80

Company or country issuing the bond

Fixed interest rate that the issuer pays the lender

Date on which the borrowers will pay the investors their principal back

Price someone is willing to pay for the bond. Always quoted in relation to 100.

The annual return until the bond matures

Source: www.investopedia.com/university/bonds/bonds5.asp.

EXAMPLE 2 ▶ Finding the Cost and Returns for Bonds

Mile wants to invest $10,000 for his children's education in a 5-year Toys Corporation bond. If the coupon rate of the bond is 7.5% paid semiannually,

(a) how much will Mile receive semiannually?

(b) what is the total return on the investment?

Solution

(a) The formula for simple interest is $I = Prt$. In this example, $P = \$10,000$, $r = 7.5\% = 0.075$, and $t = \frac{1}{2} = 0.5$ year. Therefore,

$$I = \$10,000 \times 0.075 \times 0.5$$
$$= \$375$$

Mile will receive a check of $375 semiannually (twice a year) for 5 years.

(b) To find the total return, we let $t = 5$ in the formula $I = Prt$. Thus,

$$I = \$10,000 \times 0.075 \times 5$$
$$= \$3750$$

Mile's total return, assuming that he kept the bond for the 5 years, is $3750. ∎

Now that we have seen how stocks and bonds work, would you invest in stocks or in bonds? Here are some advantages and disadvantages to help you decide.

Stocks

Advantages	Long historical track record of outperforming bank savings, money-market funds, CDs, and real estate investments. You have voting rights! (Bondholders and depositors do not.)
Disadvantages	Stock prices fluctuate (go up and down). If the stock price drops, you may lose part or all of your money

Bonds

Advantages	Higher interest rates compared with short-term investments. Less risk (compared to stocks).
Disadvantages	If you sell the bond before **maturity,** you may incur a loss (politely called a **discount**). If the issuer declares bankruptcy, you may lose money.

So what is your decision? You may decide to invest in **mutual funds** instead.

C. Investing in Mutual Funds

A **mutual fund** is a pool of many investors who pay a professional manager to purchase a variety of investments. There are literally thousands of mutual funds to suit a variety of needs and objectives. You can find tables of mutual funds similar to Table 13.8 on page 934 in the newspaper.

The most common type of mutual fund is the **open-ended** fund, in which an unlimited number of shares can be issued. The more money investors put into the fund, the more shares it issues (the fund company absorbs the shares of sellers). The price of each share, called the **net asset value (NAV),** is calculated every day (that's where the name "open-ended fund" comes from) by using the formula

$$NAV = \frac{A - L}{N}$$

where A = total fund assets
L = total fund liabilities
N = total number of outstanding shares

EXAMPLE 3 ▶ Finding the Cost of a Share in the ABC Mutual Fund

The ABC Mutual Fund has $400 million worth of stock, $200 million in bonds, and $500,000 in other assets. Total liabilities amount to $10 million, and there are 20 million shares outstanding.

(a) Find the NAV of one share.

(b) How many shares can you buy with $100,000?

TABLE 13.8

52W high	52W low	Fund	Spec.	Fri. NAVPS $chg	Fri. NAVPS %chg	Wkly NAVPS high	Wkly NAVPS low	Wkly NAVPS cls	Wkly NAVPS $chg	Wkly NAVPS %chg
Montrusco Bolton Funds										
11.71	10.12	Bal Plus	*N	-0.08	-0.76	10.58	10.50	10.50	0.02	0.15
12.50	10.25	Growth Plus	*N	-0.10	-0.96	10.89	10.78	10.78	0.02	0.22
31.39	24.78	Quebec Growth	*FR	0.05	0.17	26.97	26.75	26.97	0.43	1.61
13.78	7.24	RSP Intl Growth	*N	-0.08	-1.01	7.45	7.36	7.36	-0.03	-0.41
11.16	9.09	Value Plus	*N	-0.07	-0.75	9.39	9.32	9.32	0.01	0.14
9.65	8.90	World Inc	*N	-0.04	-0.40	9.52	9.39	9.48	0.04	0.43
Montrusco Select Funds C$(n)										
12.87	10.49	Balanced	*N	-0.04	-0.37	10.85	10.80	10.81	0.05	0.45
16.32	12.11	Balanced +	*N	-0.05	-0.43	12.57	12.52	12.52	0.06	0.45
10.36	9.86	Bond Index +	X*N	-0.03	-0.32	10.35	10.30	10.30	0.04	0.37

Highest and lowest prices for the fund during the last 52 weeks

Name of the mutual fund.

Name of the company that manages the fund is written above in **bold type**.

Fund specifics. * Retirement eligible, N no load, F front load, B both front and back end fees

Dollar change from previous day

Percent change from previous day

Highest and lowest prices the fund traded at during the last week

The last price at which the fund traded

The dollar change in price over the previous week

The percent change in price over the previous week

Source: www.investopedia.com/university/tables/tables4.asp.

Solution

(a) Here,

$$A = \$400,000,000 + \$200,000,000 + \$500,000 = \$600,500,000$$
$$L = \$10,000,000$$
$$N = 20,000,000 \text{ shares}$$

Thus,

$$
\begin{aligned}
NAV &= \frac{A - L}{N} \\
&= \frac{600,500,000 - 10,000,000}{20,000,000} \\
&= \frac{590,500,000}{20,000,000} \\
&= \$29.525
\end{aligned}
$$

(b) The number of shares that can be bought with $100,000 at $29.525 per share is

$$\frac{\$100,000}{\$29.525} = 3386.96, \text{ or } 3386 \text{ shares}$$

We have already learned how to make money in stocks and bonds, but how do you make money with mutual funds? There are at least three ways.

1. By **capital appreciation,** an increase in the NAV of the fund

2. By receiving a portion of the **dividends** paid by the stocks in the fund

3. By receiving a **capital gain distribution,** the profits made by selling stocks or bonds that have gone up in price in the fund

For convenience in tracking fund performance, we express each outcome as an annual percent by using the following formulas:

1. Capital appreciation (%) $= \dfrac{\text{change in NAV}}{\text{number of shares}}$

2. Yield (%) $= \dfrac{\text{distribution per share}}{\text{NAV (at time of sale)}}$

3. % Return $= \dfrac{\text{increase in value} + \text{distribution}}{\text{cost of initial investment}}$

(Source: *Wall Street Journal Guide to Understanding Money and Investing,* www.wachoviasec.com/education/MutualFunds03.shtml.)

EXAMPLE 4 ▶ Finding Annual Returns on a Mutual Fund

An investor buys 100 shares of the Math Fund at $80 per share. A year later she sells them at $90 per share. If she received an income distribution of $2.50 per share, find

(a) the capital appreciation.　　　(b) the yield.　　　(c) the return.

Solution

(a) Capital appreciation (%) $= \dfrac{\text{change of NAV}}{\text{number of shares}} = \dfrac{\$90 - \$80}{100} = \dfrac{10}{100} = 10\%$

(b) Yield (%) $= \dfrac{\text{distribution per share}}{\text{NAV (at time of sale)}} = \dfrac{\$2.50}{\$90.00} = 2.78\%$

(c) The increase in value is ($90 − $80) × 100 = $1000. The distribution is $2.50 × 100 = $250. Thus,

$$\text{Return (\%)} = \frac{\text{increase in value} + \text{distribution}}{\text{cost of initial investment}}$$

$$= \frac{\$1000 + \$250}{\$8000} = 15.63\%　■$$

We have calculated **annual** returns on a mutual fund. How can we calculate returns, in general, for a mutual fund? A good answer can be found by accessing links 13.5.4 and 13.5.5 on this textbook's Online Study Center.

"It would be easy to calculate your returns on mutual funds if you invested only on the first business day of every year and you sold on the last business day of the year. But calculating a compound rate of return when distributions are reinvested is considerably more complicated than that because most people buy and sell mutual funds at other times." In addition, when calculating the return of your mutual funds, you should consider the fees charged when buying and selling the fund, as well as the taxes involved. If you want a calculator to do it for you, you can access link 13.5.6 on this textbook's Online Study Center.

EXAMPLE 5 ▶ Finding the Return on a Mutual Fund When Reinvesting Dividends

Suppose the investor of Example 4 bought 100 shares of the Math Fund at $80 per share and that, for the sake of simplicity, we only consider the returns the fund earns and automatically reinvests at the rate of 1% per month. Find

(a) the beginning value of the fund.

(b) the monthly return.

(c) the annual rate of return.

Solution

(a) The beginning value of the fund is

NAV $\times$ number of shares = $80 $\times$ 100 = $8000

(b) The monthly return is 1% of $8000 = $80.

(c) Since the money is reinvested (compounded) at 1% per month, we need to find the future value A_n of 1 compounded n times at interest rate r. In this case, $n = 12$ and $r = 1\% = 0.01$. From page 891, this value is

$$A_n = (1 + r)^n = (1 + 0.01)^{12} = (1.01)^{12} \approx 1.13$$

Since we started with 1, the return is $1.13 - 1 = 0.13 = 13\%$. Thus, the annual rate of return is 13%. ■

Can you see the advantages of mutual funds? Here are some advantages and disadvantages:

Mutual Funds

Advantages	*Professional management* of your money.
	Diversification—your risk is spread out, and losses in any particular investment may be minimized by gains in other investments.
	Liquidity—you can request that your shares be converted to cash at any time.
Disadvantages	*Professional management* costs you money.
	Cost—sometimes hidden and hard to understand.
	Taxes—when making decisions, managers don't consider your personal tax situation. Sometimes you incur unwanted tax liabilities.

(Source: www.investopedia.com/university/mutualfunds/mutualfunds.asp.)

EXERCISES 13.5

A Finding the Cost and Returns for Stocks

The following lists the six New York Stock Exchange volume leaders on June 24:

NYSE VOLUME LEADERS

Symbol	Name	Last Trade	Change	Dividend
GE	GEN ELECTRIC CO	34.78 Jun 24	↑ 0.12 (0.35%)	$1.06
NWS-A	NEWS CORP CL A	16.90 Jun 24	↑ 0.26 (1.56%)	$0
PFE	PFIZER INC	28.52 Jun 24	↓ 0.38 (1.31%)	$0.72
GDT	GUIDANT CORP	63.90 Jun 24	↓ 4.70 (6.85%)	$0.40
LU	LUCENT TECH INC	3.02 Jun 24	↓ 0.09 (2.89%)	$0
XOM	EXXON MOBIL CP	58.15 Jun 24	↓ 0.87 (1.47%)	$1.10

In problems 1–6, find
  a. the total cost.
  b. the total dividend.
  c. the commission when selling the shares.
  d. the capital gain when selling the shares.
  e. the total return for the year.
  f. the percent of return (round your answer to two decimal places).

1. An investor buys 100 shares of GE at the Last Trade price, pays a $50 commission on the purchase, and sells the 100 shares a year later for $36.78 per share with a 2% commission on the sale.

2. An investor buys 100 shares of NWS-A at the Last Trade price, pays a $50 commission on the purchase, and sells the 100 shares a year later for $18.40 per share with a 2% commission on the sale.

3. An investor buys 200 shares of PFE at the Last Trade price, pays a $50 commission on the purchase, and sells the 200 shares a year later for $30 per share with a 2% commission on the sale.

4. An investor buys 200 shares of GDT at the Last Trade price, pays a $50 commission on the purchase, and sells the 200 shares a year later for $64 per share with a 2% commission on the sale.

5. An investor buys 200 shares of LU at the Last Trade price, pays a 2% commission on the purchase, and sells the 200 shares a year later for $3.00 per share with a $50 commission on the sale.

6. An investor buys 200 shares of XOM at the Last Trade price, pays a 2% commission on the purchase, and sells the 200 shares a year later for $68 per share with a $50 commission on the sale.

Problems 7–12 refer to the following chart. An investor (a day trader) always buys 1000 shares of stock at the market close price and sells them at the last sale price, paying a $25 commission per transaction. Find
  a. the total cost.
  b. the return for the day.
  c. the percent of return if the stock the trader bought is

7. MSLV.

8. PWOD.

9. NEON.

10. SDIX.

11. XXIA.

12. MKTY.

NASDAQ - After Hours Ten Most Advanced

Symbol	Company Name	Market Close	Last Sale (after hours)	% Change (after hours)	Share Volume (after hours)
MSLV	MetaSolv Inc.	$2.21	$2.60	17.65%	500
PWOD	Penns Woods Bancorp, Inc.	$41.17	$44.62	8.38%	1,500
NEON	NEON Systems, Inc.	$3.01	$3.25	7.97%	1,000
SDIX	Strategic Diagnostics Inc.	$3.50	$3.70	5.71%	7,200
XXIA	Ixia	$18.96	$19.90	4.96%	57,965
MKTY	Mechanical Technology Incorporated	$3.61	$3.78	4.59%	100

Source:http://dynamic.nasdaq.com/dynamic/afterhourma.stm.

Order	Ratings	Qty	Min	Ticker	Description	Coupon	Maturity	YTC/YTM	Price
Buy	Baa3/BB	100		GM	General Mtrs Corp 370442BB0 Global Nt Make-Whole	7.200	01-15-2011	9.254	91.242
Buy	A3/A−	100		DOW	Dow Chem Co 260543BL6 Nt Non-Callable	6.125	02-01-2011	4.257	109.195
Buy	Baa3/BBB	100		TYC	Tyco Intl Group S A 902118AY4 Nt Make-Whole	6.750	02-15-2011	4.405	111.564
Buy	A1/AA	100		ABT	Abbott Labs 002824AP5 Nt Make-Whole	3.750	03-15-2011	4.145	98.005
Buy	Aa3/A+	100		KO	Coca Cola Co 191216AH3 Nt Make-Whole	5.750	03-15-2011	4.120	108.209
Buy	Baa1/BBB+	100		K	Kellogg Co 487836AS7 Nt Make-Whole	6.600	04-01-2011	4.359	111.285

Source: www.bondtrac.com/zions/corporate/search/advanced.

B **Finding the Cost and Returns for Bonds**

The table above is to be used in problems 13–18 and 35–40.

In problems 13–18 use the information in the table above to find
   a. the amount you will receive semiannually
   b. the total return on the investment
if you bought

13. $5000 in General Motors bonds and kept them for 5 years until maturity.

14. $5000 in Dow Chem bonds and kept them for 5 years until maturity.

15. $10,000 in Tyco Intl Group bonds and kept them for 2 years until maturity.

16. $10,000 in Abbott Labs bonds and kept them for 2 years until maturity.

17. $20,000 in Coca-Cola bonds and kept them for 5 years until maturity.

18. $20,000 in Kellogg Co bonds and kept them for 5 years until maturity.

C **Finding the Cost and Returns of a Mutual Fund**

In problems 19–24, find
   a. the NAV of one share.
   b. the number of shares you can buy when you invest the amount of money shown in the last column of the table below.

	Name	Stocks	Bonds	Other Assets	Liabilities	Outstanding Shares	Amount to Invest
19.	ABC	$200 million	$100 million	$250,000	$10 million	10 million	$ 50,000
20.	XYZ	$100 million	$ 50 million	$200,000	$10 million	20 million	$ 40,000
21.	Grand	$500 million	$250 million	$1.5 million	$25 million	25 million	$100,000
22.	MathRUs	$100 million	$ 50 million	$100,000	$ 5 million	5 million	$ 25,000
23.	Liberty	$400 million	$200 million	$1 million	$20 million	20 million	$100,000
24.	Freedom	$250 million	$125 million	$750,000	$10 million	10 million	$ 75,000

In problems 25–30 find

    a. the capital appreciation.

    b. the yield.

    c. the return for the indicated mutual fund when the given number of shares is bought and sold 1 year later after receiving the distribution shown in the following table.

	Name	Shares Bought	Shares Sold	Distribution
25.	Bolting	100 at $70	100 at $80	$3 per share
26.	TrustCo	200 at $90	200 at $100	$2.50 per share
27.	NorthCo	300 at $75	300 at $80	$1.25 per share
28.	SouthCo	500 at $30	500 at 40	$0.50 per share
29.	EastCo	400 at $40	400 at $50	$4.00 per share
30.	WestCo	1000 at $20	1000 at $25	$0.75 per share

In Other Words

31. Describe in your own words what stocks, bonds, and mutual funds are.

32. Explain the difference between stocks, bonds, and mutual funds.

33. Explain in your own words what a dividend is.

34. Explain in your own words what liquidity means to you.

Using Your Knowledge

The **current yield** for an investment can be defined in many ways. Here is a formula for the current yield of a bond.

$$\text{Current yield} = \frac{\text{coupon rate}}{\text{market price}} \times 100$$

Find the current yield of the following bonds using the information in the table accompanying problems 13–18 at the top of page 938:

35. The General Motors bond

36. The Dow Chemical bond

37. The Tyco International bond

38. The Abbot Labs bond

39. The Coca Cola bond

40. The Kellogg Co bond

Premium and Discount Prices

If you sell a bond prematurely, the bond price will change depending on the **current** interest rate and the **fixed coupon payment.** For example, suppose you have a bond with a face value of $1000, a 10% coupon rate, and a $100 fixed coupon payment (10% × 1000 = $100). If interest rates go down, say, to 8%, your bond is worth more (because it pays 10%, but the interest rate is now only 8%). You must pay a **premium** for the bond. How much?

Suppose the premium price of the bond is P. You have to pay the fixed coupon payment of $100 at the new rate of 8%. Thus,

$$8\% \text{ of } P = 100$$
$$0.08\,P = 100$$
$$P = \frac{100}{0.08} = \$1250$$

The **premium price** of the bond is $1250 (you make a **premium** of $1250 − $1000 = $250).

On the other hand, if interest rates go up, say, to 12%, your bond is worth less (because it pays only 10%, but interest rates went up to 12%). You must **discount** your bond. How much?

Suppose that the discount price of the bond is D. You have to pay the fixed coupon payment of $100 at the new rate of 12%. Thus,

$$12\% \text{ of } D = 100$$
$$0.12D = 100$$
$$D = \frac{100}{0.12} = \$833.33$$

The **discount price** of the bond is $833.33 (the **discount** is $1000 − $833.33 = $166.66.

In problems 41–46, find the price of a $1000 bond with a 10% coupon rate when

41. interest rates go down to 7%.

42. interest rates go down to 6%.

43. interest rates go down to 5%.

44. interest rates go up to 11%.

45. interest rates go up to 12%.

46. interest rates go up to 13%.

 Web It Exercises

Problems 47–63 require access to the Internet. For help with the problems, access links 13.5.7 to 13.5.14 in the textbook's Online Study Center.

Investing in Bonds

47. What different types of bonds are there?

48. What are some factors that determine the price of a bond?

49. What are some of the possible credit ratings for bonds?

50. How can you evaluate a broker?

How to Read Bond Tables

51. What does **bid** and **ask** price mean?

52. What does a bid of 105:12 mean? What do the numbers after the colon represent?

Trading Stocks

53. What is a market order? A limit order? A stop order?

54. What are online, after hours, day, and active traders?

55. What is a common stock?

56. What is a preferred stock?

57. What is a bull market? What is a bear market?

How to Read Stock Tables in Detail

58. What is the P/E ratio?

59. What does the YLD column mean, and how can the value be approximated?

Fundamental Analysis

60. What is fundamental analysis?

61. What is PEG?

62. What is ROE?

Chapter 13 Summary

Section	Item	Meaning	Example
13.1A	$I = Prt$	Simple interest equals Prt, where P is the principal, r is the rate, and t is the time in years.	The interest on a \$500 2-year loan at 12% is $I = \$500 \cdot 0.12 \cdot 2 = \$120$.
13.1A	$A = P + I$	Amount equals principal plus interest.	The amount A you have to pay on a \$500, 2-year loan at 12% simple interest is $A = \$500 + \$120 = \$620$.
13.1D	$A_n = P(1 + r)^n$	Compound amount equals $P(1 + r)^n$, where P is the principal, r is the rate per period, and n is the number of periods.	The compound amount A_2 paid on a 2-year loan of \$500 compounded annually at 12% is $A_2 = \$500(1 + 0.12)^2$ $= \$500(1.12)^2$ $= \$627.20$
13.3A	APR	Annual percentage rate	

Section	Item	Meaning	Example
13.3B	Rule of 78	The unearned interest rate on a loan of n periods with r remaining periods is $\dfrac{r(r+1)}{n(n+1)} \times F$, where F is the finance charge.	The unearned interest rate on a 15-month loan with a \$120 finance charge and monthly payments of \$50 at the end of the eighth month is $\dfrac{7 \times 8}{15 \times 16} \times \$120$ $= \dfrac{7}{30} \times \$120 = \$28$
13.5	Stock	A share in the ownership of a company	You may own stock (shares) of Coca Cola or Dell Computer.
13.5	Dividend	The profits that the company allocates to the owners	The dividend of company ABC may be \$2.50 per quarter.
13.5	Broker	An individual or firm that charges a commission for executing orders submitted by investors	The commission on a transaction may be a percent of the value of the transaction or a fixed fee.
13.5	Bond	An IOU from a company or government to a lender	Treasury bonds, corporate bonds, municipal bonds
13.5	Mutual fund	A pool of many investors who pay a professional manager to purchase investments	Merrill Lynch, Smith Barney, and T. Rowe Price have mutual funds.
13.5	NAV	Net asset value	$\text{NAV} = \dfrac{A - L}{N}$ where A = assets, L = liabilities, and N = number of outstanding shares
13.5	Capital appreciation	An increase of the NAV of a fund	The capital appreciation of a fund may be 10%.
13.5	Yield	The amount the fund produces	$\dfrac{\text{Distribution per share}}{\text{NAV}}$
13.5	Percent of return	The quotient of the sum of the increase in value plus distributions and the cost of the initial investment	$\dfrac{\text{Increase in value } + \text{ distribution}}{\text{Cost of initial investment}}$

Research Questions

Sources of information for these questions can be found in the Bibliography at the end of the book.

1. Use an encyclopedia and write a report about the origin of coins, citing places, peoples, and dates.

2. Write a report about the origin of paper money, citing places, peoples, and dates.

3. Write a brief report about Continental currency.

4. Find out and report about the Federal Reserve System and its relationship to the printing and control of money in the United States.

5. Go to a bank or a savings and loan and research its requirements for getting a mortgage loan. Write out all of these requirements.

Chapter 13 Practice Test

1. The Ready-Money Loan Company charges 28% simple interest (annual) for a 2-year, $800 loan. Find the following:
   a. The total interest on this loan   **b.** The interest for 3 months
   c. The total amount to be paid to the loan company at the end of 2 years

2. A state has a 6% sales tax.
   a. What is the sales tax on a microwave oven priced at $360?
   b. What is the total cost of this oven?

3. In a sale, a store offers a 20% discount on a freezer chest that is normally priced at $390.
   a. How much is the discount?   **b.** What is the sale price of the freezer?

4. The table below is a portion of a compound interest table to use in this problem. Find the accumulated amount and the interest earned for the following:
   a. $100 at 8% compounded semiannually for 2 years
   b. $100 at 8% compounded quarterly for 2 years

Amount (in dollars) to Which $1 Will Grow in _n_ Periods Under Compound Interest

_n_	2%	4%	6%	8%	10%
1	1.0200	1.0400	1.0600	1.0800	1.1000
2	1.0404	1.0609	1.1236	1.1664	1.2100
3	1.0612	1.1249	1.1910	1.2597	1.3310
4	1.0824	1.1699	1.2625	1.3605	1.4641
5	1.1041	1.2167	1.3382	1.4693	1.6105
6	1.1262	1.2653	1.4185	1.5869	1.7716
7	1.1487	1.3159	1.5036	1.7138	1.9487
8	1.1717	1.3686	1.5938	1.8509	2.1436

5. A credit card holder is obligated to pay the balance in full if it is less than $10. Otherwise, the minimum payment is $10 or 5% of the balance, whichever is more. Suppose a customer received a statement listing the balance as $185.76.
   a. Find the minimum payment due.
   b. The finance charge is 1.5% per month. What will be the amount of this charge on the next statement if the customer makes only the minimum payment?

6. JoAnn Jones received a statement showing that she owed a balance of $179.64 to a department store where she had a revolving charge account. JoAnn made a payment of $50 and charged an additional $23.50. If the store charges 1.5% per month on the unpaid balance, find the following:
   a. The finance charge for the month
   b. The new balance

7. A car costing $6500 can be bought with $1500 down and a 12% add-on interest to be paid in 48 equal monthly installments.
   a. What is the total interest charge?
   b. What is the monthly payment?

8. The following is a table for you to use in this problem:

   True Annual Interest Rate for a 12-Payment Plan

Finance Charge	14%	14%	15%	$15\frac{1}{2}$%	16%
Finance charge (per $100 of the amount financed)	7.74	8.03	8.31	8.59	8.88

   Sam Bear borrows $200 and agrees to pay $18.10 per month for 12 months.
   a. What is the APR for this transaction?
   b. If Sam decided to pay off the balance of the loan after 5 months (with 7 payments remaining), use the rule of 78 to find the amount of the interest refund.
   c. Find the amount needed to pay off the loan.

9. The Mendoza family wants to buy a $50,000 house in New Mexico.
   a. If a bank was willing to loan the family 75% of the price of the house, what would be the amount of the loan?
   b. What would be the down payment for this house?
   c. If the family decided to obtain an FHA loan instead, what would be the minimum cash investment?
   d. What would be the maximum FHA loan the family could get?

10. Refer to problem 9. Suppose the Mendoza family contracted for a 15-year mortgage at 12% with the bank that loaned the family 75% of the price of the house. What is the family's monthly payment for principal and interest? (Use the table below.)

    Monthly Payment (in dollars) for Each $1000 Borrowed

Rate	10 years	15 years	20 years
11%	13.78	11.37	10.32
12%	14.35	12.00	11.01
13%	14.93	12.65	11.72

11. Scott McWaters bought 100 shares of Kiwi-Cola at $52.50 per share and paid a $50 commission on the transaction. A year later he sold the 100 shares at $57.75 per share and paid a 2% commission on the sale. If Kiwi-Cola paid $1.22 per share in dividends, find
    a. the total cost of the 100 shares.
    b. the total dividend.
    c. the commission when selling the 100 shares.
    d. the capital gain (profit) after he sold the 100 shares.
    e. the total return for the year.
    f. the percent of return.

12. Fernando wants to invest $10,000 for his children's education in a 5-year Kiddies Corporation bond. If the coupon rate of the bond is 6.5% paid semiannually,
    a. how much will Fernando receive semiannually?
    b. what is the total return of the investment?

13. The XYZ Mutual Fund has $200 million worth of stock, $100 million in bonds, and $250,000 in other assets. Total liabilities amount to $10 million, and there are 20 million shares outstanding.
    a. Find the NAV of one share.
    b. How many shares can you buy with $100,000?

14. An investor buys 100 shares of the Geo Fund at $70 per share. A year later she sells them at $80 per share. If she received an income distribution of $1.50 per share, find
    a. the capital appreciation.
    b. the yield.
    c. the return.

15. Suppose that the investor of problem 14 bought 100 shares of the Geo Fund at $70 per share and that, for the sake of simplicity, we only consider the returns the fund earns and automatically reinvests at the rate of 2% per month. Find
    a. the beginning value of the fund.
    b. the monthly return.
    c. the annual rate of return.

Answers to Practice Test

ANSWER	IF YOU MISSED	REVIEW		
	Question	Section	Example(s)	Page(s)
1. a. $448 **b.** $56 **c.** $1248	1	13.1	1	882–883
2. a. $21.60 **b.** $381.60	2	13.1	2	883
3. a. $78 **b.** $312	3	13.1	3	884
4. a. $116.99; $16.99 **b.** $117.17; $17.17	4	13.1	4	886–887
5. a. $10 **b.** $2.64	5	13.2	2, 3	894, 895
6. a. $1.94 **b.** $155.08	6	13.2	4, 5	896–897
7. a. $2400 **b.** $154.17	7	13.2	6	898
8. a. $15\frac{1}{2}\%$ **b.** $6.17 **c.** $120.53	8	13.3	1, 2	903, 905
9. a. $37,500 **b.** $12,500 **c.** $1500 **d.** $48,500	9	13.4	2	912
10. $450	10	13.4	3	913–914
11. a. $5300 **b.** $122 **c.** $115.50 **d.** $359.50 **e.** $481.50 **f.** 8.89%	11	13.5	1	930–931
12. a. $325 **b.** $3250	12	13.5	2	932–933
13. a. $14.5125 ≈ $14.51 **b.** 6891.8 or 6891 shares	13	13.5	3	933–934
14. a. 10% **b.** 1.875% **c.** $16.43%	14	13.5	4	935
15. a. $7000 **b.** $140 **c.** 27%	15	13.5	5	936

CHAPTER 1

Exercises 1.1

1. Step 1. Understand the problem.
   Step 2. Devise a plan.
   Step 3. Carry out the plan.
   Step 4. Look back.
3. What does the problem ask for? What is the unknown?
5. $15 \times \$.20 = \$3$
7. The Light Use Plan; it is less expensive.
9. After 63 calls, the Standard Use Plan is less expensive.
11. Add n to the nth term. The next three terms are 11, 16, and 22.
13. The odd-numbered terms are 1s. The even-numbered terms start with 5 and add 5 for each additional such term. The next three terms are 1, 20, and 1.
15. Going clockwise, move the shaded region one place, then two places, then three places, and so on.

17. Each term is half the preceding term. The next three terms are $\frac{1}{16}, \frac{1}{32}$, and $\frac{1}{64}$.
19. The odd-numbered terms are 1, 2, 3, 4, 5, 6, ... The even-numbered terms are 5, 6, 7, 8, 9, ... The next three terms are 7, 4, and 8.
21. a.

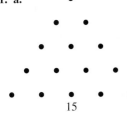

15

   b. At each step add one row, one dot longer, at the bottom. The next three triangular numbers are 15, 21, and 28.
   c. The tenth triangular number is 55.
23. a. $1 + 2 + 3 + 4 + 5 + 6 + 7 + 8 = \boxed{36}$
   b. 36
   c. $1 + 2 + 3 + \cdots + 12 = \boxed{78}$
   d. 78
   e. $1 + 2 + 3 + \cdots + (n - 1) + n = \dfrac{n(n + 1)}{2}$
   f. $50 \cdot 101 = 5050$

25. a.

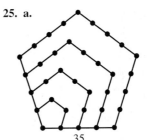

35

   b. At each step, increase the length of the bottom and the left lower side of the pentagon by 1 unit. The number of dots on each side is increased by 1 unit.
   c. The sixth pentagonal number is 51.
27. The number of diagonals is three less than the number of sides of the polygon. Thus, seven diagonals can be drawn from one vertex of a decagon.
29. a.

5	10	20	100
12	17	27	107
36	51	81	321
30	45	75	315
10	15	25	105
5	5	5	5

   The final result is always 5.
   b. n
   $n + 7$
   $3n + 21$
   $3n + 15$
   $n + 5$
   5
31. a.

5	10	20	100
10	15	25	105
40	60	100	420
20	30	50	210
10	10	10	10

   The final result is always 10.
   b. n
   $n + 5$
   $4n + 20$
   $2n + 10$
   10
33. a. 4
   b. *Hint:* Try the numbers from 0 to 10 to see what happens.
35. a. $(1 + 2 + 3 + 4)^2 = 1^3 + 2^3 + 3^3 + 4^3$
   $(1 + 2 + 3 + 4 + 5)^2 = 1^3 + 2^3 + 3^3 + 4^3 + 5^3$
   $(1 + 2 + 3 + 4 + 5 + 6)^2 = 1^3 + 2^3 + 3^3 + 4^3 + 5^3 + 6^3$
   b. The square of the sum of the first n counting numbers equals the sum of the cubes of these numbers.

37. The number of units of length of the pendulum is the square of the number of seconds in the time of the swing.

39. a. 12, 15, 18   **b.** $9\frac{2}{3}$ in.

45. $1 + 1 + 2 + 3 + 5 = 12$. The sixth term is 8, so the seventh term is 13, which is 1 more than the sum of the first five terms.

47. The fourteenth term is 377, so the sum of the first 12 terms is 376.

Exercises 1.2

1. $12,000

3. $8 + $2 + $4 + $2 + $3 = $19

5. 900 gal

7. a. $\frac{4256}{14,053} \approx 0.303$

    b. $\frac{4300}{14,100} \approx 0.305$

9. ERA $= \frac{9 \times 14}{140} = 0.900$

11. 400 to 800   **13.** 33 lb

15. 25 lb

17. a. 5182   **b.** 80 kWh   **c.** $6.40   **d.** $192

19. a. 7001   **b.** 50 kWh   **c.** $4   **d.** $120

21. $4090.00 + 0.25($10,300) = $6665.00

23. a. 71.16 in.   **b.** $73 - 71.16 = 1.84$ in.

25. About 22.87. The person's BMI is normal.

27. 980 lb. 10 lb of hay, 5 lb of grain, and 6 gal of water

29. $1655

31. a. $24 + 12 = 36$ years old

    b. $24 + 32 = 56$ years old

33. 15 mi   **35.** 15 mi

37. $22\frac{1}{2}$ mi   **39.** $2.70

41. $2.25   **43.** 8933.33 lb

45. 466.67 lb   **47.** 1800

51. 1,335,840 ft^2

Exercises 1.3

1. a. Bus   **b.** Bike   **c.** 71%

3. a. Cheddar   **b.** Swiss

    c. Mozzarella

5. a. Bathing   **b.** 150 gal

    c. Toilet leak   **d.** 20 gal

7. a. Paper   **b.** Yard trimmings

    c. 20 lb; 9 lb

9. a. Oil   **b.** Nuclear

    c. Natural gas

11. a. 114.4   **b.** 114   **c.** 111.2

    d. 2–<3/day; 110

    e. >3/day; almost 120

13. a. 39   **b.** 29

    c. 0.20–0.29; 14

15. a. 20–29   **b.** 13–15   **c.** Less than 50 years old

    d. 90$^+$; answers vary.

17. a. $75,000$^+$; 77.7; 60.3

    b. Under $15,000; 12.7; 7.1

    c. $60,000

19. a. $560 - 300 = 260$ cal

    b. $1120 - 600 = 520$ cal

21. a. Cuban toast   **b.** Cheese toast

    c. 105; about 27 (rounded from $26\frac{1}{4}$)

23. a. African American   **b.** White   **c.** 20%; 15%

    d. Answers vary.

25. a. 24%; 82%; 58%

    b. 4%; 22%; 18%

    c. "Less than 12 years" is decreasing; the other two are increasing.

27. a. <18   **b.** 65+   **c.** About 30%

    d. 65+   **e.** 18–64   **f.** <18

29. a. 60   **b.** About 10   **c.** About 2

31. a. 4 kg   **b.** 10 kg   **c.** About 10 mo

33. About $570   **35.** $300

41. a. Years 1–7   **b.** Years 2–7

    c. Years 1–7   **d.** Years 4–7

    e. Year 7 of either the breast cancer group or the stroke group; about 0.005

    f. Breast cancer; years 0–4

CHAPTER 2

Exercises 2.1

1. Not a set   **3.** A set

5. A set   **7.** Not a set

9. a. Incorrect   **b.** Correct   **c.** Incorrect

    d. Correct   **e.** Incorrect

11. $\in$ **13.** $\notin$

15. The set consisting of the first and the last letters of the English alphabet

17. The set consisting of the names of the first biblical man and woman

19. The set of counting numbers from 1 to 7

21. The set of odd counting numbers from 1 to 51

23. The set of counting numbers starting with 1 and then adding 3 successively until the number 25 is obtained

25. {Dioxin, Xylene}

27. {1, 2, 3, 4, 5, 6, 7}   **29.** {0, 1, 2, 3, 4, 5, 6, 7}

31. {4, 5, 6, 7}   **33.** $\varnothing$ or { }

35. {4, 5, 6, . . . }   **37.** {8, 9, 10}

39. {1, 2, 3, 5, 8, 10}

41. {1, 4}   **43.** {1, 2}

45. Sets A and B are not equal.

47. Sets A and B are equal.

49. a. =   **b.** $\neq$ **c.** $\neq$

51. $\varnothing$, {a}, {b}, {a, b}. The first three are proper subsets.

53. $\varnothing$, {1}, {2}, {3}, {4}, {1, 2}, {1, 3}, {1, 4}, {2, 3}, {2, 4}, {3, 4}, {1, 2, 3}, {1, 2, 4}, {1, 3, 4}, {2, 3, 4}, {1, 2, 3, 4}. All but the last of these are proper subsets.

55. $\varnothing$, {1}, {2}, {1, 2}. The first three are proper subsets.

57. 2^4, or 16   **59.** 2^{10}, or 1024

61. 5   **63.** 6

65. Yes. Every set is a subset of itself.

67. $B \subseteq A$
69. a. 5　**b.** 10　**c.** 10
71. 8
77. a. If $g \in S$, then Gepetto shaves himself; this contradicts the statement that Gepetto shaves all those men and only those men who do not shave themselves. Therefore, $g \notin S$.
b. If $g \in D$, then Gepetto does not shave himself, and so by the same statement he does shave himself. Thus, there is a contradiction and $g \notin D$.
79. The word *non-self-descriptive* cannot be classified either way without having a contradiction.

Exercises 2.2

1. a. $\{1, 3, 4\}$　**b.** $\{1\}$　**c.** $\{1, 6\}$
3. a. $\{1, 3, 4\}$　**b.** $\{1, 2, 3, 4, 5, 6\}$
5. $\{1, 2, 3, 4, 5, 6, 7\}$　**7.** $\{1\}$
9. a. $\{c\}$　**b.** $\varnothing$
11. a. Correct　**b.** Incorrect
13. a. Correct　**b.** Correct
15. a. $\{b, d, f\}$　**b.** $\{a, c\}$
17. a. $\varnothing$　**b.** $\{a, b, c, d, f\}$
19. a. $\{c, e\}$　**b.** $\{a, b, c, d, f\}$
21. a. $\{b, d, f\}$　**b.** $\{a, c\}$
23. a. $\{a, b, c, d, e, f\}$　**b.** $\{c, e\}$
25. a. $\{b, d, f\}$　**b.** $\{a, c\}$
27. a. $\{2, 3\}$　**b.** $\{2, 3\}$
29. $\mathcal{U}$　**31.** $\varnothing$
33. A　**35.** $\varnothing$
37. A　**39.** $\{1, 2, 3, 4, 5\}$
41. {Beauty, Consideration, Kindliness, Friendliness, Helpfulness, Loyalty}
43. {Intelligence, Cheerfulness, Congeniality}
45. {Intelligence, Cheerfulness}
47. {Is aware of others, Follows up on action}
49. {Follows up on action}
51. a. F　**b.** M
53. a. Male employees who work in the data processing department
b. Female employees who are under 21
55. $D \cap S$　**57.** $M \cap D$
59. Male employees or employees who are 21 or over
61. a. The set of full-time employees who do shop work, $\{04, 08\}$
b. The set of part-time employees who do outdoor field work or indoor office work, $\{02, 05, 07\}$
63. {Jimi Hendrix, Led Zeppelin, The Who}
65. 3　**67.** $\{4, 7\}$　**69.** $\{3\}$
71. $\{8\}$　**73.** $\{1, 2, 3, 6, 7, 8, 9\}$
75. $\{1, 2, 3, \ldots, 10\}$
77. A and B have no elements in common.
79. All the elements of A are elements of B, and all the elements of B are elements of A ($A = B$).

81. a.–b. The set of characteristics that are in both columns of the table {long tongue, skin-covered horns, native to Africa}
c. The set of characteristics that appear in either column of the table
d. $G' = \{\text{short, short neck}\}$
e. $O' = \{\text{tall, long neck}\}$
83. 685,000
85. 12- to 17-year-old females; $F \cap A$
87. This set is empty. There are no persons who are both male and female.
89. \$41,339　**91.** \$28,403
93. Average earnings of males with a high school degree; \$32,521

Exercises 2.3

1.
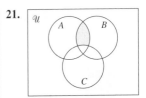

3.

5.

7.

9. Region 1　**11.** $\varnothing$
13. Regions 4, 5, and 7
15. Regions 1, 4, 5, 6, and 7
17. Region 8
19.
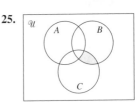

21.

23.
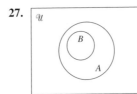

25.

27.

29.

31. a. Both $A \cup (B \cup C)$ and $(A \cup B) \cup C$ correspond to regions 1, 2, 3, 4, 5, 6, and 7. This verifies the given equality.
   b. Both $A \cap (B \cap C)$ and $(A \cap B) \cap C$ correspond to region 7. This verifies the given equality.
33. a. $A \cup A'$ corresponds to regions 1, 2, 3, 4, 5, 6, 7, and 8. Therefore, $A \cup A' = \mathcal{U}$.
   b. Since A and A' have no region in common, $A \cap A' = \varnothing$.
   c. $A - B$ corresponds to regions 1 and 5, and $A \cap B'$ also corresponds to regions 1 and 5. Thus, $A - B = A \cap B'$.
35. a. $A \cap B$ is represented by regions 3 and 7.
37. a. $A = \{a, b, c, e\}$, $B = \{a, b, g, h\}$, and $\mathcal{U} = \{a, b, c, d, e, f, g, h\}$
   b. $A \cup B = \{a, b, c, e, g, h\}$
   c. $(A \cap B)' = \{c, d, e, f, g, h\}$
39.

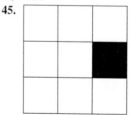

41. Arizona, California, Florida, Texas
43.

45.

47.

53. The set of elements common to A and B
55. The set of elements in $\mathcal{U}$ and not in either A or C
57. False
59. False
61. AB^+, the blood type of a person who has all three antigens and thus may receive blood from any person
63. No, because the B^- person does not have the A antigen
65. No, because the O^- person does not have the Rh antigen   **67.** 44%   **69.** 45%
71. 16, or 2^4
73. a. Region 11   **b.** Regions 8 and 16

Exercises 2.4

1. 30   **3.** 20
5. 40 families subscribe to both.
7. a. None   **b.** 10   **c.** 10
9. a. 22   **b.** 36   **c.** 6
11. a. $120,000   **b.** $510,000   **c.** $1,305,000
13. 200
15. a. 5   **b.** 30   **c.** 20
17. $450   **19.** 28
21. a. 120   **b.** 80   **c.** 50
23. a. 80   **b.** 120   **c.** 50
25. a. 73   **b.** 55   **c.** 91   **d.** 38   **e.** 5
27. 31%   **29.**

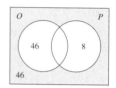

31. False. A counterexample is $A = \{1, 2\}$, $B = \{m, n\}$.
33. False. A counterexample is $A = \{1, 2\}$, $B = \{1, 2, 3\}$.
35. The Venn diagram shows that with the added information, the statistics in the cartoon are possible.

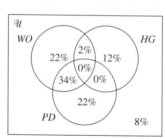

37. $2^4 = 16$ different subsets

Exercises 2.5

1. The following correspondence shows that sets N and O are equivalent:

1	2	3	$\cdots$	n	$\cdots$
$\updownarrow$	$\updownarrow$	$\updownarrow$		$\updownarrow$	
1	3	5	$\cdots$	$2n - 1$	$\cdots$

3. The following correspondence shows that sets E and G are equivalent:

2	4	6	$\cdots$	$2n$	$\cdots$
$\updownarrow$	$\updownarrow$	$\updownarrow$		$\updownarrow$	
102	104	106	$\cdots$	$100 + 2n$	$\cdots$

5. The following correspondence shows that sets G and T are equivalent:

202	204	206	$\cdots$	$200 + 2n$	$\cdots$
$\updownarrow$	$\updownarrow$	$\updownarrow$		$\updownarrow$	
302	304	306	$\cdots$	$300 + 2n$	$\cdots$

7. The following correspondence shows that sets P and Q are equivalent:

2	4	8	12
$\updownarrow$	$\updownarrow$	$\updownarrow$	$\updownarrow$
6	12	24	36

9. The following correspondence shows that sets I^- and N are equivalent:

-1	-2	-3	$\cdots$	$-n$	$\cdots$
$\updownarrow$	$\updownarrow$	$\updownarrow$		$\updownarrow$	
1	2	3	$\cdots$	n	$\cdots$

11. $n(A) = 26$
13. $n(C) = 50$
15. $n(E) = \aleph_0$
17. The set $\{100, 200, 300, \ldots\}$ can be put into a one-to-one correspondence with a subset of itself, $\{200, 300, 400, \ldots\}$. This shows that the set is infinite.
19. The set $\{\frac{1}{3}, \frac{2}{3}, \frac{3}{3}, \ldots\}$ can be put into a one-to-one correspondence with a subset of itself, $\{\frac{2}{3}, \frac{3}{3}, \frac{4}{3}, \ldots\}$. This shows that the set is infinite.
21. Sets B and D are equal and equivalent.
23. Set A is neither equal nor equivalent to any of the other sets.
25. $\aleph_0$
27. $\aleph_0$
29. a. $\frac{7}{9}$ and $\frac{8}{9}$
   b. $\frac{1}{3} + \frac{2}{9} + \frac{4}{27} + \frac{8}{81} + \cdots$; the sum gets closer and closer to 1.
31. To room 223
33. Rooms $1, 3, 5, \ldots, 2n + 1, \ldots$
35. To room 666

CHAPTER 3

Exercises 3.1

1. Not a statement
3. A compound statement with the following components: Jane is taking an English course. She has four themes to write.
5. Not a statement
7. A compound statement with the following components: Students at Ohio State University are required to take a course in history. Students at Ohio State University are required to take a course in economics.

9. $a \wedge f$
11. $d \vee f$
13. $b \wedge p$
15. $a \vee m$
17. $p \wedge q$
19. $\sim p \wedge \sim q$
21. Ricky loves Lucy, or Lucy does not love Ricky.
23. Ricky loves Lucy, but Lucy does not love Ricky.
25. It is not the case that Ricky and Lucy love each other.
27. Bill's store is not making a good profit.
29. My dog is not a spaniel.
31. I like to work overtime.
33. These two are negations of each other.
35. Some men are not mortal.
37. All basketball players are 6 ft tall.
39. He is not bald and he does not have a 10-in. forehead.
41. No circles are round.
43. Nobody up there loves me.
45. Somebody does not like to go on a trip.
47. Some persons occupying your covered auto are not insured.
49. All expenses are subject to the 2% limit.
51. Statement (**d**)
53. $(d \wedge p) \vee r$
55. $r \wedge (t \vee g)$
57. a. The diagram is neither a rectangle nor a square.
   b. The diagram is a square or not a rectangle.
   c. The diagram is a square and a rectangle.
59. Sentence (1) is true. Sentence (2) is false. If we assume that sentence (3) is true, there will be two sentences, (1) and (3), that are true; thus, sentence (3) is false!
61. F   **63.** F   **65.** T   **67.** T   **69.** T

Exercises 3.2

1. Today is Friday or Monday.
3. Today is not Friday.
5. He is a gentleman or a scholar.
7. He is a gentleman and a scholar.
9. $g \wedge s$
11. a. $p \wedge q$ **b.** $p \vee q$
   c. Statement in (**a**) is false.
      Statement in (**b**) is true.
13. $\sim q \wedge p$ or $\sim (q \vee \sim p)$
15. $p \vee q$
17. $p \vee q$, true
19. $\sim p \wedge \sim q$, false.
21. $\sim q \wedge \sim p$, false
23. $g \vee \sim j$, false
25. $(g \vee j) \wedge \sim (g \wedge j)$, true

27.

1	2	4	3
p	q	$p \vee$	$\sim q$
T	T	T	F
T	F	T	T
F	T	F	F
F	F	T	T

29.

1	2	3	4
p	q	$\sim p$	$\wedge q$
T	T	F	**F**
T	F	F	**F**
F	T	T	**T**
F	F	T	**F**

31.

1	2	5	4	3
p	q	$\sim$	$(p \vee$	$\sim q)$
T	T	**F**	T	F
T	F	**F**	T	T
F	T	**T**	F	F
F	F	**F**	T	T

33.

1	2	6	3	5	4
p	q	$\sim$	$(\sim p$	$\wedge$	$\sim q)$
T	T	**T**	F	F	F
T	F	**T**	F	F	T
F	T	**T**	T	F	F
F	F	**F**	T	T	T

35.

1	2	3	6	4	5
p	q	$(p \wedge q)$	$\vee$	$(\sim p$	$\wedge q)$
T	T	T	**T**	F	F
T	F	F	**F**	F	F
F	T	F	**T**	T	T
F	F	F	**F**	T	F

37.

1	2	3	5	4
p	q	r	$p \wedge$	$(q \vee r)$
T	T	T	**T**	T
T	T	F	**T**	T
T	F	T	**T**	T
T	F	F	**F**	F
F	T	T	**F**	T
F	T	F	**F**	T
F	F	T	**F**	T
F	F	F	**F**	F

39.

1	2	3	4	7	6	5
p	q	r	$(p \vee q)$	$\vee$	$(r \wedge$	$\sim q)$
T	T	T	T	**T**	F	F
T	T	F	T	**T**	F	F
T	F	T	T	**T**	T	T
T	F	F	T	**T**	F	T
F	T	T	T	**T**	F	F
F	T	F	T	**T**	F	F
F	F	T	F	**T**	T	T
F	F	F	F	**F**	F	T

41. a. True when p and q are both true.
   b. False if at least one of p and q is false.
   c. True if at least one of p and q is true.
   d. False only if both p and q are false.

43.

1	2	3	5	4	6	8	7
p	q	r	$p \vee$	$(q \wedge r)$	$(p \vee q)$	$\wedge$	$(p \vee r)$
T	T	T	**T**	T	T	**T**	T
T	T	F	**T**	F	T	**T**	T
T	F	T	**T**	F	T	**T**	T
T	F	F	**T**	F	T	**T**	T
F	T	T	**T**	T	T	**T**	T
F	T	F	**F**	F	T	**F**	F
F	F	T	**F**	F	F	**F**	T
F	F	F	**F**	F	F	**F**	F

Columns 5 and 8 of the above truth table show that the two statements have the same truth values, so they are equivalent.

45.

1	2	4	3	5	7	6
p	q	$\sim$	$(p \vee q)$	$\sim p$	$\wedge$	$\sim q$
T	T	**F**	T	F	**F**	F
T	F	**F**	T	F	**F**	T
F	T	**F**	T	T	**F**	F
F	F	**T**	F	T	**T**	T

Columns 4 and 7 of the above table show that the two statements have the same truth values, so they are equivalent.

47.

1	2	3	5	4	6
p	q	$(p \wedge q)$	$\vee$	$\sim p$	$q \vee \sim p$
T	T	T	**T**	F	**T**
T	F	F	**F**	F	**F**
F	T	F	**T**	T	**T**
F	F	F	**T**	T	**T**

Columns 5 and 6 of the preceding table show that the two statements have the same truth values, so they are equivalent.

49. **a.** $p \wedge q$ is true only when both p and q are true; this gives the truth values *TFFF*. $p \wedge {\sim}q$ is true only when p is true and q is false; this gives the truth values *FTFF*. ${\sim}p \wedge q$ is true only when p is false and q is true; this gives the truth values *FFTF*. ${\sim}p \wedge {\sim}q$ is true only when p and q are both false; this gives the truth values *FFFT*. This verifies the table.

   b. $p \wedge q$ is true only in the first row, and ${\sim}p \wedge {\sim}q$ is true only in the last row. So $(p \wedge q) \vee ({\sim}p \wedge {\sim}q)$ has the truth values *TFFT*.

   c. $(p \wedge {\sim}q) \vee ({\sim}p \wedge q)$ has truth values *FTTF*. $(p \wedge {\sim}q) \vee ({\sim}p \wedge q) \vee ({\sim}p \wedge {\sim}q)$ has truth values *FTTT*. ${\sim}(p \wedge q)$ is a simpler statement with truth values *FTTT*.

51. None are eligible.
53. 7 is greater than or equal to 5.
55. 0 is less than or equal to 3.
57. $\frac{1}{2}$ is greater than $\frac{1}{8}$.
59. I will not go fishing or the sun is not shining. This would be true if either or both of the components, "I will not go fishing" and "The sun is not shining," were true.
61. $[(e \wedge g) \vee a] \wedge h \wedge (c \vee n \vee {\sim}t)$
63. Mr. Baker is the carpenter.

Exercises 3.3

1.

1	2	3	5	4	6
p	q	${\sim}q$	$\rightarrow$	${\sim}p$	$p \rightarrow q$
T	*T*	*F*	*T*	*F*	*T*
T	*F*	*T*	*F*	*F*	*F*
F	*T*	*F*	*T*	*T*	*T*
F	*F*	*T*	*T*	*T*	*T*

Columns 5 and 6 are identical, so ${\sim}q \rightarrow {\sim}p$ is equivalent to $p \rightarrow q$.

3.

1	2	3	4	5
p	q	${\sim}p$	${\sim}p \rightarrow q$	$p \vee q$
T	*T*	*F*	*T*	*T*
T	*F*	*F*	*T*	*T*
F	*T*	*T*	*T*	*T*
F	*F*	*T*	*F*	*F*

Since columns 4 and 5 are identical, ${\sim}p \rightarrow q$ and $p \vee q$ are equivalent.

5. *F*   7. *T*   9. *x* can be any number.
11. *x* can be any number except 4.
13. It is false whenever the antecedent, "You've got the time," is true and the consequent, "We've got the beer," is false.

15.

1	2	3	4	6	5
p	q	r	$(p \rightarrow q)$	$\Leftrightarrow$	$(p \vee r)$
T	*T*	*T*	*T*	*T*	*T*
T	*T*	*F*	*T*	*T*	*T*
T	*F*	*T*	*F*	*F*	*T*
T	*F*	*F*	*F*	*F*	*T*
F	*T*	*T*	*T*	*T*	*T*
F	*T*	*F*	*T*	*F*	*F*
F	*F*	*T*	*T*	*T*	*T*
F	*F*	*F*	*T*	*F*	*F*

17.

1	2	3	5	4
p	q	r	$p \rightarrow$	$(q \wedge r)$
T	*T*	*T*	*T*	*T*
T	*T*	*F*	*F*	*F*
T	*F*	*T*	*F*	*F*
T	*F*	*F*	*F*	*F*
F	*T*	*T*	*T*	*T*
F	*T*	*F*	*T*	*F*
F	*F*	*T*	*T*	*F*
F	*F*	*F*	*T*	*F*

19. The final columns in the tables in problems 17 and 18 are identical, so the two statements are equivalent.
21. $p \rightarrow q$ 23. ${\sim}q \rightarrow {\sim}p$
25. $q \rightarrow {\sim}p$ 27. ${\sim}b \rightarrow {\sim}s$
29. ${\sim}a \vee b$; the temperature is not above 80°, or I would go to the beach.
31. ${\sim}a \vee g$; Eva does not have a day off, or she would go to the beach.
33. You do not have the time, or we've got the beer.
35. If it is a dog, then it is a mammal.
37. If it is a man, then it is created equal.
39. If it is a rectangle with perpendicular diagonals, then it is a square.

41.

1	2	4	3	6	5
p	q	${\sim}$	$(p \rightarrow q)$	$p \wedge$	${\sim}q$
T	*T*	*F*	*T*	*F*	*F*
T	*F*	*T*	*F*	*T*	*T*
F	*T*	*F*	*T*	*F*	*F*
F	*F*	*F*	*T*	*F*	*T*

Since columns 4 and 6 are identical, $\sim(p \to q)$ and $p \wedge \sim q$ are equivalent.

43. Johnny does not play quarterback, and his team does not lose.

45. I kiss you once, but I do not kiss you again.

47. Evel Knievel is careless, but he will not lose his life.

49. If Johnny plays quarterback, then his team wins.

51. If Joe had not had an accident, then he would be able to get car insurance.

53. No   **55.** Statement (**d**)   **57.** Statement (**d**)

59. No. $p \to q$ is true if p is false and q is either true or false.

61. The student has to take the placement examination only if the student has satisfied the freshman requirements (perhaps by advanced courses in high school) and is being admitted to sophomore standing, but is entering college for the first time.

67. $r \to a$

69. No. It only says that an adjustment will be made if a report is made in 10 days.

71. *A* must see at least one black hat, or she would know that her hat is black since they are not all white. *B* also must see at least one black hat, and further, that hat had to be on *C*, otherwise she would know that her hat was black (since she knows *A* saw at least one black hat). So *C* knows that her hat is black, without even seeing the others' hats.

73. The one who fell silent, presumably the quickest of the three, reasoned that his head must be painted also. The argument goes as follows. Let's call the quick logician *Q*, and the other two *D* and *S*. Let's assume *Q*'s head is untouched. Then *D* is laughing because *S*'s head is painted, and vice versa. But eventually, *D* and *S* will realize that their head must be painted, because the other is laughing. So they will quit laughing as soon as they realize this. *Q* waits what he thinks is a reasonable amount of time for them to figure this out, and when they don't stop laughing, his worst fears are confirmed. He concludes that his assumption is invalid and he must be crowned in crimson too.

Skill Checker 3.3

p	q	$\sim p$	$p \vee q$	$p \wedge q$
T	T	F	T	T
T	F	F	T	F
F	T	T	T	F
F	F	T	F	F

Exercises 3.4

1. If n is divisible by 2, then n is an even number.

3. $q \to p$ **5.** $q \to p$

7. If one is a mathematics major, then one takes calculus.

9. If the measure gets a two-thirds vote, then it carries.

11. If we have a stable economy, then we have low unemployment.

13. If birds are of a feather, then they flock together.

15.

			Converse	Inverse
p	q	$p \to q$	$q \to p$	$\sim p \to \sim q$
T	T	T	T	T
T	F	F	T	T
F	T	T	F	F
F	F	T	T	T

The converse, $q \to p$, is true except when q is true and p is false (third row). The inverse, $\sim p \to \sim q$, is true except when $\sim p$ is true and $\sim q$ is false (third row). The converse and the inverse have the same truth values and, hence, are equivalent.

17. $u \to h$ **19.** $u \to h$

21. $u \to h$ **23.** $u \leftrightarrow h$

25. $u \to h;\ \sim h \to \sim u;\ \sim u \to \sim h$

27. $p \leftrightarrow s$

29. a. Converse: If you are not strong, then you do not eat your spinach.
Inverse: If you eat your spinach, then you are strong.
Contrapositive: If you are strong, then you eat your spinach.

b. Converse: If you are strong, then you eat your spinach.
Inverse: If you do not eat your spinach, then you are not strong.
Contrapositive: If you are not strong, then you do not eat your spinach.

c. Converse: If you eat your spinach, then you are strong.
Inverse: If you are not strong, then you do not eat your spinach.
Contrapositive: If you do not eat your spinach, then you are not strong.

31. If the square of an integer is divisible by 4, the integer is even. True.

33. If I am neat and well dressed, then I can get a date. False.

35. If you pass this course, then you get passing grades on all the tests. False.

37. If we cannot find a cure for cancer, then the research is inadequately funded.

39. If a person does not want to improve the world, then the person is not a radical.

41. Equivalence (**c**).   **43.** Statement (**b**).

45.

1	2	3	4
p	q	$(p \wedge q) \to p$	
T	T	T	T
T	F	F	T
F	T	F	T
F	F	F	T

Column 3 is the conjunction of columns 1 and 2, and so has T only in the first row, where both p and q are true. Therefore, column 4 is all T's; this shows that $(p \wedge q) \to p$ is a tautology.

47.

1	3	2
p	$p \leftrightarrow {\sim}p$	
T	F	F
F	F	T

Since column 3 is all F's, the statement $p \leftrightarrow {\sim}p$ is a contradiction.

49.

1	2	3	4	6	5
p	q	$({\sim}p \wedge q)$	$\to$		$(p \to q)$
T	T	F	F	T	T
T	F	F	F	T	F
F	T	T	T	T	T
F	F	T	F	T	T

Since column 6 is all T's, the first statement, ${\sim}p \wedge q$, implies the second, $p \to q$.

51. Equivalent
53. $p \wedge {\sim}q$ implies ${\sim}p \vee {\sim}q$.
63. a. $Q \cap R'$ **b.** $(P \cap Q) \cap R'$
65. The contrapositive of ${\sim}q \to {\sim}p$ is $p \to q$.
67. The inverse of $p \to q$ is ${\sim}p \to {\sim}q$, and the contrapositive of ${\sim}p \to {\sim}q$ is $q \to p$.
69. $({\sim}r \wedge {\sim}s) \vee (p \vee q) \Leftrightarrow (r \vee s) \to (p \vee q)$; it is true because $(r \vee s) \to (p \vee q) \Leftrightarrow {\sim}(r \vee s) \vee (p \vee q) \Leftrightarrow ({\sim}r \wedge {\sim}s) \vee (p \vee q)$, and the converse of $(r \vee s) \to (p \vee q)$ is $(p \vee q) \to (r \vee s)$.
71. The direct statement   **73.** The contrapositive

Exercises 3.5

1. Premises: "No misers are generous" and "Some old persons are not generous." Conclusion: "Some old persons are misers."
3. Premises: "All diligent students get A's" and "All lazy students are not successful." Conclusion: "All diligent students are lazy."
5. Premises: "No kitten that loves fish is unteachable" and "No kitten without a tail will play with a gorilla." Conclusion: "No unteachable kitten will play with a gorilla."
7. Valid

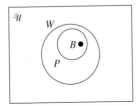

9. Invalid

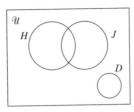

11. Valid

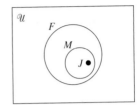

13. Invalid

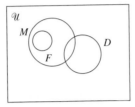

15. Invalid

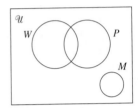

17. Invalid

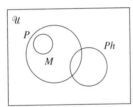

19. Valid

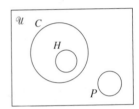

21. Invalid

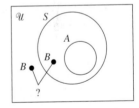

23. Invalid

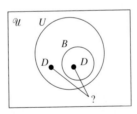

25. Invalid

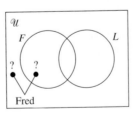

27. b.

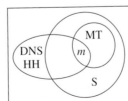

m is a salsero and does not sing hip-hop, so (**b**) is the correct conclusion

29. a.

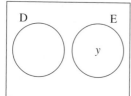

y (you) are inside Enthusiastic and outside Doctors, so you are not a doctor and the correct conclusion is (**a**).

31. b.

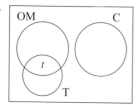

t is a thin person who is not cheerful, so (**b**) is the correct conclusion.

33. a.

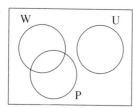

The intersection of the set of puppies P and the set of wasps, W is not necessarily empty, so no conclusion can be drawn.

35. a.

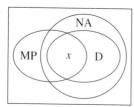

x is a difficult problem that needs attention, so the correct conclusion is (**a**).

37. b. Statement (b) can be logically deduced from the diagram.

39. a. Yes, an argument is valid if and only if the conclusion is true whenever all the premises are true.
   b. No. By the preceding statement, if the premises are all true and the conclusion is false, then the argument is invalid.

41. No. It may be that the conclusion does not follow from the premises. See Example 2 of this section.

43. The conclusion is true.

45. Valid

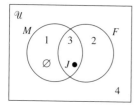

47. Invalid

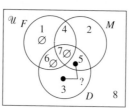

49. Invalid

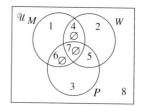

51. Invalid

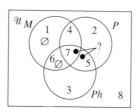

53. Valid

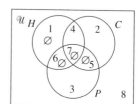

55. Only conclusion (**a**) is valid.

57. Yes. See the diagram.

59. Some z's are y's. See the diagram.

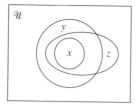

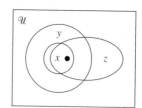

Skill Checker

1.

p	q	$\sim p$	$\sim q$	$p \rightarrow \sim q$	$p \vee \sim q$	$\sim p \wedge q$
T	T	F	F	F	T	F
T	F	F	T	T	T	F
F	T	T	F	T	F	T
F	F	T	T	T	T	F

Exercises 3.6

1. $e \rightarrow p$
$\underline{\sim e}$
$\therefore \sim p$
Invalid

3. $s \rightarrow e$
$\underline{\sim e}$
$\therefore \sim s$
Valid

5. $\underline{g}$
$\therefore g \wedge r$
Invalid

7. $w \rightarrow m$
$\underline{\sim w \rightarrow g}$
$\therefore m \vee g$
Valid

9. $t \rightarrow b$
$\underline{t}$
$\therefore b$
Valid

11. $s \rightarrow f$
$\underline{s}$
$\therefore f$
Valid

13. $m \rightarrow e$
$\underline{\sim m}$
$\therefore \sim e$
Invalid

15. $f \rightarrow s$
$\underline{\sim f}$
$\therefore \sim s$
Invalid

17. Valid

19. Invalid

21. Valid

23. $p \rightarrow r$

25. s

27. q

29. No politicians are reliable.

31. Some students are snobs.

33. All romances are well written.

35. Aardvarks do not vote.

37. c. Some college students will not pass the course.

39. b. No teacher is dumb.

41. c.

43. b.

45. b.

47. You read *X* magazine.

49. "Affirming the consequent" means that the "then" statement is affirmed and the "if" statement is taken as a valid conclusion. This is a fallacy because $p \rightarrow q$ is true if p is false and q is either true or false.

51. Kittens that will play with a gorilla do not have green eyes. (Or the equivalent: No kitten with green eyes will play with a gorilla.)

CHAPTER 4

Exercises 4.1

1. ∩∩IIIII

3. ꝃ∩∩∩∩II

5. ꝃꝃꝃꝃ∩∩∩III
ꝃꝃꝃꝃ   II

7. 113

9. 322

11. 11,232

13. ∩∩∩IIIII
+ ∩∩III
∩∩∩∩∩IIIIIIII

15.
ꝃꝃꝃ∩∩∩II →
− ꝃ∩∩∩∩III →
ꝃꝃꝃ∩∩∩∩∩∩∩∩∩∩IIIIIIIIIIIII
ꝃ∩∩∩∩   III
ꝃꝃ∩∩∩∩∩∩∩IIIIIIIII

17.
\1	40
\2	80
\4	160
\8	320
15	600

19.
1	51
\2	102
\4	204
8	408
\16	816
22	1122

21.
18	32
9	64*
4	128
2	256
1	512*
	576*

23.
12	51
6	102
3	204*
1	408*
	612*

25. ▼▼▼▼▼▼

27. ◄◄◄▼▼

29. ▼▼ ▼▼▼

31. ▼▼▼▼ ◄▼▼▼▼▼▼▼

33. ▼ ▼▼ ◄▼▼▼

35. 92

37. 192

39. 4322

41.
◄◄◄▼▼
+ ◄◄◄◄▼▼▼
◄◄◄◄◄◄◄▼▼▼▼▼ = ▼ ◄▼▼▼▼▼

43.
▼▼ ◄▼▼▼
+ ▼ ▼▼▼▼▼▼▼
▼▼▼ ◄▼▼▼▼▼▼▼▼▼▼ = ▼▼▼ ◄◄▼

45. 126

47. 42,000

49. 90,405

51. LXXII

53. CXLV

55. $\overline{\text{XXXII}}$DIII

57. 8

59. 4

61. No. C is more than two steps larger than I, so this subtraction is not allowed. I may be subtracted from V or X only.

63. The Babylonian system is a base 60 system and our decimal system is a base 10 system. Another important difference is the lack of a symbol for zero in the Babylonian system. The Babylonian system was not a good place system; it depended on spacing. The symbol for 1 was the same as that for 60, and only the spacing could show which was intended.

65. The Egyptian system was based on 10 and the Babylonian on 60. The Egyptian system was not a positional system; it depended essentially on the addition of the symbol values. The Babylonian system used spacing to change symbol values.

67. Assume the answer is 6. $6 + \left(\frac{1}{6}\right)(6) = 7$ and $21 \div 7 = 3$. Hence, the correct answer is $3 \times 6 = 18$.

69. Assume the answer is 3. $3 + \left(\frac{2}{3}\right)(3) = 5$. $5 - \left(\frac{1}{3}\right)(5) = \frac{10}{3}$ and $10 \div \frac{10}{3} = 3$. Therefore, the correct answer is $3 \times 3 = 9$.

71. $n = 8$

Exercises 4.2

1. $(4 \times 10^2) + (3 \times 10) + (2 \times 10^0)$

3. $(2 \times 10^3) + (3 \times 10^2) + (7 \times 10^0)$

5. $(1 \times 10^4) + (2 \times 10^3) + (3 \times 10^2) + (4 \times 10) + (9 \times 10^0)$

7. 1

9. 45

11. 9071

13. 748,308

15. 4,000,031

17.
23	$(2 \times 10) + (3 \times 10^0)$
+ 13	$(1 \times 10) + (3 \times 10^0)$
36	$(3 \times 10) + (6 \times 10^0)$

19.
71	$(7 \times 10) + (1 \times 10^0)$
+ 23	$(2 \times 10) + (3 \times 10^0)$
94	$(9 \times 10) + (4 \times 10^0)$

21.
76	$(7 \times 10) + (6 \times 10^0)$
− 54	$(-)(5 \times 10) + (4 \times 10^0)$
22	$(2 \times 10) + (2 \times 10^0)$

23.
$$\begin{array}{r} 84 \\ -\,31 \\ \hline 53 \end{array} \qquad \begin{array}{l} (8 \times 10) + (4 \times 10^0) \\ (-)(3 \times 10) + (1 \times 10^0) \\ \hline (5 \times 10) + (3 \times 10^0) \end{array}$$

25. 7^{11} 27. 6^{40} 29. 6^7

31. 6^{12} 33. 5^{12} 35. 10^{30}

37.
$$\begin{array}{r} 25 \\ \times\,51 \\ \hline 25 \\ 125 \\ \hline 1275 \end{array}$$

$$(2 \times 10) + (5 \times 10^0)$$
$$\times\ (5 \times 10) + (1 \times 10^0)$$
$$(2 \times 10) + (5 \times 10^0)$$
$$(10 \times 10^2) + (25 \times 10)$$
$$10^3 + (27 \times 10) + (5 \times 10^0)$$
$$= (1 \times 10^3) + (2 \times 10^2) + (7 \times 10) + (5 \times 10^0)$$
$$= 1275$$

39.
$$\begin{array}{r} 62 \\ \times\,25 \\ \hline 310 \\ 124 \\ \hline 1550 \end{array}$$

$$(6 \times 10) + (\ 2 \times 10^0)$$
$$\times\ (2 \times 10) + (\ 5 \times 10^0)$$
$$(30 \times 10) + (10 \times 10^0)$$
$$(12 \times 10^2) + (\ 4 \times 10)$$
$$(12 \times 10^2) + (34 \times 10) + (10 \times 10^0)$$
$$= (1 \times 10^3) + (5 \times 10^2) + (5 \times 10)$$
$$= 1550$$

41.
$$\begin{array}{r} 8 \\ 8\overline{)64} \\ 64 \\ \hline 0 \end{array}$$

$$(8 \times 10^0)$$
$$8 \times 10^0\overline{)(6 \times 10) + (4 \times 10^0)}$$
$$(6 \times 10) + (4 \times 10^0)$$
$$\hline 0$$

43.
$$\begin{array}{r} 12 \\ 6\overline{)72} \\ 6 \\ \hline 12 \\ 12 \\ \hline 0 \end{array}$$

$$(1 \times 10) + (2 \times 10^0)$$
$$(6 \times 10^0)\overline{)(7 \times 10) + (2 \times 10^0)}$$
$$(6 \times 10)$$
$$(1 \times 10) + (2 \times 10^0)$$
$$(1 \times 10) + (2 \times 10^0)$$
$$\hline 0$$

45. 3×10^5 47. 7.735×10^7 dogs

49. 1.6×10^6 searches   51. 5.1×10^{10} lb

53. 3.06×10^{10} lb   55. 7.2×10^9 gal

57. 5 lb   59. 31 years

61. You must add the exponents; you obtain a^{m+n}.

63. You must multiply the exponent m by the exponent n; you obtain a^{mn}.

65. There were 137,256 in all on the road to Rome.

Exercises 4.3

1. 22_{three} 3. 31_{four}

5. (*******) ******; 17_{eight}

7. (*******) (******) *; 21_{seven}

9. 22   11. 139

13. 27   15. 291

17. 30_{five} 19. 11100_{two}

21. $19_{sixteen}$ 23. 41_{six}

25. 121_{seven} 27. 46_{eight}

29. $5BB_{sixteen}$

31. $73 = 1001001_{two} = 111_{eight}$

33.
00110	01010	00101	00011	01100
3	5	2	1	6

35.
01001	00101	00011	00110	01010
4	2	1	3	5

37. ||||ıııılılılılıllılıl

39. ||ılıııl ılılılıllılılılll

41. Yes. The zip code has nine digits but the tenth digit is the checking number.

43. One of the meanings of *binary* is "based on two." The prefix *bi-* means "two."

45. *Hexadecimal* means "based on 16." The prefix *hexa-* means "six."

47. The trick works because the columns correspond to the binary digits in the number. For instance, $6 = 110_{two}$ and this corresponds to the number $6 = 2 + 4$, the numbers that head columns B and C. Note that 6 occurs in columns B and C, but not in A.

49. Use the same procedure as for the numbers from 1 to 7 but with five columns instead of three.

Calculator Corner

1. 13   3. 113   5. 1914

Exercises 4.4

1. 1001_2 3. 10011_2 5. 10010_2

7. 101_2 9. 1_2 11. 1010_2

13. 10010_2 15. 101101_2 17. 110111_2

19. 110_2 R 1_2 21. 100_2 R 10_2 23. 1011_2 R 100_2

25. HELLO. H $\leftrightarrow$ 72, E $\leftrightarrow$ 69, L $\leftrightarrow$ 76, O $\leftrightarrow$ 79

		Binary	Decimal	Hexadecimal
27.	A	01000001	65	41
29.	Q	01010001	81	51
31.	X	01011000	88	58

33.
Hexadecimal	Binary	Letter
48	01001000	H
45	01000101	E
4C	01001100	L
50	01010000	P

35. 55   37. 000001_2 39. 011111_2

Exercises 4.5

1. 600_8 3. 10112_8 5. 432_8

7. 7154_8 9. 507_8 11. 2306_8

13. 35_8 R 4_8 15. 250_8 R 5_8 17. 417_{16}

19. $9B8_{16}$ 21. $A367_{16}$ 23. $4A451_{16}$

25. Answers may vary. The main reason is that there are only two digits in the binary system contrasted with 16 digits in the hexadecimal system.

27. 2.625   29. 2.125   31. 58.75

Calculator Corner

1. Problem 13, remainder 4. Problem 15, remainder 5.

CHAPTER 5

Exercises 5.1

1. For identification only **3.** A cardinal number

5. The "First" is for identification; the "one" is an ordinal number.

7.

51	52	(53)	54	55	56	57	58	(59)	60
(61)	62	63	64	65	66	(67)	68	69	70
(71)	72	(73)	74	75	76	77	78	(79)	80
81	82	(83)	84	85	86	87	88	(89)	90
91	92	93	94	95	96	(97)	98	99	100

The primes in the table are circled.

9. 6 **11.** 4

13. a. 2 and 3

b. No. If any pair of consecutive counting numbers greater than 2 is selected, one of the pair must be an even number (divisible by 2) and, hence, not a prime.

15. a. The product part of m is exactly divisible by 2, so that m divided by 2 would have a remainder of 1.

b. The product part of m is exactly divisible by 3, so that m divided by 3 would have a remainder of 1.

c.–d. Exactly the same reasoning as in parts (**a**) and (**b**) applies here. If m is divided by any prime from 2 to P, there is a remainder of 1.

e. Because P was assumed to be the largest prime.

f. Because m is not divisible by any of the primes from 2 to P.

17. 1, 2, 5, 10, 25, 50

19. 1, 2, 4, 8, 16, 32, 64, 128

21. 1, 7, 11, 13, 77, 91, 143, 1001

23. 41 is a prime. **25.** $91 = 7 \times 13$

27. $148 = 2^2 \times 37$ **29.** 490 **31.** 1200

33. a. Divisible by 3 and by 5

b. Divisible by 2, by 3, and by 5

c. Divisible by 2, by 3, and by 5

35. Three: 2, 11, and 22 **37.** 27

39. Relatively prime **41.** 47

43. Relatively prime **45.** 20

47. $\frac{31}{44}$ **49.** $\frac{3}{14}$ **51.** $\frac{1}{4}$

53. LCM = 165; $\frac{14}{165}$ **55.** LCM = 992; $\frac{101}{992}$

57. LCM = 720; $\frac{7}{720}$ **59.** LCM = 180; $\frac{31}{180}$

61. LCM = 167,580; $\frac{281}{33,516}$ **63.** LCM = 4200; $\frac{3}{280}$

65. $\frac{1}{2}$ **67.** $\frac{3}{20}$ **69.** $\frac{3}{10}$

71. a. $100 = 3 + 97 = 11 + 89 = 17 + 83$
$= 29 + 71 = 41 + 59 = 47 + 53$

b. $200 = 3 + 197 = 7 + 193 = 19 + 181$
$= 37 + 163 = 43 + 157 = 61 + 139$
$= 73 + 127 = 97 + 103$

73. The number 1 has only one divisor: itself. It is not a prime because a prime must have exactly *two distinct* divisors, 1 and itself. It is not a composite number because it has only one divisor.

75. The largest prime that you need to try is 13, because the next prime is 17 and $17^2 = 289$, which is greater than 211.

77. All the other digits are multiples of 3, so their sum is divisible by 3. Thus, only the sum of 2 and 7 needs to be checked.

79. Since 999 and 99 and 9 are all divisible by 9, only the sum

$$2 \times 1 + 8 \times 1 + 5 \times 1 + 3$$

which is exactly the sum of the digits, needs to be checked. If this sum is divisible by 9, the original number is divisible by 9, and not otherwise.

81. a. Divisible by 4, not by 8

b. Divisible by 4 and by 8

c. Divisible by 4 and by 8

d. Divisible by 4, not by 8

83. None of the numbers 1, 2, 3, 4, or 5 is the sum of its proper divisors. Therefore, 6 is the smallest perfect number.

85. $496 = 1 + 2 + 4 + 8 + 16 + 31 + 62 + 124 + 248$ so 496 is a perfect number.

87. All primes are deficient because they have only 1 as a proper divisor.

89. They end in 6 or 28. Also, they are sums of powers of 2; $6 = 2 + 2^2$, $28 = 2^4 + 2^3 + 2^2$, and so on.

Exercises 5.2

1.

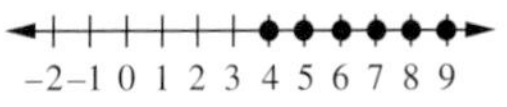

3.

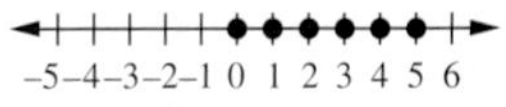

5. -3 **7.** 8

9.

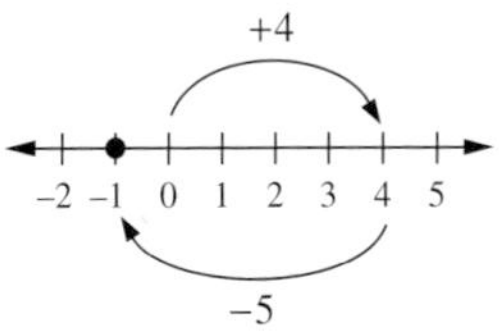

11.

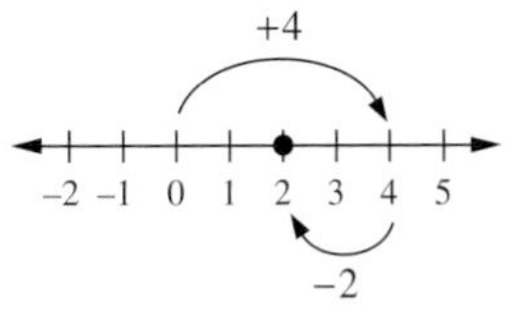

13.

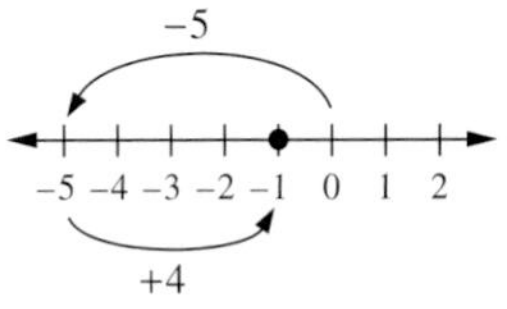

15.

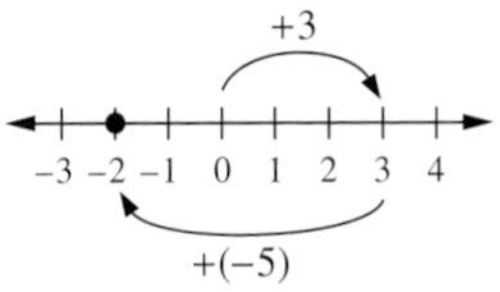

17. $3 + (-8) = -5$ **19.** $3 + (-4) = -1$

21. $-5 + (-2) = -7$ **23.** $5 + (+6) = 11$

25. $-3 + (+4) = 1$ **27.** $-5 + (+3) = -2$

29. a. −15　　**b.** −72
31. a. −20　　**b.** −39
33. a. −60　　**b.** −60
35. A negative even integer　**37.** −14; 492
39. +19; 525　　**41.** +53; 559　　**43.** 4　　**45.** 0
47. a. −27　　**b.** 4
49. a. −9　　**b.** −5　　**51.** 10　　**53.** −5
55. 11,800　　**57.** 10　　**59.** 20　　**61.** 20 km
63. Step 1. Because m is assumed to be a multiplicative identity
Step 3. Because 1 is a multiplicative identity
Step 4. Because both m and 1 equal $m \cdot 1$
65. Step 1. By the definition of subtraction
Step 3. By the associative property of addition
Step 5. Because 0 is the additive identity
67. b. adding a
c. identity
e. q (1 is the multiplicative identity)
f. identity, unique
69. Step 1. Because 0 is the additive identity
Step 3. By the distributive property
Step 5. Because the additive identity (0) is unique
71. The product of two positive numbers is a positive number.
73. The product of two negative numbers is a positive number.
75. −3　　　　**77.** 0

Exercises 5.3

1. Numerator 3, denominator 4
3. Numerator 3, denominator −5
5. $\frac{17}{41} = \frac{289}{697}$　　**7.** $\frac{11}{91} = \frac{253}{2093}$　　**9.** $\frac{5}{2}$
11. $\frac{7}{16}$　　**13.** $\frac{15}{14}$　　**15.** $\frac{2}{3}$
17. $\frac{14}{18}$　　**19.** $\frac{11}{18}$　　**21.** $\frac{16}{63}$
23. $\frac{19}{12}$　　**25.** $\frac{176}{323}$　　**27.** $\frac{2}{63}$
29. $-\frac{1}{12}$　　**31.** $\frac{62}{323}$　　**33.** $\frac{2}{3}$
35. $\frac{21}{8}$　　**37.** $\frac{56}{27}$　　**39.** $\frac{18}{77}$
41. $\frac{9}{14}$　　**43.** $-\frac{1}{6}$　　**45.** $\frac{1}{8}$
47. $\frac{1}{4}$　　**49.** $\frac{49}{80}$　　**51.** 16
53. $\frac{3}{2}$　　**55.** $\frac{2}{5}$　　**57.** $1\frac{9}{14}$
59. $\frac{2}{7}$　　**61.** $3\frac{5}{12}$　　**63.** $\frac{3}{7}$
65. $-\frac{3}{4}$　　**67.** −18　　**69.** $-\frac{4}{5}$
71. −30　　**73.** $7\frac{1}{8}$　　**75.** $2\frac{5}{8}$
77. 0　　**79.** $-\frac{2}{9}$　　**81.** $-5\frac{3}{4}$
83. $1\frac{1}{6}$　　**85.** $\frac{1}{2}$　　**87.** $\frac{7}{36}$
89. $3\frac{7}{15}$　　**91.** 3°F　　**93.** $1\frac{5}{16}$ lb
95. $\frac{1}{2}$　　**97.** $\$4\frac{9}{10}$　　**99.** $12\frac{1}{10}$ hr
101. $\frac{1}{10}$ hr　　**103.** $\frac{7}{30}$ hr　　**105.** Traffic; $\frac{1}{10}$ hr
107. No. News uses 8 min at the top of the hour and 6 min at the bottom, and traffic uses 2 min at the top of the hour and 4 min at the bottom.

109. Weather and news
111. If $\frac{0}{0} = n$, then $0 = 0 \times n = 0$ no matter what value is assigned to n. Thus, $\frac{0}{0}$ cannot be uniquely defined.
113. 126 mi　　　**115.** 3 in.
117. $\frac{5}{1} \ \frac{5}{2} \ \frac{5}{3} \ \frac{5}{4} \ \frac{5}{5} \ \frac{5}{6} \ \frac{5}{7} \ \frac{5}{8} \ \frac{5}{9} \cdots$
$\frac{6}{1} \ \frac{6}{2} \ \frac{6}{3} \ \frac{6}{4} \ \frac{6}{5} \ \frac{6}{6} \ \frac{6}{7} \ \frac{6}{8} \ \frac{6}{9} \cdots$
119. $\frac{2}{2} = 1, \frac{4}{2} = 2, \frac{3}{3} = 1$, and $\frac{2}{4} = \frac{1}{2}$; these have already been caught in the one-to-one correspondence.
121. Neither. The two sets have the same cardinal number.

Exercises 5.4

1. $(6 \times 10^2) + (9 \times 10) + (2 \times 10^0) + (8 \times 10^{-2}) + (7 \times 10^{-3})$
3. $(1 \times 10^{-3}) + (7 \times 10^{-5})$　　**5.** 5020.39
7. 0.004702　　**9.** 9.35×10^2　　**11.** 1.2×10^{-3}
13. 86,400　　**15.** 0.00671　　**17.** 2×10^{-8}
19. 6.82×10^{-1}　　**21.** 3×10^{-2}　　**23.** 4×10^{25}
25. 2×10^3 hr　　**27.** 3×10^8 m/s　　**29.** 39 years
31. a. 4.74　　**b.** −4.74
33. a. −4.158　　**b.** −5.864
35. a. 0.045　　**b.** 0.128
37. a. −0.05　　**b.** 0.02
39. $\$3831.88$ (millions)　　**41.** $\$0.21$
43. Harry lost $\$9.25$; George lost $\$16.50$.
45. $\$176$　　**47.** 3300 oz
49. 392.5 mi　　**51.** $\$14,422,500$
53. 1.26×10^{19} mi　　**55.** 59.26 mi/gal
57. 78 rpm　　**59.** Answers may vary.
61. b. $\$11$

Exercises 5.5

1. 0.9　　**3.** 1.1　　**5.** 0.17
7. 1.21　　**9.** 0.003　　**11.** 1.243
13. 0.6　　**15.** 0.5625　　**17.** 0.625
19. 0.714285 ⋯　　　　**21.** 0.266 ⋯
23. 7.142857 ⋯　　　　**25.** 0.1875
27. 0.015625　　　　**29.** 0.00992
31. $0.\overline{5}$　　**33.** $0.\overline{64}$　　**35.** $0.\overline{235}$
37. $0.21\overline{5}$　　**39.** $0.07\overline{935}$　　**41.** $5.\overline{07}$
43. $\frac{8}{9}$　　**45.** $\frac{31}{99}$　　**47.** $\frac{38}{333}$
49. $\frac{229}{99}$　　**51.** $\frac{137}{111}$　　**53.** $\frac{14}{11}$
55. $\frac{151}{330}$　　**57.** $\frac{224}{1111}$　　**59.** 0.29
61. 0.009　　**63.** 0.4569　　**65.** 0.3415
67. 0.000234　　**69.** 345%　　**71.** 56.7%
73. 900.3%　　**75.** 0.45%　　**77.** 60%
79. 83.3%　　**81.** 9.1　　**83.** $33\frac{1}{3}$%
85. 125%　　**87.** 50.4%
89. a. 64.5 lb　　**b.** 39 lb
91. $\$10.81$
93. 591,900,000 (to the nearest million)
95. $\$16,000$　　**97.** $\$1500$　　**99.** $\$3600$
101. $\frac{3000}{12,000} = 25\%$; more than the 24% in the graph

103. $\frac{2400}{12,000} = 20\%$; more than the 19% in the graph
105. 20%   **107.** 35   **109.** 29%
111. 10%   **113.** 25.3 mi/gal
115. Net loss of ¢18   **117.** $\$\frac{1}{16} = 6.25¢$
121. $\$\frac{1}{8} = 12.5¢$ **123.** $\$\frac{1}{64} = 1.5625¢$ **125.** 100

Exercises 5.6

1. Irrational   **3.** Irrational
5. Rational   **7.** Rational
9. Rational   **11.** Rational
13. Irrational   **15.** Rational
17. Irrational   **19.** Rational
21. 4   **23.** 8   **25.** 9
27. -13 **29.** 14   **31.** -9
33. <   **35.** <   **37.** =   **39.** <
41. <   **43.** =   **45.** =
47. 0.315 (Other answers are possible.)
49. 0.311212345 · · · (Other answers are possible.)
51. 0.1011 (Other answers are possible.)
53. 0.101101001000 · · · (Other answers are possible.)
55. $\frac{7}{22}$ (Other answers are possible.)
57. 0.5101001000 · · · (Other answers are possible.)
59. $\frac{11}{18}$ (Other answers are possible.)
61. $0.21 < 0.2121 < 0.21211 < 0.212112111 \cdots$
< 0.21212
63. 3.09 mi   **65.** 163 mi   **67.** 31 yd
69. A rational number can be expressed as a terminating decimal or as a nonterminating, repeating decimal. An irrational number cannot be expressed this way.
71. $(OB)^2 = 1^2 + (\sqrt{5})^2 = 1 + 5 = 6$, so $OB = \sqrt{6}$.
73. $h = 3$

Exercises 5.7

1. $3\sqrt{10}$ **3.** Simplest form   **5.** $6\sqrt{5}$
7. $10\sqrt{2}$ **9.** $8\sqrt{6}$ **11.** $14\sqrt{3}$
13. $\frac{3\sqrt{7}}{7}$ **15.** $-\frac{\sqrt{10}}{5}$ **17.** $\sqrt{2}$
19. $\frac{\sqrt{3}}{7}$ **21.** $\frac{2\sqrt{3}}{3}$ **23.** $\frac{2\sqrt{2}}{7}$
25. $\frac{3}{5}$ **27.** $\frac{4\sqrt{10}}{25}$ **29.** $5\sqrt{10}$
31. $\sqrt{14}$ **33.** $\frac{1}{5}$ **35.** $\frac{\sqrt{6}}{2}$
37. $3\sqrt{3}$ **39.** $9\sqrt{5}$ **41.** 5
43. 5   **45.** $5\sqrt{7}$ **47.** $-7\sqrt{7}$
49. $-8\sqrt{5}$ **51.** $20\sqrt{41}$ m   **53.** $\frac{5\sqrt{2}}{4}$ sec
55. 20%   **57.** 15 m/sec   **59.** 28 ft/sec
61. 3   **63.** 13   **65.** $\frac{2}{3}$
67. Check rational numbers and real numbers.
69. Check whole numbers, integers, rational numbers, and real numbers.
71. Check natural numbers, whole numbers, integers, rational numbers, and real numbers.
73. Check rational numbers and real numbers.
75. Check irrational numbers and real numbers.
79. $6\frac{4}{13}$, 6.31 (Calculator gives 6.32.)
81. $9\frac{4}{19}$, 9.21 (Calculator gives 9.22.)

Exercises 5.8

1. a. $a_1 = 7$ **b.** $d = 6$
c. $a_{10} = 61$ **d.** $a_n = 6n + 1$
3. a. $a_1 = 43$ **b.** $d = -9$
c. $a_{10} = -38$ **d.** $a_n = 52 - 9n$
5. a. $a_1 = 2$ **b.** $d = -5$
c. $a_{10} = -43$ **d.** $a_n = 7 - 5n$
7. a. $a_1 = -\frac{5}{6}$ **b.** $d = \frac{1}{2}$
c. $a_{10} = \frac{11}{3}$
d. $a_n = \frac{n}{2} - \frac{4}{3}$, or $\frac{3n - 8}{6}$
9. a. $a_1 = 0.6$ **b.** $d = -0.4$
c. $a_{10} = -3$ **d.** $a_n = 1 - 0.4n$
11. $S_{10} = 340, S_n = n(3n + 4)$
13. $S_{10} = 25, S_n = \frac{n}{2}(95 - 9n)$
15. $S_{10} = -205, S_n = \frac{n}{2}(9 - 5n)$
17. $S_{10} = 14\frac{1}{6}, S_n = \frac{n}{12}(3n - 13)$
19. $S_{10} = -12, S_n = \frac{n}{5}(4 - n)$
21. a. $a_1 = 3$ **b.** $r = 2$
c. $a_{10} = 1536$ **d.** $a_n = 3 \cdot 2^{n-1}$
23. a. $a_1 = \frac{1}{3}$ **b.** $r = 3$
c. $a_{10} = 6561$ **d.** $a_n = 3^{n-2}$ or $\frac{1}{3}(3^{n-1})$
25. a. $a_1 = 16$ **b.** $r = -\frac{1}{4}$
c. $a_{10} = -\frac{1}{16,384}$
d. $a_n = \frac{(-1)^{n-1}}{4^{n-3}}$ or $(-1)^{n-1} 4^{3-n}$
27. $S_{10} = 3(2^{10} - 1) = 3069; S_n = 3(2^n - 1)$
29. $S_{10} = \frac{1}{6}(3^{10} - 1) = 9841\frac{1}{3}; S_n = \frac{1}{6}(3^n - 1)$
31. $S_{10} = \frac{4^{10} - 1}{5 \cdot 4^7} = \frac{209,715}{16,384}; S_n = \frac{4^n - (-1)^n}{5 \cdot 4^{n-3}}$
33. $S = 12$ **35.** $S = -16$
37. $\frac{7}{9}$ **39.** $\frac{208}{99}$
41. a. $1020   **b.** $18,000
43. a. $95   **b.** $8625
45. $610.51
47. If n is an even number, there are $n/2$ pairs and the sum of each pair is $(n + 1)$. The total sum is $n(n + 1)/2$. If n is an odd number, find the sum of the first $(n - 1)$ terms. The preceding formula gives $n(n - 1)/2$. Then adding n, the omitted term, gives the sum $n(n + 1)/2$, as before.
49. In an arithmetic sequence, each term after the first is obtained by adding the constant difference d to the preceding term. In a geometric sequence, each term after the first is obtained by multiplying the preceding term by the constant ratio r.
51. It is an arithmetic sequence. The common difference is 85.
53. $127 + (n - 1)85$, or $42 + 85n$

CHAPTER 6

Exercises 6.1

1. -2 and 0 are solutions.
3. 3 and 1 are solutions.
5. c. 2 **7.** $x = 5$ **9.** $x = 3$
11. $x = 2$ **13.** $x = 9$ **15.** $x = 1$
17. $x = 2$ **19.** $n = 6$ **21.** $x = \frac{3}{2}$
23. $x = 12$ **25.** $x = \frac{15}{4}$, or $3\frac{3}{4}$ **27.** $x = \frac{10}{7}$, or $1\frac{3}{7}$
29. $x = 10$ **31.** $p = 4$ **33.** $h = V/\pi r^2$
35. $W = V/LH$ **37.** $b = P - s_1 - s_2$
39. a. $T = D/R$ **b.** 4 hours
41. a. $A = 34 - 2H$ **b.** 18 years
43. $\{x \mid x < 4\}$ **45.** $\{x \mid x > 3\}$
47. $\{x \mid x > 3\}$ **49.** $\{x \mid x \geq -2\}$
51. $\{x \mid x > -4\}$ **53.** $\{x \mid x \leq -2\}$
55. $\{x \mid x \leq -9\}$ **57.** $\{x \mid x > 3\}$
59. $\{x \mid x > -4\}$ **61.** $\{x \mid x \leq -\frac{2}{3}\}$
63. $\varnothing$ **65.** $\{x \mid x \leq 2\}$
67. $\{x \mid x > -2\}$ **69.** $\{x \mid x \leq \frac{2}{5}\}$
71. $\{x \mid x \leq -4\}$ **73.** 32
75. 10% **77.** 12.5%
79. 200 **81.** 1000 billion barrels
83. 67% **85.** The $\$100$ price
87. 2.5 years
89. a. 9.72 million; within 0.18 million
      b. 11.52 million; within 0.02 million
      c. 10.92 million
91. 11 in.   **93.** 8
95. 12 in.
97. a. 47 **b.** 48
99. When $x < 20.21$, that is, when $x < 20.21$ oz
101. When $x > 2$, that is, after 1998
107. 250 cal   **109.** 710 cal
111. 150 chirps   **113.** 1 cm/sec

Exercises 6.2

1.
3.
5.
7.
9.
11.
13.
15.
17.
19.
21. The solution set is $\varnothing$.
23.

25.
27. $\{-1, 0, 1, 2, 3, 4\}$
29. $\{3, 4, 5, 6\}$ **31.** $\{2\}$
33. $\{\ldots, -8, -7, -6, 6, 7, 8, \ldots\}$
35.   **37.** $\varnothing$
39.   **41.**
43. $\varnothing$ **45.**
47.   **49.**
51. All real numbers
53. a. $C = 4953 + 0.12\,m$
      b. $6000 < 4953 + 0.12\,m < 6500$
      c. $8725 < m < 12{,}892$. Each car is budgeted to run between 8725 and $12{,}892$ mi annually.
55. $22 < 2.2\,P_2 < 66$, or $10 < P_2 < 30$
57. a. All the real numbers between -1 and $+2$ including the 2 but not the -1
      b. All the real numbers between -1 and $+2$ including the -1 but not the 2
      c. All the real numbers between -1 and $+2$ not including the endpoints
      d. All the real numbers between -1 and $+2$ including the endpoints
59. a. The set of all the real numbers between -1 and 2
      b. The set of all the real numbers between -1 and 2 including the 2
      c. The set of all the real numbers between -1 and 2 including the -1
      d. The set of all the real numbers between -1 and 2 including both the -1 and the 2
61. $(-\infty, 5)$ **63.** $(9, +\infty)$
65. $[-4, -1)$ **67.** $[-1, 10]$
69. Let J in. be Joe's height. Then $J = 60$ in.
71. Let F in. be Frank's height and S in. be Sam's height. Then $F = S - 3$.
73. Let S be Sam's height. Then $S = 77$ in.
75. Let B be Bill's height. Then $B > 74$ in. (6 ft 2 in.).

Exercises 6.3

1. 10 **3.** $\frac{1}{8}$
5. 3 **7.** 2
9. -8
11. 2 and $\frac{5}{3}$ are solutions.
13. $\{0\}$
15. $\{-5, 5\}$

17. $\{\ldots, -3, -2, -1, 1, 2, 3, \ldots\}$

19. No interval

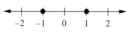

21. In interval notation, $[-4, 4]$

23. In interval notation, $(-4, 2)$

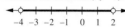

25. In interval notation, $(-\infty, -1] \cup [1, +\infty)$

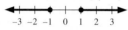

27. In interval notation, $(-\infty, -1) \cup (3, +\infty)$

29. In interval notation, $(-2, 2)$

31. In interval notation, $(-\infty, -2] \cup [2, +\infty)$

33. In interval notation, $[0, 3]$

35. In interval notation, $(-\infty, 0) \cup (3, +\infty)$

37. Any amount between \$450 and \$550, inclusive

39. Yes

41. a. $|w - 137| \le 7$
b. $-7 \le w - 137 \le 7$, or $130 \le w \le 144$

43. a. $|L - 12| \le 0.24$
b. $11.76 \le L \le 12.24$

45. a. $|s - 505| \le 4$
b. $501 \le s \le 509$

47. "The absolute value of x is less than a" is equivalent to "x is between $-a$ and a."

49. All real numbers, since the absolute value of an expression is always nonnegative. The solution of $|x| > a$ is the set of real numbers when a is negative.

Skill Checker

1. $2\sqrt{2}$ **3.** $2\sqrt{3}$ **5.** 7 **7.** $6\sqrt{3}$ **9.** 12

Exercises 6.4

1. $(x + 2)(x + 4)$ **3.** $(x - 4)(x + 3)$ **5.** $(x + 9)(x - 2)$
7. $(x - 5)^2$ **9.** $(x + 5)^2$ **11.** $(2x + 3)(x - 1)$
13. $(2x - 1)(3x - 1)$ **15.** $\{2, 4\}$ **17.** $\{-2, 3\}$
19. $\{-1, 0, 1\}$ **21.** $\{-2, \frac{1}{2}\}$ **23.** $\{-4, 4\}$
25. $\{-5, 5\}$ **27.** $\{-\frac{3}{2}, \frac{8}{5}, 2\}$ **29.** $\{-6, 6\}$
31. $\{3, 9\}$ **33.** $\{-2, 10\}$ **35.** $\{-\frac{1}{2}, -\frac{1}{5}\}$
37. $\{-\frac{5}{3}, 1\}$ **39.** $\{-\frac{5}{2}, 1\}$ **41.** $\{-\frac{7}{2}, 1\}$

43. $\left\{\dfrac{-5 - \sqrt{13}}{2}, \dfrac{-5 + \sqrt{13}}{2}\right\}$

45. $\left\{\dfrac{4 - \sqrt{6}}{5}, \dfrac{4 + \sqrt{6}}{5}\right\}$

47. $\left\{\dfrac{3 - \sqrt{2}}{7}, \dfrac{3 + \sqrt{2}}{7}\right\}$

49. $\left\{\dfrac{1 + \sqrt{3}}{3}, \dfrac{1 - \sqrt{3}}{3}\right\}$

51. $\left\{\dfrac{-1 + \sqrt{3}}{2}, \dfrac{-1 - \sqrt{3}}{2}\right\}$

53. $\{-\frac{5}{2}, \frac{1}{2}\}$

55. $(2ax + b)^2 = (2ax + b)(2ax + b)$
$= 2ax(2ax + b) + b(2ax + b)$
$= (4a^2x^2 + 2abx) + (2abx + b^2)$
$= 4a^2x^2 + 4abx + b^2$

which is the left side of the equation, as stated. If we subtract b from both sides of

$$2ax + b = \pm \sqrt{b^2 - 4ac}$$

we get

$$2ax = -b \pm \sqrt{b^2 - 4ac}$$

If we divide both sides of this last equation by $2a$, we get the quadratic formula as given.

57. 6 cm, 8 cm, and 10 cm

59. 5 in., 12 in., and 13 in.

61. 1 sec **63.** 1 sec **65.** 40 **67.** 10

69. a. $5.50 - x$ **b.** $550 + 450x - 100x^2$
c. \$0.50 or \$4 **d.** \$0.50

75. If $b^2 - 4ac = 0$, there is only one solution, $-\dfrac{b}{2a}$.

77. If $b^2 - 4ac < 0$, there are no real number solutions.

79. About 211 ft **81.** 7225

Exercises 6.5

1. $4m = m + 18$ **3.** $10x + (x - 3) = 26(x - 3)$
5. $4x + 5 = 29, x = 6$ **7.** $3x + 8 = 29, x = 7$
9. $3x - 2 = 16, x = 6$
11. $2x^2 = 2x + 12, x = -2$ or 3
13. $\frac{1}{3}x^2 - 2 = 10, x = -6$ or 6 **15.** 2.71 million lb
17. Russia has 6575 ships and Japan has 8851 ships.
19. 130 mi **21.** 14% per year
23. a. 204 mi **b.** The mileage rate
25. 10% **27.** 30 mph
29. 76.5 ft **31.** 0.6 sec
33. 20 mph **35.** 4 and 5 **37.** 36 billion

39. a. Revenue $= T + \dfrac{3N(2800)}{40} = T + 210N$ (dollars)
b. Cost $= (40)(\$100) + (40)(\$150) + 40 (\$1000) + \$3000 = \$53,000$
c. $T + 210N = 53,000$

d. $N = \dfrac{53{,}000 - T}{210}$

e. $N = 161$ is the least number of students needed to incur no loss.

41. a. See the "Total" rows in the table.

    b. Rent 5 size A units   **c.** Rent 1 size C unit

    d. Rent 1 size A unit and 1 size B unit.

Identical Row Alignment

	Unit		
	A	B	C
Cost per month	$ 25	$ 90	$128
Number per row	5	10	10
Rows	5	5	10
Layers	3	5	5
Total	75	250	500
Units needed	5	2	1
Cost for 2 months	$250	$360	$256

Staggered Row Alignment

	Unit		
	A	B	C
Cost per month	$ 25	$ 90	$128
Number per row	2–4s & 3–5s	3–10s & 2–9s	6–10s & 5–9s
Rows	5	5	11
Layers	3	5	5
Total	69	240	525
Units needed	6	2	1
Cost for 2 months	$300	$360	$256

43. Spend 3 hr on mathematics, 6 hr on science, and 3 hr on English to get your best GPA.

45. a.

Returned After	Blockbuster	Red Rabbit
2 Days		
New	$3 + $2 = $5	$3 + $3 = $6
Old	$3 + $0 = $3	$1.60 + $0 = $1.60
3 Days		
New	$3 + $4 = $7	$3 + $6 = $9
Old	$3 + $2 = $5	$1.60 + $1.50 = $3.10
4 Days		
New	$3 + $6 = $9	$3 + $9 = $12
Old	$3 + $4 = $7	$1.60 + $3 = $4.60

b. Let C_B = cost at Blockbuster, and C_R = cost at Red Rabbit.

$$C_B = 4n; \quad C_R = 4.50n - 1.40 \quad \text{(in dollars)}$$

c. $C_B = \$11; C_R = \$15$

47. a. The average stopping distance (in feet) is
$$d = 0.9v + 0.06v^2$$

    b. 162 ft   **c.** 242 ft   **d.** 3.67 sec

49. You should try to determine what is the unknown, that is, what is wanted.

51. About 5.22 years

53. If M were greater than G, then y would be negative; this is unrealistic.

Exercises 6.6

1. 7000 to 2000; 7000:2000; $\frac{7000}{2000}$; or 7 to 2; 7:2; $\frac{7}{2}$

3. 70 to 4260; 70:4260; $\frac{70}{4260}$; or 7 to 426; 7:426; $\frac{7}{426}$

5. $\frac{10}{3}$ reduced trans. ratio   **7.** 17 mi/gal

9. a. 6 cents   **b.** 5 cents   **c.** White Magic

11. $x = 12$ **13.** $x = 6$ **15.** $x = 24$

17. $\frac{9}{2} = \frac{n}{40}$ **19.** 6   **21.** 66.5 in.

23. 2.81 runs   **25.** 2650 fish

27. a. $R = kt$ **b.** $k = 45$ **c.** 2.4 min

29. a. $T = kh^3$ **b.** $k = \frac{1}{1750} \approx 0.0005714$

    c. 241 lb

31. a. $f = k/d$ **b.** $k = 4$ **c.** 16

33. 10.8 in.^{3}

35. a. $w = \frac{k}{s}$ **b.** $k = 7200$ **c.** 720

37. a. $b = \frac{k}{a}$ **b.** $k = 2970$ **c.** 90

39.   **a.** $d = ks$ **b.** $k \approx 17$

    c. The number of hours to travel the distance d at the speed s

41. a. $BAC = k(N - 1)$ **b.** 0.026

    c. 0.104   **d.** 4

43. a. $C = 4(F - 37)$ **b.** 212

45. $208.33   **47.** $C = 102.9$

49. a. $BAC = \frac{k}{w}$ **b.** 0.033

    c. 107.25 lb   **d.** Less than 0.08

CHAPTER 7

Exercises 7.1

1. Domain: $\{1, 2, 3\}$; range: $\{2, 3, 4\}$

3. Domain: $\{1, 2, 3\}$; range: $\{1, 2, 3\}$

5. Domain: $\{x \mid x$ is a real number$\}$; range: $\{y \mid y$ is a real number$\}$

7. Domain: $\{x \mid x$ is a real number$\}$; range: $\{y \mid y$ is a real number$\}$

9. Domain: $\{x \mid x$ is a real number$\}$; range: $\{y \mid y \geq 0\}$

11. Domain: $\{x \mid x \geq 0\}$; range: $\{y \mid y$ is a real number$\}$

13. Domain: $\{x \mid x \neq 0\}$; range: $\{y \mid y \neq 0\}$

15. This is a function because only one real value of y corresponds to each real value of x.

17. This is not a function because two values of y correspond to each positive value of x.

19. This is a function because only one real y value corresponds to each x value in the domain.

21. This is a function because only one real value of y corresponds to each real value of x.

23. a. 1   **b.** 7   **c.** -5

25. a. 0   **b.** 2   **c.** 5

27. a. $3x + 3h + 1$ **b.** $3h$ **c.** 3

29. $g(x) = x^2; \frac{1}{16}, 4.41, \pm 8$

31. a. 1   **b.** -5 **c.** 15

33. a. 140 beats per minute
   b. 130 beats per minute

35. a. 160 lb   **b.** 78 in.

37. a. 639 lb/ft^2 **b.** 6390 lb/ft^2

39. a. 144 ft   **b.** 400 ft

41.

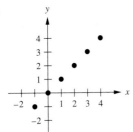

43.

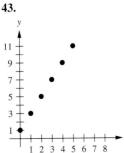

45.

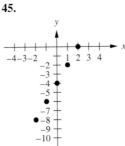

47.

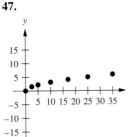

49.

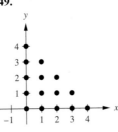

51.

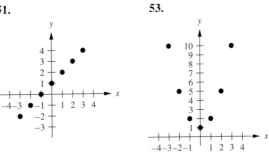

53.

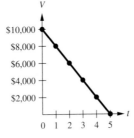

55. a. $h(x) = 2.89x + 70.64$
   b. $h(34) = 168.9$ cm, or about 169 cm

57. a. $F(x) = 10x + 20$ **b.** $F(8) = 100$; $100

59. a. $V = 10,000 - 2000t$

   b.

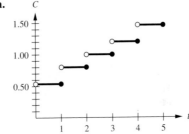

61. a.

   b. 13 min

63. $g(x) = \sqrt{x - 1}$ is real if and only if $x \geq 1$. Thus, we would exclude all values of x less than 1 if g is to have real values.

65. The graph of a function $f(x)$ is a picture of the set of points $\{(x, y) \mid y = f(x), x$ an element of the domain of $f\}$.

67. Yes. It is a relation for which there is exactly one value of y for each value of x in the domain.

69. $c = f(x) = 4(x - 40)$, x in °F

71. $f(t) = 16t^2$

73. $f(x) = \sqrt{x}$

75. 10,000 units

Exercises 7.2

1. 4

3.

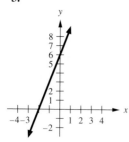

5.

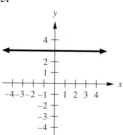

7.

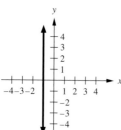

9.

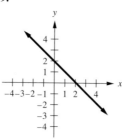

11.

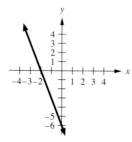

13.

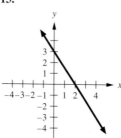

15.

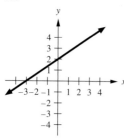

17.

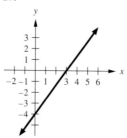

19. 5 units   **21.** $\sqrt{73} \approx 8.54$ units
23. $\sqrt{58} \approx 7.62$ units   **25.** 2 units   **27.** 4 units
29. a. $E(x) = 500 + 25x$; $S(x) = 1000 + 20x$

b.

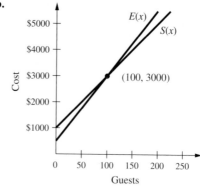

c. The cost is the same for 100 persons.

31. a.

t	$C(t)$
$0 < t \leq 1$	\$1.16
$1 < t \leq 2$	\$1.81
$2 < t \leq 3$	\$2.46
$3 < t \leq 4$	\$3.11
$4 < t \leq 5$	\$3.76

b.

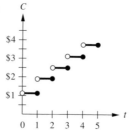

33. Yes. The distance d from $(2, -3)$ is $\sqrt{52} = 2\sqrt{13} < 10$ mi.

35. a. 50%   **b.** 5%   **c.** $2008\frac{1}{3}$, or 2009

37. a.

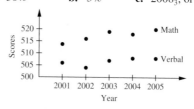

b. $V = 506.6$; $M = 517.4$
c. $y = V$, very well; $y = M$, not as well
d. Yes; vertical line test

39. Use the distance formula to find the square of the length of each side of the triangle.

$$(AB)^2 = (a_1 - b_1)^2 + (a_2 - b_2)^2$$
$$(AC)^2 = (a_1 - c_1)^2 + (a_2 - c_2)^2$$
$$(BC)^2 = (b_1 - c_1)^2 + (b_2 - c_2)^2$$

The triangle is a right triangle if and only if one of these squares equals the sum of the other two squares. You can check this.

41. See the Problem Solving procedure following Example
1 and follow it step by step for $x = c$.
43. About \$940   **45.** About \$800

Exercises 7.3

1. $m = 1$ **3.** $m = -1$
5. $m = -\frac{1}{8}$ **7.** $m = \frac{1}{4}$
9. $m = 0$ **11.** $y = \frac{1}{2}x + \frac{3}{2}$
13. $y = -x + 6$ **15.** $y = 5$
17. a. $m = 1$ **b.** $b = 2$
19. a. $m = \frac{4}{3}$ **b.** $b = 0$
21. a. $m = -1$ **b.** $b = 14$
23. a. $m = 0$ **b.** $b = 6$
25. a. The slope is not defined.
     b. The line does not intersect the y axis.
27. $3x - y = 4$ **29.** $x + y = 5$
31. $10x - y = 0$ **33.** $w = 5h - 176$
35. $w = 5h - 187$
37. $C = 1.70m + 0.30$; \$51.30
39. 6 mi
41. a. \$2   **b.** $(m - 1)$
     c. \$1.70   **d.** $1.70(m - 1)$
     e. $C = 1.70m + 0.30$; yes
43. a. $C = 2m + 55$ **b.** \$55
45. Parallel   **47.** Not parallel
49. Parallel   **51.** $4x - y = 6$
53. a. $5x - 2y = 10$ **b.** $x + 2y = 3$
     c. $2x + y = 2$ **d.** $5x - 4y = 1$
55. For the first line, $m_1 = 2$, and for the second line,
$m_2 = -\frac{415}{790}$. Since $m_2 \neq -1/m_1$, the lines are not
perpendicular.
57. a. The slope is 1. Annual sales. **b.** 27 billion
     c. $U(t) = t + 27$ **d.** $U(8) = 35$ billion
59. a. Decreasing   **b.** Decreasing
     c. Fish and shellfish
61. a. 0.15   **b.** Increasing
     c. The slope 0.15 represents the annual increase in the
     life span of American men.
63. a. 0.4   **b.** Increasing
     c. The slope 0.4 represents the annual increase in fat
     consumption.
65. a. 140g   **b.** 180g   **c.** 190g
67. a. 10 million   **b.** 24 million   **c.** 0.4 million
69. If (x_1, y_1) and (x_2, y_2) are any two distinct points on a
horizontal line, $y_1 = y_2$ and $x_1 \neq x_2$. Thus, the slope is

$$m = \frac{y_2 - y_1}{x_2 - x_1} = \frac{0}{x_2 - x_1} = 0$$

71. a. 0.79   **b.** No
73. a. 0.25   **b.** Yes
75. The slope is 0.25, which is not safe for parking.
$x = 105$ ft will give the maximum allowable slope.
77. The fixed fee is \$35. The hourly rate is \$25.

Skill Checker

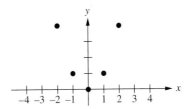

Exercises 7.4

1.

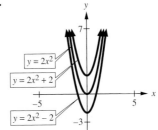

3.

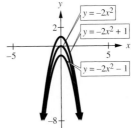

5.

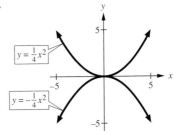

7.

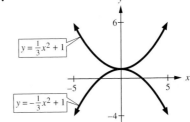

9.

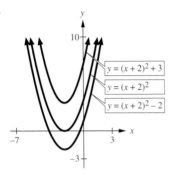

$y = (x + 2)^2 + 3$

$y = (x + 2)^2$

$y = (x + 2)^2 - 2$

11.

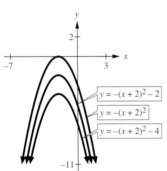

$y = -(x + 2)^2 - 2$

$y = -(x + 2)^2$

$y = -(x + 2)^2 - 4$

13.

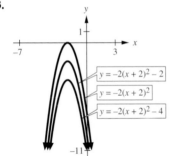

$y = -2(x + 2)^2 - 2$

$y = -2(x + 2)^2$

$y = -2(x + 2)^2 - 4$

15.

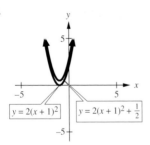

$y = 2(x + 1)^2$ $y = 2(x + 1)^2 + \dfrac{1}{2}$

17.

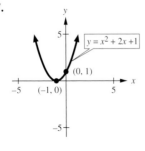

$y = x^2 + 2x + 1$

$(0, 1)$

$(-1, 0)$

19.

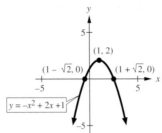

$(1, 2)$

$(1 - \sqrt{2}, 0)$ $(1 + \sqrt{2}, 0)$

$y = -x^2 + 2x + 1$

21.

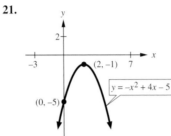

$(2, -1)$

$y = -x^2 + 4x - 5$

$(0, -5)$

23.

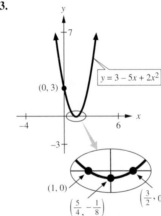

$y = 3 - 5x + 2x^2$

$(0, 3)$

$(1, 0)$ $\left(\dfrac{3}{2}, 0\right)$

$\left(\dfrac{5}{4}, -\dfrac{1}{8}\right)$

25.

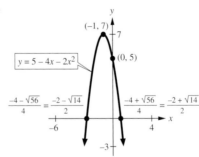

$(-1, 7)$

$y = 5 - 4x - 2x^2$

$(0, 5)$

$\dfrac{-4 - \sqrt{56}}{4} = \dfrac{-2 - \sqrt{14}}{2}$ $\dfrac{-4 + \sqrt{56}}{4} = \dfrac{-2 + \sqrt{14}}{2}$

27.

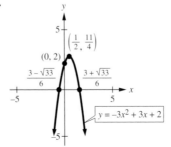

$\left(\dfrac{1}{2}, \dfrac{11}{4}\right)$

$(0, 2)$

$\dfrac{3 - \sqrt{33}}{6}$ $\dfrac{3 + \sqrt{33}}{6}$

$y = -3x^2 + 3x + 2$

29. $x = 4000$; $P = \$11,000$ **31.** \$25,000 **33.** 400 ft
35. $P = (600 + 100W)(1 - 0.10W)$; P = price, W = the weeks elapsed. The maximum for P occurs when $W = 2$ (at the end of 2 weeks).
37. a. $(42, 18)$ **b.** 18 in. **c.** 84 in.

d.

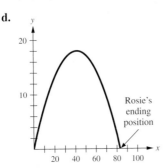

39. a. $(200, 100)$ **b.** 100 ft **c.** 400 ft
d.

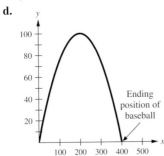

41. a. 6.3%; 27 **b.** 10.5% **c.** 22 **d.** 16
43. If $a > 1$, it is narrower. If $a < 1$, it is wider.
45. Down. If it opened up from $(1, 1)$, it would have no x intercept.
47. Negative **49.** $(1962, 520)$ (approximately)
51. $FP = \sqrt{x^2 + (y - p)^2}$ **53.** $x^2 = 4py$
55. $y^2 = 12.5x$, focus $(3.125, 0)$; or $x^2 = 12.5y$, focus $(0, 3.125)$

Exercises 7.5

1. a. $\frac{1}{5}$ **b.** 1 **c.** 5
3. a. $\frac{1}{9}$ **b.** 1 **c.** 9
5. a. $\frac{1}{10}$ **b.** 1 **c.** 10
7.

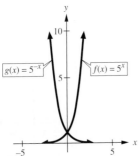

9.

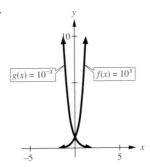

11. a. \$2459.60 **b.** \$2435.19
13. a. \$1822.12 **b.** \$1814.02
15. a. 4000 **b.** 8000 **c.** 16,000
17. a. 2000 g **b.** 699.9 g **c.** 244.9 g
19. 42,736,261
21. a. 285 million **b.** 22,654 million
  c. 1,799,492 million
23. a. 333,333 **b.** 222,222 **c.** 8671
25. a. About 4.42 lb/in.^{2} **b.** About 6.42 lb/in.^{2}
27. $f(x) = \log_5 x$

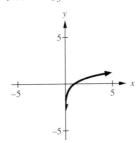

29. About 13.86 year **31.** About 10.66 year
33. About 6.88 billion **35.** About 17.3 min
37. About 80.5 min
39. a. 100,000 **b.** About 67,032
  c. About 13,534 **d.** About 1832
41. About 23,105 years **43.** About 13.3 years
45. a. 14.7 lb/in.^{2} **b.** About 11.4 lb/in.^{2} **c.** 8.92 lb/in.^{2}
47. a. About 10.57 or 11 years
  b. About 12.6 or 13 years
49. a. 5000 **b.** About 2247 **c.** 2010
51. a. 50,000 **b.** About 74,591
  c. About 166,006
53. a. 1000 **b.** About 368
55. a. 0 **b.** 39.3
57. a. -0.8055 **b.** -3.8968
59. Continuous, \$1822.12; monthly, \$1819.40. Continuous compounding gives about \$2.72 more.
61. a. $f(x) = 1$, a horizontal line **b.** Yes
63. Symmetric with respect to y axis

Skill Checker

1.

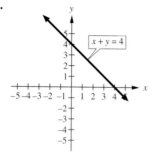

3.

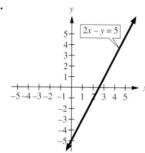

Exercises 7.6

1. $(1, 2)$ **3.** $(3, -4)$

5. $\left(2, -\frac{1}{2}\right)$ **7.** $(1, 2)$

9. $(4, 2)$ **11.** $(3, -4)$

13. No solution   **15.** $(1, 2)$

17. $\left(2, -\frac{1}{2}\right)$ **19.** $(-3, 10)$

21. $\left(-2, \frac{1}{2}\right)$ **23.** $(2, -5)$

25. $\left(\frac{2}{3}, \frac{5}{6}\right)$ **27.** No solution

29. $u = -\frac{5}{2}, v = 6;$ or $\left(\frac{-5}{2}, 6\right)$

31. a. 100

   b.

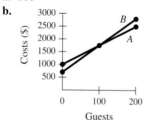

   c. Company A

33. 4   **35.** 5

37. a. $C = 20 + 35m$

   b.

m	C
6	230
12	440
18	650

   c.

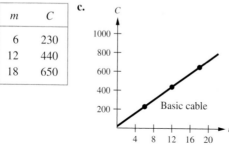

39.

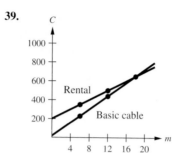

41. When you use it for more than 18 months

43. a. $W = 100 + 3t$

   b.

t	W
5	115
10	130
15	145
20	160

   c.

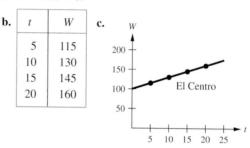

45. a. $C = 0.60m$ **b.** $C = 45 + 0.45m$

   c.

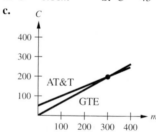

47. and 49.

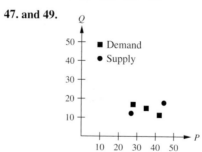

51. $(35, 15)$ **53.** 1500

Exercises 7.7

1.

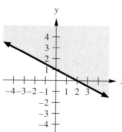

3.

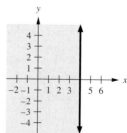

5.

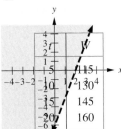

7.

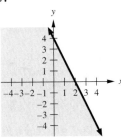

23.

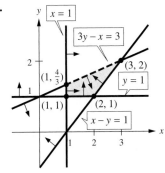

9.

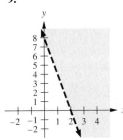

11.

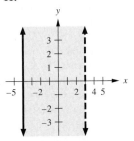

25.

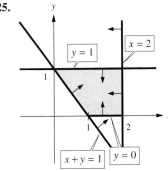

13.

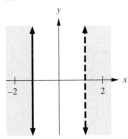

15.

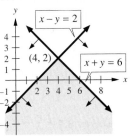

27. Conditions (**b**)
29. Conditions (**a**)
31. The region shown in (**c**)
33. $h \le 29{,}035$
35. $e \ge 2$
37. $n \ge 4 \times 10^{25}$
39. $n > 20{,}000$
41. $h < 13$

17.

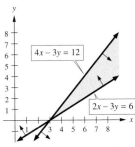

19.

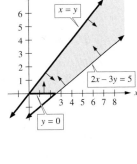

43.

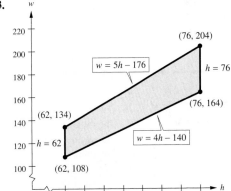

21.

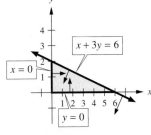

45. Suppose $c \ne 0$.
Step 1. Find the intercepts of the line $ax + by = c$.
Step 2. Draw a dashed line through these intercepts.
Step 3. Substitute $(0, 0)$ into the equation. This gives zero for the left side.
Step 4. If $c > 0$, shade the region opposite the origin. If $c < 0$, shade the region on the origin side of the line.

47. The graph would show $x = k$ as a solid line with the region to the right of this line as the shaded region.

49. The possible pairs of integers are (4, 2), (5, 2), (5, 3), (5, 4), (6, 2), (6, 3), and (7, 2).

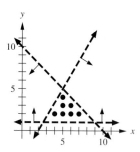

51. a. 33 mi   **b.** Rental A (It's $10 cheaper.)

53. If you plan to drive more than 33 mi, rental A is the cheaper.

55. Cutting up the 2-in. square by connecting the midpoints of the opposite sides gives four smaller squares, each 1 in. on a side. Since there are five points, at least two of them must be inside or on the perimeter of one of the four small squares. The diagonal of this small square is $\sqrt{1^2 + 1^2} = \sqrt{2}$, so the two points cannot be more than $\sqrt{2}$ in. apart.

57. Leave everything as in the given flowchart except for interchanging the shading instructions at the end, so that if the answer to "Is $c > 0$?" is no, the half-plane not containing the origin will be shaded, and if the answer is yes, then the half-plane containing the origin will be shaded.

CHAPTER 8

Exercises 8.1

1. a.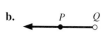

b.

c.

3. The segment $\overline{BC}$ **5.** The ray $\overrightarrow{AD}$ (or $\overrightarrow{AC}$ or $\overrightarrow{AB}$)

7. The segment $\overline{AD}$ **9.** The segment $\overleftrightarrow{AD}$

11. ∅   **13.** The point C

15. a. $\overleftrightarrow{AB}, \overleftrightarrow{AC}, \overleftrightarrow{AD}, \overleftrightarrow{BC}, \overleftrightarrow{BD}, \overleftrightarrow{CD}$
    b. $\overleftrightarrow{AB}$ and $\overleftrightarrow{CD}$; $\overleftrightarrow{AC}$ and $\overleftrightarrow{BD}$; $\overleftrightarrow{AD}$ and $\overleftrightarrow{BC}$
    c. No

17. True   **19.** True   **21.** True

23. False   **25.** False

27. a. ∠BAC (or ∠CAB)   **b.** ∠β (or ∠FAE)

29. ∠BAC, ∠CAD, ∠DAE, ∠EAF

31. ∠BAE, ∠CAE, ∠CAF

33. a. ∠DAE   **b.** ∠CAD

35. a. ∠DAF   **b.** ∠BAE

37. 35°   **39.** 15°   **41.** ∠B

43. 110°   **45.** 20°   **47.** 220°

49. a. 150°   **b.** 30°   **c.** 150°

51. ∠C, ∠E, ∠A, ∠G

53. a. 49°   **b.** 139°   **55.** $x = 16$

57. $x = 15$; 35° and 145°   **59.** $x = 10$; 30° and 60°

61. a. 30°   **b.** 180°

63. 90°   **65.** 25

67. a. $m\angle A + m\angle B = 180°$
    Angles A and B form a straight angle.
    b. $m\angle C + m\angle B = 180°$
    Angles C and B form a straight angle.
    c. $m\angle A + m\angle B = m\angle C + m\angle B$
    Both sides equal 180° (substitution).
    d. $m\angle A = m\angle C$
    Subtract $m\angle B$ from both sides.

69. Acute   **71.** Obtuse   **73.** Obtuse

75. 0°   **77.** 100°

79. Point, line, and plane

81. The ray may have only its endpoint in common with the plane, it may have only some other single point in common with the plane, or it may lie entirely in the plane.

83. a. One of the usual meanings of *acute* is "sharp" or "intense." Thus an *acute pain* means a "sharp or intense pain."
    b. One of the ordinary meanings of *obtuse* is "dull." Thus, *obtuse intelligence* means "dull intelligence" or "stupidity."

85. The error is 0.2°.

87. N 50° W

89. $360 - 40 = 320$, so the navigator's bearing would be 320°.

Exercises 8.2

1. a.   **b.**

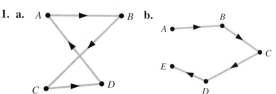

3. a. *C, D, I, J, L, M, N, O, P, S, U, V, W, Z*
    b. *B, D, O*

5. a. *D, O*   **b.** *A, E, F, G, H, K, Q, R, T, X, Y*

7. Convex   **9.** Parallelogram

11. Rectangle   **13.** Trapezoid

15. Parallelogram   **17.** Scalene, right

19. Scalene, acute   **21.** Isosceles, acute

23. Scalene, obtuse   **25.** (**a**) and (**c**)

27. $x = 5\frac{1}{3}, y = 6\frac{2}{3}$ **29.** 14

31. $1\frac{1}{3}$ **33.** $5\frac{1}{4}$ and $3\frac{1}{2}$ cm

35. 8, 12, and 16 in.   **37.** $18\frac{3}{4}$ ft

39. The two triangles are similar because the corresponding angles are equal. The tree is 25 ft tall.

41. 600 m

43. Recall that an isosceles triangle is one that has two sides of equal length. By SAS, $\triangle ABD \cong \triangle BCD$. This implies that $AB = BC$ and so $\triangle ABC$ is isosceles.

45. It is given that $\triangle ACD$ and $\triangle BCE$ have two pairs of sides with equal length. They also share the side CD. This means we can apply SSS to get $\triangle ACD \cong \triangle BCD$.

47. a. $m\angle A = m\angle B = 50°$, $m\angle C = 80°$
   b. $m\angle A = m\angle B = 40°$, $m\angle C = 100°$

49. $30°$ **51.** $2160°$ **53.** $108°$ **55.** $135°$ **57.** $144°$

59. a. $(E \cup I \cup S) = T \subset P$
   b.

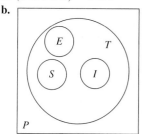

61.

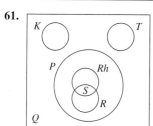

65. a.

   b.   **c.**

67. a.   **b.**

   c. Impossible

Exercises 8.3

1. 65 cm   **3.** 12.6 yd   **5.** 184.6 m
7. 57.2 m   **9.** 30 cm   **11.** 10 ft
13. 90 ft   **15.** 716 by 518 ft
17. $7\pi \approx 22.0$ m   **19.** $20\pi \approx 62.8$ ft
21. $3\pi \approx 9.42$ cm   **23.** $9\pi \approx 28.3$ yd
25. $61\pi \approx 192$ cm   **27.** $16\pi \approx 50.2$ cm
29. $8\pi \approx 25.1$ ft   **31.** $d = 15$ cm, $r = 7.5$ cm
33. $7/\pi \approx 2.23$ cm   **35.** $\dfrac{4.125}{\pi} \approx 1.31$ in.

37. About 15,500,000 (rounded from 15,520,000)
39. $16\pi \approx 50.2$ yd   **41.** 490 mi

43. 564 mi   **45.** A bicycle tire
47. The worn tire has a smaller circumference, so will turn more times per mile.
49. $\frac{3}{4}$ in.   **51.** $\frac{3}{4}$ in.

Exercises 8.4

1. 15 in.^{2} **3.** 15 cm^2
5. 24 ft^2 **7.** 30 ft^2
9. $800 + 50\pi \approx 957$ cm^2 **11.** $64 + 32\pi \approx 164$ cm^2
13. $18 - \frac{9}{2}\pi \approx 3.87$ cm^2 **15.** $\frac{25}{4}\pi \approx 19.6$ ft^2
17. 8 ft   **19.** 15 in.   **21.** \$600
23. The side opposite the base is 12 ft long and the other two sides are each 10 ft long.
25. 6400 yd^2 **27.** 18 ft wide
29. 96 ft wide   **31.** 16π cm^2
33. 12 in. wide and 18 in. long   **35.** $3\frac{1}{2}$ by $4\frac{1}{2}$ ft
37. Let ABC be an equilateral triangle of side s as in the following figure. Drop a perpendicular from C to the base $\overset{\bullet}{A}\overset{\bullet}{B}$, meeting $\overset{\bullet}{A}\overset{\bullet}{B}$ at point D. Triangles ADC and BDC are right triangles. Let the length of $\overset{\bullet}{C}\overset{\bullet}{D}$ be h. Then,

$$h^2 + \left(\frac{s}{2}\right)^2 = s^2$$

This gives

$$h^2 = \frac{3s^2}{4} \quad \text{and} \quad h = \frac{s\sqrt{3}}{2}$$

The area of the triangle is

$$A = \frac{1}{2}bh = \frac{1}{2}s\left(\frac{s\sqrt{3}}{2}\right) = \frac{s^2\sqrt{3}}{4}$$

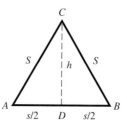

39. a. Its area is multiplied by 4.
   b. Its area is multiplied by 9.
   c. Its area is multiplied by k^2.
41. The area of the circle is larger by about 6.8 cm^2.
43. The hypotenuse of the cut-out triangle is the side x of the octagon. Thus, $x = s\sqrt{2}$. One side of the original square consists of one side of the octagon plus two sides of the cut-out triangles, so that $2s + s\sqrt{2} = a$ and

$$s = \frac{a}{2 + \sqrt{2}} = \frac{a(2 - \sqrt{2})}{(2 + \sqrt{2})(2 - \sqrt{2})} = \frac{2 - \sqrt{2}}{2}a$$

45. a. in.^{2} and cm^2 **b.** mi^2 and km^2
   c.–d. yd^2 and m^2

47. a. $2\frac{2}{3}$ gal   **b.** \$42
49. a. 15 bags   **b.** \$60
51. a. 2.49¢/in.² **b.** 2.36¢/in.² **c.** The 10-in. pie
53. 15,386 ft²   **55.** 803.84 mi²
57. 725.34 mi²   **59.** 14.13 ft²
61. 14.13 in.²
63. The area of the rectangle taken away is Wx and the area of the rectangle added on is hy. Since $hy = Wx$, the area of the new rectangle is equal to the area of the original one.

Exercises 8.5

1. a. A, B, C, D, E
    b. $\overrightarrow{AB}, \overrightarrow{AC}, \overrightarrow{AD}, \overrightarrow{AE}, \overrightarrow{BC}, \overrightarrow{BE}, \overrightarrow{CD}, \overrightarrow{CE}, \overrightarrow{DE}$
3. $ABCD$
5.

7.

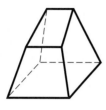

9. a. The volume is multiplied by 8.
    b. The volume is multiplied by 27.
11. a. 1600 in.³   **b.** 880 in.²
13. a. $V = 264x^3$ **b.** $S = 240x^2$
15. 50 in.³   **17.** 32 in.³   **19.** 1.5 liters
21. a. $V = 225\pi \approx 707$ in.³
        $S = 140\pi \approx 440$ in.²
    b. $V = \frac{225}{3}\pi \approx 236$ in.³
        $S = (25 + 5\sqrt{106})\pi \approx 240$ in.²
23. a. $V = 36\pi \approx 113$ ft³
        $S = 42\pi \approx 132$ ft²
    b. $V = 12\pi \approx 37.7$ ft³
        $S = 24\pi \approx 75.4$ ft²
25. $288\pi \approx 904$ in.³
27. $2,536,456\pi \approx 7,960,000$ ft³
29. $1024\pi \approx 3220$ m³
31. $r \approx 3.81$ cm, $h \approx 10.16$ cm, $V \approx 147.5\pi \approx 463$ cm³
    The can holds about 463 g.
33. a. Baseball: $S_1 = \dfrac{81}{\pi} \approx 25.8$ in.²

    $V_1 = \dfrac{243}{2\pi^2} \approx 12.3$ in.³

    Soccer ball: $S_2 = \dfrac{729}{\pi} \approx 232$ in.²

    $V_2 = \dfrac{6561}{2\pi^2} \approx 333$ in.³

    Basketball: $S_3 = \dfrac{900}{\pi} \approx 287$ in.²

    $V_3 = \dfrac{4500}{\pi^2} \approx 456$ in.³

    b. $\dfrac{S_1}{S_3} \approx \dfrac{9}{100}$ **c.** $\dfrac{V_1}{V_2} = \dfrac{1}{27}$
35. a. $\dfrac{6561}{2\pi^2} \approx 333$ in.³   **b.** About 15.66 lb
37. a. 1428 in.³   **b.** 730 in.²   **c.** 32,000 bags
39. 1760 in.³   **41.** 2288 in.³
43. 35.33 in.³   **45.** 88 in.³
47. a. 1232 ft³   **b.** 63 ft³   **c.** No, 1300 > 1295
49. About $1.41\pi \approx 4.43$ ft³   **51.** About 5.31 ft³
53. About 94.16, or 94
63. Since the sum of the face angles at a vertex must be less than 360°, three, four, or five equilateral triangles, three squares, or three pentagons can be put together at any vertex. There are no other possibilities, so only the five regular polyhedrons listed are possible.

65.

Figure	F	V	E	$F + V$
8.69B	6	8	12	$14 = E + 2$
8.69C	6	5	9	$11 = E + 2$
8.69D	7	10	15	$17 = E + 2$
8.70	5	6	9	$11 = E + 2$
8.71	6	6	10	$12 = E + 2$

Thus, Euler's formula is $F + V = E + 2$.

Exercises 8.6

1. a. 3   **b.** 0
    c. Traversable; all three vertices are possible starting points.
3. a. 3 (B, D, and E)   **b.** 2 (A and C)
    c. Traversable; start at either A or C.
5. a. 1 (A only)   **b.** 4(B, C, D, E)
    c. Not traversable; it has more than two odd vertices.
7. a. 5 (A, C, D, E, G)   **b.** 2 (B, F)
    c. Traversable; start at either B or F.
9. a. 1, the vertex of the pyramid
    b. 4, the vertices of the base
    c. Not traversable; it has more than two odd vertices.
11. Think of each region as a vertex and the individual line segments in its boundary as the number of paths to the vertex. The boundary of region A has four segments, so the corresponding vertex would be even. The boundary of region B has five segments, so the corresponding vertex would be odd. The boundary of region C has four segments, so the corresponding vertex would be even. The boundary of region D has five segments, so the corresponding vertex would be odd. The boundary of region E has ten segments, so

the corresponding vertex would be even. Thus, the network would have two odd vertices (*B* and *D*). By starting in region *B* or *D,* it is possible to draw a simple connected broken line that crosses each line segment exactly once.

13. Region *A* has three doorways, so the corresponding vertex would be odd. Regions *B, C, D,* and *E* and the outside each has two doorways, so the corresponding vertices would be even. Region *F* has three doorways, so the corresponding vertex would be odd. There are two odd vertices, *A* and *F.* By starting in either of these rooms and ending in the other, it is possible for a walk to pass through each doorway exactly once. It is not possible to start and end outside.

15. *A* and *D* have three doorways and the other rooms and the outside each have an even number of doorways. Thus, the corresponding network has two odd vertices. By starting in either *A* or *D* and ending in the other, it is possible for a walk to pass through each doorway exactly once. It is not possible to start and end outside.

17. All the rooms and the outside have an even number of doorways, so the corresponding network has no odd vertices. The walk can start in any room or outside and end in the same place and pass through each doorway exactly once.

19. Rooms *B* and *D* each have three doorways, so the corresponding network has two odd vertices. The walk can start in either *B* or *D* and end in the other one, passing through each doorway exactly once. It is not possible to start and end outside.

21. Room *A* and the outside *D* each have three doorways. Thus, the corresponding network has two odd vertices. By starting in *A* and ending in *D* (or vice versa), a walk can pass through each doorway exactly once. It is not possible to start and end in *D.* (See the traversability rules.)

23. Given a line and any point not on that line, there is one and only one line through that point that is parallel to the given line.

25. Given a line and any point not on that line, there is no line through that point that is parallel to the given line.

27. In hyperbolic geometry

29. The surface of a rectangular box

31. The surface of a sphere

33. (a)–(e) are topologically equivalent.

35. (a)–(e) are of genus 2. (f) is of genus 1.

37. Topo is correct. If you cut through a loop of the left-hand figure, you can unwind it into a single strip as you can with the circular cord.

39. Since each arc has two endpoints, the total number of endpoints must be even. An odd vertex accounts for an odd number of endpoints, whereas an even vertex accounts for an even number of endpoints. Thus, there must be an even number of odd vertices.

41. Vertices 3; regions 2; arcs 3
43. Vertices 5; regions 2; arcs 5
45. Vertices 2; regions 3; arcs 3
47. Vertices 6; regions 3; arcs 7
    Numbers 41–48 all fit Euler's formula, $V + R = A + 2$.

CHAPTER 9

Exercises 9.1

1. **a.** $\begin{bmatrix} 8 & 4 \\ 0 & -4 \end{bmatrix}$ **b.** $\begin{bmatrix} 6 & -12 \\ -9 & -3 \end{bmatrix}$
   c. $\begin{bmatrix} 5 & 1 \\ -1 & -1 \end{bmatrix}$

3. **a.** $\begin{bmatrix} 3 & 10 \\ 5 & 0 \end{bmatrix}$ **b.** $\begin{bmatrix} -3 & 0 \\ 1 & 0 \end{bmatrix}$

5. **a.** $\begin{bmatrix} 0 & 13 \\ 8 & -3 \end{bmatrix}$ **b.** $\begin{bmatrix} -1 & -21 \\ -12 & -4 \end{bmatrix}$

7. **a.** $\begin{bmatrix} 0 & 1 & 3 \\ 7 & 3 & -3 \\ 4 & 3 & 0 \end{bmatrix}$ **b.** $\begin{bmatrix} 2 & -3 & 1 \\ -1 & -3 & -1 \\ 4 & 1 & 2 \end{bmatrix}$
   c. $\begin{bmatrix} 1 & 0 & -1 \\ 2 & 2 & -6 \\ 1 & 5 & 1 \end{bmatrix}$

9. **a.** $\begin{bmatrix} -1 & 4 & 7 \\ 18 & 9 & -7 \\ 8 & 7 & -1 \end{bmatrix}$ **b.** $\begin{bmatrix} -5 & 8 & -1 \\ 6 & 9 & 1 \\ -8 & -1 & -5 \end{bmatrix}$

11. **a.** $\begin{bmatrix} 3 & -1 & 0 \\ 7 & 4 & -14 \\ 6 & 12 & 3 \end{bmatrix}$ **b.** $\begin{bmatrix} -1 & 1 & 4 \\ 5 & 1 & 3 \\ 3 & -2 & -1 \end{bmatrix}$

13. $\begin{bmatrix} 3 & 0 & 3 \\ 10 & 4 & 2 \\ 4 & -1 & 2 \end{bmatrix}$ 15. $\begin{bmatrix} 7 & -6 & 7 \\ 4 & -3 & 0 \\ 1 & -3 & 5 \end{bmatrix}$

17. $\begin{bmatrix} -19 & -3 & -2 \\ 4 & 2 & 12 \\ -7 & 3 & 6 \end{bmatrix}$ 19. $\begin{bmatrix} -15 & -9 & 2 \\ -2 & -5 & 10 \\ -10 & 1 & 9 \end{bmatrix}$

21. $I^2 = I$ 23. $BA = I$
25. $BA = I$ 27. $BA = I$

29. No. The matrices could not be conformable for both orders of multiplication.

31. **a.** $A^2 = A$ **b.** $A^2 = A$

33. **a.**

	E	*M*	*L*
Armchairs	20	15	10
Rockers	12	8	5

   b.

	E	*M*	*L*
Armchairs	120	90	60
Rockers	72	48	30

35.

	E	*M*	*L*
Armchairs	30	15	10
Rockers	12	28	0

37.

Hardware	July	Aug.	Sept.	Oct.	Nov.
Bolts	650	1300	2100	2600	2400
Clamps	400	800	1400	1600	1800
Screws	1200	2400	4100	4800	5100

39. a. $C^2 = \begin{bmatrix} 2 & 0 & 0 \\ 0 & 1 & 1 \\ 0 & 1 & 1 \end{bmatrix}$

    b. It means that Tom can communicate with himself by 2 two-step communications.

    c. Two. The 1 in the second row, third column means that Dick can communicate with Harry by a two-step communication. The 1 in the third row, second column means that Harry can communicate with Dick by a two-step communication.

41. If we do a row-column multiplication the result is the matrix

$$\begin{bmatrix} -2 & -2 & 0 \\ 0 & 4 & 4 \end{bmatrix}$$

As the figure shows, the multiplication rotated the triangle about the y axis and doubled the length of each side.

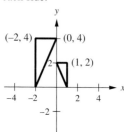

43. Reflects (a, b) across the y axis to $(-a, b)$
45. Rotates (a, b) $180°$ around the origin to $(-a, -b)$
47.

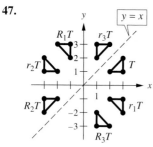

49. Yes. The sum of two 2×2 matrices is a 2×2 matrix.
51. Yes. The 2×2 zero matrix
53. Yes

55. Suppose that A has an inverse, say, $B = \begin{bmatrix} x & y \\ z & w \end{bmatrix}$. Then,

$$AB = \begin{bmatrix} 1 & 2 \\ 0 & 0 \end{bmatrix}\begin{bmatrix} x & y \\ z & w \end{bmatrix} = \begin{bmatrix} x + 2z & y + 2w \\ 0 & 0 \end{bmatrix}$$

Since there are no values of $x, y, z,$ and w to make this last matrix the identity matrix, A has no inverse.

Skill Checker

1. $(1, 1)$ **3.** $(2, -2)$

Exercises 9.2

1. $x = 1, y = 2, z = 0$ **3.** $x = -1, y = -1, z = 3$
5. $x = -2, y = -1, z = 3$ **7.** No solution
9. $x = 1 - k, y = 2, z = k,$ where k is any real number
13. 8 type I, 10 type II, 12 type III
15. 50% type I, 25% type II, 25% type III
17. a. Singular   **b.** and **c.** Nonsingular
19. $x = a, y = b,$ and $z = c$
21. The solution is $x = a, y = b,$ and z is any real number.
23. $\begin{bmatrix} 1 & 0 & 0 & | & a \\ 1 & 0 & 0 & | & b \\ 0 & 0 & 0 & | & c \end{bmatrix}$ $c \neq 0$

$\begin{bmatrix} 1 & 0 & 0 & | & a \\ 0 & 0 & 0 & | & b \\ 0 & 0 & 1 & | & c \end{bmatrix}$ $b \neq 0,$ $\begin{bmatrix} 0 & 0 & 0 & | & a \\ 1 & 0 & 0 & | & b \\ 0 & 0 & 1 & | & c \end{bmatrix}$ $a \neq 0$

Exercises 9.3

1. 4   **3.** 11   **5.** 3   **7.** 7
9. 5   **11.** 9   **13.** 1   **15.** 2
17. 11   **19.** 10   **21.** 1   **23.** 12
25. 6   **27.** 4

29.

$\otimes$	1	2	3	4	5	6	7	8	9	10	11	12
1	1	2	3	4	5	6	7	8	9	10	11	12
2	2	4	6	8	10	12	2	4	6	8	10	12
3	3	6	9	12	3	6	9	12	3	6	9	12
4	4	8	12	4	8	12	4	8	12	4	8	12
5	5	10	3	8	1	6	11	4	9	2	7	12
6	6	12	6	12	6	12	6	12	6	12	6	12
7	7	2	9	4	11	6	1	8	3	10	5	12
8	8	4	12	8	4	12	8	4	12	8	4	12
9	9	6	3	12	9	6	3	12	9	6	3	12
10	10	8	6	4	2	12	10	8	6	4	2	12
11	11	10	9	8	7	6	5	4	3	2	1	12
12	12	12	12	12	12	12	12	12	12	12	12	12

31. 3 **33.** 11 **35.** 4
37. 8 **39.** No solution **41.** False
43. False **45.** False **47.** 2 (mod 5)
49. 4 (mod 5) **51.** 3 (mod 5) **53.** 1 (mod 5)
55. 3 (mod 5) **57.** 3 (mod 5) **59.** 0
61. 0 **63.** 1 **65.** 2
67. 2 **69.** 2 **71.** 3
73. 2 **75.** Yes **77.** Yes, 1
79. 0x **81.** No; yes
83. It is in the 20s or 30s, and so on.
85. 1948, 1960, 1972, 1984, 1996, 2008, . . .
87. 24. It avoids confusion between A.M. and P.M.
91. 5 **93.** 2 **95.** 1
97. 4 **99.** 7 **101.** 3
103. 1 **105.** 7 **107.** Yes
109. No; 2 **111.** No; 4 **113.** Yes
115. No; 8 **117.** Valid number
119. Error (Sum is 18.9)

Exercises 9.4

1. a. a **b.** c **c.** b
3. a. a **b.** a **c.** Yes
5. a. c **b.** c **c.** Yes
7. Yes. If $x \in S$ and $y \in S$, then $x @ y \in S$.
9. a. Yes. If a and b are natural numbers, then $a \text{ F } b = a$, which is a natural number.
  b. Yes. $a \text{ F } (b \text{ F } c) = a \text{ F } b = a$ and $(a \text{ F } b) \text{ F } c = a \text{ F } c = a$. Thus, $a \text{ F } (b \text{ F } c) = (a \text{ F } b) \text{ F } c$.
  c. No. If $a \neq b$, then $a \text{ F } b = a$ and $b \text{ F } a = b$, so $a \text{ F } b \neq b \text{ F } a$.
11. a. No. For example, $3 + 5 = 8$, which is not an odd number
  b. Yes. The product of two odd numbers is an odd number.
c. Yes. The sum of two even numbers is an even number.
d. Yes. The product of two even numbers is an even number.

13.

$\cap$	$\varnothing$	$\{a\}$	$\{b\}$	$\{a, b\}$
$\varnothing$	$\varnothing$	$\varnothing$	$\varnothing$	$\varnothing$
$\{a\}$	$\varnothing$	$\{a\}$	$\varnothing$	$\{a\}$
$\{b\}$	$\varnothing$	$\varnothing$	$\{b\}$	$\{b\}$
$\{a, b\}$	$\varnothing$	$\{a\}$	$\{b\}$	$\{a, b\}$

15. a. $\varnothing$ **b.** $\varnothing$ **c.** Yes
17. Yes. All elements in the table are elements of S.

19. a.

L	1	2	3	4
1	1	2	3	4
2	2	2	3	4
3	3	3	3	4
4	4	4	4	4

  b. 1
21. a. No inverse **b.** No inverse
  c. No inverse **d.** 4
23. Yes; 1
25. a. The identity element is A, because for every B that is a subset of A, $A \cap B = B \cap A = B$.
  b. No, there is no other identity element.
27. Yes. The identity element is 0.
29. a. 3 **b.** 4
31. The distributive property holds. If a, b, and c are real numbers, then $a \text{ F } (b \text{ L } c) = a$ and $(a \text{ F } b) \text{ L } (a \text{ F } c) = a \text{ L } a = a$.
33. Yes **35.** Yes
37. Yes (actually, a commutative group)
39. No; no multiplicative inverses
41. No; no identity element
43. No; no multiplicative inverses
45. No; no multiplicative inverses
47. Yes
49. No; no multiplicative inverse for 0
51. No; no identity element

53. No (no multiplicative inverses)

55. No (no multiplicative inverses)

57. Check to see that $a \blacklozenge b$ is always an element of S.

59. Check the table to see whether there is an element e in S such that the column under e is identical to the column at the far left and the row opposite e is identical to the top row. If there is, then e is the identity element. If there is no such element, then there is no identity element.

61. You have to check that $x * (y * z) = (x * y) * z$ for all possible values of x, y, and z from the set $\{a, b, c\}$. If you had to check all possible cases, there would be 27 of these because each of the 3 places has 3 possible values. However, since the operation has the commutative property, the number of cases to be checked is greatly reduced. Think about it.

63. $6 \times 9999 = 6(10,000 - 1) = 60,000 - 6 = 59,994$

65. $7 \times 59 = 7(60 - 1) = 420 - 7 = 413$

67. $4 \times 9995 = 4(10,000 - 5) = 40,000 - 20 = 39,980$

69. The following steps show why the puzzle works:

Think of a number.	x
Add 3 to it.	$x + 3$
Triple the result.	$3x + 9$
Subtract 9.	$3x$
Divide by the number x with which you started.	3

Exercises 9.5

1. Strictly determined. Optimal pure strategy: Row player should play row 1, column player should play column 1. Value = 4.

3. Not strictly determined

5. Strictly determined. Optimal pure strategy: Row player should play row 1, column player should play column 3. Value = 4.

7. Strictly determined. Optimal pure strategy: Row player should play row 3, column player should play column 1 or column 3. Value = 4.

9. a. No saddle point

b. Row player should play row 1 five-sixths of the time and row 2 one-sixth of the time.

11. a. No saddle point

b. Row player should play row 1 one-sixth of the time and row 2 five-sixths of the time.

13. Optimal row strategy: Play row 1 one-fourth of the time, row 2 three-fourths of the time, and do not play row 3. Value = 3.

15. Study 2 hr half the time, and 4 hr half the time.

17. Ann's optimal strategy: Buy no bonds, buy stocks with five-sevenths and money market funds with two-sevenths of her investment. Her expected return will be $11\frac{3}{7}\%$.

19. Station R should price its gasoline at \$1 four-fifths of the time.

21. a.

	Younger	Older
Performance	70%	20%
Safety	40%	80%

b. $\frac{4}{9}$ performance, $\frac{5}{9}$ safety

23.

Freeze

		Yes	No
Water	Yes	6000	−400
	No	−4000	4000

Optimal strategy: water $\frac{5}{9}$ of the time; don't water $\frac{4}{9}$ of the time. Expected payoff $= \frac{\$14,000}{9} \approx \$1556$.

25. If row i dominates row j, this means that in the long run playing row i is more profitable than playing row j. Thus, row j can be eliminated from the row player's options.

27. Row 1

29. Send poems $\frac{5}{8}$ of the time and candy $\frac{3}{8}$ of the time. Do not send flowers.

CHAPTER 10

Exercises 10.1

1. 8 different outfits

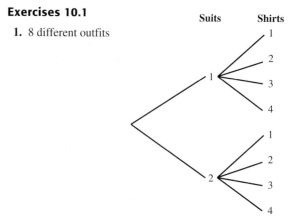

3. 8 different outcomes

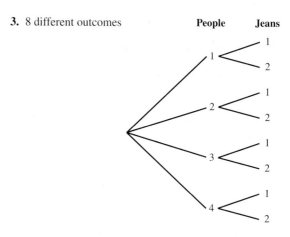

5. a. 9   **b.** 126   **c.** 25

7. a. 12   **b.** 144

9. 30

11. 17,576

13. 378

15. 16

17. 90

19. a. 900,000,000   **b.** 10^9

21. a. 59,280   **b.** $40^3 = 64,000$

23. a. 10   **b.** 2   **c.** 4   **d.** 80

25. a. 19   **b.** 152

27. 720

29. 480

31. (a, @), (a, &), (a, %), (b, @), (b, &), (b, %), (c, @), (c, &), (c, %), (@, a), (&, a), (%, a), (@, b), (&, b), (%, b), (@, c), (&, c), (%, c)

33. $2 \times 2 = 4$

35. $3 \times 2 = 6$

37.

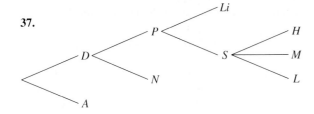

39.

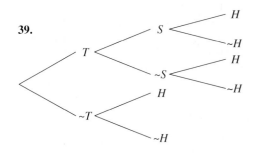

41. If a single event can occur in m ways or in n ways, then the total number of ways in which the event can occur is $m + n$ (assuming no duplications).

43. 24

45. There are not enough different sets of initials for 27,000 people, so at least 2 people must have the same set of initials.

47. 144

49. It will point to 3 if no slippage occurs.

Exercises 10.2

1. 24

3. 720

5. 120

7. 40,320

9. 362,880

11. 990

13. 126

15. 90

17. 10

19. 20,160

21. 120

23. 5040

25. 1716

27. 650

29. 60

31. 9

33. 31

35. 1

37. 57

39. 60

41. a. $10^6 = 1,000,000$ **b.** $P(10, 6) = 151,200$
    c. 151,200 are not enough for a population of 608,827.

43. $2 \cdot P(25, 3) = 27,600$

45. $n! = n \times (n - 1)!$ for $n > 1$. This formula holds for $n = 1$ only if 0! is defined to be 1.

The formula $P(n, r) = \dfrac{n!}{(n - r)!}$ holds for $r = n$, only if 0! is defined to be 1.

47. The number of elements in the union of two sets is the sum of the number of elements in each of the sets diminished by the number of elements common to the two sets.

49. 120

51. 3, including a possible tie for fourth place

53. $(n - 1)!$

55. 4

Exercises 10.3

1. $C(5, 2) = 10, P(5, 2) = 20$

3. $C(7, 3) = 35, P(7, 3) = 210$

5. $C(9, 6) = 84, P(9, 6) = 60,480$

7. $C(5, 4) = 5$

9. $C(10, 2) = 45$

11. $C(8, 2) = 28$

13. $C(12, 8) = 495$

15. a. 120   **b.** 56   **c.** 1   **d.** 1

17. $C(5, 4) = 5$

19. a. $C(5, 3) = 10$ **b.** 26

21. $C(24, 3) = 2024$

23. $C(10, 3) = 120$

25. $C(52, 5) = 2,598,960$

27. $C(100, 5) = 75,287,520$

29. 15

31. 63

33. $C(8, 4) = 70$

35. Answers will vary; the important difference is that permutations take account of order and combinations do not.

37. The order of the numbers is important, so a permutation is being used.

39. $n = 6$: 1   6   15   20   15   6   1
    $n = 7$: 1   7   21   35   35   21   7   1

41. a. $(a + b)^4 = a^4 + 4a^3b + 6a^2b^2 + 4ab^3 + b^4$
    b. $(a + b)^5 = a^5 + 5a^4b + 10a^3b^2 + 10a^2b^3 + 5ab^4 + b^5$

43. $C(5, 0) = 1$

45. $C(5, 3) = 10$

47. $C(5, 5) = 1$

49. The left side is the sum of the number of ways in which there could be 0 heads and n tails, 1 head and $n - 1$ tails, 2 heads and $n - 2$ tails, and so on, to n heads and 0 tails. The right side is exactly the number of ways in which n coins can fall either heads or tails. Thus, the two sides are equal.

Exercises 10.4

1. a. $P(52, 3) = 132,600$
    b. $C(52, 3) = 22,100$

3. a. $50 \times 50 \times 50 = 125,000$
    b. $C(50, 3) = 19,600$

5. $C(5, 3) = 10$

7. a. $C(7, 3) = 35$ **b.** $P(7, 4) = 840$

9. a. $P(10, 2)P(14, 2) = 16,380$
    b. $P(10, 2)P(12, 2) = 11,880$

11. a. $C(7, 3)C(8, 1) = 280$
b. $C(5, 3)C(2, 1) = 20$
13. 831,600   **15.** 22,680
17. $C(7, 2)C(5, 3)C(2, 2) = 210$
19. $C(n, 5) \geq 21$ for $n \geq 7$
21. $C(4, 2)P(3, 2)C(2, 1) = 72$

23. a.

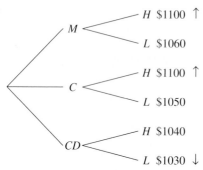

tree:
M — H \$1100 ↑, L \$1060
C — H \$1100 ↑, L \$1050
CD — H \$1040, L \$1030 ↓

b. \$1100, mutual fund (M), high (H) or management company (C), high (H)
c. \$1030, CD, low (L)

25. a.

tree:
EH — H \$2000 √, M \$1000, L \$750
LO — H \$750, M \$500, L \$300 √

b. \$2000, market size extremely high (EH), sales high (H)
c. \$300, market size low (LO), sales low (L)
27. Answers will vary.   **29.** 15
31. $(a + 1)(b + 1)(c + 1)(d + 1)$

CHAPTER 11

Exercises 11.1

1. $\frac{1}{6}$ **3.** $\frac{1}{3}$ **5.** $\frac{1}{10}$ **7.** $\frac{9}{10}$
9. 0   **11.** $\frac{1}{13}$ **13.** $\frac{1}{4}$ **15.** $\frac{11}{26}$
17. a. $\frac{1}{5}$ **b.** $\frac{3}{5}$ **c.** $\frac{4}{5}$ **19. a.** $\frac{1}{3}$ **b.** $\frac{1}{3}$
21. $\frac{1}{8}$ **23.** $\frac{3}{8}$ **25.** $\frac{1}{6}$ **27.** $\frac{1}{6}$
29. $\frac{1}{2}$
31. a. $\frac{1}{4}$ **b.** $\frac{7}{25}$ **c.** $\frac{6}{25}$
d. $\frac{23}{100}$ **e.** $\frac{1}{4}$ **f.** $P(B)$
33. a. $\frac{6}{25}; \frac{1}{6}$ **b.** $\frac{9}{100}; \frac{1}{9}$
c. $\frac{29}{100}; \frac{5}{18}$ **d.** $\frac{1}{5}; \frac{2}{9}$ **e.** $\frac{9}{50}; \frac{2}{9}$
35. a. $\frac{3}{25}$ **b.** $\frac{9}{50}$ **c.** $\frac{13}{25}$ **d.** $\frac{12}{25}$
e. Getting a sum of 8

f. Getting a sum of 2 (or 4 or 12)
g. Getting a sum of 2, 4, or 12
h. No
37. a. $\frac{86}{251}$ **b.** $\frac{73}{502}$
c. Having exactly 4 credit cards
d. Having no credit cards
e. $\frac{165}{251}$
39. a. $\frac{3}{20}$ **b.** $\frac{37}{50}$ **c.** $\frac{11}{100}$
41. a. $\frac{89}{239}$ **b.** $\frac{33}{239}$ **c.** $\frac{206}{239}$ **d.** $\frac{24}{239}$
43. a. $\frac{140}{5543}$ **b.** $\frac{2804}{5543}$ **c.** $\frac{2739}{5543}$
45. a. $\frac{13}{17}$ **b.** $\frac{4}{17}$ **47. a.** $\frac{11}{25}$ **b.** $\frac{14}{25}$
49. a. $\frac{8}{75}$ **b.** $\frac{1}{5}$
53. The probability formula does not apply if the events are not all equally likely to occur (see Belgian one-euro coin). If a die is weighted so that a 6 is twice as likely to come up as any other number, then to calculate the probability that an even number comes up, it would be wrong to use the fact that 3 of the 6 faces are even so the probability is $\frac{1}{2}$. Instead, the 6 face must be given a weight of 2 and the other faces weights of 1. Then the weight of the even faces is $1 + 1 + 2 = 4$ and the weight of all the faces is 7. Thus, the probability that an even number comes up is $\frac{4}{7}$.
57. $\frac{4653}{4720}$, or about 0.986   **59.** $\frac{253}{254}$, or about 0.996

Exercises 11.2

1. $\frac{1}{18}$ **3.** $\frac{1}{8}$ **5.** $\frac{C(2, 2)C(23, 2)}{C(25, 4)} = \frac{1}{50}$
7. a. $\frac{1}{5}$ **b.** $\frac{4}{5}$
9. a. $\frac{4}{21}$ **b.** $\frac{1}{21}$
11. $\frac{P(4, 2)}{P(52, 2)} = \frac{1}{221}$ **13.** $\frac{13 \times 3}{52 \times 51} = \frac{1}{68}$
15. $\frac{P(26, 2)}{P(52, 2)} = \frac{25}{102}$
17. a. 15   **b.** $\frac{5 \times 3}{15 \times 14} = \frac{1}{14}$ **c.** $\frac{1}{7}$
19. 0.0396   **21.** 0.005
23. a. $\frac{C(4, 2)C(4, 2)C(44, 1)}{C(52, 5)}$ **b.** $\frac{C(4, 3)C(4, 2)}{C(52, 5)}$
25. $\frac{2}{5}$ **27.** $\frac{4}{C(52, 5)} = \frac{1}{649,740} \approx 0.0000015$
29. $\frac{13 \times 48}{C(52, 5)} = \frac{1}{4165} \approx 0.00024$
31. $\frac{4 \times [C(13, 5) - 10]}{C(52, 5)} = \frac{5148 - 40}{C(52, 5)} = \frac{1277}{649,740} \approx 0.0020$
33. 0.2   **35.** Dome structure, dry
37.

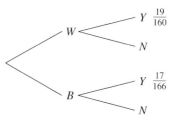

tree:
W — Y $\frac{19}{160}$, N
B — Y $\frac{17}{166}$, N

39. $\frac{141 + 149}{326} = \frac{290}{326} = \frac{145}{163}$
41. No. The coin is probably weighted to come up heads. Bet on heads.

43. No. The coin is probably weighted to come up heads. Bet on heads.

45. $\frac{C(40,000,150)C(36,000,150)}{C(76,000,300)}$

47. $\frac{C(16,6)C(14,6)}{C(30,12)}$

Exercises 11.3

1. 0; property 1   **3.** 1; property 2   **5.** $\frac{3}{5}$
7. 1   **9.** $\frac{7}{26}$ **11.** $\frac{1}{4}$
13. $\frac{3}{13}$ **15.** $\frac{41}{50}$ **17.** 0.2

19. 0.45   **21. a.** $\frac{13,776}{98,447}$ **b.** $\frac{8,135}{98,937}$

23. a. $\frac{27,381}{97,100}$ **b.** $\frac{18,180}{98,432}$ **25. a.** 1   **b.** 1

27. $\frac{11}{20}$ **29.** $\frac{4}{5}$ **31.** $\frac{13}{17}$
33. If the probability of an event is 0, then the event cannot occur.
35. If $P(A \cap B) = 0$, that is, A and B have no common elements, then $P(A \cup B) = P(A) + P(B)$.
37. About 0.83   **39.** Answers will vary.   **41.** $\frac{9}{10}$
43. $\frac{2}{5}$ **45.** $\frac{1}{5}$ **47.** $\frac{1}{20}$

Exercises 11.4

1. 0
3. a. $\frac{1}{6}$ **b.** 0
   c. 1   **d.** 0
5. $\frac{1}{2}$ **7.** $\frac{1}{2}$
9. a. $\frac{1}{5}$ **b.** $\frac{1}{3}$
11. a. $\frac{11}{20}$ **b.** 80   **c.** $\frac{3}{5}$
13. $\frac{1}{17}$ **15.** $\frac{1}{5}$
17. $\frac{P(A \cap B)}{P(A)} = \frac{2}{3}$ **19. a.** $\frac{3}{10}$ **b.** $\frac{39}{100}$
21. $\frac{1}{3}$ **23. a.** $\frac{27}{40}$ **b.** $\frac{7}{10}$
25. $\frac{19}{160} \approx 0.119$ **27.** No (only about 0.017)
29. $\frac{11}{63} \approx 0.175$ **31.** $\frac{6}{103} \approx 0.058$
33. Answers will vary.
35. a. $\frac{75}{128}$ **b.** $\frac{53}{128}$ **c.** Male
37. $\frac{9}{53}$

Exercises 11.5

1. Yes
3. a. $\frac{1}{96}$ **b.** $\frac{1}{48}$ **c.** $\frac{7}{16}$
5. a. $\frac{1}{8}$ **b.** $\frac{1}{2}$ **c.** $\frac{7}{8}$
7. a. $\frac{1}{4}$ **b.** $\frac{1}{2}$ **c.** $\frac{1}{16}$ **d.** $\frac{9}{16}$
9. a. $\frac{1}{2}$ **b.** $\frac{3}{4}$ **c.** $\frac{3}{8}$
   d. They are independent.
11. $\frac{5}{108}$ **13.** $\frac{25}{324}$ **15.** $\frac{25}{36}$
17. $\frac{9}{76}$ **19.** $\frac{9}{380}$ **21.** $\frac{28}{625}$
23. $\frac{289}{625}$ **25.** $\frac{301}{625}$
27. 0.205379   **29.** 0.636056
31. a. $\frac{1}{12}$ **b.** $\frac{1}{8}$
33. a. $\frac{1}{4}$ **b.** $\frac{1}{12}$ **c.** $\frac{1}{4}$
35. About 0.503   **37.** About 0.59

39. a. $\frac{1}{8000}$ **b.** $\frac{57}{8000}$
41. 0.189   **43.** $\frac{1}{2}$
45. $\frac{3}{5}$ (*Hint:* Draw a tree and use the final probabilities as weights for the two possibilities.)
47. a. The probability of one of the events does not depend on the probability of the other event, or the occurrence of one of the events does not affect the occurrence of the other event.
   b. Find the product $P(A)P(B)$.
49. $\frac{C(50,25)}{2^{50}}$
51. $1 - \left[C(6,0)\left(\frac{1}{3}\right)^6 + C(6,1)\left(\frac{1}{3}\right)^5\left(\frac{2}{3}\right) + C(6,2)\left(\frac{1}{3}\right)^4\left(\frac{2}{3}\right)^2 \right] = \frac{656}{729}$
53. $C(5,2)\left(\frac{1}{6}\right)^2\left(\frac{5}{6}\right)^3 = \frac{625}{3888}$
55. $\frac{1}{8}$ **57.** 0.02   **59.** 0.000008

Exercises 11.6

1. 1 to 5   **3.** 1 to 12   **5.** 1 to 3
7. 5 to 21   **9.** 1 to 1   **11.** 10 to 3
13. $C(6,3) \cdot C(47,3)/C(53,6) = \frac{32430}{2295748}$; 32,430 to 2,263,318
15. $\frac{3}{5}$ **17.** $\frac{63}{10,000}$; 63 to 9937
19. 1 to 1   **21.** $\frac{10}{4877}$ **23.** $\frac{1}{8}$
25. 33 to 67   **27.** $-\$5$ **29.** \$2.15
31. No. The probability of getting a sum of 7 is only $\frac{1}{6}$.
33. Build at the first location and make \$50,000.
35. a. $-\$4000$ **b.** \$10,000
   c. Discontinue the campaign.
37. a. \$200   **b.** \$450   **c.** B
39. The mathematical expectation is that you lose \$.58 per \$2 bet.
41. a. About 0.81   **b.** About 0.19
43. Odds against you are 9 to 1, so put up \$1 for each \$9. Thus, you should bet $\left(\frac{1}{9}\right)(\$10)$, or about \$1.11, to make the bet fair.
45. If the ball stops on red or black, you break even. If the ball stops on 0 or 00 and then on red or black, you lose 50¢. Thus, the expected value is $-(50)\left(\frac{2}{38}\right)\left(\frac{36}{38}\right) = -\frac{\$9}{361} \approx -2.5¢$.

CHAPTER 12

Exercises 12.1

1. Descriptive statistics is the science of collecting, organizing, and summarizing data.
3. A sample population usually consists of just a part of the target population.
5. a. All households in the United States
   b. The 1006 households surveyed
7. a. No
   b. The sample includes only those viewers who are willing to pay for the call.
9. a. Make a card for each student, number the cards, and mix them up. Then draw ten cards at random and select the corresponding students.

b. Make a card for each student, number the cards, and do any of the following:
   i. Pick only even-numbered cards.
   ii. Pick only odd-numbered cards.
   iii. Don't mix the cards and pick the first ten.

11. a. All TV watchers
  b. People who visit the Web site and are willing to participate in the survey
  c. No. Respondents do not necessarily represent the target population (all TV watchers).

13. Although both surveys ultimately had 50 responses, the second survey should produce more accurate results because $\frac{50}{75} \approx 67\%$ of the people responded compared with only $\frac{50}{100} \approx 50\%$ for the first survey.

15. a. Attended a football game last year?

Y	720
N	160
NR	120
Total	1000

  b. So the total will add to 1000

17. a.

Number of Hours	Tally Marks	Frequency
0	|	1
1	|	1
2	|||	3
3	|||	3
4	|||	3
5	|||	3
6	||||	4
7	||	2

  b. 6   **c.** 6; $\frac{6}{20} = 30\%$ **d.** 14; $\frac{14}{20} = 70\%$

19. a.

Outcome	Tally Marks	Frequency
1	|||| |||| |	11
2	||	2
3	|||| |	6
4	||	2
5	|||| |||	8
6	|	1

  b. The 1; 11   **c.** The 6; 1

21. a.

Number of Hours	Tally Marks	Frequency
0	||	2
1	|||	3
2	||	2
3	|||	3
4	|||	3
5	||	2
6	||	2
7	||	2
8	||||	4
9	|	1
10	||	2
11		0
12	||	2
13		0
14	|	1
15	|	1

  b. 8   **c.** 4   **d.** 15   **e.** About 36.7%

23. a.

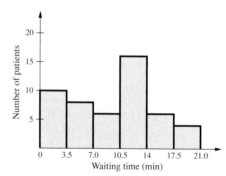

  b. 36%   **c.** 52%

25. a.

Age	Tally Marks	Frequency
6	|	1
7	|	1
8	|||	3
9	|||	3
10	||	2

b.

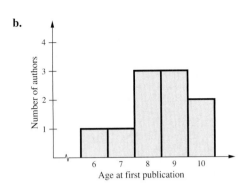

c. 20%

27. a.

Price	Tally Marks	Frequency
$0 < P \le 10$		0
$10 < P \le 20$		0
$20 < P \le 30$	‖‖‖ ‖	7
$30 < P \le 40$	‖‖‖ ‖‖‖	8
$40 < P \le 50$	‖‖‖	4
$50 < P \le 60$	‖‖‖	3
$60 < P \le 70$	‖‖‖	5
$70 < P \le 80$	‖	2
$80 < P \le 90$	‖	1

b. $30 < P \le 40$ **c.** 15 **d.** 7

e. $\dfrac{12}{30} \approx 40\%$ **f.** $\dfrac{7}{30} \approx 23\%$

29.

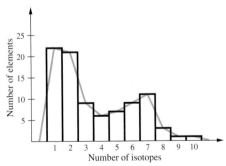

31. a.

Letter	Frequency	Letter	Frequency
a	10	n	13
b	4	o	14
c	3	p	2
d	4	q	0
e	18	r	7
f	2	s	10
g	3	t	17
h	9	u	5
i	12	v	2
j	0	w	5
k	1	x	0
l	4	y	2
m	4	z	0

b. e **c.** About 39.1%

33.

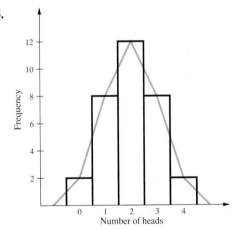

35. a.

Concentration	Tally Marks	Frequency
0.00–0.04	‖‖‖ ‖	6
0.05–0.09	‖‖‖ ‖‖‖ ‖‖‖	13
0.10–0.14	‖‖‖	5
0.15–0.19	‖‖‖	4
0.20–0.24	‖	2

b. 20%

37. a.

Weekly Salary	Tally Marks	Frequency
$ 600–$2400	ⅢⅢ ⅢⅢ	9
$2400–$4200		0
$4200–$ 6000	‖	2
$6000–$ 7800			1

b.

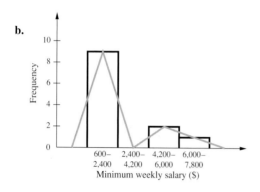

39. a.

Salary (× 1000)	Tally Marks	Frequency
$20–$34	ⅢⅢ ⅢⅢ ⅢⅢ ⅢⅢ	20
$34–$48			1
$48–$62			1
$62–$76	‖	2
$76–$90			1

b.

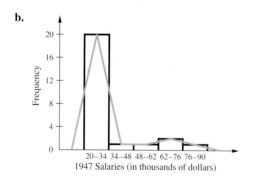

41. a. $74

b.

Price	Tally Marks	Frequency
$180–$254	ⅢⅢ	5
$254–$328	ⅢⅢ ‖	7
$328–$402		0
$402–$476			1
$476–$550	‖	2

c.

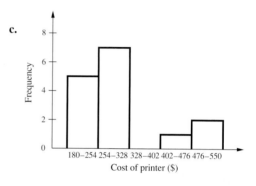

43. a. 3 years

b.

Age	Tally Marks	Frequency
109–111	ⅢⅢ ⅢⅢ ‖	12
112–114	ⅢⅢ ⅢⅢ	10
115–117	‖	2
118–120		0
121–123			1

c.

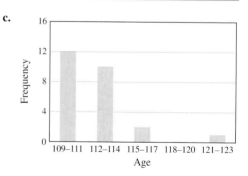

45. Tract 1304   **47.** Tract 1305

49. No. The women were not selected at random.

51. No. Not all members of the population had the same chance of being chosen.

53. The upper and lower class limits, respectively, are the least and the greatest values in that class. Each class boundary is the midpoint between the upper limit of the respective class and the lower limit of the next class.

55.

Digit	Frequency
0	1
1	4
2	5
3	6
4	4
5	4
6	3
7	3
8	5
9	5

57. Yes. 100,271
59. Yes. In each case, most of the bar is omitted.

Exercises 12.2

1. a. Mean = 9; median = 9
 b. Mean = 24.2; median = 9
 c. Mean = 11; median = 9
 d. Mean = median for part (**a**) only. Median = 9 for all three. None has a mode.
3. Mean = 6.05, median = 6.5, and mode = 8. The mode is the least representative.
5. Betty is correct. She took account of the number of students making each score, and Agnes did not.

7.

Number of Letters	Frequency
2	1
3	5
4	4
5	5
6	0
7	3
8	2
9	0
10	0
11	1

 a. 3 and 5   **b.** 5   **c.** 5.05
 d. No. There is too much repetition.

9. b.–c. Answers will vary, but should be approx. 8.
11. 60   **13.** $84 per week
15. a. 0.34   **b.** $20,000–$24,999
 c. $39,444   **d.** $20,000
17. Statements (**b**) and (**c**) are both true.
19. 70   **21.** About 577.25; 622.5
23. 20.6   **25.** About 22.2
27. 28   **29.** Internet
31. 13.425 million; 14.2 million
33. 9.5 million; 9.75 million
35. 66.5 million
37. a. 43,216.40; $40,827; none   **b.** Median
39. The median of a set of scores is the middle number (if there is one) when the scores are arranged in order of magnitude. If there is no middle number, the median is the average of the two middle numbers. The median is not a good measure of a set of scores as it gives no indication of how the scores are spread.
41. a. 2.4   **b.** 2
43. The mean
45. The mean of a secretary's and a worker's salaries

Exercises 12.3

1. a. 18   **b.** 7.21   **3. a.** 20   **b.** 7.91
5. a. 4   **b.** 1.58   **7. a.** 8   **b.** 3.11
9. a. 6   **b.** 2.16   **11. a.** 8   **b.** 6.5
 c. 6   **d.** 3.23   **e.** 70%   **f.** 100%
13. a. 110   **b.** 110   **c.** 108
 d. 3.71   **e.** 102, 103, 113; 30%
15. Range 15, mean 7, standard deviation $\sqrt{30} \approx 5.48$
17. Range 9, mean 7, standard deviation $\sqrt{10} \approx 3.16$
19. $80, $390, 0, $89, $77
21. $167   **23.** $14   **25.** $6.30
27. The numbers are all the same. If zero is the standard deviation, then $(x - \bar{x})^2 = 0$ for all x in the set.
29. Choose the numbers 1, 1, 5, and 5 (remember, repetitions are allowed). The mean of these numbers is 3, and 1, 1, 5, and 5 are as far from the mean 3 (2 units) as possible.
31. 5   **33.** About 8.66
35. If you multiply each data point by 1.1 (100% + 10% = 1.1), it increases the spread of the points by a factor of 1.1. For example, two employees making $50,000 and $30,000 are now $20,000 apart. When each gets a 10% raise, they are $1.1 \times 50,000 - 1.1 \times 30,000 = 1.1 \times (50,000 - 30,000)$, or $22,000, apart.

Exercises 12.4

1. a. 100 in.   **b.** 10 in.   **c.** 68%   **d.** 2700
3. a. 25   **b.** 25   **c.** 680
5. A, 12 or 13; B, 67 or 68; C, 340; D, 67 or 68; F, 12 or 13
7. 5   **9.** 19 ft 10.5 in. and 20 ft 1.5 in.
11. The purchasing director will decide to buy. If the lifetimes are normally distributed, $2\frac{1}{2}\%$ will last 40 or

more days. Thus, of the 8000, about 200 will last 40 or more days.

13. The student who scored 570 on the S.A.T. has the higher score relative to the test.

15. a. 10   **b.** 5   **c.** 40
d. 65   **e.** 60; 30   **f.** 75; 60

17. 16.15 oz   **19.** 90

21. a. 0.8   **b.** 1.6   **c.** 2

23. The German test score

25. 47.7%   **27.** 62.9%

29. a. 0.249   **b.** 0.477

31. a. 0.998   **b.** 0.004

33. 0.228

35. Taller than 69.4 inches or about 5 ft $9\frac{1}{2}$ in.

37. 1 indicates that $4400 is 1 standard deviation above the mean, $4200.

39. Between 7.25 and 8.75 oz

41. 44.5%   **43.** 100 and 120 gal

45. 50%   **47.** 34%

49. 0.067, or 6.7%

51. Yes. See Figure 12.14 on page 818. In this figure both curves have the mean 0. The standard deviation for A is $\frac{1}{3}$ unit while that for B is 1 unit.

53. No. If curve A in Figure 12.14 on page 818 is moved 2 units to the right, this would be an example. The standard deviation for A would be $\frac{1}{3}$ unit and that for B would be 1 unit, while the mean for A would be 2 and that for B would be 0.

55. 80th

57. a. 20th   **b.** 60th   **c.** 90th   **d.** 0th

59. a. 75   **b.** 89   **c.** 49

61. $\bar{x} = 6$; $s \approx 4.77$. Three lie within 1 standard deviation from the mean, and all nine lie within 2 standard deviations from the mean. The theorem makes no prediction for 1 standard deviation; it predicts 75% for 2 standard deviations.

63. No. It cannot exceed 25%.

Exercises 12.5

1.

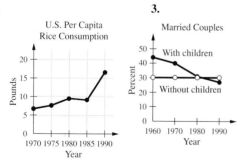

3.

5.

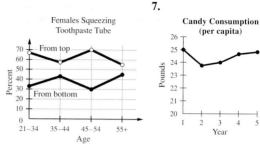

7.

9.

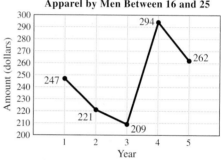

11.

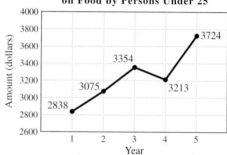

13.

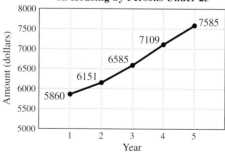

15.

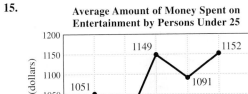

Average Amount of Money Spent on Entertainment by Persons Under 25

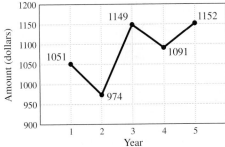

17.

Average Amount of Money Spent on Health Care by Persons Under 25

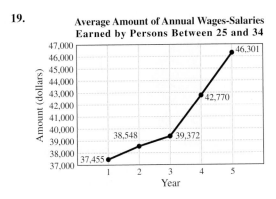

19.

Average Amount of Annual Wages-Salaries Earned by Persons Between 25 and 34

21.

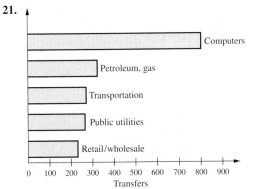

23. a.

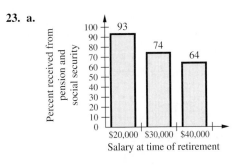

b. $18,600

25. a.

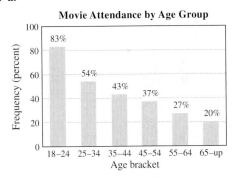

b. At specialty stores   **c.** Department stores
d. Specialty stores

27. a.

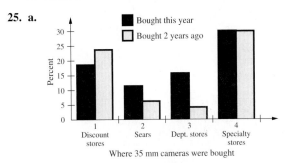

b. 18–24   **c.** 65 and up

29. a.

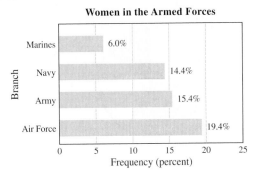

b. Air Force   **c.** Marines

d. No; we need to know the total numbers, not just the percents

31.

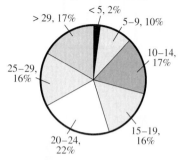

Meeting Hours Per Week

> 29, 17% < 5, 2% 5–9, 10% 10–14, 17% 15–19, 16% 20–24, 22% 25–29, 16%

33.

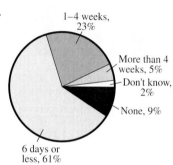

1–4 weeks, 23%

More than 4 weeks, 5%

Don't know, 2%

None, 9%

6 days or less, 61%

35.

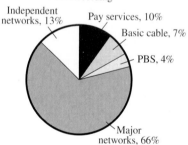

VCR Recording

Independent networks, 13%

Pay services, 10%

Basic cable, 7%

PBS, 4%

Major networks, 66%

37. a.

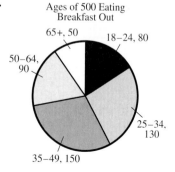

Ages of 500 Eating Breakfast Out

65+, 50

18–24, 80

50–64, 90

25–34, 130

35–49, 150

b. 35–49

39.

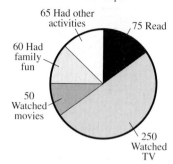

Favorite Activites of 500 People

65 Had other activities

75 Read

60 Had family fun

50 Watched movies

250 Watched TV

41.

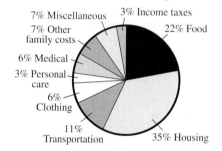

Retired Couple's Budget

7% Miscellaneous

3% Income taxes

7% Other family costs

22% Food

6% Medical

3% Personal care

6% Clothing

11% Transportation

35% Housing

43. a. Top six TVs by brand   **b.**

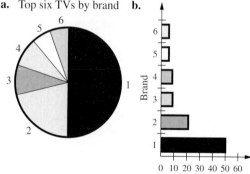

c. The circle graph. The area corresponding to brand 1 overshadows all the rest of the chart.

45. a. Yes. To give a correct visual impression, only the height should be doubled. If both the height and the radius are doubled, the volume is multiplied by 8.

b.

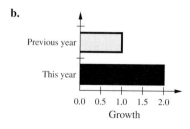

47.

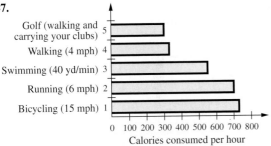

Golf (walking and carrying your clubs) 5
Walking (4 mph) 4
Swimming (40 yd/min) 3
Running (6 mph) 2
Bicycling (15 mph) 1

0 100 200 300 400 500 600 700 800
Calories consumed per hour

49.

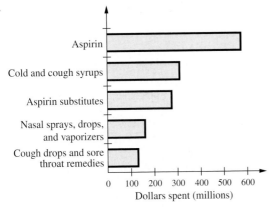

Aspirin

Cold and cough syrups

Aspirin substitutes

Nasal sprays, drops, and vaporizers

Cough drops and sore throat remedies

0 100 200 300 400 500 600
Dollars spent (millions)

51. a. $16,000,000
   b. In year 6.
53. a. Use a large scale on the vertical axis.
   b. Use a very small scale on the vertical axis.
55. The area of the bar in a bar graph indicates the amount of the item that is graphed. For this reason the bars are usually shaded. In a histogram, the height of the bar corresponds to the frequency of the item in question and there is no space between the bars.
57. The second graph has a very much compressed vertical scale which diminishes the visual effect of each increase or decrease.

Skill Checker

1. $-\dfrac{3}{2} = -1.5$ **3.** $y = -1.5x + 3$ **5.** 2

Exercises 12.6

1. a.–b.

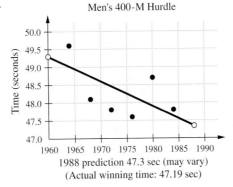

Men's 400-M Hurdle

Time (seconds)

50.0
49.5
49.0
48.5
48.0
47.5
47.0

1960 1965 1970 1975 1980 1985 1990
1988 prediction 47.3 sec (may vary)
(Actual winning time: 47.19 sec)

c. $y = -0.061x + 49.143$ **d.** *2000*   *2004*
    Prediction: 47.44 sec   46.70 sec   46.46 sec

3. a.

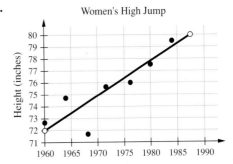

Women's High Jump

Height (inches)

80
79
78
77
76
75
74
73
72
71

1960 1965 1970 1975 1980 1985 1990

   b. 1988 prediction 80.0 in. (may vary) [actual winning height: 80.0 in., by Louise Ritter (U.S.)]
   c. $y = 0.268x + 72.197$ **d.** *2000*   *2004*
     Prediction: 79.70 in.   82.92 in.   83.99 in.

5. a.

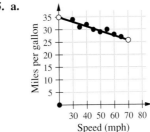

Miles per gallon

35
30
25
20
15
10
5

30 40 50 60 70 80
Speed (mph)

   b. About 26 mi/gal (may vary)
   c. $y = -0.169x + 38.155$
     Prediction: About 26.3 mi/gal

7. a.

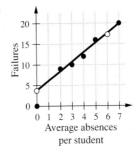

Failures

20
15
10
5
0

0 1 2 3 4 5 6 7
Average absences per student

   b. About 18
   c. $y = 2.338x + 3.581$
     Prediction: About 18 would fail.

9. a.

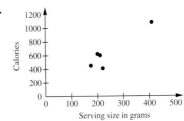

Calories

1200
1000
800
600
400
200
0

0 100 200 300 400 500
Serving size in grams

   b. About $y = 2.706x - 32.5$
   c. About 779 calories

11. a.

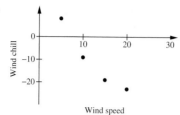

b. $y = -1.98x + 13.5$
c. -36

13. a.

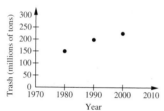

b. $y = 3.5x + 155$
c. 242.5; 260

15. a.

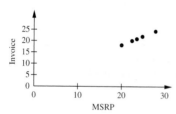

b. $y = x - 2$ (thousands)
c. \$28,000

17. a.

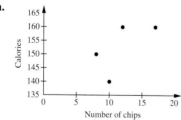

b. About $y = 2.184x + 124.66$
c. About 157 calories

19. 720   **21.** 200   **23.** 426
25. a. About 13   **b.** About 9   **c.** About 9
d. About 3   **e.** About 3   **f.** About 3
27. 100; 160

29. a.

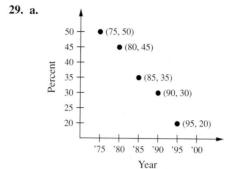

b. 13.5%   **c.** 135
31. $y = -0.872x + 55.231$
33. $y = -1.057x + 62.186$, where x is the number of the Olympics with 1960 as number 1
35. $y = -0.293x + 7.368$, where $x = \text{year} - 1965$, and $y = t - 45$
37. a. $y = -0.417x + 75$ **b.** $64.992 \approx 65$ mph
c. 50 mph   **d.** $y = 50$
39. Procedure (**d**)
41. a. The women who shop on Rodeo Drive are not necessarily a good representation of the entire population of California.
b. The same can be said of the male population of Berkeley.
c. The same can be said of the people attending an Oakland baseball game.
43. 24.7
45. $y = -0.3x + 24.7$
47. About 20.6 seconds

Exercises 12.7

1. No. Too many data points below the line.
3. Good   **5.** Positive
7. None   **9.** Negative
11. Positive   **13.** (**c**)
15. (**c**)   **17.** (**a**)
19. (**c**)   **21.** (**c**)

23. a.

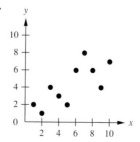

b. $y = 0.58x + 1.133$ **c.** 0.7388
d. We can be 95% confident that there is a significant positive linear correlation.

25. a.

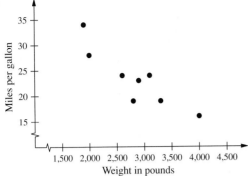

b. $y = -0.01x + 44.315$　**c.** -0.889
d. We can be 99% confident that there is a significant negative linear correlation.

27. a.

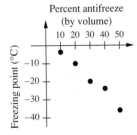

b. $y = 1.83x + 1.605$ (thousands)
c. 0.9307
d. We cannot be certain that there is a significant positive linear correlation at the 95% or 99% confidence level.
e. $19,905

29. a.

Percent antifreeze
(by volume)

b. $y = -0.78x + 4.6$　**c.** -0.99
d. We can be 99% confident that there is a significant negative linear correlation.
e. About $-14.9°C$
31. a. The seventh week　**b.** The eleventh week
　　c. After the second week　**d.** No
33. a. No
　　b. Answers may vary. It is possible that the number of cars with air bags being sold is increasing at a rate that makes them representative of the entire population of cars.
35. $r = 0.2236$. We cannot be certain that there is a significant linear correlation.

37. No. Inclusion of Ireland makes it appear as though there is no linear correlation.

CHAPTER 13

Exercises 13.1

1. $240　**3.** $540　**5.** $200
7. $62.50　**9.** $466.67
11. a. $2.41　**b.** $42.61
13. $1802.40　**15.** $4592.50
17. a. $40　**b.** $160　**c.** $168
19. a. $125　**b.** $375
21. a. $1.42　**b.** $26.98
　　Calculator answers may vary; see solutions manual.
23. $159.38; $59.38　**25.** $7154.60; $4574.60
27. $26,195; $14,195　**29.** $25,364; $5364
31. $3,880,688; $3,840,688
33. a. $2854.30　**b.** $2898.30
35. a. $3993　**b.** $993
37. $15.30　**39.** $2081.62
41. 11% semiannually yields about $51.47 more.
43. Simple interest means that the interest itself does not earn additional interest. Compound interest means that the interest earns interest at the same rate. For instance, $100 at 10% simple interest would earn $20 in 2 years, while $100 at 10% compounded annually would earn $21 in 2 years.
45. 6.09%　**47.** 8.24%　**49.** 9.31%　**51.** 19.56%

Exercises 13.2

1. $15,000 per year　**3.** $141.35
5. $185.01　**7.** $557.48
9. a. $1.28　**b.** $236.28　**c.** $11.81
11. a. $5.16　**b.** $409.16　**c.** $20.46
13. a. $1.21　**b.** $180.39　**c.** $10
15. a. $.84　**b.** $92.73　**c.** $10
17. a. $.52　**b.** $122.81　**c.** $10
19. $2.85　　　　　**21.** $1.35
23. a. $1998　**b.** $261
25. a. $450　**b.** $136
27. a. $60　**b.** $26
29. This procedure gives the longest possible time between the purchase date and the date when payment must be made to avoid a finance charge.
31. 6.4%　　　　　**33.** 10%

Exercises 13.3

1. 14%　**3.** 16%　**5.** 17%
7. $15\frac{1}{2}$%　**9.** 16%
11. a. $2　**b.** $98
13. a. $8.40　**b.** $261.60
15. a. $4.18　**b.** $45.82

17. a. $18.48   **b.** $19.88
   c. $422.52   **d.** 421.13
19. a. $104.13   **b.** $120.04
   c. $1413.39   **d.** $1397.43
21. a. $24   **b.** $14\frac{1}{2}\%$
23. a. $134   **b.** $16\frac{1}{2}\%$
25. a. $4   **b.** $196
27. a. $72   **b.** $66
   c. $9.23   **d.** $254.77
29. a. $120   **b.** $51.11
   c. $4.21   **d.** $149.12
31. a. $562   **b.** $554
33. a. $22,344   **b.** $22,637
   c. $22,932   **d.** About $300
35. $25,440
37. 14.3% (0.3% more than answer to problem 1)
39. 16.9% (0.9% more than answer to problem 3)
41. 17.7% (0.7% more than answer to problem 5)
43. 17.1% (1.6% more than answer to problem 7)
45. 17.6% (1.6% more than answer to problem 9)

Exercises 13.4

1. a. Yes   **b.** Yes   **c.** Yes
3. a. $61,600   **b.** $15,400   **c.** $2310
5. $526.27   **7.** $935.10   **9.** $1282.64
11. a. $1350   **b.** $44,437.50
13. a. $2250   **b.** $73,237.50
15. a. $46,200   **b.** $44,437.50   **c.** $374.99
17. a. $75,100   **b.** $72,850   **c.** $538.36
19. a. $3000   **b.** $667.64
21. a. $3000   **b.** $813.04
23. a. $439   **b.** 360   **c.** $158,040
   d. $108,040   **e.** Less

25. a. $11,185   **b.** $549   **c.** $563.04
27. a. $25,630   **b.** $872.80   **c.** $885.92
29. a. $10,467   **b.** $812.70
31. Answers will vary.
33. On a loan of $100,000 at 9% for 30 years, the total interest will be $189,800, which is more than the price of the house. On a $50,000 loan under the same terms, the total interest will be $94,900, which is more than the price of the house.

Section 13.5

1. a. $3528   **b.** $106   **c.** $73.56
   d. $76.44   **e.** 182.44   **f.** 5.07%
3. a. $5754   **b.** $144   **c.** $120
   d. $126   **e.** $270   **f.** 4.60%
5. a. $616.08   **b.** 0   **c.** $50
   d. -66.08 **e.** -66.08 **f.** -9.92%
7. a. $2235   **b.** $340   **c.** 15.04%
9. a. $3035   **b.** $190   **c.** 6.21%
11. a. $18,985   **b.** $890   **c.** 4.68%
13. a. $180   **b.** $1800
15. a. $337.50   **b.** $1350
17. a. $575   **b.** $5750
19. a. $29.025   **b.** 1722
21. a. $29.06   **b.** 3441
23. a. $29.05   **b.** 3442
25. a. 10%   **b.** 3.75%   **c.** 18.57%
27. a. 1.67%   **b.** 1.56%   **c.** 8.33%
29. a. 2.5%   **b.** 8%   **c.** 35%
35. 7.89%   **37.** 6.05%
39. 5.31%   **41.** $1428.57
43. $2000   **45.** $833.33

Research Bibliography

The entries in this bibliography provide a first resource for investigating the Research Questions that appear at the end of each chapter in the text. Many of these books will also contain their own bibliographies, which you can use for an even more thorough information search. Also, using your library's card catalog system, be sure to look under subject as well as title listings to gain a better idea of the resources that your library has available, even beyond the specific titles listed here.

Bell, E. T. *Men of Mathematics.* New York: Simon and Schuster, 1986.

Billstein, R., et al. *A Problem-Solving Approach to Mathematics.* 8th ed. Boston, MA: Addison-Wesley, 2003.

Borowski, E. J., and Borwein, J. M. *The Harper Collins Dictionary of Mathematics.* New York: HarperCollins, 1991.

Boyer, Carl B. *A History of Mathematics.* 2nd ed. Revised by Uta C. Mevzbach. New York: John Wiley, 1991.

Brewer, James W. "Emmy Noether, A Tribute to Her Life and Work." In *Monographs and Textbooks in Pure and Applied Mathematics,* vol. 69, ed. Martha K. Smith. New York: Marcel Dekker, 1981.

Burton, David. *The History of Math: An Introduction.* 4th ed. Dubuque, IA: Wm. C. Brown, 1999. *History of Mathematics: An Introduction,* 5th ed. McGraw-Hill, 2002.

Cajori, Florian. *A History of Mathematical Notation.* New York: Dover Publications, 1993.

Calinger, Ronald, ed. *Classics of Mathematics.* Upper Saddle River, NJ: Prentice Hall, 1994.

Copi, I. *Introduction to Logic.* 11th ed. Upper Saddle River, NJ: Prentice Hall, 2002.

Davis, Philip J., and Hersh, Reuben. *Descarte's Dream: The World According to Mathematics.* Boston, MA: Houghton Mifflin, 1986. Hardcover, 1986, Harcourt; paperback, 1987, Houghton Mifflin.

Eves, Howard. *An Introduction to the History of Mathematics.* 6th ed. New York: Saunders College, 1990. B & N, 1994, 6th ed., Saunders; Amazon, 1990, 6th ed., Harcourt Brace.

Grattan-Guinness, Ivor. *Norton History of the Mathematical Sciences,* New York: W. W. Norton and Company, B & N: 1998; Amazon: 1999.

Hogben, Lancelot. *Mathematics in the Making.* London: Galahad Books, 1960.

Kahane, Howard. *Logic and Philosophy: A Modern Introduction.* 9th ed. Belmont, CA: Wadsworth, 2001.

Katz, Victor J. *A History of Mathematics.* 2nd ed. Reading, MA: Addison-Wesley, 1998.

Kline, Morris. *Mathematical Thought: From Ancient to Modern Times.* 4 vols. New York: Oxford University Press, 1990.

Krause, Eugene. *Mathematics for Elementary Teachers.* 2nd ed. Lexington, MA: Houghton Mifflin, 1991.

Newman, James. *The World of Mathematics.* 4 vols. New York: Dover Publications, 2000.

Osen, Lynn M. *Women in Mathematics.* Cambridge, MA: MIT Press, reprinted 1994.

Pedoe, Don. *The Gentle Art of Mathematics.* New York: Dover Publications, 1990.

Perl, Teri H. *Math Equals.* Reading, MA: Addison-Wesley, 1990.

Polya, George. *Mathematical Discovery.* New York: John Wiley, 1981.

Vos Savant, Marilyn. *More Marilyn.* New York: St. Martin Press, 1994.

Acknowledgments

CHAPTER 1

Page 17, Figure 1.5, Reading electric meters, www.we-energies.com. Page 27, Figure 1.8, Circle graph of family budget, reprinted with permission of The One Heart Foundation. © 2003. Page 29, Figure 1.13, Bar graph of risks of mortality; page 36 Alcohol consumption and Risk of stroke bar graphs, reprinted by permission of Dr. David J. Hanson from www2.potsdam.edu/alcohol-info/Health/Health.html. Pages 30–31, Figure 1.14 and Figure 1.15 and page 43, monthly payment charts provided by KJE Computer Solutions, LLC. For more information please see www.dinkytown.net. Page 33, Ways of traveling to work and daily activities charts courtesy of Learnthings, Ltd. Page 42, Evolution of VCR, DVD player, and DVD recorder chart, reprinted by permission. Page 46, Introductory courses from Review of Tables, Bar Graphs and Circle Graphs, Center In Support of Teachers and Learning, Syracuse University, 1998. Page 46, Inside sales and Web site sales, line graph built using NetCharts Server by Visual Mining, Inc., www.visualmining.com. Page 48, Mortgage rates, © 2003 Bankrate.com. Used with permission.

CHAPTER 2

Pages 79 and 85, Google Search Engine screenshots, © Google, Inc. Used with permission.

CHAPTER 4

Page 188, Mayan numbers for you to translate, by Rhonda Robinson, reprinted courtesy of Rhonda Robinson. Pages 193, 211, bytes and bits table from Australian Photonics CRC, reprinted with permission.

CHAPTER 5

Page 264, Bay News 9 hourly program schedule, from Tampa Tribune TV, Sunday, September 19, 1999, p. 6, copyright © 1999 by Tampa Tribune. Page 276, bytes and bits table from Australian Photonics CRC, reprinted with permission.

CHAPTER 6

Pages 337–338, problems 31–42; page 363, example 10; pages 364–364, exercises 61–70, 78–80; pages 387–388, example 8; page 390, exercises 38–40, page 391, exercises 43–47 from *Intermediate Algebra*, First Edition, by I. Bello, copyright © 1997. Page 387, *USA Today*. Copyright © 1993. Reprinted with permission.

CHAPTER 7

Page 407, Figure 7.7, Take out takes off, from *The Wall Street Journal Almanac* © 1999 by Wall Street Journal Staff. Page 420, Texting Increases with Age, © 2006 The Pew Charitable Trusts. Page 434, Wireless Phones Outnumber Wired Lines, Source: Kaufman Brothers, "A General Flavor of Mild Decay," July 14, 2003. Page 450, Millennium Fact Box, from *The World Almanac and Book of Facts*, © 2004, World Almanac Education Group, Inc. All Rights Reserved. Page 485, Blood Alcohol Level, Fatalities, Effects of Alcohol, *USA Today*, copyright © 1993. Reprinted with permission. Pages 437–465, 478, from *Intermediate Algebra*, First Edition, by I. Bello, copyright © 1997.

CHAPTER 8

Page 568, Figure 8.92, Reprinted courtesy of Tetrad Computer Applications, Inc.

CHAPTER 9

Page 606, Figure 9.1, New York by LATA, and page 609, Housing Tenure, Reprinted courtesy of Tetrad Computer Applications, Inc.

CHAPTER 10

Pages 669, 693, Figure 10.12, tree diagrams, from TreeAge Software, data ™3.5 TreeAge Software, Inc., Williamstown, MA. Page 697, tree diagram from Haddix, Ann S. Teutch, Phaedra Shaffer, and D. Donet, eds., *Prevention Effectiveness: A Guide to Decision Analysis and Economic Evaluation*, New York: Oxford University Press.

CHAPTER 11

Page 715, Should "Intelligent design" be taught in high school biology class?, reproduced from the April 25, 2006 issue of *BusinessWeek* by special permission, copyright © 2006 by the McGraw-Hill Companies, Inc. Page 717, problems 1 and 2; page 753, excerpt from Example 8 reprinted with permission from *Parade* and Marilyn vos Savant, copyright © 1992, 1996, and 1999, respectively. Page 755, Spam categorization Breakdown, August 2005, reprinted courtesy of Clearswift Corporation.

CHAPTER 12

Page 805, Attendance at Top Five Amusement/Theme Parks in the United States; Page 808, Discrimination Complaints, from the *Wall Street Journal Almanac*, 1999 by Wall Street Journal Staff. Page 869, Figure 12.38, reprinted with permission of David C. Howell.

CHAPTER 13

Page 898, High Yield Rates for 1-Year CD, © 2006 Bankrate.com. All rights reserved. Page 900, Figure 13.1, The Amoco torch and oval is a registered trademark of BP Products North America, Inc. Pages 903–904, Figure 13.3, reprinted courtesy of Sears, Roebuck & Company. Pages 914 and 915, Figures 13.4 and 13.5, © 1995, 2004, FinanCenter, Inc. Pages 930, 932, 934, Table 13.6, Table 13.7, and Table 13.8, respectively, Copyright 1999–2006 Investopedia, Inc. www.investopedia.com. Page 938, Table for problems 13–18, © 2006 Bondtrac, Inc.

Index

Applications Index

Table I *Squares and Square Roots*

N	N^2	$\sqrt{N}$	$\sqrt{10N}$	N	N^2	$\sqrt{N}$	$\sqrt{10N}$
1	1	1.00000	3.16228	51	2601	7.14143	22.5832
2	4	1.41421	4.47214	52	2704	7.21110	22.8035
3	9	1.73205	5.47723	53	2809	7.28011	23.0217
4	16	2.00000	6.32456	54	2916	7.34847	23.2379
5	25	2.23607	7.07107	55	3025	7.41620	23.4521
6	36	2.44949	7.74597	56	3136	7.48331	23.6643
7	49	2.64575	8.36660	57	3249	7.54983	23.8747
8	64	2.82843	8.94427	58	3364	7.61577	24.0832
9	81	3.00000	9.48683	59	3481	7.68115	24.2899
10	100	3.16228	10.0000	60	3600	7.74597	24.4949
11	121	3.31662	10.4881	61	3721	7.81025	24.6982
12	144	3.46410	10.9545	62	3844	7.87401	24.8998
13	169	3.60555	11.4018	63	3969	7.93725	25.0998
14	196	3.74166	11.8322	64	4096	8.00000	25.2983
15	225	3.87298	12.2474	65	4225	8.06226	25.4951
16	256	4.00000	12.6491	66	4356	8.12404	25.6905
17	289	4.12311	13.0384	67	4489	8.18535	25.8844
18	324	4.24264	13.4164	68	4624	8.24621	26.0768
19	361	4.35890	13.7840	69	4761	8.30662	26.2679
20	400	4.47214	14.1421	70	4900	8.36660	26.4575
21	441	4.58258	14.4914	71	5041	8.42615	26.6458
22	484	4.69042	14.8324	72	5184	8.48528	26.8328
23	529	4.79583	15.1658	73	5329	8.54400	27.0185
24	576	4.89898	15.4919	74	5476	8.60233	27.2029
25	625	5.00000	15.8114	75	5625	8.66025	27.3861
26	676	5.09902	16.1245	76	5776	8.71780	27.5681
27	729	5.19615	16.4317	77	5929	8.77496	27.7489
28	784	5.29150	16.7332	78	6084	8.83176	27.9285
29	841	5.38516	17.0294	79	6241	8.88819	28.1069
30	900	5.47723	17.3205	80	6400	8.94427	28.2843
31	961	5.56776	17.6068	81	6561	9.00000	28.4605
32	1024	5.65685	17.8885	82	6724	9.05539	28.6356
33	1089	5.74456	18.1659	83	6889	9.11043	28.8097
34	1156	5.83095	18.4391	84	7056	9.16515	28.9828
35	1225	5.91608	18.7083	85	7225	9.21954	29.1548
36	1296	6.00000	18.9737	86	7396	9.27362	29.3248
37	1369	6.08276	19.2354	87	7569	9.32738	29.4958
38	1444	6.16441	19.4936	88	7744	9.38083	29.6648
39	1521	6.24500	19.7484	89	7921	9.43398	29.8329
40	1600	6.32456	20.0000	90	8100	9.48683	30.0000
41	1681	6.40312	20.2485	91	8281	9.53939	30.1662
42	1764	6.48074	20.4939	92	8464	9.59166	30.3315
43	1849	6.55744	20.7364	93	8649	9.64365	30.4959
44	1936	6.63325	20.9762	94	8836	9.69536	30.6594
45	2025	6.70820	21.2132	95	9025	9.74679	30.8221
46	2116	6.78233	21.4476	96	9216	9.79796	30.9839
47	2209	6.85565	21.6795	97	9409	9.84886	31.1448
48	2304	6.92820	21.9089	98	9604	9.89949	31.3050
49	2401	7.00000	22.1359	99	9801	9.94987	31.4643
50	2500	7.07107	22.3607	100	10000	10.00000	31.6228
N	N^2	$\sqrt{N}$	$\sqrt{10N}$	N	N^2	$\sqrt{N}$	$\sqrt{10N}$